HANDBOOK OF OCCUPATIONAL SAFETY AND HEALTH

HANDBOOK OF OCCUPATIONAL SAFETY AND HEALTH

SECOND EDITION

Edited by
Louis J. DiBerardinis
Environmental Health Service, Massachusetts Institute of Technology

A WILEY-INTERSCIENCE PUBLICATION

JOHN WILEY & SONS, INC.

New York • Chichester • Weinheim • Brisbane • Singapore • Toronto

This book is printed on acid-free paper. ♾

Published simultaneously in Canada.

Library of Congress Cataloging-in-Publication Data:

Handbook of occupational safety and health / edited by Louis J. DiBerardinis.—2nd ed.
p. cm.
Includes bibliographical references and index.
ISBN 0-471-16017-2 (cloth : alk. paper)
1. Industrial hygiene—Handbooks, manuals, etc. 2. Industrial safety—Handbooks, manuals, etc. I. DiBerardinis, Louis J., 1947–.
RC967.H26 1999
613.6′2—dc21 97-49069

Printed in the United States of America

10 9 8 7 6 5 4 3

This handbook is dedicated to William A. Burgess, Melvin F. First, ScD, and Benjamin Ferris, MD. These three individuals have influenced my professional life immensely and are well-respected pioneers in the occupational health profession. My soul mate Margie often reminds me that "if not for them you'd be driving a cement truck!"

CONTENTS

PREFACE xi

CONTRIBUTORS xiii

PART I INTRODUCTION

1. **Occupational Safety and Health Management** 1
   Thomas M. Dougherty

2. **Management's Roles and Responsibility in an Effective Safety and Health Program** 13
   Rosemary Colvin and Ray Colvin

3. **How to Locate Information Sources for Occupational Safety and Health** 33
   Robert Herrick and James Stewart

4. **Recognition of Health Hazards in the Workplace** 53
   Martin R. Horowitz and Marilyn F. Hallock

PART II SAFETY EVALUATIONS

5. **How to Conduct an Accident Investigation** 103
   Thomas M. Dougherty

6. **Risk Assessment Techniques** 127
   Thomas M. Dougherty

7. **How to Apply Computer Simulation to Occupational Safety and Health** 179
   Mahmoud A. Ayoub and Stephen D. Roberts

PART III HEALTH EVALUATIONS

8. **Evaluation of Exposure to Chemical Agents** 239
   Jerry Lynch and Charles Chelton

9. Evaluation and Management of Exposure to Infectious Agents 287
Janet M. Macher and Jon Rosenberg

10. Biological Monitoring in Occupational Health and Safety 373
Alain J. Couturier and Robert J. McCunney

11. How to Ensure Quality in a Medical Surveillance Program 415
Robert J. McCunney and Alain J. Couturier

PART IV CONTROL PRACTICES

12. How to Establish Industrial Loss Prevention and Fire Protection 429
Peter M. Bochnak

13. Hazard Communication and Worker Right-to-Know Programs 479
Lisa K. Simkins and Charlotte A. Rice

14. Health and Safety Instruction 517
Kenneth P. Martin

15. Philosophy and Management of Engineering Control 539
Pamela Greenley and William A. Burgess

16. How to Select and Use Personal Protective Equipment 601
Richard J. Nill

17. Respiratory Protective Devices 639
Howard Cohen

18. How to Apply the Microcomputer to Occupational Safety and Health 677
Brad T. Garber

PART V PHYSICAL HAZARDS AND SPECIAL TOPICS

19. Occupational Dermatoses 697
David E. Cohen

20. Indoor Air Quality in Nonindustrial Occupational Environments 743
Philip R. Morey

21. Heat Stress 791
Anne M. Venetta Richard and Ralph Collipi, Jr.

22. Occupational Noise Exposure and Hearing Conservation 825
Charles P. Lichtenwalner and Kevin Michael

23. Ergonomics: Achieving System Balance Through Ergonomic Analysis and Control **931**
Graciela Perez

24. Radiation: Nonionizing and Ionizing Sources **987**
Donald L. Haes, Jr. and Mitchell S. Galanek

PART VI LEGAL AFFAIRS

25. How to Deal with the Troubled Employee **1017**
Margie Magowan

26. Managing Workers' Compensation **1029**
James J. Paugh III

27. How to Evaluate Your Occupational Safety and Health Program **1081**
Rosemary Colvin and Ray Colvin

28. Occupational Health and Safety Regulatory Affairs **1107**
Roy Deitchman and James Degen

29. Occupational Safety and Health Legal Issues **1149**
Thea Dunmire

30. How to Provide Liability Industrial Coverage **1163**
Donald M. Weekes, Jr.

INDEX **1195**

PREFACE

As we go to press OSHA expects to propose an occupational safety and health program standard by early 1999. The basic requirement of the standard is to provide a basic framework for systematically identifying and controlling workplace hazards covered by other specific OSHA regulations and the general duty clause that mandates each employer to provide a workplace free from recognized hazards. This handbook provides assistance in accomplishing this task.

The purpose of this handbook is to provide a basic reference source that in part uses techniques and methods from various disciplines applicable to occupational safety and health. These include safety, industrial hygiene, occupational medicine, engineering, and legal issues. This handbook is not intended as an elementary text on occupational safety and health. Rather, the presumption is that the reader is a mature, intelligent person who is not necessarily a full-time safety practitioner, but may have some safety and health responsibilities or may want a basic understanding of the important concepts. It is directed to individuals who have collateral duties in safety and health, such as plant managers, engineers, human relations/personnel managers, and supervisory personnel.

It is also directed to full-time health and safety professionals with broad knowledge of the field who will use the handbook to review and update their understanding of specific topics as the need arises.

This text is designed to be a quick source of immediately useful information on a particular topic. It is meant to be a true handbook, providing tables and figures of data and examples of forms and checklists that can be immediately put to use.

The handbook provides the reader with a basic understanding of health and safety concepts and then focuses on specific, usually the most common, problems encountered in the workplace. The basic principles followed include anticipation, recognition, evaluation, and control. A good health and safety program anticipates and recognizes potential hazards by maintaining a complete knowledge of the operations performed and materials used in the workplace. Since many regulations will also drive the program, it is important to know which regulations apply to your workplace, when they apply, and how to comply. Once potential hazards are identified, a systematic method of evaluation is needed to determine the degree to which the hazard exists under current or future conditions. Finally, methods to control the existing hazards or prevent future hazards from developing must be identified and implemented.

Although each chapter is written by different contributors with, in many cases, vastly different backgrounds, the approach to each chapter is quite similar. All basically say you first need to *understand* the problem and then be able to implement *change* when necessary. Some chapters only cover one of the basic principles noted above, while others take a specific problem, i.e., noise, heat stress, ergonomics, and discuss recognition, evaluation, and control.

The handbook is divided into six major parts. The first three follow the standard health and safety approach to hazards: recognition, evaluation, and control. The intent is to provide the reader with the basic concepts, detailed steps to follow, and "helpful hints." Part IV groups special issues that don't fall into the standard approach. Part V addresses management techniques that are necessary to implement a successful health and safety program. Part VI presents legal issues, including a discussion of current regulations and methods for ensuring compliance.

The need for such a handbook is highlighted by the growing scope and diversity in the field.

I am indebted to the many individuals who have contributed to this handbook. Certainly the contributors of each chapter have worked diligently to present their sometimes quite technical expertise in an easy-to-use, quick-reference format. Two individuals deserve special acknowledgment. Margie Magowan, in addition to writing a chapter, provided the encouragement to continue and invaluable advice for keeping the project on course. Margaret Mahoney provided the administrative support needed to communicate with 36 contributors. This handbook would not have become a reality without the cooperation and patience of all of these individuals.

LOU DIBERARDINIS

Cambridge, Massachusetts

CONTRIBUTORS

Mahmoud A. Ayoub, PhD, North Carolina State University, Industrial Engineering, 328 Riddick Labs, Raleigh, NC 27695

Peter M. Bochnak, Environmental Health and Safety, Harvard University, 46 Oxford St., Cambridge, MA 02138

William A. Burgess, CIH, Massachusetts Institute of Technology, Bldg. 56-235, 77 Massachusetts Ave., Cambridge, MA 02139-4307

Charles Chelton, CIH, 30 Radtke St., Randolph, NJ 07869

David Cohen, MD, NYU Medical Center, Dermatology Dept., Rm H-100, 560 First Avenue, New York, NY 10016

Howard Cohen, PhD, CIH, University of New Haven, West Haven, CT 06516

Ralph Collipi, Jr., CIH, AT&T, 40 Elwood Rd., Londonderry, NH 03053

Ray Colvin, PE, CSP, Safety Training Dynamics, Inc. 27497 AL. Hwy 91, Hanceville, AL 35077

Rosemary Colvin, Safety Training Dynamics, Inc., 27497 AL. Hwy 91, Hanceville, AL 35077

Alain J. Couturier, MD, Occupational Health Network, 11220 County Rd. 14, Middlebury, IN 46540

James Degan, CSP

Roy Deitchman, JD, CIH, Bellcore NVC 3X283, 331 Newman Springs Rd., Red Bank, NJ 07701-5699

Louis J. DiBerardinis, CIH, CSP, Environmental Health Service, Massachusetts Institute of Technology, Bldg. 56-235, 77 Massachusetts Ave., Cambridge, MA 02139-4307

Tom Dougherty, Polaroid Corp., W2-2, 1265 Main St., Waltham, MA 02254

Thea D. Dunmire, JD, CIH, CSP, ENLAR Compliance Services, Inc., P.O. Box 3520, Clearwater, FL 33767

Mitchell S. Galanek, CHP, Massachusetts Institute of Technology, Bldg. 16-268, 77 Massachusetts Ave., Cambridge, MA 02139-4307

Brad T. Garber, PhD, CIH, University of New Haven, West Haven, CT 06516

Pamela Greenley, CIH, Massachusetts Institute of Technology, Bldg. 56-235, 77 Massachusetts Ave., Cambridge, MA 02139-4307

Donald Haes, CHP, Massachusetts Institute of Technology, Bldg. 16-268, 77 Massachusetts Ave., Cambridge, MA 02139-4307

Marilyn Hallock, CIH, Massachusetts Insitute of Technology, Bldg. 56-235, 77 Massachusetts Ave., Cambridge, MA 02139-4307

Robert Herrick, ScD, CIH, Harvard School of Public Health, Department of Environmental Health, 665 Huntington Ave., Boston, MA 02115

Martin Horowitz, CSP, CIH, Analog Devices, 21 Osborn St., Cambridge, MA 02139

Charles P. Lichtenwalner, CIH, PE, CSP, Lucent Technologies, Bell Laboratories, 600 Mountain Ave., Murray Hill, NJ 07974

Jerry Lynch, CIH

Janet Macher, ScD, CIH, Division of Environmental and Occupational Disease Control, Environmental Health Laboratory Branch, California Department of Health Services, 2151 Berkeley Way, Berkeley, CA 94704

Margie Magowan, MBA, 594 Washington St., Wellesley, MA 02482

Kenneth P. Martin, CIH

Robert J. McCunney, MD, Massachusetts Institute of Technology, Bldg. 20B-238, 77 Massachusetts Ave., Cambridge, MA 02139-4307

Kevin Michael, PhD, Michael & Associates, 246 Woodland Dr., State College, PA 16803

Philip R. Morey, PhD, CIH, AQS Services, 2235 Baltimore Pike, Gettysburg, PA 17325

Richard Nill, CIH, CSP, Genetics Institute, 1 Burtt Rd., Andover, MA 01810

James J. Paugh III, Lynch, Ryan & Paugh, Inc., 120 Front St., Suite 320 , Worcester, MA 01608-1408

Graciela M. Perez, ScD, CPE

Charlotte A. Rice, CIH

Anne M. Venetta Richard, Lucent Technologies, 1600 Osgood St., 21-2L28, North Andover, MA 01845

Stephen D. Roberts, PhD, North Carolina State University, Industrial Engineering, 328 Riddick Labs, Raleigh, NC 27695

Jon Rosenberg, MD, Division of Communicable Disease Control, Disease Investigations and Surveillance Branch, California Department of Health Services, 2151 Berkeley Way, Berkeley, CA 94704

Lisa K. Simkins, PE, CIH, Clayton Group Service Inc., P.O. Box 9019, Pleasanton, CA 94566

James Stewart, PhD, CIH, Harvard School of Public Health, Department of Environmental Health, 665 Huntington Ave., Boston, MA 02115

Donald M. Weekes, CIH, CSP, Abacus Environmental, Inc., 123 Pinney St., Ellington, CT 06029

CHAPTER 1

Occupational Safety and Health Management

THOMAS M. DOUGHERTY
Polaroid Corporation, 1265 Main St., Waltham, MA 02254

"Concern for man himself and his fate must always form the chief interest of all technical endeavor; never forget this in the midst of your diagrams and equations." Albert Einstein

1.1 INTRODUCTION
  1.1.1 Scope of This Chapter
  1.1.2 Safety and Health Resources
1.2 OCCUPATIONAL SAFETY AND HEALTH—HISTORY
  1.2.1 The Past
  1.2.2 The Present
  1.2.3 The Future
1.3 SAFETY MANAGEMENT AND THE ORGANIZATION
  1.3.1 What Types of Skills Are Required?
  1.3.2 Where in the Organization?
  1.3.3 Effects of Organizational Culture
1.4 SAFETY AND HEALTH MANAGEMENT SYSTEM ELEMENTS
  1.4.1 Introduction
  1.4.2 OSHA's Proposed Safety Management System
  1.4.3 OSHA's Process Safety Management System
  1.4.4 ISO 14001—Environmental Management System
  1.4.5 AIHA—Occupational Health and Safety Management System
  1.4.6 BS 8800—British Occupational Health and Safety Management System
REFERENCES

1.1 INTRODUCTION

1.1.1 Scope of This Chapter

This chapter provides an overview of occupational safety and health and in particular the aspects relating to the management function. The history and evolution of safety and health as it relates to the "workplace" is reviewed as well as the place of safety

Handbook of Occupational Safety and Health, Second Edition, Edited by Louis J. DiBerardinis, ISBN 0-471-16017-2

management in the organizational structure. Safety and health management elements proposed by several organizations are summarized. The focus for this chapter is primarily safety and health management aspects in the United States, but there is evidence that the benefits and opportunities to apply common management practices across country boundaries are growing. This is particularly true with the International Standards Organization (ISO) globalization efforts regarding quality management (ISO 9000 series) as well as environmental management (ISO 14000 series).

The most important aspect of occupational safety and health management performance, however, has to do with the culture of the organization. To the degree that the safety, health, and well being of each individual in the organization is a shared core value is the degree to which safety and health is or will be well managed. So, the challenge for the leadership of any organization is to create and/or embed these values and make them a "way of doing things." When this is accomplished, using and implementing the information contained in the following chapters becomes automatic and institutionalized.

A portion of the Piper Alpha oil drilling rig safety disaster report says, in part, "Safety is not an intellectual exercise to keep us in work. It is a matter of life and death. It is the sum of our contributions to safety management that determines whether the people we work with live or die."

1.1.2 Safety and Health Resources

The following organizations are only a basic listing of resources available to the safety and health professional. The explosive growth of the internet and the World Wide Web is making access to up-to-date information faster and better. Informal networking has always been a naturally occurring phenomenon among safety and health professionals as well.

- American Society of Safety Engineers—ASSE
- American Industrial Hygiene Association—AIHA
- National Safety Council—NSC
- American National Standards Institute—ANSI
- Human Factors and Ergonomics Society
- National Fire Protection Association—NFPA
- Occupational Safety and Health Administration—OSHA

1.2 OCCUPATIONAL SAFETY AND HEALTH—HISTORY

1.2.1 The Past

In 1760 BC, King Hammurabi, who belonged to the first dynasty of Babylonia, set in motion the collection of laws and edicts that came to be called the Hammurabi Code. The code, having been received by the king from the sun god, Shamash, provided the procedures regarding property rights, personal rights, and debts. It provided for, among other things, damages caused by neglect in various trades. For example, it provided that

> If a builder build a house for some one, and does not construct it properly, and the house which he built fall in and kill its owner, then that builder shall be put to death.

and

> If a shipbuilder build a boat for someone, and do not make it tight, if during that same year that boat is sent away and suffers injury, the shipbuilder shall take the boat apart and put it together tight at his own expense. The tight boat he shall give to the boat owner.

These edicts sound suspiciously close to our present-day building codes and OSHA's standards regarding Shipyard Employment as well as conceptually establishing the first Worker's Compensation requirements.

Hippocrates, the celebrated Greek physician, has been called the father of medicine. Around 400 BC he has been credited with developing tetanus, helping check a great plague around Athens, as well as prescribing treatment for injuries to the head caused by accidents.

During the early and up to the late Middle Ages a variety of hazards were identified, including the effects of lead and mercury exposure, burning of fires in confined spaces, as well as the need for personal protective equipment. However, there were no organized or established safety standards or requirements during this time. The worker was generally an independent craftsman or part of a family-run shop or farm and was solely responsible for his own safety, health, and well being.

In the early part of the eighteenth century and on the cusp of the Industrial Revolution, Beardini Ramazzini wrote his classic "Discourse on the Diseases of Workers." Regarded as the father of occupational medicine, he described causes of occupational diseases exhibited by chemists working in laboratories. His great admiration for the chemists, however, led him to believe it would be an insult to their profession if he suggested any safety interventions. He also described the pain exhibited in the hands of scribes, foretelling our modern-day interventions regarding repetitive strain injuries. As an addition to the standard patient history questionnaire, he also thought to ask: "What is your occupation?"

In the late 1700s, the factory system exposed the worker to new and unknown hazards. Textiles led the way with power looms, the cotton gin, and the spinning jenny, along with the associated risks of machinery, noise, and dust. Management was concerned with profit and loss. Deaths and injuries were accepted as part of the industrial landscape. Today, perhaps aches and pains may be considered the norm and expected as well as being accepted in some industrial occupations. The management of safety and health, then, was an unthought of concern or need. Labor was plentiful, and the workers were glad just to have a paying job.

Through the early part of the 1800s, the Industrial Revolution swept through the United States, pressures to cut costs increased, and the labor force consisted more and more of untrained immigrant labor and children. The common laws of the day favored the owners and managers, and there was virtually no compensation for occupational illnesses or injury nor any agreed upon standards for workplace safety. However, as injury rates took their toll, the first attempts at compensation started in Massachusetts with the Employer's Liability Law[1] in 1877. In most cases, however, any attempt at compensation was refuted by a variety of legal defenses if an employer could show the employee was negligent or contributed to the cause of the accident.

The twentieth century brought presidential safety interest to the political arena. In 1908, Theodore Roosevelt stated[2]: "The number of accidents which result in the death or crippling of wage earners ... is simply appalling. In a very few years it runs up a total far in excess of the aggregate dead in any major war." This was followed by the establishment of Workers Compensation requirements federally as well throughout the states. In the meantime, safety standards started to evolve regarding machine guarding, and the steel companies and railroads began the start of what we know today as safety management programs. The infamous Triangle Shirtwaist Factory fire in 1911, which resulted in the deaths of 146 garment workers, helped to galvanize these efforts. The National Safety Council was formed during this time as well.

Up until 1931, most of these safety and health intervention efforts were directed towards improving factory conditions. Then H. W. Heinrich published a book entitled *Industrial Accident Prevention.*[3] He postulated the concept that the actions of people caused many more accidents than workplace conditions. He is sometimes called the Father of Modern Safety because he proposed the first organized set of safety principles.

These principles were revolutionary for their time. They included the concepts that accidents were caused, that the majority of accidents were the result of unsafe acts of people, and that those same unsafe acts had probably been committed over 300 times on the average. He also proposed some rationales why people behaved in an unsafe manner, a basic methodology for preventing accidents, as well as a postulate that management has the responsibility for assuming the work of accident prevention.

1.2.2 The Present

In 1970, the historic Occupational Safety and Health Act (OSHA) passed and became federal law effective in 1971. This followed on the heels of the several events, including a renewed fervor regarding automobile safety with Ralph Nader's book *Unsafe at Any Speed.* Safety and occupational health had become important elements to most major manufacturing industries. Standards had been evolving, and management had realized that operating profits were directly affected when employees lost time because of a work-related injury.

Some would argue that the OSHA Act diverted management's attention from the prevention of injuries to being compliant with the law. However well intentioned, the initial safety regulations were adopted from other documents developed by standards-producing organizations. In many cases, those standards were intended to be used as guidelines. Responsible application of safety guidelines was replaced with strict "how do we comply" attitudes to some extent. Also, because the Act focused on workplace conditions, it may have inadvertently slowed the development of safety management tools based on behavioral interventions. This workplace conditions approach was in contrast to the principle proposed by Heinrich that suggested that most accidents were caused by people's actions.

In any event, the Occupational Safety and Health Act, along with its research partner, the National Institute for Occupational Safety and Health (NIOSH), and its advisory committee, the National Advisory Committee on Occupational Safety and Health (NACOSH), created a new interest and new era in safety and health. The law provides for sanctions in the event of noncompliance with the requirements of providing a workplace free from recognized hazards that may cause death or physical harm. Initial requirements tended to be specification oriented and provided in great detail what needed to be done. Much fun had been made regarding the requirements for toilet seat design as well as the height for fire extinguishers mountings. More recent legislation has moved towards a performance-based orientation, which can allow for reasoned judgments and responsible application of the requirement. An example of this approach is found in the Process Safety Management Standard, which requires risk assessments around chemical safety manufacturing.

While most major companies have incorporated the OSHA Act requirements into their health, safety, and management system, progressive companies have moved beyond just compliance. They recognize that simply being in compliance with safety regulations is not sufficient. They recognize that workplace conditions, which are the primary focus of the regulations, is only one aspect of a well-managed program. Regulations represent a minimal criteria and entry-level approach. Off-the-job safety programming by some companies takes the lessons learned in the workplace into the homes and families of their employees.

The technology of safety and health has evolved into a mature science for the most part. Standards exist and methodology has been developed for mechanical safety, chemical hazard risk assessments, electrical safety standards, fire protection standards, as well as personal protective equipment standards. Much of this technology is discussed in the following chapters. The standards around workplace ergonomics is a quickly evolving field. Although regulatory initiatives in this area have not been successful, most large corporations see the benefit of including ergonomic programs as a part of their overall safety and health efforts.

Some would argue that actual safety performance has been mixed since the establishment of OSHA. The Bureau of Labor's injury statistics show that the total recordable injury rate has fallen slightly over the past twenty years from 13.2 injuries and illnesses per 100 workers to 11.6 in the manufacturing sector. Injuries and illnesses that have resulted in lost working time have fallen in the same period from 4.4 to 2.9 percent. It is hard not to argue however, that actual workplace conditions have not substantially improved from the early days of industrialization. In the Western world, the hazards of heavy manufacturing have been replaced by other hazards that are not immediately apparent. For example, manufacturing companies now have increasing levels of chronic disorders associated with cumulative trauma, skin diseases, and respiratory ailments. The burden of immediate acute injuries is now being borne by manufacturing industries in Third World countries.

During this time, one of the more prolific writers on safety and health has been Dan Petersen. He has authored a number of books and articles on the management of safety and health. He built upon and further developed the theories of Heinrich. He espoused five principles of safety in his 1978 book *Techniques of Safety Management*,[4] the first of which was "An unsafe act, an unsafe condition, and an accident are all symptoms of something wrong in the management system." His 1993 *Challenge of Change: Creating a New Safety Culture*[5] describes a process of creating change in the management system that leads to a desired safety performance.

1.2.3 The Future

Although the future of the occupational safety and health effort is difficult to predict with certainty, some insights are beginning to emerge. Governmental intervention will probably be less. At least from a prescriptive point of view, the chances of legislating the "how to's" of safety and health are fairly remote. United States politics has taken a recent turn in trying to reinvent itself. The administration's position in the mid 1990s was "To provide adequate protection to workers without imposing unfair burdens on employers." The heretofore adversarial, often paperwork-based, inspections will probably be replaced by a system targeting the most serious hazards and dangerous workplaces. Employee participation will be stressed as well as participation in OSHA's Voluntary Protection Program, which started somewhat slowly in the early 1980s but is now gathering momentum. Third party, nongovernmental audits may play an increasing part in a health and safety assurance program. This may also offer career-enhancing opportunities for safety and health practitioners. Currently, however, the chance of having an OSHA inspection is once every 86 years. Additionally, service organizations probably will receive more attention as a result of corporate downsizing and resulting outsourcing.

The way in which companies manage their safety and health programming will evolve as well. Downsizing and outsourcing have created a need for the workers themselves to become empowered regarding their own safety and well being. Middle management levels formerly created to monitor and control employees have disappeared. Team-based management techniques and self-directed work teams have flooded the safety and health arena as well. All this comes on the heels of behavior-based safety programming. Dr. Thomas Krause has popularized this approach in recent years. His book, *The Behavior-Based Safety Process*,[6] described one way to influence behavior in the workplace. Based on the work of Harvard psychologist B. F. Skinner, it includes identifying critical behaviors, observing actual behaviors, and providing feedback that lead to changed and improved behaviors. Safety programming in the future will likely include efforts directed at individual employee behavior based on Krause's work. Others who have made significant contributions in this behavioral field include Michael Topf, Scott Geller, and Don Eckenfelder. Dr. Geller, in his book *The Psychology of Safety*,[7] has emphasized that the culture of the organization is an important influence on the behavior of the workers. Industry focus on behaviors is an important trend. These behavior measurements are indicators of potential downstream injuries and promote safe work habits upstream of the injury.

These behavioral-based safety interventions will be closely allied with renewed interest in employee assistance programs and off-the-job wellness programs. All these efforts have to do with organizational cultures and the beliefs and values held by the individual. As downsizing efforts level out, the recognition of the worker as a valuable asset will re-emerge. This will reinforce behavioral change initiatives in taking personal responsibility for one's actions, leading to improvements in health, safety, and personal well being.

Finally, probably the most important element for the future of occupational safety and health will have to do with globalization efforts. The power of international trade will create momentum for change. As the world becomes smaller, as information flows unimpeded, as world trade and commerce grow, the use of health and safety measures as a competitive advantage will become more evident. The successes of the ISO 9000 quality management standard in the international trade arena will likely be emulated in the environmental management arena by the use of the ISO 14000 standards. Although international efforts to establish similar safety and health management standards have recently been sidetracked, many organizations see value in combining occupational safety, health, and environment into one management system. Once established and certified in some fashion, it then can become a powerful marketing tool with added business value used for a competitive advantage globally. The use of the CE mark is present-day evidence of this trend. Organizations operating in Third World countries with higher levels of risk will find increasing pressures to upgrade safety and health performance in order to be successful internationally.

1.3 SAFETY MANAGEMENT AND THE ORGANIZATION

1.3.1 What Types of Skills Are Required?

The skills required to manage the safety and health effort in an organization depends on many factors. What hazards and risks are present in the organization? What are the regulatory requirements imposed by governmental agencies? What type of technology drives the organization? Does the job require the management of safety and health professionals? Does it require skills to be able to influence operating managers? Does it require technical skills for input into equipment and facility design? Are legal interpretations of regulatory issues an important part of the job?

In the past, some considered the practice of safety as simply following common sense. The early days of safety provided easily observable situations where corrections were clear and obvious. Early attempts at safety also included safety contests, safety slogans, and safety posters. This gave the early impression that safety and health was a game and that anyone could do it. Later came the three Es of safety: engineering, education, and enforcement. Some added a fourth E: enthusiasm. Applying these elements would solve any safety concern! We now know that these were simplistic approaches to establishing a robust and effective safety and health management system.

Today's safety and health practitioner has to face and solve a variety of complex problems with new and more effective tools. The skills required to apply those tools are now recognized a being multifaceted. Several professional certifications have evolved. The American Industrial Hygiene Association has adopted codes of conduct for its members. They require that their members "Practice their profession following recognized scientific principles with the realization that the lives, health, and well being of people may depend upon their professional judgment . . . ."

Similarly, the American Society of Safety Engineers has a code of ethics and allows for certification provided that the individual has a bachelor's degree in safety from an accredited institution (or other options), has four years of professional safety practice meeting selected criteria, and passes the Safety Fundamentals and Comprehensive Practice Examination. They describe the safety professional as a "person engaged in the prevention of accidents, incidents and events that harm people, property or the environment. They use qualitative and quantitative analysis of simple and complex products, systems, operations and activities to identify hazards. . . . Besides knowledge of a wide range of hazards, controls, and safety assessment methods, safety professionals must have knowledge of physical, chemical, biological and behavioral sciences, mathematics, business, training, and educational techniques, engineering concepts, and particular kinds of operations. . . . "

The degree to which the safety and health practitioner needs to develop and apply any one of these skills areas of course depends on the nature of the hazards and job in the organization. Corporate downsizing has led the safety practitioner to become less a doer and more an enabler. This requires the application of the skills of facilitation, advocacy, and being a team or group leader. The danger with this trend is the potential dilution and/or dissolution of the professional practice of safety and health. Would this be considered a sound management approach with other professional practices such as biology, accounting, or engineering?

1.3.2 Where in the Organization?

In some organizations, the Corporate CEO considers him/herself as the chief safety officer. In this case, management style has evolved to the point where it is recognized that safety starts at the top. Other large organizations have Senior Vice Presidents of Health, Safety and Environment with professional health and safety staffs working centrally or indirectly through line organizations. In many small or medium enterprises, the person handling safety and health issues wears a variety of hats such as human resources or facilities management. Evolving organizational thinking has health and safety integrated into strategic business units wherein all the organizational needs are available within the group. Some have supplemented this with an overlay of a small centralized matrix of functional experts.

Most organizational experts will advise that health and safety needs to be fully aligned with all other aspects of the organizational structure. Establishing a strong centralized health and safety function that manages in a top-down way obviously does not make sense for an organization that works in a fully decentralized business unit fashion.

In any case, wherever health and safety fits into the structure, in order to be successful, health and safety activities must be fully aligned with the business goals of the organization. The health and safety effort must provide clear and measurable business value. Thoughtful discussions with key parts of the organization to define mission as well as safety and health goals is a good first step in alignment. Providing carefully reasoned and professionally articulated health and safety advice up and down the organization are the next steps to a successful value-added effort.

1.3.3 Effects of Organizational Culture

Recent writings in the health and safety literature have postulated that the most important aspect of a successful safety and health program has to do with the organization's core values. What beliefs and values are held inviolate? When economic stress stretches the organization, which activities will continue? Those organizations that are successful in providing substance, meaning, and measurable definitions to safety and health values have ongoing successful safety results. In Tom Peters' book,[8] *A Passion for Excellence—The Leadership Difference*, he describes the DuPont experience. "DuPont's fetish provides a marvelous opportunity to see all the levers—stories, language, attention—pointed in one direction. Safety is pervasive at DuPont, defying all formal categorization. . . . Virtually even meeting, regardless of the subject matter, begins with a report on safety. . . . Any accident . . . is automatically on the chairman's desk within twenty-four hours . . . ."

Deal and Kennedy[9] suggest that values are the bedrock of any corporate culture and that they provide common direction for all employees and guidance for their day-to-day behavior. Further, shared values define the fundamental character of the organization, creating a sense of identity and making employees feel special. They become the essence of the organization's philosophy.

"The culture clearly announces every day to every worker whether safety is a key value and where it fits into the priorities," says Dan Peterson.[10] He continues, saying that "It dictates how an employee will act and how they will be treated. . . . It dictates whether elements of a safety system will work or flop."

Why do companies having the same elements in their safety programming differ so greatly in the results given the same efforts? Geller[11] says that successful organizations have "safety as an unwritten rule, a social norm, that workers follow regardless of the situation. It becomes a value that is never questioned—never compromised."

The challenge for any organization, then, is to determine what its core values are and decide what they want them to be. Changing those values involves confronting widely held beliefs throughout the organization. Is it okay to decide that safety and health is not a core organizational value? Is it okay to be less than world class in safety performance? Can the safety and health professional expect and/or accept less than safety excellence? How about top management? How about the workers in the plant? To what extent can and should the safety professional be an agent for change? These are the issues that will actually determine the degree to which the advice and recommendations provided in the following chapters is implemented and safety excellence achieved!

1.4 SAFETY AND HEALTH MANAGEMENT SYSTEM ELEMENTS

1.4.1 Introduction

The early history of health and safety management is replete with nonconnected and nonrelated activities. We still find safety programs solely managed via posters, videos, and gimmicks. A serious approach to designing a well-structured approach to safety management really starts with H. W. Heinrich in the 1930s. He proposed a rationale for why accidents happen and proposed some concepts for accident prevention. His ten axioms of industrial safety included accident prevention activities of engineering revision, personnel adjustment, persuasion, and appeal, as well as discipline. Although it was a simplistic approach, it nevertheless led to concepts that are still around today. Many will remember the three Es of safety: engineering, education, and enforcement.

Today, most safety and health professionals will advise that the management of safety and health elements needs to be directly aligned with the existing management structure of the organization to be successful. Managing this part of the operations differently can lead to a disconnect and a less than successful outcome. A fully integrated health and safety management program should be the goal of any safety practitioner. The Plan, Do, Check, Act elements popularized by Deming have evolved into a variety of approaches focused on health and safety management, as will be described below. The challenge for the safety and health professional is to ensure that all these elements are in some way integrated into the organization's management plan.

Assessing the effectiveness of the existing organization against the desired management system is the first step in determining the required next steps. Dan Petersen[12] suggests an easy way to assess a safety system: "Ask the people who know—the hourly employees. They'll tell you if it's world-class or if it sucks."

1.4.2 OSHA's Proposed Safety Management System

OSHA's 1996 draft proposed Safety and Health Program Standard contains five core elements. These are: (1) management leadership and employee participation, (2) hazard assessment, (3) hazard prevention and control, (4) training, and (5) evaluation of program effectiveness. Under management leadership, the proposal requires that the employer take responsibility for managing safety and health by identifying at least one manager to initiate corrective actions where necessary as well as ensuring employees participate. Hazard assessment requires documenting workplace inspections of certain frequencies and investigations of injuries. Hazard prevention includes the identification of workplace hazards in equipment materials and processes and controlling them, as well as training employees at certain frequencies. Finally, the proposed standard requires the employer to evaluate the effectiveness of the program.

In the 1980s, OSHA had established the Voluntary Protection Program, designed to recognize and promote safety and health management in the workplace. Most of the elements described above were included as criteria for participation in the program. Although the effort started slowly, governmental efforts towards more friendly partnerships with industry has led towards more participation.

1.4.3 OSHA's Process Safety Management System

This is an example of a management system designed to prevent catastrophic incidents in the chemical process industry. This performance-oriented standard has eleven elements:

1. Employee participation—participate in hazard reviews and have access to safety information
2. Process safety info—document hazards, technology, diagrams, and good engineering practice
3. Process hazard reviews—formal analysis such as HAZOP, fault tree, what if, job safety analysis
4. Operating procedures—written procedures, including startup, normal, shutdown, and emergencies

3. Training—in procedures with documentation initially and every 3 years
4. Contractors—Development and evaluation of contractor safe work performance
5. Prestartup review—for all new and modified facilities and equipment
6. Mechanical integrity—documented preventive maintenance procedures
7. Hot work permit—procedures for using open flames in hazardous locations
8. Management of change—formal method of management signoff for changes in process
9. Incident investigations—for all serious or potentially serious incidents and/or accidents
10. Emergency action plan—documented plan for addressing emergency situations
11. Compliance audits—Evaluation of effectiveness of management system every 3 years

1.4.4 ISO 14001—Environmental Management System

The success of the international quality management standard ISO 9000 and the impetus of the global environmental movement gave rise to the development of the ISO 14000 Environmental Management Standards. Many expect that an ISO 14000 certification will provide a competitive advantage as well as environmental improvements worldwide. The standards are closely aligned with Deming's Plan, Do, Check, Act elements. A recent movement to create a similar international health and safety management standard was defeated. Those organizations that wish to become ISO 14000 certified, however, should seriously consider adding health and safety elements into a completely integrated health, safety, and environmental management system.

The ISO 14001 standard includes six major elements:

1. General requirements—specifies the following elements
2. Environmental policy—appropriate to the organization, commitment to comply with regulations, sets, and reviews objectives and targets, and is documented
3. Planning—regarding environmental aspects, legal requirements, objectives and targets, designation of responsibilities
4. Implementation and operation—establishing requirements, reporting performance, providing training, communicating, documents, operational, and emergency procedures
5. Checking and corrective actions—for monitoring and measuring on a regular basis, evaluating compliance, acting on nonconformance, auditing, and keeping records
6. Management review—regular review of the organization's management system

1.4.5 AIHA—Occupational Health and Safety Management System

In 1996, the American Industrial Hygiene Association proposed an "Occupational Health and Safety Management System: An AIHA Guidance Document." It was developed in response to the ISO 9000 Quality Management success and potential development of an international Safety and Health standard similar to the ISO 14000 Environmental Management Standard described above. The guidance document is formatted similarly to the ISO 9000 standard and contains the following overall requirements:

4.0	General	Requiring the establishment of an occupational health and safety management system (OHSMS)
4.1	Management responsibility	Requiring the establishment of (OHSMS) policy, responsibility, authority reviews and resources
4.2	Management systems	Requiring a documented OHSMS with procedures, planning, and performance measures
4.3	Reviews	Requiring determination of conformance with laws, policies, and goals and objectives with continuous improvement
4.4	Design control	For workplace situations by qualified personnel with safety reviews
4.5	Document control	Requiring procedures for establishing and controlling OHS documents and changes
4.6	Purchasing	Requiring review of OHS aspects of purchased goods and contractors
4.7	Communication	To assure effective and relevant OHS communications, exposures, documents
4.8	Hazard trace	For feedstock used in other processes back to the point of manufacture
4.9	Process control	To measure and monitor hazards to minimize or eliminate risks to employees
4.10	Inspection	Of OHS activities and of records, resulting in corrective actions
4.11	Equipment	For control, inspection, and maintenance and calibration of test equipment
4.12	Evaluation status	Sampling data and reporting mechanisms explicitly documented

4.13 Nonconforming processes	Communicate nonconforming situations to management and employees until acceptable state can be achieved
4.14 Corrective/preventive action	Procedures for implementing corrective and preventive OHS actions
4.15 Hazardous materials	For handling, storing, packaging hazardous materials
4.16 Record control	Documenting, securing, filing, storage, and disposition of OHS records
4.17 Audits	Planning and implementing OHS management system audits
4.18 Training	Identifying needs and providing for relevant OHS training
4.19 Services	Ensure outside contractor services are included in OHS activities

1.4.6 BS 8800—British Occupational Health and Safety Management System

In 1996, the British Standards Institute issued a Guide to Occupational Health and Safety Management Systems, designated BS 8800. The elements of the standard are aligned closely with ISO 14000 and the AIHA proposal. One interesting aspect of the British Standard is the explicit requirement for root-cause analysis. As with the other ISO management documents, guidance annexes provide details on ways to actually implement, in a practical way, the standard requirements. Organizations should review these safety and health aspects if they are working towards ISO 14000 environmental certification in an integrated way.

REFERENCES

1. National Safety Council, *Accident Prevention Manual for Industrial Operations*, 1993, Ninth Edition.
2. *Ibid.*
3. Heinrich, H. W., *Industrial Accident Prevention*, McGraw-Hill, New York, 1931, First Edition.
4. Petersen, Dan, *Techniques of Safety Management*, Second Edition, McGraw-Hill, 1978.
5. Petersen, Dan, *The Challenge of Change: Creating a New Safety Culture*, Safety Training Systems, 1993.
6. Krause, Thomas R, *The Behavior-Based Safety Process*, Van Nostrand Reinhold, 1990.
7. Geller, E. Scott, *The Psychology of Safety*, Chilton Book Company, 1996.
8. Peters, Tom, *A Passion for Excellence—The Leadership Difference*, Random House, New York, 1985.
9. Deal, Terrence E., and Kennedy, Allan A., *Corporate Cultures*, Addison–Wesley, 1982.
10. Petersen, Dan, *Establishing Good "Safety Culture" Helps Mitigate Workplace Dangers*, Occupational Health and Safety, July 1993.
11. Geller, E. Scott, *The Psychology of Safety*, Chilton Book Company, Radnor, PA, 1996.
12. Sheridan, Peter J., "The Essential Elements of Safety," *Occupational Hazards*, February 1991.

CHAPTER 2

Management's Roles and Responsibilities in an Effective Safety and Health Program

ROSEMARY COLVIN and RAY COLVIN

Safety Training Dynamics, Inc., 27497 Alabama Highway, Hanceville, AL 35077

2.1 FINANCIAL IMPACT OF ACCIDENTS UPON COMPANY PROFITS
2.2 ELEMENTS OF A SAFETY AND HEALTH PROGRAM
2.2.1 Management Commitment and Employee Involvement
2.2.2 Worksite Hazards Analysis
2.2.3 Hazard Prevention and Control
2.2.4 Safety and Health Training
2.3 ACHIEVING AN EFFECTIVE SAFETY AND HEALTH PROGRAM
2.4 LEGAL RESPONSIBILITIES OF MANAGEMENT
2.5 ROOT CAUSES OF ACCIDENTS
2.6 THE HIDDEN COSTS OF ACCIDENTS
2.7 MEASURING THE EFFECTIVENESS OF THE SAFETY AND HEALTH PROGRAM
2.8 MANAGEMENT'S RESPONSIBILITY TO SELECT COMPETENT SAFETY AND HEALTH PROFESSIONALS
2.9 CONCLUDING OBSERVATIONS
APPENDIX A EXAMPLE OF A COMPANY SAFETY AND HEALTH POLICY STATEMENT
APPENDIX B ROADBLOCKS TO ACHIEVING AN EFFECTIVE SAFETY AND HEALTH PROGRAM
APPENDIX C A COMPANY'S SUMMARY OF OSHA CITATIONS AND FINES

> "We must insist that safety and health be as important to a company as earnings and market share"
>
> United States Secretary of Labor (Statement made after a serious accident with OSHA issued fines of 1.8 million dollars)

According to an advisor at *Dun & Bradstreet*, "ninety percent of businesses that fail are the result of bad management, and the failure of management to respond to change."

Every year, thousands of American workers are killed in the workplace. Tens of

Handbook of Occupational Safety and Health, Second Edition, Edited by Louis J. DiBerardinis, ISBN 0-471-16017-2

thousands of workers are permanently injured. Millions of others have their physical and mental health affected. These accidents result in lost time or reduced production, as well as disrupting personal lives.

Accidents cost American companies and the economy hundreds of billions of dollars every year. Many companies lose their market share and are even forced out of business due to accidents. Many other companies receive "bad" publicity after a serious accident. It may take years to "repair" the financial damage, especially if the accident is one that results in environmental damage. Many companies have a combination of problems due to an accident; for example, the world's largest computer chip manufacturer had its main manufacturing plant shut down due to a "fork truck mishap when an operator *accidentally* hit the main electrical power line." Additionally, electrical power was interrupted for 9,000 local residents. The chip manufacturer put the cost of production interruption into the "millions of dollars" (*Wall Street Journal*, May 25, 1994).

2.1 FINANCIAL IMPACT OF ACCIDENTS UPON COMPANY PROFITS

Years ago, DuPont and the National Safety Council did a study of both the direct and indirect costs of employee lost time injuries. In today's dollars, the average cost of each employee injury accident would exceed $20,000.

Generally, the larger property damage and production interruption accidents cost many times more than the personal injuries. Unfortunately, few employers *formally* consider controlling these types of accidents as part of their company employee safety program.

As stated above, the average direct and indirect cost of an employee lost time injury accident is $20,000. This translates into the company having to generate $200,000 in sales/services at a 10% profit margin, or $400,000 at a 5% profit margin, to pay for a $20,000 accident. In some cases insurance will cover a portion of the loss, but over time the company will pay back the loss through increased premiums. Ultimately, the company will pay for all losses directly or indirectly.

The following are typical indirect costs often overlooked when calculating accident losses:

- Lost and/or reduced production until the injured employee returns to work full time
- Long-term physical restrictions (light duty) of employee's work activities from causes such as carpal tunnel or repetitive work motion problems, back injuries, strains, etc.
- Reduced production until a replacement employee comes up to speed with the operations, if the employee's duties are done by someone else
- Increased insurance costs (insurance companies add a surcharge to the premium for the money they pay out over and above the basic insurance contract)
- The total cost of the time lost by the injured employee beyond the actual time at home away from the job. For example, after the employee returns to work while being treated, examined, filling out questionnaires, talking about injury with supervisor or accident investigation team, short-term light duty, etc.
- Overtime for other employees who must perform injured employee's duties, if a replacement cannot be obtained

- OSHA fines as a result of violations involved in the employee accident; there is a maximum of $70,000 for *each* egregious and/or flagrant violations
- Local or state government fines for violations of their codes and requirements
- Decreased production after a serious accident when workers may become fearful for their safety
- Damage to machinery, equipment, product, or facility during the accident (fork truck, materials handling, process, etc.)
- Production interruption. Accidents may cause production to stop or decrease (sometimes machines can't run until government officials investigate accident or health officials approve use of machine (e.g. blood on food processing machine)
- Delayed filling of production orders or contractual deadlines to complete work
- The cleanup of blood in machine or about the general area (this requires specially trained persons and equipment according to OSHA laws)
- Negative publicity by media can cause damage to the company's reputation. Good employees don't want to work in companies where employees are injured. Larger companies are afraid to do business with suppliers who might not be able to fill orders due to accidents. An accident can also produce negative neighborhood and community attitudes against the company. Since the serious accident to the employee made the TV and or newspapers, community leaders might ask, what about the company's environmental hazards, etc?
- Loss of contracts due to lack of an effective safety and health program, resulting in an above average accident and injury record. Many OSHA laws require the employer and the contractor to share their programs and to ensure that each of their employees is properly trained to recognize and avoid the hazards they may produce. (*Many larger companies are requiring written programs and formal training by the contractors with which they will do business.*)

Whatever the cause of a financial loss to the company, that loss comes out of *profits*. In the present national and international competitive business culture, companies with excellent safety and health programs can better control their overhead costs and be more competitive.

The safety history books are filled with catastrophic accidents that have caused businesses to shut down temporarily and sometimes permanently go out of business due to employee fatalities, fires, explosions, serious chemical reactions, spills, environmental damage, etc.

To understand the connection between employee injuries, property damage, and production interruption better, the following definition of an accident will offer insight into the totality of safety and health activities:

> "An accident is an unplanned event that results in, or suggests the possibility of, personal injury, property damage, production interruption, diminished health, or environmental damage."

This definition includes an employee having a "close call," since an incident that did not cause any injury or financial loss is considered an accident. Take, for example, an employee who slips on a stair, but does not fall down. An accident still occurred; the only thing lacking is the "result" of the accident, in this case an injury.

Most people, and even many safety people, confuse the term of an injury with an accident. An injury is the result of an accident. A fire is the result of an accident; an explosion is the result of an accident. When an employee climbs a ladder to change a light bulb, falls off the ladder, and breaks an arm, the accident was falling off the ladder. The result of the accident was the broken arm (see Fig. 2.1).

All accidents are caused! There is a logical reason why they happen and a logical solution to prevent them.

Often, company safety programs do not include the above definition in their mission statements or consider including management's responsibilities in the company safety and health policy (see Appendix A).

Companies, also, do not understand how the basic safety program relates to and involves management's roles and responsibilities, in preventing production interruption, property, and environmental damage and employee injuries.

The common mission and/or objective of most company safety programs is only to prevent employee injuries, with little effort or activity given toward controlling production interruption and property damage accidents. Typically, these safety and health programs only involve the supervisor and employees in the manufacturing/service area or "only those who can get hurt" (management's definition). Often the less obvious nonmanufacturing hazards to the other employees, such as repetitive work hazards in the office that result in ergonomic problems, including carpal tunnel, stress, back injuries, slip trips and falls, and tight building environmental hazards, are overlooked.

The *root* or basic cause of an employee having an accident that results in a personal injury is the same root/basic cause of an employee accidentally starting a fire, spilling a chemical, or tearing out a main electrical system with a fork truck.

Simply, a company cannot achieve an effective environmental program, or even a total quality management (TQM) program, without first developing a meaningful *basic* safety program that is led by management and gives equal attention to the control of personal injuries, property damage, production interruption, diminished health, and environmental damage. Companies with excellent safety programs led by management have also found it easy to achieve excellent environmental and TQM programs.

Conversely, companies with safety problems resulting in accidents to their employees, damage to their facility, products, and that interrupt production will also have problems with their products' quality, employee moral, absenteeism, meeting schedules, etc. They are all caused by the same problems in the management system. If there is a problem with the company not having an adequate program for the maintenance of machinery, the company will also have problems achieving an adequate program for maintaining employee safety equipment or the guards on the machinery.

A weakness in a management system will affect all elements of company activities. Some will be more apparent than others. An accident that happens that is obviously apparent to everyone is usually one that is costly, or results in a serious employee injury. Not as apparent is the material handler who crushes a product or materials with a fork truck, dents a storage rack, or tears out part of a wall, or a maintenance worker who gets a "small" electrical shock when he/she doesn't lockout the electrical circuit on which he/she is working.

Companies get many "early warning signs" before a serious, costly accident happens. Typically, there are a number of near-miss or minor "accidents" before the serious one occurs. There is a rule of thumb that for every 331 times a safety rule or good practice is

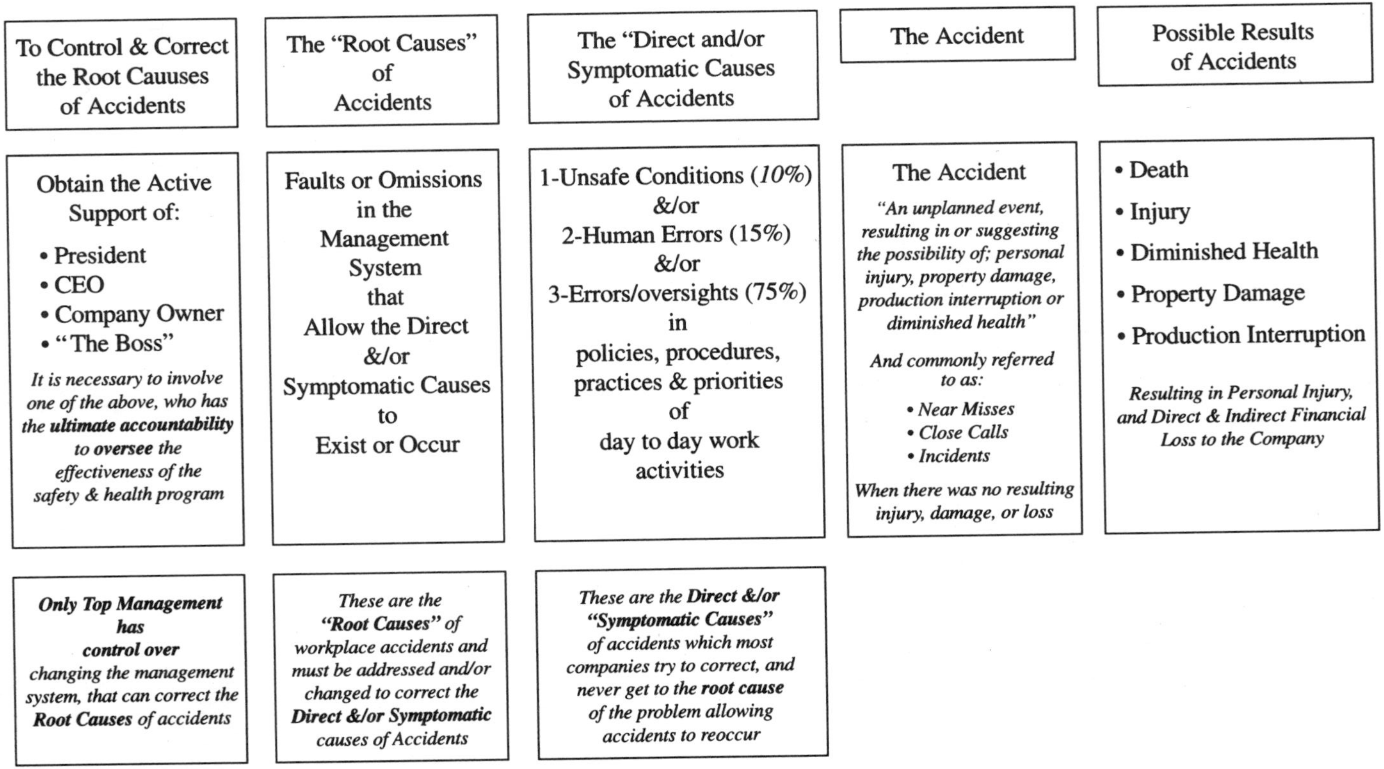

To Control & Correct the Root Cauuses of Accidents	The "Root Causes" of Accidents	The "Direct and/or Symptomatic Causes of Accidents	The Accident	Possible Results of Accidents
Obtain the Active Support of: • President • CEO • Company Owner • "The Boss" *It is necessary to involve one of the above, who has the **ultimate accountability** to **oversee** the effectiveness of the safety & health program*	Faults or Omissions in the Management System that Allow the Direct &/or Symptomatic Causes to Exist or Occur	1-Unsafe Conditions (*10%*) &/or 2-Human Errors (15%) &/or 3-Errors/oversights (75%) in policies, procedures, practices & priorities of day to day work activities	The Accident *"An unplanned event, resulting in or suggesting the possibility of; personal injury, property damage, production interruption or diminished health"* *And commonly referred to as:* *• Near Misses* *• Close Calls* *• Incidents* *When there was no resulting injury, damage, or loss*	• Death • Injury • Diminished Health • Property Damage • Production Interruption *Resulting in Personal Injury, and Direct & Indirect Financial Loss to the Company*
Only Top Management has control over *changing the management system, that can correct the **Root Causes** of accidents*	*These are the **"Root Causes"** of workplace accidents and must be addressed and/or changed to correct the **Direct &/or Symptomatic** causes of Accidents*	*These are the **Direct &/or "Symptomatic Causes"** of accidents which most companies try to correct, and never get to the **root cause** of the problem allowing accidents to reoccur*		

Fig. 2.1 The "Root Cause" of workplace accidents.

violated (such as not using a guard on a machine) 300 times there will be a "close call or near-miss" incident (the employee will get out of the danger zone just in time), 30 times there will be in a minor accident (where the employee will just get a finger scratched or "nipped" in the machine), and once there will be a serious accident (when the employee loses part of a finger in the unguarded machine). This accident ratio typically applies to every serious accident. It indicates that before every major or serious accident, there were "red flags" that were not given attention and continued to get worse over time.

In other words, one can break the safety rules most of the time and get away with it without anything happening; some of the time there will be a minor problem; and one time a serious problem will occur.

It also says that if you can read the "red flags" (the 300 near misses) early on, and you do something about their cause, the serious accident can be prevented.

2.2 ELEMENTS OF A SAFETY AND HEALTH PROGRAM

According to OSHA (OSHA Safety & Health Program Guidelines, 1989) and industry safety and health experts, there are four major elements that make up a basic safety and health program:

1. Management commitment and employee involvement
2. Worksite hazards analysis
3. Hazard prevention and control
4. Safety and health training

2.2.1 Management Commitment and Employee Involvement

This is the most important element in developing a safety program. Without management's active support, such things as safety and health budgets, training, correction of hazards, and employee support of the program would not be effective. Top management sets the pace of the program. Employees will reflect management's enthusiasm, or lack of enthusiasm (see Appendix B).

Without exception, all safety and health professionals, the courts, and government agencies agree that the most important element of the four-point program is management's commitment, without which no safety and health activities could succeed.

Some examples of how top management can support and be involved in the safety and health program are:

- Clearly makes safety and health activities a company priority equal to production, sales, research, and development
- Develops and signs a company safety and health policy statement (see Appendix A)
- Makes safety and health activities part of the company performance evaluation for all levels of employees, supervisors, and managers
- Trains all managers and supervisors about their basic safety and health roles and responsibilities
- Develops a budget for safety and health that captures the cost of accidents and

the cost of controlling accidents and charges them back to each department supervisor

- Includes the annual safety and health activities' cost and summary into the company annual report
- Requires all senior managers to participate in accident investigations within their areas of responsibility and sign off on corrective measures; identifies the costs of accidents and the impact they have on profits
- Requires all senior management to participate in a formal company audit annually to determine the effectiveness of the safety and health program
- Requires managers to participate periodically in their department safety meetings

2.2.2 Worksite Hazards Analysis

ALL workplace "recognized" hazards must be identified, to comply with federal laws. "Recognized" hazards are those hazards that are "typical" to the work being performed. Some activities and "typical" hazard exposures that should be addressed when evaluating workplace "recognized" hazards:

processes and manufacturing
fire/explosion
material handling
repetitive work
criminal activities
explosions
environment
biological
bloodborne
radioactive exposures
electromagnetic
elevated work
confined space
vehicular/traffic
powered moving equip
chemical
research and development
automation/robots
stress
machinery
electrical
human error
violence
contractors
construction
neighboring
handtools
natural disaster
overexertion
heat and cold
slips, trips, and falls
others not listed

2.2.3 Hazard Prevention and Control

Based upon OSHA standards and industry practices all workplace "recognized" hazards should be evaluated by conducting a "hazards analysis." A completed "hazards analysis" is similar to written safety instructions on how an employee will safely perform a task that has a "recognized" hazard. When completed the document will be used to:

1. Define safe work procedures for employees
2. Control and eliminate hazards
3. Justify purchase of employee personal protective equipment
4. Perform employee safety and health training
5. Used in accident investigation

2.2.4 Safety and Health Training

First, senior management should be taught about the basic safety program activities and its roles and responsibilities in the program. Management should know how to audit the program and possibly integrate the results into a "safety pay for performance" evaluation of all company employees. They should also learn the importance of their active participation, and how the success of the program will be directly related to their visible priority for safety. They should understand how the employees will reflect what they perceive is management's safety priority for the safety program. Employees are a mirror of their management.

Second, middle management must be made aware of how they will carry out senior management's priorities for a successful program. They should work with and support supervisors in carrying out their direct responsibilities to the employees. They need to learn their roles and responsibilities and how they will be measured on their support of the supervisors and the program.

Third, supervisors should be trained and motivated. Since supervisors are the people who directly oversee the employees' safety, they must be given the tools to do the job. They must be taught their *importance to the program.* They need the resources to perform all the functions of safely supervising employees, that is, how to identify various types of hazards, resolve employee communications issues, train employees, understand human relations, investigate incidents, and motivate employees to participate in the safety program.

Fourth, employee training teaches the specifics of how to avoid accidents and injuries to themselves and others. The training should include the evaluation of the employees upon completion, in order to ensure that the employee knows the safe way to perform the work.

2.3 ACHIEVING AN EFFECTIVE SAFETY AND HEALTH PROGRAM

To develop an effective basic safety program, a company must start with (1) top management wanting to have an effective safety program, and making it a priority activity equal to all other "important" activities. Its support must be obtained to do so. This is followed by (2) an educational and motivational training program for managers and supervisors, to teach them their roles and responsibilities in the safety and health program.

The next step would be for management (3) to train, and then have supervisors and employees together (4) develop a list of "recognized hazards" in the workplace, and have them (5) perform a "job hazards analysis" for each of the listed recognized hazards. After this is done, (6) management should ensure that the affected employees be taught the safe process and/or procedures to follow in order to avoid the hazards of their work to themselves, to the facility and property, to production, and to the environment. All this is accomplished by management's leadership, working within the company guidelines with the employees and setting the priorities to make it happen.

Companies should develop a list of possible accidents that can have a serious financial impact upon the company and that can be caused by employees, outside contractors, or a third party, such as a utility company, gas, electric, storage or transmission, adjoining company, traffic accident, etc. Employers should identify all potentially catastrophic hazards before a serious accident occurs, and then have an action plan of how they will control or recover from such an occurrence and how they will get back to full production with a minimum of time and expense.

The key element in achieving a successful company safety and health program that addresses all the above issues is by management making the safety and health program and its elements and activities a priority equal to all other company activities.

Unfortunately, safety and health priorities and full management support are rarely given the same importance as production in most companies. The reason for this is that the majority of managers have never been formally taught the benefits of a basic safety and health program in their college management courses and how workplace accidents are controlled. They do not understand the logic of why and how they play a major role in making the company achieve an effective safety and health program.

All levels of managers can support and directly participate in the safety and health program by:

- reviewing and participating in investigating accidents
- periodically attending and supporting the safety and health meetings
- evaluating incident trends
- establishing safety and health goals and objectives
- conducting and participating in safety and health audits and inspections
- attending and participating in safety and health training activities
- encouraging employees to report near misses and close call accidents
- encouraging employees to be proactive in safety and health activities
- creating a working environment that supports safety and health activities
- making safety and health activities a department performance goal

An excellent example of how a senior manager participated and took a leadership role in a safety and health program was when a company chemical department had to experiment with a hazardous, cancer-causing chemical in a new process. The department senior manager put on a chemical safety suit along with the employees, and worked side by side with them throughout the experiment, while the supervisor and other managers observed from a safe distance. There was no question that the senior manager had earned the respect of the department employees by putting himself in "harm's way" alongside the department employees. He is currently a senior officer of the company.

2.4 LEGAL RESPONSIBILITIES OF MANAGEMENT

Legally, the courts and government regulators require management to be responsible for the health and safety of the employees in the workplace. Unfortunately, most managers have never been made aware of their responsibilities according to the law. (Additional legal requirements may be found in Chapter 29.)

OSHA is very clear when they describe management's involvement in the company safety and health program how managers can effectively support the program's activities.

OSHA spells out everyone's role and responsibilities in many of its standards. It looks at substandard employee training, lack of proper safety equipment, or not having identified workplace hazards as part of management's responsibility. OSHA will hold management accountable if a serious accident occurs.

If an employee is seriously injured and it is later determined that the individual manager had prior knowledge about the hazard, especially if there were past accidents and/or close calls reported to the manager by the employees or there were past written safety or health inspection recommendations that were not corrected, leaving a paper trail, it is possible that the manager could be brought up on criminal negligence charges by the local district attorney or even the state attorney general. *In one state, the state Attorney General brought criminal charges against eight management people, including a Vice President, Plant Manager, Department Managers, and Supervisors since it was believed the managers were not totally supporting the safety and health program. (The safety manager was not included since it was felt that he was doing all he was expected to do.)*

(**Note:** A company can/will refuse to support the manager financially in a criminal or legal case, if the manager broke a law or a company policy, for example, by not correcting a known workplace hazard.)

There also has been a case where a manager directed a warehouse employee to use a fork truck to move some materials. The employee had not been trained to drive the fork truck. The supervisor/manager failed to check if the employee was authorized and qualified to drive a fork truck. If the employee had an accident and suffered a serious injury, or had injured another employee, it is possible that the manager/supervisor could be brought up on criminal charges. While the employee could have told the manager that he didn't know how to drive the fork truck, typically, employees will do as they are told, assuming that the manager knows what he/she is doing and would never tell him to do something that might cause an accident. *Supervisors and managers cannot assume what would appear to be obvious, in this case a warehouse employee who was not qualified to drive a fork truck.*

2.5 ROOT CAUSES OF ACCIDENTS

When management understands the basic and "root" causes of accidents, it will be in a better position permanently to prevent accidents from occurring in the workplace.

Many company safety and health programs only address the symptomatic or direct causes of accidents and never really identify the "root" causes (see Fig. 2.1).

Changing a management system that can allow accidents to happen is not difficult. For example, if an employee selects a defective ladder to change a light bulb in a ceiling lighting fixture, and falls off the ladder, breaking an arm, typically, most company corrective measures would recommend destroying or repairing the defective ladder. Some corrective measures might even "blame the employee" for the accident and recommend disciplinary measures. These measures address only the direct or symptomatic causes.

Seeking to identify the "root" cause would force the question of what had failed in the management system that allowed the accident to happen? For example:

- How did an unsafe ladder get into the company? (*What procedures or policies control the use of ladders?*)
- Wasn't there a ladder safety inspection program? (*What kind of a maintenance program is there for ladders?*)
- Was the employee properly trained to recognize an unsafe ladder? (*Were employees trained to recognize the hazards of their job?*)
- Was the employee pressured to use the unsafe ladder? (*Did management or super-*

visors "rush" the job due to production demands that forced the use of an unsafe ladder?)

Addressing the "root" cause would necessitate having a ladder safety maintenance and inspection schedule, according to OSHA standards. Ladders would be identified by a number and be periodically safety inspected and kept under lock and key to prevent unauthorized use. Employees who need to use ladders would be adequately trained and retrained to recognize the hazards and how to use the ladders safely. Employees would be medically approved to work at heights, especially as they get older.

It should be obvious that addressing the root causes is significantly better than just repairing or destroying the ladder or blaming and disciplining the employee. Blaming the employee normally results in a disgruntled employee and sends a message to other employees that management will blame them when there is an accident.

Correcting the "omissions or faults in the management system" that allow accidents to occur is typically done by using the same methods management uses to resolve production problems.

Approximately 10% of workplace accidents occur because of unsafe conditions, processes, or facilities. 15% of are due to "employee or human error," and 75% are due to "errors, oversights, or omissions in policies, procedures, practices, and especially priorities in day to day business activities.

(Data are based upon the author's 40+ years of safety and health experience, 15 years in the insurance industry as a safety consultant, and 25 years in private industry and consulting, studying accident causes and their control methods.

Since the 1930s accident causes have been placed into two categories, unsafe conditions (12%) and unsafe acts (88%). Typically, employees were identified as the cause of the accidents, even when there was an unsafe condition ("He should have known better, or he should have been aware."). Employees were disciplined, which often resulted in the employee and other employees disliking management for always blaming them for accidents. This would condition/strain employees not to report future accidents or even minor injuries.

Over the past 60 to 70 years as safety professionals obtained more formal management skills, education, and training and have worked within the management organization, the "root" or basic causes of accidents became more focused on the management systems, policies, procedures, practices, and/or priorities that control the day to day work activities of employees.

In today's business culture, in order to correct a safety or health problem permanently, it is common to examine the company management system to identify the root cause of the problem, and then to use acceptable management techniques to correct the problem. For example, repetitive work hazards are primarily due to controlling workplace hazards and controlling employee exposures through the administrative process of limiting the time exposed to the hazards, shifting jobs, and allowing for stretch breaks during the workday. It is management's policies, procedures, practices, and priorities that control the 75% of workplace accidents.

2.6 THE HIDDEN COSTS OF ACCIDENTS

In one major company shipping and warehouse accident losses exceeded 100 million dollars a year. In another company, a fork truck operator ran the forks into a 55 gal drum of specialty chemicals that cost over $50,000 to manufacture, and also added the extra environmental cost of cleanup plus all the related administrative costs, which exceeded $100,000. In the same company, fork truck drivers were "dropping" expensive products from the storage racks and placing the product back onto the pellets for shipment to customers. The customers would only find out about the damaged product when they used the product and it didn't work. While the company paid for the returned damaged product, many customers no longer used that brand of product. How is a dollar figure placed upon a lost customer?

Many companies hire entry-level persons to work in the shipping and receiving areas, putting them in charge of expensive raw or finished materials. There was another case in which, after many years of research and development, a department had developed specialized space age equipment. This equipment was dropped and seriously damaged in transportation, because an undersized fork truck attempted to lift it onto a flatbed tractor trailer. In this case, it was not considered the employee's "fault." However, managers had ordered the employee to use the undersized fork truck to move the equipment. (*Management said they would accept "responsibility" if anything went wrong.*) Fortunately, no one was injured, but the accident was costly. The multimillion dollar project was behind schedule to start with and the equipment was needed a few thousand miles away as part of a larger military and research activity. Many companies and people were involved at all levels.

How much did this accident cost? Officially, no one really calculated the loss. (It must have been too embarrassing to even consider documenting.) It is uncertain in the management circles of the company and the other management teams involved in the project if it was ever discovered that the local management "accepted the responsibility" for directing the employee to use a undersized fork truck. Suspicion has it that the story went something like "the employee used a wrong size fork truck."

Many times management does not understand nor attempt to evaluate the potential financial impact a government regulatory agency could have upon the company and its bottom line profits. While the citations could add up to sizable amounts, the additional costs of making the recommended safety and health changes to the facility, equipment, processes, training, personal protective equipment, and public relations could be many times the original citation fine. (See Appendix C, Summary of Citations, indicating how one company received $1,803,500 in safety and health penalties.) At a 10% profit margin the company would have to generate some serious product sales to pay for this multimillion dollar "loss."

At the time OSHA cited the company whose fines are described in Appendix C for $1.8 million, which was not contested by the company, the United States Secretary of Labor said in part, "We must insist that safety and health be as important to a company as earnings and market share." In this case the cost to comply with the OSHA recommendations will run into the "millions of dollars" and have a major impact upon company profits.

A classic case of an accident ultimately resulting in a company going out of business was in an airline fast food processing and packaging company. An employee cut a finger that bled into the food processing machine. The employee required a few stitches. The machine had to be shut down and a city health inspector had to approve the restarting of the machine. The result was the machine being shut down for a few days, and contracted packed foods could not be delivered to the various airlines. As the company was operating on a tight financial margin, the single accident had a serious impact upon the company. Within a year the company had lost some of its major accounts and went out of business. The cost of the employee injury was less than $1,000. The "bottom line" cost of this accident resulted in the company going out of business.

2.7 MEASURING THE EFFECTIVENESS OF THE SAFETY AND HEALTH PROGRAM

Senior management should want to "measure" the company safety and health activities to evaluate its success (see also Chapter 27). The first and foremost tool to evaluate is the same tool used to evaluate all company activities. Money. . . . A safety budget shold be developed and followed throughout the company. Safety and health line items should be captured in each department's financial statements. The costs of accidents and incidents should be identified (both the direct and indirect costs). Corporate costs to control accidents (i.e., personal protective equipment, machine guarding, special safety and health instrumentation, equipment, etc.) should be charged back to each department to get the real cost of operating the department. (There are many departments that "get the job done," but at what cost? Has the department accomplished the job, but cost the company excessive losses paid out of "the maintenance or production department's budget" due to equipment, facility and product damage? For example, the shipping department keeps the product going out the door, but damages the shipping containers and even the product, the fork trucks, the over the road trucks, the warehouse storage racks, sprinkler system, electrical wiring, electrical control panels, etc. While they continue to get the product out the door, they have caused major financial losses to the company during the process.

An interesting exercise is to do a "damage evaluation audit" of the material handling department. Document the cost of damaged (1) warehouse storage racks, (2) fork trucks, (3) power materials moving and handling equipment, (4) damage to the doorways, (5) door frames by the loading docks, (6) electrical equipment, (7) battery charging equipment, and of course the (8) product, in whatever stage it is moved.

2.8 MANAGEMENT'S RESPONSIBILITY TO SELECT COMPETENT SAFETY AND HEALTH PROFESSIONALS

How would you answer this question if it were asked after a serious accident in your company, when you are in a courtroom or at an OSHA hearing, or possibility before a District Attorney? "What are the safety and health and management qualifications of the person you have representing you as your company safety/health person?" Can he/she possibly work effectively within your management system to institute all the governmental safety and health regulations and standards required for your company and to manage a day to day, effective company safety and health program?

In today's ever-changing business culture, prudent managers are now requiring their safety persons to obtain a *formal* safety and administrative management education and the training necessary to fulfill the requirements of the safety and health position. No longer can the "safety inspector or safety engineer" of yesteryear be effective in the legal and regulatory environment of our present day business world, which now requires management to participate actively in the safety program (Refer to OSHA Guidelines for a Safety & Health Program—1989 publication).

While the driving force of an effective safety program should be good moral business practice and not the threat of criminal negligence, we have to be realistic and understand that we are in an era of "regulatory *and* legal enforcement," where it is important for companies to document a formal "paper trail" identifying exactly what management is doing to control accidents in the workplace. Corporate management would be wise to take a good hard look at the qualifications of the person it has "managing" its safety program. While, as a corporate executive, you can delegate your safety responsibilities to someone else, you can never delegate your legal accountability for the safety and health of the persons you oversee, manage, or supervise. If a person is selected who does not have adequate qualifications to be the company safety and health person, it is management who did not hire a qualified person. It is not the unqualified person's fault. If the scope of the job is beyond the capabilities of the safety and health person, and management dose not give the person the help they need to perform the job, it is not the person's fault, it is management's for not giving him/her the resources to do the job properly.

Every safety and health person in your organization should have some formal training or education which can be used to prove that your senior management has a "competent" individual representing it. This safety and health person will carry out your management's safety and health responsibilities and will, thereby, meet your legal accountability for the health and safety of your employees. At the top of the safety and health professions are the "Certified Safety Professional" and "Certified Industrial Hygienist," who by demonstrating their knowledge and experience have achieved this designation of their qualifications. They would be considered "competent" professionals.

Management will generally search for only the best engineers to design and manufacture its products. Management also hires the best available doctors, lawyers, accountants, financial advisors, and advertising professionals to handle their particular functions of the company business. Isn't the managing of the safety and health program that controls such things as the prevention of serious accidents involving employee deaths and injuries, fire damage, production interruption, etc., including the compliance with government regulations, and all other safety and health laws and standards, just as important a position in the company?

The courts and government regulatory agencies won't think so!

2.9 CONCLUDING OBSERVATIONS

> Are we to leave our children a country ridden with accidents and their corresponding burden of human and economic loss? or can we prove that, as Americans, we can progress, both technically and morally, to provide our citizens with a life and work style worthy of the sufferings, sacrifices and expectations of our founding fathers?
>
> Our past national safety record does not show that our country is heading in a direction that will significantly reduce this needless and wasteful result of our affluence, and its impact upon ours and our children's future. Only when our business, political, and government leaders join together to make safety and health a joint priority issue will we be able give our citizens the assurance of a safe and healthful country in which to work and live. To do this we must start by addressing the root causes of all accidents and not just the symptomatic causes that allow accidents to continue to be repeated as they have over the past 100 years since the Industrial Revolution.
>
> From an address to the New York University, Center for Safety, Alumni, 1959, by Raymond J. Colvin, Sr.

Since the 1959 address at New York University:

- 5,800,000 Americans have been killed in all accidents
- 92,500,000 Americans have experienced 4,800,000,000 injuries

Note: Annually, one in three Americans (men, woman, children, young, and old) will have an injury so severe they will lose a day from work or school or go to a hospital or doctor for treatment, due to an accident at work, at home, while traveling, or at play or at school.

The total cost of accidents since 1959 has exceeded $5,700,000,000,000.

APPENDIX A

EXAMPLE OF A COMPANY SAFETY AND HEALTH POLICY STATEMENT

COMPANY SAFETY AND HEALTH POLICY

Date ________________

The health and safety of our employees, contractors employees, guests and community are a major responsibility of our management and all employees.

Also, as an organization, we have a responsibility to our families, community, nation, and stockholders to maintain a cost-effective production, by not having any accidents that can injure employees, cause fires, interrupt production, or damage the facility or the environment.

As the (CEO, President, Plant Manager, etc.) I consider myself as having the primary responsibility to make safety and health activities a priority, and will oversee the success of the total program and its activities.

Managers, supervisors, and employees will be performance evaluated on their roles and responsibilities according to their level of accountability in making the programs successful.

Chief Executive Officer—President or plant manager's signature ________________

APPENDIX B

ROADBLOCKS TO ACHIEVING AN EFFECTIVE SAFETY AND HEALTH PROGRAM

- Most companies honestly do not know or understand what needs to be done for the safety of employees, production, or the facility. (Why should they know anything about managing safety/health programs when they have never had any formal safety management education?)
- If there is a strong enough incentive, company management will want to learn what needs to be done for the health and safety of employees. (After a serious accident, management tends to "pay more attention" to safety and health activities.)
- Many times when a company's senior management changes, a good safety and health program can regress, get worse, and run into serious safety and health problems. (This is because the "new" senior management may feel too comfortable with the "good safety record." They will not give the safety and health activities the support that was developed over time by the "old" senior management. The safety person will normally have to "resell" safety and health as a priority, and in many cases will fail to do so.)
- Employees' attitudes can be changed if management changes its attitudes and priorities. (Employees are very willing to change if allowed. Supervisors are more reluctant to change. Managers, especially middle managers, are the most reluctant to change, and are normally found to be the "roadblocks" to an effective safety and health program.)
- Employees generally know the workplace hazards, once they are trained to recognize them. It is then up to management to work with the employees to identify the workplace hazards. (It is really the employees who have first-hand knowledge of where the workplace hazards are, and in many cases how to control or eliminate them. It is up to management to involve the employees in the safety program.)
- A small number of companies do not want employees to be knowledgeable about workplace hazards for fear of potential workers compensation claims. (Management will resist having employees participate in the hazards assessment process, because it might make them aware of the workplace hazards, and might use that knowledge to "fake" a claim.)
- Other company managements are afraid that if employees find workplace hazards, the employer will be forced to do something about them. (Normally, this means spending money to correct them. Another employer attitude is, "out of sight, out of mind.")
- Another group of employers actually believe that, legally, it is better not to find a hazard than to find the hazard and do nothing about it! (Unfortunately, some employers receive this suggestion from their legal counsel.)

- Many companies have not *formally* identified and assessed the workplace hazards. (They have not formally identified or assessed hazards because they do not understand the benefits of doing the hazard assessment, nor do they understand the process.)
- Employees will reflect their management's safety attitudes and priorities. (If management has a weak or bad attitude about safety, this will be reflected in the employees' safety behavior.)
- The better managed safety and health programs are ones where the safety and health person reports directly to the most senior manager. (The further down the ladder to whom the safety person reports is a reflection of top management's attitude toward the value of safety in the organization.)
- Many safety and health professionals experience vacillating support from management. They normally have a good working relationship with employees, but find support for the safety and health activities weak from supervisors and managers. Top management is generally very supportive of the safety and health programs, but they typically respond to the observations and suggestions of middle management, whose priorities do not normally include the company's safety and health programs. (This vacillating support frustrates safety and health professionals, especially when policies, procedures, and day to day work practices are affected by production quotas or production problems when managers and supervisors feel it is acceptable to "bend" the safety rules. Safety and health professionals have a very difficult position, which is to get employees to change their behavior, and management, through the supervisors, to help make it happen by paying for the changes.)

APPENDIX C

A COMPANY'S SUMMARY OF OSHA CITATIONS AND FINES

After Having a Serious Workplace Accident

- Failure to provide training to ensure that the purpose and function of the energy control program (lockout/tagout) are understood by employees authorized to do maintenance and that the knowledge and skills required for the safe application, use, and removal of the energy controls are acquired by the employees, as required by the lockout/tagout standard.
- Failure to provide retraining for other authorized and affected employees when there was a change in their job assignments, machines, equipment, or processes that presented a new hazard, or when there was a change in the energy control procedures, as required by the lockout/tagout standard.
- Failure to have lockout and tagout procedures that clearly and specifically outlined the scope, purpose, authorization, rules, and techniques to be utilized for the hazardous energy, and means to enforce compliance, as required by the lockout/tagout standard.
- Failure to conduct a periodic inspection of the energy control procedure at least annually to ensure that the procedure and the requirements of the lackout/tagout standard were being followed.
- Failure to lock out hydraulic pumps and power apparatus when maintenance was being performed on the hydraulic forging press, as required by the lockout/tagout standard and the forging standard.
- Failure to provide training to ensure that the purpose and function of the energy control program (lockout/tagout) are understood by employees working in the area or affected by the maintenance operation and that the knowledge and skills required for the safe application; use and removal of the energy controls are acquired by the employees, as required by the lackout/tagout standard,
- Employer failed to meet requirement of Section 5(a)(1), the general duty clause of the Occupational Safety and Health Act and provide a place of employment free from recognized hazards that were causing or likely to cause death or serious physical harm. Employees were exposed to the hazards of an unexpected release of nitrogen at 5,000 psi or greater pressures. The employer failed to ensure that pressure vessels were not isolated from the pressure gauge. The employer also had the pressure relief valves set too high and the pressure relief valves discharged in employee work areas.
- Failure to train production employees in emergency evacuation procedures.
- Failure to ensure the adequacy of employee owned protective equipment.

- Failure to post warnings about the existence, location, and danger of permit-required confined spaces.
- Failure to train members of the rescue service about the equipment necessary to rescue persons from confined spaces.
- Failure to train employees assigned to the rescue service about their duties in rescuing persons from confined spaces.
- Failure to provide hardware necessary to isolate, secure, or block machines from hazardous energy sources.
- Failure to meet requirements for filling out the log and summary of occupational injuries (OSHA Form 200 or its equivalent) and for providing supplementary records for each occupational injury or illness in detail (OSHA form 101).

Total penalties $1,803,500

CHAPTER 3

How to Locate Information Sources for Occupational Safety and Health

ROBERT HERRICK and JAMES STEWART

Department of Environmental Health, Harvard School of Public Health, 665 Huntington Ave., Boston, MA 02115

3.1 INTRODUCTION
3.2 INFORMATION ON HAZARDS, THEIR SOURCES, AND PROPERTIES
  3.2.1 The OSHA Hazard Communication Standard
  3.2.2 Governmental Resources
  3.3.3 Sources of Published Information
3.3 ELECTRONIC DATA SOURCES
3.4 RESOURCE HOTLINES
3.5 CONCLUSION
REFERENCES

3.1 INTRODUCTION

In order to prevent workplace disease and injury, occupational safety and health practitioners need information on potential hazards, and the work activities that bring people into contact with them. The range of occupational hazards and working conditions is extremely broad; so safety and health professionals rely upon a set of information resources that is constantly changing and expanding both in content and in means of access. This chapter summarizes these information resources, recognizing that any such list will be continuously updated.

The risk of adverse effects resulting from occupational exposures has two primary determinants: the nature (extent and intensity) of human contact, and the inherent toxicity or hazard of an agent or factor. Information on both aspects of risks is essential to programs for prevention of occupational disease and injury, as these programs are founded upon the recognition, evaluation, and control of exposures to hazardous agents and conditions. The practice of prevention has been extended to include a step that is in some ways preliminary to hazard recognition: the anticipation of hazardous exposures and conditions before they actually occur. The anticipation of potentially hazardous con-

Handbook of Occupational Safety and Health, Second Edition, Edited by Louis J. DiBerardinis, ISBN 0-471-16017-2

ditions, along with intervention to prevent exposures and resulting diseases and injuries, is a classic public health approach. Information resources are an essential part of this practice of primary prevention.

Hazard anticipation and recognition involve a systematic review of the occupational environment to identify exposures and potentially hazardous working conditions. This review should include information on the materials used and produced, the characteristics of the workplace including the equipment used, and the nature of each worker's interaction with the sources of workplace hazards. Specific information can be obtained on the raw materials used in a workplace activity, the materials produced or stored, and the byproducts that may result from the process. Hazard recognition also includes gathering information on the types of equipment used in the workplace, the cycle of operation and/or frequency of exposure, and the operational methods and work practices used.

3.2 INFORMATION ON HAZARDS, THEIR SOURCES, AND PROPERTIES

There are many sources of information on hazardous properties of materials and conditions in the workplace environment. These reviews and evaluations are prepared by private organizations as well as government agencies in the United States and internationally. Chemical hazards in the workplace are the subject of a particular regulation, the OSHA Hazard Communication Standard.

3.2.1 The OSHA Hazard Communication Standard

The OSHA Hazard Communication Standard (HCS) provides a valuable information resource for hazard identification and evaluation. This standard requires employers to: (1) develop a written hazard communication program, (2) maintain a list of all hazardous chemicals in the workplace, (3) make available to workers Material Safety Data Sheets (MSDS) for each hazardous chemical, (4) place labels on containers reflecting chemical identity and handling precautions, and (5) provide workers with education and training in the handling of hazardous material. The HCS is based upon the concept that employees have both a need and a right to know the hazards and identities of the chemicals they are exposed to when working. Chemical manufacturers and importers must evaluate the hazards of the chemicals they produce or import. Using that information, they must then prepare labels for containers, and more detailed technical bulletins, the MSDS. Chemical manufacturers, importers, and distributors of hazardous chemicals are all required to provide the appropriate labels and material safety data sheets to the employers to which they ship the chemicals. The role of MSDSs under the rule is to provide detailed information on each hazardous chemical, including its potential hazardous effects, its physical and chemical characteristics, and recommendations for appropriate protective measures.

3.2.2 Governmental Resources

The U.S. National Institute for Occupational Safety and Health (NIOSH) prepares a range of documents including criteria, reviews, and recommendations on hazardous occupational exposures and working conditions. These are not legally enforceable themselves, but NIOSH recommendations are transmitted to OSHA, where they can be used in promulgating legal standards. NIOSH resources are available from several sources: The NIOSH home page on the Internet (http://www.cdc.gov.niosh) and the NIOSH toll-free telephone number (1-800-35-NIOSH) are convenient guides to NIOSH documents.

The Agency for Toxic Substances and Disease Registries (ATSDR) of the U.S. Department of Health and Human Services has developed over 200 toxicological profiles for compounds commonly found at hazardous waste sites. Information on this Hazardous Substance Release/Health Effects Database and other ATSDR documents including Case Studies in Environmental Medicine, A Primer on Health Risk Communication Principles and Practices, Hazardous Substances and Public Health, and ToxFAQs summaries on hazardous materials is most readily obtained through the ATSDR homepage (http://www.atsdr1.atsdr.cdc.gov:8080).

The National Institute of Environmental Health Sciences (NIEHS) of the U.S. Department of Health and Human Services prepares an Annual Report on Carcinogens, which reviews and evaluates information on evidence of carcinogenicity. The report provides a listing of chemicals classified on the basis of the strength of the evidence of carcinogenic risk. This and other information is available from the NIEHS Environmental Health Clearinghouse (1-800-NIEHS94, http://ehis.niehs.nih.gov).

Several international organizations review scientific information for purposes of evaluating risks resulting from human exposure to chemicals. The International Agency for Research on Cancer (IARC) prepares critical reviews of information on evidence of carcinogenicity for chemicals (http://www.iarc.fr). The International Programme on Chemical Safety (IPCS) is a joint venture of the United Nations Environment Program, the International Labor Organization, and the World Health Organization. This program develops Environmental Health Criteria Documents, which are summaries and evaluations of the information on toxic effects of specific chemicals and groups of chemicals. These are available from the World Health Organization (telephone 41 22 791 2476, email publication@who.ch, web site http://www.who.ch). The International Occupational Hygiene Association (IOHA) Telephone 44 1332 298 101, web site http://www.ed.ac.uk/ ~robin/ioha.htm) is an international organization of 15,000 members dedicated to the practice of occupational hygiene around the world. The IOHA produces a newsletter, holds conferences, and publishes a survey of occupational hygiene certifications offered in countries around the world.

A site with extensive information on hazards associated with pesticides is the EXTension TOXicology NETwork (EXTOXNET), prepared by the University of California, Davis, Oregon State University, Michigan State University, and Cornell University. This is a comprehensive source of toxicology-related information on pesticides (http://ace.ace.orst.edu/info/extoxnet).

3.2.3 Sources of Published Information

Information on the hazards of workplace exposures and conditions is published in a number of documents. Table 3.1 summarizes the books in print, with a brief description of the contents. This listing is not a complete inventory of publications, but it includes readily available documents that provide information on hazard recognition, evaluation, and control. The descriptive information on each book is abstracted from that provided by the publisher or distributor, and is not the result of a separate review by the authors of this chapter.

TABLE 3.1 Published Documents on Occupational Safety and Health

Workplace Hazards and Toxicology

1994/1995 Chemical Substance Hazard. Assessment & Protection Guide, Joan Henehan (Ed.), 467 pp, 1994

A comprehensive, concise source of information needed to evaluate basic safety issues and make an initial selection of personal protective equipment

Agrochemicals Desk Reference: Environmental Data, J. H. Montgomery, 670 pp, 1993, ISBN: 0-87371-738-4

A reference source for information on chemicals including pesticides, herbicides, fungicides, and other agricultural chemicals

Air Toxics and Risk Assessment, E. J. Calabrese and E. M. Kenyon, 662 pp, 1991, ISBN: 0-87371-165-3

Includes decision-tree methodology for derivation of acceptable levels of ambient air contaminants. The methodology is applied to over 100 specific toxic agents

Casarett and Doull's Toxicology: The Basic Science of Poisons, 5th Ed., C. D. Klaassen (Ed.), 912 pp, 1995, ISBN: 0-07-105476-6

Covers the basic concepts and fundamental principles in modern toxicology including organ system toxicology, specific agent toxicology, and environmental toxicology

Chemical Information Manual, 3rd Ed., OSHA, 400 pp, 1995, ISBN: 0-86587-469-7

A database of chemical information used by OSHA inspectors. Includes over 1,400 regulated substances, with identification synonyms, OSHA exposure limits, a description and listing of physical properties, carcinogenic status, health effects and toxicology data, as well as sampling and analysis information

Common Sense Toxics in the Workplace, I. R. Danse, 262 pp, 1991, ISBN: 0-442-00154-1

A reference on diagnosing, treating, and preventing toxic injuries in the workplace, which explains toxic substances in terms of human health effects

Concepts in Inhalation Toxicology, 2nd Ed., R. O. McClellan and R. F. Henderson (Eds.), 672 pp, 1995, ISBN: 1-56032-368-X

Covers the basic concepts and quantitative approaches in the study of health effects of airborne materials

Dictionary of Chemical Names and Synonyms, P. H. Howard, M. Neal, 2542 pp, 1992, ISBN: 0-87371-396-6.

Includes information on approximately 20,000 chemicals of broadest interest and general use, based upon sources including EPA, National Toxicology Program, and National Library of Medicine

Dictionary of Toxicology, E. Hodgson, R. B. Mailman, J. E. Chambers, 395 pp, 1988, ISBN: 0-442-31842-1

Defines and explains significant terms and concepts in toxicology for those working in the field and in related areas

Documentation of the Threshold Limit Values and Biological Exposure Indices, 6th Ed., ACGIH, 1800 pp, 1993, ISBN: 0-936712-96-1

Provides the basic rationale for the development of TLVs for chemical substances and physical agents and of BEIs for selected chemicals, including a history of the TLV adoption process, and information on the OSHA PELs, NIOSH RELs, and NTP studies; carcinogen designations from various sources

TABLE 3.1 (*Continued*)

Workplace Hazards and Toxicology (*Continued*)
Gardner's Chemical Synonyms and Trade Names, 10th Ed., M. Ash, I. Ash (Eds.), 1309 pp, 1994, ISBN: 0-556-07491-5 Includes over 40,000 tradenames and chemicals, nearly 3,000 manufacturers listed. Entries give descriptions, classification, chemical formulas/applications, and manufacturers
Guide to Occupational Exposure Values—1996 ACGIH, 132 pp, 1996, ISBN: 1-882417-14-3 Companion document to the ACGIH Threshold Limit Values and Biological Exposure Indices Booklet—a reference for comparison of the most recently published values; 1996 TLVs from ACGIH; the OSHA Final Rule PELs, RELs from NIOSH; and MAKs. Includes listing of carcinogens
Handbook of Environmental Data on Organic Chemicals, 3rd Ed., K. Verschueren, 2071 pp, 1996, ISBN: 0-442-01916-5 Covers natural and manmade sources of substances, their uses, and various formulations. Also provides extensive details on each chemical including properties; air, water, and soil pollution factors; and biological effects
Handbook of Pesticide Toxicology, W. L. Hayes, E. R. Laws, Jr., 1576 pp, 1991, Vol. 1; ISBN: 0-12-334161-2, Vol. 2, ISBN: 0-12-334162-0, Vol. 3, ISBN: 0-12-334163-9 Detailed toxicological profiles on more than 250 insecticides, herbicides, and fungicides described by their effects on humans, animals, and the environment
Handbook of Toxic and Hazardous Chemicals and Carcinogens, 3rd Ed., M. Sittig, 1685 pp, 1991, ISBN: 0-8155-1286-4 Includes chemical, health, and safety information on nearly 1300 toxic and hazardous chemicals as defined by official recognition or by associations such as ACGIH or the German Research Society (DGF). Additional information includes storage, shipping, spill handling, and fire extinguishing
Hawley's Condensed Chemical Dictionary, 12th Ed., R. J. Lewis, Sr. (Ed.), 1275 pp, 1992, ISBN: 0-442-01131-8 A condensed chemical dictionary with technical data and descriptive information covering thousands of chemicals. Each chemical substance is identified by name, physical properties, source of occurrence, shipping regulations, Chemical Abstract Service (CAS) Registry numbers, chemical formulas, hazard, derivation, synonym, and use
Hazardous Chemicals Desk Reference, 3rd Ed., R. J. Lewis, Sr., 1760 pp, 1993, ISBN: 0-442-01408-2 A quick-access reference to key information on the hazardous properties of approximately 6000 chemicals commonly encountered in industry, the laboratory and the environment. Provides information on chemical and physical properties, toxicity, hazard ratings, and safety profiles
Hazardous Materials Handbook, R. P. Pohanish and S. A. Green, 1620 pp, 1996, ISBN: 0-442-02212-3. Contains comprehensive practical and technical data as well as chemical properties for more than 1,240 substances widely used and transported as industrial materials. An updated and expanded version of the U.S. Coast Guard Chemical Hazards Response Information System (CHRIS) Manual

TABLE 3.1 ***(Continued)***

Workplace Hazards and Toxicology (*Continued*)

Hazardous Materials Handbook for Emergency Responders, J. Varela (Ed.), 551 pp, 1996, ISBN: 0-442-02104-6

Summarizes the hazard and response priorities according to the nine classes of hazardous materials explaining the physical and chemical properties of hazardous materials; and current control and mitigation techniques for chemical emergencies with details on the necessary personal protective equipment

Industrial Toxicology: Safety and Health Applications in the Workplace, P. L. Williams, J. L. Burson, 502 pp, 1985, ISBN: 0-442-23541-0

Provides specific guidance on the safe manufacture, storage, use, and disposal of dangerous materials and provides coverage of risk assessment, carcinogenesis, and mutagenesis, including the evaluation and control of heavy metals, pesticides, and organic solvents, with a glossary of toxicological terms

Introduction to Toxicology, 2nd Ed., J. A. Timbrell, 181 pp, 1995, ISBN: 0-7484-0241-1

Provides a general introduction to toxicology with updated figures, examples, and essay questions as well as a complete bibliography

Occupational Health Guidelines for Chemical Hazards, Original Manual, NIOSH, 1922 pp, 1981

This test was originally published in 1981 to disseminate technical information assembled under the Standards Completion Project and the recommendations resulting from that effort. These recommendations reflect good industrial hygiene and medical surveillance practices

Occupational Neurology and Clinical Neurotoxicology, M. L. Bleecker (Ed.), 399 pp, 1994, ISBN: 0-683-00848-X

Provides important information on the clincial and epidemiological assessment of exposure and dose and work-related factors (ergonomic stressors) in neurological disorders of the upper extremity and spine, with the use of biomarkers to measure biological change

Occupational Toxicology, Neill H. Stacey (Ed.), 408 pp, 1993, ISBN: 0-85066-9-831-X

An introduction to the basic aspects of toxicology, including chapters on metals, pesticides, solvents, plastics, gases, and particulate matter. Discusses the types of toxicity and the organs likely to be affected, emphasizing the practical use of toxicological information, including the regulations governing the use of such chemicals

Proctor and Hughes' *Chemical Hazards of the Workplace*, 4th Ed., G. Hathaway, N. H. Proctor, J. P. Hughes, 816 pp, 1996, ISBN: 0-442-02050-3; also available on CD-ROM

Contains updated information on 542 substances, including data on acute and chronic effects; expanded information on carcinogenic, mutagenic, and other reproductive effects; as well as chemical formula, CAS number, synonyms, physical form, and exposure sources for each substance

Quantitative Risk Assessment for Environmental and Occupational Health, 2nd Ed. W. H. Hallenbeck, 224 pp, 1993, ISBN: 0-87371-801-1

Covers calculation of human dose rate and dose from experimental studies (animal and human); quantitation of response; tests of significance; calculation of excess risk; calculation of confidence limits on excess risk; individual and group excess risk; conversion of risk factor units; and acceptable concentrations

Reproductively Active Chemicals: A Reference Guide, R. J. Lewis, Sr., 841 pp, 1991, ISBN: 0-442-31878-2

TABLE 3.1 (*Continued*)

Workplace Hazards and Toxicology (*Continued*)
Contains fully documented information on more than 3300 chemical substances known or suspected to cause adverse effects on human reproductive health, featuring cross-indexes by synonym and identification number; information including chemical properties, toxicity, and synonyms; and toxic effects
Sax's Dangerous Properties of Industrial Materials, 9th Ed. R. J. Lewis, Sr., 5500 pp, 1996, ISBN: 0-442-02025-2, also available on CD-ROM Features more than 44,000 entries on chemicals and industrial materials, including recent advances in biohazards, primary irritants, organometallics, agricultural chemicals, industrial and environmental carcinogens, and air contamination
Toxic Air Pollution Handbook, D. R. Patrick (Ed.), 608 pp, 1994, ISBN: 0-442-00903-8 Covers the latest acceptable control methods and identifies EPA acceptable levels of exposure for 189 pollutants. Includes methods for assessing human exposure, technologies to control emissions; fugitive emissions; air sampling methods and emission estimation; and techniques for communicating with the public, press, and regulators about risks associated with air toxics
Toxicological Chemistry, 2nd Ed., S. E. Manahan, 464 pp, 1992, ISBN: 0-87371-621-3 Both toxicological chemistry and environmental biochemistry are addressed in this text, including basic concepts of general and organic chemistry and an overview of environmental chemistry, emphasizing the chemical aspects of toxicological phenomena
Toxicology: A Primer on Toxicology Principles and Applications, M. A. Kamrin, 144 pp, 1988, ISBN: 0-87371-133-5 Presents basic toxicology principles and applications in a nontechnical style. This book supplies information on both the strengths and limitations of the discipline of toxicology, with case studies to facilitate comprehension
World-Wide Limits for Toxic and Hazardous Chemicals in Air, Water, and Soil, M. Sittig, 827 pp, 1994, ISBN: 0-8155-1344-5 Summarizes allowable domestic and international limits for over 1100 chemicals in workplace air, ambient air, water of various types, and in soils
Industrial Hygiene
Aerosol Measurement: Principles, Techniques, and Applications, K. Willeke and P. A. Baron (Eds.), 894 pp, 1993, ISBN: 0-442-00486-9 Presents measurement fundamentals and practices in a wide variety of aerosol applications including optical direct-reading techniques, bioaerosol sampling, indoor air applications, industrial aerosol processing, and measurement in semiconductor clean rooms
Aerosol Science for Industrial Hygienists, J. H. Vincent, 428 pp, 1995, ISBN: 0-08-042029-X Links aerosol science and occupational health, focusing on the worker, addressing the nature and effects of exposure to airborne particles and offering the framework for standards, measurement, and control
Air Monitoring, C. J. Maslansky and S. P. Maslansky, 304 pp, 1993, ISBN: 0-442-00973-9 Gives easy-to-understand, step-by-step instruction on the function, use, operation, and limitations of air monitoring instruments, teaching proper use of many different types of instruments and providing information on properly recording and interpreting readings

TABLE 3.1 (*Continued*)

Industrial Hygiene (*Continued*)
Air Monitoring for Toxic Exposures, S. A. Ness, 534 pp, 1991, ISBN: 0-442-20639-9 Provides guidance on evaluating potentially harmful exposures to chemicals, radon, and bioaerosols, offering practical information on how to perform air sampling, collect biological and bulk samples, evaluate dermal exposures, and make judgments on the advantages and limitations of a given method
Air Sampling Instruments, 8th Ed., ACGIH, 628 pp, 1995, ISBN: 1-882417-08-9 Presents state-of-the-art information on contemporary air sampling practices and procedures, including theory, detailed descriptions of instruments, sampling strategies in the workplace and the community, particle and gas-phase interactions, size-selective health hazard sampling, and calibration of gas and vapor samplers and aerosol samplers
Basic Guide to Industrial Hygiene, J. W. Vincoli, 383 pp, 1995, ISBN: 0-442-01960-2 Explains what industrial hygiene is, how it evolved to its present professional level, and how to establish an industrial hygiene program. It also contains related information on human anatomy and industrial hazards. A reference for safety professionals with limited industrial hygiene experience or those initiating careers in industrial hygiene
Bioaerosols Handbook, C. S. Cox and C. M. Wathes (Eds.), 639 pp, 1995, ISBN: 0-87371-615-9 Provides up-to-date knowledge and practical advice covering the principles and practices of bioaerosol sampling, descriptions and comparisons of bioaerosol samplers, calibration methods, and assay techniques, with an emphasis on practicalities
Biological Monitoring: An Introduction, S. Que Hee (Ed.), 670 pp, 1993, ISBN: 0-442-23677-8 Describes the relationship between environmental exposures to particular chemicals and concentrations of markers in body tissues and fluids, monitoring for harmful substances in the workplace, the benefits and limitations of testing for critical levels of toxic materials in body tissues and fluids
Building Air Quality, U.S. EPA and NIOSH, 229 pp, 1991 A guide for building owners and facility managers providing indoor air quality guidance from the U.S. EPA and NIOSH. Contains information on developing an IAQ building profile and management plan, identifying causes and solutions to problems and appropriate control strategies, and deciding whether outside technical assistance is needed
Definitions, Conversions, and Calculations for Occupational Safety and Health Professionals, E. W. Finucane, 352 pp, 1993, ISBN: 0-87371-863-1 A reference information source and example problems workbook containing virtually every mathematical relationship, formula, definition, and conversion factor that any professional in occupational safety and health will need or encounter
Design of Industrial Ventilation Systems, 5th Ed., J. L. Alden and J. M. Kane, 281 pp, 1982, ISBN: 0-8311-1138-0 Local exhaust systems, the interrelated areas of general exhaust ventilation and makeup air supply, and the need for energy conservation are among the topics in this text. The manual contains information required to design and build, purchase, and operate an exhaust system that will adequately and economically perform
Fundamentals of Industrial Hygiene, 4th Ed., National Safety Council, 1000 pp, 1995, ISBN: 0-87912-171-8

TABLE 3.1 (*Continued*)

Industrial Hygiene (*Continued*)
This reference covers monitoring, recognition, evaluation and control of workplace health hazards with updated OSHA regulations, professional standards, permissible exposures, and worker's right to know with chapters on an industrial hygiene overview, methods of control, the safety professional, computerizing an industrial hygiene program, and governmental regulations
Guide to Industrial Respiratory Protection, NIOSH, 296 pp, 1987 This NIOSH Technical Guide covers the selection, use, and maintenance of respiratory protective devices. Detailed information is provided on types of respirators, including respiratory inlet coverings, air-purifying respirators, and atmosphere-supplying respirators. Other major topics include respirator selection, respirator use, and special conditions
Guidelines for Selection of Chemical Protective Clothing, 3rd Ed., A. D. Schwope, P. P. Costas, J. O. Jackson, and D. J. Weitzman, 330 pp, 1987, ISBN 0-936712-73-2 Chemical protective clothing (CPC) is a key element in minimizing the potential for worker exposure to chemicals. This two-volume set brings virtually all CPC performance information to one location; provides the basic data required to select, order, and intelligently use CPC
Handbook of Health Hazard Control in the Chemical Process Industry, S. Lipton and J. Lynch, 1015 pp, 1994, ISBN: 0-471-55464-2 A state-of-the-art look at the tools and procedures for monitoring and controlling worker exposure to environmental hazards, examining the impact of the Clean Air Act Amendments of 1990, including the new allowable release rates for chemicals, technological innovations in exposure control, basic procedures for exposure evaluation, emissions measurement and estimation, sampling, and exposure assessment
Illustrated Dictionary of Environmental Health & Occupational Safety, H. Koren, 415 pp, 1996, ISBN: 0-87371-420-2 Includes over 7,500 terms encompassing all important areas of environmental health and occupational safety with more than 600 detailed illustrations. Terms are drawn from varied specialized and technical fields including biology, chemistry, medicine, epidemiology, computer science, toxicology, risk assessment, occupational disease, air toxics, and hazardous waste
Industrial Chemical Exposure: Guidelines for Biological Monitoring, 2nd Ed., R. R. Lauwerys and P. Hoet, 336 pp, 1993, ISBN: 0-87371-650-7 A practical guide to biological monitoring for industrial chemical exposure assessment, discussing the objectives of biological monitoring, the types of biological monitoring methods, their advantages and limitations, and practical aspects to be considered before initiating a biological monitoring program
Industrial Health, 2nd Ed., J. E. Peterson, 350 pp, 1991, ISBN: 0-936712-91-0 Focuses on the useful fundamentals and principles behind the evaluation and control of hazards in the working environment. Occupational hazards are stressed, ranging from chemical toxins to various energy forms. Sixteen chapters cover the broad spectrum of critical topics in industrial health
Industrial Ventilation: A Manual of Recommended Practice, 22nd Ed., ACGIH, 470 pp, 1995, ISBN: 1-882417-09-7 This manual features a compilation of research data and information on the design, maintenance, and evaluation of industrial exhaust ventilation systems. Basic ventilation principles and sample calculations are presented in a clear and simplified manner

TABLE 3.1 (*Continued*)

Industrial Hygiene (*Continued*)
Internet User's Guide For Safety & Health Professionals, M. Blotzer, 160 pp, 1995, ISBN: 0-931690-82-X This provides a guide to the use of the information available on the Internet for safety and health professionals. Information ranges from the equipment, services, tools, and techniques necessary to access the Internet to the many locations that offer valuable information
Modern Industrial Hygiene: Recognition and Evaluation of Chemical Agents, J. Perkins, 1996, Vol. 1, 1st Ed., Van Nostrand Rheinhold, ISBN: 0-44020-210-54 A guide to industrial hygiene as it relates to chemical agents
NIOSH Manual of Analytical Methods, 4th Ed. Government Printing Office, Washington, D.C. (revised with Supplements through 1997) A reference guide for more than 260 analytical methods for monitoring occupational exposures to toxic substances in air and biological samples
NIOSH/OSHA Occupational Health Guidelines for Chemical Hazards, 1922 pp, Government Printing Office, Washington, D.C. (revised with supplements through year 1995) Originally published in 1981 this document has fact sheets on a wide range of chemicals. The fact sheets cover health effects, monitoring, physical properties and advice on personal protective equipment. The document is updated through the use of supplements. The 1995 supplement was 500 pages
NIOSH Pocket Guide to Chemical Hazards, 1996, Government Printing Office, Washington, D.C., 245 pp Pocket guide listing of chemicals, their TLVs, IDLHs, recommended protective equipment, target organs and effects. This guide is very useful and its physical shape allows for placement in a large shirt or pants pocket
OSHA Technical Manual, 4th Ed., OSHA, 428 pp, 1996, ISBN: 0-86587-511-1 The manual OSHA's Compliance Safety and Health Officers use to monitor industry compliance. The wide-ranging text facilitates hazard recognition and the development of safety and health programs
Patty's Industrial Hygiene and Toxicology, Vol. I, Part A and Part B: General Principles, 4th Ed., G. D. Clayton and F. E. Clayton (Eds.) 1091 pp, 1991, ISBN: 0-471-50196-4 The classic guide to the concepts of industrial hygiene and toxicology, with focus on environmental safety and hazard control including conditions beyond the industrial workplace
Patty's Industrial Hygiene and Toxicology, Vol. II, G. D. Clayton, F. E. Clayton (Eds.) Part A, 973 pp, 1993; Part B, 846 pp, 1993; Part C, 763 pp, 1994; Part D, 907 pp, 1994; Part E, 1041 pp, 1994; Part F, 900 pp, 1994; ISBN: 0-471-54724-7, 0-471-54725-5, 0-471-54726-3, 0-471-57947-5, 0-471-01282-3, 0-471-01280-7 (Parts A–F) The Fourth Edition of Volume II offers a comprehensive guide to the toxins commonly found in industrial settings. This edition includes significant toxicological information on new products, new research trends, new computerized data and recordkeeping, and accelerated public awareness
Patty's Industrial Hygiene and Toxicology, Vol. III, Parts A and B: Theory and Rationale of Industrial Hygiene Practice R. L. Harris, L. J. Cralley, L. V. Cralley, J. S. Bus, Part A: 861 pp, 1994, ISBN: 0-471-53066-2; Part B: 800 pp, 1995, ISBN: 0-471-53065-4 This two-part set is a comprehensive guide to the practice of industrial hygiene and the theory that underlies it. The two parts address the most important aspects of the evaluation of the work environment, and the biological responses to exposures encountered in the workplace

TABLE 3.1 (*Continued*)

Industrial Hygiene (*Continued*)

Recognition of Health Hazards in Industry: A Review of Materials Processes, 2nd Ed., W. A. Burgess, 608 pp, 1995, ISBN: 0-471-57716-2

The authoritative and practical guide to the major health issues in the workplace containing detailed surveys of work tasks in a wide range of industries, enabling readers to recognize health problems in facility design and operation and to relate medical symptoms to job exposure

The Work Environment, Vol. I: Occupational Health Fundamentals D. J. Hansen (Ed.), 512 pp, 1991, ISBN: 0-87371-303-6

An introduction to workplace health and safety issues, concerned with recognition, evaluation, and control of workplace hazards. Major topic areas include: evaluating workplace hazards; occupational health standards; controlling hazards in the work environment; personal protection; and training to optimize occupational health. A glossary and an index are also provided

The Work Environment, Vol. II: Healthcare, Laboratories, and Biosafety, D. J. Hansen (Ed.) 320 pp, 1992, ISBN: 0-87371-392-3

Includes topics such as the bloodborne pathogens standards and how to comply, the resurgence of tuberculosis and how to protect against it, good work practices in any laboratory (including biosafety concepts, levels, and controls), how to respond to spills in the laboratory, medical waste disposal, and how to comply with the Laboratory Safety Standards

The Work Environment, Vol. III: Indoor Health Hazards, D. J. Hansen (Ed.) 232 pp, 1994, ISBN: 0-87371-393-1

Defines a wide range of indoor air quality (IAQ) problems and solutions, discussing common symptoms and potential environmental and chemical causes, health hazards from arts and crafts and from common household products, and the impact of common building ventilation problems and how to solve them. An expert summary of methods to conduct an IAQ survey

Ventilation for Control of the Work Environment, W. A. Burgess, M. J. Ellenbecker, and R. T. Treitman, 476 pp, 1989, ISBN: 0-471-89219-X

This source book for the specification, design, installation, and maintenance of industrial ventilation systems provides a theoretical background to the plant engineer, brings the concepts and principles of ventilation into an industrial setting, and illustrates practical aspects for students

The Occupational Environment—Its Evaluation and Control, S. R. DiNardi (Ed.), 1365 pp, 1997, ISBN: 0-932627-82-X

This is a complete update of the classic "White Book." It is an essential reference for occupational safety and health professionals

Safety and Ergonomics

A Guide to Manual Materials Handling, A. Mital, A. S. Nicholson, and M. M. Ayoub, 114 pp, 1993, ISBN: 0-85066-801-8

Lifting, pushing, pulling, carrying, and holding are addressed in this *Guide*, including recommendations in the form of design data that can be used to design different MMH work activities. The *Guide* outlines the scope of the problem, discusses the factors that influence a person's capacity to perform MMH activities, and reviews the various design approaches to solving the MMH problem

TABLE 3.1 (*Continued*)

Safety and Ergonomics (*Continued*)
Accident Prevention Manual for Business and Industry, Vol. I, 10th Ed., National Safety Council, 564 pp, 1992, ISBN: 0-87912-155-6 This three-volume set contains all the information needed to develop, maintain, and improve occupational safety programs. Includes four parts: Introduction to Safety and Health, Program Organization, Hazard Information and Analysis, and Program Implementation
Accident Prevention Manual for Business and Industry, 10th Ed., Vol. II, National Safety Council, 825 pp, 1992, ISBN: 0-87912-156-4 Emphasizing information related to specific businesses and industries, this volume contains four sections: Facilities and Workstations, Material Handling, Workplace Exposures and Protection, and Production Operations. Includes chapters on lockout/tagout, confined space, lifting recommendations, and ergonomics
Accident Prevention Manual for Business and Industry, Vol. III, 10th Ed., National Safety Council, 600 pp, 1994, ISBN: 0-87912-170-X Topics covered include history and development, economic and ethical issues, principles of U.S. and international legal and legislative framework, managing environmental resources, environmental audits and site assessments, principles of environmental science, training requirements, hazardous waste management, transportation of hazardous materials, pollution prevention technologies, public health, and indoor air quality
Applied Ergonomics Handbook, M. Burke, 270 pp, 1992, ISBN: 0-87371-367-2 This procedural guide includes forms, protocols, and suggestions for preventing musculoskeletal trauma in the workplace. The procedures described are the least cumbersome, most practical, and most objective. This book provides the tools to generate a report that comprehensively describes specific risk factors to suggest interventions, to form an ergonomic committee, and to establish a corporate or companywide ergonomics process
Analyzing Safety System Effectiveness, 3rd Ed., D. Petersen, 282 pp, 1996, ISBN: 0-442-02180-1 Provides information on developing plans by analyzing the physical, managerial, and behavioral aspects of existing safety programs. It describes how to effectively analyze safety performance; evaluate skills and behavior of management and staff; and interpret results, devise plans, and implement change
Basic Guide to System Safety, J. W. Vincoli, 207 pp, 1993, ISBN: 0-442-01275-6 This text addresses the basics of system safety. Detailed sections survey the interaction between occupational safety and system safety; the integration of system safety into equipment design; OSHA compliance; and system safety in aerospace, manufacturing, and service industries
Complete Manual of Industrial Safety, S. Z. Mansdorf, 397 pp, 1993, ISBN: 0-13-159633-0 This manual facilitates the development or improvement of industrial safety and health programs. With its emphasis on the practical application of proven techniques, it contributes to better employee relations and morale, improved production and efficiency, reduced workers' compensation costs, better product quality, and an improved bottom line
Cumulative Trauma Disorders, V. Putz-Anderson (Ed.), 151 pp, 1988, ISBN: 0-85066-405-5 This manual defines cumulative trauma disorders (CTDs) in the workplace, to enable nonmedical personnel to recognize them, and to present strategies for preventing their occurrence. Defines the cumulative trauma category of musculoskeletal disorders, presents methods for determining how many workers at a worksite have CTDs or early symptoms of CTDs, and strategies used to control or prevent the occurrence of CTDs

TABLE 3.1 ***(Continued)***

Safety and Ergonomics (*Continued*)
Cumulative Trauma Disorders, Current Issues and Ergonomic Solutions, K. G. Parker and H. R. Imbus, 144 pp, 1992, ISBN: 0-87371-322-2
Addresses both the medical and ergonomic aspects of cumulative trauma. Specific topics addressed include CTD etiology, in-plant control programs, return-to-work concepts, ergonomic stressors and their root causes, and basic guidelines for ergonomic workstation design. The text also discusses the rationale and value of implementing program components in the OSHA guidelines
Emergency Incident Risk Management: A Safety & Health Perspective, J. D. Kipp and M. E. Loflin, 329 pp, 1996, ISBN: 0-442-01926-2
Demonstrates how to analyze accident, injury, and illness data; identify and evaluate risk; establish risk management priorities; implement sound risk control measures; monitor a risk management program; incorporate risk management into an incident management system; and use proper personal protective equipment. Includes regulatory information on bloodborne pathogens, confined spaces, respiratory protection, and hazardous waste
Encyclopedia of Occupational Health and Safety, International Labour Organisation (ILO), Geneva, 4 Vols., 1997, ISBN: 92-2-109203-8
Comprehensive coverage of occupational safety and hygiene from a technical and international perspective. Practical and theoretical aspects are covered. Many graphics throughout
Ergonomic Design for People at Work, Vol. One, Eastman Kodak Company 406 pp, 1983, ISBN: 0-442-23972-6
A practical discussion of workplace equipment, environmental design, and the transfer of information in the workplace, summarizing the published literature, internal research, and observation by the members of the Human Factors Section. The guidelines and examples of approaches to design problems are most often drawn from case studies
Ergonomic Design for People at Work, Vol. Two, Eastman Kodak Company 603 pp, 1986, ISBN: 0-442-22103-7
A complement to Volume One, this book draws on physiology, psychology, engineering, medicine, and environmental sciences to provide practical information for the design of jobs and work tasks. Guidelines and procedures are based on ergonomic approaches that have proven to be effective within Eastman Kodak
Ergonomics: A Practical Guide, 2nd Ed., National Safety Council, 128 pp, 1993, ISBN: 0-87912-168-8
This manual is designed to help safety and health professionals identify and correct workplace ergonomic problems. It discusses how to analyze work methods and workstations and how to identify and resolve ergonomic problems in such areas as seated and standing operations; manual materials handling; upper extremity disorders; and the use of tools, controls and displays
Ergonomics for Beginners: A Quick Reference Guide, J. Dul and B. Weerdmeester, 146 pp, 1993, ISBN: 0-7484-0079-6
Embracing the concepts of designing tasks and environments for human comfort and satisfaction as well as optimum performance, the book shows, in an accessible and easily understandable fashion, the steps by which managers, workers, and users can achieve an appropriate balance

TABLE 3.1 (*Continued*)

Safety and Ergonomics (*Continued*)
Fundamentals of Occupational Safety and Health, J. P. Kohn, M. A. Friend, and C. A. Winterberger, 452 pp, 1996, ISBN: 0-86587-539-1 Includes basic information safety and health professionals need to control hazards and losses and protect the health and lives of workers, including: industrial hygiene, system safety, psychology and safety, safety management, ergonomics, workers' compensation, accident causation and investigation, safety regulations, record keeping, fire science, hazardous materials, workplace violence, and training requirements
Human Error Reduction and Safety Management, 3rd Ed., D. Petersen, 414 pp, 1996, ISBN: 0-442-02183-6 Illustrates how managers can modify employees' behavior to reduce error, accidents, and consequently on-the-job injuries and illnesses. The book includes a revised model of accident causation that exemplifies the procedures of today's safety technology; expanded treatment of managerial sources of error; a discussion of ergonomics; insight on how to reduce psychological overload; and new material on risk-assessment techniques
Occupational Ergonomics: Theory and Applications, A. Bhattacharya and J. D. McGlothlin (Eds.), 846 pp, 1996, ISBN: 0-8247-9419-2 An introduction to the fundamental principles of ergonomics with practical application of ergonomic principles in solving actual problems in the workplace. Reviews ergonomic case studies; examines legal implications of applying ergonomic science to workers; describes design of a medical surveillance program for ergonomic problems and considers cumulative trauma disorder, vibration white-fingers, and back disorders
Occupational Health & Safety, 2nd Ed., J. LaDou (Ed.), 950 pp, 1993, ISBN: 0-87912-154-8 Focuses on establishing a quality occupational health program in the workplace. Covers a broad range of important issues related to workplace management of an occupational health and safety program
On the Practice of Safety, F. A. Manuele, 288 pp, 1993, ISBN: 0-442-01401-5 This text provides an understanding of vital issues including a basic definition of the safety profession, incident investigation, the significance of ergonomics, the critical role of design processes, total quality management, and causation models
Safety and Health for Engineers, R. L. Brauer, 671 pp, 1995, ISBN: 0-442-01856-8 This text addresses the fundamentals of safety; legal aspects, hazard recognition, human elements, and techniques for managing safety in engineering decisions. In-depth coverage includes the engineer's duties and legal responsibilities; an exploration of all types of hazards and their engineering controls; the latest safety regulations and the agencies responsible for their enforcement
Safety Engineering, J. CoVan, 248 pp. 1995, ISBN: 0-471-55612-2 A presentation of guidelines, checklists, and safety data for safety engineers and technicians to institute a well-planned safety program. The need for professionalism and scientific analysis of risks and safety measures are stressed, with both an overview of the fundamentals and insight into the subtleties of a rapidly growing field
Work Related Musculoskeletal Disorders (WMSDs): A Reference Book for Prevention, M. Hagberg, B. Silverstein, R. Wells, M. J. Smith, H. W. Hendrick, P. Carayon, and M. Perusse, 428 pp, 1995, ISBN: 0-7484-0132-6

TABLE 3.1 (*Continued*)

Safety and Ergonomics (*Continued*)

Prevention of work-related musculoskeletal disorders (WMSDs) is the focus of this book, which examines the work-relatedness of WMSDs, and explores and synthesizes information and techniques that seek to prevent WMSDs

Medicine

A Practical Approach to Occupational and Environmental Medicine, 2nd Ed., R. J. McCunney, 856 pp, 1994, ISBN: 0-316-55534-7

This text includes chapters in 5 major sections: Occupational Medical Services; Occupational Related Illnesses; Evaluating a Health Hazard or Work Environment; Challenges in Occupational and Environmental Medicine; and Environmental Medicine

Emergency Care for Hazardous Materials Exposure, 2nd Ed., A. C. Bronstein and P. L. Currance, 649 pp, 1994, ISBN: 0-8016-7813-7

Provides field recognition and management guidelines for hazardous material exposures and associated medical emergencies, including emergency care of exposed and contaminated patients. A field and educational training reference for use by emergency medical respondents, hazmat response teams, physicians and nurses, and health and safety officers

Environmental and Occupational Medicine, 2nd Ed., W. N. Rom (Ed.), 1493 pp, 1992, ISBN: 0-316-75567-2

This textbook serves the interests of medical students as they enter their clinical years, residents and practitioners as an extensive resource, and the public health community as a guide to controlling and preventing disease. Includes sections on environmental and occupational disease by organ systems; toxicants in the workplace and the environment; and control strategies for both diseases and toxicants

Environmental Medicine, S. Brooks, M. Gochfeld, J. Herzstein, R. J. Jackson, M. B. Schenker, 799 pp, 1995, ISBN: 0-8016-6469-1

Provides health professionals with a comprehensive resource of scientific information on environmental health. Developed for health professionals who must face issues of environmental and occupational disease and especially for the practitioner who takes care of patients on a regular basis

Hunter's Diseases of Occupations, 8th Ed., P. A. B. Raffle, P. H. Adams, P. J. Baxter, W. R. Lee (Eds.) 815 pp, 1994, ISBN 0-340-55173-9

This includes chapters on diseases associated with chemical agents, physical agents, and microbiological agents; muscular and skeletal problems; mental ill-health at work; occupational cancer; occupational diseases of the skin; reproduction and work, and an appendix from IARC on evaluation of carcinogenic risks to humans

Mosby's Pocket Dictionary of Medicine, Nursing, and Allied Health, 2nd Ed., K. N. Anderson and L. E. Anderson, 1152 pp, 1994, ISBN: 0-8016-7226-0

An abridgment of *Mosby's Medical, Nursing, and Allied Health Dictionary* reflecting new developments in many facets of health care. Printed tabs on the edge of each page assist in quickly locating definitions

TABLE 3.1 (*Continued*)

Medicine (*Continued*)
Occupational Health Nursing: Concepts and Practice, B. Rogers, 544 pp, 1994, ISBN: 0-7216-7588-3 This comprehensive reference provides a framework for occupational health nursing practice and relevant discussion of the practical application of concepts. Discusses factors affecting occupational health nursing practice related to individual and collective workforce health as well as the promotion of health and quality living
Occupational Health: Recognizing and Preventing Work-Related Disease, 3rd Ed., B. S. Levy and D. H. Wegman, 791 pp, 1995, ISBN: 0-316-52271-6 This text provides the information necessary to recognize, prevent, and treat work-related disease and injury. Updated to include numerous case studies, detailed photographs, line drawings, graphs, and tables, addressing a wide range of occupational medicine issues including work and health, recognition and prevention of occupational diseases, hazardous workplace exposure, and occupational disorders by system
Occupational Medicine, 3rd Ed., C. Zenz, O. B. Dickerson, and E. P. Horvath, Jr. (Eds.), 1336 pp, 1994, ISBN: 0-8016-6676-7 *Occupational Medicine* stresses occupational environments, including the workplace and also psychosocial/cultural factors. Eight major sections are devoted to clinical factors, occupational pulmonary diseases, physical occupational environment, selected work categories of concern, behavioral considerations, fundamental disciplines and related activities for prevention and control, and other special activities within the occupational health setting
Preventing Occupational Disease and Injury, J. L. Weeks, B. S. Levy, and G. R. Wagner (Eds.), 750 pp, 1991, ISBN: 0-87553-172-5 Presents an integrated and multidisciplinary approach to prevention, describing anticipation, surveillance, analysis, and control in the workplace. Also provides a compendium of adverse health outcomes caused in whole or in part by work and overviews of occupational musculoskeletal and infectious diseases as well as of occupational cancers
Textbook of Clinical Occupational and Environmental Medicine, L. Rosenstock and Mark R. Cullen (Eds.), 927 pp, 1994, ISBN: 0-7216-3482-6 This publication provides a broad overview of the specialized skills central to the successful practice of occupational and environmental medicine, encompassing toxicology, epidemiology, and industrial hygiene; three core disciplines that are necessary complements to the diagnosis, treatment, and prevention of occupational and environmental diseases and an organ-system approach to occupational and environmental diseases

3.3 ELECTRONIC DATA SOURCES

A number of information sources are now available on CD-ROM, the Internet, and the World Wide Web. For example, the OSHA Standards, Letters of Interpretation, EPA Standards, Hazardous Substances Databanks from the National Library of Medicine, Medline, Toxline, etc. can all be accessed directly from a personal computer. The OSHA-CD, as it is called, can be obtained via OSHA's web site (www.osha.gov) or by contacting the Government Printing Office (GPO). The Internet is currently a great information and communication asset, and in future it will be central to occupational safety and health practice. For users of the Internet, the Internet User's Guide to the Internet for Health and Safety Professionals (described in the table of Industrial Hygiene publications) is an excellent introduction to the Internet and its uses. It also includes 79 pages of site listings for safety and health professionals. Methods for opening an Internet account, logging onto an Internet server, finding sites, etc. are all described. The following list is a sample of some of the readily available information on the World Wide Web:

MSDSs: hundreds of thousands of material safety data sheets

Toxicological profiles of thousands of chemicals available from EPA, ATSDR, NTP

Literature searches from the National Library of Medicine, Medline, Toxline, Chemline, Hazardous Substances Database

Exposure information from NIOSHTIC, National Cancer Institute Surveillance Exposure and Epidemiology Reports (SEER) Data

OSHA Compliance Reports, injury and illness rates by industry, etc.

Training courses, e.g., ATSDR training for lead and radon; notices for upcoming courses, conferences, etc. from the American Industrial Hygiene Association, the American Conference of Governmental Industrial Hygienists, and many other sources

New Groups comprised of groups of Internet users with similar interests, e.g., ergonomics, MSDSs, chemical safety

NIOSH Manual of Analytical Methods

Selection guides for personal protective equipment including permeability data

Complete Federal Regulations and Guidelines (EPA, OSHA, DOT, NIOSH, CDC)

Governmental and nongovernmental organizations throughout the world, including National Institutes of Safety and Health for many countries, World Health Organization, and universities throughout the world

International Standards Organization (www.iso.ch)

ANSI Standards (www.ansi.org)

To illustrate how pervasive (and useful) the World Wide Web is in industrial hygiene, a samll sample of web sites has been assembled below. Table 3.2 is a listing of organizations with web pages (sites on the World Wide Web) of interest to industrial hygienists. There are many thousands of sites that pertain to industrial hygiene; finding needed information is a challenge. This list may be helpful in providing beginning points in the search. All of the listed sites have links (connections) to other sites on the World Wide Web.

Table 3.3 illustrates how World Wide Web sites can be organized around a common interest, such as ergonomics or cancer. As in Table 3.2, all these sites also have links to other related sites.

To access sites identified in Tables 3.2 and 3.3, you must first have an internet account with an internet service provider. Setting up an internet account is discussed extensively in Blotzer (1996). Once an account has been set up, enter the address given in the tables and you will be connected to the site. Methods for downloading information are provided at each site.

TABLE 3.2 Sampling of Organizations With Sites on The World Wide Web That May Be Useful to Industrial Hygienists. The Addresses are All Accessible With Web Browsers From Microsoft and Netscape

Organization	Internet Address and Short Description of Content
Occupational Health and Safety Administration	WWW.OSHA-SLC.GOV Source of compliance and rates of injury/illness information
Environmental Proection Agency	WWW.EPA.GOV Environmental regulations, research activities, toxic substances control, chemical specific information
Agency for Toxic Substances and Disease Registry	WWW.ATSDR1.ATSDR.CDC.GOV:8080/ Health effects information concerning chemicals, chemicals released from hazardous waste disposal sites, health study design issues, physician case studies
National Institutes of Health	WWW.NIH.GOV Ongoing research, information on cancer and causation
National Toxicology Program	WWW.NTP-SERVER.NIEHS.NIH.GOV Extensive information on chemicals, reactivity, LD50, flammability, long-term and short-term effects.
Worksafe Australia	WWW.WORKSAFE.GOV.AU/WORKSAFE/HOME.HTM Chemical search capability. Review of programs in Australia
National Institute for Occupational Health and Safety	WWW.CDC.GOV/NIOSH Research studies, health hazard evaluations, extensive links to occupational safety and health resources on the internet
World Health Organization	WWW.WHO.CH Information on environmental as well as occupational health with a global perspective. Currently, largely nonoccupational health issues
International Agency for Research on Cancer	WWW.IARC.FR Cancer classifications, ongoing research, publications and ordering info
American Conference of Governmental Industrial Hygienists	WWW.ACGIH.ORG Source of information on TLVs, BEIs (biological exposure indices), chemicals under study and proposed revisions to TLVs
National Library of Medicine	WWW.NLM.NIH.GOV Over 40 databases containing more than 20 million references to a wide range of health studies. Hazardous substances databank, toxline, and medline are particularly noteworthy. Databases on genetic toxicology, reproductive toxicology, toxic release inventory, aids, cancer are also available. Search software can be downloaded from this site
American Chemical Society	WWW.ACS.ORG Chemical database searching, chemistry resources, publications
Centers for Disease Control and Prevention	WWW.CDC.GOV Information on environmental chemical risks, biosafety, emergency planning, and programs at the National Center for Environmental Health

TABLE 3.3 Sample of Internet Sites Based on Particular Health Issue. Sites Can be Accessed With Browsers From Microsoft and Netscape

Issue	Internet Address and Brief Description
EMF (electromagnetic fields)	HTTP://INFOVENTURES.COM Extensive current research and findings relating to health and electromagnetic fields
Ergonomics	WWW.HFES.VT.EDU/HFES Human Factors and Ergonomics Society homepage. Links to other ergonomics sites
MSDSs (material safety datasheets)	WWW.MSC.CORNELL.EDU A site with links to University of Virginia and University of Utah MSDS master files. Extensive access to MSDSs

3.4 RESOURCE HOTLINES

A number of emergency response services are in operation, some of which are primarily intended to provide information on environmental aspects of chemical hazards. These services are good sources of information on the toxicity and risk of exposure to a wide range of chemicals, regardless of whether exposure takes place in an environmental or an occupational setting. NIOSH operates a toll-free technical service to provide information on workplace hazards. The service is staffed by technical information specialists who can provide information on NIOSH activities, recommendations and services, or any aspect of occupational safety and health. The number is not a hotline for medical emergencies, but is a source of information and referrals on occupational hazards. The NIOSH toll-free number is 800-35-NIOSH (800-356-4674).

CHEMTREC is a 24-hour hotline to the Chemical Transportation Emergency Center operated by the Chemical Manufacturers Association (800-424-9300). CHEMTREC assists in the identification of unknown chemicals, and provides advice on proper emergency response methods and procedures. It does not provide emergency treatment information other than basic first aid, however. CHEMTREC also facilitates contact with chemical manufacturers when further information is required.

The National Pesticides Telecommunications Network Hotline is operated through Oregon State University (800-858-7378). The hotline provides information on pesticide-related health effects on approximately 600 active ingredients contained in over 50,000 products manufactured in the United States since 1947. It is also a source of information on pesticide product formulations, basic safety practices, health and environmental effects, and cleanup and disposal procedures.

Several hotline and information lines are available for response to information requests on toxic materials and environmental issues. The Toxic Substances Control Act (TSCA) Assistance Information Service (TAIS) provides information and publications about toxic substances, including asbestos (202-554-1404). The EPA also operates an Emergency Response Notification System (ERNS), which is a source of information on oil discharges, and releases of hazardous substances (202-260-2342). In addition, each of the ten USEPA Regional Offices has a hotline telephone number.

3.5 CONCLUSION

Any summary of information sources for occupational safety and health is quickly rendered out of date by the rapidly changing nature of the field. The sources described in this chapter will be updated as new information is published, and internet-based resources expand. The central role of information resources in the recognition, evaluation, and control of workplace hazards will not change, however. By effectively using the information resources available to them, occupational safety and health practitioners can prevent occupational disease and injury over the entire range of working conditions.

REFERENCES

Blotzer, M.: Internet User's Guide for Safety and Health Professionals. Schenectady, NY: Genium Publishing, 1996.

CHAPTER 4

Recognition of Health Hazards in the Workplace

MARTIN R. HOROWITZ

Analog Devices, 21 Osborn St., Cambridge, MA 02139

and

MARILYN F. HALLOCK

Massachusetts Institute of Technology, 77 Massachusetts Ave., Cambridge, MA 02139

4.1 INTRODUCTION
4.2 ABRASIVE BLASTING
 4.2.1 Application and Hazards
 4.2.2 Control
4.3 ACID AND ALKALI CLEANING OF METALS
 4.3.1 Acid Pickling and Bright Dip
 4.3.2 Alkaline Treatment
4.4 DEGREASING
 4.4.1 Cold Degreasing
 4.4.2 Vapor Degreasing
4.5 ELECTROPLATING
 4.5.1 Electroplating Techniques
 4.5.2 Air Contaminants
 4.5.3 Control
4.6 GRINDING, POLISHING, BUFFING
 4.6.1 Processes and Material
 4.6.2 Exposures and Control
4.7 HEAT TREATING
 4.7.1 Surface Hardening
 4.7.2 Annealing
 4.7.3 Quenching
 4.7.4 Hazard Potential

Adapted from "Potential Exposures in the Manufacturing Industry—Their Recognition and Control" by William A. Burgess, Chapter 18 in Clayton & Clayton, *Patty's Industrial Hygiene and Toxicology, 4th ed., Vol. IA.* New York: John Wiley & Sons, Inc., 1991, pp. 595–674.

Handbook of Occupational Safety and Health, Second Edition, Edited by Louis J. DiBerardinis,
ISBN 0-471-16017-2

4.8 METAL MACHINING
4.8.1 Conventional Metal Machining
4.8.2 Electrochemical Machining
4.8.3 Electrical Discharge Machining

4.9 NONDESTRUCTIVE TESTING
4.9.1 Industrial Radiography
4.9.2 Magnetic Particle Inspection
4.9.3 Liquid Penetrant
4.9.4 Ultrasound

4.10 METAL THERMAL SPRAYING
4.10.1 Spraying Methods
4.10.2 Hazards and Controls

4.11 PAINTING
4.11.1 Types of Paints
4.11.2 Composition of Paint
4.11.3 Operations and Exposures
4.11.4 Controls

4.12 SOLDERING AND BRAZING
4.12.1 Soldering
4.12.2 Brazing

4.13 WELDING
4.13.1 Shielded Metal Arc Welding
4.13.2 Gas Tungsten Arc Welding
4.13.3 Gas Metal Arc Welding
4.13.4 Gas Welding
4.13.5 Control of Exposure

REFERENCES

4.1 INTRODUCTION

Although employment in the United States is shifting from manufacturing to the service sector, manufacturing continues to employ in excess of 20 million workers in workplaces that present both traditional and new occupational health hazards. To understand the nature of these hazards, the occupational health professional must understand not only the toxicology of industrial materials but the manufacturing technology that defines how contaminants are released from the process, the physical form of the contaminants, and the route of exposure. Physical stresses including noise, vibration, heat, and ionizing and nonionizing radiation must also be evaluated. Twelve specific unit operations representing both large employment and potential health hazards to the worker have been chosen for discussion in this chapter; these unit operations occur in many different industrial settings. This chapter is based on the previous reviews of the subject (Burgess, 1991; 1995) as well as original scientific literature. The purpose of this chapter is to help the reader recognize potential health hazards that may exist in specific operations and industries. Other chapters in this text cover the evaluation and control of the recognized hazards.

4.2 ABRASIVE BLASTING

Abrasive blasting is practiced in a number of occupational settings, including bridge and building construction, shipbuilding and repair, foundries, and metal finishing in a variety of industries. The process is used in heavy industry as an initial cleaning step to remove surface coatings and scale, rust, or fused sand in preparation for finishing operations. Abrasive blasting is used in intermediate finishing operations to remove flashing, tooling marks, or burrs from cast, welded, or machined fabrications, and to provide a matte finish to enhance bonding of paint or other coatings.

Various abrasives are used in blast cleaning operations. The most commonly heavy-duty abrasives for metal surfaces are silica sand, metal shot and grit, coal and metallurgical slags, and synthetic abrasives such as aluminum oxide and silicon carbide. For light-duty cleaning of plastic and metal parts where erosion of the workpiece is of concern, a number of ground organic products based on corn, oat, and fruit pits are available as well as baking soda, glass beads, plastic chips, and solid carbon dioxide pellets.

Three major methods of blasting are used to deliver the abrasive to the workpiece. In pressure blasting, compressed air is used to either aspirate or pressurize and deliver the abrasive from a storage "pot" to a nozzle where it is directed to the workpiece at high velocity by the operator. This type of process may be used in either open-air blasting or in blasting enclosures. In hydroblasting, a high-pressure stream of water conveys the abrasive to the work surface; this process is used principally for outdoor work. Finally, in the centrifugal wheel system, a high-speed centrifugal impeller projects the abrasive at the workpiece; this method is used primarily in some types of blasting enclosures.

4.2.1 Application and Hazards

Blasting operations may be performed either in a variety of enclosures or in open-air operations such as bridge and ship construction. The types of enclosures used in industrial applications include blasting cabinets, automatic tumble or barrel and rotating table units where the operator controls the process from outside the operation, and exhausted rooms, where the operator in inside the enclosure. In most cases, these industrial units have integral local exhaust ventilation systems and dust collectors.

In open-air blasting in construction and shipyard applications, general area contamination occurs unless isolation of the work area can be achieved. As a result, the blasting crew, including the operator, "pot man," and cleanup personnel as well as adjacent workers may be exposed to high dust concentrations depending on the existing wind conditions.

The most obvious health hazard of abrasive blasting operations is airborne dust contamination. Dust exposures may include the abrasive in use, the base metal being blasted, and the surface coating or contamination being removed. In the United States, the widespread use of sand containing high concentrations of crystalline quartz is still a major hazard to workers. Though elevated rates of silicosis were identified in blasters by the U.S. Public Health Service in the 1930s, a NIOSH alert published as recently as 1992 identified 99 cases of silicosis in abrasive blasters with 14 deaths or severe impairment (NIOSH, 1992). In the United Kingdom, the use of sand was prohibited for in-plant abrasive blasting in 1949; a ban on the use of sand for both inside and outside applications has been adopted by the European Community. The use of abrasives based on metallurgical slags warrants attention owing to the presence of heavy metal contamination.

In most cases, the base metal being blasted is iron or steel, and the resulting exposure to iron dust presents a limited hazard. If blasting is carried out on metal alloys containing such materials as nickel, manganese, lead, or chromium, the hazard should be evaluated by air sampling. The surface coating or contamination on the workpiece frequently presents a major inhalation hazard. In the foundry it may be fused silica sand; in ship repair, the surface coating may be a lead-based paint or an organic mercury biocide. Abrasives used to remove lead paint from a bridge structure may contain up to 1 percent lead by weight. Used abrasives will often need to be disposed of as hazardous waste depending upon the level of heavy metal contamination.

Physical hazards of abrasive blasting include noise exposure and safety hazards from high-velocity nozzle discharge. The release of air by a blasting nozzle generates a wide-band noise that frequently exceeds 110 dBA. This noise is greatly attenuated in a properly designed blasting enclosure, but open-air blasting requires the implementation of a hearing conservation program and frequently the use of both ear plugs and muffs (NIOSH, 1975).

4.2.2 Control

The ventilation requirements for abrasive blasting enclosures have evolved over several decades, and effective design criteria are now available (ANSI, 1996; ACGIH, 1995). The minimum exhaust volumes, based on seals and curtains in good condition, include 20 air changes per minute for cabinets with a minimum of 500 fpm through all baffled inlets. Abrasive blasting rooms require 60 to 100 cfm/sq ft of floor for downdraft exhaust. Dust control on abrasive blasting equipment depends to a large degree on the integrity of the enclosure. All units should be inspected periodically, including baffle plates at air inlets, gaskets around doors and windows, gloves and sleeves on cabinets, gaskets at hose inlets, and the major structural seams of the enclosure. For more information on control techniques, see Chapter 14.

Operators directly exposed to the blasting operation, as in blasting room or open-air operations, must be provided with NIOSH-approved type C air-supplied abrasive blasting helmets. NIOSH has assigned a protection factor of 25 for Type C hooded respirators operating in a continuous flow mode and 2000 operating in the positive-pressure mode. The latter respirator is recommended for open-air blasting with crystalline silica containing sand (NIOSH, 1992). The intake for the air compressor providing the respirable air supply should be located in an area free from air contamination, and the quality of the air delivered to the respirator must be checked periodically. The respirator must be used in the context of a full respirator program (see Chapter 16).

4.3 ACID AND ALKALI CLEANING OF METALS

After the removal of major soils and oils by degreasing, metal parts are often treated in acid and alkaline baths to condition the parts for electroplating or other finishes. The principal hazard in this series of operations is exposure to acid and alkaline mist released by heating, air agitation, gassing from electrolytic operation, or cross-contamination between tanks.

4.3.1 Acid Pickling and Bright Dip

Pickling or descaling is a technique used to remove oxide scales formed from heat treatment, welding, and hot forming operations prior to surface finishing. On low- and high-carbon steels, the scale is iron oxide, whereas on stainless steels it is composed of oxides of iron, chromium, nickel, and other alloying metals (Spring, 1974). The term *pickling* is derived from the early practice of cleaning metal parts by dipping them in vinegar.

Scale and rust are commonly removed from low and medium alloy steels using a nonelectrolytic immersion bath of 5 to 15 percent sulfuric acid at a temperature of 60 to 82°C (140 to 180°F) or a 10 to 25 percent hydrochloric acid bath at room temperature. Nitric acid is frequently used in pickling stainless steel, often in conjunction with hydrofluoric, sulfuric, and hydrochloric acids. The most common stainless steel descaling process uses nitric acid in the concentration range of 5 to 25 percent in conjunction with hydrochloric acid at 0.5 to 3 percent. For light scale removal, the concentrations are 12 to 15 percent nitric acid and 1 percent of hydrofluoric acid by volume at a bath temperature of 120 to 140°F; for heavy oxides, the concentration of hydrofluoric is increased to 2 to 3 percent. Pickling operations on nonferrous metals such as aluminum, magnesium, zinc, and lead each have specific recommended acid concentrations.

Acid bright dips are usually mixtures of nitric and sulfuric acids employed to provide a mirrorlike surface on cadmium, magnesium, copper, copper alloys, silver, and in some cases, stainless steel.

The air contaminants released from pickling and bright dips include not only the mists of the acids used in the process but nitrogen oxides, if nitric acid is employed, and hydrogen chloride gas from processes using hydrochloric acid. Extensive information is available on the effects of exposure to inorganic acids. Accidental contact with skin and eyes produces burns, ulcers, and necrosis. For most acids, the acute effects of contact exposure are rapid and detected immediately by the affected individual. However, hydrogen fluoride penetrates the skin and the onset of symptoms may be delayed for hours, permitting deep tissue burns and severe pain. Airborne acid mists produce upper and lower respiratory irritation. Chronic exposure to nitrogen oxides can produce pulmonary edema. One series of epidemiological studies found excess laryngeal cancer in steel workers who conducted pickling operations using sulfuric and other acids (IARC, 1992).

An extensive listing of bath components and potential air contaminants for all common pickling and bright dip baths is provided in American Conference of Governmental Industrial Hygienists (ACGIH) Ventilation Manual (ACGIH, 1995). The extent of exposure will depend on bath temperature, surface area of work, current density (if bath is electrolytic), and whether the bath contains inhibitors that produce a foam blanket on the bath or that lower the surface tension of the bath and thereby reduce misting. It is a general rule that local exhaust ventilation is required for pickling and acid dip tanks operating at elevated temperature and for electrolytic processes. Extensive guidelines for local exhaust ventilation of acid dip tanks have been developed (ACGIH, 1995; ANSI, 1996; Burgess et al., 1989).

Minimum safe practices for pickling and bright dip operators have been proposed by Spring (1974): (1) Hands and faces should be washed before eating, smoking, or leaving plant. Eating and smoking should not be permitted at the work location. (2) Only authorized employees should be permitted to make additions of chemicals to baths. (3) Face shields, chemical handlers' goggles, rubber gloves, rubber aprons, and rubber platers' boots should be worn when adding chemicals to baths and when cleaning or repairing tanks. (4) Chemicals contacting the body should be washed off immediately and medical assistance obtained. (5) Supervisor should be notified of any change in procedures or unusual occurrences. Because the symptoms of hydrofluroic acid exposure can be delayed and the consequences so severe, any suspected exposure should be reported to a responsible medical authority immediately.

4.3.2 Alkaline Treatment

4.3.2.1 Alkaline Immersion Cleaning Acid and alkaline cleaning techniques are complementary in terms of the cleaning tasks that can be accomplished. Alkaline soak, spray, and electrolytic cleaning systems are superior to acid cleaning for removal of oil, gases, buffing compounds, certain soils, and paint. A range of alkaline cleansers including sodium hydroxide, potassium hydroxide, sodium carbonate, sodium meta- or orthosilicate, trisodium phosphate, borax, and tetrasodium pyrophosphate are used for both soak and electrolytic alkaline cleaning solutions.

The composition of the alkaline bath may be complex, with a number of additives to handle specific tasks (Spring, 1974). In nonelectrolytic cleaning of rust from steels, the bath may contain 50 to 80 percent caustic soda in addition to chelating and sequestering agents. The parts are immersed for 10 to 15 min and rinsed with a spray. The usual temperature range for these baths is 160–210°F, and alkaline mist and steam are potential air contaminants. Guidelines for local exhaust ventilation of these baths have been provided (ACGIH, 1995).

Electrolytic alkaline cleaning is an aggressive cleaning method. The bath is an electrolytic cell powered by direct current with the workpiece conventionally the cathode, and an inert electrode as the anode. The water dissociates; oxygen is released at the anode and hydrogen at the cathode. The hydrogen gas generated at the workpiece causes agitation of the surface soils with excellent soil removal. The gases released at the electrodes by the dissociation of water may result in the release of caustic mist and steam at the surface of the bath.

Surfactants and additives that provide a foam blanket are important to the proper operation of the bath. Ideally, the foam blanket should be 5 to 8 cm thick to trap the released gas bubbles and thereby minimize misting (NIOSH, 1985). If the foam blanket is too thin, the gas may escape, causing a significant alkaline mist to become airborne; if too thick, the blanket may trap hydrogen and oxygen with resulting minor explosions ignited by sparking electrodes.

4.3.2.2 Salt Baths A bath of molten caustic at 370 to 540° (700 to 1000°F) can be used for initial cleaning and descaling of cast iron, copper, aluminum, and nickel with subsequent quenching and acid pickling. The advantages claimed for this type of cleaning are that precleaning is not required and the process provides a good bond surface for a subsequent finish.

Molten sodium hydroxide is used at 430 to 540°C (800 to 1000°F) for general-purpose descaling and removal of sand on castings. A reducing process utilizes sodium hydride in the bath at 370°C (700°F) to reduce oxides to their metallic state. The bath utilizes fused liquid anhydrous sodium hydroxide with up to 2 percent sodium hydride, which is generated in accessory equipment by reacting metallic sodium with hydrogen. All molten baths require subsequent quenching, pickling, and rinsing. The quenching operation dislodges the scale through steam generation and thermal shock.

These baths require well-defined operating procedures owing to the hazard from molten caustic as well as safety hazards from the reaction of metallic sodium with hydrogen. Local exhaust ventilation is necessary, and the tank must be equipped with a complete enclosure to protect the operator from violent splashing as the part is immersed in the bath. Quenching tanks and pickling tanks must also be provided with local exhaust ventilation. Ventilation standards have been proposed for these operations (ACGIH, 1996; ANSI, 1996).

4.4 DEGREASING

For many decades the principal application of degreasing technology has been in the metalworking industry for the removal of machining oils, grease, drawing oils, chips, and other soils from metal parts. The technology has expanded greatly prompted by advances in both degreasing equipment and solvents. At this time it is probably the most common industrial process extending across all industries, including jewelry, electronics, special optics, electrical, machining, instrumentation, and even rubber and plastic goods. The significant occupational health problems associated with cold and vapor-phase degreasing processes are described below.

4.4.1 Cold Degreasing

The term *cold degreasing* identifies the use of a solvent at room temperature in which parts are dipped, sprayed, brushed, wiped, or agitated for removal of oil and grease. It is the simplest of all degreasing processes, requiring only a simple container with the solvent of choice. It is widely used in small production shops, maintenance and repair shops, and automotive garages. The solvents used have traditionally included low-volatility, high-flash petroleum distillates such as mineral spirits and Stoddard Solvent to solvents of high volatility including aromatic hydrocarbons, chlorinated hydrocarbons, and ketones. The choice of degreasing agent has changed significantly since the 1987 Montreal Protocol. The ozone depletion potential (ODP) of a given degreaser has become an important variable in choice.

Skin contact with cold degreasing materials should be avoided by work practices and the use of protective clothing. Glasses and a face shield should be used to protect the eyes and face from accidental splashing during dipping and spraying. Solvent tanks should be provided with a cover and, if volatile solvents are used, the dip tank and drain station should be provided with ventilation control.

Spraying with high-flash petroleum distillates such as Stoddard Solvent, mineral spirits, or kerosene is a widely used method of cleaning oils and grease from metals. This operation should be provided with suitable local exhaust ventilation (ACGIH, 1995). The hood may be a conventional spray booth type and may be fitted with a fire door and automatic extinguishers. The fire hazard in spraying a high-flash petroleum solvent is comparable to spraying many lacquers and paints.

4.4.2 Vapor Degreasing

A vapor degreaser is a tank containing a quantity of solvent heated to its boiling point. The solvent vapor rises and fills the tank to an elevation determined by the location of a condenser. The vapor condenses and returns to the liquid sump. The tank has a freeboard that extends above the condenser to minimize air currents inside the tank (Fig. 4.1). As the parts are lowered into the hot vapor, the vapor condenses on the cold part and dissolves the surface oils and greases. This oily condensate drops back into the liquid solvent at the base of the tank. The solvent is continuously evaporated to form the vapor blanket. Because the oils are not vaporized, they remain to form a sludge in the bottom of the tank. The scrubbing action of the condensing vapor continues until the temperature of the part reaches the temperature of the vapor, whereupon condensation stops, the part appears dry, and it is removed from the degreaser. The time required to reach this point depends on the particular solvent, the temperature of the vapor, the

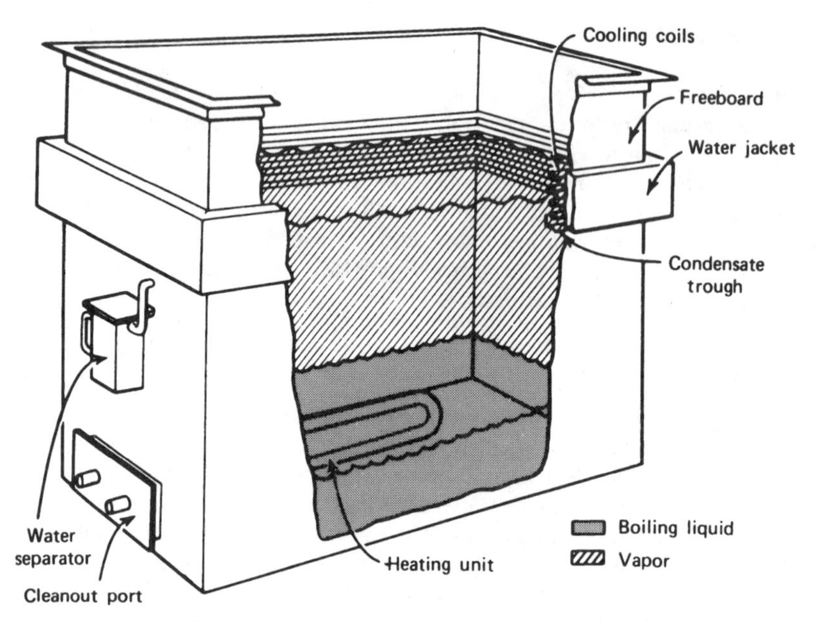

Fig. 4.1 Major components of a vapor-phase degreaser.

weight of the part, and its specific heat. The vapor-phase degreaser does an excellent job of drying parts after aqueous cleaning and prior to plating; it is frequently used for this purpose in the jewelry industry.

4.4.2.1 Types of Vapor-Phase Degreasers and Solvents The simplest form of vapor-phase degreaser, shown in Fig. 4.1, utilizes only the vapor for cleaning. The straight vapor-cycle degreaser is not effective on small, light work because the part reaches the temperature of the vapor before the condensing action has cleaned the part. Also, the straight vapor cycle does not remove insoluble surface soils. For such applications, the vapor-spray-cycle degreaser is frequently used. The part to be cleaned is first placed in the vapor zone as in the straight vapor-cycle degreaser. A portion of the vapor is condensed by a cooling coil and fills a liquid solvent reservoir. This warm liquid solvent is pumped to a spray lance, which can be used to direct the solvent on the part, washing off surface oils and cooling the part, thereby permitting final cleaning by vapor condensation (Fig. 4.2).

A third degreaser design has two compartments, one with warm liquid solvent and a second compartment with a vapor zone. The work sequence is vapor, liquid, and vapor. This degreaser is used for heavily soiled parts with involved geometry or to clean a basket of small parts that nest together. Finally, a three-compartment degreaser has vapor, boiling and warm liquid compartments with a vapor, boiling liquid, warm liquid, and vapor work sequence. Other specialty degreasers encountered in industry include enclosed conveyorized units for continuous production cleaning.

Ultrasonic cleaning modules installed in vapor degreasers have found broad application for critical cleaning jobs. In an ultrasonic degreaser, a transducer operating in the range of 20 to 40 kHz is mounted at the base of a liquid immersion solvent tank. The

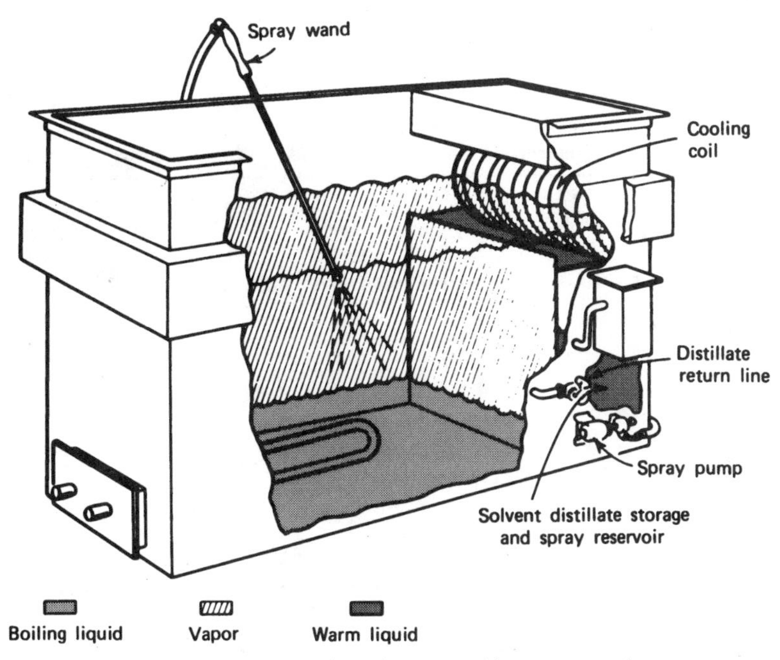

Fig. 4.2 A vapor-phase degreaser with a spray wand.

transducer alternately compresses and expands the solvent forming small bubbles that cavitate or collapse at the surface of the workpiece. The cavitation phenomenon disrupts the adhering soils and cleans the part. Ultrasonic degreasers use chlorinated solvents at 32 to 49°C (90 to 120°F) and aqueous solutions at 43 to 71°C (110 to 160°F). These degreasers commonly employ refrigerated or water-chilled coils for control of solvent vapors; the manufacturers claim that local exhaust ventilation is not needed in this configuration.

The solvents commonly used with vapor-phase degreasers have traditionally included trichloroethylene, perchloroethylene, 1,1,1-trichloroethane, methylene chloride, and a series of Freon® solvents (Table 4.1) (DuPont, 1987; Dow, 1978). The degreaser must be designed for and used with a specific solvent. Most chlorinated degreasing solvents sold under trade names contain a stabilizer present in a concentration of less than 5 percent. The purpose of the stabilizer is to neutralize any free acid that might result from oxidation of the degreasing liquid in the presence of air, hydrolysis in the presence of water, or pyrolysis under the influence of high temperatures. The stabilizer is not a critical issue in establishing health risk to the worker owing to its low concentration; the solvent itself is usually the predictor of risk.

The emphasis on using non-ozone-depleting chemicals in degreasing operations has greatly influenced the choice of degreasers. The use of 1,1,1-trichloroethane and many of the Freons® has diminished, while trichloroethylene is being used more frequently. The flammability and toxicity of the new generation of materials requires careful review before use. New materials, such as D-limonene, have been found to have their own set of issues including dermatitis and low odor thresholds. Semiaqueous processes using terpenes, dibasic esters, *n*-methyl pyrrolidone, and other materials have been used in

TABLE 4.1 Properties of Vapor Degreasing Solvents

	Trichloro-ethylene	Perchloro-ethylene	Methylene chloride	Trichloro-trifluoroethane[a]	Methyl chloroform (1,1,1-trichloroethane)
Boiling point	87	121	40	48	74
°C	188	250	104	118	165
°F					
Flammability	Nonflammable under vapor degreasing conditions				
Latent heat of vaporization (b.p.). Btu/lb	103	90	142	63	105
Specific gravity					
Vapor (air = 1.00)	4.53	5.72	2.93	6.75	4.60
Liquid (water = 1.00)	1.464	1.623	1.326	1.514	1.327

[a]Binary azeotropes are also available with ethyl alcohol, isopropyl alcohol, acetone, and methylene chloride.

conjunction with surfactants to provide effective systems. In many cases, nonchemical cleaning substitutes have been chosen, such as bead blasting with plastic pellets or frozen carbon dioxide pellets.

The loss of degreaser solvent to the workplace obviously depends on a number of operating conditions, including the type and properties of contaminants removed, cleaning cycle, volume of material processed, design of parts being cleaned, and most important, work practices. The effort to dry parts quickly with an air hose, for example, will lead to higher exposures. Maintenance and cleaning of degreasers can yield high acute exposures and may require the use of respiratory protection.

4.4.2.2 Control The vapor level in the degreaser is controlled by a nest of water-fed condensing coils located on the inside perimeter of the tank (Fig. 4.1). In addition, a water jacket positioned on the outside of the tank keeps the freeboard cool. The vertical distance between the lowest point at which vapors can escape from the degreaser machine and the highest normal vapor level is called the *freeboard*. The freeboard should be at least 15 in. and not less than one-half to three-fifths the width of the machine. The effluent water from the coils and water jacket should be regulated to 32 to 49°C (90 to 120°F); a temperature indicator or control is desirable.

Properly designed vapor degreasers have a thermostat located a few inches above the normal vapor level to shut off the source of heat if the vapor rises above the condensing surface. A thermostat is also immersed in the boiling liquid; if overheating occurs, the heat source is turned off.

There is a difference of opinion on the need for local exhaust ventilation on vapor-phase degreasers. Authorities frequently cite the room volume as a guide in determining if ventilation is needed: Local exhaust ventilation is needed if there is less than 200 ft^3 in the room for each square foot of solvent surface, or if the room is smaller than 25,000 ft^3. In fact, ventilation control requirements depend on the degreaser design, location, maintenance, and operating practices. Local exhaust increases solvent loss, and the installation may require solvent recovery before discharging to outdoors. To ensure effective use of local exhaust, the units should be installed away from drafts from open windows, spray booths, space heaters, supply air grilles, and fans. Equally important is the parts loading and unloading station. When baskets of small parts are degreased, it is not possible to eliminate drag out completely, and the unloading station usually requires local exhaust.

Later in this chapter under the discussion of welding, reference is made to the decomposition of chlorinated solvent under thermal and UV stress with the formation of chlorine, hydrogen chloride, and phosgene. Because degreasers using such solvents are frequently located near welding operations, this problem warrants attention. In a laboratory study of the decomposition potential of methyl chloride, methylene chloride, carbon tetrachloride, ethylene dichloride, 1,1,1-trichloroethane, *o*-dichlorobenzene, trichloroethylene, and perchloroethylene, only the latter two solvents decomposed in the welding environment to form dangerous levels of phosgene, chlorine, and hydrogen chloride (Dow, 1987). All chlorinated materials thermally degrade if introduced to direct-fired combustion units commonly used in industry. If a highly corroded heater is noted in the degreaser area, it may indicate that toxic and corrosive air contaminants are being generated.

Installation instructions and operating precautions for the use of conventional vapor-phase degreasers have been proposed by various authorities (ASTM, 1989). The following minimum instructions should be observed at all installations:

1. If the unit is equipped with a water condenser, the water should be turned on before the solvent is heated.
2. Water temperature should be maintained between 27 and 43°C (81 and 110°F).
3. Work should not be placed in and removed from the vapor faster than 11 fpm (0.055 m/sec). If a hoist is not available, a support should be positioned to hold the work in the vapor. This minimizes the time the operator must spend in the high exposure zone.
4. The part must be kept in the vapor until it reaches vapor temperature and is visually dry.
5. Parts should be loaded to minimize pullout. For example, cup-shaped parts should be inverted.
6. Overloading should be avoided because it will cause displacement of vapor into the workroom.
7. The work should be sprayed with the lance below the vapor level.
8. Proper heat input must be available to ensure vapor level recovery when large loads are placed in the degreaser.
9. A thermostat should be installed in the boiling solvent to prevent overheating of the solvent.
10. A thermostat vapor level control must be installed above the vapor level inside the degreaser and set for the particular solvent in use.
11. The degreaser tank should be covered when not in use.
12. Hot solvent should not be removed from the degreaser for other degreasing applications, nor should garments be cleaned in the degreaser.
13. An emergency eyewash station would be located near the degreaser for prompt irrigation of the eye in case of an accidental splash.

To ensure efficient and safe operation, vapor-phase degreasers should be cleaned when the contamination level reaches 25 percent. The solvent should be distilled off until the heating surface or element is 1-1/2 in. below the solvent level or until the solvent vapors fail to rise to the collecting trough. After cooling, the oil and solvent should be drained off and the sludge removed. It is important that the solvent be cooled prior to draining. In addition to placing the operators at risk, removing hot solvent causes serious air contamination and frequently requires the evacuation of plant personnel from the building. A fire hazard may exist during the cleaning of machines heated by gas or electricity because the flash point of the residual oil may be reached and because trichloroethylene itself is flammable at elevated temperatures. After sludge and solvent removal, the degreaser must be mechanically ventilated before any maintenance work is undertaken. A person should not be permitted to enter a degreaser or place his or her head in one until all controls for entry into a confined space have been put in place. Anyone entering a degreaser should wear a respirator suitable for conditions immediately hazardous to life, as well as a lifeline held by an attendant. Anesthetic concentrations of vapor may be encountered, and oxygen concentrations may be insufficient. Such an atmosphere may cause unconsciousness with little or no warning. Deaths occur each year because of failure to observe these precautions.

The substitution of one degreaser solvent for another as a control technique must be done with caution. Such a decision should not be based solely on the relative exposure standards but must consider the type of toxic effect, the photochemical properties, the physical properties of the solvents including vapor pressure, and other parameters describing occupational and environmental risk.

4.5 ELECTROPLATING

Metal, plastic, and rubber parts are plated to prevent rusting and corrosion, for appearance, to reduce electrical contact resistance, to provide electrical insulation, as a base for soldering operations, and to improve wearability. The common plating metals include cadmium, chromium, copper, gold, nickel, silver, and zinc. Prior to electroplating, the parts must be cleaned and the surfaces treated as described in Sections 4.2 and 4.3.

There are approximately 160,000 electroplating workers in the United States; independent job shops average 10 workers, and the captive electroplating shops have twice that number. A number of epidemiologic studies have identified a series of health effects in electroplaters ranging from dermatitis to elevated mortality for a series of cancers.

4.5.1 Electroplating Techniques

The basic electroplating system is shown in Fig. 4.3. The plating tank contains an electrolyte consisting of a metal salt of the metal to be applied dissolved in water. Two electrodes powered by a low-voltage dc power supply are immersed in the electrolyte. The cathode is the workpiece to be plated, and the anode is either an inert electrode or, most frequently, a slab or a basket of spheres of the metal to be deposited. When power is applied, the metal ions deposit out of the bath on the cathode or workpiece. Water is dissociated, releasing hydrogen at the cathode and oxygen at the anode. The anode may be designed to replenish the metallic ion concentration in the bath. Current density expressed in amperes per unit area of workpiece surface varies depending on the operation. In addition to the dissolved salt containing the metallic ion, the plating bath may contain additives to adjust the electrical conductivity of the bath, define the type of plating deposit, and buffer the pH of the bath.

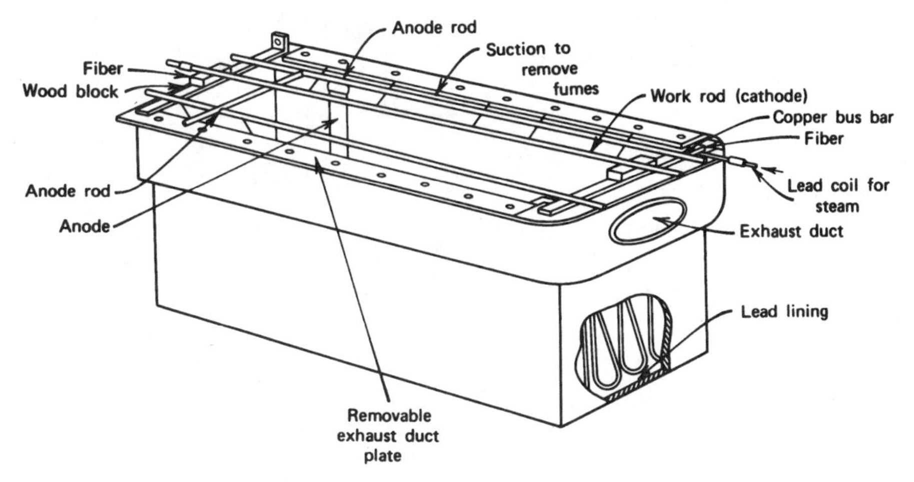

Fig. 4.3 An electroplating tank.

Anodizing, a common surface treatment for decoration, corrosion resistance, and electrical insulation on such metals as magnesium, aluminum, and titanium, operates in a different fashion. The workpiece is the anode, and the cathode is a lead bar. The oxygen formed at the workpiece causes a controlled surface oxidation. The process is conducted in a sulfuric or chromic acid bath with high current density, and because its efficiency is quite low, the amount of misting is high.

In conventional plating operations, individual parts on a hanger or a rack of small parts are manually hung from the cathode bar. If many small pieces are to be plated, the parts may be placed in a perforated plastic barrel in electrical contact with the cathode bar, and the barrel is immersed in the bath. The parts are tumbled to achieve a uniform plating.

In a small job-shop operation, the parts are transferred manually from tank to tank as dictated by the type of plating operation. The series of steps necessary in one plating operation (Figure 4.4) illustrates the complexity of the operation. After surface cleaning and preparation, the electroplate steps are completed with a water rinse tank isolating each tank from contamination. In high production shops, an automatic transfer unit is programmed to cycle the parts from tank to tank, and the worker is required only to load and unload the racks or baskets. Automatic plating operations may permit exhaust hood enclosures on the tanks and therefore more effective control of air contaminants. Exposure is also limited because the worker is stationed at one loading position and is not directly exposed to air contaminants released at the tanks.

4.5.2 Air Contaminants

The principal source of air contamination in electroplating operations is the release of the bath electrolyte to the air by the gassing of the bath. As mentioned above, the path operates as an electrolytic cell; so water is dissociated and hydrogen is released at the cathode and oxygen at the anode. The gases released at the electrodes rise to the surface of the bath and burst, generating a respirable mist that becomes airborne. The mist generation rate depends on the bath efficiency. In copper plating, the efficiency of the plating bath is nearly 100 percent; that is, essentially all the energy goes into the plating

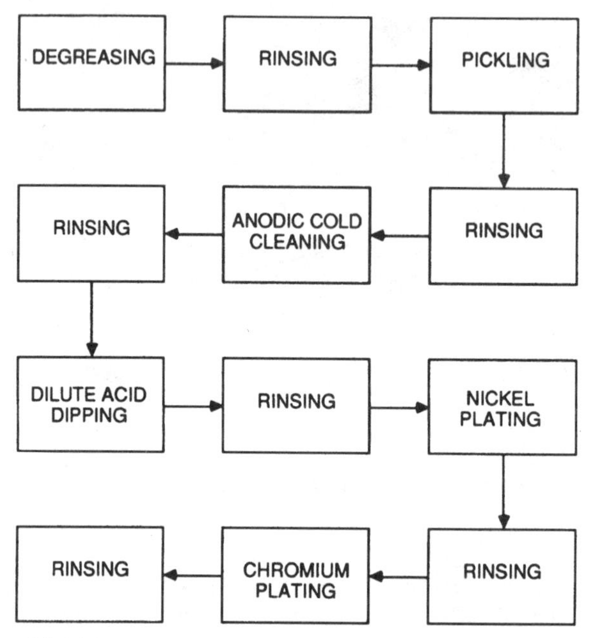

Fig. 4.4 The steps involved in chromium plating of steel.

operation and little into the electrolytic dissociation of water (Blair, 1972). Nickel plating baths operate at 95 percent efficiency; so only 5 percent of the energy is directed to dissociation of water, and misting is minimal. However, chromium plating operations are quite inefficient, and up to 90 percent of the total energy may be devoted to dissociation of the bath with resulting severe gassing, and resulting potential exposure of the operator to chromic acid mist. Although the contamination generation rate of the bath is governed principally by the efficiency of the bath, it also varies with the metallic ion concentration in the bath, the current density, the nature of the bath additives, and bath temperature. Air or mechanical agitation of the bath used to improve plating quality may also release the bath as droplets.

The health significance of the mist generated by electroplating processes depends, of course, on the contents of the bath. The electrolyte mist released from the bath is alkaline or acidic depending on the specific electroplating process. A majority of the alkaline baths based on cyanide salts solutions are used for cadmium, copper, silver, brass, and bronze plating. Acidic solutions are used for chromium, copper, nickel, and tin. The exact composition of the baths can be obtained from an electroplater's handbook or, if proprietary, from the supplier. An inventory of the nature of the chemicals in the common electroplating baths, the form in which they are released to the air, and the rate of gassing has evolved over the past several decades, drawing heavily on the experience of the state industrial hygiene programs in New York and Michigan (Burgess et al., 1989). These data are useful in defining the nature of the contaminant and the air sampling procedure necessary to define the worker exposure. As noted below, these data are also valuable in defining the ventilation requirements for various plating operations.

4.5.3 Control

Proprietary bath additives are available to reduce the surface tension of the electrolyte and therefore reduce misting. Another additive provides a thick foam that traps the mist released from the bath. This agent is best used for tanks that operate continuously. A layer of plastic chips, beads, or balls on the surface of the bath also trap the mist and permit it to drain back into the bath. Where possible, tanks should be provided with covers to reduce bath loss.

Although the use of the above mist suppressants are helpful, they will not alone control airborne contaminants from plating tanks at an acceptable level. Local exhaust ventilation in the form of lateral slot or upward plenum slotted hoods is the principal control measure.

The design approach described in the Industrial Ventilation Manual (ACGIH, 1996) provides a firm basis for the control of electroplating air contaminants. This procedure permits one to determine a minimum capture velocity based on the hazard potential of the bath and the rate of contaminant generation. The exhaust volume is based on the capture velocity and the tank measurements and geometry. Owing to the severe corrosion of duct work, periodic checks of the exhaust systems in plating shops are necessary. Qualitative assessment of the ventilation is possible using smoke tubes or other tracers. In many cases, the use of partitions to minimize the disruptive effects of drafts may greatly improve the installed ventilation.

In addition to the proper design and installation of good local exhaust ventilation, one must provide adequate replacement air, backflow dampers on any combustion devices to prevent carbon monoxide contamination of the workplace, and suitable air cleaning.

Because a low-voltage dc power supply is used, an electrical hazard does not exist at the plating tanks. A fire and explosion risk may result from solvent degreasing and spray painting conducted in areas contiguous with the plating area. The major chemical safety hazards are due to handling concentrated acids and alkalis and the accidental mixing of acids with cyanides and sulfides during plating, bath preparation, and waste disposal with the formation of hydrogen cyanide and hydrogen sulfide.

The educated use of protective equipment by electroplaters is extremely important in preventing contact with the various sensitizers and corrosive materials encountered in the plating shop. The minimum protective clothing should include rubber gloves, aprons, boots, and chemical handler's goggles. Aprons should come below the top of the boots. All personnel should have a change of clothing available at the workplace. If solutions are splashed on the work clothing, they should be removed, the skin washed, and the worker should change to clean garments. A shower and eyewash station serviced with tempered water should be available at the workplace. The wide range of chemicals presents a major dermatitis hazard to the plater, and skin contact must be avoided. Nickel is a skin sensitizer and may cause nickel itch, developing into a rash with skin ulcerations.

A summary of the health hazards encountered in electroplating shops and the available controls is presented in Table 4.2. More information on controls can be found in the references (American Electroplaters, 1989; NIOSH, 1985).

4.6 GRINDING, POLISHING, BUFFING

These operations are grouped together for discussion because they all involve controlled use of bonded abrasives for metal finishing operations; in many cases, the operations are

TABLE 4.2 Summary of Major Electroplating Health Hazards

Exposure	How contamination occurs
	Inhalation
Mist, gases, and vapors	
Hydrogen cyanide	Accidental mixing of cyanide solutions and acids
Chromic acid	Released as a mist during chrome plating and anodizing
Hydrogen sulfide	Accidental mixing of sulfide solutions and acids
Nitrogen oxides	Released from pickling baths containing nitric acid
Dust	Released during weighing and transferring of solid bath materials, including cyanides and cadmium salts
Fumes	Generated during on-site repair of lead-lined tanks using torch-burning techniques
	Ingestion
Workplace particles	Accidental ingestion during smoking and eating at workplace
	Skin contact
Cyanide compounds	Absorption through the skin
Solvents	Defatting by solvents
Irritants	Primary irritants contacting the skin
Contact allergens	Sensitization
	Control technology
Local exhaust ventilation	
Mist reduction	
Reduce surface tension	
Coat surface	
Tank covers	
Isolation of stored chemicals	

conducted in the sequence noted. This discussion covers the nonprecision applications of these techniques.

4.6.1 Processes and Materials

Nondimensional application of grinding techniques includes cutoff operations in foundries, rough grinding of forgings and castings, and grinding out major surface imperfections in metal fabrications. Grinding is frequently done with wheels and disks of various geometries made up of selected abrasives in different bonding structural matrices. The common abrasives are aluminum oxide and silicon carbide; less common are diamond and cubic boron nitride. A variety of bonding materials are available to provide mechanical strength and yet release the spent abrasive granules to renew the cutting sur-

TABLE 4.3 Grinding Wheel Specification Nomenclature

Abrasives (First Letter in Specification)	
A	Aluminum oxide
C	Silicon carbide
D or ND	Natural diamonds
SD	Synthetic diamonds
CB or CBN	Cubic boron nitride
Bond (Last Letter in Specification)	
V	Vitrified
B	Resinoid
R	Rubber
S	Silicate

face. Vitrified glass is the most common bonding agent. The grinding wheel is made by mixing clay and feldspar with the abrasive, pressing it in shape, and firing it at high temperature to form a glass coating to bond for the abrasive grains. Resinoid wheel bonds, based on thermosetting resins such as phenol-formaldehyde, are used for diamond and boron nitride wheels and are reinforced with metal or fiber glass for heavy-duty applications including cutoff wheels. Other bonding agents are sodium silicate and rubber based agents.

The abrasive industry utilizes a standard labeling nomenclature to identify the grinding wheel design (ANSI, 1978) that includes the identification of the abrasive and bonding agent as well as grain size and structure and other useful information. Table 4.3 lists grinding wheel specification nomenclature.

Exact information on the generation rate of grinding wheel debris for various applications is not available. However, the wheel components normally make up a small fraction of the total airborne particles released during grinding; the bulk of airborne particles are released from the workpiece. After use, the grinding wheel may load or plug, and the wheel must be "dressed" with a diamond tool or "crushed dressed" with a steel roller. During this brief period, a significant amount of the wheel is removed and a small quantity may become airborne.

Polishing techniques are used to remove workpiece surface imperfections such as tool marks. This technique may remove as much as 0.1 mm of stock from the workpiece. The abrasive, again usually aluminum oxide or silicon dioxide, is bonded to the surface of a belt, disk, or wheel structure in a closely governed geometry, and the workpiece is commonly applied to the moving abrasive carrier by hand.

The buffing process differs from grinding and polishing in that little metal is removed from the workpiece. The process merely provides a high luster surface by smearing any surface roughness with a lightweight abrasive. Red rouge (ferric oxide) and green rouge (chromium oxide) are used for soft metals, aluminum oxide for harder metals. The abrasive is blended in a grease or wax carrier that is packaged in a bar or tube form. The buffing wheel is made from cotton or wool disks sewn together to form a wheel or "buff." The abrasive is applied to the perimeter of the wheel, and the workpiece is then pressed against the rotating wheel mounted on a buffing lathe.

4.6.2 Exposures and Control

The hazard potential from grinding, polishing, and buffing operations depends on the specific operation, the workpiece metal and its surface coating, and the type of abrasive system in use. A NIOSH-sponsored study of the ventilation requirements for grinding, polishing, and buffing operations showed that the major source of airborne particles in grinding and polishing is the workpiece, whereas the abrasive and the wheel textile material represent the principal sources of contamination in buffing (NIOSH, 1975).

The health status of grinders, polishers, and buffers has not been extensively evaluated. One study did not find elevated cancer mortality in metal polishers (Blair, 1980), while two other studies did (Sparks and Wegman, 1980; Jarvholm et al., 1982). However, each of the polishing operations studied was unique; so it is difficult to generalize to the industry as a whole. A listing of the metal and alloys worked and information on the nature of the materials released from the abrasive system are needed in order to evaluate the exposure of a particular operator. In many cases, the exposure to total dust can be evaluated by means of personal air sampling with gravimetric analysis. If dusts of toxic metals are released, then specific analysis for these contaminants is necessary.

The need for local exhaust ventilation on grinding operations has been addressed by British authorities (United Kingdom Department of Employment, 1974), who state that control is required if one is grinding toxic metals and alloys, ferrous and nonferrous castings produced by sand molding, and metal surfaces coated with toxic material.

The same guidelines can be applied to polishing operations. The ventilation requirements for buffing, however, are based on the large amount of debris released from the wheel, which may be a housekeeping problem and potential fire risk.

The conventional ventilation control techniques for fixed location grinding, polishing, and buffing are well described in the ACGIH and ANSI publications (ACGIH, 1995; ANSI, 1996). It is more difficult to provide effective ventilation controls for portable grinders used on large castings and forgings. Flexible exterior hoods positioned by the worker may be effective. High-velocity, low-volume exhaust systems with the exhaust integral to the grinder also are suitable for some applications (Fletcher, 1988). The performance of the conventional hoods on grinding, polishing, and buffing operations should be checked periodically.

The hazard from bursting wheels operated at high speed and the fire hazard from handling certain metals such as aluminum and magnesium are not covered here, but these problems affect the design of hoods and the design of wet dust collection systems.

The hazard from vibration from hand tools leading to VWF can result from portable grinders and also pedestal mounted wheels. A strategy for the protection of workers from VWF has been developed by NIOSH (1989) and includes tool redesign, using protective equipment, and monitoring exposure and health.

4.7 HEAT TREATING

A range of heat treating methods for metal alloys is available to improve the strength, impact resistance, hardness, durability, and corrosion resistance of the workpiece (Heat Treaters Guide, 1982). In the most common procedures, metals are hardened by heating the workpiece to a high temperature with subsequent rapid cooling. Softening processes normally involve only heating, or heating with low cooling.

4.7.1 Surface Hardening

Case hardening, the production of a hard surface or case to the workpiece, is normally accomplished by diffusing carbon or nitrogen into the metal surface to a given depth to achieve the hardening of the alloy. This process may be accomplished in air, in atmospheric furnaces, or in immersion baths by one of the following methods.

4.7.1.1 Carburizing In this process, the workpiece is heated in a gaseous or liquid environment containing high concentrations of a carbon-bearing material that is the source of the diffused carbon. In gas carburizing, the parts are heated in a furnace containing hydrocarbon gases or carbon monoxide; in pack carburizing, the part is covered with carbonaceous material that burns to produce the carbon-bearing gas blanket. In liquid carburizing, the workpiece is immersed in a molten bath that is the source of carbon.

In gas carburizing, the furnace atmosphere is supplied by an atmosphere generator. In its simplest form, the generator burns a fuel such as natural gas under controlled conditions to produce the correct concentration of carbon monoxide, which is supplied to the furnace. Because carbon monoxide concentrations up to 40 percent may be used, small leaks may result in significant workroom exposure. To control emission from gas carburizing operations, the combustion processes should be closely controlled, furnaces maintained in tight condition, dilution ventilation installed to remove fugitive leaks, furnaces provided with flame curtains at doors to control escaping gases, and self-contained breathing apparatus available for escape and repair operations. In liquid carburizing, a molten bath of sodium cyanide and sodium carbonate provides a limited amount of nitrogen and the necessary carbon for surface hardening.

4.7.1.2 Cyaniding The conventional method of liquid carbonitriding is immersion in a cyanide bath with a subsequent quench. The part is commonly held in a sodium cyanide bath at temperatures above 870°C (1600°F) for 30 to 60 min. The air contaminant released from this process is sodium carbonate; cyanide compounds are not released, although there are no citations in the open literature demonstrating this. Local or dilution ventilation is frequently applied to this process. The handling of cyanide salts requires strict precautions including secure and dry storage, isolation from acids, and planned disposal of waste. Care must be taken in handling quench liquids because the cyanide salt residue on the part will in time contaminate the quench liquid.

4.7.1.3 Gas Nitriding Gas nitriding is a common means of achieving hardening by the diffusion of nitrogen into the metal. This process utilizes a furnace atmosphere of ammonia operating at 510 to 570°C (950 to 1050°F). The handling of ammonia in this operation is hazardous in terms of fire, explosion, and toxicity.

4.7.2 Annealing

Annealing is a general term used to describe many heating or cooling cycles that change the metallurgical properties of the workpiece. The process varies depending on the alloy and the use of the part, but in all cases it involves heating at a given temperature for a specific time and then cooling at a desired rate, frequently at slow rates. Different types of salt baths may be used, including a blend of potassium nitrate, sodium nitrate, and nitrite for low temperatures or a blend of sodium and potassium chloride and barium nitrite for high temperatures baths. Careful handling and storage procedures must be used for nitrate salts, because of their reactivity and explosivity.

4.7.3 Quenching

The quench baths may be water, oil, molten salt, liquid air, or brine. The potential problems range from a nuisance problem due to release of steam from a water bath to acrolein or other thermal degradation products from oil. Local exhaust ventilation may be necessary on oil quench tanks.

4.7.4 Hazard Potential

The principal problems in heat treating operations are due to the special furnace environments, especially carbon monoxide, and the special hazards from handling bath materials. Although the hazard potential is significant from these operations, few data are available.

The fire and safety hazards of these operations are considerable and have been extensively reviewed (FM, 1991; NFPA, 1991). Salt bath temperature controls must be reliable, and the baths must be equipped with automatic shutdowns. Venting rods need to be inserted in baths before shutdown and reheating to release gases when the bath is again brought up to temperature. If this is not done and gas is occluded in the bath, explosions or blowouts may occur. Parts must be clean and dry before immersion in baths for residual grease, paint, and oil may cause explosions. Where sprinkler systems are used, canopies should be erected above all oil, salt, and metal baths to prevent water from cascading into them. Dilution ventilation for fugitive emissions as well as local exhaust ventilation for baths are often provided, but specific standards have not been developed.

The physical hazards for employees include heat stress from furnace processes and noise from mechanical equipment and combustion air. Personal protective equipment may include goggles, face shields, heavy gloves and guantlets, reflective clothing, and protective screens. The Wolfson Health Treatment Centre reports (1981, 1983, 1984) provide detailed information on hazards and safe practices.

4.8 METAL MACHINING

The fabrication of metal parts is done with a variety of machine tools, the most common of which are the lathe, drill press, miller, shaper, planer, and surface grinder. The occupational health hazards from these operations are similar; so they are grouped together under conventional machining. Two rapidly expanding techniques, electrochemical and electrical discharge machining, are also discussed in this section.

4.8.1 Conventional Metal Machining

The major machining operations of turning, milling, and drilling utilize cutting tools that shear metal from the workpiece as either the workpiece or the tool rotates. A thin running coil is formed that normally breaks at the tool to form small chips. Extremes of temperature and pressure occur at the interface between the cutting tool and the work. To cool this point, provide an interface lubricant, and help flush away the chips, a coolant or cutting oil is directed on the cutting tool in a solid stream (flood) or a mist.

The airborne particles generated by these machining operations depend on the type of base metal and cutting tool, the dust-forming characteristics of the metal, the machining technique, and the coolant and the manner in which it is applied. Each of these concerns is addressed briefly in this section.

The type of metal being machined is of course of concern. The metals range from mild steel with no potential health hazard as a result of conventional machining to various high-temperature and stainless alloys incorporating known toxic metals including lead, chromium, nickel, and cobalt, which may present low airborne exposures to toxic metals depending on the machining technique. Finally, highly toxic metals, such as beryllium, do present significant exposures that require vigorous control in any machining operation. Under normal machining operations, excluding dimensional grinding, the airborne dust concentration from conventional metals and alloys is minimal.

A range of specialized alloys has been developed for use in the manufacture of cutting tools. These materials include (a) high carbon steels with alloying elements of vanadium, chromium, and manganese; (b) high-speed steels containing manganese and tungsten; (c) special cobalt steels; (d) cast alloys of tungsten, chromium, and cobalt; and (e) tungsten carbide. The loss of material from the cutting tool is insignificant during conventional machining, and therefore airborne dust concentrations from the tool do not represent a potential hazard. However, preparing the cutting tools may involve a significant exposure to toxic metal dusts during grinding and sharpening, and such operations should be provided with local exhaust ventilation.

Coolants and cutting fluids are designed to cool and lubricate the point of the cutting tool and flush away chips. These fluids are currently available in the form of (a) soluble (emulsified) cutting oils based on mineral oil emulsified in water with soaps or sulfonates; (b) straight cutting oils based on a complex mixture of paraffinic, naphthenic, and aromatic mineral oils with the addition of fatty acids; (c) synthetic oils of varying composition; and a mixture of (a) and (c). A description of the composition of the three types of cutting fluids including special additives is shown in Table 4.4 (NIOSH, 1978).

4.8.1.2 Health Effects Cutting fluids present two potential health problems: extensive skin contact with the cutting fluids and the inhalation of respirable oil mist. It has been estimated that over 400,000 cases of dermatitis occur in the United States each year from contact with coolants and cutting fluids. Soluble oils frequently cause eczematous dermatitis whereas the straight oils (insoluble) cause folliculitis. One also notes occasional sensitization to coolants.

There continues to be a difference of opinion on the role of bacterial contamination of fluids in dermatitis; however, there is agreement that maintenance of coolants is of hygienic significance. A coolant sampling procedure is now available that permits evaluation of aerobic bacteria, yeasts, and fungi concentration in coolants using simple dip slides.

The application of the cutting fluid to hot, rotating parts releases an oil mist that causes the characteristic smell in the machine shop. The health effects from extended exposure to airborne mists of mineral oil and synthetic coolants is not clear. The association between mineral oil based fluids and squamous cell carcinoma was observed in the United Kingdom in the 1800s. Unrefined or partially refined mineral oil has been classified as a carcinogen by IARC (1987); one suspected causative agent are the polycyclic aromatic hydrocarbons (PAHs) that are not removed by mild refining techniques. A study conducted in the automotive parts fabrication industry found a twofold increase

TABLE 4.4 Composition of Cutting Fluids

I. Mineral oil
   1. Base 60–100%, paraffinic or naphthenic
   2. Polar additives
      a. Animal and vegetable oils, fats, and waxes to wet and penetrate the chip/tool interface
      b. Synthetic boundary lubricants: esters, fatty oils and acids, poly or complex alcohols
   3. Extreme pressure (EP) lubricants
      a. Sulfur-free, or combined as sulfurized mineral oil or sulfurized fat
      b. Chlorine, as long-chain chlorinated wax or chlorinated ester
      c. Combination: sulfo-chlorinated mineral oil or sulfo-chlorinated fatty oil
      d. Phosphorus, as organic phosphate or metallic phosphate
   4. Germicides

II. Emulsified oil (soluble oil)—opaque, milky appearance
   1. Base: mineral oil, comprising 50–90% of the concentrate; in use the concentrate is diluted with water in ratios of 1:5 to 1:50
   2. Emulsifiers: petroleum sulfonates, amine soaps, rosin soaps, naphthenic acids
   3. Polar additives: sperm oil, lard oil, and esters
   4. Extreme pressure (EP) lubricants
   5. Corrosion inhibitors: polar organics, e.g., hydroxylamines
   6. Germicides
   7. Dyes

III. Synthetics (transparent)
   1. Base: water, comprising 50–80% of the concentrate; in use the concentrate is diluted with water in ratios of 1:10 to 1:200. True synthetics contain no oil. Semisynthetics are available that contain mineral oil present in amounts of 5–25% of the concentrate
   2. Corrosion inhibitors
      a. Inorganics: borates, nitrites, nitrates, phosphates
      b. Organics: amines, nitrites (amines and nitrites are typical and cheap)
   3. Surfactants
   4. Lubricants: esters
   5. Dyes
   6. Germicides

Source: Ref. 36.

of larynx cancer for straight oil exposure (Eisen *et al.*, 1994). Other studies have found increased digestive tract and other malignancies (Decoufle *et al.*, 1978; Jarvholm *et al.*, 1982; Vena *et al.*, 1985). The exact agent(s) responsible for the increased cancer rates have not been determined. OSHA has recommended the use of severely refined mineral oils since the mid-1980s in order to minimize PAH exposure. Also the use of synthetic fluids containing both ethanolamines and nitrites has been eliminated. These mixtures have been shown to contain nitrosamines, known animal carcinogens. Bronchitis and asthma may also be associated with machining fluid exposure; cross-shift changes in pulmonary function were demonstrated by Kennedy et al. (1989).

4.8.1.3 Control An excellent pamphlet on lubricating and coolant oils prepared by Esso outlines procedures to minimize exposure to cutting fluids (Esso, 1979). Selected work practice recommendations are the following: (1) Avoid all unnecessary contact with mineral or synthetic oils. Minimize contact by using splash guards, protective gloves, and protective aprons, etc; (2) encourage workers to wear clean work clothes, since oil-soaked clothing may hold the oil in contact with the skin longer than would otherwise occur; (3) remove oil from the skin as soon as possible if contact does occur; this means the installation of easily accessible wash basins and the provision of mild soap and clean towels in adequate supply; (4) do not allow solvents to be used for cleansing the skin. Use only warm water, mild soap, and a soft brush or in combination with a mild proprietary skin cleanser.

In many cases, the machining operations on such metals as magnesium and titanium may generate explosive concentrations of dust. Frequently, these operations must be segregated, and the operations must be conducted with suitable ventilation control and air cleaning. High-density layout of machine tools frequently results in workplace noise exposures above 85 dBA, the level that triggers the OSHA Hearing Conservation Program.

Since the 1950s, exposures to machining fluid particulates in large production facilities has decreased due to the installation of enclosures, local exhaust ventilation, general air cleaning, and probably the increased use of soluble and synthetic fluids (Hallock et al., 1994). Today most high-volume machine tools are designed with complete enclosures and local exhaust ventilation. As of this writing, the PEL for mineral oil remains 5 mg/m^3 total particulate, but OSHA has announced its plan to re-examine this value.

4.8.2 Electrochemical Machining

The electrochemical machining (ECM) process utilizes a dc electrolytic bath operating at low voltage and high current density. The workpiece is the anode; and the cutting tool, or cathode, is machined to reflect the geometry of the hole to be cut in the workpiece. As electrolyte is pumped through the space between the tool and the workpiece, metal ions are removed from the workpiece and are swept away by the electrolyte. The tool is fed into the workpiece to complete the cut. The electrolyte varies with the operation; one manufacturer states the electrolyte is 10 percent sulfuric acid.

Because the ECM method is fast, produces an excellent surface finish, does not produce burrs, and produces little tool wear, it is widely used for cutting irregular-shaped holes in hard, tough metals.

In the operation, the electrolyte is dissociated and hydrogen is released at the cathode. A dense mist or smoke is released from the electrolyte bath. Local exhaust ventilation must be provided to remove this mist and ensure hydrogen concentrations do not approach the lower flammability limit.

4.8.3 Electrical Discharge Machining

A spark-gap technique is the basis for the electrical discharge machining (EDM) procedure, which is a popular machining technique for large precise work such as die sinking or the drilling of small holes in complex parts. In one system, a graphite tool is machined to the precise size and shape of the hole to be cut. The workpiece (anode) and the tool (cathode) are immersed in a dielectric oil bath and powered by a low-voltage dc power supply. The voltage across the gap increases until breakdown occurs and there is a spark discharge across the gap, which produces a high temperature at the discharge point. This spark erodes a small quantity of the metal from the workpiece. The cycle is repeated at a frequency of 200 to 500 Hz with rather slow, accurate cutting of the workpiece. In more advanced systems, the cutting element is a wire that is programmed to track the cutting profile. The wire systems can also be used to drill small holes.

The hazards from this process are minimal and are principally associated with the oil. In light cutting jobs, a petroleum distillate such as Stoddard Solvent is commonly used, whereas in large work a mineral oil is the dielectric. When heavy oil mists are encountered, local exhaust ventilation is needed. The oil gradually becomes contaminated with small hollow spheres of metal eroded from the part. As in the case of conventional machining, these metals may dissolve in the oil and present a dermatitis problem. An ultra-high-efficiency filter should be placed in the oil recirculating line to remove metal particles.

4.9 NONDESTRUCTIVE TESTING

With the increase in manufacturing technology in the last decade of the twentieth century, a need has developed for in-plant inspection techniques. The most common procedures now in use in the metalworking industries are discussed in this section.

4.9.1 Industrial Radiography

Radiography is used principally in industry for the examination of metal fabrications such as weldments, castings, and forgings in a variety of settings. Specially designed shielded cabinets may be located in manufacturing areas for in-process examination of parts. Large components may be transported to shielded rooms for examination. Radiography may be performed in open shop areas, on construction sites, on board ships, and along pipelines.

The process of radiography consists of exposing the object to be examined to X-rays or gamma rays from one side and measuring the amount of radiation that emerges from the opposite side. This measurement is usually made with film or a fluoroscopic screen to provide a visual, two-dimensional display of the radiation distribution and any subsurface porosities.

The principal potential hazard in industrial radiography is exposure to ionizing radiation. This section deals with the minimum safety precautions designed to minimize worker exposure to radiation (X-rays and gamma rays) sources.

4.9.1.1 X-Ray Sources X-rays used in industrial radiography are produced electrically and therefore fall into the category of "electronic product radiation." For this reason the design and manufacture of industrial X-ray generators are regulated by the Food and Drug Administration, Center for Devices and Radiological Health. ANSI has developed a standard for the design and manufacture of these devices (ANSI, 1976). These standards specify maximum allowable radiation intensities outside the useful beam. They require warning lights on both the control panel and the tube head to indicate when X-rays are being generated.

The use of industrial X-ray generators is regulated by OSHA. Additionally, many states regulate the use of industrial X-ray generators within their own jurisdictions. Areas in which radiography is performed must be posted with signs that bear the radiation caution symbol and a warning statement. Access to these areas must be secured against unauthorized entry. When radiography is being performed in open manufacturing areas, it is essential to instruct other workers in the identification and meaning of these warning signs in order to minimize unnecessary exposures.

Radiographic operators are required to wear personal monitoring devices to measure the magnitude of their exposure to radiation. Typical devices are film badges, thermoluminescent dosimeters, and direct-reading pocket dosimeters. It is also advisable for radiographic operators to use audible alarm dosimeters or "chirpers." These devices emit an audible signal or "chirp" when exposed to radiation. The frequency of the signal is proportional to the radiation intensity. They are useful in warning operators who unknowingly enter a field of radiation (Isotope Radiography Safety Handbook, 1981).

The operators must be trained in the use of radiation survey instruments to monitor radiation levels to which they are exposed and to assure that the X-ray source is turned off at the conclusion of the operation. A wide variety of radiation survey instruments is available for use in industrial radiography. When using industrial X-ray generators, especially relatively low-energy generators, it is imperative that the instrument has an appropriate energy range for measuring the energy of radiation used.

4.9.1.2 Gamma Ray Sources Gamma rays used in industrial radiography are produced as a result of the decay of radioactive nuclei. The principal radioisotopes used in industrial radiography are iridium-192 and cobalt-60. Other radioisotopes, such as ytterbium-169 and thulium-170, are much less common but have some limited applicability. Radioisotope sources produce gamma rays with discrete energies, as opposed to the continuous spectrum of energies produced by X-ray generators. Owing to radioactive decay, the activity of radioisotope sources decreases exponentially with time.

Unlike X-ray generators, radioisotope sources require no external source of energy, which makes their use attractive in performing radiography in remote locations such as pipelines. However, because they are not energized by an external power supply, these sources cannot be "turned off," and they continuously emit gamma rays. For this reason, certain additional safety precautions must be exercised.

In industrial radiographic sources, the radioisotope is sealed inside a source capsule that is usually fabricated from stainless steel. The radioisotope source capsule is stored inside a shielded container or "pig" when not in use, to reduce the radiation intensities in the surrounding areas. In practice, the film is positioned and the radioactive source is then moved from the pig to the desired exposure position through a flexible tube by means of a mechanical actuator. At the end of the exposure, the operator retracts the source to the storage position in the pig.

The design, manufacture, and use of radioisotope sources and exposure devices for industrial radiography are regulated by the U.S. Nuclear Regulatory Commission (NRL). However, the Nuclear Regulatory Commission has entered into an agreement with "agreement states" for the latter to regulate radioisotope radiography within their jurisdictions. Organizations that wish to perform radioisotope radiography must obtain a license from either the Nuclear Regulatory Commission or the agreement state. In order to obtain this license, they must describe their safety procedures and equipment to the licensing authority. Check with your state agency or the NRC to see if your state is an agreement state.

Radiographic operators must receive training as required by the regulatory bodies including instruction in their own organization's safety procedures, a formal radiation safety training course, and a period of on-the-job training when the trainees work under the direct personal supervision of a qualified radiographer. At the conclusion of this training, operators must demonstrate their knowledge and competence to the licensee's management.

Areas in which radiography is being performed must be posted with radiation warning signs, and access to these areas must be secured as described earlier. An operator performing radioisotope radiography must wear both a direct-reading pocket dosimeter and either a film badge or a thermoluminescent dosimeter. Additionally, the operator must use a calibrated radiation survey instrument during all radiographic operations. In order to reduce the radiation intensity in the area after an exposure, the operator must retract the source to a shielded position within the exposure device. The only method available to the operator for assuring that the source is shielded properly is a radiation survey of the area. The operator should survey the entire perimeter of the exposure device and the entire length of guide tube and source stop after each radiographic operation to assure that the source has been fully and properly shielded. Some current types of radiographic exposure devices incorporate source position indicators that provide a visual signal if the source is not stored properly. Use of these devices may also reduce the frequency of radiation exposure incidents.

Occasional radiation exposure incidents have occurred in industrial radioisotope radiography. These incidents generally result because the operator fails to return the source properly to a shielded position in the exposure device and then approaches the exposure device or source stop without making a proper radiation survey. The importance of making a proper radiation survey at the conclusion of each radiographic exposure cannot be overemphasized. The estimated mean annual dose equivalent for radiography workers is 290 mRem/year.

4.9.2 Magnetic Particle Inspection

This procedure is suitable for detecting surface discontinuities, especially cracks, in magnetic materials. The procedure is simple and of relatively low cost; as a result, it is widely applied in metal fabrication plants.

The parts to be inspected first undergo vigorous cleaning and then are magnetized. Magnetic particles are applied as a powder or a suspension of particles in a carrier liquid. The powders are available in color for daylight viewing or as fluorescent particles for more rigorous inspection with a UV lamp. Surface imperfections, such as cracks, result in leakage of the magnetic field with resulting adherence of the particles. The defined geometry of the imperfection stands out in a properly illuminated environment or with a hand-held UV lamp.

Worker exposures using this technique are minor, consisting of skin contact and minimal air contamination from the suspension field, usually an aliphatic hydrocarbon; exposure to the magnetic field during the inspection/process; and exposure to the lamp output. General exhaust ventilation usually is adequate for vapors from the liquid carrier, and gloves, apron, and eye protection should be used to minimize skin and eye contact. A proper filter on the UV lamp minimizes exposure to UV B radiation. The hazard from magnetic fields is now under investigation.

4.9.3 Liquid Penetrant

This procedure complements magnetic particle inspection for surface cracks and weldment failures because it can be used on nonmagnetic materials. A colored or fluorescent liquid used as the penetrant is flowed or sprayed on the surface, or the part is dipped. The excess penetrant is removed, in some systems by adding an emulsifier and removing with water, or by wiping off the excess. A developer is applied to intensify the definition of the crack, and the inspection is carried out by daylight or UV lamp depending on the penetrant system of choice. The penetrant is then removed from the workpiece by either water or solvent such as mineral spirits.

The inspectors' exposures are to the solvent or carrier used with the penetrant, emulsifier, developer, and the final cleaning agent. The solvent is water, mineral spirits, or in some cases, halogenated compounds.

4.9.4 Ultrasound

Pulse echo and transmission-type ultrasonic inspection have a wide range of application for both surface and subsurface flaws. This procedure resolves voids much smaller than all other methods, including radiographic procedures. The procedure commonly involves immersing the part in water to improve coupling between the ultrasound transmitter/receiver and the workpiece. The reported response on other applications of higher-energy ultrasound systems include local heating and subjective effects. Health effects have not been reported on its application in nondestructive testing of metals.

4.10 METAL THERMAL SPRAYING

A practical technique for spraying molten metal was invented in the 1920s and has found application for applying metals, ceramics, and plastic powder to workpieces for corrosion protection, to build up worn or corroded parts, to improve wear resistance, to reduce production costs, and as a decorative surface. The health hazards from thermal spraying of metals was first observed in 1922 in the United Kingdom when operators suffered from lead poisoning (Ballard, 1972).

The application head for thermal spraying takes many forms and dictates the range of hazards one may encounter from the operation. The four principal techniques currently in use in industry are described below.

4.10.1 Spraying Methods

In the combustion spraying or flame method, the coating material in wire form is fed to a gun operated with air/oxygen and a combustible gas such as acetylene, propane, or natural gas. The wire is melted in the oxygen-fuel flame and propelled from the torch at velocities up to 240 m/sec (48,000 fpm) with compressed air. The material bonds to the workpiece by a combination of mechanical interlocking of the molten, platelet-form particles and a cementation of partially oxidized material.

In the thermal arc spraying or wire method, two consumable metal wire electrodes are made of the metal to be sprayed. As the wires feed into the gun, they establish an arc as in a conventional arc welding unit; the molten metal is disintegrated by compressed air, and the molten particles are projected to the workpiece at high velocity.

In the plasma method, an electric arc is established in the controlled atmosphere of a special nozzle. Argon is passed through the arc, where it ionizes to form a plasma that continues through the nozzle and recombines to create temperatures as high as 16,700°C (30,000°F). Metal alloy, ceramic, and carbide powders are melted in the stream and are released from the gun at a velocity of 300 to 600 m/sec (60,000 to 120,000 fpm).

In the detonation method, the gases are fed to a combustion chamber, where they are ignited by a spark plug. Metal powder is fed to the chamber, and the explosions drive the melted powder to the workpiece at velocities of approximately 760 m/sec (150,000 fpm). Tungsten carbide, chromium carbide, and aluminum oxide are applied with this technique, which is used on a limited basis.

4.10.2 Hazards and Controls

A major hazard common to all techniques is the potential exposure to toxic metal fumes. If the vapor pressure of the metal at the application temperature is high, this will be reflected in high air concentrations of metal fume. The deposition efficiency, that is, the percent of the metal sprayed deposited on the workpiece, varies between application techniques and the coating metal and has a major impact on air concentrations in the workplace. Because the airborne level varies with the type of metal sprayed, the application technique, the fuel in use, and the tool-workpiece geometry, it is impossible to estimate the level of contamination that may occur from a given operation. For this reason, if a toxic metal is sprayed, air samples should be taken to define the worker exposure.

The principal control of air contaminants generated during flame and arc spraying is local exhaust ventilation such as an open or enclosing hood (Hagopian and Bastress, 1976). The operator should also wear a positive-pressure air-supplied respirator while metallizing with toxic metals. One must also consider the exposure of other workers in the area, because systems with 50 percent overspray obviously represent a significant source of general contamination.

All thermal spraying procedures result in significant noise exposures and require effective hearing conservation programs. Appreciable noise reduction can be achieved by changing the operating characteristics of the spray equipment, which may or may not affect the quality of the surface coating. If the resulting changes are unacceptable, the normal approach to noise control, including isolation, the use of hearing protectors, and various work practices, must be used.

The thermal arc and plasma procedures present additional hazards reflecting their welding ancestry, including UV exposure and the generation of ozone and nitrogen dioxide. If ventilation is effective in controlling metal fumes, it will probably control ozone and nitrogen dioxide, although this fact should be established by air sampling. A NIOSH study demonstrated excess visible, IR, and UV radiation from thermal arc spraying (NIOSH, 1989).

In addition to air-supplied respirators, operators handling toxic metals such as cadmium and lead should wear protective gloves and coveralls and be required to strip, bathe, and change to clean clothes prior to leaving the plant.

Special fire protection problems are encountered in routine metallizing operations. Unless ventilation control is effective, particles will deposit on various plant structures and may become a potential fire hazard. If one uses a scrubber to collect the particles, hydrogen may form in the sludge, resulting in a potential fire and explosion hazard.

4.11 PAINTING

Paint products are used widely in industry to provide a surface coating for protection against corrosion, for appearance, as electrical insulation, for fire retardation, and for other special purposes. The widespread application of this technology from small job shops to highly automated painting of automobiles includes more than half a million workers. Although the health status of painters has not been well defined, the studies that have been completed suggests acute and chronic central nervous system effects, hematologic disorders, and excess mortality from cancer. In addition, respiratory sensitization may occur from two-part urethane systems, and the amine catalyst in spray paints may cause skin sensitization. The hazards associated with the industrial application of paint products are discussed here.

4.11.1 Types of Paints

The term *paint* is commonly used to identify a range of organic coatings including paints, varnishes, enamels, and lacquers. Conventional paint is an inorganic pigment dispersed in a vehicle consisting of a binder and a solvent with selected fillers and additives. The conventional varnish is a nonpigmented product based on oil and natural resin in a solvent that dries first by the evaporation of the solvent and then by the oxidation of the resin binder. Varnishes are also available based on synthetic resins. A pigmented varnish is called an enamel. Lacquers are coatings in which the solvent evaporates leaving a film that can be redissolved in the original solvent.

In the past decade, paint formulations have been influenced by environmental regulations limiting the release of volatile organic solvents. Although conventional solvent-based paints will continue to see wide application, the major thrust is to convert to formulations with low solvent content. Such systems include solvent-based paints with high solids content (>70 percent by weight), nonaqueous dispersions, powder, two-part catalyzed systems, and water-borne paints. The use of UV-curable paints may aid efforts to use powder coatings at lower temperatures for a greater variety of substrates.

4.11.2 Composition of Paint

The conventional solvent-based paints consist of the vehicle, filler, and additives. The vehicle represents the total liquid content of the paint and includes the binder and solvent. The binder, which is the film-forming ingredient, may be a naturally occurring oil or resin including linseed oil and oleoresinous materials or synthetic material such as alkyd resins.

The fillers include pigments and extenders, which historically have presented a major hazard in painting. The common white pigments include bentonite and kaolin clay, talc, titanium dioxide, and zinc oxide. Mineral dust used as extenders to control viscosity, texture, and gloss include talc, clay, calcium carbonate, barite, and both crystalline and amorphous silica. The pigments and extenders do not represent an exposure during brush or roller application, but when sanded or removed preparatory to repainting, these materials are released to the air and may present significant exposures. A group of pigments including lead carbonate, cadmium red, and chrome green and yellow do represent a critical exposure, both during spray application and surface preparation.

Additives to hasten drying, reduce skin formation in the can, and control fungus are added to paints. The fungicides warrant special attention because the ingredients, including copper and zinc naphthenate, copper oxide, and tributyltin oxide, are biologically active.

The solvent systems are varied and complex. The most common *organic solvents* include aliphatic and aromatic hydrocarbons, ketones, alcohols, glycols, and glycol ether/esters. These solvents have high vapor pressure and represent the critical worker exposure component in most painting techniques. High-solids paint systems (low solvent) represent a major contribution to reduced solvent exposure.

The ultimate high-solids, low-solvent formulation is a dry powder paint. This coating technique provides a high-quality job while eliminating solvent exposure for the worker. The application technique, to be described later, utilizes a dry powder formulated to contain a resin, pigment, and additives. Thermosetting resins, used for decorative and protective coatings, are based on epoxy, polyester, and acrylic resins. Thermoplastic resins are applied in thick coatings for critical applications.

Water-borne paints now represent 15 to 20 percent of construction and industry paints, and expanded use is anticipated. The present systems use binders based on copolymers of several monomers including acrylic acid, butyl acrylate, ethyl hexyl acrylate, and styrene. The other major ingredients and their maximum concentrations include pigments (54 percent), coalescing solvents (15 percent), surfactants (5 percent), biocides (1.1 percent), plasticizers (2.2 percent), and driers in lesser amounts. The pigments include the conventional materials mentioned under solvent-based paints. Coalescing solvents, including hydrocarbons, alcohols, esters, glycol, and glycol ether/esters, although present in low concentrations, may present an inhalation hazard. The proprietary biocides in use frequently result in exposure to significant airborne formaldehyde concentrations. The surfactants in use include known skin irritants and sensitizers (Hansen et al., 1987).

A single study of water-borne paints and worker exposure confirms that these systems represent an improvement over solvent-based systems. The authors of this study recommend that paint formulations be designed to eliminate formaldehyde, minimize the unreacted monomer content in the various polymers, eliminate ethylene glycol ethers because of their reproductive hazard, and minimize the ammonia content (Hansen et al., 1987).

Special attention must be given two-component epoxy and urethane paint systems. The urethanes consist of a polyurethane prepolymer containing a reactive isocyanate; the second component is the polyester. Mixing of the two materials initiates a reaction with the formation of a chemically resistant coating. In the early formulations, the unreacted toluene diisocyanate (TDI) resulted in significant airborne exposures to TDI and resulting respiratory sensitization. Formulations have been changed to include TDI adducts and derivatives that are stated to have minimized the respiratory hazard from these systems. For ease of application, single package systems have been developed using a blocked isocyanate. Unblocking takes place when the finish is baked and the isocyanate is released to react with the polyester.

Epoxy paint systems offer excellent adhesive properties, resistance to abrasion and chemicals, and stability at high temperatures. The conventional epoxy system is a two-component system consisting of a resin based on the reaction products of bisphenol A and epichlorohydrin. The resin may be modified by reactive diluents such as glycidyl ethers. The second component, the hardener or curing agent, was initially based on low-molecular-weight, highly reactive amines.

Epoxy resins can be divided into three grades. The solid grades are felt to be innocuous; however, skin irritation may occur from solvents used to take up the resin. The liquid grades are mild to moderate skin irritants. The low-viscosity glycidyl ether modifiers are skin irritants and sensitizers and have systemic toxicity. The use of low-molecular-weight aliphatic amines such as diethylenetetramine and triethylenetetramine, both strong skin irritants and sensitizers, presented major health hazards during the early use of epoxy systems. This has been overcome to a degree by the use of low-volatility amine adducts and high-molecular-weight amine curing agents.

4.11.3 Operations and Exposures

In the industrial setting, paints can be applied to parts by a myriad of processes including brush, roller, dip, flow, curtain, tumbling, conventional air spray, airless spray, heated systems, disk spraying, and powder coating. Conventional air spraying is the most common method encountered in industry and presents the principal hazards owing to overspray. The use of airless and hot spray techniques minimizes mist and solvent exposure to the operator, as does electrostatic spraying. The electrostatic technique, now commonly used with many installations, places a charge on the paint mist particle so it is attracted to the part to be painted, thereby reducing rebound and overspray.

In powder coating, a powder, which represents the paint formulation including the resin, pigment, and additives, is conveyed from a powder reservoir to the spray gun. In the gun, a charge is imparted to the individual paint particles. This dry powder is sprayed on the electrically grounded workpiece, and the parts are baked to fuse the powder film into a continuous coating.

The operator is exposed to the solvent or thinner in processes in which the paint is flowed on, as in brushing and dipping, and during drying of the parts. However, during atomization techniques, the exposure is to both the solvent and the paint mist. The level of exposure reflects the overspray and rebound that occur during spraying.

A common exposure to organic vapors occurs when spray operators place the freshly sprayed parts on a rack directly behind them. The air movement to the spray booth sweeps over the drying parts and past the breathing zone of the operator, resulting in an exposure to solvent vapors. Drying stations and baking ovens must therefore be exhausted. The choice of exhaust control on tumbling and roll applications depends on the surface area of the parts and the nature of the solvent.

4.11.4 Controls

Most industrial flow and spray painting operations utilizing solvent-based paints require exhaust ventilation for control of solvent vapors at the point of application and during drying and baking operations (O'Brien and Hurley, 1981; NIOSH, 1984). Water-based paints may require ventilation only when spray application is utilized. Flow application of solvent-based paints requires local exhaust ventilation depending on the application technique. The usual ventilation control for spray application of solvent-based paints is a spray booth, room, or tunnel provided with some type of paint spray mist arrestor before the effluent is exhausted outdoors. The degree of control of paint mist and solvent varies with the application method, that is, whether it is air atomization, airless, or electrostatic painting. The latter two techniques call for somewhat lower exhaust volumes. The advent of robotic spray painting permits the design of exhausted enclosures and minimizes worker exposure.

The design of the conventional paint spray booth is shown in Fig. 4.5. The booths are commonly equipped with a water curtain or a throwaway dry filter to provide paint mist removal. The efficiency of these devices against paint mist has not been evaluated. Neither of these systems, of course, removes solvent vapors from the air stream.

In industrial spray painting of parts, the simple instructions contained in several state codes on spray painting should be observed:

1. Do not spray toward a person.
2. Automate spray booth operations where possible to reduce exposure.

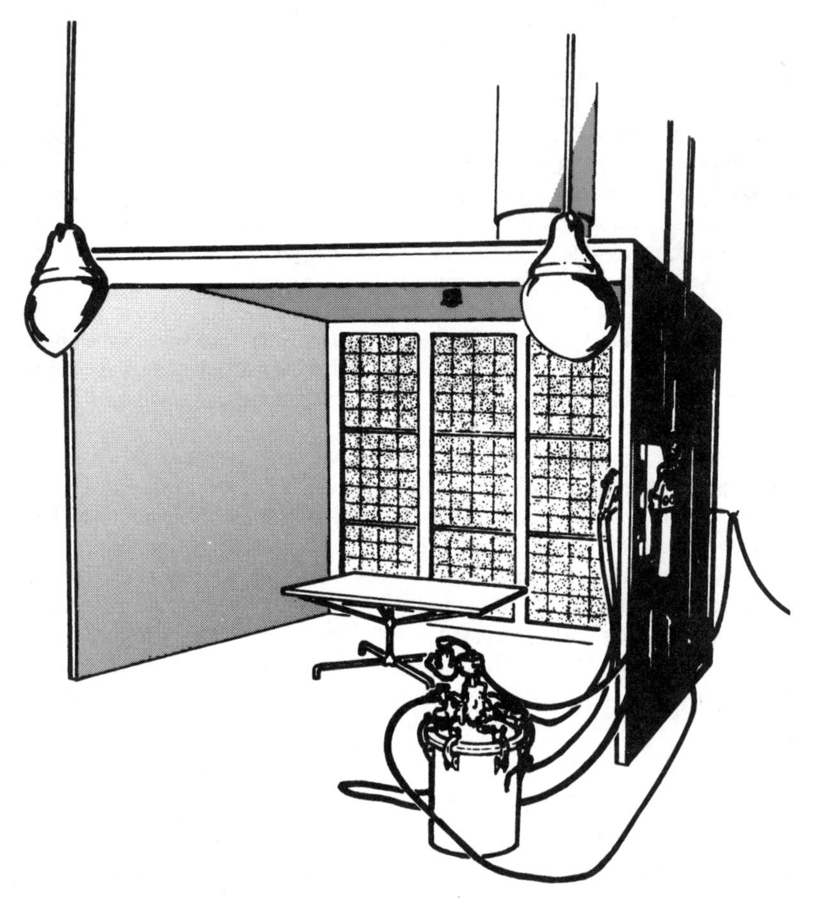

Fig. 4.5 Paint spray booth.

3. Maintain 2-ft clearance between the sides of the booth and large flat surfaces to be sprayed.
4. Keep the distance between the nozzle and the part to be sprayed to less than 12 in.
5. Do not position work so that the operator is between the exhaust and the spray gun or disk.
6. Locate the drying room so that air does not pass over drying objects to exhaust food past the breathing zone of the operator.

Controls in the application of two-component urethane and epoxy paint systems must include excellent housekeeping, effective ventilation control, and protective clothing; moreover, in applications not effectively controlled by ventilation, the operators should wear air-supplied respirators. Adequate washing facilities should be available, and eating, drinking, and smoking should be prohibited in the work area.

Dermatitis due to primary irritation and defatting from solvents or thinners as well as sensitization from epoxy systems is not uncommon. Skin contact must be minimized, rigorous personal cleanliness encouraged, and suitable protective equipment used by the operator.

4.12 SOLDERING AND BRAZING

Soft soldering is the joining of metal by surface adhesion without melting the base metal. This technique uses a filler metal (solder) with a melting point less than 316°C (600°F); hard solder is used in the range of 316 to 427°C (600 to 800°F). These temperature ranges differentiate soldering from brazing, which utilizes a filler metal with a melting point greater than 427°C (800°F).

4.12.1 Soldering

To understand the potential health hazards from soldering operations, one must be familiar with the composition of the solder and fluxes in use and the applicable production techniques.

4.12.1.1 Flux All metals including the noble metals have a film of tarnish that must be removed in order to wet the metal with solder effectively and accomplish a good mechanical bond. The tarnish takes the form of oxides, sulfides, carbonates, and other corrosion products (Manko, 1979). The flux, which may be a solid, liquid, or gas, is designed to remove any adsorbed gases and tarnish from the surface of the base metal and keep it clean until the solder is applied. The molten solder displaces the residual flux and wets the base metal to accomplish the bond.

A range of organic and inorganic materials is used in the design of soldering fluxes as shown in Table 4.5. The flux is usually a corrosive cleaner frequently used with a volatile solvent or vehicle. Rosin, which is a common base for organic fluxes, contains abietic acid as the active material.

The three major flux bases are inorganic, organic nonrosin, and organic rosin, as shown in Table 4.5. A proprietary flux may contain several of these materials. Organic nonrosin is less corrosive and thereby slower acting, and in general, these fluxes do not present as severe a handling hazard as the inorganic acids. This is not true of the organic halides whose degradation products are very corrosive and warrant careful handling and ventilation control. The amines and amides in this second class of flux materials are common ingredients whose degradation products are very corrosive.

The rosin base fluxes are inactive at room temperature but at soldering temperature are activated to remove tarnish. This material continues to be a popular flux compound because the residual material is chemically and electrically benign and need not be removed as vigorously as the residual products of the corrosive fluxes.

The cleaning of the soldered parts may range from a simple hot water or detergent rinse to a degreasing technique using a range of organic solvents. These cleaning techniques may require local exhaust ventilation.

4.12.1.2 Solder The most common solder contains 65 percent tin and 35 percent lead. Traces of other metals including cadmium, bismuth, copper, indium, iron, aluminum, nickel, zinc, and arsenic are present. A number of special solders contain antimony in concentrations up to 5 percent. The melting point of these solders is quite low;

TABLE 4.5 Common Flux Materials

Type	Typical Fluxes	Vehicle
	Inorganic	
Acids	Hydrochloric, hydrofluoric, orthophosphoric	Water, petrolatum paste
Salts	Zinc chloride, ammonium chloride, tin chloride	Water, petrolatum paste, polyethylene glycol
Gases	Hydrogen-forming gas: dry HCI	None
	Organic, nonrosin base	
Acids	Lactic, oleic, stearic, glutamic, phthalic	Water, organic solvents, petrolatum paste, polyethylene glycol
Halogens	Aniline hydrochloride, glutamic acid hydrochloride, bromide derivatives of palmitic acid, hydrazine hydrochloride or hydrobromide	Water, organic solvents, polyethylene glycol
Amines and amides	Urea, ethylenediamine, mono- and triethanolamine	Water, organic solvents, petrolatum paste, polyethylene glycol
	Organic, rosin base	
Superactivated	Rosin or resin with strong activators	Alcohols, organic solvents, glycols
Activated (RA)	Rosin or resin with activator	Alcohols, organic solvents, glycols
Mildly activated (RMA)	Rosin with activator	Alcohols, organic solvents, glycols
Nonactivated (water-white rosin) (R)	Rosin only	Alcohols, organic solvents, glycols

Source: Ref. 47.

and at these temperatures, the vapor pressure of lead and antimony usually do not result in significant air concentrations of metal fume. The composition of the common solders encountered in industry can be obtained from the manufacturers.

4.12.1.3 Application Techniques Soldering is a fastening technique used in a wide range of products from simple mechanical assemblies to complex electronic systems. The soldering process includes cleaning the base metal and other components for soldering, fluxing, the actual soldering, and postsoldering cleaning.

4.12.1.4 Initial Cleaning of Base Metals Prior to fluxing and soldering, the base metal must be cleaned to remove oil, grease, wax, and other surface debris. Unless this is done, the flux will not be able to attack and remove the metal surface tarnish. The procedures include cold solvent degreasing, vapor degreasing, and ultrasonic degreasing. If the base metal has been heat treated, the resulting surface scale must be removed.

In many cases, mild abrasive blasting techniques are utilized to remove heavy tarnish before fluxing. In electrical soldering, the insulation on the wire must be stripped back to permit soldering. Stripping is accomplished by mechanical techniques such as cutters or wire brushes, chemical strippers, and thermal techniques. Mechanical stripping of asbestos-based insulation obviously presents a potential health hazard, and this operation must be controlled by local exhaust ventilation. The hazard from chemical stripping depends on the chemical used to strip the insulation; however, at a minimum, the stripper will be a very corrosive agent. Thermal stripping of wire at high production rates may present a problem owing to the thermal degradation products of the insulation. Hot wire stripping of fluorocarbon insulation such as Teflon may cause polymer fume fever if operations are not controlled by ventilation. Other plastic insulation such as polyvinyl chloride may produce irritating and toxic thermal degradation products.

4.12.1.5 Fluxing Operations Proprietary fluxes are available in solid, paste, and liquid form for various applications and may be cut with volatile vehicles such as alcohols to vary their viscosity. The flux may be applied by one of 10 techniques shown in Table 18.8 depending on the workpiece and the production rate. Because flux is corrosive, skin contact must be minimized by specific work practices and good housekeeping. The two techniques that utilize spray application of the flux require local exhaust ventilation to prevent air contamination from the flux mist.

4.12.1.6 Soldering and Cleaning In two of the fluxing techniques listed in Table 4.6 the flux and solder are applied together. One need merely apply the necessary heat to bring the system first to the melting point of the flux and then to the melting point of the solder. In most applications, however, the flux and solder are applied separately, although the two operations may be closely integrated and frequently are in a continuous operation.

The soldering techniques used for manual soldering operations on parts that have been fluxed are the soldering iron and solder pot. A number of variations of the solder pot have been introduced to handle high production soldering of printed circuit boards (PCB) in the electronic industry. In the drag solder technique, the PCB is positioned horizontally and pulled along the surface of a shallow molten solder bath behind a skimmer plate that removes the dross. This process is usually integrated with a cleaning and fluxing station in a single automated unit. Another system designed for automation is wave soldering. In this technique, a standing wave is formed by pumping the solder through a spout. Again, the conveyorized PCBs are pulled through the flowing solder.

The potential health hazards to the soldering operators are minimal. A thermostatically controlled lead–tin solder pot operates at temperatures too low to generate significant concentrations of lead fume. The use of activated rosin fluxing agents may result in the release of thermal degradation products that may require control by local exhaust ventilation. The handling of solder dross during cleanup and maintenance may result in exposure to lead dust.

After soldering operations are carried out, some flux residue and its degradation products remain on the base metal. Both water-soluble and solvent-soluble materials normally exist so that it is necessary to clean with both systems using detergents and saponifiers in one case and common chlorinated and fluorinated hydrocarbons in the other. CFCs are being replaced with cleaners such as limonene, which are less threatening to the environment. The processing equipment may include ultrasonic cleaners and vapor degreasers.

TABLE 4.6 Soldering Flux Application Techniques

Method	Application technique	Use
Brushing	Applied manually by paint, acid, or rotary brushes	Copper pipe, job shop printed cricuit board, large structural parts
Rolling	Paint roller application	Precision soldering, printed circuit board, suitable for automation
Spraying	Spray painting equipment	Automatic soldering operations, not effective for selective application
Rotary screen	Liquid flux picked up by screen and air directs it to part	Printed circuit board application
Foaming	Work passes over air agitated foam at surface of flux tank	Selective fluxing, automatic printed circuit board lines
Dipping	Simple dip tank	Wide application for manual and automatic operations on all parts
Wave fluxing	Liquid flux pumped through trough forming wave through which work is dipped	High-speed automated operations
Floating	Solid flux on surface melts providing liquid layer	Tinning of wire and strips of material
Cored solder	Flux inside solder wire melts, flows to surface, and fluxes before solder melts	Wide range of manual operations
Solder paste	Solder blended with flux, applied manually	Component and hybrid microelectric soldering

Source: Ref. 47.

4.12.1.7 Controls The fluxes may represent the most significant hazard from soldering. The conventional pure rosin fluxes are not difficult to handle, but highly activated fluxes warrant special handling instruction. A range of alcohols, including methanol, ethanol, and isopropyl alcohol, are used as volatile vehicles for fluxes. The special fluxes should be evaluated under conditions of use to determine worker exposure. Rosin is a common cause of allergic contact dermatitis, and the fume causes an allergic asthma.

4.12.2 Brazing

Brazing techniques are widely used in the manufacture of refrigerators, electronics, jewelry, and aerospace components to join both similar and dissimilar metals. Although the final joint looks similar to a soft solder bond, it is much stronger and the joint requires little finish. As mentioned earlier in this section, brazing is defined as a technique for joining metals that are heated above 430°C (800°F), whereas soldering is conducted below 430°C. The temperature of the operation is of major importance because it determines the vapor pressure of the metals that are heated and therefore the concentration of metal fumes to which the operator is exposed (Handy and Harman, 1985).

4.12.2.1 Flux and Filler Metals Flux is frequently used; however, certain metals may be joined without flux. The flux is chosen to prevent oxidation of the base metal and not to prepare the surface as is the case with soldering. The common fluxes are based on fluorine, chlorine, and phosphorus compounds and present the same health hazards as fluxes used in soldering; that is, they are corrosive to the skin and may cause respiratory irritation. A range of filler metals used in brazing rods or wire may include phosphorus, silver, zinc, copper, cadmium, nickel, chromium, beryllium, magnesium, and lithium. The selection of the proper filler metal is the key to quality brazing.

4.12.2.2 Application Techniques Brazing of small job lots that do not require close temperature control are routinely done with a torch. More critical, high production operations are accomplished by dip techniques in a molten bath, by brazing furnaces using either an ammonia or hydrogen atmosphere, or by induction heating.

The brazing temperatures define the relative hazard from the various operations. As an example, the melting point of cadmium is approximately 1400°C. The vapor pressure of cadmium and the resulting airborne fume concentrations increase dramatically with temperature. The filler metals with the higher brazing temperatures will therefore present the most severe exposure to cadmium.

The exposure to fresh cadmium fume during brazing of low-alloy steels, stainless steels, and nickel alloys has resulted in documented cases of occupational disease and represents the major hazard from these operations. This is especially true of torch brazing, where temperature extremes may occur. On the other hand, the temperatures of furnace and induction heating operations may be controlled to ±5°C.

One study of brazing in a pipe shop and on board ship included air sampling for cadmium while brazing with a filler rod containing from 10 to 24 percent cadmium. The mean concentrations during shipboard operation was 0.45 mg/m^3 with a maximum of 1.40 mg/m^3.

4.12.2.3 Controls Controls on brazing operations must obviously be based on the identification of the composition of the filler rod. Local ventilation control is necessary is operations where toxic metal fumes may be generated from the brazing components or from parts plated with cadmium or other toxic metals. In the use of common fluxes, one should minimize skin contact owing to their corrosiveness and provide exhaust ventilation to control airborne thermal degradation products released during the brazing operation.

4.13 WELDING

Welding is a process for joining metals in which coalescence is produced by heating the metals to a suitable temperature. Over a dozen welding procedures are commonly encountered in industry. Four welding techniques have been chosen for discussion because they represent 80 to 90 percent of all manufacturing and maintenance welding. These procedures are used by more than half a million welders and helpers in the United States.

In the nonpressure welding techniques to be discussed in this section, metal is vaporized and then condenses to form initially a fume in the 0.01 to 0.1 μm particle size range that rapidly agglomerates. The source of this respirable metal fume is the base metal, the metal coating on the workpiece, the electrode, and the fluxing agents associated with the particular welding system. A range of gases and vapors including carbon monoxide, ozone, and nitrogen dioxide may be generated depending on the welding process. The wavelength and intensity of the electromagnetic radiation emitted from the arc depends on the welding procedure, inerting gas, and the base metal.

The common welding techniques are reviewed assuming that the welding is done on unalloyed steel, or so-called mild steel. The nature of the base metal is important in evaluating the metal fume exposure, and occasionally it has impact in other areas. When such impact is important, it is discussed.

4.13.1 Shielded Metal Arc Welding

Sheilded metal arc (SMA) welding (Fig. 4.6) is commonly called stick or electrode welding. An electric arc is drawn between a welding rod and the workpiece, melting the metal analog a seam or a surface. The molten metal from the workpiece and the electrode form a common puddle and cool to form the bead and its slag cover. Either dc or ac power is used in straight (electrode negative, work positive) or reverse polarity. The most common technique involves dc voltages of 10 to 50 and a wide range of current up to 2000 A. Although operating voltages are low, under certain conditions an electrical hazard may exist.

The welding rod or electrode may have significant occupational health implications. Initially a bare electrode was used to establish the arc and act as filler metal. Now the electrode covering may contain 20 to 30 organic and inorganic compounds and perform several functions. The principal function of the electrode coating is to release a shielding gas such as carbon dioxide to ensure that air does not enter the arc puddle and thereby cause failure of the weld. In addition, the covering stabilizes the arc, provides a flux and slag producer to remove oxygen from the weldment, adds alloying metal, and controls the viscosity of the metal. A complete occupational health survey of shielded metal arc welding requires the identification of the rod and its covering. The composition of the electrode can be obtained from the American Welding Society (AWS) classification number stamped on the electrode.

The electrode mass that appears as airborne fume may include iron oxides, man-

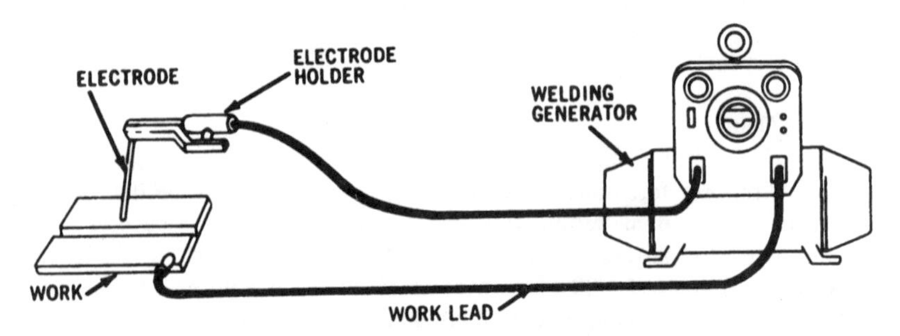

Fig. 4.6 Shielded metal arc welding (SMAW).

ganese oxide, fluorides, silicon dioxide, and compounds of titanium, nickel, chromium, molybdenum, vanadium, tungsten, copper, cobalt, lead, and zinc. Silicon dioxide is routinely reported as present in welding fume in an amorphous form or as silicates, not in the highly toxic crystalline form.

4.13.1.1 Metal Fume Exposure The potential health hazards from exposure to metal fume during shielded metal arc welding obviously depend on the metal being welded and the composition of the welding electrode. The principal component of the fume generated from mild steel is iron oxide. The hazard from exposure to iron oxide fume appears to be limited. The deposition of iron oxide particles in the lung does cause a benign pneumoconiosis known as siderosis. There is no functional impairment of the lung, nor is there fibrous tissue proliferation. In a comprehensive review of conflicting data, Stokinger has shown that iron oxide is not carcinogenic to man (Stokinger, 1984).

The concentration of metal fume to which the welder is exposed depends not only on the alloy composition but on the welding conditions, including the current density (amperes per unit area of electrode), wire feed rate, the arc time, which may vary from 10 to 30 percent, the power configuration, that is, dc or ac supply, and straight or reverse polarity. The work environment also defines the level of exposure to welding fume and includes the type and quality of exhaust ventilation and whether the welding is done in an open, enclosed, or confined space.

Many of the data on metal fume exposures have been generated from shipyard studies. The air concentration of welding fume in several studies ranged from less than 5 to over 100 mg/m^3 depending on the welding process, ventilation, and the degree of enclosure. In these early studies, the air samples were normally taken outside the welding helmet. Recent studies have shown the concentration at the breathing zone inside the helmet ranges from one-half to one-tenth the outside concentrations. If SMA welding is conducted on stainless steel, the chromium concentrations may exceed the TLV (0.5 mg/m^3); in the case of alloys with greater than 50 percent nickel, the concentration of nickel fume may exceed 1 mg/m^3 (VanderWal, 1990).

4.13.1.2 Gases and Vapors Shielded metal arc welding has the potential to produce nitrogen oxides; however, this is not normally a problem in open shop welding. In over 100 samples of SMA welding in a shipyard, Burgess et al. (1989) did not identify an exposure to nitrogen dioxide in excess of 0.5 ppm under a wide range of operating conditions. Ozone is also fixed by the arc, but again this is not a significant contaminant in SMA welding operations. Carbon monoxide and carbon dioxide and produced from the electrode cover, but air concentrations are usually minimal.

Low-hydrogen electrodes are used with conventional arc welding systems to maintain a hydrogen-free arc environment for critical welding tasks on certain steels. The electrode coating is a calcium carbonate–calcium fluoride system with various deoxidizers and alloying elements such as carbon, manganese, silicon, chromium, nickel, molybdenum, and vanadium. A large part of this coating and of all electrode coatings becomes airborne during welding (Pantucek, 1971). In addition to hydrogen fluoride, sodium, potassium, and calcium fluorides are present as particles in the fume cone.

Exposure to fumes from low hydrogen welding under conditions of poor ventilation may prompt complaints of nose and throat irritation and chronic nosebleeds. There has been no evidence of systemic fluorosis from this exposure. Monitoring can be accomplished by both air sampling and urinary fluoride measurements.

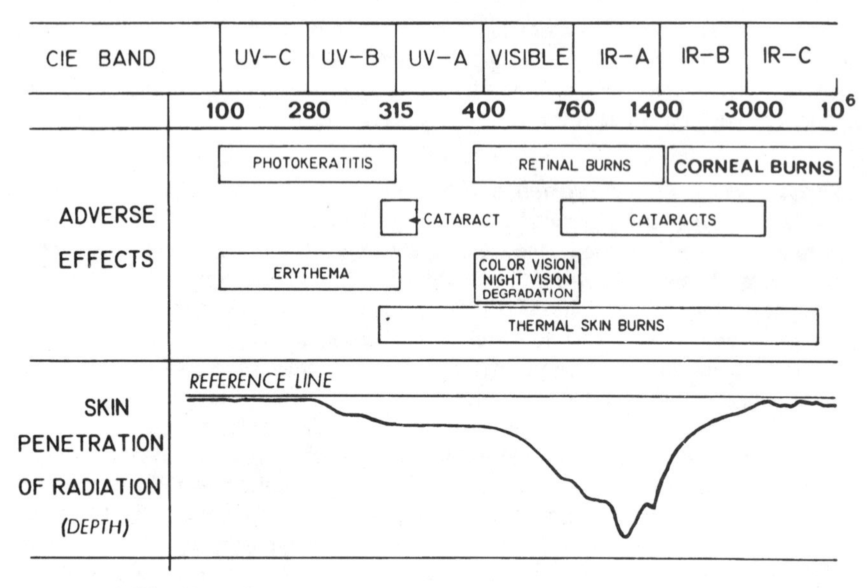

Fig. 4.7 Effects of electromagnetic radiation on the eye and skin.

4.13.1.3 Radiation The radiation generated by SMA welding covers the spectrum from the infrared C range of wavelengths to the UV C range. The acute condition known to the welder as "arc eye," "sand in the eye," or "flash burn" is due to exposures in the UV B range (see Fig. 4.7). The radiation in this range is completely absorbed in the corneal epithelium of the eye and causes a severe photokeratitis. Severe pain occurs 5 to 6 hr after exposure to the arc, and the condition clears within 24 hr. Welders experience this condition usually only once and then protect themselves against a recurrence by the use of a welding helmet with a proper filter. Skin erythema or reddening may also be induced by exposure to UV C and UV B as shown in Figure 18.11 (Sliney and Wolbarsht, 1980).

4.13.2 Gas Tungsten Arc Welding

Although shielded metal arc welding using coated electrodes is an effective way to weld many ferrous metals, it is not practical for welding aluminum, magnesium, and other reactive metals. The introduction of inert gas in the 1930s to blanket the arc environment and prevent the intrusion of oxygen and hydrogen into the weld provided a solution to this problem. In gas tungsten arc welding (GTA) (Fig. 4.8) also known as tungsten inert gas and Heliarc welding, the arc is established between a nonconsumable tungsten electrode and the workpiece producing the heat to melt the abutting edges of the metal to be joined. Argon or helium is fed to the annular space around the electrode to maintain the inert environment. A manually fed filler rod is commonly used. The GTA technique is routinely used on low-hazard materials such as aluminum and magnesium in addition to a number of alloys including stainless steel, nickel alloys, copper-nickel, brasses, silver, bronze, and a variety of low-alloy steels that may have industrial hygiene significance.

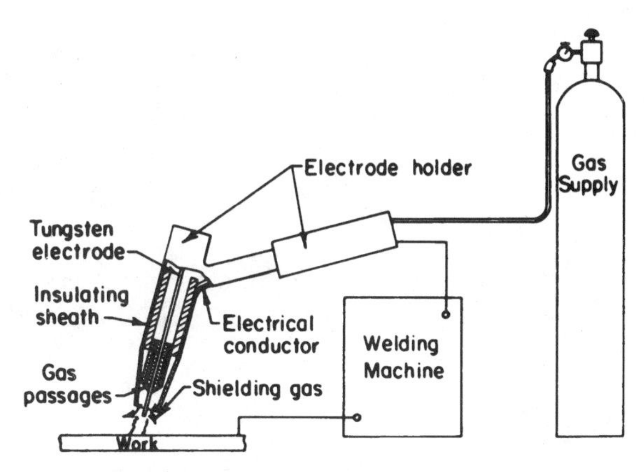

Fig. 4.8 Gas tungsten arc welding (GTAW).

The welding fume concentrations in GTA welding are lower than in manual stick welding and gas metal arc welding. High-energy GTA arc produces nitrogen dioxide concentrations at the welder's position. Argon results in higher concentrations of nitrogen dioxide than helium (Ferry and Ginter, 1953).

The inert gas technique introduced a new dimension in the welder's exposure to electromagnetic radiation from the arc with energies an order of magnitude greater than SMA welding. The energy in the UV B range, especially in the region of 290 nm, is the most biologically effective radiation and will produce skin erythema and photokeratitis. The energy concentrated in the wavelengths below 200 nm (UV C) is most important in fixing oxygen as ozone. The GTA procedure produces a rich, broad spectral distribution with important energies in these wavelengths. The ozone concentration is higher when welding on aluminum than on steel, and argon produces higher concentrations of ozone than helium owing to its stronger spectral emission. The spectral energy depends on current density. Ozone concentrations may exceed the TLV under conditions of poor ventilation.

4.13.3 Gas Metal Arc Welding

In the 1940s, a consumable wire electrode was developed to replace the nonconsumable tungsten electrode used in the GTA system. Originally developed to weld thick, thermally conductive plate, the gas metal arc welding (GMA) process (also known as manual inert gas welding) now has widespread application for aluminum, copper, magnesium, nickel alloys, and titanium, as well as steel alloys.

In this system (Fig. 4.9) the welding torch has a center consumable wire that maintains the arc as it melts into the weld puddle. Around this electrode is an annular passage for the flow of helium, argon, carbon dioxide, nitrogen, or a blend of these gases. The wire usually has a composition the same as or similar to the base metal with a flash coating of copper to ensure electrical contact in the gun and to prevent rusting.

An improvement in GMA welding is the use of a flux-cored consumable electrode. The electrode is a hollow wire with the core filled with various deoxifiers, fluxing agents, and metal powders. The arc may be shielded with carbon dioxide or the inert gas may be generated by the flux core.

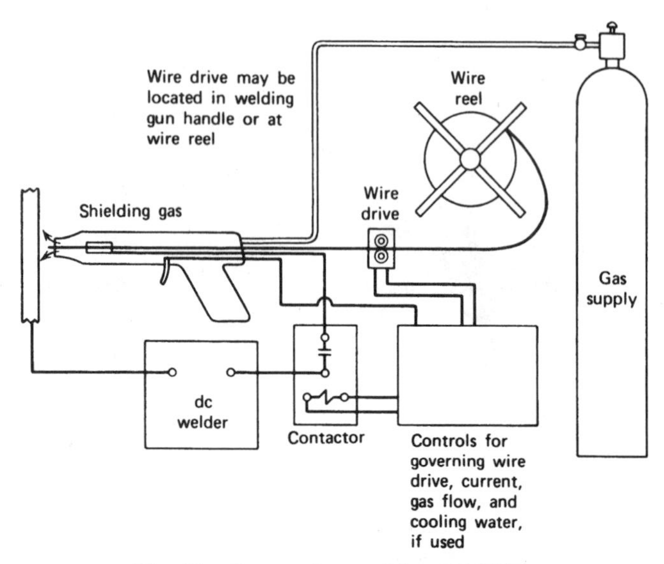

Fig. 4.9 Gas metal arc welding (GMAW).

Metal fume concentrations from GMA welding frequently exceed the TLV on mild steel. If stainless steel is welded the chromium concentrations may exceed the TLV and a significant percent may be in the form of hexavalent chromium. As in the case of SMA, welding on nickel alloys produces significant concentrations of nickel fume. Nitrogen dioxide concentrations are on the same order of magnitude as SMA welding; however, ozone concentrations are much higher with the GMA technique. The ozone generation rate increases with an increase in current density but plateaus rapidly. The arc length and the inert gas flow rate do not have a significant impact on the ozone generation rate.

Carbon dioxide is widely used in GMA welding because of its attractive price; argon and helium cost approximately 15 times as much as carbon dioxide. The carbon dioxide process is similar to other inert gas arc welding shielding techniques, and one encounters the usual problem of metal fume, ozone, oxides of nitrogen, decomposition of chlorinated hydrocarbons solvents, and UV radiation. In addition, the carbon dioxide gas is reduced to form carbon monoxide. The generation rate of carbon monoxide depends on current density, gas flow rate, and the base metal being welded. Although the concentrations of carbon monoxide may exceed 100 ppm in the fume cone, the concentration drops off rapidly with distance, and with reasonable ventilation, hazardous concentrations should not exist at the breathing zone (note the 1996 TLV for carbon monoxide is 25 ppm).

The intensity of the radiation emitted from the arc is, as in the case of GTA welding, an order of magnitude greater than that noted with shielded metal arc welding. The impact of such a rich radiation source in the UV B and UV C wavelengths has been covered in the GTA discussion. With both GTA and GMA procedures trichloroethylene and other chlorinated hydrocarbon vapors are decomposed by UV from the arc-forming chlorine, hydrogen chloride, phosgene, and other compounds. The solvent is degraded in the UV field and not directly in the arc. Studies by Dahlberg showed that hazardous concentrations of phosgene could occur with trichloroethylene and 1,1,1-trichloroethane even though the solvent vapor concentrations were below the appropriate TLV for the solvent (Dahlberg, 1971). The GMA technique produces higher concentrations of phosgene than GTA welding under comparable operating conditions. Dahlberg also identified dichloroacetyl chloride as a principal product of decomposition that could act as a warning agent because of its lacrimatory action. In a followup study, perchloroethylene was shown to be less stable than other chlorinated hydrocarbons. When this solvent was degraded, phosgene was formed rapidly, and dichloroacetyl chloride was also formed in the UV field.

In summation, perchloroethylene, trichloroethylene, and 1,1,1-trichloroethane vapors in the UV field of high-energy arcs such as those produced by GTA and GMA welding may generate hazardous concentrations of toxic air contaminants. Other chlorinated solvents may present a problem depending on the operating conditions; therefore, diagnostic air sampling should be performed. The principal effort should be the control of the vapors to ensure that they do not appear in the UV field. Delivery of parts directly from degreasing to the welding area has, in the authors' experience, presented major problems owing to pullout and trapping of solvent in the geometry of the part.

4.13.4 Gas Welding

In the gas welding process (Fig. 4.10) the heat of fusion is obtained from the combustion of oxygen and one of several gases including acetylene, methylacetylene propadiene (MAPP), propane, butane, and hydrogen. The flame melts the workpiece, and a filter rod is manually fed into the joint. Gas welding is used widely for light sheet metal and

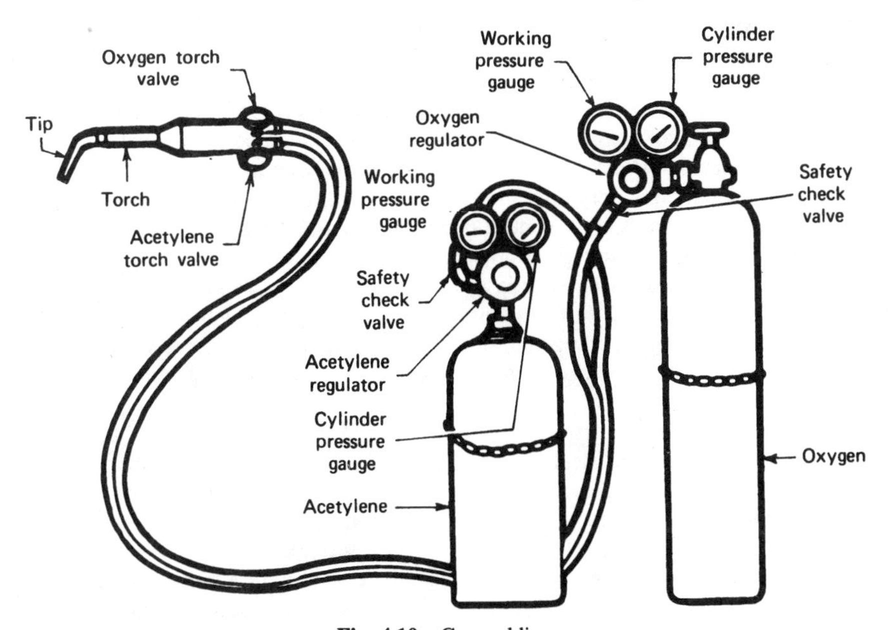

Fig. 4.10 Gas welding.

repair work. The hazards from gas welding are minimal compared to those from arc welding techniques.

The uncoated filler rod is usually of the same composition as the metal being welded except on iron where a bronze rod is used. Paste flux is applied by first dipping the rod into the flux. Fluxes are used on cast iron, some steel alloys, and nonferrous work to remove oxides or assist in fusion. Borax-based fluxes are used widely on nonferrous work, whereas chlorine, fluorine, and bromine compounds of lithium, potassium, sodium, and magnesium are used on gas welding of aluminum and magnesium. The metal fume originates from the base metal, filler metal, and the flux. The fume concentration encountered in field welding operations depends principally on the degree of enclosure in the work area and the quality of ventilation.

The principal hazard in gas welding in confined spaces is due to the formation of nitrogen dioxide. Higher concentrations occurred when the torch was burning without active welding. Striekevskia found concentrations of nitrogen dioxide of 150 ppm in a space without ventilation and of 26 ppm in a space with ventilation (Striekeskia, 1961) (note the 1996 TLV is 3 ppm). These investigators caution that phosphine may be present as a contaminant in acetylene and that carbon monoxide may be generated during heating of cold metal with gas burner.

The radiation from gas welding in quite different from arc welding. The principal emissions are in the visible and IR A, B, C wavelengths and require the use of light-tinted goggles for work. Ultraviolet radiation from gas welding is negligible.

4.13.5 Control of Exposure

In addition to the fume exposures that occur while welding steel alloys, nonferrous metal, and copper alloys, metal coatings and the welding electrodes also contribute to the metal fume exposure. Table 4.7 shows the wide range of metal fumes encountered in welding. Recent interest in the possible carcinogenicity of chromium and nickel fume has prompted special attention to these exposures during welding. Stern has suggested that welders who work on stainless and alloy steels containing chromium and nickel may be at elevated risk from respiratory cancer (Stern et al., 1986). Lead has been used as an alloy in steels to improve its machinability, and welding on such material requires rigorous control. This is also the case with manganese used in steel alloys to improve metallurgical properties. Beryllium, probably the most toxic alloying metal, is alloyed with copper and warrants close control during welding and brazing operations.

Welding or cutting on workpieces that have metallic coatings may be especially hazardous. Lead-based paints have been used commonly to paint marine and structural members. Welding on these surfaces during repair and shipbreaking generates high concentrations of lead fume. Cutting and welding on structural steel covered with lead paint can result in concentrations exceeding 1.0 mg/m^3 in well-ventilated conditions out of doors.

Steel is galvanized by dipping in molten zinc. Air concentrations of zinc during the welding of galvanized steel and steel painted with zinc silicate range from 3 to 12 times the zinc TLV under conditions of poor ventilation. Concentrations are lower with oxygen–acetylene torch work than with arc cutting methods. When local exhaust ventilation is established, the TLV for zinc is seldom exceeded.

The hazard from burning and welding of pipe coated with a zinc-rich silicate can be minimized in two ways. Because pipes are joined end to end, the principal recommendation is to mask the pipe ends with tape before painting (Venable, 1979). If it is

TABLE 4.7 Contaminants from Welding Operations

Contaminant	Source
Metal fumes	
Iron	Parent iron or steel metal, electrode
Chromium	Stainless steel, electrode, plating, chrome-primed metal
Nickel	Stainless steel, nickel-clad steel
Zinc	Galvanized or zinc-primed steel
Copper	Coating on filler wire, sheaths on air-carbon arc gouging electrodes, nonferrous alloys
Vanadium, manganese, and molybdenum	Welding rod, alloys in steel
Tin	Tin-coated steel
Cadmium	Plating
Lead	Lead paint, electrode coating
Fluorides	Flux on electrodes
Gases and vapors	
Carbon monoxide	CO_2 shielded, GMA, carbon arc gouging, oxy-gas
Ozone	GTA, GMA, carbon arc gouging
Nitrogen dioxide	GMA, all flame processes

Source: Ref. 1.

necessary to weld pipe after it is painted, the first step before welding is to remove the paint by hand filing, power brush or grinding, scratching, or abrasive blasting. If the material cannot be removed, suitable respiratory protection is required; in many cases, air-supplied respirators may be required.

A variety of techniques to reduce ozone from GMA welding, including the use of magnesium wire, local exhaust ventilation, and addition of nitric oxide, have been attempted without success. A stainless steel mesh shroud positioned on the welding head controlled concentrations to show below the TLV-ceiling level of 0.1 ppm (Faggetter et al., 1983; Tinkler, 1980).

Hazardous concentrations of nitrogen dioxide may be generated in enclosed spaces in short periods of time and therefore require effective exhaust ventilation.

4.13.5.1 Radiant Energy Eye protection from exposure to UV B and UV C wavelengths is obtained with filter glasses in the welding helmet of the correct shade, as recommended by the AWS. The shade choice may be as low as 8 to 10 for light manual electrode welding, or as high as 14 for plasma welding. Eye protection must also be afforded other workers in the area. To minimize the hazard to nonwelders in the area, flash screens or barriers should be installed. Semitransparent welding curtains provide effective protection against the hazards of UV and infrared from welding operations. Sliney et al. (1981) have provided a format for choice of the most effective curtain for protection while permitting adequate visibility. Plano goggles or lightly tinted safety glasses may be adequate if one is some distance from the operation. If one is 30 to 40 ft away, eye protection is probably not needed for conventional welding; however, high-current-density GMA welding may require that the individual be 100 ft away before direct viewing is possible without eye injury. On gas welding and cutting operations, infrared wavelengths must be attenuated by proper eye protection for both worker health and comfort.

The attenuation afforded by tinted lens is available in National Bureau of Standards reports, and the quality of welding lens has been evaluated by NIOSH. Glass safety goggles provide some protection from UV; however, plastic safety glasses may not.

The UV radiation from inert gas–shielded operations causes skin erythema or reddening; therefore, welders must adequately protect their faces, necks, and arms. Heavy chrome leather vest armlets and gloves must be used with such high-energy arc operations.

4.13.5.2 Decomposition of Chlorinated Hydrocarbon Solvents The decomposition of chlorinated hydrocarbon solvents occurs in the UV field around the arc and not in the arc itself. The most effective control is to prevent the solvent vapors from entering the welding area in detectable concentrations. Merely maintaining the concentration of solvent below the TLV is not satisfactory in itself. If vapors cannot be excluded from the workplace, the UV field should be reduced to a minimum by shielding the arc. Pyrex glass is an effective shield that permits the welder to view his or her work. Rigorous shielding of the arc is frequently possible at fixed station work locations, but it may not be feasible in field welding operations.

4.13.5.3 System Design The advent of robotics has permitted remote operation of welding equipment, thereby minimizing direct operator exposure to metal fumes, toxic gases, and radiation.

REFERENCES

American Conference of Governmental Industrial Hygienists, *Industrial Ventilation: A Manual of Recommended Practice*, 22nd ed., ACGIH, Lansing, MI, 1995.

American Electroplaters and Surface Finishers Society, *Safety Manual for Electroplating and Finishing Shops*, Orlando, FL, 1989.

American Foundrymen's Society, Inc., *Industrial Noise Control—An Engineering Guide*, Des Plaines, IL, 1985.

American National Standards Institute, "Practices for Ventilation and Operation of Open-Surface Tanks," ANSI Z9.1-1977, New York, 1977.

American National Standards Institute, "Safety Requirements for the Use, Care, and Protection of Abrasive Wheels," B7.1-1978, ANSI, New York, 1978.

American National Standards Institute, "Ventilation Control of Grinding, Polishing, Buffing Operations," Z 43.1-1966, ANSI, New York, 1966.

American National Standards Institute, "Radiological Safety Standard for the Design of Radiographic and Fluoroscopic Industrial X-ray Equipment," ANSI/NBS 123-1976, ANSI, New York, 1976.

American Society for Metals, *Heat Treaters Guide*, New York, 1982.

American Society for Testing and Materials, *Manual on Vapor Degreasing*, 3rd ed., Philadelphia, 1989.

Ballard, W. E., *Ann. Occup. Hyg.* **15,** 101 (1972).

Bastress, E. K., et al., "Ventilation Requirements for Grinding, Polishing, and Buffing Operations," Report No. 0213, IKOR Inc., Burlington, MA, June 1973.

Bingham, E., P. Trosset, and D. Warshawsky, *J. Environ. Pathol. Toxicol.* **3,** 483 (1979).

Blair, A., *J. Occ. Med.*, **22:**158–162, 1980.

Blair, D. R., *Principles of Metal Surface Treatment and Protection*, Pergamon Press, Oxford, 1972.

Burge, P. S., M. G. Harries, I. M. O'Brien, et al., *Clin. Allergy* **8,** 1 (1978).

Burgess, W., *Recognition of Health Hazards in Industry: A Review of Materials and Processes*, John Wiley, New York, 1995.

Burgess, W. A., "Potential exposures in the manufacturing industry—their recognition and control." In: *Patty's Industrial Hygiene and Toxicology*. Vol. I. Part A. George D. Clayton, Florence E. Clayton, eds., New York, John Wiley and Sons, 1991.

Burgess, W. A., Ellenbecker, M. J., and Treitman, R. D., *Ventilation for Control of the Work Environment*, John Wiley, New York, 1989.

Dahlberg, *J. Ann. Occup. Hyg.* **14,** 259 (1971).

Danielson, J., Ed., Air Pollution Engineering Manual, 2nd ed., Publication No. AP-40, U.S. Government Printing Office, Washington, DC, 1973.

Decoulfe, P., *J. Nat. Cancer Inst.* **61,** 1025 (1978).

Decoufle, P. *J. Natl. Cancer Inst.*, **61:**1025–1030, 1978.

Dreger, D. R., *Machine Design* **47** (Nov. 25, 1982).

Dow Chemical Co., "How to Select a Vapor Degreasing Solvent," Bulletin Form 100-5321-78, Midland, MI (1978).

Dow Chemical Co., "Modern Vapor Degreasing and Dow Chlorinated Solvents," Dow Bulletin, Form No. 100-5185-77, Midland, MI (1977).

Dow Chemical Co., "How to Select a Vapor Degreasing Solvent," Bulletin Form 100-5321-78, Midland, MI (1978).

DuPont, "Freon Solvent Data," Bulletin No. FST-1, Wilmington, DE, 1987.

Eisen, E., Tolbert, P. E., Hallock, M. F., Monson, R., Smith, T. J., Woskie, S. R., "Mortality studies of machining fluid exposures in the automotive industry. II. Risks associated with specific fluid types," *Am. J. Ind. Med.* **24:**123 (1994).

Elkins, H. B., *The Chemistry of Industrial Toxicology*, John Wiley, New York, 1959.

Esso Petroleum Company, "This is About Health, Esso Lubricating Oils and Cutting Fluids," Ltd., 1979.

Faggetter, A. G., V. E. Freeman, and H. R. Hosein, *Am. Ind. Hyg. Assoc. Y.* **44,** 316 (1983).

Ferry, J., and G. Ginter, *Welding J.* **32,** 396 (1953).

Fletcher, B. "Low-Volume-High-Velocity Extraction Systems," Report No. 16, Bootle, UK: Health and Safety Executive, Technology Division, 1988.

Forging Industry Association, *Forging Handbook*, Cleveland, OH, 1985.

Goldsmith, A. H., K. W. Vorpahl, K. A. French, P. T. Jordan, and N. B. Jurinski, *Am. Ind. Hyg. Assoc. J.* **37,** 217–26, 1976.

Gutow, B. S., *Environ. Sci. Technol.* **6,** 790 (1972).

Hagopian, J. H. and Bastress, E. K. Recommended Industrial Ventilation Guidelines, Publ. No. (NIOSH) 76-162, National Institute for Occupational Safety and Health, Cincinnati, OH, 1976.

Hallock, M. F., Smith, T. J., Woskie, S. R., Hammond, S. K., "Estimation of historical exposures to machining fluids in the automotive industry," *Am. J. Ind. Med.* **26:**621–634 (1994).

Handy, and Harman, *The Brazing Book*, New York, 1985.

Hansen, M. K., M. Larsen, and K.-H. Cohr, *Scand. J. Work Environ. Health* **13,** 473 (1987).

Health and Safety Executive, Portable Grinding Machines: Control of Dust, Health and Safety Series Booklet HS(g) 18, H. M. Stationary Office, London, England, 1982.

Heat Treaters Guide, American Society for Metals, New York, 1982.

HM Factory Inspectorate, "Dust Control: The Low Volume–High Velocity System," Technical Data Note 1 (2nd rev.), Department of Employment, London, England.

Hogapian, J. H., and E. K. Bastress, "Recommended Industrial Ventilation Guidelines," Department of Health, Education and Welfare, NIOSH Publication No. 76-162, Cincinnati, OH, 1976.

IARC, (International Agency for Research on Cancer) "Occupational Exposure to Mists and Vapours from Strong Inorganic Acids and Other Industrial Chemicals," *IARC Monographs on the Evaluation of Carcinogenic Risks to Humans*, Vol. 54, pp. 44–130. IARC, Lyon, France (1992).

International Molders and Allied Workers Union (AFL-CIO-CLC), No-Bakes and Others: Common Chemical Binders and Their Hazards, undated.

Isotope Radiography Safety Handbook, Technical Operations, Inc., Burlington, MA, 1981.

Jarvolm, B., Thuringer, G., Axelson, O., *Brt. J. Occ. Med.* **39:**197–197, 1982.

Kennedy, S. M., Greaves, I. A., Kriebel, D., Eisen, E. A., Smith, T. J., Woskie, S. R. (1989) *Am. J. Ind Med.* **15:**627–641, 1989.

Knecht, U., H. J. Elliehausen, and H. J. Woitowitz, *Br. J. Ind. Med.* **43,** 834 (1986).

Manko, H. H., *Solders and Soldering*, McGraw–Hill, New York, 1979.

NIOSH, "Ventilation Requirements for Grinding, Buffing, and Polishing Operations," NIOSH Publ. No. 75-105, National Institute for Occupational Safety and Health, Cincinnati, OH, 1975.

NIOSH, "Industrial Health and Safety Criteria for Abrasive Blasting Cleaning Operations," DHHS (NIOSH) Publ. No. 75-122, National Institute for Occupational Safety and Health, Cincinnati, OH, 1975.

NIOSH, "Recommendations for Control of Occupational Safety and Health Hazards Manufacture of Paint and Allied Coating Products," DHHS Publication No. 84-115, CDC, NIOSH, 1984.

NIOSH, "Recommendations for Control of Occupational Safety and Health Hazards Foundries," DHHS Publication No. 85-116, Cincinnati, OH, 1985a.

NIOSH, "Control Technology Assessment: Metal Plating and Cleaning Operations," DHHS Publication No. 85-102, Cincinnati, OH, 1985b.

NIOSH, "Criteria for a Recommended Standard: Occupational Exposure to Hand-Arm Vibration," NIOSH Publ. No 89-106, National Institute for Occupational Safety and Health, Cincinnati, OH, 1989.

NIOSH, Health Hazard Evaluation Report 88-136-1945, Miller Thermal Technologies, Inc., Appleton, WI, National Institute for Occupational Safety and Health, Cincinnati, OH, 1989.

NIOSH, "Guidelines for the Control of Exposures to Metalworking Fluids," NIOSH Publ. No. 78-165, National Institute for Occupational Safety and Health, Cincinnati, OH, 1978.

NIOSH, "NIOSH Alert, Request for Assistance in Preventing Silicosis and Deaths from Sandblasting," DHHS (NIOSH) Publ. No. 92-102, National Institute for Occupational Safety and Health, Cincinnati, OH, 1992.

NFPA, NFPA86C, Standard for Industrial Furnaces Using a Special Processing Atmosphere, National Fire Protection Association, Quincy, MA, 1991.

O'Brien, D. M., and D. E. Hurley, "An Evaluation of Engineering Control Technology for Spray Painting," DHHS (NIOSH) Publication No. 81-121, USDHHS, CDS, Cincinnati, OH (1981).

O'Brien, D., and J. C. Frede, "Guidelines for the Control of Exposure to Metalworking Fluids," Department of Health, Education and Welfare, NIOSH Publication No. 78-165, Cincinnati, OH, 1978.

Oudiz, J., A. Brown, H. A. Ayer, and S. Samuels, *Am. Ind. Hyg. Assoc. J.* **44,** 374 (1983).

Pantucek, M., *Am. Ind. Hyg. Assoc. J.* **32,** 687 (1971).

Pelmear, P. L., and R. Kitchener, "The Effects and Measurement of Vibration," in *Proceedings of the Working Environment in Iron Foundries*, March 22–24, 1977, British Cast Iron Research Association, Birmingham, England, 1977.

Sliney, D. H., C. E. Moss, C. G. Miller, and J. B. Stephens, *Appl. Optics* **20,** 2352 (1981).

Sliney, D., and M. Wolbarsht, *Safety with Lasers and Other Optical Sources*, Plenum, New York, 1980.

Sparks, P. J., Wegman, D. H., *J. Occ. Med.* **22:**733–736, 1980.

Spring, S., *Industrial Cleaning*, Prism Press, Melbourne, 1974.

Starek, J., M. Farkkila, S. Aatola, I. Pyykko, and O. Korhonen, *Br. J. Ind. Med.* **40,** 426 (1983).

Stern, R. M., A. Berlin, A. C. Fletcher, and J. Jarvisalo, *Health Hazards and Biological Effects of Welding Fumes and Gases*, Excerpta Medica ICS 676, Elsevier, Amsterdam, 1986.

Stokinger, H., *Am. Ind. Hyg. Assoc. J.* **45,** 127 (1984).

Strizerskiy, I., *Welding Prod.* **7,** 40 (1961).

Tinkler, M. J., "Measurement and Control of Ozone Evolution during Aluminum GMA Welding," Colloquium on Welding and Health, July, 1980, Brazil.

Toeniskoetter, R. H., and R. J. Schafer, "Industrial Hygiene Aspects of the Use of Sand Binders and Additives," in *Proceedings of the Working Environment in Iron Foundries*, March 22–24, 1977, British Cast Iron Research Association, Birmingham, England, 1977.

United Kingdom Department of Employment (1974), Control of Dust from Portable Power Operated Grinding Machines (Code of Practice), London.

VanDerWal, J. F., *Ann. Occup. Hyg.* **34,** 45 (1990).

Vena, J. E., Sultz, H. A., Fiedler, R. C., Barnes, R. E., *Brit. J. Ind. Med.*, **42:**85–93, 1985.

Venable, F. *Esso Med. Bull.* **39,** 129 (1979).

Verma, D. C., F. Muir, S. Cunliffe, J. A. Julian, J. H. Vogt, and J. Rosenfeld, *Ann. Occup. Hyg.* **25,** 17 (1982).

WHTC, "Guidelines for Safety in Heat Treatment, Part 1, Use of Molten Salt Baths," Wolfson Heat Treatment Centre, University of Aston, Gosta Green, Birmingham, England, 1981.

WHTC, "Guidelines for Safety in Heat Treatment, Part 2, Health and Personal Protection, "Wolfson Heat Treatment Centre, University of Aston, Gosta Green, Birmingham, England, 1983.

WHTC, "Guidelines for Safety in Heat Treatment, Part 3, Quenching, Degreasing, and Fire Safety," Wolfson Heat Treatment Centre, University of Aston, Gosta Green, Birmingham, England, 1984.

CHAPTER 5

How to Conduct an Accident Investigation

THOMAS M. DOUGHERTY

Polaroid Corporation, 1265 Main St., Waltham, MA 02254

When someone gets hurt in an accident, telling the person to "Be more careful" is not very helpful for either the individual or the organization.

5.1 INTRODUCTION
- 5.1.1 The Benefits of Investigating Accidents
- 5.1.2 The Need to Report Accidents
- 5.1.3 The Need to Investigate Accidents
- 5.1.4 Who Participates in the Investigation

5.2 THE ACCIDENT INVESTIGATION—AN OVERVIEW
- 5.2.1 Getting Medical Help
- 5.2.2 When to Start the Investigation
- 5.2.3 An Overview of the Process

5.3 THE ACCIDENT INVESTIGATION—STEP BY STEP
- 5.3.1 Step 1—Getting the Facts
- 5.3.2 Step 2—Determining the Immediate Causes—Acts and Conditions
- 5.3.3 Step 3—The Underlying Factors—Why the Acts and Conditions
- 5.3.4 Step 4—Organizational Prevention Measures—Underlying Factors and Interventions

5.4 Management Systems
- 5.4.1 Implementing the Prevention Measures—Investigation Is Not Enough!
- 5.4.2 Analyses of Consolidated Injury Investigation Data

5.5 SUMMARY

REFERENCES

"Jim," Bill said haltingly, "we've just had a major accident. ... Carl had two of his fingers smashed! there was blood everywhere ... !!!! It happened when he reached into the assembly machine to clear a jam. The machine has been a dog the last two shifts. ... I just can't get Carl's screams out of my mind. It was really awful."

Handbook of Occupational Safety and Health, Second Edition, Edited by Louis J. DiBerardinis, ISBN 0-471-16017-2

5.1 INTRODUCTION

5.1.1 The Benefits of Investigating Accidents

Conducting an effective accident investigation provides the solution to several different requirements. Organizational safety policies, as well as good safety practice, requires an investigation to determine how it happened, why it happened and what's needed to prevent it in the future. State Worker's Compensation laws require an investigation to determine if the injury is compensible. Federal Occupational Safety and Health laws require an investigation to ascertain if it should be recorded.

Enlightened organizations also understand that effective injury investigations provide clues to creating an error-free workplace. Total quality management (TQM) cultures require and demand understanding of the causes and effects that lead to errors, defects, and injuries. Dan Petersen[1] says "The goal . . . is to change your company's culture until safety becomes an internal value that is incorporated in every plan, decision, and work activity. In the new safety culture, every employee will know that the only way to do anything is *safely*."

There is no doubt that basic organizational beliefs and values have a direct effect on the outcomes of the injury reporting and investigation process. If the organization believes that injured individuals "need to be more careful" and that the organization has little or no responsibility in this effort, then a passive, ineffective, and frustrating injury investigating exercise is the likely result. The individual is blamed for making a "stupid error," and the organization washes its collective hands of the incident.

On the other hand, organizations that recognize that they have a responsibility for providing an environment embracing continuous learning, skill enhancement, and placing high value on personal well being find effective and satisfying results. The injured individual will take personal responsibility for acting in a different and better way and the organization also learns what it needs to do differently and better.

The purpose of this chapter is not only to provide some basic, traditional safety approaches for investigating accidents but also to provide some guidance on how to probe for the root causes of those incidents and the most effective organizational interventions. These elements of the accident investigation program then can become an essential element of an effective health and safety management system. This accident reporting and investigation system could in fact become the foundation of the program. This CHECK (checking and corrective action) part of the management system, however, needs to be supported by the PLAN, the DO (implementation and operation) and the ACT (management review). Reacting to accidents, incidents, and near-misses needs to be supported by the proactive process of identifying hazards, evaluating the associated risks, and installing effective controls.

The National Safety Council[2] reports 93,300 fatal accidents in 1995, of which 5,300 were work related. They also report over 60 million injuries, of which over 8 million occurred at work. Nearly 50% of the fatalities were attributable to motor vehicle accidents. Accidental death rates in the United States peak in the teenage years through the early twenties primarily due to automobile accidents. Deaths from falls increase dramatically after reaching the age of 70.

At work, backs are the part of the body most often injured (25%) and having the highest percent of worker's compensation costs (31%). Arms and hands follow at 17% and 13%, respectively. Estimates of OSHA recordable injury rates for all private industry is about 8.9%, with the construction industry at 13.1%, all manufacturing at 12.5%, and finance/insurance industry at 2.9%.

The following table is adapted from the National Safety Council's "Accident Facts."

	Deaths	Deaths per 100,000 persons	Disabling injuries
Total	93,300	35.5	19,300,000
Motor vehicle	43,900	16.7	2,300,000
Work	5,300	2.0	3,600,000
Home	26,400	10.0	7,300,000
Public	20,100	7.6	6,200,000

Worldwide facility and businesses losses to insureds of Allendale Insurance, Arkwright, and Protection Mutual Insurance were attributed[3] to five leading causes of shut sprinkler valves, arson, hot work, poor housekeeping, and smoking. Gross amounts of losses ranged from over $100 million caused by smoking to nearly $400 million caused by arson.

5.1.2 The Need to Report Accidents

Accidents must be reported for a variety of reasons. An accident is an incident that results in personal injury or property damage. An incident is an unplanned, unwanted event that results in, or under slightly different circumstances could have resulted in, personal injury or property damage. So more correctly stated, "incidents" must be reported for a variety of reasons.

The reporting of *all* incidents provides the organization with the opportunity to prevent the more serious accidents. Remembering the accident triangle (Fig. 5.1), the greater the number of near misses (incidents), the greater the likelihood of minor injuries, therefore the greater the likelihood of more serious accidents and ultimately a higher frequency of fatalities. Reporting and investigating on the near misses and *all*

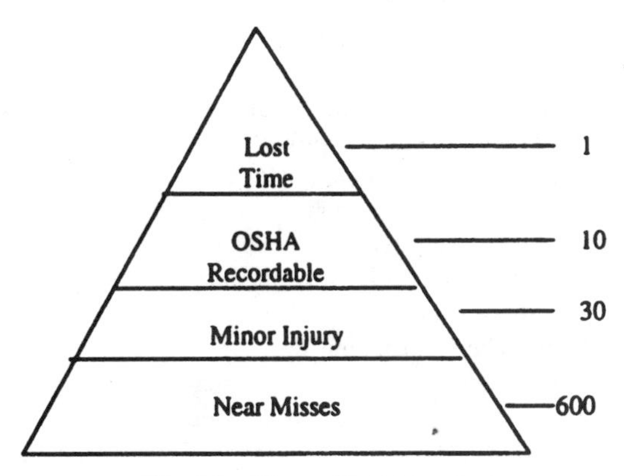

Fig. 5.1 The accident triangle.

incidents reduces the chances that they will reoccur. This leads to fewer serious accidents and a much lower chance of someone being killed in the workplace.

Several investigators have estimated the ratios of incidents to minor injuries to serious injuries. Bird and Germain[4] have reported a ratio of 90,000 incidents to 500 property damage accidents to 100 minor injuries to one disabling injury. International Safety Academy[5] reports studies finding a ratio of 600 accidents with no visible injury or damage to 30 property damage of all types to 10 minor injuries to one serious or disabling injury. DuPont[6] suggests an incident pyramid that is shaped with 1 fatality to 30 lost-time injuries to 300 recordable injuries to 30,000 near misses and having a base of 300,000 at-risk behaviors and/or conditions. Eastman[7] reports studies indicating that before each accident occurs, the worker may have performed nearly 750,000 unsafe behaviors!

The important thing to know is not the exact ratios. The important thing to notice is that the key to reducing the serious injuries is to work at the bottom of the triangle. Becoming knowledgeable of all incidents and near misses provides valuable insights that lead to a broader more deeper understanding of what is causing the more serious accidents. This approach helps focus the organization on the required safety activities that are needed before the accident happens. This is a proactive approach to injury reduction rather than a reactive after-the-fact activity.

Although both activities are important and necessary, major safety improvements cannot occur without a focus at the bottom of the accident triangle. This chapter will show ways to get even deeper into the foundation of the accident triangle with a greater appreciation for the root causes of the accidents that can lead directly to the best organizational preventive measures.

Most organizations have policies requiring the reporting and investigation of injuries that occur in the workplace. In addition to the organization's safety and health policy, there are several regulatory requirements for reporting occupational injuries and even near misses in some cases. State Worker's Compensation laws require employers to report industrial injuries so that compensation benefits can be provided for a personal injury resulting from a workplace accident. The Federal Occupational Safety and Health Act requires employers to "record" occupational injuries that have a certain level of severity. A summary record of these "recordable" injuries also must be posted in the workplace annually using the OSHA 200 log form.

Additionally, if a work-related fatality occurs or a multiple hospitalization occurs, then OSHA regulations requires that these accidents be reported to them within 8 hours. The Process Safety Management requirement of OSHA also requires affected employers to document incidents that resulted in or could reasonably have resulted in a catastrophic release of a highly hazardous chemical. Local, state, and federal environmental laws also require the reporting of certain chemical spills and releases.

The reporting of accidents and incidents then, is the first step in meeting organizational policy requirements as well as regulatory needs. It is also an important first step in preventing future accidents and incidents.

A typical accident/injury reporting form is illustrated in Fig. 5.2. Obtaining information regarding the facts of the injury in a timely and complete manner is an important part of injury prevention. The information developed will be useful in analyzing the specific injury under investigation as well as in analyzing an organizations accident record/history over a period of time. Most organizations collect this information in an electronic database. The data can then be summarized, manipulated, and reported out in a variety of ways.

ACCIDENT / INJURY REPORTING FORM

Name ____________________ SS# ______________ Age ______________
Address ____________________ Phone # ____________ Job Title ______________
City, State ____________________ Dept ____________ Supvr ______________
Date of Birth __________ Sex M/F Seniority Date __________ Shift ______________

Accident Location ____________________ Date __________ Time ______________
Description of Accident ____
____
____
____
____
Witnesses: ____
Employee Signature ____________________ Date ______________

Type of Incident	Type of Injury	Part of Body	Treatment
• Hit By	• Contusion, laceration	• Eye R / L	• Cleaned/Dressed
• Hit On	• Fracture, dislocation	• Face / Head	• Hot / Cold Pack
• Slip, Trip, Fall	• Sprain, strain	• Torso / Back	• Bandage/Butterfly
• Lifting, Pulling	• Burn ______	• Hands R / L	• Cast/Sutures
• Cumulative Trauma	• Foreign Body	• Arms R / L	• Crutches / Rest
• Transportation	• Exposure ______	• Feet / Legs R / L	• Medication
• Other ______	• Other ______	• Other ______	• Other ______

Medical Comments ____
____
____
____
____
Medical Signature ____________________ Date ______________

Report Initiator ____________________ Date ______________
Copies: Employee, Supervisor, Medical, Safety, Manager

Fig. 5.2 Accident/injury reporting form.

5.1.3 The Need to Investigate Accidents

Once the incident or accident is reported, the next step is conducting an effective investigation. Why did it happen? ... and what can be done to prevent it from happening in the future? The culture of the organization will play an important part in this process. In fact, the safety environment will also impact on the actual reporting of workplace injuries. If the prevailing atmosphere is to affix blame and find fault, many if not most minor injuries and incidents will go unreported. The opportunity to intervene and prevent future serious injuries will be lost.

World-class organizations recognize that effective injury investigations brings added value to the business. Effective investigations can result in understanding of the basic root causes of the errors leading to those "accident/incident" situations. Those root causes can then be viewed as symptoms of organizational difficulties that can lead to other areas of concern such as quality defects, cost overruns, or even a stressed and unproductive workforce. Minter[8] says "... if we really want to improve a safety program, we look at the management system that is producing injuries and seek out the root causes."

The valuable insights gained through effective accident investigations can lead to prevention strategies that can be applied to the entire management system. The lessons learned on how to prevent injuries are not much different from those learned on how to prevent other types of quality defects.

The challenge for the organization is to move from a position of preventing accidents and injuries to a culture of ensuring safety and well being for its workforce. Most safety management experts endorse the proactive approach of positive reinforcement of safe behaviors rather than the reactive message of discipline for unsafe behaviors. Although both are necessary, the accident investigation process can lead to affixing blame and finding fault if not carefully structured. Learning where the greatest opportunities to improve for both the organization and the individual is the goal. An injury-free workplace is the reward.

5.1.4 Who Participates in the Investigation

Traditionally, the supervisor has been considered the key person in the accident investigation process. First-line supervisors have had a unique perspective and also a special responsibility for ensuring the safety and well being of people working for them. They generally are familiar with the situation, including the hazards of the job tasks, the necessary controls, and the particular attributes of their workers.

As organizations reengineer, restructure, redesign, and manage differently, the responsibility for conducting accident investigations is also changing. Coaches and team leaders are now taking a more influential role in the investigation process. Many organizations are also forming teams independent of the department where the accident occurred to conduct the investigation so as to minimize the fault finding and blame assessment. Verespej[9] reports that Rhone-Poulenc has put workers into teams because "... Things are simply too complex to be done by specialists. You need to have people who know different parts of an issue get together to solve problems as a team." Teams comprising individuals who have good analytical skills, who know how things can go wrong and are not trying to affix blame can be objective and dispassionate in their findings. The accident fault or human error is recognized to be the result of an imperfect element of the system, which has allowed or may even have influenced an individual to behave inappropriately. The responsibility to do things differently and better is recognized as being an opportunity for both the organization and the injured individual.

The investigation team and/or the supervisor generally has access to support personnel who can help assess causal factors and determine the best prevention measures. Engineering personnel can provide insights into equipment design standards, trades personnel can provide recommended preventive maintenance practices, safety profession-

TABLE 5.1 Incident Investigation Resources

Supporting resources	Expertise, assistance, and insights
Design/production engineers	Can provide design basis information, equipment standards, and engineering calculations
Supervisors/team leaders	Can provide standard operating procedures and approved methods
Trades/maintenance personnel	Can provide information on maintenance needed or performed relating to the facility or equipment
Safety professional	Can provide relevant safety standards and help with judgments regarding risks
Industrial hygiene professional	Can provide specialized information on exposure level potentials and appropriate controls
Ergonomic professional	Can provide expertise and analysis of work station and work task designs
Medical personnel	Can provide information related to the type, nature, and level of injury or illness
First aiders/emergency response personnel	Can provide information relating to initial situation and possible response improvements
Equipment/material vendors	Can provide expertise relating to their equipment, materials, or process
Trade associations/similar facilities	Can provide experts, standards, or best practices relating to the situation
Professional H/S organizations	Can provide experts or standards knowledgeable regarding the situation

als can supply safety standards as well as training on how to investigate effectively. An ergonomic specialist may have clues as to why the person is hurting based on the review of the work station and job tasks. Medical personnel can provide diagnoses that can be helpful in assessing the facts surrounding the nature of the injury.

Table 5.1 summarizes some of the expertise that may be needed. Note that one individual may provide assistance in several different areas. Some of the resources may need to come from outside the organization.

Although the direct-line management has the ultimate responsibility for the effective investigation of accidents and the implementation of the appropriate prevention measures, the organization can and should derive maximum benefit from all its supporting resources.

In our case where Carl had two fingers severely injured while clearing a jam on an assembly machine, Bill, the supervisor, asked the machine design engineer, the trades mechanic, and a co-worker of Carl to join in the accident investigation. Carl would participate later when he was given medical clearance.

5.2 THE ACCIDENT INVESTIGATION—AN OVERVIEW

5.2.1 Getting Medical Help

When someone gets hurt in the workplace, the most immediate need is to be sure that medical help is quickly summoned. Most organizations have basic policies and procedures on how to accomplish this. Simple rules are the easiest to remember and generally the most effective.

Specific situations and circumstances will determine the best procedures. A high-rise building complex of office workers located in a major urban area will have different medical response capability needs compared to a large chemical manufacturing complex located in a remote outlying area. In both situations, however, employees need to know the basic procedure for summoning medical assistance. This is generally accomplished via the use of an emergency telephone number. This number can connect at local community response agencies or the organization's own emergency response crew.

For those organizations that do not have on-site medical coverage, some choose to provide training for selected employees in basic first aid and cardiopulmonary resuscitation techniques (CPR). Local hospitals, health care organizations, as well as the Red Cross and Heart Association provide excellent training resources. These first aid trained individuals can then provide a stabilizing environment and immediate support until professional assistance arrives. They can also provide basic treatment to fellow employees for minor cuts and abrasions. Additionally, this approach supports the notion of a caring environment for employees and the reinforcement of organizational safety values.

Basic training on the OSHA requirement of the hazards of blood-borne pathogens must also be provided to these individuals. The organization's written program needs to include training on the use of personal protective equipment and methods of cleaning up of blood as well as disposal of blood-soiled materials.

The organization needs to consider carefully medical transport practices and decide policy only after consultation with qualified medical experts. Transport of injured personnel to hospitals or clinics by employees in their own personal cars is not advisable. Providing stabilizing support by trained personnel until the ambulance or rescue squad arrives is usually the best choice.

Some organizations provide equipment, develop procedures, and train personnel to respond to accidents and incidents. Typical equipment includes medical supplies, oxygen bottles, machinery extrication equipment, wheel chairs, stretchers, and other similar equipment. Periodic training reviews and routine equipment inspections need to be established. Many organizations include these aspects as part of their overall emergency response procedures.

Carl was transported by ambulance to the local hospital. A co-worker who happened to be trained as an Emergency Medical Technician (EMT) stabilized the bleeding while the security personnel called the listed emergency number. Emergency response drills had been conducted semiannually at the plant which included a medical response component.

5.2.2 When to Start the Investigation

Once the injured person has been provided with the immediate medical assistance or the incident situation has been stabilized, the investigation should begin as soon as possible. Delays can easily result in a change in the physical situation and surrounding circumstances. Lighting levels change, noise levels change, temperatures change, liquids evaporate, and in general, the situation changes. Additionally, the recall ability of witnesses to the event will change. Even under ideal circumstances, memory quickly fades with time. In a traumatic accident situation where a co-worker gets seriously hurt, it is likely that there will be a variety of answers to the question "What happened?" If there is an immediate continuing hazard, actions may need to be taken to secure unsafe situations to prevent injury to the investigators or stabilize an unsafe situation for co-workers.

For accidents resulting in serious injuries it may not be possible to interview the injured person for several days or longer. For this and other reasons, it is essential that the scene of the accident be preserved until the essential facts around the situation are determined. Immediate steps should be taken to secure the area, videotape the scene, take photographs, sketch the situation, take statements from co-workers and witnesses, and otherwise document the evidence at the site. Actions not taken here will result in difficulties in accurately recreating the situation days and weeks later.

After Bill was assured that Carl had medical attention, he asked each of Carl's shift co-workers to write down the facts surrounding the accident. He took photographs at the scene and copied relevant documents. He then called key members of the organization to join him in an early morning meeting.

5.2.3 An Overview of the Process

Consider the following four steps in conducting an effective investigation. These steps form a logical progression of thinking regarding the accident causes, the resulting effects, and potential prevention opportunities. An expanded accident triangle (Fig. 5.3) forms the conceptual basis for the evaluation steps. Every near miss or injury (step 1) can have an immediate cause (step 2) of an inappropriate action or inappropriate workplace condition. Those immediate causes have associated underlying factors (step 3), and those factors, in turn, have their roots in the management system (step 4). The expanded accident triangle shows those cause and effect relationships visually.

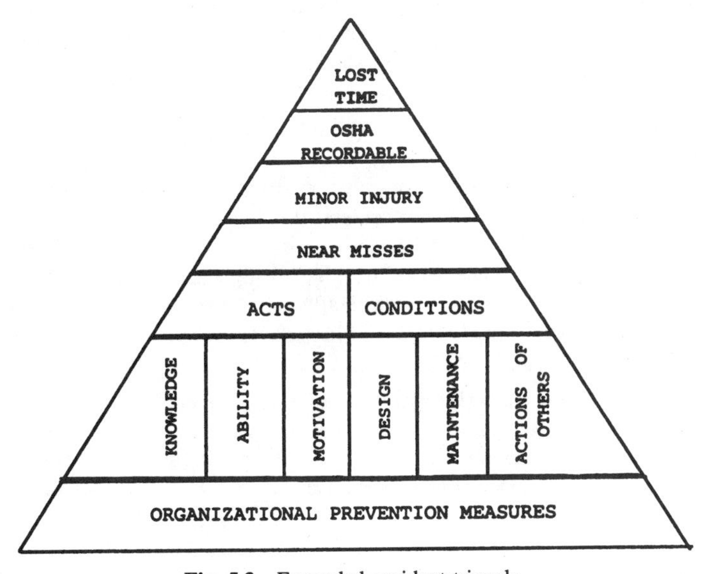

Fig. 5.3 Expanded accident triangle.

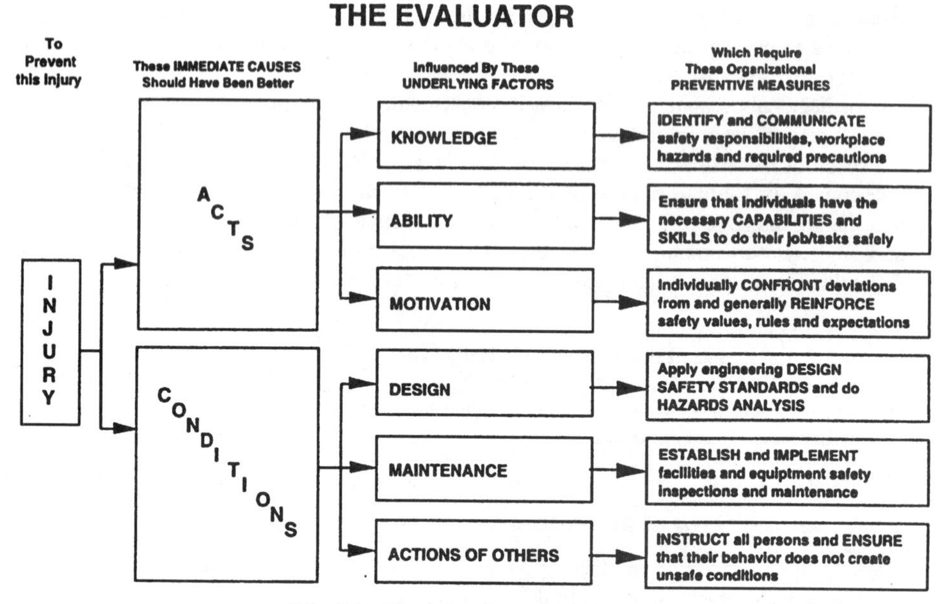

Fig. 5.4 The injury investigation tool.

The injury investigation tool, called the Evaluator and based on the extended accident triangle (Fig. 5.4), illustrates the process. Step 1, obtain the facts around the injury/accident; step 2, determine the immediate causes; step 3, ascertain the underlying factors; and step 4, identify the organizational preventive measures.

1. **"Get the Facts."** It is virtually impossible to determine the root causes, underlying factors, and ultimately the best prevention measures without first getting accurate facts about what actually did happen. So spending the right amount of time and effort in getting the facts straight starts the investigation down the right path.

2. **"What Acts and/or Conditions Should Have Been Different?"** This is the beginning of the analyses. Unfortunately, it has also been the ending of some analyses! Based on the facts of the situation developed above, judgments need to be made regarding the immediate causes of the accident. What behaviors of the injured employee could have been different or better. These have traditionally been called Unsafe Acts. Also, judgments need to be made regarding what workplace conditions could have been different or better. These have been traditionally called Unsafe Conditions. Notice how asking "What should have been different or better?" leads to a different place than asking "What were the Unsafe Acts?"

3. **"What Were the Underlying Factors That Led to Those Acts and Conditions?"** This brings the analyses to a place beyond the "be more careful" stage. Understanding why the injured person acted that way and why the workplace conditions were that way leads directly to preventive actions that the person needs to do as well as what the organization needs to do.

Regarding Acts, for example, did the person know the rules or procedures? Did the person have the ability to follow the procedures? . . . or was it a situation of the person not being motivated to follow the procedures? The answer to these questions leads to dramatically different prevention measures.

Similarly regarding the inappropriate conditions, was the condition a result of inadequate design? Was the condition a result of poor maintenance? or was the condition the result of an action of another person? Determining why the condition was that way really depends on the underlying factors and leads directly to the most effective prevention measures.

4. **"What Are the Best Prevention Measures?"** Knowing now what underlying factors influenced the injured person's actions or the work place conditions leads directly into determining the best injury prevention measures. For example, if the person did not know the rules or procedures, a training program is indicated. If the person did not have adequate ability to follow the procedures, then on-the-job skills enhancement may be the answer. Or if the person had the knowledge and the skills, then a motivational effort is indicated. Depending on the circumstances, it could require the installation of a recognition program or a discipline program or both.

This tool is then used to develop a incident investigation form, which is illustrated in Fig. 5.5. This form is then used by the accident investigators to determine the root causes and the best prevention measures.

This approach, then, leads to the best actions for the injured individual to take as well as the best things for the organization to do. It also supports the organization that is trying to establish and reinforce beliefs that accidents are caused and don't just "happen."

In summary, a logical, sequential investigation of the facts, the causes, and the associated effects can lead to a new and deeper understanding of the root causes and needed prevention measures for those accidents and injuries. When the investigation is conducted in a blame-free environment, the results can lead to a world-class safety and health program. In commenting on accidents, John Thirion[10] says, "We know that safety is a clearcut barometer of organizational excellence. You cannot have an excellent organization that has a lot of accidents. It's an oxymoron."

5.3 THE ACCIDENT INVESTIGATION—STEP BY STEP

5.3.1 Step 1—Getting the Facts

Probably the most important step in the injury investigation process is getting the facts. Without accurate facts, the assessments and judgments that follow are probably not going to be very useful. Restricting access to the area where the accident occurred until photographs are taken, videotapes are shot, sketches are made, and initial statements are gathered is a good idea to preserve the evidence. Consider having the injured person and witnesses to the accident write down a detailed statement of what occurred from their perspective before the end of their shift. A visit to the scene of the accident as soon as possible is a must. Conducting an accident investigation from an office is an impossible task.

Gather relevant facts in the following areas without making judgments about them.

ACCIDENT / INCIDENT INVESTIGATION REPORT FORM

THE EVALUATOR

To Prevent this Injury	These IMMEDIATE CAUSES Should Have Been Better	Influenced By These UNDERLYING FACTORS	Which Require These Organizational PREVENTIVE MEASURES
INJURY	ACTS	KNOWLEDGE	IDENTIFY and COMMUNICATE safety responsibilities, workplace hazards and required precautions
		ABILITY	Ensure that individuals have the necessary CAPABILITIES and SKILLS to do their job/tasks safely
		MOTIVATION	Individually CONFRONT deviations from and generally REINFORCE safety values, rules and expectations
	CONDITIONS	DESIGN	Apply engineering DESIGN SAFETY STANDARDS and do HAZARDS ANALYSIS
		MAINTENANCE	ESTABLISH and IMPLEMENT facilities and equipment safety inspections and maintenance
		ACTIONS OF OTHERS	INSTRUCT all persons and ENSURE that their behavior does not create unsafe conditions

Step 1-The Facts: Describe what happened. ____________________

____

____

____

Step 2-Immediate Causes: Describe what acts & conditions should have been different or better.

A. Acts: ____________________

____

B. Conditions: ____________________

____

Step 3-Underlying Factors: Describe what factors should have been better for those acts & conditions.

A. Knowledge,Ability &/or Motivation: ____________________

____

____

B. Design, Maintenance &/or Other's Actions: ____________________

____

____

Step 4- Preventive Measures: Describe specific actions required to influence the factors identified above

A. Influencing Behaviors: ____________________

B. Influencing Conditions: ____________________

Signed/Date : ____________________ Approved/Date: ____________________

Fig. 5.5 Accident/incident investigation report form.

1. *The injured person*
   - Nature of the injury: What part(s) of the body was affected? How serious? Medical input? Cumulative versus instantaneous? New injury versus aggravation of existing injury?
   - Work experience: How much job experience? New employee? New work assignment? Co-worker's experience?
   - Extenuating circumstances: Stress? Long working hours? Family pressures? Job pressure?

2. *The facility conditions/situation*
   - Environmental conditions: Lighting levels, noise levels, temperature level, walking surface, spills, inside/outside, obstructions, recognized hazards, facility damage
   - Time, day, shift, witnesses
   - Weights, measurements, samples, photographs, video recordings

3. *The job/task equipment condition/situation*
   - *Job/Task:* What happened? What was the sequence of events? Were written operating procedures available? Were operating procedures verbally given? What was the actual procedure? How long was the person doing the task? How much training was provided? What safety controls were expected? What actual safety controls were used? What personal protective equipment was expected? What was the actual practice?
   - *Equipment:* Design conditions, maintenance conditions, safety conditions, equipment damage, and to what extent
   - *Situation reenactment:* Walk through the accident situation to corroborate other facts and information, pull the story together

At this point in the investigation, the facts should pull together and tell a consistent story. If there is a major discrepancy that cannot be explained or if there are conflicting "facts," it is important to resolve them before making judgments around them in the next steps.

In our case where Carl had two fingers severely injured while clearing a jam on an assembly machine, the following facts could include (1) noticing that the safety guard on a modified knife cutoff device had been bent out of position and taking a photograph of it; (2) copying the operating log sheet from the past two days indicating increased levels of jams since the modification was made; and (3) obtaining a copy of the most recent written operating procedure and related job safety analysis (JSA).

5.3.2 Step 2—Determining the Immediate Causes—Acts and Conditions

Now that the facts of the situation have been determined, the next step is to ascertain the immediate causes. In the past they have been called the Unsafe Acts and Unsafe Conditions. Using the approach outlined in this chapter, the immediate causes would be characterized as actions or behaviors that need to be better or different and as workplace conditions that need to be better or different.

This is the first step in making judgments about the acceptability of certain risks associated with these human behaviors and workplace conditions. Making sound, realistic assessments is a key part of the process. What actions or behaviors should have been better, and what workplace conditions should have been better? The best approach in making these decisions is to be knowledgeable, practical, and realistic.

To help make these decisions, the person or team conducting the investigation should know or have access to relevant safety standards. The organization's safety and health policy is an excellent source of information to assist in making these assessments. Regulatory agencies also specify certain safety and health requirements that have the force of law. Voluntary safety and health guidelines have also been established for certain manufacturing sectors and provide good direction for making these judgments. Engineering safety standards, division and department policies, as well as written operating procedures and rules are also great sources of information (see Chapter 2).

For complex cases, consider using risk analyses techniques (see Chapter 6). This involves making judgments about the likelihood of people's behaviors and workplace conditions leading to an accident and if an accident does occur what would be the possible consequences, that is, how serious could it be? For an accident to occur, there usually is a combination of factors that lead to the situation. In many accidents there are at least three to five immediate causes. In almost all accidents there is at least one act and one condition that should have been different. Some studies have shown that the average ratio of inappropriate acts to inappropriate conditions is 4 to 1. H. Heinrich, who some consider as the father of safety, suggested that as many as 88 percent of accidents are caused by the unsafe act of persons.[11] Howard[12] reports a ratio of 5 to 1. Petersen[13] suggests that these inappropriate acts are often rooted in the management system.

With the facts of the investigation in mind, now start making judgments regarding the person's *acts* that needed to be different or better. Consider the following areas:

- **Do certain things better**—Does the person need to follow procedures better? need to be more alert? need to wear the proper personal protective equipment? need to do a better job in planning the task? need to focus more attention on the job task?
- **Stop doing certain things**—Does the person need to stop wearing inappropriate clothing? need to stop rushing or hurrying? need to stop using unsafe equipment? need to stop using equipment unsafely? need to stop creating unsafe situations?

Similarly with the facts of the situation in mind, now start making judgments regarding workplace *conditions* that needed to be different or better. Consider the following areas:

- **Improve the equipment**—Does the equipment need to be better guarded? need to be better maintained? need to have sharp edges removed? need better safety interlocks? need to be redesigned? need to be more reliable?
- **Improve the procedure**—Does a procedure need to be developed? need to have a Job Safety Analyses conducted? need to require less reaching? less lifting? better ergonomically designed?
- **Improve the environment**—Does the workplace need to be less cluttered? need to be better lighted? less noisy? warmer or colder?

Determining what actions or behaviors could have been better is not always an easy task. Human nature being what it is sometimes prevents the admission of acting inappropriately or doing stupid things. It is much easier to target the workplace conditions as an immediate cause of the accident. If the culture of the organization is threatening, an individual's defenses go up and determining what actually needs to be better is a very difficult challenge! Developing a positive organizational safety culture that points out opportunities for improvement rather than pointing a finger of blame is the key.

In our case, it appeared that Carl acted improperly. The inappropriate act of clearing a jam while the machine was operating was one of the primary immediate causes of the accident. The poor condition of the knife cutoff safety guard as well as the high rate of machine jams were also determined to be other immediate causes of Carl's injury.

Many injury investigations would stop at this point. The findings would indicate that the person needs to be more careful and the machine needs to be fixed. The immediate causes of the situation would be corrected, but the underlying reasons as to why those situations existed would remain the same. Doing the same safety things then would lead to the same results in the future. Actively influencing and changing the most important underlying reasons leads to different and better results. According to Pritchett and Pound,[14] "The change effort needs to become a cause, a crusade, and your job is to champion the vision." So with the safety vision is mind, the next task is to determine why those situations existed. Notice the logical and sequential nature of the process.

5.3.3 Step 3—The Underlying Factors—Why the Acts and Conditions

Now that the organization has made some judgments around what behaviors and workplace conditions needed to have been better or different, the challenge is to determine why the person acted that way and why those conditions existed in the first place. As the investigation progresses deeper into the causes and effects of the accident, notice that the focus will be on the most important prevention measures. Not only will the investigation go beyond "Be more careful," but also specific, actionable items can be identified and implemented. The cause and effect relationship of accidents is now understood, and the realization of the fact that accidents "just don't happen" is reinforced.

In determining the underlying factors, refer to the Evaluator, the injury investigation tool (Fig. 5.4). This is the tool that provides a sequential and logical framework for understanding what happened, why it happened, and what needs to be done to prevent it from happening again. The next steps in conducting the investigation then will be making judgments regarding the reasons behind the injured person's behavior and why the inappropriate workplace conditions existed. Generally, there are several different underlying factors. The challenge is to determine the most relevant and the most important ones!

A. Regarding *the actions of the person*, what factors would have been influential? Was it a question of not knowing the procedures or expectations? or not having the appropriate ability? or of not being properly motivated to do the job safely? For example:

- **Knowledge:** Did the individual clearly know of the safety expectations? ... of the rules and safety procedures? ... of the hazards in the workplace? ... of the necessary precautions?

- **Ability:** Was the individual able to do the job/task? Were the physical and mental requirements of the job determined? Were skill development requirements satisfactorily completed? Was the person temporarily disabled either physically or mentally?
- **Motivation:** Was there a competing interest to following the safety rules? Were personal difficulties interfering? Were production schedules causing pressures to "bypass" the rules or to take shortcuts?

B. Regarding the *workplace conditions*, what factors would have been influential? Was it a question of inadequate design, poor maintenance, or was the condition created by another person? For example:

- **Design:** Did the equipment, facility, or process need to be better safeguarded? . . . or need to be better specified? . . . or better engineered?
- **Maintenance:** Did the equipment, facility, or process need to be inspected better? . . . or need to have a formal written inspection procedure? . . . or be inspected at more frequent intervals?
- **Actions of another:** Was the condition created by someone other than the person injured? . . . to what extent? . . . control of situation exercised by injured person?

At this point in the investigation, there are generally several factors that can provide rich and fertile material for analysis. Some factors will also become apparent as being the most important. The power of using this approach lies in the fact that it leads to a review of all the possible underlying factors as well as being able to identify the most important interventions. The investigator(s) needs to make judgments regarding the degree to which each of these factors could have been better or could have been more influential. A third party or an investigation team with no ax to grind may be in a better position to provide the most objective judgments at this point. If you, as the supervisor, are responsible for training and motivating the worker who was just injured, it is sometimes difficult to see clearly what you could have done better.

Also, the individual who was injured needs to be upfront and as forthright as possible regarding his or her behaviors. A claim of ignorance to the rules is somewhat easier rather than admitting to rushing and hurrying because of being motivated by external pressures or stress.

In our case, the underlying factors that led to Carl's actions included the pressure he felt to make the production schedules, and as a result he was rushing and hurrying. He did not take the time to shut down and lockout the machine when jams had to be cleared. Even though the written procedure had not been updated, Carl and his coworkers knew the rules and had enough on-the-job experience to be able to follow the procedure. The unwritten rule lately, however, had become making the schedule and not letting the other shifts "outdo" your shift. The underlying Act factor was therefore a Motivation issue that needed to be different. Knowledge or Ability were not significant factors in this particular situation.

Regarding the workplace conditions, the investigation found that the underlying reason for the bent guard on Carl's machine was an inadequate maintenance program. The retaining bracket had not been replaced after removal for servicing. This resulted in the safety guard becoming bent out of position. This allowed the machine operators including Carl to put their hands in a dangerous location while the machine was running. The underlying Condition factor was therefore a Maintenance issue that needed to be better, which also reduced the rate of machine jamming. Design or Actions of Others were associated but not primary issues.

The difficult part of the investigation has now been completed. Once the underlying factors have been accurately determined, the prevention measures are usually self-evident. The organization can also use the power of this approach to determine if the underlying factors related this specific incident are also important in other similar situations throughout the workplace.

5.3.4 Step 4—Organizational Prevention Measures—Underlying Factors and Interventions

Referring to the injury investigation tool (Figure 4.4), notice that the organization has specific things to do based on what underlying factors were judged to be the most important. Instead of saying "Be more careful" or "stop rushing and hurrying", the logic model focuses the interventions towards influencing those factors that lead to the injury in the first place. Such as making sure that the person knows the rules and standards, ensuring that the person has the skills, or a motivational intervention. Similarly, the critical workplace conditions of proper design, necessary maintenance as well as situations created by other individuals are examined.

The organizational prevention measures associated with the injured person's *Acts* can be understood in the following way.

A. If the person didn't *know*: Consider if the organization meeds to do a better job in ensuring that the individuals know their personal safety responsibilities, that the workplace hazards and associated precautions have been identified, and then more effectively communicating those expectations by:
   - establishing safety rules, procedures, and expectations for job-related, common everyday tasks and emergency situations;
   - broadcasting those rules and expectations on a continual, consistent basis.

B. If the person didn't have the necessary *ability:* Consider if the organization needs to establish or change the requirements of the job and/or task, determine the physical and mental requirements of the job, and develop on-the-job skills by:
   - determining the job requirements with respect to physical, mental, and cognitive capabilities;
   - ensuring that the job and/or task does not exceed the capability of the individual;
   - reviewing the requirements and capabilities with respect to a temporary or permanent disability;
   - ensuring that the individual has developed on-the-job capabilities through applied skills training.

C. If the person wasn't properly *motivated:* Consider if the organization needs to establish a motivational intervention effort to discipline individuals for inappropriate behavior and/or reward individuals for following the rules by:
   - confronting individuals whose behaviors are not consistent with the rules;

- recognizing and rewarding those individuals who are following the rules;
- ensuring that the intervention occurs continuously and before any inappropriate behaviors lead to an injury or an incident.

The organizational prevention measures associated with the workplace *conditions* can be understood in the following way.

A. If the workplace was not *designed* appropriately: Consider if the organization needs to improve safety design standards or apply risk assessments by:
   - ensuring that engineering safety standards are understood and applied;
   - applying regulatory, consensus, and best practice safety standards;
   - conducting hazard assessments and applying the results where the situation is complex or where standards alone are not sufficient.

B. If the workplace was not *maintained* appropriately: Consider if the organization needs to establish maintenance safety procedures or fully implement those safety inspection procedures by:
   - ensuring that preventive maintenance safety procedures are developed;
   - ensuring that the safety inspection procedures are conducted;
   - improving the quality and/or frequency of the safety inspections.

C. If inappropriate workplace conditions were created by the *actions of others:* Consider if the organization needs to establish safety procedures for others or to develop a motivational program to ensure that others such as contractors follow the rules by:
   - ensuring that others, including outside contractors and visitors, know the safety rules;
   - determining if other individuals are following the rules; and/or
   - establishing adequate oversight and supervision of contractors and others.

Regarding Carl's injury while clearing the machine jam, we have now (1) ascertained the facts; (2) determined the immediate causes of inappropriate acts and workplace conditions; (3) made judgments about the underlying factors or root causes; and (4) can now identify the most effective organizational prevention measures.

Jim and Bill must recognize that the pressures they had created for shifts "not to be outdone" had led to the operators taking chances and not following the lockout/tagout rules. The focus on "meeting schedule" to the exclusion of responsibly enforcing the safety rules had led directly to the injury. Jim and Bill have to decide on a course of action that will motivate Carl and his co-workers to follow the rules. This probably will include a balance of reward and recognition efforts as well as an effort for establishing a culture for fair and responsible discipline.

Although not judged to be a primary factor that led to the injury, the out-of-date operating procedures need to be rewritten. Additionally, Jim and Bill could decide to update and recommunicate all the procedures in the plant. Some of the newer operators probably didn't know the best or proper way of operating the machines.

The poor condition of the machine cutoff safety guard and increased frequency of jams was the result of an inadequate maintenance program. No procedures had been developed and communicated regarding the restoration of guards after servicing. Some mechanics did a good job and some a poor job. Bill and Jim recognized that a formal maintenance and inspection program was needed. In addition to this effort, they decided to institute a shift safety inspection checklist for the operators' use. Bill and Jim also committed to check periodically on how well the shift inspections were being conducted. They also saw the value in establishing this program for the rest of the manufacturing facility.

Jim, Bill, and the accident review team then reviewed their findings with the rest of plant personnel and asked for further comments and recommendations. The plan and schedule for implementing the decision was then established and posted on the operations bulletin board.

The investigative phase of the process has now been completed. The underlying factors have been determined and the best prevention measures identified. Notice how the learnings from this specific injury investigation can then be applied to other similar situations for the organization. The "Be more careful" fix has been supplanted by a organizational effort towards reestablishing safety behavior expectations, and the "Fix the guard" approach has been supplanted by establishing a safety preventive maintenance inspection program. The solution has been directed at the bottom of the expanded accident triangle (Fig. 5.3), the very foundation that the organization needs to strengthen.

This portion of the chapter has described a process for reporting and investigating specific accidents and injuries so that effective root causes are determined and long-term fixes identified. Ensuring that those organizational preventive fixes are actually implemented requires a well-developed management system. As described in the introduction of this chapter, this CHECK part (checking and corrective action) of the system (i.e., the injury investigation process) needs to be supported by the PLAN, the DO, and ACT parts of the system.

5.4 MANAGEMENT SYSTEMS

5.4.1 Implementing the Prevention Measures—Investigation Is Not Enough!

Now that the investigation has been completed and the organization has determined what things need to be better, the next challenge is to implement the desired fixes effectively. The strategy for implementing the fixes is in large measures determined by how well developed is the organization's management system. Ezell[15] has suggested that "... we have paid too little attention to the effects on the human psyche of technology and mismanagement, which causes operational errors and results in accidents." Using our example of Carl's finger injury, the desired fix was to install a motivation system to ensure that employees follow the safety rules and improve the maintenance inspection function. If the existing management systems for ensuring that employees follow other rules and procedures are weak and ineffective, it will be difficult, if not impossible, to install a separate system that deals solely with enforcing safety procedures. For example, if absentee rates are high, or quality rates are low or operating mistakes abound, it should not be surprising to find that getting people to follow the safety rules is difficult. In this situation, the entire management system needs to be addressed, since the safety issues are only a symptom of a larger problem.

On the other hand, if reviewing performance is an integral, effective part of the management system and the safety portion has just been overlooked, adding the safety performance review component is a relatively easy task. Similarly, if a safety preventive maintenance inspection program is deemed necessary to assure that machines guards are properly maintained, it is likely that a broader reaching entire facility preventive maintenance program could also be needed. The learnings and rich insights gained from using this investigative tool can be quite powerful. If the appropriate organizational beliefs and values are in place, a world-class operation is quite possible.

Recently, international voluntary standards have been developed that provide guidance regarding the necessary elements of management systems, sometimes characterized by the words PLAN, DO, CHECK, ACT. These standards include the ISO 9000 series on Quality Management and the ISO 14000 series on Environmental Management Systems.[16] The preceding portion of this chapter has been dealing with the CHECK part of the ISO 14001 standard, that is, the checking and corrective actions required. The DO part of the system deals with the implementation and operational requirements.

With respect to the injury reporting and investigation process, the DO part means having certain elements in place so that the fixes determined by the CHECK part become effectively implemented. Using the ISO 14001 standard as a framework, the following DO elements need to be established.

1. Roles and responsibilities need to be established for those individuals who need to act upon the prevention measures.
2. Training needs to be provided to those individuals so that they have the required skills.
3. Communication of the necessary fixes needs to be done throughout the organization.
4. Documentation of the requirements and the implementation of the prevention measures need to be completed.

In practical terms, this means that a individual in the organization should be responsible for ensuring that the corrective actions identified as being necessary to prevent that injury are implemented and that the necessary facts around that situation are properly documented and communicated as necessary through the organization. The ACT part (Management Review) of the management system then requires a periodic check to ensure that the implementation of the prevention measures is in fact happening.

5.4.2 Analyses of Consolidated Injury Investigation Data

The Evaluator approach to injury investigation leads not only to the best and most effective fix for the specific accident under review but also can lead to activities that could prevent an entire class or type of injuries or uncover organizational management issues that need attention. Analyses of the data and information developed at each stage of the injury investigation process can lead to many different levels and types of important organizational intervention efforts. ANSI has a proposed standard out that would provide guidance on ways to compute basic incidence rates and other comparison measures.[17] Computer databases can provide a quick and convenient access to these analyses. An illustration of various ways of describing consolidated injury data and investigation analyses data follows. A document for collecting and displaying the data is illustrated in Figs. 5.6 and 5.7.

Injury Type Summary Data

Department / Division: ____________ Reporting Period: ______________

This table summarizes the types of incidents, injury types and parts of the body injured.

Incident Type	#	%	Injury Type	#	%	Part of Body	#	%
• Hit By			• Contusion, Laceration			• Eye		
• Hit On			• Fracture, Dislocation			• Face / Head		
• Slip, Trip, Fall			• Sprain, Strain			• Torso / Back		
• Lift, Push, Pull			• Burn			• Hands		
• Cumulative Trama			• Foreign Body			• Arms		
• Transportation			• Exposure			• Feet / Legs		
• Other			• Other			• Other		
Totals			Totals			Totals		

Fig. 5.6 Injury type summary data.

Evaluator Summary Data

This table summarizes the primary injury causes, underlying factors and prevention measures.

Immediate Causes	#	%	Underlying Factors	#	%	Preventive Measures	#	%
Inappropriate Acts			Knowledge			• Establish Procedure		
						• Communicate Procedure		
			Ability			• Ensure Capability		
						• Ensure Skills		
			Motivation			• Discipline		
						• Recognition		
Inappropriate Conditions			Design			• Safety Stanards		
						• Hazards Analysis		
			Maintenance			• Est. Maint. Procedure		
						• Impl. Maint. Procedure		
			Actions of Others			• Est. Contract Rules		
						• Impl. Contract Rules		
Totals			Totals			Totals		

- number of injuries in that category % - percent of injuries in that category

Fig. 5.7 Evaluator summary data.

- **Step 1 Data—The Facts:** Analysis of the consolidated data developed at this stage of the investigation can indicate opportunities for improvement regarding (1) part of the body injured such as hand, arm, back and (2) type of injury such as slip, trip, or fall and lifting, reaching, pulling and (3) part of the organization at risk such as shift, department, or facility and (4) job function of injured individual such as tradesperson, machine operator, or material control person. The American National Standard for Information Management for Occupational Safety and Health has prepared a recent revision of "Method of Recording Basic Facts Relating to the Nature and Occurrence of Work Injuries," ANSI Z16.2.[18] This document provides guidance on ways to collect information on the (a) nature of injury/illness, (b) part of body affected, (c) source of injury/illness, (d) event or exposure, (e) secondary source of injury/illness, (f) occupation of worker, and (g) industry of worker. It also provides guidance on ways to analyze the data.
- **Step 2 Data—The Immediate Causes:** Analyses of the consolidated data developed at this stage of the investigation can indicate opportunities for improvement regarding (1) unsafe acts such as not doing the right things or doing inappropriate things or (2) unsafe conditions such as facilities, equipment, environment. What percentage of the accidents are occurring because of inappropriate acts or inappropriate conditions?
- **Step 3 Data—The Underlying Factors:** Analyses of the consolidated data developed at this stage of the investigation can indicate opportunities for improvement regarding the underlying factors leading to injuries such as (1) knowledge, (2) ability, and (3) motivation as well as (4) design, (5) maintenance, and (6) actions of others.
- **Step 4—Organizational Prevention Measures:** Analyses of the consolidated data developed at this stage of the investigation can indicate opportunities for improvement regarding (1) training/setting expectations, (2) on-the-job skills enhancement, (3) reward, recognition, or discipline programs, as well as (4) hazard reviews/standards development, (5) preventive maintenance programs, and (6) contractor safety expectations.

In practical terms, analysis of this data could result in learning that 25% of the *B-Shift operators* have had *hand injuries* over the past three years. Of these, 70% were the result of *unsafe actions* while they were *maintaining* their own machines because there was no shift trades coverage. They *knew* they should wait for trades personnel to be called in but were generally *motivated* to keep the machines running. The lesson for the organization could be to determine if they should (1) provide B-Shift trades coverage or (2) train the operators to perform the required minor maintenance activities or (3) reinforce the expectations so that only authorized trades personnel can perform this maintenance on a call-in basis.

5.5 SUMMARY

The accident reporting and investigating process is considered by many as the minimum requirement of any health and safety program. Some suggest that it can be the very foundation of organizational safety excellence. Combined with the proactive element of an accident and injury prevention effort, it can be part of a world-class safety program. In any event, accident reporting and investigation is a regulatory requirement. To be truly successful, the aim and goal of the process, however, needs to reach beyond regulatory boundaries of compliance. The investigation and resulting analyses can and should reveal opportunities for the organization as well as the injured individual to improve and get better.

After ensuring that the injured person has received the proper medical attention, the investigation needs to proceed in a logical, sequential, and fact-finding way. Fault finding will lead to blocks in the investigation and will ultimately lead to to wrong conclusions with the wrong prevention measures. The organization needs to consider carefully accidents are to be treated as willful and deliberate acts. Are accidents expected to be promptly reported? The basic safety values of the organization provides the framework for the actions that follow.

Determining the basic facts surrounding the injury or accident is the key first step. What happened? Prompt review of the accident scene and interviews of co-workers and witnesses to the accident is essential. Next, ascertain the immediate causes of the accident. What actions of the individual need to be different or better and what work place conditions needed to be different or better? Those conducting the investigation should recognize that there are usually several things that go wrong. Uncovering all these things leads to a more complete analysis.

Now knowing what behaviors needed to be better, determine what would have been the best influence. Was it a question that the safety rules weren't known? or that the person didn't have the appropriate ability? or was it a question of being better motivated? There may have been several underlying factors at work influencing the person's behavior.

Likewise, knowing what workplace conditions needed to be better, determine what would have been the best influence. Was it a question of basic design? or an issue of preventive maintenance? or was the condition created by another person or outside contractor?

Now knowing what underlying factors lead to the injury, the organizational prevention measures become quite clear. The most important ones need to be assigned to individuals with responsibility for followup and implementation. In some instances, basic management organizational interventions need to be conducted. This could lead to a paradigm shift in safety and health thinking and result in a major improvement in safety and health performance.

"Jim," Bill said with feeling, "we've just completed a year without a serious accident . . . even Carl was talking about it in the lunch room!!! You know, I still remember that night and his screams!

"It sure has made a difference since we began that shift inspection program and especially since you and I have made our feelings known regarding safety and enforced the rules. Well, see you at the safety meeting tomorrow."

REFERENCES

1. Petersen, Dan, *The Challange of Change, Creating a New Safety Culture*, Creative Media Development, 1993.
2. National Safety Council, *Accident Facts*, 1996 Edition.

3. Factory Mutual System, *Record, The Magazine of Property Conservation*, Second Quarter, 1996. Vol. 73, No. 2.
4. Bird, F. E., Jr., and Germain, G. L., *Damage Control*, American Management Association, 1966.
5. O'Shell, Harold E., *Modern Principles of Loss Prevention and Control*, International Safety Academy.
6. DuPont, *Executive Safety News*, Fall 1995.
7. Eastman, Martin, *What Does It Take To Change Unsafe Behavior?*, Safety & Health; July, 1990.
8. Minter, Stephen G., "Luck Is Not the Issue," *Occupational Hazards*, October, 1996.
9. Verespej, Michael A., "Lead, Don't Manage," *Industry Week*, March 4, 1996.
10. Thirion, John: Safety Director, Johnson and Johnson, *Occupational Hazards*, August, 1991.
11. Heinrich, H. W., *Indstrl. Accident Prevention.* 4th ed. New York, McGraw–Hill, 1959.
12. Howard, P., *The Death of Common Sense*, New York, Random House, 1994.
13. Petersen, Dan, *Safety Management*, 2nd Edition, New York, Aloray, 1988.
14. Pritchett, Price, and Pound, Ron, *High-Velocity Culture Change, A Handbook for Managers*, Pritchett Publishing Company, Dallas, TX, 1993.
15. Ezell, Charles W., "The Pursuit of Error-Free Performance," *Occupational Hazards*, June 1996.
16. International Organization for Standardization, ISO 14001, *Environmental Management Systems—Specification with Guidance for Use.*
17. *American National Standard for Occupational Safety and Health Incident Surveillance*, Z16.5, 7th Draft, November, 1996, Itasca, IL, National Safety Council.
18. *American National Standard for Information Management for Occupational Safety and Health*, ANSI Z16.2-1995, Itasca, IL, National Safety Council.

CHAPTER 6

Risk Assessment Techniques

THOMAS M. DOUGHERTY
Polaroid Corporation, 1265 Main St., Waltham, MA 02254

"A thing is safe if its risks are judged to be acceptable."

William W. Lowrance[1]

6.1 INTRODUCTION
- 6.1.1 Scope of This Chapter
- 6.1.2 Benefits of Assessing Risks

6.2 RISK ASSESSMENT—AN OVERVIEW
- 6.2.1 What Is It?
- 6.2.2 When Is It Done?
- 6.2.3 What Risk Assessment Technique to Apply

6.3 SAFETY CHECKLISTS
- 6.3.1 Overview of Checklists
- 6.3.2 Developing Checklists
- 6.3.3 When to Use Safety Checklists
- 6.3.4 Pros and Cons of Safety Checklists
- 6.3.5 Some Examples of Safety Checklists

6.4 JOB SAFETY ANALYSIS
- 6.4.1 Job Safety Analysis Overview
- 6.4.2 The Benefits of Job Safety Analysis
- 6.4.3 Step One—Selecting the Job or Task
- 6.4.4 Step Two—Breaking the Task into Sequential Steps
- 6.4.5 Step Three—Identify the Hazards Associated with Each Step
- 6.4.6 Step Four—Develop the Safety Controls for the Hazards at Each Step
- 6.4.7 Summary—Making JSAs Effective

6.5 "WHAT IF" ANALYSIS
- 6.5.1 "What If" Analysis Overview
- 6.5.2 Getting Started—What Is Needed
- 6.5.3 Conducting the Review—How Is It Done?
- 6.5.4 Reporting the Results—To Whom and How?
- 6.5.5 "What If" Summary—Pros and Cons

6.6 HAZARD AND OPERABILITY ANALYSIS
- 6.6.1 HAZOP Analysis Overview

Handbook of Occupational Safety and Health, Second Edition, Edited by Louis J. DiBerardinis,
ISBN 0-471-16017-2

6.6.2 HAZOP—Getting Started
6.6.3 Conducting the HAZOP—How Is It Done?
6.6.4 Reporting the Results
6.6.5 HAZOP Summary—Pros and Cons
6.7 FAILURE MODE AND EFFECT ANALYSIS
6.7.1 FMEA Overview
6.7.2 FMEA—Getting Started
6.7.3 Conducting the FMEA Review—How Is It Done?
6.7.4 Reporting the Results
6.7.5 FMEA Summary—Pros and Cons
6.8 FAULT TREE ANALYSIS
6.8.1 FTA Overview
6.8.2 FTA—Getting Started
6.8.3 Conducting the FTA—How Is It Done?
6.8.4 Reporting the Results
6.8.5 FTA Summary—Pros and Cons
6.9 APPENDIX
REFERENCES

6.1 INTRODUCTION

6.1.1 Scope of This Chapter

This chapter provides a basic introduction to the traditional qualitative risk assessment techniques available to the safety and health professional. These techniques include the safety checklist, job safety analysis, "what-if" analysis, hazard and operability analysis, failure mode and effect analysis, and fault tree analysis. Quantitative risk analysis techniques are also available to the safety and health practitioner, but they are used fairly infrequently and in complex situations.

There are also a variety of methods that are used on consequence analysis, that is, the severity potentials. For example, models are available that assist the analyst in determining fire thermal radiation effects, explosion effects, chronic exposure effects, and contaminant dispersion possibilities. These methods, however, are beyond the scope of this chapter. For more information on these subjects, interested readers are directed to the series of guidebooks published by the Center for Chemical Process Safety (CCPS) under the auspices of the American Institute of Chemical Engineers (AIChE).[2] The endnotes also direct interested readers to a rich and varied source of literature on the subjects of hazard identification, risk analysis, risk assessment, and risk management.

6.1.2 Benefits of Assessing Risks

The primary benefit of conducting a risk assessment is to evaluate the hazards of a situation prior to the occurrence of an incident or accident. Having this knowledge, the organization or the affected person can then make an informed decision on whether or not to accept that risk. A risk assessment program is but one of the many necessary elements for a successful risk management program. Once assessed, the organization needs to be prepared to act on the findings and resulting recommendations. This PLAN (identification of safety and health aspects) part of the management system needs to be supported by the DO (implementation and operation), the CHECK (checking and corrective action), and the ACT (management review). Chapter 5, on conducting an effective accident investigation, describes one portion of the checking and corrective action management elements. The DO and ACT elements are discussed in many chapters throughout this book. This chapter describes the elements involved in the identification and evaluation of risks.

Once identified and evaluated, the risks can then be prioritized in an informed way. Recognizing that risks can never be zero, the assessment process provides organizations with information so that resources can be applied most productively so as to provide the most added value. Virtually every organization that has excellent safety and health performance also conducts risk assessments. These organizations see reduced injury rates, improved operating profits, lowered property loss insurance rates, and greater productivity improvements. The risk assessment techniques described in this chapter are valuable tools that can be used to help build this performance. They can bridge the gaps left when traditional safety approaches are not sufficient for understanding and controlling the hazards associated with evolving, innovative, and complex equipment and facilities. Also, new health, safety, and environmental regulations are now requiring the application of risk assessment techniques as a matter of law.

For example, OSHA'S Process Safety Management Standard[3] requires certain facilities handing highly hazardous chemicals to apply formal risk assessment techniques. Similarly, OSHA's personal protective equipment revision[4] requires employers to conduct and document a hazard assessment and equipment selection program. Also, EPA's Risk Management Program[5] requires certain facilities handling chemicals to conduct worst-case hazard assessments.

6.2 RISK ASSESSMENT—AN OVERVIEW

6.2.1 What Is It?

In its simplest form, a risk assessment is the process of making a determination of how safe a situation is. William Lowrance[6] defines safety as "a judgment of the acceptability of risk" and risk as "a measure of the probability and severity of harm to human health." So the work of risk assessment involves determining (1) how often a certain unwanted event could occur (the probability), (2) how serious could be the consequences of that unwanted event (the severity), and then (3) judging the acceptability of that risk.

This notion of determining the likelihood and the severity of unwanted events or situations can be visually illustrated. The risk matrix (Fig. 6.1) is one example of how this risk information can be displayed and communicated.

Embodied in all the formal risk assessment methodologies described later in the chapter are the ides of determining "how likely" and "how serious." These methodologies provide a wide variety of approaches to analyze risk situations systematically and formally. Understanding this simple risk matrix relationship provides the basis for not only determining the risk but also for controlling and reducing the risks, that is, how to reduce the likelihood and the severity of potential accidents.

For example, to reduce the risk of exposure to chemicals, noise, or radiation, the use of personal protective equipment will reduce the severity of the *consequences* of the incident. Using safety goggles could prevent a severe eye injury as a result of a chemical splash but would not reduce the possibility of the splash happening in the first place. Likewise, the use of safety belts in an automobile will reduce the injury

Likelihood of Incident	DEGREE OF CONSEQUENCE		
	Catastrophic	Serious - Critical	Minor - Marginal
Extremely Likely			
Quite Possible			
Unusual But Possible			
Unlikely			
Remote			
Extremely Remote			

Fig. 6.1 The risk matrix.

severity to occupants of a car during a crash. The safety belts, however, would not lessen the possibility or frequency of the accident.

Similarly, driving defensively and being alert could reduce the *likelihood* of an accident but would not reduce the severity of the injury. The use of machinery lockout/tagout procedures would reduce the chances of a machine inadvertently starting up with someone in the danger zone. Those practices, however, would not reduce the severity of the injury if the machine did start up. Some actions will reduce both the likelihood and the severity. By driving slower, the chances of an accident are reduced and the severity of the accident is lessened.

Typical hazard control measures include the development of operating procedures, the use of personal protective equipment, and the installation of engineering controls such as ventilation, guarding, and safety interlocks. Procedures can include the use of safety permits such as confined space entries, hazardous work permits, and chemical line openings. Other procedures can include periodic preventive maintenance requirements on safety interlocks, sprinkler systems, or hoisting equipment. Personal protective equipment requirements could be established at job task exposure hazards such as wearing of safety shoes during the handling of heavy equipment. Similarly, improvements to the facility or equipment could be needed.

Once the severity and likelihood is determined, the next step in assessing risks is to make judgments regarding the "acceptability." The greater the likelihood and the more severe the accident can be, the higher the risk and, therefore, the more unacceptable it is, given the benefits. For example, if it is determined that having a severe fire while using solvents is quite possible and could result in several fatalities, then that risk would be very high and unacceptable. On the other hand, if the assessment resulted in the determination that the likelihood of the fire is remote because of the existing safety precautions and would only result in some moderate property loss because of the sprinkler system, then it may be judged as an acceptable risk. Using these concepts, the risk matrix can then be displayed in an expanded form as illustrated in Fig. 6.2.

The risk matrix can also be used in a quantitative manner. For example, in any one

Likelihood of Incident	DEGREE OF CONSEQUENCE		
	Catastrophic	Serious - Critical	Minor - Marginal
Extremely Likely			
Quite Possible			
Unusual But Possible			
Unlikely			
Remote			
Extremely Remote			

UNACCEPTABLE RISK

Fig. 6.2 Risk matrix with acceptability criteria.

year, the chances of a situation occurring as extremely likely could be defined as one in ten, quite possible as one in a hundred, unusual as one in a thousand, unlikely as one in ten thousand, remote as one in a hundred thousand, and extremely remote as one in a million. Similarly, catastrophic consequences could be defined as a fatality or multiple fatalities, serious as permanent disabilities, and minor as temporary disabilities or less. A quantitative analysis could then be displayed numerically on the matrix.

The question arises as to who is deciding what level of risk is acceptable. This can be a complicated business. The perspective of each of the involved stakeholders is quite different. Stakeholders can include the plant workers, the plant management, the corporate officers, the stockholders, the regulatory agencies, and the local neighboring community. The acceptance by a worker personally exposed to the risk of a plant fire is probably different from that of a corporate officer to that of a regulatory agency to that of the surrounding community. The nature of the risk also directly affects the perception of its acceptability. A risk assumed voluntarily such as rock climbing is perceived differently than one borne involuntarily, such as having a nuclear power plant built next to your house. Additionally, a common hazard such as driving a car is viewed differently than a "dread" hazard such as cancer. The perceived or actual benefits, such as worker pay, taxes to the community, and social responsibility also play an important part in risk acceptance decisions.

Kaplan and Garrick[7] put it well: ". . . risk cannot be spoken of as acceptable or not in isolation, but only in combination with the cost and benefits that are attendant to that risk. Considered in isolation, no risk is acceptable! A rational person would not accept any risk at all except possibly in return for the benefits that come along with it."

The value and beauty of the risk matrix concept is to allow the various stakeholders to come to some agreement on what the risk actually is and then have the argument on the acceptability issue alone. Based on an organization's or individual's values and beliefs, there can be true differences on what is acceptable. Use of the risk matrix, however, can at least facilitate a discussion on "how likely" and "how serious" between interested stakeholders.

6.2.2 When Is It Done?

The nature and complexity of the situation will determine when to apply formal risk assessment techniques. In many cases, conforming to well-known and well-developed work practices will suffice. For example, a confined space permit guides the user through a series of steps designed to assess the likelihood of an incident occurring such as lack of adequate oxygen and the potential severity of the situation such as a fatality. Use of the permit additionally provides guidance and direction for actions necessary to control those risks. Flame permits and LockOut/TagOut procedures are other examples of work practice approaches designed to address specific, reoccurring occupational risks. These are examples of risk assessments that have already been formalized into specific safe work practices. There is no need to develop a separate risk assessment approach.

Furthermore, Kletz[8] suggests that ". . . hazard analysis is a waste of time . . . given . . . poor permit-to-work system, lack of instructions . . ." and that "It should not be used until the basic management is satisfactory." In other words, get your basic safety procedures and traditional safety programming up to speed and then you will be better able to take advantage of the power of formalized risk assessments.

Similarly, many safety and health standards have evolved to the point where they are accepted as part of design standards, industry best practices, or consensus standards and indeed regulatory requirements. These standards and regulations need to be applied as an ongoing integral part of any risk management program. The formal risk assessment methodologies described in this chapter are designed to complement and enhance this ongoing effort. They are not designed to supplant or replace these well-developed safety and health practices.

On the other hand, complex operations involving many hazards, having the potential for many fatalities, or perhaps impact on the community, may require the application of a very sophisticated quantitative risk assessment technique. For example, a large chemical processing facility may need to evaluate the risk of fire, explosion, or toxic exposures to their workers and the surrounding community neighbors. Also, evolving regulations for certain operations legally require the application of formal risk assessment techniques.[2–4]

Safety and health risk assessments can and should be conducted (1) from the conceptual stage of a program or project proposal, (2) throughout the design stage, from preliminary to final design, (3) to the construction stage, (4) during the startup and debug stage, (5) throughout the life of the operations, (6) during maintenance activities, and (7) finally to the shutdown, dismantling, and disposal operations. Traditional safety practices, design reviews, and permit systems will be sufficient for many of the safety concerns during these situations. Risk assessment techniques should be applied for new, different, or complex situations.

The National Safety Council[9] suggests the following benefits of using formal risk assessments.

1. It can uncover hazards that have been overlooked in the original design, mockup, or setup of a particular process, operation, or task.
2. It can locate hazards that developed after a particular process, operation, or task was instituted.
3. It can determine the essential factors in and requirements for specific job processes, operations, and tasks. It can indicate what qualifications are prerequisites to safe and productive work performance.

4. It can indicate the need for modifying processes, operations, and tasks.
5. It can identify situational hazards in facilities, equipment, tools, materials, and operational events (for example, unsafe conditions).
6. It can identify human factors responsible for accident situations (for example, deviations from standard procedures).
7. It can identify exposure factors that contribute to injury and illness (such as contact with hazardous substances, materials, or physical agents).
8. It can identify physical factors that contribute to accident situations (noise, vibration, insufficient illumination).
9. It can determine appropriate monitoring methods and maintenance standards needed for safety.

6.2.3 What Risk Assessment Technique to Apply

Six risk assessment techniques are described in this chapter and include (1) safety checklist, (2) job safety analysis, (3) what if analysis, (4) hazard and operability analysis, (5) failure mode and effect analysis, and (6) fault tree analysis. Several factors enter in the decision as to what methodology to apply. These factors include the complexity of the situation, the experience of the decision makers, the perception of the consequences, the regulatory requirements, and capabilities of the risk assessors.

Table 6.1 provides a basic summary guide to the application of the above methodologies.

6.3 SAFETY CHECKLISTS

6.3.1 Overview of Checklists

A safety checklist is generally considered as the first pass or preliminary review of the safety aspects of a situation. The checklist can be applied at any time of the review. It can be used during the evaluation of a piece of equipment, an entire facility, a design concept, or an operating procedure. The results of the safety checklist review should be considered as forming the basis for later more extensive design and operation reviews. It can also be used during the summary or wrapup as a reminder to ensure that all the initial safety concerns were addressed.

Checklists are generally a list of items or questions related to the situation. The main purpose is to ensure that key safety aspects of that situation are identified so that further discussion and analysis will take place. It provides assurance that key safety elements have not been overlooked or forgotten. Each area of concern can be checked physically, can be reviewed for compliance with regulatory requirements, can be analyzed to determine if it meets best industry practice, or can be set aside for a more rigorous hazard review. Checklists represent the most basic and simplest method for identifying hazards that need to be controlled.

6.3.2 Developing Checklists

Checklists are developed by individuals based on their past experience and knowledge of engineering and design codes as well as regulatory requirements and company policy.

TABLE 6.1 Risk Assessment Techniques Summary[10]

Name	Purpose	When to Use	Procedure	Type of Results	Nature of Results	Data Requirements	Limitations, Comments
Safety checklist	Identification of safety issues and concerns that need to be addressed	Early in conceptual or preliminary design phase	Check off applicable safety items on predesigned list	Checklisk of items or concern	Qualitative only	Gross knowledge of system and applicable safety standards	Success limited to the experience of the users and breadth of list
Job safety analysis	Provide safety requirements for simple job tasks	For existing job procedures with annual update	Step-by-step review of job tasks	List of specific requirements to do tasks safely	Qualitative only	Written job instructions are helpful	Only good for well-defined, noncomplex job tasks
What if analysis	Identification of likely things that could go wrong and possible controls	Popular approach that can be used in most situations, as system changes	Asking "What if" questions at each step of the process	List of potential problems and recommended controls	Qualitative only	Operating instructions, flow diagrams	Depends on team members' experience with similar situations
Hazard and operability studies	Identification of problems that could compromise a system's ability to achieve intended productivity	Late design phase when design is nearly firm; also for an existing system when a major redesign is planned	Examine inst. diagrams, flowchart at each critical node identify operational deviations, causes, and consequences	List of hazards and operating problems, deviations from intended functions, consequence, cause, and suggest change	Qualitative with quantitative potential	Detailed system descriptions, flow charts, procedures, knowledge of instruments and operation	Depends heavily for its success on data completeness and accuracy of drawings

Name	Purpose	When to Use	Procedure	Type of Results	Nature of Results	Data Requirements	Limitations, Comments
Failure mode and effect analysis	Identification of all the ways a piece of equipment can fail, and each failure mode's effect(s) on the system	At design, construction, or operation and reviewed every 3 to 5 years	Collect up-to-date design data on equipment and relationship to the rest of the system; list all conceivable malfunctions; describe effects	List of identified failure modes, potential effects, and needed controls	Qualitative, although can be quantified if failure probabilities for components are known	System equipment list; knowledge of equipment function; knowledge of system function	Poor at showing interactive sets of equipment failures that lead to events, not useful for errors or common-cause failures
Fault tree analysis	Deduction of causes of unwanted event via knowing of combinations of malfunctions	At design, operation and updated as significant changes are made in the process	Construct a diagram with logic symbols to show the logical relationships between situations	List of sets of equipment and human errors that can result in a specific unwanted event	Qualitative with quantitative potential with probabilistic data on components and subsystems	Complete understanding of the system's functions	Enables ID and quantitative examination of critical factors and interrupt modes for chains of failures

Traditional checklists vary widely in level of detail and purpose. Some organizations have well-developed and -conceived checklists that must be completed before a conceptual project moves forward to initial design. Others have checklists used by developers of operating procedures to ensure that key elements of safely operating a piece of equipment are not neglected. Some organizations require a formalized signoff of safety checklists before new capital costs are authorized to ensure that safety costs have been considered in the request for project funding.

The level of detail within the checklist needs to be related to the potential risks associated with the situation being reviewed as well as overall complexity. The effective development and use of a checklist is directly related to the experience and skills of the preparer and user. Completing the checklist assumes the user has the knowledge of the underlying questions and answers implied in the checklist structure. Experienced individuals can play an important role in the development of checklists focused on the organization's specific need and guide the inexperienced users in how they can be effectively applied.

6.3.3 When to Use Safety Checklists

A simple overview checklist can be an excellent safety reminder at the conceptual stage of developing a new facility or designing a new produce or modifying an existing facility or piece of equipment. Before committing large sums of money to establish a new manufacturing site or designing a new piece of equipment, a properly formatted checklist can identify major safety items that need to be considered. This could include things as simple as adequate water supply for fire protection to items as complex as dealing with community acceptance of a new neighborhood risk.

While developing proposals to request funds, a safety checklist can ensure that major safety and health elements have been identified and have been included in the estimated cost to build or modify a new product or facility or piece of equipment. For example, overlooking the cost of conducting product safety testing could result in a substantial financial overrun.

Applying a detailed safety checklist at the design stage ensures that safety standards, best industry practices, and regulatory requirements are identified. Even the best of designers cannot keep in mind all the requirements. This is an opportunity to make sure that major design mistakes are minimized and the cost of reengineering safety into the equipment after the fact is reduced.

During the construction phase, a safety checklist ensures that hazards associated with a specific construction project have been identified. A well-developed check at this point can reduce the chance of construction errors and resultant injuries. Contractors can also be instructed on the safety rules and expectations during this phase of the project.

Finally, an operational safety checklist can ensure that basic safety concerns have been identified prior to commissioning the new product or facility or equipment. The list would establish the need for items such as maintenance procedures, operational procedures, job safety analyses, and debugging safety checks prior to starting formal operations.

6.3.4 Pros and Cons of Safety Checklists

Safety checklists are the most basic and simplest of hazard identification techniques. They provide the impetus for further and more detailed safety analyses. They minimize the chances of major safety concerns being forgotten or neglected. The results of using a safety checklist are only as good as the experience and skills of the user. Not having a full appreciation of how certain safety controls or certain safety regulations apply to the situation may result in the checking off these items as not applicable or not required.

Individuals must be aware of the limitations in using simple safety checklists. The use of the list can become so narrow so that other associated hazards are ignored or neglected. Therefore, the preparer of the checklist has a challenge to make it inclusive but not overwhelming. Additionally, the checklist results in generalized action items rather than specific things to do. A well-organized risk assessment program will keep track of status of followup items. The checklist review can result in identifying the need to conduct a more focused and formalized assessment such as "What if" hazard analysis. In any case, the safety checklist is a simple, versatile, and highly resource-effective method to identify normally encountered hazards.

6.3.5 Some Examples of Safety Checklists

Checklists need to be developed for the specific situation under review. The examples that follow are intended to be illustrative in nature. Organizational experience and skills will dictate the specific character and shape of the safety checklists that evolve for their specific needs.

Figure 6.3 illustrates a checklist that can be used to identify the hazards associated with a new piece of mechanical equipment undergoing a conceptual design review. For example, an engineer developing a conceptual design for a new assembly machine would ensure that all of the potential hazards have been identified through the use of the checklist. Then as the detailed design proceeds, appropriate accommodations for guards, interlocks, and the associated costs would have been made.

Likewise, Fig. 6.4 illustrates a checklist that could be used to identify potential construction hazards and the actions needed to control those hazards. For example, an organization could use this checklist with an outside contractor to review the hazards associated with a construction project. The checklist would provide a way for documenting the agreements on how those hazards would be controlled.

Some checklists are designed to lead a safety review of specific pieces of equipment. Figure 6.5 illustrates a checklist for pumps.[11]

Finally, most equipment vendors have developed checklists of one form or another that provide safety guidelines on operations, maintenance, or debugging for their specific equipment.

6.4 JOB SAFETY ANALYSIS

6.4.1 Job Safety Analysis Overview

A job safety analysis (JSA) is a simple four-step hazard analysis technique used to identify the hazards associated with individual job tasks and to develop the best controls to minimize those risks. *This approach is generally used for simple, well-defined job tasks that contain injury risks.* An individual experienced in safely doing the task along with a supervisor, a team coach, or a safety representative can best conduct the analysis. A JSA is generally not appropriate for conducting design reviews or understanding the hazards of a complex process. In fact, a job safety analysis may be the result of a

MECHANICAL EQUIPMENT HAZARDS CHECKLIST			
Project / Equipment Description: ______			
Potential Hazards	Applies	Does Not Apply	Actions Required
• Rotating / Moving Parts			
• Nip / Pinch Points			
• High Loads			
• Hi / Lo Pressures			
• Hi / Lo Temperatures			
• Noise Exposures			
• Chemical Exposures			
• Electrical Exposures			
• Radiation Exposures			
• Laser Exposures			
• Fire Exposures			
• Ergonomic Exposures			
• Operating Procedures			
• Testing / Debugging Procedures			
• Maintenance Procedures			
• Applicable Standards ANSI, NFPA, ASME			
• Other / Miscellaneous			
Reviewed By: ______ Date: ______			
CC: Operations, Maintenance, Safety, Equipment File			

Fig. 6.3 Mechanical safety checklist example.

Construction Safety Checklist

Project Title / Scope		**Actions Required**
Area / Facility:	✓	
A. Contractor Safety Management: • Competent Person/Construction Supervisor - registered, regulatory requirement, identification, availability • Employee Training/ Orientation - Written program, training frequency, Company rules • Subcontractor Expectation/Orientation - Program, implementation, oversight role • Workplace Inspection Frequency/Documentation - Frequency, documentation, corrective actions • Safety Rules - availability, distribution, posting of "construction area" signs with key information.		
B. General Safety: • Job Site - local rules, barricading & warning signs, roads, material staging, vehicle parking, trailers, utility lines, radios / walkman prohibitions • Housekeeping - cleanup and disposal of debris, workplace conditions, egress routes • Emergencies - Phone #; first aid/medical, eye wash & safety showers and injury reporting; fire alarms, evacuation and reporting; spill response and reporting • Personal Protective Equipment - safety glasses & goggles, hard hats, safety shoes, gloves, respirators, life lines, ear protection, special clothing • Tools - condition, inspection, double insulated if powered, use, guarding, training • Loaning of Company Tools & Equipment - only on rare occasions and with a written permit • Environmental - PCB, asbestos, lead paint, refrigerants, ballast's, hydraulics, cleaners, spray cans, oily rags, spills		
C. Fire Safety: • Welding & Cutting - Open flames & Company fire permit requirements as well as local fire dept. permit • Flammable Liquids - handling & storage of solvents, gasoline, other fuels • Smoking - not allowed inside or on Company buildings • Sprinklers - storage of combustibles, permit for shutdown of sprinklers & fire alarm systems • Emergency Response - alarm locations, use of fire extinguishers, fire hose, exits, emergency marshals		

Fig. 6.4 Construction safety checklist example.

recommendation from a more detailed process hazard review. As such, it is ideal for analyzing *job tasks* such as clearing jams in machinery, operating a lathe in the shop, or moving a drum in the factory.

The results of a job safety analysis should be incorporated into the formal, written procedures for that task. For example, the analysis may result in the recognition of the need to wear safety glasses with side shields while operating a certain machine. Therefore, in addition to the other requirements for quality control and production specifications, the personal protective equipment requirements for safety glasses would be included as part of the operating procedures. Similarly, defining the needed safety controls, such as locking it out, in written maintenance procedures is another ideal end result of performing a job safety analysis.

The job safety analysis can be conducted using a form similar to the one shown in

D. Electrical Safety: • Locking Out - de-energizing, locking and tagging out equipt. prior to servicing, no hot work • Ground Fault - GFCI protection in outside construction and potentially wet areas, custodial work • Tools - Three wired, grounded, non-defective and double insulated powered portable hand tools • Temporary Wiring - Permit required for temporary power and lighting in Company occupied buildings • License - Electricians need for state license		
E. Chemical Safety: • Toxic/Hazardous Materials - Hazard Communication, Material Safety Data Sheets (MSDS), labeling, use • Poisons, Explosives, Pesticides - State License and registration, local safety office approval • Process Piping - Line breaking permit required to work on high temperature/pressure and process piping		
F. Confined Spaces: Permit required for entry into vessels, boilers, manholes and other confined spaces		
G. Ladders/ Staging/Scaffolding: • Ladders - Conditions, storage, non-conducting, securing • Staging/ Scaffolding - Guardrails & toe boards, scaffold grade planking, loading, stability, competent person • Lifts - powered vs manual, stability, condition, travel		
H. Excavation & Trenching: Dig Safe, egress, slopes, soil class, barricading & shoring, competent person		
I. Floor / Wall Openings & Roof Work: • Roof Work - Life lines, safety nets, motion stopping safety system, competent person monitoring system, melt pots, training, other • Floor/Wall Openings - Size, guarding, toe boards, coverings, location		
J. Rigging/Hoisting/Cranes: • Cranes - Licensed operator, inspection certificate, barricading, boom travel, load limit, overhead lines • Hoists - Load limit, training, load swing, training • Rigging - Plan, load limits, equipment condition		
K. City/State Permits: Building, Wiring, Plumbing, etc.		
L. Other Special Situations: • Powder Actuated Tools - Prohibited in occupied company bldgs • Blasting/Explosives/Demolition - local safety approval • Gas Cylinders - storage, handling, use, disposal • Radiation/Lasers - power, precautions, training		
Reviewed By/With: ____________________		Date: ____________
CC: Project File, Contractor, Construction Engineer, Safety Department		

Fig. 6.4 (*Continued.*)

Fig. 6.6. The first column shows the various steps of the job, the next column records the potential hazards of that step, and the last column identifies the required safety controls. Notice that this technique moves beyond the safety checklist approach of just hazard identification. Job safety analysis also involves the selection of the proper safe work practices.

The four basic steps in conducting a job safety analysis then are:

1. Selecting the job or task to be reviewed.
2. Breaking the job or task into sequential or successive steps.
3. Identifying the potential hazards at each step.
4. Deciding on the required action or procedure to minimize each potential hazard.

Pump Safety Checklist

Pump Identification #:		**Comments**
Process Requirements:	✓	
A. Centrifugal Pump:		
1. Can casing design pressure be exceeded?		
2. Is downstream piping / equipment adequately rated?		
3. Is backflow prevented?		
4. Suction piping overpressure (single pumps)		
5. Suction piping overpressure (parallel pumps)		
6. Is damage from low flow prevented?		
7. Can fire be limited?		
B. Positive Displacement: Pump:		
1. Can casing design pressure be exceeded?		
• Pressure relief valve in discharge		
• Set pressure = casing DP minus maximum suction pressure		
• Pressure relief valve discharge location (viscous materials)		
C. General Requirements:		
1. Guarding Requirements - rotating parts		
2. Environmental Requirements - seal design		
3. Noise Emissions - meets standards		
4. Electrical Requirements - x-proof, dust tite, water tite		

Fig. 6.5 Pump safety checklist.

The results of the documented analysis are then incorporated into the actual written job procedure. The procedure can then be used to assist in the training of those individuals who also perform the task. A periodic review of the written procedure should be conducted to assure that the safety requirements are still appropriate. Additionally, an occasional audit helps determine if the required safety controls are actually being used and incorporated into the day-to-day operations.

6.4.2 The Benefits of Job Safety Analysis

There are several major benefits that derive from conducting job safety analyses. The job safety analysis approach itself is easy to understand, does not require a great deal of training, and can be quickly completed by experienced individuals. The process also

JOB SAFETY ANALYSIS		
Division:	Machine/Operation:	
1. Job Task Description:		
2. Task Steps	3. Potential Hazards	4. Safety Controls
Date:	Review due date:	Approved:

Fig. 6.6 Job safety analysis form.

provides an opportunity for an individual to be recognized for his/her knowledge of the operation. The results of the review can provide a written document that can be used to train new employees, a consistent method of operation that can reduce process variables, as well as a common set of safety expectations and reduced injury rates. Additionally, the job safety analysis document can be reviewed as part of planned safety audits as well as providing a starting point for reviewing job procedures if an accident does occur.

6.4.3 Step One—Selecting the Job or Task

Jobs or tasks that can be simply described are best suited for job safety analysis. The task must be defined fairly specifically to take most advantage of the power of JSAs. For example, broadly defined jobs such as constructing a building, making chemicals, or working in a laboratory are not suitable for JSAs. On the other hand, tasks such as threading a coating machine, assembling scaffolding, or charging a drum of chemicals to a reactor are suitable subjects for safety analysis. Some traditional permit-type safety practices are based on the job safety analysis approach, such as confined space entry permits, chemical line opening permits, and temporary wiring permits.

In selecting and prioritizing, the jobs to be reviewed, several factors need to be considered. Analysis of past injury data (see Chapter 5) may indicate that individuals did not know the safe procedure for accomplishing certain tasks. Perhaps up-to-date written procedures need to be established. The data could illustrate the existence of a low frequency of very serious injuries or a high frequency of minor injuries. This would be considered a high-risk situation on the risk matrix (Fig. 6.2). Employees new to a department may need to be trained on doing the job, and therefore up-to-date written operating instructions may be needed. Likewise, a new process or new piece of equipment may need a job safety analysis so that the initial operating procedures can be developed. Perhaps a more advanced risk assessment technique has uncovered the fact that the existing operating procedures need improvement.

Recording the specific task to be analyzed on the JSA form is the first step in a productive job safety analysis. This brief written description of the task also helps further to define the boundaries of the analysis.

6.4.4 Step Two—Breaking the Task into Sequential Steps

After the task or job for review has been selected, the next step is to list all the discrete steps in performing the task. An experienced operator with the help of a co-worker, team leader, supervisor, or safety representative is usually the best choice to accomplish this. The operator can walk through the task and describe what is being done at each step. The supervisor or team leader can then record the steps on the first column of the job safety analysis form. Identifying the hazards and associated safety controls needed at each step should be held until all the steps in the job task are listed.

To determine the job steps, a basic question such as "What is the first thing that you do?" is a helpful starting point. Each of these steps should accomplish some discrete identifiable task. Selection of steps that are too broad in nature such as "charging the chemicals to the reactor" precludes understanding of the hazards presented at each step of the task, whereas selecting steps such as (1) close reactor charge port, (2) connect clean 2-in. solvent hose to reactor manifold, (3) open manifold valve at reactor level B, (4) start raw material pump, (4) etc., provides the basis for proper hazard identification.

Notice that each step starts with an action word. The intent at this point is to determine what is being done.

An overly detailed procedure that provides a lost of descriptive information but little in the way of additional hazard opportunities leads to boredom and needless paperwork. A proper balance is the challenge for selecting the steps in a job safety analysis. An experienced operator may find that the JSA job steps selected are too basic. On the other hand, a new operator may find that the same information developed from a job safety analysis is exactly what is needed to do the job safely. Consider combining jobs with only few steps together as well as breaking up jobs with too many or too complex tasks. Be sure that the work is actually observed so that key steps are not taken for granted, especially at the beginning or the end of the task.

6.4.5 Step Three—Identify the Hazards Associated with Each Step

The next step is to identify all of the hazards associated with each step. The hazards should then be recorded on the second column of the job safety analysis form. Identifying the proper safety controls needed at each step can be done at this point or at the next step after all the hazards are identified. If solutions are generated at this point, there is some chance of missing some hazard potentials. In any event, an experienced operator is generally able to provide excellent insights as to what can go and has gone wrong at each step of the job or task.

Consider using the injury hazard potentials listed on the injury reporting form in Chapter 5. A portion of it is reproduced as Table 6.2.

At each step of the task, the hazard potentials would then be identified and recorded. For instance, if the particular task is pouring a caustic liquid from one container to another, the potential hazards could be chemical burns to various parts of the body such as the eye, face, hands, or torso depending on the size of the container and nature of the procedure. Additionally, another potential hazard that could be associated with this task could be the strain of the back or arm depending on the size and weight of the container.

Be sure to observe and review the actual job task. Recalling the task from memory or not including elements that begin or start the task may lead to an incomplete analysis. The "sometime quirks" of the task that do not get reviewed can result in the operator not knowing the full story and the required safe procedures. The experienced

TABLE 6.2 JSA Hazard Potentials

Type of incident potential	Type of injury potential	Part of body potential
Hit by	Contusion, laceration	Eye R/L
Hit on	Fracture, dislocation	Face/head
Slip, trip, fall	Sprain, strain	Torso/back
Lifting, pushing, pulling	Burn________	Hands R/L
Cumulative trauma	Foreign Body	Arms R/L
Transportation	Exposure____	Feet/legs R/L
Other________	Other_____	Other______

analyst can maximize the chances of ensuring a complete review by asking open-ended questions such as "Does the procedure ever change?" or "Is this step ever done differently?" Having another experienced operator review the completed JSA or videotaping the procedure are other excellent approaches.

6.4.6 Step Four—Develop the Safety Controls for the Hazards at Each Step

Now that the hazards have been identified, the next step is to develop the necessary safety controls. Remembering the risk matrix (Fig. 6.2), the idea is to move the high risks out of the unacceptable range. The risks can be reduced by lowering the possibility that something will go wrong or by reducing the severity of the consequences if it does go wrong or by doing both. In some cases, judgments must be made about the potential risks of alternate procedures that could be proposed. Minimizing risks at each step in the job is the goal.

Using the JSA form and the hazards identified at each step, now develop and document the required safety controls. There are generally several controls that could do the job. The challenge is to determine the best one that is both cost effective and that, in fact, will be implemented.

If the task appears to be highly hazardous and easy solutions to reducing the risks do not appear obvious, consider the possibilities of accomplishing the task in an entirely new way. Brainstorming ideas with operators, designers, and trades personnel can often result in cost-saving, less hazardous ways of accomplishing the task. Ensure that the proposed solutions do not result in riskier situations.

Typical hazard control measures include the development of procedures, engineering controls such as equipment design and ventilation, and the requirements to use personal protective equipment. Procedures can include the use of safety permits such as confined space entries, hazardous work, and chemical line openings. Other procedures can include periodic preventive maintenance inspections on safety interlocks, fire protection systems, or ladder conditions. Procedures could be developed to check periodically on compliance with the established safety rules. The use of personal protective equipment requirements could be established, such as wearing safety shoes during movement of materials. Similarly, improvements to the facility or equipment can be made to reduce the chance of accidents, such as improved lighting, interlocked safety guards, and improved material storage layouts.

6.4.7 Summary—Making JSA's Effective

Job safety analysis is an easy-to-use and results-oriented tool for identifying and controlling the hazards of everyday tasks. The National Safety Council[12] suggests that it is an excellent starting point for questioning the established way of doing a job. Conducting the analysis does not require extensive training. The results are most effective when it is done by experienced individuals who are actually at risk from doing the task. The JSAs can be used for training new employees, agreeing on safety procedures with experienced employees, and can be the safety basis for the preparation of well-written operating procedures. The completed JSA document needs to be incorporated into the overall management system with appropriate approvals as well as periodic reviews and updates. When a new employee needs to be trained to do the task safely, it is also an excellent time to review the existing JSA and written job procedure. It is an opportunity for a knowledgeable individual to demonstrate his or her skills as well as to ensure that the JSA is actually up to date. In any event, an annual review is the best choice. An example of a completed JSA is illustrated in Fig. 6.7.

JOB SAFETY ANALYSIS		
Division: VCR Assembly	Machine/Operation: VCR Viewing Mirror Carrier Machine	
1. Job Task Description: At Mirror Carrier Machine manually insert bellows and mirrors; actuate glue application and curing process, stack completed assemblies and package and transport boxed assemblies.		
2. Task Steps	**3. Potential Hazards**	**4. Safety Controls**
1. Insert bellows into assembly machine fixture	• Possible repetitive motion problems • Danger if machine cycles while hand is in the danger zone • Danger of finger / hand being cut while inserting bellows	• Request an ergonomic evaluation • Safety interlocks are designed into machine operation and interlock reliability checks are integral to the machine operation • Risk is minor and no additional controls are needed
2. Stack mirrors inside the viewing mirror carrier machine	• Danger of cuts to the fingers from the sharp edges	• Present mirror thickness does not present unreasonable hazard however - • Provide a "sweep up" device to allow for the pick up of sharp pieces of broken mirrors
3. Place mirror onto the nest	• Danger of cuts to the fingers from the sharp edges • Danger if machine cycles while hand is in the danger zone	• Same as above - provide sweep-up device • Same as above - safety interlocks provided
4. Wipe glue tips	• Exposure of RTV epoxy glue to fingers, hand and possibly to the face and eyes inadvertently	• Re-design fixture to eliminate the need to wipe the glue • Provide MSDS & review irritant properties • Ensure that eyes, skin is washed immediately upon exposure
5. Close VCR assembly doors	• No apparent hazards	• None required
6. Activate cycle timer button	• Machine cycles while hand is in the danger zone • Potential UV light exposure to eyes &/or skin	• Same as above - safety interlocks provided • Re-do UV level measurements especially at spaces around door and provide preventive maintenance checks on mounting
7. Open VCR assembly doors	• None apparent	• None required
8. Release mechanism and remove completed assembly	• Machine mis-cycles • Sharp edges	• Same as above - safety interlocks provided • Same as above - provide sweep-up device
9. Put completed assembly into tray	• None apparent	• None required
10. Other associated hazards	• General area eye hazards • General housekeeping / slipping hazards	• Require the use of safety glasses • Ensure floor kept clear of parts
Date: June 27, 1996	Review due date: July 1997	Approved: Susan Supervisor

Fig. 6.7 Example of completed JSA.

6.5 "WHAT IF" ANALYSIS

6.5.1 "What If" Analysis Overview

"What If" analysis is a structured brainstorming method of determining what things can go wrong, judging the risks of those situations, and recommending corrective actions where appropriate. With little experience in methodology, a review team experienced in the process, equipment, or system under review can effectively and productively uncover the major safety and health issues. Led by an energetic and focused facilitator, the review team assesses step by step what can go wrong based on their past experiences and knowledge of similar situations.

Using an operating procedure and/or a piping and instrument diagram (P&ID), the team reviews the operation or process using a form similar to one illustrated in Fig. 6.8. Team members usually include operating and maintenance personnel, design and/or operating engineers, specific knowledgeable people as needed (chemist, structural engineer, radiation expert, etc.), and a safety representative. At each step in the procedure or process, "What If" questions are asked and answers generated. To minimize the chances that potential problems are not overlooked, moving to recommendations is generally held until all the potential hazards are identified.

The review team then makes judgments regarding the likelihood and severity of the "What If" answers. If the risk indicated by those judgments is unacceptable, then a recommendation is made by the team for further action. The completed analysis is then summarized, prioritized, and responsibilities assigned.

6.5.2 Getting Started—What Is Needed

The first steps in conducting an effective analysis include picking the boundaries of the review, involving the right individuals, and having the right information. The boundaries of the review may be a single piece of equipment, a collection of related equipment, or an entire facility. A narrow focus results in an analysis that is detailed and explicit in defining the hazards and specific recommended controls. As the review boundaries expand to include a large complex process or even an entire facility, the findings and recommendations can become more overview in nature. The analysis can include the various stages of a construction project, the procedural steps involved in the operation of the equipment or facility, or the written maintenance procedures for a piece of equipment. A clear understanding of the boundaries of the analysis starts the review off in an effective manner.

Assembling an experienced, knowledgeable team is probably the single most important element in conducting a successful "What If" analysis. Individuals experienced in the design, operation, and servicing of similar equipment or facilities is essential. Their knowledge of design standards, regulatory codes, past and potential operational errors, as well as maintenance difficulties brings a sense of practical reality to the review. On the other hand, including new designers and new operators in the review team mix presents an excellent learning opportunity for subjects that are not usually taught in design school or in operating classes.

The next most important step is gathering the needed information. One important way to gather information for an existing process or piece of equipment is for each review team member to visit and walk through the operation. Video tapes of the operation or maintenance procedures or still photographs are important and often underutilized excel-

"What-If" Hazard Analysis

Division:	Desc. of Operation:			By: Date:
What If ?	Answer	Likeli-hood	Conse-quences	Recommendations

Fig. 6.8 "What-If" hazard analysis form.

lent sources of information. Additionally, design documents, operational procedures, or maintenance procedures are essential information for the review team. If these documents are not available, the first recommendation for the review team becomes clear: Develop the supporting documentation! Effective reviews cannot be conducted without up-to-date, reliable documentation. An experienced team can provide an overview analysis, but nuances of specific issues such as interlocks, pressure reliefs, or code requirements are not likely to be found.

6.5.3 Conducting the Review—How Is It Done?

Now that the team has had an opportunity to review the information package, the next step is to conduct the analysis. Generally, an experienced hazards review facilitator will lead the group through a series of "What If" questions. A focused, energetic, and knowledgeable facilitator can keep the review moving productively and effectively. A scribe is usually assigned to take notes of the review. Recent advances in software as well as laptop computers can provide online data collection possibilities by the scribe. That is, as hazards are identified, judgments made, and responsibilities assigned, the scribe can input the data and agreements live! Scheduling more than four hours at a time, however, can result in the team members losing energy and being eager to finish the analysis rather than to probe deeper. Generally, in a well-designed or well-operated system the participants in the review will need to work hard to find major issues of concern. It is the job of the facilitator to keep the effort productively moving.

Step 1. **Developing the "What If" Questions.** Using the available documents and the experience and knowledge of the review team, "What If" questions can be formulated around human errors, process upsets, and equipment failures. These errors and failures can be considered during normal production operations, during construction, during maintenance activities, as well as during debug situations. The questions could address any of the following situations.

- Failure to follow procedures or procedures followed incorrectly
- Procedures incorrect or latest procedures not used
- Operator inattentive or operator not trained
- Procedures modified due to operational upset
- Processing conditions upsets
- Equipment failure
- Instrumentation miscalibrated
- Debugging errors
- Utility failures such as power, steam, gas
- External influences such as weather, vandalism, fire
- Combination of events such as multiple equipment failures

Experienced personnel are knowledgeable of past failures and likely sources of errors. That experience should be used to generate credible "What If" questions.

For example, consider a chemical manufacturing process that includes the charging of a granularlike material from a 55-gallon drum to a 1000-gallon mix vessel containing a highly caustic liquid. Some typical questions are shown in Fig. 6.9 for illustration purposes only.

As the "What If" questions are being generated, the facilitator should ensure that each member of the team has an opportunity to input potential errors or failures. Determining the answer to each question as it is generated creates the danger of closing too soon on all the possible upsets. The facilitator needs to be sure that the team has

"What-If" Hazard Analysis

Division: Chemical Mix	Desc. of Operation: Manufacturing B Mix / Drum Charging Operations - Page 2 of 4			By: Review Team Date: 9/97
What If ?	Answer	Likelihood	Consequences	Recommendations
1. Granular powder is not free flowing? 2. Drum is mis-labeled? 3. Wrong powder in the drum 4. Drum hoist is not used? 5. Two drums are added? 6. Drum is mis-weighed? 7. Drum hoist fails? 8. Drum is corroded? 9. Ventilation at Mix Tank is not operating? 10. Granular powder becomes dusty? 11. Powder gets on operator's skin? 12. Tank mix liquid level too high?				

Fig. 6.9 Example of completed step 1, "What-If" analysis form.

really probed into all the possibilities before going to the next step of answering the questions. The analysis can be divided into smaller pieces if there is a danger of just developing questions and not having the value of them fresh in mind while answering those questions.

Step 2. **Determining the Answers.** After being assured that the review team has exhausted the most credible "What If" scenarios, the facilitator then has the team answer the questions. What would be the result of that situation occurring? For example, consider the answers illustrated in Fig. 6.10 to the "What If" questions in our previous example.

If done correctly, reviewing the potential equipment failures and human errors can point out the possibilities for not only safety and health improvements but also for the opportunity to minimize operating and quality problems. Including the operators and trades personnel in the review can bring a practical reality to the conclusions that are reached.

Step 3. **Assessing the Risk and Making Recommendations.** Now having considered the answers to the "What If" questions, the next task is to make judgments regarding the likelihood and severity of that situation. In other words what is risk? Remembering the risk matrix (Fig. 6.2), the review team needs to make judgments regarding the level and its acceptability. For example, consider the risk judgments and recommendations based on answers in our example and illustrated in Fig. 6.11.

Notice that the team has not only assessed the risk at each situation but has also made related safety recommendations. The discussion of each "What If" situation leads naturally to the recommendation. The team will then continue the analysis, question

"What-If" Hazard Analysis

Division: Chemical Ops	Desc. of Operation: Manufacturing B Mix / Drum Charging Operations - Page 2 of 4			By: Review Team Date: 9/97
What If ?	Answer	Likeli-hood	Conse-quences	Recommendations
1. Granular powder is not free flowing?	1. Back injury potential when breaking up clumps			
2. Drum is mis-labeled?	2. Quality issue only			
3. Wrong powder in the drum	3. If wet, could cause chemical exothermic reaction			
4. Drum hoist is not used?	4. Back injury potential			
5. Two drums are added?	5. Quality issue only			
6. Drum is mis-weighed?	6. Quality issue only			
7. Drum hoist fails?	7. Leg, foot, back arm injury			
8. Drum is corroded?	8. Iron contamination as well as drum failure & injury			
9. Ventilation at Mix Tank is not operating?	9. Dusting & potential operator exposure			
10. Granular powder becomes dusty?	10. same as above			
11. Powder gets on operator's skin?	11. Possible burn			
12. Tank mix liquid level too high?	12. Possible caustic splash as well as quality issue			

Fig. 6.10 Example of completed steps 1 and 2, "What-If" analysis form.

by question, until the entire process or operation has been assessed. At this point, the facilitator should have the team step back and review the "big picture" and determine if they have inadvertently missed anything.

6.5.4 Reporting the Results—To Whom and How?

The hard work of conducting the analysis has been completed. The important work of documenting and reporting the results still remains. The makeup of the organization generally determines to whom and how the results get reported. Usually, the department or plant manager is the customer of the review. The leader of the review team will generate a cover memo that details the scope of the review as well as the major findings and recommendations. In some organizations, the report recommendations will also assign responsibilities and a time frame for actions. In other cases, a separate staff or function will review the recommendations and determine the actions required. A periodic report is then generated to summarize the present status of each of the recommendations. Those organizations that have a well-developed hazard review program require followup assessments every three to five years based on the associated hazard levels.

6.5.5 "What If" Summary—Pros and Cons

The "What If" analysis technique is simple to use and has been effectively applied to a variety of processes. It can be useful with mechanical systems such as production

"What-If" Hazard Analysis

Division: Chemical Ops	Desc. of Operation: Manufacturing B Mix / Drum Charging Operations - Page 2 of 4			By: Review Team Date: 9/97
What If ?	**Answer**	**Likelihood**	**Consequences**	**Recommendations**
1. Granular powder is not free flowing?	1. Back injury potential to break up clumps	Quite Possible	Serious	Design automated de-lumping equipment
2. Drum is mis-labeled?	2. Quality issue only	Remote	Serious	Improved label from vendor
3. Wrong powder in the drum	3. If wet, could cause chemical exothermic reaction	Unlikely	Minor	Include inspection in procedure
4. Drum hoist is not used?	4. Back injury potential	Possible	Serious	Train personnel & ensure use
5. Two drums are added?	5. Quality issue only	Remote	Minor	None
6. Drum is mis-weighed?	6. Quality issue only	Possible	Serious	Require 2nd check on weight
7. Drum hoist fails?	7. Leg, foot, back arm injury	Remote	Serious	Ensure hoist on PM program
8. Drum is corroded?	8. Iron contamination as well as drum failure & injury	Remote	Serious	None
9. Ventilation at Mix Tank is not operating?	9. Dusting & potential operator exposure	Unlikely	Minor	Include ventilation check in operating procedure
10. Granular powder becomes dusty?	10. same as #9 above	Unlikely	Minor	None beyond existing procedure
11. Powder gets on operator's skin?	11. Possible burn	Quite Possible	Serious	Use dust suit & gloves
12. Tank mix liquid level too high?	12. Possible caustic splash as well as quality issue	Remote	Very Serious	Use goggles and apron

Fig. 6.11 Example of completed "What-If" analysis form.

machines, with simple task analysis such as assembly jobs, as well as with reviewing complex tasks in chemical processing. No specialized skills are needed to use the methodology. Individuals with minimal hazard analysis training can participate in a full and meaningful way. It can be applied at any time such as during construction, during debugging, during operations, or during maintenance. The results of the analysis are immediately available and usually can be applied quickly. This is especially true if the review team members also operate or maintain the system being assessed.

On the other hand, the technique does rely heavily on the operational and hands-on experience and intuition of the review team. It is somewhat more subjective than other methods such as HAZOP (see Section 6.6), which involve a more formal and systematized approach. If all the appropriate "What If" questions are not asked, this technique can be incomplete and miss some hazard potentials. It may also be appropriate to assign those more complex portions of the system to a more rigorous review such as HAZOP.

6.6 HAZARD AND OPERABILITY ANALYSIS

6.6.1 HAZOP Analysis Overview

A hazard and operability (HAZOP) Analysis is the systematic identification of every credible deviation in a system or process, usually a chemical manufacturing process, from the design intent. Resultant adverse consequences from those deviations are then identified as well as the initiating causes. The risk of those deviations are then assessed, and if deemed unacceptable, then a set of recommended actions determined. It requires rigorous adherence to the methodology to be sure that no potential hazards are missed. This method has its roots in the United Kingdom with Imperial Chemical Industries (ICI) in the 1960s.[13] The analysis requires individuals who are expert in the design requirements and the design intent of the facility, up-to-date piping and instrument diagrams, a well-defined system, and a hazard review facilitator who is knowledgeable of the HAZOP technique.

The team reviews the plant section by section, line by line, and item by item using key guidewords to initiate discussion. The guidewords prompt the team members to consider deviations from the plant design intent such as more of or less of, none, reverse, and other. Those guidewords are then applied to the relevant plant operating parameters under review such as flow, pressure, temperature, materials, etc. The causes and consequences of those deviations are then assessed, and the need for added risk controls is determined.

6.6.2 HAZOP—Getting Started

Assembling the right people to conduct the analysis is a critical step. The detailed and rigorous nature of the process probing is not something that everyone enjoys. On the other hand, the logical, sequential, and ordered methodology of step-by-step review of a process can be enlightening to those who have been heavily involved in the design or operation of the process. It provides an opportunity to test and expand their skills and knowledge. There are generally five to seven review participants whose functions are illustrated in Table 6.3.

The review team leader needs to keep the analysis focused on identifying problems and prevent setting sidetracked on solving those problems. Solutions to problems are not the primary goal of the analysis. The leader also needs to keep the energy level high, ensure that everyone participates, and be sure not to get bogged down.

To be successful, the review team needs to have the process information organized into a suitable and usable form after the scope of the hazard review is set. Typically, the information consists of line drawings, flowsheets, plant layouts, as well as piping and instrument diagrams (P&ID). Additionally, operating manuals, maintenance manuals, and equipment manuals are usually available. A critical step in conducting effective reviews is dividing the process into individual nodes, sections, or operating steps for analysis. The team leader will usually define the nodes prior to the meetings. The nodes

TABLE 6.3 HAZOP Team Members

Review team leader	Experience in HAZOP methodology is essential and experience in the process is helpful
Process engineer/chemist	Familiarity with the process chemistry and operations
Design engineer	Knowledge of the piping and instrument (P&ID) as well as equipment design requirements
Maintenance engineer/supervisor/leader	Familiar with equipment deviations
Operations/supervisor/leader	Knowledge of operating unit and deviations
Safety/fire/industrial hygiene leader	Experience with health, safety, and industrial hygiene standards and guidelines
Scribe	Note taker familiar with HAZOP techniques

can be highlighted on piping and instrument diagrams where the process parameters have an identified process design intent. These can be pipe sections where pressure, temperature, and flow conditions have been established. Processing equipment components such as pumps, valves, vessels, and heat exchangers are points between nodes that could cause changes in these parameters. Key guidewords associated with the process parameters are also established.

Once the team has been selected and the information gathered and distributed to the team members, the review meetings can be scheduled. Ideally, meetings should not be scheduled for more than four hours with breaks every 1 1/2 to 2 hours. The number of meetings will depend on the depth of preparation of the leader, the knowledge of the participants and the complexity of the process. The number of hours to complete a HAZOP has been estimated by several authors. The CCPS Guidelines[14] has suggested that simple, small systems will require 8 to 12 hours preparation time, 1 to 3 days for review and 2 to 6 days for the documentation. Freeman et al.[15] have suggested that the time to complete a HAZOP analysis is a function of the skill level of the team leader, the number of P&ID's, and their complexity. A suggested formula provides weighting factors. For example, a HAZOP consisting of one P&ID, three nodes and one major piece of equipment requires about 18 hours of team meeting time. Fifteen P&ID's, 81 nodes and 13 major equipment items requires about 100 hours of team meeting time. Arco Chemical Company[16] suggests that a "two to four week study is likely to produce between 50 to 200 findings."

For large, complex processes, it may be necessary to use several teams and team leaders to complete the review within a reasonable timeframe.

6.6.3 Conducting the HAZOP—How Is It Done?

Now that the team has been assembled, the information has been gathered, and the meeting scheduled, the analysis can proceed. The analysis is systematic and includes the use of the following terms:

- Nodes—The locations on piping and instrument diagrams (P&ID) at which the process parameters are analyzed for possible deviations.
- Deviations—Departures from the design intentions that are uncovered by systematically applying appropriate guidewords to the process parameters (i.e., no flow, high pressure, low temperature)
- Intention—How the plant is expected to operate in the absence of deviations at nodes (i.e., pressures, rates, levels, conditions)
- Causes—Reasons why deviations from intentions may occur. A credible cause should be considered as meaningful and included as part of the analysis (i.e., equipment failure, human error, power failure, unanticipated situation)
- Consequences—Results of the deviations if they should occur (i.e., injury, spill, fire, explosion, release to atmosphere)
- Risk—The likelihood of the deviation occurring and the severity of the consequences (i.e., the concepts embodied in the risk matrix discussed in Section 6.2)
- Guidewords—Used to discover or derive the potential deviations from design intentions; common guidewords and meanings are listed in Table 6.4.

TABLE 6.4 HAZOP Guidewords and Meanings

Guideword	Meaning
No	Negation of the design intent—No part of the design intentions is achieved and nothing else happens
Less	Quantitative decrease—Refers to quantities and properties such as flow, temperature, and pressure
More	Quantitative increase—Same as above including quantities and properties such as flow, temperature, and pressure
Part of	Qualitative decrease—Only some of the design intentions are achieved; some are not
As well as	Qualitative increase—All the design intentions are achieved together with some additional items
Reverse	Logical opposite of intent—Applicable to activities such as flow
Other than	Complete substitution—No part of the design intention is achieved, and something different occurs

- Parameters—Characteristics of the process that when deviated from the design intent could result in an injury, environmental upset, or business loss. Common chemical process parameters are:

Flow	Mixing
Temperature	Addition
Pressure	Substitution
Level	Reaction
Composition	pH
Frequency	Time
Viscosity	Information
Voltage	Speed

Now, using the terms and definitions described as well as their knowledge of the process, the HAZOP review team conducts the systematic and structured analysis. Although described as sequential, the actual review steps are closely connected. The results are documented by the assigned scribe on a form similar to Fig. 6.12. The collected data include the causes, consequences, risk judgments, and recommended actions for potential deviations at each node in the process.

If the review is conducted on an existing process, the team members need to tour the facility in order to become familiar with the size and the inter-relationships of the key components. Operations that are spread out over five acres at one level in a rural area provide a different sense of risk and hazards than those that are located in one five-story building in an urban location.

The steps in the HAZOP method are:

1. **Select a line or vessel (node).** Using the P&ID and selected nodes, the team starts with the beginning or front end of the process. The team leader or hazard review facilitator usually has made a preliminary breakout of the nodes prior to the first meeting. The team can revise or modify the selections based on new information as the review proceeds. An operating manual can also be used as the basis for the review. At each operating step (node), an analysis equivalent to that to be described would be conducted.

Hazard & Operability Analysis

Process Reviewed: ______________________ Drawing #: ____________ Review Date: ____________

Node Reviewed: ____________ Process Parameter: ____________ Reviewers: ______________________

Design Intention: ____

Guide Word Deviation	Causes	Consequences	Risk	Recommended Action

Fig. 6.12 Hazard and operability analysis form.

2. **Describe the design intention of the line or vessel in the process.** At this point, the most knowledgeable person on the team would define the design intention. For example, this could mean a transfer line designed to transfer caustic solution from the discharge of pump A to the inlet of vessel B at 100°F and 70 psig and at a rate of 125 gallons per minute.

3. **Select appropriate process parameter(s) associated with the line or vessel for analysis.** The team would then select process parameter guides and the associated guidewords. Review the list above for potential process parameters to investigate. In this case, the parameters of *flow rate*, fluid *temperature*, and fluid *pressure* are important operating parameters to investigate for possible deviations.

4. **Apply the guideword deviations to the parameter(s) of interest).** The team would select the most important and applicable deviations that would occur with that process parameter. See the list above for typical process deviations of interest. In the example concerning caustic flow rates, the team would review the deviation possibilities of no flow, more flow, reduced flow, reverse flow, as well as other flows.

5. **List credible consequences and causes of those deviations.** As the team moves through possible deviations from the intended design, a list of scenarios including credible causes leading to unwanted consequences will be developed. For example, no flow could result in a off-specification production batch. Several causal factors were considered, such as closed valve on the pump discharge, or an empty line feeding the supply side of the pump. Similarly, reverse flow could result in catastrophic introduction of caustic into another holding tank and could have been caused by power failure and gravity siphoning effects. In some cases, there may be gaps in the process information design intention or knowledge of the team members. This could result in delaying or deferring that portion of the analysis in order to obtain more information.

6. **Judge the risk.** Given the potential deviations and the potential consequences of those deviations, the team will then make judgments on the level of risk and the acceptability of that risk. The concepts embodied in the risk matrix (Fig. 6.2) should be used in that determination. Some organizations[17] have developed a semiquantitative approach. That is, frequency rates and severity consequences are assigned numerical rankings and when combined add up to risks that are low enough to be either acceptable or marginal, requiring additional study, or unacceptably high, requiring aggressive interventions.

7. **Recommended actions.** If the risk was judged to be unacceptable and a solution was apparent to the team, a recommendation for action would be appropriate. If, on the other hand, a solution was not apparent, a recommendation for further study such as an engineering analysis or thermal calorimeter work may be the right choice. Some organizations will include in the report the assignment of recommendations to specific individuals with specific followup time frames. In some cases, an action may be recommended even though the risk may be low since the benefit of is obvious and is obtained at a low cost. Typical risk reduction actions include (a) changing the design, (b) changing the operating procedure, (c) changing the processing conditions, or (d) changing the process itself. These actions reduce the risk by reducing the severity of the resulting consequences or reducing the likelihood of the deviation or both.

8. **Repeat previous steps.** The team would then sequentially review each of the selected nodes for the entire process, analyzing each possible deviation and the resultant consequences and make appropriate recommendations. The particular section of

the P&ID under review is usually marked off as that portion of the analysis has been completed.

An example of a partially completed HAZOP worksheet is illustrated in Fig. 6.13. As the team works its way through the process, a series of potential deviations, consequences, risks, and recommendations are developed. To expedite the analysis some organizations use a predetermined list (such as Table 6.5) of relevant deviations for process section types.[18] That is, when reviewing pipe lines the process parameters of flow, pressure, temperature, concentration, leak, and rupture would apply. For large and complex facilities with many processes, the hazard review data can be collected on spreadsheet software or on specifically designed programs for HAZOP analysis. This helps make the job of managing the results of a large hazard analysis effort somewhat easier. Software programs are being developed that take information from a computer-generated P&ID directly into a HAZOP application program. In this case the HAZOP information generated "stays with" the process design information.

6.5.4 Reporting the Results

By this time the review team has generated a great deal of good information. As with other types of hazard reviews, the information needs to be communicated in an effective manner. An executive summary that highlights the important findings is directed towards the facility or department manager. Highly risky situations need to be communicated verbally as they are identified and not wait until a written report is drafted, reviewed, and issued. The person(s) who will act on the recommendations needs to know the specifics of the findings included in the report details. In some organizations a periodic report is issued that summarizes the percent of recommendations acted upon by process, by plant, or by business unit. Well-disciplined organizations will include timetables and responsibilities. This gives the top management a sense of how well the hazard review process is being managed overall. The management system should ensure that the quality of the responses are monitored or audited to ensure that the recommendations are being acted upon in a responsible and credible manner.

6.6.5 HAZOP Summary—Pros and Cons

The HAZOP methodology is used extensively in the chemical processing industry as an effective technique to conduct hazard reviews. One of the keys to the successful application of the technique is careful preparation by the team leader in selection of nodes and ensuring that design intentions have been established. This technique provides a way to probe exhaustively into all of the potential process deviations and upsets. The technique may become laborious and tedious to an inexperienced review team but it does ensure a complete and thorough analysis.

The HAZOP method is an advanced design review that is structured, systematic and complete. Assuring that the design is aligned with engineering codes alone may be a technique that is simpler to understand, more manageable and easier to apply. On the other hand, codes and best practices are based on past experience with little predictive value, are usually minimum consensus documents, and do not consider the interrelationships of process risks. Incomplete information regarding the design intent will prove to be a frustration to completing the analysis. For this reason, using the HAZOP technique early in the design stage should be avoided.

Hazard & Operability Analysis

Process Reviewed: **Caustic Treatment System** P&ID#: **D-2346 Rev. 3** Review Date: **August 9, 1996**

Node Reviewed: **Transfer Section "A"** Process Parameter: **Flow & Pressure** Reviewers: **Susan F., Frank B., John P., Mary Y.**

Design Intention: **Transfer Caustic Solution from pump discharge A to inlet of vessel B at 100°F, 70 psig & 124 gpm**

Guide Word Deviation	Causes	Consequences	Risk	Recommended Action
1A. No Flow	1. Pump not turned on 2. Pump impeller corroded. 3. In-line valve closed 4. Pipe ruptured 5. Computer does not start pump	1. No production 2. No production / quality 3. Pump motor burns out 4. Environmental release & possible injury 5. No production	1. Moderate - acceptable 2. Moderate - acceptable 3. Low - Acceptable 4. High - Unacceptable 5. Moderate - acceptable	1. Put checks in procedure 2. Do maintenance checks 3. None 4. Install collection berm & detection alarm 5. Test pump startup by input simulation
1B. Reverse Flow	1. Power failure & check valve failure 2. Over filling of Tank B and valve A open	1. Caustic flows into hold tank & large reaction. 2. Same as #1	1. Very High - unacceptable 2. Same as #1	1. Provide aux. power & valve maintenance 2. Provide level alarm on tank B & double check on Valve A in operating procedure.
1C. As Well As	1. City water back flow	1. Same as 1B. above	1. Same as 1B. above	1. Install backflow preventer
1D. High Flow	1. No credible cause determined			
1E. Low Flow	1. Same as 1A. above			
2A. Low Pressure	1. Same as 1A above	1. Same as 1A above	1. Same as 1A above	1. Same as 1A above
2B. High Pressure	1. No credible cause determined			

Fig. 6.13 Example of a partially completed hazard and operability analysis.

TABLE 6.5 HAZOP Predetermined Chemical Processing Deviation Types

	Process section type				
Deviation	Distillation column	Mix vessel	Line	Heat exchanger	Pump
High flow			×		
Low/no flow			×		
High level	×	×			
Low level	×	×			
High interface		×			
Low interface		×			
High pressure	×	×	×		
Low pressure	×	×	×		
High temperature	×	×	×		
Low temperature	×	×	×		
High concentrat'n	×	×	×		
Low concentrat'n	×	×	×		
Reverse flow			×		
Leak	×	×	×	×	×
Rupture	×	×	×	×	×

The methodology is not designed to fix problems only to uncover problems. Due to the skills, knowledge and experience of the review team some solutions may be recommended that are intuitively obvious and perhaps the best solution. Recommending solutions, however, can quickly become inefficient and time consuming resulting in a re-design effort by the team.

6.7 FAILURE MODE AND EFFECT ANALYSIS

6.7.1 FMEA Overview

A failure mode and effect analysis (FMEA) is a type of hazard review that probes the failures of components within a process or system and the resultant effects of those failures. For example, a failure of a shutdown interlock may result in a machine continuing to run when a safety guard is moved, which could further result in an injury. This methodology is largely equipment oriented and is generally not used when human failures could be major contributors to the process risks. The technique is useful in analyzing single failure modes that result in unwanted situations but is not efficient for reviewing a large number of multiple combinations of component failures. The results of the FMEA analysis generally lead to improved equipment reliability.

Once the boundaries of the analysis have been set, the review starts with updated process drawings such as piping and instrument diagrams (P&IDs). Reviewers who are familiar with equipment functions and common breakdowns then list the major system components and the possible ways in which they can fail. The effects of those failures are then determined, as well as some likely known combinations of failures. The review team then assesses the risks and makes judgments regarding the need for additional safety controls. The findings are then documented in a failure mode and effect tabular form, major recommendations determined, and final report issued.

6.7.2 FMEA—Getting Started

The experience and skills of the review team determines to a large degree the thoroughness, the completeness, and the accuracy of the resulting analysis. Individuals skilled in equipment and component failures are especially important to this methodology. Trades personnel and plant engineering personnel who fix and analyze plant problems can therefore by a very helpful resource to the effort. Also, those systems that have electrical and instrument controls with associated safety interlocks can benefit from the inclusion of instrumentation engineers as part of the review team.

In addition to piping and instrument diagrams, logic or ladder diagrams, instrument loop control diagrams, and wiring diagrams are also helpful to the effort. If the drawings are not current, they must be marked up after a field review or walk through. The facilitator could mark the key components that the team will be reviewing prior to the first review meeting and use this listing as part of the opening overview. Scheduling of meetings follows along the lines indicated for HAZOP-type reviews, although the time requirements are slightly reduced because the analysis is focused on failures that individuals are experienced with. For instance, CCPS[19] suggests that for simple systems it may take 2 to 6 hours of leader preparation, 1 to 3 days for the team review, and 1 to 3 days for the writeup incorporating team comments. On the other hand, large and complex systems may take 1 to 3 days preparation time, 1 to 3 weeks to evaluate, and 2 to 4 weeks to document.

Prior to the actual review, or as part of the start of the review, the team members should physically "walk through" the facility to become familiar with the actual layout of the equipment and its relationship with other equipment. If reviewing a new design, the design engineer should mentally "walk" the team through the process describing the system and the functions of each major piece of equipment in the system.

6.7.3 Conducting the FMEA—How Is It Done?

To conduct the analysis the facilitator will lead the team through a series of steps starting with the identification of each major piece of equipment, the possible failure modes, the effects of that failure, and the associated risks as well as recommending action items as appropriate. The results of the analysis is documented on a form similar to Fig. 6.14.

Step 1. **Identify and Describe the Equipment.** Either prior to the meeting or as the first step in the review the major equipment items are listed in the order in which they appear on the P&ID or process flow chart. As the review progresses, each piece of equipment can be "checked off" as each of its failure modes have been evaluated. The description usually characterizes the equipment functionality such as air valve safety solenoid. Equipment numbers are sometime available and can provide an easy additional identifying reference, such as steam safety solenoid valve V-431. The key is to ensure that the equipment is identified in ways that the reviewers can understand and is usable later by those who must act upon the recommendations of the FMEA.

Step 2. **Determine the Failure Modes.** For each component or piece of equipment listed, determine all of the possible failure modes that are consistent with the equipment operation. The nature of the failure should be realistic and credible. For instance, a typical failure mode for a solenoid valve could be failure to open when signaled. Common types of failure modes are illustrated in Table 6.6.

Failure Mode and Effects Analysis

Process/ System Reviewed: ____________ Facility: ________ Review Date: ________

P&ID#: ________ Reviewers: ____________ Page: of

Component Description	Failure Mode	Effects	Risk Probability/Severity	Recommended Action

Fig. 6.14 Example of failure mode and effects analysis form.

TABLE 6.6 Typical Failure Modes

Type of failure mode	Typical example
Failure during operation	Valve, pump, wire, computer, utility fails while operating
Failure to operate when signaled	Equipment fails to start
Failure to stop when signaled	Equipment fails to stop
Premature operation	Equipment starts unexpectedly
Specification failure	Equipment operates outside design specifications

A solenoid valve, relief valve, and water pump in a water tempering system may fail in several different ways, as illustrated in Fig. 6.15.

For relatively simple systems, all of the failure modes can be determined prior to analyzing the effects of those failures. For large, complex systems it is probably best to consider each component and its associated failures sequentially so as not to lose the meanings of the discussions, associations, and connections. Some organizations list the cause of those failures to describe more fully those situations and provide a basis for reviewing and understanding a later incident that has occurred that was not predicted by the analysts.

Step 3. **Determine the Effects of the Failures.** Now having a list of what failures can occur, the review team then determines in a careful and considered way the effects of those failures. The credible downstream effects should be determined as well as immediate effects. This reality check keeps the analysis within manageable bounds as well as having the practical effect of keeping the review team focused on the most likely and credible scenarios.

Using the example started above, the failures of the valves and pump could credibly result in effects illustrated in Fig. 6.16.

The review can continue sequentially along the P&ID path of component failures and effects until all these scenarios are considered. Judgments around the associated risks can be made concurrently for complex processes or the team may decide to do this analysis after all the failure effects have been determined.

Step 4. **Judge the Risk.** At this point, the review team then makes a judgment about the risks around each of the equipment failures and the associated effects of those failures. The concepts embodied in the risk matrix (Fig. 6.2) again are used. What is the likelihood of that failure? How severe is the resulting effect? The team will be using their experience to develop qualitative responses to each of the scenarios. Having known or estimated equipment failure rates can lead to a quantified assessment where needed. Common definitions can help the team members calibrate and align along these judgments (see risk matrix section for further discussion).

Although the FMEA is not ideally suited to analyze combinations of failures, the team can selectively approach scenarios that have happened in the past or the team expects could easily happen in the future. Attempting analysis on many of these combinations leads to an enormous number of scenarios and a task difficult to manage.

Failure Mode and Effects Analysis

Process/ System Reviewed: Tempering System for Eye Wash Water Supply System   Facility: Mix Plant C   Review Date: July 8, 1997

P&ID#: D- 4598 Rev. B   Reviewers: John P., Claire S., Tony M., & Frances T.   Page: 4 of 7

Component Description	Failure Mode	Effects	Risk Probability/Severity	Recommended Action
C. V-431: Steam Valve Safety Solenoid - Normally Open	1. Closes unexpectedly while operating 2. Fails to open when signaled (sticks open) 3. Fails to close when signaled (sticks closed) 4. Valve leaks to surroundings			
D. RV- 36: Relief Valve on Heater set for 90 psig	1. Fails to operate when needed (i.e. heater exceeds 90 psig & valve sticks closed) 2. Opens unexpectedly (i.e. heater at less than 90 psig)			
E. P-236: Water Supply Pump - Normally Operating	1. Fails unexpectedly while operating 2. Fails to stop when signaled (keeps pumping) 3. Fails to start when signaled (doesn't pump)			

Fig. 6.15 Example of a partially completed failure mode and effect analysis.

Failure Mode and Effects Analysis

Process/ System Reviewed: Tempering System for Eye Wash Water Supply System Facility: Mix Plant C Review Date: July 8, 1997

P&ID#: D- 4598 Rev. B Reviewers: John P., Claire S., Tony M., & Frances T. Page: 4 of 7

Component Description	Failure Mode	Effects	Risk Probability/Severity	Recommended Action
C. V-431: Steam Valve Safety Solenoid - Normally Open	1. Closes unexpectedly while operating 2. Fails to open when signaled (sticks closed) 3. Fails to close when signaled (sticks open) 4. Valve leaks to surroundings	1. Loss of tempering for incoming water. Cold water in system. 2. Same as above. 3. Potential of scalding eye wash water 4. Steam leak in small enclosed area.		
D. RV- 36: Relief Valve on Heater set for 90 psig	1. Fails to operate when needed (i.e. heater exceeds 90 psig & valve sticks closed) 2. Opens unexpectedly (i.e. heater at less than 90 psig)	1. Potential rupture of heater. 2. Release of hot water in small enclosed area.		
E. P-236: Water Supply Pump - Normally Operating	1. Fails unexpectedly while operating 2. Fails to stop when signaled (keeps pumping) 3. Fails to start when signaled (doesn't pump)	1. No water at all to eye wash system. 2. Potential injury when attempting repairs. 3. same as 1. above		

Fig. 6.16 Example of a partially completed failure mode and effects analysis.

Step 5. **Recommendations for Action.** Based on the level of risks determined for each of the situations, the team then makes recommendations. The discussions and related brainstorming at each step has usually generated a list of rich, ripe, cogent suggestions. Again, as with each of the hazard review methodologies, the team needs to guard against re-engineering the process or solving each of the problems. On the other hand, the experience and shared knowledge of the group can and does lead to excellent recommendations for solution.

In the failure mode and effect analysis method, the nature of the review generally leads to ways to improve equipment reliability. This can take the form of a better preventive maintenance program, equipment redundancy, more robust equipment selection, or additional alarms/shutdown features.

The following FMEA tabular report (Fig. 6.17), continued from above, illustrates some examples of risk judgments and recommended actions.

6.7.4 Reporting the Results

The information developed during the review needs to be communicated in a manner similar to other hazard analyses. An executive summary that highlights the important findings is directed towards the facility or department manager. The person(s) who will act on the detailed recommendations need to know the specifics. The tabular nature of the worksheets with associated equipment numbers assists in clarifying who needs to do what regarding what equipment. A periodic status report is usually issued until all of the findings are considered and resolved.

6.7.5 FMEA Summary—Pros and Cons

The failure mode and effects methodology is particularly well suited for equipment-oriented systems that have little or no human interface. The analysis is generally limited to single failure modes that directly result in an incident or accident. A complex mechanical or electrical system that has multiple operator interface opportunities is beyond the scope of FMEA methodology. The method is largely equipment oriented, and having good records and experience with the equipment failure rates specific to the system under review is necessary. Trades and maintenance engineering personnel can be quite helpful in this regard. The HAZOP approach provides a systematic review of all potential equipment failure modes throughout the system. The effects of each failure is considered separately and then in selected combinations of other failure modes.

6.8 FAULT TREE ANALYSIS

6.8.1 FTA Overview

A fault tree analysis (FTA) is a deductive approach to risk assessment. Whereas failure mode and effects analysis describes what can happen using inductive reasoning, the fault tree methodology, starting with the unwanted event, deduces how it can happen. Some authors make the distinction between hazards *identification* techniques such as HAZOP and FMEA and hazard *analysis* methods such as fault tree. The fault tree analysis technique was reported[20] to have been developed by H. A. Watson of the Bell

Failure Mode and Effects Analysis

Process/ System Reviewed: Tempering System for Eye Wash Water Supply System   Facility: Mix Plant C   Review Date: July 8, 1997

P&ID#: D- 4598 Rev. B   Reviewers: John P., Claire S., Tony M., & Frances T.   Page: 4 of 7

Component Description	Failure Mode	Effects	Risk Probability/Severity	Recommended Action
C. V-431: Steam Valve Safety Solenoid - Normally Open	1. Closes unexpectedly while operating 2. Fails to open when signaled (sticks closed) 3. Fails to close when signaled (sticks open) 4. Valve leaks to surroundings	1. Loss of tempering for incoming water. Cold water in system. 2. Same as above. 3. Potential of scalding eye wash water 4. Steam leak in small enclosed area.	1. Unlikely / Minor 2. Remote / Minor 3. Unlikely / Very Serious 4. Unusual / Minor	1. Ensure Preventive Maintenance program is fully implemented 2. Same as above 3. Install Hi Temp Alarm 4. Same as 1 above
D. RV- 36: Relief Valve on Heater set for 90 psig	1. Fails to operate when needed (i.e. heater exceeds 90 psig & valve sticks closed) 2. Opens unexpectedly (i.e. heater at less than 90 psig)	1. Potential rupture of heater. 2. Release of hot water in small enclosed area.	1. Remote / Catastrophic 2. Unlikely / Serious	1. Increase safety valve maintenance frequency 2. Same as above - safety valve maintenance frequency
E. P-236: Water Supply Pump - Normally Operating	1. Fails unexpectedly while operating 2. Fails to stop when signaled (keeps pumping) 3. Fails to start when signaled (doesn't pump)	1. No water at all to eye wash system. 2. Potential injury when attempting repairs. 3. same as 1. above	1. Remote / Very Serious 2. Extremely Remote / Very Serious 3. Remote / Very Serious	1. Install Lo pressure Alarm 2. None required 3. Install Lo pressure Alarm as stated above

Fig. 6.17 Example of a partially completed failure mode and effects analysis.

Telephone Labs in the early 1960s. The fault tree itself is a graphical representation of the various combinations of events that can result in a single selected accident or incident. The fault tree analysis starts with the hypothetical unwanted event such as a fire, an amputation, or a chemical release. All the possibilities that can contribute to that event are described in the form of a tree. The branches of the tree are continued until independent initiating events are reached. The strength of the fault tree methodology lies in its visual representation of the combinations of basic equipment and human failure modes that can lead to the unwanted event. If the failure modes have known or estimated failure rates and probability data, then the data can be applied to the tree using Boolean algebra, resulting in a quantitative determination of the top event probability.

The fault tree itself is a graphical representation of Boolean algebra using logic symbols (i.e. AND gates, OR gates, INHIBIT gates) to break down the causes of the Top unwanted event into basic or primary failures and errors. These basic events generally have known equipment failure rates (i.e. pump break downs per hour of use, switch cycle rate failures, weld failure rates) and human failure rates (i.e. errors per attempt, trained vs. non-trained, high stress vs. low stress, complex vs. simple tasks). The analysis begins with the unwanted or Top event. The immediate causes of that event are then identified. The relationships of these immediate causes are shown on the tree through the use of the appropriate connecting gates and events. The basic gates are "OR" and "AND". These gates denotes a relationship of the state of one or more other events. Each of the immediate causes is then analyzed in a similar fashion until the basic initiating events have been determined. The relationships between these basic events and the unwanted Top event are then displayed by the resulting Fault Tree.

The tree may then be evaluated qualitatively or quantitatively. A minimal cut set (combination of failures) is the smallest set of primary events which must occur in order for the Top unwanted event to happen. They are important because they represent the errors or faults which must be changed in order for the Top event to change. Using actual or estimated failure rate data allows the analyst to quantify the minimal cut sets. Computer simulation can assist with the more complex trees.

6.8.2 FTA—Getting Started

The fault tree analysis methodology requires analysts skilled in the technique as well as being generally familiar with the process to be reviewed. Construction of the tree does not easily lend itself to the committee approach, but the results are reviewed by individuals who have unique knowledge of the system, process, and equipment being analyzed. In practice the analyst, with knowledge of the system, constructs the tree, the experts review the tree and associated logic, the tree is modified and reviewed until the primary faults and associated logic is agreed upon. In complex situations, the hazard review team members are assigned to develop a particular tree or section of a tree based on their expertise. Those developed tree sections are then reviewed for correctness and logic by the entire team members.

In order to conduct an analysis, an unwanted top event is selected. This is usually accomplished by using other hazard analysis or identification techniques or by the experienced, intuitive knowledge of the process hazards. The top event analysis is usually focused on a particular piece of equipment or certain part of the process. The design parameters and operating procedures at that part of the process must be well known in order to construct a credible tree. On the other hand, a well-constructed tree can reveal or underscore the fact that there is insufficient knowledge to make a credible risk determination. More data would need to be developed to proceed further. If a quantitative analysis is desired, then failure rate data for the equipment under review need to be available or realistically estimated.

The construction of a fault tree which focuses on one out of many possible system hazards can be time consuming and resource dependent. Powers and Lapp[21] suggest that a single chemical process can generate over 50 hazardous events. Each event would require two to three man-days to analyze. They report that the U.S. Atomic Energy Commission study (WASH-1400) required over 25 man-years to complete fault trees for one boiling water reactor and one pressurized water reactor. The Guidelines for Hazard Evaluation Procedures[22] estimates that modeling of a single Top event of a simple process with an experienced team could be done in a day while complex systems could require weeks or months. For simple/small systems they further suggest times of 1 to 3 days of preparation, 3 to 6 days of tree generation, 2 to 4 days of qualitative evaluation and 3 to 5 days for documentation. For complex/large systems, equivalent requirements would be 4 to 6 days, 2 to 3 weeks, 1 to 4 weeks and 3 to 5 weeks.

6.8.3 Conducting the FTA—How Is It Done?

To generate a Fault Tree the analyst must know and understand the symbols and conventions associated with this method of hazard analysis.

Basic Fault Tree Symbology. Events and gates: Table 6.7 shows some of the basic symbols and associated meanings used in the construction of fault trees.

Basic Fault Tree Terminology. The analyst should be familiar with the following terminology and definitions. The terms have different meanings and are used to prompt different actions during the generation and analysis of the fault tree.

Fault A term indicating a human or equipment malfunction that can be "self-correcting" once the situation(s) causing the malfunction is corrected. The component operates at the wrong time due to an upstream command error. For example, an electrical switch in the wrong position because of power inadvertently applied has sustained a fault.

Failure A term indicating a human or equipment malfunction that needs to be repaired before the component can successfully operate again. The component does not operate properly when called upon to do so. For example, an electrical switch mechanically stuck in the wrong position has sustained a failure.

Categories of Failures and Faults Faults and failures are grouped into three classes: primary, secondary, or command, to assist in assuring that the analysis drives to basic, primary faults and failures.

- *Primary* is a component fault in an environment for which the component was designed. Usually attributable to a defect in the failed component. For example, failure of a compressed air receiving tank designed for 200 psig failing at 100 psig or a failure of an electrical switch, in a wet environment, designed for that situation.
- *Secondary* is a component fault or failure in an environment for which it was not designed. This is usually attributable to an external force or conditions. Similar

TABLE 6.7 Gate and Event Symbols

Event symbol	Event name	Event symbol meaning
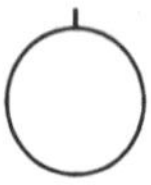	Basic event	The circle indicates a basic, initiating event or fault. One that needs no further development. Failure or error rate data can be determined or estimated.
	Undeveloped event	The diamond indicates a fault that is deliberately not developed further because of low probability or low consequence or little information. Also represents human error.
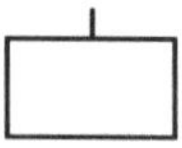	Intermediate event	The rectangle indicates a fault that is to be developed further.
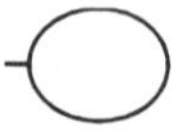	Conditional event	The oval indicates restrictions or specific conditions when used with an inhibit gate.
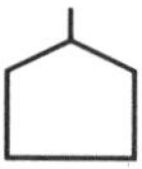	House event	The house represents an event that is occurring or not occurring. It is either on or off. An assumed boundary condition for the fault tree.

Gate symbol	Gate name	Gate symbol meaning
output  inputs	AND gate	For an AND gate, the output occurs only if all of the input events occur.
output  inputs	OR gate	For an OR gate, the output occurs if any one of the input events occur.
output 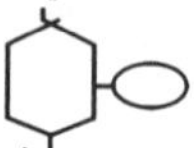 input	INHIBIT gate	For an INHIBIT gate, the input produces output when a conditional event occurs.

to the discussion above, examples would be failure of the 200 psig designed air receiver at 300 psig or an electrical switch not designed for a wet environment failing during a water wash down.

- *Command* is a fault involving the proper functioning of a component although at the wrong time. Usually attributable to a fault in commanding component. An example could be a machine operator starting a piece of equipment at the wrong time.

Fault Tree Construction. In addition to being familiar with the symbology and terminology conventions, the analysis also includes the determination of the immediate and necessary causes for the event to occur. The logic-driven nature of the analysis requires that the immediate causes of the unwanted effects be determined in a sequential and stepwise nature. The immediate causes for the unwanted event become intermediate faults that are then sequentially analyzed until the events become basic or primary failures. Early identification of basic events could indicate that the analysis jumped over several intermediate causes or that the needed complexity of the fault tree approach was not warranted.

There are generally three major steps in the fault tree methodology. The first step is the determination of the top unwanted event and the boundaries of the analysis; the second step is the construction of the fault tree itself; and the third step is the evaluation of the tree. Managing the documentation, review, and followup actions are similar to other hazard evaluation techniques.

Step 1. **Top Unwanted Event and Analysis Boundaries.** Determination of the top unwanted event usually comes out of another inductive hazard evaluation technique such as HAZOP or What If methods. Those approaches will help uncover *what* can happen, whereas the fault tree approach can help specify *how* that event can occur. Experienced analysts can also intuitively determine those situations for which a more focused review is necessary. One danger in using the fault tree method exclusively lies in the possibility of not analyzing potential unwanted events because they were not identified in the first place.

The top unwanted event must be sufficiently defined so that the evaluation can be conducted efficiently and results credible. A top event of "multiple plant fatalities" starts the analysis in a broadly scoped, poorly defined manner. On the other hand, an event defined such as "fatality due to explosion in the nitration reactor during startup operations" allows the analysis to become focused quickly and the team can productively apply their knowledge and use their time efficiently.

Clearly defining the top event also helps clarify and establish the analysis boundaries. Limiting the analysis to the "fatality" described above would exclude analysis of situations like those during unattended operations, minor upsets that would not result in an explosion or shut down/maintenance operations. Defining the physical and operational boundaries as well as the level of detail and assumptions also allows the team to be clear on the actual bounds and scope of the analysis. Does the analysis go beyond the physical bounds of the reactor? beyond normal operations? beyond simple reactor failure into modes of failure such as weld failure, corrosion effects, jacket limits, brittle failure, etc.? How about utility failures, reactor batch size changes, or status of agitator, pumps, valves? How about mitigation possibilities such as emergency response person-

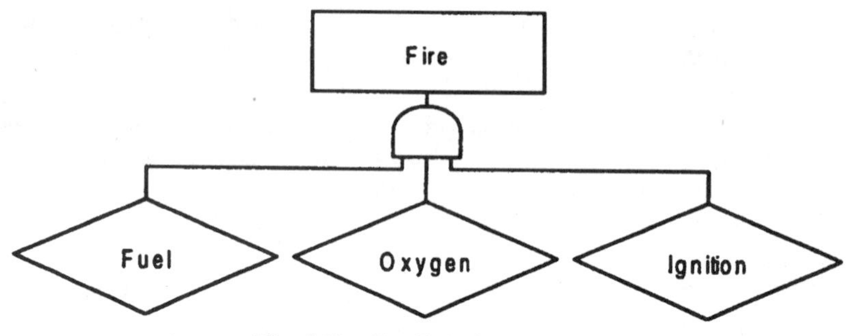

Fig. 6.18 Simplistic fire fault tree.

nel? The clearer the definition of the review, the clearer the results of the analysis. On the other hand, these clarifications may point out all the "other" situations that also may require analysis.

Step 2. **Fault Tree Construction.** Starting with the top unwanted event, the analyst then constructs in a logical, sequential fashion all the contributing fault events until the basic or primary events are uncovered. When completed, the tree should show a cause and effect relationship and trace the primary events to the top unwanted event through a series of intermediate events or faults. The construction of the tree commences in a deductive, cause and effect fashion. Simplistic immediate causes of a top unwanted event such as a "fire" would be (1) fuel and (2) oxygen and (3) ignition source. The fire triangle! (Fig. 6.18)

Another simple fault tree would be the *evaluator* described in Chapter 5 and shown in Fig. 6.19. The unwanted event, the injury or the accident, is a combination of (1) an unsafe act and/or (2) an unsafe condition. Those immediate causes are the result of six (6) underlying factors. For improper acts they are inadequate knowledge, insufficient ability, or improper motivation. For inappropriate workplace conditions they are improper design, inadequate maintenance, or improper actions of another person. The model does not explicitly show the AND or OR gates, but they are implied in its use. For example, in many situations, an injury is the result of several factors, and therefore the use of an AND gate would be appropriate.

If any *one* of the immediate causes results directly in the top event (or any of the intermediate events for that matter), they are connected via an OR logic gate. On the other hand, if *all* the immediate causes are required for the event to occur, they are connected via an AND gate. In the fire triangle example, all three factors are required for a fire (i.e., ignition source, fuel, and oxygen); therefore, they are connected to the top event via an AND gate. On the other hand, there could be several sources of ignition, such as smoking, static, welding and nonrated electrical equipment. These would be connected via OR gates, as illustrated in Fig. 6.20.

Other symbols allow the analyst to show conditions, such as the INHIBIT gate and CONDITIONAL event. The analyst proceeds until all the intermediate events have been developed into the basic or primary faults that need no further breakdowns. The construction should be methodical, logical, and systematic.

An excerpt from a fault tree analysis is shown in Figs. 6.21 and 6.22 as described by Bass.[23] The tree reflects a portion of the analysis of a high-pressure oxidation system used

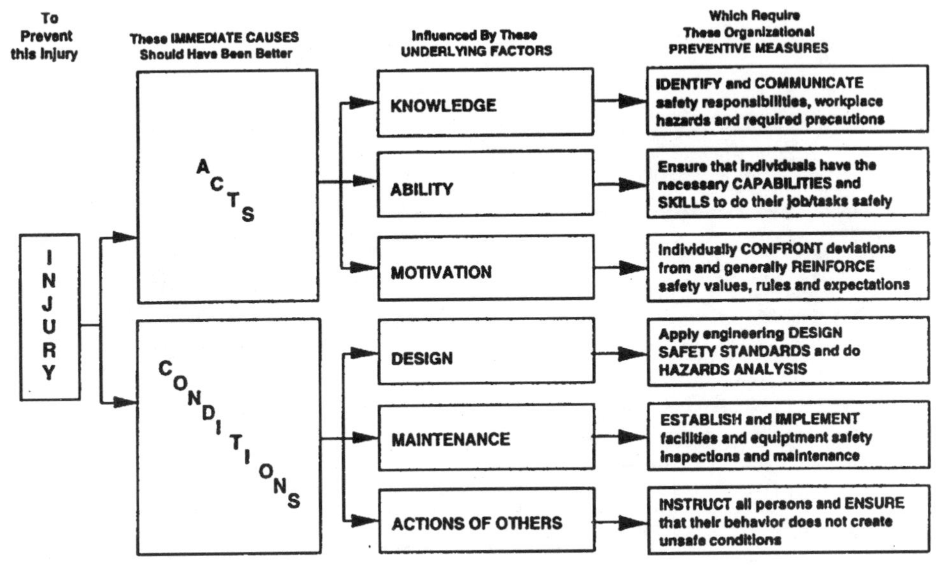

Fig. 6.19 The evaluator.

to perform high-volume wafer oxidation. Because of hydrogen and oxygen are used in this system, the potential for an explosion was evaluated. To guard against a detonation capable of causing the pressure vessel to rupture, several safety devices are used.

Step 3. **Analysis of Fault Tree.** With the completed fault tree, information is visually displayed showing how the individual failures and specific faults can combine and

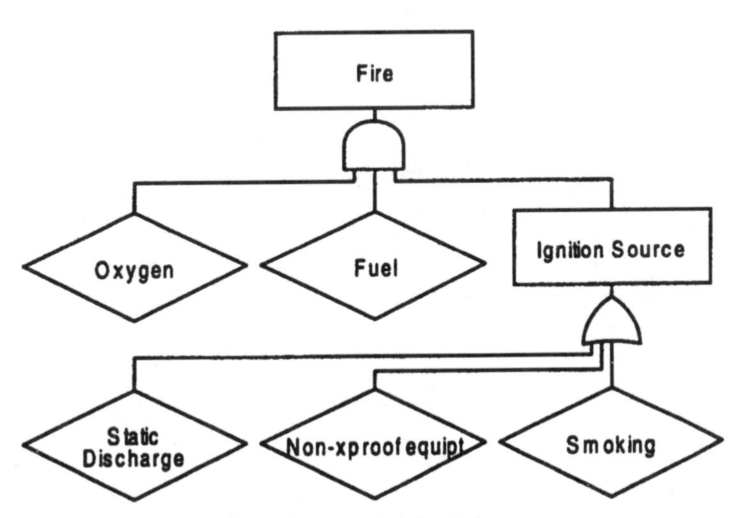

Fig. 6.20 Expanded fire fault tree.

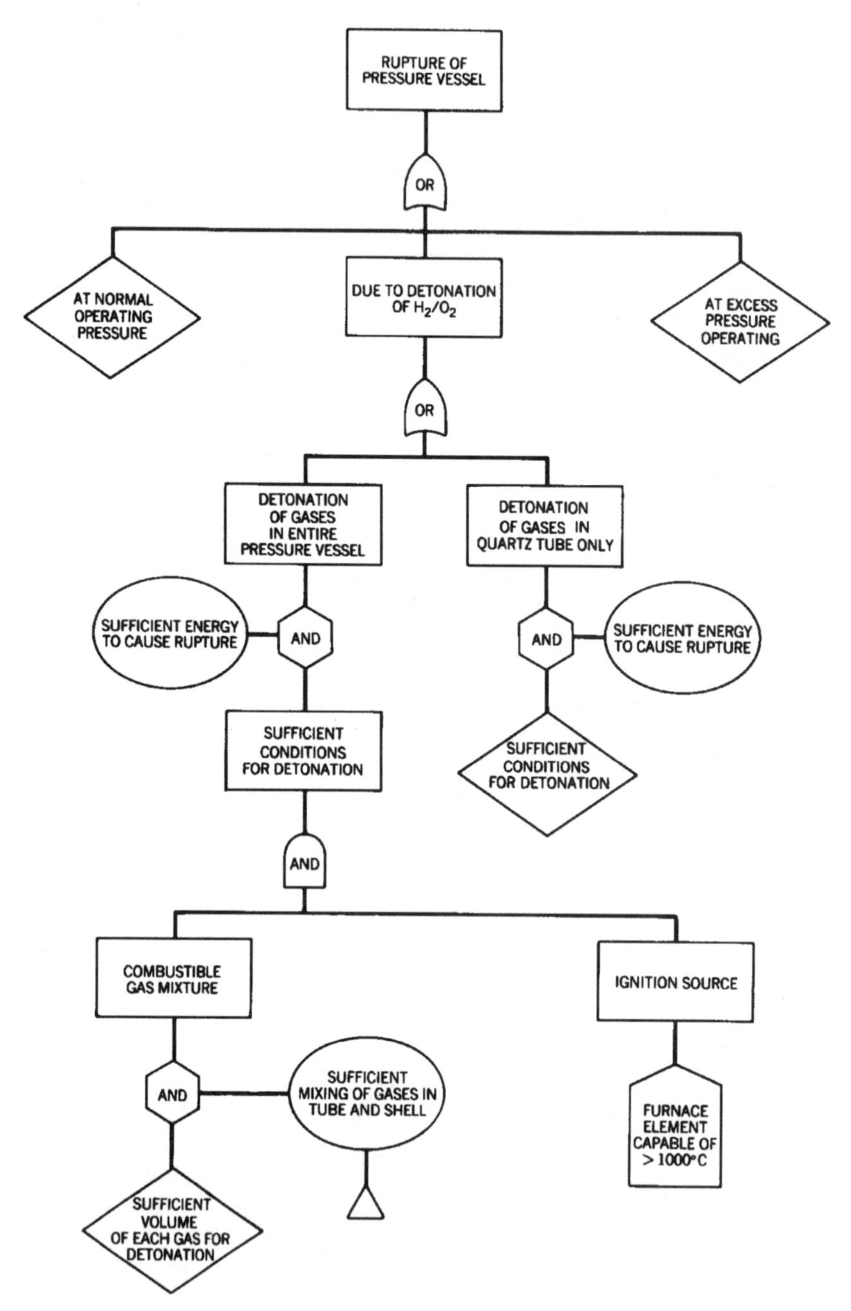

Fig. 6.21 Fault tree analysis—high-level pressure oxidation system.

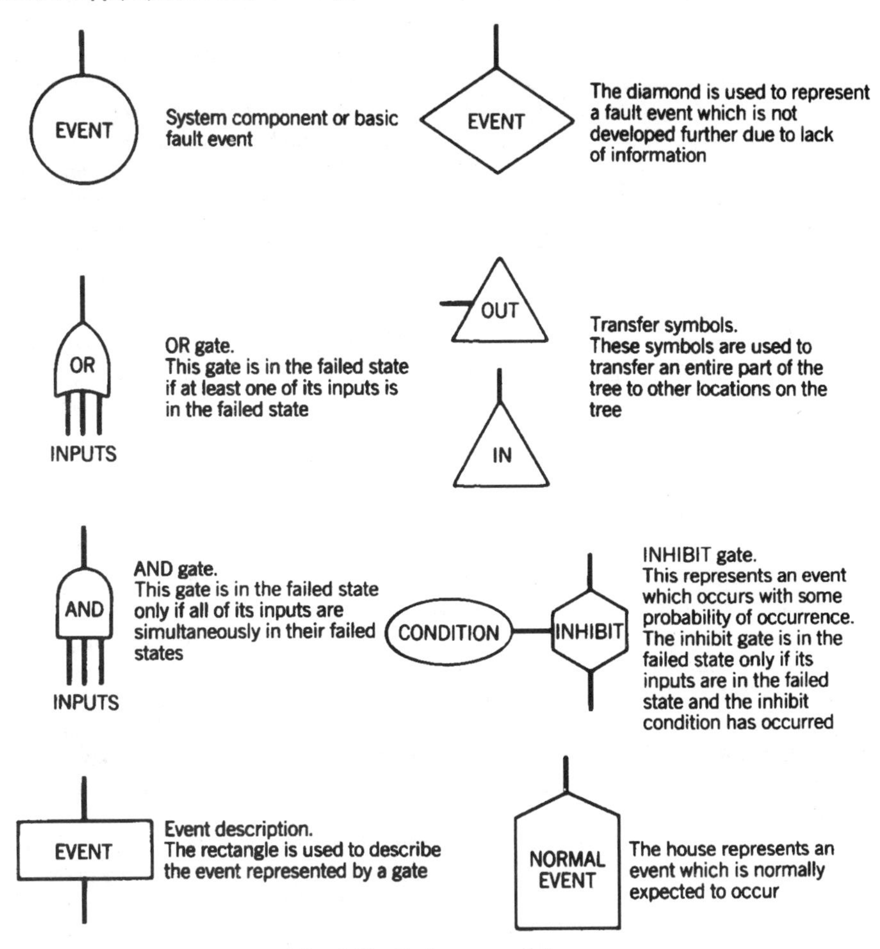
How to Apply System Safety to Occupational Safety and Health
EVENT
System component or basic fault event
EVENT
The diamond is used to represent a fault event which is not developed further due to lack of information
OR
INPUTS
OR gate.
This gate is in the failed state if at least one of its inputs is in the failed state
OUT
IN
Transfer symbols.
These symbols are used to transfer an entire part of the tree to other locations on the tree
AND
INPUTS
AND gate.
This gate is in the failed state only if all of its inputs are simultaneously in their failed states
CONDITION
INHIBIT
INHIBIT gate.
This represents an event which occurs with some probability of occurrence. The inhibit gate is in the failed state only if its inputs are in the failed state and the inhibit condition has occurred
EVENT
Event description.
The rectangle is used to describe the event represented by a gate
NORMAL EVENT
The house represents an event which is normally expected to occur

Fig. 6.22 Fault tree symbology.

cause the incident or accident, the unwanted event. Determining all these combinations, called the *minimal cut sets*, that will result in the top unwanted event is the next step. If the minimal cut sets are readily apparent, the use of the fault three methodology is probably not warranted. The minimal cut sets are useful for listing the various ways in which the event can happen. For large fault tree analysis, minimal cut sets are best determined by computer programs. Computer programs also can provide a convenient method of quantifying the tree.

There are many sources that provide detailed guidance on the creating of minimal cut sets and quantitative analysis. In general, the methods[24] involve the (1) identification of all of the logic gates and basic events, (2) resolution of all gates into the sets of basic events, (3) removing all duplicate events within the sets, and (4) deletion of all sets within sets.

The minimal cut sets then are evaluated to determine where the failures and faults can most readily result in the top unwanted event. If the basic or primary failure rates are known or can be estimated, they can then be used to quantify the resulting analysis. Typical human and equipment failure and fault probability data are available in the literature[25] and include failure rate per hour and failure rate per cycle. Presenting the resulting quantitative analysis as an absolute number should be avoided unless presented in context and used to compare with other quantitative analyses that were generated in a similar fashion. The quantitative results have their greatest value in comparing the relative risk of different options for risk reduction measures.[26] Additionally, the analyst needs to ensure that the fault events are independent in nature. In some cases, a "common cause" will increase the likelihood of the top event occurring. That is, if all events in a minimal cutset can occur due to the same event, then the top event will occur with this single cause. For example, if all the pumps in a cut set have a common power source, they are subject to the "common cause" of power failure, or similarly if all instruments in a cut set have been (mis)calibrated, they are subject to the same "common cause" failure.

6.8.4 Reporting the Results

An executive summary that highlights the important findings of the analysis is directed towards the facility or department manager. Complex trees will need an explanation of the major contributors to the unwanted event including the lists of the minimal cut sets and their significance. The basic tree and associated cut sets usually does not present self-evident findings like the tabular form of What If analysis (WIA) or hazard and operability (HAZOP) studies. The associated explanatory information therefore needs to be carefully constructed so that the results can be understood, evaluated, and critiqued. When the results are well documented, including the assumptions, boundaries, and explanatory background, the analysis can be more easily updated as the hazards and associated risks change.

Recommendations for risk reductions based on the results of the analysis can be shown in the form of a redrawn fault tree showing the proposed controls. The fault tree can then be used visually and numerically to display the effects of the recommendations on the risk.

6.8.5 FTA Summary—Pros and Cons

The fault tree analysis process provides a rigorous, methodical, structured way to evaluate how a specified accident or incident can occur. through deductive reasoning, the analyst can identify the causes and the resulting effects capable of creating the undesired event. The visual nature of the analysis provides the vehicle for communicating a large amount of complex information in a fairly efficient fashion. In addition to allowing for qualitative analysis, the logical, connected structure provides an ideal situation for quantification of results if failure rates and fault data are known.

The tree approach allows for the analysis to focus on a very specific, identified hazard. Virtually all the ways in which the unwanted event can occur can be identified, analyzed, quantified, and eventually controlled. It has a important side benefit of improved understanding of how a particular system works and how it fails.

On the other hand, depending on the complexity of the system and level of detail needed, this approach can be very time consuming. It also requires an experienced analyst who works alone for much of the time until the results are periodically reviewed. Keeping a team energized to construct and analyze a tree as a committee is very hard work.

Additionally, the validity of failure rate data must be established if the tree is to be quantified. Although there are many data available, the analyst must use care in applying the data directly because the equipment use, surrounding conditions, and preventive maintenance programs vary widely.

REFERENCES

1. Lowrance, William W., *Of Acceptable Risk*, William Kaufman, Los Altos, CA, 1976.
2. Center for Chemical Process Safety of the American Institute of Chemical Engineers, 345 East 47th Street, New York, NY 10017.
3. 29 CFR 1910.119, *Process Safety Management of Highly Hazardous Chemicals*, Occupational Safety and Health Administration, Washington, DC, 1992.
4. 29 CFR 1910 Subpart I, *Personal Protective Equipment Standards for General Industry*, Occupational Safety and Health Administration, Washington, DC 1994.
5. 40 CFR 68, *Accidental Release Prevention Requirements: Risk Management Programs*, Environmental Protection Administration, Washington, DC, 1996.
6. Lowrance, William W., *Of Acceptable Risk*, William Kaufman, 1976.
7. Kaplan, Stanley, and Garrick, B. John, "On the Quantitative Definition of Risk," *Risk Analysis*, **1**(1), 1981.
8. Kletz, Trevor A., *Hazop & Hazan*, Institution of Chemical Engineers, Second Edition, 1986.
9. National Safety Council, *Accident Prevention Manual for Industrial Operations*, 1993, Ninth Edition.
10. Adapted from Bell, Trudy E., "Managing Murphy's Law: Engineering a Minimum Risk System," *IEEE Spectrum*, June, 1989.
11. Adapted from *Guidelines for Hazard Evaluation Procedures*, The Center for Chemical Process Safety, AIChE, 1985.
12. National Safety Council, *Accident Prevention Manual for Industrial Operations*, 1993, Ninth Edition.
13. Knowlton, R. Ellis, *An Introduction to Hazard and Operability Studies, The Guide Word Approach*, 10th printing, 1992.
14. *Guidelines for Hazard Evaluation Procedures*, Second Edition with Worked Examples, Center for Chemical Process Safety of the American Institute of Chemical Engineers, New York, 1992.
15. Freeman, Raymond A., Lee, Roberto, and McNamara, Timothy P., Monsanto Company, "Plan HAZOP Studies with an Expert System," *Chemical Engineering Progress*, August, 1992.
16. Sweeney, Joseph C., Arco Chemical Company, "ARCO Chemical's Hazop Experience," *Process Safety Progress*, April 1993.
17. Greenburg, Harris R., and Cramer, Joseph J., Stone & Webster Engineering Corporation, *Risk Assessment and Risk Management for the Chemical Process Industry*, Van Nostrand Reinhold, 1991.
18. *Guidelines for Hazard Evaluation Procedures*, Second Edition with Worked Examples, Center for Chemical Process Safety, American Institute of Chemical Engineers, New York, 1992.
19. *Guidelines for Hazard Evaluation Procedures*, Second Edition with Worked Examples, Center for Chemical Process Safety, American Institute of Chemical Engineers, New York, 1992.

20. Henley, E. J. and Kumamoto, H., *Reliability Engineering and Risk Assessment*, Prentice–Hall, Englewood Cliffs, NJ, 1981.

21. Powers, G. J., and Lapp, S. A., "Computer-Aided Fault Tree Synthesis," *Chemical Engineering Progress*, April 1976.

22. *Guidelines for Hazard Evaluation Procedures*, Second Edition with Worked Examples, Center for Chemical Process Safety of the American Institute of Chemical Engineers, New York, 1992.

23. Bass, Lewis, *Handbook of Occupational Safety and Health*, Chapter 18, John Wiley & Sons, New York, 1987.

24. After *Guidelines for Hazard Evaluation Procedures*, Second Edition with Worked Examples, Center for Chemical Process Safety of the American Institute of Chemical Engineers, New York, 1992.

25. Lees, Frank P., *Loss Prevention in the Process Industries*, Butterworths, London, England, 1980.

26. Ref. 17.

CHAPTER 7

How to Apply Computer Simulation to Occupational Safety and Health

MAHMOUD A. AYOUB and STEPHEN D. ROBERTS
North Carolina State University, Raleigh, NC 27695

7.1 INTRODUCTION
7.2 SIMULATION AS A PROBLEM-SOLVING METHODOLOGY
7.2.1 The Role of Management
7.2.2 The Simulation Project Team
7.2.3 The System Specification
7.2.4 The Cost of Simulation
7.3 SIMULATION MODELING AND ANALYSIS
7.3.1 Problem Definition and Simulation Objectives
7.3.2 Model Design and Implementation
7.3.3 Data for the Model—Input Modeling
7.3.4 Model Testing, Verification, and Validation
7.3.5 Analysis of Simulation Output
7.3.6 Experimentation with the Simulation Model
7.3.7 Model Documentation and User Manuals
7.3.8 Results Implementation
7.4 SIMULATION FUNDAMENTALS
7.4.1 Monte Carlo Sampling
7.4.2 Generation of Random Numbers
7.4.3 Data Categorization: Cumulative Probability Distribution Function
7.4.4 Drawing Sample Values
7.4.5 Computer Implementation of Distribution Sampling
7.4.6 Computing the Number of Replications
7.4.7 Sampling from Standard Distributions
7.4.8 Spreadsheet Functions Useful in Computer Simulation
7.4.9 Simulation Algorithms
7.4.10 Computer Simulation Languages
7.4.11 Available Languages
7.4.12 Games and Gaming
7.5 APPLICATIONS
7.5.1 Effectiveness of the Safety Organization

Handbook of Occupational Safety and Health, Second Edition, Edited by Louis J. DiBerardinis, ISBN 0-471-16017-2

7.5.2 Evaluation of Maintenance Manuals
7.5.3 A Troubleshooting Model

REFERENCES

7.1 INTRODUCTION

A common approach to safety management is to impose hazard control wherever it can be achieved. This view is based on the questionable assumption that hazard control should be achieved at any cost and by any means, which clearly ignores the compelling limitations of resources and technology. In an era marked by competitiveness, there is great concern that all resources be used efficiently and effectively. Thus hazard control has to be predicated on an optimum allocation of resources. Methods and techniques associated with industrial engineering, operations research, and management science can be used to deal with the complex problems of allocation and scheduling of resources among competitive alternatives.

Some problems of allocating safety resources can be formulated as mathematical models of one type or another—linear and nonlinear optimization models, inventory models, queuing models, economic decision models, and so on. Every class of mathematical models has a general form that is independent of the problem or situation being studied. Each class of mathematical models not only has a prespecified form, but an analytical "solution." The generality of the model class is that the content (decision variables, parameters, constants, etc.) of each model may vary according to the application. Mathematical models have been in use in a variety of application for many years, and there is now available a collection of efficient, computerized algorithms that implement the model solutions. Furthermore, there is commercial software that can be used by those whose interests lie in the utilization of the approaches. For illustrative applications of some of these models, see Ayoub (1979).

The counterpart of mathematical models, computer simulation, is an experimental approach to modeling using computing. However, unlike mathematical models, simulation models do not have a *solution*, and thus solutions are obtained via experimentation with the model. Nonetheless, simulation models have greater applicability than mathematical models, since they make fewer assumptions about the form or structure of the problem being studied. The user builds a computer model of the system that needs study and exercises ("simulates") the computer model by asking "what if" questions. A computer simulation model is highly dependent on the purpose of the simulation study, and the availability and quality of the data used in building the model. Simulation studies use models for the analysis of present systems, but more importantly for making predictions about potential new systems. (Chapter 6 demonstrates a parallel "manual" method for performing similar exercises.)

Engineers have long used physical simulation in many engineering design and research projects. For example, cars and airplanes have been tested in wind tunnels, detailed models of dams show how the dam controls water runoff, and architectural models show how buildings look in three dimensions. Just as these physical models permit experimentation with various design parameters, computer simulation allows designers to vary parameters of the system being modeled on the computer. The growth and success of computer simulation and simulation modeling parallels the development of computer technology. The basic simulation methodology has existed for some time, but it has become a viable tool through the availability of increased computer power, especially at the desktop, and with simulation languages that are "easy to use" by many kinds of users. Also with new advances in computer graphics and pictorial representations, simulation models can now incorporate visual interpretations or animations of the system under study.

Simulation models are composed of entities that employ decision rules through a prescribed course of action, which are undertaken upon the occurrence of certain events. In some cases, simulation models can be built in such a way as to utilize mathematical models, or portions thereof, as an integral part of the model. Simulation, by its nature, is more amenable to online interactive modeling approaches, and many commercial simulation environments take full advantage of windows and graphical representations to ease the burden of modeling and render the output more understandable.

A limiting factor is using modeling to study occupational health and safety systems that are heavily dominated by behavioral components rests with the difficulty of quantifying the decision-making process that produces the behavior. Many worthwhile modeling studies were aborted because the cognitive components of the problems could not be described. Existing simulation technology makes it possible to overcome some of these problems by incorporating the user as an active component of the model, with no attempt to describe what a person (user) will do under any given condition. Instead, the person is presented with a summary or global view of the prevailing conditions in the system, and then is asked to suggest a course of action or actions to be followed during the next time period. In this fashion, several decision-makers can be incorporated in the simulation model to gain insight into system behavior.

In the remainder of this chapter we will present in Section 7.2 the use of simulation as a general problem solving methodology. In Section 7.3, the simulation modeling and analysis method is discussed. Basic simulation techniques are presented in Section 7.4, and in Section 7.5 we present several simulation applications. Throughout, we relate simulation to the general area of occupational health and safety.

7.2 SIMULATION AS A PROBLEM-SOLVING METHODOLOGY

Simulation is both a modeling technique and a problem-solving methodology. Before discussing the simulation technique, it is important to understand the general problem-solving process. Many people who have used simulation take the view that more is gained from the problem-solving approach than from the generation of numbers from the technique. Both are important, however. The general problem-solving methodology used in simulation has its roots in the general area of *systems analysis and systems engineering*. In systems work, a simulation model encompasses the study of a system for the purpose of understanding and improving that system. The focus is on the system, rather than its components, since it is the system impact that is more critical. For the purpose of this survey, a system is a process that promotes occupational health and safety. It might be an accident prevention program, an occupational nurse's office in a factory, a system of guards in the stamping department, an OSHA inspection program, a new production line, or a safety-training program.

At issue will be the analysis of a specific system and its subsequent improvement. This central concern about some system is sometimes referred to as "the design question." Thus it is important that the design question be fully understood by everyone concerned—management, workers, ergonomists, systems analysts, etc. Achieving the full understanding of the basic design question can be an exercise that lasts during the entire lifetime of a simulation project. However, there are some key aspects that must be considered. These include (1) the role of management, (2) the project team, (3) the system specification, and (4) the cost of simulation.

7.2.1 The Role of Management

A simulation model, regardless of its level of sophistication, will ultimately be used to serve management; thus management's support and needs have to be established at the outset and, furthermore, nursed throughout the simulation project. Many excellent and worthwhile models have been shelved because management failed to identify with them, or to perceive their true value and potential.

Management can be activated in the process of model development through a number of mechanisms. Personal interviews can be used to learn various points of view about the issues. Reviews and surveys can be made of typical management decision problems. Workshops can be given where illustrative examples and successful simulation applications are presented and managers' reactions and comments are solicited. Finally, model development can be examined via review of simulation and its animations. Several iterations may be required before management can reach a consensus on what is expected from the model and on the operational requirements for future maintenance and expansion of the model. Management involvement should not end with the definition of model objectives and limitations, but rather should be extended and applied at various levels to other development activities.

7.2.2 The Simulation Project Team

The various tasks associated with a simulation project have to be coordinated and scheduled in a manner that will best utilize the organization resources and assure completion within a given time frame. Furthermore, the simulation project must be relevant to the organization and be done so that the results are meaningful. The failure of many simulation projects is due to the fact that they tend to be carried out by only a few people. Thus personal biases tend to color the direction of the project (the tasks to be started, the time allocated to any given task, sequencing of different tasks, etc.).

For industry or large-scale studies, a team approach is highly recommended. A project leader may be assigned the responsibility of overseeing and coordinating the work of a number of cross-functional team members (system designers, programmers, manual writers and workshop instructors, outside consultants, etc.) assigned to the different tasks. The project team should also contain a representation of various potential user groups within the organization. Having such a comprehensive team will assure that user interests are protected throughout the various phases of model development and implementation. Furthermore, it will eliminate the climate of apprehension and suspicion usually accorded new projects or undertakings by making the users part of the process itself. Upon implementation, model maintenance and upkeep should be assigned to an individual who is thoroughly familiar with the intricacies and details of the project. The selected individual will, in effect, act as the model manager and will be in a position to respond to future user demand and needs.

For small applications and feasibility studies, other project management approaches can be used. However, regardless of the approach used, some degree of control and accountability over the project should be exercised; otherwise, the outcome is likely to be less than acceptable, if not disastrous. Furthermore, it is critical that potential users be involved in every step of the modeling process, regardless of whether the project is done by one person or a team of people.

Finally, an organization may employ a consultant or group of consultants to carry out portions or the entire simulation project. In this context, the organization assumes the role of an observer. Accordingly, when the model is developed and implemented, the organization may not have the in-house capability to work with model expansion and modification—needs that are certain to occur. If staff expertise to respond quickly to management requests is lacking, the credibility and usefulness of the simulation model will be affected, and the model may be abandoned. To overcome some of these difficulties, the organization may need to build in-house expertise in simulation modeling. In some cases, consultants may augment in-house expertise to assure that the model will be developed objectively and in conformance with the best technology currently available. The point to remember is that a simulation model is not developed once and for all. Rather, simulation models are growth assets that are continually modified and changed to meet demands of the changing environment and the characteristics of the systems they portray. For building and maintaining simulation models on a continuous basis, an in-house capability to maintain and modify the system is a must.

7.2.3 The System Specification

While the focus of a simulation study takes a system perspective, the scope of that system must be bounded if the study is to be completed in a reasonable amount of time. There is a tendency to expect too much at the beginning of the study. Often, the project will create an omni-focus, owing in part to the systems orientation. Furthermore, some on the project team will possibly not understand the inherent limitations of any modeling approach like simulation that fundamentally depends on detailed knowledge and data. Instead they may believe that models can be built without information about component interactions present in the system and without data to parametrize the model. Thus persons new to simulation need to be cautioned: (1) to set fairly narrow boundaries on the system model, (2) to be aware that any modeling effort will require detailed knowledge and information, and (3) that all expectations will change and be modified as the project progresses.

It is important to remember that a simulation model is only a *representation* of a system. It is not a substitute for the system. The model cannot, and should not, represent all the details of the system. Great care should be taken in determining what portions of the system need to be studied and why. Without a specific purpose for the model, the modeling activity never ends. Someone can always think of some detail that is missing in the model. Therefore, at the onset of the simulation study (and throughout the study), there must be constant focus on what is the purpose of the simulation. A simulation model should always be designed for a purpose. A simulation project that has no special focus or purpose will rarely produce any results of interest. Furthermore, without a focus, it is unlikely that various users and decision-makers will invest themselves into making the simulation project a success.

There are three specific issues that should be considered early in the project. First are the system boundaries. For example, it is impractical, and unnecessary, to model an entire factory just to improve the safety of the assembly operations. While system boundaries may seem easy to state initially, as the simulation project proceeds, there will be a tendency to include more and more. The entire project team should be cautioned to be alert to "detail creep."

A second area of concern that needs to be considered as early as possible are the "what if" questions to be asked of the model. These will provide the "experimental conditions" for the simulation. The content of the simulation must be such that execution of the simulation provides answers to these questions. Listing the "what if" questions in advance will help determine the system boundaries (or scope) for the model and establish concrete objectives for the simulation study.

Third, in specifying the problem to be modeled by simulation, it is important to establish, in advance of the modeling activity, the criteria by which each "what if" question will be evaluated. Will the primary measures of effectiveness be the number of lost time days, the utilization of inspectors, the production of parts, the risk of accident, etc.? Clearly, the computation of measures of effectiveness will determine what the model does, since it must, at least, produce the measures of effectiveness as output.

Finally, in many applications, a single simulation model may be too restrictive to meet the demands of varying degrees of details and exactness required. It is, therefore, not uncommon to develop a set of models to deal with various scenarios, or "what if" situations. For example, top management interested in defining general policies and procedures would need the model to represent general characteristics of the organization. On the other hand, for an operational manager concerned about meeting schedules and targeted delivery dates, a model reflecting the detailed movement of products, status of machines, and work schedules would be required. While different in focus, both models would reflect the same organizational structure and characteristics and draw upon similar databases.

How many models are needed? The answer can be easy to determine in cases where the decision needs are well defined and clearly stated. In cases where it is difficult to delineate the role for the potential model or models, developers may be forced to perform a feasibility study and develop some prototype models. Such models would then be used to arrive at the specific set of models suitable for the overall needs of the organization. In any event, in the course of defining the architecture of the simulation model or models, the developer should design models that can meet existing demands and yet be potentially flexible enough to meet future demands as well.

7.2.4 The Cost of Simulation

Large-scale simulation models can be quite expensive, since they require a large financial commitment, first for their development and then for their maintenance. Before commencing a simulation study, its cost should be clearly defined. The estimated cost of the study should include development of software (model specification, model design, programming, and testing); hardware requirements (acquisition of special equipment, modification of existing systems, etc.); and staffing for model maintenance and user services.

The total cost of the simulation model should be contrasted with the cost the organization is willing to incur in finding out the implications and consequences of various strategic and planning decisions. These costs include the value the organization places on the information to be generated by the model for a given decision scenario. In almost all cases, the benefits of having the simulation model offset its cost many times over.

Starting a simulation study does not necessarily mean it will be concluded successfully—the model fully developed and implemented. The risk associated with the development process must be appreciated and accepted. It is possible that after the specifications are defined, it may be found infeasible (for technical, political, or other reasons) to pursue the simulation further. The chance of this happening is quite good, especially in the early stages of development. Fortunately, at that point the cost is typically not very great and can be sustained without sending shock waves through an organization's accounting system. However, it takes a disciplined and experienced modeler to terminate the simulation at the right time; waiting too long may make it difficult to scrap the simulation project and to write it off as a small loss to the organization. Therefore, as pointed out previously, the need for detailed and periodic reviews of all aspects of the simulation study cannot be overemphasized.

7.3 SIMULATION MODELING AND ANALYSIS

The development and implementation of a simulation model is a process that encompasses a set of well-defined, but interrelated activities: (1) problem definition and simulation objectives, (2) model design and implementation, (3) data gathering and model parametrization, (4) model testing, verification, and validation, (5) analysis of simulation output, (6) experimentation with the simulation model, (7) model documentation, and (8) results implementation. Each of these is briefly described below. While these activities will typically be engaged in the order presented, they are revisited often during the conduct of any given simulation project. For more detailed coverage of these simulation activities as well as other related topics, see Emshoff and Sisson (1970), Shannon (1975), Fishman (1978), Bratley, Fox, and Schrage (1987), Law and Kelton (1991), and Banks, Carson, and Nelson (1996).

7.3.1 Problem Definition and Simulation Objectives

At this stage we assume that the project team has been established and that there is general agreement on the project focus. Some time should be spent in carefully listing the "what if" questions that need to be addressed. Based on the scope of these questions, there should develop some notion of system boundaries, and measures of effectiveness should be prescribed.

A problem statement should now be developed that states exactly what the simulation project is to produce. The problem statement will establish a clear purpose for the simulation project. The clearer the problem statement is, the better will be the simulation study. A clear problem statement will make clear the scope of the simulation and will serve to determine just how much detail to include in the simulation model.

Simulation objectives should accompany the problem statement by showing how the simulation will deal with the "what if" questions. The objectives set the intent of the simulation study. When the simulation model is exercised, the objectives will establish what experiments to run. Furthermore, objectives will determine what output from the simulation will be analyzed.

7.3.2 Model Design and Implementation

The development of a simulation model should be the next activity. People who are new to simulation often believe that data should be collected before the simulation model is developed. However, if the simulation model is developed immediately after or during the process of defining the simulation project, then the model can be used immediately to help with many other steps in the simulation project. For instance, if the model is developed in advance of data collection, then the model can be used to determine what data are needed and how important the data are to the synthesis of the system being studied. Furthermore, if the model is developed early, it can be constantly revised as more information is gathered and as clearer insight is obtained into the system under study through other simulation activities.

The simulation model should be an expression of the system representing the operations and activities observed. Care should be taken so that the model reflects what actually happens in the system, rather than what should happen. In addition to performance information, several other aspects of the system will be required to define the model completely. These include work and information flow procedures, physical and technological requirements for various production functions, and overall regulatory standards that must be met by the organization.

The representation can take different forms, depending on the computer implementation choice. Computer implementations can range from programming a simulation in a general programming language to employing a special purpose simulation package to a simple application of a spreadsheet (many spreadhseets now have some rudimentary simulation functions). In general, the tradeoff over the range of implementation alternatives is between modeling flexibility against ease of use. Using a programming language like BASIC, FORTRAN, PASCAL, and C++ gives the simulation modeler full control over all aspects of the model. However, that flexibility requires significant programming competency and detailed knowledge of simulation mechanics. On the other hand, if a simulation package can be found that does the job, its use will usually mean the model can be constructed by simply providing some simple structural information and data. Furthermore, the simulation package will automatically provide all the simulation functions. Thus ease of use is greatly enhanced by the "higher-level" simulation implementation alternatives.

Unless the organization possesses exceptional programming and simulation capability, the use of general programming languages to implement a simulation is not a serious choice. More likely, simulation languages and packages that have greater user convenience will be a preferred choice. While it is beyond the scope of this chapter to discuss the choice of simulation languages and packages in detail, there are some general guidelines. Law and Kelton (1991) and Banks, Carson, and Nelson (1996) provide more discussion of various languages.

General *simulation programming languages* are simulation implementation alternatives that require considerable programming skill from the user, but still relieve the user of coding common simulation functions. Common simulation functionality provided by general simulation programming languages would include facilities for managing events, creating and destroying entities and objects, assigning attributes, controlling processes, generating random variates, statistics collection, and run-time monitoring. However, the user still has the responsibility to invoke this functionality, usually through fairly extensive programming or programming-like constructs. General simulation programming languages include SIMSCRIPT (Russell, 1983), SIMULA (Birtwistle et al., 1973), and SmallTalk (Goldberg and Robson, 1983).

Rather than depending on programming to employ simulation functionality, another category of simulation implementation alternatives is the *simulation modeling languages*. These languages are unlike the simulation programming languages because they promote a special modeling view of the systems being considered. For example, a very widespread vehicle is a "network of queues." This structure is used in GPSS (Schriber, 1991), SLAM II (Pritsker, 1995), and SIMAN (Pegden et al., 1995). In this paradigm, active entities flow through a network that consist of service stations where they may wait to be served by a finite set of resources. In general the entities are used to represent demands for service or production, while the resources are used to used to model the service capacity. Queuing results when the demand exceeds supply. Common questions that are posed are how to configure the supply of resources, moderate the demand by entities, and control the flow in the network. Many problems in occupational health and safety are satisfactory modeled using this perspective. Programming inserts and additions to enhance the capability of the language can augment simulation modeling languages. However, these languages have been developed to represent so many modeling concepts that programming is less necessary.

Further there are *simulation packages* that promote even greater structure by containing specific modeling concepts that correspond directly to their modeling domain. For example, ProModel (Benson, 1996) has been developed to make the modeling of manufacturing and production systems direct. For instance, the inclusion of conveyors and milling machines can be directly specified. The concept of "domain-specific" packages has been extended to MedModel (Carroll, 1996) for the modeling of medical and health care systems. Recent interest in *business process reengineering* has sparked a large number of simulations directed at that interest, including Lazzari and Crosslin (1996) and Binun (1996).

In recent years there is a blurring of the lines between all simulation languages and packages and simulation implementations now provide *simulation environments*. Simulation implementations that had a strong modeling perspective now have incorporated programming concepts that yield features similar to simulation programming languages while at the same time offering modeling "addons" that promote special modeling functionality for important areas like manufacturing and communications. Furthermore, simulation packages that were once directed at specific domains have discovered that their technology can be used in other domains and have expanded their concepts to look more like modeling languages.

Finally, the addition of visual model construction and animation has forced greater commonality among all the simulation implementation alternatives. Most simulation vendors now have visual "front ends" to their simulation environments that construct graphical models through windows and dialog boxes, so that users can easily use and learn to use their simulation functionality. Further, most now provide some form of animation visualization so that the execution of the model can be visualized, either by dynamic changing pictures that represent the system or by visual interpretation of the statistics [as examples, see Arena (Markovitch and Profozich, 1996) and AweSim (Pritsker and O'Reilly, 1996).

7.3.3 Data for the Model—Input Modeling

Whatever the simulation implementation chosen, there should be a working version of the simulation model or models that represent the system under study and that produces information that could answer the what if questions posed in the simulation objectives. However, the model should, at this point, be constructed with strictly hypothetical data; so its numerical output should not be seriously regarded. However, there is a lot to be learned about the data from the simulation models.

Notice what data are needed to construct the models. Virtually every simulation project will require data collection, and data collection, especially in the occupational health and safety areas, and it is difficult at best. Therefore, it is important to collect only what data are actually needed. Nothing is more discouraging than to mount an extensive data collection compaign only to find out that the wrong data are collected or there are insufficient data. However, now that the simulation model exists, you can see what data are needed.

Also you can test the importance of the data through the model or models being developed and determine which data are most important. For example, if the time it takes to inspect a part has only a distant bearing on the risk of injury on the job, then do not outfit for an extensive collection of information on the time it takes to inspect the part. Why collect information that allows you to fit an entire distribution when all you need is a rough estimate of the mean?

Some people with limited simulation experience believe that simulation modeling is crippled if the organization does not have extensive data holdings. That may or may not be true. It depends or what data are needed and how critical those data are to the decision-making process imbedded in the simulation model.

When data are available, there are methods of extracting information through informal and formal interview and professional assessment. For instance, it is quite easy to "fit" a beta distribution with only three estimates (optimistic, pessimistic, and most likely) from interviews with experienced personnel. Obviously data collected from databases or from the direct observation of key operations and functions over a reasonable time period are greatly preferred. These data will typically be a more accurate and objective measure of information than those obtained otherwise.

Data provide input to the simulation and supply values for important parameters to the simulation. Very often these data represent quantities that are best described by random variables, and thus the data need to be organized as a probability distribution. The process of finding the proper representation of simulation input is called "input modeling" and provides another modeling challenge to the simulation project. This input is itself a representation of some more complex underlying process that is not being explicitly modeled but represented typically by a random variable with a given distribution.

The area of input modeling has now advanced to the point were there exist a number of software packages, such as ExpertFit (Law and McComas, 1996) and BestFit (Jankauskas and McLafferty, 1996), to especially address this important topic in simulation. In addition, simulation-modeling environments like Arena also have input modeling features. These packages help the simulation modeler identify the input model that best represents the needed input.

7.3.4 Model Testing, Verification, and Validation

A large-scale simulation model encompasses many components, data files, and decision rules. It represents the designer's abstraction of the important and key aspects of the system under consideration. The value of such a model lies in its ability to predict accurately the behavior of the system it portrays. Errors in prediction may arise from two sources: model correctness and model validity.

The analysis of model correctness is referred to as *verification*—determining whether the model is behaving as it should. There are a number of methods for determining if it is functioning correctly. Certain components of the model can be suppressed to see if the model reacts as expected. For instance, reducing variation should force queues to either become very long or nonexistent. Making arrivals very large should produce very high resource utilizations. When the behavior of the model is explained to people who actually work in the system, do they recognize the behavior observed? If a resource is removed or if an unexpected large demand is realized as it may occur in the real system, does the model behave in accordance with the real system?

Testing a simulation model against the performance of the real system is called *validation*. In many ways validation is the most important aspect of system testing because the validation process must be sufficiently convincing to establish the model as a reasonable representation of the real system. Validation is usually accomplished by examining model performance in dealing with diverse decision problems encountered in the past. Testing the model in its entirety—components, data elements, structure, and so on—is usually presented in the literature as a criterion-related validity test. It measures how accurately the model predicts the performance, as well as the response of the system for a given set of input data and policy conditions. In using criterion-related validity, no attempt is made to establish or define the accuracy of any specific component of the model, but rather to define the accuracy of total performance and output.

Other types of tests include content validity tests and face validity tests. A content test seeks to establish that the model components, structure, relationships, and so on are valid and accurate presentations of the system modeled. On the other hand, face validity tests deal with the question of relevancy: Do the model and its output show a high degree of association with the system and its operation? For the model to have validity, the potential user must be able to associate model performance with expected system performance. To define content and criterion-related validity tests, a host of statistical tests can be used, test of means, test of variances, chi-square tests, and so forth.

An important new development in simulation is animation, in which a dynamic pictorial representation of the simulation is viewed. Animations can be constructed from separate models or can be constructed as a part of the main simulation models. Animations can be very helpful in verification and validation to see that the model behaves as it should and to see that it behaves as the real system.

A common pitfall in simulation verification and validation is the notion that the model must be a faithful representation of *all* parts of a system. It cannot, and should not! The model should be verified only against the simulation project objectives and measures of effectiveness determined at the outset of the study. The model need only be sufficiently valid to provide accurate performance measures for the "what if" questions that will be posed.

The processes of verification and validation may require the model to be reworked and revised to correct flaws. The model should not be validated against the same data that were used to provide input parameters. If the data that are used to provide input are also used to validate the model, then all the model does is to replicate its input. Different data should be used for verification and validation than for input modeling. For example, flow time of parts through a factory and the utilization of resources will typically *not* serve as input to the simulation. Thus this kind of information becomes excellent sources of information for verification and validation processes.

Finally, most simulation implementation systems have various facilities to revise and modify simulations. For example, many systems provide for system *tracing* of various simulation entities and objects. Some offer full *interactive run control* or debugging options that provide a means for users to interact with an executing simulation to view and change various specifications. Using these facilities, the model can continuously be revised until its meets the verification and validation standards.

7.3.5 Analysis of Simulation Output

A fully verified and validated simulation model offers the decision-maker a laboratory by which a host of studies can be performed. A decision contemplated by the organization might be carried out by changes in use of resources, adoption of new policies and procedures, or restructuring some functions within the organization. All these can be manipulated and defined by the decision-maker; the question is which one (or ones) would have a significant impact on the overall performance.

However, before the simulation can be used for decision making, its output needs to be carefully examined. Remember that simulation is an experimental technique and there are certain tactical issues present. For instance, how long do you run the simulation to get proper data (this is the sample size question in traditional experiments)? A very important complicating factor in simulation experiments is that the data observed are often correlated. Therefore, the computation of the variance of the means, needed for almost all statistical analysis, is confounded by the correlation. Therefore, it is not sufficient simply to translate the standard deviations to standard errors, as required by the central limit theorem.

To resolve the problems inherent in the output requires that some decisions be made regarding the fundamental statistical collection. First, and most important, is whether the output from the simulation should ignore its startup behavior and focus instead on the steady-state behavior of the system. A *steady-state simulation* is indicated if the system under study does not open and/or close in an orderly fashion and, instead, the system is expected to run indefinably. If the steady-state behavior is desired, then the output must be collected so as to minimize the effect of the startup. Startup information can simply be ignored until steady state obtains, as determined by plots over time of the statistics of interest. The system may also be run long enough so that startup information is swamped by the shear amount of steady-state information. Or the system may be initialized close to steady state so that there is relatively little startup behavior. Assuming startup handled, the steady-state output may be divided into "batches" that are relatively independent. Thus using the batch means as observations the results are independent and identically distributed observations from which legitimate standard errors can be computed.

Some simulation environments like Arena contain software to plot data and their averages over time to help determine when the startup period terminates. Furthermore its statistical collection facilities are sufficiently flexible to capture statistics in batches and to employ the batch means as observations. Correlation analysis of the batch means ensures that they are relatively independent.

It is more likely in occupational health and safety analysis that analysis does not require steady-state methods. For these cases, the systems have natural starting and stopping conditions. These are referred to as *terminating simulations*. Now each simulation run provides an observation of the behavior of the system under the experimental conditions. Thus the complications of steady-state simulations are not relevant in these simulations and the statistical analysis is simplified.

The central output issue remaining is the number of replications for a terminating simulation and the number of batches for a steady-state simulation. Both issues are statistically equivalent, and the number in each case has the same formulation. To determine that number of replications or the number of batches, a pilot study or preliminary simulation is needed so that there exists some estimate of the variance of the mean associated with the measures of effectiveness. Using this known variance of the output and a need for precision in the estimate, the number of replications can be determined statistically.

7.3.6 Experimentation with the Simulation Model

There are two common approaches to experimentation with the simulation model. The most common is *scenario analysis*. Here each of the "what if" questions is posed individually to the simulation model or models. Characteristically each "what if" question indicates a fundamental shift in a policy, a procedure, a technique, etc. that the organization might consider. It will evaluate the value of the proposed change by examining the measures of effectiveness. This kind of experimentation might also be described as "one at a time" changes to the present system. Thus there are two set of results to consider—the results from the present system and the results from the proposed system. The differences can be analyzed statistically to determine if the differences observed have statistical evidence (for example using paired *t* tests). However, more important that statistical differences (which can be obtained by longer or more runs) are the practical differences observed between the present and proposed systems. On that basis should decisions rest.

Scenario analysis allows the project team to explore completely all those "what if" questions posed at the beginning of the study. Through an analysis of each, the project team should gain sufficient insight into the operations to recommend only those changes that can substantially benefit the organization. Further, it is possible to examine several scenarios simultaneously through finding one or more "best" alternatives, using multiple comparisons and *ranking and selection techniques* (Law and Kelton, 1991).

Another form of experimentation that is less common in applied simulation studies is the traditional form of statistically *designed experiments* (for example, see Law and Kelton, 1991). In these experiments, a set of experimental variables is established, typically from the set of "what if" questions posed in the simulation objectives. These experimental variables will be varied in a consistent fashion, and a simulation will be run for each combination (i.e., a *factorial experiment*). Based on an analysis of the designed experiments (actually *analysis of variance*), a set of values for the experimental variables will be chosen as those to be recommended. Designed experiments are more often used in research studies employing simulation where it is necessary to make some general statement about the values of the experimental variables.

Sometimes designed experiments are used in *exploratory* simulation studies. Suppose that the performance of the system (organization) is lagging behind expectation in one or more areas. Relying on knowledge of and past experience with the system, a set of corrective actions might be defined separately or collectively to bring the system back to an acceptable performance level. Now the questions are: What corrective actions should be implemented? At what level should those selected be implemented? To answer these effectively, a designed experiment is needed to identify the action(s) that will significantly improve the system performance.

When the number of experimental variables is large and there are many possible settings for each variable, then the number of simulation experiments can become prohibitive. In many cases where the measure of effectiveness is well understood, it is possible to determine through factor screening that some variables are of less importance and can either be omitted from experimental interest or their range of possible settings greatly reduced. Also depending on the kind of interest in the results, it may not be necessary to consider all possible combinations of experimental settings (i.e., a full factorial experiment). Instead, a *fractional factorial experiment* may provide all needed information on the main effects of the variables and on the interactions among them that are of interest.

Because of the experimental nature of the simulation experiments, it is possible to use statistical methods to identify optimal settings of experimental variables through *response surface analysis*. Other optimization procedures designed for simulation experiments are being developed (Glover, Kelly, and Laguna, 1996).

7.3.7 Model Documentation and User Manuals

An often overlooked task of a simulation project is that of model documentation and development of user manuals. Unfortunately, in many studies, it usually receives little attention and, in many cases, is done in haste and as an afterthought. If a simulation model is to be successful, it has to be understood and has to be available to the potential users. For such reasons, proper and sufficient documentation has to be provided. Two types of documentation accompany a simulation model: (1) design manuals to cover model architecture, algorithms, general data, and so on; (2) manuals to allow potential users to utilize the model effectively, to detail and interpret the output reports, and to provide the user with a range of illustrative examples of possible model applications.

These manuals should be designed and developed as the simulation project unfolds, and certainly should not be worked on until the model is developed and tested. For user manuals, the background and skill of the potential users will definitely dictate the level and type of coverage, the languages used, and so on. By planning for the manuals early in the development process, user needs will be properly accounted for and reflected in model specification and design, thus eliminating last minute surprises and perhaps confusion and disappointment.

7.3.8 Results Implementation

With the model fully developed, documented, and proved accurate, implementation is the next and final task of the simulation project. Implementation encompasses two major activities: (1) introducing the model and its supporting manuals to the potential users, and (2) monitoring model performance and user reaction. The model can best be introduced through workshops and demonstration projects geared to the needs of various departments or management groups within the organization. A model can be excellent in its design and documentation, but it may fail to gain user acceptance and support. Records should be kept on user interaction with the model, problems encountered either in preparation of input data or interpretation of output reports, and circumstances where the model failed to respond to specific needs. Examination of these records will yield some insight as to what changes and modifications, if any, might be necessary to enhance the user's interface with the model.

Further, animation is a very potent vehicle to show the importance of the successful systems that are proposed. If people can visualize the proposals, even if they are not yet implemented, the animation can demonstrate their value. Animation is especially effective for people who are familiar with the systems but not knowledgeable about simulation. In an age of video games and movie "special effects" the animation demonstrates visually what is often very difficult to present numerically. Therefore, animation should be regarded as a very important contribution to the simulation project.

Finally, in cases where more than one model is required, attempts should be made to have the users interface with the models through a simple input routine, perhaps as an addition to the animation. This routine, usually implemented in a dialog window, would aid the user in selecting the appropriate model and prepare the user's data (given unstructured) to match the input specifications of the selected model. To facilitate the production of custom-tailored reports and one-of-a-kind analyses, the user should have a special output routine. This can simply be a report writer that can extract from standard simulation output data the special reports and analyses required by the user. Some special output data the special reports and analyses required by the user. Some special applications may call for the use of a decision model in concert with, as well as in support of, the simulation model. In such cases, plans should be drawn up specifying how the models will be linked together, how the information will flow between the models, and what control the user would exercise over the two models.

7.4 SIMULATION FUNDAMENTALS

7.4.1 Monte Carlo Sampling

Simulation can be defined as the process of recreating data. It has also been said that a simulation model is a collection of objects that interact with each other over time. Therefore, as the simulation proceeds, the simulation will need data. Individually, specific input has known characteristics (usually a well-defined probability distribution). However, at any time and for a given distribution, the specific data to be recreated cannot be predicted (i.e., their occurrence is totally random) and their impact on the simulation changes the interaction of objects. Yet when sufficiently sampled, the overall pattern of interactions provides a prediction of system behavior. When changes are made to the system (via "what if" questions), this process of system manipulation in the face of uncertain information provides a behavior from which the value of the change can be judged.

Recreation of stochastic data is important in simulation and is an important feature in fully developed simulation languages. Traditionally, the process of generating observations from random variables is given the catchall name of "Monte Carlo." In less exotic words, *it* can be simply called "random sampling" (simulated from theoretical and empirical distributions). The process is rather simple and can be easily implemented.

7.4.2 Generation of Random Numbers

A sequence of numbers, each of which has an equal chance of occurring at any place in the sequence, is called random. In addition, successive numbers in the sequence are independent since the occurrence of one number does not influence the occurrence of another. The roulette wheel in a gambling casino produces such numbers, assuming of course, that the wheel in honest. Each number on the wheel has the same chance of being selected as the winning number; the occurrence of any number does not change or influence the outcome of successive spins. Random numbers have many uses and can be found in practically every textbook on statistics, management science, and operations research. Table 7.1 is an example of a collection of four-digit random numbers, listing uniformly distributed random numbers. The table can be entered at any point. However, once an entry point is selected, it is recommended that successive numbers (especially for the same distribution) be selected in the order given.

For simulation models, random numbers are generated by employing mathematical

TABLE 7.1 Four-Digit Random Numbers

5269	6867	3781	0884
5173	0108	4089	3558
1169	3049	7772	9190
1644	0384	7509	1596
9747	2344	6337	6919
6403	3730	4748	4918
7840	6510	8497	2388
4341	6063	7311	9291
5716	7602	4171	6603
0979	5202	2397	7564

expressions known as pseudorandom numbers (Law and Kelton, 1991 and Banks et al., 1996). These are generated by starting with a first number (usually referred to as a random number seed) and then through mathematical manipulation (squaring, a combination of multiplication and addition, etc.) a next random number is generated. The new number is then used as its predecessor to generate the next random number. The process continues in this fashion; each repetition yields a new number. Because pseudorandom numbers look just like real random numbers, the pseudo is often dropped and the numbers are just referred to as random numbers. It is important, however, to realize that these numbers are not true random numbers, in a philosophical sense. Thus it is important for users of random numbers to be sure that they are using random numbers that pass all the tests of randomness.

7.4.3 Data Categorization: Cumulative Probability Distribution Function

Data describing various aspects of a system and its components have to be categorized somewhat descriptively. Histograms and frequency tables are two examples of data categorization and summarization. If feasible, another type of data categorization is achieved through the use of such standard probability distributions as normal, Poisson, Erlang, triangular. Choosing from among these distributions depends on the data and its characteristics. Goodness of fit tests can be used to aid in selecting the proper distributions to represent the stochastic phenomena (Law and Kelton, 1991).

However, with simple frequency data, cumulative distributions are easily developed. Each cumulative distribution is a plot of the random variable X versus the probability of the variable being x or less. The cumulative distribution function is usually denoted $F_X(x) = P\{X \leq x\}$.

Example 7.1: Assume that for a random variable x we have

a	$P(X = a)$
1	.1
2	.2
3	.3
4	.15
5	.05

Cumulative probability $F_X(x)$ is obtained by successively adding the probability figures $P(X = a)$ preceding each of the given values. The result would be as follows:

a	$P(X = a)$	$F_X(x) = P(X \leq a)$
1	.10	.10
2	.20	.30
3	.30	.60
4	.15	.75
5	.25	1.0

A plot of the cumulative distribution above is shown in Fig. 7.1. The cumulative function in Fig. 7.1 represents a discrete distribution, for the values of the random variable X compose a set of fixed (discrete) values. The counterparts of discrete random variable are the continuous random variables. The probability function of a normal distribution (see Fig. 7.2) illustrates a continuous random variable. Notice that there is a continuous number (infinite) of potential values. Also there is no probability mass for any given value; so when we refer to the probability for a continuous random variable, we compute it for a range of values of the random variable.

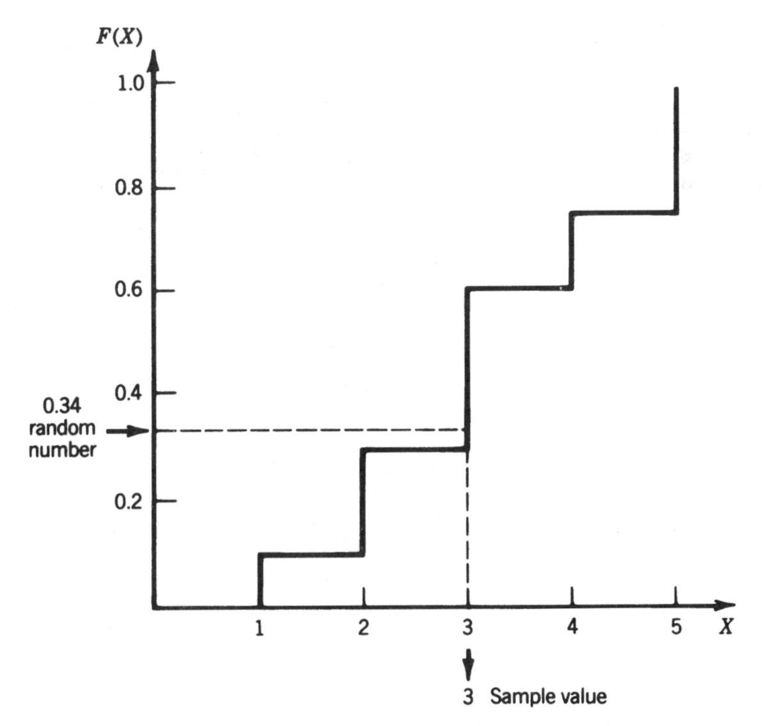

Fig. 7.1 Cumulative distribution of a discrete function.

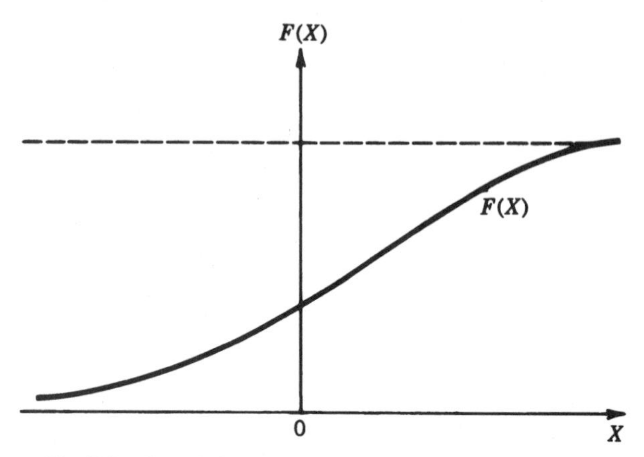

Fig. 7.2 Cumulative distribution of a continuous function.

7.4.4 Drawing Sample Values

Having a method to obtain (called sampling) random numbers and knowing how to develop the cumulative distribution function, we can now proceed to draw samples of the random variable. This is done by finding the actual values that correspond to the sampled random numbers (treated as a uniformly distributed value between 0 and 1). The value random variable can be obtained graphically by projecting a line horizontally at the point on the ordinate $F_X(x)$ corresponding to the value of the random number until the line intersects the curve (for continuous distributions) or one of the staircase segments (for discrete distributions). The value of the abscissa corresponds to this intersection. This method is called *inversion*, and it inverts the usual method of building a cumulative distribution function since it starts with the value of the distribution function to obtain the value.

Example 7.2: Suppose we have generated the sequence of random numbers

0.34 0.24 0.23 0.38 0.64 0.36 0.35

Entering each of these numbers on the ordinate of Fig. 7.1, we generate the following set of values for the random variable X:

Random number	Sample values
0.34	3
0.24	2
0.23	2
0.38	3
0.64	4
0.36	3
0.35	3

The same set of sample values can also be obtained by simply writing some decision rules specifying the value (V) that corresponds to each range of random numbers (RN). Specially, for the example at hand, we can write

$$\begin{array}{lll} \text{If} & & RN \leq 0.10, V = 1 \\ \text{If} & .10 < & RN \leq 0.30, V = 2 \\ \text{If} & .30 < & RN \leq 0.60, V = 3 \\ \text{If} & .60 < & RN \leq 0.75, V = 4 \\ \text{If} & & RN > 0.75, V = 5 \end{array}$$

Again, the two methods (graphical and decision rules) are the same; however, the latter is rather easy to implement and quite suitable for computer algorithms or spreadsheets. Simulated sampling, or Monte Carlo sampling, from probability distributions can be extended to encompass more than one distribution. Below, we consider a situation where there is a need to sample from several distributions; however, the sample obtained from one will determine how a sample is to be drawn from the others.

Example 7.3: Consider the case of simulating accidents and determining the corresponding costs. Accidents occur with the following probability:

Accident type	Probability
First aid	0.6
Lost time	0.1
Equipment damage	0.3

The costs associated with the various accident types are given by

1. First Aid		2. Lost Time		3. Equipment Damage	
Cost $	Probability	Cost $	Probability	Cost $	Probability
25	0.3	500	0.2	100	0.5
50	0.5	1000	0.5	200	0.3
100	0.2	2500	0.3	500	0.2

Based on past history, we anticipate that a total of 10 accidents will occur during the next period. Now, what would be the total cost for all accidents? To answer this question we proceed as before and use Monte Carlo sampling as follows:

1. Obtain a sample from the accident-type distribution. This gives a particular accident type (i.e., first aid, lost time, or equipment damage).
2. For the type of accident determined in (1), compute an associated cost by sampling from the appropriate distribution.
3. Repeat (1) and (2) until a total of 10 accidents has been generated. Adding the individual costs of all accidents, obtain the total accident cost for the next period.

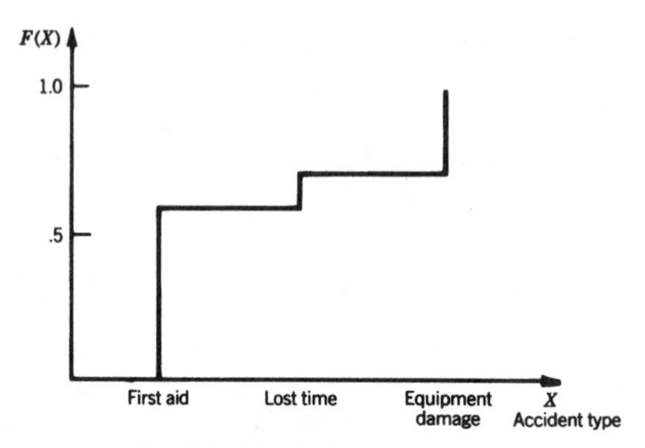

Fig. 7.3 Distribution of accidents.

4. Obtaining the total cost of 10 accidents constitutes but one sample point so that the sum of the costs is, by the central limit theorem, normally distributed. To have a more realistic estimate, repeat the process (Steps 1–3) several times. From the estimate obtained, the mean and standard deviation of the total accident cost can be computed.

Figures 7.3, 7.4, 7.5, and 7.6 give the cumulative distributions for the accident types and the three cost categories. Results of sampling from the distributions are summarized in Table 7.2. Table 7.3 shows the estimated cost for twenty trials (repetitions). From the data presented, we determine that mean cost = \$2385 and standard deviation = \$1522.19.

7.4.5 Computer Implementation of Distribution Sampling

It should be clear from the example presented that the power of computers could be used to accept the computational burden of a simulation. Many more simulations can be run using the computer. In fact, it is hard to imagine not doing simulation without a computer. Furthermore, modern spreadsheet software usually contains the facilities to obtain random numbers.

The following Visual Basic program can be added as a user-defined function to any Microsoft Excel spreadsheet. This function titled `inverseDiscrete` implements the

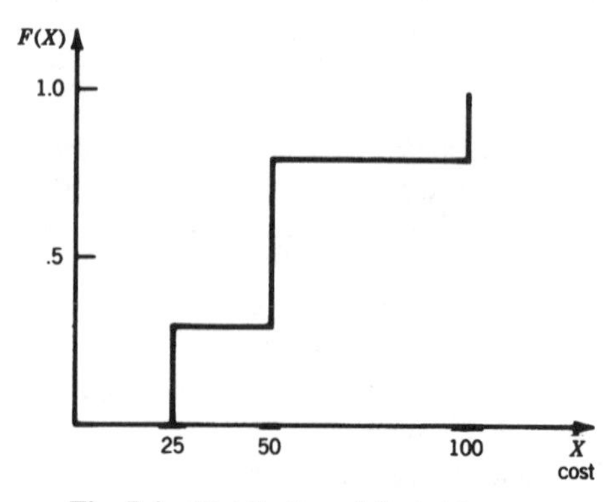

Fig. 7.4 Distribution of first-aid costs.

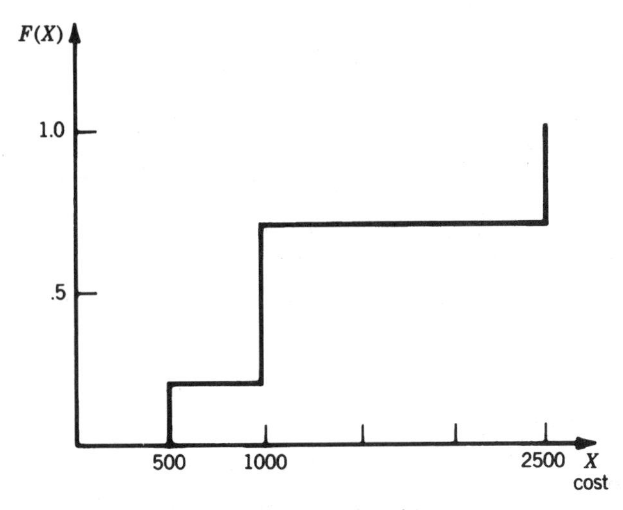

Fig. 7.5 Distribution of lost time costs.

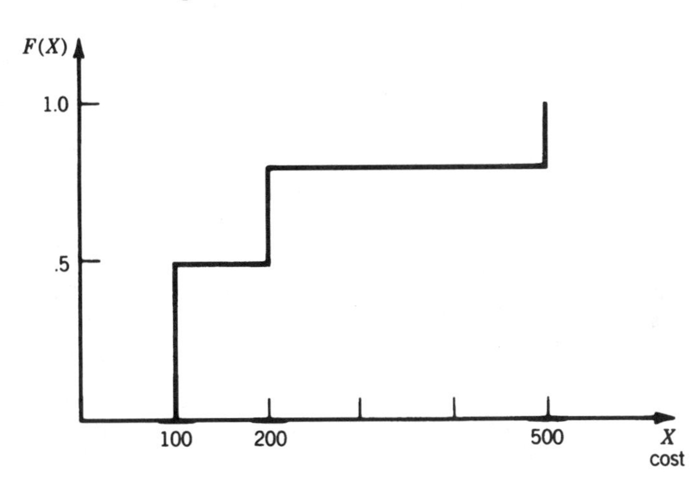

Fig. 7.6 Distribution of equipment damage costs.

TABLE 7.2 Accident Costs for the First Simulation

	1	2	3	4	5	6	7	8	9	10
First random number	.510	.426	.092	.461	.613	.895	.215	.626	.622	.800
Accident type[a]	1	1	1	1	2	3	1	2	2	3
Second random number	.758	.878	.921	.445	.942	.903	.460	.115	.362	.202
Cost ($)[b]	50	100	100	50	2500	500	50	500	1000	100

[a] 1 = First aid; 2 = lost time; 3 = equipment only.
[b] Total cost = $4950.00; average cost = $495.00; standard deviation = $730.56.

TABLE 7.3 Total Accident Costs for 20 Simulations[a]

1	2	3	4	5	6	7	8	9	10	11	12	13	14	15	16	17	18	19	20
4950	1875	3175	675	4500	1525	3425	650	1925	1025	1575	4200	1800	5750	800	1350	1275	3650	1300	1275

[a] Average total cost = $2385.00; standard deviation = $1522.19.

Monte Carlo sampling procedure from a discrete distribution described earlier. The user need only provide a spreadsheet range for the probabilities and a corresponding range for the values returned from the sample.

```
Public Function inverseDiscrete (prob As Object, val As Object) As Integer
Dim randNum As Single, cum As Single, ind As Integer
randNum = Rnd
cum = 0
For ind = 1 To prob.Rows.Count
  cum = cum + prob.Cells(ind).value
  If (randNum < cum) Then
    inverseDiscrete = val.Cells(ind).value
    Exit Function
  End If
Next ind
End Function
```

Note that the function `Rnd` is a builtin function that obtains random numbers in the range (0, 1).

7.4.6 Computing the Number of Replications

Now Example 7.3 can be replicated an arbitrary number of times, with an arbitrary number of values in its sum. Remember that the number of terms in the sum is simply enough for normality to be present, which is probably a value around 10–15.

Therefore, since we can now obtain simulation runs of almost arbitrary lengths, the natural question to ask is "how many replications should be performed in a simulation?" Clearly, the more replications we perform, the more confident we are in the result. Thus the answer to the question requires us to state how precise we want our answers. Generally, we will state a precision as a function of the estimate. For example, we will say we want an estimate that is not off by more than 10 percent. In other words, if we are estimating a mean cost, we want to be sure that we have enough replications so that the "true" mean cost is within a range ±10% of the estimate. Suppose our estimate of the cost is $1000, then we may want to be 95% confident that true mean cost is in the range [900, 1100]. The 95% confidence means that 95% of the time we compute this range from the simulation, it will contain the true mean.

There is a formula to compute the number of replications based on the previous reasoning. It does, however, require some additional information, which can be obtained from a "pilot" simulation run. A "pilot" simulation is done with an arbitrary number of simulations. Lets use the results in Table 7.3 as our pilot run and, from that, determine the number of replications. Recall that the sample mean = 2385.00 and the standard deviation of the sample = 1522.19. Therefore the half-width, given by h, of an exact $1 - \alpha$ confidence interval for the true mean, μ, centered at the sample mean, $\bar{x}$, whose observations have a sample standard deviation of $s(x)$ is computed from

$$h = t_{n-1, 1-\alpha/2} s(x)/\sqrt{n}$$

where $t_{n-1, 1-\alpha/2}$ is obtained from a table of t values and the upper $\alpha/2$ point of the student-t distribution is with $n - 1$ degrees of freedom. Applying the previous values yields the following computations:

$$\bar{x} = 2385.00$$
$$s(x) = 1522.19$$
$$n = 20$$
$$t_{19,.975} = 2.09$$
$$h = 711.38$$

This means that we are 95 percent confident that our estimate of total cost for ten accidents is between 1673.62 and 3096.38. But if we want a precision that is within 10 percent, then the half-interval should be (.10)(2385) = 238.5. This result means we need to make more simulation runs. We compute number needed from the following formula:

$$n^* = n(h/h^*)^2$$

where the n^* is the correct number of replications and h^* is the desired half-interval. Now these calculations are made

$$n = 20$$
$$h = 711.38$$
$$h^* = 238.5$$
$$n^* = 177.9$$

Now round up the n^* to 178 as the correct number of replications to obtain the desired precision. Notice that all these calculations are easily done using a spreadsheet.

7.4.7 Sampling from Standard Distributions

In a similar manner, Monte Carlo sampling can be extended to sample from many standard distributions. By virtue of their well-defined functions, standard distributions are well suited for producing sample values through direct computation instead of graphically.

For example, to obtain a sample X from a normal distribution with mean μ and standard deviation σ, we use

$$X = \mu + \sigma(\text{RNN})$$

The random number, denoted RNN, should be sampled from a table of unit normals (see Table 7.4). A unit normal is a normal distribution with a mean of 0 and a standard deviation of 1.

A random sample X from an exponential distribution whose mean is μ is given by

$$X = -\mu \ln(\text{RN})$$

where u = mean. Here, RN, the random number, is uniformly distributed between 0 and

TABLE 7.4 Random Normal Numbers $\mu = 0$, $\sigma = 1$.

−0.98	0.76	−1.22	0.99	−1.29
−0.91	0.43	−1.11	0.04	−1.96
−0.67	0.92	−0.72	−0.47	0.15
−0.79	0.66	1.70	−1.02	1.82
−0.55	−0.07	−2.08	−0.35	−0.17
−0.42	0.46	−0.61	−0.78	0.77
0.10	−0.33	−0.52	0.18	0.38
0.11	−0.65	0.30	−0.36	0.54
−1.42	1.01	0.10	−0.20	0.71
0.17	1.75	−0.86	0.93	−0.98

1 (see Table 7.1). Similar expressions can be developed for other distributions (Law and Kelton, 1991).

The procedure just described is Monte Carlo sampling from arbitrary distributions. Reflection on the method shows that we have done nothing more than use data categorization in reverse. In other words, when studying a phenomenon such as the number of accidents versus day of week, we usually start by determining (through observations, surveys, etc.) the frequency with which a certain number of accidents occurs. From this information, an appropriate probability distribution is developed.

In a simulated sampling, we reverse the roles and assume that we have a population of random numbers from which we draw a sample. To every number in the sample we assign a value. The nice thing about this procedure is that to draw a sample of random numbers, we need only a table or a mathematical expression! Below are four additional examples that demonstrate how useful Monte Carlo simulation can be in dealing with some typical safety management problems.

Example 7.4: An OSHA compliance officer is scheduled to inspect the Buoya Manufacturing Company for possible violations of the noise standard. The overall noise level in the Buoya Plant varies from hour to hour and from day to day. In general, the noise level varies with the number of machines operating simultaneously and with the type of materials being processed. From previous sound surveys, the following data are given:

Overall noise level (dBA)	Probability
88	0.4
90	0.3
92	0.2
95	0.1

What is the probability that the inspector finds the company in compliance? The inspector will base his (her) decision on the results of ten sound level readings taken throughout the plant.

Table 7.5 is used to simulate the OSHA inspector's ten noise-level readings. The assumption here is that the sound readings will be taken randomly with respect to both

TABLE 7.5 Basis for Noise-Level Determination

Noise Level dBA	Probability	Random Numbers
88	0.4	0–39
90	0.3	40–69
92	0.2	70–89
95	0.1	90–99

time and space. To be in compliance, the noise level must average less than or equal to 90 dBA.

Results of the first simulation run are shown in Table 7.6. The average values for the first five runs are 89.3, 90.1, 90.5, 90.1, and 89.5. Since two of the values are 90.0 or less, the company stands a 40% chance of being found in compliance. The standard deviation after five runs is 0.49.

After five more runs are completed, values of 91.2, 90.4, 90.5, 89.8, and 91.0 are obtained. Now there are three figures below compliance; the company's chances are now estimated at 30%. The standard deviation for the ten runs is 0.61. The next five runs produce values of 89.7, 90.5, 89.8, 91.0, and 91.5. With a total of five values out of fifteen runs indicating compliance, our probability is now 33%. The standard deviation is roughly 0.65. Therefore, we can estimate that the company will be found in compliance only about one-third of time.

Example 7.5: A state OSHA director would like to determine an appropriate level of staffing for his inspection and compliance force. He gives the following estimates for inspection time per plant:

Time (h)	Probability
6	0.4
8	0.3
16	0.2
24	0.1

TABLE 7.6 Results of the First Simulation

Sample	Random Number	Noise Level dBA
1	50	90
2	48	90
3	28	88
4	27	88
5	34	88
6	62	90
7	27	88
8	17	88
9	94	95
10	33	88
		Average = 89.3

In some cases, the safety inspection is followed by an industrial hygiene survey. The probability of a plant getting both safety and hygiene inspections is 0.25. The time required to perform an industrial hygiene inspection is assumed to be normally distributed with mean = 5 hours and standard deviation = 1.5 hours.

The OSHA director is planning to inspect 200 plants in the next six months. Existing policies limit each inspector to a maximum of 20 hours of fieldwork per week. How many safety inspectors should be hired? How many industrial hygienists would be needed? What is the average workload per inspector?

Since the problem calls for answers based on a weekly workload, let us assume that of the 200 plants to be inspected, no more than 8 plants will be attempted in the 26 weeks. We can simulate this 8-plant workweek until we feel we have a range of values for the required manpower.

Eight simulations were made in accordance with the procedure shown in Fig. 7.7. Details of the first two runs are given in Table 7.7. The overall results of eight simulations are:

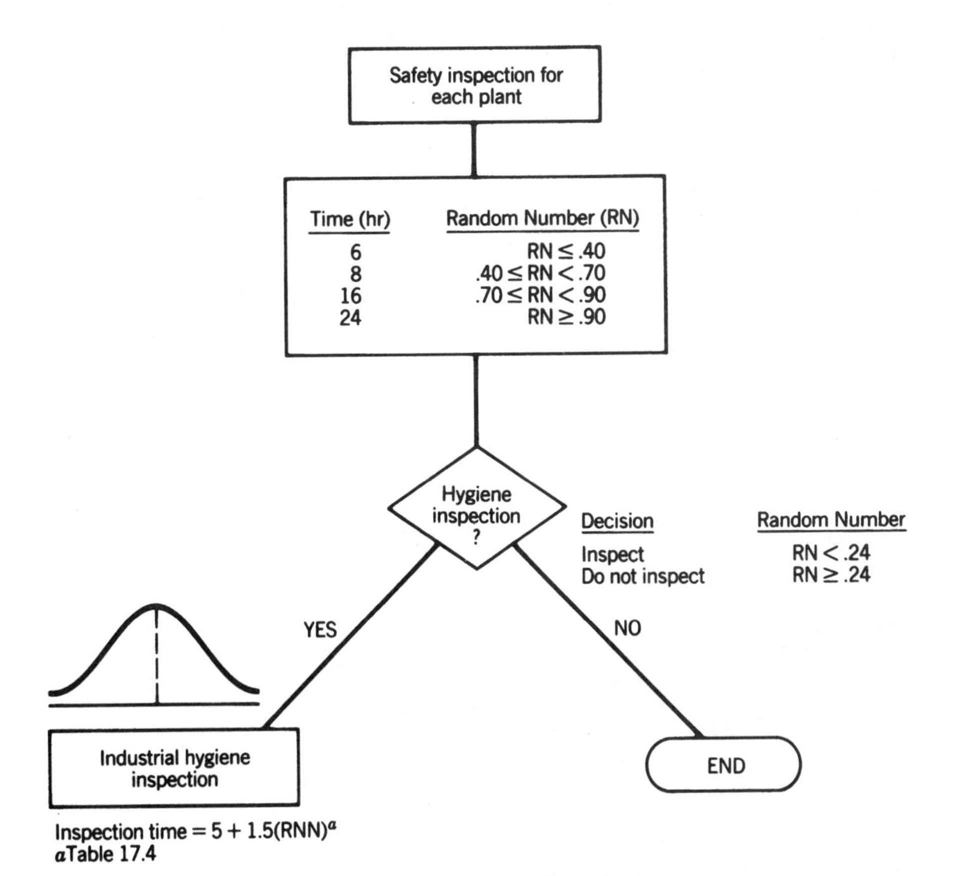

Fig. 7.7 OSHA manpower determination.

Hours/week for safety inspections	Hours/week for industrial hygiene inspections
70	0
82	0
96	14
94	14
52	13
82	17
60	21
106	5

Even without further simulation runs, it seems that five safety inspections and one industrial hygiene inspector will be sufficient at 20 hours/week each. In only one of either cases does the workload exceed this manpower level. This deficit can be covered by spare time remaining in a slack week. With this manpower allocation, the average workload can be found by averaging the simulated weekly requirements and dividing by the number of inspectors. For safety inspectors, the load is 80.25/5 = 16.05 hours/week; the industrial hygienist will put in 11.62 inspection hours/week. The slack time could be

TABLE 7.7 Results of the First Two Simulations

Safety Inspection Plants	Random Number	Time (h)	Random Number	Decision	Random Number	Industrial Hygiene Inspection Time (h)
First Simulation Run						
1	.63	8	.44	No		
2	.58	8	.63	No		
3	.65	8	.76	No		
4	.12	6	.50	No		
5	.47	8	.49	No		
6	.40	8	.63	No		
7	.69	8	.13	Yes	0	5
8	.82	16	.14	Yes	0.66	4
Totals		70				9
Second Simulation Run						
1	.72	16	.33	No		
2	.39	6	.86	No		
3	.85	16	.37	No		
4	.30	6	.48	No		
5	.31	6	.97	No		
6	.40	8	.91	No		
7	.73	16	.33	No		
8	.53	8	.55	No		
		82				0

lessened if one inspector certified for both safety and industrial hygiene was employed, leaving only five total inspectors.

Example 7.6: Figure 7.8 gives distribution of noise levels throughout a manufacturing plant. Because of various work assignments, some employees are exposed to a wide variety of noise levels during any given work shift. A time study of those employees over several days provides the following:

Area time (h)		
I		2
II		1
III		3
IV		1
	Total	9

Based on the given data, what is the probability that the company can be found in compliance with the OSHA noise standard?

From the data provided, we can see how long, on the average, a worker spends in each of the areas each day. Within each area the noise levels are different. Probabilities can be assigned to each noise level in proportion to its occurrence in each of the four areas. If we assume that over a period of time a worker will stay in each noise-level area for a length of time proportional to its fractional floor space, we can perform a simulation of expected noise exposure. Estimated percentages of each noise level and their associated random numbers are shown in Table 7.8.

A worker's daily exposure to noise may be simulated by generating two random numbers each for Areas I and II, three numbers for Area III and one for Area IV. In this manner, the noise levels are determined at intervals of one hour each. Results of the

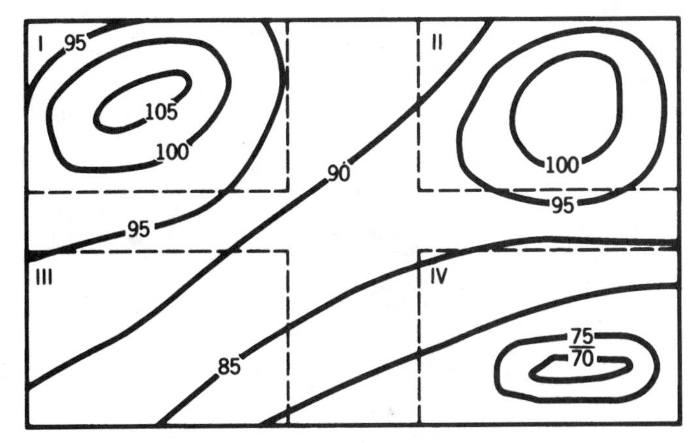

Fig. 7.8 Noise levels in the plant, dBA.

TABLE 7.8 Estimated Occurrence of Noise Levels Within Each Area

Noise Level dBA	%	Random Number	Noise Level dBA	%	Random Number
	Area I			Area II	
90–95	20	0–19	90–95	25	0–24
95–100	30	20–49	95–100	40	25–64
100–105	35	50–84	100–105	35	65–99
> 105	15	85–99			
	Area III			Area IV	
80–85	25	0–24	< 70	10	0–9
85–90	40	25–64	70–75	15	10–24
90–95	35	65–99	75–80	50	25–74
			80–85	25	75–99

first simulation are given in Table 7.9. For a state of compliance to exist, the following condition should be met:

$$\Sigma(C_i/T_i) \leq 1$$

where C_i = actual exposure time at the ith noise level and T_i = the maximum permissible exposure time for the ith noise level. Table 7.10 defines the limiting times for various noise levels.

From Table 7.9, it is clear that the time-weighted exposure is indeed greater than 1; thus compliance is lacking. Of a total of 20 simulations, only 2 runs were found to yield a weighted exposure of less than 1. Therefore, it is safe to conclude that the probability that the company will be found in a state of compliance is 0.10—a very low figure, which should entice management into taking some drastic steps for instituting corrective measures.

TABLE 7.9 Results of the First Simulation

Random Number	Noise Level Range dBA	Assumed Value dBA
	Area I	
99	> 105	105
07	90–95	92
	Area II	
26	95–100	97
82	100–105	102
	Area III	
84	90–95	92
41	85–90	87
48	85–90	87
	Area IV	
46	75–80	77

TABLE 7.10 Permissible Noise Exposure[a]

Duration per day (h)	Noise Level dBA
8	90
6	92
4	95
3	97
2	100
1 1/2	102
1	105
1/2	110
1/4 or less	115

[a]OSHA general industry standards (1910.95).

Example 7.7: An insurance company is interested in defining the premium it should charge its industrial customers for workers' compensation coverage. The company estimates that the value of all the claims to be processed in a year in normally distributed with a mean of $750,000 and standard deviation of $50,000. The company will insure 1000 clients. The overhead cost (service charge) per claim is as follows:

Cost per claim	Probability
50% of the claim	0.60
25% of the claim	0.40

What insurance premium should the company charge each of its clients?

The basic solution approach for the problem is shown in Fig. 7.9. We will demonstrate the procedure by carrying out several simulation runs of ten trials each. Results of the first run are given in Fig. 7.10. The total claims value would be approximately $1,053,000 annually. Each of the 1000 clients should be charged $1053.

Example 7.8: OSHA estimates that the canning industry will incur $150 per plant in order to comply with a new standard regulating conveyors. However, because some degree of compliance with the standard can be attributed to standard industry practices, the estimated cost has to be somewhat modified. From industry data and known practices, the following can be assumed:

Existing level of compliance	Probability
10%	0.35
25%	0.25
75%	0.25
80%	0.25
90%	0.10

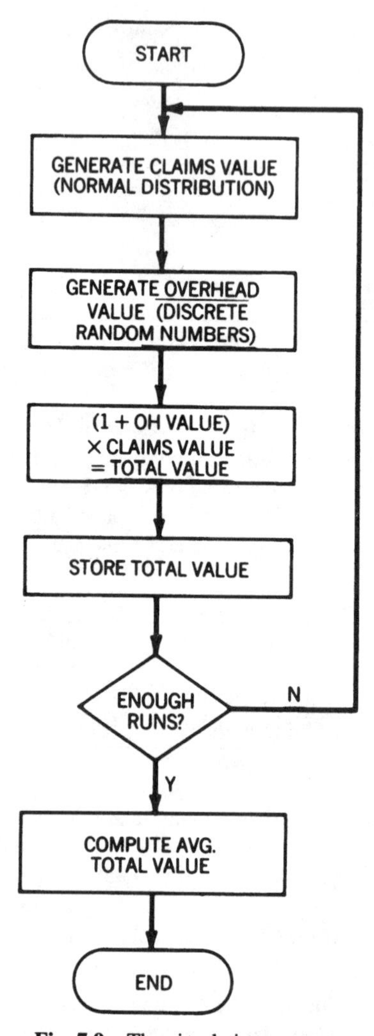

Fig. 7.9 The simulation process.

In computing the annualized compliance cost per plant, OSHA assumes an interest charge (i) of 10% with an economic life (n) of 15 years. There are many reasons for i and n to vary, for instance, the state of the economy (growth, decline, etc.). Economic forecasts provide the following:

n (years)	Probability	i (%)	Probability
10	0.5	10	0.2
15	0.3	15	0.5
20	0.2	20	0.3

Normal Random Number (*RNN*)	Claims Value[a] (dollars)	Random Number (*RN*)	OH Value	Total Value
.5	775,000	.96	.25	968,750
0	750,000	.72	.25	937,500
−1.0	700,000	.38	.50	1,050,000
2.5	825,000	.01	.50	1,237,500
− .5	725,000	.36	.50	1,087,500
.5	775,000	.97	.25	968,750
−2.0	650,000	.94	.25	812,500
1.0	800,000	.28	.50	1,200,000
.5	775,000	.67	.25	968,750
− .5	725,000	.38	.50	1,087,500

[a] = 750,000 + 50,000 (*RNN*) Average = 1,031,875

The simulation was repeated 14 more times; the corresponding average values of the claims are:

$1,031,875	$1,137,500	$1,049,375
1,100,750	1,007,500	1,086,250
1,072,000	1,002,962	1,051,250
1,050,875	1,056,250	1,023,750
1,063,750	1,070,625	1,000,000

Fig. 7.10 Average cost of claims.

What is the most likely annualized cost for the industry?

The cost of compliance depends on the existing level of compliance and its corresponding percentage of occurrence. Hence, the cost will be .90($150) = $135 in 35% of the plants; .75($150) = $112.50 in 25% of the plants, and so on.

The annualized cost is determined from the nine different combinations of i and n, which yield the capital recovery factor $(A/P, i\%, n)$ and the cost of compliance (present cost, P):

$$A = P(A/P, i\%, n)$$

The three stochastic functions (cost of compliance, i, and n) can be represented as random number strings, as in the preceding examples. Three random numbers will be required to generate one value for the annualized cost of compliance. The problem model is presented in Fig. 7.11. A few manual runs were performed to illustrate typical simulation results. The average annualized cost after 15 runs of $19.62 may be used as a rough estimate of the probable yearly cost to the industry.

Example 7.9: In the maintenance department, a utility worker is assigned an area that is about 20% of the total shop area. The worker performs a variety of jobs in the assigned area: loading and unloading, cleaning, helping other workers, and so on. Throughout the shop, an overhead (bridge) crane travels from end to end for handling heavy pieces of equipment. Movement of the crane is random and in a sense cannot be predicted, for its position is controlled by the needs of the different workstations in the maintenance area. The crane is rather old and has not been kept in good working order due to poor preventive maintenance. The probability of the crane dropping its load due to chain breakage or improper loading is 0.10. What is the probability that the utility worker will be involved in a crane accident in a period of eight hours?

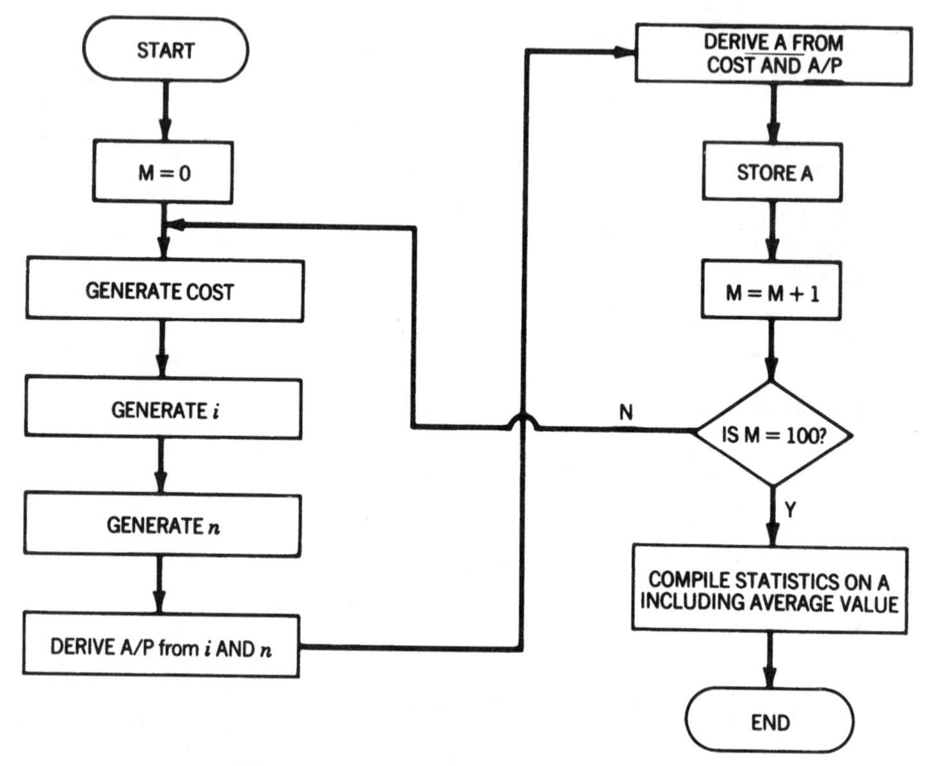

Fig. 7.11 Flow chart for OSHA compliance cost.

In order to work this example, three assumptions must be made:

1. If the crane drops its load over the worker's assigned area (20% of the total shop area), he will be involved in an accident.
2. The worker's area will be arbitrarily assigned random numbers 30–49 (20% of the random numbers).
3. The action of the crane dropping its load will be assigned to random numbers 0–9.

Since the crane is above the worker's area 20% of the day, the appearance of any of the random numbers 30–49 will represent the hazardous condition. Then, if any of the random numbers 0–9 appears next, a simulated accident occurs.

Three hundred trials were performed, and five accidents occurred. Therefore, the probability that the worker will be involved in an accident is 5/300 = 1.7%.

Example 7.10: A large company is evaluating a pre-employment screening program for strength, which is claimed to have been developed in accordance with the Equal Employment and Opportunity Commission (EEOC) guidelines and recommendations. The company implemented the program for a pilot test at one of its plants. The strength scores for those persons examined are given below:

Men: Scores are normally distributed with mean = 80 and standard deviation = 15.

Women: Scores are similarly distributed with mean = 60 and standard deviation = 20.

The company will hire a labor market equally divided between men and women. The company plans to hire 30 workers per year for the next five years. Minimum passing score for both men and women is 70.

What will be the composition of the company labor force by the end of the fifth year? Will the company be in compliance with the EEOC rule that demands that the workforce should be representative of the labor market accessible to the company?

A simulation model, shown in Fig. 7.12, is designed to count the number of males

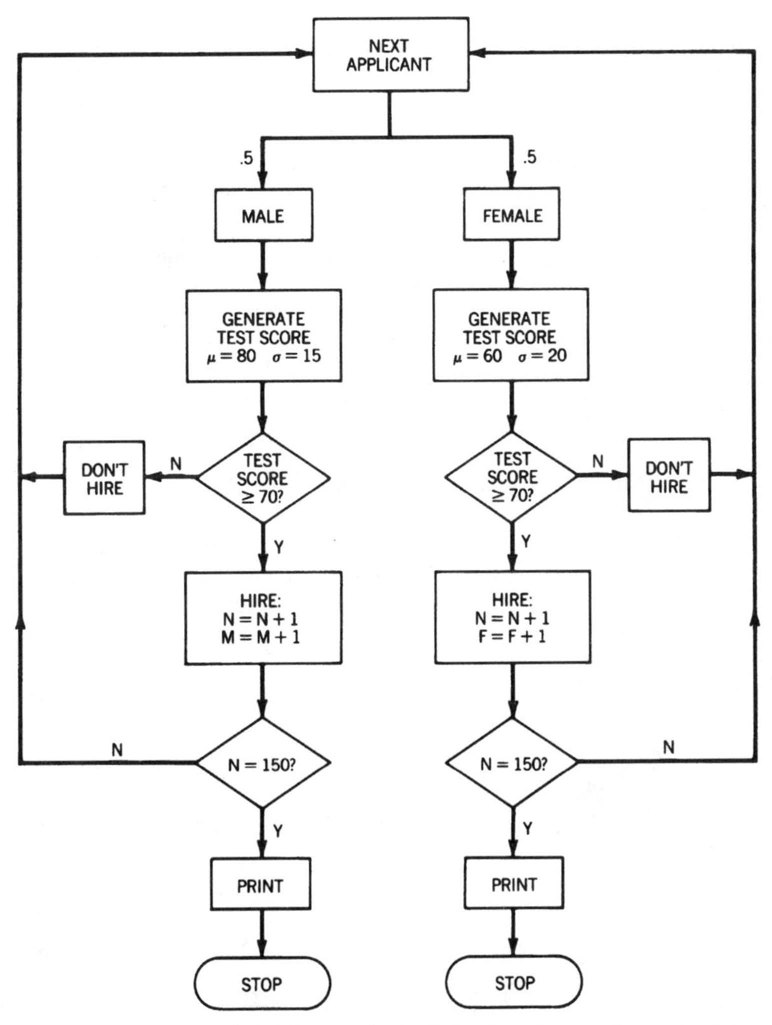

Fig. 7.12 Flow chart of EEOC strength discrimination rule compliance simulation.

(M) and females (F) that would be employed out of the next 150 applicants (N). The history of test scores indicates that there will be some favor given to males under the screening program. The simulation will show how much difference this bias will make as far as the total workforce is concerned. If the company cannot comply with the EEOC rule, experiments may be conducted with the model to find the minimum reduction in job requirements (strength) to make the job more accessible to females.

Example 7.11: An operator works on a small punch press, feeding the press with small parts to be stamped. Operating the press involves two distinct steps. First, the operator loads the metal piece into the press. Second, the press moves the die downward (down cycle) to complete stamping. Therefore, in this situation, we have two patterns synchronized with each other in such a way that the operator manages to pull his hands out of the die area before the press die reaches the metal piece. The two patterns can be depicted as shown in Fig. 7.13.

The first observation about this problem might be that working on the press in question should certainly be avoided. However, the high accident rate is caused by a couple of biases in the solution. For simplicity, an accident was assumed to occur if the cumulative loading time (CUMTO) equaled or exceeded the press cycle time (TM) when, in fact, an accident could not occur unless the two times coincided. Also, the fact that negative (and some low) loading times cannot be used suggests that the normal distribution should be altered slightly in order to reflect the loading time function accurately. Correction of these two factors will result in a more accurate simulation.

Timing of press action (cycle) is uniform and can be determined in advance. In contrast, for many reasons (fatigue, distraction, poor eye–hand coordination, etc.) the operator action can occur at somewhat irregular intervals. An accident results when the sequence of loading actions is disrupted to the extent that it coincides with the press cycle: The operator's hand will be caught in the press die.

Suppose that the press cycles once very minute, $t_m = 1$. The time interval between successive loadings (t_0) is normally distributed with mean (μ) = 1 and standard deviation (σ_x) = 0.5 min. What is the probability that a press-related injury will occur in an eight-hour period?

We may safety assume that the operator will not try to load a part before the preceding part has been stamped. Let us also assume that the operator begins the sequence at $t = 0.5$ min. Therefore, the first possible accident would occur at $t = 2$ if $t_0 \geq 1.5$ min. In the computer simulation model (Fig. 7.14), TO will be generated from a normal distribution random number routine.

The execution of the simulation runs produced an accident in 4 of 13 cycles.

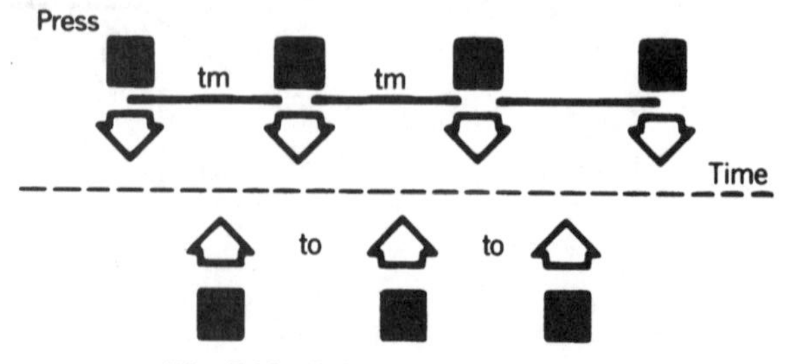

Fig. 7.13 Press operation sequence.

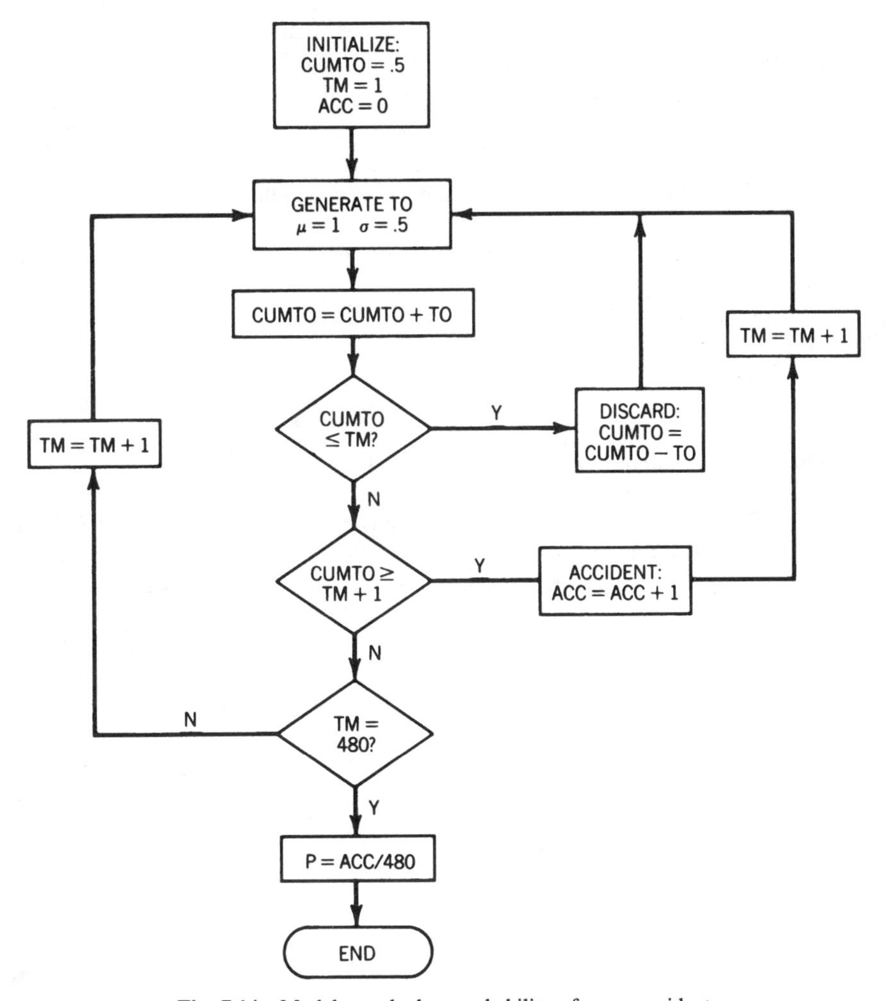

Fig. 7.14 Model to calculate probability of press accident.

7.4.8 Spreadsheet Functions Useful in Computer Simulation

Modern spreadsheet software usually contains several sampling functions. For example, in Microsoft Excel, we have already used the function that obtains a random number in the range (0, 1) within Visual Basic. However, there are many simple spreadsheet functions that can be directly used. For example the function RAND is used to obtain a random number in the range (0, 1), and RANDBETWEEN can be used to obtain a random number within a given range. The following is a list of spreadsheet functions from Excel that can be used to directly obtain samples from standard distributions:

BETAINV: obtains a sample from a beta distribution

CHIINV: obtains a sample from a chi-square distribution

LOGINV: obtains a sample from a lognormal distribution

NORMINV: obtains a sample from a normal distribution
NORMSINV: obtains a sample from a unit normal distribution
TINV: obtains a sample from a Students-*t* distribution

Thus to obtain a value corresponding to a sample from a unit normal, similar to that obtained in Table 7.4, you would use the spreadsheet formula:

NORMSINV(RAND())

To obtain a value from the normal distribution with mean of 15.6 and standard deviation of 2.7, you would use:

NORMINV(RAND(), 15.6, 2.7)

Therefore, arbitrary numbers of replications can be obtained with computer implementation of the procedures outlined.

In Microsoft Excel, there is a "sampling" analysis tool that is located under "Tools" in the main menu and within the "Data Analysis" submenu. This tool allows the sampling from an input range as a population. You can restrict the sample in various ways.

Also, it should be noted that spreadsheets contain various financial functions. Such functions would be especially useful in doing Example 7.8. If you were using Excel, then the annual cost for $10,000 present cost with an annual 8% interest that must be paid in 10 years would be obtained from the spreadsheet formula:

PMT(8%/12, 10, –10000)

Similar functions exist for other kinds of interest computations.

7.4.9 Simulation Algorithms

Monte Carlo sampling makes no provisions for expenditure of time or scheduling of activities in cases where limitations of resources may introduce delays and the formation of queues. In this context, Monte Carlo implies that events proceed (occur) with no time lost between successive events. This assumption is not always true, especially when one is modeling complex systems and their related activities. With queues, issues concerning the selection of which activities (tasks) to be started next, how resources will be assigned to competing activities, and a host of other issues, have to be dealt with. Dealing with all these issues requires an algorithm that offers more than just sampling. The algorithm, among other things, should be able to sample from theoretical as well as empirical distributions and should maintain files of model activities and allocation of resources in accordance with some prescribed rules. In summary, these requirements describe the characteristics of a simulation algorithm. It follows that we view simulation as general Monte Carlo sampling, but allowing for the expenditure of time and the occurrence of delays. However, because the expenditure of time and the movement of entities within a simulation can be complicated enough to warrant computer simulation, even if *no* Monte Carlo sampling takes place.

The following is an example of how one simulation can recognize the passage of time.

Example 7.12: A state OSHA program has a compliance officer assigned on a full-time basis to conduct industrial hygiene investigations. Requests for such investigations are made on the basis of the results of the general OSHA inspections and compliance surveys. In any given week, the demand for industrial hygiene investigations can be given by

Number of requested investigations	Probability
0	0.5
1	0.25
2	0.25

Time per investigation is assumed to be normally distributed with mean = 2.5 days and standard deviation = 0.5 days. Investigations are carried out on a first-come first-served basis. Accordingly, requests received while the officer is busy are placed on the list of future investigations. From this list, requests are handled in the order in which they were filed. All requests are assumed to be made at the beginning of the week. The officer works only five days a week, with no provision for overtime. Therefore, investigations that cannot be finished in the week in which they started are carried forward to the following week. Determine the number of investigations the officer would be able to handle in a ten-week period.

To deal with the problem at hand, we would proceed as follows:

1. Generate the requests for a week. The number of industrial hygiene investigations to be conducted is obtained by sampling, using the demand data given.
2. If the compliance officer is busy, the requests have to be added to the others that have not yet been processed. Otherwise, the officer is immediately assigned to start a requested investigation. Time (days) of the investigation is assigned using the formula for sampling from normal distribution: 2.5 = 0.5 (RNN); RNN = random number obtained from Table 7.4; 2.5 and 0.5 are the mean and standard deviation of the distribution, respectively. The officer will remain busy for this amount of time.
3. When the officer becomes free, the requests, if any, at the top of the waiting list will be considered next.
4. The process of creating requests and carrying out the corresponding investigations continue until the total time assigned to the simulation expires—ten weeks, in our example.

The steps above are carried out in a format similar to that of Table 7.11. Solid lines in Table 7.11 represent investigation times, while dashed lines portray time before the investigation can be commenced for a given request. Circled numbers are successive requests generated during the simulation period. Dashed lines in Table 7.11 give the total waiting times before investigations can be started. This gives an average of two days of waiting per request for investigation. From the table we also can determine the utilization level of the industrial hygienist, the longest number of periods he remained busy, remained idle, and so forth.

TABLE 7.11 Simulation Data Sheet

	Week									
	1	2	3	4	5	6	7	8	9	10
Random number for generating requests	.82	.12	.84	.98	.46	.70	.08	.76	.77	.82
Number of requests	2	0	2	2	0	1	0	2	2	2
Random numbers for investigation times	1.57 −0.23	0	1.44 0.05	−2.56 2.23	—	0.81	—	0.17 −1.06	2.39 0.25	−1.17 0.08
Investigation time 2.5 ± 0.5 (*RNN*)	3.28 2.38	0	3.72 2.75	1.22 3.66	0	2.9	0	2.56 1.97	3.6 2.62	1.92 2.54

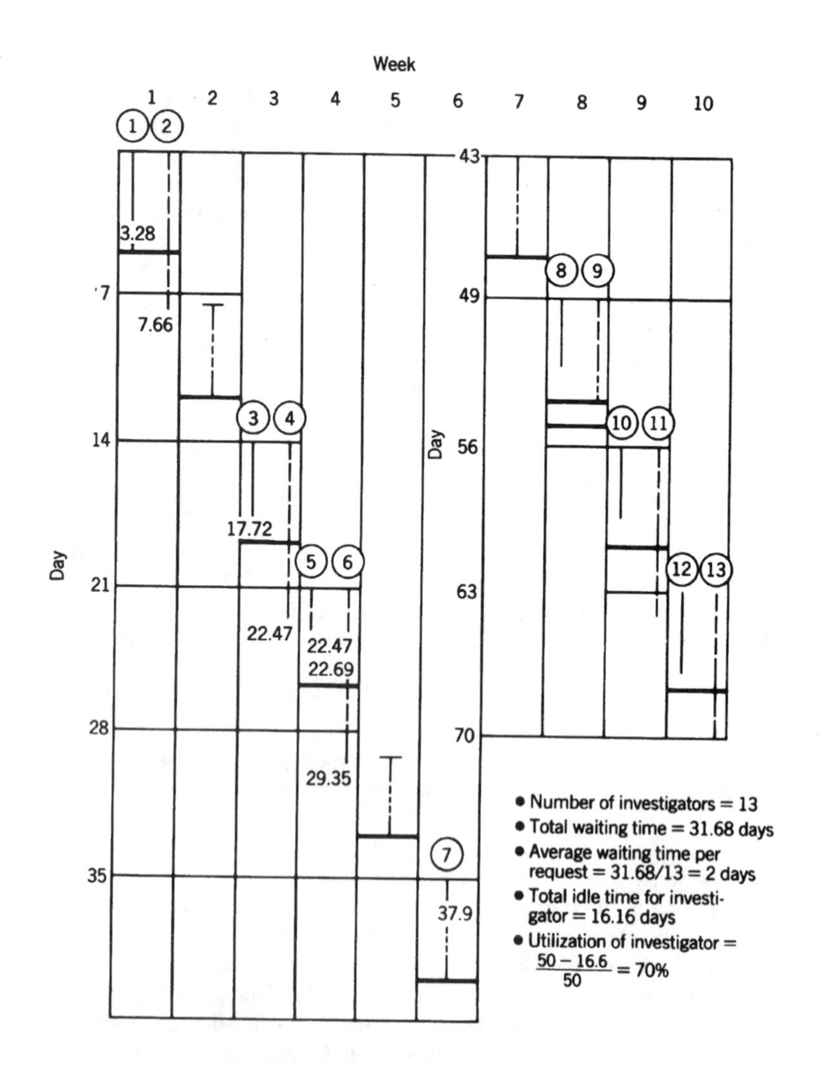

The preceding adaptation to handle simulations of models with resources and time constraints is a simple exercise that is characterized by very few events and activities. It should not be difficult, however, to appreciate how cumbersome and time consuming the procedure gets when we start dealing with several events coupled with a hierarchy of activities. In other words, it would no longer be the case of scheduling one activity, such as hygiene investigation, but rather the case that the start or completion of one or more activities will dictate the events to be scheduled. This brings us a bit closer to requiring a computerized method for handling our simulation.

7.4.10 Computer Simulation Languages

The two general classes of simulation systems are discrete and continuous. *Discrete* means that we are interested in monitoring the system of interest only at some finite or discrete points in time. Accordingly, the simulation is a sequence of "snapshots," each of which is taken at a time the system changes state. Each of these points in time changes information on the status of the system and its activities. Because of these changes, simulation time is incremented in steps (see Fig. 7.15), each of which corresponds to duration of an activity, delay time spent in a queue, and so forth.

Discrete simulations are applicable to the class of simulation models where decisions concerning starting of activities or assignment of resources are made at the occurrence of some events. During the time between events we assume that the status of the system being modeled remains in its state.

As an example, consider the OSHA inspection situation of the preceding example. To have an inspection, we need (1) an inspector who is free and (2) a site where the inspection will be carried out. After the inspection is commenced, there is no change in the system until the inspector becomes free or a new request for inspection is received. Either of these (free inspector or request for inspection) constitutes an event that will cause us to change the state of the system. That is precisely the way we carried out the simulation for the OSHA example.

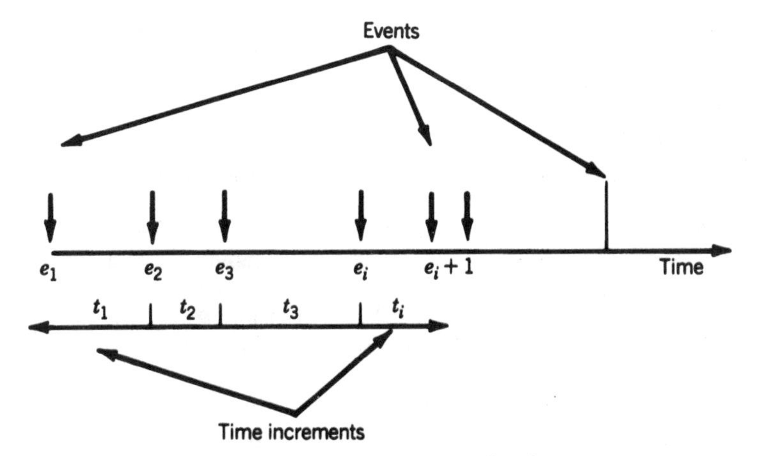

Fig. 7.15 Increments of simulation time.

Continuous algorithms are utilized for systems whose variables vary in a continuous fashion. At any point, the value of a variable depends on, among other things, its previous value and the current time. It is customary for continuous algorithms to represent the relationships among successive values by using difference equations and/or differential equations. Usually simulation time is represented as a sequence of equal intervals, each of a relatively short duration so as to include any changes. Economic, biological, and political simulation models are highly suited for continuous algorithms. An example of a continuous simulation language is the continuous systems modeling program (CSMP) that has been used to simulate the cardiovascular system. As another example, consider the safety management model developed by Diehi (1974). Utilizing the approach of systems dynamics, Diehi's model is an attempt at defining the impact of (1) organizational financial posture, (2) management attitude, (3) type of production technology utilized, and (4) investment in accident (disease) prevention measures on the total safety and health performance of the firm. The system dynamics models use complex feedback loops and relationships that involve people, materials, equipment, money, and production orders combined with expressions of their variables and changes in variables by rates and levels. Systems dynamics was originally developed by Forrester (1971) but remains an active language.

Short of programming our own simulation completely, we may consider the use of one of the many existing languages for simulation. A simulation language will implement the detailed requirements of a simulation and provide a convenient interface to construct models. Most newer simulation languages include animation capabilities. For a discussion of the most recent commercial (and noncommercial) simulation languages, see the latest *Proceedings of the Winter Simulation Conference*, which is held every year. This conference is a must for anyone interested in simulation and appeals to both the beginner and the expert.

7.4.11 Available Languages

Discrete simulation languages may be classified into two distinct groups: modeling and programming languages. *Modeling languages* present the modeler with a set of concepts and features that can be used directly in modeling. For example, a common approach is to provide the model as a flowchart consisting of building blocks and functional components. Later, or simultaneously, the software can be used to create code for computer processing. GPSS, SIMAN, and SLAM are languages that are primarily used for modeling systems characterized by queues and limited resources.

In contrast to the modeling languages are the *programming languages*, which present the modeler with a collection of routines that can be used for carrying out many of the standard tasks typical of simulation languages (scheduling of events and activities, collection of statistics, maintaining the simulation clock, etc.). These languages give the user the sole responsibility for programming the simulation model into logically and functionally correct computer codes. SIMSCRIPT, SIMULA, and GASP fall into the category of programming languages. Languages such as these either have their own syntax and semantics or use those of a general purpose programming language, such as Fortran or C.

The fundamental difference between modeling and programming languages lies in the level of programming involvement expected from the modeler. Modeling languages require little or no experience with computer programming or with the mechanics of the simulation process itself. On the other hand, the use of programming languages mandates proficiency on the part of the modeler in both programming, and simulation algorithms.

In general, modeling languages offer the following advantages:

1. They are self-documenting. Through the use of standard symbols and graphics, the structure and basis of the simulation model (components, data, relationships, etc.) are easily communicated to others who may not have been involved with the model development process.

2. They free the modeler from many of the time-consuming and error-prone tasks needed to structure and translate simulation models into appropriate computer codes. In this respect, the modeler is involved only with the modeling process.

3. The modeler does not need to be proficient in any programming language (e.g., FORTRAN or SIMSCRIPT). He or she only needs to understand how to translate the model into the input data (building blocks) defined by the language.

4. Modeling languages often have a graphical interface that make modeling very easy by simply identifying modeling elements and responding to dialogs requested by the program processor.

Modeling languages, however, have the drawback of requiring a somewhat rigid structure, which makes modeling certain phenomena a chore. Indeed, for some large-scale applications, the use of a programming language is a necessary. However, to the disappointment of many transient simulation modelers, mastering one of the programming languages requires time and hands-on experience—two requisites that cannot be justified for minor one-time applications.

To illustrate the use of a modeling language, we use the Arena product to build a simulation and animation model in the following example.

Example 7.13: Workers in a plant periodically drop by the Nurse's Office to seek various treatments. Generally the nurse's assistant sees the arriving worker first. Most problems are of a minor nature and can be handled by the nurse's assistant. However, some of the workers need to be seen by the nurse. After being seen by the nurse, the worker goes back to the nurse's assistant to complete paperwork and is then released. The Nurse's Office has been the subject of some concern, and reorganization of it is being contemplated. Therefore, a simulation model is needed to explore some of the potential changes.

Data show that workers arrive randomly, at about 4 per hour (from this we can assume that the interarrival time is 12 min, distributed exponentially). The time spent with the nurse's assistant is best described by a lognormal with a mean of 9 min and a standard deviation of 3.6 min. About fifty percent of the workers need to see the nurse, and the time spent with the nurse can be described by an Erlang distribution of order 4 with an exponential mean of 3.5 min.

The following shows the Arena simulation and animation model of the Nurse's Office

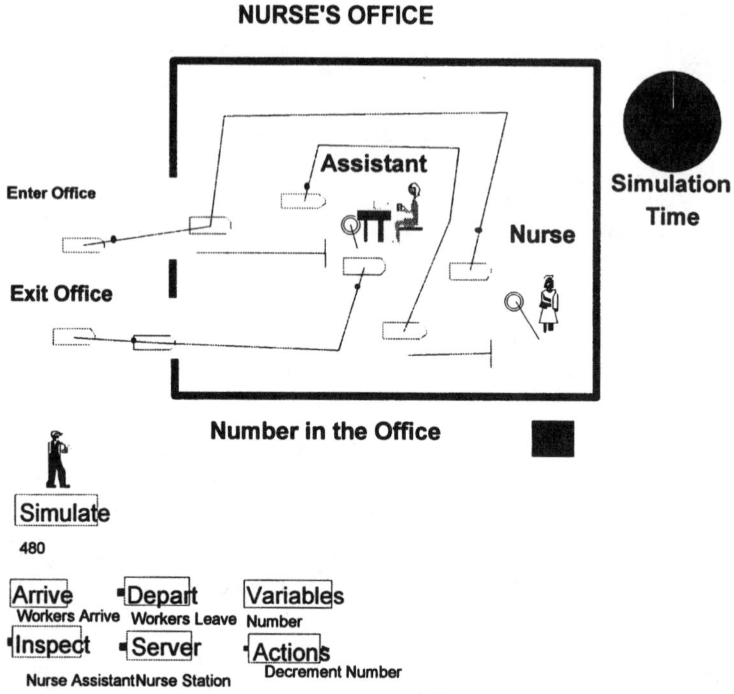

Ex. 7.13.1

The graphical part of the model is used to create the animation. The lower part of the model contains the model components, which include the provisions for the arrivals, services, departures, variables, and actions that take place in the model. The specification for these are seen in a dialog box when a modeler clicks on the component. Thus people building or using the model only need to know how the components are positioned together and how to fill out the needed information. In the previous model, we added "walking times" to permit the animation to show the various people moving about the office.

If the model is run for 480 min, then output is automatically produced as illustrated in Ex. 7.13.2. Therefore, you can see that both the nurse and nurse assistant are busy about 75 percent of the time. Workers spend about 5.3 min waiting for the nurse assistant and about 8.1 min waiting for the nurse. On the average workers spend about 52 min in the nurse's office from arrival to departure. However there were some extremes. For example, one worker spent as much as 165 min.

This simulation can be used to evaluate other organizational options. Perhaps a new kind of person should be added that can do some of what the nurse does but accepts workers directly? Maybe new equipment could help the office run more efficiently? All these questions can be evaluated using the simulation. Notice also, because of the animation, people who might not understand the numerical values from a simulation could look at the animation. They could "see" for themselves just what an impact each change would have.

```
                    ARENA Simulation Results
                    NC State - License #9400000

                  Summary for Replication 1 of 1

Project:                                  Run execution date :    2/23/1997
Analyst:                                  Model revision date:    2/23/1997

Replication ended at time       : 480.0

                            TALLY VARIABLES

Identifier              Average   Variation   Minimum    Maximum   Observations
_______________________________________________________________________________

Time in Office          52.058     .82025     8.1526     164.75        29
NurseQ Queue Time       8.0598     1.2482     .00000     32.436        26
AssistantQ Queue Time   5.2923     1.1190     .00000     22.002        56

                        DISCRETE-CHANGE VARIABLES

Identifier              Average   Variation   Minimum    Maximum   Final Value
_______________________________________________________________________________

Nurse Available         1.0000     .00000     1.0000     1.0000     1.0000
Assistant Available     1.0000     .00000     1.0000     1.0000     1.0000
Assistant Busy          .75341     .57210     .00000     1.0000     1.0000
# in NurseQ             .43657     1.5235     .00000     2.0000     .00000
# in AssistantQ         .61744     1.2893     .00000     3.0000     .00000
Nurse Busy              .75012     .57717     .00000     1.0000     .00000


Simulation run time: 0.03 minutes.
Simulation run complete.
```

Ex. 7.13.2

7.4.12 Games and Gaming

Simulation is traditionally used to evaluate a given system design or action plan. As discussed previously, simulation gives the user a paper-and-pencil laboratory to study various "what if" questions concerning a system and its performance. Another use of simulation is in the area of management gaming. A gaming situation results when (1) a decision has to be made by a person and (2) a consequence of this decision results. The decision outcome may be positive in that the person has perceived the correct situation and, accordingly, the system has performed as expected. Conversely, failure of the decision-maker to comprehend the prevailing conditions of the system can lead to erroneous decisions that may seriously affect system performance. By way of feedback, a person can be trained to make decisions, to integrate several pieces of information in order to define a prevailing condition, and to make use of some "rules of thumb" gained from continuous experimentation. Such training is usually given by on-the-job experience, a process that may consume years before an acceptable level of training is reached. To reduce somewhat the normal time span required for management decision-making, carefully structured computer games are widely utilized. The essence of any of the games is simple: A person is presented with a scenario and then asked to make a decision based on the information provided and his or her understanding of the situation. Next, the person is given some kind of feedback on the behavior of the system following implementation of the decision. At this point, the person may introduce some changes to the system or may elect to do nothing. The specific action will depend on the feedback and the judgment of the person involved. The process repeats itself for many cycles. As the person goes through the game cycles, his or her ability to make and reach the right decisions should be improved. In some situations, the person may capture the characteristics of the situation under consideration to the extent of developing some rules for making decisions. These rules can vary from defining some simple conditions for each decision to rules that would involve the use of highly developed decision models such as linear programming. In any event, the outcome of using a game, computerized or otherwise, will have its profound impact (in a positive sense) on the person's ability to make decisions and to "size up" given situations.

7.5 APPLICATIONS

The potential of simulation as a tool for safety and health research and applications is basically unlimited. It is readily applicable to a host of safety and health problems; the following is a partial list:

1. Study of the accident phenomenon, its underlying mechanism, its contributing variables (human, machine, and environment).
2. Evaluation of the effectiveness of various hazard control measures: control of physical hazards (alternatives for safeguarding), control of environmental hazards (selection of noise control approach), control of human behavior (training selection, enforcement).
3. Manpower planning for the safety and health organization based on organizational structure and available resources.
4. Inflationary impact of safety and health regulations given current state of compliance in industry, compliance alternatives, and potential compliance costs.
5. Evaluation of various modes for safety and health training—on-the-job retraining and retrofitting versus formal (college) training.
6. Analysis of the standard-setting process and its driving and restraining forces—politics, media campaigns, pressure groups, degree of public awareness, and comprehension of industrial hazards and economy.
7. Evaluation of alternative sites for storage of nuclear waste based on site topography, population at risk, and quality of transportation links between sites and waste producers.
8. Evaluation of land, sea, and air transportation systems for handling hazardous materials. Considerations should be given to characteristics (physical, chemical, etc.) of materials, reliability of transportation system employed, population exposed, feasibility and effectiveness of hazard containments following spills and regulatory constraints.
9. Effectiveness of safety communication within multiplant corporations. The simulation should consider: degree and extent of safety inspection and accident investigation practiced, the place of the safety and health departments within the organization, characteristics of data handling system, training and expertise of the safety and health staff.

10. Design of safety performance aids (safety signs, safe operating procedures, accident investigation manuals, etc.).
11. Evaluations of different patterns of materials flow within a given plant, taking into account characteristics of materials handled, materials handling equipment used, plant layout, regulatory requirements, and variations in product demands.
12. Impact of various work schedules (shift work, four-day workweek) on safety performance.
13. Development of management games for training safety and health professionals to handle effectively issues concerning the allocation and scheduling of resources; formulation of policies and procedures.
14. Assessment of the impact of various pre-employment screening program on safety performance and the regulatory requirements of Equal Employment Opportunity Commission (EEOC).
15. Prediction of health profiles of employees exposed to a mix of industrial hazards (with varying intensities and durations) throughout ten or more years of employment.

Two examples given below will illustrate some of the problem areas outlined.

7.5.1 Effectiveness of the Safety Organization

To develop a computerized approach for studying the effects of different management systems on the overall safety performance of the industrial firm, management gaming and simulation are used in the following model. (For a good review of gaming literature, see Belch, 1973, Gibbs, 1974.) The proposed model simulates the interaction between the accident process and the management process. The *accident* process describes the steps involved in producing accidents: hazard generation and transformation of the generated hazards into accident-producing situations. The *management* process deals with managerial decisions and acts that span the three phases of the accident control process: hazard recognition, hazard analysis, and hazard control.

Management and accident processes are time dependent. For instance, in the case of the accident process, time will elapse before a hazard is generated; following this, the hazard may be queued for some time before it produces an accident. On the other hand, the time behavior of the management process is dependent upon such items as the structure of the organization, existing safety policies, and the availability of the safety program. Using time as the control variable, the model simulates a race against time by management to eliminate hazards capable of causing either an accident or near accident. Figure 7.16 is intended to portray the basic philosophy behind the model: An accident or an incident occurs when protection normally provided that a safety program is absent. This protection could be in the form of personal protective equipment, machine guarding, isolation devices, substitution devices, or other equipment normally used to protect the worker from the hazards of the industrial environment. The concept, exemplified in Fig. 7.17, follows from what Kibbee et al. (1972) calls the "teeter-totter" principle. Kibbee relates his explanation for this principle to the field of economics, postulating the counterbalancing effect of a firm that distributes stock profits versus its working capital. Worker protection costs management an initial capital outlay that either may

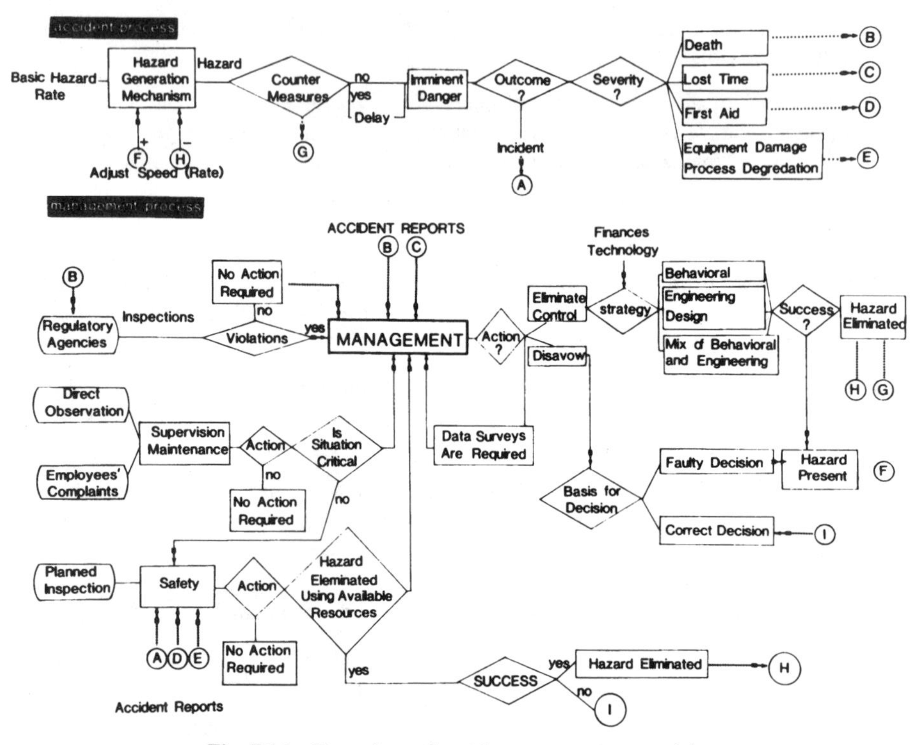

Fig. 7.16 Flow chart of accident management model.

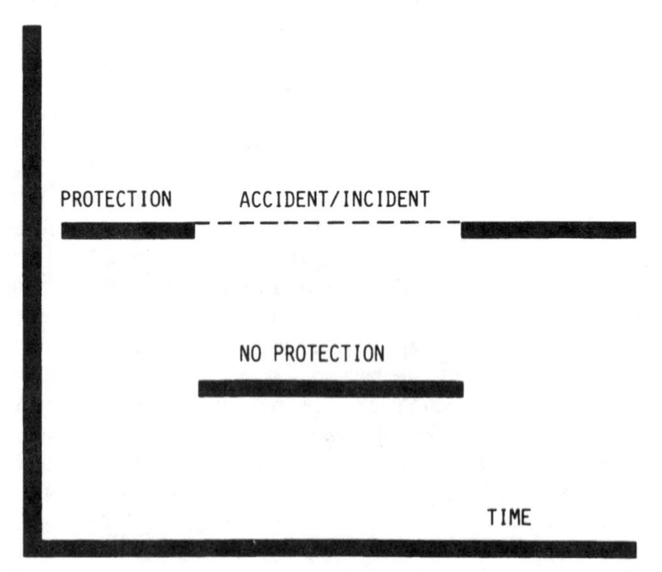

Fig. 7.17 Control sequence for avoidance or generation of an accident.

outweigh its usefulness, or prove a wise investment; however, the unprotected worker (or system) will be involved in an accident or an incident only while unprotected (the bottom section of the figure).

Hazards are assumed to be generated on a time basis; their occurrence is a discrete event, but their lingering presence is a continuous phenomenon. Each firm, institution, or others using the model must express the hazard generation rate at a daily function of its (OSHA) incidence rate. One way to be obtain this note is by using Heinrich's 1–29–300 classic pyramidal formula and extending the base by approximately the same proportion for one level. The ratio is then 1 : 29–300 = 3000, with the latter (Fig. 7.18) representing the number of hazards contributing to the top parts of the pyramid. By the use of proportion, the mean time between hazardous occurrences can be estimated for a given incidence rate—a rate based on the data included in the first two levels of the pyramid. As shown in Fig. 7.18, other formulae may also be used (Kann, 1974).

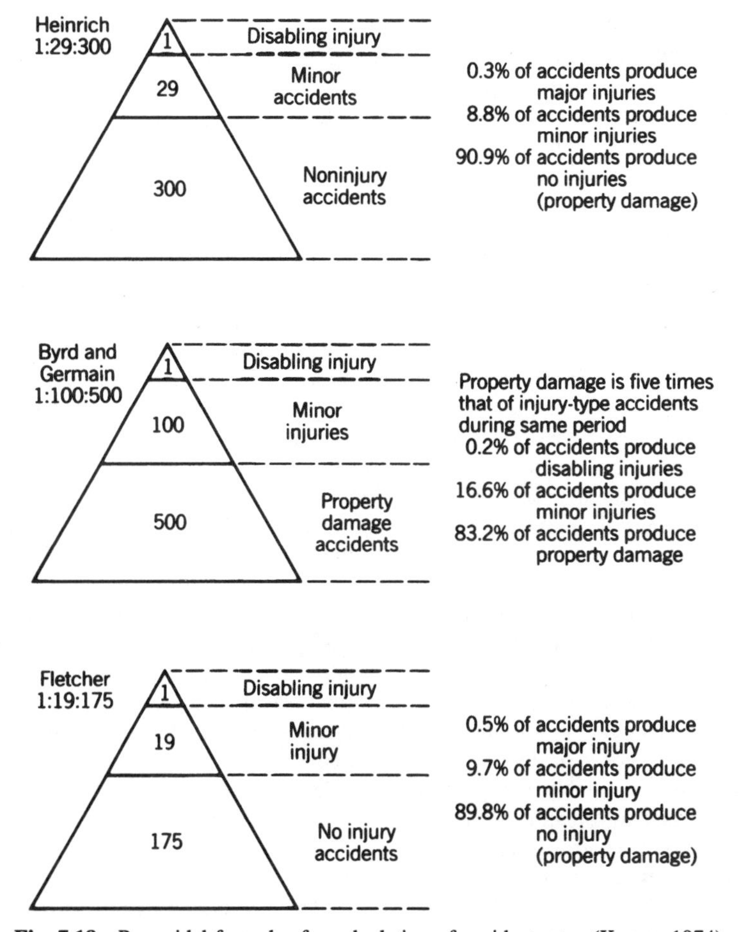

Fig. 7.18 Pyramidal formulas for calculation of accident rates (Kanon, 1974).

As the hazards are generated and subsequently queued for potential accident/incident cases, some fade and do not actually cause accidents or incidents (personnel obviously being keyed to avoid the hazard). Hazards culminating in either an accident or an incident can lead to one of four events; first aid, lost time, death, or process or equipment degradation, the latter being a result in which personnel are not injured. Based on Heinrich's formula, the probability of each of the four possible events can be derived. While a host of information inputs and events trigger the communication processes that ultimately lead to hazard elimination, the enumeration of hazards in terms of specific type, nature, and level is considered to be superfluous to the intent of the model. This philosophy is similar to that of Howard (1968); then indicate a growing consensus of safety specialists is that "it is infeasible to initiate system development by modeling a total organization's information system." Howard reports that such models have invariably failed, whereas others attempting to model smaller segments of the organization have been successful.

Communication is shown to begin either with the various personnel within the organization or from those outside the organization. It follows seven possible channels: inspections by regulatory agencies; direct observation by management or supervision; employee complaints; a rash of first aid cases; substantial losses due to equipment or process degradation; routine inspections from within the organization; and an accumulation of near accidents. Regulatory agency inspections are communicated directly to management, direct observations and employee complaints are passed on to supervision or maintenance, and all other information inputs (primarily generated from accident reports) are communicated to the safety function. Each communication channel is given a time value to indicate how often the flow of information can be expected to occur and also is given a regenerating loop to indicate the time interval between successive inputs.

Management, made aware of the hazardous condition, can be expected to do one of three things: (1) recognize the condition as hazardous and initiate hazard elimination procedures; (2) ask for further surveys and analyses before making a decision; or (3) conclude that the condition or practice reported to be hazardous can be accepted as a fair risk. Further surveys could result in no action or implementation of hazard control measures. No action, after a selected time interval, could lead to reevaluation by management, maintenance, or supervision. Probabilities are assigned for the three possible outcomes of the management decisions; time values are also assigned to indicate the amount of time required to perform the indicated activities and are expressed as slow, usual, and fast responses. If management decides to provide interim protection (such as personal protective equipment), then the accident/incident process will be slowed down. If none is provided, the process is speeded up or continues to operate at the same speed. The increase or decrease in speed of the model is represented by the section of the model under the accident process, labeled "countermeasures."

To summarize, the intensity of management repsonse will depend upon the value of the information received. For instance, management receiving an OSHA citation will undoubtedly move quickly and, perhaps, at considerable expense, to correct the hazards cited. In contrast, information contained in some inspection reports may not produce any immediate hazard-control actions.

Hazard elimination requires the availability of technology, finances, and time. First, we should be able to define an approach that is technically and otherwise feasible for controlling the recognized hazards. Second, cost of the selected control approach must be within reach, considering the organization resources and financial posture. Finally, it takes time to implement the selected control approach and to bring the recognized hazards effectively under control.

In general, regardless of hazard type and characteristics, control is achieved by using either behavioral or engineering approaches; in some situations, using a mix of the two can be justified. Behavioral approaches encompass such things as training, selection, and enforcement. On the other hand, engineering emphasizes the technical changes to be introduced in the workplace and work practices, for example, safeguarding of machines. Behavioral approaches can be implemented in a relatively short time. In contrast, their reliability (i.e., remaining effective in the long run) is rather low. Behavioral approaches usually cause a surge in a safety and health activities, the result being the elimination or control of most workplace hazards. However, this does not last long, and with time, the sharpened interest in safety activities will dissipate, and eventually control of the hazards will be all but lost. Nor are engineering approaches without shortcomings. Due to a host of factors (poor maintenance, wear and tear, etc.), hazard control through engineering approaches may not be successful under all circumstances and for all situations.

Therefore, any type of hazard elimination (control) is subject to either success or failure. If success (abatement) is obtained, then the accident/incident process is slowed down proportionately: failure increases the speed of the process. Information concerning failure to control the hazards—after a time interval necessary for detection—can trigger either a reevaluation through maintenance and supervision or an initiation of a new hazard-control attempt. As pointed out by Firenze (1973), hazard control is neither total failure nor total success. Stated differently and in light of the discussion above, hazard control approaches (behavioral or engineering) do have a limited service life, after which their effects will be lost and hazards will start to accumulate. When this point is reached, the system is changed once more to reflect the new speed of the process.

The steps involved in the simulation (running the model) are as follows:

Step 1. Define performance and decision parameters for various model components. In each case, three pieces of data are sought: the time it takes before a decision is reached, the type and number of inputs required to trigger a response, and the frequency with which each input is received. Whenever applicable, time parameters should be described using appropriate probability distributions.

Step 2. According to the organizational structure and existing policies and regulations, determine the different activities of the management process that will be active during the model simulation. For example, if the particular company under consideration does not have an active safety program, all information inputs to be generated by the safety department will not be considered during simulation.

Step 3. Simulate the model using input data generated from Steps 1 and 2.

Step 4. After examining the simulation results of Step 3, some policy and/or procedure changes may be introduced. The effects of these changes upon the overall safety performance of the organization can be ascertained by simulating the model several more times. This process of introducing or making changes, and then simulating, can be repeated until all possible policies have been assessed.

The model described above was coded and then tested by studying the effects of three management policies on the overall safety performance of a hypothetical firm. The three policies used were:

TABLE 7.12 Result of Executed Simulations

Condition	First aid	Process or equipment lost		
		Degradation	Time	Death
Safety program	36	405	0	0
No safety program	41	414	0	1
No safety program; further surveys required	54	500	2	1

1. An ongoing safety program does exist.
2. No safety program is available.
3. No safety program is available; management is not always supportive of safety activities and expenditures.

Data for the various components of the model were estimated by conducting a survey of ten furniture plants in North Carolina. A summary of ten simulations (each representing a full year of performance) is shown in Table 7.12. The results were in the expected vein; that is, the firm with the safety program and the firm with no safety program fared better than the firm with no safety program and in which management always asked for further surveys (considering the lowest number of observances as the better performance).

To test accurately the impact of various organizational structures and policies, one needs to go a little bit further than what has been presented thus far. That is, a full-fledged experiment should be designed and carried out using the proposed model; such an experiment in itself would constitute a worthwhile study.

7.5.2 Evaluation of Maintenance Manuals

Preventive and corrective maintenance activities do constitute an important and integral part of any trustworthy hazard-control as well as accident-prevention program. Any practical system can fail from time to time, no matter how remote the chance is kept through system design (reliability) and preventive maintenance. To fix a system, the failing components have to be identified, then repaired or replaced with equivalent components (spares). The process of identifying the failing components is called *troubleshooting*—an important concept in corrective maintenance.

Successful troubleshooting requires knowing the system and its operation; this was relatively easy when systems were simple and small. However, as systems grew larger and more complex, it became difficult, time consuming, and sometimes even impossible for a single person to acquire all the knowledge needed to maintain such systems (Elliott, 1967). To overcome this difficulty, fully procedural troubleshooting guides are used. These are step-by-step procedures that guide maintenance personnel through the troubleshooting process. A guide tells the maintenance person where to check first, how to interpret the check result, and then what to do *next*. The purpose of using fully procedural guides is to minimize the number of decisions to be made by the maintenance person.

Several studies have demonstrated the effectiveness of fully proceduralized guides in performing *maintenance* activities (Elliott, 1967; Post and Price, 1973; Foley and Camm, 1972; Joyce et al. 1973; Smillie and Ayoub, 1977). Fewer errors and shorter maintenance time can be expected when troubleshooting guides are used. Furthermore, the use of guides can practically eliminate the need for training or reduce it to as little as a few dozen hours. Troubleshooting guides, however, have some limitations. Some of these are:

1. In designing a troubleshooting guide, it is practically impossible to deal with every potential system problem; for example, a conductive metal can drop into electronic circuits and short adjacent circuits. This could cause a malfunction of the system. Since the short can occur in any place, it would be impossible to design a guide for isolating and repairing this type of fault.

2. Fully proceduralized guides do not promote opportunities for logical reasoning and learning. This can have a serious impact on the attitude and motivation of maintenance personnel.

3. The guides are designed on the implicit assumption that when a system fails, the initiating causes (troubles) will remain in a permanent state—the trouble will appear every time a check is made.

This is not always the case for many systems. Indeed, troubles are known to appear intermittently. Therefore, using fully procedural guides to deal with intermittent troubles may not be very effective, and in some instances, may lead to erroneous conclusions.

The degree to which trouble intermittency appears varies a great deal within and among system components. Some troubles may appear once in three or four tests; some may appear once in several hours of operation or perhaps every month or two. As long as the trouble appears relatively frequently, it will not too severely impact the effectiveness of the guide. On the other hand, if the trouble occurs rarely, say, once a month, it would be almost impossible to pin down the failing component on the first attempt. The issue then becomes one of identifying under what conditions of trouble intermittency a fully proceduralized guide would lose its effectiveness. For this and similar issues, the following model can provide a starting point.

7.5.3 A Troubleshooting Model

We consider any troubleshooting process as a sequence of the following steps (see Figure 7.19):

1. Intermittent trouble occurs and a maintenance person is called in.
2. He makes preparation for troubleshooting (gets symptom information from the operator, removes cover, etc.).
3. He runs a test.
4. If the trouble appears, he proceeds to the next step. If not, he skips to Step 6.
5. He follows the subsequent instruction steps until he is directed to the failing component or comes to the next test run. When the failing component is found, the troubleshooting process is completed. Otherwise, he goes back to Step 3.

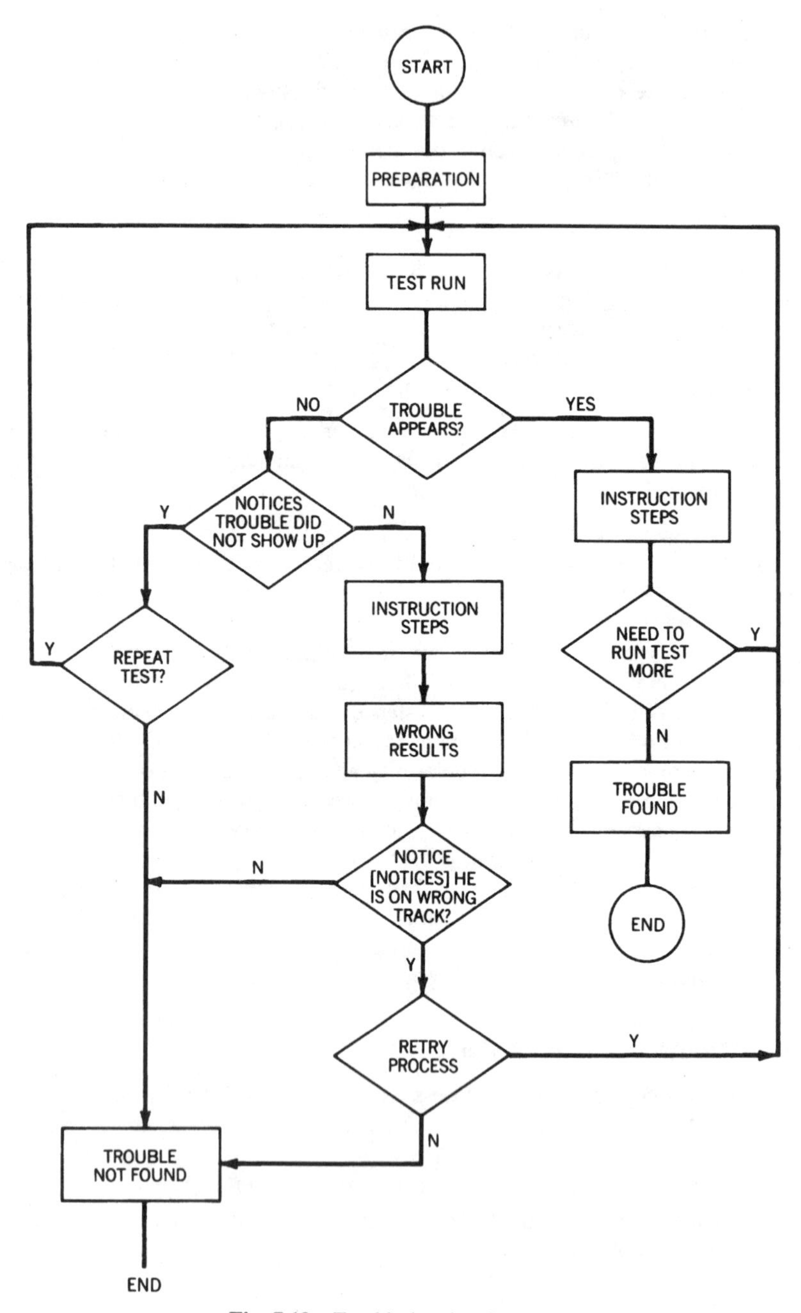

Fig. 7.19 Troubleshooting flowchart.

6. He may or may not notice that the trouble did not show up during the test. If he notices, he would repeat the test within some time limits or for a certain number of trials. In this case, he goes back to Step 3. If repeating the test does not identify the failing component, he would quit further troubleshooting, assuming that the trouble is too intermittent. In this case, the system is not fixed and most likely will fail later.
7. If he fails to recognize that the trouble did not occur, then he will follow a wrong path in the guide. And after following some instruction steps, he will conclude the test without locating the malfunctioning component.
8. He may or may not notice that he was on a wrong track. If he does not, the trouble is not really fixed. If he does, there are two possible consequences. One is to quit right there, while the other is to repeat the whole process from the beginning, that is, go back to Step 3.

The troubleshooting process described above was simulated with the following specifications and parameters.

1. Availability of a minicomputer with a builtin exerciser for logic testing is assumed. This test takes three minutes to run.
2. Time to perform instruction steps has a beta distribution.
3. Troubles appear intermittently. If the trouble is not encountered during a test run, the test will be repeated up to 5 times.
4. On the average, 2.5 decision steps will be used in any given troubleshooting session. The distribution of the number of decision steps is assumed to be normal, with standard deviation equal to 0.8.
5. If the maintenance person notices that he was not following the right sequence of instructions, he may repeat the process only once.

The results of 500 simulations are given in Table 7.13 and Fig. 7.20. As shown in Fig. 7.20, the effectiveness of the troubleshooting guide drops very rapidly as the trouble intermittency decreases (trouble appears less and less). When trouble appears very intermittently (probability less than 0.1), the troubleshooting guide is almost useless (Table 7.13). This becomes significant if we consider an extreme case:

TABLE 7.13 Guide Effectiveness

Trouble intermittency	Guide effectiveness	Average diagnostic hour
0.9	0.98	81.55
0.7	0.93	101.34
0.6	0.83	113.16
0.5	0.65	122.53
0.4	0.47	122.05
0.3	0.25	122.20
0.1	0.01	83.60

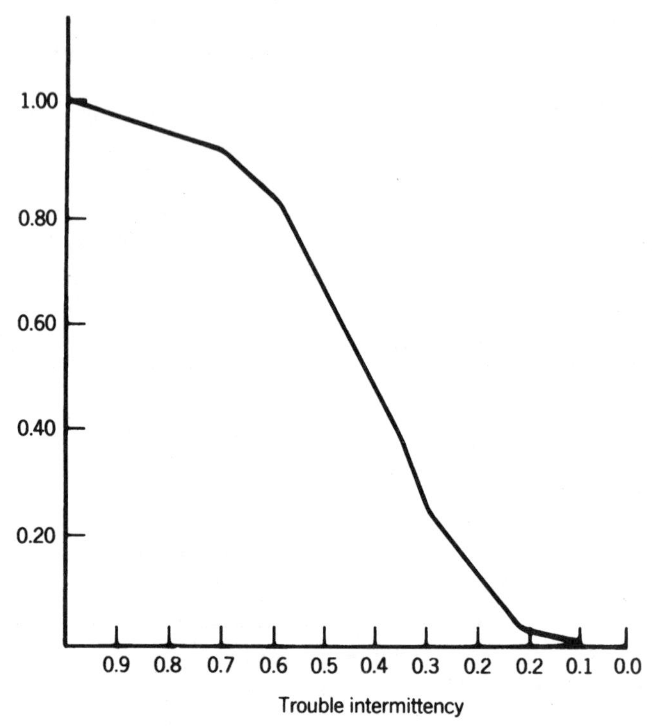

Fig. 7.20 Trouble intermittency versus guideline effectiveness.

A problem appears, say, once a month. Suppose we could tell whether the trouble appears or not, and we could see the trouble at the first decision point in the guide. To proceed further, we have to wait at the next decision point until the trouble appears again, which may be a month ahead.

Generally speaking, by increasing the number of test repetitions, the effectiveness (the probability of locating the troubles) will be improved. However, it will be at the expense of more troubleshooting hours. The fully procedural troubleshooting guides can be very ineffective on intermittent failures. It will be necessary to improve the guides in this respect or to develop a new maintenance concept to deal with intermittent failures. Post and Price (1973) conducted a study on maintenance documentation comparing the traditional maintenance documents (deductive maintenance approach) and the proceduralized troubleshooting guide (directive maintenance approach). They suggest an integrated approach that takes advantage of both approaches.

The model just described may be used to explain how, under certain conditions, accident investigation can be misleading, even with the best intentions and availability of seasoned investigators. For example, consider the following hypothetical situation.

1. A traffic intersection was controlled with four-way stop signs. Following a rash of serious accidents, it was decided to replace the stop signs by a traffic light. Here most of the accidents at the intersection were attributable to poor traffic control. Indeed, after the traffic light was installed, serious accidents at the same intersection were eliminated.

In other words, reviewing some of the accident cases led the investigator to define the absence of a traffic light as the primary cause of the problem. Since the intersection had no traffic light, that is, the trouble appeared permanently; accordingly, it was not difficult for the investigator to reach the proper conclusion.

2. Consider the same traffic intersection, now equipped with a traffic light. A few months after the installation of the traffic light, a sequence of serious accidents occurred at the intersection. Careful analysis of these accidents led to the inevitable conclusion that identified human error was the primary cause, for every time an investigation was conducted, each aspect of the intersection, including the traffic light, was according to standards and specifications. However, the number of accidents did not decrease and persisted even after additional warning signs were posted at the intersection. With this as background, a special study of the intersection and its traffic was undertaken for a period of several months. This lengthy investigation revealed that the traffic light did not function properly all the time, and in many instances it gave the right of way to both traffic directions simultaneously—an open invitation to accidents.

In this case, the trouble (malfunctioning traffic light) was intermittent to the extent that every time an investigation was undertaken (following an accident) it did not appear; every investigation concluded that the traffic light was functioning properly. When it was discovered to be otherwise, it was too late as well as being costly.

The essence of the traffic case just described can be found (at least conceptually) in many accident-producing situations, for the simple reason that we deal with people who can change their predictable responses as the characteristics and nature of their working environments change. These changes can be periodic and might be difficult to trace following an accident—an important prerequisite for accident reconstruction.

REFERENCES

Anderson, V. L., and McLean, R. A., *Design of Experiments: A Realistic Approach*, Marcel Dekker, New York, 1974.

Ayoub, M. A., Integrated Safety Management Information System; Part II: Allocation of Resources, *J. Occupational Accidents* **2:** 135–57, 1979.

Banks, J., Carson II, J. S., and Nelson, B. L., *Discrete-Event System Simulation*, 2nd ed., Prentice Hall, Upper Saddle River, New Jersey, 1996.

Belch, J., *Contemporary Games: A Directory and Bibliography Covering Games and Play Situations Used for Instruction and Training by Schools, Colleges and Universities, Government, Business and Management*, Gale Research Company, Detroit, 1973.

Benson, D., Simulation Modeling and Optimization using ProModel, in *Proceedings of the 1996 Winter Simulation Conference*, Ed. J. M. Charnes, D. J. Morrice, D. T. Brunner, and J. J. Swain, IEEE, Piscataway, NJ, pp. 447–52, 1996.

Binun, M., Business Process Modeling with SIMPROCESS, in *Proceedings of the 1996 Winter Simulation Conference*, Ed. J. M. Charnes, D. J. Morrice, D. T. Brunner, and J. J. Swain, IEEE, Piscataway, NJ, pp. 530–41, 1996.

Birtwistle, G. M., Dahl, O., Myhrhaug, B., and Nygaard, K., *Simula BEGIN*, Petrocelli/Charter, New York, 1973.

Bratley, P., Fox, B. L., and Schrage, L. E., *A Guide to Simulation*, 2nd ed., Springer–Verlag, New York, 1987.

Carroll, D., MedModel—Healthcare Simulation Software, in *Proceedings of the 1996 Winter Sim-*

ulation Conference, Ed. J. M. Charnes, D. J. Morrice, D. T. Brunner, and J. J. Swain, IEEE, Piscataway, NJ, pp. 441–46, 1996.

Diehi, A. E., *A System Dynamics Simulation Model of Occupational Safety and Health Phenomena.* Unpublished Ph.D. Dissertation, North Carolina State University, Raleigh, NC, 1974.

Elliott, T. K., *Development of Fully Proceduralized Troubleshooting Routines*, Aerospace Medical Research Laboratories Technical Report 152, November, 1967.

Emshoff, J. R., and Sisson, R. L., *Design and Use of Computer Simulation Models*, Macmillan, New York, 1970.

Firenze, R. J., The Logic of Hazard Control Management, in: Widner, J. T., *Selected Readings in Safety*, Academy Press, Macon, GA, 1973.

Fishman, G. S., *Principles of Discrete Event Simulation*, Wiley, New York, 1978.

Foley, J. P., and Camm, W. B., *Job Performance Aids Research Summary*, Air Force Systems Command, August, 1972.

Forrester, J. W., *Principles of Systems*, Wright-Allen Press, Cambridge, MA, 1971.

Gane, C., and Sarson, T., *Structured Systems Analysis: Tools and Techniques*, Prentice-Hall, Englewood Cliffs, NJ, 1979.

Gibbs, G. I., Eds., *Handbook of Games and Simulation Exercises*, Sage Publications, Beverly Hills, California, 1974.

Glover, F., Kelly, J. P., and Laguna, M., New Advances and Applications of Combining Simulation and Optimization, in *Proceedings of the 1996 Winter Simulation Conference* Ed. J. M. Charnes, D. J. Morrice, D. T. Brunner, and J. J. Swain, IEEE, Piscataway, NJ, pp. 144–151, 1996.

Goldberg, A., and Robson, D., *Smalltalk-80: The Language and Its Implementation*, Addison-Wesley, Reading, MA, 1983.

Graybeal, W., and Pooch, V. W., *Simulation: Principles and Methods*, Winthrop Publishers, Cambridge, MA, 1980.

Howard, J. A., and Morgenroth, W. M., Information Processing Model of Executive Decision, *Management Science* **14** (March 1968): 416–28.

Jankauskas, L., and McLafferty, S., BestFit, Distribution Fitting Software by Palisade Corporation, in *Proceedings of the 1996 Winter Simulation Conference*, Ed. J. M. Charnes, D. J. Morrice, D. T. Brunner, and J. J. Swain, IEEE, Piscataway, NJ, pp. 551–55, 1996.

Joyce, R. P., Chenzoff, A. P., Mulligan, J. F., and Mallory, W. J., *Fully Proceduralized Job Performance Aids* (Three Books), Air Force Systems Command, December, 1973.

Kanon, J. C., Safety Standards—The Systematical Approach and Methodical Basis for Personal Safety and Damage Control, in *Symposium on Working Place Safety*, 22–26 July, 1974, Bad Grund, FRG, 1974.

Kibbee, J., Kraft, C., and Nanua, B., Management Games, in: Carlson, J. G. H., and Misshauk, M. J., *Introduction to Gaming: Management Decision Simulations*, John Wiley, New York, 1972.

Law, A. M., and Kelton, W. D., *Simulation Modeling and Analysis*, 2nd ed., McGraw–Hill, New York, NY, 1991.

Law, A. M., and McComas, M. G., ExpertFit: Total Support for Simulation Input Modeling, in *Proceedings of the 1996 Winter Simulation Conference*, Ed. J. M. Charnes, D. J. Morrice, D. T. Brunner, and J. J. Swain, IEEE, Piscataway, NJ, pp. 588–593, 1996.

Lazzari, D., and Crosslin, R., Introduction to Work Flow Modeling with BPSimulator, in *Proceedings of the 1996 Winter Simulation Conference*, Ed. J. M. Charnes, D. J. Morrice, D. T. Brunner, and J. J. Swain, IEEE, Piscataway, NJ, pp. 429–31, 1996.

Linstone, H. A., and Turnoff, M., *The Delphi Method: Techniques and Application*, Addison–Wesley, Reading, MA, 1975.

Markovitch, N. A., and Profozich, D. M., Arena Software Tutorial, in *Proceedings of the 1996*

Winter Simulation Conference, Ed. J. M. Charnes, D. J. Morrice, D. T. Brunner, and J. J. Swain, IEEE, Piscataway, NJ, pp. 437–40, 1996.

Pegden, C. D., Shannon, R. E., and Sadowski, R. P., *Introduction to Simulation Using SIMAN*, 2nd ed., McGraw–Hill, 1995.

Phillips, D. T., *Applied Goodness of Fit Testing*, AIIE Monograph Series, AIIE-OR-72-1, Atlanta, GA, 1972.

Post, J. T., and Price, H. E., *Development of Optimum Performance Aids for Troubleshooting*, BioTechnology, Virginia, April 1973.

Pritsker, A. A. B., *Introduction to Simulation and SLAM II*, 4th ed., John Wiley, New York, 1995.

Pritsker, A. A. B., and O'Reilly, J. J., AweSim: The Integrated Simulation System, in *Proceedings of the 1996 Winter Simulation Conference*, Ed. J. M. Charnes, D. J. Morrice, D. T. Brunner, and J. J. Swain, IEEE, Piscataway, NJ, pp. 447–52, 1996.

Pulat, B. M., *Computer Aided Panel Design and Evaluation System—CAPADES*, Ph.D. Dissertation, Dept. of Industrial Engineering, North Carolina State University, Raleigh, NC, 1980.

Rider, K., Hausner, J., Shortell, R., Bligh, J., and Candeloro, T., *An Analysis of the Deployment of Fire-Fighting Resources in Jersey City, New Jersey*, The Rand Corporation, R-1566/4-HUD, August 1975.

Russell, E. C., *Building Simulation Models with SIMSCRIPT II.5*, CACI Products Company, LaJolla, CA, 1983.

Schriber, T. J., *An Introduction to Simulation Using GPSS/H*, John Wiley, New York, 1991.

Shannon, R. E., *Systems Simulation: The Art and Science*, Prentice–Hall, Englewood Cliffs, NJ, 1975.

SAS User's Guide, SAS Institute, Inc., Raleigh, NC, 1979.

Smillie, R. J., and Ayoub, M. A., Accident Causation Theories: A Simulation Approach, *Journal of Occupational Accidents*, **1** (1976): 47–68. SPSS for the IBM PC, Chicago, 1984.

CHAPTER 8

EVALUATION OF EXPOSURE TO CHEMICAL AGENTS

JERRY LYNCH and CHARLES CHELTON

8.1 SAMPLING STRATEGY
- 8.1.1 Purpose of Measurement
- 8.1.2 Environmental Variability
- 8.1.3 Location
- 8.1.4 Sampling Period
- 8.1.5 Frequency
- 8.1.6 Method Selection

8.2 METHODS FOR GASES AND VAPORS
- 8.2.1 Absorbent Tubes
- 8.2.2 Passive Badges
- 8.2.3 Impingers and Liquid Traps
- 8.2.4 Length of Stain Tubes
- 8.2.5 Evacuated Containers and Bags
- 8.2.6 Direct Reading Instruments

8.3 METHODS FOR AEROSOLS
- 8.3.1 Method Selection
- 8.3.2 Filter Sampling Methods
- 8.3.3 Impactors
- 8.3.4 Impingers
- 8.3.5 Direct Reading Instruments

8.4 GENERAL CONSIDERATIONS
- 8.4.1 Planning the Collection of a Sample
- 8.4.2 Analytical Lab Services and Chain of Custody
- 8.4.3 Instrument Calibration, Verification, and Maintenance
- 8.4.4 Sampling Equipment and Instrument Certification
- 8.4.5 Radio Frequency Effects
- 8.4.6 Shipping
- 8.4.7 Records
- 8.4.8 Data Logging

Adapted from "Measurement of Worker Exposure" by Jeremiah R. Lynch, Chapter 2 in Harris, Cralley, and Cralley, *Patty's Industrial Hygiene and Toxicology, 3rd ed., Vol. IIIA*. New York: John Wiley & Sons, Inc., 1994, pp. 27–80.

Handbook of Occupational Safety and Health, Second Edition, Edited by Louis J. DiBerardinis, ISBN 0-471-16017-2

8.5 SURFACE SAMPLING
8.5.1 General
8.5.2 Dermal Exposure Assessment
REFERENCES
APPENDIX A CONSIDERATIONS IN ESTABLISHING AN INDUSTRIAL HYGIENE FIELD OFFICE
APPENDIX B INDUSTRIAL HYGIENE FIELD OFFICE INVENTORY CHECKLIST
APPENDIX C DIRECT READING INSTRUMENT INVENTORY
APPENDIX D PRELIMINARY INDUSTRIAL HYGIENE EVALUATION REPORT
APPENDIX E FIELD SAMPLING REPORT
APPENDIX F INDUSTRIAL HYGIENE MONITORING RESULTS

8.1 SAMPLING STRATEGY

This chapter explains why workplace measurements of air contaminants are made, discusses the options available in terms of number, time, and location, and relates these options to the criteria that govern their selection and the consequences of various choices.

A person at work may be exposed to many potentially harmful agents for as long as a working lifetime, upward of 40 years in some cases. These agents occur singly and in mixtures, and their concentration varies with time. Exposure may occur continuously, at regular intervals, or in altogether irregular spurts. The worker may inhale the agent or be exposed by skin contact or ingestion. As a result of exposure to these agents, they come in contact with or enter the body of the worker, and depending on the magnitude of the dose, some harmful effects may occur. All measurements in industrial hygiene ultimately relate to the dose received by the worker and the harm it might do.

Changes in working conditions, in technology, and in society have changed old methods of measurement.

- With few exceptions, workplace exposure to toxic chemicals is much below what is commonly accepted as a safe level.
- As a consequence of the reduction of exposure, frank occupational disease is rarely seen. Much of the disease now present results from multiple factors, of which occupation is only one.
- Workers have the right to know how much toxic chemical exposure they receive, and this often results in a need to document the absence of exposure.
- Technology provides enormously improved sampling equipment that is rugged and flexible. This equipment, used with analytical instruments of great specificity and sensitivity, has largely replaced the old "wet" chemical methods.

As a consequence of these changes in the workplace and advances in technology, it is now both necessary and possible to examine in far more detail the way in which workers are exposed to harmful chemicals. Personal sampling pumps permit collection of contaminants in the breathing zone of a mobile worker. Pump–collector combinations are available for long and short sampling periods. Passive dosimeters, which do not require pumps, are available for a wide range of gases and vapors. Systems that do not require the continual attention of the sample taker permit the simultaneous collection of multiple samples. Data loggers can continuously record instrument readings in a form easily transferable to a computer. Automated sampling and analytical systems can collect data continuously. Sorbent-gas chromatography techniques permit the simultaneous sampling and analysis of mixtures and, when coupled with mass spectrometers, the identification of obscure unknowns. Analytical sensitivities have improved to the degree that tens and hundreds of ubiquitous trace materials begin to be noticeable.

As a starting point for a complete assessment of the risk to health posed by an occupational environment, it is necessary to know the substances to which workers are exposed. Systematic recognition of all possible hazards requires a review of inventories of the materials brought into the workplace, descriptions of production processes, and identification of any new substances, byproducts, or wastes. However, these sources of information may not be enough to identify all substances, particularly those present as trace contaminants or substances generated by production process, either inadvertently or as unknown byproducts. To complete the identification of all substances present, before going to the next step of evaluating exposure and risk, it may be necessary to make some substance recognition measurements. Since these measurements, which are typically made by such techniques as gas chromatography–mass spectrometry (GC-MS), are not intended to evaluate exposure, they may be area rather than personal samples and may be large-volume samples for maximum sensitivity.

The term *sampling strategy* as used here means the analytical reasoning used to decide how to make a set of measurements to represent exposure for a particular purpose. The measurements should yield data that are logically and statistically adequate to provide information to describe the environment under assessment (see Table 8.1). An optimum strategy is that selection of options under the control of the exposure assessor, which efficiently achieves the objective given the physical circumstances and environmental variability (HSE, 1989). The occupational exposure assessment charts in Figs. 8.1 and 8.2 provide tools that can be used to accomplish this end.

TABLE 8.1 Objectives of Exposure Assessment

Hazard recognition	Identification of the presence of trace byproduct or waste stream hazards that are not identifiable through review of materials inventories and process descriptions
Exposure evaluation	Comparing measured exposure to reference exposure levels to asses risks to employee health
Assessment of control method effectiveness	Evaluating changes in exposure levels resulting from changes to process operations work products or modifications to engineering controls
Model validation	Comparing field measurements with predicted values to evaluate the powers and limitations of a model to predict responses
Method research	Validation of new and alternative techniques of sample collection and analysis
Source evaluation	Point source identification and determination of magnitude of emissions
Operational evaluation	Identification of sources and tasks that contribute to employee exposure
Epidemiology	Provide employee exposure data for epidemiological investigation into temporal and population trends in occupational illness

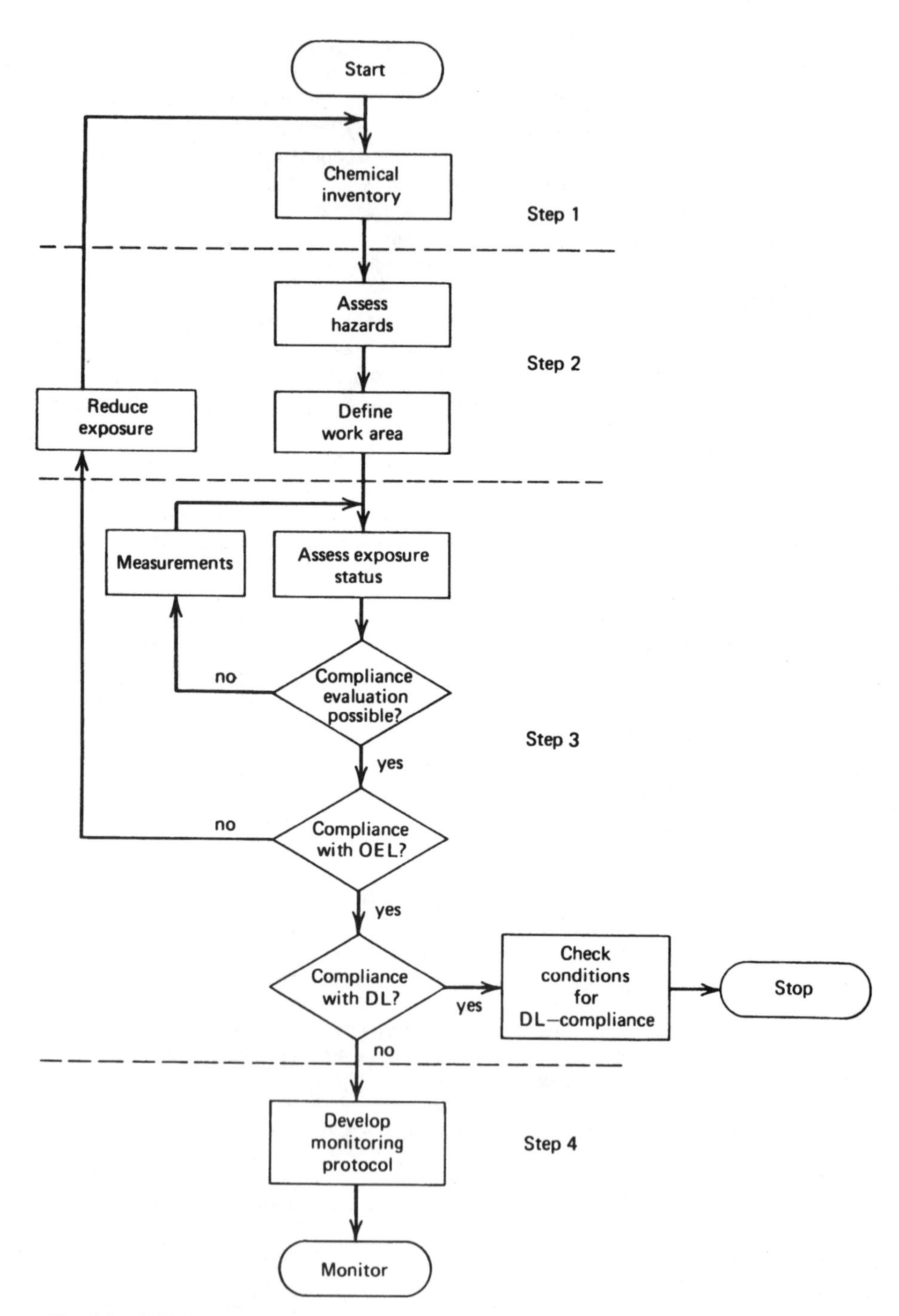

Fig. 8.1 CEFIC occupational exposure analysis chart. The occupational exposure level (OEL) and decision level (DL) used depends on the substance being evaluated.

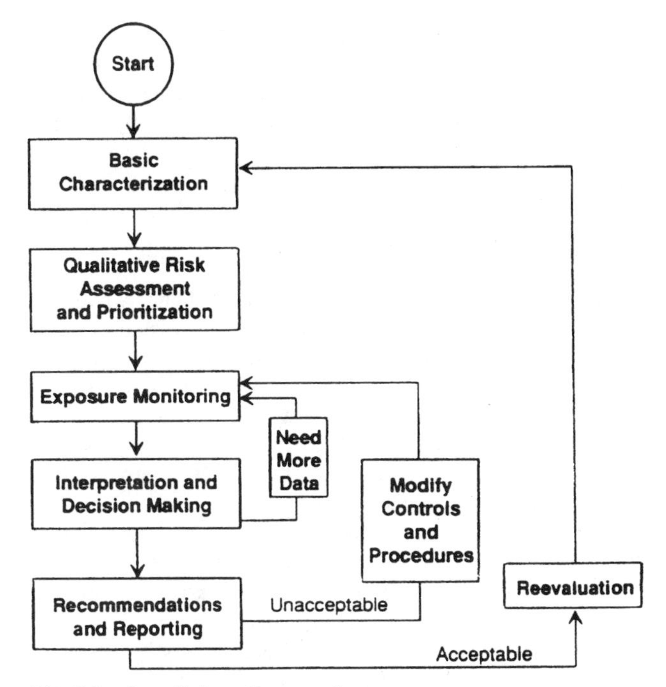

Fig. 8.2 Overall flow diagram of an exposure assessment strategy.

8.1.1 Purpose of Measurement

The development of a sampling strategy requires a clear understanding of the purpose of the measurement. Rarely are data collected purely for their own sake. Even when data are collected because of a demand by others, such as the government or employees, the use of the data should be considered.

> What questions will be answered by the data? What decisions could depend on those answers? For example, do we want to know whether a group of workers are overexposed? And if so, will a decision to take action to control the exposure follow? Is the control likely to be a minor change in a work practice or an expensive engineering modification? Or are the numbers to be assembled to answer the question: What level of control is currently being achieved in this industry? And may this answer lead to new decisions regarding what control is feasible? Last, will workers use the result to find out whether their health is at risk and, as a consequence, decide to change jobs or seek changes in the conditions of work?

Often there are several questions that need answers, and thus data are collected for multiple purposes. Table 8.1 summarizes the most common objectives of sample gathering. As the discussion that follows indicates, the purpose of the data determines the design of the measurement scheme. All too often data intended for multiple purposes turn out not to be suitable for any purpose. Thus it is usually necessary to focus on the prime need to be sure the strategy will meet this requirement. If possible, minor adjustment or additions can be made to meet other needs. The optimum sampling strategy is that which combines the choice of method and sampling scheme with respect to sampling location, time, frequency, and sample number so that we are confident that the data are adequate for the decisions that follow. Most employers attempt to meet the most common of purpose of exposure measurement through a simple routine monitoring program. The U.S. National Institute for Occupational Safety and Health (NIOSH) has developed a scheme for a logical stepwise analysis of data to arrive at decisions (Leidel et al., 1977).

8.1.2 Environmental Variability

An important factor in the design of any measurement scheme, whether it be a process quality control program, the measurement of an analyte in a lab or an industrial hygiene exposure assessment is to recognize the inherent patterns of the data due to natural phenomena and to match the data to an appropriate mathematical model. We are familiar with linear relationships in our daily lives where one variable increases or decreases in direct proportion to another. In our school careers, we were introduced to the use of a bell-shaped curve to model the distribution of grades in a classroom setting. In this circumstance, student performance clusters around a central value with small proportions of poor and exceptional performances, creating a tail at each end of the curve. Industrial hygiene monitoring data and environmental measurements, in general, also tend to be described by a bell curve. The data cluster toward a central value; however, high excursions from the average, which occur infrequently, create a long tail. At the same time, the extreme variability in measurements causes a very broad-shaped curve. This type of data is best characterized by a lognormal distribution. Statistical mathematics are applied, not to the data points themselves, but to the logarithmic values that represent them. Exposure data are usually described by the mean value of a group of readings collected from a similar population of workers and by the statistic that characterizes the variability in the data, the standard deviation. The range of values that are the points of the lognormal curve from two standard deviations below the mean to two standard deviations above the mean is a data set, called the 95% confidence interval. This range is reported because it is highly probable (95% probability) that the true answer for the exposure being evaluated lies somewhere within this range (Fig. 8.3).

An important factor in the design of any measurement scheme in advance of data gathering is to recognize the degree of variability likely to exist in the data set. This variability has a primary effect on the number of samples to be taken and the accuracy of the results that can be expected. Generally speaking, the fewer the number of samples, the less the accuracy. The greater the variability, the larger the number of samples required for assurance of a minimum accuracy. There is substantial documentation that the variability in employee exposure measurements tends to be quite high. During the course of a day, there are minute-to-minute variations. Daily averages vary widely from day to day.

One can speculate on the probable causes for this variability in worker exposure. The volume of space that a worker moves through in performing a task can be viewed as having an exposure zone centered on one or more point sources. Fugitive emissions from these points occur randomly like frequent small accidents rather than being the main consequence of the production process. Production rates change within a day or between days, affecting emission rates. Overlapping multiple operations within the exposure zone shift irregularly. The distribution pattern of contaminants within the zone by bulk flow, random air turbulence, and eddy diffusion is uneven in both time and space. Through all this, the target system, the worker, moves in a manner that is not altogether pre-

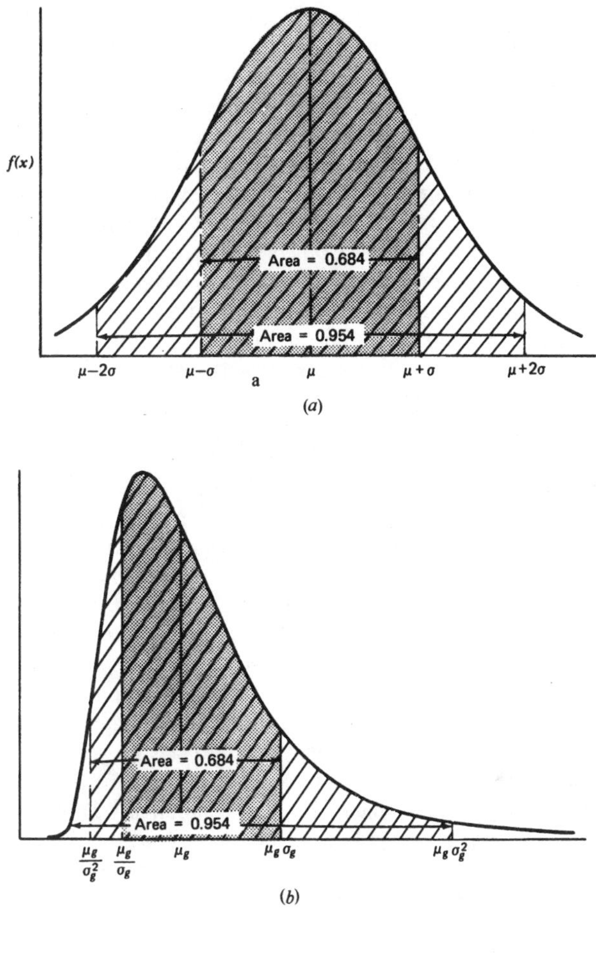

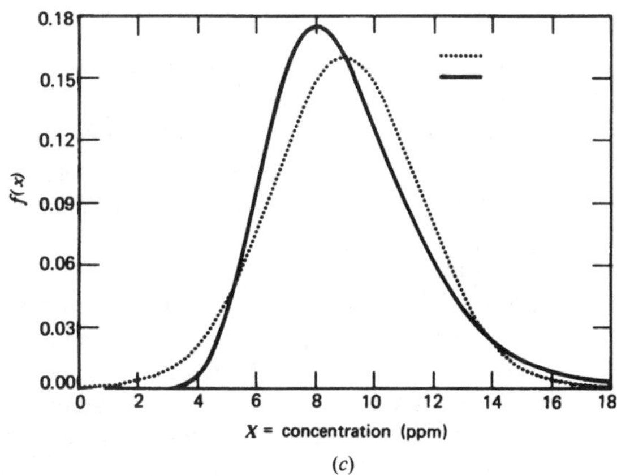

(c)

Fig. 8.3 (*a*) Normal distribution curve. (*b*) Log-normal distribution curve. (*c*) Comparison of normal (dotted) and log-normal (solid) distribution curves.

dictable. These and other uncertainties are the probable causes of the variability typically observed. While a goal of a monitoring effort is often to identify the most likely causative sources of exposure, the situation is far too complex to pinpoint each source of variation and to calculate its consequences. Variance, the statistical measure of variability, can be divided into within-worker variance and between-worker variance. Within-worker variance is a result of changes in the exposure of a worker from day to day when the tasks on each day are nominally similar; between-worker variance is a result of differences in procedures practiced by individuals who are essentially doing the same job. Between-worker variation is, at least theoretically, controllable by the way workers are grouped for sampling purposes. Task observations or review of job descriptions can allow the health and safety professional to draw some conclusions on defining nominally homogeneous groups for sampling purposes. Since it is not usually practical to measure the exposure of all the workers all the time, it is customary to sample a subset of the homogeneous group and to apply the mean exposure value to all employees who are members of the same grouping. However, when the between-worker variance is seen to be quite large with respect to the within-worker variance, it may be necessary to use information from historical sampling data or to conduct some preliminary sampling to group workers.

Sampling schemes may be designed that deal with the variability from all sources as a single pool and derive whatever accuracy is required by increasing the number of samples. Alternatively, one can postulate that a large part of the variability is due to an observable factor or factors. No hard and fast rules can be made regarding the choice of sampling schemes except that is seems logical to expect the factors that have been identified to account for a statistically significant fraction of the variance and to design a plan to collect and analyze the data in such a way that a defensible description of the exposure results. Typical factors may include rate of production, shift, season of the year, and wind velocity and can be singled out as affecting worker exposure. Even after constructing sampling plans that create nominally homogeneous exposure groupings, it is likely that the residual or error variance will still be quite large. This source of variance will normally outweigh the error contributed to the analysis from other sources such as analytical or instrumental inaccuracies.

8.1.3 Location

The most common purpose of measurement of exposure of workers is to estimate the dose so as to prevent or predict adverse health effects. These health effects result from a substance entering the body by some route. Industrial hygienists most often estimate inhaled dose by measurement of the concentration of a substance in inhaled air. Although air samples are sometimes collected from inside respirator face pieces, it is generally not possible to sample the air being inhaled directly. Therefore, the location of the sample collector inlet in relation to the subject's nose and mouth is important. We categorize sampling methods in terms of their closeness to the subject and the point of inhalation as personal samples, breathing zone or vicinity samples, and area or general air samples. A personal measurement is collected from an individual's immediate environment using a device that is worn and travels with the individual. Personal measurements are most often collected from the envelope or "breathing zone" around the worker's head, which is thought, based on observation and the nature of the operation, to have approximately the same concentration of the contaminant being measured as the air breathed by the worker. Area or general air samples are the most remote and are collected in fixed locations in the workplace. Area samples are useful for such purposes as evaluating controls or to characterize emissions, but they are at best a crude estimate of exposure.

Obviously, personal samples are the preferred method of estimating dose by the inhalation route since they most closely measure inhaled air. OSHA enforcement operations "reflect a long-standing belief that personal sampling generally provides the most accurate measure of an employee's exposure . . ." (OSHA, 1982).

If personal sampling cannot be used, some other means of estimating exposure must be accepted. Breathing zone measurements, made by a person who follows the worker while collecting a sample, can come close to measuring exposure. However, this intrusive measurement method may influence worker behavior, and the inconvenience of the measurement will limit the number of measurements and therefore reduce accuracy, as discussed below.

When fixed station samplers are used, knowledge of the quality of the relation between their measurements and the exposure of the workers is necessary if worker exposure is to be estimated. The important question in the use of general air measurements is: What confidence can be placed in the estimate of worker exposure?

In studying the relation between area and personal data with respect to asbestos, the British Occupational Hygiene Society concluded (Roach et al., 1983):

> The relationship between static and personal sampling results varies according to the characteristics of the dust emission sources and the general and individual work practices adopted in a particular work area.
>
> 1. When identical sampling instruments are deployed simultaneously at personal and static sampling points and the distances between them are reasonably small, at least two-thirds of the personal sampling results obtained in a given working location are higher than those obtained from static sampling.
> 2. The differences found between the two types of result tend to be particularly great where the static sampling points are relatively remote from dust emission points, as, for example, when "background" static testing is adopted.
> 3. In certain cases, results from personal sampling may be lower than those from static sampling, owing to factors such as the positioning of the sampling point with respect to air extraction systems.
> 4. The correlation coefficient between the personal and static measurements is statistically significant, but, even so, no consistent relationship of great practical utility could be found in the limited data available.

Although worker exposure measurements are most often used in relation to health hazards, not all measurements made for the protection of health need be measurements of exposure. When it has been established that an industrial operation does not produce unsafe conditions when it is operating within specified control limits, fixed station measurements that can detect loss of control may be the most appropriate monitoring system for workers' protection. Local increases in contaminant concentration caused by leaks, loss of cooling in a degreaser, or fan failure in a local exhaust system can be detected before important worker exposure occurs. Continuous air monitoring equipment that detects leaks or monitors area concentrations is often used in this way. All such systems should be validated for their intended purpose and a performance maintenance program established with their deployment.

8.1.4 Sampling Period

Free of all other constraints, the most biologically relevant time period over which to measure or average worker exposure should be derived from the time constants of the uptake, action, and elimination of the toxic substance in the body (Roach, 1966, 1977; Droz and Yu, 1990). These periods range from minutes in the case of fast-acting poisons such as chlorine or hydrogen sulfide, to days or months for slow systemic poisons such as lead or quartz. In the adoption of guides and standards, such as PELs or TLVs, this broad range has been narrowed, and the periods have not always been selected based on speed of effect. Measurement of the long-term average (multishift) exposure is much more efficient than measuring single shift "peaks." However, standards have been developed or have been interpreted as single shift limits. For most substances, a time-weighted average over the usual work shift of 8 hours has been accepted since it is long enough to average out extremes and short enough to be measured in one work day. Several systems have been proposed for adjusting limits to novel work shifts (Brief and Scala, 1975; Mason and Dershin, 1976; Anderson et al., 1987).

Once the time period over which exposure is to be averaged has been decided for either biological or other reasons (Calabrease, 1977; Hickey and Reist, 1977), there are several alternate sampling schemes to yield an estimate of the exposure over the averaging time. A single sample could be taken for the full period over which exposure is to be averaged (Fig. 8.4). If such a long sample is not practical, several shorter samples can be strung together to make up a set of full-period, consecutive samples. In both cases, since the full period is being measured, the only error in the estimate of the exposure for that period is the error of sampling and analytical method itself. However, when these full-period measurements are used to estimate exposure over other periods not measured, the interperiod variance will contribute to the total error. When a measurement is made of a quantity that is varying, the uncertainty in the measurement caused by the variability becomes part of the error.

It is often difficult to begin sample collection at the beginning of a work shift, or an interruption may be necessary during the period to change sample collection device. Several assumptions may be made with respect to the unsampled period. Observation of employee activities may allow the assumption that exposure was zero during this period. Alternatively it could be assumed that the exposure during the unmeasured period was the same as the average over the measured period if employee activities remained unchanged. This is the most likely assumption in the absence of information that the unsampled period was different. However, it is difficult to calculate confidence limits on the overall exposure estimate since the validity of the assumption is a source of error, there is no internal estimate of environmental variance, and the statistical situation is complex.

When only very short period or grab samples can be collected, a set of such samples can be used to calculate an exposure estimate for the full period. Such samples are usually collected at random; thus each interval in the period has the same change of being included as any other and the samples are independent. This sampling scheme of discrete measurements within a day is analogous to a set of full-period samples used to draw inferences about what is happening over a large number of days. In both cases the environmental variance, which is usually large, has a major influence on the accuracy of the results.

Short-period sampling schemes can be useful with dual standards. For example, if a toxic substance has both a short-term, say, 15-minute, limit and an 8-hour limit,

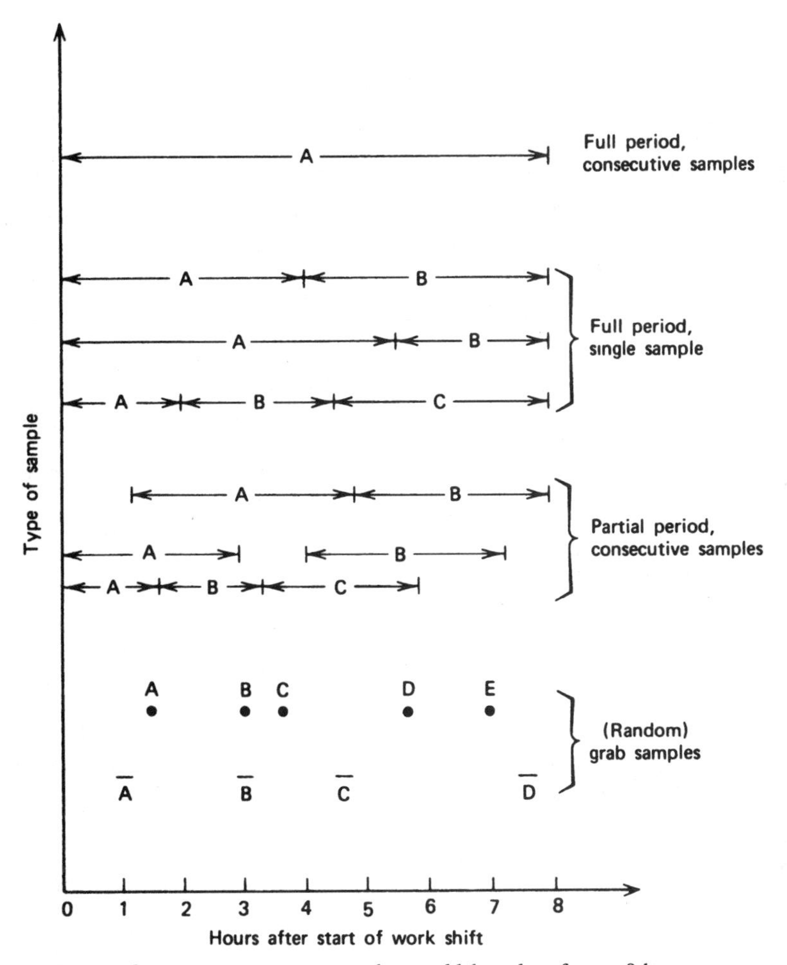

Fig. 8.4 Types of exposure measurements that could be taken for an 8-hr average exposure standard.

15-minute samples taken during the 8-hour period could be used to evaluate exposure against both standards. This involves some compromise, however, since samples taken to evaluate short-period exposure are likely to be taken when exposure is likely to be at a maximum rather than at random.

The traditional method of estimating full-period exposure is by the calculation of the "time-weighted average." In this method the workday is divided into phases based on observable changes in the process or worker location. It is assumed that concentration patterns are varying with these changes and are homogeneous with each phase. A measurement or measurements, usually shorter than the length of the phase, are made in each phase, and the exposure estimate E is calculated:

$$E = \frac{C_1 T_1 + C_2 T_2 + \ldots + C_n T_n}{8}$$

where C_n is the concentration measured in phase n; and T_n, the duration (h) of phase n ($\Sigma T = 8$).

When an averaging time longer than a full shift is needed, that long-term average (LTA) is usually calculated from some number of full-shift samples. For example, the coal mine dust standard is based on the average of five full-shift measurements. Single samples of multiple shifts, excluding nonshift periods, are not widely used. When workplace measurements are made for purposes other than the estimation of worker exposure, different considerations apply. While a single 8-hour sample may be an accurate measure of a worker's average exposure during that period, the exposure was probably not uniform, and the single sample gives no information on the time history of contaminant concentration. To find out when and where peaks occur, with the aim of knowing what to control, short-period samples or even continuous recordings are useful. Similarly, when a control system is evaluated or sampling methods are compared, measurements need be only long enough to average out system fluctuations and provide an adequate sample for accurate analysis. As in the case of the decision on location, the purpose of the measurement is a primary consideration in the selection of a time period of a measurement.

8.1.5 Frequency

By increasing the number of measurements made over a period of time or in a sampling session, the width of the error envelope as described by the confidence limits narrows and the mean result is better defined. Also the maximum number of periods in which a limit could be exceeded (the exceedance) is reduced. With narrower (tighter) confidence limits it becomes easier to arrive at a decision with a given degree of certainty or to be more confident that a decision is correct. The choice of the number of samples to be collected rests on three factors: the magnitude of the error variance associated with the measurement, the size of a difference between the result and the standard or guide that would be considered important, and the consequence of the decision based on the result.

The error variance associated with the measurement depends in most cases on the environmental variance. An exception is the rather limited instance of evaluating the exposure of a worker over a single day by means of a full-period measurement. In that case the error variance is determined by only the sampling and analytical error and confidence limits tend to be quite narrow. Usually, however, our concern is with the totality of a worker's exposure, and we wish to use the data collected to make inferences about other times not sampled. There is little choice, unless the universe of all exposure occasions is measured, we must "sample," that is, make statements about the whole based on measurement of some parts.

As discussed earlier, the universe has a large variance quite apart from the error of the sampling and analytical method. In terms of our decision-making ability, the error of the sampling and analytical method may have very little impact.

The American Industrial Hygiene Association has addressed the issue of appropriate sample size (Hawkins et al., 1991) and recommends in the range of 6 to 10 random samples per homogeneous exposure group. Fewer than 6 leaves a lot of uncertainty, and more than 10 results in only marginal improvement in accuracy. Also, it is usually possible to make a reasonable approximation of the exposure distribution with 10 samples. The difference between the mean and the standard necessary to achieve confidence in the conclusion decreases sharply as the number of samples used increases from 3 to 11. An important conclusion is that for a fixed sampling cost and level of effort, many samples by an easy but less accurate method may yield a more accurate overall result than a few samples by a difficult but more accurate method due to the effect of increasing sample numbers on the error of the mean in highly variable environments.

In selecting a sample size, it must be kept in mind that it is possible to make a difference statistically significant by increasing the number of samples even though the difference may be of small importance. Thus given enough samples, it may be possible to show that a mean of 1.02 ppm is significantly different from a TLV of 1.0 ppm, even though the difference has no importance in terms of biological consequence. Such a statistically significant difference is not useful. Therefore, in planning the sampling strategies first decide how small a difference is important in terms of the data and then select a frequency of sampling that could prove this difference significant, if it existed.

The consequences of the decision made on the basis of the data collected should be the deciding factor in selecting the level of confidence at which the results will be tested. Although the common 95 percent (1 in 20) confidence level is convenient because its bounds are two standard deviations from the mean, it is arbitrary, and other levels of confidence may be more appropriate in some situations. When measurements are made in a screening study to decide on the design of a larger study, it may be appropriate to be only 50 percent confident that an exposure is over some low trigger level. On the other hand, when a threat to life or a large amount of money may hang on the decision, confidence levels even beyond the three standard deviations common in quality control may be appropriate. To choose a confidence limit, first consider the consequence of being wrong and then decide on an acceptable level of risk.

Since sampling and analysis can be expensive, some thought should be given to ways of improving efficiency. Sequential sampling schemes in which the collection of a second or later group of samples is dependent on the results of some earlier set are a possibility. This common quality control approach results in infrequent sampling when far from decision points, but increases as a critical region is neared. Another means of economizing is to use a nonspecific, direct reading screening method, such as a total hydrocarbon meter, to obtain information on limiting maximum concentrations that will help to reduce the field of concern of exposure to a specific agent.

Few firm rules can be provided to aid in the selection of a sampling strategy because data can be put to such a wide variety of uses. However, the steps that can be followed to arrive at a strategy can be listed:

1. Decide on the purpose of the measurements in terms of what decisions are to be made. When there are multiple purposes, select the most important for design.

2. Consider the ways in which the nature of the environmental exposure and of the agent relates to measurement options.

3. Identify the methods available to measure the toxic substance as it occurs in the workplace.

4. Select an interrelated combination of sampling method, location, time, and frequency that will allow a confident decision in the event of an important difference with a minimum of effort.

8.1.6 Method Selection

The selection of a measurement method depends on the sampling strategy considerations discussed above. Should a full-shift, partial-shift, task-specific, or grab sample be taken?

TABLE 8.2 Sampling and Analysis Method Attributes

Method[a]	Sampling period[b]	Ability to concentrate contaminant[c]	Ability to measure mixtures[d]	Time to result[e]	Intrusiveness[f]	Proximity to nose and mouth[g]
Personal sampler/solid sorbent						
Sorption only	Medium to long	Yes	Yes—gases	After analysis	Medium	Very close
Sorption plus reaction	Medium to long	Yes	No	After analysis	Medium	Very close
Personal sampler/filter						
Gross gravimetric	Medium to long	Yes	Yes—particulate	After weighing or analysis	Medium	Very close
Respirable gravimetric	Long	Yes	Yes—particulate	After weighing or analysis	Medium	Very close
Count	Medium to long	Yes	Yes—particulate	After counting	Medium	Very close
Combination filter and sorbent	Medium to long limited	Yes	Yes	After analysis	Medium	Very close
Passive dosimeter	Long	Yes	Yes—gases	After analysis	Low	Very close
Breathing zone impinger/bubbler						
Analysis	Medium–limited	Yes	Yes	After analysis	High	Close
Count	Medium–limited	Yes	Yes—particulate	After counting	High	Close
Detector tubes						
Grab	Short	NA	No	Immediate	High	Close
Long period	Medium to long	NA	No	Immediate	Medium	Very close

Gas vessels						
Rigid vessel	Short to long	No	Yes—gases	After analysis	High	Medium
Gas bag	Short to long	No	Yes—gases	After analysis	High	Close
Evacuated/critical orifice	Medium to long	No	Yes—gases	After analysis	Medium	Distant
Direct reading portable meters						
Nonspecific (flame ion, combination gases)	Instantaneous or recorder	NA	Yes	Immediate	High	Slightly distant
Specific (carbon monoxide, hydrogen sulfide, ozone, sulfur dioxide, etc.)	Instantaneous or recorder	NA	No	Immediate	Medium	Slightly distant
Multiple compound (infrared, gas chromatography, etc.)	Instantaneous or recorder	Some	Yes	Almost immediate	High	Slightly distant
Mass monitor (β absorber, piezoelectric)	Short	Yes	No	Almost immediate	High	Slightly distant
Particle counters (optical, charge)	Short	No	No	Almost immediate	High	Slightly distant
Sensor with datalogger	Short or long	No	No	Hours	Medium	Close
Fixed station						
High volume	Medium to long	Yes	Yes—particulate	After analysis	Low	Remote
Horizontal or vertical elutriator	Long to short	Yes	Yes—particulate	After analysis	Low	Remote
Installed monitor	Short to long	Some	No	Almost immediate	Low	Remote
Freeze trap	Medium	Yes	Yes—vapors	After analysis	Low	Remote
FTIR	Instantaneous	No	Yes—gases	Immediate	Low	Remote

TABLE 8.2 (*Continued*)

Method	Specificity[h]	Convenience rating[i]	Sample transportability[j]	Recheck of analysis possible[k]	Accuracy[l]
Personal sampler/solid sorbent					
Sorption only	High by analysis	High	Good	Elution—yes; thermal des. no	Good
Sorption plus reaction	High by analysis	High	Good	Yes	Good
Personal sampler/filter					
Gross gravimetric	None for weight only—high by analysis	High	Fair	Yes	Good
Respirable gravimetric	High by analysis	Medium	Fair	Yes	Fair
Count	Fair—depends on particle identification	High	Good	Yes	Poor
Combination filter/sorbent	High by analysis	Medium	Good	Yes	Fair
Passive dosimeter	High by analysis	Very high	Good	Yes	Fair
Breathing zone impinger/bubbler					
Analysis	High by analysis	Low	Poor	Yes	Fair
Count	Fair—depends on particular analysis	Low	Poor	Yes	Poor
Detector tubes					
Grab	Medium—some interference	High	No sample	No	Fair
Long period	Medium—some interference	High	No sample	No	Fair
Gas vessels					
Rigid	High by analysis	Low	Fair	Yes	Good
Gas bag	High by analysis	Low	Fair	Yes	Good
Evacuated/critical orifice	High by analysis	Low	Good	Yes	Good
Direct reading portable meters					
Nonspecific (flame ion, combination gases)	None—total of measured class	High	No sample	No	Good
Specific (carbon monoxide, hydrogen sulfide, ozone, sulfur dioxide, etc.)	Medium—some interference	High	No sample	No	Good

Multiple compound (infrared, gas chromatography, etc.)	Medium—frequency overlap	Medium	No sample	No	Fair
Mass monitor (β absorber, piezoelectric)	Mass only	High	No sample	No	Fair
Particle counters (optical, charge)	Count/size only	High	No sample	No	Fair
Sensor with datalogger	Medium—some interference	High	No sample	No	Good
Fixed station					
High volume	High by analysis	Low	Fair	Yes	Good
Horizontal or vertical elutriator	High by analysis	Low	Fair	Yes	Good
Installed monitor	Medium—may be interferences	High	No sample	No	Good
Freeze trap	High by analysis	Very low	Poor	Yes	Fair
FTIR	Medium—may be interferences	High	No sample	No	Fair

This table shows that not all sampling strategies are possible, since for some strategies the sampling and analytical method with the necessary combination of attributes may not exist. The technology gaps thus revealed are fruitful areas for future research and development.

[a] The methods listed include both sampling and direct reading methods. For the sampling methods the ratings of attributes that follow assume the usual range of analytical methods that can be applied to the size and type of sample collected.

[b] By *short* is meant essentially instantaneous or grab samples, while long means 8 hours or longer in a single sample.

[c] Sampling methods that extract a contaminant from the air and collect it in a reduced area or volume are potentially able to improve analytical sensitivity by several orders of magnitude. However, the concentrating mechanism (filtration, sorption) may introduce errors.

[d] Most sampling methods provide a sample that can be analyzed for more than one gas or vapor, but usually not for both gases and vapors or particulates.

[e] Certain decisions (vessel entry) must be made immediately, while others can wait until after the sample is transferred to a laboratory and analyzed.

[f] When the method requires the presence of a person to collect the sample or the wearing of a heavy or awkward sampling apparatus, this intrusion of the sampling system into the work situation may affect worker behavior and exposure.

[g] As discussed earlier, locating a sampler inlet even a small distance from a worker's mouth may bias the exposure measurement. Samplers remote from the worker may not be measuring the air inhaled at all.

[h] Some methods give only nonspecific information like total weight of all dust particles or concentration of all combustible gases, while others measure a specific substance directly or provide a sample that can be analyzed for any species or element.

[i] These are estimatees of the amount of work or difficulty involved in collecting samples.

[j] If the sample must be transported to a distant laboratory for analysis, the ability to withstand shock, vibration, storage, and temperature and pressure changes without being altered or destroyed is important.

[k] Some samples may only be analyzed once, while others are in a form such that rechecks, reanalyses at different conditions, or analysis for other substances is possible.

[l] Given all the possibilities for error form sampler calibration, sample collection, transport, and analysis, an overall coefficient of variation (CV) of 10 percent is considered good. Some count methods are subject to such counter variability that poor accuracy is usual. Method inaccuracy should not be judged alone but should be seen in combination with the inaccuracy caused by environmental variability, which is usually larger, in making decisions whether a method is sufficiently accurate for a purpose.

Should the sample be representative of individual personal exposure, or of the area adjacent to the task being assessed? Should the source be assessed independently of the tasks that may place the employees at risk or should equipment emissions or a confined space be sampled to verify the absence of risk?

Is the contaminant of concern a gas, liquid, solid, or in some cases, in more than one phase? Does the environment contain only the contaminant of concern, or are other substances present as either additional potential hazards or potential interferences? The final consideration in developing a sampling plan is often the immediate availability of sampling equipment. This last decision is constrained by the original efforts to outfit an industrial hygiene field office with sampling equipment that anticipates the answers to the questions raised above. Appendixes A, B, and C contain information to assist in setting up this function. Also various trade magazines who publish annual buyers guides can be a source of vendors (*Ind. Hyg. News*, 1996; Chiltons, 1996).

The discussion of sampling devices that follows will reflect the most common or typical application in assessing exposures. However, each type of device has been developed and improved over time so that they can be applied interchangeably. The notable exceptions are the direct reading instruments, which are too large in most cases to be worn for a personal exposure assessment. Very often the selection of equipment begins with a choice between collecting a sample over an extended period of time (integrated sample) or performing an abbreviated instantaneous test of the environment (grab sample). The choice of sample time is often driven by the desire to compare employee exposures or source emissions to a regulatory (i.e., OSHA) or advisory (i.e., ACGIH) exposure limit. When the limits are based on exposures over 8-hour, 30- or 15-minute averaging times, it is necessary to sample the workplace in question for a similar period of time for comparison.

Grab or short-term sampling is driven most often by the need to perform quick and convenient measurements of contamination as a first response to an employee complaint; as a screening tool to identify target populations or activities for more intensive investigation; to assess engineering or administrative controls; to localize sources of contamination; to confirm the appropriate selection of respiratory protection or to access acute hazards to the general population. Many of the long-term sampling techniques discussed earlier can be modified for application as short-term sampling techniques; however, generally three techniques are most often used because of the immediacy of the data. They are (1) length of stain detector tubes; (2) portable direct reading instruments; and (3) bag or container sampling.

Integrated samples are collected in a variety of media. These are selected based on considerations of chemical interactions that capture and stabilize the contaminant. These samples most often are sent to a laboratory for analysis. For this reason, the chemistry of the contaminant, its collection method, and analysis technique are closely associated. Table 8.2 provides a general list of techniques for sampling and analysis of such contaminants. These general methods have been evaluated and developed into specific field sampling and lab analysis protocols by the National Institute of Occupational Safety and Health (NIOSH), 1984) and are updated and republished periodically. Part of planning a successful sampling program would be to select the best media to collect the contaminants of greatest concern based on the operations and activities covered by the industrial hygienists scope of responsibilities and to arrange support from a laboratory qualified to perform the required analysis. Laboratory qualification and certification are discussed in Section 8.4.2. An added issue in selecting a support lab would be to determine its capabilities in providing technical advice and sampling supplies for circumstances when the industrial hygienists must evaluate unique or unusual exposures.

8.2 METHODS FOR GASES AND VAPORS

As the terms are used in this field, gases are substances such as carbon monoxide whose vapor pressure is greater than atmospheric under normal conditions, while vapors are the gaseous phase of a substance such as benzene, whose vapor pressure is less than atmospheric under normal conditions and which can therefore be present in both a liquid and gaseous phase. Both gases and vapors are capable of mixing completely with air, although they are generally not uniformly mixed. Sampling methods for gases and vapors must have some means of capturing the contaminant for subsequent analytical evaluation. A typical sampling system consists of a portable battery-operated vacuum pump connected to a collection media by soft plastic tubing (usually 1/8 or 1/4 in. ID). The pump is designed to be clipped to the worker's belt and to be light enough to be worn a full shift with a minimum of discomfort. A glass tube containing solid sorbent (Fig. 8.5) is attached to the plastic tube and pinned to the worker's lapel.

8.2.1 Absorbent Tubes

Although Table 8.2 lists a variety of choices in methods for sample collection, the most frequently used method is the concentration of the contaminant on a solid sorbent contained in a glass tube (Fig. 8.6). This technique employs a sampler that is relatively small and easily worn by an employee, is dry and also convenient to handle, store, and ship. The glass tube contains a main section of solid sorbent followed by a porous plug then a second backup section of sorbent. The backup section, which is commonly half

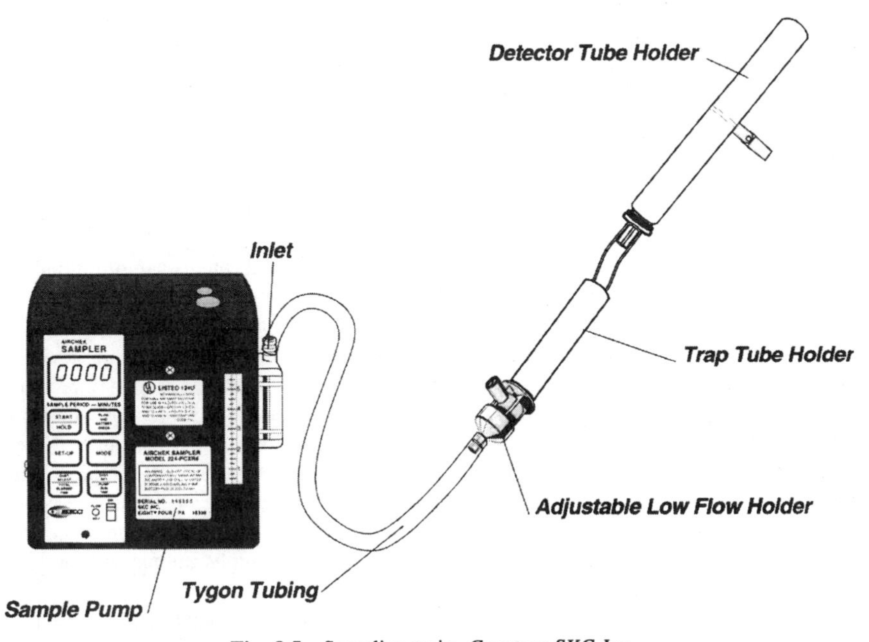

Fig. 8.5 Sampling train. *Courtesy SKC Inc.*

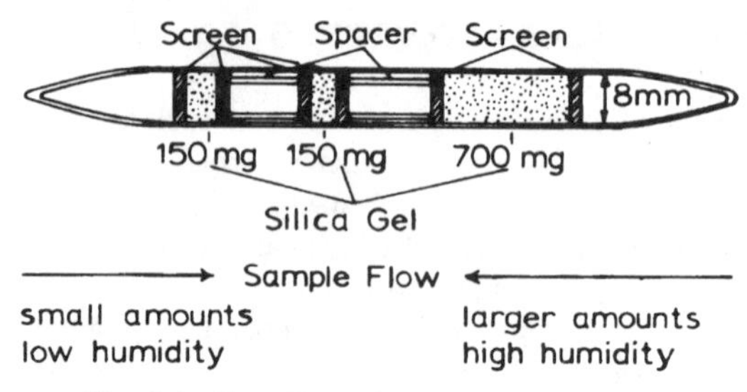

Fig. 8.6 Sampling tube for aromatic amines.

the mass of the front section, is analyzed to determine if the analyte has broken through the front section and compromised the sample results.

The main drawback to this sampling method is that the contaminated air must be drawn across the sorbent with a battery-driven portable pump. The pump and a connecting tube are usually clipped to the employee's belt, and the sampling tube is attached to a shirt collar. The pumps are designed to allow adjustments to the rate of air volume sampled, the overall volume collected, and the total amount of contaminant collected. These parameters are critical in efficiently collecting an accurate representation of the environment being assessed. Limitations of absorbent tubes include the fact that not all gases and vapors are efficiently collected (e.g., ethylene) or desorbed (e.g., PNAs). Also the volume of air sampled must be carefully selected so as to yield adequate material on the sorbent for analysis without exceeding the collection capacity of the tube and causing breakthrough. The sampling and analytical method will determine these factors.

8.2.2 Passive Badges

As the sensitivity of analytical tools in the laboratory have improved, smaller and smaller amounts of material have been required for quantification. This has made possible the development and use of lightweight passive sampling badges (Fig. 8.7). In these devices, the contaminant migrates across a narrow turbulent free space to the adsorbent material in the same way that fragrant aromas originating in a kitchen spread throughout a home. The rate of diffusion of the material represents its sampling rate and controls the total material deposited on the sorbent, which is available for analysis. Thus passive diffusion replaces the sampling pump as the means of bringing the contaminant and the sorbent into contact. Diffusion rates are determined for each material under laboratory conditions and may require adjustments if field conditions of temperature, humidity, and pressure vary greatly from the typical lab setting. Either the vendor or the analytical laboratory is the best source of information on the performance of these devices. While passive samplers offer greater convenience over pumped samplers, their effectiveness is poor in situations where significant air turbulence or the source causes rapid fluctuations in contaminant concentrations. Also the sampling time must be long enough to yield an adequate sample size. If these limitations apply, pumps and sampling tubes should be utilized.

Activated charcoal has found the widest application as an adsorbent material for collecting organic contaminants and is available packed in glass tubes and in passive

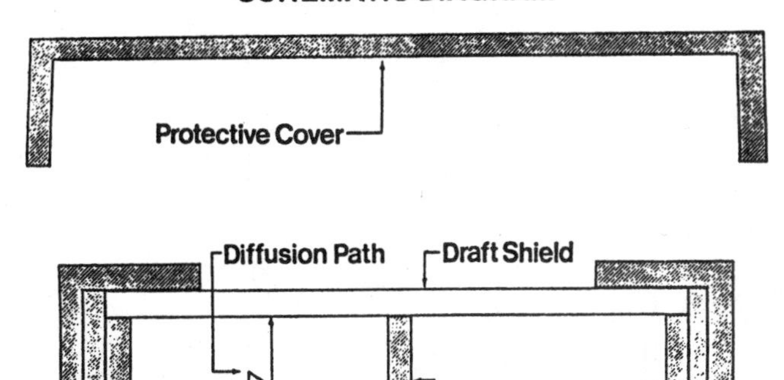

Fig. 8.7 Passive sampling dosimeter.

badges. Other sorbents such as silica gel, alumina, and molecular sieves have been developed for collecting polar compounds, and in some cases materials that have been developed for use in chromatography, such as XAD, have also been applied to sampling contaminants that are not more easily collected. The recommended absorbent sampling rate, total volume, and recommended techniques for sample preservation are summarized in NIOSH's established methods (NIOSH, 1984). Sample calculations of these variables will be discussed in Section 8.4.1.

8.2.3 Impingers and Liquid Traps

In some cases, where the contaminant is unstable, reactive, or cannot be collected on solid media, solutions contained in small bubblers are employed (Fig. 8.8). Air is drawn by a sampling pump through a hollow glass tube drawn to a fine tip into an impinger filled with a liquid to produce small bubbles. The bubbles provide a large gas to liquid surface, allowing the contaminants to migrate into the liquid phase. In the liquid they are trapped and in some cases stabilized by reacting with a reagent selected for this purpose. Some impingers have been designed to minimize spillage while they are attached to an employee's lapel. An additional concern is loss of the trapping solution during the sampling period due to evaporation.

8.2.4 Length of Stain Tubes

Length of stain detector tubes are a popular method of grab sampling because of the wide variety of tubes that have been developed for specific chemical contaminants, their ease of use, as well as their low cost and storage stability. These tubes provide an inexpensive method of anticipating the broadest possible range of demands for a quick initial exposure assessment. Detector tubes are sold as sealed glass tubes with a solid granular matrix such as silica gel, alumina, or pumice that has been impregnated with a

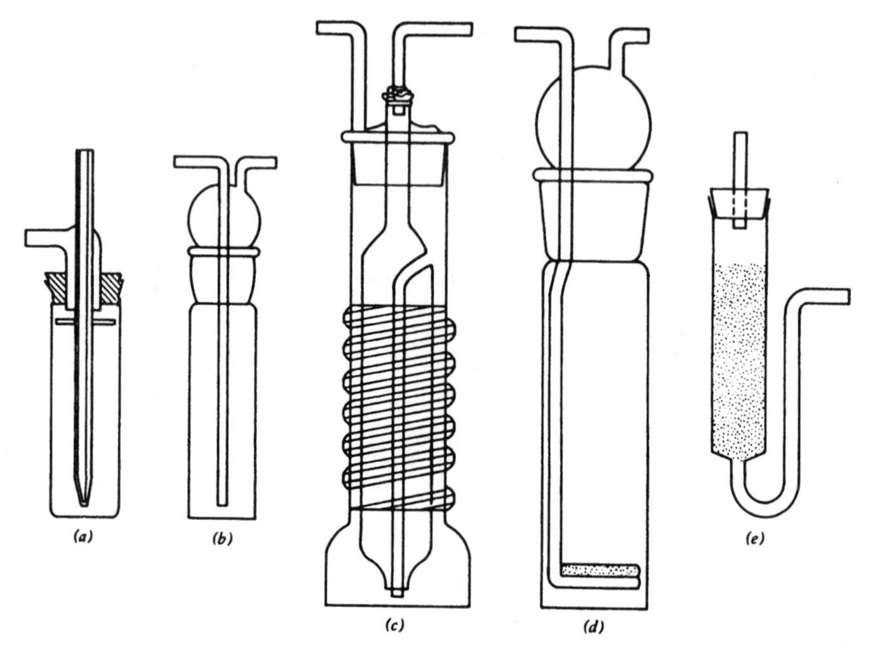

Fig. 8.8 (*a*)–(*c*) Three types of wet impingers. Smith–Greenberg impinger samples air at 28.3 lpm, δP = 3-in Hg, and terminal velocity is 100 m/sec. Water in impinger may be ~75 cc. For midget impinger, sampling rate is 2.83 lpm, δP = 12-in H_2O, terminal velocity 70 m/sec, and typical liquid = 10 cc H_2O. (*d*) Bubbler. (*e*) Packed absorber.

dye that will change color as air containing a specific contaminant is drawn through the tube. These tubes are most commonly used for short-duration grab samples using hand-operated piston or bellows pumps. The tubes are calibrated by the manufacturer such that when a fixed volume of air is drawn across the tube, the length of the coloration corresponds to the concentration in air.

Standard detector tubes are of limited value in assessing exposure relative to a full-shift standard due to their short sampling time. For certain reactive gases and vapors a length of stain indicating tube has been developed to measure full-shift exposures. In some select cases low-flow sampling pumps are used for long-period samples. Passive diffusion may be used when it allows enough contaminant to migrate onto the tube that a stain can be developed. Some applications are listed in Table 8.3.

All detector tubes are subject to limitations of specificity and sensitivity. Specificity is the ability of a method to detect and measure a desired contaminant in the presence of other chemicals. Chemicals that react similarly in a procedure are said to "interfere." Sensitivity is the smallest detectable change in analyte concentration that is measurable by the method. These issues are discussed in detail in the manufacturers' literature, and the restrictions must be strictly observed for the measurement to be considered valid.

8.2.5 Evacuated Containers and Bags

A technique that is often used for grab sampling is slow-filling containers that collect both contaminant and air without any method of separating or concentrating of the

TABLE 8.3 Long-Term Colorimetric Detector Tubes

Acetic acid
Acetone
Ammonia
Benzene
Butadione
Carbon dioxide
Carbon disulfide
Carbon monoxide
Chlorine
Ethanol
Ethyl acetate
Hydrochloric acid
Hydrocyanic acid
Hydrofluoric acid
Hydrogen sulfide
Methylene chloride
Nitrogen dioxide
Nitrous fumes
Perchloroethylene
Sulfur dioxide
Toluene
Trichloroethylene
Vinyl chloride
Water vapor

contaminant at the time of sampling. This technique can only be considered when the analytical procedure has sufficient sensitivity to detect the contaminant at ambient concentrations (usually parts per million). Gas chromatography and infrared are best suited to this purpose. Samples are most often collected using a gas bag that is slowly filled by a low-flow pump or by using an evacuated glass or metal container. The container can be opened and filled instantly, or the mouth can be equipped with a critical orifice that will allow for slower filling at a steady rate. Syringe needles obtained from a chromatography lab can be used as critical orifices. Needles of different diameter openings can be used to vary the sampling rate. The evacuated container is sealed with a one-hole stopper with a hollow glass tube passing through it. The tube and needle are connected by a short run of plastic tubing that is tightly clamped. In some instances where a transportable field gas chromatograph is nearby, the gas-tight syringe normally used to inject samples into the instrument can become the grab sampling device itself.

With these techniques the loss of contaminant to the surface of the container is a common concern. Bags made of inert materials such as Mylar, Tedlar, and PTFE are available to minimize absorption or permeation through the bag walls. Prior to application it is important to monitor the rate of decay under controlled conditions as part of the preuse selection and calibration. Cross-contamination and memory from surface residue are also of concern. Even liquid washing may only result in the wash liquid itself leaving a residue. Gentle heating under a fresh air or inert gas flow followed by testing of the container provides the best means of avoiding this problem.

8.2.6 Direct Reading Instruments

Direct reading instruments provide a number of advantages over sampling and analytical methods and detector tubes, but these devices tend to be more expensive than most other equipment. They are most often purchased with specific applications in mind. A disadvantage is that they usually cannot be worn by the worker; so they are not suitable for personal sampling. These instruments are capable of "instantaneous" measurements, and they can also make integrated measurements when used in connection with a data logger (see Section 8.4.8). These instruments have evolved over the past 25 years from larger bench-scale models originally developed for the analytical chemistry lab to small hand-held devices. The drive towards miniaturization and portability in the electronics industry has resulted in weight and size reductions in instrumentation that have been remarkable. Instruments often provide a quicker response, greater accuracy, and greater specificity than detector tubes, although they are not as accurate or specific as analytical methods. With the proper technical support direct reading instruments are amenable to fairly rapid development of field methods for analyzing unusual materials where published protocols do not exist. Table 8.4 summarizes the most common sensing technologies employed in today's instruments. Manufacturers combine the detectors with electronics and other hardware to provide for many applications.

1. Point and measure;
2. Continuous area or process monitors;
3. Personal dosimeter;
4. Complex environment analyzers.

Point and measured devices are used to identify point sources of contaminant release (leak meter), to assess general background conditions, or to perform a preliminary identification and assessment of an exposed population. They also serve a critical purpose in identifying environments that contain acute hazards such as carbon monoxide, hydrogen sulfide, or explosive gases. Permitted entry into confined spaced that may contain such hazards relies heavily on measurements by these types of instruments.

Stationary area monitoring instruments that identify process leaks are often equipped with alarms or are linked to emergency shut down systems to control the release of acutely hazardous materials such as flammable vapors, hydride gases, and acutely toxic gases (i.e., hydrogen fluoride). They can be positioned in areas adjacent to the processes or within process enclosures to monitor for upsets.

Personal dosimeters consist of a sensor that produces a continuous signal whose output fluctuates in real time with concentration. These are either equipped with alarms that activate at hazardous concentrations and/or dosimeters that record and store a time-history profile of the exposure.

Many instruments are available that employ chromatographic techniques to separate complex environments into individual components and quantify a single toxic substance within a group of similarly toxic materials.

The wide range of selection and breadth of applications of direct reading instruments make it unlikely that one would be able to afford or anticipate all the possible applications of these devices. However, in the past few years an equipment rental busi-

TABLE 8.4 Direct Reading Instrument Sensors and Their Most Common Applications

Detector type	Measurement principle	Most common applications
Solid-state pellister	Detects changes of heat of reaction on the surface of a catalytic solid (pellister), which alters the pellister's sensitivity	Explosive atmosphere; high concentration of total hydrocarbon
Semiconductor (1)	Detects a change in conductivity of the semiconductor material produced by trapping or release of charge carriers at the surface. The shift in conductivity is proportional to concentration	Explosive atmospheres; high concentration of total hydrocarbon
Semiconductor (2)	Semiconductor materials made of *n*-type metal oxides are doped or mixed with other metal oxides to react selectively with contaminants that donate electrons or remove adsorbed oxygen from the solid matrix. An electric current is produced that is proportional to concentration	Inorganic reactive gases, such as oxygen, carbon monoxide, hydrogen sulfide, sulfer oxides and nitrogen oxides, hydride gases
Infrared analyzers	Molecules that vibrate at the atomic level absorb a characteristic infrared wavelength in proportion to the total concentration of material in the beam path	Wide variety of gases, principally organic compounds
Ultraviolet analyzer	Measurement principle is the same as in infrared; however, only mercury vapor has a specific absorption wavelength in the UV range	Mercury
Electrochemical	Gases diffuse into an electrochemical cell consisting of solid, liquid, or gel electrolyte, an anode, and a cathode. The contaminant undergoes an electrochemical reaction, which produces a current proportional to concentration	Inorganic reactive gases as above and oxygen
Impregnated paper tape sulfide, isocyanites	Contaminant is drawn through a paper tape impregnated with a specific color developing reagent. The intensity of color is proportional to concentration	Hydride gases, TDI, hydrogen phosgene, chlorine organic ammonia
Photoionization (PID)	Ultraviolet light impinging on contaminant molecules causes the species to ionize. The ions migrate across an applied electric field to a collecting electrode when a current is produced that is proportional to concentration	Many hydrocarbons
Flame ionization (FID)	Organic compounds are ionized in a hydrogen/air flame, which reduces the electrical resistance of the flame in proportion to the concentration of hydrocarbon	Many hydrocarbons

ness has developed that will supply an instrument calibrated and ready to use. Several companies provide just-in-time delivery of industrial hygiene equipment either to supplement existing equipment such as pumps or calibrators or to serve as a source of more expensive specialty equipment. This resource is helpful when a specific project or a high-visibility exposure assessment justifies a one-time rental fee but not the purchase of the equipment. Vendors should be selected who demonstrate sufficient understanding of both the sampling problem and instruments to be capable of configuring the device to the application (i.e., selecting proper columns and conditions). The supplier should also be capable of calibrating the instrument in the working range anticipated for the project since the capability to generate challenge concentrations can be quite difficult to achieve in a field office for specialty assessments. Finally, the supplier should be prepared to supply a replacement instrument in the case of equipment failure during a project.

8.3 METHODS FOR AEROSOLS

Aerosols are solid particles or liquid droplets suspended in air. Sampling methods based on filtration, impaction, and impingement were some of the earliest techniques developed to aid in evaluating exposures in the dusty trades and over the past 75 years have evolved to be applicable as both area samplers as well as personal samplers. To understand their use it is necessary to consider the properties of aerosols that affect sampling.

Sampling for aerosols differs in three significant ways from gas/vapor sampling:

1. Unlike gas/vapor environments that quickly become homogeneous after leaving the source, materials suspended in air tend to vary in concentration because of large particle settling, small particle agglomeration, or turbulence in the air itself. Droplets may evaporate while in suspension or condense as the vapor cools. Particulates may become resuspended as a result of worker activities or air turbulence.
2. Particle size is a factor in determining health hazard. Particles above 10 μm are all trapped in the nasal passages and have little probability of penetration to the lung and would not be of interest if the lung were the only target organ. However, particles such as toxic metal that are swallowed after being trapped in the throat, larynx, and upper bronchia can be dissolved in the stomach and migrate to internal target organs or, as is the case with acid mist and some allergens, cause harm at the point of deposition.
3. Finally, the dynamic interrelationship between the particles and the air that keeps them in suspension requires sampling methods that take into account the physical behavior of both. For example, it is critical in sampling to attempt to maintain nonturbulent air flow patterns at the point of interface between the sampler and the ambient environment to avoid collecting a sample that is not representative of the exposure. Therefore, sampling rate must be more specifically defined for particulates than for gases (see Chapter 2).

It is well recognized that some particulate material exerts its effects at the locus of deposition in the lung. Therefore, regulatory standards set by OSHA for substances such as quartz, christobalite, tridimite, coal dust, and cotton dust specify exposure limits expressed as the mass concentration or total particle count of the sampled fraction of dust that is respirable (OSHA, 1970). The ACGIH's Chemical Substances TLVs are currently examining particulate substances to better define the size fraction most closely associated with the health effect of concern. Future TLVs for these chemicals will be based on the mass concentration of the specified fraction. The "Particle Size-Selection" TLVs are expressed in three forms (ACGIH, 1996):

1. Inhalable particulate mass (IPM-TLVs) will be applied to those materials that are hazardous when deposited anywhere in the respiratory tract.
2. Thoracic particulate mass (TPM-TLVs) will be applied to those materials that are hazardous when deposited anywhere within the lung airways and the gas-exchange region.
3. Respirable particulate mass (RPM-TLVs) will be applied to those materials that are hazardous when deposited in the gas-exchange region.

Airborne fibrous particles such as asbestos are collected on open-face filters also to provide a homogeneous sample for microscopic analysis. Health standards are reported in terms of fiber concentration.

8.3.1 Method Selection

The equipment that has been developed to measure aerosol concentrations falls into three categories:

1. Devices that are small enough that they can be worn as personal samplers either for full-period integrated samples or as grab samples;
2. Devices that can be used as either full-period or grab samples but whose size or bulk limits their application to fixed area monitoring;
3. Hand-held portable and transportable instruments that can be used for grab or full-period sampling as either area or personal samplers.

The most common sampling methods and least expensive in categories 1 and 2 depend either on filtration, impaction, or impingement of the particles as the means of collection.

8.3.2 Filter Sampling Methods

Personal sampling is most often accomplished by passing the contaminated environment through filtering media that capture the solids. Table 8.5 lists the most commonly used filters and typical applications. The sampling assembly will consist of an air moving pump, filter cassette, and connecting tubing. Where a respirable sample is to be collected, the cassette is attached to a cyclone size selector.

The *filter cassette assembly* (Fig. 8.9) consists of three pieces: a base upon which a cellulose pad (sometimes a metal screen) is placed to provide structural support for the filter, a center section that holds the filter against the support pad for "open-faced" sampling (sampling in which the entire top of the cassette is removed), and a top cover that protects the filter and seals the cassette for storage and shipment. The base and top cover have sampling ports in the center with plastic plugs. In most sampling operations the center section is not used. The top cover fits against the filter, and the sample is taken

TABLE 8.5 Common Applications of Filters

Filter matrix	Common applications	Most common pore size
Cellulose ester	Asbestos counting, particle sizing, metallic fumes, acid mists	0.8 μm
Fibrous glass	Total particulate, all mists, coal tar pitch volatiles	—
Paper	Total particulate, metals, pesticides	—
Polycarbonate	Total particulate, crystalline silica	—
Polyvinyl chloride	Total particulate, crystalline silica, all mists, chromates	5.0 μm
Silver	Total particulate, coal tar pitch volatiles, crystalline silica, PNAs	0.8 μm
Teflon	Special applications (high temp)	—

by drawing air through the sampling ports. The joints where the base, center section, and/or top cover of the cassette join should be sealed when sampling (plastic electrical tape works fine), and the center section of top cover should be pressed firmly against the filter to ensure that air passes through the filter and not around the edges.

Many occupational health limits for particulates are based on measurements of total dust. Therefore, the filter cassette is most frequently used with either the top lid in place (closed faced) or removed (open faced) for more uniform distribution of dust particles. Open-faced sampling is advisable when the method of analysis is either qualitative identification of the materials by microscopy or when the concentration is being determined by particle counting. Asbestos fibers are collected using open-faced cassettes. These cassettes are specifically designed to eliminate undersampling due to fiber adhesion to the walls of the cassette caused by static charge.

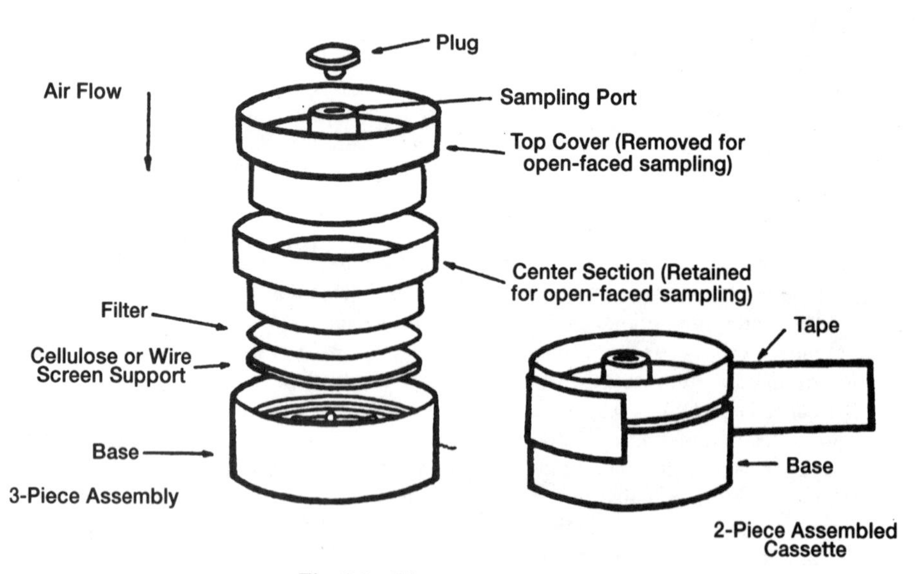

Fig. 8.9 Filter cassette assembly.

The cyclone separator is used when only the respirable fraction (the particles less than 10 μm in aerodynamic diameter) is to be collected. Since this size range represents the particles likely to be deposited deep in the lungs, this respirable dust sample is required to assess the hazard to employees exposed to particulates that are capable of producing lung damage using the ACGIH or OSHA limits for respirable dust (ACGIH, 1996). The cyclone works by directing the stream of sampled air into a vortex. The larger particles are unable to make the turn with the curved air stream and hit the walls of the cyclone. The result is that the particles fall out of the stream before reaching the filter. The smaller particles remain in the air stream and are collected on the filter. The cyclone (Fig. 8.10) consists of a tube for collecting the particles that have fallen out of the stream (grit pot), a section that fits into the top of that tube and converts the incoming air stream to a vortex, and a supporting framework for holding the cassette in place.

Because of the construction of the cyclone samplers, the calibration (see Section 8.4.3) setup is not quite the same as either cassettes or charcoal tubes because there is no easy way to attach plastic tubing to the cyclone. Figure 8.11 shows the cyclone and cassette assembly configured to allow calibration.

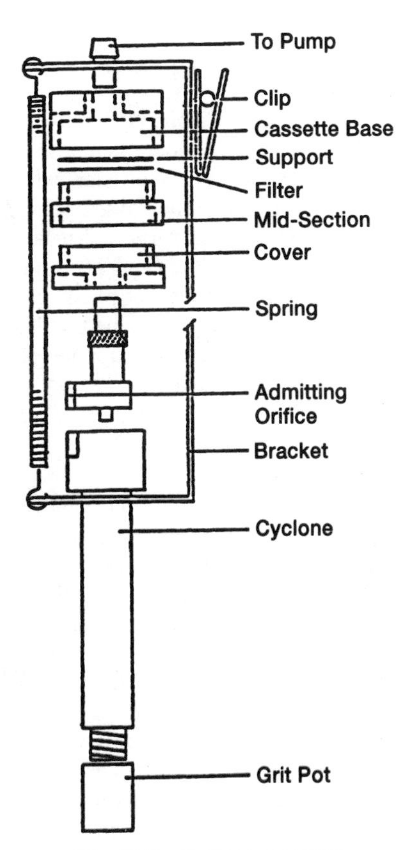

Fig. 8.10 Cyclone separator.

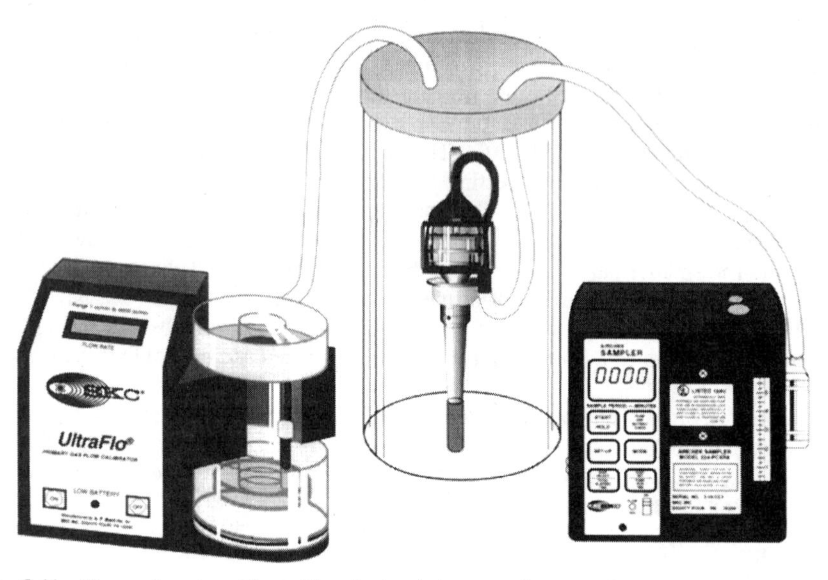

Fig. 8.11 The cyclone is calibrated by placing it in a one liter vessel attached to an electronic bubble meter. *Courtesy SKC Inc.*

Another technique, elutriation, has been applied to fixed-area sampling for dust. In an elutriator the air speed and kinetic energy are controlled in such a way that nonturbulent laminar air flow is produced. In a horizontal elutriator, which is used for coal dust sampling in the United Kingdom but not in the United States, the larger particles are drawn by gravity downward and settle out at varying distances from the entrance as a function of their mass. Large particles drop out quickly, and small particles continue to be carried along with the air stream and are collected on the filter at the end. In a vertical elutriator, used for dust measurement in cotton textile mills, air enters the base of the sampler and small particles are carried upward in the air stream to a filter, while heavier debris settles to the bottom of the sampler.

8.3.3 Impactors

Impactors were initially designed to collect and characterize the size distribution of dusty environments. Impactors are designed to create sudden change in direction in air flow and the momentum of dust particles so that the particles impact against a flat plate. Particles are retained on the plate's surface by an adhesive coating that minimizes bounce. Small particles traveling in the air stream contain relatively less momentum than larger particles; therefore, as the air stream is redirected at a right angle to its flow, the larger particles cross the air stream and strike the flat plate, where they are captured. In a multistage impactor, a series of plates are configured in a stack and air speed is reduced in stages. In this way, particles of progressively smaller size ranges are captured on separate plates. Devices of this design can be fairly large and were originally deployed as area samplers. However, devices such as the Anderson Mini-Sampler (Fig. 8.12) have been developed that are small enough to be worn on an employee's lapel to permit personal sampling to characterize inhalable particle size.

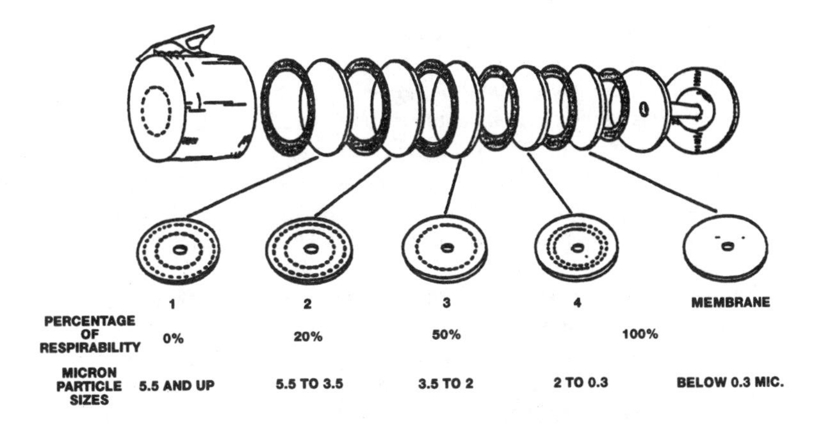

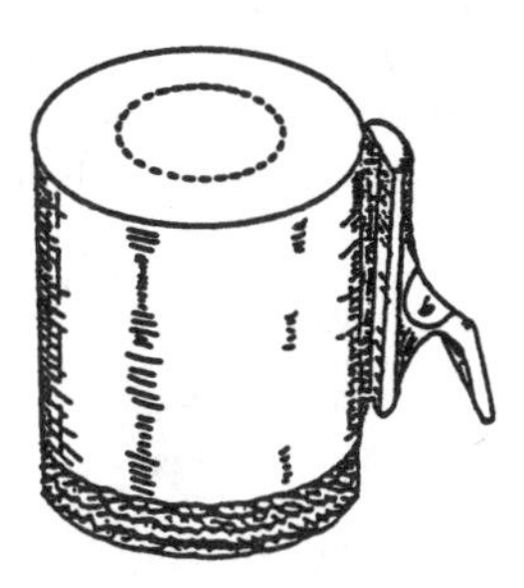

Fig. 8.12 Anderson mini-sampler.

8.3.4 Impingers

Impingers are apparatus that accelerate the dust-laden air stream to a high velocity through a small orifice. Impingers were used in the early part of this century to capture dust particles. The particles would strike a plate immersed in a liquid and be trapped. Particles were then counted by microscope. The most familiar impingers are the Greenberg–Smith impinger and the midget impinger (Fig. 8.8). These methods, however, have largely been replaced by the use of filter and impactor methods today.

8.3.5 Direct Reading Instruments

Instruments that can provide real-time data on particulate concentrations have continued to become available in both hand-held and transportable versions and have improved in parallel with the evolution in illumination sources and microelectronics. Many bench-scale instruments, which have been developed to support air pollution monitoring and aerosol research efforts, will not be discussed here. An in-depth summary of these devices can be found in the most current edition of *Air Sampling Instruments for Evaluation of Atmospheric Contaminants*, published by the ACGIH (1978).

Instruments that are both lightweight and battery operated are available for workplace monitoring either as held-held instruments or transportable area monitors (Table

8.6). These instruments rely on indirect methods of sensing aerosols, and frequently the calibration of signal response is developed using an artificial dust. As a result field measurements with these instruments should be viewed as approximations of exposure because the materials in the environment being monitored may differ in signal response characteristics and size distribution from the calibration dust.

Devices are available that measure total particle count and/or total mass concentration. Some of the same devices came with fractionating cyclones that allow for determining respirable and total particulate concentration. Other instruments incorporate size separation with particulate measurement to allow detailed real-time measurement of size distribution. The devices most often used measure a parameter of the aerosol over its entire size range. A commonly used instrument measures the concentration of the total number of particles by counting condensation nuclei. In other instruments total mass concentration is measured by the alteration in frequency of a vibrating quartz crystal caused by mass loading of particulates or by attenuating a β radiation source. Forward scattering photometers have been developed that measure both mass and number concentration. In one application of this method, an instrument has been miniaturized and data logging electronics have been incorporated to introduce a personal real-time monitor that also records a time history exposure profile (see Section 8.6).

In another instrument particles are deposited by electrostatic precipitation onto an oscillating quartz crystal, thereby increasing its mass and decreasing its frequency. The

TABLE 8.6 Direct Reading Aerosol Instruments

Detector	Flow rate (L/min)	Size range (μm)	Lower conc. limit
	1. Number concentration measurement		
Condensation nucleii (particle) counter, (CNC or CPC)	0.3–1.0	0.01–2.0	<0.001 particle/cm^3
Condensation nuclei counter ultrafine (UCNC or UCFC)	0.03–0.30	0.003–3.0	<0.001 particle/cm^3
Optical particle counter, white light and laser	0.01–28	<1.0–2.0	0.001 particle/cm^3
Cloud condensation nucleii counter	1–5	0.08–1	1 particle/cm^3
Ice nuclei counter		10	0.001 particle/cm^3
	2. Mass concentration measurement		
Quartz crystal microbalance	1–5	0.01–20 (electrostatic) 0.3–20 (impactor)	0.01 mg/m^3
Vibrating sensor	1–5	0.01–20	<1 mg/m^3
β-attenuation sensor	1–12	0.01–20 (filtration) 0.3–20 (impactor)	0.01 mg/m^3
Photometer, nephlometer	1–100	0.1–2	0.01 mg/m^3
Electrical aerosol detector	0.5–20	0.1–2	5/cm^3 @ 1 μm

difference in frequency between this crystal and a reference crystal is sensed and converted into mass concentration, which is automatically displayed. The device is sold with a variety of size-selecting inlets that can be chosen to measure particulates below a certain cutoff size, including the respirable fraction. The measurement range is from 10 micrograms to 10 milligrams per cubic meter. The vibrating crystal is somewhat sensitive to temperature and humidity fluctuations and should be allowed to equilibrate to the surrounding environment before actual measurement begins. The instrument's accuracy is believed to improve as the target dust becomes stickier. A β attenuating analyzer operates by drawing a known quantity of air through a critical orifice and depositing its particulate contents on a surface. This surface can be a rotatable glass disc. The particulate deposits are small pinpoints on the outer periphery of the glass discs. After a sampling cycle is completed, the instrument activates a radiation emitter beneath the glass disc. Radioactive particles then travel through the pinpoint particulate deposits. A sensor on the other side of the glass disc determines how much radiation passes through the deposit and relates the degree of radiation penetration to mass of particulate collected. By placing a cyclone over the instrument inlet, it is possible to obtain a readout of mass respirable particulates.

Instruments that employ incandescent light or lasers as sources and dark-field microscopy optics have been developed. In these instruments a narrow beam of light is focused on an aerosol cloud and light falling on photoreceptor in the near forward field is measured. These instruments can measure aerosol concentration either as total mass or particle concentration. Measurements using these devices are less sensitive to variations in the refractive index and size range of the target aerosol relative to the calibration dust than photometers. These types of instruments are available as a hand-held lightweight device and also as a personal time history monitor.

Fibrous aerosols can be measured using an instrument that continuously draws air into a horizontal cylindrical cell, where it is illuminated by a helium–neon laser source. The fibers are subjected to a rotating electrical field that allows their variable scattering to be detected. Fibers that are 2 to 3 μm long and 0.2 μm in diameter can be measured in concentrations of 10 to 25 f/cm^3.

To determine particle size distribution, size-selecting devices are attached upstream of the sensing chambers as modifications to the instruments discussed above. Instruments that can provide real-time particle size analyses are generally large lab models requiring greater power sources, which place them in the category of transportable, fixed-area monitors that are not frequently used in routine industrial hygiene evaluations. These types of instruments are well described and cataloged in the ACGIH publication mentioned earlier (1978).

8.4 GENERAL CONSIDERATIONS

8.4.1 Planning the Collection of a Sample

To successfully collect a representative sample, three criteria must be considered.

1. An adequate amount of contaminant must be collected to provide the lab with a large enough quantity to be analyzed. Normally a total quantity that allows for analysis in the midpoint of the sensitivity range of the analytical method is ideal.

2. The maximum capacity of either air volume or contaminant mass that the sampling matrix can retain must not be exceeded. Exceedance of this factor is usually recognizable by a lab report indicating breakthrough to the backup section of the sampler. (See Section 8.2.1 for a discussion of sampling tube design.)

3. Sampling rate must be chosen such that (1) and (2) are satisfied while sample collection takes place over the desired period of monitoring. This period closely approximates the 8 hours and 15 minutes exposure limits or the duration of the task if it is less.

Information on criteria (1) and (2) will be provided in the sampling and analytical method (NIOSH, 1984). Minimum sampling times and volumes can be calculated as follows:

$$\text{minimum volume of air sampled} = \frac{10 \times \text{analytical detection limit}}{\text{hygiene standard}}$$

$$\text{minimum duration of sample} = \frac{\text{minimum volume}}{\text{flow rate}}$$

The analytical sensitivity is normally discussed in units of mass such as milligrams or micrograms; therefore, parts per million should not be used to express the hygiene standard but rather milligrams per cubic meter. Normally one begins by calculating the two minimum values and then adjusting the parameters to the particulars of the situation.

Example: Collect an 8 hour sample for benzene to detect at 50% of the hygiene standard of 1 ppm (3 mg/m^3) using a laboratory that can detect 0.005 mg and a sampling tube with breakthrough capacity of 100 liters.

$$\begin{aligned}
\text{minimum volume of air sample} &= \frac{10 \times 0.005\,\text{mg}}{0.5 \times 3\,\text{mg/m}^3} \\
&= 0.033\,\text{m}^3 \text{ of air} \\
\text{converting m}^3 \text{to liters} &= 1000 \times 0.033 \\
&= 33 \text{ liters of air} \\
\text{minimum duration of sample} &= 8\,\text{h} = 480\,\text{min} = \frac{33\,\text{liters}}{\text{flow rate}} \\
\text{flow rate} &= 3.3\,\text{liter/480 min} \\
\text{converts units to common cc/min} &= \frac{33\,\text{liter} \times 1000\,\text{cm}^3/\text{liter}}{480\,\text{min}} \\
&= 70\,\text{cm}^3/\text{min}
\end{aligned}$$

The calculation indicates approximately 70 cm^3/min as required for the example conditions. Typically a sample such as benzene will be collected between 100 and 200 cm^3/min without adversely affecting the outcome of the results. The total volume required to cause breakthrough is roughly three times the volume to be sampled. The risk of breakthrough only exists if either of two circumstances exist alone or in combination. First, if sampling at the highest rate (200 cm^3/min) exceeds a full shift or if the concentration of benzene in the environment is substantially above the collection capacity of the charcoal. Normally total air volume to breakthrough at a given concentration is determined as part of the evolution of a sampling method, while mass loading is not. As a rule of thumb, a flow rate closest to the minimum specified range should be selected when sampling environments are suspected of high concentrations. Alternatively, the sampling period can be divided into segments and tubes can be replaced at the end of each period (e.g., 4 tubes sampled consecutively for 2 hours each during an 8 hour operation) and the results averaged to give a full period exposure. The maximum sampling time can be calculated as shown below:

$$\text{maximum sampling time} = \frac{1}{70\,\text{cm}^3/\text{min} \times \dfrac{\text{liter}}{1000\,\text{cm}^3} \times \dfrac{1}{100\,\text{liter}}}$$
$$= 1430\,\text{min or } 24\text{ h}$$

8.4.2 Analytical Lab Services and Chain of Custody

Industrial hygienists most often rely on support from laboratories who are capable of analyzing samples collected in the field. These labs can be small, independent service providers, part of larger environmental testing labs, or captive, internal company labs; or, in some cases where unique tests are required, university or research labs. In selecting a lab, the health and safety professional should be convinced that the laboratory has the technical expertise to support the problems most likely encountered in the field and can demonstrate competence. Labs may have available a sampling protocols manual, which assures that the techniques used to collect material are compatible with their established analytical procedures. They may also provide collecting media. The selected laboratory should demonstrate general competence by participating in appropriate external quality assurance and certification programs. The most familiar program is the American Industrial Hygiene Association Proficiency Analytical Testing program. Laboratories that demonstrate analytical and quality assurance competency are accredited by AIHA. A list of accredited labs is published biannually in the AIHA journal (usually April and September issues). The selected laboratory should be capable of producing a set of standard operating procedures, chain of custody practices, sample handling and control procedures, instrument calibration and maintenance records, internal and external testing programs, and quality control of individual methods. Frequently, these laboratories will follow protocols published by the National Institute for Occupational Safety and Health (NIOSH, 1984); however, on some occasions, novel sampling and analytical procedures will be developed on request. In these cases, the client health and safety professional should review the methodology used to verify the accuracy, precision, sensitivity, temperature, humidity, and storage effects of the procedure.

Both the health and safety professional and the laboratory management should be concerned with maintaining the integrity of the sample through a chain of custody procedure. This practice assures that the analyzed samples are not confused with each other and that the sample was not damaged or in some way altered as it passes through various individuals' or organization's areas of responsibility. In the field, samples should be dated, labeled, and, if necessary, stored to remain fresh as soon as possible after collection. Sampling data sheets should be updated continuously with pre- and postsampling flow rates, date and duration of sampling, type of sample, individual collecting the sample, and critical information such as the individual location or operation being investigated as well as pertinent field notes. Sample and data sheets should be linked by a common identification number. The laboratory and field personnel should maintain sample logs. Examples of forms that can be used for this purpose can be found in Appendixes D–F.

Frequently, bulk samples are requested by the service lab. These often need to be shipped separately to avoid leakage and contamination of the field samples. The health and safety professional should also submit blank and spiked samples to determine if unanticipated circumstances have compromised the sample. Spiked samples are most easily prepared by injecting a minute volume of the pure contaminant onto the sampling media. This can be in the form of a saturated vapor obtained with a gas-tight syringe from the space above the liquid in a shaker bottle. In some cases, the liquid material itself can be injected. The health and safety professional should calculate the mass loading to be approximately in the midrange of the anticipated exposure measurements.

There are two types of blank samples. In one case, a sealed tube or filter is analyzed to identify any positive interferences due to contamination of lab agents or zero drift in the lab procedure. In a second case, a blank tube is broken or filter cassette uncovered in the field but not employed to actively collect a sample. At the completion of sampling this "field blank" is treated as all the other samples. This field blank identifies potential sampling errors due to loading from the environmental conditions and/or shipping and handling procedures. For example, solvents from magic markers used to identify individual passive badges have been reported to be evaporating during shipment and collecting on the charcoal of blanks and samples to produce false positive data.

8.4.3 Instrument Calibration, Verification, and Maintenance

There are two degrees of calibration. A laboratory grade calibration is designed to determine within specified limits the true values associated with scale readings on an instrument under critical challenge conditions that reflect field applications. A field grade calibration tests the instrument's ability to indicate a correct reading. The laboratory calibration is most often performed by the manufacturer or a third party lab and is discussed in Section 8.4.4. Field calibration is performed prior to a monitoring activity to verify instrument performance. The simplest field calibration is to challenge an instrument with a prepared gas mixture purchased from a vendor. The challenge contaminant is certified by the supplier to be at a given concentration and the span setting of the instrument is adjusted to conform to this value. A second source of calibration mixture can be produced by evaporating calculated quantities of challenge liquid into clean static containers such as a bag, bottle, or drum in such a way that a uniform environment is produced. In some cases, the headspace above a liquid stored in its bottle serves as a most convenient source of this type of a challenge vapor. A third technique bubbles clean, dry air through a heated liquid in order to create a saturated vapor stream. This stream is then diluted in a mixing chamber with a second source of clean or humidified air to produce a challenge concentration. Each of these techniques has advantages and disadvantages and are usually chosen based on convenience, space limitations, and access to materials. References discussing calculations and setup of calibration systems can be found in the bibliography (Chelton, 1993).

Field verification should also include a test to confirm that sampling tubes are not leaking. Sample collecting pumps should be calibrated for proper flow using bubblemeters or flow sensors. This is critical for sampling pumps connected to tubes, impingers, and filters, and pumps internal to direct reading devices should be checked periodically. In addition, the performance of batteries, filters, flashback arrestors, sensor life, and electronics performance should also be confirmed before beginning a sampling exercise. All these evaluations should be recorded and maintained for field quality control purposes. Example forms can be found in Appendixes D–F.

The above evaluation discussed can be performed either prior to use and as part of an ongoing scheduled maintenance program. The most common equipment failures are loss of battery rechargeability, leaking sample tubes, clogged prefilters, failed or clogged flashback arrestors (this should be tested and serviced by the vendor), sensor aging, electronics drift, or oxidation of contacts. Scheduled servicing or field parts replacement should be discussed with each equipment supplier on an individual basis.

8.4.4 Sampling Equipment and Instrument Certification

It is not uncommon for the health and safety professional to be expected to assess environmental conditions in a setting where the possibility of an explosive or flammable atmosphere of a vapor, gas, or dust exists. For this reason a majority of the equipment discussed in this chapter is designed and certified by their companies as intrinsically safe. The National Fire Protection Agency (NFPA) in Article 500 of the National Electric Code describes intrinsically safe equipment as being incapable of releasing sufficient electrical or thermal energy under normal and abnormal conditions to cause ignition of specific flammable or combustible atmospheric mixtures in their most easily ignitable condition. NFPA has developed a classification scheme of hazardous environments consisting of seven groupings of gases, vapor, and dusts according to their explosive/flammable potential (NFPA, 1987). Manufacturers may seek intrinsic safety certification by having their product tested in all or selected challenge atmospheres that represent these groupings. Independent testing laboratories that are listed by OSHA as Nationally Recognized Testing Laboratories are qualified to perform challenge tests in the United States applying either the ANSI/UL 913-1988 (ANSI, 1988) standard or similar protocol. These labs also test and certify instrument performance applying the procedures found in ANSI/ISA-S12-13.1-1986 (NESVIG, 1986). In this procedure instruments are evaluated for accuracy, temperature effects, response to concentration change, humidity effects, and ambient air velocity effects. Instruments that have successfully passed an evaluation are permitted to carry the certifying labs label. Greater details on label information can be obtained by contacting the certifying lab directly. Samples of certification labels are shown in Fig. 8.13.

8.4.5 Radio Frequency Effects

Radio frequency interference from two-way radios and other sources has been reported to affect the readings obtained from electronic instruments. (Cook and Huggins, 1984). The greatest concern is interference to direct reading devices from two-way radios during field monitoring. Devices with simple circuity and no amplification are less likely to be affected. Some vendors shield critical circuits to prevent this problem.

8.4.6 Shipping

After collection, the samples are packaged and shipped to the lab for analysis. Shipment of liquid samples will require proper Department of Transportation labeling. Most

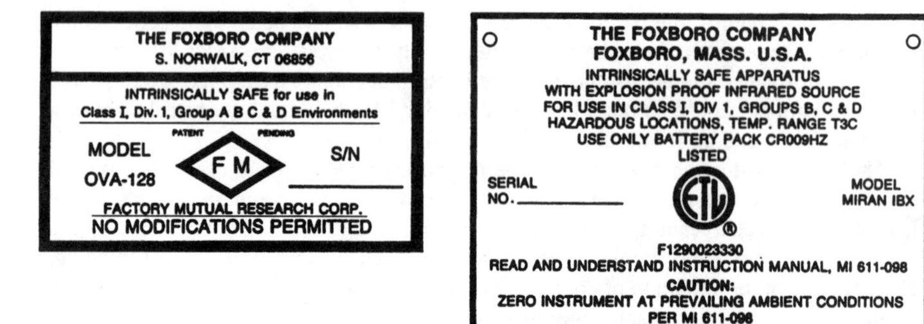

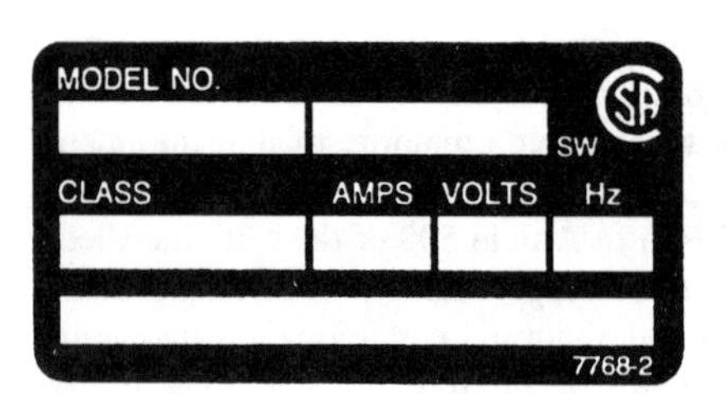

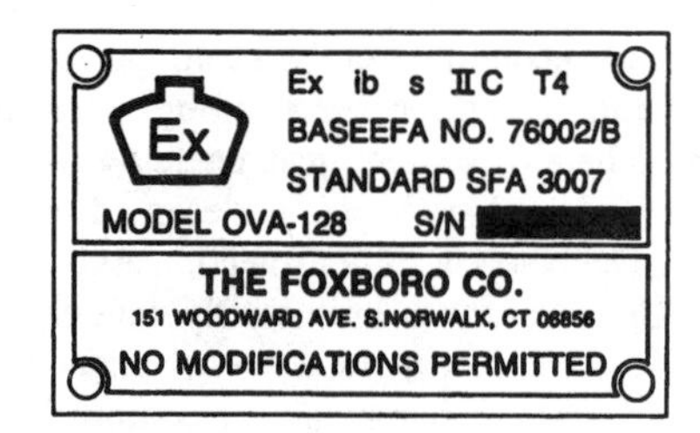

Fig. 8.13 Equipment certification labels.

samples will be classified as flammable, poisonous, and/or corrosive. Information on proper labeling and packaging can be found by contacting the lab chosen for the analysis. Another resource may be the supplier of reagents to a company's research, quality control, or process lab since the collecting solutions are common lab reagents. A review of the Material Safety Data Sheet that accompanies the reagent bottle will also prove to be a useful source of shipping information.

8.4.7 Records

The creation and maintenance of a good records is a critical element of any industrial hygiene program. The records on employee exposure can be the basis for epidemiological studies; as the source of predicting proper respiratory selection when sampling is not feasible; as a response to employee or union concerns in support of medical surveillance programs; and/or to determine the need or effectiveness of engineering controls. They can also be used to advise management on the impact of proposed regulations and in litigation. Finally, a good record keeping system exhibits a commitment to quality and professional performance. Discussions of this topic can be found in many references (Wrench, 1990; Confer, 1994). Ideally, records should include a standard operating procedures manual that includes procedures for calibrating, testing, and maintaining equipment, routine sampling protocols, example calculations of flow determination, and data analysis. Equipment maintenance, repair, and calibration logs should also be included. Chain of custody records also verify the validity of exposure data. Finally, data on exposure assessments should be maintained that define the nature of the characteristics that make a population homogeneous (Tait, 1997): the number of sampling sessions, the details of sampling protocol (i.e., specified flow rate, time, and volume), analytical lab results, and the results of calculations. Included in this database should be a narrative of individual or group performance during the sampling session, the types of personal protection or administrative practices in practice, and influencing environmental conditions. Many databases have been developed and discussed over time, and individuals or organizations have created their own customized systems over time. A joint effort of the ACGIH and AIHA is underway to publish guidelines discussing occupational databases in greater detail (ACGIH/AIHA, 1997).

8.4.8 Data Logging

Data loggers are electronic devices that record the output of an instrument for later analysis. The output is linked to a data logging record, which both averages the signal over brief periods of time and records the average as data points over the duration of the sample. This produces a time history of the exposure that identifies peak periods of exposure as well as the integrated overall average exposure. Data logging capabilities are often included in the overall instrument package; however, standalone data logging devices are also available on the market. In many cases, data loggers have added another level of convenience to the field application of direct reading instruments by replacing the larger, more inconvenient, strip chart recorder.

8.5 SURFACE SAMPLING

8.5.1 General

The terms *wipe sampling*, *swipe sampling*, and *smear sampling* are synonyms that describe the techniques used to assess surface contamination. Wipe sampling is most often used to determine the presence of materials such as asbestos, lead and other metals, aromatic amines, and PCBs. Wipe sampling techniques are applied to:

1. Evaluate potential contact with surface contaminants by wiping surfaces that workers can touch.
2. Surfaces that may come into contact with food or other materials that are ingested or placed in the mouth (e.g., chewing tobacco, gum, cigarettes) may be wipe sampled (including hands and fingers) to detect contamination.
3. Effectiveness of personal protective gear (e.g., gloves, aprons, respirators) may sometimes be evaluated by wipe sampling the inner surfaces of the protective gear (and protected skin).
4. Effectiveness of decontamination of surfaces and protective gear (e.g., respirators) can sometimes be evaluated by wipe sampling.

When accompanied by close observation of the operation in question, wipe sampling can help identify sources of contamination and poor work practices.

Generally, two types of filters are recommended for taking wipe samples. Glass fiber filters (GFF) (37 mm) are usually used for materials that are analyzed by high-performance liquid chromatography (HPLC), and often for substances analyzed by gas chromatography (GC). Paper filters are generally used for metals. For convenience, the

Whatman smear tab (or its equivalent) or polyvinyl chloride filters for substances that are unstable on paper-type filters are commonly used. Normally a 100 cm^2 area of the surface is wiped while wearing gloves. The filter can be dry or wetted with distilled water; however, organic solvents are not normally used (OSHA, 1995). Collected filters are normally stored in glass vials for shipment.

8.5.2 Dermal Exposure Assessment

Many substances, especially fat-soluble hydrocarbons and other solvents, can enter the body and cause systemic damage directly through the skin when the skin has become wet with the substance by splashing, immersion of hands or limbs or exposure to a mist or liquid aerosol. Some substances, such as amines and nitriles, pass through the skin so rapidly that the rate at which they enter the body is like that of substances inhaled or ingested. The prevention of skin contact to phenol is as important as preventing inhalation of airborne concentrations. A few drops of dimethyl formamide on the skin can contribute to a body burden similar to inhaling air at the TLV for 8 hours. The "skin" notation in the TLV list identifies substances for which skin absorption is potentially a significant contribution to overall exposure. However, the TLV list offers no precise definition of a "significant" contribution.

Note that the skin notation relates to absorption or toxicity via route and not to the potential to cause skin damage and dermatitis. These later effects are important, however, and should be indicated on such hazard data sources as material safety data sheets (MSDSs).

Dermal exposure assessment is a complex matter that has not received much attention in the practice of industrial hygiene. Estimation of dermal exposure involves exposure scenarios and pathways, contact duration and frequency, body surface area in contact, and substance adherence. The amount of contaminant that crosses the skin barrier and enters the body is influenced by the properties of the substance and the properties and condition of the skin at the exposed site. Models of dermal absorption have been developed (Berner and Cooper, 1987; Flynn, 1990). These models start with the partition coefficients of the substance, its mole weight and solubility, and the diffusivity of the compound in the lipid and protein phases of the skin. Complex theoretical models have been simplified and combined with empirical models to yield an estimate of the chemical specific dermal permeability constant. The dermally absorbed dose (mg/kg day) can then be calculated from this constant and the skin contact area, exposure time, frequency and duration, and body weight. Since these quantities are rarely known, they must be assumed. When conservative assumptions are used, the conservatisms may compound to yield a very conservative estimate of the dermal dose.

REFERENCES

ACGIH (1978). "Air Sampling Instruments for Evaluation of Atmospheric Contaminants. 5th edition," ACGIH, P. O. Box 1937, Cincinnati, Ohio 45201.

ACGIH (1996). American Conference of Government Industrial Hygienists, "Threshold Limit Values for Chemical Substances and Physical Agents in the Work Environment with Intended Changes for 1996–1997," ACGIH, P. O. Box 1937, Cincinnati, Ohio 45201.

ACGIH/AIHA (1997). "Data Elements for Occupational Exposure Databases," M. Lippman et al, Joint ACGIH/AIHA Task Group on Occupational Exposure Databases.

Anderson, M. E., M. G. McNaughton, H. J. Clewell, and D. P. Paustenbach (1987). *Am. Ind. Hyg. Assoc. J.* **48,** 335–43.

ANSI (1988). ANSI/UL 913-1988, Intrinsically Safe Apparatus and Associated Apparatus for Use in Class I, Class II and III, Division I, Hazardous (Classified) Locations, New York, NY.

Brief, R. S., and R. A. Scala (1975). *Am. Ind. Hyg. Assoc. J.* **36,** 467–69.

Calabrease, E. J. (1977). *Am. Ind. Hyg. Assoc. J.* **38,** 443–46.

Chelton, C. F. (1993). "Manual of Recommended Practice for Combustible Gas Indicators and Portable Direct-Reading Hydrocarbon Detectors," Am. Ind. Hyg. Assoc., Fairfax, VA.

Confer, R. G. (1994) Workplace Exposure Protection: Industrial Hygiene Program Guide. Lewis Publishers, CRC Press, Boca Raton, FL.

Cook, C. F., P. A. Huggins (1984). *Am. Ind. Hyg. Assoc. J.* **45,** 740–44.

Droz, P., and M. Yu (1990). Biological Monitoring Strategies, in *Exposure Assessment for Occupational Epidemiology and Hazard Controls*, S. M. Rapport and T. J. Smith, Eds., Lewis Publishing Co., Chelsea, MA.

Hawkins, N. C., S. K. Norward, and J. C. Rock, Eds. (1991). A Strategy for Occupational Exposure Assessment, Am. Ind. Hyg. Assoc., Akron, Ohio.

Hickey, J. L. S., and P. C. Reist (1977). *Am. Ind. Hyg. Assoc. J.* **38,** 613–21.

HSE (1989). Health and Safety Executive Monitoring Strategies for Toxic Substances, Guidance Note, EH42, HSE, Bottle, Merseyside, UK.

Ind. Hyg. News (1996). Buyers Guide, *Industrial Hygiene News* **19**(3), Rimbach Publishing Co.

Leidel, N. A., K. A. Busch, and J. R. Lynch (1977). Occupational Exposure Sampling Strategy Manual, Department of Health, Education and Welfare, Publication (NIOSH) 77-173, Cincinnati, OH.

Mason, J. W., and H. Dershin (1976). *J. Occup. Med.* **18,** 603-60ISA.

NESVIG (1986). Performance Requirements, Combustible Gas Detectors (ISA-S12.13.1), Instrument Society of America, Pittsburgh, Pa.

NFPA70 (1985). National Electrical Code, National Fire Protection Association, Quincy, MA.

NIOSH (1984). National Institute of Occupational Safety and Health, Manual of Analytical Methods, 4th ed., U.S. Dept of Health and Human Services, Washington, DC.

Occup. Safety and Health (1996). Medical Publications, Inc., Waco, TX.

OSHA (1982). Use of Personal Sampling Devices During Inspection, 47 Fed. Reg. 55478.

OSHA (1991). Analytical Methods Manual, Pub. 1985 with updates through 1991.

OSHA (1995). Technical Manual, Chapter 2, Sampling for Surface Contaminants, Occupational Safety and Health Administration, Washington, DC.

Roach, S. A. (1966). *Am. Ind. Hyg. Assoc. J.* **27,** 1–12.

Roach, S. A. (1977). *Ann. Occup. Hyg.* **20,** 65–84.

Roach, S. A. et al (1983). *Am. Occup. Hyg.* **27,** 1–13.

Tait, K. The Workplace Exposure Assessment Workbook (Workbook). Appl. Occup. and Env. Hyg. J. **8**(1) 55–68. 1993.

Wrench, C. (1989) Data Management for Occupational Health and Safety: A User's Guide to Integrating Software. Van Nostrand Reinhold, New York, NY.

APPENDIX A

CONSIDERATIONS IN ESTABLISHING AN INDUSTRIAL HYGIENE FIELD OFFICE

- Review the operations and activities within the hygienist's scope of responsibilities to determine the most likely hazardous materials exposures to be evaluated.
- Assess the emerging concerns in occupational risk that may impact the employee population in the near term.
- Determine if integrated and/or grab sampling will be necessary.
- Select the most probable sampling procedures to be used.
- Select a support industrial hygiene laboratory service based on analytical capabilities, certification of performance, and backup technical support.
- Develop an inventory of equipment to be maintained on site (see Apendix B).
- Obtain or develop sampling protocols and gather them into a centralized file or binder.
- Collect and centralize equipment manuals.
- Develop field calibration capabilities for sampling pumps direct reading instruments and other equipment as needed or arrange for external support.
- Establish an equipment calibration and repair log.
- Develop exposure sampling data.
- Develop an exposure data management system.
- Allocate space for a minimum of simple field laboratory functions (evaluate the need for lab hood and flammable materials storage cabinet).
- Collect a file of relevant MSDS.

APPENDIX B

INDUSTRIAL HYGIENE FIELD OFFICE INVENTORY CHECKLIST

Sampling pumps

- Lo flow
- High flow
- Combination hi/lo flow

Adsorbent tubes

- Charcoal
- Silica gel
- Alumina
- Chromatographic packings

Passive badges
Colorimetric detector tubes
35/27 mm filter cassettes

- Cellulose acetate
- PVC
- Gold film
- PTFE

Microscope
Asbestos ID kit
Sound level meter
Noise dosimeters
Radiation survey meter
Microwave survey meter
Flow calibrator/bubblemeter

APPENDIX C

DIRECT READING INSTRUMENT INVENTORY

Instrument/device	Application	Common brands
Hand pumps	Detector tube sampling	SKC, MSA, Gastech, National Draeger
Pumps low flow	Charcoal tube sampling	Gilian, SKC, Du Pont, MSA
Pumps medium flow	Filter sampling	Same
Flow calibrator	Pump calibration	Gilian
Fibrous aerosol monitors	Asbestos/fibers	MIE, MDA, PPM, TSI
Double range meters	Oxygen, combustible gas	Gastech, MSA
Triple range meters	Carbon monoxide, oxygen combustible gas, hydrogen sulfide	Many brands
Carbon monoxide dosimeter	Indoor air quality	Interscan, Scott, Draeger
Carbon dioxide meter	Indoor air quality	Gastech
Infrared analyzers	Organics, carbon monoxide carbon dioxide	Foxboro
Hydrogen cyanide monitors	Hydrogen cyanide	MDA, Monitox
Hydrogen sulfide monitors	Hydrogen sulfide	Draeger, Interscan
Mercury vapor	Mercury	Jerome

APPENDIX D

PRELIMINARY INDUSTRIAL HYGIENE EVALUATION REPORT

DATE ________ TIME CONTACTED ________ UNIT/AREA ________

WHO REQUESTED INVESTIGATION/SURVEY: ________________________________

____

____

____

DATE IH RESPONDED: ______________________ TIME: ______________________

RESPONDING IH NAME(S): ____

OTHER PEOPLE INVOLVED IN INVESTIGATION/SURVEY: ______________________

FINDINGS AT TIME OF FIELD INVESTIGATION/SURVEY: ______________________

____

____

____

____

FOLLOW UP NEEDED: ____ YES ____ NO IF YES, WHAT AND WHEN TO BE DONE:

____

____

WAS FOLLOW UP COMPLETED: ____ YES ____ NO IF NOT, WHY: ______________

____

____

RESULTS OF INVESTIGATION/SURVEY: ________________________________

____

____

____

____

Report By: ______________________

APPENDIX E

FIELD SAMPLING FORM

1. Employee and company information

Name ______________________ SS# ____________ Company ID# ____________

Company Name ______________________ Billing Department ____________________

Address __________________________ Supervisor __________________________

__________________________________ Report Sent _________________________

__________________________________ To _________________________________

2. Task Description

Identifying task name _______________________ Shift _______________________

Job Title ____

Equipment or location identified ____

Operations (circle one) routine, maintenance, upset conditions, other

Task length (circle one) continuous, intermittent, extension, partial shift, other.

Task duration ____________

Narrative description (engineering controls, task performance & characteristics)

Signature _______________ Date ______

3. Personal Protective Equipment

Respiratory protection (circle one): None, supplied air, SCBA: ______ Air Purifying: Cartridge ______

Body protection: Nomex coveralls, disposable coveralls, slicker suit, nitrile suit, acid suit, encapsulated suit

Cotton: Long sleeve Short sleeve

Gloves Y or N Type ____ Boots Y or N Type ____

Eyewear Y or N Type ____ Hearing Y or N Type ____

4. Sampling Strategy

Sample type (circle one) typical; worst case; random

Sample period (circle one) full shift; partial period; short-term grab

Representation (circle one) breathing zone; employee area; point source, area

Relative humidity ____ Barometric pressure ____ Other relevant weather conditions ____

5. Measurement

Sample ID Start time Stop time Elapsed time Sample rate Total value Sampling media

APPENDIX F

INDUSTRIAL HYGIENE MONITORING RESULTS

Report to Supervisor: ____________________

Department Name and #: ____________________

Date Sent: ____________________

IH Log #: ____________________

Date Sampled: ____________________

Process/Equipment: ____________________

____

____

____

Employee Name: ____________________

Employee #: ____________________

Exposure: ____________________

Permissible Exposure Limit (PEL): ________

Threshold Limit Value (TLV): ________

Sample Results: ____________________

Comments: ____________________

____

____

Inform employee and return to Safety Manager

by ______________________________

____

Safety Manager

____

______________________________ was informed of the above information

(Employee Name)

____

(Date)

Employee/supervisor's comments: ______________________________

____

____

____

____

Supervisor's Signature

CHAPTER 9

Evaluation and Management of Exposure to Infectious Agents

JANET M. MACHER

Division of Environmental and Occupational Disease Control, Environmental Health Laboratory Branch, California Department of Health Services, 2151 Berkeley Way, Berkeley, CA 94704

JON ROSENBERG

Division of Communicable Disease Control, Disease Investigations and Surveillance Branch, California Department of Health Services, 2151 Berkeley Way, Berkeley, CA 94704

9.0 INTRODUCTION
- 9.0.1 Health Care
- 9.0.2 Child Care
- 9.0.3 Agriculture, Food Processing, and Animal-Associated Occupations
- 9.0.4 Laboratories
- 9.0.5 Occupational Travel
- 9.0.6 Personal Service Work
- 9.0.7 Wastewater and Sewage Treatment
- 9.0.8 Unlisted Occupations

9.1 INFECTIOUS DISEASES—BASIC PRINCIPLES
- 9.1.1 Infectious Microorganisms and Their Sources
- 9.1.2 Transmission of Infectious Agents
- 9.1.3 The Infectious Disease Process
- 9.1.4 Modeling Infectious Disease Transmission

9.2 EVALUATING EXPOSURE TO BIOLOGICAL AGENTS
- 9.2.1 Regulations, Recommendations, and Guidelines
- 9.2.2 Medical and Biological Evaluation of Exposure
- 9.2.3 Laboratory Diagnosis of Infectious Diseases
- 9.2.4 Infectious Disease Surveillance
- 9.2.5 Inspections and Environmental Testing

9.3 INFECTIOUS DISEASE PREVENTION
- 9.3.1 Immunizations
- 9.3.2 Substitution and Engineering Controls

Handbook of Occupational Safety and Health, Second Edition, Edited by Louis J. DiBerardinis, ISBN 0-471-16017-2

9.3.3 Education, Training, and Administrative Controls
9.3.4 Personal Protective Equipment
9.3.5 Postexposure Prophylaxis
9.3.6 Workers at Increased Risk from Infectious Diseases
CONCLUSIONS
REFERENCES

Abbreviations

AIDS	acquired immunodeficiency disease
AOAC	Association of Official Analytical Chemists
ARDS	adult respiratory distress syndrome
BCCV	Black Creek Canal virus
BCG	Bacille Calmette Guérin
BSL	biosafety level
ca.	circa, approximately
CDC	Centers for Disease Control and Prevention
CFR	Code of Federal Regulation
CMV	cytomegalovirus
CNS	central nervous system
DNA	deoxyribonucleic acid
EEE	eastern equine encephalomyelitis
FDA	Food and Drug Administration
GI	gastrointestinal
HAV	hepatitis A virus
HBV	hepatitis B virus
HCV	hepatitis C virus
HEPA	high-efficiency particulate air
HIV	human immunodeficiency virus
HPV	human papillomavirus
HSV	herpes simplex virus
HTLV-I	human T-lymphotropic virus type I
Ig	immunoglobulin
IgA	immunoglobulin class A
IgG	immunoglobulin class G
IgM	immunoglobulin class M
IG	immune globulin (serum)
ID50	infectious dose (50 percent)
ISG	immune serum globulin
LD	Legionnaires' disease
MSDS	material safety data sheet
MMWR	Morbidity and Mortality Weekly Report
NPRM	Notice of Proposed Rulemaking
NSF	National Sanitation Foundation
ODTS	organic dust toxic syndrome
OSHA	Occupational Safety and Health Administration
PCR	polymerase chain reaction
PEP	postexposure prophylaxis

PF	Pontiac fever
PPE	personal protective equipment
RNA	ribonucleic acid
SLE	St. Louis encephalomyelitis
SNV	Sin Nombre virus
TB	tuberculosis
Td	tetanus-diphtheria
TLV	threshold limit value
USEPA	U.S. Environmental Protection Agency
USDOL	U.S. Department of Labor
USDOT	U.S. Department of Transportation
UVGI	ultraviolet germicidal irradiation
WEE	western equine encephalomyelitis
WHO	World Health Organization
VZ	varicella-zoster
VZV	varicella-zoster virus

Agencies

AAP	American Academy of Pediatrics
ACGIH	American Conference of Governmental Industrial Hygienists
AIHA	American Industrial Hygiene Association
APHA	American Public Health Association
ASM	American Society for Microbiology
ASTM	American Society for Testing and Materials
CDC	Centers for Disease Control and Prevention
NIH	National Institutes of Health
NIOSH	National Institute for Occupational Safety and Health
NRC	National Research Council
NSF	National Sanitation Foundation
USDHHS	U.S. Department of Health and Human Services
USDOL	U.S. Department of Labor
USEPA	U.S. Environmental Protection Agency

9.0 INTRODUCTION

The evaluation and management of workplace exposures to infectious agents have received increasing attention in recent years. This interest is due in part to (a) the emergence of new organisms, from the human immunodeficiency virus (HIV, the agent of HIV infection and acquired immunodeficiency syndrome or AIDS) to hantaviruses; (b) the resurgence of previously controlled diseases (e.g., tuberculosis); and (c) better recognition of existing hazards (e.g., hepatitis B virus, HBV, infection), all of which present risks to workers. Biological hazards in the workplace may cause acute or chronic infectious diseases as well as toxic and hypersensitivity reactions. This chapter covers the major infectious diseases seen in the United States as a result of work-related exposures (Gantz, 1988; Garibaldi and Janis, 1992; Ellis and Symington, 1994; Gerberding and Holmes, 1994; Wald and Stave, 1994). In some cases, sport and recreational activities overlap with work-related ones, because many of the former involve full-time or part-time workers as well as volunteers (Collins et al., 1997). Biological hazards not covered here include bites or attacks by domestic or wild animals as well as hypersensitivity reactions (e.g., hypersensitivity pneumonitis and allergic dermatitis, rhinitis, sinusitis, and asthma) and organic dust toxic syndrome (ODTS) for which the reader is referred elsewhere (NIOSH, 1994; Rosenstock and Cullen, 1994; Rylander and Jacobs, 1994; Wald and Stave, 1994; Harber et al., 1996).

Health and safety professionals are playing an expanding role in anticipating, recognizing, evaluating, and preventing occupational infections. This is so even though the role these professionals play in infectious disease control often differs, for substantive and historical reasons, from the role they play in preventing other illnesses. Many occupational diseases related to chemical or physical agents are chronic, progressive processes, whereas infectious diseases are often acute and follow shortly after exposure. Understandably, few heath and safety professionals have extensive training in microbiology, infection control, or epidemiology. Many readers may primarily encounter biological hazards as bioaerosols in problem-building investigations (see Chapter 20). Rather than health and safety professionals, infection control practitioners generally handle worker surveillance for human-borne diseases (i.e., those transmitted from person to person), veterinarians and agricultural healthcare workers manage zoonoses (i.e., those animal-borne infections that affect humans), and biosafety specialists monitor laboratory-acquired infections. However, some of these duties may fall to occupational health and safety professionals for a variety of reasons.

The traditional approach to health and safety, besides primary prevention, is often workplace monitoring (to detect excessive exposures and correct them) and worker surveillance (to identify affected workers early). However, exposure limits do not exist for infectious agents, and testing for pathogenic microorganisms in the work environment is seldom required, even in outbreak investigations. The presence of microorganisms in a workplace may easily be missed because of their small size and the frequent absence of warning properties, such as odor or taste, to alert workers to an agents' presence. While all workers can be assumed to be susceptible to chemical or physical hazards, workers who have been immunized (vaccinated) or naturally infected may be protected from infectious diseases.

Occupational infectious diseases range from the familiar and relatively benign (e.g., the common cold and warts) to the rare and usually or frequently fatal (e.g., rabies and B-virus infection). Besides the damage they cause, these diseases may require treatment with toxic drugs, such as antibiotics for multidrug-resistant tuberculosis. Even infections that are commonplace and mild in children (e.g., chickenpox, measles, and mumps) may result in lost time and be more serious when contracted by adult workers. Affected workers can also become sources of infection for their co-workers, clients, families, and the public, and job restrictions may be necessary to prevent this. Job restrictions may also be prudent for workers with underlying risk factors (e.g., immunodeficiency or pregnancy) that may make them unusually susceptible to infectious diseases or make the consequences of infection more serious.

Infectious agents in the workplace include a wide range of microorganisms that affect a diverse variety of occupations. Occupational health and safety professionals need to recognize when infectious diseases are or may become a concern and should acquire the necessary training to handle these hazards. This instruction should enable health and safety staff to identify (a) infectious diseases possibly related to a job; (b) the causative agents of these infections; (c) workplace sources or reservoirs of the agents; (d) the transmission mechanisms from these sources or reservoirs to workers; (e) available surveillance and monitoring methods; and (f) appropriate preventive measures.

Tables 9.1–9.3 organize and cross-index information on occupational infectious diseases in different ways. Table 9.1 lists a variety of occupations and work-related infections that have been reported or are considered possible. Respectively, Tables 9.1a–c address infectious diseases transmitted by human contact, animal or arthropod contact, and environmental (soil, water, or plant) contact. In each section, infections are grouped

TABLE 9.1a Infectious Diseases and Occupations—Human Contact

Transmission Mode[a] Disease	Child care	Teaching	Health care	Public safety	Personal service
Viral infections					
Contact transmission					
Acute respiratory disease[a]	×	×	×	×	
Conjunctivitis (viral)	×				
Cytomegalovirus disease	×				
Herpetic whitlow			×		
Warts			×		
Contact (bloodborne) transmission					
AIDS/HIV infection			×	×	×
Hepatitis B, C, D			×	×	×
Contact (fecal-oral/ingestion) transmission					
Hepatitis A	×				
Droplet transmission					
Rubella (German measles)[a]	×		×		
Airborne transmission					
Chickenpox	×		×		
Influenza[a]		×	×	×	
Measles (rubeola)[a]	×	×	×		
Protozoal and helminth infections—Contact (fecal-oral/ingestion) transmission					
Cryptosporidiosis	×		×		
Giardiasis	×		×		
Fungal infections—Contact transmission					
Dermatophytosis					×
Bacterial infections					
Contact transmission					
Conjunctivitis (bacterial)	×				
Streptococcal skin disease			×		
Contact (fecal-oral/ingestion) transmission					
Campylobacter enteritis	×				
Shigellosis	×				
Droplet transmission					
Pertussis[a]	(×)[b]	(×)	×		
Airborne transmission					
Tuberculosis			×	×	

[a]Only the most common mode of transmission is identified for infectious diseases transmitted by more than one mechanism (see Tables 9.2a–c for other transmission modes).

[b](×) = work-association unusual or uncertain.

TABLE 9.1b Infectious Diseases and Occupations—Animal or Arthropod Contact

Transmission mode[a] Disease	Abattoirs (slaughter houses)	Animal hide, hair, wool, bone processing	Animal handling/ trapping	Bird/ poultry handling/ breeding	Farming	Fishing/ fish handling	Food handling	Forestry	Meat processing/ packing	Public safety	Travel	Veterinary care
Viral infections												
Contact transmission												
B virus infection												
Orf virus disease	×				×							
Rabies[a]			×		×			×		×	(×)	
Warts									×			
Airborne transmission												
Hantavirus pulmonary syndrome			×		×							
Vectorborne transmission												
Arthropod-borne viral disease[a]			(×)		×			×			×	(×)
Protozoal and helminth infections												
Contact transmission												
Hookworm disease			×		×						×	
Contact (fecal-oral/ ingestion) transmission												
Cryptosporidiosis					×						×	×
Echinococcosis					×						×	
Toxoplasmosis					×				×			
Vectorborne transmission												
Malaria											×	
Fungal infections—Contact transmission												
Dermatophytoses	×	×	×		×							

Bacterial infections

Contact transmission										
Anthrax[a]	×	×			×				×	×
Brucellosis[a]	×				×				×	×
Cat-scratch disease[a]			×							×
Erysipeloid	×			×		×			×	×
Leptospirosis	×		×		×	×			×	×
Mycobacterium marinum infection						×				
Tetanus					×			×		
Tularemia[a]		×	×		×			×		×
Contact (fecal-oral/ Ingestion) transmission										
Campylobacter enteritis				×	×				×	×
Salmonellosis	×		×	×	×		×		×	×
Airborne transmission										
Psittacosis[a]				×					×	×
Q fever[a]	×	×	×		×				×	×
Vectorborne transmission										
Lyme disease			×		×			×		
Plague[a]			×		×			×		×
Rocky Mt. spotted fever			×		×			×		

[a]Only the most common mode of transmission is identified for infectious diseases transmitted by more than one mechanism (see Tables 9.2a–c for other transmission modes).
[b](×) = work-association unusual or uncertain.

TABLE 9.1c Infectious Diseases and Occupations—Environmental (Soil, Water, or Plant) Contact

Transmission mode[a] / Disease	Construction	Farming	Fishing/ fish handling	Forestry	Gardening/ horticulture	Mining	Travel	Utility/ maintenance work	Wastewater/ sewage treatment
Viral infections									
Contact transmission									
Rabies[a]						×			
Contact (fecal-oral/ ingestion) transmission									
Hepatitis A							×		
Airborne transmission									
Hantavirus pulmonary syndrome		×		×				×	
Protozoal and helminth infections									
Contact transmission									
Cutaneous larva migrans		×			×			×	
Hookworm disease		×				×	×		×
Contact (fecal-oral/ infestion) transmission									
Amebiasis[a]							×		(×)
Cryptosporidiosis		×							×
Giardiasis				×					(×)

					Fungal infections				
Contact transmission									
Dermatophytosis						×			
Sporotrichosis[a]		×			×	×			
Airborne transmission									
Aspergillosis		×							
Blastomycosis	×	×		×					
Coccidioidomycosis	×	×							
Cryptococcosis	×							×	
Histoplasmosis	×	×						×	
					Bacterial infections				
Contact transmission									
Leptospirosis		×				×			×
Mycobacterium marinum infection			×						
Tetanus	×	×		×	×		×		×
Airborne transmission									
Legionellosis							×	×	

[a]Only the most common mode of transmission is identified for infectious diseases transmitted by more than one mechanism (see Tables 9.2a–c for other transmission modes).
[b](×) = work-association unusual or uncertain.

TABLE 9.2a Basic Information on Infectious Diseases Transmitted by Human Contact or Direct Agent Handling

Disease	Infectious agent	Health effects	Sources	Transmission	Surveillance	Prevention
			Viral infections			
Acquired Immunodeficiency Syndrome (AIDS, HIV infection)	Human immunodeficiency virus (HIV)	AIDS: progressive damage to immune and other organ systems, many opportunistic infections	Human blood, serum-derived body fluids (semen, vaginal fluids)	Parenteral; mucous membrane contact; direct contact nonintact skin	Postexposure serology	Standard precautions (needlestick prevention, barriers, disinfection) PEP: antiretroviral agents
Acute viral respiratory disease excluding influenza (acute viral rhinitis, pharyngitis, laryngitis, the common cold, coryza)	Many (rhinoviruses, parainfluenza viruses, respiratory syncytial virus, adenovirus, coronaviruses, coxsackieviruses, echoviruses)	Acute upper and lower respiratory tract infections	Human nose, throat discharges	Droplet nuclei; direct and indirect contact (hands, articles freshly soiled with nose/thorat discharges)	For illness only	Hygiene (handwashing, covering of coughs and sneezes, sanitary disposal of discharges); avoid crowding
Chickenpox (varicella)	Varicella-zoster virus (VZV) [human (alpha) herpesvirus 3, V-Z virus], a member of the *Herpesvirus* group	Varicella: rash, fever Congenital varicella: malformations	Human respiratory secretions, vesicular fluid from chickenpox rash or herpes zoster (shingles)	Droplet nuclei; direct contact; indirect contact with freshly soiled articles; droplets	Serology for evidence of past infection or immunization	Immunize susceptible healthcare workers; restrictions for infected workers PEP: immune globulin for pregnant/immunodeficient workers

Conjunctivitis (keratitis)	Adenoviruses and picornaviruses (see also Bacterial Infections)	Acute inflammation of eyes	Human eye secretions	Direct and indirect contact with eyes	Not indicated	Hygiene, sanitation
Cytomegalovirus disease	Human (beta) herpesvirus 5, human cytomegalovirus (CMV)	CMV mononucleosis; congential CMV	Human respiratory secretions, urine	Mucosal contact	Possibly serology to identify susceptible childcare workers prior to pregnancy	Hygiene (handwashing); possibly barrier precautions (gloves); reassignment during pregnancy
Hantavirus pulmonary syndrome	See Environmental Contact (Below)					
Hepatitis A (infectious hepatitis, HAV infection)	Hepatitis A virus virus (HAV) (*Hepatovirus*)	Abrupt onset of fever, malaise, anorexia, nausea, abdominal discomfort, followed by jaundice	Human feces, rarely other primate feces or human blood	Fecal-oral route; ingestion of contaminated water or food	For illness only	Immunization; hygiene (handwashing), sanitation PEP: immune globulin
Hepatitis B (HBV infection)	Hepatitis B virus (HBV)	Acute, systemic hepatitis (fever, malaise, jaundice), 10% chronic carriers, ca. 2% fatal	Human blood, serum-derived body fluids (saliva, semen, vaginal fluids)	Direct and indirect contact; parenteral, mucous membrane contact; nonintact skin	Serology for evidence of infection or immunization	Immunization; standard precautions PEP: immune globulin, vaccine
Hepatitis C (parenterally transmitted non-A non-B hepatitis, HCV infection)	Hepatitis C virus (HCV) (*Hepacavirus*)	Rarely acute hepatitis, 50% chronic liver disease	Human blood, serum-derived fluids (only sources documented)	Percutaneous (only route documented)	Serology for evidence of past infection	Standard precautions
Delta Hepatitis (viral hepatitis D)	Hepatitis D virus (Delta agent), virus inactive except with HBV	Acute, fulminant hepatitis, more severe than HBV alone	Same as HBV	Same as HBV; percutaneous (only route documented)	Serology for evidence of infection	Same as HBV
Herpetic whitlow	Herpes simplex virus 1 & 2 (HSV)	Finger pain, vesicles	Human body fluids (oral HSV-1)	Direct contact of nonintact skin with saliva or skin lesions	Not indicated	Gloves

TABLE 9.2a (*Continued*)

Disease	Infectious agent	Health effects	Sources	Transmission	Surveillance	Prevention
			Viral infections (Continued)			
Influenza	Influenza viruses type A, B, C	Acute respiratory tract infection with fever, headache, myalgia, prostration, cough, sore throat	Human respiratory secretions	Airborne; possibly direct contact	For illness only (viral isolation and antigen detection for outbreaks)	Annual immunization of workers at high risk of exposure/ complications; hygiene, avoid crowding; antiviral (type A) PEP: amantadine, rimantadine
Measles (rubeola)	Measles virus (*Morbillivirus*)	Acute: rash, fever, encephalitis, occasionally pneumonia Pregnancy: fetal death, prematurity, maternal complications	Human respiratory secretions	Airborne; droplets or direct contact; less commonly by indirect contact with freshly soiled articles	History of 2-dose vaccination, serologic evidence of immunity, confirmed measles, born before 1957	Immunization PEP: vaccine if ≤3 days of exposure, immune globulin if ≤6 days of exposure or vaccination contraindicated
Rubella (German measles)	Rubella virus (*Rubivirus*)	Acute: rash, adenopathy, arthritis (25–50% subclinical) Congenital: anomalies in developing fetus	Human respiratory secretions	Droplets or direct contact	Serology to determine susceptibility (15–25% of young adults susceptible)	Immunize all susceptible workers (require proof of immunity or vaccination); identify exposed pregnant females
Warts (verruca vulgaris, comon wart)	Human papillomavirus (HPV)	Cutaneous warts	Human and animal (type 7 virus) tissue	Direct contact; droplets from laser surgery	None	Gloves; hygiene; local exhaust ventilation during laser surgery

			Protozoal and helminth infections			
Cryptosporidiosis	*Cryptosporidium parvum*	Diarrhea, asymptomatic infection common; prolonged, fulminant course in immunodeficient people (especially with AIDS)	Cattle, human, domestic animal feces	Fecal–oral route; waterborne; foodborne	Not indicated	Sanitation and hygiene
Giardiasis	*Giardia lamblia*	Usually asymptomatic, various intestinal symptoms: bloating, intermittent loose stools	Human, animal feces	Fecal–oral route (ingestion of cysts in contaminated water or food)	For illness only	Hygiene (handwashing); water treatment
			Fungal infections			
Dermatophytosis (tinea, ringworm, athlete's foot)	Various fungi known as dermatophytes (*Epidermophyton*, *Microsporum*, *Trichophyton* spp.)	Scaly or papular rash of keratinized areas of the body (hair, skin, nails)	Humans; cats, dogs, cattle, horses, rodents wild animals; environment	Direct skin-to-skin or indirect contact (contaminated floors, shower stalls, toilet articles, clothing)	Not Indicated	Awareness; hygiene
			Bacterial infections			
Campylobacter enteritis	See Animal Contact (below)					
Conjunctivitis (keratitis)	*Haemophilus*, *Streptococcus*, and other genera (see also Viral Infections)	Acute inflammation of eyes	Human eye secretions	Direct and indirect contact with eyes	Not indicated	Hygiene, sanitation

TABLE 9.2a ***(Continued)***

Disease	Infectious agent	Health effects	Sources	Transmission	Surveillance	Prevention
			Bacterial infections (Continued)			
Pertussis (whooping cough)	*Bordetella pertussis*	Prolonged (>2 weeks) cough, paroxysmal (often not in adults)	Human respiratory secretions	Droplet or direct contact	Suspect diagnosis in patients with prolonged cough	Prompt diagnosis, isolate suspected patients PEP: 14-day course of erythromycin, vaccine booster during outbreaks
Shigellosis (bacillary dysentery)	*Shigella dysenteriae, Shigella flexneri, Shigella boydii, Shigella sonnei*	Acute disease of large/small intestines with diarrhea (blood, mucus in stools); fever, nausea, sometimes vomiting/cramps	Human feces	Direct or indirect fecal–oral route; contact and ingestion of contaminated food or fluid	For illness only	Sanitation, hygiene (handwashing)
Streptococcal skin disease (pyoderma, impetigo)	Group A (beta hemolytic) *Streptococcus* spp.	Superficial skin infection often with vesicular, pustular, and encrusted stages	Humans	Direct or indirect contact of nonintact skin with lesion or contaminated item	For infection only, wound recognition	Hygiene; treat injuries
Tuberculosis (TB, consumption)	*Mycobacterium tuberculosis* complex	Pulmonary disease (ca. 5% lifetime risk for those infected; higher rate for immunocompromised persons); extrapulmonary granulomas less common than pulmonary	Human respiratory secretions	Droplet nuclei; contamination of abraded skin (prosector's wart)	Tuberculin skin test, chest X-ray	Identify active cases, droplet nuclei control PEP: one or more antimicrobial drugs; immunization (BCG) rarely used in U.S.

TABLE 9.2b Basic Information on Infectious Diseases Transmitted by Animal or Arthropod Contact or Direct Agent Handling

Disease	Infectious agent	Health effects	Sources	Transmission	Surveillance	Prevention
			Viral infections			
Arthropod-borne viral disease	Many viruses: St. Louis encephalitis (SLE), eastern and western equine encephalomyelitis (EEE, WEE), California group encephalitis, yellow fever viruses	Abrupt onset of fever, chills, headache, malaise, myalgia, arthralgia, nausea, rash	Reservoirs: animals (birds, rodents/ other mammals, reptiles, amphibians) Vectors: arthropods (mosquitos, ticks, sand flies)	Arthropod bite; droplet nuclei (laboratory)	For illness only	Arthropod control; personal protection to avoid arthropod bites
B-virus infection (simian B virus disease; herpes B-virus infection)	Cercopithecine herpesvirus 1 (B virus) (formerly herpes simiae or *Herpesvirus simiae*)	Rare, frequently fatal encephalomyelitis, no evidence of asymptomatic infection	Old World monkey (macaques) mucosal tissues	Bites, scratches from infected monkeys; direct contact with infected animal's mucous membranes; mucous membrane contact with body fluid from infected animal	Followup injuries	Standard precautions; medical attention for injuries PEP: acyclovir
Hantavirus pulmonary syndrome	See Environmental Contact (below)					
Orf virus disease (milker's nodule, contagious pustular dermatitis, human orf, ecthyma contagiosum)	Orf virus (*Parapoxvirus*)	Blistered/red nodule/pustule on hands, also arms, face; usually solitary	Sheep, goats, reindeer, musk oxen; important occupational disease in New Zealand	Direct contact with infected animal's mucous membranes/ulcer; indirect contact through contaminated knives, stalls, clothing	Not indicated	Hygiene (wash exposed areas); barrier precautions (gloves)

TABLE 9.2b ***(Continued)***

Disease	Infectious agent	Health effects	Sources	Transmission	Surveillance	Prevention
			Viral infections (Continued)			
Rabies (hydrophobia, lyssa)	Rabies virus (*Lyssavirus*)	Almost invariably fatal, acute encephalomyelitis	Animal (raccoons, dogs, cats, skunks, bats, foxes) saliva, tissues (brain, cerebrospinal fluid)	Bite, tissue or saliva contact; airborne (rare—caves, laboratory work)	Animal diagnosis	Highrisk workers: preexposure immunization, 2-yr serologic testing/booster vaccine Laboratories: biosafety practices, 6-mo serologic testing/booster vaccine PEP: prompt, thorough wound care; immune globulin; vaccine
Warts	See Human Contact (above)					
			Protozoal and helminth infections			
Cryptosporidiosis	See Human Contact (above)					
Echinococcosis (hyatid disease)	*Echinococcus granulosus* (Utah, Arizona, New Mexico, California); *Echinococcus multilocularis* (Canada, Alaska, northern Asia, Europe)	Cysts form in the liver, lung, other sites	Canine feces; *E. granulosus* (dog/sheep hosts); *E. multilocularis* (foxes, wolves, dogs)	Fecal–oral route	Not indicated	Proper disposal of carcasses and entrails; hygiene; limit dog populations on farms and ranches; parasite control in dogs
Hookworm disease	See Environmental Contact (below)					

Malaria	*Plasmodium vivax, Plasmodium malariae, Plasmodium falciparum, Plasmodium ovale*	Cycles of fever, chills, headache, nausea, sweats; more severe with *P. falciparum*	Human blood	Bite of infective female anopheline mosquito	For illness only	Mosquito control (clothes, repellents, netting); chemoprophylaxis (specific for area)
Toxoplasmosis	*Toxoplasma gondii*	Asymptomatic or cervical lymphadenopathy, flulike illness Congenital: stillbirth, severe infection (CNS, ocular)	Cat feces, sheep, goat, cattle meat	Fecal–oral route, ingestion of cyst-contaminated food	Not indicated	Avoidance; hygiene (handwashing); gloves; cook food completely; dispose of cat litter daily
			Fungal infections			
Dermatophytoses	See Human Contact (above)					
			Bacterial infections			
Anthrax (malignant pustule, malignant edema, woolsorter's disease, ragpicker's disease, charbon)	*Bacillus anthracis*	Cutaneous: papules, ulcers (95%), necrotic eschar Pulmonary: hemorrhagic mediastinitis; regional lymphadenopathy	Tissue, hide, hair, wool, bones of non-U.S. herbivores (cattle, sheep, goat, horse)	Direct contact (skin lesions); ingestion (meat); droplet nuclei (occasionally)	For illness only	Infected animal disposal; dust control; wool cleaning; hygiene; animal immunization PEP: vaccine
Brucellosis (undulant fever, Malta fever, Mediterranean fever)	*Brucella abortus, Brucella suis, Brucella melitensis, Brucella canis*	Flulike illness with continued, intermittent, or irregular fever; localized infections especially bone;	Cattle, swine, goat, sheep, dog tissues, blood, urine, placentas, milk	Direct contact; droplet nuclei; ingestion (unpasteurized milk, milk products); inoculation	For illness only	Livestock: immunization, treatment, slaughter Abattoir workers (endemic areas): PPE, work

TABLE 9.2b (*Continued*)

Disease	Infectious agent	Health effects	Sources	Transmission	Surveillance	Prevention
		Bacterial infections (Continued)				
		U.S. cases only in abattoir and laboratory workers		during vaccine administration		restrictions based on immunity; laboratories: biosafety practices
						Pasteurize milk
Campylobacter enteritis (vibrionic enteritis)	*Campylobacter jejuni*, other species	Acute diarrhea, abdominal pain, malaise, fever, nausea, vomiting; usually self-limited (≤10 days)	Animal feces, meat (primarily poultry, cattle, also puppy, kitten, other pet, swine, sheep, rodent, bird); human feces	Fecal–oral contact with infected pets, farm animals, infants; ingestion of contaminated food (undercooked meat), water, raw milk	For illness only	Avoid raw, undercooked, unpasteurized foods; control infection in animals; sanitation, hygiene (handwashing)
Cat-scratch disease (cat-scratch fever, benign lymphoreticulosis)	*Bartonella* (formerly *Rochalimaea*) *henselae*	Subacute, self-limited infection with malaise, lymphadenitis, fever	Domestic cats; possibly cat fleas as vectors	Scratch, bite, lick, other cat (usually young and healthy) exposure; possibly flea bite	For illness only	Thorough cleaning of cat scratches/bites may be helpful; flea control
Erysipeloid (Rosenbach's disease, fish handler's disease)	*Erysipelothrix rhusiopathiae*	Localized purple-red skin lesions; generalized disease; septic (endocarditis)	Fish (slime layer), other wild/domestic animals (swine, rabbits, cattle, birds, turkeys, rats)	Contact (usually hands)	For illness only	Gloves; handwashing; animal immunization
Leptospirosis (Weil disease, canicola fever, hemorrhagic jaundice,	*Leptospira interrogans*, ca. 23 serogroups with >200 serovars	Flulike illness of sudden onset, severe myalgia (calves, thighs), without jaundice	Dog, cat, livestock, rodent, wild mammal urine	Contact with urine-contaminated water, soil; direct animal	For illness only	Animal immunization; rodent control; hygiene; disinfection;

mud fever, swineherd disease, Fort Bragg fever)		(90%), jaundice (10%), skin rash, meningitis, mental confusion		contact, bites		antibiotic during high exposure
Lyme disease	*Borrelia burgdorferi*	Early: erythema chronicum migrans Delayed: meningitis, encephalitis, myocarditis, arthritis	Ixodes ticks from deer, wild rodents	Tick bites	For attached ticks (ticks may be tested)	Tick protection; tick removal; early antibiotic treatment; vaccine licensed in 1998
Mycobacterium marinum infection (swimming pool granuloma, fish tank granuloma)	*Mycobacterium marinum*	Nodular inflammatory skin lesion (elbow, knee, foot, toe, finger), abscess (possibly deep infection)	Fresh/salt water with infected fish, tropical fish	Direct contact of nonintact skin with contaminated water	Not indicated	Gloves and other protection
Plague (pestis)	*Yersinia pestis*	Bubonic: fever, regional lymph-adenopathy (85–90%) Septicemic: bloodstream dissemination Pneumonic: primary and secondary All three forms rare and sporadic in U.S.	Fleas/secretions/tissues from wild rodents (ground squirrels), lagomorphs (rabbits, hares), wild carnivores (coyotes), domestic cats	Flea bits; direct contact with infected tissues; airborne respiratory droplets from animal (or person) with plague pharyngitis or pneumonia	For illness only; surveillance for plague activity in animals in endemic areas	Avoid animal handling; rodent and flea control; hygiene (clothing); immunization (formalin-killed plague) Laboratories: biosafety practices PEP: chemoprophylaxis

TABLE 9.2b ***(Continued)***

Disease	Infectious agent	Health effects	Sources	Transmission	Surveillance	Prevention
			Bacterial infections (Continued)			
Psittacosis (ornithosis, parrot fever, avian chlamydiosis)	*Chlamydia psittaci*	Flulike, acute, respiratory symptoms	Dried feces, scretions, feathers, tissues from psittacine (parakeets, parrots), other birds	Airborne; direct contact, mouth-to-beak inoculation	For illness in pet shops, aviaries, farms, processing plants, cases linked to birds	Tetracycline-treated poultry feed; quarantine/treatment of imported psittacine birds; domestic parakeet breeding
Q fever (query fever)	*Coxiella burnetii*	Acute flulike illness; rarely pneumonia, hepatitis, endocarditis	Sheep, cattle, goat, cat, dog placental tissue; birds, ticks	Airborne dust from dried placental tissue; direct contact with infected tissue; indirect contact with contaminated articles	Skin test or serology for evidence of infection or immunization	Work with live agent: immunization with investigational vaccine; control animal movement Laboratories: biosafety practices
Rocky Mountain spotted fever (New World spotted fever, tickborne typhus, Sao Paulo fever)	*Rickettsia rickettsii*	Fever, headache, rash	Ticks in south Altantic and western south-central U.S.	Tick bite	Not indicated	Avoid ticks; vaccine under development

Salmonellosis	*Salmonella typhimurium*, *Salmonella enteritidis*, >2000 other serotypes	Acute enterocolitis with sudden onset of headache, fever, abdominal pain, diarrhea, sometimes vomiting	Domestic, wild animals (poultry, swine, cattle, rodents; pet birds, reptiles, mammals); human, animal feces	Ingestion of food from infected animal or of feces-contaminated food; fecal–oral route; patient contact	For illness only	Educate food handlers; avoid raw, undercooked, unpasteurized animal foods (eggs, meat, milk); control infection in animals
Tetanus	See Environmental Contact (below)					
Tularemia (deer fly fever, rabbit fever)	*Francisella tularensis*	Dependent on transmission route Ulceroglandular: flulike (75–85%) Glandular (5–10%) Typhoidal (5–15%) Oculoglandular (1–2%) Oropharyngeal: abdominal pain, diarrhea, vomiting	Wild mammals (rodents, lagomorphs), hard ticks	Direct animal tissue, fluid contact; arthropod bite (tick, deer fly), animal bites; airborne; ingestion	For illness only	Immunization (investigational live attenuated bacteria); barriers (gloves); repellents; tick removal Laboratories: biosafety practices

TABLE 9.2c Basic Information on Infectious Diseases Transmitted by Environmental (Soil, Water, or Plant) Contact or Direct Agent Handling

Disease	Infectious agent	Health effects	Sources	Transmission	Surveillance	Prevention
Hantavirus pulmonary syndrome (*Hantavirus* adult respiratory distress syndrome—ARDS)	Two or more hantaviruses in the Americas: Sin Nombre virus (SNV—Southwest U.S.), Black Creek Canal virus (BCCV—Florida)	Fever, myalgia, GI complaints followed by abrupt onset of respiratory distress and hypotension, then respiratory failure, cardiogenic shock	SNV: deer mouse (*Peromyscus maniculatus*), pack rats, chipmunks, other rodents BCCV: cotton rat (*Sigmodon* spp.)	Airborne rodent excreta; person-to-person transmission reported from Argentina but not seen in the U.S.	For illness only; surveillance of animal infection in endemic areas; banked serum for baseline serology	Rodent avoidance; rodent control; standard precautions
Hepatitis A	See Human Contact (above)					
Rabies	See Animal Contact (above)					
			Protozoal and helminth infections			
Amebiasis (amoebiasis)	*Entamoeba histolytica*	Asymptomatic, intestinal disease, extraintestinal (liver)	Human feces, contaminated food, water	Ingestion; direct oral–fecal contact; droplets	Not indicated	Barriers; hygiene (handwashing), sanitation
Cryptosporidiosis	See Human Contact (above)					
Cutaneous larva migrans (creeping eruption, ground itch)	*Ancylostoma braziliense* (Gulf Coast) *Ancylostoma caninum* (less often seen)	Tunnel-like migrating rash with intense itching	Dog, cat feces	Direct contact with contaminated soil	Not indicated	Nematode control of pets; barrier precautions
Giardiasis	See Human Contact (above)					
Hookworm disease (ancylostomiasis, uncinariasis, necatoriasis)	*Necator americanus*, *Ancylostoma duodenlae* (southeastern U.S.)	Various in proportion to degree of anemia, occasional severe acute pulmonary and GI reactions	Human feces	Skin penetration (usually foot) by infective larvae	Not indicated	Proper disposal of human, cat, dog feces; wear shoes, gloves

			Fungal infections			
Aspergillosis	*Aspergillus* spp.	Ear, sinus, lung infections, central nervous system infection (immunodeficient people)	Ubiquitous saprophyte, but higher in compost, hay, grain	Inhalation of airborne conidia	Not indicated	Dust control; air sampling to monitor control efficacy; respirators
Blastomycosis	*Blastomyces dermatitidis* (*Ajellomyces dermatitidis*)	Pneumonia, granulomatous lesions in skin, lungs	Moist soil in central, southeastern U.S., other countries	Inhalation of airborne conidia	Not indicated	Unknown
Coccidioidomycosis (valley fever, San Joaquin fever, desert fever, desert rheumatism, coccidioidal granuloma)	*Coccidioides immitis*	Asymptomatic, upper respiratory infection (60%); flulike illness (40%); pulmonary nodule/cavity (5%); disseminated (0.5%)	Soil in Lower Sonoran Life Zone (southwestern states–California to southern Texas)	Inhalation of airborne arthroconidia from soil, laboratory culture	Skin test high-risk workers to identify those susceptible	Dust control; respirators; immunization (future); for dusty jobs (road work), recruit skin-test positive workers or local residents Laboratories: biosafety practices
Cryptococcosis	*Cryptococcus neoformans* var. *neoformans*	Deep mycosis presenting as subacute/chronic meningitis; infection of lungs, kidneys, prostate, bone may occur	Old pigeon nests, droppings	Airborne	Not indicated	Dust control (wet pigeon droppings with disinfectant before removing); respirators
Dermatophytosis	See Human Contact (above)					

TABLE 9.2c ***(Continued)***

Disease	Infectious agent	Health effects	Sources	Transmission	Surveillance	Prevention
		Fungal infections (Continued)				
Histoplasmosis (histoplasmosis capsulati, American histoplasmosis)	*Histoplasma capsulatum* var. *capsulatum* (*Ajellomyces capsulatus*)	Asymptomatic pneumonitis, acute pulmonary infection, flulike illness, chest pain, mediastinal lymphadenopathy	Soil, bird/bat excrement, central U.S.	Inhalation of airborne conidia	Not indicated	Dust control; selectively disinfect contaminated foci; respirators
Sporotrichosis (rose thorn fever)	*Sporothrix schenckii*	Skin (usually extremities), nodule, spreads through lymphatics forming a series nodules that may ulcerate; rarely extracutaneous	Thorny plants, wood splinters, soil, sphagnum moss	Direct skin inoculation; rarely inhalation of airborne conidia	Not indicated	Gloves, long sleeves (particularly when in contact with sphagnum moss); some lumber may warrant fungicide treatment
		Bacterial infections				
Legionellosis (Legionnaires' disease, LD; Pontiac fever, PF)	*Legionella pneumophila*, other species	LD: fever pneumonia PF: flulike illness (possibly hypersensitivity response)	Heat-transfer equipment, water supplies; natural waters	Airborne; possibly aspiration	For illness only	Source control (limit bacterial growth); avoid aerosolizing contaminated water
Leptospirosis	See Animal Contact (above)					
Mycobacterium marinum infection	See Animal Contact (above)					
Tetanus (lockjaw)	*Clostridium tetani* exotoxin	Local, cephalic, generalized (>80%)	Animal (horse), human feces, soil, ubiquitous	Spore entry into wounds	Wound recognition	Booster immunization every 10 years PEP: immune globulin

TABLE 9.3 Potential Occupational Infectious Diseases, Their Agent Categories, and Sources

Disease	Agent category	Source
AIDS	Virus	Humans
Acute respiratory infections	Virus	Humans
Amebiasis	Protozoan	Environment[a]
Anthrax	Bacterium	Animals
Arthropod-borne virual diseases	Virus	Animals/arthropod
Aspergillosis	Fungus	Environment
B-virus infection	Virus	Animals
Blastomycosis	Fungus	Environment
Brucellosis	Bacterium	Animals
Campylobacter enteritis	Bacterium	Animals, humans
Cat-scratch disease	Bacterium	Animals
Chickenpox	Virus	Humans
Coccidioidomycosis	Fungus	Environment
Conjunctivitis	Virus, bacterium	Humans
Cryptococcosis	Fungus	Environment
Cryptosporidiosis	Protozoan	Animals, humans, environment
Cutaneous larva migrans	Helminth	Environment
Cytomegalovirus disease	Virus	Humans
Dermatophytosis	Fungus	Animals, humans, environment
Echinococcosis	Helminth	Animals
Erysipeloid	Bacterium	Animals
Giardiasis	Protozoan	Humans, environment
Hantavirus pulmonary syndrome	Virus	Animals, environment
Hepatitis A	Virus	Humans, environment
Hepatitis B, C, D	Virus	Humans
Herpetic whitlow	Virus	Humans
Histoplasmosis	Fungus	Environment
Hookworm disease	Helminth	Animals, environment
Influenza	Virus	Humans
Legionellosis	Bacterium	Environment
Leptospirosis	Bacterium	Animals, enviroment
Lyme disease	Bacterium	Animals/arthropods
Malaria	Protozoan	Animals/arthropods
Measles	Virus	Humans
Mycobacterium marinum infection	Bacterium	Animals, environment
Orf virus disease	Virus	Animals
Pertussis	Bacterium	Humans
Plague	Bacterium	Animals/arthropods
Psittacosis	Bacterium	Animals
Q-fever	Bacterium	Animals
Rabies	Virus	Animals, environment
Rocky Mountain spotted fever	Bacterium	Animals/arthropods
Rubella	Virus	Humans
Salmonellosis	Bacterium	Animals
Shigellosis	Bacterium	Humans

TABLE 9.3 (*Continued*)

Disease	Agent category	Source
Sporotrichosis	Fungus	Environment
Streptococcal skin disease	Bacterium	Humans
Tetanus	Bacterium	Animals, environment
Toxoplasmosis	Protozoan	Animals
Tuberculosis	Bacterium	Humans
Tularemia	Bacterium	Animals
Warts	Virus	Animals, humans

[a]Environment = soil, water, or plants.

by type of agent (i.e., virus, protozoan or helminth, fungus, or bacterium) and mode of transmission (e.g., contact, droplet, airborne, or vectorborne transmission). Tables 9.2a–c list infectious diseases alphabetically (within type of contact and agent group) and summarize information of occupational health importance, that is, infectious agent, health effects, source, transmission mode, surveillance recommendations, and preventive measures. Readers responsible for workers in particular occupations can determine the infectious diseases of potential importance by reviewing Tables 9.1a–c and then can identify items of importance for recognition and control of these diseases in Tables 9.2a–c. Table 9.3 lists all diseases alphabetically, identifying the category of infectious agent and type of contact, to help readers locate the diseases in the appropriate sections of Tables 9.1 and 9.2.

Table 9.1 shows that certain occupations carry greater risks of infection than others, for example, health care; child care; agriculture, food processing, and animal-associated work; occupational travel; personal service work; and wastewater and sewage treatment. Laboratory work could be included in Table 9.1, but is not listed because laboratory handling of any infectious agent or potentially contaminated material carries some risk of infection. Sections 9.0.1 to 9.0.7 discuss key issues specific to work-related infections in these high-risk occupations. Readers unfamiliar with infectious diseases and related terminology may wish to review first Section 9.1 on the basic principles of infectious diseases and Section 9.2.3 on laboratory diagnosis of infectious diseases.

9.0.1 Health Care

Historically, hospitals and the healthcare professions have been recognized as posing a risk of infection to both workers and patients (Sepkowitz, 1996). Health-care workers may be employed at hospitals, physicians' and dentists' offices, blood banks, and outpatient clinics and similar facilities as well as in home health care. In 1847, Semmelweiss discovered the transmission mode (and means of prevention through handwashing) of puerperal fever, an often fatal streptococcal infection following childbirth. This discovery came only after a professor died of the disease from a scalpel wound received during an autopsy. Following the pioneering efforts of Florence Nightingale and Joseph Lister, along with the advent of the antibiotic era, the public perception of hospitals as hazardous places decreased, as did the frequency of nosocomial (literally, hospital-acquired) infections. However, even through the 1970s, 5 percent of patients in the United States became infected while hospitalized. Although incidence data for employee infections are unavailable, healthcare workers attending these patients were also recognized as being at risk for a variety of pathogens present in their workplaces.

The AIDS epidemic has resulted in an increased focus on risks to healthcare workers. This attention includes the overdue recognition that each year (prior to effective immunization) an estimated 2000 U.S. healthcare workers were infected with the HBV. In recent years, a system utilizing barriers and other precautions to minimize or prevent exposure to blood (and certain body fluids and tissues) of all patients was devised: "universal precautions" or "body substance isolation." Reasons for these precautions include the following (a) HBV and HIV are present in the blood of a significant percentage of patients; (b) carriage of these bloodborne viruses is often silent; (c) screening for risk factors of infection (e.g., injecting-drug use, high-risk sexual behavior, and hemophilia) will identify only a fraction of infected patients in advance of providing them care; (d) the viruses are present only in blood and a limited number of other blood-derived or blood-contaminated fluids; and (e) additional bloodborne agents may suddenly emerge (as did HIV). Implementation of these precautions may be particularly needed but difficult for certain types of emergency care or rescue work (Carrillo et al. 1997).

These voluntary guidelines were incorporated into the *OSHA Bloodborne Pathogens Standard* [USDOL, 29 Code of Federal Regulation (CFR) Part 1910.1030], the first OSHA regulation pertaining to infectious agents (see Section 9.2.1). This standard was unusual in applying to any workplace where workers were "reasonably anticipated" to be occupationally exposed. Also, employers were required to make available, free of charge, hepatitis B vaccine to all such covered employees. In the event of an exposure incident, employers were required to make immediately available a postexposure evaluation and to provide postexposure prophylaxis if medically indicated. Because application of the standard was not confined to hospitals or even traditional healthcare facilities, it resulted in attention to occupational exposure to bloodborne and other pathogens in nursing homes and other long-term care facilities, outpatient facilities, patients' homes, emergency care delivery (including public safety work), schools and childcare facilities, correctional facilities, and laboratories.

The AIDS epidemic was followed shortly by the tuberculosis epidemic and the emergence of multidrug-resistant *Mycobacterium tuberculosis*. As a group, healthcare workers have a tuberculosis rate similar to the general population. However, elevated rates have been observed for inhalation therapists and lower-paid healthcare workers (McKenna et al., 1996). The Centers for Disease Control and Prevention (CDC, Atlanta, GA) has published comprehensive guidelines for preventing tuberculosis in healthcare facilities (CDC, 1994a). In 1997, OSHA released a proposed rule and notice of public hearing on occupational exposure to tuberculosis (USDOL, 1997a). This document includes a summary of studies of occupational tuberculosis. The latest information on this and other actions is available at http://www.osha.gov/index.html or through OSHA offices.

In 1997, CDC published a draft guide for infection control in healthcare personnel (CDC, 1997h; http://www.cdc.gov.ncidod/hip/draft_gu/phgwww.htm). General recommendations for immunization of health-care workers have also been published (CDC, 1997g). In particular, influenza, measles, and varicella are airborne infections that are important in healthcare settings. All healthcare workers with patient contact should be vaccinated annually for influenza to prevent transmission to persons at high risk of influenza-related complications (CDC, 1997de). Specific recommendations for immunization and management for measles (CDC, 1989) and varicella (chickenpox) (CDC,

1996c) are also available. Susceptible workers exposed to patients with chickenpox, measles, mumps, or rubella must often be restricted from patient care for long periods because the agents can be transmitted before an infected worker showed signs of disease. Appropriate use of vaccines can often eliminate these restricted periods and thus may be cost-effective (see Section 9.3.1).

Besides the bloodborne and airborne routes, droplet exposure and contact can transmit infectious agents to healthcare workers. It is important for healthcare workers and occupational health professionals responsible for them to appreciate the many transmission modes and the wide range of pathogenicity of the infectious agents that may be found in healthcare settings (see Sections 9.1.2 and 9.1.3). Fortunately, many of the agents that patients carry (including many antibiotic-resistant bacteria and fungi) are pathogenic only for persons with compromised immune function or underlying medical conditions.

Infection control committees at hospitals and healthcare centers oversee the development and implementation of mandated infection control programs. Institutional health and safety professionals may contribute to these activities in the areas of facility operation and maintenance (e.g., of heating, ventilating, and air-conditioning systems; water supplies; and waste disposal). Additional areas where health and safety professionals may play a role in infection control are in implementing engineering controls to prevent the spread of infectious agents, conducting environmental inspections and testing, and ensuring general safety and security (AIA, 1996; JCAHO, 1997).

9.0.2 Child Care

Childcare settings tend to promote infectious disease transmission because such operations closely group together young children who lack appropriate hygienic behavior and who have not yet developed immunity to common infectious agents (Holmes et al., 1996). In the United States and many other industrialized countries, the demand for and utilization of out-of-home child care has expanded as the number of working mothers with young children has increased and fewer extended families are available to provide child care. In 1990 in the United States, over 50 percent of mothers of preschool children worked either part or full time, and as many as 10 million children ≤5 years attended childcare facilities (Holmes et al., 1996). Compared to children cared for at home, youngsters in childcare settings are more likely to be associated with illness outbreaks and to experience increased infectious disease rates and severity. Table 9.4 supplements Table 9.1a and lists infectious diseases associated with child care, their causative agents, and the degree of risk for childcare workers.

Many gastrointestinal and respiratory tract diseases affect both childcare staff and children. Some infections affect primarily children but not adults, such as *Haemophilus influenzae* type b disease (which is now vaccine preventable), otitis media (inflammation of the middle ear), and chickenpox. A few infectious agents, notably the hepatitis A virus (HAV), cause disease in adults but primarily inapparent infection in children. Other agents, principally cytomegalovirus (CMV) and to a lesser extent parvovirus B19, cause mild or inapparent infections in children but may have serious consequences for the fetus of a pregnant employee (or parent) or for immunocompromised persons (Osterholm, 1994). Besides those infections transmitted from person to person, some diseases have been contracted from handling pets and other animals that may be kept at childcare centers and in primary school classrooms, such as salmonellosis from handling infected chicks, ducks, and reptiles.

TABLE 9.4 Infectious Diseases in Child Care[a]

Disease	Agents	Increased incidence with child care[b]	Risk to childcare workers
Respiratory infection			
"Colds"	Rhinoviruses, other viruses	Yes	Yes
Otitis media	*Streptococcus pneumoniae*, other bacteria, viruses	Yes	No
Pharyngitis	Group A streptococcus, other bacteria viruses	Probably	Yes
Pneumonia	*S. pneumoniae*, other bacteria, viruses	Probably	Probably
Influenza	Influenza virus	Yes	Yes
Diarrhea	Norwalk virus, Astrovirus, other viruses	Yes	Yes
	Shigella and *Salmonella* spp., other bacteria	Yes	Yes
	Giardia and *Cryptosporidium* spp., other parasites	Yes	Yes
Vaccine-preventable diseases	Measles, mumps, and rubella viruses *Bordetella pertussis*	Not established	Yes
Invasive bacterial disease	*Haemophilus influenzae* type b (vaccine preventable)	Yes	No
	Neisseria meningitidis	Probably	Probably
	S. pneumoniae	Yes	Low
Hepatitis A	Hepatitis A virus (vaccine available)	Yes	Yes
Hepatitis B	Hepatitis B virus (vaccine preventable)	Few reports	Low
Hepatitis C	Hepatitis C virus	None known	Low
Skin infections			
Impetigo	*Staphylococcus aureus*, Group A streptococcus	Yes	Little
Pediculosis	Head lice (*Pediculus humanis*)	Yes	Yes
Scabies	Scabies mite (*Sarcoptes scabiei*)	Yes	Yes
Herpes	Herpes simplex virus	Yes	Probably
Other infections			
"Fifth" Disease	Parvovirus B19	Yes	Yes
Chickenpox	Varicella-zoster virus (vaccine preventable)	Yes	Yes
Cytomegalovirus infection	Cytomegalovirus	Yes	Yes

[a]Adapted from Holmes et al. (1996) and Osterholm (1994).
[b]From Holmes et al. (1996).

Common sense measures can minimize infectious agent transmission and reduce the burden of illness in childcare facilities. A national health and safety performance standard has been published with specific recommendations for personal hygiene (handwashing), disinfection, exclusion policies, management of ill children, requirements for immunization, exposure to blood, and reporting of infectious disease outbreaks to a health agency (APHA/AAP, 1992). Following these recommendations (particularly in regard to personal hygiene and disinfection) and the recommended childhood immunization schedule has been shown (albeit in limited studies) to be effective at preventing and controlling infectious diseases in childcare, including among childcare workers.

9.0.3 Agriculture, Food Processing, and Animal-Associated Occupations

Agricultural employment (including farming), food processing (e.g., slaughterhouse or abattoir work and butchering), and veterinary care are among the most hazardous occupations. Although less frequent, less serious, and less studied than injuries, infectious diseases are common in these professions. Animals are the most common sources of exposure to infectious agents for such workers (Table 9.1b), but soil, water, and plants are also important (Table 9.1c).

Field workers are often at increased risk from enteric pathogens due to lack of sanitation and clean drinking water. In the United States, migrant farm workers have enterically transmitted viral and bacterial diseases at ca. 10 times the rate of the general population. Social conditions also place migrant farm workers at increased risk of interpersonally transmitted diseases [e.g., tuberculosis (McKenna et al., 1997) and sexually transmitted diseases]. Workers with direct soil contact, particularly barefoot farmers in less developed countries, are at risk for parasitic infections including hookworm, ascariasis, cutaneous larva migrans, and visceral larva migrans.

Operations that involve the moving of soil place workers at risk for fungal infections, which often have specific geographic distributions, such as histoplasmosis in the Ohio and Mississippi River Valleys and coccidioidomycosis in the southwest (Lower Sonoran Life Zone). Disturbing compost and moldy grain or other plant material may also places workers at risk for fungal (e.g., *Aspergillus* spp.) or bacterial (e.g., actinomycete) infections as well as for hypersensitivity diseases and ODTS.

Farm and ranch workers along with field researchers spend much of their time outdoors, placing them at risk for vectorborne diseases, such as Lyme disease, erlichiosis, plague, and Rocky Mountain spotted fever (in the U.S.) as well as malaria, viral encephalitis, and other diseases (elsewhere). Farmers (and their guests) often acquire campylobacteriosis, salmonellosis, and listeriosis from drinking unpasteurized milk. Food handlers (and cooks) may taste uncooked items such as meat, placing themselves at risk for bacterial and parasitic infections.

Underdiagnosis of illness is common in these populations due to the acceptance of illness, limited medical care access (particularly for farm workers), and the need for serologic (blood) tests to diagnosis many zoonoses. Infectious disease outbreaks due to organisms not previously identified as human pathogens (e.g., *Streptococcus iniae* in persons handling aquacultured fish from fish farms) point out the evolving nature of work and workplaces and the continuing need to look for emerging pathogens. Some previously common zoonoses (e.g., brucellosis, anthrax, leptospirosis, and psittacosis) have been controlled to a large extent in many developed countries through vaccination, chemoprophylaxis, and identification of infected animals. Vaccination of workers with animal contact is recommended primarily for veterinarians and animal control personnel at risk for rabies. Often, clinically inapparent infections in young, healthy workers (e.g., Q fever and coccidioidomycosis) or frequent low-dose exposures (e.g., possibly to enterohemorrhagic *Escherichia coli*) protect workers against the development of more serious disease later in life. In less developed countries and for migrant workers, the principal means of preventing infectious diseases involves wearing protective clothing and shoes, better education, and measures to improve the living standard.

9.0.4 Laboratories

Microbiology laboratories are obvious workplaces in which persons may be exposed to infectious agents because laboratory workers handle pathogens and infected animals. Laboratories do not appear in Table 9.1 because workers in medical and research laboratories may be considered to be at risk of infection from any of the microorganisms they may encounter in the materials they handle. Thus laboratory work would be listed for all the infections shown as well as for many others. Some of the same concerns apply to workers in biotechnology laboratories. These operations involve the large-scale production of microorganisms for many uses, such as in the manufacture of vaccines, antibiotics, and microbial products (such as enzymes and hormones); as animal feed; as pesticides; for environmental management of oil spills; and as biodegradation agents at hazardous waste sites (NIH, 1991, 1994; Cottam, 1994; McGarrity and Hoerner, 1995). Biosafety practices aim (a) to ensure the health and safety of laboratory workers and the surrounding community; and (b) to protect the environment (NRC, 1989; USDHHS, 1993; WHO, 1993; AIHA, 1995; ASM, 1995a). Laboratory workers handling infectious agents are clearly at increased risk of laboratory-associated infections. However, other laboratory staff (e.g., glassware washers, janitors, maintenance personnel, and clerical workers) as well as nonlaboratory occupants of the same building may also be at greater risk than the public.

Accurate data on laboratory-associated infections are difficult to obtain, but clearly such infections do occur. Adding to the difficulty of there being no systematic reporting system for laboratory-acquired infections, such infections are often subclinical and have atypical exposure routes as well as incubation periods (Sewell, 1995). Gantz (1988) wrote that over 6000 infections and 250 fatalities associated with laboratory exposures had been reported in the literature. The most frequently identified laboratory-associated infections worldwide (in decreasing order of reporting) are brucellosis, Q fever, typhoid fever, hepatitis, tularemia, tuberculosis, dermatomycosis, Venezuelan equine encephalitis, typhus, psittacosis, coccidioidomycosis, streptococcal infections, histoplasmosis, leptospirosis, salmonellosis, and shigellosis (Gantz, 1988; USDHHS, 1993; Sewell, 1995; Harding et al., 1996). Many laboratory-associated infections result from obvious accidents, such as spills, splashes, needlesticks, cuts, and animal bites. However, the incident that led to laboratory worker exposure is unclear in many cases (often the affected person is only known to have handled an agent or worked nearby). In these cases of unidentified exposure, transmission is assumed to be by droplet or droplet nuclei exposure (see Section 9.1.2).

Practices have been developed over the years that allow the safe laboratory handling of even the most hazardous agents as well as the large-scale production of microorganisms and cells (NRC, 1989; NIH, 1991, 1994; USDHHS, 1993, 1994; WHO, 1993; AIHA, 1995; ASM, 1995a, 1995b). Basic precautions include posted biohazard warning signs; limited laboratory access; careful manipulation of infectious fluids; biological safety cabinets to contain accidentally generated aerosols; hand-pipetting devices for transferring infectious materials; proper disposal of needles, syringes, culture material, animal carcasses, and so forth; handwashing following laboratory activities; decontaminating work surfaces before and after use and immediately after spills; and prohibiting eating, drinking, smoking, and food storage in laboratories.

Other texts discuss regulations and guidelines related to laboratory biosafety practices as well as the safe use of pathogenic and oncogenic microorganisms (NRC, 1989; Hansen, 1993; WHO, 1993; AIHA, 1995; ASM, 1995a; Sewell, 1995; Harding et al., 1996). These references also provide lists of standard setting or credentialing groups as well as professional associations in these fields. The National Sanitation Foundation (NSF, Ann Arbor, MI) Standard 49 describes requirements for design, construction, and performance of Class II biological safety cabinets, which are widely used in microbiology laboratories to contain aerosols (NSF, 1992). Increased attention has recently been placed on the safe transport of biohazardous agents and verification that a receiving laboratory is qualified to handle them properly. This concern has resulted in more stringent shipment regulations in the United States (USDHHS, 1996) (see Section 9.2.1.5).

Infectious agents may receive the most attention in microbiology laboratories, but typically they are not the only biological hazards present. Workers in microbiology and biotechnology laboratories may also need engineering controls and training in proper work practices to avoid exposure to endotoxin from Gram-negative bacteria, mycotoxins from fungi, biologically active pharmaceutical products, and proteins and other allergenic materials from whole cells, cell fragments, and cell products that may cause hypersensitivity diseases, such as hypersensitivity pneumonitis and asthma. Laboratory work with oncogenic (tumor-producing) viruses and some recombinant DNA molecules may carry a risk of cancer induction. Appropriate training and supervision in chemical, electrical, and radiation safety are also necessary.

Immunizations for some infectious agents (e.g., the rabies virus and the Q fever agent) may be recommended for laboratory workers at high risk of exposure even though these immunizations are not recommended for adults in general. Certain poxviruses (e.g., variola major and minor, vaccinia, and monkeypox) are infectious for humans, and laboratory workers handling these viruses should be considered for smallpox vaccination (Ellis and Symington, 1994). Engineering controls can reduce laboratory-associated infections, but the individual behavior of laboratory workers plays a large role in their safety and that of their co-workers. Prevention of laboratory-associated infections depends on how each worker perceives the risks and consequences of exposure to pathogenic agents as well as on workers' awareness of, belief in, and adherence to good practices (Gershon and Zirkin, 1995).

9.0.5 Occupational Travel

Military personnel are often required to travel worldwide and may spend considerable time abroad. With the globalization of industry, job-required travel is increasing in other occupations and may involve extended residence, often in areas outside usual tourist routes. Fortunately, this trend was preceded by years of increasing tourist and other travel (including visits to exotic locales). With this travel came the related growth of the travel medicine industry and the 1991 establishment of the International Society of Travel Medicine. Occupational health staff for large corporations with overseas locations or customers likely are familiar with the infectious disease risks of those regions. Local health departments and the CDC may be able to offer recommendations on locating the nearest travel clinic. However, a survey revealed that North American travel and health clinics often gave inaccurate advice; that is, as many as 75 percent of immunizations were inappropriate and 60 percent of antimalarial drugs were incorrect (Kozarsky et al., 1991). Hopefully this situation has improved. The CDC issues a periodic publication, *Health Information for International Travel*, and maintains an automated travelers' hotline accessible from a touchtone phone at (404)332-4559 and by facsimile at (404)332-4565. Travel medicine information may also be available on the CDC website (http://www.cdc.gov). It is not unreasonable to verify advice received from a travel clinic with these sources.

Some workers in travel-related and import industries have close contact with passengers and cargo. However, these workers do not appear to be at greater risk for infectious diseases than others. Tuberculosis transmission has occurred on aircraft in which people may share air for many hours at close quarters (Driver et al., 1994; CDC, 1995). However, the risk does not appear to be greater on aircraft than in other confined spaces. Practices are under development to deal with infectious diseases on aircraft (Grainger et al., 1995). Border patrol and coast guard workers occasionally have close, possibly physical, contact with legal and illegal travelers who may have communicable diseases. The infectious agents of concern are those related to food, water, and sanitation; respiratory pathogens; blood-borne pathogens; and vector-borne agents (USDDT, 1994).

Precautions that may be indicated for travelers include (a) vaccine or immune serum globulin for hepatitis A; (b) vaccines or chemoprophylaxis for mosquito-borne diseases such as yellow fever, malaria, and Japanese encephalitis; and (c) insect repellents and general protective measures to avoid dengue and other mosquito-borne diseases for which prophylaxis is not available. As needed, travelers should request information on (a) reliable drinking water sources, proper food selection, and other measures to prevent and treat traveler's diarrhea; (b) the need for wading and swimming precautions (to avoid schistosomiasis and leptospirosis); and (c) the possible need for rabies and other prophylaxis. Although not directly work-related, travelers should be reminded of the risk for acquiring sexually transmitted diseases and AIDS from unprotected sex, particularly in areas where HIV seroprevalence is high or there is a legal sex trade.

Immunizations recommended for travelers may depend on whether a person will make only a short visit or reside in an area. Travelers should obtain those vaccinations recommended for all adults as well as those recommended for their particular occupations (see Section 9.3.1). Readministration of measles, mumps, and rubella vaccines has been advised for young adults who travel. Travelers should ensure protection against diphtheria (which recently reemerged in Russia and has spread to other countries) and tetanus by receiving a tetanus-diphtheria toxoid (Td) booster whenever ≥10 years have elapsed since previous administration. Rabies vaccine is recommended for travelers staying ≥30 days in countries with ineffective control of animal rabies. The uniformed services have recommendations for active service members, which typically include immunization against diphtheria, tetanus, and polio with documented protection against measles, mumps, and rubella (USDOT, 1994). In addition, immunization against influenza, hepatitis A, and hepatitis B may be recommended or required. Some nations require immunizations for entry into the country and other precautions are recommended for travel to certain regions.

For entry, some countries *require* an official certificate of vaccination for certain diseases. Yellow fever is the only disease for which the World Health Organization (WHO, Geneva, Switzerland) prepared a certificate in 1998. Such requirements are intended to safeguard the residents of a host country. The WHO, CDC, and other agencies *recommend* certain immunizations and preventive treatments to protect travelers. Malaria is an example of a disease for which prophylaxis is recommended but no country requires visitors to use such protection. Vaccination against meningococcal disease is recommended for travelers to countries recognized as having epidemic meningococcal disease, necessitating up-to-date information on disease occurrence, particularly in Africa. Cholera vaccines available in 1998 were only partially effective for older *Vibrio cholerae* strains and were probably ineffective for the strain causing outbreaks at that time in a number of countries. Typhoid fever vaccine is recommended for travelers to areas where there is a recognized risk of infection. Japanese encephalitis vaccine is recommended for travelers staying ≥30 days in specific risk areas in all of Asia, the Indian Subcontinent, and the Western Pacific.

9.0.6 Personal Service Work

Personal service work includes cosmetology, barbering, hair dressing, manicure, pedicure, massage therapy, acupuncture, electrology, tattooing, and body piercing. All of these activities involve hands-on contact with clients, therefore, transmission of bacterial, fungal, and viral skin infections may occur, as may parasitic infestations. Some work-related activities involve needles or may result in skin puncture in the performance of services (e.g., HBV has been transmitted in some of these activities, notably tattooing.) State or local authorities regulate these occupations, but enforcement of such regulations may be irregular. The U.S. Food and Drug Administration (FDA) has developed resource and curriculum materials, including slides, for a one-day workshop on preventing disease transmission in personal service work (USDHHS, 1990).

9.0.7 Wastewater and Sewage Treatment

Garbage handling and work at sanitary landfills has been associated with worker eye irritation, dermatitis, and sore throats, but not infectious diseases (Gelberg, 1997). Proximity to waste water and sewage might be assumed to carry a risk for occupational infections because human feces are a major component of domestic waste water in the United States. Bacteria (e.g., *E. coli*) that are pathogenic to body sites outside the gastrointestinal tract are a normal constituent of human feces, which may also contain viral, protozoal, fungal, and helminthic pathogens. Disease outcomes associated with waterborne infections include mild to life-threatening gastroenteritis, hepatitis, skin infections, wound infections, conjunctivitis, and respiratory infections. Exposure may occur through aerosol inhalation or hand-to-mouth contamination (Moe, 1997). Some diseases (e.g., cholera, typhoid fever, hepatitis A, and amebiasis) are of such low frequency in the United States, in contrast to many developing countries, that they pose at most a low risk of infection from wastewater exposure. However, definitive studies on the risk of occupational infections in wastewater and sewage-treatment workers are lacking. McCunney (1986) published a review of available information on the health effects of work at wastewater treatment plants with guidelines for medical surveillance and personal hygiene. The NSF has standards on drinking water and wastewater treatment as well as food equipment, related products, and swimming pools, spas, and hot tubs.

9.0.8 Unlisted Occupations

General terms for occupations are used throughout this chapter in the hope that readers will recognize the category encompassing the job in which they are interested. For example, a reader needing to know what infectious diseases might affect firefighters, law enforcement officers, or animal control workers would check the listings for public safety workers in Table 9.1a and for public safety workers and animal handlers/trappers in Table 9.1b. Someone concerned about emergency aid workers exposed to flood water could also review the listing for wastewater/sewage treatment workers in Table 9.1c. Laboratory work is missing from these tables as discussed in Section 9.0.4 on laboratories. Certain common occupations neither appear in the tables nor are discussed in the text (e.g., office work and commerce). This is not because infectious disease transmission does not occur in these settings, but because transmission is not directly due to the work performed there. Outbreaks of airborne infections (e.g., tuberculosis) and dropletborne ones (e.g., mumps) have occurred in office buildings and crowded workplaces, respectively. Foodborne diseases (e.g., shigellosis and salmonellosis) have been acquired from workplace cafeterias, and infections transmitted by indirect contact (e.g., conjunctivitis) have been spread by sharing office equipment (e.g., computer keyboards). Legionnaires' disease may be acquired in nonindustrial workplaces, and influenza may be contracted in crowded buses or shops. However, these could be considered building-related or community-acquired infections rather than work-related ones. State and Federal OSHA offices can help employers determine what infections may be considered work related (see Section 9.2.1).

9.1 INFECTIOUS DISEASES—BASIC PRINCIPLES

The following section describes fundamental terms and concepts related to infectious diseases, their transmission, and prevention. *Infection* is the entry (invasion) and multiplication or growth of microorganisms in body tissues so that a response, at least an immune response, occurs. In contrast, *colonization* is the presence of a microorganism in or on a host (e.g., a person or animal) with multiplication or growth but without evidence of an effect or an immune response. The following four elements are required for an illness to be considered an *infectious disease*: an illness due to a specific *infectious microorganism* or its toxin that arises through *transmission* of the agent from a *source* (i.e., an infected person, animal, or inanimate reservoir) to a susceptible *host* (APHA, 1995).

Only a small fraction of all microorganisms on earth have been identified but they are found wherever environmental conditions allow them to survive. The majority of microorganisms inhabit soil or water where they often have important ecological functions, such as nutrient break down and recycling. Only a small portion of known microorganisms are *pathogenic* (disease causing) for humans, animals, or plants. Certain microorganisms are associated with plants, animals, or humans as normal flora, such as on plant leaves and roots or on animal skin and mucosal membranes as well as in digestive tracts.

A *communicable disease* is one that can be transmitted from an infected person or animal or an inanimate reservoir to another person or animal. Infestations may also be communicated from person to person or between animals and humans. An *infestation* of a person or animal involves the lodgement, development, and reproduction of arthropods (e.g., body lice or scabies mites) on a body surface or in clothing. Infested articles or premises are those that harbor or give shelter to animals, especially arthropods and rodents, for example, a building may be described as infested with house flies or mice (APHA, 1995).

Some microorganisms are obligate pathogens (*parasites*) and require a living host, but many are *facultative* or *opportunistic* pathogens, that is, they can grow either as parasites in a living host or as *saprophytes* on decaying organic matter. Saprophytic microorganisms that generally do not harm healthy persons may cause disease in those who are immune compromised. Many bacteria and fungi can be isolated from clinical and environmental samples and grown in laboratory culture. Some obligate pathogens can only be grown in laboratory animals or in cell or tissue cultures.

Identifying an infectious disease's causative *agent* may be critical for disease control and prevention. In the late 1800s, Koch and Henle developed a series of criteria that a microorganism must meet to be confirmed as the agent of an infectious disease. Known as *Koch's postulates*, these criteria require (a) that a specific microorganism always be found in association with a given disease; (b) that the microorganism be isolated from an infected animal or person and grown in pure culture (i.e., separate from all other microorganisms); (c) that the pure culture reproduce the disease when inoculated into a susceptible animal; and (d) that the microorganism be reisolated from the experimentally infected animal. These criteria form the foundation of medical microbiology even though some accepted agent/disease associations do not meet them, for example, difficult-to-culture agents may be identified using molecular criteria rather than culture isolation (Podzorski and Persing, 1995).

9.1.1 Infectious Microorganisms and Their Sources

9.1.1.1 Microorganisms *Microorganisms* are so small that the individual units cannot be seen by eye. Table 9.5 lists physical characteristics for a variety of microorganisms including their shapes and sizes. Microorganisms can be divided into two groups for classification or taxonomic purposes (a) prokaryotes, which lack nuclei to contain their chromosomes (e.g., bacteria); and (b) eukaryotes, true cells with

TABLE 9.5 Physical Characteristics of Microorganisms

Organism	Dimensions (μm) Width	Dimensions (μm) Length	Shape	Reproductive or resistant form
Viruses	0.020–0.25		Geometric, complex, helical, ovoid, spherical	—
Bacteria	0.3–1	1–10	Coccoid, spiral, rod-shaped: branched, unbranched	Endospores
Fungi	1–10	1–100	Spores/conidia: spherical, ellipsoid, fusiform Hyphae: branched, unbranched	Spores/conidia
Protozoa	5–20	5–75	Spherical, elongated, changing	Cysts
Helminths				
eggs or larvae,	15–75	25–150	Spherical, ellipsoid	Eggs
adult worms	<1–>5 mm	5–>10 mm	Elongated and cylindrical, flattened, elongated and segmented	

membrane-bound nuclei (e.g., fungi, protozoa, and helminths). Most prokaryotes and eukaryotes contain all the enzymes required for their own replication and have the necessary equipment to obtain or produce energy. Exceptions are certain bacteria (e.g., the chlamydia and rickettsia) which are obligate intracellular parasites.

Viruses fall outside the above classification system because they are neither truly alive nor true cells, that is, they consist only of nucleic acid (RNA or DNA) in a protein coat. Viruses can multiply only within host cells and generally do not retain their activity very long outside a host. Host-virus interactions are often highly specific; for example, viral pathogens of animals may also be human pathogens, but plant viruses are usually harmless for humans. Viruses are responsible for the majority of acute respiratory infections in humans (e.g., influenza and the common cold), which show seasonal incidence patterns related to increased indoor transmission in winter.

Prions are proteinaceous particles smaller even than viruses but with some similar properties. These rare and unusual infectious agents are believed to be responsible for degenerative neurological diseases in humans and animals that are known collectively as transmissible spongiform encephalopathies (e.g., Creutzfeldt-Jakob disease, a form of human dementia), Kuru (a human disease seen in Papua New Guinea), scrapie (a nervous system disease of sheep), and bovine spongiform encephalopathy ("mad-cow disease"). If prions are responsible for transmissible diseases, they are the only agents without nucleic acid that can do so. Consequently, the prion theory is not universally accepted.

Bacteria are grouped according to their Gram-stain reaction (positive or negative) and their cell shape (e.g., cocci or rods). Bacteria multiply by binary fission with generation times as short as 20 min. Chlamydia and rickettsia are bacteria that, like viruses, are intracellular parasites. Bacteria are found in a wide range of ecological sites in the environment (usually fairly damp ones) and also in association with animals, humans, and plants both as normal flora and, less often, as pathogens. Some bacteria (e.g., *Bacillus* and *Clostridium* spp.) produce endospores (internal dormant bodies) that enable them to survive adverse environmental conditions.

Fungi come in many forms, from unicellular yeasts to multicellular molds and mushrooms. Fungi can reproduce by budding or forming (often in large numbers) sexual or asexual spores or asexual conidia (i.e., reproductive bodies formed, respectively, in saclike structures or externally). Many fungi are found in soil and on leaf surfaces where they help decompose organic matter and recycle nutrients. Many humans suffer superficial fungal infections of the skin, nails, hair, and mucous membranes. Some airborne fungal infections are fairly common in certain parts of the country (e.g., histoplasmosis and coccidioidomycosis), others are somewhat rare (e.g., blastomycosis and cryptococcosis), and those caused by opportunistic fungal pathogens (e.g., aspergillosis and mucormycosis) usually only cause invasive disease in immunocompromised individuals or when a person receives an overwhelming exposure. Dimorphic fungi (e.g., the agents of blastomycosis and histoplasmosis) grow as yeasts in the body and when incubated at 37°C in laboratory culture but as molds in the environment or when incubated at 25 to 30°C.

Protozoa are mostly microscopic animals, made up of a single cell or a group of more-or-less identical cells. Protozoa live chiefly in water, although some are parasitic in humans and other animals. The majority of pathogenic protozoa cause hematogenous diseases transmitted by insects (e.g., malaria and babesiosis) or gastrointestinal diseases transmitted via ingestion (e.g., amebiasis, giardiasis, cryptosporidiosis, and cyclosporidiosis). Toxoplasmosis is also acquired by accidental ingestion of cysts shed in cat feces or by eating raw contaminated meats. Toxoplasmosis is of particular concern for immunocompromised persons and the offspring of women infected during pregnancy. People have become infected with free-living amebae (e.g., *Naegleria fowleri* and *Acanthamoeba* spp.) by swimming in contaminated waters, being sprayed or splashed in the eyes or on the mucous membranes of the nose or mouth with contaminated water, or wearing improperly decontaminated contact lenses.

Helminths are parasitic worms including cestodes (tapeworms), nematodes (round worms), and trematodes (flatworms). Some helminths have complex life cycles, with one or more host, and cause intestinal disease or live in host tissues. Many human helminth infections arise from contact with animal or human feces or with feces-contaminated soil or other material, resulting in intestinal or tissue diseases.

9.1.1.2 Sources of Microorganisms A *source* is the person, animal, object, or substance from which an infectious agent was passed to a host; sources may also be reservoirs. A *reservoir* is any person, animal, arthropod (insect, e.g., mosquito or flea, or arachnid, e.g., tick or mite), plant, soil, or substance, or combination of these (a) in which an infectious agent normally lives and multiplies; (b) on which it primarily depends for survival; and (c) where it reproduces and from which it can be transmitted to a susceptible host (APHA, 1995). For example, the placental tissue of an ewe that recently kidded (gave birth) or dust from dried placental tissue may be the source for the agent that caused Q fever in a laboratory researcher. However, the reservoir for the Q fever agent most likely is the flock from which the ewe was purchased. Identifying infectious agent reservoirs (not just their immediate sources) may be important to achieve long-term control of workplace transmission.

Sources and reservoirs of infectious microorganisms in the workplace include *humans*, most often patients in healthcare facilities or children in childcare centers and schools, but also the public, clients, and co-workers (Tables 9.1a and 9.2a). Infectious persons may have apparent disease or be asymptomatic (without symptoms) but still able to transmit infection; for example, they may be in the incubation period of a disease, be colonized by an infectious agent without being themselves affected, or be chronic carriers of an agent (see Section 9.1.3). *Animals* (fish, reptiles, birds, mammals, and arthropods) are also often asymptomatic carriers of infectious agents and serve as exposure sources for agricultural, food production, and veterinary care workers as well as animal handlers/trappers (Tables 9.1b and 9.2b). Agents with human or animal reservoirs seldom harm their hosts severely because the microorganisms evolved as parasites and depend on their host for survival. Finally, the *environment* (water, soil, and plants) can be a source or reservoir of infectious microorganisms (Tables 9.1c and 9.2c). Inanimate objects (medical devices, toys, tools, and work materials) may become contaminated and also serve as sources of infectious agents or means of agent transmission.

9.1.1.3 Nomenclature

Microorganism names. The taxonomic system that classifies similar organisms into increasingly restrictive groups uses Latin or Greek names, of which the *genus* and *species* designations are the most familiar. A name may describe an organism, commemorate a distinguished scientist, or identify the geographic source of the first isolate. For example, the name *Francisella tularensis* recognizes Francis' work and Tulare, the California county where the bacterium was first found. A microorganism may be reclassified and renamed as more is learned about it. For example, the family *Legionellaceae* is generally considered to have only one genus (*Legionella*), and *Fluorobacter dumof-*

fii and *Tatlockia micdadei* are now generally referred to as *Legionella dumoffii* and *Legionella micdadei*, respectively.

The first letter of a genus name is capitalized and is followed by the species name in lower case, for example, *Bacillus subtilis*, a rod-shaped, slender bacterium, and *Bacillus anthracis*, a related rod-shaped bacterium that produces a dark carbuncle. Only the first initial of a genus name is written when used repeatedly or when several species from the same genus are named, for example, *B. subtilis* and *B. anthracis*. Italicized genus names are followed by the word "species" (abbreviated "sp." for singular and "spp." for plural) when no species is named, such as a *Bacillus* sp. or several *Bacillus* spp. Identification to the species level is often important when evaluating a microorganism's significance; for example, *B. subtilis* is an essentially harmless soil bacterium frequently used in experimental work, whereas *B. anthracis* causes anthrax, a potentially serious zoonotic infection. The sexual and asexual forms of a fungus may be known by different genus names (e.g., *Histoplasma* and *Ajellomyces*, which, in this case are also the names of the mold and yeast forms of this dimorphic fungus). A genus name used as a noun or adjective is neither capitalized nor italicized, for example, the streptococci or a streptococcal infection.

Isolates within a species may be further separated into subspecies based on minor differences. Serologic distinctions among closely related microorganisms are based on detecting different antigenic determinants on their surfaces (see Section 9.2.3). Serologic tests are used primarily in epidemiologic investigations to compare multiple isolates of the same species, such as from several persons or environments. This comparison can help investigators determine if the isolates represent a single strain (possibly identifying a person or environment as a common source) or multiple strains (possibly indicating that they arose from different sources). Serologic techniques allow division of species into serogroups and serotypes (often referred to by number or letter, e.g., serogroup 1 or serogroup A) and within these into serovars (often referred to by the place of origin). However, the naming of serologically identified subspecies is not as consistent as that for genera and species.

Viruses are grouped into 73 families according to (a) the type and form of the nucleic acid present (i.e., RNA or DNA, single or double stranded); and (b) the size, shape, structure, and mode of viral replication. Within these families, viruses are placed into genera and species. Family and genus names for viruses are italicized but the common species names are not, for example, the mumps and measles viruses (the common species names) are both in the family *Paramyxoviridae* but belong, respectively, to the genera *Paramyxovirus* and *Morbillivirus*.

Disease names. Names for microorganisms are chosen in many ways, but they must comply with international codes of nomenclature or taxonomy. However, there is no official, international nomenclature for diseases, although the WHO is a good resource. A disease may be named for the microorganism that causes it (e.g., brucellosis from *Brucella* spp.), or an infectious agent may be known for the disease it produces (e.g., the rubella virus). Such naming usually depends on whether the disease or the agent was recognized first. Diseases may also be known by more than one name, for example, viral hepatitis B or serum hepatitis.

Some diseases may be caused by more than one agent (e.g., viral or bacterial conjunctivitis or pneumonia) or by more than one species of a genus (e.g., shigellosis and salmonellosis). In addition, some infectious agents can cause more than one disease (e.g., the varicella-zoster virus, VZV, causes chickenpox/varicella and herpes zoster/shingles). Disease names that use a microorganisms' genus or species name end in -osis (singular) or -oses (plural) (e.g., psittacosis from *Chlamydia psittaci* or the shigelloses caused by *Shigella* spp.). Use of a disease name (e.g., salmonellosis or tuberculosis) implies that specific symptoms were present (e.g., diarrhea or cough and weight loss, respectively). In the absence of symptoms, a person with a stool culture positive for a *Salmonella* sp. would be reported as having a salmonella infection rather than salmonellosis. Likewise, an asymptomatic person with a positive tuberculosis skin test is said to have tuberculosis infection rather than tuberculosis.

9.1.2 Transmission of Infectious Agents

Transmission describes the mechanism by which an infectious agent spreads from a source or reservoir to a person. Infectious agents are transmitted via contact (direct or indirect), droplets, air (droplet nuclei, dust, or spores), common vehicle, or vector (Table 9.6). The *portal of entry* for an infectious agent into the human body may be the respiratory tract (via inhalation); the mouth (via ingestion); the mucous membranes, conjunctivae, and broken skin (via contact); and again the skin (via subcutaneous or intravenous injection or penetration). Intact skin is believed to be an effective barrier to microorganism entry. Infectious agents are transmitted to these entry sites by a variety of routes (Tables 9.1 and 9.2). Some microorganisms are transmitted by more than one means, possibly leading to distinct diseases. For example, a veterinarian in a plague-endemic area may acquire infection through a flea bite (resulting in bubonic plague) or by inhaling droplets generated while caring for an animal with plague pneumonia (resulting in pneumonic plague). Likewise, the agent of tularemia (a zoonotic infection) can cause a skin ulcer on a worker's hand where contact occurred, systemic infection if inhaled, or pharyngitis (an inflamed throat) if ingested. In addition, a common, nonoccupational means of acquiring an infection (e.g., ingesting *Brucella* spp. in unpasteurized milk or cheese) may differ from an uncommon, occupational transmission mode (e.g., inhaling airborne bacteria from brucella cultures in a microbiology laboratory). Nevertheless, the same clinical illness, brucellosis, would result in either case. Viruses tend to cause the same manifestations of infection regardless of their transmission mode. For example, the hepatitis viruses cause hepatitis whether transmitted via ingestion or percutaneous exposure.

Contact transmission is the most important and frequent mode of spread for occupational infections. Contact may be either direct (e.g., by touching or being bitten) or indirect (e.g., via contaminated tools or utensils, clothing or bedding, or surgical instruments or dressings). *Direct-contact* transmission requires personal contact between a worker and a source, such as a patient in a healthcare setting, a fellow worker, a child in a childcare or school setting, or an animal in an agricultural or food production setting. Direct contact involves touching of a part of a susceptible worker's body to another infected or colonized body surface, with physical transfer of an infectious agent between the two. Hepatitis, herpes, rabies, and warts are transmitted in this way. *Indirect-contact* transmission involves contact of a susceptible worker with a contaminated intermediate object, usually inanimate (e.g., an instrument, glove, or wound dressing; a laboratory culture; or animal bedding or waste). Examples of diseases transmitted by indirect contact are bacterial infections (e.g., streptococcal skin diseases, salmonellosis, shigellosis, and Q fever), viral and bacterial conjunctivitis, and dermatophytosis (athlete's foot). *Bloodborne* transmission involves contact transfer of infectious agents found in blood, blood-derived fluids, or blood-contaminated items.

TABLE 9.6 Infectious Agent Transmission

Mode	Mechanism	Examples
Contact	Essentially immediate transfer of an infectious agent to a receptive site; includes percutaneous (through the skin) transmission and the mouth as a receptive site for fecal-oral transmission	Direct contact: by touching or being bitten (broken skin); hepatitis, herpes, HIV infection, rabies, warts Indirect contact: via contaminated tools, handkerchiefs, clothing, bedding, cooking or eating utensils, and surgical instruments or dressings; HBV, HCV, HIV infection, conjunctivitis
Droplets	Projection into the eyes, nose, or mouth (usually within a distance of ≤1 m)	Acute viral and bacterial respiratory infections, rubella, pertussis
Airborne	Dissemination of infectious aerosols to a suitable portal of entry (usually the respiratory tract via inhalation); particles 1–5 μm reach the alveoli	Droplet nuclei: the residue that remains after liquid evaporates from droplets (such as those produced by infected people or aerosol-generating equipment) containing an infectious agent; chickenpox, influenza, measles, tuberculosis Dust: small particles from soil, plant or animal material, clothes or bedding, and floors; anthrax, hantavirus pulmonary syndrome Spores: products of fungal sexual or asexual reproduction often designed for air transport; coccidioidomycosis, cryptococcosis, histoplasmosis
Common-vehicle	Ingestion of contaminated food, water or medication	Salmonellosis, shigellosis, hepatitis A
	Via equipment or medical device	Conjunctivitis
Vector-borne	Mechanical carriage of infectious agents by crawling or flying insects and transfer through insect bites	Tickborne: Lyme disease, Rocky Mountain spotted fever, tularemia Fleaborne: murine typhus fever, plague Mosquito-borne: arthropod-borne viral disease, malaria

Droplet transmission can be considered a form of contact transmission (and it shares some features with airborne transmission), but the mechanism of pathogen transfer is distinct. Droplets (usually of respiratory secretions and often large enough to be seen or felt) are generated when a person coughs, sneezes, or speaks or during medical procedures such as suctioning and bronchoscopy (visually examining the tracheobronchial tree using a flexible tube). Droplets may also be generated when liquids (e.g., process waters such as a metalworking fluid, body fluids such as blood or sputum, or laboratory cultures) are spilled, splashed, or sprayed. Transmission occurs when droplets containing microorganisms are propelled a short distance (usually ≤1 m) through the air and deposit on a worker's conjunctivae, nasal mucosa, or mouth. Many acute viral and bacterial infections (e.g., the common cold and streptococcal infections, respectively) are transmitted in this way.

True *airborne transmission* results from inhalation of droplet nuclei, dust particles carrying microorganisms, or fungal or bacterial spores. Droplet nuclei are the solid residue of dried droplets, usually ≤5 μm. Few droplet nuclei contain microorganisms, depending on the original source of the droplets (e.g., coughs from infected persons or animals or liquid aerosols). Small particles can remain airborne for some time, and workers may be exposed near a source or at some distance from one.

Common-vehicle transmission is similar to indirect contact transmission but applies to microorganisms transmitted by contaminated items to many persons rather than to just one. Examples of contaminated items involved in common-vehicle transmission are food, water, medications, medical devices, and equipment or tools. Outbreaks of food-borne or waterborne disease may occur in workplaces. However, such infections usually do not result from work-related activities, except perhaps in food handlers, agricultural workers, and field crews.

Vectorborne transmission occurs when an animal serves as an intermediary, carrier, or vector and transmits an infectious agent from one host (or reservoir) to another. Examples of vectors are mosquitoes, fleas, ticks, flies, rats, and other vermin. Some vectors serve simply as a mechanical means of transmitting a microorganism (e.g., a housefly carrying bacteria from one place to another). Other vectors play a biological role and serve as a host in which a microorganism multiplies or develops (e.g., mosquitos and *Plasmodium* spp., protozoa that cause malaria). A vector may also be an intermediate host for an agent (e.g., a tick carrying *Borrelia burgdorferi*, the bacterial agent of Lyme disease). An intermediate host transports an infectious agent between a definitive host (e.g., a rodent, deer, or other mammal) and a human who may serve as an additional reservoir. Outdoor workers (including vector and animal controllers) are at risk of vectorborne diseases as are indoor workers if they are exposed to animals that serve as reservoirs of infectious agents transmitted by vectors.

9.1.3 The Infectious Disease Process

An *infection* occurs when a microorganism invades the body, multiplies, and causes the body to react. To produce infectious disease, a microorganism must be (a) pathogenic (i.e., disease causing); (b) viable (alive and able to multiply, e.g., bacteria, fungi, protozoa, and helminths) or active (able to initiate replication by a host cell, e.g., viruses and perhaps prions); (c) present in sufficient numbers; and (d) successfully transmitted to a susceptible host at a suitable entry site. Both living and dead microorganisms and microbial fragments can elicit hypersensitivity (allergic) reactions in sensitive persons, but only viable or active microorganisms can cause infection and then only in susceptible individuals. The term *virulence* describes a microorganism's degree of pathogenicity or capacity to cause disease. Virulence is determined by an agent's ability to invade and damage tissues and by the case-fatality ratio for an infection, that is, the percentage of infected persons who experience serious illness or die (APHA, 1995). This term is often confused with infectivity. *Infectivity* describes the tendency of an organism to be transmitted and cause infection. For example, rhinoviruses (which cause the common cold) have high infectivity but low virulence, that is, they attack many people, but the resulting illness is never itself fatal. In contrast, influenza viruses may be both highly infectious and virulent.

Some of the effects of an infection are due to toxins that an invading microorganism produces, such as the tetanus toxin of *Clostridium tetani*. Some foodborne diseases are due to preformed toxins present as a result of microbial multiplication in a food before consumption (e.g., the botulinum toxin of *Clostridium botulinum*). Such foodborne diseases are referred to as intoxications rather than infections. Gram-negative bacteria produce endotoxins and some fungi produce mycotoxins (Morey et al., 1990; Rylander and Jacobs, 1994; Wald and Stave, 1994; Burge, 1995; ASM, 1997). However, the health problems these toxins may cause (e.g., asthma and inhalation fevers) are not considered communicable diseases and are not addressed here. See Chapter 20 for additional information on these.

A person must receive a threshold number of organisms or amount of toxin, that is, an *infectious* or *effective dose*, to contract a disease. Reliable information on infectious doses for humans is available for only a limited number of agents (Sewell, 1995; Harding et al., 1996). Infectious doses vary widely among agents and may depend on the exposure route. For example, one *M. tuberculosis* cell deposited in the gas exchange region of the lungs may initiate infection. However, the tissues of the upper airways are not susceptible to this bacterium, and even large numbers of bacteria deposited there appear to be harmless. This observation illustrates the importance of particle size for airborne infections as well as the distinction between transmission by droplets and droplet nuclei. Infectious doses are often reported as ID50, the number of organisms required to infect 50 percent of a test population. The infectious dose for an individual worker depends on the person's general health and may be higher or lower than the ID50. In general, larger infectious doses are required by the oral route than by inhalation or mucous membrane contact. Doses that greatly exceed the minimum may result in more severe infection or more rapid onset.

Worker exposure to infectious agents varies, and with it a worker's likelihood of infection. However, individual resistance is the principal distinction among exposed workers that determines their infection risk. *Resistance* describes the degree to which a person can eliminate an organism without developing disease or is able to limit the severity of illness following infection. Resistance varies greatly from person to person. Immunization or past infection may contribute to resistance. Host factors may render some workers more susceptible to infection. Examples of such host factors are age, race, gender, underlying disease, antimicrobial treatment, administration of corticosteroids or other immunosuppressive agents, irradiation, and breaks in the first line of defense, such as damage to the skin or respiratory tract. In addition, such risk factors may predispose individuals to more severe illness should they become infected. Other than immunizations, no medications, treatments, or dietary supplements have been demonstrated to effectively increase a person's resistance to infection, despite popular claims.

Exposure of a susceptible person to an infectious dose of an agent may result in asymptomatic infection, subclinical infection, or clinical disease. *Asymptomatic infections* occur when an organism enters the body and multiplies, but by virtue of a healthy immune response, the person has no clinical symptoms or outward signs of illness. Asymptomatic infection is the most common form of infection for some agents (e.g., HBV). Such infections may only be recognized through specific serologic testing (see Sections 9.2.2.1 and 9.2.3). *Subclinical infections* are generally mild (e.g., a slight fever, minor aches, or transient fatigue) and are usually of short duration.

Clinical disease is associated with signs or symptoms of infection that may aid diagnosis, such as the fever, chills, night sweats, fatigue, weight loss, and hemoptysis (coughing blood) of tuberculosis. However, tuberculosis is an example of an infection that seldom leads to clinical disease in healthy infected adults and is one in which infected individuals only transmit the agent (i.e., become infectious) when they have active disease. In the United States, only ca. 1 to 5 percent of persons infected with

M. tuberculosis develop disease in the first year following infection identified by a skin test conversion. For persons infected as children, the lifetime risk of tuberculosis disease approaches 10 percent (APHA, 1995). In contrast, for individuals coinfected with HIV, the annual tuberculosis disease risk has been estimated at 2 to 7 percent and the cumulative risk at ca. 70 percent. *Carriers* harbor infectious agents, temporarily or chronically, and may transmit them to others. Carriers may be colonized, infected but asymptomatic, or have chronic or relapsing disease (e.g., 0.2 to 0.9 percent of North American adults chronically carry HBV).

Infectious agents typically have characteristic *incubation periods*, the time interval between initial contact and the first appearance of symptoms or immunologic evidence of infection. Incubation periods are usually a few days to a few weeks. However, incubation periods may be only a few hours (e.g., salmonellosis), extend to months or years (e.g., warts and rabies, respectively), or are unknown (e.g., dermatophytoses). During this period, an infectious agent is multiplying and the body typically begins to mount an immune response. An *immune response* is the development of antibodies against specific protein or carbohydrate antigens on infectious agents. For many agents, the larger the dose received, the shorter the incubation period.

The *communicable period* is the time during which an infected person or animal can transmit an infectious agent. This period may begin before signs or symptoms of infection are evident. For many infectious agents (particularly viruses), the time of greatest communicability is during the incubation period, which is when a microorganism is multiplying at the greatest rate. Consequently, the period of greatest risk of transmitting infection is often past by the time disease is recognized.

Antibodies belong to a group of molecules called *immunoglobulins* (Ig), of which there are several classes represented by letters (e.g., M or IgM, G or IgG, A or IgA). Temporally, there are two antibody responses to an infection. The first response consists of IgM antibodies. IgM appears as early as 4 days following infection, peaks at ca. 7 to 10 days thereafter, and then declines rapidly. The second response usually involves IgG antibodies. IgG appears within 2 to 3 weeks of exposure and begins declining 4 to 6 weeks after infection or may persist for many years. An antibody name may identify an infectious agent, for example, "anti-"hepatitis A virus (anti-HAV), or a specific antigen of the agent, such as "anti-"hepatitis B core antigen (anti-HBc). Laboratory serologic reports often omit the term "anti-." For example, a laboratory result may report detection of legionella antibodies rather than detection of anti-legionella antibodies.

9.1.4 Modeling Infectious Disease Transmission

Modeling disease transmission can help investigators identify significant contributing factors and select and evaluate control measures. Epidemiologists have used this approach to estimate unknown terms in transmission equations (Catanzaro, 1982) and to predict and compare the benefits of interventions (Nardell et al., 1991; Gammaitoni and Nucci, 1997). Health and safety professionals, familiar with exposure assessment for other workplace hazards, may recognize similarities for infectious disease prevention if they can separate and individually consider the factors involved in the process. Such risk modeling was used to develop a process to select respirators for protection against infectious aerosols (Nicas, 1994, 1996). The model was based on estimating an agent's infectious dose and air concentration, a worker's number of exposures, and the acceptable risk of infection. A similar exercise is presented here.

Equation 1 predicts the number of new cases of infectious disease, C, that would result if a given number of susceptible persons, S, were exposed at a particular rate, r (Burge, 1990; Nardell et al., 1991):

$$C = S(1 - e^{-r}) \tag{1}$$

Clearly, the higher the exposure rate, the greater the number of new cases for a given number of susceptible individuals. The exposure term, r, varies with the mechanism of disease transmission, but may contain familiar terms used to assess other work-related exposures. For example, the airborne transmission of an infectious agent spread from person to person or animal to person depends directly on (a) the number of infectious persons or animals present, I; (b) the rate at which these sources release microorganisms, q (infectious doses h^{-1}, which depends on the concentration of infectious agents at a source and the rate at which it generates infectious droplets); (c) the breathing rates of susceptible workers, p (m^3 h^{-1}); and (d) the exposure times for these potential new cases, t (h). Transmission depends inversely on the rate, Q (m^3 h^{-1}), at which dilution ventilation or other particle removal processes reduce airborne contaminant concentration. Substituting these terms for r, Eq. 1 becomes:

$$C = S\left(1 - e^{-I\frac{qpt}{Q}}\right) \tag{2}$$

This presentation illustrates that person-to-person or animal-to-person disease transmission can be reduced by

1. Minimizing S, the number of individuals susceptible to an infectious disease, for example, by reducing the number of persons exposed, immunizing them, or selecting workers who have been immunized or have evidence of previous infection with an agent;
2. Minimizing I, the number of infectious persons or animals present, for example, by preventing them from becoming infected or by screening to exclude them from the workplace;
3. Reducing q, the number of organisms infected sources release, for eaxample, (a) by treating infected persons or animals to render them noninfectious or reduce the load of infectious agents they carry, or (b) by having people cover their noses and mouths or wear masks when coughing or sneezing or by using local exhaust ventilation, isolation, or enclosure to contain aerosols;
4. Reducing exposures for susceptible persons, for example, by minimizing the amount of contaminated air they breathe, (p t), through the use of respirators or by limiting the time, t, they spend in a contaminated environment,
5. Increasing Q, the contaminant removal rate, for example, by increasing the dilution ventilation rate, using particle air cleaners, or irradiating the air.

Similar models can be developed for other transmission mechanisms (e.g., contact or ingestion). Likewise, other sources may be substituted for infected persons or animals used in the example (e.g., aerosol-generating equipment, activities, or processes). Models are best used to identify factors contributing to disease transmission and prevention rather than to predict infectious disease risk. Other factors to consider when estimating the degree of hazard an agent poses are (a) the agent's infectivity and virulence; (b) the exposure route and target organ system; (c) a worker's susceptibility and ability to resist infection; (d) the potential severity of an infection should it occur; and (e) the availability of effective therapeutic measures and postexposure prophylaxis (see Section 9.2.2.3).

9.2 EVALUATING EXPOSURE TO BIOLOGICAL AGENTS

Health and safety programs for control of infectious agents in the workplace should (a) identify potential work-related infections (Table 9.1); (b) identify applicable regulations or guidelines; (c) develop an appropriate medical surveillance program (Chapter 10), including accident reporting and investigation (Chapter 5); (d) develop control and prevention strategies, including emergency procedures; (e) educate workers and train them to recognize and avoid exposures and to respond correctly to emergencies (Chapter 14); and (f) inspect the workplace for compliance with health and safety rules. The first item has already been discussed; comments on the other components follow.

9.2.1 Regulations, Recommendations, and Guidelines

The control of communicable diseases is one of the oldest fields of public health. Not surprisingly, there are many laws designed to prevent infectious disease transmission. Federal, state, and local regulations on food and water protect the public from many infectious agents. Similarly, vector control programs and many of the restrictions on immigration and travel are intended to prevent infectious diseases. These regulations focus control on the environment (e.g., food preparation and sanitary regulations) or on human conduct (e.g., compulsory examinations, immunization, and detention/isolation). Local and federal public health officials can exercise considerable authority in these areas when they choose to do so. However, health officers prefer to work cooperatively with employers, employees, employee representatives, and state and federal labor departments. There are no workplace or ambient exposure limits for infectious agents, and it is unlikely there will soon be any (ACGIH, 1997). Following are summaries of some work-related regulations and guidelines that may apply to occupational infectious disease control.

9.2.1.1 General Federal, State, and Local Regulations The handling of agricultural items that will become food products shipped through interstate commerce is regulated by the U.S. Department of Agriculture and the FDA. Food preparation facilities and food handlers are regulated by state departments of agriculture, food, and drugs as well as by local jurisdictions. State and local air quality, water, and waste management agencies govern aspects of pollution control that impact infectious disease transmission.

Healthcare and childcare facilities are subject to a variety of state and local regulations to control communicable diseases. Personal service work is regulated by state and local authorities. Some professional licenses and credentials require evidence of immunization as well as training and examination in infectious disease control. However, these regulations and those covering the handling of food products as well as liquid and solid wastes are intended primarily to protect consumers, clients, and the public, not necessarily the workers in these industries.

Many state laws specify that a local health officer must clear an infected worker before that person is allowed to return to work. However, these laws generally leave to the health officer's discretion the conditions that must be met before return to work.

Section 9.2.2.4 addresses return-to-work requirements for healthcare workers and food handlers. Local, state, and federal public health authorities and professional associations are the best sources of information on regulations and recommendations that apply to particular occupations. The Federal Register and CFR are accessible in a searchable format at http://law.house.gov/cfr.htm. Further descriptions of government regulations of general safety and health matters are available elsewhere (Gillotti, 1996).

9.2.1.2 CDC, WHO, and Professional Associations CDC, WHO, public health associations, and professional groups have authored many guidelines and recommendations referenced throughout this chapter. These recommendations do not have the impact of regulations or requirements but are considered accepted current standards of practice. Activities contrary to these recommendations are generally considered unacceptable.

9.2.1.3 The Occupational Safety and Health Administration OSHA commonly addresses workplace hazards associated with infectious agents through enforcement actions using the General Duty Clause of the Occupational Safety and Health Act of 1970 Section 5(a)(1) (USDOL, 1970). This section requires employers to provide a workplace free of recognized hazards that cause or are likely to cause death or serious physical harm.

Other codes that OSHA may invoke relative to communicable disease transmission in workplaces include the following. Standard 29 CFR 1910.22 (housekeeping) requires that workplaces be kept clean, orderly, and in a sanitary condition. Standard 29 CFR 1910.141 (sanitation) requires that employers provide potable (drinking quality) water; disposal of putresible (decomposable) wastes; control of vermin (e.g., rodents and insects); change rooms for employees required to wear protective clothing; lavatories exclusively for washing hands, arms, faces, and heads; and toilet facilities. Standard 29 CFR 1910.142 (temporary labor camps) requires vector control and sanitation, openable windows and avoidance of overcrowding in sleeping quarters, and reporting of communicable diseases to local health officers. Standard 29 CFR 1928.110 (field sanitation) requires employers of ≥11 field hand-laborers to provide toilets, potable drinking water, and handwashing facilities in the field.

Standard 29 CFR 1910.132 (general requirements for personal protective equipment, PPE) requires that protective equipment (including PPE for eyes, face, head, and extremities; protective clothing; respiratory devices; and protective shields and barriers) be provided, used, and maintained in a sanitary and reliable condition. Standard 29 CFR 1910.133 (eye and face protection) requires use of protective eye and face equipment where there is a reasonable probability of injury that can be prevented by such equipment. Standard 29 CFR 1910.134 (respiratory protection) requires that respirators applicable and suitable for the purpose intended be provided when such equipment is necessary to protect the health of an employee.

Standard 29 CFR 1910.145 (specifications for accident prevention signs and tags) requires biohazard postings to identify the actual or potential presence of a biological hazard and to identify equipment, containers, rooms, materials, experimental animals, or combinations of these that contain or are contaminated with hazardous biological agents. This standard applies primarily to laboratory and healthcare settings.

In 1991, OSHA promulgated a standard that addressed occupational exposure to bloodborne agents (codified under 29 CFR 1910.1030, Occupational Exposure to Bloodborne Pathogens), which included, but was not limited to, HBV and HIV. Table 9.7 summarizes major elements of the standard and its implementation (Gomez, 1993). The

TABLE 9.7 Key Provisions of the Bloodborne Pathogens Standard

Exposure Control Plan

Each employer must establish a written exposure control plan designed to eliminate or minimize employee exposure to bloodborne pathogens. The plan must include at least the following three elements

(a) Exposure determination to identify potentially exposed employees;
(b) A schedule and method for implementing provisions of the standard, such as hepatitis B vaccination, postexposure evaluation and followup, communications of hazards to employees, and recordkeeping;
(c) A procedure for evaluating the circumstances surrounding exposure incidents.

Methods of Compliance

Engineering and work practice controls are designated as the primary means of eliminating or minimizing employee exposures. The standard defines *engineering controls* as "controls (e.g., sharps disposal containers, self-sheathing needles) that isolate or remove the bloodborne pathogens hazard from the workplace." The standard defines *work practice controls* as "controls that reduce the likelihood of exposure by altering the manner in which a task is performed (e.g., prohibiting recapping of needles by a two-handed technique)."

The standard states that where occupational exposure remains after institution of engineering and work practice controls, *personal protective equipment* must also be used. Personal protective equipment includes, but is not limited to, "gloves, gowns, laboratory coats, face shields or masks and eye protection, and mouthpieces, resuscitation bags, pocket masks, or other ventilation devices."

Housekeeping

Employers must ensure that worksites are maintained in a clean and sanitary condition. All equipment and environmental and work surfaces must be cleaned and decontaminated after contact with blood or other potentially infections materials.

Hepatitis B Vaccination and Postexposure Evaluation and Followup

Employers must make hepatitis B vaccination available at no cost to employees who have potential occupational exposure. Following a report of an exposure incident, employers must make a confidential evaluation and followup immediately available to the exposed worker.

Communication of Hazards to Employees

Warning labels are required on all containers of regulated waste, refrigerators, and freezers containing blood or other potentially infectious materials and on other containers used to store, transport, or ship them. Employers must post signs at the entrances to specified work areas identifying the infectious agent, special requirements for entering the area, and a contact name and telephone number.

Information and Training

All employees with occupational exposure must participate in a training program provided at no cost to employees and during regular working hours. Training must be provided at the time of initial assignment and at least annually thereafter.

Recordkeeping

Each employer must establish and maintain employee medical records that include name, social security number, hepatitis B vaccination status, and results of examinations, tests, and followups.

other bloodborne infections named in the standard are arboviral infections; babesiosis; brucellosis; Creutzfeldt-Jakob disease; cytomegalovirus infection and disease; hepatitis A, C, D, and E; human T-lymphotropic virus type I (HTLV-I); leptospirosis; malaria; relapsing fever; syphilis; and viral hemorrhagic fever.

In the absence of standards, OSHA may reference CDC guidelines to substantiate a citation using the General Duty Clause of the OSHA Act Section 5(a)(1), for example, recommendations to prevent exposure to *M. tuberculosis* (CDC, 1992, 1994a, 1996b) and hantavirus (CDC, 1993c). The proposed rule on occupational tuberculosis includes several components from the CDC guidelines: written exposure control plans, procedures for early identification of individuals with suspected or confirmed infectious tuberculosis, procedures for initiating isolation of individuals with suspected or confirmed infectious tuberculosis or for referring those individuals to facilities with appropriate isolation capabilities, procedures for investigating employee skin test conversions, and education and training for employees (USDOL, 1997a). In addition, OSHA has incorporated CDC recommendations for engineering control measures and CDC's standard performance criteria for the selection of respiratory protective devices appropriate for use against exposure to *M. tuberculosis*. The latest information on this and other actions is available at http://www.osha.gov/index.html or through OSHA offices. The *OSHA Technical Manual* addresses Legionnaires' disease in Section II, Chapter 7, and tuberculosis in Section V, Chapter 1—Hospital Investigations: Health Hazards (USDOL, 1997b).

Standard 29 CFR 1904.2 (log and summary of occupational injuries and illnesses) requires employers to maintain a log and summary of all occupational injuries and illnesses (including those involving infectious agents) that result in death, lost workdays, transfer to another job or termination of employment; require medical treatment other than first aid; or involve loss of consciousness or restriction of work or motion. Standard 29 CFR 1904.8 (reporting of fatality or multiple hospitalization accidents) requires reporting of work-related injuries and illnesses that result in death or the hospitalization of ≥ 5 employees. OSHA has published a Notice of Proposed Rulemaking (NPRM) for Occupational Injury and Illness Recording and Reporting Requirements. This NPRM discusses proposed revisions to 29 CFR 1904 (recording and reporting occupational injuries and illnesses), supplemental recordkeeping instructions, and replacement of the OSHA 200 form with an OSHA 300 form (see Chapter 29 and Section 9.2.4).

9.2.1.4 Medical Waste Handling, Disinfectants, and Medical Devices Various groups offer guidelines on the handling and disposal of waste material that may contain infectious agents, for example, the CDC Hospital Infections Program, the Joint Commission for the Accreditation of Hospitals, and the College of American Pathologists. The Medical Waste Tracking Act (PL 100–582; 40 CFR 259) requires the U.S. Environmental Protection Agency (USEPA) to track medical waste in ten states (New York, New Jersey, Connecticut, and the seven states bordering the Great Lakes). Medical waste is any solid waste generated in the diagnosis, treatment (e.g., provision of medical services), or immunization of human beings or animals, in research pertaining thereto, or in the production or testing of biological reagents. Most states have passed medical waste legislation or promulgated rules to address medical waste disposal. The

USEPA (1986) has also published advice on infectious waste management. The FDA sets standards for disinfectants as well as medical devices and equipment. The FDA registers only those products sold for the purpose of sterilizing critical devices such as surgical instruments that penetrate the blood barrier and semicritical devices that contact mucous membranes (see also Section 9.3.2.1).

9.2.1.5 Transfer of Hazardous Agents The U.S. Department of Agriculture (Animal and Plant Health Inspection Service and Veterinary Services, Riverdale, MD) regulates the importation of animals and animal-related materials to ensure that animal and poultry diseases are not introduced into the United States. Although some animal diseases also affect humans, these regulations are designed primarily to protect domestic and wild animals. The U.S. Public Health Service (CDC, Atlanta, GA) has jurisdiction over importation of human and nonhuman primate materials.

The recent threat of illegitimate use of infectious agents (e.g., for bioterrorism) has led to new provisions to regulate the transfer of hazardous agents. Regulations address the packaging, labelling, and transport of select agents shipped in interstate commerce. The CDC issued a regulation that places additional shipping and handling requirements on facilities that transfer or receive select agents that are capable of causing substantial harm to human health (USDHHS, 1996). The rule was designed to (a) establish a system of safeguards to be followed when specific agents are transported; (b) collect and provide information concerning the location where certain potentially hazardous agents are transferred; (c) track the acquisition and transfer of these specific agents; and (d) establish a process for alerting the appropriate authorities if an unauthorized attempt is made to acquire these agents. The rule has the following components: (a) a comprehensive list of select agents (including viruses, bacteria and rickettsiae, fungi, toxins, and recombinant organisms and molecules); (b) a registration of facilities transferring these agents; (c) transfer requirements; (d) verification procedures including audit, quality control, and accountability mechanisms; (e) agent disposal requirements; and (f) research and clinical exemptions.

9.2.2 Medical and Biological Evaluation of Exposure

Medical evaluation of worker exposure to biological agents can occur in a variety of contexts, for example, before job placement, periodically after placement, following exposure, and for diagnosis of possible illness. Chapters 10 and 11 describe medical surveillance for chemical exposures, which has some relevance to infectious disease surveillance in terms of general approaches.

9.2.2.1 Preplacement Examinations The goal of a *preplacement examination* is to assign an employee to an appropriate job with any reasonable accommodations that may be necessary. A preplacement examination for a worker potentially exposed to infectious agents should include (a) collection of baseline data for evaluation of any abnormalities that may later be noted; (b) medical and occupational histories; (c) a history of immunizations; (d) a history of past infections; and (e) a physical examination. Laboratory tests should be based on the specific agents to which an employee may be exposed. Intradermal skin tests can be used to establish probable immunity as a result of previous infection for tuberculosis, coccidioidomycosis, and Q fever.

Often *blood samples* are drawn and stored as sera (*sing.*, serum, the fluid portion of blood remaining after the solid portion clots). Such samples are collected at the beginning of work (and perhaps periodically thereafter) to provide baselines for comparison with subsequent antibody levels (USDHHS, 1993). A portion of a stored serum may be tested concurrently with a recent sample to detect a change in antibody concentration (see Section 9.2.3). Recent and stored sera are tested at the same time to reduce analytic variability due to the method or reagents used. This "serum banking" is particularly useful for occupations in which new pathogens may be uncovered in the future. Asymptomatic or unrecognized past infections may be detected or ruled out through comparison of current and prior antibody levels. Such testing was performed for many workers exposed to rodents following the discovery of hantavirus pulmonary syndrome (an often fatal infection first seen in 1993 in southwestern United States) (Zeitz et al., 1997). These tests confirmed that asymptomatic or unrecognized infections had not occurred in the past, that is, stored sera did not contain antibodies to this virus. Similar findings were observed among workers exposed to B virus from monkeys.

Antibody levels may be determined at the time of worker placement if it depends on knowing an employee's susceptibility/immunity status. The possible future need to perform a postexposure evaluation (described below) presents a basis for conducting a preplacement test or for banking a serum sample. Without the information such tests can provide, it may not be possible to distinguish a recent, occupationally acquired infection from one that predated employment. Preplacement tests can also be useful for worker's compensation purposes, such as to establish that a bloodborne viral infection such as HBV, hepatitis C virus (HCV), or HIV infection was a consequence of exposure prior to employment. However, such testing may raise legal questions. An employer may not be able to require testing because of privacy, confidentiality, or discrimination issues although medical test results are generally confidential (i.e., accessible only to a worker and medical personnel). Storage of a serum sample for possible future testing may temporarily avoid these issues (i.e., tests will only be run if a need arises). Nevertheless, collection of a blood sample is an invasive procedure that must be voluntary. In addition, it can be expensive to draw and store these samples. Therefore, preplacement examinations and tests should be recommended for bona fide occupational needs.

9.2.2.2 Periodic Examinations *Periodic examinations* may be performed for purposes of (a) screening and surveillance of asymptomatic workers; (b) early detection and treatment of work-related disease; (c) determining the adequacy of workplace controls by detecting worker exposure; and (d) general medical screening of conditions unrelated to specific occupational exposures, for example, hypertension or breast cancer (Ehrenberg and Frumkin, 1995). Examples of periodic examinations include skin tests for tuberculosis and serologic tests to detect antibodies developed against specific agents known or suspected to be present in a workplace (see Sections 9.2.2.1 and 9.2.3). Serologic or skin test conversions in these cases will usually not indicate the time of infection, other than that it occurred between performance of the current and previous tests. Serologic testing will also identify immunized workers whose antibody titers require boosting to remain protective (e.g., anti-rabies and anti-polio titers; also possibly hepatitis B along with hepatitis A and varicella in the future).

9.2.2.3 Postexposure Examinations *Postexposure examinations* are medical evaluations following known or suspected acute exposure incidents (Ehrenberg and Frumkin, 1995). These exams take place in one of two time frames (a) while a worker

TABLE 9.8a Postexposure Medical Evaluation: Worker Asymptomatic or within Incubation Period

Step	Action
1	Document route and circumstances of exposure.
2	Identify source of exposure (e.g., human, animal, environment), if known, and conduct appropriate testing, if possible.
3	Establish or re-establish baseline tests (e.g., serologic antibody titers).
4	Update history: past infections, immunizations, medical history, family history.
5	Physical examination.
6	Laboratory tests: based on suspected agent Blood—draw and store serum sample for baseline antibody testing; Other—as indicated.
7	Consider postexposure prophylaxis or treatment, if appropriate (see Table 9.13).

is within a disease's incubation period and asymptomatic (Table 9.8a); or (b) after a worker has developed symptoms or has passed through the incubation period without developing them (Table 9.8b). The primary distinction between these exams is the focus on (a) establishing or re-establishing baseline test results during the incubation period and the possible provision of postexposure prophylaxis (PEP) (see Section 9.3.5); and (b) diagnosing and possibly treating infection that does occur.

9.2.2.4 Return-to-Work Examinations *Return-to-work examinations* are medical evaluations required to be given before a worker with a communicable disease can resume work. Local or state regulations may require such exams for healthcare workers or food handlers (see also Section 9.2.1). For some intestinal infections, one or more negative stool cultures (e.g., for shigella or salmonella) may be required before a worker is given clearance. For other infectious diseases, documentation of treatment or passage of the communicable period may suffice. Healthcare workers infected with certain agents should be precluded from caring for extremely vulnerable patients. However, workers who use PPE (e.g., surgical masks for respiratory infections and gloves for skin and nail infections) may be permitted to provide patient care in selected circumstances.

9.2.2.5 Chronically Infected Workers Sections 9.2.2.3 and 9.2.2.4 discussed postexposure examinations and return-to-work criteria for certain workers acutely

TABLE 9.8b Postexposure Medical Evaluation: Worker Symptomatic or beyond Incubation Period

Step	Action
1	Update history: past infections, immunizations, medical history, family history.
2	Symptoms: nature and pattern.
3	Current illness: medical diagnosis; treatment, if any.
4	Physical examination: vital signs; other, as appropriate.
5	Laboratory tests: consult a laboratory early on specimens and specimen collection appropriate for the suspected agent.
6	Consider postexposure prophylaxis or treatment, if appropriate (see Table 9.13).

infected with communicable diseases transmitted by direct contact or the fecal/oral route. Concern about workers with chronic infections has been confined primarily to healthcare workers. All workers (especially healthcare, childcare, and emergency workers as well as those with public contact) should be aware that they may transmit communicable diseases to others. In addition, some facilities have policies that healthcare workers performing exposure-prone invasive procedures should know their HIV, HBV, and HCV status. Defining "exposure-prone invasive procedures" has been difficult and controversial, although some practices obviously qualify (e.g., blind suturing, using a fingertip to feel for a needlepoint during surgical stitching).

Policies for dealing with chronically infected workers seek to protect the workers and their co-workers along with patients, other contacts, and the public. Programs generally do not actively attempt to identify infected workers, except perhaps as followup on exposure incidents. Instead, such programs rely on voluntary reporting by infected individuals. HIV-, HBV-, or HCV-positive persons in high-risk occupations are advised to report their status to an appropriate individual or panel, which decides if work restrictions or reassignment is required. These decisions may be difficult to reach due to a lack of quantifiable data on patient risk and a lack of consensus on acceptable risks. Additionally, there is controversy over requiring informed consent from patients and the related issue of confidentiality for infected healthcare workers. Some countries (e.g., the United Kingdom) have explicit restrictive regulations for specific conditions, such as healthcare workers who are HBV carriers and are positive for hepatitis B e antigen (HBeAg).

9.2.3 Laboratory Diagnosis of Infectious Diseases

A discussion of infectious disease diagnosis is beyond the scope of this text, and other thorough references are available (Wald and Stave, 1994; ASM, 1995b; Mandell et al., 1995). The following are brief descriptions of the most common laboratory tests used specifically in the diagnosis of infection. Other diagnostic tests may indicate disease but are not specific for an infectious etiology; for example, pneumonia can be diagnosed by chest X-ray, but this condition has many infectious and noninfectious causes.

The *sensitivity* of a test is defined as the frequency of a positive reaction in persons with an infection (i.e., the percentage of true positives among all positive patients). Test *specificity* is the frequency of a negative test in uninfected persons (i.e., the percentage of true negatives among all negative patients). The terms *sensitive* and *specific* are also used in other ways to describe tests. The *predictive value* of a test may be a more useful criterion to rule an infection in or out. The predictive value of a positive test is the probability that a positive result accurately indicates that a person has an infection. The predictive value of a negative test is the probability that a negative result correctly indicates that a person does not have an infection (Herrmann, 1995; Sewell and Schifman, 1995).

Clinical microbiology laboratories use a number of methods to detect infectious agents in human specimens (e.g., throat and wound swabs as well as blood, sputum, and urine samples). Laboratories use many of the same methods to detect, identify, and quantify microorganisms in environmental samples (e.g., air, surface, liquid, or bulk material samples) (see Section 9.2.5.2). The primary techniques for microorganism detection and identification are (a) direct visualization by microscopic examination; and (b) isolation by culture.

Microscopic examination is usually done on a portion of a specimen submitted for culture. Visual examination is rapid and suitable for organisms with distinctive physical features, such as fungal spores, protozoa, and helminths. However, the sensitivity of this method (i.e., its ability to detect small numbers of microorganisms) may be low (Chapin, 1995; Lawrence et al., 1997). Microorganisms in some samples can be concentrated by centrifugation or filtration to increase sensitivity. The specificity, or accuracy, of organism identification by microscopic examination is limited for many bacteria and yeasts without the use of stains (e.g., dyes and fluorescent-antibody stains) or gene probes.

Many infectious agents can be grown in the laboratory and identified using microscopy, bioassays, or immunoassays (ASM, 1995b; Payment, 1997; Tanner, 1997; Warren et al., 1997). Agent detection by *culture isolation* can be very sensitive and subsequent identification of a cultured microorganism can be very precise. Unfortunately, test results may not be available until several days or weeks after sample collection. Culturing is routine for many clinical specimens and environmental samples when looking for bacterial or fungal pathogens. However, laboratory growth is not as often attempted to detect viruses, protozoa, or helminths. Any body fluid likely to contain a suspected infectious agent should be cultured. If a worker has a fever, a blood culture should be obtained as soon as possible. Some workplace-acquired organisms (e.g., *Brucella* spp., *Bordetella pertussis*, and *Leptospira* spp.) can usually be isolated only during the first days of illness. Only culture isolation requires that microorganisms be able to multiply under laboratory conditions for detection. The other diagnostic tests will generally detect both live and dead microorganisms as well as active and inactive viruses.

Antibody tests may be for a specific antibody class (e.g., IgM or IgG), or for total antibodies, not distinguishing among them (Herrmann, 1995). A *titer* is the highest serum dilution at which a test clearly detects an antibody. Dilutions are often done in twofold steps (e.g., 1:2, 1:4, 1:8, and so forth). A test may be considered positive only if it exceeds a specific titer (e.g., 1:64 or 1:128), the value varying from test to test (APHA, 1995; ASM, 1995b, 1996). These cutpoints are needed because nonspecific low-level reactivity may occur; that is, reactivity seen at low serum dilutions may disappear at higher dilutions. Also, judgement is involved in determining titers.

A person who previously tested negative on a serologic examination who later tests positive is said to have *seroconverted*. A significant rise in titer, usually a four-fold or greater increase, between two blood samples is evidence of recent infection. A *baseline* specimen (one collected before exposure) or an *acute* specimen (one collected ≤7 days since exposure or illness onset) should be tested at the same time as a *convalescent* specimen (one collected 3 to 4 weeks after exposure or illness onset). A positive test for specific IgM antibody (e.g., IgM anti-hepatitis A virus, IgM anti-HAV), is also indicative of recent infection. However, IgM antibodies to some agents may persist for several months. An IgG or total antibody titer that is positive acutely but does not rise at least fourfold in a convalescent specimen is consistent with an infection in the distant past (>3 or 4 months). A single, positive convalescent titer for IgG or total antibody that is very high (e.g., >1:1024) may reflect recent infection. However, all that can often be concluded from a single positive test is that the person has been infected, but not when this occurred.

The diagnostic sensitivity and specificity of antibody tests vary among infectious diseases and are generally quite high. Analytic tests that detect low antibody concentrations are described as sensitive. Specificity describes the degree to which an antibody binds a target antigen and not others. False-positive antibody tests may occur due to sample contamination or if an antibody cross reacts with a related antigen on another organism or tissue. Occasionally, medications or drugs can cause the formation of reactive antibodies, also leading to false-positive antibody reactions. Such cross-reactivity decreases the value of a test for diagnostic purposes.

A test's predictive value depends not only on test sensitivity and specificity but also on the likelihood (prior probability) that infection is present. A significant number of false-positive results are seen with some diagnostic tests used on extremely large numbers of persons (e.g., tests for HIV and HCV infection). *Confirmatory tests* are required for all positive screening tests with these diseases because of the nontrivial consequences of diagnosing them. Confirmatory tests are particularly important for individuals with a low risk of infection (e.g., volunteer blood donors). A positive initial HIV or HCV test in a person with no risk factors for infection with these agents is often a false positive and will not be confirmed with a more definitive test (in this situation, the predictive value of the positive result was low). However, a positive initial antibody test in a person with multiple risk factors for infection (e.g., injecting-drug use, multiple sexual partners, or blood transfusions that predated screening for these viruses) is likely to be due to infection and generally will be confirmed (in this case, the predictive value was high).

There are a number of different methods to test for antibodies, including neutralization, agglutination, precipitation, complement fixation, immunofluorescence, and immunoassay (ASM, 1995b; Herrmann, 1995). Although more than one method may be available for any agent, a clinical laboratory will probably use only one test and not always the most sensitive or specific one. The results of antibody tests performed with nonstandardized antigens may be difficult to interpret (e.g., tests for *Legionella* spp. other than *Legionella pneumophila* and *L. pneumophila* serogroups other than 1 to 6). Reference laboratories (e.g., commercial laboratories and local, state, or federal public health laboratories) often can confirm test results and may offer additional tests not ordinarily available.

Antigen tests can also be used to diagnose infection and are run similarly to tests for antibody. However, these procedures offer the advantage of testing directly for the presence of an infectious agent (as do microscopic examination and culture isolation) or a microbial toxin. Direct detection of antigen in clinical specimens is particularly effective for identification of respiratory viral infections and permits diagnosis within hours rather than days or weeks. Antigen tests are positive as soon as an organism is present and, therefore, will be positive before most antibody tests. Unfortunately, antigen tests may be relatively insensitive, the availability of immunoreagents is limited, and the tests do not distinguish viable from nonviable organisms. For example, antigen from *Legionella* spp. can be detected in the urine of infected persons early in acute illness (weeks before antibodies develop). Antigens can also be detected in spite of early antibiotic treatment, which may result in negative cultures and poor antibody development. This test has a sensitivity of ≥80 percent and very high specificity. Unfortunately, a urine antigen test is available only for *L. pneumophila* serogroup 1.

There are a number of *molecular methods* to detect the presence of infectious agents. By far, the most common technique is nucleic acid amplification through *polymerase chain reaction* (PCR), which mimics natural DNA replication (Podzorski and Persing, 1995). Double-stranded microbial DNA is repeatedly separated into single-stranded DNA templates using heat. The templates are reacted with (a) primers that flank a target DNA sequence (one that is more or less specific to the microorganism of interest); (b) nucleotides (the building blocks needed to copy the target sequence); and (c) DNA polymerase (an enzyme that facilitates the copying). Repeated separation and replication generate an exponential increase in the number of target sequences to amounts that are easily detected. PCR techniques may be able to detect fewer numbers of organisms than culture methods (i.e., they are more sensitive) and can often be performed more rapidly. In addition, PCR-based genomic sequence analysis has been used to detect pathogens that cannot currently be recovered by culture techniques. For example, PCR was used in the initial identification of the viruses causing hantavirus pulmonary syndrome. However, the risk of false-positive results (from sample contamination), test failure on environmental samples (due to interfering compounds in the samples), and the additional complexity and cost involved have limited the availability of PCR to research laboratories.

Past infection with some agents can be detected through skin testing. *Intradermal skin tests* detect a cell-mediated delayed hypersensitivity response to an antigen. To perform this test, a small amount of test solution is injected between layers of the skin (usually on the forearm) with a fine hypodermic needle. The injection site is marked and examined 48 to 72 hours later for evidence of a raised, hardened (indurated) area. The diameter of the raised area may be used to judge the strength of a person's response and to categorize the results (i.e., a person may be described as "skin-test positive" or "skin-test negative"). Factors other than degree of cell-mediated immunity can affect a person's skin test response. *Anergy* describes the false-negative reaction of individuals unable to mount a cellular response due to immunosuppression. Such a condition can be the result of medication (corticosteroids or cancer chemotherapy), overwhelming infection, or advanced age. When suspected, anergy can be detected by demonstrating a person's inability to react to agents for which humans are universally responsive (e.g., mumps, trichophyton, or candida antigens). *Two-step testing* is a form of baseline skin testing used to identify a boosted skin test reaction from that of a new infection. The procedure involves placing a second skin test 1 to 3 weeks after an initial negative test. Boosting is seen in infected persons whose sensitivity has waned since initial exposure but is raised and becomes detectable upon subsequent testing. Repeated skin tests are used to track the status of skin-test-negative workers and to identify recent infections. Skin testing does not lead to the development of antibodies in uninfected persons who are repeatedly tested. Skin tests are most often used for workers potentially exposed to tuberculosis, but may also be useful to track exposure to the agents of coccidioidomycosis, paracoccidioidomycosis, histoplasmosis, leishmaniasis, and leprosy.

9.2.4 Infectious Disease Surveillance

Infectious disease surveillance involves the epidemiologic collection and analysis of medical data about a group of workers. Surveillance is used to measure the occurrence of illness as well as to identify changes in trends or distributions of cases. Infectious disease surveillance may be used to direct investigative and control (preventive) measures. The information collected may not necessarily benefit individual workers, in contrast to screening activities such as those described above. However, a group of workers may benefit from surveillance through resultant prevention of future disease or injury (Ehrenberg and Frumkin, 1995). Chapter 26 discusses the use of injury and illness records for surveillance.

Strategies for evaluating and surveying occupational infections may include (a) identifying the potential presence of infectious organisms; (b) establishing workers' baseline (pre-exposure) status; (c) identifying susceptible workers (i.e., those at risk because of an absence of immunity); (d) identifying workers with conditions that may increase their risk of infection or disease; (e) diagnosing occupational infections in individual workers, characterizing epidemiologic features of infections (e.g., locations, times, persons, and activities involved), and verifying a work association; (e) comparing current attack rates with usual baseline rates; and (f) checking the effectiveness of existing control measures and identifying workplace operations and sites needing better controls. Surveillance for certain zoonotic infectious diseases may also be conducted in animals in workplaces. Animal surveillance programs may monitor animals that are part of a process (e.g., farm animals, animals at abattoirs, and laboratory animals) as well as those incidentally present in a workplace (e.g., wild rodents).

9.2.4.1 Infectious Disease Investigation and Reporting Individual workers may become infected in isolated events, or more than one worker may be involved in an episode. An outbreak of an infectious disease (i.e., an *epidemic* if in humans, an *epizootic* if in animals) is a greater-than-expected number of cases of a specific infection. The expected case number is determined relative to (a) an endemic infection rate (i.e., the constant or usual prevalence for a given area, if applicable); or (b) the baseline rate for a workplace or occupation. A single infection is usually not considered an outbreak.

The first recognized case in an initial outbreak investigation is termed the *index case*, the case that brought the outbreak to someone's attention. However, an investigation may uncover earlier infections; that is, an index case may not have been the first or *primary case* in an outbreak. There may be more than one primary case in common-source outbreaks, such as foodborne infections. Subsequent infections contracted by exposure to a primary case are termed *secondary* and *tertiary* cases, and so forth. Subsequent infections with agents transmitted from person to person may be found among co-workers, customers, and clients of cases along with family members and other contacts.

All states and territories in the United States participate in a national morbidity reporting system. The CDC establishes a list of infectious diseases and related conditions reportable by physicians and other healthcare providers (CDC, 1996f). CDC publishes data on the incidence of notifiable diseases in the *Morbidity and Mortality Weekly Report* (MMWR). MMWR is listed under Publications, Products, and Subscription Services at the CDC website (with a searchable index at http://www.cdc.gov/cgi-bin/mmwrsearch.pl/). Information on rates of notifiable infections can be obtained from CDC publications and from state epidemiologists. Table 9.9 lists the 52 infectious diseases notifiable in the United States as of December, 1996. Table 9.10 lists the U.S. incidence of selected reportable diseases from 1987 to 1996.

9.2.4.2 Notifiable Diseases and Case Definitions *Notifiable diseases* are those for which regular, frequent, and timely information on individual cases is considered necessary for prevention and control (CDC, 1996f). The authority to require case notification resides in respective state legislatures. State health departments receive these reports and provide the data to CDC. States vary in the following: how the authority is enumerated (e.g., by statute or regulation), conditions and diseases to be reported (states and local authorities can make additional diseases or conditions reportable), time frames for reporting, agencies receiving reports, persons required to report, and conditions under which reports are required. In many states, healthcare providers are required or encouraged to report diseases directly to local rather than state health department. This is because investigation and intervention is often provided at the local level. Increasingly, clinical laboratories report diagnosis of some diseases (e.g., tuberculosis) directly, often electronically, to a health department. Occupational infections may also

TABLE 9.9 Nationally Notifiable Infectious Diseases in the U.S.—1997[a] (not all of the following diseases are seen as work-related infections)

Anthrax	Measles (Rubeola)
Botulism	Mumps
Brucellosis	*Mycobacterium leprae* (leprosy, Hansens disease)
Chlamydia psittaci (psittacosis)	*Mycobacterium tuberculosis* (tuberculosis)[(7)]
Chlamydia trachomitis (gential)[(1)]	*Neisseria gonorrhoeae* (genital)[(2)]
Cholera	*Neisseria meningitidis* (meningococcal disease)
Coccidioidomycosis	Pertussis
Cryptosporidiosis	Plague
Diphtheria	Poliomyelitis (paralytic)
Encephalitis, California	Rabies (animal)
Encephalitis, easter equine	Rabies (human)
Encephalitis, St. Louis	Rocky Mountain spotted fever
Encephalitis, western equine	Rubella (congenital)
Escherichia coli O157:H7	Rubella (German measles)
Haemophilus ducreyi (chancroid)	Salmonellosis[(4)]
Haemophilus influenzae (invasive)	Shigellosis[(6)]
Hantavirus pulmonary syndrome	Streptococcal disease (group A, invasive)
Hemolytic uremic syndrome (post-diarrheal)	*Streptococcus pneumoniae* (drug resistant)
Hepatitis A[(5)]	Syphilis (congenital)
Heaptitis B[(10)]	Syphilis (all other stages)[(8)]
Hepatitis, C/non-A, non-B	Tetanus
Human immunodeficiency virus infection (AIDS)[(3)]	Toxic shock syndrome (staphylococcal)
Human immunodeficiency virus infection (pediatric)	Toxic shock syndrome (streptococcal)
Legionellosis	Trichinosis
Lyme disease[(9)]	Typhoid fever
Malaria	Yellow fever

[a]The ten most common infections noted by superscript ranking (ASM, 1996; CDC, 1996f, 1997a).

be reportable to state or federal OSHA programs (see Section 9.2.1.3) and to facility licensing agencies, particularly in the case of healthcare facilities.

Case definitions are used to decide when an illness qualifies for a particular diagnosis and should be reported (CDC, 1997e). Case definitions are useful in occupational surveillance and outbreak investigations and are available at http://www.cdc.gov/epo/mmwr/other/case_def. A case of infectious disease is described as laboratory confirmed if one or more listed diagnostic tests are positive. Case definitions are included for some infectious conditions that are no longer considered nationally notifiable or which may become so in the future [i.e., amebiasis, aseptic meningitis, bacterial meningitis, *Campylobacter* infection, *Cyclospora* infection, dengue fever, ehrlichiosis, genital herpes (HSV), genital warts, giardiasis, granuloma inguinale, leptospirosis, listeriosis, lymphogranuloma venereum, mucopurulent cervicitis, nongonococcal urethritis, pelvic inflammatory disease, rheumatic fever, tularemia, and varicella].

A case may be described as clinically compatible if the syndrome was generally similar to a disease but could not be confirmed by a laboratory test. Case definitions for surveillance work are deliberately biased toward increased sensitivity (to identify all

TABLE 9.10 Incidence of Reportable Diseases of Occupational Significance, 1987–1996 (total cases reported, not only work-related cases)

	Number of cases reported in year									
Disease	1996	1995	1994	1993	1992	1991	1990	1989	1988	1987
AIDS	66,885	71,547	78,279	103,533	45,472	43,672	41,595	33,722	31,001	21,070
Amebiasis	[a]	[a]	2,983	2,970	2,942	2,989	3,328	3,217	2,860	3,123
Anthrax	—	—	—	—	1	—	—	—	2	1
Brucellosis	112	98	119	120	105	104	85	95	96	129
Hepatitis A	31,032	31,582	29,796	24,238	23,112	24,379	31,441	35,821	28,507	25,280
Hepatitis B	10,637	10,805	12,517	13,361	16,126	18,003	21,102	23,419	23,177	25,916
Legionellosis	1,198	1,241	1,615	1,280	1,339	1,317	1,370	1,190	1,085	1,038
Leptospirosis	[a]	[a]	38	51	54	58	77	93	54	43
Lyme disease	16,455	11,700	13,043	8,257	9,895	9,465	—	Not previously nationally notifiable		
Malaria	1,800	1,419	1,229	1,411	1,087	1,278	1,292	1,277	1,099	944
Pertussis	7,796	5,137	4,617	6,586	4,083	2,719	4,570	4,157	3,450	2,823
Plague	5	9	17	10	13	11	2	4	15	12
Psittacosis	42	64	38	60	92	94	113	116	114	98
Rabies	3	5	6	3	1	3	1	1	—	1
Rocky Mt. spotted fever	831	590	465	456	502	628	651	623	609	604
Tetanus	36	41	51	48	45	57	64	53	53	48
Tuberculosis	21,337	22,860	24,361	25,313	26,673	26,283	25,701	23,495	22,436	22,517
Tularemia	[a]	[a]	96	132	159	193	152	152	201	214

[a] No longer nationally notifiable (some states continue to collect case reports).

Source: CDC, 1997a.

possible cases), perhaps at the expense of specificity (i.e., some false-positive reporting may occur). Therefore, surveillance case definitions should be used with caution to interpret an individual worker's diagnosis. However, surveillance data most often are used to monitor trends. Thus a relatively high false-positive reporting rate is of little consequence if the bias remains stable over time.

9.2.5 Inspections and Environmental Testing

Infectious diseases are primarily monitored in workplaces through serologic and skin testing of workers and disease surveillance rather than by environmental sampling for infectious agents. Investigators use various criteria to establish cause and effect relationships between workplace exposures and infections. Environmental testing may help establish some of these points and confirm or fail to confirm the work relatedness of an infection. For example, (a) an infectious agent must be known, or strongly suspected, to be or have been present in a workplace; (b) there must be a plausible means of worker exposure; (c) an infected worker must be shown, or strongly suspected, to have been exposed in the workplace; and (d) potential exposures (places and routes) other than the workplace must be ruled out (Burge, 1989, 1990).

A work association is easier to accept for a disease that workers are unlikely to encounter outside their jobs (e.g., brucellosis) than for infections that are fairly common in the community (e.g., tuberculosis, HAV infection, influenza, or the common cold). On the other hand, detecting an infectious agent in a workplace by environmental sampling does not necessarily mean that workers are or were at risk. Rather, risk depends on the agent, the concentration found, and possible exposure routes. For example, many of the infectious agents in Table 9.2c are common in indoor and outdoor reservoirs, but these agents do not necessarily present an occupational hazard when found. Rather than test for exposure to infectious agents, health and safety professionals may assess worker safety indirectly by auditing worker training, work practices, and worker supervision. Evaluation of control equipment is also an indirect assessment of safety (e.g., checks of containment devices, ventilation systems, and PPE performance).

Healthcare professionals handle most infectious disease evaluations because they consist of medical examinations and clinical laboratory tests. However, health and safety professionals may need to understand the interpretation of medical findings to help healthcare personnel design environmental inspection and testing programs and implement appropriate controls. For example, ≥2 confirmed Legionnaires' disease cases in workers without likely nonworkplace exposures should trigger an inspection for possible legionella sources at a jobsite. However, misdiagnosis of Legionnaires' disease, due to misinterpretation of a serologic test, is not uncommon. This is understandable because laypersons (and even physicians) are unfamiliar with these tests, and laboratories often do not interpret the findings they report. As many as 20 percent of the general population in certain areas have serum antibodies to legionella reflecting prior exposure (usually unrecognized) or a nonspecific cross-reaction. Therefore, in any workplace, someone may test positive for legionella antibody even though their exposure was not job related and possibly occurred years in the past. Environmental sampling for legionellae often finds the bacteria because they are frequently present in water supplies, often without causing apparent illness (Fields, 1997). Thus an incorrect interpretation of a serum test could lead to environmental sampling that finds legionella, which in turn could lead to expensive (and likely unnecessary) remediation. Health and safety professionals may be placed in a difficult situation if they have concerns about or question a medical diagnosis or laboratory finding. Local public health authorities and infectious disease specialists may be able to provide advice and consultation in such situations.

9.2.5.1 Inspections In some workplaces, walkthrough inspections are conducted routinely to identify infectious agent reservoirs and dissemination routes. However, in most workplaces, inspections take place following an exposure episode (e.g., a laboratory accident) or a report of a potentially work-related infection (e.g., a confirmed case of plague in a field researcher or utility worker). Initial walkthrough inspections in these cases are similar to inspections conducted to identify other workplace hazards or evaluate indoor environmental quality (Chapter 20). During a walkthrough, health and safety professionals (a) identify potential reservoirs or sources of infectious agents as well as possible exposure routes; and (b) formulate plans for further investigation (including environmental testing if indicated) and for remedial actions on noted problems. Inspections may identify potentially hazardous situations that merit mitigation even if these situations have not been linked to worker infections. For example, rodent infestation, poor equipment maintenance, or improper waste handling should be noted even if not related to the issues that prompted the inspection. Management and employee representatives should be included in these inspections and subsequent deliberations.

9.2.5.2 Environmental Testing Previous sections have described workplace monitoring (through medical surveillance) for the *effects* of infectious agents to identify their presence and measure their impact. This section briefly describes sampling for the *presence* of infectious agents. Air, surface, liquid, and bulk material sampling to detect, quantify, and identify infectious agents in workplaces is a special case of environmental testing. Environmental sampling in hospitals and laboratories (where infectious agents are known to be present) formerly was routine. However, even in these facilities such testing has been almost completely abandoned. Environmental sampling is relatively rare because testing is often expensive and time consuming when conducted correctly, for example, many samples may be needed to detect infectious agents that are present only sporadically or in low concentrations. In addition, certain organisms can be expected to be found in some types of samples, for example, *Aspergillus fumigatus* at a compost operation. Therefore, air concentration measurements or other test results cannot be interpreted without a database of expected concentrations and information on infectious doses.

Investigators can often obtain the information they need by means other than environmental sampling. For example, it may be more productive to perform periodic visual inspections, measure water temperatures and biocide concentrations, and initiate an employee medical surveillance program than it is to sample for infectious agents in a workplace. Environmental testing most often is undertaken (a) to identify or test potential sources or reservoirs of pathogenic microorganisms (e.g., raw materials, manufacturing supplies, or cooling fluids); (b) to determine the ability of certain organisms to survive in a given environment; (c) to test the efficacy of a cleaning or disinfecting method; or (d) to check hypotheses generated by epidemiologic or medical surveillance.

Environmental sampling for infectious agents generally focuses on one or a few microorganisms (Table 9.2), which simplifies the selection of suitable collection and detection techniques. Culture methods are frequently used on environmental samples because it is assumed that agents must be viable or active (able to multiply) to cause infection and, therefore, should be culturable (able to grow on laboratory media). It is also necessary to culture organisms to compare environmental and clinical (worker or patient) isolates to help determine if an infection was work related or a cluster of infections originated from one source. Detection methods other than culture are more convenient or appropriate for some organisms, for example, immunologic methods or methods based on detection of nucleic acids (see Section 9.2.3).

Sample collection methods should be selected in consultation with an experienced microbiologist who can anticipate how sample collection, transport, and storage may affect test outcomes. For example, *air samples* to identify airborne infectious agents are generally collected directly by impaction onto culture media (agar plates) or by impingement into a suitable liquid collection medium (ACGIH, 1989; Burge, 1995; Macher et al., 1995a; AIHA, 1996; Buttner et al., 1997). Filters are used less often because infectious agent isolation may be low from filters due to drying and loss of cell viability. However, filters may be compatible with assays not based on organism culturability. Air sampling has proven useful in hospitals to test the effectiveness of physical barriers separating construction sites from patient areas and to determine when patient rooms closed for construction were sufficiently clean to return patients.

Surfaces may be sampled by washing a measured area with a swab wetted with sterile water or other liquid or by using special contact plates containing agar medium. Surfaces occasionally tested for the presence of infectious agents include work benches, instruments and tools, and human and animal skin. *Personal samples* (specimens from the hands, skin, nose, or other body sites) are occasionally collected from healthcare workers to identify the person or persons who could have transmitted an infectious agent to a patient. However, this type of testing should only be done when epidemiologic data links certain workers with infections.

Liquid and *bulk material samples* from potential sources of airborne infectious agents (e.g., cooling-tower water, metalworking fluid, animal droppings, or soil) are often more useful than air samples because the agents are concentrated in the source material, increasing the chances of detecting them (Macher et al, 1995b; Harding et al., 1996; Fields, 1997). Such samples are also appropriate for identifying sources of infectious agents transmitted by contact (e.g., amebae in eye-wash stations) and by ingestion (e.g., foodborne pathogens in consumable items).

Microbiologists experienced in handling environmental samples can advise health and safety professionals about appropriate sample collection methods and can choose suitable laboratory techniques to obtain the information required from the samples. All samples should be protected from contamination and extreme temperatures to ensure that the samples are still representative of the original source when they reach a laboratory. Other texts discuss environmental microbial sampling outdoors as well as in laboratories, industrial workplaces, animal houses, and problem buildings (ACGIH, 1989; Morey et al. 1990; Nevalainen et al., 1993; Lighthart and Mohr, 1994; Rylander and Jacobs, 1994; AIHA, 1995, 1996; Burge, 1995; Cox and Wathes, 1995; Macher, 1995a,b; Harding et al., 1996; ASM, 1997).

9.3 INFECTIOUS DISEASE PREVENTION

Preventing infectious diseases in the workplace can decrease absenteeism and the costs associated with disability, sick leave, and health insurance even if the primary source of infection is nonoccupational (Gerberding and Holmes, 1994). Infectious disease prevention should be part of a facility's health and safety program and begins with identifying possible infectious agents along with their reservoirs, modes of transmission, and risk factors (Tables 9.1 and 9.2). The American Public Health Association (APHA, 1995) publishes a pocket reference, *Control of Communicable Diseases Manual*, that lists infectious diseases alphabetically and serves as a comprehensive but concise general reference. The manual provides information in each of the following categories: disease identification, infectious agent, occurrence, reservoir, transmission mode, incubation period, communicable period, susceptibility and control, and control methods. Each section on control measures identifies applicable preventive measures; recommended control of patients, contacts, and the immediate environment; epidemic measures; disaster implications; and international measures.

Some of the primary tools for preventing work-related infections include immunization, standard (universal) precautions for blood and body fluids, and good hygiene (e.g., frequent handwashing, wound covering, and proper food storage and handling). Prevention can be approached through the three classic strategies shown in Table 9.11 (i.e., primary, secondary, and tertiary prevention). Respectively, these aim to prevent exposure, intervene after exposure or when signs or symptoms of infection are first detected, and limit the consequences of clinical illness once it occurs. Occupational health and safety professionals can use their training in recognizing sources, pathways, and receptors and the approach outlined in Section 9.1.4 to identify where and how transmis-

TABLE 9.11 Infectious Disease Prevention Strategies

Strategy	Definition	Example
Primary prevention	Prevent exposure	Preplacement examinations and immunization of susceptible workers Worker education and training Engineering controls: Substitute safer materials or processes—choose uncontaminated materials or disinfect materials before handling Isolate or enclose sources or processes Ventilation—local exhaust and dilution ventilation Inspections and environmental monitoring Personal protective equipment—hand protection, skin and clothing protection, eye and face protection, respiratory protection
Secondary prevention	Intervene postexposure, postinjury, or when signs or symptoms of infection are detected	Periodic medical examinations and surveillance Postexposure or -injury prophylaxis or treatment
Tertiary prevention	Limit consequences of clinical illness once it occurs	Medical treatment Work restrictions Worker removal

sion may be prevented or interrupted. Infectious agents may move from sources (e.g., humans, animals, or the environment) to workers and back again. Biosafety practices in laboratories and healthcare settings are often designed to protect both workers and the materials they handle (e.g., laboratory cultures and patients).

Laboratories. Microorganisms are not equally hazardous, and the CDC Office of Biosafety has classified agents on the basis of hazard (USDHHS, 1994). Agents in microbiology and biomedical laboratories are handled at one of four *biosafety levels* (BSLs). The necessary BSL is based on the risk associated with an agent, the volume and concentration of infectious material being handled, and the laboratory activities being conducted (USDHHS, 1993). At the extremes are BSL-1 agents (the least hazardous with the fewest restrictions on their handling) and BSL-4 agents (the most hazardous, which are only handled in high-containment facilities). Decisions on containment levels for laboratory work at BSL-1 through BSL-3 can be made at the institutional level. Very few facilities have BSL-4 capability, and those that do need to consult experienced biosafety experts.

The BSL classification system applies primarily to work in laboratories. Transmission modes may differ in laboratory settings (compared to other workplaces), and higher organism concentrations may be encountered there. Nevertheless, the recommended BSLs reflect the relative hazards of infectious agents. Health and safety professionals can use an agent's classification or rating as a starting point to assess its hazard. For each BSL, recommendations are given for facility design, safety equipment, and laboratory practices and techniques. Determining an agent's BSL can help health and safety professionals identify appropriate precautions (a) to prevent routine exposure (e.g., the need to avoid or contain aerosols, to prevent ingestion, to prevent skin or mucous-membrane exposure, or to protect workers from arthropod vectors); and (b) to handle emergency spills or releases.

The Office of Biosafety of Health Canada (Laboratory Centre for Disease Control, Ottawa, Ontario) has prepared Material Safety Data Sheets (MSDSs) for over 150 infectious substances. Each MSDS provides basic facts about an agent and information on the following: (a) the microorganism's antimicrobial susceptibility, susceptibility to disinfectants and physical inactivation, and ability to survive outside a host; (b) recommended first aid or treatment, immunizations, and prophylaxis; (c) recommended precautions (e.g., containment requirements, suitable protective clothing, and other precautions); and (d) information on agent handling for addressing spills, disposal, and storage.

Health Care. Infection control in healthcare settings involves identifying the type of contact that workers may have with potentially infectious body substances (e.g., blood and body fluids) and contaminated materials (e.g., wound dressings and linens) (Gerberding and Holmes, 1994). In 1996, hospital precautions to protect both patients and workers from infectious agent transmission were revised (CDC, 1996a). Table 9.12 summarizes these precautions, showing the categories of infection control practices, diseases or conditions to which they apply, and the specific precautions recommended for each.

Standard precautions are the equivalent of what have been called universal precautions (Table 9.12). Compliance with these precautions and the OSHA Bloodborne Pathogens Standard (Section 9.2.1.3) will reduce workers' risk of exposure to bloodborne agents. The effective application of specific precautions to patients and clients infected with organisms transmitted by modes other than blood should also reduce work-

TABLE 9.12 Infection Control Precautions for Healthcare Settings

Type of precaution Diseases or conditions requiring precaution	Summary of precaution
Standard (universal) precautions All patients	Handwashing: after touching blood or other body fluid whether or not gloves worn; between patient contacts or certain tasks and procedures Gloves: when touching blood or other body fluids Other protective equipment: as indicated (when splash or splatter anticipated) Environmental control: adequate procedures for routine care, cleaning, and disinfection Occupational health and bloodborne pathogens: needles and other sharp instruments are not manipulated after use and are disposed of immediately in puncture-resistant containers Patient management: place in private rooms patients who contaminate environment or do not assist in hygiene
Airborne precautions[a] Tuberculosis, measles, varicella (chickenpox and disseminated zoster)	Patient placement: private room under negative pressure, 6 to 12 air changes per hour, discharge air to outdoors or filter, door closed, patient in room Respiratory protection: whenever entering room (tuberculosis isolation) or if worker not immune (measles, varicella exposures) Patient transport: limit movement, mask patient
Droplet precautions[a] Bacterial: invasive *Haemophilus influenzae* type b, invasive *Neisseria meningitidis*, pertussis, pneumonic plague, strepto-coccal pharyngitis, pneumonia, scarlet fever in infants or children Viral: adenovirus, influenza, mumps, parvovirus B19 infection, rubella	Patient placement: private room if possible Mask: when working within 1 m (3 ft) of patient Patient transport: limit movement, mask patient
Contact precautions[a] Infections or colonizations with multidrug-resistant bacteria Enteric infections with a low infectious dose or prolonged environmental survival, including *Clostridium difficile*, enterohemorrhagic	Patient placement: private room if possible Gloves and handwashing: gloves when entering room, change as indicated, remove before leaving environment, wash hands immediately with antimicrobial agent, leave room without contaminating hands Gown: when entering if substantial contact with patient or environmental surfaces anticipated

TABLE 9.12 (*Continued*)

Type of precaution Diseases or conditions requiring precaution	Summary of precaution
Contact precautions[a] (*Continued*)	
Escherichia coli O157:H7, *Shigella* spp., hepatitis A virus, rotavirus infection	Patient transport: limit movement, prevent contamination
Respiratory syncytial virus, parainfluenza virus, enteroviral infections in infants and children	
Highly contagious skin infections on dry skin: herpes simplex virus (neonatal or mucocutaneous), impetigo, pediculosis, scabies, zoster, major abscesses, cellulitis	
Viral/hemorrhagic conjunctivitis	

[a]In addition to Standard Precautions.

Source: CDC, 1996a.

ers' risks (Tables 9.2 and 9.12). Compliance with three practices (i.e., handwashing, vaccination, and appropriate isolation of infected patients) has been found effective at controlling infectious disease transmission in hospital settings (Sepkowitz, 1996). Guidelines for specific healthcare settings are also available, such as dentistry (CDC, 1993b; Gomez and Gomez, 1993), as are precautions for other occupations that involve person-to-person contact, such as child care (APHA/AAP, 1992; Osterholm, 1994), personal service work (USDHHS, 1990), and public safety workers (ARC, 1993).

9.3.1 Immunizations

Immunization against infectious agents should be provided to workers when such protection is available, safe, effective, and the benefits clearly exceed the risks (e.g., side effects such as local or systemic reactions). Immunization generally is preferred over engineering controls and PPE or is used in addition to these measures. Immunity acquired as a response to vaccination or natural infection is called *active* immunity. That acquired by inoculation with sera containing specific protective antibodies is called *passive* immunity (see Section 9.3.5). Active immunity may last from several years to a lifetime, whereas passive immunity may last for only a few days or months. The U.S. Public Health Service provides recommendations on immunizations appropriate for healthcare workers as well as healthy and immunodeficient adults in the United States and its territories (Table 9.13) (CDC, 1991, 1993a, 1994b; USDHHS, 1993). The WHO may have different recommendations for regions other than North America (see Section 9.0.5).

One of the reasons for low compliance with recommended immunizations among workers is their fear of adverse events. In fact, vaccine safety has improved in recent years, and the occurrence of serious side effects for the most widely recommended vaccines is quite low (CDC, 1996d). Possible contraindications to immunization include the taking of immunosuppressive agents, coincidental severe illness, and pregnancy. Recommendations for giving workers less efficacious vaccines, those associated with high rates of local or systemic reactions, and those that produce increasingly severe

TABLE 9.13 Immunoprophylaxis for Workers Exposed to Infectious Agents[a]

Vaccine or immunotherapy	Occupations
Anthrax vaccine	Selected occupations, (e.g., those who work with imported animal hides, fur, wool, hair, or bristles) and military personnel potentially exposed to biological warfare/terrorism agents PEP:[b] *Bacillus anthracis* exposure
Diphtheria vaccine	See tetanus/diphtheria vaccine
Hepatitis A immune globulin	PEP: HAV exposure Pre-exposure prophylaxis: travellers to highly endemic areas (Africa, the Middle East, Asia, and Central South America
Hepatitis B vaccine and immune globulin	Vaccine for healthcare workers and others who handle human blood or body fluids (e.g., laboratory workers, animal handlers, public safety workers, barbers, tattoosits, cosmetologists, and staff of institutions for the developmentally disabled) PEP: immune globulin for HBV exposure
Influenza vaccine	Adults with cardiopulmonary disease; workers with public contact (e.g., community service workers, healthcare workers, military personnel, office workers, public safety workers, teachers); workers older than 65 years; some healthy workers to minimize worktime lost time due to illness
Lyme disease	Vaccine to be licensed in 1998; recommendations on what workers should receive vaccine not yet available
Measles vaccine and immune globulin	Vaccine for all susceptible adults, especially healthcare and childcare workers and travelers PEP: measles virus exposure; vaccine if $\leq$3 days since exposure, immune globulin if $\leq$6 days since exposure or vaccine contraindicated
Mumps vaccine	All susceptible adults, especially healthcare and childcare workers and travelers
Pertussis vaccine	PEP: *Bordetella pertussis* exposure; vaccine boosters during outbreaks
Plague vaccine	Veterinary-care workers, animal handlers, and others with high risk of exposure to possibly infected animals (e.g., selected field personnel)
Polio vaccine	Not routinely recommended for unimmunized adults (>18 years) unless traveling to endemic areas or potentially exposed to the polio virus (e.g., healthcare and laboratory workers)
Rabies vaccine and immune globulin	Vaccine for veterinary-care workers and others with high risk of exposure (e.g., animal handlers, laboratory workers, and selected field personnel): pre-exposure immunization and serologic testing with administration of booster vaccine, if indicated, every 2 years PEP: vaccine and immune globulin for exposure to rabies virus/rabid animal
Rubella vaccine	All susceptible adults, especially healthcare and childcare workers and travelers

TABLE 9.13 ***(Continued)***

Vaccine or immunotherapy	Occupations
Tetanus/diphtheria (Td) vaccine	All adults, booster vaccine every 10 years
Tetanus toxoid and immune globulin	PEP: *Clostridium tetani* exposure/wound exposure; minor, uncontaminated wound, person immunized within past 10 years—no booster; major/contaminated wound, no booster within past 5 years —single injection of tetanus toxoid (preferably Td, see above); no primary immunization series, wound major/contaminated —tetanus toxoid and tetanus immune globulin
Tuberculosis vaccine (BCG)	Not recommended in the U.S. except for healthcare workers in areas where multidrug-resistant tuberculosis is prevalent, a strong likelihood of infection exists, and where comprehensive infection control precautions have failed to prevent TB transmission to healthcare workers
Varicella vaccine and immune globulin	Vaccine for healthcare workers, laboratory workers PEP: immune globulin for VZV exposure for pregnant and immunodeficient workers

[a]See also recommendations in Section 9.0.5, Occupational Travel.
[b]PEP = postexposure prophylaxis.

reactions with repeated use (e.g., cholera, tularemia, and typhoid vaccines) should be carefully considered (CDC, 1993a). Employers should maintain complete records of immunoprophylaxis that workers receive on the basis of occupational requirements or recommendations.

9.3.2 Substitution and Engineering Controls

Substitution and engineering controls are preferred over PPE and administrative controls to prevent exposure to infectious agents just as these control measures are preferred to control other workplace hazards (see also Chapter 15). For example, legionellosis prevention is often addressed by controlling bacterial multiplication in water systems and reducing aerosol exposures. Bacterial multiplication can be minimized by proper water temperature control, equipment cleaning and maintenance, and appropriate biocide use. Aerosol exposures can be reduced by minimizing mist generation and dispersal. Maintenance procedures to decrease the survival and multiplication of *Legionella* spp. in potable-water distribution systems and procedures for cleaning cooling towers and related equipment are available for healthcare facilities (Freije, 1996; CDC, 1997c) and other settings (ASTM, 1996). CDC developed recommendations for maintenance and operation of whirlpool spas following a Legionnaires' disease outbreak on a cruise ship (CDC, 1996g).

9.3.2.1 Substitution: Changing the Material or Process Whenever possible, potentially infectious materials that are part of a work process should be treated before workers handle the items. For example, plant and animal materials used in textile production and in animal-product manufacturing should be selected carefully or disinfected before use (as appropriate for the infectious agents that may be present). Likewise,

TABLE 9.14 Decontaminants and Their Use in Research and Clinical Laboratories[a]

	Use parameter				Effective against					Important characteristics								Applications																							
Decontaminant	Concentration of active ingredients	Temperature (°C)	Relative humidity (%)	Contact time (min)	Lipid viruses	Hydrophilic viruses	Vegetative bacteria	Bacterial spores	*Mycobacterium tuberculosis*	Inactivated by organic matter	Residual	Corrosive	Skin irritant	Eye irritant	Respiratory irritant	Toxic (absorbed or ingested)	Work surface maintenance	Floor maintenance	Floor drains	Biohazard spill, floor surfaces	Safety cabinet surface maintenance	Safety cabinet biohazard spill	Safety cabinet total decontamination	Equipment surfaces	Electrical instruments	Lensed instruments[b]	Equipment total decontamination	Large area & air system decontamination	Water baths	Contaminated glassware	Contaminated instruments	Contaminated liquid—discard	Infectiouse laboratory waste	Contaminated animal bedding	Infected animal carcasses	Books, paper, shoes
Autoclave, 15 lb in^{-2}	Saturated steam	121	—	50–90	+	+	+	+	+	–	–	–	–	–	–	–	–	–	–	–	–	–	–	–	–	–	–	–	–	+	±	+	+	+	±	–
27 lb in^{-2}	Saturated steam	132	—	10–20	+	+	+	+	+	–	–	–	–	–	–	–	–	–	–	–	–	–	–	–	–	–	–	–	–	+	±	+	+	+	+	–
Dry-heat oven	—	160–180	—	180–240	+	+	+	+	+	–	–	–	–	–	–	–	–	–	–	–	–	–	–	–	–	–	–	–	–	±	±	–	–	±	–	–
Incinerator	Heat	649–926	—	10–60	+	+	+	+	+	–	–	–	–	–	–	–	–	–	–	–	–	–	–	–	–	–	–	–	–	–	–	–	+	+	+	–
UV radiation (253.7 nm)[c]	40 $\mu W\ cm^{-2}$	—	—	10–30	+	±	+	–	+	+[c]	–	–	+	+	–	–	–	–	–	–	±	–	–	–	–	–	–	–	–	–	–	–	–	–	–	–
Ethylene oxide	400–800 mg L^{-1}	35–60	30–60	105–240	+	+	+	+	+	–	+	–	+	+	+	+	–	–	–	–	–	–	–	–	+	–	±	–	–	±	±	–	–	–	–	+
Paraformaldehyde (gas)	10.6 g m^{-3}	>23	>60	60–180	+	+	+	+	+	–	+	–	–	+	+	+	–	–	–	–	–	–	+	–	+	–	+	+	–	–	–	–	–	–	–	–
Vaporized hydrogen peroxide	2.4 mg L^{-1}	4–50	<30	8–60	+	+	+	+	+	–	–	+	+	+	+	–	–	–	–	–	–	–	+	–	–	–	+	–	–	–	–	–	–	–	–	–
Quaternary ammonium cpds	0.1–2%	–	–	10–30	+	–	+	–	–	+	–	–	–	–	–	+	+	+	+	±	+	±	–	+	–	–	–	–	±	+	–	–	–	–	–	–
Phenolic compounds	0.2–3%	–	–	10–30	+	±	+	–	+	±	+	+	+	+	±	+	+	+	+	+	–	+	–	+	–	–	–	–	+	+	+	+	+	–	–	–
Chlorine compounds	0.01–5%	–	–	10–30	+	+	+	±	+	+	±	+	+	+	+	+	–	–	+	+	–	+	–	+	–	–	–	–	±	+	+	–	–	–	–	–
Iodophor compounds	0.47%	–	–	10–30	+	±	+	–	+	+	+	+	+	+	–	+	+	–	–	±	+	±	–	+	–	–	–	–	+	+	–	–	–	–	–	–
Ethyl and isopropyl alcohol	70–85%	–	–	10–30	+	±	+	–	+	+	–	–	–	+	–	+	+	–	–	±	+	–	–	+	–	–	–	–	±	–	–	–	–	–	–	–
Formaldehyde (liquid)[d]	4–8%	–	–	10–30	+	+	+	±	+	–	+	–	+	+	+	+	–	–	–	±[d]	–	–	–	±	–	–	–	–	–	±	±	±	±	–	–	–
Glutaraldehyde	2%	–	–	10–600	+	+	+	+	+	–	+	–	+	+	+	+	–	–	–	–	–	–	–	+	–	±	–	–	–	+	+	–	–	–	–	–
Hydrogen peroxide (liquid)	6%	–	–	10–600	+	+	+	+	+	–	+	+	+	+	–	–	–	–	–	–	–	–	–	+	–	+	–	–	–	+	+	–	–	–	–	–

[a] +, Very positive response; ±, less positive response; –, negative response or not applicable.
[b] Contact instrument manufacturer.
[c] UV radiation does not penetrate surface deposits and other materials.
[d] Formaldehyde's pungent and irritating characteristics preclude its use for biohazard spills.

Source: Modified from Vesley and Lauer, 1995; printed with permission.

some medical and laboratory wastes should be sterilized before transport and disposal (USEPA, 1986; USDOL, 1988; Rutala, 1990; Block, 1991; Marchese, 1993; Vesley and Lauer, 1995). Table 9.14 lists decontaminants and their use in research and clinical laboratories (e.g., concentration, temperature, and contact time; effectiveness against various microorganisms; possible interferences; adverse health effects; and applications). These recommendations may also be appropriate for similar applications in settings other than laboratories.

Biocide manufacturers test their products using standard methods of the Association of Official Analytical Chemists (AOAC). This standardized laboratory testing requires specified contact times and temperatures, the use of designated strains of microorganisms, and preparation of the use-concentration of a biocide as specified on the manufacturer's label. All disinfectants, sanitizers, and sterilants must be registered with the USEPA before they can be marketed. However, the agency does not verify a manufacturer's claims about a product. The USEPA maintains lists of registered sterilants (List A), tuberculocides (anti-*M. tuberculosis* agents, List B), and anti-HIV agents (List C). The FDA sets standards for disinfectants and registers products for sterilizing critical devices that penetrate the blood barrier or contact mucous membranes.

Sanitization implies achieving a significant reduction in the number of vegetative environmental microorganisms of public health importance. *Disinfection* implies elimination of many or all pathogenic microorganisms, except bacterial spores. *Sterilization* implies complete elimination or destruction of all microbial life and is the surest means of rendering material noninfectious. Sterility can be achieved by incineration or steam or ethylene oxide autoclaving. Waste materials that may warrant or require special processing or disposal because they may carry a risk of infectious disease transmission include (a) laboratory cultures and stocks; (b) human and animal pathological wastes (e.g., tissues, organs, carcasses and body parts; body fluids and their containers; and discarded bedding and materials saturated with body fluids other than urine); (c) blood and blood products; and (d) sharps such as needles and scalpel blades (Marchese, 1993; USDHHS, 1993; ASM, 1995a; Sewell, 1995; Harding et al., 1996; Turnberg, 1996).

9.3.2.2 Containing Infectious Agents Successful *containment* of infectious agents can simplify worker protection. Methods designed to prevent contact spread of infectious agents from potential sources or reservoirs include the sterile techniques used in laboratories and health care, proper food handling, and safe practices recommended for child care, personal service work, and animal handling. Aerosolization of contaminated materials should also be minimized, for example, the spraying of water and other liquids should be avoided, dust generation should be suppressed, fluids should be handled to avoid drips and splashes, and people should cover their noses and mouths when coughing and sneezing. Operations for which it is difficult to prevent aerosol generation should be performed within *primary containment* devices such as the biological safety cabinets and glove boxes used in laboratories (NSF, 1992; DHHS, 1993). Other examples of primary containment are negative-pressure booths (used in healthcare settings for sputum collection and in machine shops to enclose cutting equipment that requires metalworking fluids) and isolation rooms (used in hospitals to house potentially infectious patients) (CDC, 1994a, 1996a). Primary barriers protect workers in the immediate area of aerosol generation. These containment devices are ventilated and serve as source controls. *Secondary containment* provides another level of protection, principally for persons outside a facility. The room in which a primary containment device is located and the building surrounding a primary barrier are examples of secondary containment.

9.3.2.3 Ventilation Ventilation designs used to control airborne particulate matter generally work equally well to control infectious aerosols. However, ventilation should only be used when substitution, process change, or source isolation or enclosure is not possible or in addition to these control measures. Source control to limit contaminant dispersal and minimize worker exposure can best be achieved through local exhaust ventilation (e.g., a hood, duct, air cleaner, and fan) along with isolation or enclosure (see Section 9.3.2.2). General ventilation is used to dilute and remove contaminated air, control airflow patterns within rooms, and control the direction of air throughout a facility (CDC, 1994a).

Dilution ventilation can reduce the concentration of infectious aerosols that cannot be controlled at their sources, for example, because a source cannot be readily identified, is mobile, or is too large to enclose. Stand-alone or in-room air cleaners, which return air to a space, clean already conditioned room air rather than outdoor air. Portable air cleaners and ceiling- and wall-mounted units have been studied as economical means of increasing dilution ventilation without additional heating or cooling costs (Marier and Nelson, 1993; Rutala et al., 1995; Miller-Leiden et al., 1996). This research and that with respirators confirms that filters remove infectious aerosols as efficiently as they remove other particles with equivalent aerodynamic diameters (Chen et al., 1994; Johnson et al., 1994; Willeke et al., 1996; Brosseau et al., 1997).

Another means of cleaning the air (at least in terms of infectiousness) and providing a form of dilution ventilation is the use of ultraviolet germicidal irradiation (UVGI or UV "light") (Table 9.14). UVGI has a long history of use in laboratories and hospitals for surface and air disinfection. Specially designed UV lamps can be used to disinfect the air in occupied rooms, and UVGI use to prevent tuberculosis transmission has been encouraged (Riley and Nardell, 1989). UVGI may offer an appropriate means of protection against airborne infectious diseases provided that worker eye and skin exposures do not exceed current limits (NIOSH, 1972; Macher, 1993; CDC, 1994a).

9.3.3 Education, Training, and Administrative Controls

Among the administrative controls used to prevent occupational infections are education of workers about infectious diseases and training in good work habits (see also Chapter 14). Additional administrative controls include providing the necessary means and motivation to work safely, supervising workers to ensure implementation of proper practices, and institution of appropriate programs for medical evaluation, job placement, and infectious disease surveillance. One of the means by which employers ensure that workers are aware of biohazards is compliance with the requirement to post warnings at entrances to areas where biohazards are handled. These signs inform workers and visitors of biological hazards and outline routine and emergency precautions required in the area. Section 9.2.1.3 describes recording and reporting requirements, other sanitary and safety regulations, and requirements related to PPE and bloodborne pathogens.

Some OSHA standards mandate workplace *training* (e.g., the bloodborne pathogen and respiratory protection standards) (see Section 9.2.1.3). Standards may specify training content, frequency, recordkeeping requirements, and trainer qualifications. Worker education about infectious diseases should include a description of the biology of the infectious agents that workers may encounter and identification of symptoms that may indicate exposure. Job activities and applicable regulations should be reviewed with workers, and potential sources of infectious materials and exposure routes should be discussed. Acceptable work practices should be described clearly, and proper use of process and control equipment should be explained and demonstrated. Regular reviews and updates should be scheduled, as appropriate or required. Written injury and illness prevention programs are helpful. However, staff must understand their responsibilities, be properly trained to carry out their assignments, know how to implement emergency procedures in case of accidents, and be motivated to work safely (Gershon and Zirkins, 1995).

9.3.4 Personal Protective Equipment

PPE use to prevent infectious diseases is a last line of defense when risks cannot be sufficiently reduced through engineering and other controls (Kuehne et al., 1995) (see also Chapters 16 and 17 and Section 9.2.1). Table 9.15 outlines the types of PPE used to protect workers against infectious agents (i.e., hand protection, skin and clothing protection, eye and face protection, and respiratory protection). Appropriate precautions, other than personal protection, should be used to prevent animal bites and scratches, arthropod bites and stings, and injury from sharp instruments and needles (see Section 9.2.1).

Respirators are the least satisfactory means of controlling inhalation exposure to airborne agents because respirators only protect workers if the equipment is properly selected, fit tested, worn, and replaced (Colton, 1996). Respirator selection should be based on the hazard to which a worker is exposed as well as factors such as work rate, mobility, and work requirements (e.g., a need to communicate with others). Selection of respiratory protection against airborne infectious agents has been hampered by

TABLE 9.15 Personal Protective Equipment

Type	Function	Examples
Hand protection	To prevent contact with infectious materials To prevent cuts and bites	Surgical gloves: laboratory work, patient care Leather: animal handling Steel mesh: shellfish and animal handling
Skin and clothing protection	To prevent skin contact and transfer of infectious materials from workplace	Laboratory coats: patient care, laboratory work, animal handling Operating gowns: human and veterinary surgery Disposable coveralls: cleaning, contaminated waste removal, animal handling, animal trapping
Eye and face protection	To prevent exposure to infectious droplets, sprays, and splashes and to prevent eye and facial injuries	Safety glasses, goggles, and face shields: dental care, laboratory work, human and veterinary surgery, animal handling, animal trapping, veterinary care
Respiratory protection	To prevent inhalation of infectious aerosols	Respirators: patient care, dental care, laboratory work, animal and veterinary surgery, animal handling, animal trapping, veterinary care

the lack of reference values (e.g., threshold limit values, permissible exposure limits, or recommended exposure limits). In addition, inhalation and deposition of a single infectious agent may be sufficient to initiate infection and sources of infectious agents may be mobile (e.g., humans and animals) or unrecognized as potential sources (infectious agents have no warning properties). Until standards on respirator selection for protection against infectious aerosols are available, health and safety professionals should follow other relevant guidelines as well as local and federal regulations on respirator selection and use (USDOL, 29 CFR 1910.134; ATS, 1996). Respirator use has been recommended to prevent exposure to the agents of tuberculosis (CDC, 1994a; USDOL, 1997a), legionellosis (CDC, 1997c), and hantavirus pulmonary syndrome (CDC, 1993c). In healthcare settings, respirator use has met considerable resistance. Nevertheless, respirators remain the surest means of worker protection from airborne infectious agents in some situations.

Lynn et al. (1996) summarized recommended control measures to reduce potential worker exposure to hantavirus. Person-to-person transmission of a hantavirus was reported from Argentina, but this mode of transmission has not been seen in the United States (Wells et al. 1997). These recommendations are similar to precautions used for emergency and rescue operations as well as possible exposure to *Histoplasma capsulatum* (the agent of histoplasmosis) (Lenhart, 1994; Lenhart et al., 1997). These recommendations would also apply to similar high-risk exposures in other workplaces and are similar to requirements for work in high-containment microbiology laboratories (USDHHS, 1993). Besides PPE, these measures for possible hantavirus exposure include hygiene practices and work practice controls, as follows:

1. A high-efficiency particulate air (HEPA)-filtered respirator properly fitted to the wearer;
2. Disposable clothing and rubber gloves;
3. Use of a HEPA-filtered vacuum, instead of compressed air, to clean work clothing;
4. Use of a HEPA vacuum or other dustless method, instead of compressed air, to clean work areas where evidence of rodent infestation is present;
5. Removal of protective clothing inside out with a respirator on;
6. Washing and cleaning of respirators and gloves; and
7. Washing hands and exposed skin with soap and water following removal of protective clothing and prior to eating, drinking, smoking, and so forth.

9.3.5 Postexposure Prophylaxis

Exposure to infectious agents is almost unique among workplace hazards in offering opportunities for *postexposure prophylaxis* (PEP) (i.e., interventions to prevent or modify the course of infection following a known exposure). The bloodborne pathogen standard defines an "exposure incident" as a specific eye, mouth, other mucous membrane, nonintact skin, or parenteral contact with blood or other potentially infectious materials that results from the performance of an employee's duties (USDOL, 29 CFR 1910.1030). Tables 9.2 and 9.13 list PEP for exposures to infectious agents (e.g., passive immunization, vaccination, administration of antimicrobial agents, and prompt and thorough wound treatment). The lack of PEP for HCV has resulted in the absence of a clear set of recommendations for follow-up of workers exposed to this virus. Recommendations for follow-up of healthcare workers after occupational exposure to HCV have been published (CDC, 1997b).

The most common and effective PEP regimens involve passive immunization through the provision of immune globulins (see Section 9.3.1). *Passive immunization* involves injection of a serum preparation containing antibodies (globulins) that confers temporary protection from antigenically related agents. In the United States, all immune globulins are prepared from human serum except botulism and diphtheria globulins, which are prepared from equine (horse) serum. In the cases of hepatitis A (for institutional and childcare workers) and measles (for healthcare workers), pooled human serum contains sufficient antibodies to confer protection, that is, immune globulin (IG) or immune serum globulin (ISG). Human sera for PEP against HBV, VZV, the rabies virus, and *C. tetani* are all hyperimmune preparations; that is, they are collected from individuals whose immunity has been boosted to increase the concentration of antibodies against a specific agent. Antibodies administered passively disappear over a matter of weeks to months. In the case of viruses with long incubation periods (e.g., HBV and the rabies virus), it is usual to administer vaccine (*active immunization*) at the same time as immune globulin to ensure adequate protection for the duration of an infectious diseases' incubation period.

PEP other than immune globulin and vaccines may be available for some infectious organisms through administration, shortly after exposure, of *antimicrobial agents*, such as antibiotics or antiviral agents. The risk of adverse effects from antimicrobial agents (and often their cost) must be balanced against their potential benefits. Antimicrobial agents must often be administered for a week or more, are more likely to cause adverse effects than immune globulins and vaccines, and are only effective during the administration period. Antimicrobial PEP is available for exposure to HIV (e.g., healthcare workers following needlesticks) (CDC, 1996e; Ippolito et al., 1997). The decision to administer PEP after potential exposure to HIV and the choices of antiretroviral regimens are difficult and complex. To assist those involved in making these decisions, a national hotline (1-888-HIV-4911) and website (http://epi-center.ucsf.edu) have been established. Antimicrobial PEP is also available for exposure to the agents of meningococcal disease (e.g., healthcare workers, teachers, and childcare workers in selected situations following contact with respiratory secretions), pertussis (e.g., susceptible healthcare workers, childcare workers, and teachers), and influenza A (e.g., susceptible healthcare workers). Recommendations for evaluation of workers exposed to B virus from monkeys, including PEP with acyclovir, are available (Holmes et al., 1995). The use of antimicrobial prophylaxis is still investigational in other situations, for example, laboratory exposure to *Brucella* spp. and institutional outbreaks of mycoplasma pneumonia.

Theoretically, antimicrobial agents could be used prophylactically after any exposure to an infectious agent. However, for diseases that can be treated effectively at illness onset, there is little benefit in treating many exposed workers to prevent illness in some fraction thereof. In contrast, antimicrobial PEP may be indicated for diseases that often follow relapsing courses with severe consequences (e.g., brucellosis). Likewise, antimicrobial PEP may be appropriate during outbreaks in which many secondary cases may occur and extensive time would consequently be lost from work (e.g., mycoplasma pneumonia and possibly shigellosis and scabies). Even in such cases, one severe adverse reaction to an antimicrobial agent may outweigh the entire potential benefit of PEP (i.e., the intervention caused more harm than it prevented). Infectious disease consultants as well as local and state health departments can provide information and assist with decisions on providing PEP in these situations.

All PEP must be administered soon after exposure to prevent infection. The maximum recommended time for IG delivery varies by infectious agent. For example, the treatment window is known to be ca. 24 hours for VZV, 24 to 48 hours for HBV, and 7 days for HAV, but is uncertain for some agents such as the rabies virus. Such protection is not absolute (e.g., efficacy may be 90 to 95 percent if PEP is administered immediately after exposure), and protection decreases incrementally with time. Some agents may still cause infection even if PEP is provided. However, the infection may be reduced to an inapparent one or cause only a mild illness. Unfortunately, ineffective PEP may prolong a disease's incubation period. In such a case, an employee may have resumed work before illness began due to an atypically long incubation period. This uncertainty may complicate the decision of when to allow an exposed worker to return to work.

9.3.6 Workers at Increased Risk from Infectious Diseases

A number of characteristics of a worker as host, noted previously, may affect the risk of becoming infected, the likelihood of infection if exposed, and the severity of any disease that does occur. These factors include age; alcoholism; coexisting, particularly chronic, diseases (infectious and noninfectious); abnormalities in the skin or respiratory tract that provide an entry site for pathogens; nutritional status; and immune status (e.g., immunity from infection or immunization; or lack thereof), immunodeficiency, and pregnancy. Extremes of age and health status are of lesser concern for the general work population. However, immunodeficient or pregnant employees may be present in any working population. Persons without functioning spleens are principally at increased risk of infection from *Babesia* spp. (protozoa transmitted by tick bite) and encapsulated bacteria (such as the streptococci) that are otherwise of limited occupational significance. Susceptibility to tuberculosis disease is greater for immunodeficient persons (e.g., those with HIV infection or other immune suppression); persons who are underweight or undernourished; persons with debilitating diseases such as chronic renal failure, cancer, silicosis, diabetes, or postgastrectomy; and substance abusers (e.g., drug or alcohol abusers) (APHA, 1995).

9.3.6.1 The Immunodeficient Worker Persons with altered immunocompetence (whose immune system is suppressed or otherwise not functioning normally) may still be able to work. Causes of immunodeficiency or immunosuppression include HIV infection, organ transplantation (due to administration of immunosuppressive drugs to prevent organ rejection), leukemia, high-dose chronic corticosteroid therapy, malignant disease (e.g., cancer, due to the disease process and administration of immunosuppressive chemotherapy or radiation), and congenital (inherited) immunodeficiency diseases (CDC, 1993a; Frazier et al. 1994). Immunodeficiencies can be antibody mediated or cell mediated and be primary or secondary.

Primary immunodeficiencies are uncommon and usually so severe as to preclude working if a victim survives to adulthood. In most cases, a deficiency mild enough to allow a person to be otherwise qualified to work will not place the person at increased risk of infection on the job. Nevertheless, health and safety professionals should make available to all workers information on infectious agents potentially present in a workplace. Immunocompromised workers can then share this information with their personal physicians and decide if the work poses an unreasonable risk of infection. In rare cases, it may be advisable for workers to avoid specific occupations, jobs, or job assignments. For example, chronic granulomatous disease is an inherited disorder of phagocytes (part of the cell-mediated immune process). This condition is characterized by an increased risk for bacterial and fungal infections (e.g., aspergillus pneumonia related to exposure while shoveling moldy wood chips). Physicians caring for such patients should recognize occupations with a potential for exposure to high concentrations of opportunistic pathogens and advise their patients accordingly.

Secondary immunodeficiencies (acquired immunodeficiencies) can result from certain malignancies as well as intentionally and secondarily immunosuppressive therapies. Medical conditions such as cancers and their treatment generally produce significant immunodeficiency for a limited time. During this period, individuals are either too ill to work or are advised to restrict their activities to limit exposure to infectious agents. Currently, HIV infection is the principal cause of acquired deficiency in cell-mediated immunity among workers. Guidelines for preventing opportunistic infections in HIV-positive individuals have been published (CDC, 1997f). The risk of opportunistic infection is highest in those who are severely immunosuppressed. However, persons affected to this degree may be able to continue working in some occupations. Many opportunistic pathogens are ubiquitous, thereby limiting immunodepressed workers' options for avoidance. Additionally, immunocompromised healthcare, childcare, and laboratory workers as well as those with animal contact face particular potential risks.

Susceptible workers in healthcare settings may be exposed to airborne agents that are difficult to control completely (e.g., *M. tuberculosis* and VZV, the latter from patients with chickenpox or disseminated herpes zoster/shingles) (see Section 9.0.1). Bacterial and parasitic enteric pathogens, which are acquired by contact, may be effectively controlled through barrier precautions and handwashing. However, the potentially devastating effects of these infections in severely immunocompromised persons makes reliance on such precautions questionable. Similar caution is advised for exposure to laboratory-acquired infectious agents; the various zoonotic pathogens encountered in agriculture and food production work; and cytomegalovirus, parvovirus B19, enteric and respiratory pathogens, and other agents present in childcare settings (see Section 9.0.2). Opportunistic infectious diseases that can be acquired from animals include toxoplasmosis, cryptosporidiosis, giardiasis, salmonellosis, and listeriosis (Glaser et al., 1994) (see Section 9.0.3).

Physicians and their immunodeficient patients should discuss the patients' susceptibility to workplace infectious agents, the advisability of continuing to work in certain occupations, and the point at which a patient should change or restrict work because of an infection risk. Institutional occupational health departments are usually not involved in these considerations due to the confidential nature of medical information. Also, workers disabled as a result of medical conditions are protected from workplace discrimination. Therefore, employers should provide information on the nature and extent of known or potential infectious hazards to all workers and to individual workers and their physicians upon request.

9.3.6.2 The Pregnant Worker Infections acquired during pregnancy may cause fetal morbidity (e.g., malformations), fetal mortality (e.g., spontaneous abortions or stillbirths), or severe maternal illness. Table 9.16 summarizes those infections most often of concern for pregnant workers, related occupations, and prevention strategies. The CDC can provide advice on immunoprophylaxis for pregnant workers who may be exposed to infectious agents (CDC, 1993a).

TABLE 9.16 Infectious Diseases of Concern for Pregnant Workers

Infection	Effect in pregnancy	Occupation	Prevention
Rubella	Fetal dealth, congenital rubella syndrome	Healthcare	Immunize susceptible workers (male and female) with patient contact
Measles	Fetal death, maternal morbidity and mortality	Healthcare	Immunize workers with patient contact; immune globulin for exposed susceptible pregnant women
Chickenpox/ varicella zoster (shingles)	Complications more frequent and severe in pregnant women; congenital varicella syndrome, premature birth, neonatal varicella	Healthcare, childcare	Immunize susceptible healthcare workers; varicella-zoster immune globulin for exposed, susceptible pregnant women
Cytomegalovirus infection	Fetal death or severe nervous system abnormalities for 25% of infants infected during pregnancy	Childcare (younger, particularly diapered, children), not healthcare	Uncertain: screen and, if susceptible, either do not work during first half of pregnancy or care only for older children; avoid "intimate contact;" frequent handwashing, glove use; disinfection of toys and surfaces has been proposed
Erythema infectiosum (human parvovirus B19 infection, "Fifth" disease, transient anemic crisis)	Fetal morbidity (hydrops fetalis) and death (<10% case fatality rate)	Healthcare, childcare	Healthcare: inform workers of risks; routine infection-control measures to minimize risks; pregnant workers should not care for patients with transient anemic crisis Childcare: greatest risk of transmission occurs before symptoms appear in children limiting options for prevention; routine exclusion policy for high-risk groups not recommended; individual should make decision for risk avoidance
Toxoplasmosis	Nervous system abnormalities if infected during pregnancy	Meat and animal handlers	Consider serologic screening before or during pregnancy; barrier precautions and handwashing, avoid mucous membrane contact; do not eat raw or undercooked meat; consider re-assignment during pregnancy; prenatal fetal diagnosis for first trimester infection if abortion considered; treatment during pregnancy; avoid cats, change cat liter daily

CONCLUSIONS

Infectious disease evaluation and management are relatively new, but growing, areas of involvement for many occupational health and safety professionals. A wide variety of microorganisms may cause work-related infections, and these agents are transmitted by many exposure routes and affect a number of organ systems. Certain workplaces are associated with obvious microbiological hazards, such as child care, human health care, animal care, and microbiology laboratories. However, transmission may occur incidentally or on rare occasions in many other workplaces. The fields of infectious disease control and biosafety are rapidly evolving. This evolution has been influenced by increasing and increasingly rapid international travel, changes in technology and land use, the emergence and reemergence of a number of infectious diseases, rising numbers of immunocompromised individuals in the workforce, and adaptation and change among infectious agents and their vectors. Environmental sampling, a routine approach for assessing many workplace health hazards, plays a smaller role in the control and prevention of infectious diseases in the workplace. The tools used in infectious disease control are primary prevention to avoid worker exposure, immunization of workers likely to be exposed, and surveillance to identify and treat infections early, prevent infections in co-workers and other contacts, and understand how exposure may occur. Applying fundamental occupational health and safety principles to the design, installation, operation, and testing of engineering controls as well as the implementation of safe work practices to reduce exposures to infectious agents are important and expanding areas of involvement for occupational health and safety professionals.

ACKNOWLEDGMENTS

The authors thank, from the California Department of Health Services, Rita Brenden, Edward Desmond, Paul Duffey, and Michael Janda of the Microbial Diseases Laboratory; Michael Hendry and David Schnurr of the Viral and Rickettsial Diseases Laboratory; S. Benson Werner of the Disease Investigations Section; Kevin Reilly and Mark Starr of the Veterinary Public Health Section; Sandra McNeel of the Environmental Health Investigations Branch; and Cynthia O'Malley of the Immunization Branch; from the University of California at Berkeley, Center for Occupational and Environmental Health, Barbara Plog; from the University of California at San Francisco, Occupational Health Clinic, Patricia Quinlan; from San Francisco General Hospital, William Charney; from the California Division of Occupational Safety and Health, Les Michael, John Howard, and William Krycia; from the U.S. Department of Labor, Occupational Safety and Health Administration, Gabriel Gillotti and Hannah-Marie Miller; and from the U.S. Coast Guard, Timothy Ungs.

REFERENCES

ACGIH, 1989. *Guidelines for the Assessment of Bioaerosols in the Indoor Environment.* Cincinnati, OH: American Conference of Governmental Industrial Hygienists.

ACGIH, 1997. *1996 TLVs® and BEIs®. Threshold Limit Values for Chemical Substances and Physical Agents. Biological Exposure Indices.* Cincinnati, OH: American Conference of Governmental Industrial Hygienists, pp. 11–14.

AIA, 1996. *Guidelines for Design and Construction of Hospital and Health Care Facilities.* Washington, D.C.: American Institute of Architects Press.

AIHA, 1995. *Biosafety Reference Manual.* Fairfax, VA: American Industrial Hygiene Association.

AIHA, 1996. *Field Guide for the Determination of Biological Contaminants in Environmental Samples*. HK Dillon, PA Heinsohn, and JD Miller, Eds. Fairfax, VA: American Industrial Hygiene Association.

APHA, 1995. *Control of Communicable Diseases Manual*. Benenson, AS, Ed. Washington, DC: American Public Health Association.

APHA/AAP, 1992. *Caring for Our Children: National Health and Safety Performance Standards: Guidelines for Out-of-Home Child Care Programs*. Washington, DC/Elk Grove Village, IL: American Public Health Association/American Academy of Pediatrics.

ARC, 1993. *Public Safety Workers and the HIV Epidemic. A Guidebook for Ambulance, Corrections, Fire Service and Law Enforcement Professionals*. 5th ed. Sacramento, CA: American Red Cross.

ASM, 1995a. *Laboratory Safety: Principles and Practices*. 2nd ed. DO Fleming, JH Richardson, JJ Tulis, D Vesley, Eds. Washington, DC: American Society for Microbiology.

ASM, 1995b. *Manual of Clinical Microbiology*. 6th ed. PR Murray, Ed. Washington, DC: American Society for Microbiology.

ASM, 1996. *Pocket Guide to Clinical Microbiology*. 6th ed. PR Murray, Ed. Washington, DC: American Society for Microbiology.

ASM, 1997. *Manual of Environmental Microbiology*. CJ Hurst, GR Knudsen, MJ McInerney, LD Stetzenbach, MV Walter, Eds. Washington, DC: American Society for Microbiology.

ASTM, 1996. *Standard Guide for Inspecting Water Systems for Legionellae and Investigating Possible Outbreaks of Legionellosis (Legionnaires' Disease or Pontiac Fever)*. D 5952-96. West Conshohocken, PA: American Society for Testing and Materials.

ATS, 1996: American Thoracic Society, Medical Section of the American Lung Association. Respiratory protection guidelines. *Am J Respir Crit Care Med*. **154:**1153–1165.

Block, SS, Ed. 1991. *Disinfection, Sterilization, and Preservation*. 4th ed. Philadelphia, PA: Lea and Febiger.

Brosseau, LM; McCullough, NV; Vesley, D: 1997. Mycobacterial aerosol collection efficiency of respirator and surgical mask filters under varying conditions of flow and humidity. *Appl Occup Environ Hyg*. **12:**435–445.

Burge, HA, 1989. Indoor air and infectious disease. *Occ Med:State of Art Rev*. **4:**713–721.

Burge, HA, 1990. Risks associated with indoor infectious aerosols. *Toxicol Ind Health*. **6:**263–274.

Burge, HA, Ed. 1995. *Bioaerosols*. Boca Raton, FL: Lewis Publishers.

Buttner, MP; Willeke, K; Grinshpun, SA: 1997. Sampling and analysis of airborne microorganisms. In: *Manual of Environmental Microbiology*. CJ Hurst, Ed. Washington, DC: American Society for Microbiology, pp. 629–640.

Carrillo, L; Fleming, LE; Lee, DJ: 1997. Bloodborne pathogens risk and precautions among urban fire-rescue workers. *J Occup Environ Med*. **38:**920–924.

Catanzaro A, 1982. Nosocomial tuberculosis. *Am Rev Respir Dis*. **125:**559–562.

CDC, 1989. Prevention and control of measles. *MMWR*. **38**(S-9).

CDC, 1990. Case definitions for public health surveillance. *MMWR*. **39**(RR-13).

CDC, 1991. Update on adult immunization: recommendations of the Immunization Practices Advisory Committee (ACIP). *MMWR*. **40**(RR-12).

CDC, 1992. Prevention and control of tuberculosis in migrant workers. *MMWR*. **41**(RR-10).

CDC, 1993a. Recommendations of the Advisory Committee on Immunization Practices (ACIP): Use of vaccine and immune globulins in persons with altered immunocompetence. *MMWR*. **42**(RR-4).

CDC, 1993b. Recommended infection-control practices for dentistry, 1993. *MMWR*. **42**(RR-8).

CDC, 1993c. Hantavirus infection—Southwestern United States: Interim recommendations for risk reduction. *MMWR*. **42**(RR-11).

CDC, 1994a. Guidelines for preventing the transmission of *Mycobacterium tuberculosis* in health-care facilities, 1994. *MMWR.* **43**(RR-13).

CDC, 1994b. Recommendations of the Advisory Committee on Immunization Practices (ACIP): General recommendations on immunization. *MMWR.* **43**(RR-1).

CDC, 1995. Exposure of passengers and flight crew to *Mycobacterium tuberculosis* on commercial aircraft, 1992–1995. MMWR **44:**137–140.

CDC, 1996a. *Guideline for Isolation Precautions in Hospitals.* Public Health Service, USDHHS, Mailstop E-69, Hospital Infections Program, National Center for Infectious Diseases, Centers for Disease Control and Prevention, Atlanta, GA 30333.

CDC, 1996b. Prevention and control of tuberculosis in correctional facilities. *MMWR.* **45**(RR-8).

CDC, 1996c. Prevention and control of varicella. *MMWR.* **45**(RR-11).

CDC, 1996d. Update: Vaccine side effects, adverse reactions, contraindications, and precautions. Recommendations of the Advisory Committee on Immunization Practices (ACIP). *MMWR.* **45**(RR-12).

CDC, 1996e. Update: provisional Public Health Service recommendations for chemoprophylaxis after occupational exposure to HIV. *MMWR.* **45:**468–80.

CDC, 1996f. Notifiable disease surveillance and notifiable disease statistics—United States, June 1946 and June 1996. *MMWR.* **45:**530–536.

CDC, 1996g. *Final Recommendations to Minimize Transmission of Legionnaires' Disease from Whirlpool Spas on Cruise Ships.* The National Center for Environmental Health. In Collaboration with the National Center for Infectious Diseases Center for Disease Control and Prevention.

CDC, 1997a. Summary of notifiable diseases, United States, 1996. *MMWR.* **45**(53).

CDC, 1997b. Recommendations for follow-up of health-care workers after occupational exposure to hepatitis C virus. *MMWR.* **46:**603–606.

CDC, 1997c. Guidelines for prevention of nosocomial pneumonia. *MMWR.* **46**(RR-1).

CDC, 1997d. Prevention and control of influenza. Recommendations of the Advisory Committee on Immunization Practices (ACIP). *MMWR.* **46**(RR-9):1–25.

CDC, 1997e. Case definitions for infectious conditions under public health surveillance. *MMWR.* **46**(RR-10):1–57.

CDC, 1997f. USPHS/IDSA Prevention of Opportunistic Infections Working Group. 1997 USPHS/IDSA Guidelines for the prevention of opportunistic infections in persons infected with human immunodeficiency virus: *MMWR.* **46**(RR-12):1–46.

CDC, 1997g. Immunization of health-care workers. Recommendations of the Advisory Committee on Immunization Practices (ACIP) and the Hospital Infection Control Practices Advisory Committee (HICPAC): *MMWR.* **46**(RR-18):1–42.

CDC, 1997h. *Draft Guidelines for Infection Control in Health Care Personnel.* Federal Register, September 8, 1997 **62**(173):1–77.

Chapin, K, 1995. Clinical microscopy. In: *Manual of Clinical Microbiology.* 6th ed. PR Murray, Ed. Washington, DC: American Society for Microbiology, pp. 33–51.

Chen, S-K; Vesley, D; Brosseau, LM; Vincent, JH, 1994. Evaluation of single-use masks and respirators for protection of health care workers against mycobacterial aerosols. *Am J Infect Control.* **22:**65–74.

Collins, CH; Aw, TC; Grange, JM, 1997. *Microbial Diseases of Occupations, Sports and Recreations.* Boston, MA: Butterworth-Heinemann.

Colton, CE, 1996. Respiratory protection. In: *Fundamentals of Industrial Hygiene.* 4th ed. BA Plog, J Niland, PJ Quinlan, Eds. Itasca, IL: National Safety Council. pp. 619–656.

Cottam, AN, 1994. Biotechnology. In: *Hunter's Diseases of Occupations*, 8th ed. PAB Raffle, PH Adams, PJ Baxter, and WR Lee, Eds. Boston, MA: Edward Arnold. pp. 577–589.

Cox, CS; Wathes, CM, Eds. 1995. *Bioaerosols Handbook*. Boca Raton, FL: Lewis Publishers.

Driver, CR; Valway, SE; Meade, M; Onorato, IM; Castro, KG, 1994. Transmission of *Mycobacterium tuberculosis* associated with air travel. *JAMA* **272:**1031–1035.

Ehrenberg, RL; Frumkin, H, 1995. Design and implementation of occupational health and safety programs. In: *Laboratory Safety: Principles and Practices*. EO Fleming, JH Richardson, JJ Tulis, and D Vesley, Eds. pp. 279–288.

Ellis, CJ; Symington, IS, 1994. Microbial disease. In: *Hunter's Diseases of Occupations*, 8th ed. PAB Raffle, PH Adams, PJ Baxter, and WR Lee, Eds. Boston, MA: Edward Arnold. pp. 545–576.

Fields, BS, 1997. Legionellae and Legionnaires' disease. In: *Manual of Environmental Microbiology*. CJ Hurst, Ed. Washington, DC: American Society for Microbiology, pp. 666–675.

Frazier, LM; Stave, GM; Tulis, JJ, 1994. Prevention of illness from biological hazards. In: *Physical and Biological Hazards of the Workplace*. Wald, PH; Stave, GM, Eds. New York, NY: Van Nostrand Reinhold, pp. 257–270.

Freije, MR, 1996. *Legionellae Control in Health Care Facilities: A Guide for Minimizing Risk*. Indianapolis, IN: HC Information Resources, Inc.

Gammaitoni, L; Nucci, MC: 1997. Using a mathematical model to evaluate the efficacy of TB control measures. *Emerg Infect Dis*. **3:**335–342.

Gantz, NM, 1988. Infectious agents. In: *Occupational Health: Recognizing and Preventing Work-Related Disease*. BS Levy and DH Wegman, Eds. Boston, MA: Little, Brown and Co., pp. 281–295.

Garibaldi, R; Janis, B, 1992. Occupational infections. In: *Environmental and Occupational Medicine*. WN Rom, Ed. Boston, MA: Little, Brown and Co., pp. 607–617.

Gelberg, KH, 1997. Health study of New York City Department of Sanitation landfill workers. *J Occup Environ Med*. **39:**1103–1110.

Gerberding, JL; Holmes, KK, 1994. Microbial agents and infectious diseases. In: *Textbook of Clinical Occupational and Environmental Medicine*. Philadelphia, PA: W.B. Saunders Co. pp. 699–716.

Gershon, RRM; Zirkin, BG, 1995. Behavioral factors in safety training. In: *Laboratory Safety: Principles and Practices*. 2nd ed. DO Fleming, JH Richardson, JJ Tulis, and D Vesley, Eds. Washington, DC: American Society for Microbiology Press, pp. 269–277.

Gillotti, G, 1996. Government regulations. Chapter 29. In: *Fundamentals of Industrial Hygiene*. 4th ed. BA Plog, J Niland, PJ Quinlan, Eds. Itasca, IL: National Safety Council. pp. 769–785.

Glaser, CA; Angulo, FJ; Rooney, JA, 1994. Animal-associated opportunistic infections among persons infected with the human immunodeficiency virus. *J Clin Dis*. **18:**14–24.

Gomez, MA, 1993. Occupational exposure to bloodborne pathogens. In: *The Work Environment. Volume Two. Healthcare, Laboratories, and Biosafety*. Hansen, DJ, Ed. Boca Raton, FL: Lewis Publishers. pp. 53–79.

Gomez, AS; Gomez, MA, 1993. Occupational health hazards in the dental office. In: *The Work Environment. Volume Two. Healthcare, Laboratories, and Biosafety*. Hansen, DJ, Ed. Boca Raton, FL: Lewis Publishers. pp. 103–124.

Grainger, CR; Young, MJF; John, HH, 1995. A code of practice on dealing with infectious diseases on aircraft. *J Roy Soc Health*. 175–177.

Hansen, DJ, Ed. 1993. *The Work Environment. Volume Two. Healthcare, Laboratories, and Biosafety*. Boca Raton, FL: Lewis Publishers.

Harber, P; Schenker, MB; Balmes, JR, Eds. 1996. *Occupational and Environmental Respiratory Disease*. St. Louis, MO: Mosby.

Harding, L; Fleming, DO; Macher, JM, 1996. Biological hazards. In: *Fundamentals of Industrial*

Hygiene. 4th ed. BA Plog, J Niland, PJ Quinlan, Eds. Itasca, IL: National Safety Council. pp. 403–449.

Harding, L; Liberman, D, 1995. Epidemiology of laboratory-associated infections. In: *Laboratory Safety: Principles and Practices.* EO Fleming, JH Richardson, JJ Tulis, and D Vesley, Eds. pp. 7–15.

Herrmann, JE, 1995. Immunoassays for the diagnosis of infectious diseases. In: *Manual of Clinical Microbiology.* 6th ed. PR Murray, Ed. Washington, DC: American Society for Microbiology, pp. 110–122.

Holmes, GP; Chapman, LE; Stewart, JA; Strus, SE; Hilliard, JK; Davenport, DS, 1995. Guidelines for the prevention and treatment of B-virus infections in exposed persons. *Clin Infect Dis.* **20:**421–439.

Holmes, SJ; Morrow, AL; Pickering, LK, 1996. Child-care practices: Effects of social change on the epidemiology of infectious diseases and antibiotic resistance. *Epidemiol Rev.* **18:**10–28.

Ippolito, G; Puro, V; Petrosillo, N; Pugliese, G; Wispelwey, B; Tereskerz, PM; Bentley, M; Jagger, J, 1997. *Prevention, Management and Chemoprophylaxis of Occupational Exposure to HIV.* Charlottesville, VA: University of Virginia.

JCAHO, 1997. *Managing Indoor Air Quality in Healthcare Organizations.* Chicago, IL: Joint Commission on Accreditation of Healthcare Organizations.

Johnson, B; Martin, DD; Resnick, IG, 1994. Efficacy of selected respiratory protective equipment challenged with *Bacillus subtilis* subsp. *niger. Appl Environ Microbiol.* **60:**2184–2186.

Kozarsky, PE; Lobel, HO; Steffen, R, 1991. Travel medicine 1991: New frontiers. *Ann Int Med.* **115:**574–575.

Kuehne, RW; Chatigny, MA; Stainbrook, BW; Runkle, RS; Stuart, DG, 1995. Primary barriers and personal protective equipment in biomedical laboratories. In: *Laboratory Safety: Principles and Practices.* 2nd ed. DO Fleming, JH Richardson, JJ Tulis, and D Vesley, Eds. Washington, DC: American Society for Microbiology Press, pp. 145–170.

Lawrence, JR; Korber, DR; Wolfaardt, GM; Caldwell, DE, 1997. Analytical imaging and microscopy techniques. In: *Manual of Environmental Microbiology.* CJ Hurst, Ed. Washington, DC: American Society for Microbiology, pp. 29–51.

Lenhart, SW, 1994. Recommendations for protecting workers from *Histoplasma capsulatum* exposure during bat guano removal from a church's attic. *Appl Occup Environ Hyg.* **9:**230–236.

Lenhart, SW.; Schafer, MP; Singal, M; Hajjeh, RA, 1997. *Histoplasmosis: Protecting Workers at Risk.* Cincinnati, OH: Centers for Disease Control and Prevention. National Institutes for Occupational Safety and Health. Publication No. 97–146.

Lighthart, B; Mohr, AJ, 1994. *Atmospheric Microbial Aerosols: Theory and Applications.* New York, NY: Chapman and Hall.

Lynn, M; Vaughn, D; Kauchak, S; England, M, 1996. An OSHA perspective on Hantavirus. *Appl Occup Environ Hyg.* **11:**225–227.

Macher, JM, 1993. The use of germicidal lamps to control tuberculosis in health-care facilities. *Infect Control Hosp Epidemiol.* **14**(12):723–729.

Macher, JM; Chatigny, MA; Burge, HA, 1995a. Sampling airborne microorganisms and aeroallergens. In: *Air Sampling Instruments.* 8th ed. Cincinnati, OH: American Conference of Government Industrial Hygienists.

Macher, JM; Streifel, AJ; Vesley, D, 1995b. Problem buildings, laboratories, and hospitals. In: *Bioaerosols—Handbook of Samplers and Sampling.* Chelsea, MI: Lewis Publishers, Inc.

Mandell, GL; Bennett, JE; Dolin, R, Eds. 1995. *Principles and Practice of Infectious Diseases.* 4th ed. New York, NY: Churchill Livingstone.

Marchese, J, 1993. Medical waste disposal. In: *The Work Environment. Volume Two. Healthcare, Laboratories, and Biosafety.* Hansen, DJ, Ed. Boca Raton, FL: Lewis Publishers. pp. 151–182.

Marier, RL; Nelson, T, 1993. A ventilation-filtration unit for respiratory isolation. *Infect Control Hosp Epidemiol.* **14:**700–705.

McCunney, RJ, 1986. Waste water/sewage treatment: Health effects of work at waste water treatment plants: A review of the literature with guidelines for medical surveillance. *Am J Indust Med.* **9:**271–279.

McGarrity, GJ; Hoerner, CL, 1995. Biological safety in the biotechnology industry. In: *Laboratory Safety: Principles and Practices.* 2nd ed. DO Fleming, JH Richardson, JJ Tulis, D Vesley, Eds. Washington, DC: American Society for Microbiology, pp. 119–131.

McKenna, MT; Hutton, M; Cauthen, G; Onorato, IM: 1997. The association between occupation and tuberculosis: A population-based survey. *Am J Respir Crit Care Med.* **154:**587–593.

Miller-Leiden, S; Lobascio, C; Macher, JM; Nazaroff, WW, 1996. Effectiveness of in-room air filtration for tuberculosis infection control in healthcare settings. *J Air Waste Manage Assoc.* **46:**869–882.

Moe, CL, 1997. Waterborne transmission of infectious agents. In: *Manual of Environmental Microbiology.* CJ Hurst, Ed. Washington, DC: American Society for Microbiology, pp. 136–152.

Morey, PR; Feeley, JC; Otten, JA, Eds. 1990. *Biological Contaminants in Indoor Environments.* Philadelphia, PA: American Society for Testing and Materials.

Nardell, EA; Keegan, J; Cheney, SA; Etkind, SC, 1991. Airborne infection: Theoretical limits of protection achievable by building ventilation. *Am Rev Respir Dis.* **144:**302–306.

Nevalainen, A; Willeke, K; Liebhaber, F; Pastuszka, J; Burge, H; Henningson, E, 1993. Bioaerosol sampling. In: *Aerosol Measurement: Principles, Techniques, and Applications.* K Willeke and PA Baron, Eds. New York, NY: Van Nostrand Reinhold, pp. 471–492.

Nicas, M, 1994. Modeling respirator penetration values with the beta distribution: An application to occupational tuberculosis transmission. *Am Ind Hyg Assoc J.* **55:**515–524.

Nicas, M, 1996. Refining a risk model for occupational tuberculosis transmission. *Am Ind Hyg Assoc J.* **57:**16–22.

NIH, 1991. *Guidelines for Research Involving Recombinant DNA Molecules. FR* **59:**34496–34547, July 18, 1991.

NIH, 1994. *Actions under the Guidelines, Guidelines for Research Involving Recombinant DNA Molecules. FR* **55:**33174–33183, July 5, 1994.

NIOSH, 1972. *Criteria for a Recommended Standard...Occupational Exposure to Ultraviolet Radiation.* Washington, DC: U.S. Department of Health, Education and Welfare. Public Health Service, Publication No. (HSM)73–110009.

NIOSH. 1994. *NIOSH Alert. Request for Assistance in Preventing Organic Dust Toxic Syndrome.* Department of Health and Human Services, Public Health Service, Centers for Disease Control and Prevention, National Institute for Occupational Safety and Health, DHHS (NIOSH) Publication No. 94–102.

NRC, 1989. *Biosafety in the Laboratory: Prudent Practices for the Handling and Disposal of Infectious Materials.* Washington, DC: National Academy Press.

NSF, 1992. *NSF 49: Class II (Laminar Flow) Biohazard Cabinetry.* Ann Arbor, MI.

Osterholm, MT, 1994. Infectious disease in child day care: An overview. *Pediatrics* **94:**987–990.

Payment, P, 1997. Cultivation and assay of viruses. In: *Manual of Environmental Microbiology.* CJ Hurst, Ed. Washington, DC: American Society for Microbiology, pp. 72–78.

Podzorski, RP; Persing, DH, 1995. Molecular detection and identification of microorganisms. In: *Manual of Clinical Microbiology.* 6th ed. PR Murray, Ed. Washington, DC: American Society for Microbiology, pp. 130–157.

Raffle, PAB; Adams, PH; Baxter, PJ; Lee, WR, Eds. 1994. *Hunter's Diseases of Occupations.* Boston, MA: Edward Arnold.

Riley, RL; Nardell, EA, 1989. Clearing the air: The theory and application of ultraviolet air disinfection. *Am Rev Respir Dis.* **139:**1286–1294.

Rosenberg, J; Clever, LH, 1990. Medical surveillance of infectious disease endpoints. In: *Occupational Medicine: Medical Surveillance in the Workplace.* D Rempel, Ed. *Occupational Medicine: State of the Art Reviews.* **5:**583–605.

Rosenstock, L; Cullen, MR, 1994. *Textbook of Clinical Occupational and Environmental Medicine.* Philadelphia, PA: W.B. Saunders Company.

Rutala, WA, 1990. *APIC Guideline for Selection and Use of Disinfectants.* Mundelien, IL: Association for Practitioners in Infection Control, Inc.

Rutala, WA; Jones, SM; Worthington, JM; Reist, PC; Weber, DJ, 1995. Efficacy of portable filtration units in reducing aerosolized particles in the size range of *Mycobacterium tuberculosis. Infect Control Hosp Epidemiol.* **16:**391–398.

Rylander, R; Jacobs, RR, Eds. 1994. *Organic Dusts: Exposure, Effects, and Prevention.* Boca Raton, FL: Lewis Publishers.

Sepkowitz, KA, 1996. Occupationally acquired infections in health care workers. Parts I and II. *Ann Int Med* **125:**826–834, 917–928.

Sewell, DL, 1995. Laboratory-associated infections and biosafety. *Microbiol Rev.* **8:**389–405.

Sewell, DL; Schifman, RB, 1995. Quality assurance: Quality improvement, quality control, and test validation. In: *Manual of Clinical Microbiology.* 6th ed. PR Murray, Ed. Washington, DC: American Society for Microbiology, pp. 55–66.

Tanner, RS, 1997. Cultivation of bacteria and fungi. In: *Manual of Environmental Microbiology.* CJ Hurst, Ed. Washington, DC: American Society for Microbiology, pp. 52–60.

Turnberg, WL, 1996. *Biohazardous Waste: Risk Assessment, Policy and Management.* New York, NY: John Wiley & Sons.

USDHHS, 1990. *Preventing Disease Transmission in Personal Service Worker Occupations. Resource and Curriculum Materials for a One-Day Workshop.* Public Health Service, Food and Drug Administration, Center for Devices and Radiological Health, Rockville, Maryland 20857.

USDHHS, 1993. *Biosafety in Microbiological and Biomedical Laboratories.* 3rd ed. HHS Publication no. (CDC)93–8395, Public Health Service, Centers for Disease Control and Prevention, and National Institutes of Health. CDC, Washington, DC: USGPO (Stock 017-040-00523).

USDHHS, National Institutes of Health. 1994. Appendix B. Classification of etiologic agents and oncogenic viruses on the basis of hazard. In: *Guidelines for Research Involving Recombinant DNA Molecules (NIH Guidelines).* Federal Register, July 5, 1994, Separate Part IV. pp. 20–24.

USDHHS, Centers for Disease Control and Prevention, 1996. *Additional Requirements for Facilities Transferring or Receiving Select Agents.* 42 CFR Part 72, RIN 0905-AE70. Federal Register, October 24, 1996.

USDOL, Occupational Safety and Health Act, 1970. Public Law 91–596. 91st Congress, S. 2193. (29 USC 651).

USDOL, Occupational Safety and Health Administration, 1987. Title 29, Code of Federal Regulations, Field Sanitation; Final Rule Vol 52, No. 84, Washington, DC: U.S. Government Printing Office.

USDOL, Occupational Safety and Health Administration, 1988. Title 40, Code of Federal Regulations, Part 259. Washington, DC: U.S. Government Printing Office (http://law.house.gov/cfr.htm). Standards for the Tracking and Management of Medical Waste.

USDOL, Occupational Safety and Health Administration, 1994. Title 29, Code of Federal Regulations, Part 1910. Washington, DC: U.S. Government Printing Office (http://law.house.gov/cfr.htm).

Part 1904.2: Recording and Reporting Occupational Injuries and Illnesses. Log and Summary of Occupational Injuries and Illnesses.

Part 1904.8: Recording and Reporting Occupational Injuries and Illnesses. Reporting of fatality or multiple hospitalization accidents.

Part 1910.22: Walking-Working Surfaces. General Requirements.

Part 1910.132: Personal Protective Equipment. General Requirements.

Part 1910.133: Personal Protective Equipment. Eye and Face Protection.

Part 1910.134: Personal Protective Equipment. Respiratory Protection.

Part 1910.141: General Environmental Controls. Sanitation.

Part 1910.142: General Environmental Controls. Temporary Labor Camps.

Part 1910.145: General Environmental Controls. Specifications for Accident Prevention Signs and Tags.

Part 1910.1030: Occupational Exposure to Bloodborne Pathogens.

Part 1928.110: Field Sanitation.

USDOL, Occupational Safety and Health Administration, 1997. *OSHA Technical Manual.* Office of Science and Technology Assessment, Washington, DC.

USDOL, Occupational Safety and Health Administration, 1997. *19 CFR, Part 1910, Occupational Exposure to Tuberculosis: Proposed Rule.* Federal Register, October 17, 1997.

USDOT, U.S. Coast Guard, 1994. *Public Health and Communicable Disease Concerns Related to Alien Migrant Interdiction Operations (AMIO).* COMDTINST M6220.9, Washington, DC.

USEPA, 1986. *EPA Guide for Infectious Waste Management.* Washington, DC: EPA, Report EPA/530-SW-86-014.

Vesley, D; Lauer, JL, 1995. Decontamination, sterilization, disinfection, and antisepsis. In: *Laboratory Safety: Principles and Practices.* 2nd ed. DO Fleming, JH Richardson, JJ Tulis, D Vesley, Eds. Washington, DC: American Society for Microbiology.

Wald, PH; Stave, GM, Eds. 1994. *Physical and Biological Hazards of the Workplace.* New York, NY: Van Nostrand Reinhold.

Warren, A; Day, JG; Brown, S, 1997. Cultivation of algae and protozoa. In: *Manual of Environmental Microbiology.* CJ Hurst, Ed. Washington, DC: American Society for Microbiology, pp. 61–71.

Wells, RM; Estani, SS; Yadon, ZE; Enria, D; Padula, P; Pini, N; Mills, JN; Peters, CJ; Segura, EL, 1997. An unusual hantavirus outbreak in Southern Argentina: Person-to-person transmission? *Emerg Infect Dis.* **2:**171–174.

WHO, 1993. *Laboratory Safety Manual.* 2nd edition. Geneva, Switzerland: World Health Organization.

Willeke, K; Qian, Y; Donnelly, J; Grinshpun, S; Ulevicius, V, 1996. Penetration of airborne microorganisms through a surgical mask and a dust/mist respirator. *Am Ind Hyg Assoc J.* **57:**348–355.

Zeitz, PS; Graber, JM; Voorhees, RA; Kioski, C; Shands, LA; Ksiazek, TG; Jenison, S; Khabbaz, RF, 1997. Assessment of occupational risk for hantavirus infection in Arizona and New Mexico. *J Occ Environ Med.* **39:**463–467.

CHAPTER 10

Biological Monitoring in Occupational Health and Safety

ALAIN J. COUTURIER

Occupational Health Network, 11220 County Rd. 14, Middlebury, IN 46540

ROBERT J. McCUNNEY

Massachusetts Institute of Technology, 77 Massachusetts Ave., Cambridge, MA 02139

10.1 INTRODUCTION

10.2 SCOPE AND PURPOSE OF BIOLOGICAL MONITORING
- 10.2.1 Types of Monitoring

10.3 PRACTICAL CONSIDERATIONS FOR BIOLOGICAL MONITORING
- 10.3.1 Sampling Strategies
- 10.3.2 Who Should Collect Biological Samples
- 10.3.3 Correspondence of Biological Monitoring Results

10.4 TOXICOLOGIC AND PHARMACOKINETIC CONSIDERATIONS
- 10.4.1 Routes of Entry
- 10.4.2 Dose
- 10.4.3 Duration
- 10.4.4 Distribution
- 10.4.5 Metabolism
- 10.4.6 Excretion

10.5 CLINICAL APPLICATION OF BIOMARKERS IN OCCUPATIONAL MEDICINE
- 10.5.1 Definitions
- 10.5.2 Role of the Occupational Physician
- 10.5.3 Role of Biomarkers in Illness Evaluation
- 10.5.4 Summary

10.6 BIOLOGICAL EXPOSURE INDICES
- 10.6.1 Implementation of Biological Exposure Indices
- 10.6.2 Interpretation of Biological Exposure Indices
- 10.6.3 When to Conduct Sampling

Adapted from "Biological Markers of Chemical Exposure, Dosage, and Burden" by Richard S. Waritz, Ph.D., Chapter 3 in Cralley, Cralley, and Bus, *Patty's Industrial Hygiene and Toxicology, 3rd ed., Vol. IIIB*. New York: John Wiley & Sons, Inc., 1995, pp. 79–158.

Handbook of Occupational Safety and Health, Second Edition, Edited by Louis J. DiBerardinis, ISBN 0-471-16017-2

10.7 SUMMARY
10.8 CASE STUDY ON ARSENIC TOXICITY
10.8.1 Direct Biological Indicators
10.8.2 Indirect Biological Indicators
10.8.3 Conclusion
10.9 CASE STUDY ON CHOLINESTERASE-INHIBITING PESTICIDE TOXICITY
10.9.1 Direct Biological Indicators
10.9.2 Conclusion
10.10 CASE STUDY ON TOLUENE TOXICITY
10.10.1 Population at Risk
10.10.2 Biological Fate in Humans
10.10.3 Laboratory Tests
10.10.4 Direct Biological Indicators
10.10.5 Indirect Biological Indicators
10.10.6 Conclusion
10.11 CASE STUDY ON TETRACHLOROETHYLENE TOXICITY
10.11.1 Biological Fate in Humans
10.11.2 Direct Biological Indicators
10.11.3 Indirect Biological Indicators
10.11.4 Conclusion
REFERENCES
SUGGESTED READINGS

10.1 INTRODUCTION

Historically, the disciplines of industrial hygiene and occupational medicine have mutually embraced the concept of preventing health impairment that may result from exposures to chemicals in the workplace. The practical extension of this principle involves the development and implementation of comprehensive chemical exposure control programs that may include health hazard assessment, occupational health history and physical examination, administrative and engineering controls, protective clothing and equipment, exposure monitoring of contaminants in work room air, as well as biological monitoring.

"Biological monitoring has been defined as the systematic or repetitive measurement and assessment of agents or their metabolites, either in tissues, secreta, excreta, expired air, or any combinations of these media to evaluate exposure and health risk in comparison with an appropriate reference" (Zenz, 1994). Long traditions in biological monitoring date back to the 1920s, when elevated urinary lead concentrations were reported in individuals working in the automotive industry (Baselt, 1988). Elkins, in the 1950s, is credited with being the first to advocate the adoption of biological monitoring as an essential element of an industrial hygiene program, and has done pioneering work on the correlation between exposure concentrations of industrial chemicals and their concentrations in body fluids. While biological monitoring has been used since the advent of modern medical practice to determine the relationship between occupational hazards and ill health, it is only more recently that it has been used to prevent the onset of ill health. Biological monitoring has seen a rapid expansion in the past few years from the measurement of lead, mercury, cadmium, phenol, and hippuric acid to the measurement of more than 100 chemicals or groups of chemicals (Alessio, 1990).

In this chapter, important aspects of biological monitoring will be reviewed from the perspective of its application in the routine practice of occupational health and safety.

Included in this presentation will be a brief overview of the fundamental principles of toxicology and pharmacokinetics that apply to biological monitoring, as well as a discussion of clinical applications of biomarkers in occupational medicine. Finally, several examples from the published literature will be presented for illustration of these concepts.

10.2 SCOPE AND PURPOSE OF BIOLOGICAL MONITORING

The main purpose of biological monitoring is to prevent chemically induced health effects through the assessment of internal exposure or uptake of chemicals in the workplace and the associated health risks. Biological monitoring provides an estimate of the amount of a chemical that has actually entered the body (the so-called *dose*). This concept is likely to yield a better prediction of the potential toxic effects than mere measurements of the concentrations in the ambient air.

Biological monitoring may be used in the following ways (ACGIH, 1995):

1. To evaluate dermal or oral absorption, or any other route of entry of chemicals that is not typically addressed by ambient air monitoring.
2. To monitor the accumulation of a substance in the body.
3. To test the efficacy of personal protective equipment or engineering controls and industrial hygiene methods.
4. To evaluate the effects of overexposure to a given chemical retrospectively.
5. To evaluate workers' symptoms that are not substantiated by ambient air concentrations measured in the workplace.
6. To determine nonoccupational exposures to chemicals of interest.

10.2.1 Types of Monitoring

Three complementary forms of monitoring programs exist: Ambient air monitoring assesses health risks through the measurement of inhalation exposure to airborne concentrations of chemicals. Depending on the type of sampling method, area or personal, an estimate of risk may be determined on a group or individual basis. A health hazard may be estimated by comparison with standard reference values, such as the threshold limit value (TLV) of the American Conference of Governmental Industrial Hygienists (ACGIH) or to Occupational Safety and Health Act (OSHA) standards (Lauwerys, 1993).

With biological monitoring of exposures, the actual levels of the chemicals or their metabolic breakdown products (metabolites) are measured in various biological specimens, such as blood, urine, or expired air. Examples include the analysis of blood lead levels in workers exposed to lead dust or the measurement of trichloracetic acid in the urine (metabolite of trichloroethylene) of people working near degreaser tanks (McCunney, 1994).

In contrast, biological monitoring of health effects measures various biochemical or physiological changes in the body as a result of exposures to different chemicals. An illustration of this would be the measurement of the activity level of the enzyme, acetyl cholinesterase, in the blood of workers exposed to organophosphate pesticides.

Both types of biomonitoring account for factors affecting the uptake of chemicals into the body (see Table 10.1) as well as differences in host susceptibility, such as state of health, age, and gender. In addition, there is consideration of the absorption, distribution, biotransformation, and excretion of the chemicals.

TABLE 10.1 Factors Affecting the Uptake of Chemicals in the Body

1. Variation in the concentration of the chemical at different locations and at different times
2. Particle size and aerodynamic properties of particles
3. Solubility characteristics of the chemical
4. Physical form or state of the chemical, i.e., liquid, gas, fumes, aerosols, dusts, etc.
5. Variation in environmental factors, i.e., temperature, humidity, barometric pressure
6. Alternative absorption routes (skin, gastrointestinal tract)
7. Industrial hygiene controls, protective devices, and their efficiency
8. Respiratory volumes (workload—job energy demands)
9. Personal habits (hygiene—hand washing, etc.)
10. Nonoccupational exposures
11. Accumulation of chemical in the body

Source: ACGIH, 1995a.

Biological monitoring may be employed in a medical surveillance program to evaluate the overall health status of a person in the context of exposure to a hazard. One of the goals of medical surveillance is to identify individuals with early signs of adverse health effects, to prevent progress to disease states or that may be ameliorated by improvement of the exposure conditions.

Biological monitoring of effects must be distinguished from the actual diagnosis of occupational illness. Diagnosis is based on many factors, including occupational history, physical examination, as well as the judgement of the effects of biological exposure. An example would be the development of methemoglobinemia from exposure to aniline in workers at a dye factory. The measurement of methemoglobin in the blood of workers represents the biological monitoring of the effect of aniline on the hemoglobin. The presence of a health risk is determined by reference to permissible levels in biological medicine (biological exposure indices, BEI) (ACGIH, 1995b). If enough hemoglobin is reduced into methemoglobin, the workers may experience symptoms varying from headache, lightheadedness, and dizziness, to loss of consciousness and possibly death. A cardiac event, such as a heart attack, may result from exposure to aniline, because the methemoglobinemia leads to the loss of oxygen delivered to the heart tissue. The diagnosis of a heart attack, however, is reached by analysis of all laboratory data: EKGs, cardiac enzyme studies, as well as the clinical evaluation of the patient.

In a well-designed workplace, *ambient monitoring* can ascertain the effectiveness of primary preventive measures, such as engineering controls, and determine whether potentially hazardous worker exposure is taking place (ACGIH, 1995c). Although ambient monitoring is a necessary component of primary prevention, the measured concentrations of a chemical agent may not necessarily correlate with the quantity actually absorbed by exposed workers. Unless monitoring is continuous, or performed frequently, dangerous surges in ambient levels may be missed. In addition, interindividual variation exists in personal hygiene habits and work practice, such as handwashing, use of personal protective equipment, and smoking on the work site. Each factor could lead to potentially toxic body burdens of a workplace agent, despite acceptable ambient measurements (McCunney, 1994).

The use of personal protective equipment, such as respirators, or special clothes or gloves, may decrease a worker's intake of toxic agents. However, individuals may not use them properly because they are uncomfortable or inconvenient. Therefore, even when the workplace air levels are considered safe, these workers may be exposed to potentially toxic amounts of chemicals.

Biological monitoring should be employed to enhance the effectiveness of a monitoring program, because it can account for factors that are not otherwise addressed. For instance, although air monitoring is useful for estimating toxic chemicals that may be inhaled by the pulmonary system, it ignores other routes of exposure, such as skin or gastrointestinal exposure. Biological monitoring thus provides a more comprehensive picture of potential toxic exposures.

The presence of a substance in either the ambient environment or inside the body, however, does not necessarily imply toxicity. Depending upon the specimen selected (i.e., blood, urine, exhaled air, nails, or hair), the amount recovered may not accurately reflect damage to the target organ because some material may be bound to proteins and not totally excreted. Sophisticated pharmacokinetic models have been developed to help address the issue of protein binding of toxicants (Shaikh, 1987).

10.3 PRACTICAL CONSIDERATIONS FOR BIOLOGICAL MONITORING

"In considering biologic monitoring as a method of exposure assessment, four underlying conditions should be satisfied:

1. The chemical, and/or its metabolites, must be present in some tissue, body fluid, or excreta suitable for sampling.
2. Valid and practical analytical methods of sampling must be available.
3. The measurement strategy must be adequate (the samples are representative).
4. The result can be interpreted in a meaningful way; that is, the concentrations noted should be correlated with certain health effects" (Zenz, 1994).

10.3.1 Sampling Strategies

In the context of mercury exposure, most studies report biological levels in terms of *blood* and *urine* mercury, HgB or HgU (ACGIH, 1995d). Urine is commonly used because of the ease in obtaining the sample. Urine samples may be a spot sample or 24-h collection of urine. It is recommended that the spot urine samples be reported in mg of Hg/g creatinine to enhance reproducibility of spot samples compared to 24 hour urine collection. In addition, it is recommended that repeated samples be collected at the same time of day, preferably preshift, because of significant fluctuations in urine mercury levels, with the highest concentrations occurring in the morning (Clarkson, 1988).

For chemicals with long half-lives, such as lead, mercury, and cadmium, the concentration in the blood or in the urine usually reaches a plateau, which reflects the equilibrium between daily intake and excretion. In a stable exposure situation, the daily variation of concentration is small, and an accurate picture may be determined by a single determination of a blood or urine level (Clarkson, 1988).

The situation is quite different for chemicals with short half-lives (see Fig. 10.1). The concentration (especially in blood) changes rapidly with time, and the concentrations only reflect recent exposure. Strict standardization of the specimen's sampling should be used to obtain meaningful results. Another consideration is the *frequency* of sampling in relation to the fluctuation in the air concentrations of the index chemical. The

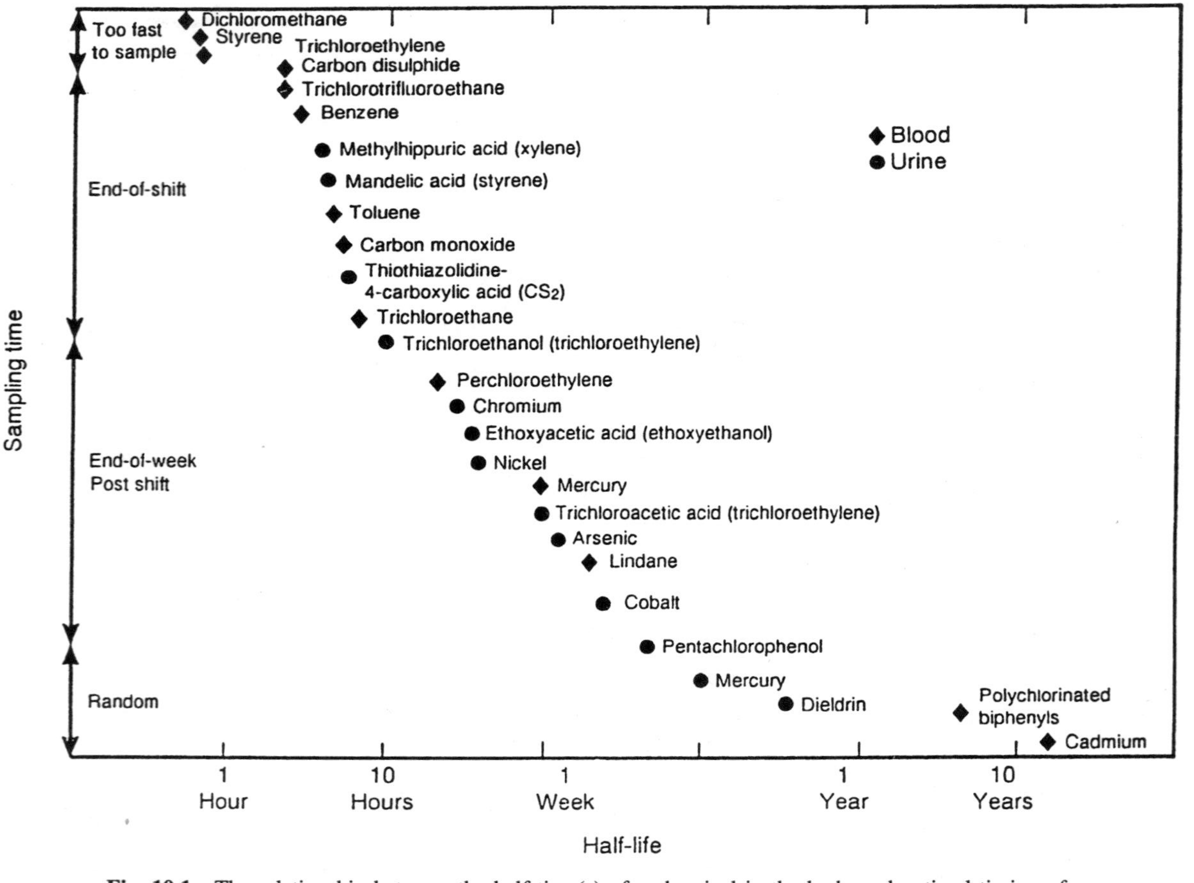

Fig. 10.1 The relationship between the half-time(s) of a chemical in the body and optimal timing of samples for biologic monitoring. Reproduced from the Guidance Note EH 56 from the Health and Safety Executive, with permission of the Controller of Her Britannic Majesty's Stationary Office.

concentration of chemicals in the workplace is seldom stable, and biological monitoring may have to be done frequently to obtain representative results (Aito, 1994).

The ideal biological measure of chemical absorbed is a *noninvasive* technique that accurately reflects the degree of toxic insult of the chemical to the worker and is not disruptive or repugnant to the worker. Most commonly, blood, urine, and rarely exhaled air are used in biological monitoring. Urine is optimal because of its ease of collection, analysis, and interpretation. Use of blood and exhaled air has also been adopted as specimens for biological exposure indices. Other specimens, such as hair and nails, are subject to contamination and are not very reliable on a consistent basis (Zielhuis, 1986).

10.3.2 Who Should Collect Biological Samples

Occupational health professionals who are involved in conducting biological monitoring must have adequate training in how to collect biological samples. They should be prepared to take all the necessary precautions during collecting, handling, and transport of samples and disposal of equipment used. In addition, they should explain the rationale for the procedure, inform the workers what tests will be done on the samples, and reassure them on tests not performed, such as HIV testing on blood samples. Those who obtain biological samples should also have the means to review the tests and be cognizant of extraneous factors that can affect these results. Ideally, biological monitoring should be performed by competent occupational health professionals (Zielhuis, 1986).

10.3.3 Communication of Biological Monitoring Results

The occupational health physician who oversees the monitoring is responsible for informing individual workers of their results. In some instances, the results may be communicated to the person's private physician for inclusion into health records. Management and unions can be provided with *grouped data*, taking the necessary precautions to maintain confidentiality by removing specific identifiers. It would be beneficial to inform safety directors and industrial hygienists to coordinate results from ambient monitoring and to confirm satisfactory control measures for reduction of exposure (Zielhuis, 1986).

10.4 TOXICOLOGIC AND PHARMACOKINETIC CONSIDERATIONS

"Toxicology is concerned with the study of the mechanisms of actions and adverse effects of chemical agents on living organisms" (McCunney, 1994). Its primary objective is to describe the nature of deleterious health effects produced by toxicants, as well as the dose range over which these effects occur. This information provides a foundation for predicting acceptable levels of exposure for working populations. Biological monitoring must adhere to the principles of toxicology in order to be effective.

Toxic effects of a chemical agent in humans cannot exist unless the agent or its metabolites reach appropriate receptors in the body at a concentration and for a length of time sufficient to initiate toxic manifestations. The critical factors influencing this event are the route of contact of the chemical, the dose, and the duration and frequency of exposure (McCunney, 1994).

10.4.1 Routes of Entry

Toxic agents enter into the body by three major routes. Some chemicals are inhaled through the lungs, for example, beryllium dust, while others are absorbed through the skin or ingested through the gastrointestinal tract. For many chemicals, such as organic solvents, multiple routes of exposure exist. Numerous factors affect the uptake of chemicals in the body and, in turn, the internal concentration. Some factors relate to the chemical itself, such as the physical form (gas, liquids, solids), while others relate to the environment, such as temperature and humidity. Host factors also play an important role, as illustrated by the fact that strenuous activity may increase the absorption of toxicants through the lungs (see Table 10.2).

10.4.2 Dose

The concentration or dose of exposure is another key determinant of a material's potential toxicity. A dose–response relationship implies that toxic manifestations typically occur with greater frequency and severity as the dose increases (see Fig. 10.2).

A key concept in the dose–response relationship is a threshold dose for observable biological effects. Typically toxicology studies are performed on a population of animals to determine the dose of a given chemical that will cause a particular effect. This effect may be adverse, such as lethargy. The lowest concentration whereby an observable adverse effect occurs is known as the *lowest observable adverse effect level*, or the LOAEL. As the dose is increased, a certain percentage of the animals in the study will perish. The dose at which 50% of the animals die is called the *lethal dose 50*, or LD50. Similarly, the lowest concentration of chemical that leads to the death of all the animals is called the *lethal dose 100*, or LD100. These results are usually extrap-

TABLE 10.2 Factors that Modify Toxicity

Host
Age
Gender
Infectious/immunological history
Behavioral stress history
Activity level/fitness
Nutritional status
Toxicant exposure history
Environment
Temperature
Light: cycle, intensity, spectral distribution
Air: Flow rate, ion content, humidity
Toxicant
Matrix/bioavailability
Physical form: liquid, solid, gas, fume, dust, etc.
Chemical form

Source: ACGIH, 1995a.

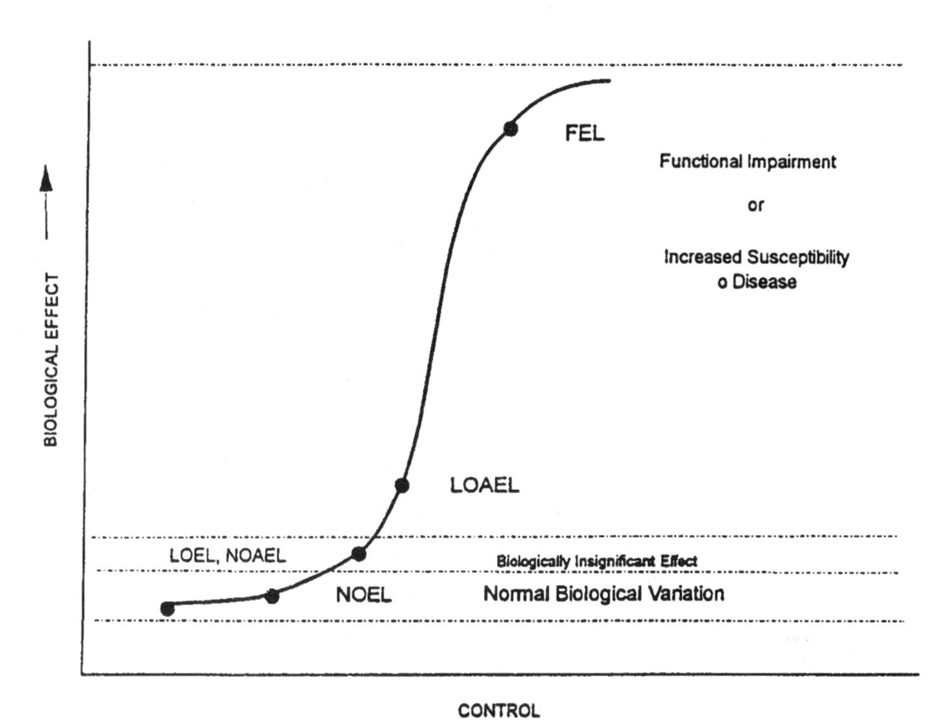

Fig. 10.2 Dose–response curve. FEL = Frank effect level. LOAEL = lowest observable adverse effect level. LOEL = lowest observable effect level. NOAEL = no observable adverse effect level. NOEL = no observable effect level. *Source*: Adapted from Derelanko, 1995.

olated to human populations by taking into account such factors as size and different metabolisms (Derelanko, 1995).

Although this principle applies to most situations, when immunological mechanisms are involved, this model does not work. In the case of beryllium exposure, certain individuals may be susceptible to trivial doses and develop a hypersensitivity reaction in the lungs. Here, the dose of immunogenic material needed to elicit symptoms in a susceptible worker can be very low due to prior sensitization (McCunney, 1994).

10.4.3 Duration

Toxicity is also determined by the duration of exposure. In some cases, a continuous long-term exposure at a lower dose may be as toxic as a large dose over a shorter period. The assumption here is that there may be a cumulative effect of chronic low-level exposures in some instances. Both immediate and late effects can occur with both acute and chronic exposures. An excellent illustration of this concept is occupational lead exposure. For lead poisoning to develop, major acute exposures to lead may not occur. The body accumulates this metal over a lifetime and releases it slowly; so even small doses, over time, can cause lead poisoning. The total body burden of lead appears related to the risk of the adverse effects. If acute lead exposures do occur, people may suffer various gastrointestinal and nervous system effects, and this may be reflected in blood lead levels (McCunney, 1994).

Several other factors influence whether a sufficient concentration of a toxic material reaches receptor sites within the body to initiate an adverse effect. These include the rate and the amount of the material absorbed, the distribution of the toxicant within the body, the rate of metabolism, or biochemical transformation, if any, and the rate of excretion of the toxicant, or its metabolites (Rappaport, 1985).

Respiratory absorption is primarily affected by the solubility of the chemical in the respiratory tract and interaction with the surface of the respiratory airways. Interindividual variation may alter solubility of toxicants in the blood according to the quantity of food intake and the quality of the diet. Dermal absorption depends on the exposure site, the condition of the skin (i.e., intact versus abraded), and humidity and temperature in the workplace (Fiserova-Bergerova, 1990). Gastrointestinal absorption depends on the stomach content; for example, there is less ethanol absorption when the stomach is full.

10.4.4 Distribution

Chemicals are distributed in the body based primarily on body size and composition. For example, lipid-soluble chemicals, such as toluene, distribute differently in slim versus obese individuals. The site of accumulation of a material may determine where the material exerts its major toxicity. For example, asbestos and other respirable dusts accumulate in the lungs, and this organ is where the toxicity is most observed. Some chemicals, such as inorganic lead, are found in storage sites, such as bones, such that levels measured in the blood will not accurately reflect total body burden (Derelenko, 1995).

Some toxicants bind to plasma proteins, thereby limiting the amount available for biomonitoring. Cadmium binding to metalloprotein in the kidneys is an example of how the saturation of these binding sites alters its measurement in urine (Shaikh, 1987). In addition, other toxicants or drugs may simultaneously compete for plasma protein and alter measurement of the index chemical (Table 10.3).

Essentially, the extent of protein binding of the chemical in the bloodstream determines the concentration of free chemical available for measurement. Therefore, it becomes necessary to understand the protein-binding characteristics of toxicants for meaningful biological monitoring to occur (Que Hee, 1993).

10.4.5 Metabolism

Biotransformation, or metabolism, is probably the most important source of pharmacokinetic variability (Table 10.4). Cytochrome P450 enzymes in the liver play a predominant role in processing chemicals in the body; however, extrahepatic metabolism can also be significant through the lungs and skin (Que Hee, 1993).

Most substances are detoxified by the liver into less harmful metabolites; however, some chemicals, such as methanol, are chemically transformed by the liver into toxic metabolites like formic acid. In this instance, it is the formic acid that is more toxic and causes damage to the optic nerve, leading to blindness. Measuring levels of methanol in the blood would not be an appropriate biological marker of toxicity (Sedivec, 1981).

The rate of metabolism of chemicals in the body can be enhanced (induced) or inhibited depending upon the presence of other chemical exposures. For example, xylene has been shown to induce the metabolism of both benzene and toluene in the liver. Thus, when present in a mixture of other solvents, xylene can increase the adverse effects of

TABLE 10.3 Medication that Potentially Interferes with Biological Monitoring

Mechanism medications	Chemicals affected
Nontoxicokinetic Effects	
Phenol-containing antiseptic lotion, lozenges, antacids	Benzene, phenol
Sedative/hypnotic: chloral hydrate	Trichloroethylene, perchloroethylene, methyl chloroform
Analgesics: Phenacetin, acetanilide, phenazopyridine	Aniline
Disulfiram (Antabuse)	Carbon disulfide
Inorganic fluoride-containing dental products, fluorine-containing anesthetics, fluorosteroids	Fluoride
Aluminum-containing antacids	Aluminum
Vitamins with essential metals supplements	Mercury and other metals
Homeopathic and patent medicines	Arsenic, lead, mercury, other metals
Effects on Uptake	
Adrenergic stimulants or blockers (pulmonary)	None known
Antacids (gastrointestinal)	Possibly ingested metals
	Organic chemicals metabolized by microsomal enzymes
Effects on Metabolism	
INHIBITION	
Sedative/hypnotics: barbiturates (allobarbital, secobarbital)	
Antihistamine: diphenhydramine	
Antibiotic: chloramphenicol	
Antiviral: amantidine	
Anti-inflammatory: indomethacin	
Miscellaneous: Propanolol, aspirin, cimetidine, disulfiram	*m*-Xylene
INDUCTION	
Anticonvulsants: phenobarbital, phenytoin, carbamazepine	
Sedative/hypnotics: barbiturates (barbital, amobarbital, phenobarbital), glutethimide, meprobamate	
Analgesic: antipyrine	
Anti-inflammatory: phenylbutazone	
Steroid: testosterone	
Antibiotic: rifampin	
Antifungal: griseofulvin	
Antimalarial: quinine	

TABLE 10.3 (*Continued*)

Mechanism medications	Chemicals affected
Effects on Hepatic Elimination	
Decreased hepatic blood flow: adrenergic blockers (propanolol), cimetidine Increased hepatic blood flow: Andrenergic stimulants (epinephrine, nicotine, caffeine)	Chemicals with high intrinsic hepatic clearance
Effects on Renal Elimination	
EFFECTS ON URINE pH	Determinants excreted as weak organic acids or bases (e.g., 2,4-dichlorophenoxyacetic acid)
Ammonium chloride, sodium bicarbonate, thiazide diurectics, acetazolamide	
EFFECTS ON CREATININE EXCRETION	Chemicals with determinants in urine corrected for creatinine concentration
Cimetidine	
EFFECTS ON URINE FLOW	Chemical excreted by glomerrular filtration
Diuretics administered acutely	
Effects on Binding	
Salicylates, sulfonamides, phenylbutazone Metals in vitamin supplements	Metals
Pharmacodynamic Effects	
Cholinesterase inhibitors: antiglaucoma (isofluorophate, demecarium, echothiophate) antimyasthenia gravis (neostigmine, pyridostigmine)	Cholinesterase inhibitors (organophosphate, carbamate pesticides) monitored by cholinesterase activity
Methemoglobin inducers: nitrates, nitroglycerin, benozocaine, prilocaine, silver nitrate, sulfonamides, phenacetin, acetanilide, bismuth subnitrate	Methemoglobin inducers monitored by methemoglobin levels in blood

Source: ACGIH, 1995a.

those compounds, which exert their toxicity through more toxic metabolites (phenol for benzene, hippuric acid for toluene) (Tardif, 1991).

Special consideration should be given to alcohol because of its frequent consumption and its unique ability to affect the metabolism of many workplace chemicals. An example of this is the inhibition of the metabolism of trichlorethylene following the ingestion of ethanol (Muller, 1975). It has been shown by Sato et al. (1993) that ingestion of moderate amounts of ethanol before the start of work or at lunch time produces marked increases in blood trichloroethylene concentrations and decreases in the urinary excretion rates of its metabolites (trichloracetic acid, etc.). This effect is due to the fact that when ethanol is present in high concentrations, it competes with trichloroethylene

TABLE 10.4 Sources of Pharmacokinetic Variability

Absorption	Distribution	Metabolism
Route	Body size	Genetic factors
Physical form	Body composition	Age and sex
Solubility	Protein binding	Environment (pollution, diet)
Physical workload	Physical workload	Chemical intake (alcohol, medication)
Exposure concentration	Exposure concentration	Physical activity (pulmonary ventilation), blood flow
Exposure duration	Exposure duration	Protein binding lifestyle (smoking)
Skin characteristics		Exposure level

Source: ACGIH, 1995a.

for metabolism by the liver enzymes. This factor can significantly affect the results of biological monitoring of exposure to organic solvents, such as trichloroethylene (Sato, 1993).

Chemicals that are poorly metabolized have more of the parent chemical present in biological media for measuring purposes. Thus the parent chemical is a better indicator of exposure than the metabolite. An example would be tetrachloroethylene, an organic solvent that is slowly metabolized and more readily detectable in expired air as the unchanged parent chemical (Skender, 1991).

For highly metabolized chemicals, the opposite effect is true. The biological level of the metabolic product is a better indicator of exposure than the level of the parent chemical. For example, only a small proportion of inhaled styrene (5%) is eliminated unchanged with expired air or excreted unchanged ($<2\%$) in urine. The large majority of absorbed styrene is oxidized to mandelic acid, which is excreted in the urine. Thus it would be more appropriate to monitor biologically for styrene exposure by measuring urinary mandelic acid level (Ong, 1995).

The same metabolite may be produced by more than one parent chemical; for example, both styrene and ethylbenzene may be metabolized into mandelic acid. Also, trichloroethylene, tetrachloroethylene, methylchloroform, and pentachloroethanol may be transformed into trichloracetic acid via metabolic pathways in the liver. This fact must be considered when selecting the best indicator of exposure for a given situation (Ducatman, 1984).

10.4.6 Excretion

Foreign substances, or *xenobiotics*, are eliminated from the body through a variety of different routes depending upon their physical and chemical properties and the body's ability to handle them. Although most chemicals are predominantly excreted in urine, feces, and expired air, smaller quantities are removed through other routes, such as perspiration and tears. One of the basic considerations in biological monitoring is which biological specimen will contain the index chemical at a level that will most accurately reflect exposure (Klassen, 1986).

Some volatile organic solvents, such as tetrachloroethylene, are mainly excreted unchanged in exhaled air (97%). On the other hand, water-soluble compounds, such as cadmium and mercury, are more favorably excreted by the kidneys in the urine (Klassen, 1986).

Other factors influence the amount of xenobiotic excreted. With regard to kidney excretion, the following factors may influence the amount of index chemicals found in the urine:

1. The degree of protein binding of the chemical in the blood.
2. Urine pH.
3. Fluid intake (increased fluid intake tends to dilute the concentration of the index chemical in the urine, whereas dehydration has the opposite effect).
4. State of health (kidney disease can impair the body's ability to eliminate xenobiotic effectively) (Klassen, 1986).

10.5 CLINICAL APPLICATION OF BIOMARKERS IN OCCUPATIONAL MEDICINE

The occupational medicine physician utilizes biological monitoring to help guide decision-making in evaluating patients who have been exposed to toxic chemicals. It is used in conjunction with the occupational history and physical examination to determine a possible chemical cause for various disease states. "It must be noted, however, that occupational disease is often indistinguishable from non-occupational disease" (McCunney, 1995). A primary example is oat cell carcinoma of the lung, a cancer associated with exposure to bischloromethyl ether, but also the same histologic type that develops in cigarette smokers (McCunney, 1995). The resultant prognosis and treatment is the same for both conditions, despite the fact that they arise from different causes.

Other examples include asthma associated with nonoccupational substances, such as dust and mites, and asthma, which is due to exposure to isocyanates. Once again both forms of asthma are indistinguishable clinically, may result in the same prognosis, and require the same treatment.

The occupational history is extremely important in deciding whether occupational factors are involved in the development of disease. However, subjective responses inherent in such a history may lead to false conclusions. In addition, although a physical is extremely useful in guiding the physician to a diagnosis, the physical examination may provide no useful information in determining the cause of an illness. Biological monitoring can serve as an useful adjunct to the occupational history and physical examination in determining the cause of a particular disease condition (McCunney, 1995).

10.5.1 Definitions

The terms *biomonitoring*, *biomarkers*, *medical monitoring*, and *medical surveillance* have been erroneously used interchangeably in the literature. The term *biomarker* refers to clinical tests that are specific to an occupational or environmental exposure or illness. Biomarkers were defined by the National Research Council in 1982, as follows:

1. Biological markers of *exposure* reflect an exogenous substance, or its metabolite, that is measured in a compartment of the body (i.e., blood lead levels).

2. Biological markers of *effect* refer to measurable alterations that can be recognized as an established or potential health impairment (i.e., carboxyhemoglobin levels in the blood as a measure of carbon monoxide exposure).
3. Biological markers of *susceptibility* are indicators of a person's ability to respond to a chemical challenge. (i.e., liver function tests as a marker of a person's ability to detoxify organic solvents).

10.5.2 Role of the Occupational Physician

The occupational physician is frequently asked to determine causality between a chemical exposure and a disease. These situations are usually broken down into two exposure scenarios: acute and chronic, and biomarkers may be implemented for each of these types of exposures. Examples of acute exposures include workers who are exposed to organic solvents during a spill at a factory or individuals exposed to high levels of cadmium dust as a result of a poorly functioning ventilation hood. Individuals may also suffer from the toxic effects of chronic exposures, during months or years in the workplace to such chemicals as polycyclic aromatic hydrocarbons in the steel industry, cobalt in metal refining operations, or asbestos insulation in pipes. Other examples of chronic exposures include arsenic in hazardous waste workers and molybdenum exposure from producing catalysts for the chemical industry (McCunney, 1995).

The retrospective study of exposures allows for assessment of past exposure. Examples of this would be state health officials performing epidemiologic studies on previous exposures to lead, beryllium, asbestos, and other hazardous materials (McCunney, 1995).

Occupational medicine physicians are often called upon to perform medical surveillance evaluations on populations of workers. "Medical surveillance refers to the systematic collection, analysis, and dissemination of health-related information on groups of people in order to identify occupational disease" (see Chapter 11). Biological monitoring is a component of medical surveillance. The Occupational Safety and Health Administration has in place, as of 1994, 28 standards that have medical surveillance provisions; however, only three of these (i.e., arsenic, lead, and cadmium) require specific biological monitoring (Table 10.5) (McCunney, 1994).

For effective biological monitoring, professional judgment must be exercised in determining an appropriate plan. There should be a hypothesis developed for assessing a potential exposure/disease relationship. Decisions are necessary regarding what groups of people will be tested, what parameter (the parent chemical or one of its metabolites) will be assessed, what type of biological specimen (blood, urine, exhaled air) will be analyzed, and the timing of the sampling. Other factors that need to be considered include the availability of appropriate reference limits for the substance, as well as its metabolic profile. In addition, there needs to be some consideration of the potential effects of such host factors as dietary deficiencies that can alter the absorption of some hazardous substances. An example would be the increased absorption of cadmium and lead secondary to either calcium or iron deficiency syndromes (Clarkson, 1988).

Several reliable references are available to assist in interpreting results of biological monitoring, including the American Conference of Governmental Industrial Hygienists Biological Exposure Indices (ACGIH, 1995d), and texts by Lauwreys (1993) and Baselt (1988). With reference to the Biological Exposure Indices (see BEI table), one may decide to examine a blood specimen as opposed to an exhaled air specimen, depend-

TABLE 10.5 Medical Screening Provisions in OSHA Standards: Primary Adverse Health Effects

Agent/worker group	To be detected by screening	Examination interval[a]	Examination contents[a]
2-Acetylaminofluorene	Cancer (site unspecified)	Annual	Hx, PE
Acrylonitrile	CNS effects, dermatitis, lung cancer, GI cancer	Annual	Hx PE CXR, fecal blood[b]
4-Aminodiphenyl	Cancer (side unspecified)	Annual	Hx, PE
Arsenic, inorganic	Skin lesions (including cancer), lung cancer, nasal septal perforation	Annual or semiannual[b]	Hx, PE, CXR, sputum cytology[b]
Asbestos	Pulmonary disease, lung and GI cancer, mesothelioma	Annual	Hx, questionnaire,[c] PE, CXR,[b] PFTs
Benzene	Hematological disease, including aplastic anemia and leukemia	Annual	Hx, CBC, PFTs
Benzidine	Cancer (site unspecified)	Annual	Hx, PE
bis-Chloromethyl ether	Cancer (site unspecified)	Annual	Hx, PE
Coke oven emissions	Lung cancer, kidney cancer, skin cancer	Annual or seminnual[b]	Hx, PE, CXR, PFTs, U/A sputum and urine cytology[b]
Cotton dust	Byssinosis	Biennial, annual, or semiannual[d]	Hx, questionnaire, PFTs
1,2-Dibromo-3-chloropropane	Male and female germ cell toxicity, cancer (site unspecified)	Annual	Hx, PE, reproductive tests[a]
3,3-Dichlorobenzidine	Cancer (site unspecified)	Annual	Hx, PE
4-Dimethylaminoazobenzene	Cancer (site unspecified)	Annual	Hx, PE
Ethylene oxide	Leukemia, dermatitis, reproductive toxicity, neurological toxicity, pulmonary toxicity	Annual	Hx, PE, CBC, Reproductive tests[a]
Ethylenemine	Cancer (site unspecified)	Annual	Hx, PE
Formaldehyde	Mucous membrane iriation, skin and pulmonary sensitization	Annual	Questionnaire,[c] PE, PFTs

Agent/worker group	To be detected by screening	Examination interval[a]	Examination contents[a]
Lead	Hematological effects, renal disease, reproductive, toxicity, neurological disease	Exam: at least annual[f] biomonitoring: at least semiannual[f]	Hx, PE, CBC, U/A, BUN, Cr, biomonitoring
Methyl chloromethyl ether	Cancer (site unspecified)	Annual	Hx, PE
Naphthylamine	Cancer (site unspecified)	Annual	Hx, PE
Naphthylamine	Cancer (site unspecified)	Annual	Hx, PE
4-Nitrobiphenyl	Cancer (site unspecified)	Annual	Hx, PE
N-Nitrosodimethylamine	Cancer (site unspecified)	Annual	Hx, PE
Noise	Hearing loss	Annual	Audimetry
Propiolactone	Cancer (site unspecified)	Annual	Hx, PE
Vinyl chloride	Hepatic disease, hepatic cancer	Annual or semiannual[b]	Hx, PE, LFTs
Employees using respirators	Conditions causing inability to wear respirator	"Periodically"[g]	[g,h]
Hazmat workers	[a]	Annual	Hx, PE[b]
Laboratory workers	[a]	[a]	[a]

[a]All standards require preplacement and termination as well as periodic examinations. For most standards, additional examinations are mandated in emergency situations, for symptomatic employees, or at the discretion of the examining physician.

[b]Frequency determined by age of employee, years of exposure or both.

[c]Questionnaire provided in appendix to standard.

[d]Frequency determined by exposure level and PFT results.

[e]If requested by physician or employee.

[f]Examination and biomonitoring frequency determined by previous blood lead levels and symptomatology.

[g]The formaldehyde standard requires annual PFTs for respirator users; the benzene standard requires PFTs every 3 years.

[h]Testing to be determined by physician.

Abbrevs.: Hx, medical, work, family history; PE, physical examination; CXR, chest x-ray; LFTs, routine serum liver function test; CBC, complete blood count, or other routine hematological test; U/A urinalysis; PFTs, pulmonary function tests (spirometry only), BUN, serum blood urea nitrogen, Cr, serum creatinine; CNS, central nervous system; GI, gastrointestinal.

Source: From U.S. Department of Labor, Occupational Safety and Health Administration, Washington, DC; General Industry OSHA Safety and Health Standards (29 CFR 1910), 1988 Hazardous Waste Operations Emergency Response (29 CFR 1910, 120), 1989 Occupational Exposures to Hazardous Chemicals in Laboratories (29 CFR 1910, 1450), 1990.

ing on the circumstances. The ACGIH text also assists one in determining appropriate sampling time and appropriate method of analysis for the various biological specimens and index chemicals concerned. These guidelines refer to the levels of the agent itself, or its metabolites, in urine, blood, or exhaled air that are associated with exposures to each substance at the threshold limit value (TLV). "In certain instances, the concentrations noted in these indices represent levels of exposure and not adverse health effects" (Lowrey, 1986).

10.5.3 Role of Biomarkers in Illness Evaluation

The use of biomarkers in evaluating illness has proven to be extremely beneficial to the occupational medicine physician. Although biological monitoring per se is not used to make a clinical diagnosis, it can be instrumental in determining early effects of a chemical on individuals who may be susceptible to a variety of toxic responses. In this sense we are talking about both biological effect and susceptibility.

In some cases, biological markers of effect may assist physicians in the treatment of acute exposures to toxic chemicals. An illustration of this concept would be the treatment of a firefighter overcome by carbon monoxide while involved in fire suppression. By measuring the *carboxyhemoglobin* level in the blood, the clinician could monitor the response to oxygen treatment.

Recently, it has been shown that subtle changes in such molecules as DNA and hemoglobin may serve as biomarkers for exposure to such carcinogens as polycyclic aromatic hydrocarbons (PAHs). The formation of adducts to both DNA and hemoglobin result from the chemical binding that occurs between the carcinogen and these macromolecules. It is hoped that in the future some forms of cancer may be prevented by the use of similar biomarkers.

Other tests have been recently developed to identify susceptibility to such toxic agents as beryllium. Specifically, the lymphocyte transformation test (LTT) allows physicians to identify individuals with an immunologic sensitivity to beryllium who may go on to develop a serious and sometimes fatal lung disease known as chronic berylliosis (Kreis, 1993). Blood samples are taken from individuals exposed to beryllium, and beryllium sulfate is added in vitro. The lymphocytes (white blood cells) in sensitive individuals will transform into primitive forms and subsequently proliferate in culture. In the past, the diagnosis of this disease was only made in the advanced stages, with little chance of amelioration. Now doctors may intervene at earlier points and slow down the disease process. An even newer generation of tests looking at genetic susceptibility may allow physicians to prevent the disease entirely before exposure occurs. A blood test for this gene and other genes shows great promise for future disease prevention (Saltini, 1995).

Biomarkers currently play a small role in diagnosing common occupational diseases such as occupational asthma and the pneumoconioses (i.e., asbestosis, black lung disease, etc.). Further research is needed to aid in the early recognition and prevention of these and other maladies.

10.5.4 Summary

The modern clinical practice of occupational medicine more effectively employs biomarkers in assessing exposures rather than diseases. Although biomarkers exist for chronic exposures (i.e., evaluation of lead exposure over the past 3 months by measuring blood levels of zinc protoporphyrin (ZPP), the use of biomarkers seems more amenable to acute exposures.

The ramifications of conducting biological monitoring should be carefully considered prior to implementing a program. Patients are often concerned when advised that a toxicologic screen identifies the presence of a chemical agent in their urine or blood. It is incumbent upon the physician to realize: (1) that the mere presence of the chemical does not imply toxicity; and (2) that although the test exists, there may be no sound scientific reason to perform it. The physician must be able to understand the correlation between given exposures, the levels found in the biological specimens, and the existence of symptoms or disease (McCunney, 1995).

There are many different perspectives to biological monitoring. In some instances, labor and management may resist the use of biomarkers in their work force for different reasons. For example, labor unions may be skeptical about the adverse impact on employment if workers will be removed secondary to biomonitoring results. Similarly, management officials may view the results of biomonitoring as a source of liability. Individual employees may see themselves as "guinea pigs" and question the validity of such a program from its inception (McCunney, 1995).

Ethical issues surround controversial decisions of what to do as a result of the findings of biological monitoring. An illustration of this concept involves the former practice of chelating workers with grossly elevated lead levels and then returning them to the work force without regard to reducing the exposure with costly engineering controls. In this way, the lead exposure was being "masked" by removing the excess lead from the workers. Fortunately, this practice has been eliminated, but it serves as a reminder that only those with the highest ethical standards implement biological monitoring the way it was intended.

As newer forms of biomarkers become available, the sociopolitical and ethical demands associated with biological monitoring will increase. It becomes increasingly important for the occupational health professional to be judicious in recommending biomarkers for certain work and environmental settings.

10.6 BIOLOGICAL EXPOSURE INDICES

The American Conference of Governmental Industrial Hygienists (ACGIH) in 1982 formed a Biological Exposure Indices (BEI) Committee. The committee, which was composed of a distinguished group of scientists, developed the first six biological exposure indices, which were promulgated in 1984. In 1995 recommendations were made to add 35 chemicals, of which 32 have been adopted, and 3 are on notice of intent to establish (Table 10.6). In the development of the biological exposure indices, the ACGIH BEI Committee utilized information from multiple sources (Table 10.7).

10.6.1 Implementation of Biological Exposure Indices

Before implementing the BEIs in a biological monitoring program, one must consider conditions which may affect the values of the BEIs. It is important to recognize that a single measurement that falls below the BEI does not imply safety in the workplace. If the sampling is repeated, however, and if measurements and specimens obtained from a worker on different occasions persistently exceed the BEI, or if the majority of measurements and specimens obtained from a group of workers at the same workplace exceed

TABLE 10.6 Adopted Biological Exposure Determinants

Chemical (CAS #) determinant	Sampling time	BEI	Notation
Acetone (67-43-1) (1994)			
Acetone in urine	End of shift	100 mg/L	B Ns
Aniline (62-53-3) (1991)			
Total *p*-aminophenol in urine	End of shift	50 mg/g creatinine	Ns
Methemoglobin in blood	During or end of shift	1.5% of hemoglobin	B, Ns
Arsenic and soluble compounds including arsine (7784-42-1) (1993)			
Inorganic arsenic metabolities in urine	End of worksheet	50 μg/g creatinine	B
Benzene (71-43-2) (1987) (see note below)			
Total phenol in urine	End of shift	50 mg/g creatinine	B, Ns
Benzene in exhaled air	Prior to next shift		
mixed-exhaled		0.08 ppm	Sq
end-exhaled		0.12 ppm	Sq
Cadium and inorganic compounds (1993)			
Cadmium in urine	Not critical	5 μg/g creatinine	B
Cadmium in blood	Not critical	5 μg/L	B
Carbon disulfide (75-15-0) (1988)			
2-Thiothiazolidine-4-carboxylic acid (TTCA) in urine	End of shift	5 mg/g creatinine	
Carbon monoxide (630-08-0) (1993)			
Carboxyhemoglobin in blood	End of shift	3.5% of hemoglobin	B, Ns
Carbon monoxide in end-exhaled air	End of shift	20 ppm	
Chlorobenzene (108-90-71) (1992)			
Total 4-chlorocatechol in urine	End of shift	150 mg/g creatinine	Ns
Total *p*-chlorophenol in urine	End of shift	25 mg/g creatinine	Ns
Chromium (VI), water-soluble fume (1990)			
Total chromium in urine	Increase during shift	10 μg/g creatinine	B
	End of shift at end of workweek	30 μg/g creatinine	B
* Cobalt (7740-48-4) (1995)			
Cobalt in urine	End of shift at end of workweek	15 μg/L	B
Cobalt in blood	End of shift at end of workweek	1 μg/L	B, Sc

* N,N-Dimethylacetamide (127-19-5) (1995)			
N-Methylacetamide in urine	End of shift at end of workweek	30 mg/g creatinine	
+ N,N-Dimethylformamide (DMF) (68-12-2) (1988)			
N-Methylformamide in urine	End of shift	(40 mg/g creatinine)	
2-Ethoxyethanol (EGEE) (110-80-5) and 2-Ethoxyethyl acetate (EGEEA) (111-15-9) (1994)			
2-Ethoxyacetic acid in urine	End of shift at end of workweek	100 mg/g creatinine	
Ethyl bezene (100-41-4) (1986)			
Mandelic acid in urine	End of shift at end of workweek	1.5 g/g creatinine	Ns
Ethyl benzene in end-exhaled air			Sq
Fluorides (1990)			
Fluorides in urine	Prior to shift	3 mg/g creatinine	B, Ns
	End of shift	10 mg/g creatinine	B, Ns
Furfural (98-01-1) (1991)			
Total furoic acid in urine	End of shift	200 mg/g creatinine	B, Ns
n-Hexane (110-54-3) (1987)			
2,5-Hexanedione in urine	End of shift	5 mg/g creatinine	Ns
n-Hexane in end-exhaled air			Sq
* Lead (1995)[b]			
Lead in blood	Not critical	30 μg/100 ml	B
Mercury (1993)			
Total inorganic mercury in urine	Preshift	35 μg/g creatinine	B
Total inorganic mercury in blood	End of shift at end of workweek	15 μg/L	B
Methanol (67-56-1) (1995)			
Methanol in urine	End of shift	15 mg/L	B, Ns
Methemoglobin inducers (1990)			
Methemoglobin in blood	During or end of shift	1.5% of hemoglobin	B, Ns
Methyl chloroform (71-55-6) (1989)			
Methyl chloroform in end-exhaled air	Prior to the last shift of workweek	40 ppm	
Trichloroacetic acid in urine	End of workweek	10 mg/L	Ns, S
Total trichloroethanol in urine	End of shift at end of workweek	30 mg/L	Ns, S
Total trichloroethanol in blood	End of shift at end of workweek	1 mg/L	Ns
Methylethyl ketone (MEK) (78-93-3) (1988)			
MEK in urine	End of shift	2 mg/L	

TABLE 10.6 ***(Continued)***

Chemical (CAS #) determinant	Sampling time	BEI	Notation
Methylisobutyl ketone (MIBK) (108-10-1) (1993)			
MIBK in urine	End of shift	2 mg/L	
Nitrobenzene (98-95-3) (1991)			
Total *p*-nitrophenol in urine	End of shift at end of workweek	5 mg/g creatinine	Ns
Methemoglobin in blood	End of shift	1.5% of hemoglobin	B, Ns
Organophosphorus cholinesterase inhibitors (1989)			
Cholinesterase activity in red cells	Discretionary	70% of individual's baseline	B, Ns
Parathion (56-38-2) (1989)			
Total *p*-nitrophenol in urine	End of shift	0.5 mg/g creatinine	Ns
Cholinesterase activity in red cells	Discretionary	70% of individual's baseline	B, Ns
Pentachlorophenol (PCP) (87-86-4) (1989)			
Total PCP in urine	Prior to the last shift of workweek	2 mg/g creatinine	B
Free PCP in plasma	End of shift	5 mg/L	B
+ Perchloroethylene (127-18-4) (1989)			
Perchloroethylene in end-exhaled air	Prior to the last shift of workweek	(10 ppm)	
Perchloroethylene in blood	Prior to the last shift of workweek	(1 mg/L)	
Trichloroacetic acid in urine	End of workweek	(7 mg/L)	Ns
Phenol (108-95-2) (1987)			
Total phenol in urine	End of shift	250 mg/g creatinine	B, Ns
Styrene (100-42-5) (1986)			
Mandelic acid in urine	End of shift	800 mg/g creatinine	Ns
	Prior to next shift	300 mg/g creatinine	Ns
Phenylglyoxylic acid in urine	End of shift	240 mg/g creatinine	Ns
	Prior to next shift	100 mg/g creatinine	
Styrene in venous blood	End of shift	0.55 mg/L	Sq
	Prior to next shift	0.02 mg/L	Sq
+(Toluene (108-88-3) (1986)			
(Hippuric acid in urine)	(End of shift)	(2.5 g/g creatinine)	(B)
	(Last 4 h of shift)		
(Toluene in venous blood)	(End of shift)	(1 mg/L)	(Sq)
(Toluene in end-exhaled air)			

Trichloroethylene (79-01-6) (1986)			
Trichloroacetic acid in urine	End of workweek	100 mg/g creatinine	Ns
Trichloroacetic and trichloroethanol in urine	End of shift at end of workweek	300 mg/g creatinine	Ns
Free trichloroethanol in blood	End of shift at end of workweek	4 mg/L	Ns
Trichloroethylene in blood (1993)			Sq
Trichloroethylene in end-exhaled air			Sq
Vanadium pentoxide (1314-62-1) (1995)			
Vanadium in urine	End of shift at end of workweek	50 μg/g creatinine	Sq
Xylenes (13307) (Technical Grade) (1986)			
Methylhippuric acids in urine	End of shift	1.5 g/g creatinine	
	Notice of Intent to Establish or Change		
N,*N*-Dimethylformamide (DMF) (68-12-2) (1993)			
N-Methylformamide in urine	End of shift	20 mg/g creatine	
+2-Methoxyethanol (EGME (109-86-4) and 2 Methoxyethyl acetate (EGMEA) (110-49-6)			
2-Methoxyacetic acid in urine	End of shift at end of workweek		Nq
Perchloroethylene (127-18-4) (1993)			
Perchloroethylene in end-exhaled air	Prior to last shift of workweek	5 ppm	
Perchloroethylene in blood	Prior to last shift of workweek	0.5 mg/L	
Trichloroacetic acid in urine	End of workweek	3.5 mg/L	Ns, S
Toluene (108-88-3) (1995)			
Toluene in venous blood	Prior to last shift of workweek		Nq
Hippuric acid in urine	End of shift		B, Nc
o-Cresol in urine	End of shift		B, Nc
Toluene in urine	Prior to last shift of workweek		Nq

[a]Note: The Chemical Substances TLV Committee has proposed a revision of the TLV for benzene. See Chemical Substances Notice of Intended Changes and TLV Documentation.

[b]Women of child bearing potential whose blood Pb exceeds 10 μg/dl are at risk of delivering a child with a blood Pb over the current Centers for Disease Control guideline of 5 μg/dl. If the blood Pb of such children of such remains elevated, they may be at increased risk cognitive deficits. The blood Pb of these children should be closely monitored and appropriate should be taken to minimize the child's exposure to environmental lead (CDC: Preventing Lead Poisoning in Young Children, October 1991: See BEI and TLV Documentations).

Source: ACGIH, 1995b.

TABLE 10.7 Practical Points Table

1. The presence of a chemical does not necessarily imply toxicity. The dose makes the poison.
2. Four checkpoints for effective biological monitoring are:
   a. Appropriate specimen selection;
   b. Selection of correct sampling time;
   c. Use of sensitive and selective analytical method;
   d. Interpretation based on a full knowledge of the metabolism.
3. Biological monitoring better assesses total chemical exposure reaching the biological target.
4. Biological monitoring of effects, i.e.; cholinesterase inhibitors in organophosphate exposure, usually give greater insight into the toxicity of different chemicals.
5. Correct compliance decisions are more likely to arise from biological monitoring than from traditional air monitoring.
6. Static (area) samples usually correlate poorly with biological exposure indices.
7. Biological monitoring is useful in assessing effects of personal protection equipment and environmental controls.
8. Advantages of biological monitoring
   a. Attempt to assess the parameter most directly related to potential health effects.
   b. Results can aid in formulating a more precise estimate of risk of illness secondary to exposure.
   c. Attempts to account for total exposure by factoring in:
      i. Breathing capacity;
      ii. Work effort;
      iii. Underlying medical conditions.
   d. Nonoccupational exposures and individual variability are assessed.
   e. Multiple exposures and other routes of exposure, such as dermal and oral can be evaluated.
9. Disadvantages of biological monitoring
   a. Effectiveness is dependent on adequate toxicologic data (for most substances used in industry today, this is not available).
   b. Test results can be affected by other factors such as alcohol and pregnancy (at the same blood lead levels women have higher levels of zinc protoporhyrinthanmen).
   c. Cigarette smoking may increase levels of cadmium in blood and urine.
   d. Dietary deficiencies can enhance the toxicity of some chemicals.
   e. For some substances, relatively short biologic half-lives affect the monitoring. In monitoring for dimethylformamide, a solvent used in the production of adhesives, 24-hour urinary samples. Within the time of last exposure can indicate the level of exposure, a worker has experienced. Beyond 48 hours from the time of last exposure, the major metabolite, *n*-methylformamide (NMF) is not detectable
   f. Monitoring is ineffective for surface acting agents such as sulfur dioxide and ammonia.
10. Biologic monitoring data are used to assess exposure to a hazard rather than to make a "clinical" diagnosis.
11. The biologic half-life of the agent will determine whether it will be detectable following exposure and, therefore, affects sampling time decision.
12. Timing of sample acquisition is most critical in substances that have short half-lives, i.e., obtaining a blood lead specimen after a 24-hour absence from work might be appropriate. However, obtaining a carbon monoxide analysis of exhaled air after such a time lapse would be invalid.
13. Urine is usually more suitable for monitoring hydrophilic chemicals, metals, and metabolites than for monitoring chemicals poorly soluble in water.

TABLE 10.7 *(Continued)*

14. Expired air may be more suitable for monitoring volatile organic solvents, which are poorly metabolized such as tetrachloroethylene.
15. It is not appropriate to use the human as an instrument for exposure assessment.
16. The pressure of measured chemicals or metabolites in body fluids or tissue indicates exposure and does not necessarily indicate disease.
17. Negative information is useful to confirm the adequacy of protective measure.
18. Positive information is useful to attempt to quantify delivered dose in circumstances in which protective measures have been breached or have failed.
19. The use of a qualified laboratory is essential.

the BEI, the cause of the excessive values should be evaluated and proper action taken (ACGIH, 1995d). Biological monitoring should be considered complementary to air monitoring in testing the efficacy of personal protective equipment and in determining the potential for absorption via the skin and the gastrointestinal system and in detecting nonoccupational exposure. "The existence of a BEI does not indicate the need for conducting biological monitoring" (ACGIH, 1995b). Once again, occupational health specialists must exercise professional judgement in designing appropriate monitoring protocols.

10.6.2 Interpretation of Biological Exposure Indices

In the interpretation of the results of biological monitoring, one must consider intra- and interindividual variability. These differences may arise as a result of variations in breathing rate, body composition, blood flow, efficacy of excretory organs, and activity of enzyme systems that metabolize a chemical. Multiple sampling is essential to reduce the effects of variable factors. Biological monitoring may supplement the results of air monitoring, but where there is discrepancy between the results, the entire exposure situation should be carefully reviewed, and an explanation found (Lowry, 1986). The biological exposure indices are reference values intended as *guidelines* for the evaluation of potential health hazards. They represent concentrations likely to be observed in specimens collected from a healthy worker exposed to chemicals to the same extent as a worker with inhalation exposure to the threshold limit value (TLV). Due to biological variability, the BEIs do not indicate a sharp distinction between hazardous and nonhazardous exposures. For example, it is possible for an individual's measurements to exceed the BEI without incurring a symptom or health problem. In general, the BEIs apply to 8 hour exposures, 5 days a week; however, BEIs for altered work schedules can be extrapolated. The BEIs are not intended for use as a measure of adverse effects or for diagnosis of occupational illness (Lowry, 1986).

10.6.3 When to Conduct Sampling

One of the most important considerations in developing an effective biological monitoring program is *timing* of the sampling. Timing must be carefully addressed because distribution and elimination of a chemical or its metabolic products, as well as biochemical changes induced by exposure to the chemical, are modified over time (Mutli, 1993). The BEIs are applicable only if collection is conducted at the specified time (Table 10.8). In many instances, when the level of the determinant changes rapidly or

TABLE 10.8 Methodological Considerations in Sampling, Storage, and Analysis of Exhaled Air, Blood, and Urine (ACGIH B)

	Exhaled air	Blood	Urine
		Sampling	
Suitable determinants	Volatile, stable, hydrophobic	Any	Any (most convenient for polar determinants)
Specimen characteristics	End-exhaled or mixed-exhaled air; mode of respiration (nose-mouth breathing	Whole blood, plasma, serum, cells, clotted blood	Spot speicmen, timed specimen
Technique	Noninvasive	Invasive	Noninvasive
Collection period	Instantaneous (single breath) or short-term (multibreath sample)	Instantaneous	Short-term (2–4 h)
Special qualification of health personnel	Well informed on technique	Medical	Informed
Infection protection	Sterile mouthpiece	Sterile needle	Clean container
Suitable containers	Air tight		
Material	Made of material that does not react or absorb determinant	Made of material that does not react or absorb determinant	Made of material that does not react or absorb determinant
Volume	20–50 ml for end-exhaled air; more than 1 L for mixed-exhaled air	Depends on method	50 ml or more
Precautions	Proper timing Normal pulmonary function, normal breathing (avoid hyperventilation; use low-resistance apparatus); sampling apparatus made from nonabsorbing material; condensation	Proper timing Venous blood (limited use of capillary blood): proper anticoagulant dry syringe	Proper timing Normal renal function
Source sof contamination	Ambient air	Skin exposure, cleaning fluid, syringe, needle anticoagulant	Skin exposure (hands, hair, cloth) (sampling after shower and clean cloths)
Hazards	Respiratory infection	Hepatitis, AIDS	

	Exhaled air	Blood	Urine
Transportation and Storage			
Source of contamination	Ambient air or container	Container	Container
Sources of deterioration	Temperature changes (leaks, condensation on surface of container)	Hemolysis, bacterial decomposition	Bacterial decomposition
Inconvenience in transporting	Avoid temperature changes	Low temperature required	Large volume and weight
Storing temperature	Room temperature	Refrigerated or frozen (after separation of serum and plasma)	Refrigerated or frozen
Analytical			
Cleanup procedure	None	Complex	Some
Possible interference	Condensation	Protein binding, conjugation, and chelation (result dependent on analytical method) Distribution in blood constituents; hemolysis	Protein binding, conjugation, and chelation (result dependent on analytical method) pH (for weak electrolytes)
Method	Sensitive and specific on-site analysis desirable	Sensitive and specific	Sensitive and specific
Determinants Requiring Special Considerations			
Parenteral chemical	Contamination during sampling procedure	Contamination during sampling procedure	Contamination during sampling procedure
With dermal absorption	None	Contamination during sampling procedure Representative sample	Contamination during sampling procedure Contamination
Volatile	Air tight containers avoid condensation	Airtight containers and controlled head space Anaerobic collection	Airtight containers and controlled head space Rapid collection
Solvents	Avoid contact with rubber and some plastics Mouthpieces, container, stopcock	Avoid contact with rubber and some plastics Container, stoppers	Airtight containers and controlled head space Container, stoppers
Metals	NA	Contamination free needles	Avoid contamination
Enzymes	NA	Low temperature	NA
Photosensitive	Dark containers		Dark containers

Sources: ACGIH, 1995a,c.

when there is potential accumulation of a substance, the sampling time is critical and must be carefully observed. The sampling time is specified in Table 10.7 according to the differences in the uptake and elimination rates of chemicals and their metabolites and according to the persistence of induced biochemical changes. Refer to the ACGIH documents for the BEIs for a more comprehensive discussion of sampling time, as well as methods of analysis, and storage of biological specimens. Where applicable, notations are provided to indicate increased susceptibility for some individuals, the presence of background levels in nonoccupationally exposed groups, and the nonspecificity of some findings. Some of the determinants are direct measures of exposure, such as the blood lead levels, while other determinants are indirect measures of effect, such as the carboxyhemoglobin levels secondary to exposure to methylenechloride, or the cholinesterase activity secondary to exposure to organophosphate pesticides. How the specific specimen should be obtained from urine, blood, or exhaled air, and the methods of analysis are described in Table 10.8.

10.7 SUMMARY

For effective biological monitoring, one must utilize the biological exposure indices as intended, and consider the variables affecting levels of determinants in biological samples (Table 10.9). Because of all the variables involved, BEIs are recommended only as *guidelines* for evaluation of data obtained in biological monitoring of the exposure in the workplace. Occupational health specialists must consider pertinent circumstances and exercise professional judgement when using BEIs. The documentations for the BEIs published by the ACGIH provide basic information to be considered when biological monitoring of the workplace is instituted and the BEI is applied (ACGIH, 1995c).

Biological monitoring can play an important role in the assessment of exposure to toxic chemicals. In a properly designed program, it complements ambient air monitoring and can identify health risks by assessing the overall effectiveness of a hazard control program. As part of a medical surveillance effort, it can help prevent the development of adverse health effects by early detection of biochemical changes that may herald subclinical disease. Biological monitoring's main focus is the primary prevention of toxic effects of chemical agents.

Although thousands of chemicals are used in occupational settings, standard reference values for interpreting the results from biomonitoring exists for only about forty chemicals. It remains a great challenge to determine safe or no effect levels of toxicants in biological media. For an assay to be useful, sufficient toxicological information must be available on the human absorption and metabolic fate of the agent. In addition, pre-exposure data or population norms should be available. Research will be needed to bridge the gap between current methods of analysis of biological specimens and the successful interpretation of the corresponding results. The mere presence of the technology to perform biological monitoring is not reason enough for its implementation in all cases. Occupational health professionals must weigh ethical, moral, and legal implications before implementing a program. Authorities tend to agree, however, that as biological monitoring techniques improve, occupational health specialists will detect lower levels of determinants in biological specimens. The implications of these developments will ideally aid in understanding illnesses associated with exposure to occupational and environmental hazards.

TABLE 10.9 Variable Affecting Levels of Determinants in Biological Samples (ACGIH A)

	Exhaled air	Blood	Urine
		Time	
Exposure duration[a]	Levels increase with exposure duration	Levels increase with exposure duration	Levels increase with exposure duration
Repetitious exposure	Levels increase if elimination half-time is greater than 5 h	Levels increase if elimination half-time is greater than 5 h	Levels increase if elimination half-time is greater than 5 h
Sampling time[a]	Levels increase (or decrease) with length of interval between start (or end) of exposure and sampling	Levels increase (or decrease) with length of interval between start (or end) of exposure and sampling	Levels increase (or decrease) with length of interval between start (or end) of exposure and sampling
Sampling period[b]	Momentary or mean levels of the sampling period; duration can affect the measurement	Momentary levels	Mean levels of the sampling period; duration can affect the measurement
		Physiological Parameters	
Pulmonary and cardiovascular[a]	Affect sampling		
	Physical activity during exposure increases absorption rate	Physical activity during exposure increases absorption rate	Physical activity during exposure increases absorption rate
	Decrease concentration	Increase levels of parental chemical; small increase of metabolites	Increase levels of parental chemical; small increase of metabolites
	Postexposure physical activity accelerates elimination	Postexposure physical activity accelerates elimination	Postexposure physical activity accelerates elimination
	Decrease concentration	Decrease levels of parental chemical; small effect on metabolites	Decrease levels of parental chemical; small effect on metabolites

TABLE 10.9 (*Continued*)

	Exhaled air	Blood	Urine
		Physiological Parameters (Continued)	
Metabolism of chemical	Decrease concentration	Decrease the concentration of parental chemical; increase the concentration of metabolite	Decrease the concentration of parental chemical; increase the concentration of metabolite
Binding to plasma proteins	Decrease levels	Levels of free determinant decrease Level of total determinant increase	Decrease levels
		Nutrition	
Body fat	Increase concentrations on lipophilic substances in samples collected on the following day[a]	Effects depends on determinant	Little effect
Fasting	Affects levels of lipid-soluble determinants	Affects levels of lipid-soluble determinants	Affects levels of lipid-soluble determinants
Diet	Levels rarely affected	May be source of determinant (increase background level); may change pH and thus levels of weak electrolytes; may affect binding sites of determinant	May be source of determinant (increase background level); may change pH and thus levels of weak electrolytes; may affect binding sites of determinant
Water intake	Insignificant effect	Insignificant effect	Usually affects concentration (adjustment for solids recommended)[a], extreme may affect excretion mechanism

	Exhaled air	Blood	Urine
		Effect of Disease	
Pulmonary disease	Significant	Significant	Indirect
Renal disease	Usually none	Significant	Significant
Liver disease	Significant for all chemicals undergoing metabolism	Significant for all chemicals undergoing metabolism	Significant for all chemialcs undergoing metabolism
Medication	May affect elimination either by altering metabolism or protein binding of determinant or be a source of determinant	May affect elimination either by altering metabolism or protein binding of determinant or be a source of determinant	May affect elimination either by altering metabolism or protein binding of determinant or be a source of determinant
		Environment	
Home exposure and workplace coexposure	Nonoccupational exposure to the chemical increases level of determinants; exposure to other chemical can alter the levels of determinants by altering activity of metabolizing enzymes or availability of binding sites.		
Temperature and altitude	Some	Some	Some

[a]Indicates variables with considerable effects.

[b]Measurements reflect the time-weighted average level of determinant during the sampling period. Sampling period is critical for those determinants whose biological levels rapidly respond to changes in the concentration of the chemical in ambient air.

10.8 CASE STUDY ON ARSENIC TOXICITY

You are the Safety Director at a well-known microelectronics company that manufactures semiconductors (ATSDR, 1990). Recently an OSHA inspection revealed that ambient levels of arsenic were well below the permissible exposures limits of 10 mcg/m^3. However, you know that some employees have been nonchalant about using respirators because of inconvenience or uncomfortableness. In addition, you have noted that one of these individuals has been uncharacteristically absent from work and was recently evaluated by his physician for pain and numbness in both feet and hands.

1. What biological monitoring measures could be employed to detect the internal dose of arsenic that these workers are receiving?

10.8.1 Direct Biological Indicators

The best indicator of recent arsenic exposure is a total urinary arsenic measurement in a 24-hour urine collection at the end of the work week (ATSDR, 1990). Spot urine collection may be useful in emergency situations if corrected to grams of creatinine in the urine. Most urinary arsenic levels in unexposed populations are below 50 micrograms/liter, and values above 200 micrograms/liter are considered abnormal (ACGIH, 1995c).

A dietary history is critical, as ingestion of seafood in the 48–72 hours preceding a test may significantly increase urinary arsenic concentrations. In one study, volunteers with an average pretest urinary arsenic level of 30 micrograms/liter were given a lobster tail for lunch. Four hours after eating, they had an average urinary level of 1300 micrograms per liter (ATSDR, 1990).

The BEI for arsenic is 50 micrograms per gram of creatinine and correlates well with exposures to arsenic compounds at air concentrations of 10 micrograms/m^3 (ACGIH, 1995c). Arsenic blood levels do not correlate well with acute environmental exposures and hair and nail analyses, although helpful in chronic exposures, are usually confounded by external arsenic contamination (ATSDR, 1990).

10.8.2 Indirect Biological Indicators

A complete blood count is useful as arsenic exposures may lead to anemia, low white cell count, and diminished platelets. Liver enzymes may be elevated in both acute and chronic arsenic neurological disorders. When arsenic neuropathy is suspected, nerve conduction velocity testing may reveal reduced amplitude and velocity. If skin lesions are present, a biopsy is recommended to rule out skin cancer, a well-known complication of arsenic exposure (ATSDR, 1990).

10.8.3 Conclusion

Arsenic, ubiquitous in the environment, has nonoccupational sources of exposure that must be considered when conducting biological monitoring. In the past, biological monitoring of workers chronically exposed to inorganic arsenic in industry has been carried out by measuring the total amount of arsenic present in the urine, either preshift or postshift. It is now well-established that the determination of inorganic arsenic metabolites (monomethyl arsenic acid and cacodylic acid) in the urine is the method of choice for biomonitoring workers exposed to inorganic arsenic, since these determinations are not influenced by the presence of organoarsenics from marine origin.

10.9 CASE STUDY ON CHOLINESTERASE-INHIBITING PESTICIDE TOXICITY

You are the plant manager at a organophosphate pesticide manufacturing facility that has apparently effective hygiene controls without a medical surveillance program. One day you receive a call from a foreman that a 30-year-old chemist is found unresponsive at his work station. While CPR is initiated, an ambulance arrives to transfer him to an emergency room for definitive care. The physician calls you to confirm that the employee is now recovering from parathion poisoning. You decide to consult an occupational medicine physician to devise and implement a medical surveillance program.

1. From the perspective of biological monitoring, what specific tests could be implemented in this medical surveillance program?

10.9.1 Direct Biological Indicators

The parent organophosphate compound remains in the blood for only a few minutes to hours after acute exposure unless great quantities are absorbed, or liver enzymes are inhibited. The urinary metabolites may be detected for up to 48 hours. Their presence may reflect lower exposures than those required to depress cholinesterase activity or produce symptoms. Determination of cholinesterase activity is useful in routine biologic monitoring of workers chronically exposed to cholinesterase-inhibiting pesticides, as well as in diagnosing acute poisoning. Cholinesterase depression is usually apparent within a few minutes or hours after significant absorption. Acetylcholinesterase enzymes measured in red blood cells, referred to as true cholinesterase, and in plasma or serum, referred to as pseudocholinesterase, are surrogates for activity occurring at the neural receptor sites. *Both plasma and RBC cholinesterase* should be determined.

Plasma cholinesterase is more labile than RBC cholinesterase, and is therefore less reliable in reflecting depression of enzyme activity at neural receptor sites. Plasma cholinesterase is more rapidly inactivated by exposure to organophosphates, and may be slightly depressed by infection, alcohol, hepatic disease, birth control pills, and pregnancy, among other factors. Because plasma cholinesterase is produced in the liver, it can be regenerated quickly. RBC cholinesterase is the same compound found in the nervous system. It is depressed more slowly than plasma cholinesterase, but because the depression is attributable to organophosphates, its level more accurately reflects the actual enzyme and activation at neural receptor sites. RBC cholinesterase activity is restored only as new red blood cells are formed (regeneration of red blood cells takes place at a rate of about 1% per day). RBC cholinesterase activity is slightly affected by rare conditions that damage cell membranes, such as hemolytic anemias.

It is difficult to interpret cholinesterase inhibition without baseline values, because normal human cholinesterase levels vary widely. The laboratory normal range is not useful because upper and lower limits may differ by a factor of 4 with some common laboratory methods. Inhibitions from 25 to 50% of a person's baseline enzyme activity are regarded as evidence of toxicity. In both acute and chronic exposures, however, symptoms have been absent even when cholinesterase levels were inhibited by 50% of baseline. The development of signs and symptoms is related to both the rate of decline in enzyme activity and the absolute level of enzyme activity.

Because of marked variation among different levels of analysis and among laboratories using the same method, cholinesterase testing, including baseline and followups, should be performed by the same laboratory using the same methods. "It is misleading to extrapolate from one method to another or from one laboratory's results to another's" (ATSDR, 1993a). A laboratory performing cholinesterase testing should have rigorous quality control and an accurate and reproducible method of analysis. Other laboratory evaluations where patients are seriously exposed to organophosphate or carbamate pesticides may include CBC and determination of electrolytes, glucose, BUN, creatinine, and liver enzymes. Arterial blood gases and chest radiography are useful in cases of inhalation exposure or respiratory compromise.

10.9.2 Conclusion

In a medical surveillance program, acetylcholinesterase activity would be a more appropriate measure of biological effect than direct measurement of the parent organophosphate or carbamate compound. The ACGIH Biological Exposure Index is 70% of baseline acetylcholinesterase activity levels (ATSDR, 1993a).

10.10 CASE STUDY ON TOLUENE TOXICITY

A pregnant 28-year old with cough and dyspnea (ATSDR, 1993b): A 28-year old pregnant female comes to the office in the late afternoon with complaints of coughing spasm, chest tightness, and a sensation of being unable to breathe. These symptoms began about 6 hours earlier, while she was repainting a disassembled bicycle with an acrylic lacquer spray paint in a small, poorly vented area. The painting took about 2 hours to complete. The patient also experienced nausea, headache, dizziness, and lightheadedness, which cleared within an hour after leaving the basement. Chest complaints, however, have persisted, prompting the office visit. She is concerned that her symptoms are related to the paint spraying and that it may affect her pregnancy (ATSDR, 1993b).

On questioning the patient further, you discover that 2 years ago, she was exposed to fumes of toluene di-isocyanate from an accidental spill at a research laboratory. The patient had eye and upper airway irritation at the time of the accident, but developed severe shortness of breath and coughing 4 hours later. She was hospitalized for several days, but recovered (ATSDR, 1993b).

10.10.1 Population at Risk

People who work in the following areas have potential exposure to toluene:

1. Adhesives and coatings manufacturers and applicators;
2. Chemical industry workers;
3. Coke oven workers;
4. Fabric manufacturers (fabric coating);
5. Hazardous waste site personnel;
6. Linoleum manufacturers;

7. Pharmaceutical manufacturers;
8. Shoe manufacturers;
9. Styrene producers.

An estimated 4–5 million workers are occupationally exposed to toluene. Automobile mechanics, gasoline manufacturers, shippers and retailers, dye and ink makers, and painters are at greatest risk (ATSDR, 1993b).

10.10.2 Biologic Fate in Humans

Inhalation is the primary route of toluene exposure, but significant amounts can be absorbed through ingestion and dermal contact. Peak blood concentrations occur 15–30 minutes after inhalation. The amount of toluene absorbed by inhalation depends on the respiratory minute volume; thus exercise affects the absorption rate of toluene. At rest, the lungs absorb 50% of an inhaled dose (ATSDR, 1993b).

The rate of absorption after oral intake is slower than after inhalation. Nevertheless, gastrointestinal absorption is nearly complete and blood toluene levels peak 1–2 hours after ingestion. Percutaneous absorption is slow through intact skin and rarely produces toxicity (ATSDR, 1993b).

Toluene is lipophilic and has little water solubility. It is distributed quickly to highly perfuse tissues, such as brain, liver, and kidney. It passes readily through cellular membranes and accumulates primarily in adipose and other tissues with high fat content. In the body, the half-life of toluene ranges from several minutes in highly vascular organs to slightly over 1 hour in fatty tissues. Toluene's affinity for lipid-rich structures of the nervous system tissue results in central nervous system toxicity effects within minutes (ATSDR, 1993b).

About 80% of absorbed toluene is oxidized in the liver to benzoic acid, which is then conjugated with glycine to form hippuric acid or with glucoronic acid to form benzylglucuronate. A small amount of toluene undergoes aromatic ring oxidation to form ortho and paracresols. Most inhaled or ingested toluene is eliminated in urine within 12 hours after exposure; a small amount (up to 20%) is eliminated as free toluene in expired air. Less than 2% of the total toluene metabolites are excreted in the bile (ATSDR, 1993b).

10.10.3 Laboratory Tests

If toluene exposure is suspected, baseline studies may include the following, depending on the circumstances: electrolytes, BUN and creatinine, liver enzymes, urinalysis, electrocardiogram, and rhythm monitoring. Repeat baseline tests in 3–6 months to detect delayed hepatic, renal, or neuropsychiatric effects may be beneficial. For patients with substantial chronic exposures, annual assessments may be valuable. Referral for detailed neuropsychological assessment is indicated only if the patient abnormal mental status or behavioral changes persist after exposure ceases (ATSDR, 1993b).

10.10.4 Direct Biological Indicators

Because excretion of toluene and its metabolites is rapid (essentially complete within 12–24 hours), biologic samples for analysis must be obtained soon after exposure. A venous blood sample taken within a day after exposure can be used to confirm toluene exposure (normal for unexposed population is 0.1 mg/deciliter); however, the toluene level obtained will not correlate well with the degree of exposure or to symptoms. Analysis of exhaled air for toluene is experimental only (ATSDR, 1993b).

10.10.5 Indirect Biological Indicators

Hippuric acid, a metabolite of toluene, may also result from the metabolism of other chemicals, including common food additives, and is typically found in significant amounts in the urine from unexposed persons. Hippuric acid levels higher than 2.5 g/g creatinine suggest toluene exposure (ATSDR, 1993b).

10.10.6 Conclusion

The patient's transient nausea, headache, dizziness, and lightheadedness correlate well with acute exposure to toluene but not with exposure to toluene diisocyanate. Although toluene can be irritating to the airways, the degree of wheezing experienced by this patient and the persistence for several hours after exposure has ceased suggests the presence of an underlying pulmonary disorder, such as asthma.

One biological exposure index for toluene is 1 mg/L toluene measured in venous blood taken at the end of a work shift (which may be useful in monitoring recent exposure). However, specimen collection and handling are important issues due to the volatile nature of toluene (Gill, 1988).

The BEI that appears most useful for assessing moderate to heavy toluene exposure is 2.5 g/g creatinine of hippuric acid measured as a spot sample at the end of the shift (Muelenbelt, 1990).

Toluene diisocyanate, a strong respiratory tract irritant, currently has no established biological monitoring indices.

10.11 CASE STUDY ON TETRACHLOROETHYLENE TOXICITY

Headache, decreased concentration, and irritability in a 37-year-old silk screener (ATSDR, 1990b): A 37-year-old woman who was 4 months postpartum presents with headache, irritability, and difficulty concentrating. She states that she has become impatient and short-tempered with her husband and child; minor things make her angry. These findings began about one month ago. She is most aware of them in the evenings when they are sometimes accompanied by a throbbing frontal headache. She has no psychiatric history, but admits to drinking 3 ounces of alcohol a day since being married 4 years ago. She did not drink during her pregnancy and denies using any other drugs or medications. She has no trouble sleeping (ATSDR, 1990b).

Two weeks ago, the patient and her family visited her parents for a week. During that time she felt well; the irritability and headaches subsided. Since coming home last week, however, symptoms have returned (ATSDR, 1990b).

The patient is worried that something in the home is causing her symptoms. She reports that the house was sprayed for termites 2 year ago, but she does not remember the name of the fumigant used. Her husband feels fine and has not been ill. Her infant daughter's delivery was uneventful, and the baby appears to be developing normally, but has been very fussy lately. The infant is still breast-feeding (ATSDR, 1990b).

A month ago, the patient returned to her job as a word processor, working mornings and relaxing with her hobby, silk screening, in the afternoon. She gets along well with her employer and fellow employees, and the job is not generally stressful; however, she is concerned that a loss in typing accuracy and a decreased ability to concentrate may lead to conflict with her supervisor. The patient has no symptoms of postpartum depression, and has had no history of headaches before she returned to these activities.

1. Could this patient's symptoms be related to her silk-screening hobby?
2. What type of biologic monitoring, if any, would be appropriate in this case?

10.11.1 Biological Fate in Humans

About 70% of an inhaled tetrachloroethethylene dose is absorbed by the lungs and about 80% of an oral dose is absorbed by the gut. Tetrachloroethylene penetrates human skin slowly. Once tetrachloroethylene is absorbed, it is readily distributed to all body tissues. Because it is highly lipid soluble, it tends to concentrate primarily in adipose tissue (ATSDR, 1990b).

More than 80% of inhaled tetrachloroethylene is eliminated unchanged by the lungs. With minimal physical activity, elimination of tetrachloroethylene from blood occurs in a biphasic pattern. In one study, the average half-life of each phase was 2.6 and 33 hours, respectively. The average half-life of tetrachloroethylene in adipose tissues is about 72 hours. Less than 2% of absorbed tetrachloroethylene is metabolized in the liver to trichloracetic acid and trichloroethanol, which are then excreted in the urine. The rate of urinary elimination is slower than exhalation rate, with urinary biological half-lives ranging from 12 to 55 hours for the first phase, and 100–200 hours for the second phase. Studies of dry cleaning shop workers have shown that urinary metabolite levels increase linearly with air concentrations up to 100 ppm tetrachloroethylene, then level off at higher concentration. This indicates the saturability of the tetrachloroethylene metabolic pathways (ATSDR, 1990b).

10.11.2 Direct Biological Indicators

In exposed persons, tetrachloroethylene may be measured in expired air and blood. Its metabolite, trichloracetic acid, may be measured in blood and urine. The cause of symptoms is questionable. Direct biologic testing may be warranted. However, other chemical exposures, such as 1,1,1-trichloroethane and trichloroethylene, can also result in trichloracetic acid in blood and urine. Trichloroethanol, another metabolite of tetrachloroethylene and trichloroethylene, has been reported, but not consistently, in urine of tetrachloroethylene-exposed workers. Trichloroethanol and trichloracetic acid can be found in those patients taking chloral hydrate (ATSDR, 1990b).

To measure tetrachloroethylene in blood or expired air, samples should be collected within *16* hours after exposure; urine tests will remain positive up to 5 days after exposure, depending on the dose. Few laboratories perform these specialized tests; regional poison control centers may be able to identify such facilities. The method of sampling and sample storage must be coordinated with the laboratory to ensure proper specimen collection and processing. The laboratory should provide reference values appropriate for the analytical method used. Recording the time of sample collection relative to the last exposure is critical to proper interpretation of results (ATSDR, 1990b).

Expired air and blood tetrachloroethylene levels and urine trichloracetic acid levels have been linearly correlated with ambient air concentrations, up to 100 parts/million. In workers, trichloracetic acid levels of 7 mg/liter in urine obtained at the end of the work week was found to correlate with exposure to an average of 50 parts/million of tetrachloroethylene for one week. The same exposure level will result in approximately 100 mcg/deciliter tetrachloroethylene in the blood drawn 16 hours after the last work shift of the week. Increased physical activity during exposure can result in higher levels (ATSDR, 1990b).

10.11.3 Indirect Biological Indicators

Although tetrachloroethylene may cause upper airway irritation and coughing, the chest x-ray, and the pulmonary function tests will probably be normal. In general, results of routine laboratory tests, including renal and liver function tests, will also be normal, unless the patient has had a significant exposure and has concurrent neurologic symptoms (ATSDR, 1990b).

Transient elevation of liver enzyme levels has been reported in tetrachlorethylene exposures, but frank hepatic necrosis has not been documented. If a known acute exposure to tetrachlorethylene results in CNS symptoms, such as syncope, then liver function tests, kidney function tests, serum creatinine, and urinalysis should be obtained immediately to establish a baseline. Testing should be repeated after several days to monitor for possible effects. Liver function tests should include SGOT, SGPT, lactic dehydrogenase, bilirubin, and alkaline phosphatase. If levels are mildly elevated, tests should be repeated in several weeks to document return to baseline. If levels remain elevated, other causes of hepatic dysfunction should be investigated (ATSDR, 1990b).

The value of a neurophysiologic evaluation for differentiating between organic and functional impairment is controversial, especially when no baseline evaluation is available. The tests, however, may be useful when comparing an exposed occupational population to a control group. Although neurologic tests provide data, they may be used to raise suspicion of cognitive impairments that are not evident on mental status testing or as a baseline for followup (ATSDR, 1990b).

10.11.4 Conclusion

The loss of typing ability and decreased ability to concentrate may indeed result from tetrachloroethylene exposure during silk screening. Some rags may be permeated with this material, and if there is poor ventilation in the room where the silk screening is done, significant exposure may result (ATSDR, 1990b).

The assessment of tetrachloroethylene exposure can be accomplished through measurement of the parent chemical in exhaled air. However, due to the technical difficulty of collecting samples, it may be easier to measure either the parent chemical in the blood (BEI 0.5 mg/ml) or the main metabolite, trichloracetic acid, in the urine (BEI is 7.0 mg/L). One confounding factor is that trichloracetic acid in the urine may also be generated from exposure to 1,1,1-trichloroethane or trichloroethylene (Ducatman, 1984).

REFERENCES

ACGIH, 1995a, "Topics in Biological Monitoring: A Compendium of Essays by Members of the ACGIH Biological Exposure Committee," ACGIH, Cincinnati, Ohio.

ACGIH, 1995b, American Conference Governmental Industrial Hygienists: 1995–1996 Threshold Limit Values for Chemical Substances and Physical Agents and Biological Exposure Indices, Cincinnati, Ohio.

ACGIH, 1995c, American Conference of Governmental Industrial Hygienists: Documentation of the Threshold Limit Values and Biological Exposure Indices, 6th Edition and Supplement, Cincinnati, Ohio.

Aito, A., 1994, "Biological Monitoring, Today and Tomorrow," *Scandinavian Journal of Work and Environmental Health* **20:** Special Issue, pp. 46–58.

Alessio, L., 1990, "History and Concept of Biological Monitoring in the European Communities." In *Biological Monitoring of Exposure to Industrial Chemicals V*, (Fiseroua-Bergeroua and Ogatam, Editors), ACGIH, Cincinnati, Ohio, 1990.

ATSDR, 1990, Case Studies in Environmental Medicine: Arsenic Toxicity, Agency for Toxic Substances and Disease Registry, Delima Associates, San Raphael, California.

ATSDR, 1990b, Case Studies in Environmental Medicine: Tetrachloroethylene Toxicity, Agency for Toxic Substances and Disease Registry, Delima Associates, San Raphael, California.

ATSDR, 1993a, Case Studies in Environmental Medicine: Cholinesterase-Inhibiting Pesticide Toxicity, Agency for Toxic Substances and Disease Registry, Delima Associates, San Raphael, California.

ATSDR, 1993b, Case Studies in Environmental Medicine: Toluene Toxicity, Agency for Toxic Substances and Disease Registry, Delima Associates, San Raphael, California.

Baselt, R., 1980, "Biological Monitoring Methods for Industrial Chemicals," 2nd Edition, Chicago, Yearbook Medical Publishers.

Clarkson, T., Friberg, L., et al., 1988, "Biological Monitoring of Toxic Metals," Plenum Press, New York, London.

Derelanko, M., Hollinger, M., 1995, *CRC Handbook of Toxicology*, CRC Press.

Ducatman, A., Moyer, T., 1984, "Environmental Exposure to Common Industrial Solvents," *American Association for Clinical Chemistry* **5**(11), May.

Fiserova-Bergerova, V., Pierce, et al., 1990, "Dermal Absorption of Industrial Chemicals: Criteria for Skin Notation," *American Journal of Industrial Medicine* **17:**617–35.

Gill, R., Hatchetl, S. E., et al., 1988, "Sample Handling and Storage for the Quantitative Analysis of Volatile Compounds in the Blood: The Determination of Toluene by Headspace Gas Chromatography," *Journal of Analytical Toxicology* **12**(3):141–46.

Klassen, C. D., 1986, "Distribution, Excretion, and Absorption of Toxicants in Cassarett and Doull's Toxicology. The Basic Science of Poisons," pp. 33–63, C. D. Klassen, M. O. Amdur, and J. Doull, Eds., Macmillan, New York, New York.

Koh, H., Prout, M., et al., 1989, "Carcinogen Monitoring by DNA Adduct Methodology in Humans: Clinical Perspectives for the 1990's," *Advances in Internal Medicine* **34:**243–64.

Kreis, K., et al., 1993, "Beryllium Disease Screening in the Ceramics Industry: Blood Lymphocyte Test Performance and Exposure Disease Relations," *Journal of Occupational and Environmental Medicine* **35**(3):267–74, March.

Lauwerys, R., Hoet, P., 1993, "Industrial Chemical Exposure: Guidelines for Biological Monitoring," 2nd Edition, Lewis, Boca Raton, Florida.

Lowrey, L., 1986, "Biological Exposure Index as a Complement to the TLV," *Journal of Occupational Medicine* **28**(8), 578–83.

McCunney, R. J., 1994, "Practical Approach to Occupational Medicine," 2nd edition, Little, Brown & Co., Boston, Massachusetts, 1994.

McCunney, R. J., 1995, "Clinical Applications of Biomarkers in Occupational Medicine," pp. 148–60. In *Biomarkers and Occupational Health: Progress and Perspectives*, Mendelsohn, M., Peeters, J., Normandy, M. J. (Eds.), Joseph Henry Press, Washington, D.C.

Muelenbelt, J., deGroot, G., et al., 1990, "Two Cases of Acute Toluene Intoxication," *British Journal of Industrial Medicine* **47**(6):417–20.

Muller, G., Sparsowski, et al., 1975. "Metabolism of Trichloroethylene and Ethanol," *Archives of Toxicology* **33:**173–89.

Mutli, A., Bargamaschi, E., Ghittori, S., et al., 1993, "On the Need of a Sampling Strategy in Biological Monitoring: The Example of Hexane Exposure," *International Archives of Occupational and Environmental Health* **65** (Supplement 1):5171–75.

Ong, C. N., Shi, C. Y., et al., 1995, "Biological Monitoring of Exposure to Low Concentrations of Styrene," *American Journal of Industrial Medicine* **28**(1):143–49.

Que Hue, S., 1993, *Biological Monitoring: An Introduction*, 1st Edition, VanNostrand Reinhold, New York, New York.

Rappaport, S. M., 1985, "Smoothing of Exposure Variability at the Receptor: Implications for Health Standards, *Annals of Occupational Hygiene* **29:**201.

Saltini, C., 1995, "A Genetic Marker for Chronic Beryllium Disease," in *Biomarkers and Occupational Health: Progress and Perspectives*, Joseph Henry Press, Washington, D.C.

Sato, A., 1993, "Confounding Factors in Biological Monitoring of Exposures to Organic Solvents," *International Archives of Occupational and Environmental Health* **65:**561–62.

Sedivec, V., Miaz, M., et al., 1981, "Biological Monitoring of Persons Exposed to Methanol Vapors," *International Archives of Occupational and Environmental Health* **48:**257–71.

Shaikh, Z. A., Tohyama, C., et al., 1987, "Occupational Exposure to Cadmium: Effect on Metallothiomin and Other Biological Indices of Exposure and Renal Function," Archives of Toxicology, **59:**360–64.

Skender, L., Karalic, V., 1991, "A comparative Study of Human Levels of Trichloroethylene and Tetrachlorethylene After Occupational Exposure," *Archives of Environmental Health* **46:**174–78.

Tardif, R., Lapare, S., et al., 1991, "Effect of Simultaneous Exposure to Toluene and Xylene on Their Respective Biological Exposure Indices in Humans," *International Archives of Occupational and Environmental Health* **63:**279–84.

Zenz, C., 1994, "Occupational Medicine," 3rd Edition, Mosby Year Book Inc., St. Louis, Missouri.

Zielhuis, R., Henderson, P., 1986, "Definitions of Monitoring Activities and Their Relevance for the Practice of Occupational Health," International Archives, *Occupational and Environmental Health* **57:**249.

SUGGESTED READINGS

Aito, A., Riihimaki, V., Vainio, H., Eds., Biological Monitoring and Surveillance of Workers Exposed to Chemicals. New York: Hemisphere Publishing Corp., 1984.

Ashford, N. A., Spadafor, C. J., Hattis, D. B., Caldart, C. C., Monitoring the Worker for Exposure and Disease. Scientific, Legal, and Ethical Considerations in the Use of Biomarkers. Baltimore, MD: The Johns Hopkins University Press, 1990.

Dillon, H. K., Ho, M. H., Eds., Biological Monitoring of Exposure to Chemicals, Metals. New York: John Wiley & Sons, 1991.

Droz, P. O., Wu, M. N., in: Rappaport, S. M., Smith, T. J., Eds., Exposure Assessment for Epidemiology and Hazard Control. American Conference of Governmental Industrial Hygienists. Chelsea, MI. Lewis Publications, 1991.

Hathaway, G. J., Proctor, N. H., Hughs, J. P., Fischman, M. L., Eds., Chemical Hazards of the Workplace, 3rd Edition. New York: Van Nostrand Reinhold, 1991.

Ho, M. H., Dillon, H. K., Eds., Biological Monitoring of Exposure to Chemicals, Organic Compounds. New York: John Wiley & Sons, 1987.

Kneip, T. J., Crable, J. R., Eds., Methods for Biological Monitoring. A Manual for Assessing Human Exposure to Hazardous Substances. Washington, D.C., American Public Health Association, 1988.

Rempel, D. M., Biological Monitoring. In: LaDou, J., Ed., *Occupational Medicine*, East Norwalk, CT: Appleton & Lange, 1990, 459–66.

U.S. Dept. of Health and Human Services, Public Health Service, Centers for Disease Control and Prevention, National Institute for Occupational Safety and Health Administration, NIOSH/OSHA Occupational Health Guidelines for Chemical Hazards (1981, 1988, 1989), Mackison, F. W., Stricoff, R. S., Patridge, L. J., Jr., Little A. D., Eds., DHHS (NJOSH). Publications Nos. 81-123; 88-118, Supplement I-OHG; and 89-104, Supplement II-OHG, U.S. Dept. of Health and Human Services, 1981, 1988, 1989.

U.S. Dept. of Labor, Occupational Safety and Health Administration, Occupational Safety and Health Standards. (29 CFR Part 1910. Air Contaminants. Final Rule). Federal Register, Volume 58, No. 124, pp. 35337–51. U.S. Dept. of Labor, Washington, D.C., U.S. Government Printing Office, Wednesday, June 30, 1993. (Revised as of July 1, 1993.)

CHAPTER 11

How to Ensure Quality in a Medical Surveillance Program

ROBERT J. McCUNNEY

Massachusetts Institute of Technology, 77 Massachusetts Ave., Cambridge, MA 02139

and

ALAIN J. COUTURIER

Occupational Health Network, 11220 County Rd. 14, Middlebury, IN 46540

11.1 INTRODUCTION

11.2 ASSESSING THE NEED FOR MEDICAL SURVEILLANCE
- 11.2.1 Background
- 11.2.2 Worksite Assessment
- 11.2.3 Decision Making
- 11.2.4 Selection of a Physician or a Medical Facility

11.3 IMPLEMENTATION OF THE PROGRAM
- 11.3.1 Development of a Medical Protocol
- 11.3.2 Notification of Personnel and Responsible Authorities
- 11.3.3 The Medical Evaluation
- 11.3.4 Interval Evaluations

11.4 ASSESSING THE EFFECTIVENESS OF A MEDICAL SURVEILLANCE PROGRAM

11.5 SUMMARY

REFERENCES

11.1 INTRODUCTION

The purpose of this chapter is to provide practical guidance on the establishment and operation of a medical surveillance program. Medical surveillance is broadly defined as the evaluation of workers and other people potentially exposed to hazardous materials that may affect their health in order to identify disease sufficiently early to prevent or reduce morbidity or mortality. Medical surveillance is based on the fundamental principles of medical screening, that is, the administration of a test designed to uncover an illness in its asymptomatic phase, such that intervention slows, halts, or reverses

Handbook of Occupational Safety and Health, Second Edition, Edited by Louis J. DiBerardinis, ISBN 0-471-16017-2

the condition (U.S., 1996). Medical surveillance has played a major role in preventing a variety of occupational and environmental illnesses (Halprin, 1986; Millar, 1986). Its effectiveness is greatly dependent upon a proper assessment of the need for such a program, followed by periodic reviews of aggregate data and abnormalities uncovered during medical surveillance examinations.

The ensurance of quality in a medical surveillance program is directly dependent upon appropriate scientific rigor in the establishment of such an effort. This chapter is designed to acquaint the reader with fundamental principles in assessing the need for medical surveillance, how to implement such a program, and how to evaluate its effectiveness. Specific attention is given to well-known hazards in the occupational setting for which medical surveillance has been effective.

11.2 ASSESSING THE NEED FOR MEDICAL SURVEILLANCE

11.2.1 Background

Although the term *surveillance* may conjure unsavory connotations, the use of the term in the medical setting represents medical oversight of people who may be exposed to hazardous materials that can adversely affect their health. A variety of motivations stimulate the development of medical surveillance programs, including regulatory requirements, product stewardship, and safe work practices. In the United States, the Occupational Safety and Health Administration (OSHA) has established a variety of standards, for which medical surveillance is either required or recommended. It is important to distinguish medical surveillance from biological monitoring, which is a test designed to assess exposure to a hazard itself, such as a blood lead measurement or a metabolite of the hazard such as the level of urinary mandelic acid in the context of exposure to styrene. Medical surveillance connotes the overall evaluation of people exposed to a hazard, including questionnaires, physical examinations, and laboratory testing, whereas biological monitoring is a component of medical surveillance. The term *biomarkers* is also used to refer to a specific test used to evaluate exposure to a hazard; examples include, among others, DNA-adducts and various genetic tests (see Table 11.1). Examples of compounds used in the workplace for which medical surveillance examinations may be required include lead, cadmium, and benzene, among others. In these three examples, periodic administration of tests for the substance itself, a metabolite, or end-organ damage can help indicate early signs of serious illness, such that intervention, either in the form of treatment or removal from exposure, can benefit the person exposed.

Medical surveillance may be recommended for people accidentally exposed to hazardous materials that may cause long-term consequences. In other cases, such as in the need to oversee product stewardship, especially for hazards with unknown long-term consequences, medical surveillance can offer reassurance to those exposed to the substance that such contact has not adversely affected their health. Medical surveillance may also be recommended based on Biological Exposure guidelines of the American Conference of Governmental Industrial Hygienists (ACGIH) or on Indices (BEI) of the ACGIH also (see Table 11.2). Biological Exposure Indices are based on *biological monitoring*, a method designed to uncover the presence of the hazard itself, or a metabolite, usually either in the blood or urine. The use of biological monitoring in the context of medical surveillance programs will be described later in this chapter and in Chapter 10 in this text.

TABLE 11.1 21 OSHA Standards as of 1996

Standard Name	Standard Number
Acrylonitrile	29CFR 1910.1045
Arsenic, inorganic	29CFR 1910.1018
Asbestos	29CFR 1910.1001
Benzene	29CFR 1910.1028
Bloodborne pathogens	29CFR 1910.1030
Cadmium	29CFR 1910.1027
Carcinogens[a]	29CFR 1910.1003 to 1016
Coke ovens	29CFR 1910.1029
Cotton dust	29CFR 1910.1043
Dibromochloropropane	29CFR 1910.1044
Ethylene oxide	29CFR 1910.1047
Field sanitation	29CFR 1910.142
Fire protection	29CFR 1910.156
Formaldehyde	29CFR 1910.1048
Hazardous waste operations	29CFR 1910.120
Hearing conservation	29CFR 1910.95
Lead	29CFR 1910.1025
Laboratories	29CFR 1910.1450
Methylenedianiline	29CFR 1910.1050
Respirator program	29CFR 1910.134
Vinyl chloride	29CFR 1910.1017
	Mean Standard deviation

[a]OSHA carcinogen standard covers 4-nitrobiphenyl, a-naphthylamine, methyl chloromethyl ether, 3,3′-dichlorobenzidine and its salts, bis-chloromethyl ether, B-naphtylamine, benzidine, 4-aminodiphenyl, ethyleneimine, B-propiolactone, 2-acetylaminofluorene, 4-dimethyl-aminoazobenzene, N-nitrosodimethylamine.

11.2.2 Worksite Assessment

Prior to the establishment of a medical surveillance program, it is critical to understand not only the potential toxicity of the hazard, but also the circumstances upon which a person may be exposed to such a hazard. With exceedingly rare examples, such as in exposure to beryllium and cobalt alloys, the risk of suffering an adverse health effect secondary to exposure to a hazardous substance is directly dependent upon the level and degree of exposure to the hazard. In short, the greater the exposure, the greater is the risk of adverse health effects.

The type of professional who performs the *worksite assessment* can vary; however, in general, those trained in industrial hygiene have the greatest appeal. Industrial hygiene, a discipline focused on the anticipation, evaluation, and control of occupational and environmental hazards, can offer valuable information in the decision-making process regarding the need for medical surveillance. A worksite review conducted by a properly trained occupational health professional can address the nature of the hazard, whether physical, chemical, or biological, and the corresponding level of control, whether ventilation, personal protective equipment, or administrative maneuvers, such as job rotation.

In the worksite assessment, it is essential to observe people in the context of potential exposure, such as in the handling of materials, to understand fully the toxicity of a substance, such as whether its route of exposure to the body is through inhalation,

TABLE 11.2 Clinical Applications of Biomarkers; Airborne Chemicals for which Biological Exposure Indices (BEI) Have Been Adopted

Airborne chemical	Determinant
Aniline	Total *p*-aminophol in urine
	Methemoglobin in blood
Benzene	Total phenol in urine
	Benzene in exhaled air
Cadmium	Cadmium in urine
	Cadmium in blood
Carbon disulfide	2-Thiothiazolidine-4-carboxylic acid in urine
	CO in end-exhaled air
Chlorobenzene	Total 4-chlorcatechol in urine
	Total *p*-chlorophenol in urine
Chromium (VI)	Total chromium in urine
N,N-Dimethylformamide	*N*-Methylformamide in urine
Ethylbenzene	Mandelic acid in urine
	Ethylbenzene in end-exhaled air
Flourides	Fluorides in urine
Furfural	Total fuoric acid in urine
n-Hexane	2,5-Hexanedione in urine
	n-Hexane in end-exhaled air
Lead	Lead in blood
	Lead in urine
	Zinc protoporphyrin in blood
Methanol	Methanol in urine
	Formic acid in urine
Methemoglobin inducers	Methemoglobin in blood
Methyl chloroform	Methyl chloroform in end-exhaled air
	Trichloroacetic acid in urine
	Total trichloroethanol in urine
	Total trichloroethanol in blood
Methyl ethyl ketone	MEK in urine
Nitrobenzene	Total *p*-nitrophenol in urine
	Methemoglobin in blood
Organophosphorus cholinesterase inhibitor	Cholinesterase activity in red cells
Parathion	Total *p*-nitrophenol in urine
	Cholinesterase activity in red cells
Pentachlorophenol (PCP)	Total PCP in urine
	Free PCP in plasma
Phenol	Total phenol in urine
Styrene	Mandelic acid in urine
	Phenylglyoxylic acid in urine
	Styrene in venous blood

TABLE 11.2 ***(Continued)***

Airborne chemical	Determinant
Toluene	Hippuric acid in urine
	Toluene in venous blood
	Toluene in end-exhaled air
Trichloroethylene	Trichloroacetic acid in urine
	Trichloroacetic acid and trichloroethanol in urine
	Free trichloroethanol in blood
	Trichloroethylene in end-exhaled air
Xylenes	Methylhippuric acids in urine

ingestion, or skin contact, or all the above. Prior to the recommendation of any medical surveillance program, it is highly recommended and in some cases required by law to conduct *exposure monitoring* to assess the concentrations of the hazards to which people may be exposed. For most OSHA standards, exposure monitoring is usually required at periodic intervals. For substances for which OSHA standards have not been promulgated or in environmental settings, a review of exposure data is a critical step in determining the value of medical surveillance. In these latter settings, the exposure data should be compared to guidelines proposed by authoritative sources, such as the ACGIH, which publishes an annual review of numerous materials used in commerce or present in the environment and their corresponding levels of proposed safety. Professional judgement related to many fundamental aspects of the hazard, including its solubility, half-life, and route of exposure, among others, is an important component to the assessment process.

Following a fundamental review of the toxicity of the substance, the manner in which the substance is used, and exposure monitoring, a decision can be made as to the potential value of a medical surveillance program. In most cases where OSHA standards have been established, medical surveillance is recommended when a person is exposed at or above the action level for 30 days per year. The action level is usually 1/2 of the permissible exposure limit (PEL) for OSHA standards and the threshold limit value (TLV) for substances for which OSHA standards have not been promulgated, but have been addressed by the ACGIH.

11.2.3 Decision Making

The fundamental purpose of any medical surveillance program is to uncover disease in its asymptomatic stage such that treatment or other interventions favorably affect the outcome of the disease. For many illnesses, medical screening has not been effective. Notable examples include lung cancer, a serious illness for which even the most aggressive form of monitoring, such as chest films and sputum cytology administered every 4 months in high-risk groups (that is two-pack-a-day smokers over the age of 40) has not been shown to be effective (Marfin, 1991). To properly establish a medical surveillance program, it is not only important to conduct a proper workplace assessment, but also to understand the natural course of the disease that may occur as a result of the exposure to the hazard. A careful review of the disease that may result from exposure to the hazard is critical to the selection of appropriate medical tests to be included in a program.

Professionals who oversee medical surveillance programs are advised to seek the advice of physicians both trained and experienced in the recognition of diseases due to exposure to environmental and occupational hazards. Unfortunately, the average physician receives little, if any, formal education in either medical school or postgraduate training in occupational and environmental medicine. (See Section 11.3.4 for more discussion on this topic.) The medical protocol should address the specific hazard through either an assessment of potential end-organ damage that may occur from exposure to the hazard, the concentration of the hazard itself, usually in either the blood or urine, or one of the metabolites of the material. For example, in monitoring for heavy metal exposure, one can determine the level of cadmium, lead, and mercury, among others, in the blood and/or urine itself. In the context of exposure to certain types of pesticides, such as organophosphates, serial measurements of serum cholinesterase can be helpful. In other cases, however, one cannot reliably measure either the substance or its metabolite, and as a result, end-organ damage must be assessed through determination of liver, kidney, or lung function.

11.2.4 Selection of a Physician or a Medical Facility

Too often when a medical surveillance program is recommended or suggested, scant attention is given to the selection of a proper physician or medical facility to oversee the effort. In general, physicians with certification and/or training and interest in occupational medicine are advisable. Such physicians usually have a more firm grasp on the relationship between exposure to a hazard and its potential corresponding health effect. The American College of Occupational and Environmental Medicine (ACOEM) publishes an annual directory of physicians who practice various aspects of occupational medicine (55 West Seeges Road, Arlington Heights, IL 60005. Tel: 847-228-6850). In addition, ACOEM publishes a directory of physicians engaged in either private practice or in the direction of hospital-based programs in occupational medicine. Since not all ACOEM members are board certified, one can contact either the American Board of Preventive Medicine (ABPM) or inquire from the physician about board certification status. Although certification does not guarantee quality, it does demonstrate a commitment to the specialty and a minimum level of formal training in the field and in related disciplines important in medical surveillance, such as epidemiology, toxicology, and health services administration. Such physicians also tend to be more adept in the multidisciplinary functions overseeing medical surveillance, such as coordinating medical results with exposed data obtained by industrial hygienists and other professionals.

To become board certified in occupational medicine through a residency program, physicians must spend a concentrated 2-year period following internship, during which they earn a Master's Degree in Public Health and engage in a variety of clinical, research, and administrative activities in occupational medicine. The crucial role such physicians play in understanding the occupational and environmental context in which exposure to a hazard may occur can be a valuable asset to any medical surveillance program.

Since the early 1980s, many hospitals and private facilities have focused on occupational medicine, the specialty of the American Board of Preventive Medicine concerned with the prevention of occupational and environmental illnesses (McCunney, 1984). Such facilities are more apt to have the necessary protocols and systems in place, not only to conduct the examinations, but also to ensure confidentiality of specific diagnostic information and to prepare appropriate reports (in the context of confidentiality restraints) to both management and labor.

11.3 IMPLEMENTATION OF THE PROGRAM

11.3.1 Development of a Medical Protocol

Once the decision is made regarding the potential value for performing periodic medical evaluations, the specific protocol needs to be established. The selection of appropriate tests, consistent with the hazard, is an essential element in the development of a quality program.

In general, the examination should be specific, not only to the hazard, but to the job itself. In some cases, attention must be directed towards legal requirements, such as those of OSHA standards, in which medical surveillance and the corresponding battery of tests may be prescribed with little room to exercise independent judgment (Silverstein, 1994).

Biological monitoring may be an important component to a medical surveillance program. This form of testing refers to the analysis of either the substance itself, one of its metabolites, or physiological effect of the exposure (Lauwerys, 1993). Unfortunately, biological monitoring has limited value in most settings, largely because of the few substances for which specific indies of exposure have been established (McCunney, 1995) (see Table 11.2). See Chapter 10 for a more detailed discussion of biological monitoring. These substances and their corresponding concentrations are guidelines designed to indicate possible overexposure to a hazard. Clearly, judgment is necessary whenever interpreting the results of a medical surveillance examination, including biological exposure indices. For example, in monitoring exposure to arsenic and mercury, one must consider ingestion of fish, of which some types may affect urinary arsenic levels (see Chapter 9 for a case study).

In situations where no particular biological monitoring method may be available or for which specific standards have yet to be prescribed, judgement may need to be exercised regarding the selection of tests design to uncover "end-organ damage." For example, specific tests that monitor liver or kidney function can be performed to assess potential damage to these organs as a result of exposure to some hazardous materials, such as solvents. Other examples include the use of pulmonary function testing to monitor lung function damage from inhalation of a variety of dusts (Committee, 1992).

11.3.2 Notification of Personnel And Responsible Authorities

Once a decision has been made to implement a medical surveillance program, proper notification of responsible authorities, such as workers and supervisors, is essential. Such notification can take many forms, but should focus on the purpose of the testing, relative components of the evaluation, how results will be reviewed, records stored, and who may have access to these records. Proper notification of the intentions of this effort will enhance the likelihood for success by increasing participation rates.

11.3.3 The Medical Evaluation

Once a physician and a medical facility have been selected to oversee the medical surveillance effort, fundamental administrative matters must be addressed. For example, if a program is designed in accordance with epidemiological studies, it is necessary to ensure that the examinations are performed within a prescribed period of time. Clearly, the efficiency of such an effort depends upon the number of examinations to be performed as well as the resources of the corresponding medical facility, whether an in-house operation or through contracted services. For examinations performed outside of a sponsoring organization, it is especially important to have a letter of agreement that specifies the respective roles and responsibilities of corresponding parties (McCunney, 1996). For example, decisions need to be made as to who bears the responsibility for informing workers of the need for the evaluation and the ensurance that such examinations actually take place. Personnel decisions related to those who do not complete the examination should be considered. Although workers may, in turn, refuse to undergo medical surveillance examinations, the organization may restrict exposure to a hazard or a job duty that necessitate medical evaluations.

Fundamental components of a medical surveillance examination that can be tailored to the specific job or hazard often include a number of the following elements.

11.3.3.1 Medical History Computerized forms are readily available to tabulate a person's past medical history, particularly as it relates to current or future job responsibilities. A thorough medical history also addresses prior and current medication use, past surgical history, and hospitalizations. It is important, however, for the examinee to be assured that specific medical information will be maintained in a confidential manner in accordance with principles of medical confidentiality as put forth in the Code of Ethical Conduct of the American College of Occupational and Environmental Medicine (CEC, 1994). Without the assurance of confidentiality of personal medical information that bears no bearing on a person's ability to work, the integrity and the success of a program may be compromised (McCunney, 1996).

11.3.3.2 Occupational History Many different forms are available from a variety of different organizations that can be helpful in tabulating a person's work history. In light of the latency (the interval between exposure and disease) associated with carcinogens and other hazards, a well-done occupational history can be instrumental in assessing causality of various health ailments. Moreover, the occupational history can document past exposures to a variety of hazards.

11.3.3.3 Physical Examination The extent of a physical examination in the context of a medical surveillance examination will vary depending on its purpose. For example, in performing a respirator clearance evaluation, the examination ought to focus on the heart and lungs as well as vision and hearing. In other settings, and depending on the person's age, a more focused evaluation, such as that directed towards the prostate or colon, may be appropriate depending on routine guidelines, especially those of the American Cancer Society (U.S., 1996).

11.3.3.4 Laboratory Testing The extent of laboratory testing performed at a medical surveillance examination depends upon the hazard(s) as well as the job duties. For example, the OSHA standard for occupational exposure to benzene specifies a detailed examination of the blood system, including a complete blood count, hemoglobin and hematocrit, and an assessment of the platelet count for those workers exposed above the action level of benzene for at least 30 days per year (OSHA). Care must be exercised in interpreting the results of any laboratory test because at least 5% of all tests will be abnormal, because 95% confidence intervals are used by labs to establish "normal" values. Ideally, abnormalities should be initially reevaluated with a repeat test, before proceeding to more sophisticated assessments. In the context of medical evaluations, a variety of blood studies have become routine and are often used in medical surveillance examinations. Examples include a complete blood count or screening studies for liver and kidney function, among other parameters. In other situations, depending on the symptoms or a patient's physical findings, assessments of thyroid function or bone metabolism may be performed. Interpretation of these results must be conducted in the context of a person's work and medical history as well as the limitations of the tests themselves. A full discussion of these types of studies, however, is beyond the scope of this chapter.

11.3.3.4.1 Chest Films. A chest film is required for many OSHA standards, especially in the context of occupational exposure to asbestos. In these cases, the chest film ought to be performed according to recommended techniques of the International Labor Organization (ILO, 1980). In many cases, the chest film must be read by a B reader, a physician certified by NIOSH in the review of chest films according to the ILO (Wagner, 1992). The ILO method for chest film review, originally designed in 1950, are intended to ensure uniformity in assessing chest films to reduce interobserver variability. Although challenges have been raised regarding the quality of the B reader program (Ducatman, 1988; Balmes, 1992), it is nonetheless essential to have chest films reviewed by a physician knowledgeable about this method.

11.3.3.4.2 Pulmonary Function Tests. Pulmonary function studies, an assessment of one's breathing capacity, can be performed both on-site and off-site. NIOSH has specified criteria for technicians who administer the pulmonary function studies to ensure that quality control measures are addressed. In interpreting the results of pulmonary function testing, one's assessment should be based on recommended guidelines of the American Thoracic Society (ATS, 1991; Harber, 1991).

Customarily, the evaluation of pulmonary function testing considers a number of parameters, including the forced expiratory volume (FEV_1) in the first record of the 6-second test and the forced vital capacity (FVC). Both volumes are measured in liters, reflect the competency primarily of the large airways, and are based on a person's age, height, and sex. Other parameters designed to assess the small airways have wide confidence intervals; thus, their value in routine surveillance is unclear.

Pulmonary function declines with age. As a result, periodic testing may indicate that a person's FEV_1, for example, is deteriorating at a faster clip than expected (approximately 30 cm^3 per year). Such an assessment, however, must ensure proper training of the technician, calibration of the equipment, and performance by the subject. It is necessary to have the person perform three maneuvers, all of which must be within 5% of one another in total volume exhaled and have slopes on the flow volume curve that become asymptomatic.

In occupational medical practice, lung function testing is also used to assess a person's medical suitability for wearing a respirator and to determine whether asthma or some other allergic disorder is causing breathing impairment. Although there are no standards for lower limits of lung function for safe use of a respirator, levels below 75% of that predicted warrant careful screening of the person's job, potential for exposure to a hazardous substance and the circumstances of respirator use—in addition to the clinical assessment. It can also be used to monitor lung function before and after a work shift to assess an occupational induced asthma.

11.3.3.4.3 Biological Monitoring. As noted earlier, biological monitoring refers to the assessment of a hazard or one of its metabolites, usually by testing blood or urine.

Other techniques, such as hair and nail analyses, have been performed, but are not routinely available and not recommended for general use. Other forms of biological monitoring include breath testing, but this form of testing other than for alcohol testing, has not been used regularly for over 15 years. As a result, most medical surveillance programs, when biological monitoring is appropriate, include testing of the blood or urine. As noted earlier in this chapter, the ACGIH has proposed biological exposure indices for a variety of substances for which biological monitoring may be appropriate (see Table 11.2). Notable OSHA standards that include biological monitoring components are those referable to cadmium and lead, for which specific measures of these compounds is required for workers exposed to these materials above certain concentrations for designated periods of time. Standard references are available to assist in the selection of the appropriate biological monitoring method and an interpretation of the results (Murphy, 1995).

11.3.3.4.4 Review of the Results of Medical Surveillance Examinations. In the view of results obtained during medical surveillance examinations, it is essential to understand the purpose of the examination, that is, to uncover a disease in its asymptomatic stage, such that intervention, in the form of treatment or removal from exposure, may prevent or slow the progression of the disease. In general, it is most appropriate to compare the results of the examination to the previous evaluation or to the baseline, a process that is also part of the OSHA standard on occupational exposure to noise (McCunney, 1992). All results need to be interpreted in light of the job duties and potential exposure to hazardous materials.

When abnormalities are noted during a medical surveillance examination, the physician has the responsibility of determining whether the change may be related to the work environment or have some nonoccupational explanation. This assessment, however, can be complicated. For example, in evaluating liver function abnormalities, in the context of work in the hazardous waste industry, one needs to be aware that the most common explanation for minor abnormalities in liver function studies is alcohol ingestion and not exposure to solvents or other hepatotoxins (Helzberg, 1986; Hodgson, 1990). Moreover, most cases of noninfectious hepatitis are usually related to medications and not solvents and other workplace hazards. As a result, great care needs to be exercised in the interpretation of results, even to the point of removing a worker from exposure to determine whether the abnormalities resolve.

Once a medical surveillance examination indicates that abnormalities are occupationally related, a followup worksite assessment is essential to evaluate the working conditions to reduce the exposure and to ensure that other workers are not similarly affected. In some cases, job modification may be necessary in accordance with the Americans With Disabilities Act (Peterson, 1994). The degree of the worksite assessment appropriate in the context of the followup of abnormalities depends on individual situations; it may entail full exposure evaluations performed by an industrial hygienist and corresponding engineering changes. It is essential, however, for the physician who oversees the medical surveillance program to ensure that the abnormalities are work related because of the necessary responsibilities borne by the employer for recording work-related illnesses according to OSHA requirements.

11.3.3.4.5 Dissemination of Results. Following the performance of a medical surveillance examination, the physician has the responsibility to inform the individual about the results and their implications. In some cases, the physician may need to recommend that the individual be evaluated by his/her personal physician for abnormalities not caused by work and that may have no bearing on the workplace.

The employer and designated representatives need to be informed about the results related to work-related conditions. Management, however, does not need to be informed about non-work-related medical conditions. Unless there is an abnormality uncovered that can be related to the workplace, only *aggregate* data should be provided. For example, a physician should inform management that an individual has been evaluated and that no conditions were uncovered that could be related to exposure to a hazard at work. Specific medical information is usually not provided, except in situations mandated by OSHA standards.

Physicians who provide occupational health services often look to the Code of Ethical Conduct of the American College of Occupational and Environmental Medicine regarding principles of confidentiality. Individuals who undergo medical surveillance examinations may request that results be forwarded to their personal physician upon written request.

Review of *aggregate data* obtained during medical surveillance examinations can provide valuable information about the effectiveness of workplace control measures. General information that ought to be provided includes number of people evaluated, the number with abnormalities related to work, and the extent of those abnormalities. For example, in conducting medical surveillance examinations on a hepatotoxin (a substance that can damage the liver), the physician should indicate the number of people with liver function abnormalities related to exposure to this solvent. Such information can be of great help in preventing future exposures to other employees. Appropriate reassurance that the control measures are effective, especially if no abnormalities have been noted, is another benefit of such an approach. Review of aggregate data can be enhanced by computer software programs that integrate medical and industrial hygiene data. Information on these programs can be obtained from the "Computers Section" of the American College of Occupational and Environmental Medicine (847-228-6850).

11.3.3.4.6 Recordkeeping Responsibilities. In performing medical surveillance examinations, questions invariably surface as to where the medical records ought to be stored. In general, records should be maintained at a medical facility under the responsibility of a licensed physician or registered nurse. The length of storage time varies considerably, and in some cases, where OSHA standards prevail, records need to be maintained for the length of employment *plus* 30 years. Access to these records, including the medical evaluations and exposure assessments are dictated by the OSHA Medical Record Access Standard (Peterson, 1994). Care needs to be exercised in storage of chest films, since many health care facilities do not keep chest films longer than 5 years. Although guidelines regarding medical confidentiality may not be understood by business managers, it is wise for the employer to have a firm knowledge of the location of the records and to advise the physician and/or medical facility of the legal requirement for maintaining the records for lengthy periods of time (up to 30 years or longer for some OSHA standards). It is naive to assume that all physicians who participate in medical surveillance programs understand their responsibilities for record storage and retention.

11.3.4 Interval Evaluations

The frequency and composition of interval examinations, after the initial examination, depend on a variety of factors. Clearly, where an OSHA standard prevails, the interval frequency should be based on that which is required. In other cases, the frequency depends on the toxicity of the hazard, including other specific matters such as latency, the age of the individual, as well as any abnormalities that may have been uncovered in earlier examinations. Annual examinations are performed in most cases; however, such a frequency may not be essential for all situations. Judgement needs to be exercised in those cases where OSHA standards do not specify the interval frequency or the composition of such an examination. For example, in monitoring laboratory workers who may be exposed to Macaque monkeys, for which there may be a risk of the transmission of the Herpes B Simian virus, a baseline serum sample is obtained at the time of placement and then after an incident where a monkey bite may occur; annual or a periodic serum testing is not usually conducted.

11.4 ASSESSING THE EFFECTIVENESS OF A MEDICAL SURVEILLANCE PROGRAM

The fundamental approach to assessing the effectiveness of a medical surveillance program is to ensure that people do not become ill, injured, or disabled from a work duty or a workplace hazard. In the context of medical surveillance, an effective means of performing such an analysis is to review aggregate data through a variety of methods. One might, for example, tabulate the number of people who have either illnesses or laboratory abnormalities that can be related to the work in comparison to the number of people who have participated in the program. Questions one might pose include: Has anyone suffered chest film changes *secondary* to exposure to asbestos, silica, or some other pulmonary hazard? Do people involved in lead-removal operations have blood levels above the OSHA recommended level? Is the participation rate sufficiently high to draw definitive opinions regarding the success of the program?

Ideally, the results of medical evaluations ought to be coordinated with industrial hygiene assessments. If the data are properly tabulated, there is always the potential for epidemiological assessments of the relationship between exposure data and medical abnormalities. Bear in mind, however, that the ideal epidemilogical program is designed de novo, and not "retrofitted" to an existing medical surveillance program. In some cases, valuable medical and industrial hygiene data, if available, may serve valid epidemiological purposes.

The success of any medical surveillance program depends upon proper communication of the results to management and workers. This communication can be formal in the sense of a written report or as an oral presentation. Specific recommendations as to measures that need to be employed to prevent illnesses in the future should be considered.

Following the review of results obtained from a medical surveillance program, worksite changes should be implemented where appropriate. In some cases, job site modification may be necessary based on the results of individual evaluations. Care needs to be exercised to ensure that such recommendations pay heed to the dictates of the Americans with Disabilities Act and other measures designed to protect those with disabilities.

11.5 SUMMARY

A well-run medical surveillance program can be effective in preventing occupational and environmental illnesses. It is important to realize, however, that not all situations whereby exposure to a hazard may occur actually warrant medical surveillance. One needs to keep in mind that, with exceedingly rare exceptions, it is the dose that makes the poison. In other words, simply because a person may be exposed to a hazard, a medical surveillance program may not be necessary. One needs to exercise judgement in the consideration of the level and degree of exposure to the hazard, as well as the circumstances upon which exposure to the hazard may occur.

Although most regulations assume a linear dose–response relationship between exposure to a hazard and an adverse health effect, recent empirical data suggest that effects at low levels of exposure may not be as severe as expected by the extrapolation of results from acute high-dose exposures. In fact, despite the regulatory paradigm that assumes there is no safe level exposure to a carcinogen, recent epidemiology reports suggest that a threshold exists at low levels of exposure to indicate that certain concentrations of a hazard may not pose a deleterious effect. Notable examples include recent studies of benzene exposed and radiation-exposed workers.

In the context of medical surveillance examinations, one should avoid unrealistic expectations. For example, certain medical conditions do not lend themselves well to screening, such as lung cancer. In these settings, the physician has a role in informing the worker who undergoes an evaluation regarding the purpose and scope of a medical surveillance program. Moreover, the physician also has a responsibility to ensure that nonoccupational abnormalities uncovered during such a program are referred to the family physician.

Despite inherent limitations in any program designed to uncover adverse health effects secondary to exposure to a hazard, medical surveillance has been helpful in many settings by pinpointing early signs of illness, as well as providing valuable reassurance to both management and labor regarding the effectiveness of workplace control programs.

REFERENCES

American Thoracic Society, 1991, Lung Function Testing: Selection of Reference Values and Interpretative Strategies. *American Review of Respiratory Diseases* **144:**1202.

Balmes, J. R., 1992, To B Read or Not to B Read (Editorial). *Journal of Occupational Medicine* **34:**885.

Carois, E., Gilbert, E. S., Carpenter, L., et al., 1995, Effects of Low Dose Rates of External Ionizing Radiation: Cancer Mortality Among Nuclear Industry Workers in Three Countries. *Radiat Research* **142:**117–32.

Code of Ethical Conduct of the American College of Occupational and Environmental Medicine, 1994, *J. Occup. Med.* **36:**27–30.

Committee on Occupational Lung Disorders of the American College of Occupational and Environmental Medicine, 1992, Spirometry in the Occupational Setting: Notes for Guidance. *Journal of Occupational Medicine* **34:**559.

Ducatman, A. M., Yang, W. N., Forman, S. A., 1988, B-Readers in Asbestos Medical Surveillance. *Journal of Occupational Medicine* **30:**644.

Halperin, W. E., Ratcliff, J., Frazier, T. M., et al., 1986, Medical Screening in the Workplace: Proposed Principles. *J. Occup. Med.* **23:**547–52.

Harber, P., 1991, Interpretation of Lung Function Tests. *Curr. Pulmonol.* **12:**261.

Helzberg, J. H., Spiro, H. M., 1986, LFT's Tests More Than the Liver. *JAMA* **256:**21.

Hodgson, M. J., Goodman-Klein, B. M., Van Thiel, D. H., 1990, Evaluating the Liver in Hazardous Waste Workers. *Occup. Med. State-of-the-Art Review* **5:**67.

ILO, 1980, Guidelines for the Use of the ILO International Classification of Pneumonoconiosis, Geneva, International Labor Office.

Lauwerys, R. R., Hoet, P. (Eds.), 1993, Industrial Chemical Exposure: Guidelines for Biological Monitoring, Boca Raton, FL, Lewis Publishers, CRC Press, 2nd Edition.

Marfin, A. A., Schenker, M., 1991, Screening for Lung Cancer: Effective Tests Awaiting Effective Treatment. *Occup. Med. State-of-the-Art Reviews* **6:**111.

McCunney, R. J., 1984, A Hospital Based Occupational Health Service. *J. Occup. Med.* **26.**

McCunney, R. J., 1995, Clinical Applications of Biomarkers in Occupational Medicine, in *Biomarkers and Occupational Health: Progress and Perspectives* Eds. Mendelsohn, M. M., Peeters, J., Normandy, M. J., John Henry Press, Washington DC, pp. 148–60.

McCunney, R. J. (Ed.), 1996, *A Manager's Guide to Occupational Health Services*, OEM Press, Beverly, MA.

McCunney, R. J., 1996, Preserving Confidentiality in Occupational Medicine Practice. *Am. Fam. Phys.* **53:**1751–56.

McCunney, P. J., 1992, Occupational Noise Exposure, in *Environmental and Occupational Medicine*, Rom, W. N. (Ed.) Little Brown, Boston.

Millar, D., 1986, Screening and Monitoring: Tools for Prevention. *J. Occup. Med.* **28:**544–46.

Murphy, L. I., Halperin, W. E., 1995, Medical Screening and Biological Monitoring. A Guide to the Literature for Physicians. *J. Occup. Med.* **37:**170–84.

Occupational Safety and Health Administration Access to Employee Exposure and Medical Records. 29 CFR 1910.20.

Occupational Safety and Health Administration. Exposure to Benzene 29 CFR 1910, 1028.

Peterson, K., 1994, The American's with Disability Act, in *A Practical Approach to Occupational and Environmental Medicine*, McCunney, R. J. (Ed.), Little Brown, Boston.

Silverstein, M., 1994, Analysis of Medical Screening and Surveillance in 21 Occupational Safety and Health Administration Standards: Support for a Generic Medical Surveillance Standard. *Am. J. Ind. Med.* **26:**283–95.

U.S. Preventive Services Task Force, 1996, Guide to Clinical Preventive Services and Assessment of the Effectiveness of 169 interventions Report of the U.S. Services Task Force, Baltimore, Williams & Wilkins, Second Edition.

Wagner, G. R., et al., 1992, The NIOSH B-Reader Certification Program: An Update Report. *Journal of Occupational Medicine* **34:**879.

Wong, O., Raabe, G. K., 1996, Cell-Type Specific Leukemia Analyses in a Combined Cohort of more than 208,000 Petroleum Workers in the United States and the United Kingdom (1937–1989). *Reg. Tox. Pharm.*

CHAPTER 12

How to Establish Industrial Loss Prevention and Fire Protection

PETER M. BOCHNAK

Environmental Health and Safety, Harvard University, 46 Oxford Street
Cambridge, MA 02138

12.1 INTRODUCTION
- 12.1.1 Basic Process Safety
- 12.1.2 Equipment Safety
- 12.1.3 Operator Safety
- 12.1.4 Safety Audits and Training Programs

12.2 RISK
- 12.2.1 Definition
- 12.2.2 Quantification
- 12.2.3 Methods for Risk Quantification

12.3 HAZARD IMPACT
- 12.3.1 Identifying Potential Hazards
- 12.3.2 Consequence Analysis

12.4 ELEMENTS OF A LOSS CONTROL PROGRAM
- 12.4.1 Policy Statement
- 12.4.2 Loss Control Organization
- 12.4.3 Identification
- 12.4.4 Evaluation
- 12.4.5 Control

12.5 FIRE PROTECTION
- 12.5.1 Fire Prevention Plan
- 12.5.2 Elements of Emergency Action Plan
- 12.5.3 Evacuation Drills
- 12.5.4 Life Safety Principles

12.6 EMERGENCY PLANNING
- 12.6.1 Elements of Emergency Planning

The information provided in this chapter was current and reliable at the time of publication. Regulatory and technical requirements are subject to change, hence, the information detailed herein is only in compliance with current regulations in effect at the time of publication. Consult OSHA regulations or seek legal counsel for accurate information on current requirements. The writer disclaims any liability resulting from reliance on obsolete or other inaccurate information presented in this chapter.

Handbook of Occupational Safety and Health, Second Edition, Edited by Louis J. DiBerardinis,
ISBN 0-471-16017-2

12.7 AUDITS
12.7.1 Self-Inspection Program
12.8 PLANT SITE SELECTION
12.8.1 Systems Approach
12.8.2 Site Selection
12.8.3 Plant Layout
12.9 PROCESS EQUIPMENT AND FACILITIES DESIGN
12.10 TRAINING
REFERENCES

12.1 INTRODUCTION

Today, with the real presence of the Occupational Safety and Health Administration (OSHA) and the employee's "right" to safety and health, coupled with the "age of risk management," captive insurance companies, and large self-retention of risk by major corporate entities, there is a greater need for management to familiarize itself with the general field of fire protection. In the past, there appears to have been a general feeling that the protection of plants and facilities from fire and the maintenance of their integrity could be safely left to insurance underwriter personnel or perhaps to a very small staff or an individual who might have something less than a heavy background in this area. Now, with greater retention of risk by corporations and the tremendous investments involved in many facilities, there is a need for many more people, at the very least, to have more than a superficial knowledge of the world of loss prevention.

This need has impacted not only upon plant management, but also upon architects and engineers. Many architectural engineering firms have found it prudent to employ specialists in fire protection on their staffs to handle what should be considered a very important part of the overall plant design. It is important for these technical people to develop some knowledge of fire protection, since many firms of this type are not only providing fire protection concepts, but in many cases are doing the actual design, although they may not possess a high degree of fire protection expertise.

Risk management, with its attempts to bring a degree of sophistication to decisions on handling and preserving corporate assets, must by necessity begin with personnel who have the knowledge of safety and fire protection necessary to evaluate the risks facing a corporation.

Where does the risk manager turn—or, for that matter, the plant manager, technician, architect, or engineer who needs to know something about fire protection? Many publications are available on this very broad subject that, for the most part, are directed at those having more than a causal interest in this field. Many of these publications are voluminous and represent complete works either on the entire scope of fire protection or on specific areas. For the professional, a library of these publications is indispensable.

However, turning back to the question as to where the busy manager should turn for some background, there appear to be few publications that give a relatively short treatment or an overview on this subject. So we now come to the objective of this chapter, which is to provide the manager, the architect, the plant engineer, the technician, and others with a background in safety and fire protection in as brief and concise a manner as possible and then point out other sources available for more in-depth information.

This chapter is, therefore, specifically directed to management-oriented personnel with an involvement in loss prevention; to technicians who are not engineering oriented, but who nevertheless must have a relatively strong background in safety, and to the architects and engineers involved with overall plant design.

12.1.1 Basic Process Safety

The integration of safety into any plant, especially on processing toxic or flammable liquids or gases, involves several disciplines. The initial requirement is to determine the toxicological and flammability properties of the products. The next is to determine whether the products can be processed safely. These determinations may require review by a toxicologist, industrial hygienist, process engineer, equipment design engineers, an instrument engineer, and a safety engineer, among others.

12.1.2 Equipment Safety

After the process has been approved as safe, then safety must be built into the plant hardware. Pressure vessels, pumps, fired heaters, and other equipment must be designed to withstand the operating conditions, which include abnormal upsets such as exothermic reactions, failure of instrument controls, and human failures. In anticipation of such failures, safety alarms and shutdown controls must be designed into the process. Wherever possible, automatic safety shutdowns should fail in the safe position. Where practical, backup manual systems should be provided.

12.1.3 Operator Safety

Next, operators must be trained in the safe operation of the unit, in how to detect incipient equipment problems, and finally in what action to take when an emergency arises. Procedures must be written for normal plant operation and also for any anticipated emergency. These procedures should be available to all affected personnel and periodically reviewed and updated to reflect changing conditions. Management should publicize to plant personnel that they support these procedures and take the necessary steps to assure that the procedures are being practiced.

12.1.4 Safety Audits and Training Programs

In addition to day-to-day safe plant operations, periodic safety audits should be made and firefighters be kept trained to respond to any emergency (see Chapter 6 for a discussion of hazard analysis and Chapter 28 for a discussion on program audits). An emergency response plan should be developed utilizing plant and municipal medical staff, plant security staff, and mutual aid provided by other local industries. The emergency response plan should be tested periodically to ensure effectiveness. Safety cannot be initially designed into a plant and then forgotten. Maintaining safety is an ongoing daily struggle. Even when it is maintained to the highest degree, there is no guarantee that accidents will not occur. However, the converse is true: *If safety is not observed, accidents are inevitable.* Plants for the most part are still run by people. Even if plants become completely automated with robots and controlled by computers, people will still be required to assure safety.

12.2 RISK

12.2.1 Definition

A risk is the possibility of an undesirable occurrence. The only justification for taking a particular risk is that the reward clearly exceeds the penalty if the associated accident takes place. In order to decide whether to take a particular risk, it must be quantified carefully.

12.2.2 Quantification

To quantify a risk, it is necessary first to determine what the risk is and the extent of damage in a worst-case scenario. Next it is necessary to determine the probability of the accident occurring. In evaluating the probability of an event, there are *two current popular hypotheses:* (1) A disaster that never happened is about to happen. (2) A disaster that never happened won't happen. Both of these hypotheses are false. Three Mile Island proved that a nuclear reactor meltdown was possible, and Chernobyl made it a reality. A likely risk should not be accepted unless the penalty of the incident is also accepted. If the loss of property and production is acceptable, then the risk is acceptable. Of course, this also assumes that all reasonable precautions have been taken to eliminate or mitigate the penalty. Pure chance taking is never good business practice, and when the risk involves the possible loss of life, then if should not be accepted.

12.2.3 Methods for Risk Quantification

In order to quantify a risk, some of the following tools can be used:

PAST EXPERIENCE. What has caused an incident, how often it can be expected to occur, and what is the expected extent of damage. To make this determination, a database should be constructed from experience within the plant or company and throughout identical and similar industries. This is a costly, time-consuming procedure, and where very few incidents have occurred, the validity of the information may be questionable. Some data on the reliability of instruments appear in Anyakora, Engel, and Lees (1981). Probabilities for equipment failure can be found in Browning (1969).

LOGIC MODELS. These are concerned primarily with the "on/off" connection between variables and incidents and are useful for analyzing complex systems.

Two common techniques used in logic models are failure modes and effects analysis (FMEA) and fault tree analysis (FTA).

FMEA breaks down a system into its simplest components and asks "what if" questions regarding all possible failure modes for each component. The probability of a failure occurring in the system is the sum of the failure probabilities for each component. The advantages of FMEA are that it highlights components with the highest probability of failure and permits improving these components. FMEA results in a systematic piece-by-piece review of each component and permits establishing inspection and test programs to monitor the performance of high-failure components.

Fault tree analysis, which goes further than FMEA in that it identifies failures in subsystems, is often used in conjunction with FMEA. It determines the consequence of two or more component failures occurring simultaneously.

A more detailed discussion of these forms of hazard analysis is provided in Chapter 6.

12.3 HAZARD IMPACT

12.3.1 Identifying Potential Hazards

As stated previously, the risk-taking decision should be based not only on the probability of an incident occurring, but also on the potential impact or consequences of the incident. *Consequence analysis* can be the most difficult phase of overall hazard analysis. Some of the most serious hazards involve the formation of toxic and flammable vapors. It is often extremely difficult to determine the maximum quantity of flammable vapor that may be involved in an explosion or the maximum quantity of a toxic release that could be expected. As an example, the percentage of a propane gas release that can be in the flammable range at any given time is 2.15 to 9.60 percent by volume in air. Thus the assumed amount of explosives could be incorrect by a factor of 4.5. For this example, one method of calculating peak overpressure of an explosion is to determine TNT equivalent based on the heat of combustion per pound of the chemical, related to TNT. This is a rough correlation, but is used because more is known about the pressure waves and destruction caused by the detonation of TNT than that of any other explosive.

12.3.2 Consequence Analysis

From the calculated extent and intensity of potential damage, the impact on both plant/personnel and the surrounding community can be determined. Using these data, plans can be formulated to cope with the results. The determination of the impact on plant and personnel should consider the following factors:

- The number of people exposed in the plant onsite areas and in administration buildings, shops, warehouses, laboratories, etc., both during the day and at night.
- The extent of property damage within the plant.
- In-plant and outside first aid response.
- The time required for cleanup and repair or replacement of damaged equipment and the lost production cost.
- The effect on employees who may lose work if a portion of the plant is shut down for an extended period.
- Backlash legislation, that is, the passing of local, state, or federal laws that would increase the cost of doing business. An example is mandating additional safety features that are not necessary, forcing the plant to relocate or shut down.

In determining the impact of an incident on the surrounding community, the following factors should be considered:

- The maximum number of people outside the plant exposed to a hazard during the day and night.
- The type of people exposed and the probability of panic, i.e., adults, children, senior citizens, handicapped.

- Medical response: the time required to evacuate injured as casualties increase.
- Extent of property damage outside plant, including lost production and lost work time.
- Damage to the company's overall public image.
- Passage of laws that could cause considerable added cost without appreciable added safety.
- Permanent loss of some markets from boycotts and lost production.
- Personal injury and damage suits.

12.4 ELEMENTS OF A LOSS CONTROL PROGRAM

As new technologies are developed, hazards may be introduced, and when industry and business expand, existing hazards may increase. To remain prosperous, a corporation must minimize the possibility of casualties, interrupted production, and property losses caused by fires and explosions. Corporations are also required by law (OSHA 29 CFR 1910) to provide a reasonably safe workplace for their employees. To satisfy these responsibilities and also remain competitive, corporations must eliminate hazards where possible and minimize those that cannot be eliminated.

12.4.1 Policy Statement

A program of identification, evaluation, and control of hazards is known as a *loss control program.* For this program or any program involving the welfare of a corporation or its employees to be effective, the sincere and total support and the sustained interest of top management are essential. This is particularly true for all corporate loss control programs. In addition to top management support, this policy of controlling losses must be communicated to lower levels of management and to all employees. This can be accomplished during the formation of the loss control program by a written policy statement from the chief executive officer demonstrating his or her complete support of the program; outlining the procedures, objectives, responsibilities, and accountabilities; and further indicating his or her desire that all employees of the organization support those responsible for the formation and implementation of the loss control organization and function. This shows that management cares, that management is directly involved.

The objective of the fire loss control program should be fully stated, with emphasis on the protection of employees against injury and the conservation of corporate assets. Both management and labor benefit from the responsible safeguarding of profit centers and the dependable continuity of operations. This policy should emphasize to all employees that major damage to a facility by fire or explosion in most cases caused temporary or total loss of jobs.

12.4.2 Loss Control Organization

Some form of a loss control organization should be developed to implement any corporate loss control program. No definitive guidelines can be established for the simple reason that such an organization could vary from a very informal structure to one that is highly structured.

Under any loss control program it is important to have some control of these func-

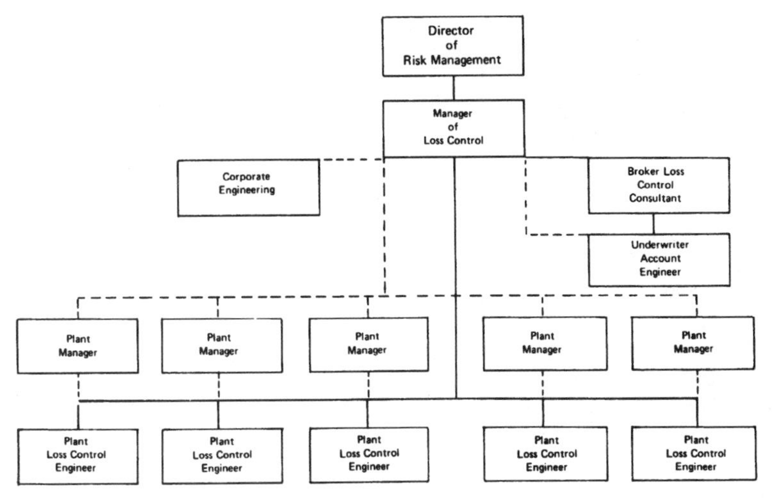

Fig. 12.1 A suggested loss control organization for a large multidivision corporation with a corporate loss prevention staff.

tions at the corporate level and to assign line responsibilities at the division and plant levels. Figures 12.1 and 12.2 indicate suggested loss control organizational functions for (1) a large, multidivision, multiplant facility with a corporate loss prevention staff; and (2) a similar organization not employing fire protection personnel, but depending upon outside sources, principally the technical services of insurance brokerage firms.

The loss control manager could be known by a variety of other names, depending on the industry—for example, fire loss control manager, fire loss prevention and control manager, or property conservation director. In some industries, the loss control manager reports to the facility manager and is responsible for employee safety and health management, fire protection management, property insurance management, and security management.

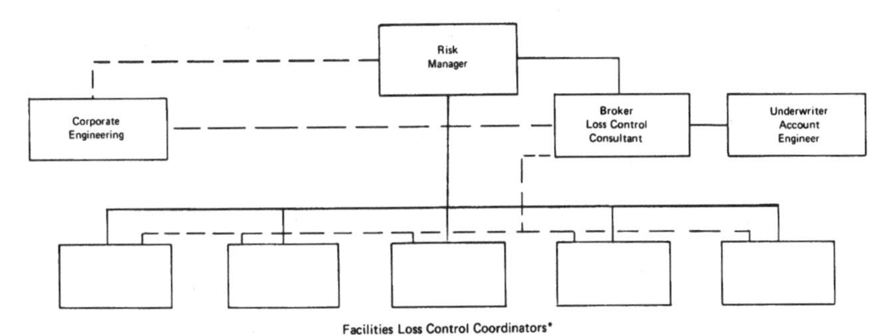

Fig. 12.2 A suggested loss control organization for a large corporation depending principally upon others for loss control efforts. The coordinators (*) will normally have principal responsibilities other than loss prevention. Ideally, coordinators will be members of facilities managers' staffs.

Basically, as a minimum, there should be some in-house loss control organization coupled with the assistance of outside sources, such as consulting fire protection and safety engineers and industrial hygienists, insurance brokerage personnel, and insurance company inspectors and engineers.

12.4.3 Identification

Once a loss prevention organization has been developed, or even if such an organization is not yet in effect, an audit of the facilities should be conducted. This audit is suggested as a first step in identifying hazards so that they can be evaluated and so that priorities can be established for the concentration of efforts in the loss control program. Quite often, the corporate staff will require assistance in the evaluation of their facilities from outside sources. This assistance can be from the insurance personnel, either in a detailed review of property loss prevention reports received from insurance sources with some special visits to particular sites or from technical services personnel of insurance brokers. Brokers will supply this service using either their reports or the services of their control personnel accompanying corporate personnel in the physical inspection of various sites. An audit form should be developed to be used in the loss control program. A suggested form is shown in Fig. 12.3. In addition to identifying the problems at the facility, the audit form can be of considerable value when kept on file at a corporate level for answering questions that frequently develop concerning construction, utilities, various occupancy details, etc.

Identification of hazards and their probability is facilitated by:

EXPERIENCE WITH SIMILAR UNITS. Some chemical processes have an inherent hazard such as high pressure or temperature, an exothermic, or "runaway," type of reaction, involve the formation of a toxic material, or use hazardous reactants such as corrosive acids.

PLANT ACCIDENT REPORTS. Most companies maintain plant accident reports, which may provide a good reference for anticipating hazards and estimating probability. Many companies are computerizing these accident reports to facilitate access and compile probability indices (see Chapters 4 and 24).

STEP-BY-STEP REVIEW OF PROCESS. A detailed review of each step of the process should be made by following through process flow and piping and instrument drawings. This review can be more effective if made with the process engineer familiar with the process and the operator familiar with operations.

SEMINARS AND MEETINGS. Seminars and meetings periodically conducted by the trade organizations and professional societies such as the American Petroleum Institute, The National Petroleum Refiners Association, the American Society of Mechanical Engineers, American Institute of Chemical Engineers, and the Society of Fire Protection Engineers (SFPE) and American Society of Safety Engineers (ASSE). These organizations provide access to a cross-section of experiences and practices throughout similar industries and each has a journal such as *Professional Safety* from ASSE and *Journal of Fire Protection Engineering* from SFPE.

PUBLISHED PAPERS. Papers published in journals often identify problems at the incep-

XYZ Corporation
Facilities Audit Form*

Location: Date:

Construction

Walls: Roof:

Floors: Partitions:

Unusual features (combustible interior finishing, insulation, etc.):

Boilers

Description and rating:

Fuel:

Combustion controls (including details on interlocks, purge cycle, flame safeguard, etc.):

Controls testing program:

Electrical

Power supply (including capacity, no. of feeders, etc.):

Transformers: No. Capacity

Major motors: HP Spares?

Hazardous electrical equipment (Class I or II):

Emergency generators:

Data processing: Description

Functions

Location and envelope construction

Tape storage

Air conditioning

Protection

Detectors

Protection:

Computer area

Underfloor

Tape library

Hazardous operations (flammable liquids, dust, etc.):

Plant protection

Water supplies (include water test data):

Underground mains and valving:

Automatic sprinklers: Full Partial Type

Design

Alarms

Special protective systems (CO_2, halon, etc.):

Fig. 12.3 A form suggested for use in an audit of corporate facilities.

Security

Watchman service: Alarms:

Environs

Flood: Seismic zone:

Human-element programs (give date of programs)

Fire brigade:

Self-inspections:

Emergency planning:

Cutting and welding (hot work):

---

*Attach plan of facility.

Fig. 12.3 (*Continued*)

tion stage. This permits correction of problems without having to repeat the "learning experience."

INSURERS. The insurer of a particular plant usually has compiled a history of losses with documentation of event causing these losses of all the sites insured by the company. Thus the insurer's experience may point out similar problems that occurred in other plants.

12.4.4 Evaluation

Once the information has been collected through the use of the audit program identifying potential fire hazard areas, it is important to develop and improve the recommendations list based upon the following priorities in descending order of importance:

1. Life safety exposures: The potential for fire casualties in industrial types of occupancies is directly related to the hazard of the operations or processes. Most of the multiple fatalities are a result of (a) flash fires in highly combustible materials; or (b) explosions of dusts, flammable liquids and vapors, and gases.
2. Continuity of operations: This includes hazards in areas that will have a significant effect upon the continuity of operations in the event of their loss, thus affecting possible corporate income and loss of corporate resources. Management is extremely conscious of financial return on investment, and those locations that affect production, such as computer facilities controlling production processes, should be highlighted.
3. Other hazard areas: This includes problems of protection from hazards not necessarily of a business continuity nature such as woodworking or machine shops.

Frequently, large firms will need numerous and similar facilities such as distribution locations, theaters, warehouses, etc. Where these situations are encountered, it may be advisable to develop corporate loss control guidelines outlining structural, building services, and protection requirements as related to property loss. The desirability of incorporating guidelines of this nature may be highlighted by an examination of past loss experience, which may show a need for improvement.

While the prime responsibility for the development of the suggested guidelines will rest with the loss control personnel, the project should be coordinated with both operating and facilities personnel. Input and approval should be obtained from all those involved, particularly those having architectural or engineering responsibilities, so that the guidelines will carry the necessary weight and will command the respect and attention of both corporate employees and outside firms who may have some interface with the guidelines.

The components of protection guidelines should include:

1. Construction details: The construction details include those having a bearing on the structure's ability to withstand a fire or to provide fuel in a fire. This may involve the specific design of a desired fire resistance rating that may be important guidelines for locations; for example, specifications for insulated metal-deck roofs so that the roof might be classified as noncombustible rather than combustible; specifying mechanical fastening of the rigid insulation of a metal-deck roof at the roof perimeter to avoid windstorm losses.

2. Safe installation of building services: This will include the location and enclosure of switch gear; the location, types, and protection of transformers; the designation of wiring means; the designation of combustion controls for heating and processing equipment and the enclosure of this equipment; the proper installation of flues; the proper installation of air-conditioning systems for sensitive areas such as those housing computers.

3. Required fire protection features: This could take the form of specifying the hydraulic design requirements for a hydraulically calculated sprinkler system and engineering requirements for similar installations; for theaters, the protection of storage areas and projection rooms.

When formulating loss control recommendations and guidelines, an important question to ask is: "What loss expectancy can be tolerated?" This is particularly significant when the trend is toward higher insurance deductibles and self-retention of losses. When considering loss expectancy, particularly for lesser value but numerous locations such as supermarkets or theaters, the desirability of the formulation of loss control guidelines takes on significant importance.

12.4.5 Control

12.4.5.1 Human Elements Despite excellent engineering and a good loss control program at the corporate level, problems at facilities and poor loss experience can still exist if the facility manager or first-line supervisor is not involved in the overall effort. The human elements of a loss control program refer to the continuing actions of personnel at the local level in the loss prevention and control effort in the formulation and updating of emergency planning procedures; in the constant readiness of the emergency organization (fire brigade); in the conduct of self-inspection programs; in the exercise of an established impairment notification system or fire protection system shutdowns; in the use of a cutting, welding, and a hot-work permit system; in the limiting of smoking to designated areas; and in good housekeeping procedures. The implementation of human-element programs does not involve extensive expenditures. Such programs provide for greater safety and have the further benefit of making local employees more aware of their environment, coupled with a feeling of being a part of the action.

12.4.5.2 Hazard Elimination

After a hazard has been identified, the potential results should be assessed to determine which risks are acceptable and which are not. Risks involving human life and health are not acceptable. Some risks involving property only may be acceptable to a limited degree. Providing no fire protection for a plant and letting it burn down on the premise that it is fully insured and can be rebuilt is *not* an acceptable risk. This is true even when economics are against fire protection.

Some methods used to eliminate or control hazards are:

1. **Changing the process conditions where possible.** For instance, reducing the pressure or temperature, substituting a less hazardous material for one or more of the reactants or using dilution ventilation or local exhaust ventilation may be possible.
2. **Changing the design of process equipment.** Using pumps with double seals or can-type pumps for hazardous materials; using storage tanks with no bottom piping connections for refrigerated hydrocarbon storage, such as LNG; eliminating piping sliding and bellows-type expansion joints and proprietary coupling shaving soft elastomeric ring gaskets; using self-reinforcing forged nozzles rather than nozzle reinforcing pads for high-pressure vessels and exchangers; upgrading equipment materials, such as using higher alloys for high-temperature equipment and equipment exposed to highly corrosive liquids and gases.
3. **Installing remote operated safety shutoff valves or interlocks that automatically operate such valves when an emergency occurs.** A few examples are remote operated valves on bottom outlets of vessels and storage tanks containing large volumes of flammable or toxic liquids; check valve in large-diameter piping to prevent backflow if a break occurs (only when flow is always in the same direction); excess flow valves that automatically close at high flow velocity resulting from a pipe rupture; automatic valves interlocked with a pressure sensor to shut in a line when the pressure drops because of a piping failure; air-operated or spring-loaded valves equipped with a fusible plug or link to close when exposed to fire.
4. **Installing fixed automatic firefighting equipment to control or extinguish a fire and minimize losses.** Some protective systems are: automatic fixed halon extinguishing systems in computer rooms or inside gas turbine generator housings; sprinkler systems in buildings; high-density fixed water sprays designed for fire intensity control; low-density water sprays to cool shells of pressure vessels when exposed to fire to prevent rupture due to overheating the metal.

12.5 FIRE PROTECTION

The overall plant fire protection system should be designed to meet the requirements to contain and extinguish the most serious fires that could occur. Where there is adequate manpower and a well-trained plant fire department is maintained, firefighting systems may be manual. Where the manpower is limited and high hazards exist, automatic and remote systems are prudent. The basic elements of firefighting systems are as follows:

FIRE WATER SUPPLY. The first step is to provide a reliable fire water supply. For small low-hazard chemical plants the municipal water supply may be adequate. The installation of a fire water storage tank is usually advisable. The quantity of water stored should be determined by the design flow rate to the fire main and the length of time expected to fight a major fire. This time varies from 4 to 8 h, depending upon the size of the plant and the exiting hazards. About 3000 gal/min (680 m^3/h) normally would be adequate for a small chemical plant, and a minimum of 4000–5000 gal/min (909–1135 m^3 h) would be required for a petrochemical plant or refinery. The design flow should be based on the assumption that only one process unit at a time will be on fire. Although there is a possibility of a fire spreading from one unit to another, the probability is low if units are adequately spaced. It is also unlikely that adequate manpower and firefighting equipment would be available to fight two process unit fires of major magnitude. In determining the flow rate, the total of fixed monitors, water sprays, and hose streams should be determined for each process unit. However, this flow should be based on the equipment expected to be required for any single type fire, and not the total firefighting equipment installed onsite.

FIRE PUMPS. Fire water pumps should be selected on the basis of reliability and meet the requirements of NFPA 20, *Standard for the Installation of Centrifugal Fire Pumps* (NFPA 20, 1993). Electric-motor-driven pumps require little maintenance and are easily started, but should not be relied upon where power failures can occur. Diesel-driven pumps are very reliable, but cost more than electric driven and require more maintenance. Steam-turbine-driven pumps should be used only where there is a reliable source of steam, especially during a fire when steam requirements for spare pumps may be a high.

It is good practice to install more than one fire pump as backup. Often one electric pump and one or more diesel pumps are installed. Typical pump sizes vary between 1500 and 3000 gal/min (340–680 m^3/h).

When a fire water system is required to be pressured constantly for immediate use, then a jockey pump (which maintains the pressure in the system) should be installed. When the fire main pressure cannot be maintained by the jockey pump, then the main fire pumps should start automatically and sequentially upon demand. A constantly running jockey pump is preferred to an intermediate running pump with an air pressure tank. Jockey pumps should be electric with a minimum capacity of 100–200 gal/min (23–45 m^3 h) to avoid frequent startup of fire pumps. Fire pumps should be installed in a fire safe area at least 200 ft from any fire hazard.

FIRE MAINS AND HYDRANTS. The fire main system is usually laid out in a grid pattern. With this design, water supply is from two or more directions, which permits smaller diameter pipe. Reliability is also improved by installing diversion valves, which can be used to isolate a main break at any location and continue fighting a fire.

Steel pipe is the preferred material for resistance to mechanical damage. Pipe sections should be welded together and flanges used at valves, hydrants, and other fittings. Friction-type flexible couplings with elastomeric ring gaskets should be avoided where possible. Underground mains should have exterior surfaces coated and wrapped for corrosion protection. In addition, cathodic protection is usually provided. When salt water is used, the pipe interior should be cement or epoxy lined. Plastic pipe should be used with caution. Use a good quality plastic pipe and make sure by careful inspection that all joints are sound and do not leak. Fire mains around process units preferably should be underground to protect them against fires, explosions, and mechanical damage. In less hazardous offsite areas, fire mains are often above grade. Mains generally are 8 in.

minimum size except that in low-hazard offsite areas 6-in. mains may be used when the required flow rate is well known.

Fire hydrants usually have two $2\frac{1}{2}$-in. hose connections and a $4\frac{1}{2}$ or 5-in. fire truck pumper connection. They should be located on fire mains looped around process units and spaced 150–200 ft (45–60 m) apart. They should also be located on mains looped around tank storage areas on about 200–300 ft (60–90 m) spacing. Hydrants around buildings and other offsite areas should be located as the hazard demands. They should be located near plant roads to be accessible to fire trucks. Hose threads should be compatible with municipal fire department and mutual aid services. When this is not feasible, hose coupling adapters should be available. Wet barrel hydrants, without underground shutoff valves, should have a valve on each hose connections.

FIRE MONITORS. Fixed water monitor nozzles (water nozzle with separate shutoff valve) should be installed around process units to protect specific pieces of equipment such as heaters, exchanger banks, vertical and horizontal pressure vessels, and compressors. This can best be accomplished by installing permanent monitor nozzles around such equipment at grade level, on special trestles, or on the roofs of buildings. They are also used in offsite areas to protect marine tankers and equipment on marine wharfs, at truck terminals to protect the loading facilities and trucks, tank cars at tank car loading racks, in-tank storage farms to protect LPG tanks, at gasoline treating plants, around TEL (tetraethyl lead) facilities and cooling towers.

Around process units it is preferable to locate monitors outside the battery limits, where firefighters are in a less exposed location. Where the process area is congested, it may be necessary to install some monitors inside the unit and also in elevated locations. Elevated monitors usually are operable from grades so that firefighters can readily escape.

The commonly used monitor size is 500 gal/min (113 m^3/h). This size is adequate for cooling most equipment and does not drain the water supply from other firefighting equipment. Larger monitors, 1000 gal/min (227 m^3/h), are sometimes used to protect such high hazards as marine wharfs and tankers. For flammable liquid and gas storage areas, foam monitor nozzles are used.

Two-wheeled portable foam/water monitors are useful for towing to a fire as needed. The 500 and 100 gal/min sizes are common, and are usually kept in the plant firehouse when not needed. The monitor generally is supplied with the foam/water solution properly proportioned by a fire truck.

FIXED WATER SPRAYS. Fixed water sprays have the following uses:

- Cooling metal to prevent distortion, rupture, or buckling when fire exposed.
- Controlling fire intensity.
- Dispersing combustible and toxic vapor clouds.
- Maintaining the integrity of electrical cable exposed to fire.
- Cooling Class A materials below their ignition temperature to extinguish a fire.

Water sprays should be designed in accordance with NFPA 15, Standard for Water Spray Fixed Systems for the Fire Protection (NFPA 15, 1990).

Water sprays for cooling metal are applied to:

- Uninsulated pressure vessels containing flammable and toxic liquids.
- Uninsulated pressure storage vessels containing light ends such as propane and butane.
- Compressors.

Water sprays to control fire intensity are applied to:

- Pumps.
- Hot piping manifold, particularly when the liquid is above autoignition temperature.

Vapor cloud dispersal applications are:

- Toxic liquids or gases that could escape from a vessel, pump, compressor, manifold, or other equipment.
- Flammable liquids under very high pressure, which could leak out of flanges and pump seals to form a large vapor or mist cloud.

Electrical cable and transformer protection are used to:

- Limit the loss of grouped electrical cable runs in tunnels or other areas difficult to reach for firefighting.
- Maintain the operability of grouped electrical cables during a fire.
- Maintain integrity of transformers.

Fire extinguishment applications are:

- Extinguish fires on conveyor belts carrying combustible materials such as coal.
- Protect long conveyor belts from being destroyed by fire caused by hot roller bearings.
- Extinguish fires in wooden cooling towers.

Activation of water sprays may be manual or automatic. Manual activation is usually by a quarter-turn ball valve located in a fire safe area. Automatic activation may be by heat or smoke detectors.

Advantages of manual operation:

- Less expensive than automatic.
- More reliable.
- Less maintenance required.
- Permits operator to exercise judgment and prevents false alarms.

Advantages of automatic operation:

- Limit losses by quick activation.
- Frees operator to shut down the unit.
- Protects areas normally unattended or during off hours.

When considering water spray protection, keep in mind that sufficient water must be available so that the fire water supply will not be depleted in an emergency. Also, drainage of spray water must be considered; otherwise a hydrocarbon fire could be spread throughout the area.

LIVE HOSE REELS. Live hose reels hold 50–100 ft (15–30 m) of noncollapsible hard rubber hose, usually $1\frac{1}{2}$ in. diameter. The water line is pressured up to the hose inlet, and to change the hose all that is required is to open the valve at the hose inlet. With hard rubber hose, not all of the hose needs to be removed from the reel in order to use it. Live hose reels have a flow rate of only about 100 gal/min (23 m^3/h), and are not suitable for fighting large fires. Their primary purpose is in fighting small incipient fires to provide holding action until the fire department arrives.

Some applications for live hose reels are:

- At process units, where they are usually installed under the unit pipe rack for easy access and can be deployed on either side of the rack.
- On marine wharfs, to protect loading areas and piping manifolds.
- Inside buildings, such as warehouses and storehouse, where immediate fire hose response is desired.
- At truck loading racks, where manpower is limited and quick response is desired.
- At areas where small gas containers, such as propane, are filled.

In cold climates winterizing is required, usually by electric heat tracing the aboveground portion of the water supply piping.

PORTABLE FIRE EXTINGUISHERS. Portable fire extinguishers for industrial use range from the small 5 lb (3.2 kg) hand-held unit to the 350 lb (160 kg) wheeled dry chemical unit. NFPA 10, Standard for Portable Fire Extinguishers (NFPA 10, 1994), covers selection, installation, inspections, maintenance, and testing. The frequently used types of extinguishing materials are:

- Potassium and sodium bicarbonate dry chemical, used for Class B fires such as most flammable liquids and gases.
- Multipurpose dry chemical, used for Class A, B, and C fires.
- Halon 121 (BCF), used for electrical equipment.
- CO_2, used for electrical equipment.
- Pressurized water, used for Class A fires, usually in office areas containing paper and other combustible materials.
- AFFF (aqueous film forming foam), used where securement of small flammable liquid fire is desired.

UL tests and classifies fire extinguishers regarding the types of fires and fire area that can be extinguished. A UL classification, such as 120-BC, indicates that a type B or C fire, 120 ft in area, can be put out by a trained operator using the extinguisher. For a more complete explanation of the UL rating, see NFPA 10, Paragraph A-1-42 (NFPA 10, 1994).

Some plant applications for extinguishers are:

- **Office buildings:** Five pound (2.3 kg) multipurpose dry chemical extinguishers located a maximum of about 75 ft (23 m) from any hazard. Pressurized water extinguishers, $2\frac{1}{2}$ gallons (10 L), also may be used, but are not as effective as dry chemical.
- **Warehouses and maintenance buildings:** Thirty pound (9 kg) multipurpose dry chemical extinguishers. Where Class B materials are stored and handled, use 30 lb potassium bicarbonate. Locate extinguishers not more than 50 ft (15 m) from any hazard.
- **Laboratories:** At least 9 lb BCF and one 30 lb potassium dry chemical extinguisher of more, depending upon the size of the laboratory.
- **Process areas:** Locate 30 lb potassium bicarbonate extinguishers so that at least one extinguisher is not more than 50 ft (15 m) from all hazards. Also provide one or two 350 lb potassium bicarbonate wheeled extinguishers in each process unit. Extinguishers mounted on pipe rack columns are accessible and easily located. Also locate extinguishers at each fired heater, compressor, in pump areas and at transformers. At least one 22 lb BCF extinguisher for electrical fires should be located in control rooms and outside substations.
- **Offsite areas:** One 30 lb multipurpose extinguisher at top deck stairway landing of cooling towers.
- Locate one or more 30 lb potassium bicarbonate extinguishers at boilers and at offsite pump areas.
- **Marine terminals:** One or two 30 lb potassium bicarbonate extinguishers at each cargo loading area, plus one 30 lb unit in the control room and near piping manifolds. Also locate one 350 lb potassium bicarbonate wheeled extinguisher near cargo holes and loading arms. Large capacity 2000 lb (909 kg) skid-mounted dry chemical units are frequently used on wharfs, particularly where access for mobile firefighting equipment is poor.

The use of dry chemical extinguishers in areas where computers and other electrical equipment is located should be avoided. because the cleanup required may be more expensive than fire damage.

Extinguishers must be properly mounted and the location clearly marked for quick identification. They should be periodically inspected and tagged. After using, they should be immediately recharged and replaced.

FIRE HOSE. Soft fire hose is used for hand hose line, to supply water from hydrants or afire truck to portable firefighting equipment such as wheeled monitors, and to supply water from hydrants to fire trucks. The basic hose sizes are $1\frac{1}{2}$, $2\frac{1}{2}$, and 3 in. Three-inch hose is used primarily to supply water from hydrants to fire trucks and is equipped with $2\frac{1}{2}$ in. couplings. Fire hoses should meet the requirements of NFPA 1961, Standard for Fire Hose (NFPA 1961, 1992). Hose couplings should be in accordance with NFPA 1963, Standard for Fire House Connections (NFPA 1963, 1993).

Single jacket hose consisting of all synthetic fiber with rubber tube liner is commonly used. Some hose is impregnated with polyvinyl chloride, Hypalon, or other plastic coating to reduce wear.

Hose is often stored in hose houses or cars in the plant area where it is expected to be used. The disadvantage of this practice is that a considerable inventory of hose may be stored throughout the plant, and often is not maintained, due to lack of inspection. In large plants with a well-trained fire department and one or more fire trucks, all hose is carried on fire trucks. This is the preferred practice, because the hose is tested and maintained in good condition and is available wherever it is needed.

FIRE FOAM. Fire foams are used to extinguish and secure hydrocarbon pool fires such as spills and atmospheric storage tank fires. Types of foam concentrates are:

- **Regular protein foam:** Used for spill fires and storage tank fires for top application only. Protein foam is less used since the development of synthetic foams with superior properties.
- **Fluoroprotein foam:** This foam serves the same uses as protein foam, but is generally more effective and can be used for subsurface application for cone roof storage tank fires.
- **AFFF (Aqueous Film Forming Foam):** This foam is a combination of fluorocarbon surfactants and synthetic foaming agents that quickly spread across the surface of a hydrocarbon pool fire. It has quicker control and extinguishment than fluoroprotein foam, but does not have as good sealing properties against hot metal, such as the shell of a storage tank on fire.
- **Alcohol-type foam:** Used for fires involving alcohol and other polar solvent liquids (such as acetone, methylethyl ketone, enamel, and lacquer thinners) that break down other foams. It can also be used against hydrocarbon fires.
- **High-expansion foam:** This foam is used at expansion rates of 800 to 1. It is intended primarily for use on Class A materials in confined areas such as warehouses and other buildings. Because of its low density, it can be blown away easily when applied outdoors.

Foam concentrate is supplied in two concentrations, 3% and 6%. The quality and cost of both foams is equivalent, but 3% foam has the advantage of making double the quantity of foam when stored in equal-size containers. This becomes important on mobile firefighting equipment where allowable weight is limited. NFPA 11, Standard for Low Expansion Foam and Combined Agent Systems (NFPA 11, 1994), covers the characteristics of foam producing agents and the requirements for design, installation, and operation of foam extinguishing systems.

Foam solution can be proportioned and generated by:

- **Fixed proportioning systems:** One such fixed system, balanced pressure proportioning, consists of a foam concentrate storage tank, a foam concentrate pump, and a control valve to proportion the foam. Another fixed system, pressure proportioning, consists of a foam concentrate storage tank with an internal membrane or bladder and a venturi for proportioning foam. Water pressure on the bladder is used to force foam out of the liquid storage tank.
- **Mobile proportioning systems:** These are installed on fire trucks and have the advantage of being able to proportion foam solution wherever there is a water supply such as from hydrants or a pond. Balanced pressure proportioning is usually used on fire trucks.
- **Fixed and portable line proportioners:** This is an inexpensive proportioning

method, designed for a predetermined water discharge at a predetermined water pressure. Water flowing through a venturi creates a vacuum that picks up foam concentrate through a connection in the side of the venturi. In a fixed installation, a foam concentrate tank supplies one or more venturis. In the portable system, a tube connect to the venturi is inserted into a drum of foam concentrate. The portable pickup tube is useful for converting a hose water line to a foam line by using a foam-making nozzle on the end of the hose. This arrangement is useful for small hydrocarbon spill fires.

Some applications for foam are:

- **Cone roof and covered floating roof storage tanks:** Foam chambers installed at the top of the shell expand foam solution to extinguish tank fires. The system must be designed to cover the entire liquid surface at the prescribed rate. A fixed proportioning system or a fire truck may be used to supply foam solution to the proportioners.
- **Subsurface foam:** A high-back-pressure foam maker may be used to generate foam and inject it into the bottom of a cone roof storage tank. Fluoroprotein foam is preferred for this application. Subsurface application has the advantage of being simpler to install, because a product line to the tank may be used without having to weld a new connection onto the tank. Foam may be proportioned by a fixed system or a fire truck.
- **Catenary and over-the-top foam systems:** These systems are used to extinguish rim fires in open-top floating roof tanks. In the catenary system, foam is deposited at the floating roof seal by rigid pipe and flexible metal hose to adjust for movement of the floating roof. The over-the-top system deposits foam at the top of the shell, on the inside, and the foam runs down the shell and around the roof seal. Foam may be supplied by fixed system or a fire truck.
- **Foam/water sprinkler systems:** These systems are used mainly beneath marine wharfs and at truck loading racks. They are usually supplied by a fixed foam proportioning system.
- **Fire trucks:** Trucks are used to proportion foam and apply it to fires in process units or in storage tanks, marine terminals, and other plant offsite areas. Foam is generated and applied by hose lines with air aspirating foam nozzles or from turret monitors mounted on the truck. In this capacity, the foam fire truck becomes a very effective fire-fighting apparatus.

HALON AND CARBON DIOXIDE EXTINGUISHING SYSTEMS. Halon and carbon dioxide total flooding systems are used where valuable equipment is located, in unattended indoor locations, and where a clean agent is required for extinguishment.

The two basic extinguishing agents used are Halon 1301 and carbon dioxide. Halon 1211 is also used sometimes, but it has a greater toxicity and should be avoided wherever personnel could be present. Halon 1301 is nontoxic and is preferred to CO_2, but is considerably more expensive.

NFPA 12A, *Standard for Halon 1301 Fire Extinguishing Systems* (NFPA 12A, 1992), contains minimum requirements for halogenated agent extinguishing systems.

NFPA 12, *Standard on Carbon Dioxide Extinguishing Systems* (NFPA 12, 1993), contains requirements for CO_2 extinguishing systems.

Halon and carbon dioxide flooding systems are used for:

- Computer rooms, where a clean agent is required.
- Gas turbine enclosures, which would not be safe to enter during a fire.
- Storage vaults.

FIRE DETECTORS. Fire detectors are used to give warning that a fire has occurred in unattended places and also to activate automatic firefighting equipment. They are usually located in remote high-risk areas or in high-investment areas. Typical locations are computer rooms, grouped shipping pumps, tank truck loading terminals at automatic water spray installations, warehouses and other storage areas, and gas compressors of combustion. Types of detectors and characteristics are as follows:

HEAT DETECTORS

- **Fusible link or plug:** Very reliable and needs to no power source, but is slow to respond.
- **Fixed temperature:** Reliable and low cost but slow and affected by wind.
- **Rate of rise:** Will detect slow temperature rises and rapid rise faster than fixed temperature detectors. Affected by wind and should be confined to indoor use.

SMOKE DETECTORS

- **Ionization:** Early warning, will detect smoldering fires, but subject to false alarms where internal combustion equipment is used or personnel are smoking; limited to indoor use.
- **Photoelectric:** Early warning, will detect smoldering fires, subject to false alarms from dusts; limited to indoors.
- **Ultraviolet (UV):** Fast response, high sensitivity, self-testing, but response retarded by thick smoke and false alarms from welding, lighting, etc.
- **Infrared (IR):** Fast response, but very prone to false alarms from hot surfaces such as furnace firebrick and hot engine manifolds.
- **Combined UV/IR:** High-speed response and sensitivity, low false alarms, but thick smoke obscures range, is expensive.

PLANT FIRE ALARMS. A plant fire alarm system is necessary to notify all personnel that a fire has occurred and to report the location. Methods used to report a fire are:

- **Telephone:** The preferred system utilizes a dedicated number to report a fire. The call is usually received in the control building and sometimes also the gatehouse. The coded number dialed also activates audible fire horns or sirens in the plant.
- **Radio:** Walkie-talkie radios may be used to report a fire to the control room and also activate the plant fire alarm.
- **Manual fire alarm stations:** Pull box or pushbutton alarms located throughout the plant can be used to report a fire and sound the audible alarm. In large plants the location of the activated alarm station often appears on a dedicated computer CRT.

FIRE TRUCKS. Medium-sized and large plants often have one or more fire trucks and a well-trained fire fighting crew. The designated fire fighters usually perform other duties in the plant and respond to the firehouse when the plant fire alarm is sounded. The design of plant fire trucks varies with the types of fires expected. Some types of trucks used are:

- **Foam truck:** Used primarily to fight hydrocarbon spill fires and atmospheric storage tank fires, it carries a large volume, up to 1000 gal (3800 L), of foam concentrate. Water is supplied to the truck from hydrants, and the pressure is boosted by a fire water pump (usually 1000 gal/min) driven by the truck engine. Foam solution is proportioned by a foam concentrate pump and proportioning system.
- **Dry chemical truck:** Used primarily for rapid fire knockdown and extinguishment of process unit fires. A large capacity of potassium bicarbonate dry chemical is carried on the truck, up to 2000 lb (900 kg). The truck also usually contains a fire water pump to supply cooling water streams.
- **Combination foam/dry chemical truck:** Combines the advantages of both the foam and the dry chemical trucks. It contains 200 lb of dry chemical and 100 gal of foam concentrate and is effective for fighting both process unit and tank fires.
- **Triple agent truck:** Contains a large quantity of dry chemical, foam concentrate, and premixed AFFF solution. This truck is very effective in fighting various types of fires, but because of weight limitations, some compromises have to be made. The dry chemical capacity is 1800–2000 lb, the foam concentrate tank holds about 500 gal (1900 L), and the premix tank holds about 200 gal (760 L) of AFFF/water solution, which is pressurized for immediate use by nitrogen cylinders when activated. This truck can quickly extinguish a spill fire followed by securing with premixed AFFF. It can also be used to fight a tank fire when an outside backup source of foam concentrate is provided by a foam trailer or nurse truck.

FIREPROOFING. Fireproofing consists of a passive insulating coating over steel support elements and pressure-containing components to protect them from high-temperature exposure by retarding heat transfer to the protected member.

The objectives of fireproofing are:

- To prevent the collapse of structures and equipment and limit the spread of fire.
- To prevent the release of flammable or toxic liquids from failed equipment.
- To secure escape routes for personnel.

Fireproofing is more critical in fire exposed areas where flammable liquids are processed, stored, or shipped and where a fire could occur from a leak or spill. In determining the need for fireproofing, it is necessary to consider the characteristics of the materials being handled, the severity of operating conditions, the quantity of fuel contained in equipment and piping, and the replacement value of plant facilities and business interruption.

Some types of fireproofing materials available are:

- **Dense concrete:** This consists of Portland cement, sand, and aggregate having a dried density of 140–150 lb/ft^3 (2240–2400 kg/m^3). Concrete may be formed, troweled, or gun applied. Concrete requires no unusual skill to apply and has high resistance to impact and other abuse. Its high density makes it an inefficient insulator and requires more structural support than lighter-weight materials.

- **Lightweight concrete:** Also called cementitious materials, they consist of a lightweight aggregate such as perlite or vermiculite and a cement binder. The density dry is from 25 to 80 lb/ft^3 (400–1280 kg/m^3). Since they are lightweight, they are very good insulators. However, the low density also results in reduced strength, less resistance to impact than dense concrete, and high porosity, which leads to moisture penetration. These materials usually require a weather protective coat to prevent corrosion of substrate and spalling in freezing climates. In selecting the density of the mix, the properties of fire protection versus strength must be considered.
- **Intumescent and subliming mastics:** These mastics provide heat barriers through one or more of the following mechanisms:

- Intumescence: Materials expand to several times their volume when exposed to heat and form a protective insulating char at the barrier facing the fire.
- Subliming: Materials that absorb large amounts of heat while transferring directly from a solid to a gaseous state.
- Mastics are sprayed on the substrate to form a thin coat, usually $\frac{3}{16}$–$\frac{1}{2}$ in. (5–13 mm). The main advantages are light weight, thin coats, and speed of application. Since proper bonding to the substrate and careful control of coat thickness are very important, only vendor-approved and experienced applicators should be used. Some mastics contain a flammable solvent and should not be applied near such ignition sources as fired heaters and boilers, during operation.

- **Preformed inorganic panels:** Precast or compressed fire-resistant panels composed of a lightweight aggregate and a cement binder or a compressed inorganic insulating material such as calcium silicate. Panels are attached to the substrate by mechanical fasteners designed to withstand fire exposure without appreciable loss of strength. When used outdoors, and external weather coating system may be required. Also, all joints should be caulked with a mastic. These materials have the advantage of requiring no time-consuming curing or drying cycle. However, they are labor intensive when applied to existing units having instruments or other equipment supported on structural steel columns.

The thickness of fireproofing required for protection of a substrate is expressed in hours of protection, that is, the time until the steel reaches a temperature of 1000°F (538°C). Failure occurs when the average temperature reaches 1000°F (538°C). This temperature is perceived as the critical temperature for structural steel. Until recently, the fire test used to determine the hours of protection was ASTM E119 (ASTM, 1983). Several years ago tests conducted by some oil companies determined that the E119 time–temperature curve rate of rise was too slow to predict accurately the hours of protection in a high-temperature-rise hydrocarbon fire (Warren and Corona, 1975). Underwriters Laboratories has recently developed a high-rise fire test, UL 1709, which gives a more realistic evaluation of fireproofing materials in a hydrocarbon type fire (UL 1709, 1984).

In determining fireproofing requirements, the following factors should be considered:

- Volume of flammable or combustible liquid likely to be involved;
- Characteristics of material spilled;
- Ability of drainage system to remove spill;

- Congestion of equipment;
- Severity of operating conditions;
- Importance of the facility;
- Ability to isolate the leak with safety shutoff valves.

Some applications of fireproofing in fire exposed areas are:

- **Multilevel equipment structures:** Structures supporting equipment containing hazardous materials should be fireproofed, usually up to the equipment support level. Structures supporting nonhazardous equipment are usually fireproofed from grade to a height of 30–40 ft (9–12 m).
- **Pipe racks:** Process unit pipe racks generally should have columns and at least the first level of beams supporting pipe fireproofed. When a quantity of large-diameter pipe (10 in. and larger) is carried on a second or third level, then consideration should be given to extending fireproofing to upper levels.
- **Equipment supports:** Pressure vessel skirts and legs, air cooler legs, legs of fired heaters, and high saddled supports over 3 ft high (1 m) are often fireproofed.
- **Grouped instrument cable:** Preferably, this should run underground. When located on the pipe rack, fireproofing should be considered.
- **Emergency valves:** The preferred design is to fail in the safe position. When this cannot be done, the valve operator and power supply should be fireproofed. Steel valve bodies do not require fireproofing. In designing fireproofing for electrical cable, it should be kept in mind that the insulation deteriorates at about 320°F (160°C).

12.5.1 Fire Prevention Plan

With any discussion of a loss control program, it is important to ensure that any local, state, or federal regulations in the areas of fire prevention and protection are followed. OSHA has established a number of general industry regulations, including fire protection requirements covering means of egress, fire brigades, fixed and portables fire suppression equipment, fire detection systems, fire prevention plans, and emergency action plans.

This section discusses a model plan to prepare for emergencies and fire prevention and outlines the specific OSHA criteria to follow (including references to OSHA regulations). This plan was adapted from the model plan outlined in the North Carolina Department of Labor's *A Guide to Occupational Fire Prevention and Protection* (NC, 1989). By auditing the work area, by training the employees, by acquiring and maintaining the necessary equipment, and by assigning responsibilities and preparing for an emergency, human life and facility resources will be preserved. At the beginning, this model plan requires management decisions as to whether employees will be employed to fight fires. Management's selection of a course of action regarding employees and fire protection depends on the requirements and the needs of each individual facility. This decision, usually made by top management, requires careful consideration. The most important factors in providing adequate safety in a fire are the availability of proper exit facilities to ensure ready access to safe areas and the proper education of employees as to the actions to be taken in a fire.

There are two basic options available to management: Option A: Employees will fight fires; and Option B: Employees will not fight fires.

OPTION A: EMPLOYEES WILL FIGHT FIRES. The selection of Option A entails two additional decisions. First, who will fight fires: (1) all employees, (2) designated or selected employees, (3) fire brigade/emergency organization, or (4) any combination of these? Second, what types of fires will be fought: (1) incipient stage fires only, or (2) interior structural fires.

Once these decisions have been made, management should follow one of the following plans:

Plan 1. All employees. Provide education in fire extinguisher use and hazards involved with incipient stage fire fighting upon initial employment and annually thereafter as required by OSHA in 29 CFR 1910.157{g} (OSHA, 1996). Should an employer except all employees to fight an incipient stage fire in their immediate work areas, as a designated group (such as a fire squad), those employees should receive hands-on training with the appropriate firefighting equipment.

Plan 2. Designated or selected employees. Provide education in fire extinguisher use and hazards involved with incipient stage fire fighting upon initial employment and annually thereafter. Provide an emergency action plan that designates specific employees to use fire-fighting equipment (29CFR 1910.38(a)) (WHA, 1996). Provide annual training with fire-fighting equipment.

Plan 3. Fire brigades (emergency organization, incipient-stage fires only). Prepare a fire brigade organizational statement that establishes the existence of a fire brigade (organizational structure, training, number of members, functions) as required by OSHA 29 CFR 1910.156 (OSHA, 1996). Provide training for duties designated in organizational statement. Provide hands on training annually in the use of extinguishers, $1\frac{1}{2}$ in. hose lines (such as Class II standpipe system), and small hose lines ($\frac{5}{8}$ to $1\frac{1}{2}$ in.). Train and educate members in special hazards and provide standard operational procedures. Provide a higher level of training and education for leaders and instructors.

[Or] Fire brigade (emergency organization, interior structural fires). Prepare a fire brigade organizational statement. Ensure physical capability of members. Provide training for duties assigned in fire brigade organizational statement. Provide educational or training sessions at least quarterly and handson training with appropriate firefighting equipment at least annually. Train members in special hazards and provide standard operational procedures. Provide higher level of training and education for leaders and instructors. Provide required protective clothing and breathing apparatus.

OPTION B: EMPLOYEES WILL NOT FIGHT FIRES. The selection of Option B entails one additional decision—whether to have portable fire extinguishers and hoses in the facility. Note that specific regulation may require that extinguishers be provided.

One of the following plans should be followed:

Plan 1. Provide fire extinguishers and hoses. Provide fire extinguishers and hoses required by regulation or insurance carrier. Provide emergency action plan and fire prevention plan as required by OSHA 29 CFR 1910.38{a} (OSHA, 1996). Maintain and test this equipment. Provide for critical operations shutdown and evacuation training.

Plan 2. Fire extinguishers and hoses not provided. Provide emergency action plan and fire prevention plan. Provide for critical operations shutdown and evacuation training.

Fire Protection and Prevention Assignments

Name	Work Location	Job Title	Assignment
________	________	________	Responsible for maintenance of equipment and systems installed to prevent or control ignition of fires.
________	________	________	Responsible for control of fuel source hazards.
________	________	________	Responsible for regular and proper maintenance of equipment and systems installed on heat-producing equipment to prevent fires.

Emergency Plan and Fire Protection Plan Coordinator:

Name: ________________ Date: ________________

Fig. 12.4 Example of a form for personnel assignments. (Courtesy of North Carolina Department of Labor.)

12.5.1.1 Elements of Fire Prevention Plan

As was seen earlier, depending on what option is chose, a written fire prevention plan may be required. Obviously, if a good fire loss control program is established, the required elements of the fire prevention plan will be included automatically. If there is no fire loss control program, then at a minimum a fire prevention plan should be developed. Facility managers and/or loss control managers should be responsible for developing this plan and keeping it current. This section discusses the necessary elements of the plan.

Element 1. Names of persons responsible for control of fire protection and maintenance equipment and ignition sources. Persons who are responsible for the control and maintenance of equipment related to fire control or for the control of particular hazards should be clearly identified. An example form for personnel assignments is shown in Fig. 12.4.

Element 2. Control of major workplace fire hazards. Major fire hazards peculiar to the facility and a plan for control of such hazards should be identified. The plan should include proper handling and storage procedures, potential ignition sources and their control procedures, and the type of fire protection equipment available. An example of a type of form that could be used is shown in Fig. 12.5.

Element 3. Housekeeping procedures established. Proper housekeeping is an element in ensuring effective fire prevention.

Element 4. Fire prevention for heat-producing equipment. Heat-producing equip-

Control of Major Workplace Fire Hazards

The following major fire hazards are present in the work area, and proper control procedures are described.

Major Fire Hazard	Location	Company Controls
1. ______	______	______
2. ______	______	______

Fig. 12.5 Example of a form for listing fire hazards and control procedures. (Courtesy of North Carolina Department of Labor.)

ment, such as heaters, furnaces, and temperature controllers for such equipment, need maintenance to ensure proper operation. An example form for keeping track of heat-producing equipment and controls is shown in Figure 12.6.

Element 5. Review plan with employees. Once established, the plan should be reviewed by all employees covered by the plan to ensure that they are aware to the types of fire hazards of the materials and processes to which they might be exposed.

12.5.2 Elements of Emergency Action Plan

An emergency action plan (EAP) should be an integral part of the life safety element in a fire loss control program. Many emergencies at facilities—such as fire, explosion, bomb threats, chemical releases, and natural disasters—require that employees and people evacuate a building. History has shown that an EAP and adequate employee and occupant familiar with a building can prevent disasters. A written EAP is recognized as the best way to plan evacuation and, in most occupational situations, is required by OSHA in 29 CFR 1910.38, "Employee emergency plans and fire prevention plans" (OSHA, 1996). Facility managers and/or loss control managers are responsible for developing an EAP and keeping it up to date. This section discusses the necessary elements and development of an EAP.

Fire Prevention for Heat-Producing Equipment

Heat-Producing Equipment and Controls	Routine Maintenance	Frequency of Maintenance	Assigned Responsibility
______	______	______	______

Fig. 12.6 Example of a form for listing heat-producing equipment and controls and maintenance procedures. (Courtesy of North Carolina Department of Labor.)

12.5.2.1 Responsibilities

Loss Control Manager The loss control manager (or if there is no loss control manager, the facility manager) should be responsible for (1) overseeing the development, implementation, and maintenance of the overall facility EAP; (2) designating and training evacuation wardens (or fire wardens, floor safety officers, or similar designation); (3) reviewing the EAP with employees and building occupants, arranging to train new personnel, and notifying employees of plan changes; and (4) relaying applicable information to public fire department personnel, emergency organization personnel, employees, and evacuation wardens in event of an emergency.

Evacuation Wardens The evacuation wardens are normally appointed by the loss control manger to ensure evacuation, together with other emergency duties. The evacuation wardens may also be members of the emergency organization. They should be trained in the complete facility layout and the various evacuation routes. Responsibilities include: (1) checking for complete evacuation of their designated area and notifying the public fire department and emergency organization of missing persons or location of the fire and of the location of any trapped persons; (2) performing emergency operations as needed and detailed in the EAP, such as closing windows and doors or turning off certain electrical appliances; (3) reporting any malfunctioning alarms; and (4) assisting any disabled persons with emergency evacuation procedures.

Elements of Emergency Action Plan The EAPs are specific to each building and should be coordinated among the various departments by the loss control manager. The following elements should be included in an EAP:

1. Preferred means of reporting fires and other emergencies such as chemical spills or personnel injury;
2. Emergency evacuation procedures and emergency evacuation route assignments;
3. Procedures to be followed by employees who remain to perform critical plant operations before they evacuate;
4. Procedures to account for all building occupants after emergency evacuation has been completed;
5. Rescue and medical duties for those employees who are to perform them;
6. Names or regular job titles of persons or departments who can be contacted for further information or explanation of duties under the plan;

Element 1. The EAP should direct employees to activate the nearest alarm box or other alerting mechanism in the event of a fire. It should state that it may be necessary to activate additional boxes or shout the alarm if people are still in the building and the alarm has stopped sounding or if the alarm does not sound. Any unusual alarm notification procedures should be explained. Examples might include areas that have no audible alarm systems, noisy areas, areas with deaf employees, and places of assembly where a public address (PA) system should be used. The plan should instruct any person discovering a fire to notify the public fire department and the emergency organization stating the location of the fire. The plan should direct employees and occupants to call security and the emergency organization to report all other emergencies. It should direct people to give their name, location of the building, and nature of emergency and should advise people to mention specific features of the building, such as toxic substances, critical operations, disabled persons, etc.

Element 2. All employees and occupants should know where the primary and alternate exits are located and be familiar with the various evacuation routes available. Primary and alternate evacuation routes and exit locations should be described in the EAP. If a floor plan is posted, it should display the various evacuation routes and exits. The EAP should direct people *not* to used elevators as an evacuation route in the event of a fire. In buildings where departments share mutual areas and/or evacuation routes, the department EAPs should be coordinated. It may be necessary to combine the department EAPs into an overall building EAP. Personnel who are regularly transient because of the nature of their jobs (e.g., facilities maintenance personnel) should be trained in general evacuation procedures.

Element 3. Critical operation shutdown is dependent on the building use. It may be as complicated as dealing with pressure vessels or fuel supply or as simple as turning off bunsen burners, hot plates, or electrical equipment. Persons assigned to these duties should be either facility management and/or evacuation wardens or those who cannot leave immediately because of danger to others if the area is abandoned, such a chemists who are conducting potential runaway reactions. These responsible people should be identified in the EAP. Ventilation system shutdown in some facilities will be controlled by automatic protection systems or at a central control station. If facility management or evacuation wardens are required to perform necessary shutdown operations, this should be identified in the EAP.

Element 4. The EAP should direct groups working together in the same area to congregate at a prearranged location identified in the EAP. It is a good idea to have an alternate location for inclement weather, if an outside location, or an alternate safe refuge area. A department organization list should be developed to provide a roster of personnel to ensure that everyone has evacuated.

Element 5. Those personnel who will perform rescue and medical duties should be identified in the EAP.

Element 6. An EAP organization list should contain the names of employees, managers, or other personnel and their job titles, positions, and relative EAP collateral duties. The EAP should include appropriate numbers to call for emergency: fire, police, medical rescue, chemical, biological or radioactive release, water service, fire protection systems, utilities, etc.

Once established, the plan should be reviewed by each employees and occupant covered by the plan to ensure that they know what it expected of them in all emergency possibilities that have been planned, to ensure their safety. A well-designed EAP will anticipate possible impediments to quick and complete evacuation and will provide solutions and options for these problems. Management should make every effort to develop an EAP that includes necessary contingency planning.

12.5.3 Evacuation Drills

Just providing exits and an emergency action plan is not enough to ensure life safety during a fire. Evacuation drills are essential so that employees and occupants of the building will know how to make an efficient and orderly evacuation.

The frequency of drills should be determined by the amount and type of hazardous operations present in the facility and by the complexity of shutdown or evacuation procedures. All elements of the emergency action plan should be practiced during the drill.

After each drill, the loss control manager, facility managers, and evacuation wardens should discuss and evaluate the drill to fine tune the emergency action plan and solve any problems that may have occurred.

12.5.4 Life Safety Principles

Generally, life safety requires the following principles which are also covered in the Life Safety Codes (NFPA 101, 1997):

- Provide sufficient number of unobstructed exits;
- Protect exits from growth;
- Provide alternate exists;
- Subdivide areas and construct to provide safe areas of refuge in occupancies where evacuation is last resort;
- Protect vertical openings to limit spread of fire;
- Ensure early warning of fire;
- Provide adequate lighting;
- Ensure exists and routes of escape are clearly marked;
- Provide emergency evacuation procedures;
- Ensure construction is adequate to provide structural integrity during a fire while people are evacuating.

The *Life Safety Code* deals with occupancy groups according to their life safety hazard. These groups are: (1) assembly, (2) educational, (3) health care and penal, (4) residential, (5) mercantile, (6) business, (7) industrial, (8) storage, and (9) unusual structures. Each group has life safety requirements specific to that occupancy.

When discussing the hazards of contents the three classifications are low, ordinary, or high. *Low hazard* covers contents with low combustibility such that no self-propagating fire can occur, with the only probable danger being from panic, fumes, smoke, or fire from an external source. *Ordinary hazard* covers contents that are liable to burn with moderate rapidity or develop a considerable volume of smoke, but with no poisonous fumes or threat of explosion. *High hazard* covers contents that are liable to burn with extreme rapidity with poisonous fumes and the threat of explosion.

EXITING. Reference should be made to the Life Safety Code (NFPA 101) for detailed requirements on exiting and life safety features. Also state and local building codes should be queried for any requirements peculiar to the local area. Basically the requirements in the Life Safety Code for means of egress in general industrial occupancies include many of the features required in any structure. The travel distance to an exit is 100 ft except with a complete automatic sprinkler system the distance is increased to 150 ft. In most large industrial complexes this travel distance is hard to maintain due to the vast open areas in some plants. In some cases exit tunnels, overhead passageways or travel through fire walls with horizontal exits may be necessary. In most cases, though, it is not practical or economical to build exit tunnels or overhead passageways. In these cases travel distances of 400 ft may be permitted if approved by the local fire officials and meet the following:

- Limit to one-story buildings;
- Limit interior finish to Class A or B (flame spread less than 75 and smoke developed less than 450) in accordance with NFPA Std. 255, *Method of Test of Surface Burning Characteristics of Building Materials*;
- Provide emergency lighting;
- Provide automatic sprinkler or other automatic fire extinguishing system that is supervised for malfunctions, closed valves, and water flow;
- Provide smoke and heat venting designed so employees will not be overcome by smoke and heat within 6 ft of the floor before reaching exits.

For high-hazard industrial facilities, the travel distance to an exit is limited to 75 ft, with no common path of travel. All high-hazard facilities are required to have at least two separate exits from each high-hazard area. In general industry facilities a 50 ft common path of travel is allowed to the separate exits.

12.6 EMERGENCY PLANNING

While the need for emergency planning may vary somewhat with different facilities, there should be general agreement on the importance of providing at least some level of planning for emergencies that may be encountered at a particular facility. Previously, the value of an audit of the hazards or problems that could possibly be encountered by local facility managers was discussed. This is the first step in emergency planning: the recognition of the nature of possible emergencies. The second step is determining the impact of these emergencies. Finally, it is important to evaluate the effect that these possible emergencies may have on the interruption or continuity of operations.

Often, reviews of large loss situations indicate confusion and lack of effective actions by local facility personnel at the time of a fire, explosion, or other emergency, which may have added considerably to the adverse impact of disasters. These catastrophes may include, but will not necessarily be limited to, the following:

Fire	Earthquake
Explosion	Civil disturbances
Water damage	Bomb threats
Storms/high wind	Hazardous material releases

Once the type of emergencies have been recognized, emergency planning programs should be established to provide a way to address these problem areas properly by emergency response and thus eliminate or lessen any potential loss. The initial actions will involve the emergency organization and outside response. The second action will be the plans that are put into effect after the emergency to lessen its impact by providing ways to continue to provide safety for the employees and occupants and to continue production, maintain the integrity of the facility, etc., by instituting contingency plans.

The third part of emergency planning is to ensure that there is an emergency organization available to implement the plans that have been formulated. The formation of emergency plans must be specific to the individual facility.

While the emphasis for the formation for these plans should be at a local level, corporate management should establish policies requiring the formation of plans and guidelines containing a general outline and areas that should be included in the planning. It may be possible for corporate staff members to provide a general emergency plan if there is a great repetitiveness in locations, which could conceivably be encountered in distribution facilities, a theater chain, or similar types of facilities. The need for emergency planning will, to a great extent, depend upon whether large loss factors exist and upon an analysis of what type of response might be anticipated at the time of a specific emergency.

The emergency plan should be coordinated with any emergency response procedures and hazardous chemical facility profiles that may be required by community emergency planning commissions under the Environmental Protection Agency (EPA) community right to know regulations (EPA, 1996). A well-thought-out emergency plan will help plant management to meet the emergency response procedures necessary under the regulations and to ensure that the plant personnel will be protected during a community-wide hazardous chemical disaster.

12.6.1 Elements of Emergency Planning

Elements of emergency planning will include:

- Assessment of risk;
- Facility organization for dealing with emergencies;
- Assessment of resources;
- Training;
- Emergency headquarters;
- Security;
- Public relations.

12.6.1.1 Risk Assessment This assessment must enumerate all the potential problems that may be encountered at a facility, including those that could occur in the surrounding community and affect the plan. The magnitude of the potential problem must be established as well as the probability of the event occurring, enabling management to prioritize its effort in plan development.

12.6.1.2 Facility Organization A management team must be established to cope with emergencies. Ideally, it should include staff members from production, utilities, fire and safety, security, medical, and maintenance. In small facilities some organization members may have multiple roles. In any event, roles and authority must be clearly defined. In addition, the organization listing must include contact phone numbers when staff members are not on site. More importantly, each emergency organization member must have a staff alternate in the event that the primary team member cannot be contacted or is otherwise unaccessible.

12.6.1.3 Assessment of Resources A factual current knowledge of plan emergency response equipment is essential to any emergency preparedness program. Not only must the capability and utility be known, but plant personnel must know how to use it. In addition, to the extent possible, plant personnel must access the capability, availability, and utility of the emergency organization that may be called upon to assist in an emergency. Following this, a meeting involving all potential respondents to an emergency should be held. The objectives of the meeting are to familiarize them with the facility, resources at the facility, and resources in the community, and to define the role of various groups.

12.6.1.4 Training Plant personnel must be proficient in several areas, including knowledge of the emergency alerting system, familiarity with respect to the emergency organization, shut-down procedures, firefighting, first aid, etc. Each of these must be outlined in concise clear terms. Plant personnel must be thoroughly trained in the use of emergency equipment and the carrying out of the emergency response procedures. Community response personnel should also be informed of these plant operations that may have an impact on their role in the response effort. Chapter 14 provides advice on conducting training programs.

12.6.1.5 Headquarters In large facilities there should be at least two widely separated areas that would serve as group headquarters for dealing with emergencies. The locations should be equipped with all the necessary equipment to enable the staff to be in contact with plant personnel as well as mutual aid organizations. These access lines should be dedicated, so that essential communication will be readily available to the response group.

Once the above elements are in place, it is essential that joint exercises be held to evaluate the effectiveness of the plan and to correct any shortcomings that may be present.

12.6.1.6 Security Considerations Provisions should be made to have a security component to the emergency response team. Typically, security deals with such hazards as sabotage, terrorism, and so on; however, more typically the primary function in most emergencies is to limit access to those people and that equipment that will assist in coping with and resolving the emergency. To accomplish this requires:

- Identifying key plant and community personnel who can be activated instantaneously.
- Establishing a group having security responsibility or other plant personnel who assume security roles in an emergency.

12.6.1.7 Public Relations An important element of any emergency plan is appointing an individual who is responsible for informing both plant and community personnel about aspects of the emergency. A person must be designated to deal with the public/media. To accomplish this effectively, he or she must have access to the highest levels of management dealing with the emergency.

12.7 AUDITS

Periodic audits are recommended to assure that engineering controls are being maintained and also to determine if any upgrading of safety controls is advisable. Preferably an operator familiar with unit operation, and sometimes a process design engineer, should accompany the safety or loss prevention engineer during the audit. The safety audit should include:

- Piping and instrument drawings to verify that safety shutdown systems have been installed and maintained in accordance with the latest drawings. Process unit plot plans and equipment lists to identify equipment.
- Checklist of emergency safety shutdown systems, so that an inspection of these facilities can determine if they are maintained in good operating condition.
- Review of plant safety procedures to assure that they are kept current with any changes in plant operations. Visual inspection and testing of plant fire protection systems, such as fixed water sprays, water monitors, fire pumps, fire mains, etc.
- Review of plant accident and fire reports to determine what types of accidents occur most frequently.

12.7.1 Self-Inspection Program

Another important part of a fire loss control program is the facility self-inspection. This is recognized by insurance underwriters as being extremely valuable and can help satisfy the OSHA fire protection requirements. The purpose of the self-inspection is basically to ensure that the facility fire protection systems are operable and therefore capable of performing their function; and to detect hazard areas that may create fires, such as poor housekeeping, the unsafe handling of flammable liquids, poorly maintained electrical equipment, and others.

The first item to consider is the assignment of the inspection task. The loss control manager should establish inspection schedules, the types of inspection to be conducted, and the routing of inspection reports. It is also the manager's responsibility to see that these inspections are conducted properly. In a large facility, it is important that the individual delegated with this responsibility have a good mechanical knowledge and, specifically, a knowledge of the facility fire protection equipment. The most logical individual will be maintenance oriented and, in addition, may be involved in the emergency organization operations. This individual should be reliable, possess the necessary knowledge, and have the confidence of the loss control manager and the upper management.

Next, the inspection frequency should be determined. Normally, where sprinkler protection is involved, the recommended frequency is weekly, with particular emphasis on ensuring that water is available and sprinkler control valves are open. Where the facility is of small or moderate size and there are not fixed protection systems, monthly inspections are usually sufficient, since unusual conditions should be readily apparent to facility personnel.

Facility management should make sufficient time available to the inspector to complete the task. This inspector must have confidence that management will correct the deficiencies that he or she finds and give them proper attention on a priority basis.

Fire insurance underwriters usually furnish forms that are generally all inclusive and can be used in the performance of self-inspections. On the other hand, the loss control managers at many facilities have found it advisable to adapt a form that is specific to their facility. As an example, a self-inspection form from one of the major insurance underwriters is shown in Fig. 12.7. This type of form is used extensively throughout industry. An example of an inspection form that is specific to a facility is shown in Fig. 12.8.

While this section does not cover all possible elements that may be included in an inspection program, some of the items highlighted here are sprinkler system control valves, extinguishers, fire pumps, electrical deficiencies, and the handling of flammable liquids.

OVERVIEW FORMS PACKET
(See Section 12 in the OVERVIEW Manual)

Overview
A Total Management Program for Loss Prevention and Control.

FIRE PROTECTION EQUIPMENT INSPECTION REPORT

Facility: ______________________ Inspector: ______________________

Location: ______________________ Date: ______________________

The Following Items Should Be Checked At Least Weekly.

Any "No" response should be explained.

WATER SUPPLY, SECTIONAL, AND SPRINKLER SYSTEM CONTROL VALVES

Valve ID	Open	Shut	Sealed	Valve ID	Open	Shut	Sealed	Valve ID	Open	Shut	Sealed	Valve ID	Open	Shut	Sealed

PUBLIC WATER

Public water supply in service? ☐ Yes ☐ No ______________________ Pressure: ________ psi

Fire department connection accessible, caps in place, couplings free to rotate? ☐ Yes ☐ No ______________________

FIRE PUMPS

Pump ID	Type	Set For Auto.?		Operated Today?		Checklist Completed?		Comments
		Yes	No	Yes	No	Yes	No	

WATER SUPPLY TANKS

Tank ID	Tank Full?		Heater Working?		Water Temp.	Comments
	Yes	No	Yes	No		

AUTOMOTIVE FIRE APPARATUS

Each fully in service? ☐ Yes ☐ No ______________________

Checklist completed? ☐ Yes ☐ No ______________________

SPECIAL EXTINGUISHING SYSTEMS

System ID	Type	In Service?		Date Last Serviced	Date Last Tested	Comments
		Yes	No			

The Following Items Should Be Inspected At Least Monthly.

Any "No" response should be explained.

WET PIPE, DRY PIPE, DELUGE, AND PRE-ACTION SPRINKLER SYSTEMS

System ID	Alarm Tested?		Water Pressure			Heat Adequate?		Air/ Supv. Press.	Comments
	Yes	No	Static	Flow	Differ-ential	Yes	No		

 (over) Printed in USA.

Fig. 12.7 Example of a self-inspection form. (Reproduced with permission of Industrial Risk Insurers.)

FIRE EXTINGUISHERS, INSIDE HOSE CONNECTIONS, AND STANDPIPES

Each unit in service? ☐ Yes ☐ No ____________

Checklist completed? ☐ Yes ☐ No ____________

HYDRANTS, HOSE HOUSES, AND MONITOR NOZZLES

Monitor Nozzle/ Hydrant ID	Accessible?		Drained?		Equipment				Comments
					Adequate?		Cond. OK?		
	Yes	No	Yes	No	Yes	No	Yes	No	

FIRE DOORS

Fire doors and shutters in good condition? ☐ Yes ☐ No ____________

Automatic closing devices operable? ☐ Yes ☐ No ____________

SMOKE AND HEAT, AND EXPLOSION-RELIEF VENTS

Vents operable? ☐ Yes ☐ No ____________

Areas around vents unobstructed? ☐ Yes ☐ No ____________

PROTECTIVE SIGNALING SYSTEMS

All systems been tested satisfactorily? ☐ Yes ☐ No ____________

OTHER PROTECTION DEFICIENCIES FOUND DURING THE COURSE OF EACH INSPECTION SHOULD BE REPORTED BELOW:

	Yes	No	If "Yes," note location.
Stock within 36 inches of sprinkler heads?	☐	☐	
Sprinkler heads or piping bent?	☐	☐	
Sprinkler heads painted?	☐	☐	
Sprinkler heads or piping corroded?	☐	☐	
Sprinkler heads loaded with debris?	☐	☐	
Items hanging from, or supported by sprinkler heads?	☐	☐	
Sprinkler heads obstructed by partitions?	☐	☐	
Signs of internal sprinkler piping obstruction?	☐	☐	
Fire doors blocked by materials?	☐	☐	

ADDITIONAL COMMENTS AND RECOMMENDATIONS

____

____

____

Report reviewed by: ____________ (signed) Position: ____________

Has prompt action been initiated? ☐ Yes ☐ No ____________

FILE FOR REVIEW BY IRI REPRESENTATIVE

Fig. 12.7 *(Continued)*

DISTRIBUTION: ORIGINAL TO: FACILITIES SERVICES
COPY TO: INSURANCE DEPARTMENT

If condition is satisfactory, enter a check [✓]. If unsatisfactory enter [O] and explain on back of this form.

		FLOOR						
		B	1	2	3	4	5	6
Housekeeping	General orderliness & cleanliness							
Portable Fire Extinguishers	Water units tagged within 12 mos.							
	CO_2 & dry chem. units tagged within 6 mos.							
	Gage pressure in operating range							
	Properly mounted & sealed							
	Accessible							
Exits	All aisles & doors clear							
	Doors open freely							
	No storage in stairways							
	No stairway doors blocked open							
	All exit signs & stairway lights lit							
Fire Hose	In rack							
	Connected							
	In good condition							
Fire Doors	None blocked open							
	No damaged doors							
	All operate freely							
	Fusible links missing or painted							
Electrical	Junction & panel boxes closed							
	No temporary wiring							
	No frayed or unsafe wiring							

Sprinkler Control Valves	Open	Locked
2½" O.S. & Y.'s in 1st floor rear passageway		

Prepared by: ____________ Name ____________ Title Mgmt. review by: ____________ Name ____________ Title Deficiencies Corrected ☐ Yes ☐ No

*IF "NO", EXPLAIN ON BACK OF SHEET

Fig. 12.8 A self-inspection form specific to a facility.

Every control valve should be recorded, showing whether each valve is open or closed and sealed or unsealed and including notes about conditions to be remedied. When the seal on a post indicator valve (PIV) is found to be broken or missing, it is advisable to make a drain test at the nearest sprinkler riser to ensure that the valve is open. Even though a target or supervisory switch indicates that the valve is open,

PIVs should be tested by turning the valve wide open or until the "spring" is felt, thus ensuring that the valve is in an open position. The outside screw and yoke (OS&Y) valve does not require more than a visual inspection that the valve is open; however, if a seal is broken, the reason for the broken seal should be investigated. With dry pipe valves, checks should be made to ensure that the proper air pressure is provided in the system and, where "quick opening" devices (accelerators and exhausters) are provided, that the equipment is operable. An examination of the sprinkler piping and heads will indicate whether any heads are damaged, corroded, or loaded with deposits, dust, or lint and whether or not sprinklers are obstructed.

Self-inspection programs should include drain tests of sprinkler systems when weather conditions permit. The inspector should know the normal pressure and record it on the inspection form. This will indicate a problem such as a partially shut valve or other obstructions when the pressure is considerably lower than normal.

Where water supply depends upon fire pumps, the pumps should be visually checked to determine that power is supplied to the pump where electrical drives are provided, that steam is available where steam is the driving force, and that fuel is available and that batteries are in operation with internal combustion engine drives.

Other parts of the overall facility protection that are also involved in the inspection program include water tanks (properly heated during extremely cold weather) hydrants (not obstructed), hydrant wrenches (available), and hose houses (readily accessible). Accessibility also applies to extinguishers and hose stations within the facility. A check of extinguishers will also indicate whether pressure is satisfactory and whether extinguishers are properly located and serviced.

Part of the inspection program should be to check that proper containers are being used for the handling of flammable liquids, that electrical grounds and bonding are in place, and that bulk quantities of flammable liquids are stored in the flammable-liquids vault.

Unsafe electrical items include overloading and improper overcurrent protection of circuits, poorly maintained electrical cords that are frayed or cracked, combustible storage located within switch-gear rooms, etc.

Other areas involved in the self-inspection program are safe welding and cutting practices, housekeeping practices, fire door, storage practices, control of smoking, heating, water supplies, special hazards, and special types of protection.

12.8 PLANT SITE SELECTION

Fire protection input plant design should be provided as early as possible and preferably at preengineering conferences. Although fire protection generally represents a small portion of the overall project cost, this expense can be substantial. Early involvement of the fire protection engineer can possibly avoid the necessity of costly change orders or addenda. While the fire protection engineer is, or should be, cost conscious, factors in addition to cost–benefit ratios and cost effectiveness must be included in the decision-making process to determine whether or not to provide certain levels of protection or structural features. One major consideration is the uniqueness of a facility or operation and the importance of maintaining continuity of operation of a profit center. It is suggested that this type of consideration is often of greater importance than a cost–benefit ratio or immediate cost effectiveness.

12.8.1 Systems Approach

Fire safety can be incorporated into facility design in several different ways. One way is to require that the building be designed strictly to specification codes, such as building codes or military and government specifications. These are normally very stringent and have very little flexibility. They often do not take into account overall facility designs and often need to be interpreted because of confusion in certain specifications and lack of coverage of certain design features.

Another way to incorporate fire safety is to use performance codes. These codes try to overcome the inflexibility of specification codes by defining the expected fire safety performance of a separate component of the design and allowing alternative ways to meet the design. This approach has problems when one looks at the building as a whole design, since some components have a better chance of successful performance than do others.

The final approach is to look at fire safety as an integrated subsystem of the building, along with the functional, structural, electrical, or mechanical subsystems. This method relies on the use of the best engineering methodology rather than on strict compliance with codes. In other words, this approach provides an "equivalent" alternative to code requirements that provide equivalent fire safety. In the long run, this is the best way to incorporate fire safety into the overall design.

To effectively incorporate the facility's fire protection considerations into the design, the fire safety objectives must be identified. One descriptive tool to help designers identify the fire safety objectives in a systems approach is the systems tree of fire protection. This concept was first developed by the NFPA in its "Fire Safety Concepts Tree" (NFPA 550, 1995) and by the General Services Administration in its systems tree (GSA, undated). A trimmed-down version developed by the Architecture Life Safety Group, University of California at Berkeley, for the U.S. Fire Administration, Department of Commerce, is shown in Fig. 12.9. Reference should also be made to the *Guide to the Fire Safety Concepts Tree*, NFPA 550, for a more detailed fire safety concepts tree and explanation and examples on how to use it.

By moving through the various elements of the tree, the facility can be analyzed or designed for fire safety. As an example, the "self-termination" objective (Fig. 12.9) can be easily controlled by the designers. Geometry, fuel, and ventilation can be manipulated through the design process. The height and volume of the room; the amount, volume, and distribution of furnishings; the flame spread factor; and the type of wall finishes will have effects on a fire.

Design decisions have an effect on the facility's fire protection, which creates a compelling argument for the early introduction of fire protection engineering into the design process. Conversely, the "prevent" section has an operational component that stresses workplace procedures such as storage, maintenance, etc.

12.8.2 Site Selection

Site features should be considered in the overall fire safety design, as well as in the other design subsystems. The site characteristics can play a major roll in the type of fire defenses used for the facility. The major site features follow.

12.8.2.1 Water Supplies From the viewpoint of a fire protection engineer, one of the main considerations in plant site selection is the availability and strength of the water supply. Although water supply information is only one of many inputs that must go into

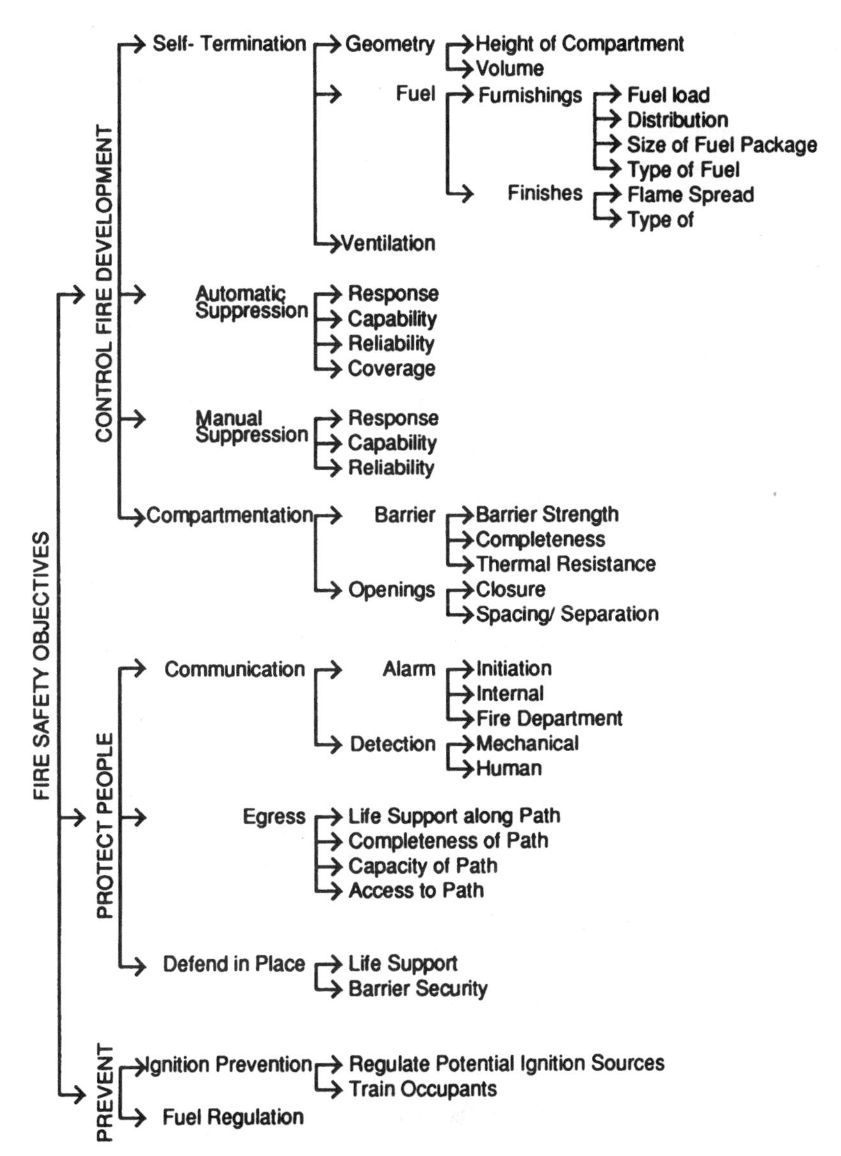

Fig. 12.9 Fire protection tree. (Courtesy of the Department of Commerce.)

site selection, this input can be of utmost importance because costly expenditures may be required if an adequate municipal supply is unavailable.

12.8.2.2 Traffic and Transportation Time plays a major role in the response of the public fire department to the facility. Traffic and the arrangement of the streets in the community have an effect on the response time. Long fire department response times may signal a need for a fully equipped internal fire brigade.

12.8.2.3 Public Fire Protection The type of protection, whether it is paid, part-paid, or volunteer; the location of the fire stations and the anticipated response time; and the nature and condition of the firefighting equipment and the alarm system need to be considered. Access to the site is very important. Built-up, congested areas may include other properties, limiting access to the site. In addition, the topography of the site may limit fire department access.

12.8.2.4 Exposures to the Site The facility could be exposed to conflagration hazards from a large concentration of combustibles within close proximity, such as hazardous chemical risks, oil refineries, large quantities of piled lumber, etc. During a fire, exposure can occur from horizontal radiation and from flames coming from the roof or top of a lower burning building. A number of factors—such as the intensity of the exposing fire, the duration, the total heat produced, the features of exposed and exposing buildings, the air temperature and humidity, the wind velocity and direction, and fire department access—significantly influences the danger of an exposing fire. A number of detailed guides have been established to quantify the exposure hazard. One of these is the *Recommended Practice for Protection of Buildings from Exterior Fire Exposures* (NFPA 80A, 1993).

12.8.2.5 Exposures to the Community As stated earlier, the EPA has established regulations concerning the community's right to know of the processing, storing, manufacturing, and use of hazardous materials by industry. In addition, the community may also have noise control ordinances, explosives ordinances, and other types of environmental regulations. The site should be planned to take into account the total effect on the surrounding community.

12.8.2.6 Flood Plain Fire hazards are increased during flooding, and fire department access is severely restricted. Careful attention should be directed to whether the site is in a known flood plain. For instance, being in a 100-year flood zone may seem fairly secure, but, in fact, works out to be a 63% chance of being flooded within the 100-year period.

Information on flood-plain areas can be obtained from the U.S. Geological Survey flood maps, U.S. Army Corps of Engineers reports, Tennessee Valley Authority reports, and Federal Insurance Administration Flood Hazard boundary Maps and Flood Insurance Studies. Tied in with the general subject is the nature of surface drainage in the area, which could possibly create flash-flooding conditions.

12.8.2.7 Earthquake Zone A large amount of the total damage caused by earthquakes may be due to the fires and explosions that follow. The site should be assessed for its location in a possible earthquake zone. A number of design modifications would be necessary if the site is located in such a zone. In a recent earthquake in the San Francisco Bay area (October, 1989) that measured 7.0 on the Richter scale, the buildings designed to be earthquake resistant had little or no damage, particularly the high-rise buildings, while the conventionally designed older buildings had considerable damage.

Fault locations can be identified from fault and seismic risk maps available from the U.S. Geological Survey, *Uniform Building Code*, National Science Foundation, and the American National Standards Institute.

12.8.3 Plant Layout Fire protection generally requires compromises. Compromise, which may seem inappropriate, is based upon the necessity of being able to operate a plant efficiently incorporating engineering design features that, at the same time, will make the facility as fire safe as practical. For example, fire protection engineers often must be content with having a hazardous operation located in the center of a plant building, where explosion-relief venting can be obtained only through the roof, rather than at the periphery of a building, where explosion relief can be more readily obtained through side walls and the process can be more readily enclosed. Other factors such as process flow my dictate the exact location of the hazardous operation.

While the interior layout of a single-building plant can present some problems, the layout is relatively simple and clearcut from a fire safety standpoint. Layouts become somewhat more complicated, again from a fire safety standpoint, when one is involved with complex, multibuilding plants involving hazardous processing, such as petrochemical facilities. Here, consideration must be given to the concentration of these hazardous processes; the accessibility; the location and protection of the vital utilities and control rooms; and the location of employee amenity areas, such as offices, locker rooms, lunch rooms, etc.

Important considerations involved in plant layout are as follows:

1. The separation of hazardous operations or processing by distance or building separation. If explosion hazards are involved, it must be determined that separation by distance is adequate, or as a less desirable alternative, protective measures should be considered, including the construction of substantial explosion-resistant barricades. The adequacy of the distance must be determined by considering the nature of the hazard involved from a fire and explosion standpoint (possibly a vapor cloud analysis) and also by considering the exposure to loss of unusual value concentrations and specific problems involving the possibility of disruption to the continuity of operations, that is, the criticality of a part or unit of the production chain. The NFPA Handbook (NFPA, 1997) is a good source for additional information on this topic.

2. Accessibility for manual fire suppression. This will determine the adequacy of roadways or other means of access by fire equipment to production units. The inability of fire equipment to reach certain locations and provide effective firefighting will often dictate the need for further automatic protection, such as automatic water-spray systems and monitor nozzles either remotely or manually controlled.

3. The separation of important and vital plant utilities such as boiler rooms, electrical substations, and control rooms, from exposure to fire or explosion. In multibuilding operations, it is generally considered that the safest location for such facilities will be on the periphery of the plant site.

4. Location of employee amenity areas, as mentioned earlier, such as locker rooms, cafeteria, office areas, etc. From a standpoint of life safety these areas should be located on the periphery of the facility.

5. The extent and concentration of valuable areas. While it is generally desirable to group more hazardous operations together, consideration should be given the concentration of values and separation to avoid the possibility of unacceptable losses. In single-building locations, this will usually involve the separation by fire walls of high-valued storage areas from processing areas.

6. Topography. This is an extremely important consideration where plant operations involve the use of extensive volumes of flammable liquids. In these instances, consid-

eration must be given to locating vital production units and utilities so that they will not be exposed by possible flows from the release of flammable liquids under emergency conditions. Where is it impractical to locate these units at elevations not subject to flammable liquid flows, the use of diversionary curbs or walls to direct flammable liquid flows to safe impounding areas or confinement of flammable liquids by diking is advisable.

Large plants, and indeed every facility, require careful consideration of the plant layout, taking into account the elements discussed above. At the same time, one must recognize the problems brought about by modern technology; the increasing tendency to a greater concentration of values and hazards; the need for a high degree of employee training in handling complex processing, and the very important need, particularly in complicated, high-valued chemical plants, for preventative maintenance programs of the highest quality.

12.9 PROCESS EQUIPMENT AND FACILITIES DESIGN

DESIGNING SAFETY INTO PROCESS EQUIPMENT OPERATION. The safe design of most process equipment is usually assured by recognized codes such as the ASME Boiler and Pressure Vessel Code, ASTM and ANSI standards, and Tubular Exchanger Manufacturers Association (TEMA). The finished product is hydrostatically tested, inspected by a registered inspector, and stamped. However, after the equipment is installed, it is the responsibility of the user to assure safe operation. This section discusses design features to protect equipment from poor operation, faulty fabrication, inadequate maintenance, and circumstances beyond the control of the user.

FIRED HEATERS. Fired heaters are either (1) direct (flue gas, kilns) or (2) indirect (coils and jackets) and are used in chemical plant processing equipment.

Most heater firebox explosions occur during lightoff, usually due to the failure of the operator to follow the recommended lightoff procedure. After an unscheduled shutdown, the temptation is great to take a "shortcut" to make up lost production. Therefore, it is mandatory that the company establish a clear operating procedure and see to its enforcement. A "prover"-type system to assure that all pilot and main burner fuel valves are completely closed may be installed. This involves a 1-in. valved bypass around the pilot control valve. When the bypass valve is opened, the pilot control valve cannot be opened unless the pressure in the bypass reaches operating pressure. After about 30 sec, a timer automatically closes the fuel bypass valve. Such systems can improve safety on lightoff, but are also no substitute for a good procedure enforced.

Other safety features normally recommended are:

Low-Fuel-Pressure Shutdown. A valve on both the pilot and main burner fuel lines can be interlocked to shut down sequentially on main burner fuel low pressure and low pilot fuel. These valves, commonly called "chopper valves," require resetting in the field so that an operator must go to the site and determine what caused the valve to trip.

Coil Backflow Prevention. The most effective way to extinguish a heater fire resulting from a process coil rupture is to shut off flow in the coil. This can be done by the use of a check valve on the coil outlet and a remote operated valve on the inlet to the coil. In fouling service, where deposits such as coke are laid down inside the coil, it is necessary to install a remote operated valve on the coil outlet, because a check valve may fail to close. Ability to close valves remotely, that is, from the control room or in the field (at least 75 ft, or 23 m, away from the heater) may be provided.

Snuffing Steam. A provision for injecting steam for purging flammable vapors in the fire box prior to lightoff should be present. This steam can be used also for snuffing out a fire resulting from a ruptured process tube. Valves for injecting steam may be manual and are usually located a minimum of 50 ft (15 m) from the heater to permit access during a fire. On large heaters it may be advisable to install remotely operated valves for quick steam injection.

Remote Damper Control. Remote control of the heater stack damper should be considered, at least for large-duty heaters (over 20 million BTU/h) and especially for heaters sharing a common stack. Besides helping to control the fire and limit damage inside ducting, improved operating economy will result.

Stack Thermocouple. A thermocouple with high-temperature alarm may be used to detect a fire resulting from a ruptured coil. A thermocouple becomes more important with a windowless control room and little or no manpower in the field, because it provides quick fire detection.

PRESSURE VESSELS. Some pressure vessels contain large quantities of flammable or toxic liquids that would create a serious hazard if released to the atmosphere. Although pressure vessels are inherently safe, precautions must be taken to prevent leaks and spills. Some protective systems are:

Shutoff Valves. A shutoff valve is usually provided on all bottom connections so that the vessel contents can be controlled when piping has holed through due to corrosion or damaged by an accident. A shutoff valve should seriously be considered when the vessel contents of flammable or combustible liquid exceeds 50 barrels (8 m^3). The valve is usually operated manually, but in the case of large-diameter piping a remote operated valve should be considered.

Depressuring Valves. A depressuring valve is often used to reduce pressure inside an uninsulated vessel when it is exposed to fire. The vessel relief valve, set to open at the vessel maximum allowable working pressure, will not protect the vessel during a fire of long duration. If the vessel metal becomes overheated (above 1000°F, 540°C) the shell may rupture at the normal operating pressure. The valve is usually remotely operated and discharges to the flare header (where gases are allowed to burn off). When the valve is located close to the vessel, the valve operator may require fireproofing to assure that it functions properly when exposed to fire. Guidelines for designing depressuring systems are provided in API Recommended Practice 521 (API, 1982).

Unless the normal operating pressure of the vessel is over 100 psig (690 kPa), it becomes difficult to reduce pressure effectively before the metal temperature rises to a high level; otherwise, the valve and depressuring line would have to be of very large diameter.

Pressure Relieving Systems. To prevent pressure vessels, exchangers, pumps, and other pressure-containing equipment from being overpressured and damaged, relief valves are installed. These valves are generally spring-loaded devices that to open when the pressure of the system reaches the maximum allowable working pressure. Relief valves should be located as specified in the ASME *Boiler and Pressure Vessel Code* (ASME, 1996), API *Recommend Practice* 521 and API *Recommend Practice* 520.

The relieving load for each valve should be calculated based on upset conditions such as electrical or steam power failure, cooling water failure, blocked outlet, or fire. The valve is sized for a single maximum load. The total release from valves in the process unit its collected in the unit flare header, which then goes offsite to the flare knockout drum and is burned at the flare.

It is desirable to release only clean, nontoxic flammable vapors into the atmosphere. This can be done safely only if the vapors are:

- Lighter than air;
- Heavier than air but with a molecular weight not exceeding 80 and with a discharge velocity of 500 ft/s (150 m/s) to disperse vapors

Water Draws. Water accumulates in the bottom of some pressure vessels and periodically must be drawn off to an open sewer. This operation may be performed by manually operating a valve or by a control valve set to open automatically on high water level and close on low level. Both methods can be hazardous. The manual system requires that an operator remain at the valve whenever water is being drawn, to prevent the escape of flammable liquids or gases when the water level falls too low. If the operator is not alert, a combustible vapor cloud can form, when the vessel contains a light-vapor-pressure material, and expose both the operator and equipment. This hazard can be reduced by keeping the water draw line small, preferably 1 in.; installing a low-level alarm to warn the operator; and keeping the water draw valve as far as practical from the point of emergence (usually an open drain hub going to the sewer). When the product is a high-vapor-pressure material that auto-refrigerates (freezes when released), there should be two water draw valves at least 2 ft apart. The downstream valve is opened first, and then the upstream valve is opened. If the downstream valve freezes in the open position, the upstream valve can be used to shut off flow.

While the precautions mentioned above enhance safety during manual water drawing, the responsibility for safety lies with management. The following items may be utilized to assure the required protection:

- Training personnel;
- Warning signs;
- Written procedures;
- Fresh-air breathing equipment;
- Color coding piping to identify hazardous materials;
- Plugging or capping drains;
- Toxic or combustible gas analyzers with visual and audible alarms.

When water is drawn automatically from vessels to an open sewer, the control valve may fail in the open position, permitting combustible vapors to escape. When automatic water draws are used, the vessel usually should have a low-water-level gauge to alarm both in the field and in the control room.

Whenever a vessel contains a toxic liquid material, the water draw should discharge to a closed system, but never to the flare header, which is generally designed for vapor releases only. Even with other less hazardous materials, it is preferable to discharge to a closed system when feasible.

COMPRESSORS. Compressors generally require protection from fire exposure from more hazardous equipment. However, some inherent hazards should be guarded against.

The most serious is a seal leak that could cause the formation of a vapor cloud, resulting in a fire or deflagration. The severity of the event can be minimized by installing a remote shutoff valve on the compressor inlet an a check valve on the outlet to stop the flow of process gas.

If possible, compressors should not be installed inside closed buildings. When a shelter is advisable, the sides should be open to facilitate ventilation. Sometimes it becomes necessary to install compressors indoors in very cold climates or to install enclosures around them for noise suppression. In these cases, the building or enclosure should be ventilated, sometimes using fans, to prevent the accumulation of combustible vapors. The installation of combustible gas analyzers should be considered, to detect any accumulation in the early stages and permit shutting down the compressor. An inerting and fire extinguishing system using halon or CO_2 may be advisable inside noise-suppression enclosures.

PUMPS. Most pump fires result from seal leaks, which usually are a result of lack of proper maintenance. An experienced operator or maintenance person can detect a bad seal or bearing and shut down the pump before a failure occurs.

Pumps should be located outdoors where possible, rather than indoors, where combustible vapors can accumulate.

All pumps handling flammable liquids should have steel casings. Cast iron can crack readily when exposed to fire, releasing additional fuel.

AIR COOLERS. Fin fan air coolers are prone to damage from any nearby fire, because the fan will induce the hot combustion gases, exposing the tubes. A clearly identified fan shutoff switch located at grade, about 50 ft (15 m) from the cooler, will permit shutting down in an emergency.

EXCHANGERS. Shell and tube heat exchangers sometimes require cleaning while the process unit remains on stream. If this is anticipated, provisions should be made to isolate the exchanger using blinds (barriers) to protect personnel. A valve alone should never be relied on for tight shutoff.

PIPING. Piping should be adequately supported, guided, and anchored. The practice of hanging pipe from rods should be avoided, since such supports are vulnerable to failure in a fire. The use of sliding and bellows expansion joints and proprietary couplings using rubber rings should also be avoided, because they are prone to failure. Preferably, expansion loops should be used. Small-size piping can be damaged easily by vibration and impact and should be adequately supported.

EMERGENCY SHUTDOWN VALVES. Emergency shutdown (ESD) valves may be manual, solenoid, pneumatic, or motor operated. Solenoid and pneumatic-operated valves should be designed to fail in the safe position. The safe position is not always obvious. Should a valve for depressuring a pressure vessel fail in the open or closed position? The safety engineer prefers the valve to fail *open* in case it becomes inoperable in a fire, but the operator wants the valve to fail *closed*, to prevent all depressuring valves from opening upon loss of instrument air. Both conditions create a hazard. This situation may be resolved by installing backup air cylinders or installing depressuring valves in non-fire-exposed areas.

Where these valves cannot be made to fail safe, they should be fireproofed to assure operation during a fire. Fireproofing should include electrical power and valve operator, but not the steel valve body. When stainless steel instrument air tubing is used, it usually requires no fireproofing. Electric-motor-operated valves fail in the last position and must be fireproofed to assure operation.

ESDs are used to isolate systems where a hazard exists, such as fired heaters, compressors, vessels containing large quantities of flammable materials or toxic materials, or long pipelines where a break could occur, and so forth.

BATTERY LIMIT VALVES. Battery limit valves are used to isolate a process unit from the main pipe rack and other process units to permit safe shutdown. In some applications they are used to shut in the entire plant. They can also be used in an emergency to shut off feed to a process unit during a fire. The valves usually are located on top of the unit pipe rack before joining the main pipe rack. Since they are used infrequently, they are generally manually operated and should be located in a fire-safe area to be tenable in an emergency.

DRAINAGE. Grading within the process unit is important, not only for the removal of storm water, but also for the removal of flammable and combustible liquid spills. The preferred design of the concrete slab is to have the high point of paving beneath the centerline of the process unit pipe rack with catch basins located on both sides of (but not under the rack). This design will cause any piping spill to run from beneath the rack and also prevent a spill from some other source from running beneath the rack. No catch basin should be located beneath equipment, especially fired heaters. Catch basins should be fire stopped by a turned down elbow or over–under weir, both having a 6-in. liquid seal. These liquid seals will prevent a fire from going through the entire sewer and also stop an explosion at the first seal. The seal or fire stop is usually located at the inlet to the catch basin, with a straight-through outlet, so that flammable and combustible liquids will not be tapped in the catch basin.

Storm water and oily water may be combined in a single sewer system, but it is becoming common to have separate storm and oily water sewers. With secondary and tertiary water treatment systems being installed, economics dictate segregating the two types of water to reduce the quantity of water being treated.

The storm water system should be designed to handle both the maximum rate of rainfall expected and the fire water rate anticipated, but not simultaneously. The fire water-drainage requirement should take into account the fixed monitor, fixed water sprays, and hose lines that are likely to be utilized. This does not include the total fire water that could be used by activating all fire-water systems simultaneously, but the equipment expected to be required to fight a fire. A total of 4000 gal/min (908 m^3/h) is often used to fight a fire in one process unit. Catch basins are often designed to handle 500 gal/min and the maximum drainage area for one catch basin usually does not exceed 3000 ft^2 (279 m^2).

ELECTRICAL CLASSIFICATION OF AREAS. The classification of hazardous locations for electrical equipment is defined in NFPA 30, Flammable and Combustible Liquids Code (NFPA 30, 1993) and NFPA 70, National Electric Code, Article 500 (NFPA 70, 1996), and API 500A, Classification of Locations for Electrical Installations in Petroleum Refineries (API, 1982). Once an area has been classified, the National Electric code specifies the type of electrical equipment that can be installed safely.

Most areas inside plants handling flammable liquids will be classified as Division II. Division I areas are those where heavier-than-air gases are present in below-grade areas such as sumps, open trenches, and sewers; enclosed areas containing equipment handling flammable such as compressor enclosures and pump houses; and immediately around vents. With lighter-than-air gases, Division I areas might be found in the roof portion of an inadequately ventilated compressor house.

Electrical switches and other sparking devices located in Division I and II areas should be explosions proof or intrinsically safe. Motors located in Division II areas should be nonsparking type rated for Division II service; motors in Division I areas should be explosion proof.

"Explosion proof" does not mean that an explosion cannot occur. If flammable vapors penetrate into the electrical housing, an explosion will occur, but it will be contained within the housing and not ignite vapors outside. However, the electrical equipment inside the housing may be destroyed.

CONTROL BUILDINGS. Since control buildings contain numerous electrical sparking devices, it is important to exclude flammable gases. This is accomplished by keeping the inside of the building at a slightly higher pressure (usually $\frac{1}{4}$–$\frac{1}{2}$ in. of water). The pressurization air must be taken from a gas-free area, and the intake is usually at least 25 ft (7.6 m) above grade. It is preferable to use two air blowers with an interlock so that if the operating blower fails the other blower will operate. An alarm should warn the operator when a blower has failed so that it may be repaired. A vestible at each doorway is preferred to maintain pressure when a door is opened.

Pressurization is preferable to installing explosion-proof electrical equipment in the control room. Not only is explosion-proof equipment expensive, but the explosion proof rating of such equipment is difficult to maintain over a period of time.

Many central control buildings constructed today are designed to be blast resistant. The objective of blast-resistant construction is to protect personnel in the building from an explosion in a process unit and also to permit an orderly shutdown of all other units. Such buildings are designed for the usual maximum anticipated wind and snow loads and also a short-time peak blast overpressure. Determining the design peak overpressure may be difficult. The pressure anticipated will depend upon the distance to the control building from the process units and a realistic determination of what flammable material and how much should be involved. A hazards analysis of all involved units may be advisable in order to make a realistic determination of the possible blast severity. When the critical material processed has a tendency to form a high-velocity (above the speed of sound) detonation and the control building is close to the source (50–100 ft). Then a blast-resistant design may not be practical. The only solution in this is to locate the control building further from the blast source (200 ft or more).

Blast-resistant construction usually requires the elimination of windows. Although some operators object to running a plant they cannot see, it has been proven that an operator can "see" almost anything almost anything that goes wrong in the process unit from instrument data. With the use of computerized data processing, instantaneous information on what is happening in the unit has been greatly expanded.

12.10 TRAINING

Continuous training of plant personnel is very important to assure that proper actions are take in any emergency. A good training program will teach personnel to take the correct action instictively in any emergency, because all emergencies that can arise have been anticipated and carefully thought out prior to their occurrence. The following types of preplanning and training are necessary:

- Training of process unit operators regarding corrective measures to take in all unit upset conditions and in emergency shutdown of the unit of plant. The unit operation manual should contain detailed operational procedures for all emergency conditions.
- Training of the plant firefighting crew in the operation of all plant firefighting equipment. A fire training ground should be established where various types of fires can be started and extinguished by the plant firefighting crew. It is usually beneficial to invite the local municipal fire department for training exercises, in order to familiarize them with the types of hazards in the plant and the use of plant firefighting equipment. Other plant personnel should be trained in the use of first aid firefighting equipment, such as fire extinguishers.
- Training of the plant security force regarding what action to take in emergencies, such as making the accident area accessible to municipal and mutual aid fire companies, and keeping the area free of people who have no useful task to perform in the emergency.
- Training personnel in first aid so that they can administer the proper minor treatment until trained medical help arrives. The is training serves as a useful purpose not only in the plant, but also in the home and elsewhere outside the plant.
- Instructing the medical department (nurse, doctor and ambulance crew) in actions to be taken in emergencies and accidents. They should be familiar with the effects of all toxic and hazardous materials used in the plant and the recommended treatment.
- The affected plant personnel should be trained in the proper use of all plant safety equipment such as air masks, safety showers, and fire blankets.

REFERENCES

API (American Petroleum Institute), 500A, 1982, *Classification of Locations for Electrical Installations in Petroleum Refineries.*

API RP 2003, undated, *Protection Against Ignitions Arising Out of Static, Lightning, and Stray Currents.*

API RP 520, undated, *Recommended Practice for the Design and Installation of Pressure-Relieving Systems in Refineries.*

API RP 521, 1982, *Guide for Pressure Relieving and Depressuring Systems.*

Anyakora, S. N., Engel, G. F. M., and Lees, F. P., 1981, Some Data on the Reliability of Instruments in the chemical Plant Environment, *Chemical Engineer.* November.

ASME (American Society of Mechanical Engineers), 1996, *Boiler and Pressure Vessel Code,* Section VII, Pressure Vessels, Division 1 and 2.

ASTM (American Society of Testing and Materials), 1983, E119, *Fire Tests or Building Construction and Materials.*

Bochnak, Peter M., 1991, *Fire Loss Control: A Management Guide.* Marcel Dekker, Inc. New York, NY.

Bochnak, P. M., NIOSH Member and Project Officer, 1982, *Classification of Gases, Liquids, and Volatile Solids Relative to Explosion-Proof Electrical Equipment,* NMAB 353-5. Committee on Evaluation of Industrial Hazards. Performed by National Academy of Sciences under sponsorship of NIOSH. August.

Bochnak, P. M., NIOSH Member and Project Officer, 1982, *Rationale for Classification of Combustible Gases, Vapors, and Dusts with Reference to the National Electrical Code*, NMAB 353-6. Committee on Evaluation of Industrial Hazards Performed by National Academy of Sciences under sponsorship of NIOSH. July.

Bochnak, P. M., *NIOSH ALERT: Request for Assistance in Prevention Fatalities Due to Fires and Explosion Oxygen-Limiting Silos*. DHHS (NIOSH) Publication No. 86-118. National Institute for Occupational Safety and Health.

Bochnak, P. M. and Pettit, T. A., 1983, *Occupational Safety in Grain Elevators and Feed Mills*. NIOSH, Cincinnati, OH, DHHS (NIOSH) No. 83-126.

Bochnak, P. M. and Pettit, T. A., 1983, *Comprehensive Safety Recommendations for Land-Based Oil & Gas Well Drilling*. NIOSH, Cincinnati, OH, DHHS (NIOSH) 83-127.

Bochnak, Peter M., and Moll, Michael B., 1984, "Relationships Between Worker Casualties, Other Fire Loss Indicators, and Fire Protection Strategies" (internal report), U.S. Department of Health and Human Services, National Institute for Occupational Safety and Health, Morgantown, W. Va.

Bochnak, Peter M., 1980, "State of the Art: Fire Alarm Technology Related to Protecting Life in Work places," Proceedings from the 52nd Fire Department Instructors Conference, the International Society of Fire Service Instructors, Memphis, Tenn.

Bochnak, Peter M., 1976, "Developments in Fire Detection," Proceedings from the Mutual Engineers' Conference, American Mutual Insurance Alliance, April 5–7, Atlanta, Ga.

Bochnak, Peter M., 1977, "Smoke and Fire Detectors for Wood Heating Systems," Proceedings from the Wood Heating Seminar I, Wood Energy Institute, April 20–21, Cambridge, Mass.

Bochnak, Peter M., Pizatella, Timothy, J., and Lark, Joseph J., NIOSH, 1981 "LP-Gas Emergencies—A Review and Appraisal of Selected Ignited Leaks during Liquid Transfer Operations," National Institute for Occupational Safety and Health, Morgantown, W. Va.

Browning, R., 1969, Reactive Probabilities of Loss Incidents, *Chemical Engineering*, p. 134, December 15.

Cote, Ron, Ed., 1994, *Life Safety Code Handbook*, 6th Edition, National Fire Protection Association, Quincy, Mass.

Cote, Arthur E., Ed., 1997, *Fire Protection Handbook*, 18th Edition, National Fire Protection Association, Quincy, Mass.

Cotte, Arthur E., Ed., 1991, *Fire Protection Handbook*, 17th Edition, National Fire Protection Association, Quincy, Mass.

"Earthquake Concerns," 1988, P8805, Factory Mutual Engineering Corporation, Norwood, Mass.

EPA, 1996, *Hazardous Chemical Reporting: Community Right to Know*. 40 CFR 370, Environmental Protection Agency, Washington, D.C.

Factory Mutual Engineering Corporation, 1996, *Approval Guide*, Norwood, Mass.

Factory Mutual Engineering Corporation, 1993, *The Handbook of Property Conservation*, Norwood, Mass.

"Flood! And Whey You're Not as Safe as You Think," 1980, P8001, Factory Mutual Engineering Corporation, Norwood, Mass.

Lerup, Lars, Cronwrath, David, and Liu, John Koh Chiang, 1977, *Learning from Fire: A Fire Protection Primer for Architects*, U.S. Fire Administration, Washington, D.C.

McCoy, C. S., and Hanly, F. J., 1985, Fire-Resistant Lubes Reduce Danger of Refinery Explosions, *Oil and Gas Journal*, November 10, pp. 191–197.

"National Disasters," 1988, P8812, Factory Mutual Engineering Corporation Norwood, Mass.

National Fire Protection Association, 1998, *National Fire Codes*, Quincy Mass.

NFPA, 1990, *Industrial Fire Hazards Handbook*, 3rd Edition, National Fire Protection Association, Quincy, Mass.

NFPA, 1994, *NFPA Inspection Manual*, 7th Edition, National Fire Protection Association, Quincy, Mass.

NFPA (National Fire Protection Association) 10, 1994, Standard for Portable Fire Extinguishers.

NFPA 11, 1994, *Standard for Low Expansion Foam and Combined Agent Systems.*

NFPA 12, 1993, *Standard on Carbon Dioxide Extinguishing Systems.*

NFPA 12A, 1992, *Standard for Halon 1301 Fire Extinguishing Systems.*

NFPA 15, 1990, *Standard for Water Spray Fixed Systems for Fire Protection.*

NFPA 1961, 1992, *Standard for Fire Hose.*

NFPA 1963, 1993, *Standard for Fire Hose Connections.*

NFPA 20, 1993, *Standard for the Installation of Centrifugal Fire Pumps.*

NFPA 214, 1992, *Standard on Water Cooling Towers.*

NFPA 30, 1993, *Flammable and Combustible Liquids Code.*

NFPA 321, 1991, *Standard on Basic Classification of Flammable and Combustible Liquids.*

NFPA 550, 1995, Guide to Fire Safety Concepts Tree.

NFPA 70, 1996, *National Electric Code.*

NFPA 80A, 1993, *Recommended Practice for Protection of Buildings from Exterior Fire Exposures.*

NFPA 8502, 1995, Prevention of Furnace Explosions/Implosions in Multple Burner Boilers.

NFPA 86, 1995, Ovens and Furnaces.

Nelson, Harold E., undated, *The Application of Systems Analysis to Building Fire Safety Design*, Accident and Fire Prevention Division, General Services Administration.

OSHA, 1996, General Industry Standards—29 CFR 1910, Occupational Safety and Health Administration, Washington, D.C.

Smith, Michael R., Ed., 1989, A Guide to Occupational Fire Prevention and Protection, NC-OSHA Industry Guide #4, North Carolina Department of Labor, Raleigh, N.C.

SFPE, 1995, *Handbook of Fire Protection Engineering*, 2nd Edition, National Fire Protection Association, Quincy, Mass.

UL, 1996, *Fire Protection Equipment Directory*, Underwriters Laboratories, Inc., Northbrook, IL.

Underwriters Laboratories, 1996, Publications Stock, Fire Resistance Directory.

Underwriters Laboratories, 1984, UL 1709, *Structural Steel Protected for Resistance to Rapid Temperature Rise Fires.*

Warren, J. H., and Corona, A. A., 1975, This Method Test Fire Protective Coatings, *Hydrocarbon Processing.*

CHAPTER 13

Hazard Communication and Worker Right-to-Know Programs

LISA K. SIMKINS
Clayton Group Service Inc., 1252 Quarry Lane, Pleasanton, CA 94566

CHARLOTTE A. RICE

13.1 INTRODUCTION
13.2 REQUIREMENTS OF A HAZARD COMMUNICATION PROGRAM
- 13.2.1 Application of Hazard Communication
- 13.2.2 Program Elements—Chemical Manufacturer, Supplier, Importer Responsibilities
- 13.2.3 Program Elements—Employer Responsibilities
- 13.2.4 Trade Secrets

13.3 PROGRAM MANAGEMENT
- 13.3.1 Policies and Procedures
- 13.3.2 Program Responsibilities
- 13.3.3 Flow of Communication
- 13.3.4 Quality Control and Program Improvement
- 13.3.5 Record Keeping

13.4 TOOLS FOR PROGRAM IMPLEMENTATION
- 13.4.1 Hazardous Material Information Systems
- 13.4.2 Training Tools
- 13.4.3 Material Labeling

13.5 PROGRAM AUDIT
- 13.5.1 Program Appraisal
- 13.5.2 Record Review
- 13.5.3 Performance

13.6 RELATED OSHA REGULATIONS
- 13.6.1 Access to Employee Exposure and Medical Records (20 CFR 1910.20)
- 13.6.2 Air Contaminants (29 CFR 1910.1000)
- 13.6.3 Occupational Exposure to Hazardous Chemicals in Laboratories (29 CFR 1910.1450)

Adapted from "Hazard Communication and Worker Right-to-Know Programs" by Heideman and Simkins, Chapter 6 in Clayton & Clayton, *Patty's Industrial Hygiene and Toxicology, 4th ed., Vol. IA*. New York: John Wiley & Sons, Inc., 1991, pp. 123–177.

Handbook of Occupational Safety and Health, Second Edition, Edited by Louis J. DiBerardinis, ISBN 0-471-16017-2

13.6.4 Occupational Exposure to Benzene (1910.1028)
13.6.5 Process Safety Management of Highly Hazardous Chemicals (1910.119)
13.6.6 Hazardous Waste Operations and Emergency Response (1910.120)
REFERENCES

13.1 INTRODUCTION

Millions of natural and synthesized chemicals are in use today with approximately 25,000 new chemicals introduced every year. Many of these chemicals find significant commercial application, exposing a growing number of workers to potentially hazardous chemicals. Although many of these chemicals may not pose a serious threat, some can cause harm to human health and the environment, if handled improperly. Providing information on chemical hazards is a critical factor in protecting workers. This is reflected in the following quotation from the Occupational Safety and Health Act of 1970[1]:

> "Any standard promulgated under this subsection shall prescribe the use of labels or other appropriate forms of warning as are necessary to insure that employees are apprised of all hazards to which they are exposed, relevant symptoms and appropriate emergency treatment, and proper conditions and precautions of safe use or exposure."[2]

The Occupational Safety and Health Administration (OSHA) met these requirements in the standards promulgated for specific chemicals. However, since only a handful of chemicals have been regulated by a specific standard, OSHA identified the need for a "generic" chemical information standard, and proposed the Hazard Communication Standard (HCS) in March of 1982.

Chemical manufacturers, labor unions, health and safety professionals, and various governmental agencies supported the idea for a national, generic chemical information and labeling standard. This support arose in part because of the proliferation of state and community right-to-know laws that were creating a confusing patchwork of overlapping and, in some cases, contradictory regulations. However, despite this broad support, the rule underwent numerous court battles and revisions after its introduction. With approximately 25 changes, modifications, amendments, and clarifications to the HCS, it is no wonder that there is confusion regarding the requirements and application of the standard. This confusion may, in part, explain why the HCS has the distinction of being the most frequently cited OSHA standard.

The HCS (29 CFR 1910.1200), as originally proposed, applied only to manufacturing business sectors. In 1987 it was extended to virtually all employees. However, OSHA was prevented from enforcing the rule in construction. OSHA was also prevented from enforcing requirements dealing with Material Safety Data Sheets (MSDSs) on multiemployer worksites, coverage of consumer products, and drugs in the nonmanufacturing sectors. These issues were subsequently resolved. A hazard communication standard, identical to the general industry HCS, is now part of the OSHA construction standards.[3] The other issues were addressed in the final rule that was published in the federal register in 1994. The HCS covers approximately 32 million workers, in 3.5 million locations.

The HCS preempts state right-to-know regulations, except in states with approved OSHA plans. This ensures similar minimum requirements in all states. Some local regulations that exceed, but do not conflict with the HCS are still applicable. Employers in a state or municipality with "right-to-know" regulations should contact local authorities for interpretation on how federal, state, and local regulations apply.

Since its introduction, the basic provisions have not changed. A written program document must explain how compliance is achieved. Material Safety Data Sheets (MSDSs) must be developed by chemical manufacturers and maintained by users for all hazardous chemicals. A list of chemicals must be maintained at the worksite, and employees must receive information and training regarding the hazards of the chemicals. These seem like simple requirements. However, changes in chemicals, materials, processes, hazard determinations, and/or employees can make ongoing compliance a challenge.

To add to the challenge, the HCS is a performance standard. The requirements are specific, but the implementation methods are left to the employer. This allows the employer a great deal of flexibility. However, the yardstick used to measure the effectiveness of a hazard communication program is whether employees recognize the hazards posed by chemicals and understand how to protect themselves. It is not always easy to determine if this is true.

The relationship between the HCS and other OSHA standards can be complex. The standards covering specific chemicals, for example, have requirements for labeling, training, information, and controls. In general, the requirements of these more specific standards take precedence over the generic approach of the HCS. However, it is wise to study the standards carefully to determine which requirements actually do apply.

Three years after OSHA proposed the HCS, the U.S. Congress passed the Emergency Planning and Community Right-to-Know Act of 1986 (EPCRA). This law, also known as SARA (Superfund Amendments and Reauthorization Act), Title III, attempts to address community concern regarding hazardous materials. It requires the inventory and reporting of hazardous materials for the purpose of providing information to the community and planning for emergencies. Although this law will not be covered in any depth in this chapter, the relationship between it and the HCS will be discussed briefly. A brief discussion, however, does not imply that it is less important than the HCS. A thorough understanding of EPCRA is also important. The U.S. Environmental Protection Agency (USEPA) has published informational pamphlets on specific sections of SARA Title III (EPCRA) regulations.

Although it may appear difficult to achieve continuing compliance to these laws, the consequences of noncompliance can be significant. Citations, penalties, adverse publicity, increased liability, and, in extreme cases, criminal prosecutions are among the possibilities. Failure to warn employees of a potential health hazard may be interpreted as a negligent violation of the workers' rights. This may permit civil actions against the employer outside the protection of the workers' compensation system.

The purpose of this chapter is to help you set up a Hazard Communication program, if you do not have one yet. It will help you determine if your existing program is effective and will also give you ideas for fine tuning and managing your program. This chapter is designed primarily for a company using, rather than manufacturing, importing, or distributing chemicals. It does not include the requirements and control for conducting hazard determinations and distributing MSDSs to downstream users. In Section 13.2 the essential elements of a Hazard Communication program are reviewed and explained. Program management strategies are described in Section 13.3. The materials and systems used to implement a program are discussed in Section 13.4. Section 13.5 contains recommendations for auditing a program. Some related regulations are summarized in Section 13.6.

13.2 REQUIREMENTS OF A HAZARD COMMUNICATION PROGRAM

13.2.1 Application of Hazard Communication

The Hazard Communication Standard applies to chemical manufacturers, importers, and distributors of hazardous chemicals, as well as all employers who use hazardous chemicals. The standard requires chemical manufacturers or importers to assess the hazards of chemicals they import or produce. Distributors of hazardous chemicals must provide required hazard information to employers. The standard requires all employers to provide information to their employees about the hazardous chemicals used in the workplace.

A "hazardous chemical" is broadly defined in the HCS as any chemical that is a physical hazard or a health hazard. However, there are a few exceptions to the standard including:

- Any hazardous waste subject to the Resource Conservation and Recovery Act (RCRA);
- Any hazardous substance subject to the Comprehensive Environmental Response, Compensation, and Liability Act (CERCLA);
- Tobacco or tobacco products;
- Wood or wood products where the only hazard they pose to employees is the potential for flammability or combustibility;
- Articles (defined as "a manufactured item other than a fluid or particle: (i) which is formed to a specific shape or design during manufacture; (ii) which has end use function(s) dependent in whole or in part upon its shape or design during end use; and (iii) which under normal conditions of use does not release more than very small quantities, e.g., minute or trace amounts of a hazardous chemical, and does not pose a physical hazard or health risk to employees;")
- Food or alcoholic beverages that are sold, used, or prepared in a retail establishment and foods intended for personal consumption by employees while in the workplace;
- Cosmetics that are packaged for sale to consumers in a retail establishment and cosmetics intended for personal consumption by employees while in the workplace;
- Any consumer product or hazardous substance where the employer can show that it is used in the workplace for the purpose intended by the manufacturer or importer and the use results in a duration and frequency of exposure that is not greater than that expected for consumers;
- Nuisance particulates that do not pose any physical or health hazard;
- Ionizing and nonionizing radiation;
- Biological hazards.

Some chemicals are subject to labeling requirements of other laws or regulations and are, therefore, exempt from hazard communication labeling requirements. Chemicals exempt from labeling requirements include:

- Pesticides subject to the labeling requirements of the Federal Insecticide, Fungicide, and Rodenticide Act or pursuant regulations;
- Chemical substances and mixtures subject to the labeling requirements of the Toxic Substances Control Act (TSCA) or pursuant regulations;

- Any food, food additive, color additive, drug, cosmetic or medical or veterinary device, or product subject to the labeling requirements of the Federal Food, Drug and Cosmetic Act or the Virus-Serum-Toxin Act or pursuant regulations;
- Distilled spirits (beverage alcohols), wine, or malt beverage intended for nonindustrial use that are subject to the labeling requirements of the Federal Alcohol Administration Act or pursuant regulations;
- Consumer products or hazardous substances as defined in the Consumer Product Safety Act and Federal Hazardous Substances Act, respectively, when subject to a consumer product safety standard or labeling requirement of these Acts or pursuant regulations;
- Agricultural or vegetable seed treated with pesticides and labeled in accordance with the Federal Seed Act or pursuant regulations.

The standard also includes a provision addressing work situations in which employees handle only sealed containers, such as retail sales, warehousing, and cargo handling. Requirements under the regulation are limited for these operations to:

- Ensuring labels on incoming containers are not removed or defaced;
- Maintaining copies of MSDSs received with incoming shipments of sealed containers;
- Requesting MSDSs from the manufacturer if an employee asks for one;
- Ensuring that MSDSs as described above are accessible;
- Providing information and training to employees as required by the HCS (except for location and availability of written hazard communication program) to the extent necessary to protect them in the event of a leak or spill.

Laboratories as defined by the HCS are exempt from some provisions. The provisions of the standard that do apply to laboratories include the following:

- Labels on incoming containers of hazardous chemicals must not be removed or defaced;
- Employers must maintain MSDSs that are received with incoming shipments of hazardous chemicals and ensure that they are readily accessible to laboratory employees;
- Employers must ensure that laboratory employees are apprised of the hazards of chemicals in their workplaces in accordance with the information and training provisions of the standard, except for the location and availability of the written hazard communication program;
- Laboratory employers that ship hazardous chemicals are considered to be either a chemical manufacturer or a distributor and therefore must ensure that containers of hazardous chemicals leaving the laboratory are labeled and that a material safety data sheet (MSDS) is provided to distributors or employers.

13.2.2 Program Elements—Chemical Manufacturer, Supplier, Importer Responsibilities

The basic responsibilities of chemical manufacturers, suppliers, or importers of hazardous substances in a Hazard Communication program are described as follows:

13.2.2.1 Hazard Determination Hazard determination is the responsibility of chemical manufacturers and importers. When evaluating chemicals, the chemical manufacturer or importer must consider the available scientific evidence concerning such hazards. For health hazards, hazard determination criteria are specified in Appendix B of the HCS. This appendix specifies criteria for (a) carcinogenicity, (b) human data, (c) animal data, and (d) adequacy and reporting of data.

Determination of whether a chemical is a carcinogen or a potential carcinogen is based on findings by the National Toxicology Program (NTP), the International Agency for Research on Cancer (IARC), and/or OSHA. Positive determinations on carcinogenicity by any of these groups is considered conclusive.

When human data such as epidemiological studies and case reports of adverse health effects are available, they must be considered in a hazard evaluation. Since human evidence of health effects is not generally available for many chemicals, animal data must often be relied upon. The results of toxicology testing in animals must often be used to predict the health effects that may occur in exposed workers. This reliance is particularly evident in the definition of some acute hazards that refer to specific animal data, such as lethal dose (LD_{50}) or lethal concentration (LC_{50}).

The results of any positive studies that are designed and conducted according to established scientific principles, and have statistically significant conclusions regarding health effects of a chemical, are considered sufficient basis for hazard determination and reporting on a Material Safety Data Sheet (MSDS).

The HCS specifies the following categories for hazard determination in its mandatory Appendix A.

- Carcinogen—a cancer-causing substance based on research by International Agency for Research on Cancer (IARC), National Toxicology Program (NTP), or OSHA;
- Corrosive—a chemical that causes visible destruction of, or irreversible alterations in, living tissue by chemical action at the site of contact;
- Highly toxic—a chemical with:
  (a) median LD_{50} less than or equal to 50 mg/kg by oral ingestion,
  (b) median LD_{50} less than or equal to 200 mg/kg by contact, or
  (c) median LC_{50} less than or equal to 200 ppm;
- Irritant—a chemical that causes reversible inflammatory effect on living tissue by chemical action at the site of contact;
- Sensitizer—a chemical that causes a substantial proportion of exposed people or animals to develop an allergic reaction in normal tissue after repeated exposure;
- Toxic—a chemical with:
  (a) median LD_{50} of more than 50 mg/kg but no more than 500 mg/kg by oral ingestion,

(b) median LD_{50} of more than 200 mg/kg but no more than 1000 mg/kg by contact, or

(c) medial LC_{50} of more than 200 ppm but no more than 2000 ppm.

- Target organ effects—chemicals that affect specific organs. Examples of target organ effect categories include: (a) hepatotoxins (liver), (b) nephrotoxins (kidney), (c) neurotoxins (nervous system), (d) agents that act on the blood or hematopoietic system, (e) agents that damage the lung, (f) reproductive toxins, (g) cutaneous hazards (skin), and (h) eye hazards.

13.2.2.2 Material Safety Data Sheets Chemical manufacturers and importers are required to obtain or develop a Material Safety Data Sheet (MSDS) for each hazardous chemical they produce or import. Chemical manufacturers and importers must then ensure that distributors and purchasers of hazardous chemicals are provided with an MSDS with the initial shipment of each hazardous chemical and with the first shipment following an MSDS update. Distributors must ensure that MSDSs and updated information are provided to other distributors and purchasers of hazardous chemicals.

The preparer of the MSDS must ensure that the information reported is accurate and reflects scientific evidence used in making the hazard determination. If the MSDS preparer becomes newly aware of significant information regarding the hazards of a chemical or ways to protect against the hazards, the new information must be added to the MSDS by the manufacturer within 3 months. The updated MSDS must be sent to customers with the next shipment of the chemical.

The MSDS may be in any format, but must be in English (although the employer may maintain copies in other languages as well). The MSDS must provide specific information regarding the hazardous chemical. Required information on an MSDS includes:

a. *Identity of the material*—The identity on the MSDS must match the identity on the label. The MSDS must also provide the chemical and common name(s) of the hazardous chemical. In the case of a mixture, the chemical and common name(s) of the ingredients must be listed. If the mixture has been tested as a whole to determine its hazards, the chemical and common name(s) of any ingredients contributing to the known hazards must be listed. If the hazardous chemical has not been tested as a whole, the chemical and common name(s) of all hazardous ingredients that comprise 1% or greater of the composition must be listed, except for carcinogens, which must be listed if they comprise 0.1% or greater of the composition. In all cases, the chemical and common name(s) of ingredients that present a physical hazard when present in the mixture must be listed. In some states, the Chemical Abstract Services (CAS) number is also required for each component. The standard does include provisions for protecting trade secrets of chemical manufacturers. The specific identity of hazardous chemicals can be withheld from the MSDS if this information is a trade secret; however, all other information regarding the properties and effects of the chemical must be provided.

b. *Physical and chemical characteristics*—This information includes characteristics such as vapor pressure, flash point, etc.

c. *Physical hazards*—This includes information regarding potential for fire, explosion, and reactivity.

d. *Health hazards*—Health hazard information must include signs and symptoms of exposure, and any medical conditions that are generally aggravated by exposure to the chemical.

e. *Primary route(s) of entry into the body*—Typical routes of entry are inhalation, ingestion, skin contact, and skin absorption.

f. *Exposure limits*—These limits include one OSHA permissible exposure limit (PEL), the ACGIH Threshold Limit Value (TLV), and any other exposure limit used or recommended.

g. *Carcinogenicity*—The MSDS must state whether the chemical has been determined to be a carcinogen or potential carcinogen by the National Toxicology Program (NTP) Annual Report on Carcinogens, the International Agency for Research on Cancer (IARC) Monographs, or OSHA.

h. *Precautions for safe handling and use*—This includes precautions that are generally applicable including hygienic practices, protective measures during repair and maintenance of contaminated equipment, and procedures for cleanup of spills and leaks.

i. *Control measures*—These include engineering controls, work practices, or personal protective equipment.

j. *Emergency and first aid procedures.*

k. *Date of preparation or latest change.*

l. *Name, address and telephone number*—This is required for the chemical manufacturer, importer, employer, or other responsible party preparing or distributing the MSDS. This is provided in case the user needs additional information on the hazardous chemical and emergency procedures.

If the MSDS preparer cannot find any information for one of these categories to include on the MSDS, then that section should be marked as not applicable or a notation that no information was found should be made. There must be no blanks on an MSDS.

OSHA provides an optional form to use for MSDSs in the HCS. A copy of the OSHA form is shown as Fig. 13.1.

The American National Standards Institute (ANSI) has developed a detailed standard, ANSI Z400.1-1993, which provides further guidance on the preparation of MSDSs.[4] This publication is available from the American National Standards Institute, 11 West 42nd Street, New York, New York, 10036.

13.2.2.3 Labels The chemical manufacturers, importers, or distributors must ensure that each container of hazardous chemicals leaving their workplaces is labeled. The label on a "shipped" container must include:

- Identity of the hazardous chemical
- Appropriate hazard warning;
- Name and address of the chemical manufacturer, importer, or other responsible party.

The label is meant to be an immediate warning. The appropriate hazard warning should cover the major effects of exposure; however, it will not include all the detailed information provided by the MSDS.

The identity of the chemical must match the MSDS chemical identity so that an employee seeking additional information can link the material in a container with an MSDS. The label must provide an immediate warning of specific acute and chronic health hazards as well as physical hazards. A precautionary statement such as "caution,"

MATERIAL SAFETY DATA SHEET

Material Safety Data Sheet May be used to comply with OSHA's Hazard Communication Standard, 29 CFR 1910.1200. Standard must be consulted for specific requirements.	**U.S. Department of Labor** Occupational Safety and Health Administration (Non-Mandatory Form) Form Approved OMB No. 1218-0072
IDENTITY *(As Used on Label and List)*	*Note: Blank spaces are not permitted. If any item is not applicable, or no information is available, the space must be marked to indicate that.*

Section I

Manufacturer's Name	Emergency Telephone Number
Address *(Number, Street, City, State, and ZIP Code)*	Telephone Number for Information
	Date Prepared
	Signature of Preparer *(optional)*

Section II — Hazardous Ingredients/Identity Information

Hazardous Components (Specific Chemical Identity; Common Name(s))	OSHA PEL	ACGIH TLV	Other Limits Recommended	% *(optional)*

Section III — Physical/Chemical Characteristics

Boiling Point		Specific Gravity (H_2O = 1)	
Vapor Pressure (mm Hg.)		Melting Point	
Vapor Density (AIR = 1)		Evaporation Rate (Butyl Acetate = 1)	
Solubility in Water			
Appearance and Odor			

Section IV — Fire and Explosion Hazard Data

Flash Point (Method Used)	Flammable Limits	LEL	UEL
Extinguishing Media			
Special Fire Fighting Procedures			
Unusual Fire and Explosion Hazards			

(Reproduce locally) OSHA 174, Sept. 1985

Fig. 13.1 Material safety data sheet form.

31:9204

Section V — Reactivity Data

Stability	Unstable		Conditions to Avoid
	Stable		

Incompatibility *(Materials to Avoid)*

Hazardous Decomposition or Byproducts

Hazardous Polymerization	May Occur		Conditions to Avoid
	Will Not Occur		

Section VI — Health Hazard Data

Route(s) of Entry:	Inhalation?	Skin?	Ingestion?

Health Hazards *(Acute and Chronic)*

Carcinogenicity:	NTP?	IARC Monographs?	OSHA Regulated?

Signs and Symptoms of Exposure

Medical Conditions
Generally Aggravated by Exposure

Emergency and First Aid Procedures

Section VII — Precautions for Safe Handling and Use

Steps to Be Taken in Case Material Is Released or Spilled

Waste Disposal Method

Precautions to Be Taken in Handling and Storing

Other Precautions

Section VIII — Control Measures

Respiratory Protection *(Specify Type)*

Ventilation	Local Exhaust	Special
	Mechanical *(General)*	Other

Protective Gloves	Eye Protection

Other Protective Clothing or Equipment

Work/Hygienic Practices

Page 2

• USGPO 1986-491-529/45775

Fig. 13.1 *(Continued)*

"harmful," or "harmful if inhaled" does not provide information on the actual hazard. The hazard warning should include target organ effects information, such as "causes lung damage when inhaled," when a specific target organ effect is known. Some manufacturers include additional information, such as emergency first aid procedures, that can prove useful. These, however, are not mandatory for compliance.

13.2.3 Program Elements—Employer Responsibilities

An employer that uses hazardous substances must have a Hazard Communication program that includes the components described as follows.

13.2.3.1 Written Hazard Communication Program The employer must develop, implement, and maintain in the workplace a written Hazard Communication program that explains how compliance with the standard is achieved. The written program must include a description of how the employer complies with labeling, MSDS, and employee information and training requirements. When OSHA inspects a workplace, the OSHA compliance officer will ask to see the written program and will look for specific items to ensure that each of the elements are properly addressed.

For the labeling, the following items should be described in the written program:

- The person(s) responsible for ensuring labeling of in-plant containers;
- The person(s) responsible for ensuring labeling of any shipped containers;
- Description of the labeling system used;
- Description of written alternatives to labeling of in-plant containers, if used;
- Procedures to review and update label information when necessary.

For MSDSs, the following items should be described in the written program:

- The person(s) responsible for obtaining and maintaining MSDSs;
- How MSDSs are maintained in the workplace and how employees can obtain access to them in their work areas during their work shift;
- Procedures to follow when an MSDS is not received at the time of the first shipment of a hazardous chemical;
- For producers, procedures to update the MSDS when new and significant health information is found;
- Description of alternatives to actual MSDSs in the workplace, if used.

For employee information and training, the following elements should be described in the written program:

- The person responsible for conducting training;
- Format of the program to be used;
- Elements of the training program;
- Procedure to train new employees at the time of their initial assignment to work with a hazardous chemical and to train employees when a new hazard is introduced.

The written program must also include the following:

a. A list of hazardous chemicals known to be present at the workplace. The identity on the list must match the identity used on the MSDS and label. The list can be for the entire facility or for individual work areas.

b. The methods the employer will use to inform employees of the hazards of nonroutine tasks, such as cleaning reactor vessels, and the hazards associated with chemicals contained in unlabeled pipes in their work areas.

For multiemployer workplaces where employers produce, use, or store hazardous chemical at a workplace in a way that other employees may be exposed (e.g., contractors working in another employer's workplace), the Hazard Communication program must also include the following:

a. Methods to be used to provide the other employer(s) on-site access to MSDSs for hazardous chemicals the other employer(s)' employees may be exposed to while working;
b. Methods to be used to inform the other employer(s) of any precautionary measures that need to be taken to protect employees; and
c. Methods to be used to inform the other employer(s) of the labeling system used in the workplace.

Employers must make the written program available, upon request, to employees and their representatives. If employees travel between workplaces during a workshift, the written program can be kept at the primary workplace.

13.2.3.2 Material Safety Data Sheets The employer may rely on the chemical manufacturer, importer, or distributor for the hazard determination and MSDS preparation. The employer is responsible for maintaining copies of the MSDSs in the workplace and ensuring that MSDSs are readily accessible during each work shift to employees when they are in their work areas. MSDSs can be available through electronic access, microfiche, and other alternatives to paper copies provided that this does not prevent immediate employee access in each workplace. When employees travel between workplaces during a workshift, the MSDSs may be kept in the primary workplace facility; however, employees need to be able to immediately obtain required information in an emergency.

MSDSs may be kept in any form, including operating procedures, and may be designed to cover groups of hazardous chemicals in a work area where it may be more appropriate to address the hazards of a process rather than individual hazardous chemicals. However, the employer shall ensure that in all cases the required information is provided for each hazardous chemical, and is readily accessible during each work shift to employees when they are in their work area(s).

13.2.3.3 Labels The employer must ensure that each container of hazardous chemicals in the workplace is labeled, tagged or marked with (a) the identity of the hazardous chemical and (b) an appropriate hazard warning.

The labels provided by suppliers on original containers are typically adequate for those containers. The employer must label in-plant containers into which hazardous chemicals are transferred from original containers. Labels similar to those used on original containers can be used, or an alternate method can be employed. Signs, placards, process sheets, batch tickets, operating procedures, or other such written materials may be used in lieu of actual labels on individual stationary process containers provided the method conveys the required label information.

Employers are not required to label portable containers into which hazardous chemicals are transferred from labeled containers, and which are intended for immediate use by the employee who performs the transfer. Immediate use is typically interpreted to mean within the same work shift that the transfer was done.

Existing labels on original containers of hazardous chemicals must not be defaced or removed, unless the required information is immediately marked on the container again.

Labels must always be present or available in English. It is acceptable and often advantageous to supplement the English labels with hazard information in other languages, where employees speak other languages.

13.2.3.4 Employee Information and Training Employers must provide specific information and training to employees to comply with the HCS. Employees must be trained on hazardous chemicals in their work area at the time of their initial assignment and whenever a new hazard is introduced into the work area. Information and training may be designed to cover categories of hazards (e.g., flammability, carcinogenicity) or specific chemicals. Labels and MSDSs must always be available to provide chemical specific information.

Information that must be provided to employees includes:

- Requirements of the HCS;
- Operations in the work area where hazardous chemicals are present;
- Location and availability of the written Hazard Communication program, including the list(s) of hazardous chemicals and MSDSs.

Employee training must include, at a minimum, the following:

- Methods and observations that may be used to detect the presence or release of a hazardous chemical in the work area;
- Physical and health hazards of chemicals in the work area;
- Measures employees can take to protect themselves from hazards;
- Procedures the employer has implemented to protect employees from exposure to hazardous chemicals, such as work practices, emergency procedures, and personal protective equipment;
- Details of the Hazard Communication program developed for the facility, including an explanation of the labeling system, the MSDS, and how employees can obtain and use appropriate hazard information

13.2.4 Trade Secrets

A trade secret is defined in the HCS as follows:

> "A trade secret may consist of any formula, pattern, device or compilation of information which is used in one's business, and which gives him an opportunity to obtain an advantage over competitors who do not know or use it. It may be a formula for a chemical compound, a process of manufacturing, treating or preserving materials, a pattern for a machine or other device, or a list of customers."

Chemical manufacturers, importers, or employers may withhold the specific chemical identity from a material safety data sheet if the information is a trade secret provided that:

- The trade secret claim can be supported;
- The MSDS contains information regarding the properties and effects of the hazardous chemical;
- The MSDS indicates that the specific chemical identity is being withheld as a trade secret; and
- The specific chemical identity is made available to health professionals, employees, and designated representatives as required by the standard.

The specific chemical identity must be immediately disclosed to a treating physician or nurse during a medical emergency if the information is necessary for emergency or first-aid treatment. This disclosure must be made regardless of the existence of a written statement of need or a confidentiality agreement, which may be required later as soon as circumstances permit.

Trade secret chemical identity information must also be disclosed in nonemergency situations provided it is requested in writing and meets a specific occupational health need as detailed in the standard. A written confidentiality agreement is typically required by the manufacturer, importer, or employer.

13.3 PROGRAM MANAGEMENT

The objective of hazard communication is to provide information to employees on the chemical hazards in the workplace. If employees understand this information, they will be better able to protect themselves. Assembling the essential elements, such as MSDSs, labels, a written program, and training, although vital to the program, will not, in itself, accomplish this objective.

Even if you have developed a compliant program, it can quickly become stagnant with ever-growing collections of MSDSs clogging file cabinets, and canned training programs shown over and over again. *A Hazard Communication program must be a dynamic process that includes continuous updating of MSDSs, chemical lists, and training.* This will not happen without ongoing management commitment to ensure that resources are available. However, immature programs that require too much professional staff input can exhaust resources. Management commitment can wither over time with changes in philosophy and leadership.

This section will provide some suggestions on designing and managing an effective Hazard Communication program for a chemical user. The philosophy presented in this section is that mature programs are most effective. A mature program is one that has become, simply, the way the company does business. Program costs tend to be stable or decline, and professional staff time is limited. A mature program will be flexible and require minimal maintenance because essential elements have been integrated into existing processes. Stable systems support essential functions. The program is less vulnerable to loss of management commitment because support for, and ownership of, the program is spread through many parts of the company, not focused in one or two.

13.3.1 Policies and Procedures

Company policies and procedures reflect the support structure for programs, assuming that the company is large enough to have written procedures addressing most key functions. The philosophical approach to health and safety is usually stated in one or more company policies. Either a general or specific statement of top management support for Hazard Communication is very desirable. The tactical program support is in the procedures. Effective procedures clearly state the:

- Responsibility for each key task and the overall program;
- Interaction of responsible parties;
- Systems that will be used to maintain information and records;
- Associated job tasks needed to support those functions;
- Mechanisms to determine that tasks are completed;
- Methods to measure the effectiveness of the program;

Responsibility for specific HCS tasks may be covered in procedures, work instructions, within the written HCS program itself, or all three. Early in the program development, the primary concern will be to "get compliant." At this stage, considerable professional time will probably be needed to organize and maintain the program. The drive and responsibility for the program will probably be focused within one group, possibly one or two individuals.

Once this plateau is achieved, begin to plan the maturation process. Review existing procedures and look for opportunities to capture and use existing systems. Identify all the departments or individuals within the company that have a stake in, or a contribution to, Hazard Communication. Consider individuals or groups within the company that may be able to assume some or all of the following tasks:

- Develop and manage the Hazard Communication program;
- Develop and maintain the chemical list;
- Acquire, review, and distribute MSDSs;
- Verify readability and completeness of labels;
- Develop and maintain in-house labeling program for secondary containers, pipes, and vessels;
- Develop content and conduct employee training;
- Verify contractor(s) compliance with HCS;
- Track the purchase and use of chemicals within the company;
- Select personal protective equipment;
- Audit Hazard Communication program;
- Keep program records.

Some suggestions for the organization of these responsibilities are provided in Section 13.3.2 below.

A word of caution: Although ownership of the program should be shared, a program that is too diffused within the organization may disappear altogether. Make sure that where there is responsibility there is also ownership, accountability, and support.

13.3.2 Program Responsibilities

As mentioned in the previous paragraphs, it is beneficial to distribute the responsibilities for some program elements across organizational boundaries. Only in smaller companies can one or two people effectively drive or manage the entire program. In most cases, programs become more effective as tasks become integrated into existing jobs. Table 13.1 contains examples of typical departments and HCS functions that they might perform.

TABLE 13.1 Possible HCS Functions for Typical Departments

Department	Possible HCS functions
Purchasing	Request MSDS
	Develop purchasing procedures that incorporate HCS requirements
	Collect documentation of contractor(s) compliance with the HCS
Receiving	Verify that an MSDS accompanies incoming chemicals, or MSDS is on file
	Check containers for complete, readable labels
	Hold material and contact the vendor if labels are defective
	Apply internal labels, if needed
Operations, production, maintenance, construction groups	Order chemicals according to company procedures
	Check labels of containers as received in area and periodically as used
	Check chemicals against chemical list and MSDS file when received
	Update list of chemicals, if appropriate
	Apply additional labeling for transfer and secondary containers
	Ensure new employees and transferred employees are trained
	Ensure training is received by all employees when a new hazard is introduced
	Respond to employee questions and concerns about chemical usage or safety
	Maintain records for department or group
	Ensure that employees have and use proper personal protective equipment
	Conduct periodic self-audits of work areas to make sure that containers are labeled, MSDSs are available for material in use, employees are using personal protective equipment, and training records are complete
	Conduct periodic audits of employees' knowledge of safe handling of chemicals
Health and safety professionals	Participate in review of new and reformulated chemicals
	Participate in the review of chemical usage; recommend safer alternatives where possible
	Conduct HCS compliance reviews or audits of the workplace
	Review incoming MSDSs for completeness and accuracy
	Establish labeling practices and participate in the selection of secondary containter labeling system
	Develop, conduct, or review content and effectiveness of training
	Conduct assessments, including air sampling to determine potential employee exposure

TABLE 13.1 (*Continued*)

Department	Possible HCS functions
Health and safety professionals	Assess, then develop procedures for safe handling of chemicals in routine and nonroutine tasks Provide guidance in the selection of personal protection equipment Review accident and injury records related to chemical usage to assess improvements needed in personal protective equipment usage and training
Laboratories	Review test data from manufacturer on testing of mixtures Research potential substitute chemicals
Engineering	Ensure that MSDSs for new chemicals are received and reviewed with users before processes are changed or new designs are implemented Ensure that a review of employee safety, including chemical handling safety is part of project review in major redesigns or new projects Ensure that on-site contractors have fulfilled safety requirements, including HCS, before beginning work
Legal	Review initial HCS written program; conduct periodic review Provide legal interpretation of regulations Assist in the negotiation of confidentiality agreements
Medical	Ensure ready access of MSDS for emergency care providers Contact manufacturers for additional information when required in cases of trade secret formulations
Training	Assist in the selection and development of training programs Support training efforts with equipment and technical advice Conduct training or train the trainers

Table 13.1 is not comprehensive, and is not meant to imply that certain department or functional groups within a company should perform the duties indicated. Use it as a guide to identify groups within the company that may participate in some aspect of the program.

Early in the program development process assess the financial impact of the additional work load, equipment needs, and support. This analysis will help to determine whether the program responsibilities are workable, or whether additional staffing is needed. Estimate both startup and ongoing program costs. These costs will include staff time, equipment costs, and outside services. Figure 13.2 may serve as a guide for this analysis.

13.3.3 Flow of Communication

The smooth flow of information from the source (the chemical manufacturer, importer, or distributor) to the employee is critical to the program. This transfer of information utilizes MSDSs and labels as sources of information and training as the means of communication. The program must facilitate, not impede, this flow. Training, although it is the key step in this transfer of information, is not the final step. Evaluate how infor-

KEY ELEMENTS	TASK	START-UP COSTS			PROGRAM MAINTENANCE COSTS		
		Hours (Direct/ Indirect)	Equipment (Purchase/ installation)	Outside Services	Hours (Direct/ Indirect)	Equipment (Maintenance)	Outside Services
Written Program	Develop written program						
	Review/revise existing policies						
	Develop policies/procedures						
	Meet with key personnel to plan strategy						
Chemical List	Conduct inspection to identify chemicals used						
	Compile list of chemical						
	Input into data management system						
	Check against chemical purchasing records						
MSDS	Prepare MSDSs for all products*						
	Collect and organize MSDSs						
	Contact manufacturer for missing MSDSs						
	Match MSDSs with chemical list						
	Copy and distribute MSDSs						
	Develop data management system for MSDS						
	Establish process for obtaining and distributing new MSDSs						
	Review MSDSs for completeness						

Fig. 13.2 Program cost analysis.

KEY ELEMENTS	TASK	START-UP COSTS			PROGRAM MAINTENANCE COSTS		
LABELING	Develop labels for all products*						
	Establish procedure to check labels of incoming containers, isolate unlabeled containers at receiving stations, and obtain labels from supplier						
	Develop in-house labeling system for transfer and secondary containers						
	Apply labels to secondary containers, transfer containers, vessels, etc.						
	Review piping systems, vessels and information that may be used in lieu of labels						
TRAINING	Assemble or develop training materials						
	Train the trainers						
	Train current employees						
	Train new hires/transfer employees						
	Train upon introduction of new hazard/changes in processes						
DOCUMENTATION	Maintain training records						
	Maintain list of chemicals						
	Maintain MSDS documentation						
CONTRACTORS	Provide information to contractors on hazardous materials at worksite						
	Request information on contractor's materials						
	Obtain documentation that contractor complies with HCS						

* These are tasks to be performed by chemical manufacturers and distributors.

Fig. 13.2 (*Continued*)

mation will flow back from the employee, between key individuals in the program and between departments. The paragraphs that follow may help you avoid communication "sinks" and "traps" within the program.

Put a mechanism in place to field employee questions regarding chemical hazards. Response should be quick, accurate, and consistent, but need not be given on the spot. It is often better to admit that the answer is not immediately available than to "shoot from the hip." Assure the employee that someone will obtain the answer and get the requested information. Then follow up to make sure it happens.

Questions may be addressed in a number of ways. Refer them to in-house expert staff. Obtain additional information from the manufacturer or distributor; they often have very knowledgeable technical personnel. Make additional information sources available to the employee. Failure to address employee questions will result in loss of confidence in the program and possibly more serious employee relation problems. Inconsistency or inaccuracy in responses will also result in loss of confidence. Investigating these questions can often provide fresh insight into hazards or controls that can then be incorporated into training.

Employees often have concerns regarding the level of exposure to chemicals, the efficiency of personal protective equipment, or the effectiveness of engineering controls. Such concerns may be expressed in training classes, safety committees, employee participation groups, directly to area supervisors, or to safety and health professionals. Regardless of where the concern is voiced, key people will need training on how to respond, and how to address these concerns. Employee concerns should be taken seriously. Act quickly to determine the best course of action. Inform the employee(s) of the actions that will be taken to address his or her concerns. An approach that involves the participation of the employees, their representatives, and management is often effective in obtaining resolution.

Open communication between departments can improve the program, but more important, can improve safety and save money. The following example illustrates a situation in which interaction between departments works toward both goals.

As a matter of convenience and cost savings, the purchasing department recommends that a production department buy a chemical in a more concentrated form, and in larger volume container. The cost savings appear to be substantial, but are they? Considering only direct savings may lead to erroneous conclusions. In this example, significant hidden costs could be associated with any one of the following factors:

- Inadequate storage facilities or inadequate leak containment for the larger storage volume;
- Increase in risk to employee health, public health, and the environment in the event of a release or spill;
- Inadequate leak or spill response capabilities for the new process, increased volume, or more concentrated chemical;
- Inadequate distribution system(s);
- Increased health and safety risk associated with distribution and handling (including dilution of concentrated chemical);
- Increased waste disposal costs;
- Additional regulatory requirements;
- Additional environmental reporting responsibilities;

- Additional personal protective equipment and spill response equipment;
- Additional health, safety, and environmental training.

Anticipating the costs associated with this change can help to present a more accurate financial picture. Through this process, risks will be identified. It should not be used as a strategy to resist change. Changes such as those in the example above can and should be made when the risks are identified and found to be acceptable, given appropriate management and controls.

The efficiently functioning program thrives on feedback from all active participants. Collecting and managing this information helps to maintain compliance. More important, it can improve employee safety and health. The following examples show contributions that can be made by various personnel, and how communication of information can benefit the program.

- Health and safety professionals evaluate employee exposure, the effectiveness of personal protective equipment, and engineering controls. Include the results of these evaluations in training.
- Supervisors or lead persons will know how personal protective equipment or control devices are used (or misused) in the workplace. In addition, they will be aware of accidents or injuries involving chemicals. All this information should influence decisions regarding the purchase and use of chemicals. Real-life examples from supervisors can make training more relevant to employees.
- Training specialists should be aware of whether employees are putting hazard communication training to work. Change the content, emphasis, or methods of training if accidents, near-misses, or misuse of controls indicate a lack of hazard awareness. (See Chapter 14 on training.)
- Legal staff tracking court cases and decisions may recommend changes to the program in light of new cases or based on regulatory trends.
- Physicians and nurses will know of medical cases involving chemicals. This can help to evaluate the effectiveness of personal protective equipment, level of employee knowledge, and handling procedures.
- Engineering personnel will be aware of planned changes in processes, and changes that may affect the effectiveness of controls.

Provide a forum in which these sources of information can be used to fine tune and redirect the program.

13.3.4 Quality Control and Program Improvement

Once a program is up and running, look beyond the essentials toward program quality control and continuous improvement. Develop a quality control strategy that utilizes the principles of continuous quality improvement. Design quality into the program. Although auditing is important, do not rely on it exclusively to uncover flaws. Use your product quality control program as a guide.

Reviewing your program is an essential part of quality control. However, introducing upgrades helps to stimulate interest and contributes to overall effectiveness. Table 13.2

TABLE 13.2 HCS Program Upgrade Suggestions

KEY ELEMENTS	SUGGESTION FOR PROGRAM UPGRADES
Written Program	Review and analyze applicability of HCS to workplace. For example • Areas in which employees handle only sealed containers • Laboratories covered under HCS or Lab Standard Review written program annually, include • Compliance audit • Review by key personnel Revision, where needed
Chemical List	Conduct periodic, documented inspection of usage areas. Review process to verify that the hazardous materials currently used and generated are in the program and covered in training. Check chemical list periodically against • Chemical purchasing records • MSDS file • Materials in work areas Establish procedures for continuous updating of the chemical list. Check the chemical list against lists kept for EPA SARA reporting. Evaluate the potential benefits of using a single chemical tracking system. Establish multidisciplinary team(s) to • Review chemical usage • Review chemical purchasing • Substitute less hazardous materials • Evaluate benefits of reducing the number of different chemicals purchased and different sources used. Develop an "approved chemical" list to help control purchases • Limit off-the-shelf purchases Identify ways chemicals may enter workplace outside purchasing procedures (e.g. blanket POs direct employee purchase and cash reimbursement). Develop procedures to control these chemical purchases.
MSDS	Purge MSDS files of duplicates and materials no longer onsite. Compare MSDSs from various manufacturers supplying similar products. Identify and research major differences. Address discrepancies in training. Test employee accessibility in usage areas. Evaluate the potential benefit of electronic MSDS systems. Conduct a spot check of MSDSs for completeness of information.

provides some useful ideas for improvements. The main Hazard Communication program elements and tasks are listed in the first two columns. Program upgrade suggestions are in the third column. These suggestions go beyond the minimal requirements of the HCS. Therefore, choose carefully and implement them a few at a time as resources permit.

TABLE 13.2 *(Continued)*

LABELING	Conduct routine spot checks of containers for legibility and completeness of labels. Report consistent problems to supplier and purchasing agent. Conduct routine spot checks of secondary containers for correct labeling. Evaluate the potential benefits of using a computer program to generate secondary labels. Consider pre-labeling frequently used transfer containers If batch tickets or process flow diagrams are used in lieu of labels, test accessibility of this information.
TRAINING	Review training materials at least annually and include some or all of the following activities: • Review technical content • Compare training content with current hazards • Evaluate language skill level and accessibility to sight and hearing impaired employees • Review and upgrade A/V materials • Insert real life examples • Utilize information from employees, accident/injury experience • Review accident/injury reports related to chemical usage. Use results to alter emphasis in training if indicated • Request employee training evaluation • Review test scores, if comprehension tests are used. Provide skills development for trainers. Consider using area employees as trainers. Evaluate alternative training methods, such as interactive computer-based training. Vary training methods to fit group (e.g. operations groups using limited numbers of chemicals vs. maintenance groups using a large variety) Interview employees for knowledge of chemical hazards and appropriate handling. Routinely spot check use of PPE. Use observations of lack of use or misuse to re-direct training, if needed. Consider integrating portions of Hazard Communication training with other required safety and environmental training, such as RCRA or "HazWhopper." Integrate Hazard Communication training with on-the-job training and work instructions. Provide periodic refresher training for all employees. Spot check training records against employee lists where HCS training is required.
CONTRACTORS	Obtain acknowledgment that contractors receive the information on worksite hazardous materials. Review service contract agreements to ensure that compliance with safety and health regulations is part of the document.

13.3.5 Record Keeping

Record keeping is one of the most challenging aspects of Hazard Communication. Early in the planning stages, decide how and where the records will be kept, and how files will be updated and purged. The major recordkeeping needs are outlined below.

MSDSS AND CHEMICAL LISTS Current MSDSs, the written Hazard Communication program, and the chemical list must be immediately accessible to employees on all shifts. However, additional record keeping is required. As explained in Section 13.6.1, 29 CFR 1910.20 (Access to employee exposure and medical records) requires that exposure and medical records be maintained for 30 years. MSDSs and chemical lists are considered exposure documentation. All this information must also meet the requirements for employee access in 29 CFR 1910.20. Please note that some other OSHA standards have record retention requirements longer than 30 years. Develop a records retention schedule to manage these documents and ensure that records are kept for the correct period of time.

TRADE SECRET DISCLOSURES If trade secret information is requested from a chemical supplier, keep files that include the requests, the responses, and the outcomes. Once the information has been provided by the supplier, keep a copy of the confidentiality agreement with the trade secret information. Implement controls on access to the filing system that are appropriate for the level of confidentiality required by the agreements. It is generally not a good idea to store them in the MSDS filing system, since this may increase the risk of inadvertently violating confidentially agreements. There should be some way to identify materials for which there is trade secret information, so that it is easy for authorized persons to obtain information, if the need arises. At least some controls are needed to prevent inappropriate distribution of confidential information. Part of the confidentiality agreement will address how these records are to be safeguarded and who within the company may have access. Breach of these contractual agreements could result in legal action.

TRAINING RECORDS Although not required under the HCS, it is strongly advised to document training. These records should include:

- Initial training: Content of training (course curriculum), date, person conducting the training, materials used, sample of test administered, (if used) training evaluations, names of employees trained, department or work area, job classification or position, completed test or exam (if used);
- Followup training: hazards reviewed, persons conducting training, date, materials used, persons trained, department, or work area;
- Qualification of persons conducting the training, train-the-trainer courses;
- Records documenting employee training on the use of electronic MSDS, if applicable.

CONTRACTORS There are no requirements within the HCS for the retention of contractor records. However, information on chemicals used by another employer at a worksite may qualify as exposure records for company employees under 29 CFR 1910.20 (Access to employee medical records). It is advisable to document the information provided by the contractor. This documentation should include, at a minimum:

- A list of the chemicals used;
- Copies of MSDSs provided by contractors;
- Documentation that onsite hazardous material information was provided to the contractor;

- Concerns regarding chemicals, employee exposure, or protection raised during the work and how those concerns were resolved.

Most companies keep records of contract work. The above information may be easier to organize as part of a single set of files on the contract. However, filing all this information with the contract will make it difficult to retrieve information in response to a specific employee request.

13.4 TOOLS FOR PROGRAM IMPLEMENTATION

13.4.1 Hazardous Material Information Systems

Hazard Communication programs for medium to large manufacturers and employers require gathering and communication of large volumes of data. Since the HCS is performance oriented, the system used is left to the discretion of the individual company and a multitude of options are available.

13.4.1.1 MSDS Management Managing MSDSs can present a challenge to employers who use many different chemicals in their workplaces. As discussed in Section 13, employers are required to maintain MSDSs such that they are readily accessible to employees during each work shift.

There are several options for maintaining MSDSs. The simplest method is to maintain hard-copy MSDSs in an accessible file location at the worksite. This type of system works well for a workplace where all the employees are located in one building or within reasonable proximity to one another and MSDSs can therefore be kept in just one or two locations. A manual system is typically practical only when the types of hazardous materials are somewhat limited.

Another option used by some larger employers is to enter information from MSDSs into a database, which then provides uniformly formatted MSDS information to the user. This allows MSDS information to be available online at computer terminals located throughout a large facility. Employees must, of course, be trained to use the computer system or have access to an individual with appropriate training on all shifts. This form of total automation is very costly, since MSDSs received from manufacturers will vary in format and must be reformatted for computer input. However, over time, the cost may be offset by savings in staff time, especially in large companies with many plant locations. Employers should do a careful cost analysis to determine what benefit, if any, would be gained from a fully automated system. The major advantage to this type of system is the ease in use by the employees. The MSDS user need only understand one MSDS format to easily access the information through a computer terminal. A disadvantage is the increased potential liability the end user incurs by changing the format of, and interpreting information from, the original MSDS.

A third option used by some employers is to scan the original MSDSs into a computer system to allow for online access. This option provides the advantage of online access with much less time required to input data. One problem with this system is scanning inaccuracies due to color, light print, and other format differences. The scanned MSDSs must therefore be reviewed prior to distribution to ensure that they are accurate and readable. MSDSs will also be in a variety of formats, making their use online more difficult.

Whichever option is chosen, a method for identifying out-of-date MSDSs (e.g., over 3 years old) should be devised to allow the employer to update their files periodically. The employer should check to determine if chemicals with out-of-date MSDSs are still in use and if so, contact the chemical manufacturer to determine whether the MSDS has been updated.

13.4.1.2 Hazardous Material Tracking Employers who use hazardous chemicals are faced with the task of managing very large volumes of data related to hazardous materials. Once MSDSs are received and stored for existing hazardous chemicals, the employer is still confronted with the task of receiving, tracking, and storing MSDSs for new chemicals and updating MSDSs for existing chemicals. The employer will also typically track "approved" hazardous materials that may be purchased for use in the facility.

Many approaches are available for tracking MSDSs, ranging from completely manual system to a completely automated system. Except in cases where very few chemicals are used, some form of automated tracking is typically needed. Tracking of MSDSs by a database program allows for (a) cross-referencing and updating of hazardous chemical list, (b) tracking the date of the latest MSDS revision, (c) checking chemicals and suppliers for preapproval, and (d) tracking departments using each chemical for training purposes.

A database system facilitates communication between users, purchasing, departments, and health/safety professionals. Materials approved by either a health/safety professional or review committee can be entered into the database as "approved" so that purchasers and users of the chemical know that the material is acceptable for use. This can help avoid duplication of effort in reviewing substances and speed the purchasing process for commonly used materials. The hazardous materials tracking system will also allow tracking of departments that use each chemical. This information can be used to determine training requirements within the department. An enhancement to the system may include tracking the training (initial and update) of employees within each department. When updated MSDSs are received, the departments using the chemical can be alerted to provide updated information to their employees on that chemical.

A database tracking system can be simple or elaborate, depending on the needs of the company. Databases specifically designed for HCS compliance are available commercially. Alternatively, standard database programs can be set up to track Hazard Communication information.

13.4.1.3 Related Programs An information system set up for Hazard Communication compliance should be flexible enough to be used in complying with other related regulatory requirements, such as those discussed in Section 13.6. Additionally, other state or local regulations may require information tracking that could be easily incorporated into a hazardous material tracking system set up for Hazard Communication.

An example of another use of the Hazard Communication tracking information is the requirement of Section 311 of Superfund Amendments and Reauthorization Act (SARA) Title III to produce copies of MSDS for certain chemicals. In addition, Section 312 of SARA Title III requires the submission of an inventory of hazardous chemicals with additional information regarding quantities present at a facility, daily use, locations, and storage information. This information can be submitted either as an aggregate by OSHA categories of health and physical hazards (Tier I) or by individual chemicals (Tier II). Including the information required for producing SARA Title III reports with the Hazard Communication database helps avoid duplication of efforts. The Hazard Communication database may also be useful as a starting point for preparing a Toxic Chemical Release Form as required by Section 313 of SARA Title III. The USEPA has pamphlets available that provide details on how to comply with certain sections of SARA Title III.

If the information tracking system includes a provision for tracking training, other training programs may be added to the system for tracking purposes. This may be particularly useful when training programs are combined. For example, a Hazard Communication training program may be expanded to include training of employees who handle hazardous waste, as required by Resource Conservation and Recovery Act (RCRA) regulations. The training may be combined, but records must be kept to show compliance with both sets of regulations. An automated system for tracking this is helpful, particularly when many employees require training.

13.4.2 Training Tools

The Hazard Communication Standard allows the employer to determine "how" training will be done and specifies only "who" must be trained, "when" training must occur, and on "what" topics training must be covered. This provides employers with a great deal of flexibility to implement a training program that is both effective and practical for their organizations. Obviously, different approaches may be applicable at different work settings depending on the number of employees, number of worksites, number of chemicals, projected turnover of personnel, and the need for refresher or updated training.

There are a variety of training techniques and tools available. Typically, a combination of techniques will enhance the learning process and allow for more effective training. The following are some of the most common techniques used for Hazard Communication training:

- Group lectures (this method is used when all members of a large group need to learn the same information and the information is conducive to mass display);
- Group and individual role plays (group members play roles and make decisions based on real-life situations; this method can be used to simulate emergency evacuations and other emergency procedures that must be performed within a critical time frame);
- Emergency equipment demonstrations (people learn by watching demonstrations and practicing behavior with fellow class participants);
- Self-paced programmed instruction (the worker responds to written questions or situations at his/her own pace and receives immediate feedback);
- Interactive computer-based training programs (these systems can be used to train employees individually);
- Charts and/or diagrams (pictures and diagrams reinforce learning);
- Slides/audiotape (this media allows flexibility of site-specific photographs that are readily available for playback and review);
- Film/videotape (an alternative to lectures, with ability to use graphics and animation).

Training effectiveness can be evaluated by a number of different methods. At a minimum, training should be documented using a roster, both to document training and track each employee's need for retraining. Additionally, the instructor may test effectiveness of the training by administering either oral or written tests at the end of each session. An auditor or supervisor may later verbally quiz employees during field visits to test for retention of information. This method is often used by OSHA compliance officers. Additionally, the employee's work practices may be observed to determine whether principles taught during training are being implemented in the workplace.

13.4.3 Material Labeling

The standard allows flexibility in the format and content of labels, provided they include the required information.

13.4.3.1 Manufacturer's Labels As discussed in Section 13.3, manufacturers must ensure that each container of hazardous chemical leaving the workplace is labeled with (1) the identity of the hazardous chemical, (2) an appropriate hazard warning, and (3) the name and address of the chemical manufacturer, importer, or other responsible party. For substances shipped in a tank truck or rail car, the appropriate label or label information may either be posted on the vehicle or attached to the shipping papers.

There are numerous labeling systems in use in industry. Labeling systems that rely on numerical or alphabetic codes to define hazards typically do not provide target organ effect information and are not appropriate for shipped containers unless additional narrative information is also included on the container's label. In some cases of chemicals regulated by OSHA in a substance-specific health standard, the warning label must meet the requirement of that standard. Examples of these are asbestos, benzene, and ethylene oxide.

13.4.3.2 Labels in the Workplace As discussed in Section 13.2, employers are responsible for ensuring that all containers in their facility are properly labeled with (a) the identity of the hazardous chemical, and (b) appropriate hazard warning.

The employer has even more flexibility than the manufacturer regarding labeling containers. In the case of individual stationary process containers, employers have the option of using signs, placards, process sheets, batch tickets, operating procedures, or other written materials in lieu of affixing labels directly on the container. The alternative method used must identify the container to which it is applicable and convey the required label information. The written materials must be accessible to employees throughout their work shift.

Labeling systems that include numerical or alphabetic codes to convey hazards may be permissible for in-plant labeling provided the entire Hazard Communication program is effective. The target organ effects must be communicated to employees in some manner. Proper training on a specific labeling system is essential to help ensure its effectiveness in promoting safe handling and use of chemicals.

13.4.3.4 American National Standards Institute Labeling The American National Standards Institute (ANSI) published a voluntary labeling standard ANSI Z129.1–1988 as guidance for precautionary information on chemical container labels.[5] ANSI recommended that labels provide much more detailed precautionary information than that required by OSHA in the HCS. Additional information included by ANSI includes:

- Signal word;
- Precautionary measures;
- Instructions in case of contact or exposure;
- Antidotes;
- Notes to physicians;
- Instructions in case of fire and spill or leak;
- Instructions for container handling.

ANSI also requires the identity of the material and a statement of the hazard to be included, as does the HCS. ANSI does not include the supplier's name and address, and, therefore, would not meet the requirement for shipped containers.

The additional information provided by the ANSI labels is generally available on the MSDS. The possible dilution of the immediate warning must be considered before deciding to use the ANSI labeling system.

13.5 PROGRAM AUDIT

13.5.1 Program Appraisal

Every Hazard Communication program should be periodically reviewed to determine its effectiveness and degree of compliance with the OSHA standard. The best form of review is a formal audit, performed by an independent outside party, such as a consultant, an internal knowledgeable individual, such as an industrial hygienist, or a team of individuals.

The audit should include a review of records to determine the adequacy of the documentation of the program, as well as a determination of how well the program is being performed. These items are discussed in Sections 13.5.2 and 13.5.3.

A quick check for employer compliance can be done by checking that the following items have been completed:

_____ Obtained a copy of the HCS
_____ Read and understood the requirements
_____ Assigned responsibility for tasks
_____ Prepared an inventory of hazardous chemicals
_____ Ensured that containers of hazardous chemicals are labeled
_____ Obtained MSDS for each hazardous chemical
_____ Prepared a written program
_____ Made MSDSs accessible to workers
_____ Conducted training of workers
_____ Established procedures to maintain current program
_____ Established procedures to evaluate effectiveness

A more detailed checklist to use during a thorough onsite audit is provided as Table 13.3. This checklist may be augmented to include state and local regulations. It also can be supplemented to include related regulations. The checklist is prepared in two parts, one for employers and one for suppliers. The appropriate portions may be used for specific audits.

TABLE 13.3 Audit Checklist

PART I
HAZARD COMMUNICATION AUDIT-EMPLOYERS

WRITTEN PROGRAM

A. Statement of Purpose *(optional)*

[ ] Purpose of Hazard Communication Program

Comments: ______________________________

B. Location: ______________________________

[ ] Available for review upon request to employees and OSHA

Comments: ______________________________

C. Labeling

[ ] Who is responsible for checking and maintaining original labels?
[ ] Who is responsible for making and maintaining inplant labels?
[ ] Description of inplant labeling system for secondary containers.
[ ] Description of labeling system for bulk storage tanks *(if applicable)*.
[ ] Description of alternative systems used for batch processes *(if applicable)*.
[ ] Description of fixed labels *(if applicable)*.

Comments: ______________________________

D. MSDS

[ ] Who is responsible for requesting MSDS and keeping files up to date?
[ ] How are MSDS obtained for every Hazardous Substance?
[ ] Where are MSDS kept?
[ ] Accessibility to employees.
[ ] Review of MSDS for omissions (checklist).
[ ] How are missing MSDS requested?
[ ] MSDS alternatives *(if used)*.

Comments: ______________________________

E. Training

[ ] Who is responsible?
[ ] Initial training plan.
[ ] New employee training.
[ ] Refresher training plan.
[ ] Update of training for new hazards.
[ ] Documentation of training.

TABLE 13.3 (*Continued*)

Comments: ____________________

____

____

F. List of Hazardous Substances

[ ] Included in written program.
[ ] Identities match MSDS and labels.
[ ] Procedure to update list.
[ ] Locations listed by work area.

Comments: ____________________

____

____

G. Non-Routine Tasks and Unlabeled Pipes

[ ] How employees will be informed.
[ ] Who is responsible for informing employees?
[ ] Incorporation of existing procedures.

Comments: ____________________

____

____

H. Contractors

[ ] How they will be informed of hazards and protective measures.
[ ] How they will provide MSDS for hazardous substances they bring onsite.
[ ] Who is responsible for informing contractors and obtaining MSDS?

Comments: ____________________

____

____

I. Hazard Determination *(optional)*

[ ] How chemicals are evaluated (e.g. rely on supplier or do own evaluation).
[ ] Who is responsible for evaluating chemicals?

Comments: ____________________

____

____

LABELING

A. Labeling System

[ ] Include product identity and hazard warning (health and physical hazards).
[ ] Labels displayed and legible.
[ ] Method for replacing damaged labels.
[ ] Checking labels on incoming containers.

Comments: ____________________

____

____

TABLE 13.3 (*Continued*)

MATERIAL SAFETY DATA SHEETS

[ ] MSDS available for each hazardous substance.

[ ] Accessible location to employees on all shifts.

[ ] Requests for missing MSDS documented.

[ ] Requests for update of incomplete MSDS documented.

[ ] Review of MSDS for completeness.

[ ] MSDS in all files up-to-date.

[ ] Method for updating MSDS files.

[ ] Provisions for providing MSDS to employees, physicians, and/or representative upon request.

Comments: ____

____

____

INFORMATION AND TRAINING

A. Training Curriculum Include:

[ ] Explanation of MSDS and information it contains.

[ ] MSDS contents for each substance or class of substances used.

[ ] Explanation of requirements of Hazard Communication Standard.

[ ] Location and availability of written program.

[ ] Location of operations where hazardous substances are used.

[ ] Observation and detection methods for hazardous substance presence or release.

[ ] Physical and health hazards of substances used in work area.

[ ] Protective measures for work area.

[ ] Details of employer's hazardous communication program, including labeling and MSDS program.

Comments: ____

____

____

B. Training Update

[ ] Periodic refresher training.

[ ] Provision for informing employees within 30 days of receiving a new or revised MSDS which indicates significantly increased risks or protective measures.

Comments: ____

____

____

C. Training Documentation

[ ] Attendance lists.

[ ] Tests of comprehension (Pre/Post)

Comments: ____

____

____

TABLE 13.3 *(Continued)*

PART II
HAZARD COMMUNICATION AUDIT-SUPPLIERS

HAZARD DETERMINATION

[ ] Who is responsible for hazard determination?

[ ] List of information sources used.

[ ] Written procedure for evaluating all products.

[ ] Plan for getting updated information.

Comments: ____

____

____

MATERIAL SAFETY DATA SHEETS

[ ] Who is responsible for:

- [ ] Preparing and updating MSDS.
- [ ] Distributing MSDS.
- [ ] Replying to MSDS requests from customers.

[ ] Complete and accurate MSDS written for each product.

[ ] Copies of MSDS provided to each purchaser (even if not requested)

[ ] Method of providing MSDS: (e.g., with bill of lading, mailed to purchaser).

____

____

____

[ ] Procedure for updating MSDS when new information becomes available.

Comments: ____

____

____

LABELS

[ ] Who is responsible for:

- [ ] Determining label working.
- [ ] Updating label wording.
- [ ] Making and maintaining labels.

[ ] Product containers labeled with:

- [ ] Name and address of manufacturer or supplier.
- [ ] Product identity (same as on MSDS).
- [ ] Hazard Warning (specific physical and health hazards).

TRADE SECRETS *(if applicable)*

[ ] Who is responsible for making trade secret determination?

[ ] Justification documentation.

[ ] Procedure for applying to OSHA for trade secret status.

[ ] Procedure for providing trade secret chemical identities in emergencies.

[ ] Procedure for providing trade secret chemical identities in non-emergencies under specified conditions.

[ ] Standard Confidentiality Agreement?

13.5.2 Record Review

A key part of every OSHA compliance officer's visit is a review of an employer's written Hazard Communication Program. As discussed earlier, the written Hazard Communication program provides an explanation of all the elements of an employer's Hazard Communication program and how they are implemented. Therefore, a logical first step of any audit is a review of the written Hazard Communication program. The written program should be checked against the audit checklist elements shown in Table 13.3.

For an employer using chemicals, the availability of MSDSs to employees should be audited. Documentation to be reviewed includes (1) MSDS copies in the workplace, (2) written requests to suppliers for missing or inadequate MSDSs, and (3) tracking method for new and updated MSDSs.

Hazard Communication training should be well documented, and, therefore, a review of training records is an important part of the records review. The records should include a description of (1) what was covered at each training session, (2) who was in attendance, and (3) when the training occurred. Training of new employees and employees transferred to a department with different chemicals in use should also be audited. Documentation of followup training for new hazards in the workplace should also be reviewed. Refresher training is required at specific time intervals in some states. The documentation should be reviewed for compliance with state as well as Federal OSHA requirements.

Specific labeling systems should be audited for their compliance with OSHA's requirements. Compliance with the system discussed in the written program should be spot checked. During a walkthrough of work areas, labels should be noted for adequacy of the information as well as their legibility and condition.

13.5.3 Performance

The most important test of its effectiveness is whether a program actually results in a more aware worker. The OSHA compliance officer tests this by actually interviewing employees at random and asking them about the chemicals in their workplace. A good audit will also include this spot check. It is not feasible to interview every worker, but a few employees in several different work settings should be interviewed. When possible, include maintenance, process control, inhouse construction workers and outside contractors in the interview, in addition to production employees with routine exposure. Employees should be asked key questions about hazard communication such as:

- What are the potential hazards of that chemical?
- What protective equipment do you use when working with that chemical?
- Where would you find the MSDS?
- What would you do if some of that chemical spilled?
- Did you receive any training regarding chemicals?

Other elements of the program, such as MSDS and labeling, should be checked in the workplace for (1) compliance with OSHA standards, (2) implementation in the workplace, and (3) actual performance. This part of the audit should be through a random check, as with training. A number of chemicals should be selected in the workplace. The adequacy of the labels on the containers should be checked at the workstation. The

MSDS should then be located and checked for completeness. The MSDS must be easily accessible to the worker, not locked in a supervisor's office. The protective measures and air monitoring data should also be reviewed as appropriate.

13.6 RELATED OSHA REGULATIONS

The complex issue of managing hazardous chemicals safely from manufacture to disposal has spawned a variety of Federal regulations. Many of these regulations have requirements for training, hazard information, and labeling. Although each regulation deals with a slightly different aspect of this problem, they are related and, in some cases, overlap. This section provides a summary of some of these regulations and discusses how they relate the Hazard Communication Standard.

13.6.1 Access to Employee Exposure and Medical Records (20 CFR 1910.20)

The purpose of this regulation is to provide employees and their designated representatives access to relevant exposure and medical records. MSDSs are identified in the regulation as one of the types of information that constitute an employee exposure record. As such, they must be available to employees, their designated representatives, and OSHA.

Exposure records must be maintained for 30 years; however, you do not need to keep MSDSs for materials that are no longer used, provided you do keep a record of the identity of the chemical, the original formulation, where is was used, and when it was used. Some companies choose to archive all old MSDSs and chemical lists to assure compliance with this regulation.

You do not need to keep two lists. The requirements for maintaining a list of chemicals in the HCS can also meet the requirements for this regulation, if the list includes information on where and when (time period, not each application) each chemical is used. Purchasing records and annual chemicals inventories can be the source of this information.

OSHA has noted in their "Inspection Procedures for the Hazard Communication Standard" that employers have confused the accessibility requirements of the 1910.20 with those of the Hazard Communication Standard. For example, some employers mistakenly believe that they have 15 days to produce a copy of the written Hazard Communication program at the worksite. This in not the case. A copy of the program must be at the worksite and accessible to employees on all work shifts.

13.6.2 Air Contaminants (29 CFR 1910.1000)

When the permissible exposure limits (PELs) were revised in 1989, limits for 376 chemicals were added or changed. Currently, these changes have been stayed and are therefore not in effect. However, when PELs are changed, they must be amended on MSDSs within 3 months of the change.

13.6.3 Occupational Exposure to Hazardous Chemicals in Laboratories (29 CFR 1910.1450)

This standard takes precedence for laboratories that fit the laboratory use and scale definition in 1910.1450. However, there are laboratories, such as some quality control labs, that do not fit this definition. For these worksites, the HCS applies. Determine, first, which category your laboratories fit. If your lab is *not* a "1910.1450 lab," it may fit the definition within the HCS. Laboratories as described within the HCS are exempted from certain provisions of the HCS; however, facilities you may call laboratories may not fit this definition either. Pilot plants, dental, photofinishing, and optical labs, for example, are considered to be manufacturing operations, not labs. Your careful assessment of how your facility fits these definitions is critical because it determines exactly which requirements will apply. Note that other activities associated with research and university laboratories such as maintenance, housekeeping, food services, administration, etc. are covered by the HCS and not the laboratory standard.

13.6.4 Occupational Exposure to Benzene (1910.1028)

There are references to the HCS within this standard, under sections dealing with the communication of benzene hazards to employees. Training and information requirements must meet those of the HCS when benzene is present. However, if exposure exceeds the action level, training and information must be provided at least annually. The labeling requirements for benzene also supersede those of the HCS. Consider the HCS as a minimum requirement. If benzene exposure in your facility fits those specified within the benzene standard, the more specific standard prevails.

13.6.5 Process Safety Management of Highly Hazardous Chemicals (1910.119)

Specific highly hazardous toxic, reactive, flammable, or explosive chemicals are listed in this standard. If these materials are involved in a process at threshold quantities, the provisions of the process safety management standard apply. The requirements for providing employees with information on these materials go well beyond those of the HCS. MSDSs may be used to provide a portion of the information, but will not provide all the information required.

13.6.6 Hazardous Waste Operations and Emergency Response (1910.120)

This standard covers both hazardous waste handling and emergency response. Most of the confusion with HCS occurs in the interpretation of the training requirements for emergency procedures.

When the emergency procedures involve evacuation of the work areas and notification to emergency responders, the training required is fairly basic and should include: emergency procedures, emergency notification (alarms, etc.), and evacuation routes. When employees are required to take additional actions, the level of training required would be dictated by 1910.120. At a minimum, it wound include: emergency procedures, spill and leak cleanup procedures, personal protective equipment, decontamination procedures, shutdown procedures, recognizing and reporting incidents, and evacuation.

REFERENCES

1. The Occupational Safety and Health Act of 1970 (P.L. 91-596, 91st Cong., S. 2193, December 29, 1970) is codefined at 29 U.S.C. 651 et seq.
2. Federal Register, 47 FR 12092.
3. 29 CFR 1926.59.
4. ANSI Z400.1–1993, American National Standard for Hazardous Industrial Chemicals, Material Safety Data Sheet Preparation.
5. ANSI Z129.1–1988, American Nation Standard for Hazardous Industrial Chemicals Precautionary Labeling.

CHAPTER 14

Health and Safety Instruction

KENNETH P. MARTIN

14.0 INTRODUCTION
14.1 NEEDS ANALYSIS
    14.1.1 Conducting a Needs Analysis
    14.1.2 Needs Analysis Outcome Example
14.2 SETTING INSTRUCTIONAL GOALS
14.3 TASK ANALYSIS
14.4 LEARNING OBJECTIVES
14.5 THE LEARNING AUDIENCE
14.6 PRESENTATION
    14.6.1 Presentation Styles
    14.6.2 Presentation Materials
    14.6.3 Presentation Materials
14.7 EVALUATION OF INSTRUCTION
14.8 FEEDBACK ON THE INSTRUCTION PROCESS
14.9 REFERENCES AND READINGS

14.0 INTRODUCTION

Education is a broad term that encompasses all the experiences in which people learn. A person's education in an area may have been attained through a long process of life experience, or through instruction. Instruction is the delivery of focused educational experiences that lead toward a set of learning goals. When the instruction involves a very specific set of skills and knowledge that will be put to use almost immediately, it is generally referred to as *training*. Most health and safety instruction involves training. Our goal is to efficiently impart the skills and knowledge to perform a particular task in a prescribed manner.

The decision to provide training to personnel is one response to a need of the orga-

Handbook of Occupational Safety and Health, Second Edition, Edited by Louis J. DiBerardinis, ISBN 0-471-16017-2

nization. Training takes places in many areas, from personnel relations to job skills involving production to job skills involving health and safety. The health and safety training may be integrated into other training or provided separately.

The type of training given will depend upon such factors as the attitude and commitment of the organization to the training, the knowledge and skills of the training participants, the time and resources allocated for the training, and the skills of the trainer.

14.1 NEEDS ANALYSIS

A needs analysis for an organizational problem is usually undertaken when there is a gap between the actual and optimal performances in a specific area. The needs assessment study should have a clear description of the problem, the range and scope of the problem, and a solution to the problem. Training may constitute all, part, or none of that solution.

If training is undertaken as part of the solution, it should be directed to meet the specific requirements expressed by the needs analysis.

14.1.1 Conducting a Needs Analysis

Rossett (1987) outlined the five elements that the needs analysis would include (see Fig. 14.1).

1. A description of the organizational problem that currently exists (or will exist in the future) is the first step in the needs analysis. An examination of these problems would determine their scope and extent. Examples might be: "Workers are not wearing their respirators during a particular work operation," or "We want to perform small-scale lead abatements but have no trained workers," or "We cannot now comply with the new OSHA regulation that goes into effect next month."

2. The required level of performance of the organization must be determined. Organizational policy, acceptable health and safety work practices, standard operating procedures for the work process, and regulatory reviews may form the basis of the performance level criteria.

3. Identify the gaps between the problems of the organization and the level of performance required by the organization. This gap is the referred to as the organizational need. If there is no gap, there is no reason for change. It is generally not a good idea to train (for instance) in areas that are unnecessary. If employees already have adequate knowledge on a subject it would be boring to repeat it to them. Worse, it may set up the attitude that they are wasting their time. This could result in their not learning other subject materials that they need in their jobs. If a gap exists, it should be clearly defined.

4. The causes of the gap should be determined. Communicating with everybody involved in the work process will reveal different viewpoints and opinions on the situation.

5. Determine the solutions to the organizational problem. The result of the needs analysis should be a clear presentation of a solution to address the organizational needs. Determine which gaps are best closed by instruction and training. While our main focus may be training, it is not the case that training is the key to all solutions. Employees may already possess sufficient knowledge to perform their tasks adequately, but may lack the organizational mandate. Certain tasks may need to be restricted to fewer employees.

NEEDS ANALYSIS

NAME John Peters	DEPARTMENT OF TASK Assembly Dept. and Maintenance Dept.
DATE 05-14-97	NUMBER OF EMPLOYEES 17, Assembly 4, Maintenance Mechanics
DESCRIPTION OF TASK Clearing jams on assembly line #4, cutter.	

ORGANIZATIONAL PROBLEM OR DEFICIENCY	REQUIRED LEVEL OF PERFORMANCE	GAP TO BE CORRECTED	CAUSES OF THE DEFICIENCY	SOLUTION TO CLOSE THE GAP
Assembly workers do not recognize early symptom of jam in line.	Assembly workers should recognize problems early so they can be fixed without a line shut-down.	Workers should be able to recognize jam symptoms before jam occurs.	Workers were never told that it is part of their job. Workers don't know what symptoms to look for.	Make recognition of jam symptoms part of assembly worker's job. Train on symptoms of future jam.
Mechanics do not know how or when to shut down line to prevent jams.	Mechanics should know how to institute an assembly line shut-down.	Mechanics should know how to shut down line when they determine it is required.	Assembly Department has not authorized the Maintenance Department to institute a line shut-down.	Institute chain of command to allow the Mechanics to shut down line.
No Lock-out/Tag-out procedure being followed when line is shut down for mechanics.	Follow OSHA required Lock-out/Tag-out requirements.	Mechanics need to follow Lock-out/Tag-out requirements.	There is a Lock-out/Tag-out procedure for the Assembly Department for the line but none exists within the Maintenance Department for the mechanics.	Develop a Lock-out/Tag-out SOP for Maintenance Mechanics. Provide training on the Lock-out/Tag-out SOP for Maintenance Mechanics.

Fig. 14.1

These would involve organizational decisions that may not include training. There may be several solutions for the same problems. A range of options with different solutions may be presented for review to close the same gap.

14.1.2 Needs Analysis Outcome Example

An organization has a problem with some employees not wearing respirators during an operation that requires their use. The cause of this and the response to it could include the following.

The employees may not know that they need respiratory protection, nor how to use it. A change in the standard operating procedure for the operation to include respirator use for that operation would be needed. The employees would not be allowed to perform the operation until the change in procedure and adequate performance could be ensured.

The employees may know they should be using respirators and are not because they do not know how to use them or how to get them. Instruction may constitute the entire solution to this problem. The employees would not be allowed to perform the operation until adequate performance could be ensured.

The employees may know the why and how-to of respirator use for the operation but feel that they are too uncomfortable. A change in the behavior of the employees would be needed, not a change in their level of understanding of respirator use. This may be addressed entirely as communications and supervisory issues before the operation could be performed.

14.2 SETTING INSTRUCTIONAL GOALS

The instructional goal is a clear understanding or statement of the result of the instruction for the learners. It should directly respond to the gap described in the needs assessment that can be most efficiently solved by instruction (see Fig. 14.2).

Establishing goals is not simply an exercise for the instructional designer. The goals must meet the needs of the persons with the problem defined in the needs assessment. There may be financial and/or time restrictions on both the instruction development and availability of the learners. There may be regulatory training requirements that need to be met to meet the organizational need. Setting the instructional goals will allow for an initial assessment of the time and resources that will be required to solve the organizational problem.

A scenario may have a manager who wants the maintenance staff to be able to perform small asbestos abatements. There may be additional in-house training in hazard communication and respirator use. Those personnel may need to take a 40-hour course required by their State. Annual refresher training may be required. The course may not be given in-house because of training certification issues. Employees may need to be away from their job for an uninterrupted week of training. None of these issues may have been known to the manager. This may cause a reassessment of the solution initially proposed in the needs assessment. It may become apparent that it is more efficient and cost effective to have outside contractors perform the work.

The most fundamental aspect of the instructional goal is the description of what the learners are expected to know or do as a result of the instruction. The identity of the learners, the context in which the learned information and skills will be used, and the resources and tools available to the learners should be described. All this information is

ORGANIZATION AND INSTRUCTIONAL GOALS

NAME	DEPARTMENT OF TASK
John Peters	Assembly Dept. and Maintenance Dept.
DATE	**NUMBER OF EMPLOYEES**
05-14-97	17, Assembly 4, Maintenance Mechanics
DESCRIPTION OF TASK	
Clearing jams on assembly line #4, cutter.	

GAP TO BE CLOSED	TRAINING ACTION NEEDED TO CLOSE GAP	TRAINING APPROVALS BY?	ORGANIZATIONAL PROCEDURE NEEDED TO CLOSE GAP	ORGANIZATIONAL APPROVALS AND ACTION TO BE TAKEN BY?
Early recognition of jams.	None	None	Institute policy of checking logs for jamming.	Assembly supervisor
Lock-out/ Tag-out	Review course in LO/TO, 4 hours.	1) Assembly supervisor 2) Maintenance supervisor	Reference safety procedures for production by sups.	1) Assembly supervisor 2) Maintenance supervisor
Day-to day operations of cutters	Review of safety procedures of cutters by sups.	Assembly sups.	Review procedures to ensure guards are functioning	1) Assembly supervisor 2) Maintenance supervisor

Fig. 14.2

needed to select the most appropriate instructional strategies for learning and applying the skills learned to workplace situations. The clear statement of the goals will be useful in the future to determine whether the training materials developed can be used for other instructional programs.

An instructional program developed for refresher training for respirator use could not be used as initial training. An initial training course for frequent respirator users would not be an efficient refresher course. Hazard communication training for chemicals used by laboratory workers could be different from the training given to the persons delivering those chemicals to the laboratory. While the main topics of instruction may be the same, the emphasis on certain topics would change to more directly and efficiently meet the needs of each group and the organization. An instructional program designed in 1979 to fulfill OSHA training requirements for working with asbestos would not be adequate in 1996 (Figs. 14.3 and 14.4).

OSHA GENERAL INDUSTRY TRAINING REQUIREMENTS

TOPIC GIVEN	REGULATION	CERTIFICATION?	REQUIRED ?	DATE
MEANS OF EGRESS Personal protective equipment Employee emergency plans and fire prevention plans	 1910.32(f)(1)(i)-(v),(2),(3)(i)-(iii)&(4) 1910.38(a)(5)(i)(ii) (a)-(c),&(iii)			
POWERED PLATFORMS, MANLIFTS, AND VEHICLE-MOUNTED WORK PLATFORMS Operations - Training Care and use Appendix C, Section 1	 1910.66(i),(ii),&(ii) (A)-(E),&(iii)-(v) 1910.66(e)(v)(9)			
OCCUPATIONAL HEALTH AND ENVIRONMENTAL CONTROLS Personal protection Inspection, maintenance, and installation Hearing protection Training program Ionizing radiation Testing	 1910.94(d)(9)(vi) 1910.94(d)(11)(v) 1910.94(d)(11)(v) 1910.95(I)(4) 1910.95(k)(1)-(3) (i)-(iii) 1910.96(I)(2) 1910.96(f)(3)(viii)			
HAZARDOUS MATERIALS Flammable and combustible liquids Explosives and blasting agents Bulk delivery and mixing vehicles Storage and handling of liquefied petroleum gases Process safety management of highly hazardous chemicals Contract employer responsibilities Mechanical integrity Hazardous waste operations and emergency response New technology programs	 1910.106(b)(5)(vi)(v)(2)&(3) 1910.109(d)(3)(i)& (iii) 1910.109(g)(3)(iii) (a) 1910.109(g)(6)(ii) 1910.109(h)(3)(d) (iii) 1910.110(b)(16) 1910.110(d)(12)(i) 1910.111(b)(13)(ii) 1910.119(g)(l)(i)-(ii) 1910.119(g)(2) 1910.119(g)(3) 1910.119(h)(3)(i)-(iv) 1910.119(j)(3) 1910.120(e)(1)(i)(ii)(2)(i)-(vi),(3)(i)-(iv)(4)-(9) 1910.120(o)(i) 1910.120(p)(7)(i)-(iii) 1910.120(p)(8)(iii) (A) 1910.120(q)(4) 1910.120(q)(5) 1910.120(q)(6)(i)(A)(F);(iii)(A)-(I);(v)(A)-(F) 1910.120(q)(7) 1910.120(q)(8)(i)(ii)			

Fig. 14.3

PERSONAL PROTECTIVE EQUIPMENT				
Respiratory protection	1910.134(a)(3) 1910.134(b)(3) 1910.134(e)(2)-(4), (5)(i)			
GENERAL ENVIRONMENTAL CONTROLS				
Temporary labor camps	1910.142(k)(1),(2)			
Specifications for accident prevention signs and tags	1910.145(c)(1)(ii), (2)(ii) &(3)			
Permit required confined spaces	1910.146(g)(1),(2) (i)-(iv) (3),(4) & (k)(1)(i)-(iv) 1910.151(a),(b)			
Medical services and first aid The control of hazardous energy (lockout/tagout)	1910.147(a)(3)(ii); (4)(i)(D);(7)(i)(A)-(C);(ii)(A)-(F); (iii)(A) (C)(iv);(8) 1910.147(e)(3)			
Lockout or tagout devices removed Outside personnel	1910(f)(2)(i)			
MEDICAL AND FIRST AID				
Medical services and first aid	1910.151(b)			
FIRE PROTECTION				
Fire protection	1910.155(c)(iv)(41)			
Fire brigades	1910.156(b)(1)			
Training and education	1910.156(c)(1)-(4)			
Portable fire extinguishers	1910.157(g)(1)-(4) 1910.158(e)(2)(vi)			
Fixed extinguishing systems	1910.160(b)(10)			
Fire detection systems	1910.164(c)(4)			
MATERIALS HANDLING AND STORAGE				
Employee alarm systems	1910.165(d)(5)			
Servicing of multi-piece and single-piece rim wheels	1910.177(c)(i)-(iii);(2)(i)-(viii),(3)* 1910.177(f)(1);(2) (i),(ii) &(3)-(11) 1910.177(g)(1)-(12)			
Powered industrial trucks	1910.178(I)			
Moving the load	1910.179(n)(3)(ix) 1910.179(o)(3)			
Crawler locomotive and truck cranes	1910.180(i)(5)(ii)			
MACHINERY AND MACHINE GUARDING				
Mechanical power presses	1910.217(e)(3) 1910.217(f)(2)			
Mechanical powers presses instructions to operators	1910.217(9)(e)(2)			
Training of maintenance personnel	1910.217(9)(e)(3)			
Operator training	1910.217(H)(13)(i) (A)-(E) &(ii)			
Forging machines	1910.218(a)(2)(iii)			

Fig. 14.3 *(Continued)*

An example would be a Hazard Communications training in the area of flammability labels for workers in a chemistry laboratory. The training for the persons delivering the chemicals to the laboratory would be quite different from the Ph.D. chemists. The delivery persons would need to be able to read and understand the warning labels. They would need to understand the potential consequences of, for instance, a spill. They would then have the knowledge to take whatever action was warranted. The chemist would need to know these things as well. In addition, they may need additional information. The chemist may create mixtures of flammables in the laboratory that he/she

WELDING, CUTTING AND BRAZING				
General requirements	1910.252(a)(2)(xiii)(c)			
	1910.253(a)(4)			
Oxygen-fuel gas, welding and cutting	1910.254(a)(3)			
Arc welding and cutting	1910.255(a)(3)			
Resistance welding				
SPECIAL INDUSTRIES				
Pulp, paper, and paperboard mills	1910.261(h)(3)(ii)			
Laundry machinery and operating rules	1910.264(d)(1)(v)			
Sawmills	1910.265(c)(30)(x)			
Pulpwood logging	1910.266(c)(5)(i)-(xi)			
	1910.266(c)(6)(i)-(xxi)			
	1910.266(c)(7)			
	1910.266(e)(2)(i)(ii)			
	1910.266(i)(1)(2)(i)-			
	(iv);(3)(i)-(vi);(4), (5)(i)-			
Logging	(iv);(6),(7)(i)- (iii);(8),(9)			
	1910.268(b)(2)(i)			
	1910.268(c)(1)-(3)			
	1910.268(j)(4)(iv) (D)			
Telecommunications	1910.268(l)(1)			
	1910.268(o)(1)(ii)			
Derrick trucks	1910.268(o)(3)			
Cable fault locating	1910.268(q)(1)(ii)			
Guarding manholes	(A)-(D)			
Joint power and telecommunication	1910.268(q)(2)(ii)			
manholes	1910.268(q)(2)(iii)			
Tree trimming - electrical hazards	1910.269(b)(1)(i)(ii);(d)(vi)(			
	A)-(C);(vii); (viii)(A)-(C);(ix)			
	1910.272(e)(1)(i)(ii)& (2)			
	1910.272(g)(5)			
Electric power generation, transmission, and distribution	1910.272(h)(2)			
Grain handling facilities				
Entry into bins, silos, and tanks				
Contractors				
ELECTRICAL SAFETY-RELATED WORK PRACTICES				
Content of working	1910.332(b)(1)			
COMMERCIAL DIVING OPERATIONS				
Qualifications of dive team	1910.410(a)(1);(2) (i)-			
	(iii);(3)(4)			
	1910.410(b)(1)			
	1910.410(c)(2)			

Fig. 14.3 *(Continued)*

would be required to label. The issues of exactly how the flammability rating is applied to a mixture and even which regulatory definition of flammability to use would need to be covered. The instructional goals of these two programs would be quite different, and one would not efficiently meet the needs of the other.

The following series of questions should be answered and clarified by the instructional goals:

TOXIC AND HAZARDOUS SUBSTANCES				
Asbestos	1910.1001(j)(50(i)-(iii)(A)-(H)			
4-Nitrobiphenyl	1910.1003(e)(5)(i), (ii)			
Alpha-Naphthylamine	1910.1004(e)(5)(i), (ii)			
Methyl Chloromethyl Ether	1910.1006(e)(5)(i), (ii)			
3,3'-Dichlorobenzidine(and its Salts)	1910.1007(e)(5)(i), (ii)			
Bis-Chloromethyl Ether	1910.1008(e)(5)(i), (ii)			
Beta-Napthylamine	1910.1009(e)(5)(i), (ii)			
Benzidine	1910.1010(e)(5)(i), (ii)			
4-Aminodiphenyl	1910.1011(e)(5)(i), (ii)			
Ethyleneimine	1910.1012(e)(5)(i), (ii)			
Beta-Propiolactone	1910.1013(e)(5)(i), (ii)			
2-Acetylaminofluorene	1910.1014(e)(5)(i), (ii)			
4-Dimethylaminoazobenzene	1910.1015(e)(5)(i), (ii)			
N-Nitrosodimethylamine	1910.1016(e)(5)(i), (ii)			
Vinyl Chloride	1910.1017(j)(1)(i), (ii)			
Inorganic Arsenic	1910.1018(o)(1),(i), (ii)(A)-(F),(2)(i)(ii)			
Lead	1910.1025(l)(1)(i)-(v)(A)-(G)(2)(i)-(iii)			
Cadmium	1910.1027(m)(4)(i)-(iii)(A)-(H),(m)(iv) (A)(B)			
Benzene	1910.1028(j)(3)(i)-(iii)(A)(B)			
Coke oven emissions	1910.1029(k)(1)(i)-(iv)(a)-(e),(2)(i)& (ii)(i)			
Bloodborne pathogens	1910.1030(g)(2)(i); (ii)(A)-(C);(iii)-(vii) (A)-(N);(viii),(ix) (A)-(C)			

Fig. 14.3 *(Continued)*

- What is the goal of the instruction?
- How does the instruction solve the gap outlined in the needs assessment?
- Who are the learners that the instruction is designed for?
- How will the learned skills be used?
- What resources will the learners have when they apply their learned skills?

For example, the needs analysis reports that chemical stockroom employees do not know how to react to chemical spills for the materials that they are delivering. The company has a standard operating procedure for spill response (Spill SOP) that uses the quantity

Cotton dust	1910.1043(i)(1)(i) (A)-(F),(2)(i)(ii)			
1,2-Dibromo-3-Chloropropane	1910.1044(n)(1)(i) (ii)(a)-(d),(2)(i)(ii)			
Acrylonitrile (Vinyl Cyanide)	1910.1045(o)(1)(i) (iii)(A)-(G),(2)(i)(ii)			
Ethylene Oxide	1910.1047(j)(3)(i); (ii)(A)-(D);(iii)(A)-(D)			
Formaldehyde	1910.1048(n)(1)-(3) (i)(ii),(A)(B)(iii)-(vii)			
4,4' Methylenedianiline	1910.1050(k)(3)(i) (ii)(A)(4)(i)(ii)			
Hazard communication	1910.1200(h)(2)(i)-(iv)			
Occupational exposure to hazardous chemicals in laboratories	1910.1450(f)(1)(2); (4)(i)(A)-(C)(ii)			

* Includes single piece wheels per Federal Registrar of February 3, 1984 (pp. 4338-4352) but not automobile or truck tires marked "L.T."

Fig. 14.3 (*Continued*)

of material and hazard classification labeling to determine the appropriate response. No materials are delivered without labels, with unknowns given the highest hazard ratings.

A learning goal might be as follows: "Each and every stockroom employee who handles chemicals will be able to choose and implement the proper spill response using the Spill SOP during the delivery of chemicals." This learning goal directly addresses the needs analysis. The learners are clearly identified. Each and every stockroom employee who handles chemicals" is fairly specific in terms of the group involved and to the level of involvement of each individual in the group. The learned behavior is specific: "will be able to choose and implement the proper spill response." The resources available to the learners is outlined; "using the Spill SOP." Finally, the context in which they will be using information learned is given; "during the delivery of chemicals."

The learning goal stated here is a general one. It addresses the needs analysis and will be our focus when we develop learning materials. During the development of those materials, specific narrowly focused learning objectives will be stated. These would include such topics as understanding the chemical labeling system and the estimation of spilled quantities. In another context one of these might have been our goal. In this case they are necessary learned steps that, by themselves, do not solve the organizational problem.

The following are questions that relate to the costs and acceptability of the development of a learning program (Dick, 1996).

- Is the proposed instruction the most efficient way to solve the problem outlined in the needs assessment? (There could be other options such as a change in job design, substitution of less hazardous materials, and other industrial hygiene techniques to reduce hazard. Even for training professionals, training to cope with a hazard should not be the first option.)
- Are the instructional goals acceptable to the management of the organization? (A review by management may uncover areas that should be included and reveal where training may be redundant with other training.)

OSHA CONSTRUCTION INDUSTRY TRAINING REQUIREMENTS				
TOPIC	REGULATION	CERTIFICATION?	REQUIRED ?	DATE GIVEN
GENERAL SAFETY AND HEALTH PROVISIONS				
General safety and health provisions	1926.20(b)(2),(4)			
Safety training and education	1926.21(a)			
	1926.21(b)(1)-(6) (i),(ii)			
OCCUPATIONAL HEALTH AND ENVIRONMENTAL CONTROLS				
Medical services and first aid	1926.50(c)			
Ionizing radiation	1926.53(b)			
Nonionizing radiation	1926.54(a),(b)			
Gases,vapors, fumes, dusts, and mists	1926.55(b)			
Asbestos, tremolite, anthophyllite, and actinolite	1926.58(k)(3)(i)-(iii)(A)(E),(4)(i),(ii)			
Hazard communication, construction	1926.59(h)(2)(i)-(iv)			
Lead in construction	1926.62(l)(1)(i)-(iv);(2)(i)-(viii) & (3)(i)-(ii)			
PERSONAL PROTECTIVE AND LIFE SAVING EQUIPMENT				
Hearing protection	1926.101(b)			
Respiratory Protection	1926.103(c)(1)			
FIRE PROTECTION AND PREVENTION				
Fire protection	1926.150(a)(5)			
	1926.150(c)(1)(viii)			
SIGNS, SIGNALS, AND BARRICADES				
Signaling	1926.201(a)(2)			
TOOLS-HAND AND POWER				
Power-operated hand tools	1926.302(e)(1),(12)			
Woodworking tools	1926.304(f)			
WELDING AND CUTTING				
Gas welding and cutting	1926.350(d)(1)-(6)			
Arc welding and cutting	1926.351(d)(1)-(5)			
Fire prevention	1926.352(e)			
Welding, cutting, and heating in way of preservative coatings	1926.354(a)			
ELECTRICAL				
Ground fault protection	1926.404(b)(iii)(B)			

Fig. 14.4

- Are the proposed costs and time commitments by the learners acceptable to the management of the organization? [Sometimes the required training time and cost commitments for an operation make the operation not worth performing for a particular group. The group performing the operation may be made smaller, the job may be reassigned to a group that has already received the training for another reason, or an outside trained group may be engaged (as in hazardous materials cleanup). These are management decisions that simply cannot be made by the training personnel.]
- How long will the materials to be developed be current for and how often will they need to be updated? (If changes in the work operation or in regulations are frequent, materials may become dated quickly.)
- Are there sufficient time, skill, and financial resources available to develop the instructional materials? (A program of instruction with associated professional

SCAFFOLDING Scaffolding Guarding of low-pitched roof perimeters during the performance of built-up roofing work Fall protection	1926.451(a)(3) 1926.451(b)(16) 1926.451(c)(4),(5) 1926.451(d)(9) 1926.451(g)(3) 1926.451(h)(6)-(14) 1926.451(k)(10) 1926.500(g)(6)(i)-(ii)(a)-(f)&(iii) 1926.503(a)(1),(2)(i)-(vii)			
CRANES,DERRICKS, HOISTS, ELEVATORS, AND CONVEYERS Cranes and derricks Material hoists, personnel hoists, and elevators	1926.550(a)(1),(5),(6) 1926.550(g)(4)(i)(A) 1926.550(g)(5)(iv) 1926.552(a)(1) 1926.552(b)(7) 1926.552(c)(15)&(17)(i)			
MOTOR VEHICLES, MECHANIZED EQUIPMENT, AND MARINE OPERATIONS Material handling equipment Site clearance	1926.602(c)(1)(vi) 1926.604(a)(1)			
EXCAVATIONS General protection requirements (excavations, trenching, & shoring)	1926.651(c)(1)(i) 1926.651(h)(2)&(3) 1926.651(I)(1) 1926.651(i)(2)(iii) 1926.651(i)(2)(iv) 1926.651(k)(1)&(2)			
CONCRETE AND MASONRY CONSTRUCTION Concrete and masonry construction	1926.701(a) 1926.703(b)(8)(i)			
STEEL ERECTION Bolting, riveting, fitting-up and plumbing up	1926.752(d)(4)			

Fig. 14.4 (*Continued*)

materials takes time to develop. A rule of thumb is that it takes 10 hours of preparation for 1 hour of instruction, provided that you are already familiar with the instructional topic.)

- Are the developed learning materials useful in other instructional programs? (The development of sets of modular materials can save time in an overall program. For example, a single module that is used in different training courses would need to be updated as regulations or company policy change for that topic, rather than each training course.)
- Are the instructional materials more efficiently developed in-house or outside of the organization? [If the training topic is fairly generic, it may be more cost effective to purchase and modify materials from other sources. Some topics, such as asbestos and hazardous materials operations, are regulated by federal agencies (OSHA, EPA, DOT, etc.), state agencies (with analogous agencies), and local agencies (health, building, fire departments, etc.). These regulatory changes, with associated

UNDERGROUND CONSTRUCTION,CAISSONS, COFFERDAMS, AND COMPRESSED AIR				
Underground construction	1926.800(d) 1926.800(g)(2) 1926.800(g)(5)(iii)-(v) 1926.800(j)(1)(i)(A) 1926.800(j)(1)(vi)(A)&(B) 1926.800(o)(3)(i)(A) 1926.800(o)(3)(iv)(B) 1926.800(t)(3)(xix)&(xx)			
Compressed air	1926.803(a)(1)(2) 1926.803(b)(1),(10)(xii) 1926.803(e)(1) 1926.803(f)(2),(3) 1926.803(h)(1)			
DEMOLITION				
Preparatory operations	1926.850(a)			
Chutes	1926.852(c)			
Mechanical demolition	1926.859(g)			
BLASTING AND USE OF EXPLOSIVES				
General provisions (blasting and use of explosives)	1926.900(a) 1926.900(k)(3)(i) 1926.900(q)			
Blaster qualifications	1926.901(c),(d),(e)			
Surface transportation of explosives	1926.902(b),(i)			
Firing the blast	1926.909(a)			
POWER TRANSMISSION AND DISTRIBUTION				
General requirements (power transmission & distribution)	1926.950(d)(1)(ii)(a)-(c),(vi),(vii) 1926.950(d)(2)(ii) 1926.950(e)(1)(i)(ii)&(2)			
Overhead lines	1926.955(b)(3)(i) 1926.955(b)(8)&(d)(i) 1926.(e)(1),(4)			
Underground lines	1926.956(b)(1)			
Construction in energized substations	1926.957(a)(1) 1926.957(d)(1) 1926.957(e)(1)			
Ladders	1926.1060(a)(1)(i)-(v) &(b)			

Fig. 14.4 *(Continued)*

changes in company policy, require that training materials be updated frequently. Generic training courses can be efficiently changed with the purchase of materials. Health and safety training courses that are extensively tailored to integrate with other job skills training will require a more time-consuming review process.]

- Is the instruction more efficiently given in-house or outside of the organization? (For a small number of employees with a long training program, it may be more cost effective to have outside training. Some training programs may need state licensing or certification (asbestos and lead), a costly and time-consuming annual process.)

- What are the technical and instructional resources available to develop materials and perform instruction? (Sufficient time must be allocated for the development of materials. A training site must be available, along with presentation equipment. For highly regulated operations, instructional designers need access to time from company personnel who track those regulations.)

14.3 TASK ANALYSIS

After the decision has been made to include instruction as a solution to closing the organizational gap, the examination of the work process is begun. The task analysis is a detailed, step-by-step, analysis of the task (with the organizational gap) to be learned (see Fig. 14.5). The task is broken down for the ultimate purpose of designing instruction. At this stage each step of the task should be recorded. It will be evaluated for instructional design at a later time. A chronological description of the task is most often recorded. This can be accomplished by observing and recording the elements of the task or by having the personnel who perform the task record its elements. It is important to have learners, their supervisors, etc. review the task description for completeness. There may be scenarios that were not recorded at the time of the task analysis or critical steps omitted. Regulations may also require training on steps that are not yet incorporated into the task.

Determining the level of detail of each step of the task is not an exact process. Eventually, particularly steps will be associated with a unit of instruction. Depending upon the skill level of the group of learners, the steps that need instruction will change. One approach is to have each step be associated with a unit of instruction based upon a student with the lowest skill level expected. Another approach is to perform a skills assessment of the targeted audience.

In the chemical spill example you might have the following steps after the spill occurs:

1. Read hazard label.
2. Determine spill quantity.
3. Use Spill SOP to determine response.

A more detailed analysis of step (1) might turn it into:

1. Locate label on package;
2. Determine type of label system used;
3. Read health hazard rating;
4. Read flammability rating;
5. Read reactivity rating;
6. Read specific hazard rating.

This more detailed analysis has broken the task into segments that will allow us to analyze them individually and plan learning objectives for each.

TASK ANALYSIS

NAME John Peters	DEPARTMENT OF TASK Assembly Dept. and Maintenance Dept.
DATE 05-14-97	NUMBER OF EMPLOYEES 17, Assembly 4, Maintenance Mechanics
DESCRIPTION OF TASK Clearing jams on assembly line #4, cutter.	

DESCRIPTION OF ACTION	DESCRIBE EMPLOYEES' SKILLS AND PRIOR TRAINING	IS TRAINING REQUIRED ?	PREREQUISITE TO TRAINING ?
Recognize problem.	All current employees have at least 4 years on line. Assembly workers have been through "Production Training".	Yes, for assembly workers on recognizing potential jams.	None
Shut down line.	Assembly workers trained in shut-down procedures.	Yes, for mechanics.	None
Lock-out/Tag-out.	Assembly workers have had Lock-out/Tag-out training.	Yes, initial for mechanics and refresher for all.	None for initial training. Initial training for refresher training.
Start up line.	All trained in start up procedure.	No	None
Day-today operations.	All current employees have at least 4 years on line. Assembly workers have been through "Production Training".	Yes, review of "Production Training"	Initial "Production Training" program.

Fig. 14.5

14.4 LEARNING OBJECTIVES

The learning objective describes the skills that a learner will be able to perform when he/she completes the instruction in a particular area. The learning objective is always written from the perspective of what the learner should learn. It is not written from the perspective of what the instructor will teach or what the organization needs. There are three major components to a learning objective (see Fig. 14.6).

The first component describes the behavior or skill that the learner will perform. These behaviors are derived from the individual steps or groups of steps in the task analysis. Action verbs such as demonstrate, perform, describe, or identify should be used. Passive verbs such as learn, know or understand should not be used. The avoidance of passive verbs will force the instructional designer to clearly state the skill to be learned.

The second component describes the conditions in which the learner will demonstrate the skills learned. It will describe the stimuli for performing the task, a definite scope of the task, and resources available to the learner when performing the task.

The third component will be the objective criteria to be used to assess learner performance. This may involve the score on a written exam or the performance of the learned task. The criteria for acceptable performance is always given (i.e., better than 70% grade on an exam, estimation of quantities within 10%, perform a task with no more than one error).

A learning objective may look like: "The learner should describe the responses in the Spill SOP, without assistance, using the Spill SOP, based upon the flammability rating system used on a package given to them from among the three systems in use at the company, with no errors."

The first component that describes the behavior or skill that the learner performs: "The learner should describe the responses in the Spill SOP."

The second component that describes the conditions in which the learner will demonstrate the skills learned: "without assistance." The stimuli for performing the task is "on a package given to them." The scope of the task is given as: "from among the three systems in use at the company." The resources available to the learner when performing the task are: "using the Spill SOP."

The third component of the objective criteria is given: "with no errors."

As with the tasks in the task analysis, the learning objectives can be rather broad, including several subskills. They can also be narrowly focused. How the material will be presented can influence the scope of the learning objectives.

The learning objectives serve many functions in the learning and instructional process, and include:

- The instructional designer uses them to decide upon and prepare instructional strategy;
- The instructor uses them to determine what is to be learned from the materials;
- The learners use them to see what is expected to be learned (often a shortened, more focused, version of the learning objective is presented to the learners);
- The test designer uses them to construct examinations.

Following the form of the described learning objective forces the instructional designer to focus on the goals and performance objectives of the learners. Again, the focus is on what the learners should be learning.

LEARNING OBJECTIVES

NAME	DEPARTMENT OF TASK
John Peters	Assembly Dept. and Maintenance Dept.
DATE	**NUMBER OF EMPLOYEES**
05-14-97	17, Assembly 4, Maintenance Mechanics

DESCRIPTION OF TASK

Clearing jams on assembly line #4, cutter.

TASK	LEARNING OBJECTIVE
Early recognition of jamming problem.	All assembly workers, using maintenance logs, should be able to institute the "Cutter Jamming Policy", before it causes a stop in a production run, at least 80% of the time.
Lock-out/Tag-out.	All assembly workers and maintenance mechanics, using their respective Lock-out/Tag-out procedure for reference, should be able to demonstrate that procedure, with 100% compliance.
Day-to-day operation of cutters.	All assembly workers, from memory, should be able to recognize unsafe conditions from guards on cutters, 100% of the time.
Day-to-day operation of cutters.	All mechanics, following the maintenance procedure for cutters, should be able to recognize and correct unsafe guarding on cutters, whenever the cutters are worked on, 100% of the time.

Fig. 14.6

14.5 THE LEARNING AUDIENCE

The learners in a particular instruction must have a minimum skill level to successfully learn the materials presented. As part of the instructional design of the learning program, the prerequisite skill level needs to be determined. These would be compared to the skill levels of the proposed learners. Either the instruction may be modified to meet the skill level of the learners, or all the learners brought to the same skill level.

The motivation and attitudes of the participants will be an important factor in their learning. Determining the participant's attitudes and modifying the training to change those attitudes may be necessary. The participants must understand the reasons that they are expected to learn materials and new skills. It is difficult, at best, to successfully train participants when they are not motivated to learn.

The more that you know of the group's preferences, the better you can design the learning program. Some groups prefer a dense presentation of the fact, with few examples. Other groups prefer a less dense presentation style. Factors such as prior learning experience with the subject matter, educational level, and language ability can influence the instructional design. Care must be taken not to let the learning preferences of the instructional designer determine the presentation style for a particular group.

14.6 PRESENTATION

14.6.1 Presentation Styles

The four most commonly used presentation styles are lecture, group discussion, hands-on exercises, and self-learning. Each has a number of variations and combinations that are used in presentations.

The lecture format can present a great deal of information in a short period of time. Presentations by lecture should be relatively short. Whenever possible, presentation materials should be given to participants to allow easier note taking. Lectures work well for knowledge-based materials. Too long a lecture can overwhelm participants with information. Lectures can be used along with group discussion for longer training sessions. College graduates are more used to and accepting of this type of format. It is an area where the instructional designers (who in the Health and Safety field is more likely to be a college graduate) may inadvertently introduce their own bias and use this format too often.

Group discussion is a more time-consuming, but more engaging, process than lecture. It is effectively used when there is a base of knowledge to which new materials are being added. It can also be useful in revealing potential problems that a new piece of equipment or new procedure would have. The group may be directed by the presenter, keeping focused on the topic. It may be directed by the group, to come up with their own solutions to the problems under discussion. Anything within the range of these two extremes is possible. The important aspect to the group discussion is that it remains a discussion. It can work well in problem-solving situations, where the group can put its new knowledge together with their experience.

The hands-on approach is usually the most time consuming of the presentation formats, but can facilitate better learning in a style that employees relate to their jobs. To perform a hands-on exercise there must be adequate equipment and classroom space to perform the exercise. It is well suited to tasks that involve new psychomotor skills. An example would be in using a respirator. The classroom is the time and place for employees to familiarize themselves with using new protective equipment, not in a hazardous work situation.

Self-learning involves a participant learning materials with no instructor present. Some written materials are easily understood by some groups and will present problems for others. Newer interactive computer/video systems can present materials in an interesting way that is determined by the pace of the learner. Some systems allow for feedback into the system to ensure that the participant has understood the materials. The computer has an infinite amount of patience with the learner. The flexibility of scheduling learning is an advantage to the system. As the costs of equipment and training materials decrease, more of these systems will be put into use.

14.6.2 Presentation of Materials

The presentation of materials will depend on such factors as the learning audience, the content of materials to be presented, the time allotted for presentation, presentation equipment and environment, and the skills of the presenter.

The learners, to a large extent, will determine the style of presentation. Highly motivated groups will generally pay more attention to detail and allow for a more compact presentation. A lecture-style format can be effectively used in this situation.

The content of materials to be presented is an important factor in determining presentation styles. If the materials to be learned involve a new psychomotor skill, hands-on learning is generally the best style of learning. Lecture format is good for short, knowledge-based materials.

Some presentation methods simply take more time and equipment than others. A lecture presentation to learn a new computer program may take one-quarter of the time that it would compared to having each participant use a computer and run the program themselves. There would also be the need for more class equipment. The same is true for learning how to use a respirator or protective clothing. The time allotted for presentation is not always determined by the instructional designer. Many times a limited amount of time is allocated for training, and the materials must be fit into that time frame.

The presentation equipment and environment will have an effect on the presentation of materials. To perform a hands-on exercise there must be adequate equipment and classroom space to perform the exercise. A class demonstration may take the place of a hands-on exercise when there is a lack of equipment, time, or space. The use of overhead materials, slides, videos, and chalk or white boards will depend on the classroom configuration and the availability and those aids. The well-prepared presenter will have a backup method of presentation if their original method is not functioning (broken slide projector, etc.).

The skill level of the presenter with different presentation styles should be taken into account. While this may mean using overheads rather than slides, it could also mean using a different instructor. An instructor who does not perform well using the style that may be required for a particular class should not be instructing that segment.

14.6.3 Presentation Materials

Prepared presentation formats such as slides, overheads, videos, and computer presentations allow for the rapid and structured presentation of materials. The use of these aids will focus the learners on the important point of the instruction. They also serve as a focus for the trainer for covering all the instructional points in a specific presentation order. This is especially helpful for the trainer if the materials are presented infrequently.

Overhead transparencies offer a system of presentation that is easy to produce, easy to modify, and allows modification during presentation. The use of computers and low-cost color printers allow for the rapid and low-cost production of professional-quality overhead transparencies. Photographic-quality production of images is available for this format.

Thirty-five millimeter slides offer their greatest advantage in training in their ability to provide a relatively low-cost way to present images. Computers can be used to produce low-cost, professional-quality materials. The computer output can be sent for processing into slides. While this is a more rigid presentation format than overheads and the room needs to be darker, the advantage of presenting a large number of images can be an effective tool with some instruction.

The use of computers with overhead LCD projection panels and LCD projectors allows for a versatile system of presentation. Both still and moving images can be presented in the same presentation. Equipment costs are relatively high, but these can be offset by the low production costs of the individual presentations. In-class flexibility in presentation is between overhead transparencies and slides.

Videos can offer an effective presentation tool in some circumstances. Its advantages are consistency of presentation, less need for a highly skilled presenter (although not less of a need for somebody to answer questions about materials), and relatively low cost. A presentation can be given to many small groups to meet their schedules in a way that may not be possible with a "live" presenter. Relatively short videos with relevant moving images offer the most effective instruction. High production costs take in-house video production out of the hands of most training departments. The problems include the difficulty of inexpensively updating materials and having generic presentations fit the needs of the organization.

There are no hard and fast rules concerning the form of the presented materials. Selection of the presentation form will depend on the participants, the materials, and the instructor. The materials themselves should be presented in a way that is easily understood by the participants. A common mistake is to put too much material on overheads and slides. One of the many "rules of thumb" for overheads, slides, and videos is the "rule of five." There should be no more than five lines of information on the screen, there should be no more than five words in each line of information, and the distance from the screen to any participant should be no more than five times the size of the screen.

It is helpful to participants if presented materials are given to them for the presentation. Presentation materials can be integrated into other class materials. Often the presentation materials are reduced in size with several to a page. Participants can follow the presentation without excessive note-taking.

14.7 EVALUATION OF INSTRUCTION

The best evaluation of health and safety training is not demonstrated in the learning environment, but in the workplace. Performance audits can be undertaken to assess whether learned behaviors are being implemented. If they are not, it should be determined whether the employees have not retained what they had learned or whether there is an attitudinal or behavioral problem in the workplace.

The evaluation of the learning materials and experience can be performed in other ways. Examinations are often used to test the understanding of the materials.

Written exams are often given as the most objective method. The exam should be focused on the important aspects of the learning. Exam question should be derived from the learning objectives and should be a test of the learned behaviors. Questions should not be a test of language ability, but of materials learned. Use the same terminology that was used in the instructional process. If you say "take off the respirator" in class, don't put "doff the respirator" on the exam. Don't use convoluted logic. Double negatives in questions and answers confuse people. Keep exams up to date. Changes in regulations can change whether an answer is correct or not. Exam questions should be reviewed by others. Questions written from one perspective may have another meaning from a different perspective, causing more than one answer to be correct.

Hands-on or practical exams are best suited for hands-on skills. The best test of how to clean a respirator is, in part, to have the person clean a respirator. This puts the exam in the same context as how the learned material will be used. This type of exam is generally considered to be less objective than a written exam.

If an exam is used as part of the instructional process, the criteria for successfully passing the exam should be included in the learning objective.

14.8 FEEDBACK ON THE INSTRUCTION PROCESS

The feedback on the instruction will come from several sources. These would include the learners, the instructors, and management personnel. Each class should have some type of formal course evaluation process.

The learners would focus on how well the materials were presented. Feedback generally pertains to the relevance of the instruction, technical content of materials, quality of instruction, quality of instructional aids, and the learning environment. Evaluation is generally put into rather broad categories such as "poor," "adequate," "good," and "excellent." A general comment section should be included. While the learners are not always privy to the constraint put on the training, their input can be useful to management. For instance, if there were many comments that the time needed for instruction should be longer (or shorter), a case could be made for changing the length of instruction. Participants generally are not asked to identify themselves in their evaluation.

Instructors should evaluate their own performance for each class. In addition to the categories of comments that the learners evaluate, factors such as the adequacy of presentation time, effectiveness of presentation style, and class participation should be noted. Instructors should always evaluate whether they accomplished the organizational goal of the instruction.

The management team should evaluate the effectiveness of the instruction by the results in the workplace. Did the training meet their expectations in terms of the stated organizational goals?

REFERENCES AND READINGS

Dick, W., and Cary, L. (1996). The Systematic Design of Instruction, pp. 18–31.

Heinich, R., Molenda, M., and Russell, J. (1993). *Instructional Media and the New Technologies of Instruction.*

Mager, R. F. (1988). *Making Instruction Work.*

Richey, R. (1992). *Designing Instruction for the Adult Learner.*

Rossett, A. (1987). *Performance and Instruction* **31**(10), pp. 6–10.

CHAPTER 15

Philosophy and Management of Engineering Control

PAMELA GREENLEY and WILLIAM A. BURGESS

Massachusetts Institute of Technology, 77 Massachusetts Ave., Cambridge, MA 02139

15.1 INTRODUCTION
15.2 GENERAL GUIDELINES
15.3 CONTROL STRATEGY
  15.3.1 Toxic Materials
  15.3.2 Replace with Alternate Material
  15.3.3 Dust Control
  15.3.4 Impurities in Production Chemicals
15.4 EQUIPMENT AND PROCESSES
  15.4.1 Diagnostic Air Sampling
  15.4.2 Equipment
  15.4.3 Processes
  15.4.4 Work Task Modification, Automation, and Robotics
  15.4.5 Facility Layout
  15.4.6 Ventilation
  15.4.7 Acceptance and Startup Testing
  15.4.8 Periodic Testing and Maintenance
15.5 SUMMARY
REFERENCES
APPENDIX A NIOSH CONTROL TECHNOLOGY REPORTS

15.1 INTRODUCTION

This chapter will cover the engineering methods widely used to minimize adverse effects on health, well-being, comfort, and performance at the workplace and in the community. Major attention will be given to controls for chemical airborne contaminants, although

Adapted from "Philosophy and Management of Engineering Control" by William A. Burgess, Chapter 5 in Harris, Cralley, and Cralley, *Patty's Industrial Hygiene and Toxicology, 3rd ed., Vol. IIIA*. New York: John Wiley and Sons, Inc., 1994., pp. 129–180.

Handbook of Occupational Safety and Health, Second Edition, Edited by Louis J. DiBerardinis, ISBN 0-471-16017-2

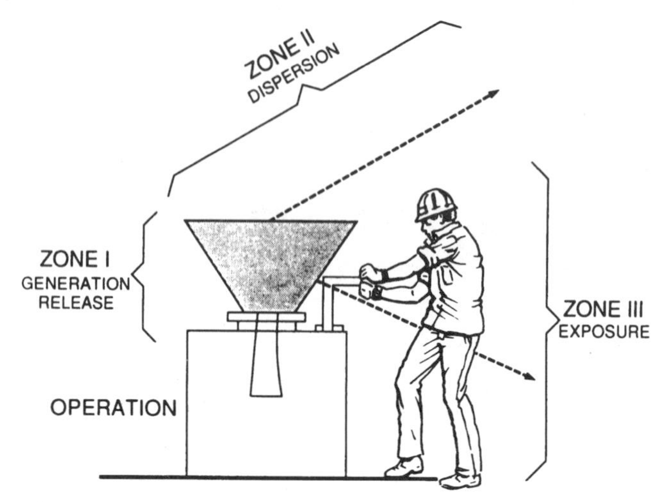

Fig. 15.1 Contaminant generation, release, and exposure zones in the workplace.

many of the control principles are also applicable to biological and physical hazards. Brief case studies will illustrate the applications of the techniques. The target for control may be a hand tool, a piece of equipment or device, an integrated manufacturing process such as an electroplating line, or a complete manufacturing facility in a dedicated building such as a foundry.

The critical zones of contaminant generation, dispersion, and exposure are shown in Fig. 15.1. Ideally the goal is to design each element of the process to eliminate contaminant generation. If it is impossible to achieve this goal and a contaminant is generated, the second defense is to prevent its dispersal in the workplace. Finally, if we fail in that defense and the material released from the operation results in worker exposure, the backup control is collection of the air containing the contaminant by exhaust ventilation. When engineering controls are not feasible or as a supplement to them, personal protective equipment such as respiratory protection and clothing are used. See Chapters 16 and 17 for guidance on their selections and use. The contaminant is frequently removed from the exhaust airstream by air cleaning before returning the air to the workplace or the general environment. Until recently, little emphasis was given to the prevention of contaminant generation and release and ventilation was relied on as a principal remedy. Since the late 1980s a cooperative effort between industry and Federal and state regulatory agencies to prevent pollution has emerged. Although the main thrust of the pollution prevention approach is to reduce environmental releases, there are frequently benefits to the occupational environment as well. In this chapter we will confirm the importance of ventilation, but emphasis will be placed on the primary controls associated with process and material design to minimize generation and release of the contaminant.

The Occupational Safety and Health Administration (OSHA) considers engineering controls as any modification of plant, equipment, processes, or materials to reduce employees' exposure to toxic materials and harmful physical agents. The zones of generation and control methods for each are listed in Fig. 15.2. Each of these approaches should be considered when designing exposure controls. These approaches provide a useful outline for discussion and act as checklists to ensure a comprehensive review.

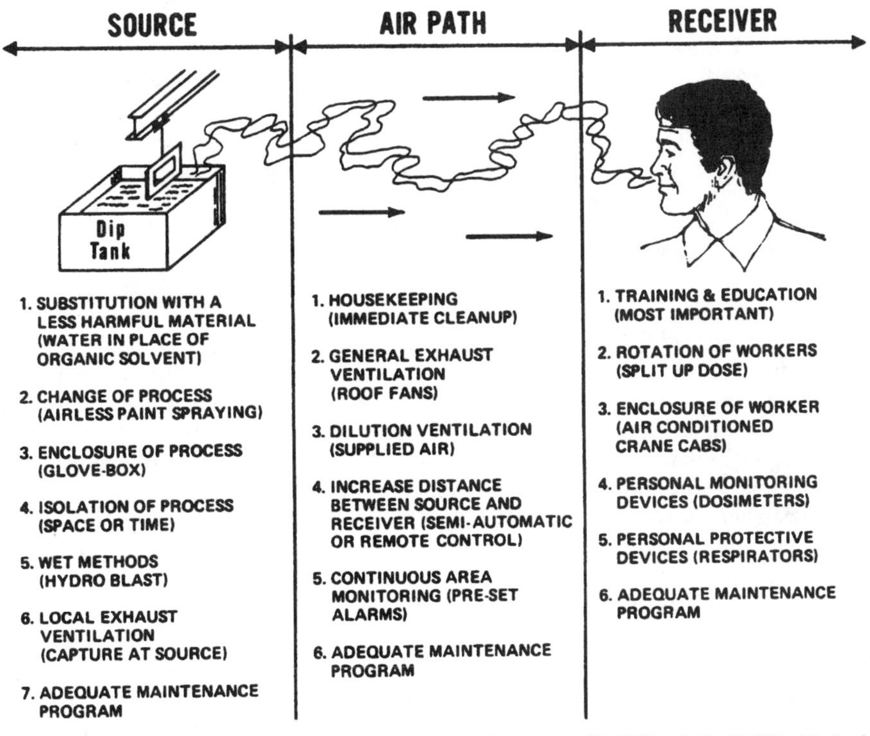

Fig. 15.2 Generalized diagram of methods of control. *Source:* Olishifski, J. B. (1988). *Methods of Control, Fundamentals of Industrial Hygiene.*

A comprehensive algorithm for control of air contaminants has been designed by Sherwood and Alsbury (1983). Brandt (1947) states: "Measures for preventing the inhalation of excessively contaminated air have been discussed by many authors, and there are as many classifications of these methods as there are papers on the subject. The principles expounded, however, are always essentially the same." Brandt also reminds the reader that: "The control of an atmospheric health hazard is rarely accomplished by a single measure. It usually involves a combination of measures."

In this chapter engineering controls will include all techniques except personal protective devices and administrative controls, those changes made in the work schedule to reduce the time-weighted average exposure of the worker.

15.2 GENERAL GUIDELINES

The strategy to be used in solving a control technology problem varies depending on the setting. As described above, the control problem may involve a specific piece of equipment, a process, or an integrated manufacturing plant. In a common situation faced by many in the 1990s, a shop repairing electric motors must consider a change in solvents in a vapor-phase degreaser. The number of persons participating in the review is small, the decision will be made in a short time, and it will probably rely heavily on vendor information. The range of options presented in this chapter probably will not be thoroughly reviewed. An industrial hygienist will not be involved unless the service shop is a support facility operated by a major company or consultation is available from an insurance carrier, consulting industrial hygienist, or state OSHA consulting program.

A far more complex situation is the scaleup of a chemical process to manufacture a photographic dye. The research to develop the new molecule and synthesize a small quantity for initial tests is completed by a research chemist at a laboratory bench. At this stage little is known about the chemical or its toxicity. However, since gram quantities are handled in a well-controlled laboratory hood, the risk is negligible. The next step involves a scaleup of the operation in a pilot plant facility designed to manufacture kilogram quantities of the material for testing purposes. This testing will provide limited data on the physical properties of the new chemical, its product potential, and toxicity data. If the results are encouraging, production scaleup is considered. A premanufacturing notification must be submitted to the Environmental Protection Agency (EPA) prior to production. This submission, which requires a description of air contaminant control technology and an estimation of worker exposure, involves a comprehensive review of the operation as noted in Table 15.1. It is decided to manufacture the chemical at an existing plant using conventional chemical processing equipment. If the industrial hygienist has been involved in the pilot plant operation, he or she is well placed to participate in a review of the control technology necessary for the full-scale plant. As noted in Table 15.1 the range of talent available to define the engineering controls is extensive and the time frame is extended. Hopefully, step-by-step review to identify the appropriate engineering controls will minimize worker exposure. If exposure problems are not anticipated, controls must be retrofitted at significant cost, whereas early integration of controls are most easily accepted and economically implemented.

The most complex engineering control package is encountered in the design and construction of a major production facility. The project is complex due to the range of issues that are encountered; however, the design process is started with a clean design sheet, and state-of-the-art controls can be included in the design. In the example in Table 15.1 a large number of specialists are included in the review team. If the design is done in-house, the industrial hygienist should be a key participant. If the design work is contracted to a design and construction firm, plant personnel are one step removed from the design process. The company will have a project engineer to interface between the outside team and company engineering group. In this case clear communication must be established by the industrial hygienist to ensure involvement in the design review.

For the health and safety professional facing his or her first process hazard review, the job may seem awesome. An initial step is to identify your responsibility. Does it include merely health issues at the workplace, or are environmental and fire protection problems to be included in your review? Once your responsibility has been defined, the issues of concern for this particular plant should be identified. Mature industries such as metallurgical, glass, and ceramics have been studied, and extensive information is available on standard operations and sources of contamination. The EPA has published emission rates for many processes that are useful in process hazard reviews. Another source of information on control technology for mature industries is the series of over 50 technical reports published by the National Institute for Occupational Safety and Health (NIOSH) (see Appendix 15.1).

If the company has similar operations elsewhere, obtain material and process data, industrial hygiene survey data, and air sampling information from that plant to assist in the

TABLE 15.1 Specialists Participating in Risk Assessment of New Chemical Manufacture

	Administrative	Legal	Sales	Marketing	Chemists	Engineers	Toxicologists	Occupational hygienists	Environmental scientists	Statisticians	Risk analyst	Transportation systems analysts	Economists	Information specialists
Part I: General information														
A. Manufacturer identification	×	×												
B. Chemical identity					×									×
C. Marketing data			×	×						×			×	
D. *Federal Register* notice	×	×		×	×	×	×	×	×					
E. Schematic flow diagram					×	×		×	×					
Part II: Risk assessment data														
A. Test data		×			×		×		×	×	×			
B. Exposure from manufacture														
1. Worker exposure					×	×	×	×		×	×			
2. Environmental release					×	×			×	×	×			
3. Disposal					×	×	×	×	×		×			
4. By-products, etc.					×	×					×			×
5. Transportation		×		×		×					×	×		
C. Exposure from operations														
1. Worker exposure					×	×	×	×		×	×			
2. Environmental release					×	×			×	×	×			
3. Disposal					×	×	×	×	×		×			
D. Exposure from consumer use	×	×	×	×		×			×		×			
Part III: Risk analysis and optional data														
A. Risk analysis and optional data	×	×			×		×		×		×			
B. Structure–activity relationships					×		×				×			
C. Industrial hygiene	×	×			×	×	×	×			×			
D. Engineered safeguards					×	×	×	×	×		×			
E. Industrial process and use restriction data					×	×								
F. Process chemistry					×	×								
G. Nonrisk factors: Economic					×	×								
and noneconomic benefits	×	×	×	×	×	×					×		×	

Source: Arthur D. Little, Inc.

review. A step-by-step review of the operations is then undertaken. This health and safety review should be as rigorous as the review of percent yield on the process and should produce a detailed process flow diagram showing the locations of contaminant loss in addition to the major physical stresses such as noise, heat, and ergonomics. It is important to also consider the ancillary systems including compressed air, refrigeration, cooling towers, and water treatment. Bulk handling of chemicals and granular minerals frequently presents a major industrial hygiene problem and should be given special attention. Finally, a detailed description of the workers' tasks including required emergency response actions are required, since work location and movement have a major effect on contaminant control. A comprehensive checklist for review of environmental and occupational health and safety issues in planning major plant construction is given by Whitehead (1987).

An element that is frequently overlooked in the design of engineering controls is the worker. A review of a specific operation conducted by several workers frequently will identify subtle differences in the way the task is performed. A worker inspired by a wish to reduce lifting, provide "cool down" in the heat, or reduce dust exposure may modify the tasks to accomplish this end. Workers frequently have insight into details of the operation that the design engineer will overlook. The video, real-time monitoring approach discussed in Section 15.4.1 is useful in defining these modifications in tasks.

It is difficult to recommend engineering controls for emerging technologies. Hopefully each industrial group will develop and publish its own engineering controls. Such is the case with advanced composite manufacturing, a technology that has seen expanding applications in recent years in the manufacture of aircraft. An association of companies involved in this work has published a volume describing the major hazards and the controls that should be applied (SACMA, 1991).

Frequently the best control cannot be achieved directly on the design board. In the 1970s a number of cases of asbestosis in shipyard workers fabricating asbestos insulating pads for steam propulsion plants prompted the plant to install an integrated workstation incorporating a number of control features. Initial designs of the downdraft work table were fabricated and evaluated by workers resulting in a number of recommendations that were incorporated in the final design improving its acceptance and performance (Fig. 15.3).

15.3 CONTROL STRATEGY

The general strategy as noted earlier is to eliminate the generation source. If that is not possible, then prevent escape of the contaminant, and if it does escape, collect and remove the contaminant before the worker is exposed. This strategy will be reviewed by considering action on toxic materials, equipment, processes, job tasks, plant layout, and exhaust ventilation.

15.3.1 Toxic Materials

The use of toxic materials range in complexity from simple wet degreasing to the synthesis of new chemicals. See Chapter 3 for a review of the hazards associated with various industries. In degreasing the exposure is associated with the use of the cleaning agent or handling its waste. In synthesis of a new chemical, exposure may occur to raw materials, intermediates, byproducts, the new chemical and waste streams. In minimizing potential worker exposure to these materials, it is necessary to consider the following options.

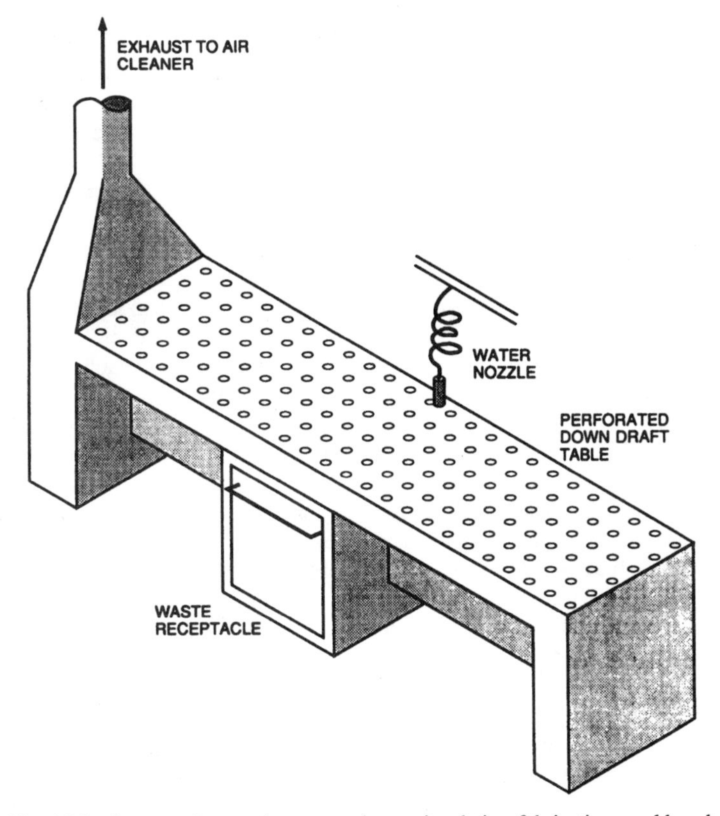

Fig. 15.3 Integrated controls on an asbestos insulation fabrication workbench.

15.3.1.1 Eliminate the Toxic Material This option refers to elimination without replacement. Frequently toxic chemicals are used in a process to improve yield or in a product to improve function or appearance. In review, it may be possible to eliminate the chemical. The pressure to do so may come from potential health effects, the cost of workplace and environmental controls to introduce the toxic chemical into the plant, or it may arise from pressure from the marketplace. An example of the latter resulted from legislation in Switzerland requiring precautionary labeling of batteries containing mercury. The conventional Leclanche dry cell battery utilizes electrodes of manganese dioxide and zinc with an electrolyte of ammonium chloride. A characteristic of the cell is zinc corrosion with subsequent power loss. A mercury compound is added to inhibit this corrosion and resulting performance degradation. At least one manufacturer has determined the performance enhancement from mercury is not necessary, and the mercury compound has been removed from the battery formulation (Ahearn, 1991). This change has eliminated exposure to mercury for several hundred workers in this company. The Montreal Protocol calling for the elimination of ozone depleting compounds has resulted in the use of water and detergents as cleaning agents in many areas. Other possible substitutes for ozone depleting compounds are discussed in Section 15.3.2.4.

15.3.2 Replace with Alternate Material

This control approach holds the greatest potential for significant reduction in worker exposure to toxic chemicals. Historically, the approach has had success in the United States. One of the most effective steps was the cooperative action taken by the U.S. Public Health Service, the State of Connecticut Department of Public Health, industry, and labor unions in 1941 to prevent mercurialism by replacing mercuric nitrate with nonmercury compounds in the carroting process in felt hat manufacture.

EPA's ability to ban the use of toxic materials, in this case asbestos, was severely curtailed in the case *Corrosion Proof Fittings v. EPA*. The court ruled that EPA must fully evaluate the relative safety of known alternatives to asbestos and choose the least burdensome regulatory measure for each application (Rossi, 1994). Another environmental concern is the exposure of children to lead from lead-based paint and potable water supplies containing lead–tin–copper sweat joints. In 1986 the Amendments to the Safe Drinking Water Act effectively banned the use of lead-based solder. The acceptable alternative solders are alloys of tin in the range of 90–95 percent, and the balance is copper, silver, zinc, or antimony (Kireta, 1988). The introduction of the alternative solder will minimize exposure of manufacturing, construction, and maintenance personnel to lead. The banning of mercury in heat manufacture demonstrates an action taken specifically for occupational health reasons; in the second and third examples the action was prompted by environmental concerns but will have a positive impact on the health of all workers.

A total of 10 materials have been banned for specific operations in the United Kingdom (HSE, 1988). An important element of the United Kingdom regulation is the prohibition of silica sand for abrasive blasting. Although there have been attempts to restrict the use of sand for this purpose in the United States, as the most frequently used blasting material it continues to present a major risk to workers.

Greater emphasis will be given to replacement control technology as we enter the next century. As discussed below, the impetus will continue to result not from the workplace health issue but from the environmental health and ecological impact. Health and safety characteristics including toxicity, smog contribution, ozone depletion, and fire potential must be evaluated. In addition to the review of the performance of the replacement material, the changes in process and product design to satisfactorily change over to the new material must be considered in the final decision.

15.3.2.1 National Toxic Use Reduction Programs Until the 1980s the EPA emphasis was on pollution control and dealt principally with "end-of-pipe" abatement technology. In 1984 the Hazardous and Solid Waste Amendments to the Resource Conservation and Recovery Act of 1976 redirected environmental quality efforts from the conventional emphasis on waste treatment and disposal to waste minimization. This approach to waste management is characterized by source reduction, recycling, and reuse. In 1986 the Office of Technology Assessment published a report that described the concept of prevention versus control (OTA, 1986). Since these environmental initiatives have great impact on the replacement of toxic chemicals in industry and, therefore, worker exposure, it is important that the health and safety professional understand the environmental management nomenclature (Table 15.2). The Pollution Prevention Act of 1990 consolidated this approach by specifying the following waste management hierarchy as national policy: first source reduction as the most desirable approach, then recycling and reuse, then treatment, and finally disposal as the least desirable approach (U.S. Congress, 1990).

To initiate this strategy, in 1989 the EPA established the Waste Reduction Innova-

TABLE 15.2 Environmental Control Approaches

Form of control	Emphasis	End rule
Pollution control	"End-of-pipe" control	Does not eliminate pollution, but merely transfers contaminant from one medium to another. Incurs an environmental risk from transporting toxic chemicals
Pollution prevention	Reduction in the production of contaminants	Direct waste reduction, convers all pollutants
Toxic use reduction	Reduction in pollutants generated in manufacturing by changes in plant operation procedures, production processes, materials, or end products Focus on a target list of chemicals	Priority setting permits control of high potential risk chemicals

tive Technology Evaluation (WRITE) Program. This program funded programs in six states and one county over a 3-year period to (1) provide engineering and economically feasible solutions to industry on specific pollution prevention problems, (2) provide performance and cost information on these techniques, and (3) promote early introduction of these pollution prevention programs into commerce and industry. A review of completed and ongoing WRITE programs that demonstrates the importance of replacement technology is shown in Table 15.3 (Harten, 1991). Although initiated for environmental quality reasons, each of these changes will have a major impact on worker health.

TABLE 15.3 Examples of Material Replacement from EPA WRITE Program

Operation	Original material	Replacement
Anodizing aluminum parts	Chromic acid anodizing, hexavalent chromium exposure	Sulfuric acid anodizing eliminates exposure to chromium
Removal of paint on defective parts	Methylene chloride-based paint stripper	Abrasive blasting with tic beads, no vapor exposure
Flexograph printing	Application of alcohol-based ink labels	Water-based inks eliminate vapor exposure
Cleaning and deburring small metal parts	Vapor degreasing and alkaline tumbling	Automatic aqueous rotation washer. Eliminates vapor and caustic mist exposure
Cleaning flexographic plates	Solvent cleaning with Stoddard solvent, acetone, toluene, and alcohol	Cleaning with terpenes aqueous cleaners

Source: Adapted from Harten and Licis (1991).

In 1991, the EPA's Office of Pollution Prevention initiated the Design for the Environment Program. This is a cooperative effort between EPA and industry where EPA uses its technical resources to help an industry segment eliminate the use of a toxic material. Many industries have recognized that to produce a product that can compete effectively on the global market, the design effort must go beyond minimizing the use of toxic materials. Areas also included in the design for the environment approach include minimizing the use of natural resources and producing a finished product that is recycleable. On a conceptual level, design for the environment evolved out of the field of Industrial Ecology.

Industrial Ecology, introduced to the general public in 1989 by Robert Frosch and Nicholas Gallopoulos (Frosch, 1989), promotes the concept that the traditional linear manufacturing model of raw materials in, produce and waste materials out, be modified to mimic the biological ecosystems. Natural resource consumptions and waste generations is minimized, wastes that are generated are used as raw materials for other processes. Even the final product, when its useful life has ended, is reused. An example of this is BMW's roadster produced in 1988, which can be quickly disassembled and the coded body plastic panels easily recycled.

Examples of the EPA–industry cooperative efforts include the printed wiring board and screen printing industries. Pollution prevention efforts with the printed wiring board focused on six areas: reducing dragout between the process baths, reducing chemical use, reducing copper buildup on plating racks, reducing chemical losses from evaporation by using polypropylene balls, using chemical substitutes that produce less sludge, and recovering metals from the wastewater treatment process for recycling. Pollutions prevention efforts must be examined closely to ensure both environmental releases and occupational exposures are controlled. The use of polypropylene balls to reduce evaporation achieves both goals. Recycling metals from wastewater, if handling of a dry filter cake is required, may not.

Plant-specific case studies are available on the EPA's Design for the Environment World Wide Web Page (http://es.epa.gov/index.html). The case studies typically involve smaller plants using low-capital-investment solutions to produce a quick rate of return on investment.

Denmark, one of the many European countries embarking on a national toxic use reduction program, base their work on the simple step model shown in Table 15.4 (Goldschmidt, 1991). The simplicity of this approach suggests that replacement technology can be done on an ad hoc basis; however, Goldschmidt warns that chemical replacement for occupational health reasons is a complex task and requires a systematic review of

TABLE 15.4 Step Model for the Substitution Process

Step	
1	Problem identification
2	Identification and development of a range of alternatives
3	Identification of consequences of the alternatives
4	Comparison of the alternatives
5	Decision
6	Implementation
7	Evaluation of the result

Source: Goldschmidt (1991).

both the existing conditions and those prevailing after the replacement action. A review of 162 individual replacement actions by the Danish Occupational Health Services confirms that replacement is successful if done in a rational manner following the step model shown in Table 15.4. Among the impressive examples of this program is the replacement with water-based systems of the solvent-based systems in all indoor paints and most outdoor paints.

The principal advances in worker protection in the next decade will probably result from replacement technology accomplished as a result of such worldwide environmental health initiatives.

15.3.2.2 State Toxic Use Reduction Programs The recent emphasis on pollution prevention and specifically on toxic use reduction in manufacturing is reflected in recent legislative action in 17 states and by the Federal government (Rossi, 1991). The Massachusetts Toxic Use Reduction Act of 1989 (Massachusetts, 1989) is described as a seminal program in that it does focus on toxic use reduction and excludes off-site recycling, off-site nonproduction unit recycling, transfer, or treatment. Rossi identifies the five toxic use reduction techniques acceptable under the Massachusetts act as:

- In-process recycling or reuse. Recycling, reuse, or extended use of toxics by using equipment or methods that become an integral part of the production unit of concern, including, but not limited to, filtration and other closed-loop methods.
- Improved operations and maintenance. Modification or addition to existing equipment or methods including, but not limited to, such techniques as improved housekeeping practices, system adjustments, product and process inspections, or production unit control equipment or methods.
- Changes in the production process. Includes both the modernization of production equipment by replacement or upgrade based on the same technology or the introduction of new production equipment of an entirely new design.
- Input design. Replacement of a chemical used in production with a chemical of a lower toxicity.
- Product reformulation. Redesign of the product to produce a product that is nontoxic or less toxic on use, release, or disposal.

Rossi proposes two divergent routes the manufacturing group may take. The route for ease of implementation includes, first, recycling and improved operations and maintenance, then production process changes, and finally replacement of toxic chemicals and product reformulation. However, for effective toxic use reduction the first step is to eliminate or replace the toxic chemical in the process or reformulate the product, then initiate production process changes, and finally recycle and apply operations and maintenance control techniques.

An institute funded by industry assessment has been established at the University of Massachusetts at Lowell with the function to provide a teaching and research facility to encourage aggressive action in toxic use reduction. Ellenbecker (1996) reviews the experiences of the Toxic Use Reduction Institute to date by presenting three case studies involving process changes. The first case study involves investigating the use of multiprocess wet cleaning to replace the dry cleaning process. Perchloroethylene, the primary solvent used by the dry cleaning industry, creates both occupational health and environmental release problems for the many small dry cleaning establishments. Wet cleaning includes a variety of techniques including steaming, immersions and gentle hand washing in soapy water, spot cleaning, and tumble drying. The more labor-intensive wet cleaning was found to be cost competitive with the capital-intensive dry cleaning.

In the second case study, blanket washes containing small percentages of volatile organic compounds (VOC) were found to remove ink as effectively as high-VOC products. Blanket washing is an open type of process where even with local exhaust ventilation, printers are exposed to high concentrations of the wash. In the final case study, the replacement of chlorinated solvents (methylene chloride, 1,1,1-trichloroethane, and trichloroethylene) used in vapor degreasers to clean oils from metal parts with an immersion tank using an aqueous cleaner cleaner was evaluated. The operating costs of the immersion tank were one-sixth the costs of the vapor-phase degreaser. Unfortunately the aqueous cleaners recommended by the immersion tank manufacturer did not work as well in cleaning all the metal parts. Research was conducted in conjunction with TURI's Surface Cleaning Laboratory to find the most effective aqueous cleaner.

In all cases where water was incorporated into the process, occupational chemical exposure potential would be reduced. In the cases where water becomes contaminated, costs of treatment must be considered in the economic evaluation of the process.

15.3.2.3 Industry Programs A number of company initiatives have been described that respond to specific legislation or broad environmental concerns. One company program has multiple goals to reduce the use of toxic chemicals, minimize toxic emissions to all environmental media, encourage recycling of chemicals, and reduce all waste (Ahearn, 1991). In the design of this program the company followed many of the recommendations of the Office of Technology Assessment report on the reduction of hazardous waste (OTA, 1986). In this company program specific goals for toxic use and waste reduction are set for all production managers, and their annual performance reviews include this issue.

Each chemical used in the company is assigned to one of five Environmental Risk Categories (ERC). The materials in ERC I pose the greatest risk to the environment and ERC II, ERC III, and ERC IV represent progressively less risk. A fifth category includes plastic, steel, paper and other waste identified as rubbish, rubble, and trash. A critical issue in the design of the program was the decision logic for assigning categories. A number of options were considered including ranking chemicals based on workplace exposure guidelines such as threshold limit values (TLVs) or permissible exposure limits (PELs), label signal words (ANSI, 1988), regulatory lists published by OSHA and EPA, and the mathematical weighting of a range of risk factors including toxicity and chemical and physical properties (Karger and Burgess, 1992).

The company chose to assign the ERCs based on a "wise person" approach based on a battery of risk factors reflecting the materials' total occupational and environmental impact from release both as an isolated incident such as a spill or the chronic release of small amounts of the chemical to air, water, and soil. The majority of risk factors are toxicity based and chosen to identify significant adverse health and environmental impact. Included in the group of risk factors are acute and chronic toxicity based on animal studies or structural analysis, human case history and epidemiological studies, carcinogenicity, mutagenicity, teratogenicity, and reproductivity effects. A limited number of physical and chemical properties associated with the "releasability" of the chemicals were also included in a second group of risk factors. The third group of risk factors reflected broad environmental impacts. The specific criteria for assignment to the five categories are shown in Table 15.5.

The company has a major chemical production facility involved in the manufacture

TABLE 15.5 Categories for a Corporate Toxic Use Reduction Program

Categories I and II: Use to be reduced
- Human or animal carcinogens, teratogens, or reproductive agents
- Highly toxic chemicals
- Chemicals with human chronic toxicity
- Chemicals for which there are adverse environmental impacts

Category III: Transportation off site to be reduced via source reduction or recovery and reuse on site
- Suspected animal carcinogens
- Moderately toxic materials
- Chemicals that cause severe irritation of eyes and respiratory tract at low concentrations
- Chemicals with limited chronic toxicity
- Corrosive chemicals
- Chemicals for which there are environmental considerations

Category IV: Disposal volume to be reduced via waste reduction or reuse following on-site recycling
- All other chemicals not included in Category V

Category V: Any material such as paper, metal parts, and so on that is identified as rubbish, rubble or trash

Source: Karger and Burgess (1992).

of photographic dyes, which require frequent changes in processes. In the design of new processes and products the evaluation of toxic use and waste minimization technology is given equal importance to considerations of cost and performance of the new process or product. The engineering group developing the new processes rigorously reviews all materials and process alternatives to ensure that the target reductions are met. In this company the "end of pipe" abatement strategy is given secondary attention relative to optimization of process and product design for toxic use reduction.

An important part of the program is the goal of a 10 percent reduction in the use of toxic materials per unit of production each year for the first 5 years following the general strategy of elimination of use, replacement with a material of lower risk, reduction in quantity used, and a hierarchical program of waste handling. The goal is to eliminate, replace, or reduce the use of category I and II materials and assure the maximum reuse of categories III and IV materials.

In describing the success of the program the authors (Ahearn, 1991) present an example of a process in which a change resulted in elimination of a carcinogenic material. The original process required an aqueous oxidation step using a hexavalent chromium compound; the new process is based on a catalyzed air oxidation step, which does not require the chromium compound (Fig. 15.4). As a result of this change, worker exposure to hexavalent chromium was eliminated, and this process change saved the company over $1 million. It is the conclusion of this company that the process that uses the least quantity of toxic chemicals is usually the most economical.

As industries seek new ways to compete on the global market, many are implementing international management consensus standards. The international Standards Organization recently published ISO 14,000, the Environmental Management Systems Standard (International Standards Organization, 1996). To become registered as an ISO 14,000 company, a corporation is visited by an outside auditor, who, through an interview process, determines whether the company has the required management system in place to make the appropriate decisions regarding their impact on the environment. An

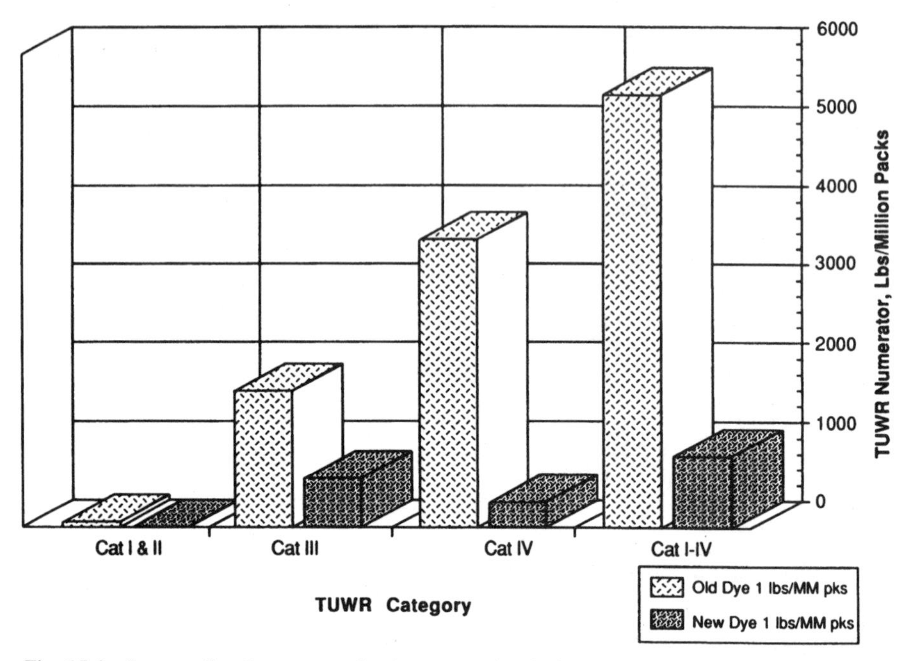

Fig. 15.4 Impact of replacement technology on a chemical process resulting in the elimination of hexavalent chromium from the workplace and its impact on toxic use and waste reduction categories (TUWR). Courtesy of Polaroid Corp.

analogous health and safety standard is being considered by ANSI International Advisory Council/Occupational Health and Safety Advisory Group and is expected within the next few years. Implementation and evaluation of engineering controls will certainly be part of this standard.

15.3.2.4 Outcome of Replacement Technology Although the replacement of toxic chemicals may provide impressive gains, the replacement of a target material must be done with great caution. If the replacement technique is to be used successfully, it must be carefully reviewed to define its impact on worker health and environmental impact, product yield, and product quality. It is important that the impact be evaluated in both pilot plant operation and field trials before making the change.

Industrial hygienists practicing from the 1950s to the 1990s devoted more attention to solvent replacement for metal cleaning than any other control problem. It is useful to track the history of such cleaning agents during those years since it shows the complexity of the replacement technology approach and indicates pitfalls that may arise in using this technique. Wolf (1991) traced the major decisions on general degreasing solvents during this period as follows:

In the 1960s and 1970s in particular, chlorinated solvents were widely substituted for flammable solvents because they better protected workers from flammability. As smog regulations were promulgated in the late 1960s and early 1970s, users moved from the photochemically reactive flammable solvents and trichloroethylene to the other "exempt" chlorinated solvents. In the 1980s there was increased scrutiny of the chlori-

TABLE 15.6 Characteristics of Generic Solvent Categories

Generic solvent category	ODP[a]	Photochemical reactivity	GWP[b]	Flash point[c]	Tested for chronic toxicity
Flammable solvents	—	Yes	—	F	
Isopropyl alcohol					Rule issue
Mineral spirits					No
Combustible solvent[d]	—	Yes	—	C	
Terpenes					Limited[e]
DBE					No
NMP					Rule issue
Alkyl acetates					No
Chlorinated solvents					
TCE	—	Yes	—	—	Yes
PERC	—	Yes[f]	—	—	Yes
METH	—	No	—	—	Yes
TCA	0.1	No	0.02	—	Yes
Chlorofluorocarbons (CFCs)		No		—	
CFC-11	1.0		1.0	—	Yes
CFC-113	0.8		1.4	—	Yes
Hydrochlorofluorocarbons (HCFCs)		No			
HCFC-123	0.02		0.02	—	In testing
HCFC-141b	0.1–0.18		0.09	—	In testing
HCFC-225	NA		NA	—	In testing
Hydrofluorocarbons (HFCs) and fluorocarbons (FCs)					
Pentafluoropropanol	—	NA[g]	NA	NA	No

Source: Wolf et al. (1991). Reprinted by permission of *Journal of the Air & Waste Management Association.*

[a]OPD: The oxygen depletion potential is the potential for ozone depletion of one kilogram of a chemical relative to the potential of one kilogram of CFC-11, which has a defined ozone depletion potential of 1.0.

[b]GWP: The global warming potential of a chemical is the potential of one kilogram of the chemical cause global warming relative to the potential of one kilogram of CFC-11 to cause global warming. CFC-111 a GWP of 1.0.

[c]F refers to flammable; C refers to combustible.

[d]DBE is dibasic esters; NMP is *N*-methyl-2-pyrrolidone.

[e]One of the terpenes, *d*-limonene, has been tested.

[f]Although PERC is not photochemically reactive, it is not exempt under the Clean Air Act.

[g]NA means not available.

nated solvents. Trichloroethylene, perchloroethylene, and methylene chloride were considered undesirable because of their suspect carcinogenicity, and trichloroethylene and freon 113 (trichlorotrifluoroethane) were being examined for their contribution to stratospheric ozone depletion.

Each change starting in the 1950s, progressing from flammable solvents to chlorinated solvents to nonphotochemical reactive materials to chlorofluorocarbon (CFC) solvents, introduced a new set of problems. The six major classes of solvents in use in 1992 or that can be considered for application as general solvents and the critical characteristics that determine their acceptability are shown in Table 15.6 (Wolf, 1991). The alternatives to the common chlorinated solvents are not encouraging. The flammable solvents require rigorous in-plant fire protection controls. Both the flammable and combustible solvents including the terpenes are smog-producing and therefore are subject to tight federal and local air district regulations. The chlorinated solvents will continue to see pressure for replacement due to the array of human toxicity and environmental effects. The increased use of CFCs in the 1970s and 1980s to the present level of 180 thousand metric tons per year is based on performance and reduced chronic toxicity; however, the impact of these chemicals on ozone depletion has resulted in the ban on manufacturing of CFC-113 and 1,1,1-trichloroethane (TCA) worldwide in 1996. The hydrochlorofluorocarbons are considered interim candidates for industrial degreasing but are considered for banning in 2020 and 2040. The hydrofluorocarbons and the other candidate solvents not containing chlorine do not present an ozone depletion risk but may present unacceptable toxicity concerns.

A vigorous research effort is being pursued by dozens of companies to develop alternatives to the cleaning agents listed in Table 15.6. Presented in Table 15.7 are the advantages and disadvantages of three new cleaning techniques compared to CFC-113 and 1,1,1-trichloroethane from the perspective of a company performing precision cleaning for the military.

A flow diagram of an aqueous and semiaqueous cleaning process is shown in Fig. 15.5. When aqueous cleaners are not sufficient, a hydrocarbon such as a terpene is used to provide cleaning. This is followed by an emulsion rinse stage to limit the amount of hydrocarbon that mixes with the water rinse. These are the systems where regula-

TABLE 15.7 Advantages and Disadvantages of Cleaning Methods Compared to CFC-113 and 1,1,1-trichloroethane

Cleaning method	Advantages	Disadvantages
Aqueous cleaning (water with an acid, alkaline, or neutral pH soap)	good for bulk processing limited health hazards water component of cleaner recycled	specialized equipment required cleaning cycle time increased over CFCs rinse cycle required drying time required
Semiaqueous and terpenes	excellent solvents that clean a wide range of materials low VOC emissions low toxicity	compatability issues terpenes have low flash point residues odors some terpenes contain *d*-limonene, an irritant
Supercritical CO_2 cleaning	good for bulk processing removes trapped fluids from parts fluid and contaminants can be recovered from SCF/CO_2 separator CO_2 inexpensive	high-pressure operation high pressure can damage delicate parts high up-front capital expenditures for the extractor

Source: Adapted from a presentation by Dr. Bill Agopovich, Draper Labs, Semiconductor Safety Association Meeting, October 28, 1993, Cambridge, MA.

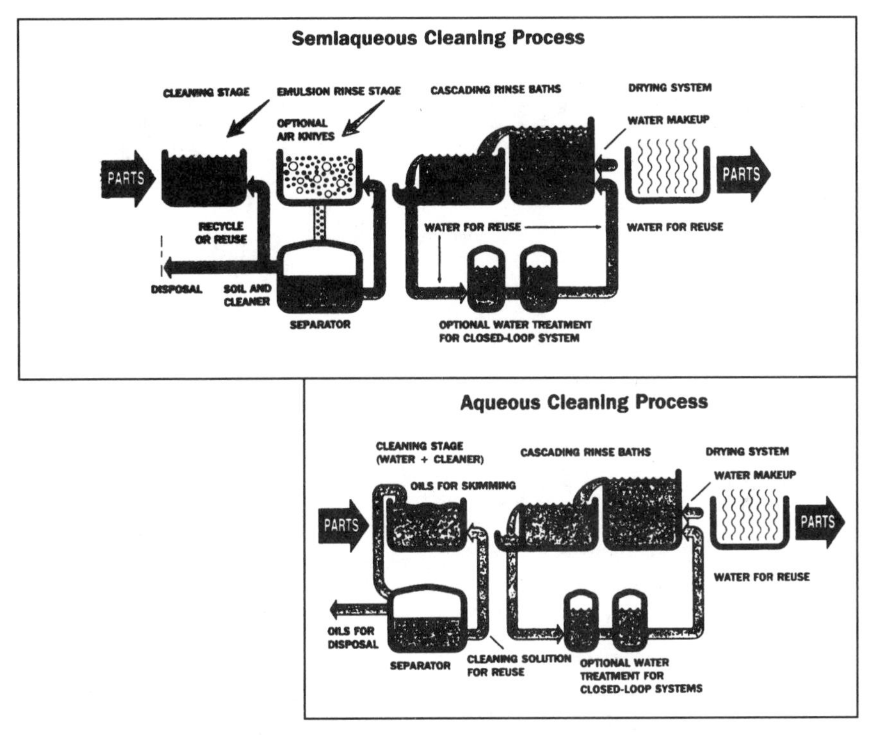

Fig. 15.5 Semiaqueous and aqueous cleaning process. Sprow, E., 1993, "How to be Solvent Free in 1993," *Manufacturing Engineering*, **110** (2 February).

tory pressure in the environmental area may decrease occupational health and safety. Flammability of the hydrocarbons and limited toxicology data are the two main areas of concern for these compounds. It is doubtful that a single agent will be found that will have the broad application of the chlorinated solvents; however, progress is being made in non-ozone-depleting cleaning techniques.

Burgess (1996) identified powder coating as an example of a new technology that significantly improves worker health and safety by reducing solvent exposure. Powder coatings are composed of a finely pulverized powder of thermoplastic or thermosetting resins with little or no solvent. Powder coatings produce a durable, high-quality finish and can be applied to a variety of materials. The powder guns have a high transfer efficiency of fifty percent. As with any replacement technology, new process hazards, in this case inhalation exposure to the powder, must be understood and controlled. The advantages and limitations of powder coatings and other alternatives to solvent-based coatings are presented in Table 15.8.

Employees affected by process changes may not always agree with industrial hygienists and process engineers that changes have improved their work environment. An interesting example of this occurred in the semiconductor industry in the substituting of a less hazardous, although more odorous, photoresist solvent. Many photoresists used in the 1980s contained ethylene glycol monomethyl ether acetate (EGMEA). OSHA reduced the permissable exposure limit for EGMEA from 50 to 5 ppm based on reproductive

TABLE 15.8 Summary of Alternatives to Solvent-Based Coatings

Alternative	Applications	Toxics use reduction benefits	Operational advantages	Operational disadvantages	Cost	Product quality	Limitations
Water-based coatings	Metal, wood, plastics, concrete, paper, leather	Reduced VOC emissions; reduced fire and explosion hazards; reduced hazardous waste; solvent not required for cleanup	Most formulations can be applied with conventional nonelectrostatic spray equipment and techniques; overspray easily recovered and reused; equipment may be cleaned with water; decreased drying time with drying oven; low odor levels	Require careful temperature and humidity control; require careful surface preparation; may require longer drying time; corrosion inhibitor may be needed on metal substrate; bacterial sensitivity reduces shelf life; may become unstable if frozen; emulsion coatings susceptible to foaming	Higher costs per gallon; special equipment and techniques needed for electrostatic application; may require special pumps and piping; may require drying oven	Reduced gloss; may cause grain raising in wood; some resins may cause water spotting; impact resistance may be reduced; some forms may have reduced corrosion resistance	Reductions in VOCs may be offset by use of solvents in surface preparation; additives to control water spotting may present worker safety hazards
High-solids coatings	Metal, wood, plastics	Reduced VOC emissions; reduced fire and explosion hazards; reduced hazardous waste	Can increase paint transfer efficiency; lower viscosity coatings compatible with conventional equipment	Narrow "time–temperature–cure" window; require careful surface preparation; generally require high cure temperatures; generally shorter pot life; may require worker retraining	Lower-viscosity coatings applied with conventional equipment; higher-viscosity coatings may require special equipment; reduced paint waste and supply needs; reduced energy use	Similar to that with solvent based coatings	Reductions in VOCs may be offset by use of solvents in surface preparation; solvents still needed for cleanup

Powder coatings	Mostly metals, but also wood, plastics, glass, and ceramics	VOC emissions and exposure eliminated or significantly reduced in application; no solvent required for cleanup; reduced fire hazard; reduced hazardous waste	High transfer efficiency; minimal solid waste; no dripping or running during application; thick coatings can be applied in one operation; overspray easily retrieved and recycled; no overspray with fluidized bed application; no mixing or stirring; requires little operator expertise	Color changes and matches can be difficult; potential for explosion must be minimized; some difficulty in applying thin coatings; requires handling of heated parts	Higher equipment and materials costs offset by savings in labor, maintenance, energy, waste, and pollution control	Durable, high-quality finish with good corrosion resistance	May present skin contact or dust inhalation hazards; good ventilation and protective equipment required; potential for explosion must be minimized; resins may still produce low VOC emissions
Radiation-cured coatings	Plastics, wood, paper, metal	VOC emissions and fire and explosion hazards eliminated or greatly reduced; reduced hazardous waste	Rapid curing; high transfer efficiency; low heat requirement for drying, useful on heat-sensitive substrates; consistent performance; low maintenance; unreacted overspray can be collected for reuse	Requires new equipment and operating procedures; curing of pigmented coatings may be difficult; may be difficult to strip	High capital investment costs—considerably higher for EB systems; lower energy requirement; lower materials use; less waste	Similar to that with solvent-based coatings	Solvent still needed for cleanup; acrylate materials in most coatings present worker safety concerns and require protective equipment
Supercritical fluid spray	Metal, plastics, wood	Reduced VOC emissions; reduced fire	Easily retrofitted into existing facilities; higher	High-pressure gas and operating temperature	Replacement of fluid handling equipment;	Thicker coatings may be	Require care in working with high pressure

TABLE 15.8 (*Continued*)

Alternative	Applications	Toxics use reduction benefits	Operational advantages	Operational disadvantages	Cost	Product quality	Limitations
Application		hazard; reduced hazardous waste	viscosity allows thicker coatings without runs and sags	requires care in operation; lower fluid delivery rates than airless or spray guns	potentially reduced operating costs	applied without runs and sags	and high temperature; still in testing phase for some industries
Surface-coating-free Materials	Metals and plastic; other substrate materials under development	Elimination of VOC emissions, fire and explosion hazards, and hazardous material use	Stripping and repainting not required throughout service life; elimination of coating operation		Initial increased cost may be offset by reduced operating costs	Surface finish appearance limited	Substrate may contain other materials of concern

Source: The Massachusetts Toxic Use Reduction Institute, Fact Sheet 5, Alternatives to Solvent-Based Coatings, March, 1994, #5, University of Massachusetts, Lowell, MA.

health effects. Photoresist manufacturers began marketing a photoresist product containing propylene glycol monomethyl ether acetate (PGMEA), which had been tested and found not to produce the reproductive health effects. Digital Equipment Corporation was eager to try the new product since the original studies investigating concerns over increased spontaneous abortion rates were conducted at their semiconductor fabrication plants. Process engineers and industrial hygienists were surprised to see OSHA-reportable illnesses increase by 30 percent when the PGMEA-based photoresist was introduced into production (Nill, 1996). The increased headache and related illnesses may have resulted from the lower odor threshold of PGMEA compared to EGMEA. The employees' perception was that health risks were increased by the substitution rather than decreased (Hallock, 1993). Engineering controls satisfactory for transfer of EGMEA were not satisfactory for PGMEA. Canister filling had to be conducted outside the fab area, and process delivery lines had to be upgraded. In this case, if production employees had been involved in the acceptance testing of the new photoresist, the odor issue may have come up and illnesses avoided due to early implementation of engineering controls.

15.3.3 Dust Control

A number of techniques have been proposed to minimize worker exposure to pneumoconiosis-producing and toxic dusts based on changes in physical form or state of the material. The common techniques to reduce dustiness in this manner are discussed below.

15.3.3.1 Moisture Content of Material The relationship between moisture content of granular materials and worker exposure to dust is well known. The use of water as a dust suppressant has been of great interest to the mining, mineral processing, and foundry industries. In the 1960s research by the British Cast Iron Research Association demonstrated that silica in air concentrations in foundries were maintained below existing exposure standards if the moisture content of foundry sand was kept above 30 percent (BCIRA, 1977). A sand handling technique that involves direct blending of new moist sand directly from the mixer with shakeout sand to add moisture to cool the shakeout sand and reduce dustiness is described by Schumacher (1978). The author also demonstrated a correlation between results of a simple laboratory dustiness test and inplant dust exposure. Goodfellow and Smith (1989) identify the importance of a dustiness index: "Dustiness testing can be a useful testing tool in establishing the type and efficiency of dust control required for different materials and materials handling systems. Further field data and verification are required to develop this procedure into a useful tool for practitioners."

The Bureau of Mines and the National Industrial Sand Association have investigated various moisture application techniques for dust control in mineral processing plants (Volkwein, 1989). The various methods investigated included the use of foams, steam, and water sprays. The type of moisture added and the percentage in dust reduction is presented in Table 15.9. Foam was found to be about 20 percent more effective than water alone or water with a surfactant. Steam experiments showed dust reductions of 65 percent. While both foam and steam are generally more effective, they are more costly to produce than water sprays. The authors also recommended processing partially wet product rather than complete drying after crushing, which is now the common practice.

The British Occupational Hygiene Society (BOHS, 1985) has also explored various

TABLE 15.9 Type of Moisture Added and Resulting Dust Reductions at One of the Study Sites

Test condition	Volume of liquid (ml)	Dust reduction (%)
Foam	1,420	91[a]
	1,300	73[a]
	764	68
Water	757	46
	1,324	58
Water with 1.5% surfactant	1,324	54
Water with 2.5% surfactant	1,324	54

[a] Average reduction.

Source: Volkwein, (1989).

dustiness estimation methods that may be useful in design of control systems. In chemical processing a solid product is frequently isolated in cake form that is wet with solvent or water. Frequently this material must be dried to ensure product quality and shelf life and to permit packing and shipping. However, if the product is to be used "in-house," one should determine if the drying operation can be omitted permitting direct handling of the wet cake, thereby saving money and reducing dustiness.

15.3.3.2 Particle Size In general, the more extensive the grinding or comminution of a granular material, the greater the dust hazard the material will present during transport, handling, and processing. When possible, purchase the most coarse form of the chemical that is suitable for the process. There are production implications that may override considerations of worker exposure. As an example, if the material must be placed in solution the large particle size will slow this process.

15.3.3.3 Dust-Controlled Forms In the past two decades the rubber, pharmaceutical, pigments, and dyestuff industries have given attention to the dustiness of the raw materials they produce and use. Dustiness testing has been extended, and significant product design changes have been adopted to minimize dustiness, worker exposure, and product acceptance. In tire manufacture at least a dozen chemicals in granular form are added in small quantities to the batch mix. In an effort to minimize worker exposure to dust from these chemicals, the British Rubber Manufacturing Association has sponsored design of low dusting forms of these common chemicals. The properties of seven dust-controlled forms of rubber chemicals is reviewed by Hammond (1980) in Table 15.10. The author notes that the disadvantages of the most effective approach, coating the chemical with a polymer, include the variability in active chemical content based on bulk weight, the reduced chemical content, and the unsuitability of the polymer in the formulation.

15.3.3.4 Slurry Form This application has limited application, but in those cases where it can be used it does have great impact on dust concentrations. In the tire industry "master batch" rubber, rubber processed with all chemicals except the vulcanizing agent,

TABLE 15.10 Dust-Controlled Forms of Rubber Chemicals—Comparative Performance[a]

		Wax or otherwise bound				Polymer Bound	
Property	Untreated powder	Soft paste	Putty	Prills	Pellets/ granules	Slab	Pellets or granular
Active content	5	3	4	5	4	4	4
Convenience of handling	3	1	2	4	5	2	5
Freedom from dust	1	5	5	4	4	5	5
General cleanliness and safety	1	1	2	3	3	5	5
Suitability for automatic weighing	3	1	1	4	4	1	5
Wastage	3	2	4	4	4	5	5
Ease of disposal of containers	3	1	3	3	4	4	4
Identification	—	—	—	—		—	—
Mill mixing behavior	3	1	3	3	3	5	4
Internal mixing behavior	5	2	3	5	5	5	5
Dispersion in rubber	5	5	4	3	4	4	4
Total	32	22	31	38	40	40	46

Source: Hammond (1980). Reprinted by permission of the British Occupational Hygiene Society.

[a]5 = excellent; 4 = good; 3 = average; 2 = below average; 1 = poor.

is processed from the Banbury to the drop mill, where it is "sheeted off" for storage. Until 1960 dry talc or limestone was dusted on the slabs of master batch material to keep it from sticking. This operation resulted in poor housekeeping and a significant exposure to talc. The present technique, adopted in the 1960s, involves dipping the stock in a slurry of talc in water before racking for storage. This simple change resulted in a significant reduction in worker exposure to talc.

15.3.4 Impurities in Production Chemicals

In low concentrations impurities or unreacted chemicals in raw or final product may represent a potential exposure that warrants attention.

15.3.4.1 Residual Monomer in Polymer In polymer manufacture there is frequently unreacted monomer in the final product. Residual monomer had not been given much attention until the early 1970s, when angiosarcoma, a rare liver cancer, noted in workers manufacturing the vinyl chloride polymer, was attributed to the monomer exposure. Investigation showed that the polymer used in subsequent fabricating operations had unreacted monomer present in concentrations as high as 0.4 percent (Braun and Druckman, 1976). This level of contamination prompted concern about the monomer exposure of workers handling the bulk polymer in plastic fabrication operations such as injection molding. As a result of this concern, the vinyl chloride manufacturers modified the manufacturing process to reduce the monomer concentration to less than 1 ppm (Berens, 1981), thereby eliminating significant worker exposure. Residual monomer is frequently present in concentrations up to 1 percent in many of the common polymers and may warrant attention. If significant air concentrations are noted when handling the polymer, engineering control is first based on the removal or reduction of the monomer content in the polymer with other controls considered later if this is not adequate.

15.3.4.2 Solvent Impurities Impurities may pose an unrecognized risk, especially in solvents of high volatility. In the manufacturing of automobile tires, the various rubber components are "laid-up" on a tire building machine. To effect good bonding between the components, the rubber is made tacky by applying a small amount of solvent to the surface with a pad. For several decades this solvent was benzene. The worker exposure, probably in the range of 1–10 ppm, may be responsible for the excess leukemia seen in older tire builders. Starting in the 1950s the industry started to replace benzene with white gasoline. In studies completed in the 1970s by the Harvard School of Public Health Joint Rubber Studies Group, the residual benzene content in white gasoline was 4–7 percent, and air sampling indicated that one-third of the air samples on tire builders exceeded 1 ppm (Treitman, 1976). The PEL for benzene at that time was 10 ppm, although it was anticipated that it would be dropped to 1 ppm (this change did occur in 1987). Technical-grade chemicals commonly have significant impurities. The level of contamination should be identified, and if sufficiently high, worker exposure should be evaluated. At that time the necessity for reduction in the impurity level can be determined.

15.4 EQUIPMENT AND PROCESSES

In Section 15.3.1 a variety of engineering control options are focused on the choice of materials to minimize the generation and release of airborne contaminants. An equally important step is a review of the various alternatives in the choice of equipment and processes.

In the discussion on dusty materials in Section 15.3.3.3, techniques to determine the relative index of dustiness are mentioned to assist in the choice of material form and the dust suppression treatment. We do not have such an index for equipment and processes; however, there are a number of operational insights that should be considered in choice of facility. Wolfson (1993) has emphasized the importance of this step in stating that the removal of the dispersal device should be a first step in the engineering control of air contamination.

15.4.1 Diagnostic Air Sampling

Contaminant control cannot be achieved until the significant operational elements of the process that generate and release the contaminant are identified. Occasionally this can be done simply by a critical review of the operations but usually diagnostic air sampling is necessary. In Chapter 7, the traditional exposure monitoring for compliance purposes is discussed. The value of short-interval, task-oriented air sampling using conventional integrated sampling with subsequent analysis has been clearly stated by Caplan (1985a). In describing this approach, illustrated in Table 15.11 Caplan states:

TABLE 15.11 Task-Oriented Air Sampling

Task	GA or BZ[a]	Minutes/ day	Concentration (mg/m^3)	Minutes/day × concentration
Charge pot	BA	40	0.12	4.8
Unload pot	BZ	80	0.50	40.0
General survey	GA	250	0.16	40.0
Lab—sample trips	GA	20	0.05	1.0
Change room	GA	30	0.08	2.4
Pump room—repack	BA	30	0.32	9.6
Lunch room	GA	30	0.07	2.1
		480		99.9

LV = 0.2 mg/m^3; Wt. avg. = 99.9/480 = 0.21 mg/m^3

Source: Caplan (1985a). Reprinted by permission of John Wiley and Sons from *Patty's Industrial Hygiene & Toxicology*, Cralley and Cralley, Eds.

[a]GA = general air; BZ = breathing zone.

> For the job analyzed in Table 5.11 presumably a single sample would have shown a concentration of 0.21 mg/m^3. The task-oriented sampling, however, reveals several interesting things. Column 5 shows that tasks B and C are the major contributors to the day's exposures and that a significant reduction in the concentration at either of those tasks would be adequate to bring the 8 hr exposure well below the TLV. This is true even though task C in itself is below the TLV. In addition, it shows that task F, well above the TLV concentration, is of such short duration that significant improvement in that part of the exposure would not have a large effect on the 8-hour exposure.

The difficulty in this approach is that frequently it is not possible to measure the air concentrations during brief individual tasks and activities due to the low air sampling rate and the limited sensitivity of the analytical procedures. To reveal important generation points in the job, it is necessary to resolve the air concentration profile in a time frame of seconds. The advent of real-time, direct-reading air sampling instruments for particles, gases, and vapors with response times of less 1 second now permit the investigator to identify these critical contaminant elements in process events and work practices. In the 1980s this air sampling technique saw expanded application with coincidental videotaping of the worker during a work cycle. In its most sophisticated form the time-coupled, real-time contaminant concentration at the workers' breathing zone is superimposed on the video display, permitting the viewer to analyze the data display to identify the specific time and location of release. Control technology is then applied to those tasks or incidents.

The video display may also be used to identify work practices that may either positively or adversely affect worker exposure. In a talc bagging operation reviewed in Vermont in the 1970s one individual consistently had the lowest dust exposure, although visual inspection did not reveal any differences in equipment or work practice. If the real-time technique were available, the worker's "secret" could have been identified and applied to the other workers. If a "correct way of doing the job" can be identified, the video concentration format is an excellent educational tool for workers.

A series of studies by NIOSH investigators describe the application of this technique to a range of in-plant tasks (Gressel, 1987: O'Brien, 1989). The air sampling instrument

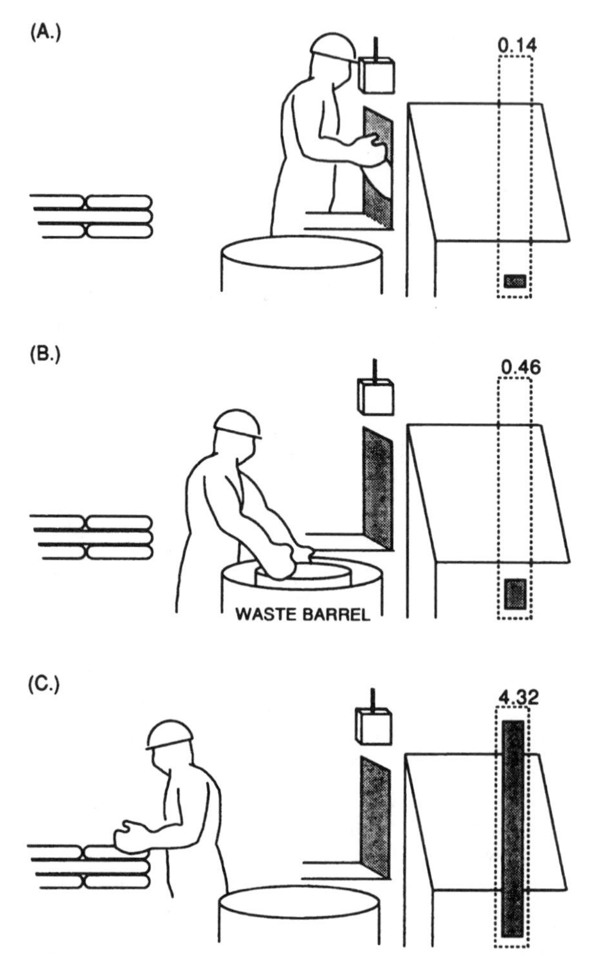

Fig. 15.6 Relative dust exposure during bag dumping as determined by video-air sampling technique: (A) Operator slits bag and dumps the granular material into a hopper equipped with an exhaust hood with minor dust exposure. (B) Operator drops the bag into the waste barrel; there is a significant increase in dust concentration. (C) As the operator pushes the bag into the waste barrel, a cloud of dust is released and the relative dust concentration increases to 4.32. [Reprinted with permission from *Applied Occupational and Environmental Hygiene* 7(4), 1992.]

is a real-time monitor with a response time much shorter than the period of the shortest worker activity or movement. The instrument is equipped with a data logger with a clock "locked-in" or synchronized with the video. The data logger is downloaded to a computer, and the data file is analyzed permitting a graphical overlay of air contamination data on the video tape. In Fig. 15.6 the video display from a study of a bag dumping operation is recreated in a line drawing for clarity to show the time-coupled concentration on the video screen display (Cooper and Gressel, 1992). Graphical representation of the air sampling data for three jobs with the concentration from the video overlay is shown in Fig. 15.7 (Gressel, 1988).

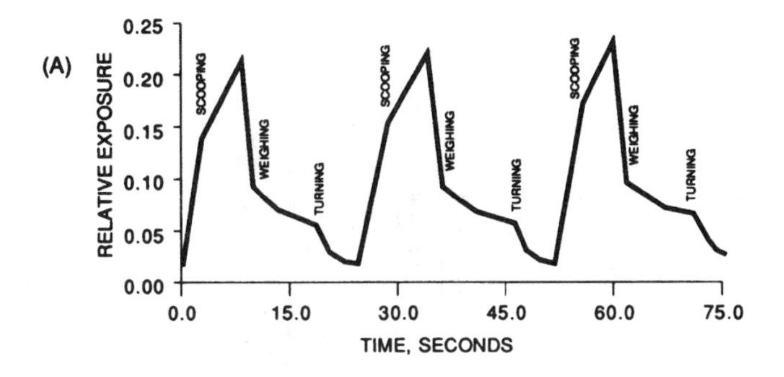

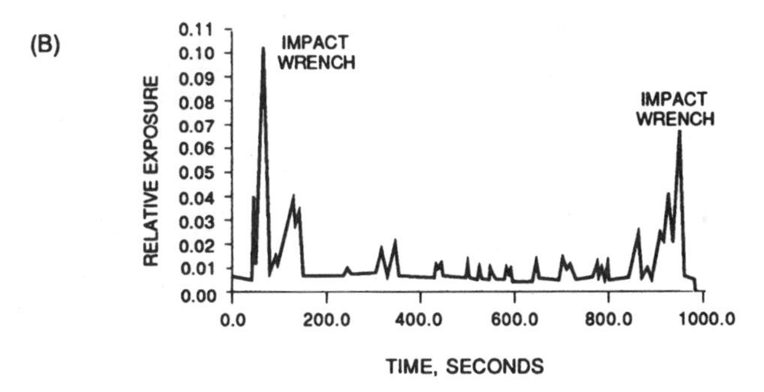

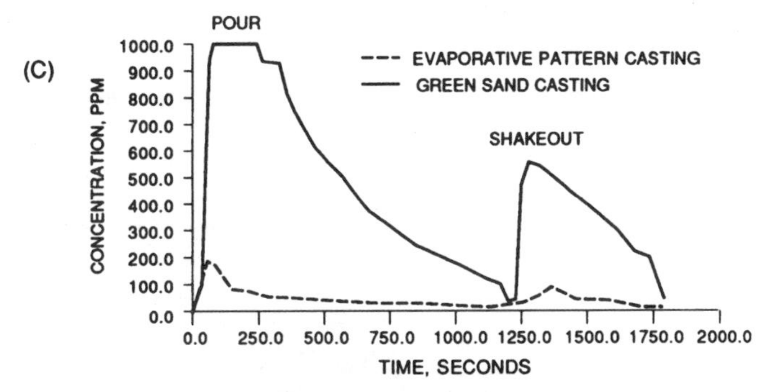

Fig. 15.7 Peak dust exposures identified by video-real-time monitoring: (A) Relative exposure versus time for three bags during manual weighout of powders. (B) Relative dust exposure during automotive brake servicing. (C) Carbon monoxide emissions from evaporative pattern casting and green sand processes. (From Gressel et al., 1988. Reproduced by permission of *Applied Industrial Hygiene.*)

The application of this technique to control technology was shown by Gressel (1989) in a study of a chemical weighing and transfer station. The information obtained permitted the investigators to redesign the workstation controls based on a perimeter exhaust hood and an air shower to eliminate eddies induced by the worker's body. Effective control of the dust exposure was obtained with one-third the airflow of the original system. In addition to improved worker protection, the cost savings of the new system resulted in a payback period of 4.5 years.

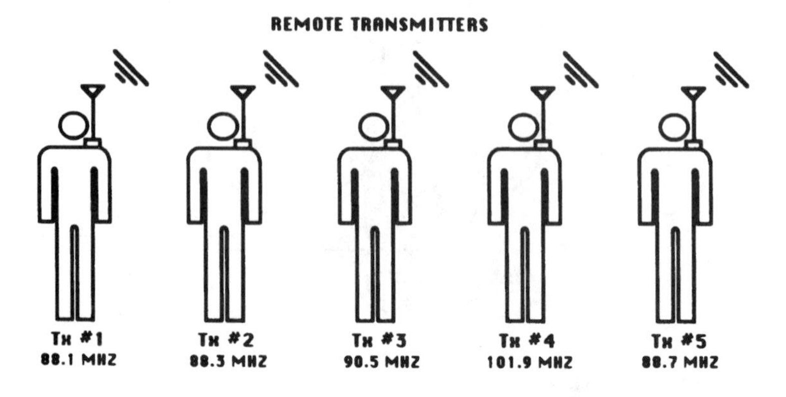

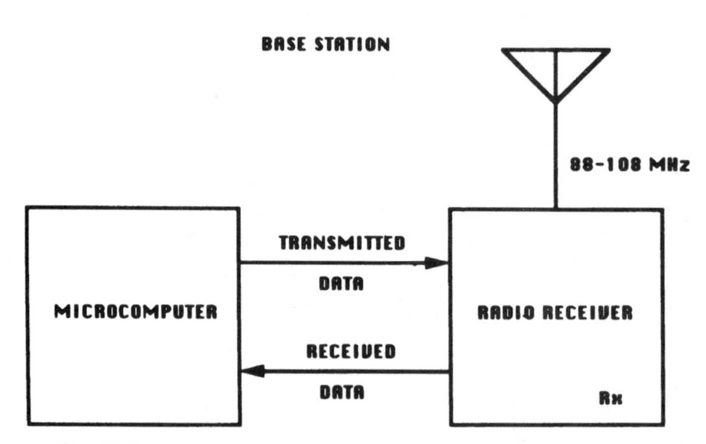

Fig. 15.8 Radio telemetry block diagram. From Kovein, 1992.

A variation of this technique first developed by Sweden's National Institute of Occupational Health and adapted for use in this country by NIOSH allows real-time mixing of the concentration and video portions by the use of radio telemetry (Kovein and Hentz, 1992). Rather than recording the concentration data on a datalogger, the direct reading instrument signal is converted to a radio frequency and broadcast with a telemetry transmitter to a receiver for immediate processing by a microcomputer (Fig. 15.8). Researchers and employees can get immediate feedback on the effect of engineering controls and their work practices on exposure levels. This allowed for immediate investigation into the cause of exposure and allowed NIOSH investigators to work with employees to reduce exposure levels.

In addition to using air sampling to determine where controls are needed, air sampling can also be used to initiate controls. This approach is common in the semiconductor industry, where continuous monitoring of toxic gases is required (Uniform Fire Code, 1988). If the gas is detected above some predetermined level in a gas storage cabinet, emergency shutoff valves for the toxic gases are activated. Through the use of sensitive, specific air monitoring devices, the leak is controlled before any adverse occupational or environmental releases occur.

These diagnostic tools will see expanded application to workstation design in the next decade since it does permit the engineer to identify the specific tasks, equipment function, or worker movement that contribute to worker exposure.

15.4.2 Equipment

15.4.2.1 Ancillary Equipment The stepwise review of the dispersal potential noted above should include all equipment, not only the major machinery. The importance of this approach is highlighted when one looks at the history of the simple *air nozzle* in industry. For most of the 1900s the widespread application of the air nozzle operating at line pressure (100 psi) was used to remove chips and cutting oil from machined parts and to dry parts after cleaning with solvent. In the 1970s OSHA required that the operating pressure for these nozzles for cleaning purposes be dropped from 100 to 30 psi. This change had many positive effects, including a reduction in noise, eye injuries, solvent mist, and vapor exposures.

Another common piece of equipment, albeit much more complex, is the *centrifugal pump*. The difficulty of equipment selection is typified by the experience of a chemical engineering team designing a modern chemical plant or refinery having hundreds of centrifugal pumps. A major issue in such plants is the impact of fugitive losses from rotating machinery such as pumps on the workplace and general environment (NIOSH, 1991; BOHS, 1984). The total loss percentage from rotating machinery varies from 8 to 24 percent, with overall uncontrolled emissions ranging from 31 to 3231 tons/year for a model chemical plant or a large refinery (Lipton and Lynch, 1987). The authors describe two possible control programs for pump fugitive losses—either a monitoring and maintenance program or engineering control by installation of dual seal pumps, seal-less pumps, and a closed exhaust hood with air cleaning.

The phaseout of 1,1,1-trichloroethane and trichlorotrifluoroethane solvents in vapor-phase degreasers in the 1990s has accelerated the search for alternative solvents, as described in Section 15.3.2.4. Initial steps to reduce emissions from this common equipment in the 1980s included the use of covers, increased freeboard height, refrigerated condenser fluids, and lower hoist speeds. Although these changes do reduce losses and therefore worker exposure, the survival of this type of equipment rests with the availability of a closed system unit as described by Mertens (1991). Such *vapor-tight degreasers* are available in Europe and are under development in the United States. In the past, closed system technology has been associated with major processing facilities. In the future individual job shop equipment will utilize this approach to meet critical workplace and environmental constraints.

15.4.2.2 Major Processing Equipment The complexity of the control technology problem becomes apparent when one moves from choices of ancillary equipment to decisions on major pieces of processing equipment. As an example, a small chemical processing plant for organic synthesis is a multifloor plant with a series of major operations staffed by six chemical technicians. The condensed version we will review (Fig. 15.9) permits discussion of equipment alternatives to minimize the generation and release of air contaminants. The three-step process includes, first, a reaction step to form

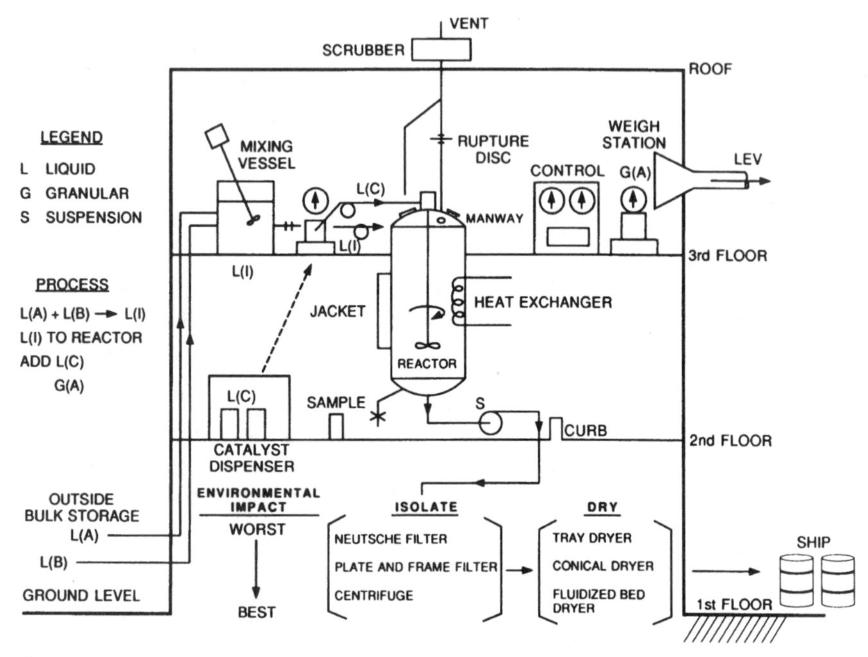

Fig. 15.9 Chemical processing facility. Liquid reagents A and B are transferred to a mixing vessel. After mixing, the resulting liquid C is transferred to the reactor by vacuum. Granular chemical X is weighed under local exhaust ventilation (LEV) and dumped into the reactor through a change port also equipped with LEV. A catalyst D is transferred to the reactor by a positive displacement pump. After the reaction is complete, the new chemical is isolated by one of three techniques—Neutsche filter, plate and frame filter, or centrifuge. The resulting wet cake is then dried by one of three methods—tray drier, rotary drier, or fluidized bed drier. The isolation and drying techniques are ranked by environmental impact with the worst process noted first.

a solid in suspension, conducted in a pressure vessel. In this case there is no alternative equipment. The reaction must be conducted in a closed pressure vessel equipped with the safety controls shown in Fig. 15.9 to minimize the possibility of a chemical accident. The second step, the isolation of the solid reaction product, can be done by at least three different techniques. The environmental contaminant potential of these processes varies considerably. However, the choice cannot be made on this basis alone, since these techniques do have specific production capabilities that may be the major determinant in the choice. Finally, the reaction product is dried and drummed for sale or use in a subsequent operation. Again there are several ways to accomplish the drying. The ranking of these isolation and drying operations based on their environmental impact is shown in Fig. 15.9.

The responsibility for the choice of specific equipment in the processing plant will rest with the project design group and will be based on production capacity, ancillary services required, maintenance history, cost and delivery, in addition to health and safety considerations. It is not expected that the health and safety professional will have complete operational and application data on all chemical unit operations, but their environmental impacts must be understood and made known to the engineering group.

It would be extremely helpful to the health and safety professional if the individual equipment were assigned an index of contaminant generation/dispersal similar to the index of dustiness or fugitive losses from a pump. This approach is not currently realistic; however, a better characterization of the performance of such equipment will emerge in the next decade.

15.4.3 Processes

A review of the production engineering literature shows a range of processes that can be considered for a manufacturing facility. The choice is usually made based on production output and cost data. Industrial processes designed to accomplish a given task have one other parameter, the ability to contaminate the workplace; this characteristic of the process should be given equal weight to production criteria (Peterson, 1985).

The development of technology in a given area such as welding is pushed by production needs, not by health concerns. Since 1900 the technology has become more complex and frequently more environmentally challenging. Infrequently, studies are conducted to influence the process design to minimize air contamination. This was the case with Gray and Hewitt (1982), who defined the welding operating parameters that influence fume generation and proposed operational configurations that minimize fume formation rate and the chemical composition of the fume, thereby permitting improved efficiency of controls.

An exercise that demonstrates the value of a ranking approach for process environmental impacts is shown in the manufacture of an electronic cabinet (Fig. 15.10). As in the case of the chemical process shown above, the fabrication steps are simplified for ease of discussion. The four enclosure components are fabricated from lightweight aluminum stock using sheet metal shear, brake, and punch. These operations may present a noise hazard but do not generate air contaminants and will not be included in our review. For a discussion of controls used for noise, see Chapter 21. The enclosure can be assembled and painted by a number of techniques; the processes are ranked from worst to best in terms of adverse effect from air contamination.

Another general approach that applies to many processes is the use of containment. This control approach was considered by an engineering control panel sponsored by the Environmental Protection Agency, and the following containment ranking was proposed (EPA, 1986).

Standard control measures may often be categorized for each process: a sealed and isolated system where airborne contaminant levels are very low (ppb range for vapors); a substantially closed system where airborne contaminant levels are still relatively low (fraction of ppm or low ppm range for vapors); a semiclosed system, which is typical of nondedicated equipment used in job shop chemical processing facilities; and an open system.

An example of a sealed and isolated system is the modern refinery, where such an approach is required both for production reasons and health and safety. A substantially closed system is typified by a semiconductor facility, and a semiclosed system by the chemical processing plant in Fig. 15.9. Most common facilities, that is, foundries, electroplating, painting, welding, and machining, are examples of open systems.

The contaminant released as a mist from electrolytic plating operations can be trapped by a layer of plastic chips or a persistent foam blanket on the surface of the bath. Also surfactants may be added to the electrolyte to reduce mist escape. The plate

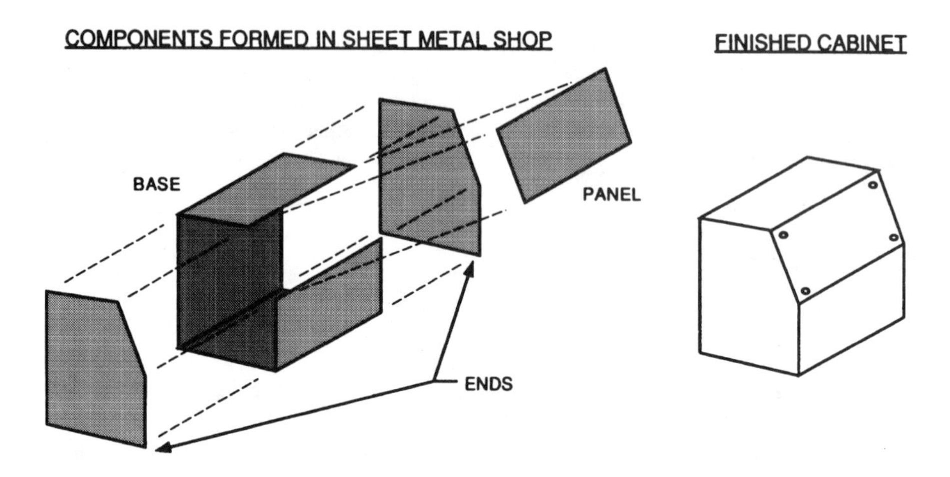

FINISHING OPTIONS

Metal cleaning	**Welding**	**Painting**
Wet degreasing with chlorinated HC	Torch welding	Air atomization
Vapor with chlorinated HC	Inert gas shielded arc welding	Airless spraying
Wet degreasing with terpenes	Submerged arc	Hot paint
Water with detergent	Resistance	Electrostatic
		Powder
		Dipping

Fig. 15.10 Manufacture of an electronic cabinet. The operational options for cleaning, abrasive blasting, and painting are ranked in terms of environmental acceptance with the worst process noted first.

and frame filter in the chemical processing industry is difficult to handle with local exhaust ventilation. Frequently a solvent wash of the cake with a highly toxic material results in high exposure when the filter is broken and the solvent-wet cake is removed. In some cases after the initial wash, the cake can be washed with isopropanol to strip out the toxic solvent. It is then washed with water so that when the filter is broken and the product is removed, exposure is nil.

A series of process changes developed in the WRITE program is shown in Table 15.12.

15.4.4 Work Task Modification, Automation, and Robotics

It is well known that simple changes in work tasks may have significant impact on job outcome. In the early 1900s, workplace time and motion studies were used to improve productivity. Later a job placement technique devised by Hanman (1968) based on a

TABLE 15.12 Examples of Process Change Technology from EPA WRITE Program

Operation	Process change
Hand mixing of paint	Proportional mixer blends paint at the gun, thereby eliminating handling paint and solvent at a mix operation
Conventional manual air spray painting	Computer-controlled robotic painting with an electrostatic spray to reduce worker exposure to paint mist and solvent vapors
Performance testing of electronic arts with CFC-based cooling system	Installed compressed air cooling system

Source: Adapted from Harten and Licis (1991).

detailed analysis of the time and effort of the job tasks was effective in reducing on-the-job injuries. More recently, the analytical tools of the ergonomic specialist permits identifying difficult tasks contributing to occupational injuries and illness.

As indicated in Section 15.4.1, the health and safety professional has techniques to investigate the source of contaminants and the generation mechanism, and to make a semiquantitative assessment of the generation rate. This approach has tremendous value in analyzing not only the critical generation points on the machine but in viewing the impact of specific worker actions and movement on air concentrations. Preliminary studies reveal that minor changes in work position and movement may offer significant reductions in exposure.

Occasionally the specific modification in work practice is dictated by knowledge of the mechanisms of generation and release of the contaminant. Such is the case with the flow of granular material at material transfer points. The generation mechanism is the airflow induced by the falling granular material (Anderson, 1964). The induced airflow can be minimized by restricting the open area of the upstream face, reducing the free-fall distance, and reducing the material flow rate as defined by Anderson's equation.

Automation is a general technique that separates the worker from the individual process. In the 1960s and the 1970s this was usually done by simple electromechanical equipment design. A good example of this procedure is the manufacture of asphalt roof shingles. In the 1970s competition resulted in the automation of all parts of the process from the dipping of the stock to the bundling of the package.

Another example of a simple automation process that reduces worker exposure is tire curing. In the early plants the worker lifted the tire out of a curing press at the end of the curing cycle and placed another tire in the mold for curing. During this period the worker was directly exposed to the emissions released from the press. By 1970 most plants had automated this process. The worker now places the uncured tire on a holding rack in front of the curing press line. When the curing cycle is completed, the tire is ejected from the press to a belt conveyor, and the next tire to be cured is transferred from the rack to the press without exposure of the worker.

The advent of robotic techniques has permitted almost all industrial procedures to be candidates for automation. The movement of the worker who buffs rubber boots can now be captured by a robotic system, as shown in Fig. 15.11. The ultimate application of robotic techniques is to spray painting. In this case the robotic system can reproduce the movement of a skilled painter. Although ventilation is still required on this job, the worker overseeing the operation is separated from the point of release of the paint mist and solvent.

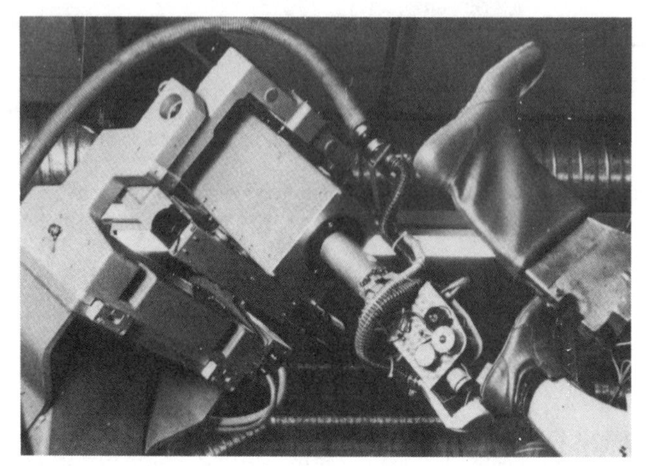

Fig. 15.11 Robotic buffing of rubber boots. (*Source:* Photograph reproduced by permission of Matti Koivumaki.)

Robotics may have been a solution to a difficult problem in the 1950s. In a large generator shop the generator coils were preformed of copper bar stock. The coils were then wrapped with insulation tape and painted with an asphalt compound. Protective clothing notwithstanding, the workers had skin exposure to the asphalt, which required aggressive cleaning at the end of the shift. Dermatitis and photosensitivity were frequent occurrences. At that time coil winding experts devoted time and money to developing machine wrapping concepts without success. This type of problem can be solved by robotics in the 1990s.

15.4.5 Facility Layout

As stated earlier, the most efficient way to do a job is probably the one that impacts the least on the workplace environment. Certainly this is true insofar as overall plant layout is concerned (Caplan, 1985b). In the shop fabricating the enclosure in Fig. 5.7 the desired flow of materials is from the incoming truck dock or railroad siding to sheet metal fabrication and then to assembly, finishing, inspection, and finally shipping. In a facility manufacturing pharmaceuticals, the input chemicals are transported to bulk storage and then to chemical processing, packaging, inspection, and shipping. In all manufacturing processes from handling metal to fine chemicals, it is important to minimize the distance the material is moved and the number of times it is picked up and transferred. If this rule is followed, worker exposure to air contaminants will be minimized.

15.4.5.1 Material Transport This issue is given major attention in industry where large quantities of raw material, intermediates, and final products are handled. Examples are injection molding of children's toys, manufacture of automobile tires, and paint manufacture. In some cases material handling alone defines the plant layout. Frequently bulk storage is in large silos located outside the plant with delivery to the workstations by mechanical or pneumatic conveyors. This is true in the manufacture of plastic tape,

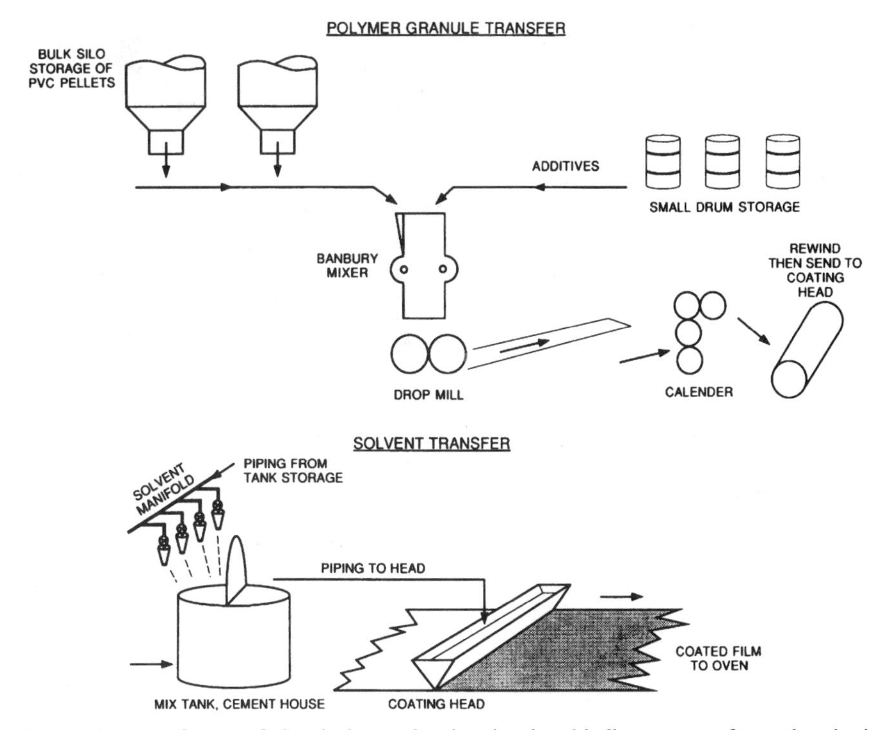

Fig. 15.12 Manufacture of electrical tape showing the closed bulk transport of granular plastic and the piping of solvents to the mix tank.

where tons of PVC granules are used each day (Fig. 15.12). This plant also requires large quantities of solvent delivered from bulk storage by piping to mix tanks and then directly to the coating heads. The intent of this system is to have all material handling to the individual coaters done in closed systems.

In organic synthesis it is common practice to have outside bulk storage of at least a dozen solvents. The solvents are transferred to an inside reaction vessel by piping to a reactor manifold with necessary valving and meters. Granular materials used in small quantities are stored in adjoining storage areas and delivered to the reactor and charged by manual dumping, with a dumping fixture, or occasionally with a transfer lock. When a large number of drums must be dumped, a bulk handling system is used with direct delivery to the reactor vessel.

15.4.5.2 General Considerations The location of the process within the facility may influence the worker's exposure. This is true of operations such as foundries, where jobs may be easily classed as clean or dirty and exposures vary greatly. The conventional iron foundry provides a good example of the importance of plant layout. It is common practice to define the optimal plant layout as one with airflow from the cleanest to the dirtiest operations, with the exhaust focal point establishing this gradient [Fig. 15.13(A)]. The layout of a similar foundry that does not follow this guideline is shown in Fig. 15.13. In the latter case the relatively clean molding line is positioned adjacent to the shakeout, an area where the control of airborne foundry dust and thermal degra-

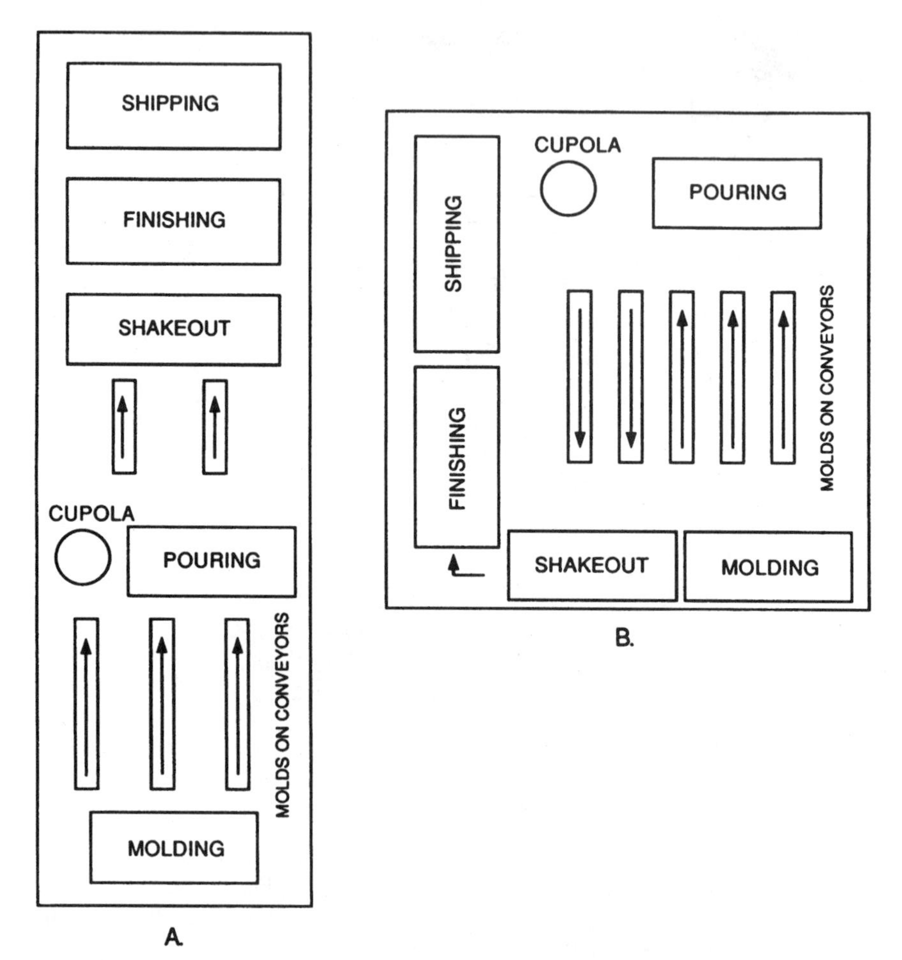

Fig. 15.13 Layout of two gray iron foundries. Foundry in (A) is designed for straight process flow and gradual transition from activities that represent limited contamination to those of greater contamination. In foundry (B) the process flow reverses and the dirty shakeout ends up adjacent to the molding line.

dation products is difficult. This poor layout results in silica and other contaminants released by shakeout moving into the molding area and exposing this work population. If the desired concentration gradient cannot be achieved by ventilation or distance, then structural walls or plastic barriers are a possibility. Although such barriers define the space, compartmentalizing complicates the design and application of local exhaust ventilation, replacement air, HVAC, and other important services.

15.4.5.3 Workstations A foundry in Finland faced problems not only of airborne contaminants, but heat, noise, and housekeeping in casting cleaning. The problems were resolved by redesigning the open work space to provide individual workstations designed with integrated services positioned for ease of work (Fig. 15.14). This lay-

(a)

(b)

Fig. 15.14 Well-integrated workstation in a foundry. (a) Overall layout of foundry finishing area; (b) individual workstation with exhaust hood. (*Source:* Photograph reproduced by permission of Matti Koivumaki.)

out resulted in improved material flow, better housekeeping, reduced air contamination, better ergonomics, and improved productivity. The success of such an installation was due, in part, to close employee–employer involvement in the design.

15.4.5.4 Service and Maintenance Frequently, close attention is given production area layout as noted in the above example, but rarely is attention given the working environment of the maintenance group. An example of an area where maintenance was not considered was tire curing presses in the plants of the 1970s. The presses were arranged in double rows back to back with limited space between rows for steam, water,

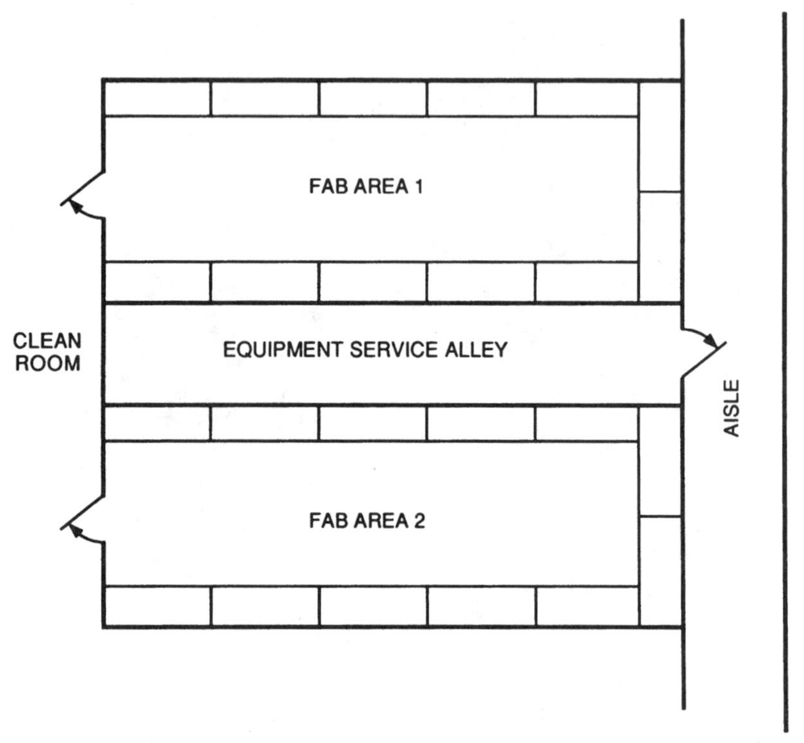

Fig. 15.15 Layout of semiconductor facility showing service aisle isolated from, but adjoining, the fabrication bays.

and electrical services. While this layout and density represented optimal use of space, repairs of steam and water piping were done in an extremely tight space and controls for the worker were difficult to set up.

The best example of a planned layout that provides proper work space for the trades is noted in the semiconductor industry (Fig. 15.15). In this layout all transfer pumps, distribution lines, and vacuum pumps are located in a service alley separated from the fabrication bay. This design reflects principal concern, not for the comfort of the maintenance worker, but rather to protect the fabrication area against contamination. Notwithstanding, it does result in adequate space for the trades to carry out their work in a safe manner.

An extension of this concern is the installation of a small field service bench with necessary tools and equipment in plant production areas where maintenance is frequently done. This arrangement permits the maintenance person to do many repairs at the site and not transport equipment back to a main facility with the potential for chemical spills.

15.4.5.5 Segregation of Operation It is frequently necessary to segregate or remove an operation from the main production area as an engineering control measure. In the semiconductor and fiber optics industry each time a new facility is designed, the engineers must choose whether to store small amounts of highly toxic gases close to the production tool or store large quantities at a segregated position some distance from the plant to be distributed by double-walled piping to the production tools. This latter segregation technique permits the plant to reduce the number of persons at risk while providing extensive controls at the work site. This option must be weighed against storing small quantities of gas directly at the tool, thereby minimizing failures in the transfer systems but requiring frequent change of gas cylinders with the entailed risks.

15.4.5.6 Isolation of the Worker The practice of isolating the worker as a control measure is placed under this section since it should be considered in conjunction with layout of the major equipment. This widely used technique is mandated by OSHA coke oven regulations, which requires enclosures with clean, conditioned air. The rail cars that travel above the coke ovens have controlled environment cab enclosures; workers can also retreat to enclosures designed to remove them from the hostile coke oven environment during available rest periods.

In other industries this approach has been chosen to eliminate worker exposure to air contaminants while providing a comfortable working environment. These applications include enclosed booths on a variety of construction equipment, front-end loaders in smelters, crane cabs in metallurgical industries, operator cabs in steel rolling mills, and pouring stations in foundries. Unfortunately off-the-shelf control booths are not available for the range of applications seen in industry, nor have engineering guidelines been published.

15.4.6 Ventilation

In the introduction to this chapter we stated that ventilation is the third step in the control hierarchy of: (1) do not generate the contaminant; (2) if you do generate a contaminant, do not allow it to be released; and (3) if it is released, collect the contaminant before it reaches the worker. The design goal of industrial ventilation is to protect the worker from airborne contamination in the workplace. To the newcomer this may suggest installing a system that will reduce exposure below the permissible exposure guidelines or an appropriate action level. This is not the case. The professional will design the system to meet the goal of "as low as reasonably practical" (BOHS, 1987).

Within this control approach the effectiveness of the major ventilation techniques is shown in Table 15.13 (BOHS, 1987). There is no agreement on the position of low-

TABLE 15.13 Ventilation Control Hierarchy

Type of ventilation[a]	Hood type	Example
LEV	Total enclosure	Glove box
LEV	Partial enclosure	Laboratory hood
LEV	Low-volume–high-velocity, tool integrated	Portable grinder
LEV	Exterior hood	Welding hood
GEV	Mechanical exhaust	Roof ventilators
GEV	Natural	Wind induced

Source: Adapted from BOHS (1987).

[a]LEV = local exhaust ventilation; GEV = general exhaust (dilution) ventilation.

TABLE 15.14 Application of Local Exhaust and General Exhaust Ventilation

Local exhaust ventilation	General exhaust ventilation
Contaminant is toxic	Contaminant has low order of toxicity
Workstation is close to contaminant release point	Contaminants are gases and vapors not particles
Contaminant generation varies over shift	Uniform contaminant release rate
Contaminant generation rate is high with few sources	Multiple generation sources, widely spaced
Contaminant source is fixed	Generation sites not close to breathing zone
	Plant located in moderate climate

Source: Adapted from Soule (1991).

volume, high-velocity systems since its effectiveness varies greatly depending on the degree of integration with the tool. When designing a hood for a particular operation, Burton (1991) recommends starting with a complete enclosure concept and then removing only those portions of the enclosure necessary to provide access to the equipment. This will ensure the most effective containment with the lowest airflow. When it is not possible to enclose the equipment or operation, an exterior hood must be used. These are hoods that direct the contaminant away from the breathing zone of the operator once the contaminant is released. Enclosing hoods are preferred over capture hoods by designers because they provide a better containment with less airflow and are less affected by crossdrafts. Users may prefer capture hoods because they tend to be less restrictive. For a capture hood, the amount of air required for effective capture is a function of the square of the distance between the release point and the hood. Thus the distance between the hood and the release point must be determined during design and the hood used accordingly once installed. Some hood designs can take advantage of the momentum of the contaminant in ensuring its capture. These are called *receiving hoods*, and the best example is a canopy hood, which receives rising hot air and gases.

Soule (1991) reviews the application of the two major ventilation control approaches, dilution and local exhaust ventilation (Table 15.14). In most industries dilution is not the primary ventilation control approach for toxic materials. It is accepted that local exhaust ventilation will not provide total capture of contaminant, and dilution ventilation is frequently applied to collect losses from such systems. In addition, it is used for multiple, dispersed, low-toxicity releases.

15.4.6.1 Design Phase The specific design methods for both general exhaust ventilation and local exhaust ventilation are presented by Soule (1991) and in other volumes dedicated to ventilation control (Burgess, 1989; Burton, 1991; ACGIH, 1996). An important predesign phase identified by Burton (1991) as problem characterization is frequently given little attention (Table 15.15). This is a topic that can be best addressed by an industrial hygienist, who can provide data on emissions, air patterns, and worker movement and actions.

If the process is new, a videotape of a similar operation with the same unit operations may be available. The best of all worlds would be the availability of a video concentration tape as discussed in Section 15.2.1. If the facility is a duplicate of one in the company, the industrial hygienist should visit the operation with the ventilation designer. Frequently the designer is an outside contractor. In this case it is important

TABLE 15.15 Information Needs for Problem Characterization

Emission source behavior
- Location of all emission sources or potential emission sources
- Which emission sources actually contribute to exposure?
- What is the relative contribution of each source to exposure?
- Characterization of each contributor: Chemical composition, temperature, rate of emission, direction of emission, initial emission velocity, continuous or intermittent, time intervals of emission

Air behavior
- Air temperature
- Air movement
- Mixing potential
- Supply and return flow conditions
- Air changes per hour
- Effects of wind speed and direction
- Effects of weather and season

Worker behavior
- Worker interaction with emission source
- Worker location
- Work practice
- Worker education, training, cooperation

Source: Burton (1989). Reprinted with permission of D. Jeff Burton.

that the problem characterization approach be followed and an information package be provided the designer.

The precautions that should be reflected in the design have been discussed in detail elsewhere, but should include worker interface, access for maintenance, and routine testing. Computer-aided manufacturing and design (CAM/CAD) technology now permits precise placement of equipment and ductwork so that ad hoc placement by the installer should be a thing of the past. As discussed earlier, it may be worthwhile to "mock up" a specific design solution prior to final design and construction. This is especially true when a large number of identical workstations are to be installed, as was the case in a shipyard asbestos insulation workroom, as noted in Fig. 15.3.

General cautions are appropriate on the use of available design data for control of industrial operations. The ACGIH Industrial Ventilation Manual provides the most comprehensive selection of design plates for general industry (ACGIH, 1996). Each of these plates provides four specific design elements: hood geometry, airflow rate, minimum duct velocity, and entry loss. The missing element is the performance of the hood in terms of percent containment. Roach (1981) has recommended such a index as a minimum performance specification for local exhaust hoods.

As noted by Burgess (1993), there has been little consolidation and publication of successful designs by industrial groups. With the exception of the Steel Mill Ventilation volume published by the American Iron and Steel Institute in the 1960s (AISI, 1965) and the Foundry Environment Control manual in 1972 (AFS, 1972), there is no evidence that companies in the United States wish to share ventilation control technology. One exception to this is the semiconductor industry which mostly through information exchange promoted by the Semiconductor Safety Association has published some information on ventilation design for their industry. Table 15.16 from the Semiconduc-

TABLE 15.16 Types of LEV Systems Commonly Used in the Semiconductor Industry

Laboratory-type hoods. Used in many locations for storage/use of chemicals, parts storage, or in QA/QC/Reliability labs for analytical procedures; ensure materials of construction are compatible with starter and intermediate chemicals, and sash adjustments are marked for minimum face velocities of 80–120 lf/min as per ASHRAE/ANSI Z9.5-1992, Lab Ventilation; crossdrafts from general HVAC and especially PCS/VLS can disperse air contaminants, as well as foot traffic in front of hoods; pay attention to compatibility of materials stored in the hood.

Wet sinks (primarily for etching or cleaning of wafers/boats). Plenum exhaust with shroud that fits over the deck surface to allow better capture at the etch/clean baths and to direct overhead vertical flow from laminar flow hood (making a push–pull system), with face velocities at shroud of 100–125 lf/min, in addition to compatibility of materials in the sink; pay attention to their compatibility with drains (e.g., cyanide solutions should not be kept in sinks with acid aspirators). Another approach used within the industry is to provide open wet sinks without shrouds with an exhaust volume of 125–150% of the laminar flow supply.

Gas cabinets. Use 150 CFM per bottle in cabinet design; UFC Article 51.107(c)3 requires gas cabinet be ventilated with 200 FPM average face velocity at the access port, with minimum velocity of 150 FPM at any point; ensure vent lines are routed to gas conditioning system or silane burnoff system; make sure exhaust from one cabinet is not intermingled with other cabinet exhausts unless chemically compatible.

Gas jungles or gas control valves. Must be exhausted; some companies have set internal standards, for example, a minimum of 5 air changes per minute (ACM) and 125 feet per minute of face velocity if accessible, or 5 ACM if not accessible.

Open surface tanks (plating or degreasing). Commonly used in "back-end" processing for adding precious metal layers; depending on the plating tank constituents may require use of push–pull systems; use of cyanides require stringent precautions with the use of acids (possible HCN formation); general HVAC system can cause crossdrafts that will dilute air contaminants into the general plating room air.

Diffusion furnaces. Local exhaust provided at "source end" jungle, and at "load end" of tube, using collector-type end caps to route tube exhaust products to vestibule exhaust opening; ensure VLF & HLF systems do not cause eddy currents and transient leads/odors.

Chemical storage cabinets. Minimum of 50 CFM per standard sized cabinet; ensure compatibility of materials stored within cabinet; use rated flammable storage vessels if flammables are transferred from original container.

Equipment cleaning hoods. Used for cleaning equipment parts that are contaminated with arsenic, antimony, or spinner/coater residues, and pumps, etc.; typically want a minimum face velocity of 125 lf/min at hood face; housekeeping in the hood is very important; also important to ensure the hoods are big enough to handle the largest parts to be cleaned.

Pump & equipment exhaust lines. Chemical compability with effluents, ability to handle pyrophoric materials, and potential for duct fires; blockage of ducting is possible from reaction products (such as CVD products from Si_3N_4); monitor exhaust duct/line pressure closely.

Portable chemical hoods. Use with nonflammable materials or as temporary exhausts for equipment or materials that are cleaned sporadically; they require preventative maintenance more than in-place chemical hoods and are more limited in the chemicals that can be effectively used in them.

Glove boxes. Used for the containment of highly toxic source materials in vials or jars (antimony, arsenic) or flammable metals (phosphorus); need LEV to provide negative pressure at load door, and HEPA filtration on LEV duct to capture fugitive particulate.

Burn-in testers. Temperature testing for ICs; need LEV for heat, and smoke detectors for ICs or boards that may start to decompose from dropping on heating elements or over-temperature shutoff malfunctioning on tester.

TABLE 15.16 (*Continued*)

Drying ovens including Blue M®-type ovens. Used primarily to drive-off moisture after cleaning of equipment parts; other volatiles can come off IC package materials and cause odors if not purged properly prior to door opening; sometimes due to temperature control problems, local exhaust is provided via a exhaust duct damper opened immediately prior to door opening; sometimes exhausted for heat only.

Welding/brazing/cutting hoods. LEV control velocities vary depending on the composition of the materials being worked on; LEV exhaust duct must be situated to avoid re-entrainment into general HVAC system.

Abrasive blasting hoods. Commonly used for cleaning contaminated parts from metal deposition areas (copper, platinum, gold, titanium/tungsten, aluminum, etc.) segregate "bead blasters" that are used for cleaning ion implanter parts (arsenic-contaminated or phosphorous-contaminated—which is a fire hazard); and other metallization equipment; ensure fine particle leakage is contained within the bead blaster by closing leakage points; establish controls/procedures for changing the contaminated bead blasting media (usually glass bead or silicon carbide).

HEPA portable and house vacuum systems. Portable HEPA vacuums are used extensively for cleaning the interiors of ion implanters or collector deposition in diffusion furnaces, cleaning up residues in parts-cleaning hoods or around bead blasters; care must be taken in using the standard HEPA vacuum without impregnated activated charcoal (designed for vacuuming up mercury) as arsenic vapors or arsine gas may be generated during implanter and molecular beam epitaxy—the impregnated activated charcoal was very effective in stripping out contaminants from the vacuum exhaust stream; house vacuum systems are very susceptible to contamination with hazardous materials in the system; they should not be used to clean surface contaminated with hazardous materials; however, because their use is difficult to control, they must be treated as being contaminated (i.e., full protective equipment and procedures).

Source: Williams, M., Baldwin D., and Manz, p. (1995). *Semiconductor Industrial Hygiene Handbook*, Noyes Publications, Park Ridge, NJ.

tor Industrial Hygiene Handbook lists the types of local exhaust ventilation systems commonly used in the semiconductor industry. Some of the control equipment is very specific to the industry, such as wet stations and diffusion furnaces. The original source of exhaust recommendations is frequently the equipment manufacturer, who may or may not have fully investigated optimal enclosure design and exhaust rates. In the case of wet stations, the concern on the part of manufacturers and process engineers is to provide chemical exposure (personnel protection) and particle contamination control (product protection) with the same device. Users of the equipment must ensure that the manufacturer's concern for product protection does not outweigh concerns for personnel protection. Verification of recommended exhaust rates by the customer or a third party (prepurchase containment testing) is frequently required.

Major sources of information on the performance of ventilation systems are the technical reports on engineering control technology published by NIOSH (Appendix A). Many of these reports couple ventilation assessment with measurement of worker exposure.

This discussion indicates that ventilation control designs on standard operations in the mature industries have been published, although performance has usually not been reported. It is important to evaluate performance by diagnostic air sampling, both to ensure the worker is protected and to prevent overdesign. The latter was shown to be the case in the design for control of a push–pull system for open surface tanks (Sciola, 1993). A mockup of one tank demonstrated that satisfactory control could be achieved at minimal airflow. Operating at the reduced airflow rate saved $100,000 in installation costs and $263,000 in annual operating costs.

A substantial portion of the operating costs for a local exhaust system are the costs to heat or cool the makeup air required by the exhaust system. These heating and cooling costs may range from three to five dollars per cfm per year. One approach that has been used to reduce these costs is to recycle the air back into the workplace once the contaminant has been removed from the exhaust stream. This technique is most successful when a relatively nontoxic material, such as wood dust, that can be reliably and efficiently cleaned from the airstream is present. Recirculation is less viable when multiple contaminants and/or high-toxicity contaminants are present. The American National Standard for the Recirculation of Air from Industrial Process Exhaust Systems (ANSI, 1996) must be followed when considering recirculation of air from a local exhaust system. This standard requires that a thorough hazard evaluation be conducted before designing a local exhaust system that may include recirculation. It also requires that a continuous monitoring device be used to ensure the exhaust air contains less that 10 percent of the acceptable level of a highly toxic substance before it is recirculated back into the building. The system must be designed to divert the air to the outside or initiate backup air cleaning if the primary air cleaner fails.

The types of air cleaning devices available depends on the physical state of the contaminant present in the local exhaust system. The most common types of air cleaning devices are for dusts, and there is a wide variety of filtration media available. For heavy loading of dusts, a baghouse that may contain upwards of a hundred large vacuum cleaner–type bags may be used. This may be followed by HEPA (high-efficiency particulate arrestor) filtration, which provides a high removal efficiency (99.97 percent at 0.3 μm) for a light dust loading. For vapors and gases, carbon adsorption, wet scrubbers, or incineration are possible choices. The air cleaning device usually represents a significant pressure drop in the local exhaust system. To add air cleaning to an existing system that does not already have it, upgrading the fan and motor is usually required. As a minimum, upgrading the fan and motor will be required if air cleaning is added onto a system.

15.4.7 Acceptance and Startup Testing

It is important that new ventilation systems be inspected and tested on completion of the installation to ensure the system conforms to the design specifications and that worker exposure be kept as low as reasonably possible. Unfortunately it is the author's experience that only 10 percent of the new systems in general industry undergo such scrutiny. The situation is baffling given the importance of contaminant control, the general practice of industry to test other services before acceptance, and the cost of ventilation systems ($10 to $30 per cfm to install and $3 to $5 per cfm per year to operate).

15.4.7.1 Construction Details The ventilation system design, completed by the plant facilities and engineering group or a consulting firm, is based on certain assumptions and specifications on hardware details such as elbows, entries, expansions, contractions, and so forth. The hood construction has been detailed by the draftsperson. If the hardware elements provided for installation are supplied by a standard manufacturer, the losses will be approximately those used by the designer. If the system is large, it may be worthwhile obtaining samples of the components to test for losses. The hoods warrant special attention. The entry losses on standard hood designs as noted in the ACGIH ventilation manual are quite reliable. However, if the hood design is not typical, it may be difficult to estimate entry loss. In such a case the best solution is to have a single hood fabricated and tested by a laboratory to define the entry loss. This is especially important when a large number of hoods are to be installed. With a sample hood the coefficient of entry can be defined for use in routine hood static suction measurements to calculate airflow. This approach will be discussed in the next section.

An initial physical inspection of the completed system should be conducted to determine if the system is built according to the design. The size and type of fittings should be checked and the hood construction reviewed. The fan type and size should be as specified and the direction of rotation, speed, and current drain should be noted. A detailed inspection sheet should be completed. This type of inspection may seem redundant, but in the author's experience it is worthwhile. In one case a mitre elbow was added in a large submain to save space; the single elbow presented high losses that caused the system to malfunction.

The air cleaning component should be inspected to determine if it is installed according to specifications. If ancillary equipment such as pressure sensing or velocity measuring equipment is included in the systems, it should also be inspected and calibrated.

All conventional ductwork and air cleaning equipment should be mounted on the suction side of the fan. If this is not done, toxic contaminants may leak to the occupied space. In one such case an installation handling a volatile and odorous organic chemical with the duct under positive pressure contaminated the workplace. Due to the complexity of the system, the cost of rectifying this problem was $50,000. This problem should have been picked up in the design review. If a duct run must be under positive pressure, special design features must be utilized by the engineer.

15.4.7.2 Airflow The general format for testing a new local exhaust ventilation system balanced without blast gates is shown in Fig. 15.16. The airflow rate through each branch servicing a hood should be evaluated by a pitot-static traverse (ACGIH, 1996). The pitot-static tube is a primary standard and, if used with an inclined manometer, it does not require calibration. Care is required in the choice of traverse location, and the device is limited at low velocities. In the past decade electronic manometers with microprocessor-based instruments are available for direct reading of pressure and velocity. When used in conjunction with a small portable recorder, the pitot traverse can be conducted by one person with ease.

One of two critical design velocities will be specified depending on the hood type. A partial enclosure such as a paint spray hood has a design velocity or control velocity at the plane of the hood face. Exterior hoods such as a simple welding hood will have a capture velocity specified at a certain working distance. The control and capture velocities may be calculated from the airflow rate and the hood dimensions. Frequently the actual velocities must be measured to satisfy plant or regulatory requirements. Face velocities at partial enclosures are usually evaluated by one of three direct-reading anemometers: rotating vane, swinging vane, or heated element. These instruments have a wide range of applications, but care should be taken to observe their limitations (Burgess, 1989). These simple direct-reading instruments can be used for the required inspections described in Section 15.2.7.

The measurement of hood static suction is also of value in periodic inspections and should be measured at the acceptance tests. The measurement, made 2–4 diameters downstream of the hood, provides a simple method of calculating airflow (Burgess, 1989).

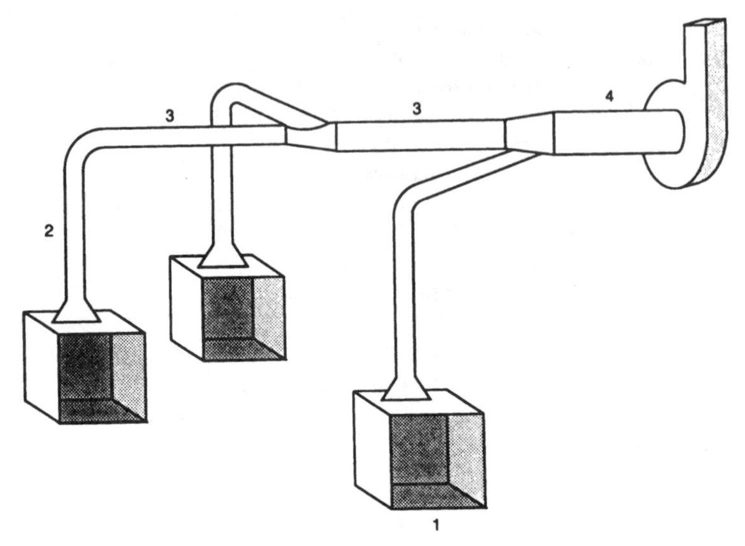

Fig. 15.16 Locations for ventilation measurements: (1) At the face of booth-type hoods, (2) just downstream of a hood at a branch location, (3) in the main to define hood exhaust rate by difference, and (4) in the main to define total system airflow.

The data collected in the acceptance tests are important baseline information for subsequent periodic inspections described in Section 15.2.8.

15.4.7.3 Performance The term *performance*, as used in this discussion, describes the ability of the local exhaust ventilation to minimize the release of air contaminants from equipment serviced by the system. This measurement can be done by qualitative, semiquantitative, and quantitative methods.

15.4.7.4 Qualitative Methods Smoke sticks, smoke candles, and theatre fog generating devices allow one to visualize air flow patterns around a hood. The use of smoke tracers released at the generation point, if done properly, provides an excellent qualitative method of exploring the performance of the hood. The control boundary for capture hoods can be established with this technique. In addition, it can be an excellent teaching tool for the worker who can view the impact of his own body and actions and that of external disturbances such as drafts from windows, doors, and traffic on the ventilation. Corrosive smoke and theater fog cannot be used in semiconductor facilities; dry ice, liquid nitrogen, and water vapor wands have been evaluated for this purpose. In Europe another qualitative technique, light scatter with back lighting, is widely used to identify the source and generation mode of particle contamination. Exquisite photographs of dust release based on this technique have been used as a design input for ceramic industry dust control.

15.4.7.5 Semiquantitative The local exhaust ventilation system is first assessed to determine if it meets good practice in terms of general design and airflow rate. If that is acceptable, the system is checked qualitatively by smoke and then semiquantitatively using a tracer (Fig. 15.17). A series of release grids are designed that model the

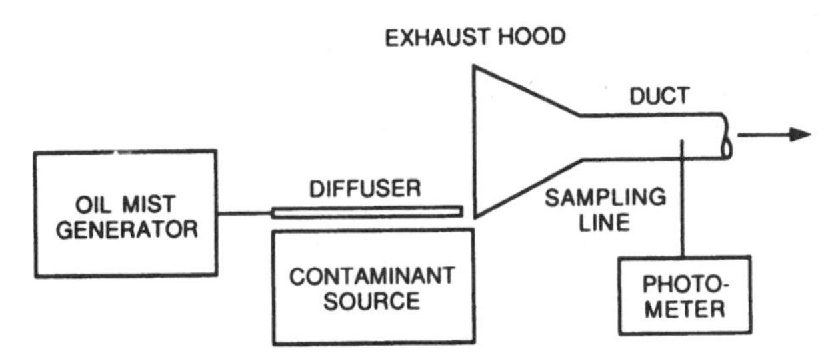

Fig. 15.17 Containment fraction for a hood can be determined in the field using a tracer introduced first inside the hood to establish the 100 percent benchmark and then at the normal contaminant release point. [From Ellenbecker et al. (1983). Reproduced with permission of the *Journal of the American Industrial Hygiene Association.*]

actual release area. An oil mist generator operating with corn oil is used to generate a submicrometer-sized aerosol. The generator is positioned to release the mist deep in the hood so that all mist is collected. The probe of a forward light scatter photometer samples the duct stream laden with the mist, and the reading is defined as 100 percent containment. The grid is then positioned at the actual work release point and a new reading is taken in the duct. If all the mist is collected, the reading will be 100 percent; if only one-half the oil mist is collected, the reading of containment is 50 percent.

This system and others now available using tracers such as sulfur hexafluoride are semiquantitative. The most common of these is the ASHRAE 110-1995 used to test performance in laboratory fume hoods (ASHRAE, 1995). This test procedure can be modified to test other types of enclosing hoods. These tests do not attempt to closely model the specific chemical or its generation rate. However, the approach provides insight into performance of specific systems and permits "tuneup" of operating parameters such as airflow rate, hood geometry, baffles, and so forth before the quantitative studies are conducted.

15.4.7.6 Quantitative The containment efficiency of a local exhaust ventilation system is usually impossible to establish in a plant using the actual chemical. To conduct such a test the system must either be modified so that all the contaminant enters the hood to establish 100 percent containment, or the fraction lost from the hood must be identified and that fraction added to the material collected by the hood. The fraction lost to the room usually cannot be measured.

In most cases, however, quantitative measurement of performance is made by air sampling on the worker. In this case the demonstrated performance is not specifically of hood containment but rather the ability of the local exhaust ventilation system to achieve control at the breathing zone of the worker. The most realistic test involves the operator using the actual chemical. Occasionally, as in the case of a premanufacturing notification (PMN) chemical, it may be useful to demonstrate performance of the controls using a substitute material. In one such test a simulant chemical was used for the actual PMN chemical. The simulant was a nontoxic blue dye that was inexpensive, reasonably dusty, and easy to sample and analyze. A later test with the actual chemical confirmed the information obtained with the tracer.

The Semiconductor Equipment and Materials International Organization published a Test Method for Enclosures Using Sulfur Hexafluoride Tracer Gas and Gas Chromatography (SEMI F15-93). The test method requires that tracer gas be released at a flow rate that would be expected during an accidental toxic gas release. This is based on the trip point of the emergency shutoff (excess flow) valve or the flow-restricting orifice located just downstream of the main cylinder valve.

While tracer gas is released, air samples are collected outside of the enclosure in polypropylene syringes and analyzed with a gas chromatograph. An equivalent release concentration for the process gas is calculated and the enclosure is considered acceptable if the TLV or PEL of the process gas is not exceeded.

15.4.8 Periodic Testing and Maintenance

The local exhaust ventilation system must be inspected periodically to ensure that the system meets performance and design standards and complies with various regulations. The inspection frequency varies and may range from a monthly to a semiannual interval. The system is inspected visually. The motor-fan components are inspected to determine the condition of the fan blades, direction of rotation, belt tension, guarding, and lubrication. The hoods and ductwork are inspected for corrosion damage, plugging, leaks, and any local modification by the plant. The airflow rate at each station is determined, preferably by the hood static suction method since it is accurate and fast. On semienclosing hoods such as paint spray booths the face velocity must be evaluated. Certain regulations may require measurement of capture velocity, that is, the velocity at the contaminant release point. This measurement is difficult to conduct accurately. The measurement of airflow rate to achieve that velocity is the more accurate measurement.

The records management system for the inspection program is of great importance especially for facilities with hundreds of hoods. The need for such a system is described by Stott and Platts (1986).

The recording and manipulation of the necessary data is time consuming, tedious and prone to inaccuracies. Most is recorded in notebooks, which must be transcribed onto official forms, which means that comparisons with previous measurements is often difficult. So at this stage, although a wealth of information is available regarding the extraction systems, the opportunity to analyze the data, initiate maintenance work, or diagnose faults is wasted.

These authors developed a computer-based system to satisfy these needs. The person conducting the field test uses a hand-held data terminal that has been loaded with a program on the ventilation system under test. Each ventilation station is named, the position to be tested is identified, and a space is provided for entry of the test results. In the plant the tester calls up the station and conducts the velocity or pressure test measurement. The data are inserted into the portable data logger. On return to the office the data logger is downloaded to a personal computer. The data are scanned, and those stations where a preset minimum value is not achieved are identified by the computer for followup. Additional measurements are requested, and based on these data a diagnosis of the ventilation system problem is identified by the computer for referral to plant engineering.

The importance of well-qualified personnel to design, install, operate, and inspect local exhaust ventilation systems has been discussed in detail (BOHS, 1987; Burton, 1991). It is especially important to develop an in-house training program for personnel who conduct the periodic inspections (Burgess, 1993; Johnson, 1993).

The maintenance problems encountered in system operation and hopefully identified by the acceptance and periodic inspection schedule have been discussed by Burton

(1991). The cost of conducting an inspection program has not been published, but in evaluating the cost-benefits of such a program, one should recall the installation cost of the system and its operating cost in addition to the important role it plays in worker health. In many locations the required inspection frequency has prompted the installation of direct-reading monitoring equipment, such as pressure sensors at a hood static suction location, a pitot-static tube in the hood branch, a heated element sensor or a swinging vane anemometer at the hood face. In a nuclear energy facility an in-line orifice meter is installed to measure flow through a glove box. This mass flow measuring technique has limited application due to energy loss, erosion of the orifice, and contamination.

In discussing the performance of local exhaust ventilation, DallaValle (1952) stated:

> "Whereas hoods are installed to eliminate a health hazard, tests should be conducted to establish their effectiveness. The ultimate criterion is not the provision of a 'strong' suction but the handling of an air volume which reduces the concentration of the contaminants in question below the MAC (maximum acceptable concentration) level."

15.5 SUMMARY

In this chapter the advances in control technology in the past decade are reviewed. The ability to identify critical exposure conditions has been enhanced by the coupling of sophisticated, direct-reading air sampling instruments with video taping of the job tasks. This technique permits the investigator to characterize the nature and origin of emissions in great detail with the subsequent application of effective controls. A second major shift during the decade is the emphasis on the reduction in the use of toxic materials with replacement by materials "kinder" to the worker and the environment. Equipment advances during this period include robotic techniques that permit the worker to be separated from the point of greatest exposure. Finally the application of ventilation control is enhanced by improved knowledge of airflow patterns into hoods, new techniques for evaluating hood containment, and management tools for ventilation systems.

REFERENCES

ACGIH (1996). Committee on Industrial Ventilation, Industrial Ventilation, *A Manual of Recommended Practice*, 25th ed., American Conference of Governmental Industrial Hygienists, Cincinnati, OH.

AFS (1972). *Foundry Environmental Control*, American Foundrymen's Society, Des Plaines, IL.

Ahearn, J., H. Fatkin, and W. Schwalm (1991). "Case Study: Polaroid Corporation's Systematic Approach to Waste Minimization," *Pollution Prevent. Rev.* **1**(13), Summer.

AISI (1965). *Steel Mill Ventilation*, American Iron and Steel Institute, New York.

Anderson, D. M. (1964). "Dust Control by Air Induction Technique," *Ind. Med. Surg.* **34,** 168.

ANSI (1988). "Precautionary Labeling of Hazardous Industrial Chemicals," ANSI Z129.1-1988, American National Standards Institute, New York.

ANSI (1996). "American National Standard for the Recirculation of Air from Industrial Process Exhaust Systems," ANSI/AIHA Z9.7-1996, American National Standards, Institute, New York.

ASHRAE (1995) ANSI/ASHRAE 110-1995, Performance Testing of Laboratory Hoods, American Society of Heating Refrigeration and Air Conditioning Engineers, 1791 Tullie Circle NE, Atlanta, GA 30329.

BCIRA (1977). *Proceedings of the Working Environment in Iron Foundries*, British Cast Iron Research Association, Birmingham, UK.

Berens, A. R. (1981). "Vinyl Chloride Monomer in PVC: From Problem to Probe," *Pure Appl. Chem.* **53,** 365–75.

BOHS (1984). British Occupational Hygiene Society, *Fugitive Emissions of Vapors from Process Equipment*, Technical Guide No. 3, Science Reviews Ltd., Northwood, Middlesex, UK.

BOHS (1985). British Occupational Hygiene Society, *Dustiness Estimation Methods for Dry Materials, Their Uses and Standardization*, Technical Guide No. 4, Science Reviews Ltd., Northwood, Middlesex, UK.

BOHS (1987). British Occupational Hygiene Society, *Controlling Airborne Contaminants in the Workplace*, BOHS Technical Guide No. 7, Science Reviews Ltd., Northwood, Middlesex, UK.

Brandt, A. D. (1947). *Industrial Health Engineering*, Wiley, New York.

Braun, P., and E. Druckman, Eds. (1976). "Public Health Rounds at the Harvard School of Public Health—Vinyl Chloride: Can the Worker Be Protected," *N. E. J. Med.* **294,** 653–57.

Burgess, W. A. (1993). "The International Ventilation Symposia and the Practitioner in Industry," *Ventilation '91, Proceedings of the Third International Symposium on Ventilation for Contaminant Control*, September 16–20, 1991, ACGIH, Cincinnati, OH.

Burgess, W. A. (1996) Cummings Award Lecture at American Industrial Hygiene Conference and Exhibition, Washington, D.C.

Burgess, W. A., M. J. Ellenbecker, and R. D. Treitman (1989). *Ventilation for Control of the Work Environment*, Wiley, New York.

Burton, D. J. (1991). *Industrial Ventilation Work Book*, IVE, Inc., Salt Lake City, UT.

Caplan, K. (1985a). "Philosophy and Management of Engineering Controls," in L. J. Cralley and L. V. Cralley, Eds., *Patty's Industrial Hygiene & Toxicology*, Wiley, New York.

Caplan, K. C. (1985b). "Building Types," in L. V. Cralley, L. J. Cralley, and K. C. Caplan, Eds., *Industrial Hygiene Aspects of Plant Operations*, Vol. 3, Macmillan, New York.

Cooper, T., and M. Gressel (1992). "Real-Time Evaluation at a Bag Emptying Operation—A Case Study," *Appl. Ind. Hyg.* **7**(4), 227–30.

DallaValle, J. M. (1952). *Exhaust Hoods*, Industrial Press, New York.

Ellenbecker, M. J. (1996) "Engineering Controls as an Intervention to Reduce Worker Exposure," *American Journal of Industrial Medicine* **29:**303–70.

Ellenbecker, M. J., R. J. Gempel, and W. A. Burgess (1983). "Capture Efficiency of Local Exhaust Ventilation Systems," *Am. Ind. Hyg. J.* **44,** 752–55.

EPA (1986). "Workshop: Predicting Workplace Exposure to New Chemicals," *Appl. Ind. Hyg.* **1**(3), R-11–R-13.

First, M. W. (1983). "Engineering Control of Occupational Health Hazards," *Am. Ind. Hyg. Assoc. J.* **44**(9), 621–26.

Frosch, R. A., and Gallopoulos, N. E. (1989). "Strategies for Manufacturing," *Scientific American*, September.

Gideon, J., E. Kennedy, D. O'Brien, and J. Talty (1979). "Controlling Occupational Exposures—Principles and Practices," National Institute for Occupational Safety and Health, Cincinnati, OH.

Goldschmidt, G. (1991). "Improvement of the Chemical Working Environment by Substitution of Harmful Substances. An Iterative Model and Tools," personal communication.

Goodfellow, H. D., and J. W. Smith (1989). "Dustiness Testing—A New Design Approach for

Dust Control," in J. H. Vincent, Ed., *Ventilation '88, Proceedings of the Second International Symposium on Ventilation for Contaminant Control*, Pergamon Press, Oxford, UK.

Gray, C. N., and P. J. Hewitt (1982). "Control of Particulate Emsisions from Electric-Arc Welding by Product Modification," *Ann. Occup. Hyg.* **25**(4), 431–38.

Gressel, M., and T. Fischbach (1989). "Workstation Design Improvements for the Reduction of Dust Exposures During Weighing of Chemical Powders," *Appl. Ind. Hyg.* **4**(9), 227–33.

Gressel, M., W. A. Heitbrink, J. McGlothlin, and T. Fischbach (1987). "Real-Time, Integrated, and Ergonomic Analysis during Manual Materials Handling," *Appl. Ind. Hyg.* **2**(3), 108–13.

Gressel, M., W. Heitbrink, J. McGlothlin, and T. Fischbach (1988). "Advantages of Real Time Data Acquisition for Exposure Assessment," *Appl. Ind. Hyg.* **3**(11), 316–20.

Hallock, M. F., Hammond, S. K., Kenyon, E., Smith, T. J., and Smith, E. R. (1993). "Assessment of Task and Peak Exposures to Solvents in the Microelectronics Fabrication Industry," *Applied Occupational and Environmental Health* **8**(11), November.

Hammond, C. M. (1980). "Dust Control Concepts in Chemical Handling and Weighing," *Ann. Occup. Hyg.* **23**(1), 95–109.

Hanman, B. (1968). *Physical Abilities to Fit the Job*, American Mutual Liability Insurance Company, Wakefield, MA.

Harten, T., and I. Licis (1991). "Waste Reduction Technology Evaluations of the U.S. EPA WRITE Program," *J. Air Waste Manag. Assoc.* **41**(8), 1122–29.

HSE (1988). Health and Safety Executive, "Control of Substances Hazardous to Health Regulations 1988, Approved Codes of Practice," HMSO, London.

International Standards Organization 14,000 (September, 1996). International Organization for Standardization, ISO Central Secretariat, Geneva, Switzerland.

Johnson, G. Q., R. Ostendorf, D. Claugherty, and C. Combs (1993). "Improving Dust Control System Reliability," *Ventilation '91, Proceedings of the Third International Symposium on Ventilation for Contaminant Control*, September 16–20, 1991, ACGIH, Cincinnati, OH.

Karger, E., and W. A. Burgess (1992). Personal communication.

Kireta, A. G. (1988). "Lead Solder Update," *Heat/Pip./Air Cond.*, 119–25.

Kovein, R. J., and Hentz, P. A. (1992). "Real-Time Personal Monitoring in the Workplace Using Radio Telemetry," *Applied Occupational and Environmental Health* **7**(3), March.

Lipton, S., and J. Lynch (1987). *Health Hazard Control in the Chemical Process Industry*, Wiley, New York.

Massachusetts General Laws (1989). "Massachusetts Toxics Use Reduction Act," Chapter 211, July 24.

Mertens, J. A. (1991). "CFCs: In Search of a Clean Solution," *Envir. Prot.*, 25–29.

Nill, R. (1996). Personal communication.

NIOSH (1991). Technical Report, Control of Emissions from Seals and Fittings in Chemical Process Industries, National Institute for Occupational Safety and Health, DHHS (NIOSH) Publication No. 81-118, Cincinnati, OH.

O'Brien, D., T. Fischbach, T. Cooper, W. Todd, M. Gressel, and K. Martinez (1989). "Acquisition and Spreadsheet Analysis of Real Time Dust Exposure Data: A Case Study," *Appl. Ind. Hyg.* **4**(9), 238–43.

Olishifski, J. B. (1988). *Methods of Control, Fundamentals of Industrial Hygiene*, Barbara Plog, Ed., 3rd ed., National Safety Council, Chicago, Illinois.

OTA (1986). Office of Technology Assessment, *Serious Reduction of Hazardous Waste for Pollution and Industrial Efficiency*, Washington, D.C., U.S. Government Printing Office.

Peterson, J. (1985). "Selection and Arrangement of Process Equipment," in L. V. Cralley, L. J. Cralley, and K. C. Caplan, Eds., *Industrial Hygiene Aspects of Plant Operations*, Vol. 3, Macmillan, New York.

Roach, S. A. (1981). "On the Role of Turbulent Diffusion in Ventilation," *Ann. Occup. Hyg.* **24**(1), 105–33.

Rossi, M., and Geiser, K. (1994). "A Proposal for Managing Chemical Restrictions at the State Level," *Pollution Prevention Review*, Spring.

Rossi, M., K. Geiser, and M. Ellenbecker (1991). "Techniques in Toxic Use Reduction: From Concept to Action," *New Solutions* **2**(2), 25–32.

SACMA (1991). Suppliers of Advanced Composite Materials Association, "Safe Handling of Advanced Composite Materials," Arlington, VA.

Schumacher, J. S. (1978). "A New Dust Control System for Foundries," *Am. Ind. Hyg. Ass. J.* **39**(1), 73–78.

Sciola, V. (1993). "The Practical Application of Reduced Flow Push Pull Plating Tank Exhaust Systems," *Ventilation '91, Proceedings of the Third International Symposium for Contaminant Control*, September 16–20, 1991, Cincinnati, OH.

SEMI F15-93 (1993). *A Test Method for Enclosures Using Sulfur Hexafluoride Tracer Gas and Gas Chromotography*, Semiconductor Equipment and Materials International Organization, 1993.

Sherwood, R. J., and R. J. Alsbury (1983). "Occupational Hygiene, Systematic Approach and Strategy of Exposure Control, in L. Parmeggiani, Ed., *Encyclopaedia of Occupational Health and Safety*, International Labor Organization, Geneva.

Soule, R. D. (1991). "Industrial Hygiene Engineering Controls," in G. Clayton and F. E. Clayton, Eds., *Patty's Industrial Hygiene and Toxicology*, Vol. 1, Part B, Wiley, New York.

Stott, M. D., and P. J. Platts (1986). "The Ventdata Ventilation Plant Monitoring and Maintenance System," in H. D. Goodfellow, Ed., *Ventilation '85, Proceedings of the First International Symposium for Contaminant Control*, Elsevier, Amsterdam.

Treitman, R. L. (1976). Personal communication.

Uniform Fire Code (1988). Article 51, Semiconductor Fabrication Facilities Using Hazardous Production Materials.

U.S. Congress (1990). Pollution Prevention Act of 1990, Congressional Record, Section 6606 of the Budget Reconciliation Act, p. 12517, October 26.

Volkwein, J. C., Cecala, A. B., Thimons, E. D. (1989). "Moisture Application for Dust Control," *Applied Industrial Hygiene* **4**(8), August.

Whitehead, L. W. (1987). "Planning Considerations for Industrial Plants Emphasizing Occupational and Environmental Health and Safety Issues," *Appl. Ind. Hyg.* **2**(2), 79–86.

Wolf, K., A. Yazdani, and P. Yates (1991). "Chlorinated Solvents: Will the Alternatives be Safer?" *J. Air Waste Manage. Assoc.* **41**(8), 1055–61.

Wolfson, H. (1993). "Is the Process Fit to Have Ventilation Applied?" in *Proceedings of the Third International Symposium on Ventilation for Contaminant Control*, September 16–20, 1991, Cincinnati, ACGIH.

APPENDIX A

NIOSH CONTROL TECHNOLOGY REPORTS*

RN: 00192317
TI: Engineering Health Hazard Control Technology for Coal Gasification and Liquefaction Process. Final Report
AU: Anonymous
SO: Division of Physical Sciences and Engineering, NIOSH, U.S. Department of Health and Human Services, Cincinnati, Ohio, Contract No. 210-78-0084, 105 pages, 78 references
PY: 1983

RN: 00092812
TI: Control Technology Assessment: The Secondary Nonferrous Smelting Industry
AU: Burton-DJ; Coleman-RT; Coltharp-WM; Hoover-JR; Vandervort-R
SO: Division of Physical Science and Engineering, NIOSH, Cincinnati, Ohio, NIOSH Contract No. 210-77-0008, 393 pages
PY: 1979

RN: 00092805
TI: An Evaluation of Occupational Health Hazard Control Technology for the Foundry Industry
AU: Scholz-RC
SO: Division of Physical Sciences and Engineering, NIOSH, Cincinnati, Ohio, DHEW (NIOSH)
Publication No. 79-114, Contract No. 210-77-0009, 436 pages, 56 references
PY: 1978

RN: 00145199
TI: Engineering and Other Health Hazard Controls in Oral Contraceptive Tablet Making Operations
AU: Anastas-MY
SO: Division of Physical Sciences and Engineering, NIOSH, U.S. Department of Health and Human Services, Cincinnati, Ohio, 94 pages

RN: 00080675
TI: Development of an Engineering Control Research and Development Plan for Carcinogenic Materials
AU: Hickey-JLS; JJ-Kearney
SO: Applied Ecology Department, Research Triangle Institute, Research Triangle Park, North Carolina NIOSH Contract No. 210-76-0147, 168 pages, 19 references
PY: 1977

**Source:* Courtesy of National Institute for Occupational Safety and Health.

RN: 00074602
TI: Engineering Control of Welding Fumes
AU: Astleford-W
SO: Division of Laboratories and Criteria Development, NIOSH, Cincinnati, Ohio, DHEW (NIOSH) Publication No. 75-115, Contract No. 099-72-0076, 122 pages, 13 references
PY: 1974

RN: 00074249
TI: Engineering Control Research and Development Plan for Carcinogenic Materials
AU: Hickey-J; Kearney-JJ
SO: Division of Physical Sciences and Engineering, NIOSH, Cincinnati, Ohio, Contract No. 210-76-0147, 167 pages, 19 references
PY: 1977

RN: 00052419
TI: Engineering Control Research Recommendations
AU: Hagoapian-JH; Bastress-EK
SO: Division of Physical Sciences and Engineering, NIOSH, Cincinnati, Ohio, Contract No. 099-74-0033, 210 pages, 181 references
PY: 1976

RN: 00182292
TI: Control Technology Assessment of Enzyme Fermentation Processes
AU: Martinez-KF; Sheehy-JW; Jones-JH
SO: Division of Physical Sciences and Engineering, NIOSH, U.S. Department of Health and Human Services, Cincinnati, Ohio, DHHS (NIOSH) Publication No. 88-114, 81 pages, 34 references
PY: 1988

RN: 00177551
TI: Minimizing Worker Exposure during Solid Sampling: A Strategy for Effective Control Technology
AU: Wang-CCK
SO: Division of Physical Sciences and Engineering, NIOSH, U.S. Department of Health and Human Services, 28 pages
PY: 1983

RN: 00177548
TI: A 3-E Quantitative Decision Model of Toxic Substance Control through Control Technology Use in the Industrial Environment
AU: Wang-CCK
SO: NIOSH, U.S. Department of Health and Human Services, Cincinnati, Ohio, 53 pages, 4 references
PY: 1982

RN: 00168700
TI: The Illuminating Engineering Research Institute and Illumination Levels Currently Being Recommended in the United States

AU: Crouch-CL
SO: The Occupational Safety and Health Effects Associated with Reduced Levels of Illumination, Proceedings of a Symposium, July 11–12, 1974, Cincinnati, Ohio, NIOSH, Division of Laboratory and Criteria Development, HEW Publication No. (NIOSH). 75-142, pages 17–27
PY: 1975

RN: 00133474
TI: Control Technology Assessment of the Pesticides Manufacturing and Formulating Industry
AU: Fowler-DP
SO: Division of Physical Sciences and Engineering, NIOSH, Cincinnati, Ohio, Contract No. 210-77-0093, 667 pages
PY: 1980

RN: 00148166
TI: NIOSH Technical Report: Control Technology Assessment: Metal Plating and Cleaning Operations
AU: Sheehy-JW; Mortimer-VD; Jones-JH; Spottswood-SE
SO: NIOSH, U.S. Department of Health and Human Services, Cincinnati, Ohio, Publication No. 85-102, 115 pages, 71 references
PY: 1984

RN: 00144935
TI: A Study of Coal Liquefaction and Gasification Plants: An Industrial Hygiene Assessment, a Control Technology Assessment, and the Development of Sampling and Analytical Techniques, Volume II
AU: Cubit-DA; Tanita-RK
SO: Division of Respiratory Disease Studies, NIOSH, U.S. Department of Health and Human Services, Morgantown, West Virginia, Contract No. 210-78-0101, 179 pages, 15 references
PY: 1983

RN: 00144934
TI: A Study of Coal Liquefaction and Gasification Plants: An Industrial Hygiene Assessment, a Control Technology Assessment, and the Development of Sampling and Analytical Techniques: Volume I
AU: Cubit-DA; Tanita-RK
SO: Division of Respiratory Disease Studies, NIOSH, U.S. Department of Health and Human Services, Morgantown, West Virginia, Contract No. 210-78-0101, 192 pages, 22 references
PY: 1983

RN: 00136362
TI: Control Technology Assessment of Selected Petroleum Refinery Operations
AU: Emmel-TE; Lee-BB; Simonson-AV

SO: Division of Physical Sciences and Engineering, NIOSH, U.S. Department of Health and Human Services, Cincinnati, Ohio, NTIS PB83-257-436, Contract No. 210-81-7102, 122 pages
PY: 1983

RN: 00136196
TI: Proceedings of the Second Engineering Control Technology Workshop, June 1981
AU: Konzen-RB
SO: Division of Training and Manpower Development, NIOSH, U.S. Department of Health and Human Services, Cincinnati, Ohio, NTIS PB-83-112-755, NIOSH Report No. 80-3794, 138 pages
PY: 1982

RN: 00133178
TI: Principles of Occupational Safety and Health Engineering, Instructor's Guide
AU: Zimmerman-NJ
SO: Division of Training and Manpower Development, NIOSH, U.S. Department of Health and Human Services, Cincinnati, Ohio, P.O. No. 81-3030, 262 pages, 17 references
PY: 1983

RN: 00135180
TI: Health Hazard Control Technology Assessment of the Silica Flour Milling Industry
AU: Caplan-PE; Reed-LD; Amendola-AA; Cooper-TC
SO: Division of Physical Sciences and Engineering, NIOSH, U.S. Department of Health and Human Services, Cincinnati, Ohio, 60 pages, 18 references
PY: 1982

RN: 00135171
TI: Engineering Noise Control Technology Demonstration for the Furniture Manufacturing Industry
AU: Hart-FD; Stewart-JS
SO: NIOSH, U.S. Department of Health, Education, and Welfare, Grant No. 1-ROH-OH-00953, 123 pages, 8 references
PY: 1982

RN: 00132081
TI: Control Technology Assessment for Chemical Processes Unit Operations
AU: Van-Wagenen-H
SO: NIOSH, Cincinnati, Ohio, Contract No. 210-80-0071, NTIS PB83-187-492, 19 pages
PY: 1983

RN: 00134239
TI: Engineering Control of Occupational Health Hazards in the Foundry Industry. Instructor's Guide
AU: Scholz-RC
SO: NIOSH, Cincinnati, Ohio, NTIS PB82-231-234, 156 pages, 49 references
PY: 1981

RN: 00133884
TI: Demonstrations of Control Technology for Secondary Lead Reprocessing
AU: Burton-DJ; Simonson-AV; Emmel-BB; Hunt-DB
SO: NIOSH, U.S. Department of Health and Human Services, Rockville, Maryland, Contract No. 210-81-7106, 291 pages
PY: 1983

RN: 00132005
TI: Pilot Control Technology Assessment of Chemical Reprocessing and Reclaiming Facilities
AU: Crandell-MS
SO: Engineering Control Technology Branch, NIOSH, Cincinnati, Ohio, NTIS PB83-197-806, 11 pages
PY: 1982

RN: 00130228
TI: Occupational Health Control Technology for the Primary Aluminum Industry
AU: Sheehy-JW
SO: Public Health Service, NIOSH, U.S. Department of Health and Human Services, Cincinnati, Ohio, DHHS Publication No. 83-115, 59 pages, 6 references
PY: 1983

RN: 00130213
TI: Control Technology Assessment in the Pulp and Paper Industry
AU: Schoultz-K; Matthews-R; Yee-J; Haner-H; Overbaugh-J; Turner-S; Kearney-J
SO: NIOSH, Public Health Service, U.S. Department of Health and Human Services, Cincinnati, Ohio, Contract No. 210-79-0008, 974 pages
PY: 1983

RN: 00106373
TI: Mechanical Power Press Safety Engineering Guide, Wilco, Inc., Stillwater, Minnesota
AU: Anonymous
SO: NIOSH, Division of Laboratories and Criteria Development, U.S. Department of H.E.W., Cincinnati, Ohio, 207 pages
PY: 1976

RN: 00123450
TI: Assessment of Engineering Control Monitoring Equipment. Volume II
AU: Anonymous
SO: Enviro Control, Inc., Rockville, Md., NIOSH, Cincinnati, Ohio, 393 pages, 23 references
PY: 1981

RN: 00123377
TI: Control Technology Assessment of Selected Process in the Textile Finishing Industry
AU: Collins-LH
SO: Bendix Launch Support Division, Cocoa Beach, Florida, NIOSH, Cincinnati, Ohio, 227 pages, 44 references
PY: 1978

RN: 00119991
TI: Phase I Report on Control Technology Assessment of Ore Beneficiation
AU: Todd-WF
SO: NIOSH, U.S. Department of Health and Human Services, Cincinnati, Ohio, 82 pages, 80 references
PY: 1980

RN: 00117613
TI: Symposium Proceedings, Control Technology in the Plastics and Resins Industry
AU: Anonymous
SO: NIOSH, U.S. Department of Health and Human Services, Cincinnati, Ohio, 333 pages, 23 references
PY: 1981

RN: 00117359
TI: Proceedings of the Symposium on Occupational Health Hazard Control Technology in the Foundry and Secondary Non-Ferrous Smelting Industries
AU: Anonymous
SO: NIOSH, U.S. Department of Health and Human Services, 401 pages, 45 references
PY: 1981

RN: 00116023
TI: Control Technology Summary Report on the Primary Nonferrous Metals Industry, Vol. IV, Appendix C
AU: Hoover-JR
SO: NIOSH, Center for Disease Control, Public Health Service, U.S. Department of Health, Education, and Welfare, 120 pages
PY: 1978

RN: 00116022
TI: Control Technology Summary Report on the Primary Nonferrous Metals Industry, Vol. 5, Appendix D
AU: Hoover-JR
SO: NIOSH, Center for Disease Control, Public Health Service, U.S. Department of Health, Education, and Welfare, 180 pages
PY: 1978

RN: 00115792
TI: Control Technology for Primary Aluminum Processing
AU: Sheehy-JW
SO: Department of Health and Human Services, Public Health Service, Center for Disease Control, NIOSH, Division of Physical Sciences and Engineering, Cincinnati, Ohio, pages 1–22
PY: 1980

RN: 00115530
TI: An Evaluation of Engineering Control Technology for Spray Painting

AU: O'Brien-DM; Hurley-DE
SO: NIOSH, Center for Disease Control, Public Health Service, U.S. Department of Health and Human Services, 117 pages, 52 references
PY: 1981

RN: 00114224
TI: Control Technology Summary Report on the Primary Nonferrous Metals Industry, Vol. II, Appendix A
AU: Coleman-RT; Hoover-JR
SO: Division of Physical Science and Engineering, NIOSH, 291 pages, 10 references
PY: 1978

RN: 00114223
TI: Control Technology Summary Report on the Primary Nonferrous Metals Industry, Vol. III, Appendix B
AU: Coleman-RT
SO: Division of Physical Science and Engineering, NIOSH, Cincinnati, Ohio, 102 pages
PY: 1978

RN: 00113373
TI: Control Technology Summary Report on the Primary Nonferrous Smelting Industry, Volume 1: Executive Summary
AU: Coleman-RT; Hoover-JR
SO: Division of Physical Science and Engineering, NIOSH, Cincinnati, Ohio, Contract No. 210-77-0008, Radian Corporation, Austin, Texas, 36 pages, 8 references
PY: 1978

RN: 00112367
TI: Control Technology Assessment of Raw Cotton Processing Operations
AU: Anonymous
SO: Envirocontrol Inc., Rockville, Md., NIOSH, Cincinnati, Ohio, Contract No. 210-78-0001, 351 pages
PY: 1980

RN: 00112366
TI: Engineering Control Technology Assessment for the Plastics and Resins Industry
AU: Anonymous
SO: Division of Physical Sciences, NIOSH, U.S. Department of Health, Education and Welfare, Contract No. 210-76-0122, Cincinnati, Ohio, 234 pages, 60 references
PY: 1977

RN: 00102819
TI: Proceedings of the Symposium on Occupational Health Hazard Control Technology in the Foundry and Secondary Non-Ferrous Smelting Industries
AU: Scholz-RC; Leazer-LD
SO: Department of Health, Education, and Welfare, Public Health Service Center for Disease Control, NIOSH, Cincinnati, Ohio, 447 pages, 10 references
PY: 1980

RN: 00094260
TI: Assessment of Selected Control Technology Techniques for Welding Fumes
AU: Van-Wagenen-HD
SO: Division of Physical Sciences and Engineering, NIOSH, Cincinnati, Ohio, NIOSH Publication No. 79-125, 29 pages, 17 references
PY: 1979

RN: 00094228
TI: Control Technology for Worker Exposure to Coke Oven Emissions
AU: Sheehy-JW
SO: Division of Physical Sciences and Engineering, NIOSH, Cincinnati, Ohio, NIOSH Publication No. 80-114, 29 pages, 26 references
PY: 1980

RN: 00092888
TI: Proceedings of NIOSH/University Occupational Health Engineering Control Technology Workshop
AU: Talty-JT
SO: Proceedings of the Workshop on Occupational Health Engineering Control Technology, May 16–17, 1979, Division of Physical Sciences and Engineering, NIOSH, Cincinnati, Ohio, 149 pages
PY: 1979

RN: 00092810
TI: Control Technology Summary Report on the Primary Nonferrous Metals Industry. Volume V. Appendix D: Review of the Testimony Presented at the 1977 OSHA Public Hearing on Sulfur Dioxide
AU: Hoover-JR
SO: Division of Physical Science and Engineering, NIOSH, Cincinnati, Ohio, NIOSH Contract No. 210-77-0008, 183 pages
PY: 1978

RN: 00092809
TI: Control Technology Summary Report on the Primary Nonferrous Metals Industry. Volume IV. Appendix C: Review of the Testimony Presented at the 1977 OSHA Public Hearings on Inorganic Lead
AU: Hoover-JR
SO: Division of Physical Science and Engineering, NIOSH, Cincinnati, Ohio, NIOSH Contract No. 210-77-0008, 123 pages
PY: 1978

RN: 00092808
TI: Control Technology Summary Report on the Primary Nonferrous Metals Industry. Volume III. Appendix B: Review of the Testimony Presented at the 1975/6 OSHA Public Hearing on Inorganic Arsenic
AU: Coleman-RT
SO: Division of Physical Science and Engineering, NIOSH, Cincinnati, Ohio, NIOSH Contract No. 210-77-0008, 105 pages
PY: 1978

RN: 00092807
TI: Control Technology Summary Report on the Primary Nonferrous Metals Industry. Volume II: Appendix A
AU: Coleman-RT; Hoover-JR
SO: Division of Physical Science and Engineering, NIOSH, Cincinnati, Ohio, NIOSH Contract No. 210-77-0008, 298 pages, 9 references
PY: 1978

The following reports are not yet listed in NIOSHTIC:

Control Technology for Ethylene Oxide Sterilization in Hospitals, V. D. Mortimer and S. L. Kercher, Division of Physical Science and Engineering, NIOSH, Cincinnati, Ohio, Pub No. 89-120, 167 pages, 82 references, 1989

Control of Asbestos Exposure During Brake Drum Service, J. W. Sheey, T. C. Cooper, D. M. O'Brien, J. D. McGlothlin, and P. A. Froelich, Division of Physical Science and Engineering, NIOSH, Cincinnati, Ohio, Pub No. 89-121, 69 pages, 48 references, 1989

CHAPTER 16

How to Select and Use Personal Protective Equipment

RICHARD J. NILL

Genetics Institute, 1 Burtt Road, Andover, MA 01810

16.1 INTRODUCTION
16.2 GENERAL CONSIDERATIONS
- 16.2.1 Goals of an Effective PPE Program
- 16.2.2 Hazard Assessment
- 16.2.3 PPE Can Be Hazardous
- 16.2.4 PPE in the Hierarchy of Hazard Control
- 16.2.5 Selecting and Acquiring Personal Protective Equipment
- 16.2.6 User Training
- 16.2.7 PPE Inspection, Maintenance, and Storage
- 16.2.8 Standards and Regulations
- 16.2.9 Who Must Pay for PPE

16.3 TYPES OF PPE
- 16.3.1 Chemical Protective Clothing
- 16.3.2 Head Protection
- 16.3.3 Eye and Face Protection
- 16.3.4 Hand Protection
- 16.3.5 Foot Protection
- 16.3.6 Other PPE

REFERENCES

16.1 INTRODUCTION

> A school crossing guard's reflective vest. An ironworker's fall-arrest system. A floor installer's kneepads. A police officer's bulletproof vest. An astronaut's spacesuit. A bridge-worker's life jacket. A firefighter's breathing apparatus. A machinist's safety glasses. An NFL quarterback's helmet. A nail-machine operator's ear plugs. A fish processor's cut-resistant gloves. A chemical worker's butyl rubber boots. A lead battery worker's dust respirator. A traveling salesman's seat belt. A chemist's faceshield. A construction worker's hard hat. A hazmat responder's encapsulating suit. . . .

Worn daily by millions of workers for protection against both the expected and the unexpected hazard, personal protective equipment ranges from the simple article to the

Handbook of Occupational Safety and Health, Second Edition, Edited by Louis J. DiBerardinis, ISBN 0-471-16017-2

complex system. It may protect the entire body, or just one part. It may be durable or disposable, low cost or expensive, fully portable or motion restricting, comfortable or annoying, fashionable or dull. Widely varied in both form and function, all personal protective equipment has one important thing in common: It is worn by individual workers to reduce the personal risk of occupational injury and illness.

This chapter will cover many of the types of personal protective equipment (PPE) in common use in general industry. It will introduce the reader to the different types available, selection and use considerations, factors affecting successful PPE use in an organizational setting, and sources of additional information.

Some PPE is covered elsewhere within this handbook, including respiratory and hearing protection (Chapters 17 and 22, respectively).

16.2 GENERAL CONSIDERATIONS

16.2.1 Goals of an Effective PPE Program

To deliver the desired protection, the use of personal protective equipment must be effectively managed by the organization, no matter how large or small. PPE is well managed when the following elements are in place:

- Hazardous tasks have been evaluated, and PPE has been selected where needed;
- The right PPE, properly sized, is always available to the worker;
- Users are knowledgeable in when, where, why, and how to use PPE;
- PPE is kept in good working condition;
- PPE is worn every time.

16.2.2 Hazard Assessment

Shortly after assuming a safety role at the operations level, the safety and health practitioner will be asked to help determine whether personal protective equipment is required for a specific job or task in the workplace. Where do you find this information? With few exceptions, the appropriate protective equipment for a task or occupation will not be found prescribed in a textbook, or in a government regulation or concensus standard. Instead, case-by-case judgments must be made using information about the workplace and job task, common sense estimates of the potential for injury, and knowledge about the types of protective equipment available.

The need for PPE in the workplace is determined using some form of formal or informal hazard assessment. The more common types of formal hazard assessments that can uncover the need for PPE are listed below.

16.2.2.1 Accident/Incident Investigation An injury should not be, but often is, the first indication that the job is not being performed as safely as it should be. Lessons from accidents are hard won, and should not be wasted.

PPE is commonly prescribed after an accident as a measure to prevent reoccurrence, or to reduce its severity. If you are lucky enough to experience a noninjury incident or "near miss," it can be a golden opportunity to examine work practices, including assessing the need for PPE, before harm to the worker actually occurs.

PPE requirements identified in an accident investigation should be shared with other parts of the organization having similar hazards. Knowledge about an actual injury experience can help motivate workers to wear protective equipment.

A more detailed discussion of accident investigation is presented in Chapter 4.

16.2.2.2 Job Safety Analysis Job safety analysis, commonly referred to as JSA or task safety analysis, is a methodical, easy to understand evaluation of a job or task. Preparation and updating of JSAs is an excellent worker-involvement safety activity.

A three-column form is used, as shown in Chapter 6. Each step of the job or task is listed in the left column. The hazards associated with each task are then listed in the center column, and protective or preventive measures, often including the use of PPE, are listed in the right column. A completed JSA is a good tool for initial and ongoing worker training.

16.2.2.3 Process Hazard Review What if/checklist, failure mode and effect, and fault tree analysis are techniques used to evaluate the hazard of complex systems, especially those carrying a risk of serious injury or death. One frequent outcome of these reviews is to identify or reinforce the need for personal protective equipment. A more detailed discussion of hazard analyses is presented in Chapter 6.

In addition to the above hazard review techniques, many organizations also establish other types of formalized procedures for safety review of changes to existing, or introduction of new, processes. For example, introduction of a new chemical may require the sponsoring engineer to obtain and circulate an MSDS to a defined list of functional organizations responsible for storing, handling, and disposing the chemical, hazmat emergency response, first aid staff, and industrial hygiene, toxicology, medical and safety advisors. The hazard review and communication that occur as part of this signoff procedure often includes specification of appropriate PPE for all those who may be at risk of exposure.

16.2.2.4 Targeted PPE Hazard Assessment PPE can be targeted as the subject for a department or organization-wide hazard assessment. A PPE hazard assessment is similar to a JSA, but is limited to identifying the different types of PPE as a preventive measure. Using a form like that in Fig. 16.1, a member or members of the organization survey the department, looking for hazards by walking around, observing work activity, and talking to workers and management. The following hazard categories are of particular interest:

1. *Impact* that could arise from sources of motion, such as machinery or processes where any movement of tools, machine elements, or particles could exist. Impact could also occur from movement of personnel that could result in collision with stationary objects (low pipes), or from objects falling from above, or from work surface to floor.
2. *Penetration* arising from sharp objects that could pierce the hands or feet, or the head if dropped from above.
3. *Compression* of feet by rolling or pinching objects.
4. *Chemical or harmful dust exposure* via inhalation or skin contact.
5. *Heat,* capable of causing contact burns or physiological heat stress.
6. *Light (optical) radiation* such as ultraviolet or laser light that could harm the eye.
7. *Noise.*

PPE Hazard Assessment and Selection Form

Bldg/Area:____________________ Survey By:____________________
(Dept Supervisor/Team Leader)

Dept: ____________________ Date:____________________

Hazard Type (Impact, penetration, chemical, heat, harmful dust, compression, light, radiation (e.g. welding, laser) electric shock, noise, etc)	**Location/Source/Task**	**Recommended PPE**

Fig. 16.1. PPE hazard assessment and selection form.

Workers should be involved throughout the hazard assessment and personal protective equipment selection process. Experienced workers are nearly always the persons most familiar with how the job is performed, and their input is key to a thorough hazard assessment. Workers will also often have to change their behavior in order to perform the work while wearing the protective equipment; involvement in hazard assessment helps provide an understanding as to why the equipment should be worn, a key to motivating individuals to make the accommodations to use it.

16.2.3 PPE Can Be Hazardous

The use of personal protective equipment can itself bring new hazards to the job that are easily overlooked. PPE is often viewed as being inherently free of hazards because, after all, it is "safety equipment." When selecting or specifying PPE, watch for hazard tradeoffs.

For example, nonbreathable chemical protection garments can lead to heat stress, which can be more of a hazard than the chemical threat. Placing hearing protection on an already hearing impaired worker can make him/her unable to hear warning sounds. Loose-fitting gloves used near in-going nip points can add to the compression hazard. Ill-fitting safety shoes can injure the foot or trip the wearer. An air-supplied respirator hooked into a nitrogen pipe can be deadly. A hard hat worn during elevated work can become the most likely falling-object hazard if worn without a chin strap. In all these cases, the benefits of the PPE should be weighed against the new risks, and the application of PPE will require that additional efforts, especially in worker training, be made to minimize the new risk.

16.2.4 PPE in the Hierarchy of Hazard Control

The use of PPE makes the most sense when the hazard cannot be fully controlled by first using others means, including the following (refer also to Chapter 15):

- Change materials or methods
- Use engineering controls to enclose or isolate the hazard from the worker using full or partial barriers. Use exhaust ventilation.
- Develop and communicate safe work practices

Personal protective equipment may appear to be an easy way to control hazards, but it should be viewed as one of the organization's last lines of defense. Why? Because *it takes a significant effort by the organization to make sure that the correct PPE is worn by everyone, all the time.*

Each user must be properly equipped, must be knowledgeable of what is required when, know how to use it, and be motivated to use and maintain it all the time. Even in a small department of four workers, that can add up to too much opportunity for noncompliance or error, especially when the equipment is uncomfortable or inhibits productivity.

There are many downsides to the use of PPE. Usually there are some workers who do not believe the equipment is necessary, and therefore will not use it, no matter how hard you try to convince them. Some individuals are never comfortable with the PPE on, while most need at least some time for familiarity and adaptation. Sizing the very large or small worker is a frequent problem. Dirty, contaminated, or damaged personal protection can be worse than using none at all. Finally, the use of inappropriate PPE can give the worker a false sense of security, possibly resulting in a more serious injury or illness.

In spite of the limitations of PPE, its own possible hazards, and the difficulty in managing its use, personal protective equipment in many cases is highly useful and essential for safe task performance.

16.2.5 Selecting and Acquiring Personal Protective Equipment

A hazard assessment of the task or workplace will lead into the selection and specification of protective equipment. Specific selection guidelines are provided later in this chapter. General factors to consider include the following:

- Protective capability. Can the available equipment provide the degree and type of protection needed for the particular hazard? What are the equipment's limitations? What new hazards will be introduced by using the equipment, and how can these be controlled?
- Impact on product quality. In some environments like manufacturing, research, and health care, some protective gloves or clothing can shed particles or elements harmful to the particular process.
- Task compatibility. Can the worker perform the task while using the equipment?
- Comfort. Is one type of equipment more comfortable than another?
- Ease of use. How difficult is it to don and operate the equipment? Is one brand more user-friendly?
- Supplier service. Is a knowledgeable supplier available to assist with selection and delivery of samples and product?
- Product identification. How are the different brands packaged? Can sizes be readily determined by labels or color coding?
- Sizes. Is a range of sizes available? Does one product provide a single size with more of a universal fit compared with another?
- Durability. Will the equipment stand up to the physical and chemical demands of the task? Is disposable equipment more practical than reusable?
- Cleaning and decontamination. If the equipment is reusable, can it be cleaned and contaminated? Who will clean it, and how often?
- Cost. Does the risk justify the more expensive PPE?
- Fashion. Will one brand be more acceptable to workers because it is more up-to-date?

The most important factor determining whether or not the equipment will be worn is *user comfort.* It is essential to allow workers to participate in the selection. *People like to choose what they wear*—even safety equipment. A variety of types, styles, colors, and sizes should be made available to workers, within practical limits. Note that some regulations, like the OSHA asbestos standard as it relates to respiratory protection, require that workers be provided with more than one type or style (*Asbestos*, 1996). Allow for a trial use of samples in the workplace before making a final selection.

Personal protective equipment is a big business, and it should not be difficult to find manufacturers and distributors to help with PPE selection for most applications. *Best's Safety Directory* (1996) is one of the most complete resource guides for PPE shoppers. *Industrial Hygiene News* (1996), *Occupational Health and Safety* (1996), *Occupational Hazards* (1996), and *Chilton's Industrial Safety & Hygiene News* (1996) are periodicals that heavily advertise PPE to industry in the form of monthly and bimonthly magazines and tabloids, and annual buyer's guides. Free subscriptions are generally available to the safety and health practitioner.

The annual conferences of the American Industrial Hygiene Association (2700 Prosperity Avenue, Suite 250, Fairfax, VA 22031 (703) 849-888), American Society of Safety Engineers (1800 E. Oakton Street, Des Plaines, IL 60018-2187, 708-692-4121), and National Safety Council (1121 Spring Lake Dr., Itasca, IL 60143-3201, 800-621-7619) include large expositions featuring a wide variety of exhibitors demonstrating PPE products. Local and national suppliers can also be found in the yellow pages under Safety Equipment & Clothing.

16.2.6 User Training

The ultimate goal of a good personal protective equipment program is *to ensure that the right PPE is correctly worn every time its needed for worker protection.* Training and motivating workers expected to wear PPE is no less important, and probably more challenging, than each of the other elements of a PPE program. See Chapter 14 for assistance in developing a training program.

User training should be provided upon initial assignment to the job, whenever the hazards or PPE type changes, or when usage indicates a lack of skill or understanding on the part of the user.

By definition, training conveys mostly "how to" as opposed to why, but a PPE training session should provide some background as to the reasons why it is needed, especially including hazard assessment information.

Training content should be based on the PPE manufacturer's recommendations and the organization's own specific requirements, and should include the following:

- A brief explanation of why the particular equipment is necessary;
- Where and when it should be used;
- Where to get it, including replacement parts;
- How to properly don, doff, adjust, and wear it;
- The equipment's limitations, especially as applied in the specific workplace;
- Proper care, maintenance, useful life, storage, cleaning, and disposal of PPE.

Each worker should demonstrate an understanding of the training, and the ability to use the equipment properly. A written quiz at the conclusion of the training session can be helpful. The opportunity to try the equipment on should be provided during the training session. For more complicated equipment such as self-contained breathing apparatus (SCBA), each user should be given a practical use test, such as correctly inspecting and donning the unit and negotiating a maze in the dark. New users of any PPE should be observed closely by supervisors and co-workers for proper technique. PPE use is also a common item included in periodic safety walkthrough audits. (Also see Chapter 30).

The organization should take advantage of multiple opportunities to communicate management's expectations regarding when PPE should be used. They may include the following:

- Signs or rules posted in the workplace. PPE may be a posted requirement for an entire area, and/or for specific tasks or workstations within the area;
- As part of new employee orientation;
- Periodic training classes on PPE;
- One-on-one safety contacts;

- Job safety analyses, or other hazard reviews, kept accessible to the workplace or posted on walls;
- As a periodic topic in department safety meetings;
- Within standard operating procedures, such as batch prep sheets;
- As part of a set of general safety rules for the organization;
- As part of department or employee safety handbooks;
- As a standalone PPE program document;
- Floor plans with PPE requirements mapped (Fig. 16.2);
- PPE Matrixes (Fig. 16.3);
- In written job postings or advertisements.

16.2.7 PPE Inspection, Maintenance, and Storage

When purchased, most PPE arrives with written instructions from the manufacturer that cover inspection, maintenance, and storage. These should be covered in the user training as discussed above.

Workers have a tendancy to extend the life of protective equipment beyond its capability, especially if replacements are not readily available. Worn or damaged equipment should be discarded and replaced. PPE should be stored in a clean, dry designated location and put away after use. It should not be draped around the work area, exposed to dirt or chemical contamination, or degredation by sunlight or weather.

While responsibility for PPE maintenance is usually left to the individual worker, some organizations will issue PPE from a central stockroom to which the worker returns the used equipment at the end of the workday or workweek. The stockroom will then perform inspection, cleaning, and maintenance as needed before returning the equipment to service.

16.2.8 Standards and Regulations

The use of personal protective equipment is regulated in general industry and construction in the United States by the U.S. Department of Labor, Occupational Safety and Health Administration. For general industry, Subpart I of 29 CFR 1910, General Industry Standards holds the bulk of the PPE standards, while the subject index under Protective Clothing & Equipment makes reference to miscellaneous PPE requirements sprinkled through the various standards. The PPE standards were updated in 1994 to require employers to select PPE based on a hazard assessment, and to provide user training. A brief section on hand protection was added, and some changes were made to sections covering eye and face, head, and foot protection, and electrical protective equipment. A provision was added specifically prohibiting the use of defective or damaged equipment. In 1994 OSHA also updated its fall protection standards for construction, Subpart M.

OSHA enforces by reference the provisions of several PPE-related national consensus standards published by the American National Standards Institute (ANSI), which specify performance requirements for the manufacture and use of eye and face, head, respiratory, and foot protection. The American Society for Testing and Materials (ASTM) publishes testing procedures and performance requirements for electrical insulating equipment, and test procedures for chemical permeation and liquid penetration of materials used to make chemical protective clothing. The National Fire Protection Association (NFPA), publishes standards applicable to firefighters clothing, including suits worn for chemical vapor and

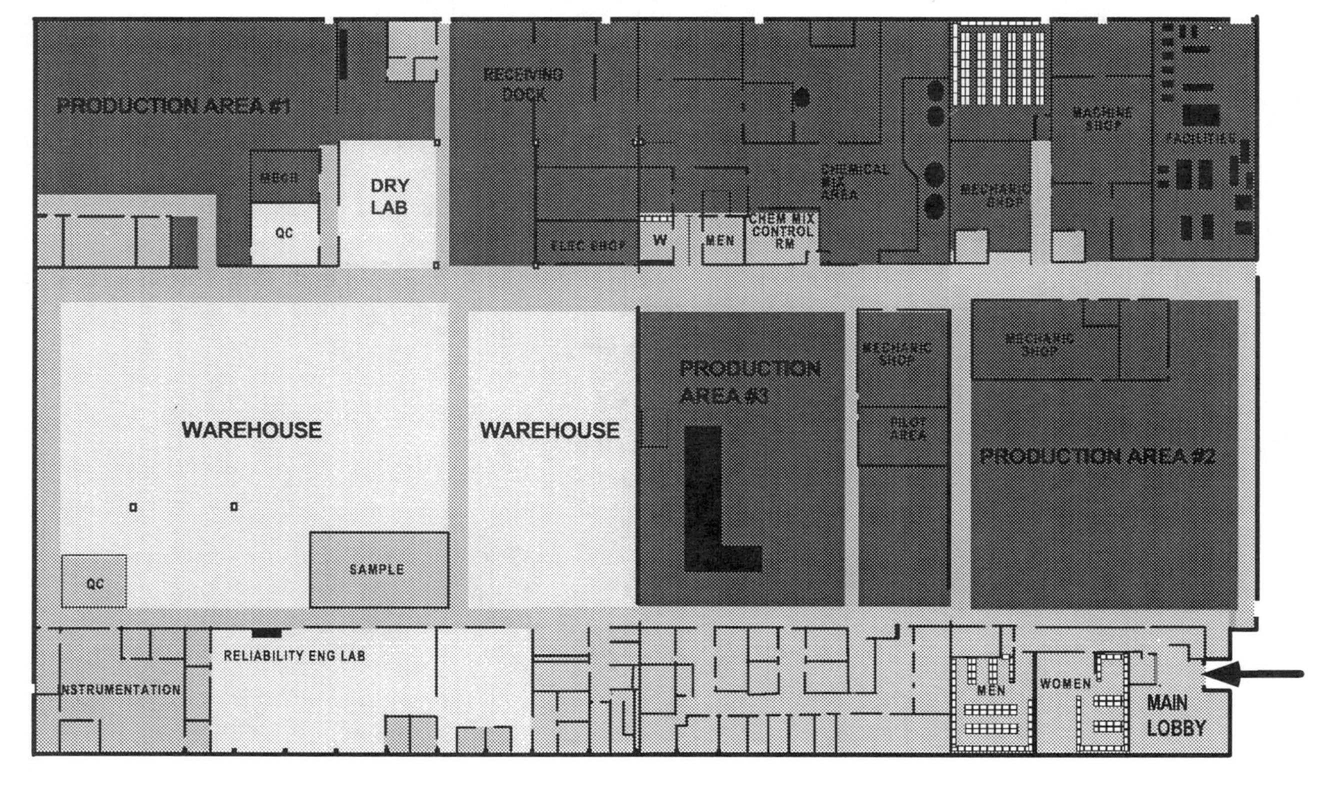
SAFETY SHOE POLICY
PRODUCTION AREA #1
DRY LAB
QC
RECEIVING DOCK
ELEC SHOP
W
MEN
CHEM MIX CONTROL RM
CHEMICAL MIX AREA
MECHANIC SHOP
MACHINE SHOP
FACILITIES
WAREHOUSE
WAREHOUSE
PRODUCTION AREA #3
MECHANIC SHOP
PILOT AREA
MECHANIC SHOP
PRODUCTION AREA #2
QC
SAMPLE
RELIABILITY ENG LAB
INSTRUMENTATION
MEN
WOMEN
MAIN LOBBY
SAFETY SHOES REQUIRED
SAFETY SHOES NOT REQUIRED
SAFETY SHOES NORMALLY NOT REQUIRED BUT MAY BE REQUIRED WHEN PERFORMING CERTAIN JOB ACTIONS.
THIS SAFETY SHOE REQUIREMENT APPLIES TO ANYONE IN THESE AREAS. THE ONLY EXCEPTIONS ARE FOR NON-WORKING VISITORS, WHERE THE RISKS OF FOOT INJURY HAVE BEEN PROPERLY CONTROLLED.
NOTE: SOME SPECIFIC JOBS OR TASKS REQUIRE THE USE OF SAFETY SHOES REGARDLESS OF LOCATION. CONSULT WRITTEN POLICY FOR DETAILS.

Fig. 16.2 PPE requirements mapped onto a floor plan.

Personal Protective Equipment (PPE) Requirements for Location or Task

Bldg/Area: ______________________ Survey By: ______________________
(Dept Supervisor/Team Leader)

Dept: ______________________ Date: ______________________

Certification of Hazard Assessment

Based upon an assessment of workplace hazards, the following is the minimum level of personal protection required for the locations and operations listed.

Minimum PPE Required		Job Task, Area, or Job Function		
Head	Hard hat			
	Bump cap			
Eye and Face	Safety glasses w/side shields			
	Chemical goggles			
	Face shield			
	Laser glasses/goggles			
	Welder's goggles or helmet *			
	Other eye and face			
Hand	Gloves, leather			
	Gloves, chemical *			
	Gloves, thermal			
	Other hand protection *			
Body	Apron *			
	Coverall *			
	Other body protection *			
Foot	Chemical boots *			
	Safety shoes			
	Other foot protection *			
Respiratory	Respiratory protection *			
Ear	Hearing protection			
Other PPE				

*Use footnotes to fully specify, i.e. optical shade, material of construction, brand and model, or stock number.

Footnotes : ______________________

____

Fig. 16.3. Personal protective equipment (PPE) requirements for location or task.

splash protection. The U.S. Center for Disease Control's National Institute for Occupational Safety and Health (NIOSH), commonly referred to as OSHA's sister agency for research, must approve all occupational respiratory protection.

A list of various PPE types with applicable standards and regulations is provided in Table 16.1.

16.2.9 Who Must Pay for PPE

It is good business to provide protective equipment that helps keep workers on the job, ready to come to work day after day, and many employers pay for PPE required by the job. In 1994, OSHA indicated in its instructions to field compliance personnel that

TABLE 16.1 Standards and Regulations Applicable to Personal Protective Equipment

PPE type	OSHA 29 CFR 1910 General Industry Standards	Consensus standards
General	1910.132 General Requirements	General information available thru the Industrial Safety Equipment Association, (703)525-1695
Chemical protective clothing (CPC)	1910.120 Hazardous Waste Operations and Emergency Response	ASTM F739, Standard Test Method for Resistance of Protective Clothing Materials to Permeation by Liquids or Gases under Conditions of Continuous Contact
		ASTM F903 Standard Test Method for Resistance of Protective Clothing Materials to Penetration by Liquids
		ASTM F1001 Standard Guide for Chemicals to Evaluate Protective Clothing Materials
		ASTM F1052 Standard Practice for Pressure Testing of Gas-Tight Totally Encapsulating Chemical-Protective Suits
		ASTM 1154 Standard Practice for Qualitatively Evaluating the Comfort, Fit, Function and Integrity of Chemical Protective Suit Ensembles
		ASTM F1342 Standard Test Method for Protective Clothing Material Resistance to Puncture
		ASTM F1359 Standard Practice for Determining the Liquid-Tight Integrity of Chemical Protective Suits or Ensembles under Static Conditions
		NFPA 1991, Standard on Vapor Protective Suits for Hazardous Chemical Emergencies
		NFPA 1992 Standard on Liquid Splash-Protective Suits for Hazardous Chemical Emergencies
		NFPA 1993 Support Function Protective Clothing for Hazardous Chemical Operations
Head	1910.135 Head Protection	ANSI Z89.1-1986 American National Standard for Personnel Protection—Protective Headwear for Industrial Workers—Requirements
Eye and face	1910.133 Eye and Face Protection	ANSI Z87.1-1989 American National Standard Practice for Occupational and Educational Eye and Face Protection
		ANSI Z136.1-1993 Safe Use of Lasers
Hand	1910.138 Hand Protection	Some information available from The International Hand Protection Association, (301)961-8680 and the National Industrial Glove Distributors Association (215)564-3484
Foot	1910.136 Foot Protection	ANSI Z41-1991 American National Standard for Personal Protection—Protective Footwear
Fall protection	Subpart M of OSHA 1926 Construction Standard	ANSI A10.14-1991 Requirements for Safety Belts, Harnesses, Lanyards, Lifelines, and Droplines in Construction and Industrial Use
		ANSI Z359.1-1992 Safety Requirements for Personal Fall Arrest Systems, Subsystems, and Components

where PPE is required to perform the job safely in compliance with OSHA standards, and the equipment is not uniquely personal in nature or usable off the job, then the employer must provide and pay for it (OSHA, 1994a). This directive was challenged by the Union Tank Car Company, and in October, 1997, the Occupational Safety and Health Review Commission ruled that while employers must ensure that PPE is provided, they need not pay for it because this was not explicitly stated in the OSHA PPE standard. (*Secretary of Labor vs. Union Tank Car Co., OSHRC. No. 96-0563, 10/16/97.*) At this writing, OSHA has stated its intention to amend the standard through a formal rulemaking requiring employers to pay.

OSHA makes an exception to the "employer must provide" rule to allow workers to provide their own equipment to accommodate work situations in which it is customary for workers in a particular trade, such as welding, to provide their own PPE. Such cases are the exception, not the norm.

16.3 TYPES OF PPE

16.3.1 Chemical Protective Clothing

Gloves and garments specifically designed for protection against chemicals are referred to as chemical protective clothing (CPC). CPC is usually intended to protect the body from direct contact with chemicals in their solid or liquid phase, but is sometimes used for protection against gases or vapors as well.

16.3.1.1 CPC Selection Considerations In addition to the general selection considerations identified in Section 16.2.5, the following specifics about the hazard should be considered during CPC selection:

- Which parts of the body are at risk of chemical contact?
- What is the anticipated duration of contact?
- How toxic or hazardous is the chemical? Is it corrosive to tissue? Does its threshold limit value (ACGIH, 1996) have a "Skin" designation, indicating that skin absorption is a potentially significant contributor to the overall dose? Is the chemical a skin sensitizer?
- Is the chemical in solid or liquid form? Will it be used at an extreme temperature?
- Does the chemical warn the body of its presence on the skin?
- Can the chemical be readily removed from skin once contaminated?

16.3.1.2 Rating Chemical Resistance A key characteristic of CPC is its ability to serve as a protective barrier to specific chemical agents as determined by its resistance to permeation, degredation, and penetration.

16.3.1.2.1 Permeation Permeation is the resistance to chemical movement through protective clothing material on a molecular level via absorption, diffusion, and then desorption. Low-level permeation is measured using ASTM Standard Test Method F739 (ASTM F739, 1991), in which a swatch of the clothing is exposed on one side to the chemical of interest, while analytical equipment measures desorption on the opposite side. Permeation as measured with this method is expressed as time (minutes) to initial breakthrough (breathrough time, BT), and also as a steady-state permeation rate following initial breakthru (mg/m^2 min). There are no quantitative health standards for permissible skin contact; hence breakthrough time is the parameter typically used for clothing selection.

One widely used field resource *Quick Selection Guide for Chemical Protective Clothing* (Forsberg and Mansdorf, 1993) is a lookup guide providing clothing material selection guidance based on permeation data for commonly used hazardous materials tested against 15 generic and proprietary barrier materials used to construct gloves, suits, and boots. The test data in the guide are those published by CPC manufacturers. In general, the guide considers a breakthrough time of >4 hours to be acceptable protection, while materials with breakthrough times of 1–4 hours should be used with caution, and those with breakthrough at <1 hour are not recommended.

Studies have shown that breakthrough time for the "same" materials, such as natural rubber used to make gloves, can vary significantly between manufacturers. Therefore, when selecting CPC, permeation data specific to the products under consideration should be obtained from the manufacturers. Figure 16.4 is an example of a typical table of permeation data provided by a CPC supplier. Another widely cited source of permeation data is *Guidelines for the Selection of Chemical Protective Clothing*, 3rd Ed. (Schwope et al., 1987).

As a general statement, the thicker the barrier material, the longer the chemical holdback will be. For example, latex gloves typically sold as "chemical gloves" will usually be at least 11 and up to 60 mils (thousandths of an inch) thick, while surgical latex gloves may be only 6 mils thickness.

16.3.1.2.2 Degradation. Another CPC performance characteristic is degradation, which is the resistance to deterioration of the material's physical properties such as weight, thickness, elongation, tear strength, and cut and puncture resistance. Significant degradation by chemical attack will render a material unsuitable for use. While degradation tests are important, interpretation of their results is somewhat subjective.

16.3.1.2.3 Penetration. A third important characteristic of CPC is resistance to penetration, which is the resistance to flow of a chemical on a nonmolecular, or gross, level through closures, seams, pinholes, or other material imperfections. A standard penetration test has been developed that exposes a portion of the garment to the challenge chemical, with the penetration measured as the time for a visible droplet to appear on the far side (ASTM F903, 1990). Penetration is especially important when considering the way garments are stitched, glued, heat sealed, and seamed.

Penetration is an important selection factor. Excellent resistance to permeation by a material is of little use if the garment is seamed with common stitches, which in effect are holes poked throughout the material.

16.3.1.4 Other Considerations In addition to permeation, degredation, and penetration, several other factors must be considered during CPC selection. These affect not only chemical resistance, but also the worker's ability to perform the required task. These include the following:

- Heat transfer—Almost without exception, CPC materials are nonbreathable materials and severely inhibit the evaporation of water from the skin, which is important

Key to Degradation and Permeation Ratings

E	Excellent	Fluid has no effect
G	Good	Fluid has minor effect
F	Fair	Fluid has moderate effect
P	Poor	Fluid has severe effect, ranging from moderate to complete destruction
ND		None detected
ID		Insufficient data, data not available or conflicting data

Permeation Chart Color Key

- Good for Total Immersion
- Good for Accidental Splash/Intermittent Exposure
- Not Recommended

	Silver Shield (4 Mil)			Viton (9 Mil)			Butyl (17 Mil)			Nitrile Latex (11 Mil)		
CHEMICAL	D	BT	PR	D	BT	PR	D	BT	PR	D	BT	PR
n-Hexane	E	> 6 hrs	ND	ID	> 11 hrs	ND	P	ID	ID	E	ID	ID
Hydrazine (70% in water)	G	> 6 hrs	ND	P	ID	ID	E	> 8 hrs	ND	G	> 8 hrs	ND
Hydrochloric Acid (37%)	E	> 6 hrs	ND	E	ID	ID	E	ID	ID	P	ID	ID
Hydrofluoric Acid (50%)	G	> 6 hrs	ND	G	ID	ID	F	ID	ID	P	ID	ID
Isobutyl Alcohol	E	ID	ID	E	> 8 hrs	ND	E	> 8 hrs	ND	G	> 8 hrs	ND
Isobutyraldehyde	E	ID	ID	P	4 min	11.5	E	> 8 hrs	ND	P	ID	ID
Methacrylic Acid	ID	ID	ID	F	> 8 hrs	ND	G	> 8 hrs	ND	P	1.7 hrs	23
Methacrylonitrile	E	ID	ID	F	4 min	462	G	6.8 hrs	0.001	P	7 min	560
Methyl Chloroform	ID	> 6 hrs	ND	E	> 15 hrs	ND	P	ID	ID	P	41 min	76.4
Methyl Cyanide	ID	> 8 hrs	ND	ID	ID	ID	E	> 8 hrs	ND	ID	ID	ID
Methyl Ethyl Ketone	E	> 24 hrs	ND	P	ID	ID	E	> 8 hrs	ND	P	ID	ID
Methyl Isocyanate	ID	ID	ID	P	4 min	121	P	1.1 hrs	9.0	P	ID	ID
Methylamine (40% in water)	F	1.9 hrs	2.0	E	> 16 hrs	ND	E	> 15 hrs	ND	G	> 8 hrs	ND
Methylene Chloride	G	> 8 hrs	ND	F	1 hr	7.32	P	24 min	133	P	4 min	766

Fig. 16.4 Example of a glove chemical resistance guide. Courtesy of North Safety Products, Hand Protection Division.

to shedding excess body heat. While some CPC may be thicker or have a higher insulation value than others, it is the impermeability to water vapor that most affects the level of heat stress to the worker. The choice of material is therefore not critical from a heat stress standpoint—all CPC increases heat stress.

- Durability—Does the material have sufficient strength to withstand the physical stress of the task at hand? Will it resist tears, punctures, and abrasions? Will it withstand repeated use after contamination/decontamination?
- Flexibility—Will the CPC, especially gloves, interfere with the worker's ability to perform the task?
- Vision—Can the worker see well enough in all directions with the equipment on?
- Temperature effects—Will the material maintain its protective integrity and flexibility under hot and cold extremes?
- Ease of decontamination—Can the material be decontaminated? Should disposable CPC be used?
- Duration of use—How does the estimated task duration compare with the permeation, degredation, and penetration data?

16.3.1.5 CPC Ensembles Chemical protective clothing is worn in a wide variety of occupations—wherever hazardous chemicals are used. Frequently gloves worn alone, or gloves worn with sleeves, apron, or chemical gown (apron with attached sleeves), eye protection, and boots may be all that is needed. In some extreme cases, however, more chemical protection may be needed.

The push to clean up hazardous waste "Superfund" disposal sites begun in the early 1980s and continuing to the present day created the need to identify PPE ensembles for workers initially approaching, and then working in, environments where chemical exposure is a risk, and where the hazard is often unknown and potentially high. In response to this risk, the Environmental Protection Agency first alone, and then in conjunction with NIOSH, OSHA, and the U.S. Coast Guard, issued recommended PPE ensembles termed "levels of protection" based on the potential hazard. These are described in Table 16.2. While the EPA levels of protection are usually applied to hazardous waste sites or hazardous materials ("hazmat") emergency response operations (OSHA 1910.120, 1996), the idea of "low to high risk" PPE ensembles which are specified based on the potential hazard has broader applicability to most any task or occupation where chemical exposure is a risk.

The National Fire Protection Association (NFPA) has issued standards covering manufacture and use of chemical protective clothing for use by firefighters (NFPA-1991, 1994, NFPA-1992, 1994, and NFPA-1993, 1994).

Figure 16.5 depicts examples of CPC ensembles.

16.3.1.6 Decontamination of CPC Much of the CPC in use today is designed to be "limited use," meaning it is intended to be disposed after a single use, or after a use that results in chemical contact and/or physical wear. The reason limited-use CPC prevails over more durable garments is the difficulty associated with cleaning chemical contamination from CPC material, and then assuring or measuring that its protective properties remain after being used and subjected to the decontamination process.

Some decontamination processes include aeration, that is, evaporation of volatile

TABLE 16.2 Sample Protective Ensembles Based on EPA Protective Ensembles

Level of protection	Equipment	Protection provided	Should be used when	Limiting criteria
A	Recommended Pressure-demand, full facepiece SCBA or pressure-demand supplied-air respirator with escape SCBA Fully encapsulating, chemical-resistant suit Inner chemical-resistant gloves Chemical-resistant safety boots/shoes Two-way radio communication Optional Cooling unit Coveralls Long cotton underwear Hard hat Disposable gloves and boot covers	The highest available level of respiratory, skin, and eye protection	The chemical substance has been identified and requires the highest level of protection for skin, eyes, and the respiratory system based on either: measured (or potential for) high concentration of atmospheric vapor, gases, or particulates or site operations and work functions involving a high potential for splash, immersion, or exposure to unexpected vapors, gases, or particulates of materials that are harmful to skin or capable of being absorbed through intact skin; substances with a degree of hazard to the skin are known or suspected to be present, and skin contact is possible; operations must be conducted in confined, poorly ventilated areas until the absence of conditions requiring Level A protection is determined	Fully encapsulating suit material must be compatible with the substances involved
B	Recommended Pressure-demand, full-facepiece SCBA or pressure-demand supplied-air respiratory with escape SCBA Chemical-resistant clothing (overalls and long-sleeved jacket; hooded,	The same level of respiratory protection but less skin protection than level A; it is the minimum level	The type and atmospheric concentration of substances have been identified and require a high level of respiratory protection, but	Use only when the vapor or gases present are not suspected of containing high concentrations of chemicals that are harmful

	one- or two-piece chemical splash suit; disposable chemical resistant one-piece suit) Inner and outer chemical-resistant gloves Chemical-resistant safety boots/shoes Hard hat Two-way radio communications Optional Coveralls Disposable boot covers Face shield Long cotton underwear	recommended for initial site entries until the hazards have been further identified	less skin protection; this involves atmospheres: with IDLH concentrations of specific substances that do not represent a severe skin hazard; or that do not meet the criteria for use of air-purifying respirators Atmosphere contains less than 19.5 percent oxygen Presence of incompletely identified vapor or gases is indicated by direct-reading instrument, but vapors and gases are not suspected of containing high levels of chemicals harmful to skin or capable of being adsorbed through the skin	to skin or capable of being absorbed through the intact skin. Use only when it is highly unlikely that the work being done will generate either high concentrations of vapor gases, or particulates or splashes of material that will affect exposed skin
C	Recommended Full-facepiece, air-purifying canister-equipped respirator Chemical-resistant clothing (overalls and long-sleeved jacket; hooded, one- or two-piece chemical splash suit; disposable chemical-resistant one-piece suit)	The same level of skin protection as Level B, but a lower level of respiratory protection	The atmospheric contaminants, liquid splashes, or other direct contact will not adversely affect any exposed skin The types of air contaminants have been identified, concentrations measured, and a canister is available that can remove the contaminant	Atmospheric concentration of chemicals must not exceed IDLH levels The atmosphere must contain at least 19.5 percent oxygen

TABLE 16.2 (Continued)

Level of protection	Equipment	Protection provided	Sould be used when	Limiting criteria
	Inner and outer chemical-resistant gloves Chemical-resistant safety boots/shoes Hard hat Two-way radio communications Optional Coveralls Disposable boot covers Face shield Escape mask Long cotton underwear		All criteria for the use of air-purifying respirators are met	
D	Recommended Coveralls Safety boots/shoes Safety glasses or chemical splash goggles Hard hat Optional Gloves Escape mask Face shield	No respiratory protection; minimal skin protection	The atmosphere contains no known hazard Work functions preclude splashes, immersion, or the potential for unexpected inhalation of or contact with hazardous levels of any chemicals	This level should not be worn in the Exclusion Zone The atmosphere must contain at least 19.5 percent oxygen

Source: NIOSH, 1985.

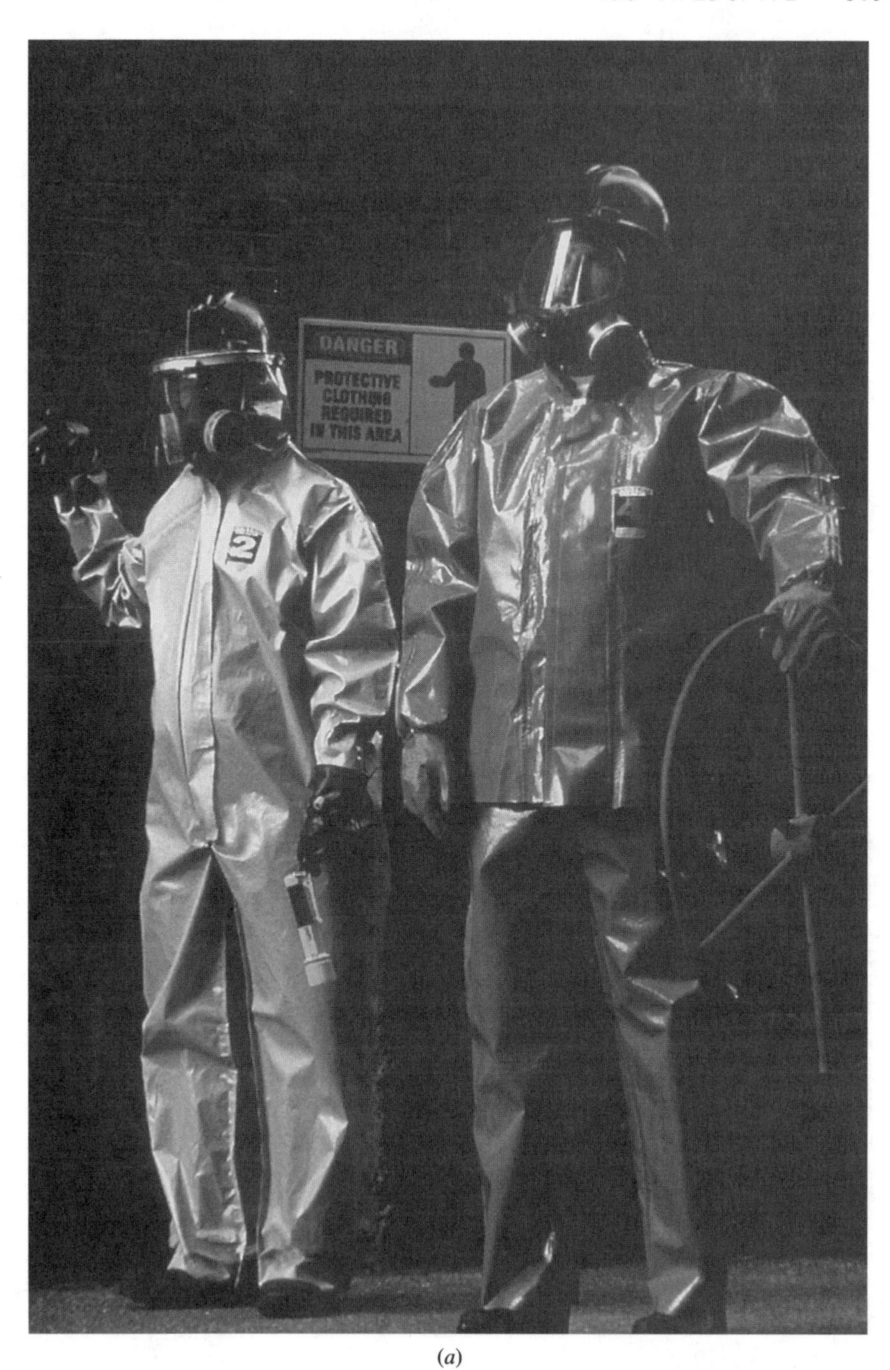

(*a*)

Fig. 16.5. Chemical protective clothing. (a) EPA Level C ensemble. Courtesy of Kappler Safety Group.

(*b*)

Fig. 16.5. (*Continued.*) (b) Chemical encapsulating suit, available as splashproof and gas tight Level A, or splashproof Level B. Courtesy of Kappler Safety Group.

(*c*)

Fig. 16.5. (*Continued.*) (c) Chemical and flash resistant chemical suit, fully compliant with NFPA 1991 (gas tight) or 1992 (splashproof) standards. Courtesy of Kappler Safety Group.

contaminants, water/detergent washing, water/bleach washing, and freon-based dry cleaning (Johnson and Anderson, 1990).

All CPC use, including use of disposables, should be accompanied by a decontamination procedure to prevent chemical exposure to the wearer while doffing the garments, contamination spread throughout the workplace, exposure to the worker's family from taking contaminated apparel home, and to prevent contamination of workers who may handle the waste garments.

16.3.2 Head Protection

16.3.2.1 Hard Hats The most common form of head protection is the hard hat. The hard hat is designed to protect against the hazard of solid objects falling from above, which can produce perforation or facture of the skull, or brain lesions from sudden displacement. (*Encyclopedia*, 1983). Head protection in the form of a hard hat or other covering can also protect against electric shock, burns from splashes of hot or corrosive liquids, or molten metal.

16.3.2.2 When Required In the United States, OSHA requires that hard hats be provided and worn where there is a risk of injury from objects falling from above, or where workers are exposed to electrical conductors that could contact the head. OSHA further requires that hard hats comply with the ANSI protective headwear standard (ANSI Z89.1-1986), which specifies physical and performance requirements and test methods for helmets.

16.3.2.3 ANSI Hard Hat Types ANSI Type 1 helmets have a full brim, while Type 2 have a baseball cap–like bill that protrudes out over the wearer's eyes. ANSI further specifies three classes of electrical insulation:

- Class A provides protection against impact and exposed low-voltage conductors (proof tested at 2200 volts);
- Class B provides protection against impact and exposed high-voltage conductors (proof tested at 20,000 volts);
- Class C provides protection against impact only.

The majority of helmets in use are Type 2, Class A or B, while Class C finds use especially in the forestry industry. The wearer should be able to identify the type of helmet by looking inside the shell for the manufacturer name, ANSI standard, and ANSI class.

16.3.2.4 How They Work Impact protection is provided through the design of two key components: the hat's hard shell and the suspension system (Fig. 16.6). The shell is designed to be smooth so as to deflect falling objects to the side. It is also designed to flex or deflect upon impact into the space between the skull and shell created by the suspension. The suspension itself is constructed to distribute the impact energy over a larger area, and to absorb energy by stretching slightly (while still preserving the suspension-to-shell space).

High-density polyethylene is the most common hard hat shell material. Other materials in use include polycarbonate, polycarbonate–glass, polyester–glass, and aluminum.

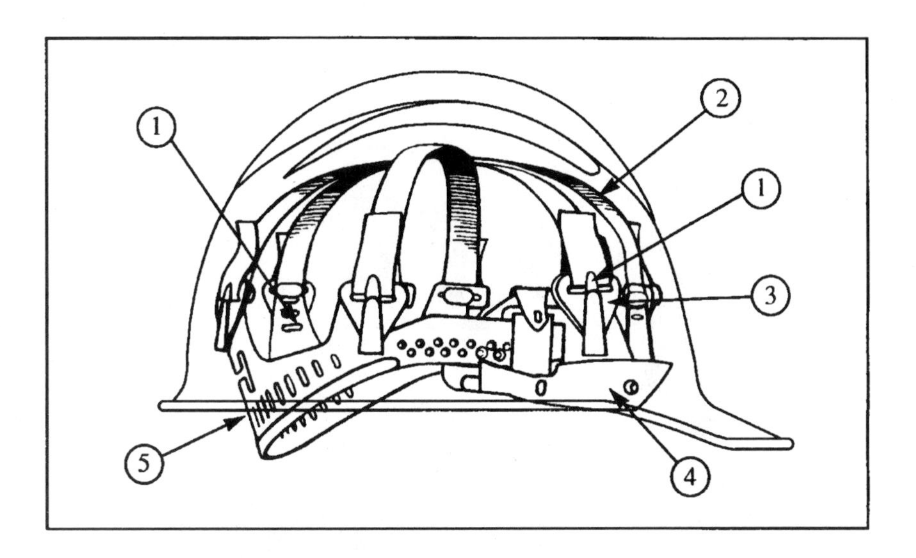

1. Vertical Adjustment *3. Hanger Key* *5. Sizing Buckle*
2. Nylon Crown Strap *4. Absorbent Brow Pad*

Fig. 16.6. Hard hat suspension design. Courtesy of E. D. Bullard Company.

For high-temperature applications, polycarbonate or polyester is more appropriate than polyethylene, and the suspension should be a woven material.

16.3.2.5 Accessories Hard hats are available that can be accessorized with hearing protection, visors, chin straps, winter liners, and miner's lamps. (Refer to Fig. 16.7.)

16.3.2.6 Use Considerations

- Persons working overhead should wear a hard hat with chin strip to secure the hat from becoming a falling object itself;
- Plastic hard hats kept in the rear window ledge of a car are subjected to damage from heat and sunlight, and can become hazardous projectiles if the car suddenly stops;
- Ventilation holes cannot be drilled in the shell because they reduce or eliminate both impact and electrical insulation protection;
- Most hard hats are not designed to provide impact protection when worn backwards;
- Some adhesives in stickers applied to plastic hard hats, and some chemicals used in the work environment can attack plastic shell material;
- Damaged hats should be replaced.

16.3.2.7 Bump Caps Bump caps (Fig. 16.8) are thin shelled caps with a front bill which are usually worn for "struck against" hazards in environments having low

Fig. 16.7. Accessorized hard hat, ANSI type 2. Courtesy of E. D. Bullard Company.

headroom. There are no design standards for bump camps. While they are not designed to protect against falling objects, bump caps may be sufficient protection for persons at risk of collision with low pipes or ducts while walking.

16.3.3 Eye and Face Protection

Eye and face protection includes safety spectacles (with or without sideshields), eyecup goggles, cover goggles, and faceshields. Eye and face protection is typically used for protection against the following hazards:

- Physical hazards such as flying objects, or collision with stationary objects. Flying or falling objects are the most common category of hazard causing eye injury (U.S. Dept. of Labor, 1980);
- Chemical hazards, including dusts, mists, and splashes;
- Radiant energy, including ultraviolet light and lasers;

Fig. 16.8. Bump cap. Courtesy of E. D. Bullard Company.

16.3.3.1 When Required In the United States, OSHA requires that eye and/or face protection be provided as determined by a hazard analysis when workers are "exposed to eye or face hazards from flying particles, molten metal, liquid chemicals, acids or caustic liquids, chemical gases or vapors, or potentially injurious light radiation." (OSHA 1910.133, 1996).

Table 16.3 provides guidance in the selection of appropriate eye and face protection. Note that spectacles and goggles are considered to be "primary" eye protection, while faceshields are termed "secondary" protection. While primary protectors may be worn alone, secondary protectors are to be worn only in combination with primary protection. Figure 16.9 depicts various types of eye and face protection.

Table 16.4 provides minimum protective shade values for protection against radiant energy.

16.3.3.2 Sideshields Sideshields should be worn on spectacles where there is a hazard from flying objects. OSHA made this an explicit requirement with their 1994 PPE standards revisions (OSHA, 1994b; OSHA, 1996). Most spectacles with plano (noncorrective, nonprescription) lenses are equipped with sideshields incorporated into the design and manufacture of the frame and temple bars. Wearers of prescription safety glasses, particularly those designed to look and be worn as streetwear, frequently add detachable sideshields when working in an eye hazard area.

Where peripheral vision is especially important, such as in driving powered industrial vehicles, glasses without side shields may be more appropriate. The risk from decreased vision should be weighed against the risk from flying objects.

16.3.3.3 ANSI Standard The ANSI occupational and educational eye and face protection standard (ANSI Z87.1-1989) provides minimum requirements for eye and face protective devices and guidance for their selection, use, and maintenance. Specta-

TABLE 16.3 Eye and Face Protection Selection Chart

Source	Assessment of hazard	Protection
Impact—chipping, grinding, machining, masonry work, woodworking, sawing, drilling, chiseling, powered fastening, riveting, and sanding	Flying fragments, objects, large chips, particles sand, dirt, etc.	Spectacles with side protection, goggles, face shields; see notes (1) (3), (5), (6), (10); for severe exposure, use faceshield
Heat—furnace operations, pouring, casting, hot dipping, and welding	Hot sparks	Faceshields, goggles, spectacles with side protection; for severe exposure use faceshield; see notes (1), (2), (3)
	Splash from molten metals	Faceshields worn over goggles; see notes (1), (2), (3)
	High-temperature exposure	Screen face shields, reflective face shields; see notes (1), (2), (3)
Chemicals—acid and chemicals handling, degreasing, plating	Splash	Goggles, eyecup and cover types; for severe exposure, use faceshield; see notes (3), (11)
	Irritating mists	Special-purpose goggles
Dust—woodworking, buffing, general dusty conditions	Nuisance dust	Goggles, eyecup and cover types; see note (8)
Light and/or radiation—welding: electric arc	Optical radiation	Welding helmets or welding shields; typical shades: 10–14; see notes (9), (12)
Welding: gas	Optical radiation	Welding goggles or welding face shield; typical shades: gas welding 4–8, cutting 3–6, brazing 3–4; see note (9)

cles, goggles, faceshields, and welding helmets worn for personal protection must be manufactured in accordance with ANSI Z-87, and each major component (other than lenses on spectacles) must be marked with a trademark identifying the manufacturer and "Z87."

Safety spectacles are tested as a complete assembly, with the frames being just as important as the lenses. Combinations of normal streetwear frames with safety lenses is not allowed.

16.3.3.4 Contact Lenses Contact lenses are considered to offer no protection to the eye from any hazards. Safety glasses or goggles must be worn over contact lenses in environments presenting a hazard from flying objects. The original hard contacts were considered a possible hazard in dusty environments by possibly trapping abrasive particles between the lens and the cornea. Until recently, contact lenses have been considered a hazard to users in chemical use areas because of the risk of chemical absorption by the lense, or "trapping" of chemicals behind the lens and against the eye tissue. While there is

TABLE 16.3 **(*Continued*)**

Source	Assessment of hazard	Protection
Cutting, torch brazing, torch soldering	Optical radiation	Spectacles or welding faceshield; typical shades, 1.5–3; see notes (3), (9)
Glare	Poor vision	Spectacles with shaded, or special-purpose lenses, as suitable; see notes (9), (10)

[1]Care should be taken to recognize the possibility of multiple and simultaneous exposure to a variety of hazards. Adequate protection against the highest level of each of the hazards should be provided. Protective devices do not provide unlimited protection.
[2]Operations involving heat may also involve light radiation. As required by the standard, protection from both hazards must be provided.
[3]Faceshields should only be worn over primary eye protection (spectacles or goggles).
[4]As required by the standard, filter lenses must meet the requirements for shade designation in 1910.133(a)(5). Tinted and shaded lenses are *not* filter lenses unless they are marked or identified as such.
[5]As required by the standard, persons whose vision requires the use of prescription (Rx) lenses must wear either protective devices fitted with prescription (Rx) lenses or protective devices designed to be worn over regular prescription (Rx) eyewear.
[6]Wearers of contact lenses must also wear appropriate eye and face protection devices in a hazardous environment. It should be recognized that dusty and/or chemical environments may represent an additional hazard to contact lens wearers.
[7]Caution should be excercised in the use of metal frame protective devices in electrical hazard areas.
[8]Atmospheric conditions and the restricted ventilation of the protector can cause lenses to fog. Frequent cleansing may be necessary.
[9]Welding helmets or faceshields should be used only over primary eye protection (spectacles or goggles).
[10]Nonsideshield spectacles are available for frontal protection only, but are not acceptable eye protection for the sources and operations listed for "impact."
[11]Ventilation should be adequate, but well protected from splash entry. Eye and face protection should be designed and used so that it provides both adequate ventilation and protects the wearer from splash entry.
[12]Protection from light radiation is directly related to filter lens density. See note (4). Select the darkest shade that allows task performance.

Source: OSHA, 1910. 133 (1996).

still agreement that the presence of a contact lens can complicate the irrigation of an eye exposed to chemical splash, the hazard of contact lenses in chemical environments is now considered to be less than once thought. Some chemical safety experts suggest allowing contact lens use in chemical environments provided that co-workers are trained to assist with lens removal in the event of a splash exposure (*Chemical Health & Safety*, 1995).

In their 1998 revision to the respirator standard, OSHA deliberately made no mention of contact lenses in the text of the standard, citing in the standard's preamble that there is no evidence of an increased hazard from contact lens use when compared with the use of conventional eyeglass inserts inside full facepiece respirators (*Fed. Reg.*, 1998). In its respirator standard, ANSI allows contact lenses with respirators provided that the individual "has previously demonstrated that he or she has had successful experience wearing contact lenses." (ANSI Z88.2-1992).

A story surfaced in the safety trade press in the late 1970s reporting a case where a contact lens user was exposed to optical radiation from welding and experienced a severe eye injury when the ultraviolet energy fused the plastic contact lens to his cornea. This widely published story was found to be untrue—an alarming rumor. This false anecdote still occasionally surfaces today, which is why it is mentioned here.

TABLE 16.4 Filter Lenses for Protection Against Radiant Energy

Operations	Electric size 1/32 in.	Arc current	Minimum[a] protective shade
Shielded metal arc welding	Less than 3	Less than 60	7
	3–5	60–160	8
	5–8	160–250	10
	More than 8	250–550	11
Gas metal arc welding and flux cored arc welding		Less than 60	7
		60–160	10
		160–250	10
		250–500	10
Gas tungsten arc welding		Less than 50	8
		50–150	8
		150–500	10
Air carbon	(Light)	Less than 500	10
Arc cutting	(Heavy)	500–1000	11
Plasma arc welding		Less than 20	6
		20–100	8
		100–400	10
		400–800	11
Plasma arc cutting	(light)[b]	Less than 300	8
	(medium)[b]	300–400	9
	(heavy)[b]	400–800	10
Torch brazing			3
Torch soldering			2
Carbon arc welding			14

Operations	Plate thickness (in.)	Plate thickness (mm)	Minimum[a] protective shade
Gas welding:			
Light	Under 1/8	Under 3.2	4
Medium	1/8 to 1/2	3.2–12.7	5
Heavy	Over 1/2	Over 12.7	6
Oxygen cutting			
Light	Under 1	Under 25	3
Medium	1–6	25–150	4
Heavy	Over 6	Over 150	5

[a]As a rule of thumb, start with a shade that is too dark to see the weld zone. Then go to a lighter shade, which gives sufficient view of the weld zone without going below the minimum. In oxyfuel gas welding or cutting where the torch produces a high yellow light, it is desirable to use a filter lens that absorbs the yellow or sodium line in the visible light of the (spectrum) operation.

[b]These values apply where the actual arc is clearly seen. Experience has shown that lighter filters may be used when the arc is hidden by the workpiece.

Source: 29 CFR1910.133.

16.3.3.5 Laser Eyewear

The ANSI laser standard (ANSI Z136.1-1993) provides guidance on selection of protective eyewear for workers potentially exposed to ANSI Class 3b or 4 lasers above the maximum permissible exposure level. Sixteen selection factors are provided in the standard. Laser eyewear is clearly labeled with the optical density and wavelength for which protection is afforded.

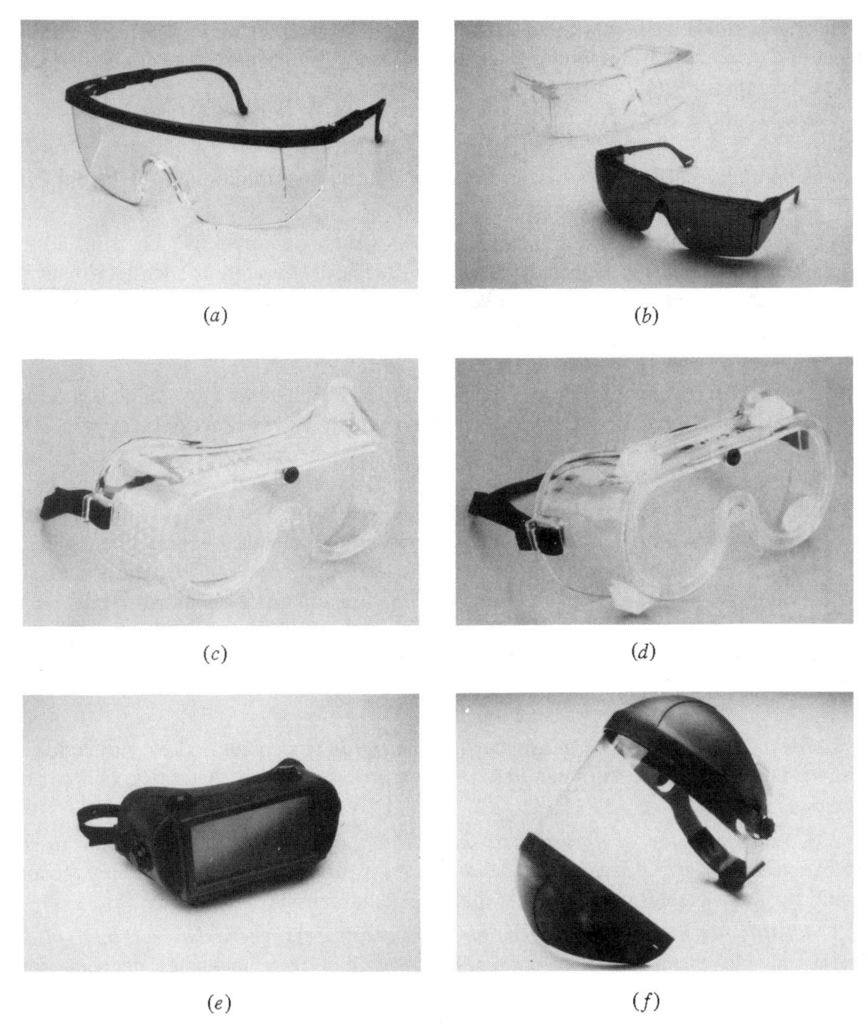

(a) (b) (c) (d) (e) (f)

Fig. 16.9. Types of Eye and Face Protection. (a) Basic spectacles w/sideshields & adjustable temple length, (b) Visitor's glasses can fit over prescription eyewear, (c) Ventilated goggle (d) Chemical goggle w/indirect vent, (e) Welder's goggle, (f) Faceshield w/neck and forehead protection. Courtesy of Aearo Company.

16.3.4 Hand Protection

Hands are the worker's primary tools for most jobs, and are at relatively high risk of injury from cuts, burns, abrasion, electrical shock, amputation, and chemical injury.

16.3.4.1 Types Available There is a wide assortment of gloves, hand pads, sleeves, and wristlets available for hand and arm protection. Each glove type has its own advantages and disadvantages relative to the protection offered and the ability of the user to perform the task while wearing the glove. Involvement of the end user in PPE selection is particularly important with gloves, and different types and manufacturers should be experimented with freely.

There are three general functional types of gloves:

- *General Purpose, Leather* from cowhide is the glove traditionally worn for protection against cuts, abrasion, and heat/cold.
- *General Purpose, Fabrics* include a wide variety of gloves woven from cotton and cotton/polyester blends, or specialty cut-resistant materials such as Kevlar by DuPont, or steel mesh. Fabric gloves are commonly coated or stippled on the palm with a polymer for better grip in specific environments, or fully dipped for greater liquid or chemical resistance.
- *Chemical-Resistant* gloves are seamless types manufactured by dipping a hand-shaped solid form into a viscous solution which polymerizes upon drying. The less expensive types most commonly encountered are made of natural latex rubber, neoprene, and nitrile rubber. Less common but having specific application are butyl rubber, Viton, PVC (vinyl), and PVA (polyvinyl alcohol). A relatively new material constructed of a PE/EVAL laminate is resistant to the widest variety of chemicals. With its relatively poor grip and finger dexterity, this type is frequently worn as an underglove for applications involving high or unknown chemical hazard.

16.3.4.2 Selection Considerations The following glove selection process is suggested:

1. *Identify the one or two most important hazards* from which glove protection is desired. Is it abrasion, cuts, punctures, temperatures, microbes, chemicals, or ionizing radiation?

2. *Consider degree of exposure to the hazards.* When a glove is to be worn for protection against two different chemical operations, for example, immersion exposure should be given greater consideration than incidental splash.

3. *Identify potential impacts of the glove on the process where they will be applied.* The wrong glove can introduce contamination with serious, unwanted consequences to the underlying objectives of the task. Examples include the bacteria-free, or sterile, gloves needed in an operating room, and the powder-free, low-particle-shedding gloves needed in a computer chip fabrication area. Inks used to mark glove sizes may dissolve and introduce contamination.

4. *For chemical protection, use permeation and degredation information* provided by the glove manufacturer, or from sources such as the *Quick Selection Guide to Chemical Protective Clothing* (1993) to identify protective materials.

In applications where protection against chemical hazards is needed, data on permeation and degredation of the *specific* glove under consideration for the particular application should be obtained from the manufacturer if possible. Although charts of chemical resistance for generic material types such as natural rubber are useful, they do not take into account important factors such as the thickness of the product under consideration, or the manufacturer-specific glove manufacturing process, which has been shown to affect chemical permeation rates.

Where test data are not available, the user can perform a simple, gross test for chemical compatibility by dipping the glove into, or filling a finger with, the challenge chemical and observing for swelling, hardening, softening, dissolving, or penetration of the material. If the chemical substance is especially hazardous, however, laboratory testing of the specific agent and glove should be either requested of the glove manufacturer, or commissioned directly with a testing company.

Not only does glove thickness affect permeation rate, but it is also a factor determining resistance to tears and abrasions. Thicker is usually better. For example, vinyl (PVC) as a material in general has excellent resistance against most concentrated acids. However, thin, amibidexterous vinyl gloves sold in 100 per pack dispenser boxes should generally not be worn when handling concentrated acids because they tear and abrade more easily, and have quicker breakthrough times than their thicker counterparts intended for use as chemical protection. The same is true of latex and nitrile gloves, which are available in very thin (e.g., 4 mil) versions.

5. *Consider the grip and dexterity demands of the task.* Tasks requiring manipulation of fine parts or tools with the fingers cannot be performed with thicker gloves. Embossed rubber gloves will grip wet objects better than smooth rubber.

6. *Consider reusability of the glove after exposure to the task.* Gloves exposed to glue or resins that are difficult to remove, for example, will often be discarded after a single use. Conversely, chemical gloves exposed to volatile organic liquids may dry out and stay contamination free for long periods of time.

7. *Consider price*, especially relative to reusability and numbers of individuals who will need protection.

8. *Consider cuff style and length.* Choices include band top, knit wrist, safety cuff, bell gauntlet, slip-on, and open cuff. For welding or metal fabrication jobs, choose a gauntlet cuff for greater protection. For cold environments, a knit-wrist will keep cold air out. Slip-on or open cuff styles make for easier donning and doffing where gloves will be frequently removed.

9. *Consider color or marking.* The correct glove for a particular task often comes to be identified within an organization by *color* because it is the simplest way to differentiate gloves. Glove colors *are not standardized*—you cannot identify the material of an unknown glove by its color. An organization can, however, create its own glove color standard, which is very useful in communicating the correct glove for the task. Color is therefore a consideration when adding a new glove type, or replacing an existing one. In addition to color, glove types and sizes can be identified by labels printed on the glove packaging or on the glove cuffs themselves.

16.3.4.3 Gloves Bring Tradeoffs All PPE, and gloves in particular, present tradeoffs or compromises to the user. Usually (but not always) protection from a physical or chemical hazard comes at the expense of dexterity. A chemistry lab or chemical plant cannot stock the most chemically resistant glove for every substance on hand; protection should be provided against the most hazardous substances and exposures. The tradeoffs need to be identified in advance, and understood by the user so that accommodation becomes easier.

In some cases two pairs of gloves may need to be worn simultaneously to address the significant hazards and/or meet the demands of the task. For example, a hazardous materials incident responder may wear a thicker, more abrasion-resistant glove over one having higher chemical resistance. A meat packer may wear a waterproof glove over one made of steel mesh.

16.3.4.4 Glove Size A poor-fitting glove is awkward to wear and can lead to clumsy, costly mistakes or accidents. Gloves are often sized numerically using a measurement of the circumference around the palm of the hand at the knuckle area, as follows:

Glove size	Hand circumference (in.)
XS	6–7
S	7–8
M	8–9
L	9–10
XL	10–11

16.3.4.5 Leak Checking Gloves should be visually inspected before each use. While a quick visual inspection may be all that is needed, sometimes a small or pinhole-type leak can create an unacceptable risk, such as applications involving protection against high electrical voltage, or high-hazard chemicals with poor skin-contact warning properties such as hydrofluoric acid. In this case a more thorough leak check should be performed, as illustrated in Fig. 16.10. Some chemical glove manufacturers will use compressed air to perform leak checking of every glove sold for high-hazard applications, if specified by the customer.

16.3.4.6 Latex Allergy Gloves made of natural latex have been identified in recent years as being responsible for causing or aggravating an allergic sensitivity in a significant percentage of wearers. The allergic reaction most commonly appears as a reddening or itching at the site of glove contact, often the wrists. In some cases it can progress into generalized hives, shortness of breath, or rarely still, full anaphylactic shock and death. Certain unidentified natural proteins in latex harvested from rubber trees are believed responsible for the allergic response. In high latex use areas such as hospitals, general contamination of the area by glove powders that have themselves been contaminated by latex protein can make the environment a hazard for those already sensitized. Hypo-allergenic latex or nonlatex alternatives are available.

16.3.5 Foot Protection

Protective footwear is available for protection against a variety of hazards. The most common hazards are impact from falling objects, and compression from rolling objects.

16.3.5.1 Types Available and When Required The American National Standards Institute specifies minimum performance for 6 types of protective footwear (ANSI-Z41-1991).

- *Impact and compression resistance* is needed where heavy objects are being moved around in the workplace by hand, or with materials handling equipment. Most safety footwear includes the steel toe box, which provides protection against falling or rolling objects.
- *Metatarsal protection* is available for environments having a higher risk of falling

Inspecting Gloves for Small Leaks

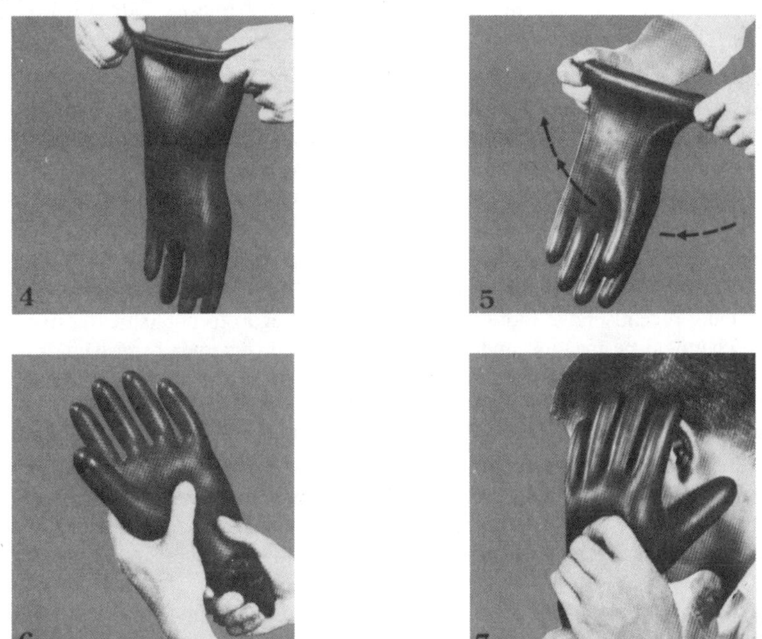

Fig. 16.10. Glove leak checking procedure. Inspecting gloves for small leaks. To air test gloves for pinholes and other damage, follow these procedures: Hold glove downward and grasp the cuff as shown in illustration 4. Twirl the glove upward toward your body to trap the air inside the glove as shown in illustration 5. Then squeeze the rolled cuff tightly into a U shape with the right hand to keep the trapped air inside. Squeeze with the other hand and look for damage exposed by inflation as shown in illustration 6. Then hold the inflated glove close to your face and ear, squeezing the glove, to feel and listen for air escaping from holes as shown in illustration 7. Important practices to follow: Inspect frequently for glove damage. Rinse gloves daily with clear water inside and out and thoroughly dry. Wear leather protectors. Replace old, worn, or damaged protectors. Photos and corresponding procedures courtesy of North Safety Products, Hand Protection Division.

object injury, and features what amounts to an extended toe box protecting much of the top of the foot.

- *Conductive footwear* can reduce the hazard of static electricity buildup, which could cause ignition of flammable atmospheres. Chemical workers handling flammable liquids or gases should wear Type 1 conductive footwear. Type 1 footwear is designed to be electrically conductive, and should not be worn near exposed electrical circuits. Type 2 conductive footwear is intended for use by electrical linemen or others needing equalization of the electrical potential between the person and the energized equipment.
- *Electrical hazard footwear* is designed to be resistant to electrical conductance, protecting the wearer from electric shock. They should be worn by persons exposed to energized electrical equipment. They should not be worn in hazardous/flammable locations where conductive footwear is required.

- *Sole-puncture-resistant footwear* is designed to protect the wearer from penetration of the bottom of the foot by sharp objects, such as protruding nails at construction sites. This is typically accomplished by a steel insole.
- *Static-dissipative footwear* is a compromise between conductive and electrical hazard footwear, providing protection against hazardous levels of static buildup while at the same time providing resistance against electric shock. Most chemical workers exposed to flammable atmospheres also have some degree of exposure to electrical hazards, making static dissipative the footwear of choice.

While chemically resistant footwear often meets one or more of the above performance standards, there is no widely used standard for protection against chemical hazards. Only recently have footwear manufacturers begun to publish extensive charts showing chemical permeation and degredation for their chemically resistant products constructed of natural or synthetic polymers. One standard referenced by footwear manufacturers is NFPA 1991 (NFPA, 1991, 1994).

In many chemical environments such as chemical laboratories or semiconductor fabrication areas, impact or compression are low risks, and while the risk of chemical splash is low enough that rubber boots are not needed, something more protective than open-toed or canvas street shoes is required. In these environments some organizations will require "shoes constructed of leather or leather-like material which completely cover the foot."

One shoe of each pair designed and constructed to meet one or more of the ANSI foot protection categories must be marked on the inside with letters and numbers coded to Section 1.5 of the ANSI standard. For example, the label can be used to confirm that the shoe is designed for impact and compression (PT), electrical hazard (EH), or to be static dissipative (SD).

16.3.5.2 Posting Foot Hazard Areas Many organizations post PPE requirements at the entrances to hazard areas, often with signs composed of a signal word like CAUTION or NOTICE, followed by a short command such as "Safety Glasses Required in this Area." In the case of safety glasses, all workers or visitors to an area can be accommodated with eye protection relatively easily because visitors glasses or goggles are relatively inexpensive to keep on hand, and one size fits all. The same is true for hard hats. With safety footwear, however, organizations serious about compliance with safety rules should think twice before posting the area unless the risk to all who enter is very high. While toe or metatarsal caps designed to provide temporary impact and compression protection over street shoes are available, they are rarely put into use. They can be noisy and uncomfortable, and worse still, can present a slipping or tripping hazard to the wearer. In most instances, posting the area as "safety shoes required" is an unreasonable rule frequently ignored, which ultimately works against the organization's safety credibility. A better alternative is to make safety footwear a job or task requirement for those who routinely work in the foot hazard area.

16.3.6 Other PPE

16.3.6.1 Seat Belts All workers will travel over the road while on business at one time or another. While sales and field service personnel are most at risk of vehicular

Fig. 16.11. Personal fall arrest system with restraint lanyard. Courtesy of Rose Manufacturing Company.

injury, all organizations, no matter what the size or type of the business, should have a policy requiring the use of seat belts while traveling on company business.

16.3.6.2 Fall Protection Falls from heights have serious consequences. In the U.S. construction industry, falls are the leading cause of worker fatalities. Each year, on average, between 150 and 200 workers are killed and more than 100,000 are injured as a result of falls at construction sites (OSHA, 1995).

Personal protective equipment employed as fall protection are of two general types: personal fall arrest systems and positioning device systems.

Personal fall arrest systems (Fig. 16.11) are used to arrest a worker in a fall from a working level, and consist of an anchorage, connectors, and a body belt or body harness and may include a deceleration device, lifeline, or suitable combinations. According to OSHA, if a personal fall arrest system is used for fall protection, it must do the following (OSHA, 1995):

- Limit maximum arresting force on a worker to 900 pounds when used with a body belt;
- Limit maximum arresting force on an employee to 1800 pounds when used with a body harness;
- Be rigged so that a worker can neither free fall more than 6 feet nor contact any lower level;

- Bring a worker to a complete stop and limit maximum deceleration distance a worker travels to 3.5 feet; and
- Have sufficient strength to withstand twice the potential impact energy of a worker free falling a distance of 6 feet or the free-fall distance permitted by the system, whichever is less.

Effective January 1, 1998, the use of a body belt for fall arrest was prohibited by OSHA. A body harness must be used.

A *positioning device system* is a body belt or harness system set up so that a worker can free fall no farther than 2 feet. It allows a worker to be supported on an elevated vertical surface, such as a wall or tower, and work with both hands free while leaning.

For more detailed information on fall protection, consult OSHA's fall protection regulations (OSHA, Subpart M, 1994), OSHA's fall protection booklet (OSHA, 1995), ANSI's fall protection standards (ANSI Z359.1-1992 and A10.14-1991), or *Introduction to Fall Protection*, by J. Nigel Ellis (1993).

16.3.6.3 Electrical Protection Electrically insulating, nonconductive garments, tools, and shielding devices should be used by those who are working near exposed energized parts that might be accidentally contacted, or where dangerous electrical heating or arcing might occur (OSHA, 1910.335, 1990). Specifications for electrical protective equipment are provided in Section 1910.137 of the OSHA general industry standards, and are based on applicable ASTM standards.

16.3.6.4 Back Belts In spite of an absence of scientific evidence that shows they prevent injury, back belts came into widespread use during the late 1980s and early 1990s as personal protection against back injury caused by materials handling. Scientific studies of Home Depot and WalMart employees who are/have been required to wear back belts are currently underway. The former study is being conducted by UCLA and California–OSHA, while the latter is by NIOSH (the National Institute for Occupational Safety and Health).

Until these and additional studies are completed, the safety and health professional community will continue to view back belts as capable only of helping with the healing process and preventing reinjury (as supported by scientific studies of injured workers), but not as personal protective equipment for the average worker. One concern is that the belts may actually increase risk by creating a false sense of security, which can lead to greater risk taking by the user, or by weakening of abdominal muscles from prolonged wearing, but an increase in injury among belt users has not been scientifically proven either.

REFERENCES

ACGIH (1996). *1996 TLV's and BEI's, Threshold Limit Values for Chemical Substances and Physical Agents, Biological Exposure Indices*, American Conference of Governmental Industrial Hygienists, 1330 Kemper Meadow Drive, Cincinnati, OH 45240-1634, (513)742-2020.

The following ANSI standards are available from the American National Standards Institute, 11 West 42nd Street, New York, NY 10036, (212)642-4900:

ANSI A10.14-1991. Requirements for Safety Belts, Harnesses, Lanyards, Lifelines, and Droplines for Construction and Industrial Use.

ANSI Z41-1991. American National Standard for Personal Protection—Protective Footwear.

ANSI Z87.1-1989. American National Standard Practice for Occupational and Educational Eye and Face Protection.

ANSI Z88.2-1992. American National Standard for Respiratory Protection, Paragraph 7.5.3.3.

ANSI Z89.1-1986. American National Standard for Personnel Protection—Protective Headwear for Industrial Workers—Requirements.

ANSI Z136.1-1993. Safe Use of Lasers.

ANSI Z359.1-1992. Safety Requirements for Personal Fall Arrest Systems, Subsystems and Components.

Asbestos (1996). 29 CFR1910.1001, Asbestos, Appendix C, U.S. Department of Labor, Occupational Safety and Health Administration.

ASTM F739, 1991. Standard Test Method for Resistance of Protective Clothing Materials to Permeation by Liquids or Gases under Conditions of Continuous Contact, American Society for Testing and Materials, Philadelphia, PA.

ASTM F903, 1990. Standard Test Method for Resistance of Protective Clothing Materials to Penetration by Liquids, American Society for Testing and Materials, Philadelphia, PA.

Best's Safety Directory (1996). A. M. Best Company, Oldwick, NJ, (908)439-2200.

Chemical Health & Safety (1995) "Contact Lens Emergencies," January/February, American Chemical Society, 1155 16th St., N. W., Washington, DC 20036.

Chilton's Industrial Safety & Hygiene News, (1996). Chilton Company, Chilton Way, Radnor, PA 19087.

Ellis, J. N. (1993). *Introduction to Fall Protection*, Second Ed. American Society of Safety Engineers, 1800 E. Oakton Street, Des Plaines, IL 60018-2187, (708)692-4121.

Encyclopedia (1983). *Encyclopedia of Occupational Health and Safety*, Third (Revised) Ed. International Labour Office, Geneva, Switzerland.

Fed. Reg. (1998). *Respiratory Protection; Final Rule*, **63**(5), January 8.

Forsberg, K. and S. Z. Mansdorf, (1993). *Quick Selection Guide to Chemical Protective Clothing*, Second Ed., Van Nostrand Reinhold.

Industrial Hygiene News, (1996). Rimbach Publishing, Inc., 8650 Babcock Road, Pittsburg, PA, 15237-5821 (800)245-3182.

Johnson, J. S., and K. J. Anderson, Eds. (1990) *Chemical Protective Clothing*, Vol. I, p. 143. Published by the American Industrial Hygiene Association, 2700 Prosperity Avenue, Suite 250, Fairfax, VA 22031.

NFPA (1991, 1994). Standard on Vapor-Protective Suits for Hazardous Chemical Emergencies, National Fire Protection Association, Quincy, Massachusetts.

NFPA (1992, 1994). Standard on Liquid Splash-Protective Suits for Hazardous Chemical Emergencies. National Fire Protection Association, Quincy, Massachusetts.

NFPA (1993, 1994). Standard on Support Function Protective Clothing for Hazardous Chemical Operations. National Fire Protection Association, Quincy, Massachusetts.

NIOSH (1985). Occupational Safety and Health Guidance Manual for Hazardous Waste Site Activities, Prepared by NIOSH, OSHA, USCG, and EPA, NIOSH Publication No. 85-115.

Occupational Hazards (1996). 1100 Superior Ave., Cleveland, OH, 44114-2543 (216)696-7000.

Occupational Health & Safety Magazine (1996). Stevens Publishing, 3700 IH-35, Waco, Tx 76706, (817)776-9000.

OSHA (1994a). Office of Information, Washington, D.C., (202)219-8151. USDL 94-520, "OSHA Clarifies Obligation of Employers to Pay for Personal Protective Equipment."

OSHA (1994b). Final Rule Revising General Industry Standards for Personal Protective Equipment for Eyes, Face, Head and Feet, 59 FR 16334, April 6.

OSHA (1995). Fall Protection in Construction, Publication 3146, U.S. Department of Labor.

OSHA (1996). Technical Amendment to OSHA's Final Rule on Personal Protective Equipment for General Industry, 61 FR 19547, May 2.

OSHA 1910.120 (1996). 29 CFR 1910.120 Hazardous Waste Operations and Emergency Response, U.S. Department of Labor.

OSHA 1910.133 (1996). 29 CFR 1910.133 Eye and Face Protection, U.S. Department of Labor.

OSHA 1910.335 (1990). 29 CFR 1910.335 Safeguards for Personnel Protection, U.S. Department of Labor.

OSHA, Subpart M (1994). 29 CFR Parts 1910 and 1926. Safety Standards for Fall Protection in the Construction Industry; Final Rule, August 9.

Schwope, A. D., P. P. Costas, J. O. Jackson, and D. J. Weitzman. (1987). Guidelines for the Selection of Chemical Protective Clothing, 3rd Ed. Available from the American Conference of Governmental Industrial Hygienists, 1330 Kemper Meadow Drive, Cincinnati, OH 45240-1634, (513)742-2020.

U.S. Dept. of Labor (1980). Accidents Involving Eye Injuries, Report No. 597, Bureau of Labor Statistics. Available from U.S. Government Printing Office.

Other useful personal protective equipment references:

NIOSH Pocket Guide to Chemical Hazards, U.S. Dept of Health and Human Services, National Institute for Occupational Safety and Health, Publication No. 94-116, June 1994.

Inspection Guidelines for 29 CFR 1910, Subpart I, The Revised Personal Protective Equipment Standards for General Industry, OSHA Instruction STD 1-6.6, June 16, 1995.

CHAPTER 17

Respiratory Protective Devices

HOWARD COHEN

University of New Haven, West Haven, CT 06516

17.1 INTRODUCTION

17.2 REGULATIONS AND GUIDANCE DOCUMENTS COVERING RESPIRATORS AND RESPIRATORY PROTECTION PROGRAMS

17.3 TYPES OF RESPIRATORS

- 17.3.1 Introduction
- 17.3.2 Respirator Facepieces
- 17.3.3 Air-Purifying Respirators
- 17.3.4 Airline Respirators
- 17.3.5 Self-Contained Breathing Apparatus
- 17.3.6 Escape-Only Respirators

17.4 RESPIRATOR PROGRAMS

- 17.4.1 Establishment of a Program Administrator
- 17.4.2 Written Standard Operating Program
- 17.4.3 Hazard Assessment
- 17.4.4 Respirator Selection
- 17.4.5 Training
- 17.4.6 Respirator Maintenance

17.5 RECORDKEEPING

17.6 SPECIAL RESPIRATOR PROBLEMS

- 17.6.1 Vision
- 17.6.2 Facial Hair
- 17.6.3 Voice Communications
- 17.6.4 Oxygen Deficiency
- 17.6.5 Cold Environments

17.7 DEFINITIONS OF RESPIRATOR TERMS

17.8 REFERENCES

APPENDIX A SAMPLE RESPIRATOR PROGRAM

A.1 INTRODUCTION

A.2 RESPIRATOR SELECTION

Adapted from "Respiratory Protective Devices" by Darrel D. Douglas, CIH, Chapter 19 in Clayton & Clayton, *Patty's Industrial Hygiene and Toxicology, 4th ed., Vol. IA.* New York: John Wiley & Sons, Inc., 1991, pp. 675–719.

Handbook of Occupational Safety and Health, Second Edition, Edited by Louis J. DiBerardinis, ISBN 0-471-16017-2

A.3 INSTRUCTION OF RESPIRATOR USERS
A.3.1 General Training
A.3.2 Specific Training
A.4 FIT TESTING OF RESPIRATOR
A.5 ASSIGNMENT OF RESPIRATORS
A.6 CLEANING AND DISINFECTING
A.6.1 Cleaning
A.6.2 Rinsing
A.6.3 Drying
A.7 INSPECTION AND MAINTENANCE
A.8 STORAGE
A.9 RESPIRATOR USE UNDER SPECIAL CONDITIONS
A.10 WORK AREA SURVEILLANCE
A.11 INSPECTION AND EVALUATION OF THE RESPIRATOR PROGRAM
A.12 MEDICAL EXAMINATIONS
A.12.1 Respirator Program Evaluation
A.12.2 Respirator Training Outline

17.1 INTRODUCTION

Respiratory protective devices (respirators) are one of many types of personal protective devices used to protect workers from a specific hazard. Some devices such as steel-toed shoes or eyeglasses are used to protect a worker from hazards that are not likely to occur on a regular basis. Respirators fall into a category of devices that are being used to protect a worker from a known and real-time hazard. If the device fails to provide the expected protection, it is likely that the wearer may suffer some immediate or delayed adverse consequence. Respirators are also unlike personal safety equipment such as shoes and glasses in that their proper selection for a specific hazard is complex and requires training and education.

Respiratory protective devices have been used as early as Roman times, and scattered mention of them occurs in reports of industrial processes during the Middle Ages. Until the nineteenth century, all respirators were air-purifying devices intended to prevent the inhalation of aerosols. They were varied in design, ranging from animal bladders or rags wrapped around the nose and mouth to full face masks made of glass with air inlets covered by particulate filters. During the 1800s masks were produced that combined aerosol filters and vapor sorbing materials. These advances were made primarily for firefighters.

Chemical warfare agents, introduced in World War I, focused attention on the need for adequate respirators. In the United States this work was successfully carried out for the army by the Bureau of Mines. After the war, misuse of surplus army gas masks by civilians highlighted the need for respiratory protection standards.[1] The Bureau of Mines developed these standards, and in 1920 approved their first respirator, a self-contained breathing apparatus (SCBA).[2] Out of this effort grew the Federal respirator approval system that is still in effect. Many of the respirator developments made during the decade following 1920 remain in use today.

To implement a comprehensive respirator program, one must have some knowledge about the regulations covering the use of respirators, the types of devices available, and elements of a respirator program critical to ensure the safe use of respirators. This chapter is organized in this manner, with a definition of respiratory protection terms located in Section 17.7.

17.2 REGULATIONS AND GUIDANCE DOCUMENTS COVERING RESPIRATORS AND RESPIRATORY PROTECTION PROGRAMS

In the United States, the Occupational Safety and Health Administration (OSHA) has established regulations covering the use of respirators in the workplace. These regulations can be found in 29 Code of Federal Regulations Part 1910.134.[3] The regulations cover the types of programs required to support the use of respirators including training, fit testing, and medical clearance. These regulations were established in 1971 and were revised by the agency in 1998.

The National Institute for Occupational Safety and Health (NIOSH) has responsibility for testing and certifying respiratory protective devices for use by employers.[4] Current regulations were originally promulgated by the Bureau of Mines prior to the establishment of NIOSH under the 1971 OSH Act. Responsibility transferred to NIOSH and the Mine Safety and Health Administration with the passage of the OSH Act. NIOSH is currently reviewing and revising existing requirements covering the certification of respirators. The first of these changes affects the design of air-purifying particulate filters and respirators. Since OSHA requires employers to provide NIOSH-approved respirators to their employees, nearly all respirators in use in workplaces are certified by NIOSH.[5]

There are several organizations that provide important reference documents for individuals with responsibility for the selection of respirators and implementation of a respirator program. The American National Standards Institute provides voluntary consensus standards for a variety of items. The ANSI Z88.2 "Practices for Respiratory Protection," issued in 1969, provided the basis of much of the original program requirements found in the current OSHA standard.[6] This standard has since been updated with the most current edition, as of the writing of this chapter, being the 1992 version, and efforts are currently underway to provide another update.[7] The reader is encouraged to obtain a current copy of the most recent version of this document, as it is specifically designed to provide guidance to individuals with responsibility for selecting and maintaining a respiratory protection program. There is also an ANSI Z88.6 standard covering recommended practices for medically evaluating respirator wearers.[8] ANSI standards covering the specific use of respirators for infectious diseases and fit testing procedures are currently in progress.

Two professional societies provide additional useful information in this area. The National Fire Protection Association publishes a variety of documents designed to help individuals in the Fire Service including a document currently in progress focusing specifically on respiratory protection. The American Industrial Hygiene Association also publishes a manual for use by occupational health professionals specifically covering the selection and use of respirators.[9] Current articles on respiratory protection can be found in the *Journal of the International Society for Respiratory Protection.*[10] The *American Industrial Hygiene Association Journal*[11] and *Applied Occupational and Environmental Hygiene.*[12]

17.3 TYPES OF RESPIRATORS

17.3.1 Introduction

There are a large variety of respirators available to protect workers. Some have very specific uses such as an abrasive blasting helmet or mercury air-purifying respirator. Others, such as a particulate respirator or airline respirator, can be used to protect workers from a wide variety of different contaminants. There is no one universal method for categorizing respirators for their selection. One important consideration is whether a respirator is being selected for emergency or routine use. In the former case, the concentration of a contaminant may be unknown or difficult to estimate, while for routine use the workplace environment may be well known and under close control. Another consideration is whether a respirator is selected to provide a safe environment from an external source of breathing air or by purifying the air for the worker. In both cases, careful selection of the device and supporting equipment is required to ensure the safety of the wearer. One scheme for classifying respirators is shown in Fig. 17.1.

17.3.2 Respirator Facepieces

Most respirators can be considered modular units. There are facepieces and then air-purifying elements or supplied air attachments. Many manufacturers make identical facepiece molds for both air purifying and supplied-air models. Facepieces from a manufacturer must be matched with specific air-purifying elements or supplied-air connections. NIOSH approval covers the entire respirator assembly, and any effort to arbitrarily use one type of filter, cartridge, or supplied-air connection other than what was specifically designed by the manufacturer will void the NIOSH approval and violate OSHA requirements.

Facepieces can be classified into tight- and loose-fitting varieties. Tight-fitting facepieces require a good seal between the surface of the device and the wearer's face; loose-fitting devices such as hoods or helmets do not. Until the 1970s respirator manufacturers usually produced tight-fitting respirator facepieces in one or two sizes only. Although each company's facepiece was different, the respirators tended to fit males better than females. This was probably a reflection of employment trends from earlier years. Today most manufacturers supply facepieces in two to three sizes and are designed to fit the facial features of both men and women. However, some individuals who have unusual facial characteristics may have difficulty finding a device that offers both comfort and proper fit, and it is recommended that more than one manufacturer be considered when selecting respirators.

17.3.2.1 Tight-Fitting Facepieces The tight-fitting facepiece is intended to adhere snugly to the skin of the wearer. It is available in three varieties: quarter mask, half mask, and full face mask. The quarter mask covers the nose and mouth; the half mask covers the nose, mouth, and chin (see Fig. 17.2); and the full face mask covers the entire face from chin to hairline and from ear to ear (see Fig. 17.3). The quarter mask is an older design and is seldom seen today due to poor comfort and fit capability.

The facepiece is usually held in place by elastic or rubber straps. Quarter and half masks may be secured by one strap attached to each side of the facepiece, that is, two-point suspension, or two straps attached at two points on each side of the facepiece, that is, four-point suspension. The four straps, instead of being attached to tabs on the

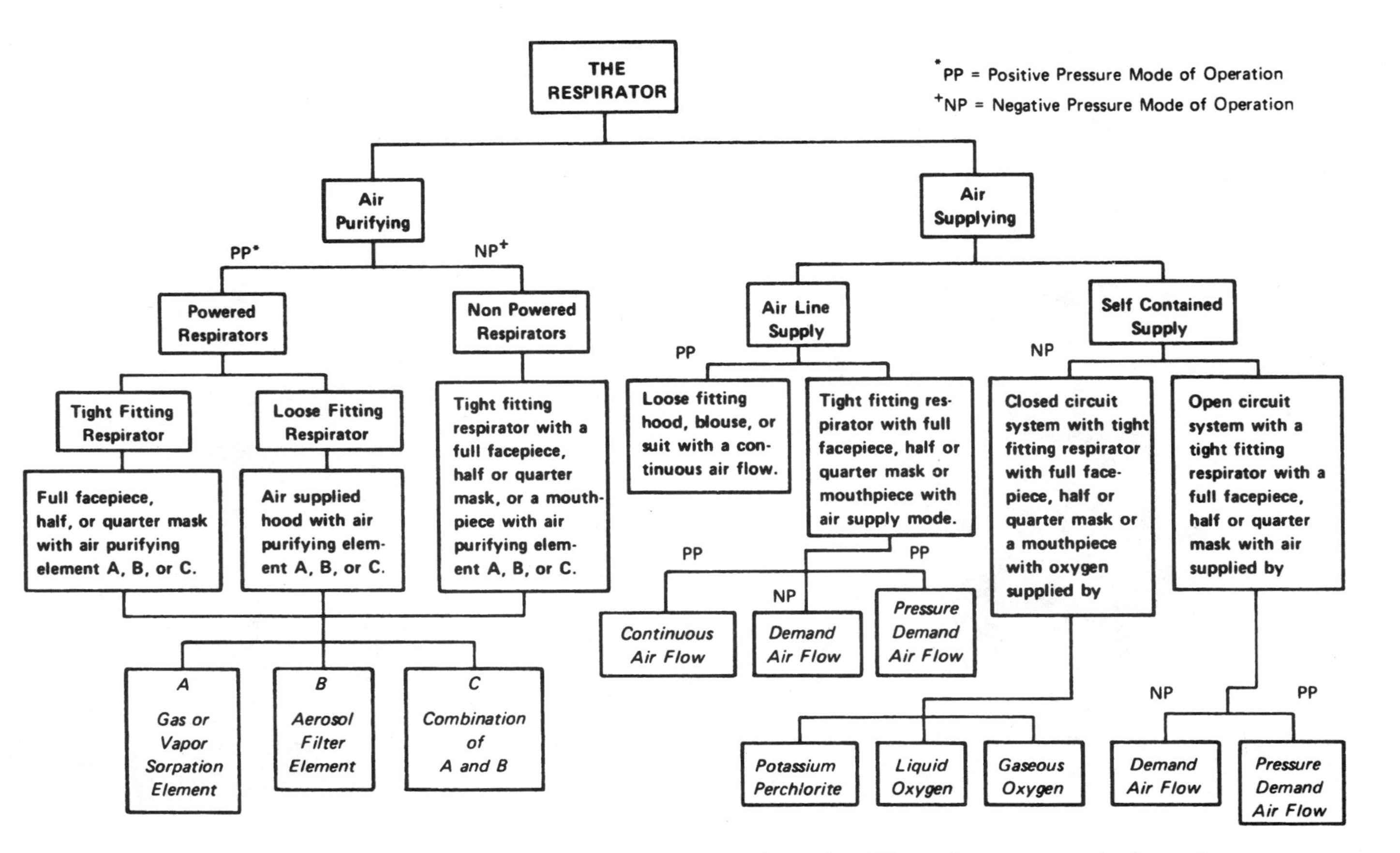

Fig. 17.1 Classification of respirators: PP, positive pressure mode of operation; NP, negative pressure mode of operation.

Fig. 17.2 Half-mask respirator in three sizes—large, medium, and small. Model 1482 Twin Cartridge Respirator, Pro-Tech Respirators.

edge of the facepiece, may be part of a yoke that is fastened to the facepiece by one or two points in the front of the facepiece. Full face masks have a head harness attached to the facepiece at four, five, or six points. The large sealing surface of the full face mask and the distribution of the headband attachment assists in maintaining a stable facepiece with less slippage than is experienced with quarter or half masks. The most

Fig. 17.3 Full-face respirator in three sizes—extra large, large, and small. Model 65 Twin Cartridge Respirator. Scott Aviation.

recent designs in half-mask respirators provide a wide sealing surface to reduce possible facepiece leakage of contaminants.

An important consideration in selecting a facepiece is that there can be nothing that interferes with the sealing surface of a tight-fitting respirator and the face. This includes facial hair and the temple bars of eye glasses. In addition, large facial scars that come into the sealing surface of a respirator can provide a place where contaminants may leak into the facepiece. Because tight-fitting respirators are designed to carefully fit the contours of a face, fit testing is required to insure that the respirator has been properly selected (see Section 17.4.5.2).

17.3.2.2 Loose-Fitting Respirators Loose-fitting respirators are designed to operate without a tight sealing surface. As such, there is less concern about selecting a proper size respirator or requiring that wearers be clean shaven. There is also no requirement for fit testing wearers of such devices. The most common of these devices is the supplied-air hood. The hood covers the head, neck, and upper torso, and usually includes a neck cuff. Air is provided through a hose leading into the hood. Because the hood is not tight fitting, it is important that sufficient quantity of air be provided to maintain an outward flow of air and prevent contaminants from entering the hood.

Supplied-air hoods can be made rugged and used as an abrasive blaster's hood (see Fig. 17.4). NIOSH testing and certification regulations have specific requirements to ensure a durable covering to withstand the rigors of the abrasive blasting atmosphere. There are lightweight plastic and paper hoods available that can be easily disposed of when used in atmospheres where decontamination may be difficult (e.g., handling pharmaceuticals).

Some of these devices incorporate a helmet with face shield. Air is supplied to the rear of the helmet and travels inside the top and comes down along the face shield. There are also supplied-air suits that extend to cover the entire body and provide whole body protection. There are no NIOSH approval regulations for supplied-air suits, but the Department of Energy does have a contract with Los Alamos National Laboratory for evaluating these devices.

17.3.3 Air-Purifying Respirators

Respirators can be categorized according to those that remove hazardous contaminants before entering the facepiece and those that supply a wearer with a clean source of breathable air. The most common forms of respirators are air-purifying devices. They tend to be of a more simple construction and are less costly. These devices can be found with any type of tight-fitting facepiece, and some loose fitting facepieces with auxiliary blowers use air-purifying elements. The two most important features of these respirators are that the air-purifying element must be carefully chosen to remove the toxic contaminant present and the element must be periodically replaced. Some respirators have the air-purifying element built directly into the respirator and are referred to as a "disposable" respirator since the entire device is discarded when the element must be replaced. Disposable respirators can be purchased as "single-use" or reusable devices. The single-use device is discarded after each use. This is a useful feature when the respirator is difficult to clean or disinfect such as when used for protection against infectious agents, or when a respirator is infrequently used and the expense and effort of having a respirator maintenance program is not cost effective.

Air-purifying respirators can be used to remove particulate contaminants, gases,

Fig. 17.4 Abrasive blasting supplied-air hood. 77 Series Supplied-Air Respirator. E. D. Bullard Company.

vapors, and combinations of these. Air-purifying respirators cannot be used when there is an insufficient concentration of oxygen in the atmosphere, when the concentration of a contaminant is so high as to present an imminent hazard to the worker, or for contaminants that cannot be effectively removed by exiting filters and sorbents.

In the 1998 revision to its respirator standard, OSHA now requires that the employer have objective data to determine when to change an air-purifying respirator for protection against gas and vapors. The change schedule will be a function of: the concentration of the contaminant(s), length of time the respirator is worn, and the efficacy of the cartridge for a specific gas, vapor or mixture. It is recommended that the respirator manufacturer be contacted to assist in obtaining the objective data required by OSHA. There is an ANSI standard, K13-1, that covers the color coding of cartridges and canisters. The color coding scheme used for filters and sorbents is summarized in Table 17.1. Although this allows for quick identification, each cartridge and canister will also have a written label indicating its approval and proper use. Respirator wearers should be trained to check the wording of the respirator and air-purifying element prior to donning the device.

TABLE 17.1 Color Code for Air-Purifying Filters, Cartridges, and Canisters

Contaminant	Color
Dust. fumes, and mists (other than radioactive particles)	Orange
Radioactive aerosols	Purple
Acid gases	White
Organic vapors	Black
Ammonia gas	Green
Carbon monoxide	Blue
Acid gases plus organic vapors	Yellow
Acid gases, ammonia, and organic vapors	Brown
Acid gases, ammonia, carbon monoxide, and organic vapors	Red
Other vapors and gases not listed above	Olive

Source: Adapted from ANSI K13.1-1973.

Air-purifying respirators are available with auxiliary blowers and are known as powered-air purifying respirators (PAPRs; see Fig. 17.5). These devices have several advantages over the more common nonpowered respirators. The blower eliminates the need for the wearer to inhale through the cartridge, which can present a significant resistance to breathing. This is a useful feature either for workers exerting substantial energy while wearing a respirator or for those individuals who have subnormal pulmonary function. These devices tend also to provide some cooling for the wearer and are well suited for work in hot environments. Loose-fitting versions of PAPRs allow employees to have facial hair or other obstacles (e.g., glasses) that would traditionally interfere with the fit of a respirator. Tight-fitting PAPRs offer some significant improvements over nonpowered respirators by reducing the likelihood of contaminants' entering the facepiece from leaks.

17.3.3.1 Particulate Filter Respirators Particulate filter respirators protect against dusts, mists, metal fumes, smokes, and biological contaminants. The filter elements remove these contaminants by mechanical filtration, electrostatic attraction, or a combination of both. Mechanical filtration is affected by particle size. Aerosols (the general term used to describe all particles) below 1 μm in aerodynamic diameter are the most difficult to filter efficiently.

NIOSH traditionally classified filters and testing protocols according to: dusts, mists, fumes, and high-efficiency particulate filters (HEPA).[4] There were different tests for each type of filter, and some devices that passed multiple tests received multiple certifications (e.g., dust/mist filters). HEPA filters can stop all aerosols and have a minimum filter efficiency of 99.97% against the most penetrating size aerosols. For mechanical filtration there is a direct relationship between filter efficiency and the pressure drop through a filter. The later translates as difficulty in breathing through a filter. Therefore, while a HEPA filter is the most protective, it often has a higher pressure drop than a dust filter. Given a high enough concentration of aerosols and/or long enough use, all filters can become clogged with particulate matter, making it impossible to breathe through the media. Employees must be trained to replace the filter media before a noticeable increase in breathing resistance has occurred.

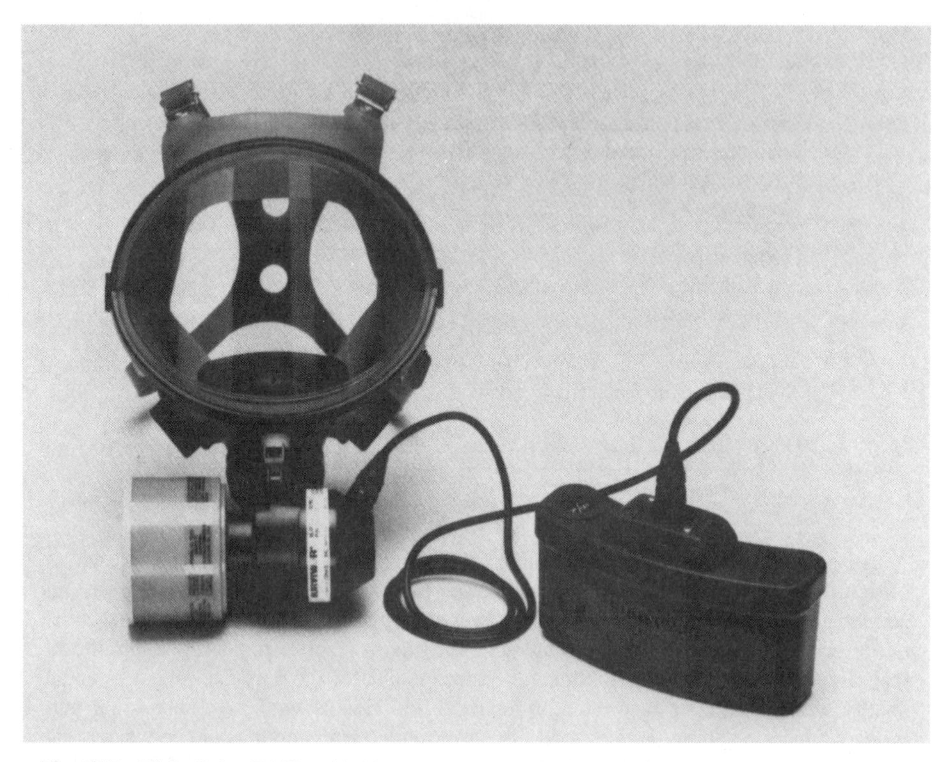

Fig. 17.5 Tight-fitting PAPR with blower, motor, and filter element mounted on facepiece. Survivair Company.

To select one of these filters properly, some knowledge about the workplace is required. Often information concerning the size of aerosols in the workplace is unknown. In 1995, NIOSH introduced a revision to their testing and certification requirements for particulate filters.[5] The new certification and description identifies filters by three-letter designation and three filter efficiencies. The letters N, R, and P are used to describe the filter's ability to function in an environment of oil mist. The letter N means that no oil mist can be present, R indicates that the filter is resistant to oil mist, and P indicates that the filter is oil proof. There are now three efficiencies used to describe filters: 95%, 99%, 100% (>99.97%). All filters are tested against a small penetrating aerosol. The traditional dust/mist filter is being replaced by an N95 filter, while HEPA filters are being referred to as P-100 filters.

The advantages of the new testing and certification scheme is that it should ensure that at least 95% of all aerosols are removed when an N95 filter is used in a non-oil mist environment. While it may take some retraining for those familiar with the previous NIOSH classification scheme, the new certification designations may prove to be simpler to select a filter with confidence.

17.3.3.2 Gas and Vapor Removing Respirators The second major category for air-purifying respirators is those that remove gases and/or vapors. Gases and vapors are captured by adsorption, chemisorption, absorption, or catalysis when passing through a sorbent. The most widely used material for removing gases and vapors is activated carbon. This material has a large internal surface area (typically 1000 m^2/g) and captures material through adsorption; physical attraction due to molecular forces (van der Waals forces of adhesion). Alone it is an excellent adsorbent of many organic vapors. Impregnating it with specific materials increases its retention efficiency for certain gases and vapors. Sorbents are packed in either cartridges or canisters. The only difference is that canisters contain more sorbent and will last for a longer time. They also weigh more and may have a higher resistance to airflow for the wearer.

There are specific cartridges and canisters (see Table 17.1) manufactured for some gases, including: ammonia for ammonia and/or methyl amine; acid gases for chlorine, hydrogen chloride, and sulfur dioxide; and carbon monoxide. The most common need for a gas and vapor air-purifying respirator is for protection against organic vapors. NIOSH has a general certification test for organic vapor cartridges and canisters using carbon tetrachloride as a test agent.[4] This test is designed to examine the quality of the activated carbon and cartridge or canister design. This test is not designed to inform the user that it can or cannot be used for protection for a specific chemical or application.

The parameters governing whether a gas or vapor will be efficiently collected include the contaminant concentration, airflow, relative humidity, and nature of the contaminant. Organic vapors with low molecular weights that are polar (e.g., methanol) tend to be poorly adsorbed onto activated carbon. Those with larger molecular weight and are nonpolar tend to be efficiently collected.

There is published literature in the area of gas and vapor respirator testing that may help in determining the appropriateness of a cartridge or canister for a specific application.[13] There are also a few testing laboratories that will test a single substance or mixture with a specific brand of respirator cartridge. Individuals selecting respirators for specific gas and vapor application are encouraged to consult with representatives from respirator manufacturers for their advice. NIOSH specifically approves organic vapor respirators only for contaminants with good warning properties. This means that such a device should only be used when a wearer can detect the contaminant by smell or taste (or some other sensory means) at a concentration below that which would represent a health hazard. This eliminates the use of air-purifying respirators for many organic vapors. A useful reference that contains odor thresholds for a number of compounds is *Odor Thresholds for Chemicals with Established Occupational Health Standards*.[14]

OSHA no longer allows air-purifying respirators for protection against organic vapors to be selected on the basis of sensory perception. OSHA now requires that objective data be used in determining if an air-purifying respirator can be used, and when gas and vapor cartridges should be changed. In addition to the published literature previously mentioned, the respirator manufacturer should be able to provide assistance in obtaining the objective information required by OSHA.

17.3.3.3 Combination Cartridges and Canisters There are a number of uses of air-purifying respirators requiring the combination of a filter and sorbent. These include, but are not limited to, pesticide application and paint spraying. Respirator manufacturers make organic vapor cartridges with either a builtin prefilter or filter that can be added to the front of a cartridge or canister. There are also combination cartridges and canisters for capturing a variety of gases and vapors. These include combination organic vapor and acid-gas (sulfur dioxide, hydrogen chloride) and a Type N canister that covers the four major gases and vapors (organic vapors, acid gases, carbon monoxide, and ammonia) plus a HEPA filter.

17.3.4 Airline Respirators

Airline respirators provide the wearer with a fresh supply of breathable air. All devices have a facepiece, a compressed air hose connecting to the facepiece, and a source of breathing air. They come either with their own supply of air directly attached to the respirator, in which case they are referred to as a self-contained breathing apparatus (SCBA), or require attachment to an auxiliary source of breathing air. This source may be compressed air cylinders or tanks, or it may be an air compressor located near the wearer or in some remote location. The facepieces may be a full facepiece, half-facepiece, hood, helmet, or even a suit.

These devices have a number of advantages over air-purifying devices. The tend to provide the wearer with a higher degree of protection because they normally maintain a positive pressure in the facepiece and reduce the amount of contaminants that can leak into the respirator. These devices can protect against gases and vapors that are not well adsorbed by air-purifying cartridges. They can provide protection for longer periods of time since there is no need to change an air-purifying element. They also can provide some significant cooling for workers operating in hot environments.

The use of airline respirators also have a number of disadvantages. They generally are more expensive to purchase as well as to operate. For fixed systems, where piping is permanently in place, careful planning must be made in designing where workers will need to work and wear respirators. With the exception of a self-contained breathing apparatus, all these devices require attachment to a compressed line that limits how far a worker can be from a source of breathable air. Airline respirators are generally not useful when employees must work on multiple levels or have far to walk with a compressed air hose.

As previously mentioned, one of the advantages of airline respirators is the ability to maintain a positive pressure inside the facepiece. Manufacturers provide either pressure-demand or continuous flow designs to accomplish this. For pressure-demand devices, the design is such that air is bled into the respirator during inhalation to try to maintain a slight positive pressure. This also requires that the wearer overcome several additional inches of water pressure during exhalation. Continuous-flow respirators, as the name implies, keep a steady stream of air coming into the facepiece. An advantage of continuous-flow respirators is the absence of breathing resistance for the wearer. The selection of these devices are very dependent on the source of breathing air. When compressed-air bottles are used, a pressure-demand tight-fitting facepiece is often required to conserve the amount of air available. When an air compressor is available, continuous-flow loose-fitting respirators are often the choice for comfort, and the quantity of breathable air available is often not a problem. NIOSH testing and certification requires a minimum of 4 cubic feet per minute (CFM) for tight-fitting airline respirators and 6 CFM for loose-fitting respirators.[5] These requirements are designed to try to maintain a positive pressure in the respirator facepiece at all times.

17.3.4.1 Breathing Air Systems The quality and dependability of breathing air for use with airline respirators is a major consideration. The quality of breathing air is covered under OSHA respirator requirements, which state that as a minimum Grade D air (as specified by the Compressed Gas Association) must be used.[15] Grade D air covers contaminants such as condensed hydrocarbons, carbon monoxide, carbon dioxide, and odor. These contaminants can result from the use of a compressor and do not take into consideration additional contaminants that may be located near the source of the compressor. It is important that the intake for a compressor used for breathing air be in a clean environment away from known sources of emissions such as diesel exhaust or exhaust lines. When oil-lubricated compressors are used for breathing air, some planning should be made for the accidental release of carbon monoxide should the compressor overheat.

Moisture is an important consideration when air compressors are used to provide breathing air. The presence of excess moisture in the air can result in the formation of water droplets downstream of the compressor and can result in frozen compressed airlines if low ambient temperatures are experienced. However, systems that remove nearly all the water vapor in the air also create problems if employees are expected to wear respirators for a substantial period of the workday. The lungs require lubrication with water vapor and extreme dry conditions can cause respiratory discomfort to workers.

In addition to the quality of breathing air, the dependability of the source of breathing air is an important factor. If a worker has entered a heavily contaminated environment and the air compressor fails, or if the compressed air hose leading to the respirator is compromised (accidentally run over by a motor vehicle), how will the worker escape from the environment? This is an obvious problem for workers entering confined spaces where the environment is not breathable, but may also be a problem when working in a work environment that may be dangerous to life and health. For these environments only an SCBA or airline with an auxiliary escape compressed air bottle is acceptable. This will automatically switch to the bottled air if the source of breathing air from the compressed airline is shut off. These bottles can provide from 15 minutes or more of air, but can only be used with tight-fitting pressure-demand facepieces. For less hazardous environments, an alternative is to have a reservoir of compressed air that allows the wearer to exit an environment should power to the compressor fail. (Note: This does not address the issue of integrity of the compressed airline. The use of bottled air systems solves some of the breathing air quality issues, but the availability of a sufficient quantity of air may be a concern for some operations.

17.3.5 Self-Contained Breathing Apparatus

A self-contained breathing apparatus is an airline respirator in which the wearer has a source of breathing air directly attached to the respirator (see Fig. 17.6). The major advantage of these devices is that the wearer is not dependent on some remote source for breathing air.

SCBAs can be classified as either open or closed-circuit devices. All SCBAs have tight-fitting facepieces and will operate in the demand (for some closed-circuit units) or pressure-demand mode. The closed-circuit device uses either a bottle of compressed oxygen or an oxygen generating source to maintain a necessary level of oxygen in the respirator. Air is rebreathed through the unit, and the wearer can generally use such devices for much longer times. These units were originally designed for mine rescue, where a long service life was required. The open-circuit units tend to provide the wearer with breathing air for between 15 and 60 minutes (a minimum of 30 minutes is required for use in IDLH environments). The exact time is dependent on the size of the compressed breathing air tank, lung (vital) capacity of the wearer, and the level of activity (breathing rate). These devices are used by firefighters and others who must spend a relatively short amount of time in a highly hazardous environment. Breathing air comes into the respirator from a compressed air tank, and the wearer exhales into the environment, hence the term *open circuit*.

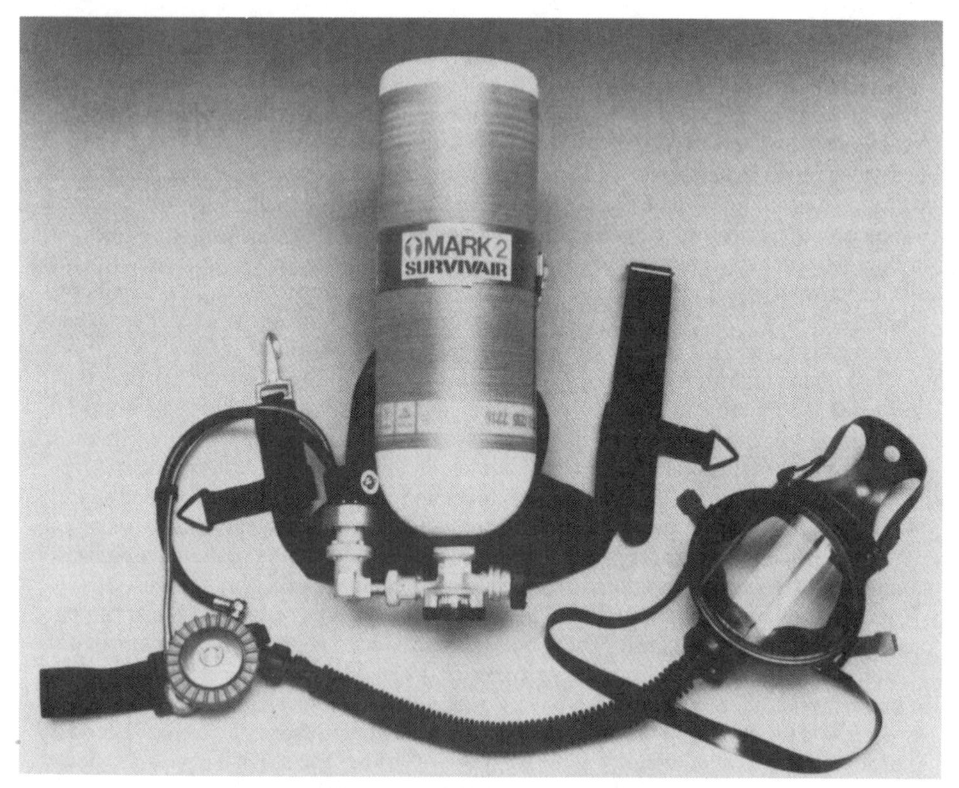

Fig. 17.6 Self-contained breathing apparatus, open circuit. Model Mark 2. Survivair Company.

Several general rules govern the use of these devices. The wearer must be physically fit to use these devices, since they are often 30 or more pounds in weight. These devices are used in highly hazardous work environments, and the wearer must be trained in their use to allow them to handle emergencies (e.g., bypass a faulty regulator on the respirator) and work in highly stressful environments using a respirator. Finally, the devices themselves must be checked at least monthly to ensure that they are in excellent condition and ready for immediate use. SCBAs are often used in response to emergencies, and the wearer may not have the luxury of carefully inspecting the device prior to having to don the respirator and perform their activity. A general checklist covering items that need to be inspected is shown in Table 17.2.

17.3.6 Escape-Only Respirators

OSHA requires that employers plan to protect their employees in the case of an emergency. For some facilities and operations this may include the use of respirators for escape purposes only. While any of the previously mentioned respirators can be used for escape purposes, there are some devices specifically designed for this purpose. These devices are easy to use and require no prior fit testing.

One air-purifying respirator that is approved by NIOSH for escape only is the mouth-

TABLE 17.2 Inspection List for Self-Contained Breathing Apparatus[a]

- Back pack and harness assembly
  - Straps: all are complete and no frayed or damaged straps
  - Buckles: locking function works
  - Back plate: no cracks or missing rivets or screws
  - Cylinder lock: fully engaged
- Cylinder and cylinder valve assembly
  - Cylinder: fastened to back plate, hydrostatic test date is current, no dents, gouges or corrosion
  - Valve assembly: cylinder valve lock is present, cylinder needle and lens in good condition, no leakage heard when cylinder valve is opened
- High-pressure hose
  - No leakage in hose or connection heard or felt
- Regulator and low-pressure alarm
  - Check that regulator reads up to 90% of rated capacity
  - Check low-pressure alarm (should sound at 450–550 PSI for 2216 PSI cylinder)
  - Check pressure-demand function by opening cylinder valve and covering regulator outlet and then quickly uncover outlet
  - Check proper function of bypass valve
- Facepiece
  - Check head harness for damage and deterioration
  - Check lens for proper seal and excess scratches or cracks
  - Check exhalation valve for deterioration or presence of any foreign material
- Breathing tube and connector
  - Check tube for deterioration and holes
  - Check connector for good condition of treads and O-ring
- Negative pressure test
  - Make sure facepiece remains negative when deep breath is taken with breathing tube covered

[a]Consult manufacturer for recommendations specific to each unit.

piece respirator. A mouthpiece is held in the wearer's mouth, and a clamp is placed over the nostrils. The lips are placed tightly around the mouthpiece, and all air comes through the filtering device or air supply attached to the mouthpiece. It has been used for mine rescue and is used at other work sites as an emergency escape device. These devices are small and very portable and for some industries (e.g., chlorine manufacture) have been routinely assigned to all individuals working in an area should an unplanned chemical release occur.

The other type of escape-only device uses a small cylinder of compressed air and usually has a plastic hood that can be quickly pulled over the wearer's head. These devices have an air source that may be rated for between 5 and 15 minutes. The advantage of this device over a mouthpiece device is that one is not dependent on being able effectively to remove whatever contaminants may be present. Therefore, it is especially well suited when the contaminants present from a catastrophic release may vary or are unknown. One disadvantage is that these devices tend to be large and, while they can be carried by a wearer, they are not as convenient as a mouthpiece respirator that can be clipped to one's belt. Another disadvantage of these devices, is that they rely on a highly compressed air source (3000–5000 psi), and the tanks must be regularly inspected (at least monthly) to ensure that they are full and have the prescribed service life.

17.4 RESPIRATOR PROGRAMS

A respirator is no better than the program designed to ensure its proper selection, training of the wearer, and maintenance of the device. A description of each critical component of a respirator program is discussed below. The reader is encouraged to consult with other reference material for additional information.[7,9] The OSHA respirator standard focuses on respirator programs required to ensure the safety of the wearer.[3] Each component of this standard is discussed below. Appendix A contains a sample copy of a respirator program.

17.4.1 Establishment of a Program Administrator

An individual (or group of individuals) must be assigned to develop and oversee the respirator program. The administrator will have responsibilities for the overall program, although additional individuals may have responsibilities for certain components of the program (e.g., wearers may have responsibility for inspection and maintenance of their device). Although there is no specific training required for program administrators, these individuals should be given some formal training in this subject. A number of private consultants and professional societies offer courses in this subject.

The administrator should participate in a formal audit of the respirator program on at least an annual basis. This is designed to ensure that every individual with responsibility for a portion of the program is successfully performing his or her tasks. The use of an outside auditor is recommended, as this individual may be more objective in assessing the strengths and weaknesses of the respirator program. All deficiencies found in an audit should be put in writing and a plan for correcting these deficiencies implemented and recorded.

17.4.2 Written Standard Operating Program

An important aspect of any respirator program is written evidence that it exists and that specific individuals have been identified with key responsibilities. The standard operating program (SOP) is primarily designed for someone outside of the facility to examine whether all the required components of a successful respirator program have been identified. The SOP will include:

- A list of each individual responsible for the program including the Program Administrator.
- The basis for selecting respirators, including the hazard assessment for each location where respirators may be worn.
- The methods used for medically clearing all individuals to wear specific respirators. Also, what methods are used to make sure that respirators are functioning as designed and that employees are not adversely affected by the contaminants they encounter.
- The training of respirator wearers.
- The methods and facilities available for maintaining respirators.
- A list of the records kept that document all required components of a respirator program. This should also include those steps taken to investigate the use of engi-

neering and/or administrative controls that could eliminate the need for respiratory protection.

17.4.3 Hazard Assessment

The need for respiratory protection begins with an assessment of a hazard. The hazard may exist on a routine basis or infrequently due to required maintenance. A respirator may be selected not because a hazard exists, but because a control method may fail to work properly. For example, a respirator may be selected when breaking into a chemical line that supposedly had previously been flushed with water. The respirator is used should the line contain some unexpected contaminant. Sometimes respirators are selected when no health hazard is present, but employees find a material to be a respiratory irritant.

Each of the situations mentioned above requires a different type of hazard assessment in determining how properly to select a respirator. Only qualified individuals, such as Certified Industrial Hygienists, should perform this hazard assessment.

Careful attention must be taken to consider the consequences should a respirator fail to perform its proper function due to the hazard present. Some contaminants are irritants, and employees will quickly know if they should leave a work area due to the malfunction or limitation of a respirator. Other contaminants can produce chronic organ damage, and employees may not be aware of the hazard of continuous overexposure until an irreversible and adverse health outcome has occurred.

The hazard assessment should include options for other controls besides the use of respirators (see Chapter 15). This may include permanent engineering solutions from the substitution of a material with a less toxic substance, reduced emissions through the use of local exhaust ventilation, or other engineering controls, along with the possibilities of reducing the contact time for an employee to a hazard through administrative or other controls. Respirators are generally uncomfortable to wear and can reduce employee productivity and means to communicate with others. Even when a respirator program has been implemented to control a hazard, efforts should continue to explore other control options.

17.4.4 Respirator Selection

The advantages and disadvantages of the variety of respirators available for selection have been mentioned in Section 17.3. There are three main items that govern the selection of a respirator. The first and by far most important is the protection afforded by the respirator. A respirator must be selected that will provide sufficient protection to ensure that a worker will not be overexposed to a contaminant. Less important are the costs associated with both the respirator and program necessary to support its safe use and comfort of the wearer.

17.4.4.1 Protection Factors of Respirators One of the more contentious issues in the field of respiratory protection concerns the protection factors assigned to respirators. The protection factor for a respirator is defined as the ambient contaminant concentration divided by the contaminant concentration inside the facepiece. A protection factor of 1 represents a respirator that allows a contaminant to enter the respirator readily and offers no protection to the wearer. A protection factor of 100 represents a respirator that allows 1% of the contaminants from the ambient environment to enter the inside of a respirator.

The protection afforded by a respirator, and its resultant protection factor, is a complex function of how well the air-purifying elements work, whether the respirator and all its parts have been properly maintained, and how well the respirator is used by the wearer.

In the early 1970s Edwin Hyatt and others at the Los Alamos National Laboratories developed a method to test the fit of a respirator to a panel of individuals with varying facial dimensions.[16] The test was referred to as a quantitative fit test (QNFT), since it could measure leakage on a real-time basis inside the facepiece. The test used a small aerosol, and by having users wear negative-pressure air-purifying respirators equipped with HEPA filters, it was assumed that all contaminants entering the facepiece were the result of a leakage from the faceseal and not from contaminants going through filters or valves. These studies were useful in determining which respirator facepiece designs offered superior protection, and resulted in better respirator designs by manufacturers.

A consequence of the development of this test was the concept that respirators could be selected for a hazard based on facepiece leakage. Prior to this time, full facepiece respirators equipped with gas canisters could be used for protection against very high concentrations of contaminants (10,000–20,000 ppm) based on the performance of the canister. With the advent of QNFT, respirators were placed into categories according to their facepiece design and given an "assigned protection factor." The assigned protection factor (APF; see Section 17.7 for definition) was a conservative estimate of the protection for a class of respirators for most wearers who passed a fit test and received adequate training. NIOSH and ANSI have their own tables with APFs, and OSHA has used APFs in certain 6b health standards for determining the minimum class of respirator to be selected. A list of NIOSH and ANSI APFs are shown in Table 17.3.

A further development in this area has been studies done in workplaces to determine workplace protection factors (WPFs; see Section 17.7 for definition). These are studies done on workers, and contaminant levels are simultaneously measured inside and outside the facepiece. Although the concept of a WPF study may seem straightforward, these studies are extremely difficult to conduct. One reason is the difficulty in finding workplaces where ambient contaminant levels are sufficiently high to provide meaningful results.

The 1992 ANSI Z88.2 standard assigned half-facepiece negative-pressure respirators (both air-purifying and airlines) an APF of 10. The same respirator in a full-facepiece configuration was given an APF of 100. Tight-fitting half-facepieces that are either PAPRs or airlines (in continuous-flow or pressure-demand modes) were assigned an APF of 50. The same respirators in a full-facepiece configuration were assigned an APF of 1000. Finally, loose-fitting respirators were assigned an APF of 25. For respirators that have been properly selected and for employees trained and fit tested, a maximum use limit can be obtained by multiplying the APF for a device by the exposure limit.

EXAMPLE The maximum use limit for protection against benzene having an OSHA permissible exposure limit of one part per million (ppm) when wearing an air-purifying half-mask respirator (APF of 10) is 10 ppm (1 ppm $\times$ 10).

The reader is cautioned when using APF values. These values refer only to facepiece leakage and do not take into consideration whether respirators are being properly worn and maintained by users. There is little evidence of how well respirators work when part of a comprehensive respirator program. Only one or two program protection

TABLE 17.3 Assigned Protection Factors

Type of respirator	ANSI Z88.2-1992	NIOSH RDL[a]
Air purifying		
Half-mask		
Single use	10	5[b]
All others	10	10
Full facepiece		
Non-HEPA filter	100	10
HEPA filter, gas/vapor cartridge	100	50
Powered air purifying		
Half-mask	50	25[c]
Full facepiece	100[d]	25[c]
Helmet or hood	100[d]	25[c]
Loose-fitting facepiece	25	25[c]
Atmosphere supplying		
Pressure-demand airline		
Half-mask	50	1000
Full facepiece	1000	2000
Continuous-flow airline		
Half-mask	50	50
Full facepiece	1000	50
Helmet or hood	1000	50
Loose-fitting facepiece	25	50
Self-contained breathing apparatus	10000[e]	10000

[a]RDL stands for respirator decision logic.
[b]10 is allowed if quantitative fit testing is performed.
[c]50 is allowed if a HEPA filter or gas/vapor cartridge is used.
[d]1000 is allowed if a HEPA filter or gas/vapor cartridge is used.
[e]10,000 is to be used for emergency planning only.

factors (see Section 17.7 for definition) studies have been published, and these suggest generally modest protection afforded by respirators (below current APF values used by ANSI and NIOSH). When respirators are used for protection at levels above ten times an acceptable exposure limit, wearers should participate in a medical surveillance program and have appropriate indices of their health routinely monitored (see Chapters 9 and 10). This may include blood chemistry, examining biological fluids for evidence of contaminants or their metabolites (e.g., blood lead), or pulmonary function testing.

17.4.4.2 Cost of Respiratory Protection The cost of implementing any control of workplace exposure is an important determinant. There has been nothing published in the peer-review literature that compares the cost to establish and maintain a respirator program to that of alternative engineering or administrative controls. In addition, there is nothing in the open literature that compares the costs of using different respiratory protection programs. Some of the more obvious costs of a respirator program are listed below.

17.4.4.2.1 Cost of Purchasing the Respirator A single-use disposable respirator will be the least expensive type of device to purchase initially, and an SCBA will be the most expensive respirator. The single-use respirator may make economic sense when the use of respirators does not warrant the establishment of a comprehensive maintenance and cleaning program or when the contaminant is highly toxic and difficult to remove from the respirator (e.g., infectious agent). A loose-fitting PAPR may be a reasonable choice when relatively few individuals need respirators, and they currently have facial hair that could interfere with the fit of a tight facepiece. This approach may not be economically viable when large numbers of individuals must wear respirators, and it is better management to require respirator users to be clean shaven. Airline respirators may require the purchase of an expensive central compressor and piping, or it may be possible to use an existing system. There are also small portable compressors that are appropriate for one or two individuals that are moderately priced.

17.4.4.2.2 Cost of Maintaining Respirators Air-purifying respirators require that cartridges and filters be regularly replaced. It may be less expensive in the long term to purchase an airline respirator system and avoid the cost of purchasing replacement cartridges and filters. Disposable respirators avoid the need for having to purchase parts and have employees take the time to clean and dry respirator components. Some disposable respirators are reusable and allow for some limited maintenance. PAPRs not only require facepiece maintenance and filter and cartridge replacement, but also have the added expense of having to maintain and replace batteries. SCBAs may have the additional expense of purchasing a special compressor to provide high pressure for recharging compressed air cylinders.

17.4.4.2.3 Cost of Training Each respirator has its own level of training associated with it. Disposable single-size respirators require the least training, while SCBAs require the most training. It may be cost effective for facilities that have historically seldom responded to emergencies to contract with an outside service to perform this function.

17.4.4.2.4 Cost of Medical Clearance All users must be qualified to wear a respirator by a licensed health care professional (LHCP) using either a questionnaire or an initial examination that includes specific information found in Appendix C of the 1998 OSHA Respirator Standard. The LHCP must be informed of the specific respirator to be worn by the user, the duration of its use, the respiratory hazard and be given a complete copy of the respirator program. The LHCP must provide a written recommendation including any limitations governing the use of the respirator, the need for any follow-up examination, and in special cases a recommendation that only a powered-air-purifying respirator be worn.

17.4.4.2.5 Worker Productivity Nearly all respirator usage will result in some loss of productivity. This loss can be the result of difficulty communicating with other coworkers while wearing a respirator, loss of some peripheral vision, some increase in the overall exertion required to complete a task, and simply the time it takes to inspect, don, and maintain a respirator. This loss can vary depending on the type of respirator. For work in hot environments, the use of a supplied air hood or suit could possibly increase productivity by reducing the heat load of a worker.

17.4.4.3 Comfort Nearly all respirator usage results in some discomfort to an employee. This may be caused by added resistance to inhalation caused by air-purifying filters or cartridges, loss of peripheral vision for full facepieces, or the weight caused by devices such as SCBAs. For a given class of respirator, different manufacturers offer various features that can be more or less acceptable to a given wearer. Some individuals may prefer cradle suspensions to four straps. Most manufacturers now have silicon facepieces that are softer and more supple than traditional neoprene facepieces. Tight-fitting facepieces vary by manufacturer, and some wearers will find a certain model and size to be more comfortable and fit better than another model. Finally, some individuals may prefer a PAPR that eliminates breathing resistance during inhalation to conventional negative-pressure air-purifying respirators. Since wearer acceptance and use are a critical part of the overall protection afforded by a respirator, employees should be given as broad a choice in the selection of a respirator as is economically and administratively feasible for the employer.

17.4.5 Training

OSHA requires that respirator wearers receive adequate training prior to wearing a device in the workplace.[3] In addition, some supervisory personnel including those with responsibility for portions of the respirator program will need to have respirator training. Training can be divided into two general categories; general classroom instruction and hands-on training that includes fit testing of the user. Annual refresher training is also required.

17.4.5.1 Classroom Training Individuals should receive training prior to being assigned a respirator. The extent of the training depends on the complexity of the respirator selected and the nature of the hazard. For hazards that are from nuisance ducts, less training is necessary than for entering atmospheres that are dangerous to life and health. The amount of training required will vary and depend on whether wearers have to maintain their own device (cleaning, disinfecting, and storage), or whether this is done by a central service or completely avoided by the use of disposable devices.

Some of the items that should be covered in most training programs include:

- The nature of the hazard resulting in the selection of respirators.
- What controls are being investigated to minimize or eliminate the future need for respiratory protection.
- What are the function, capabilities, and limitations of the respirator(s) selected for this hazard.
- What are the effects of overexposure to the contaminant or other signs that a respirator may be failing to provide adequate protection (e.g., odor detection).
- What are the applicable government regulations that the employer must follow and ensure that their employees follow.
- When must respirators be worn (both for routine and maintenance operations).
- Where can respirators and parts (e.g., filters, cartridges) be obtained, and where employees should discard spent and used respirators or respirator parts.

17.4.5.2 Hands-On Training Employees should be shown how to inspect a respirator to make sure that all the parts are present and that it is in working condition. They should have an opportunity to check for faulty inhalation valves and be shown how to detect worn or deteriorated heat straps and how to replace air-purifying elements (where appropriate). Individuals who must use SCBAs for rescue operations should have an opportunity to use these devices in environments that simulate expected situations.

Employees assigned to use tight-fitting facepieces must be fit tested to ensure that the device is capable of fitting their face. This must be done by someone specifically trained to perform this task. Acceptable respirator fit testing procedures can be found in Appendix A of the OSHA Respirator Standard (mandatory appendix).

Respirator fit tests can be categorized as qualitative that rely on a sensory response by the wearer to a challenge agent, and quantitative that provide a numerical value as to the quality of the fit of the respirator. Qualitative fit tests use isoamyl acetate (banana oil), irritant smoke, or saccharin. These are relatively simple to perform, require inexpensive equipment, and result in a positive or negative assessment of the fit of a respirator to a wearer. Quantitative fit tests either measure ambient aerosols or detect changes in facepiece pressure while subjects hold their breath. Both these tests require more expensive equipment but are able to provide a numerical assessment of the fit of a respirator to a given wearer. Fit tests are performed while wearers conduct a variety of exercises. Fit test exercises are specified by OSHA in 29CFR 1910.134 (1998). The fit factor (see Section 17.7 for definition) for each specified exercise can be averaged. A fit factor of at least 100 is required for a half-facepiece respirator, and a fit factor of at least 500 is required for a full-facepiece respirator. Many individuals performing quantitative fit tests will notice that for good-fitting negative-pressure respirators (leakage less than 0.1% corresponding to a fit factor of >1000), multiple testing of the same wearer performing standard exercises with the same respirator can provide widely different results. This suggests that these respirators are very sensitive and susceptible to minor leakage.

OSHA requires that the employer conduct fit testing on an annual basis.[3] Individuals who undergo substantial weight change or develop scars in the sealing surfaces of the respirator, or when respirator models are changes, should be retested to ensure that the respirator chosen is still appropriate. As previously mentioned, individuals with facial hair that exists at the sealing surfaces of a respirator should be prohibited from wearing tight-fitting facepieces. Do not fit test individuals with facial hair that could interfere with the fit of a respirator and use the results as a means of determining whether they can wear such a respirator. Hair growth and grooming is a dynamic issue and even should a respirator fit properly during a fit test, there is no assurance that it will continue to fit properly while facial hair is constantly growing and/or being groomed.

Respirators should also be checked each time they are put on to make sure that the straps have been properly adjusted. This is referred to as a user seal check. Detailed information on these checks can be found in the ANSI Z88.2 standard.[5]

17.4.6 Respirator Maintenance

When respirators are used repeatedly, inspection, cleaning, disinfection, and storage are normally required.

17.4.6.1 Inspection Respirators should always be inspected before they are used and prior to being placed in storage. Emergency use respirators are required by OSHA to be inspected at least monthly. Items required for inspection include: facepieces, valves, connecting tubes, air-purifying elements, and compressed air cylinders. Rubber parts should be carefully examined for oxidation and cracking, cloth straps should be examined for evidence of wear, and the lens of the facepiece should be examined for scratches that can interfere with visibility. A critical part for negative pressure air-purifying respirators is the exhalation valve. If a valve is missing or improperly seated on the respirator, contaminated air will be diverted from the cartridge, canister or filter into leakage around the exhalation, virtually eliminating the protection offered by the respirator. Respirator wearers should examine respirators for the correct size, model, and components (cartridge, filters, etc.) prior to using them at their work sites.

17.4.6.2 Respirator Cleaning and Disinfecting Respirators should be cleaned daily or after each use and disinfected when used among more than one worker. There are disposable respirators that allow for some degree of cleaning and single-use devices that are not intended to be cleaned or disinfected. Some respirators that are used in areas heavily contaminated or with highly toxic contaminants may need to be cleaned after each use, while others may require daily cleaning. Respirators should not be placed back into storage until they have been adequately cleaned and disinfected, when appropriate. All operations involving cleaning, rinsing, and drying of respirators should be done in contaminant-free environments.

Respirators must be disassembled to allow for adequate cleaning (see Fig. 17.7). There are a variety of cleaning materials available for respirators. Some care must be taken to avoid the use of materials that may cause skin irritation (e.g., quaternary ammonium compounds). Facilities that use a large number of reusable respirators daily may find that a central facility with a commercial dishwasher is the most efficient means for respirator cleaning. Care must be taken to make sure that the water used for cleaning does not exceed 43°C to avoid permanently distorting respirator facepieces. After a respirator is cleaned, it must be rinsed and then dried. Finally, the respirator components must be reassembled and the respirator placed into storage.

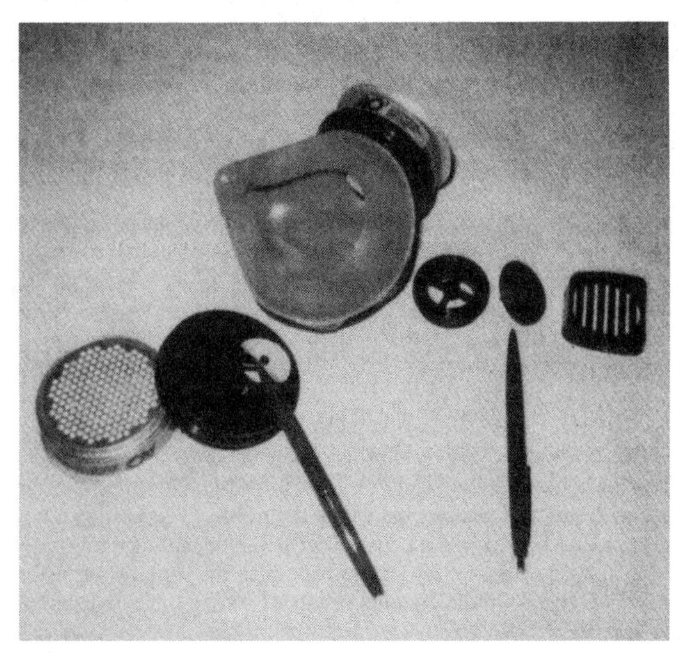

Fig. 17.7 Inhalation valve and exhalation valve assembly, air-purifying respirator. AO 5 Star Dual Element Air Purifying Respirator, AO Safety Products.

17.4.6.3 Respirator Storage Respirator storage varies depending on whether devices are used for emergency-only use or routine wear. Emergency devices such as SCBAs should be stored in areas that are clearly identified. A number of companies make appropriate signs identifying the location of such devices. Emergency-use respirators should be placed in areas of a facility quickly and easily accessible, yet are likely to remain in a contamination-free environment so that they can be safely donned.

Non-emergency-use respirators are often kept either in a central supply area or issued to employees. In the latter case, employees must be careful to store their respirator in such a manner that does not result in the distortion or damage of the facepiece or other components, degradation due to vibration, sunlight, heat, and extreme cold. Ideally, separate lockers or sections of lockers can be designated for the storage of individually assigned respirators. Most respirators should be kept in a sealed plastic bag. This protects the device from contamination and can keep important components such as gas- and vapor-removing cartridges from being contaminated with water vapor. PAPRs that use rechargeable batteries may require separate storage areas that keep batteries charged on a continuous basis until they are ready for use. Some effort is required for multiple PAPR users to distinguish batteries ready for use from those that are currently being charged.

17.5 RECORDKEEPING

There are a variety of records that should be kept documenting the respiratory protection program. Ideally, a central location will have all records kept in a logical manner to allow for program evaluation. Some records such as medical records will be confidential and will be in an employee's medical file. However, a respirator recordkeeping file can maintain a list of individuals who have been medically cleared to use a respirator. It is worthwhile to cross list some of these records, as they are useful for more than one purpose. Many records can be stored electronically, and this facilitates storage of these records in multiple file. Although there is no specific retention time for these records, it is recommended that previous records be kept in anticipation of possible future litigation.

The following are considered minimum records required to be kept:

- A copy of the standard operating procedure covering the operations of the entire respirator program and those responsible.
- Hazard evaluations of all areas and operations where the use of respirators are required for either routine use or maintenance. These should be updated whenever a reevaluation of an operation or job is completed.
- Emergency planning records indicating what emergencies have been anticipated, who will respond to them, and what respirators will be worn.
- Records for individual wearers indicating the results of fit testing, training, the respirator (size and brand) specifically selected for them, and some medical release indicating that they have been cleared to wear the specific respirator they use. Note: Medical records indicating the results of testing must be kept confidential in employee records.
- Training records indicating an outline of material used for training along with the instructor who performed the training.

- Inspection records for emergency-use respirators. It is advantageous to keep these both with the device and copies in a central location. By keeping them with the device, someone can quickly determine in an emergency that the device was inspected within a month, as required by OSHA regulations.[3]
- Documentation of any respirator problems noted and investigations and corrective action taken.
- Results of all audits (internal and external) of the respirator program and what action was taken as a result of recommendations from these audits.
- Copies of all appropriate respirator standards including OSHA CFR 1910.134 and ANSI Z88.2 (current edition).

17.6 SPECIAL RESPIRATOR PROBLEMS

There following are items that affect respirator programs that warrant some additional discussion.

17.6.1 Vision

Many individuals need corrective lenses to perform their job adequately. It should be noted that many half-facepiece respirators do not have a good interface with eyeglasses. This means that the eyeglass frame and respirator must be carefully chosen to allow the wearer to have eye protection and proper vision while wearing a respirator. Full-facepiece respirators do not allow the use of standard glasses as the temple bars will interfere with the seal of the respirator. This can be corrected with special eyeglass kits provided by the respirator manufacturer or by having a prescription ground into the glass of the respirator.

A number of individuals can successfully wear contact lenses to correct a vision problem. Some facilities prohibit the wearing of contact lenses due to fear that an individual may not quickly and thoroughly wash their eyes with water if an accident occurs while they are wearing contact lenses. OSHA originally prohibited the use of contact lenses and respirators, but changed their position following the results of a study of fire fighters that failed to find problems when these respirator wearers wore contact lenses[18].

17.6.2 Facial Hair

It has been previously mentioned that the presence of facial hair may interfere with the seal or function of a tight-fitting respirator. Tight-fitting pressure-demand airline and SCBA devices are designed to keep a positive pressure inside the facepiece at all times. A review by Stobbe et al. found that the pressure in devices examined remained positive despite the presence of facial hair.[19] However, variation among workers, the devices, and the physical exertion required by a task cannot guarantee that facial hair will not interfere with the proper function of these devices. OSHA and ANSI have taken the position that facial hair that comes into contact with the seal of a tight-fitting facepiece should not be allowed.[3,7] A number of court challenges to employers who have required employees to be clean shaven when assigned tight-fitting respirators have found in favor of employers, based on safety considerations. When relatively few employees are required to wear respirators and when loose-fitting PAPRs are a possible choice, employers should consider this option as opposed to requiring that all employees be clean shaven. This is especially true when respirator usage is infrequent.

17.6.3 Voice Communications

The use of respirators may significantly interfere with communication among other employees. Respirator facepieces can be equipped with either a special speaking diagram or with microphones to facilitate communication when this is critical to the job.

17.6.4 Oxygen Deficiency

An ambient environment should contain approximately 20.5% oxygen. When oxygen levels fall below 19.5% oxygen, the environment should be considered dangerous to life and health. Although most individuals can sustain work at levels below 19.5% oxygen with no adverse effects, it may be impossible to predict adequately how stable the oxygen level is, or how low the level may fall. Special care should also be taken when workers must operate in special subambient pressure environments. In these environments, workers may require oxygen-enriched atmospheres to be able to conduct their work.

17.6.5 Cold Environments

Cold weather may cause respirator facepieces to fog, prevent valves from seating properly, or keep facepieces from being supple and forming a good faceseal. Nose cups can be installed in full-facepiece respirators and special compounds used to prevent the facepiece lens from fogging.

It was previously mentioned that components used in airline respirators and SCBAs could freeze if there is sufficient water vapor in the compressed air being used. To prevent this, the dew point of the compressed air should be well below (10–20°C) the lowest ambient temperatures that will be encountered.

17.7 DEFINITIONS OF RESPIRATOR TERMS

Aerosol: solid or liquid particles suspended in air

Airline respirator: an atmosphere supplying respirator in which the respirable gas is not designed to be carried by the wearer

Air-purifying respirator: a respirator in which ambient air is passed through an element (filter, cartridge or canister), and the contaminant(s) of concern is removed

Assigned protection factor (APF): the level of protection expected for a class of respirators when properly fit and used by wearers

Continuous-flow respirator: an airline or other atmosphere-supplying respirator that provides for a continuous flow of air into the facepiece

Dangerous to life and health: any atmosphere that can produce either an acute or chronic irreversible debilitation effect on a worker. Please note that OSHA and others may use the term immediately dangerous to life and health (IDLH) to define conditions that can produce this effect with a relatively short exposure of 30 minutes or less. There is a vagueness to IDLH because the time frame is not specified and it is limited to short exposures only. Therefore, "dangerous to life and health" has been used not only to encompass IDLH but serious health conditions resulting from chronic conditions

Demand respirator: an airline or other atmosphere-supplying respirator that allows air into the facepiece only during inhalation and creates a negative pressure with respect to ambient conditions

Disposable respirator: a device that does not permit the changing of an air-purifying element and is intended to be discarded and not maintained

Escape-only respirator: a respirator designed for emergency use to allow the wearer to escape from a hazardous environment. It is not intended for routine use or for entering hazardous environments

Filter: a respirator component used to remove aerosols

Fit factor: a measure of the fit of a particular respirator to a specific individual obtained during quantitative fit testing

Fit test: a test designed to determine the fit of a specific respirator to a specific individual

Gas: a fluid that has neither independent shape nor volume

High-efficiency filter (HEPA): a filter than will remove at least 99.97% of small aerosols having a diameter of approximately 0.3 μm

Hood: a respiratory inlet covering the head and neck and that may cover portions of the shoulders

Loose-fitting facepiece: a respiratory inlet covering that is designed to form a partial seal with the face

Maximum use limit: the maximum airborne concentration of a contaminant for which a class of respirators is approved. The maximum use limit is obtained by multiplying the assigned protection factor for the respirator by a specific exposure limit (e.g., OSHA permissible exposure limit)

Negative-pressure respirator: a respirator that becomes negative with respect to ambient air pressure during inhalation

Positive-pressure respirator: a respirator in which the pressure inside the facepiece is designed to remain positive during inhalation. This includes both continuous-flow and pressure-demand devices

Powered air purifying respirator: an air-purifying respirator that uses a mechanical blower to force ambient air through air-purifying elements

Pressure-demand respirator: an atmosphere supplying respirator that is specially designed to provide additional air during inhalation to maintain air pressure greater than ambient conditions

Program protection factor: the protection factor of an entire program designed to protect employees. These are obtained by determining ambient exposure levels for employees and measuring the contaminant or its metabolite in body fluid (e.g., blood or urine) or other suitable media (e.g., breath or hair). The program protection factor is obtained by dividing the expected level of contaminant or metabolite in the body by the actual concentration measured

Qualitative fit test: a test designed to determine the adequacy of the fit of a respirator relying on a wearer's sensory response. The result of the test is either a pass or fail

Quantitative fit test: a test that can determine the fit of a respirator in quantitative terms. Unlike qualitative fit tests, these results can be examined and the quality of the fit can be varied to determine what represents a pass or fail

Respirator: a personal device designed to protect the wearer from the inhalation of hazardous atmospheres

Respirator inlet covering: the portion of the respirator that protects the wearer's breathing zone. This may be a facepiece, hood, suit, or mouthpiece

Self-contained breathing apparatus (SCBA): an atmosphere-supplying respirator in which the respirable air is carried by the wearer

Sorbent: a material that is contained in a cartridge or canister and that removes specific gases and vapors

Tight-fitting facepiece: a respiratory inlet covering designed to form a complete seal with the face

User seal check: an action conducted by the respirator user to determine if the respirator is properly seated to the face.

Vapor: the gaseous phase of a substance that normally is a liquid or solid at room temperature

REFERENCES

1. B. J. Held, "History of Respiratory Protective Devices in the U.S., Pre World War I." Lawrence Livermore National Laboratory, Energy Research and Development Administration, contract W-7405-Eng-48.
2. W. P. Yant, "Bureau of Mines Approved Devices for Respiratory Protection," *J. Ind. Hyg.* 473–80 (Nov. 1933).
3. Code of Federal Regulations, Title 29, Part 1910.134, "Respiratory Protection."
4. Code of Federal Regulations, Title 30, Part 11.
5. Code of Federal Regulations, Title 42, Part 84.
6. American National Standard Institute, ANSI Z88.2-1969, "Practices for Respiratory Protection," New York, NY (1969).
7. American National Standard Institute, ANSI Z88.2-1992, "Practices for Respiratory Protection," New York, NY (1992).
8. American National Standard Institute, ANSI Z88.6-1984, "Respirator Use—Physical Qualifications for Personnel," New York, NY (1984).
9. American Industrial Hygiene Association, *Respiratory Protection: A Manual and Guideline*, Second edition. Fairfax, VA (1991).
10. *Journal of the International Society for Respiratory Protection*, International Society for Respiratory Protection c/o Dr. James S. Johnson 5135 Oakdale Ct., Pleasanton, CA 94588.
11. *Journal of the American Industrial Hygiene Association*, American Industrial Hygiene Association, 2700 Prosperity Avenue, Suite 250, Fairfax, VA 22031.
12. *Applied Occupational and Environmental Hygiene*, Elsevier Science, Inc., 655 Avenue of the Americas, New York, NY 10010.
13. G. O. Nelson and C. A. Harder, "Respirator Cartridge Efficiency Studies," *Am. Ind. Hyg. Assoc. J.* **35**(7), 491–510 (1974).
14. *Odor Threshold for Chemicals with Established Occupational Health Standards*, American Industrial Hygiene Association. Fairfax, VA, 1989.

15. CGA Specification G-7.1 (ANSI Z86.1-1989), "Commodity Specification for Air," Arlington, VA: Compressed Gas Association (1989).
16. E. C. Hyatt, "Respiratory Protection Factors," Los Alamos Scientific Laboratory Report No. LA-6084-MS, January 1976.
18. Lawrence Livermore National Laboratory: "Is it Safe to Wear Contact Lenses with a Full Facepiece Respiratory? Livermore, CA: Lawrence Livermore National Laboratory (1986).
19. T. J. Stobbe, R. A. da Roza, and M. A. Watkins, "Facial Hair and Respirator Fit: A Review of the Literature," *Am. Ind. Hyg. Assoc. J.* **49**(4), 199–203 (1988).

APPENDIX A

SAMPLE RESPIRATOR PROGRAM

A.1 INTRODUCTION

"The Occupational Safety and Health Administration (OSHA) Standard on Respiratory Protection, 29 CFR 1910 : 134, requires that a Respiratory Protection program be established by the employer and that respirators be provided and be effective when such equipment is necessary to protect the health of the employee." (OSHA Industrial Hygiene Field Operation's Manual, Chapter III)

It is (COMPANY NAME) goal to control respiratory hazards at their point of generation by using engineering controls and good work practices. In keeping with this goal, the use of respirators as the primary means of protecting employees from airborne hazards is considered acceptable only in very specific situations.

When it has been determined that respiratory protection may be used, it is the responsibility of each Division to provide a formal Respiratory Protection program. The Respirator Program includes the following components:

- Respirator selection (including NIOSH approval)
- Instruction to respirator wearers
- Fit testing of respirators
- Assignment of respirators
- Cleaning and disinfecting
- Inspection and maintenance
- Storage
- Work area surveillance

- Inspection and evaluation of the program
- Medical examinations

The individual responsible for overseeing the implementation of the respiratory protection program for (division or unit name) is (name, location, number).

A.2 RESPIRATOR SELECTION

Before the proper respirator can be selected, an evaluation of the potential hazard must be carried out.

Once the airborne hazards have been evaluated and it has been determined that there exists a potential for exposure to airborne contaminants in concentrations that exceed the allowable exposure level (e.g., TLV, PEL, etc.), then a respirator should be selected. Decision logic is used to select the appropriate respirator (Fig. 17.1) and the following factors are taken into account:

- Physical state of the contaminants
- Toxicity of the contaminants
- Warning properties of contaminants
- Potential for oxygen deficiency
- Potential concentrations (immediately dangerous to life and health?)
- Polaroid exposure guidelines
- Expected respirator protection factor
- Only NIOSH/MSHA-approved respirators shall be used.

It should be noted that whenever a process change occurs that may affect the airborne concentration of the contaminant, the choice of respirator should be reviewed. The respirator and its replacement parts are available in sizes necessary to fit the user population from: (location or name of person)

A.3 INSTRUCTION OF RESPIRATOR USERS

A.3.1 General Training

All individuals who wear respirators are trained in the proper use of such equipment by qualified instructors. This training includes discussion and/or demonstration of the following elements of respirator use:

The training will be done by (name or organization) in conjunction with Division personnel, and it includes both general information and specific details of how each aspect will be handled by the Division. Training is conducted before anyone uses a respiratory protective device and anually thereafter.

A.3.2 Specific Training

In addition to the general training outlined above, respirator users must receive training in the health hazards associated with the specific chemicals to be handled, and why there is a need for respiratory protection. This includes specific information on how and why the respirator was selected, a discussion of its limitations, and the consequences of improper use of the respirator.

This specific information may be a part of the general training or may be presented separately each time a new chemical is introduced that may require the use of a respirator.

A.4 FIT TESTING OF RESPIRATOR

In order to receive the desired protection from a respirator, it is essential that it fit properly. Therefore, all employees required to wear respirators must undergo fit testing. There are two basic steps in determining the fit of a respirator:

A. Quantitative fit testing in a test atmosphere is performed to determine which particular size of a given respirator model provides the best fit for an individual user. This testing is done when the respirator is first issued. It is repeated annually thereafter or more frequently if a user has a facial change that may affect the fit of his respirator (e.g., growth or shaving of facial hair, significant gain or loss of weight, plastic surgery, or change in dentures). This testing results shall be kept by the Division as part of the documentation of a formal respirator program. These records are kept by (name and location) (or in appendix).

B. The OSHA regulations require that a respirator be checked prior to wearing a respirator each time. This requirement is usually met by performing a positive- or negative-pressure fit test. These instructions must be included as part of the formal training of respirator users. An example of these instructions is given here:

1. Positive user-seal check. Block off the exhalation valve with your hand and exhale gently into the facepiece. The facepiece fit is considered satisfactory if a slight positive pressure can be built up inside the facepiece without any evidence of outward leakage of air at the seal. For most respirators, this method of leak testing requires that the wearer first remove the exhalation valve cover and then carefully replace it after the test.
2. Negative user-seal check. Close off the inlet opening of the canister or cartridge(s) by covering it with the palm of the hand(s) or by replacing the seal(s). Inhale gently so that the facepiece collapses slightly, and hold the breath for approximately 10 seconds. If the facepiece remains slightly collapsed, the fit of the respirator is considered satisfactory.

Individuals who cannot be properly fitted with a respirator, as demonstrated by quantitative fit testing, shall not be allowed to perform jobs where respirator protection is required when an exposure level has been exceeded.

Individuals with facial hair that interferes with the respirator face seal (e.g., beard, moustache, sideburns) shall be allowed to wear only loose-fitting respirators.

A.5 ASSIGNMENT OF RESPIRATORS

Where practicable, respirators are assigned to individual workers for their exclusive use. This policy should be adhered to for all routine respirator users. For emergency and some nonroutine uses, individual assignment may not be practical. In such cases, general-use respirators may be used, but they must be made available in all sizes necessary to accommodate the user population. Respirators are assigned to individuals with the following exceptions: (emergency response personnel, confined entry, other).

Each respirator permanently assigned to an individual is durably marked to indicate to whom it was assigned. This mark shall not affect the respirator performance in any way.

The date of respirator assignment is recorded for both individual and general-use respirators. These records are kept by ...

A.6 CLEANING AND DISINFECTING

A.6.1 Cleaning

Individually assigned reusable respirators shall be cleaned at the end of each day on which they are used, or more often if necessary.

Respirators used by more than one person and emergency-use respirators shall be cleaned and disinfected after each use.

Individually assigned respirators shall be cleaned with a mild detergent, a detergent–disinfectant combination, or a respirator wipe designed specifically for this purpose.

General-use respirators, including emergency respiratory equipment, shall be claned with a detergent–disinfectant combination or wipe designed for this purpose.

A.6.2 Rinsing

The cleaned and disinfected respirators should be rinsed thoroughly in clean water (120°F maximum) to remove all traces of detergent, cleaner, sanitizer, and disinfectant. This is important to prevent dermatitis.

A.6.3 Drying

The respirators may be allowed to dry by themselves on a clean surface. They may also be hung from a horizontal wire taking care not to damage the facepieces.

Specific locations for the storage of respirator cleaning supplies and for performing the actual cleaning are as follows:

(Locations)

The products to be used, and the location of the cleaning, shall be specified in the written respirator program.

A.7 INSPECTION AND MAINTENANCE

The formal respirator training program shall include instruction on the inspection of the respirator before and after use.

All nonemergency respiratory equipment shall be inspected by the user during cleaning and immediately before and after each use.

Respiratory equipment designated for emergency use shall be thoroughly inspected during cleaning after each use, and at least once a month.

Note: Records of inspection dates and findings for respirators maintained for emergency use are located (name and location).

The following is a list of items to look for when inspecting various types of respirators and action items when appropriate:

A. Air-purifying respirators (half-mask, full facepiece, and gas mask)

1. Rubber facepiece, check for:
   Excessive dirt (clean and sanitize)
   Cracks, tears, or holes (obtain new facepiece)
   Distortion (allow facepiece to sit free from any constraints and see if distortion disappears; if not, obtain new facepiece)
   Cracked, scratched, or loose-fitting lenses (replace lens if possible; if not, obtain new facepiece)
2. Headstraps, check for:
   Breaks or tears (replace headstraps)
   Loss of elasticity (replace headstraps)
   Broken or malfunctioning buckles or attachements (obtain new buckles)
   Excessively worn serrations on the head harness that might allow the facepiece to slip (replace headstrap)
3. Inhalation valve and exhalation valve, check for:
   Detergent residue, dust particles, or dirt on valve or valve seat (remove residue with soap and water)
   Cracks, tears, or distortion in the valve material of valve seat (replace valve material if possible; contact manufacturer for instructions)
   Missing or defective valve cover (obtain new valve cover)
4. Filter elements, check for:
   Proper filter for the hazard
   NIOSH approval designation
   Missing or worn gaskets (replace with new parts)
   Worn threads—both filter thread and facepiece threads (replace filter or facepiece)
   Cracks or dents in filter housing (replace filter)
   Deterioration of gas mask canister harness (replace harness)
5. Gas mask corrugated breathing tube, check for:
   Cracks or holes (replace tube)
   Missing or loose hose clamps (obtain new clamps)
   Broken or missing end connectors (obtain new connectors)

B. Atmosphere supplying respirators

1. Check facepiece, headstrap, valves, and breathing tube as for air-purifying respirators
2. Faceshield, hood, helmet, full suit check for:
   Cracks or breaks in faceshield (replace faceshield)
   Rips and torn seams (if unable to repair, replace)
   Headgear suspension (proper adjustment)

3. Air supply system
   Breathing air quality
   Breaks or kinks in air supply hoses and end fitting attachments (replace hose and/or fittings)
   Tightness of connections
   Proper setting of valves and regulators (consult manufacturer's recommendations)
   Correct operation of air-purifying elements
4. Self-contained breathing apparatus
   Consult manufacturer's literature
   Worn or deteriorated parts shall be replaced

A.8 STORAGE

After cleaning and disinfecting, each should be placed in its storage box or bag. If not individually assigned, store each respirator in a heat-sealed or resealable plastic bag.

Respirators shall be stored in a convenient, clean, and sanitary location.

Care shall be taken to protect them from dust, sunlight, extremes of temperature, moisture, and chemicals.

Respirators in bulk storage shall be placed in a single layer, in such a way that all components (e.g., facepiece, valves, breathing tubes) will rest in a normal position. Storage in an abnormal position may cause deformation of the unit, which will impair the fit and functioning of the respirator.

A.9 RESPIRATOR USE UNDER SPECIAL CONDITIONS

A. Dangerous atmospheres
If respirator protective equipment usage in atmospheres "dangerous to life or health" is anticipated, special preparations must be made. A standard operating procedure for work in high hazard areas must be written.

The standard operating procedure must cover at least the following:

Individuals designated to enter into dangerous atmospheres must have training with the proper equipment, i.e., self-contained breathing apparatus (SCBA). These individuals must be equipped with safety harnesses and safety lines so that they can be removed from the atmosphere if necessary.

Designation and provision of a standby individual, equipped with proper rescue equipment, who must be present in a nearby safe area for possible emergency rescue.

Provision for communication between persons in the dangerous atmosphere and the standby person must be made. Communication may be visual or by voice, signal line, telephone, radio, or other suitable means. Other important data such as toxicologic information and emergency telephone numbers should also be included.

B. Confined spaces
Confined spaces are defined as enclosures that are usually difficult to get out of, such as storage tanks, tank cars, boilers, sewers, tunnels, pipelines, and vats. In many cases, confined spaces contain toxic air contaminants, are deficient in oxygen, or both. As a result, special precautions must be taken, and these are outlined in CSI-406.

When choosing the appropriate respirator for work in a confined space, the following factors should be considered:

Airline supplied-air respirators may be worn in a confined space only if the tests show that the atmosphere contains adequate oxygen and that air contaminants are well below levels immediately dangerous to life or health. While individuals wearing these types of respirators are in a confined space, the atmosphere must be monitored continuously.

If the atmosphere in a confined space is dangerous to life or health due to a high concentration of air contaminants or oxygen deficiency, those entering the space must wear a positive-pressure SCBA or a combination airline and a positive-pressure self-contained breathing respirator.

C. Low and high temperature
Use of respiratory protective equipment in low temperatures can create several problems. The lenses of the full-facepiece equipment may fog due to condensation of the water vapor in the exhaled breath. Coating the inner surface of the lens with an anti-fogging compound will reduce fogging. Nose cups that direct the warm, moist exhaled air through the exhalation valve without passing over the lens are available from the manufacturer for insertion into the full facepiece. At low temperatures, the exhalation valve can freeze onto the valve seat due to the moisture in the exhaled air. The user will be aware when this situation occurs by the increased pressure in the facepiece. When unsticking the valve, be careful not to tear the rubber diaphragm.

Respirator usage in hot environments can put additional stress on the user. The stress can be minimized by using a light-weight respirator with low breathing resistance. In this respect, an airline type atmosphere-supplying respirator equipped with a vortex tube can be used. Since the vortex tube may either cool or warm the supplied air (depending on the connection and setting), this protection scheme can be used in both hot and cold environments.

A.10 WORK AREA SURVEILLANCE

Exposure concentrations are determined in compliance with CSI #400 in order to select appropriate respiratory protective devices.

A.11 INSPECTION AND EVALUATION OF THE RESPIRATOR PROGRAM

Numerous factors affect the employee's acceptance of respirators including comfort, ability to breathe without objectionable effort, adequate visibility under all conditions, provisions for wearing prescription glasses, if necessary, ability to communicate, ability to perform all tasks without undue interference and confidence in the facepiece fit.

For a respirator program to be effective, it is important that all these factors be considered as the program is developed. Furthermore, it is essential that the respirator users be involved in the process of developing the procedures for respirator use. To this end, the (Division) Respirator Program Coordinator actively involves the users both by observing respirator use and by soliciting user comments with respect to resolution of problems associated with respirator use. This cooperation is vital to the ultimate success of any respirator program, and the user involvement is documented as part of the program.

Finally, once the program has been implemented, the (Division) Coordinator performs regular inspections to insure that respirators are properly selected, used, cleaned, and maintained. These inspections must also be documented. (See Appendix A for checklist).

A.12 MEDICAL EXAMINATIONS

The use of any type of respirator may impose some degree of physiological stress on the user. The degree of stress will be a function of the following factors:

Type of respirator to be used
Tasks to be performed while wearing it
Estimation of the energy requirements of the task
Visual and audio requirements of the task
Length of time the respirator must be worn
The nature of the hazard prompting the need for respiratory protection
Thermal environment

Consequently, it is necessary to determine that all persons expected to wear a respirator are physically able to perform the work and use the required equipment. This determination is made by the Medical Department on an annual basis. Furthermore, as part of a formal respirator program, the documentation necessary to demonstrate that all their respirator users have been approved by the Medical Department is kept by (name and location).

A.12.1 Respirator Program Evaluation

In-general, the respirator program should be evaluated at least annually, with program adjustments, as appropriate, made to reflect the evaluation results. Program function can be separated into administration and operation.

1. Program administration

   A. Is responsibility for overseeing program vested in one individual who is knowledgeable and who can coordinate all aspects of the program? (name)
   B. What is the present status of the implementation of engineering controls, if feasible, to alleviate the need of respirators? (Complete, In Progress, Needs Evaluation)
   C. Are there written procedures/statements covering the various aspects of the respirator program?

Designation of administrator;
Respirator selection;
Purchase of approved equipment;
Medical aspects of respirator usage;
Issuance of equipment;
Fitting;
Maintenance, storage, repair;
Inspection;
Use under special condition.

2. Program operation

 A. Respiratory protective equipment selection
 Are work area conditions and employee exposures properly surveyed?
 Are respirators selected on the basis of hazards to which the employee is exposed?
 Are selections made by individuals knowledgeable to selection procedures?
 B. Are only approved respirators purchased and used, and do they provide adequate protection for the specific hazard and concentration of the contaminant?
 C. Has a medical evaluation of the prospective user been made to determine their physical and psychological ability to wear respiratory protective equipment?
 D. Where practical, have respirators been issued to the users for their exclusive use, and are there records covering issuance?
 E. Respiratory protective equipment fitting

 Are the users given the opportunity to try on several respirators to determine whether the respirator they will subsequently be wearing is the best fitting one?
 Is the fit tested at appropriate intervals?
 Are those users who require corrective lenses properly fitted?
 Is the facepiece to face seal tested in a test atmosphere?

 F. Maintenance of respiratory protective equipment

Cleaning and disinfecting
 Are respirators cleaned and disinfected after each use when different people use the same device, or as frequently as necessary for devices issued to individual users?
 Are proper methods of cleaning and disinfecting utilized?

Storage
 Are respirators stored in a manner so as to protect them from dust, sunlight, heat, excessive cold or moisture, or damaging chemicals?
 Are respirators stored properly in a storage facility so as to prevent them from deforming?
 Is storage in lockers and tool boxes permitted only if the respirator is in a carrying case or carton?

Inspection
 Are respirators inspected before and after each use and during cleaning?

Are qualified individuals/users instructed in inspection techniques?
Is respiratory protective equipment designated as "emergency use" inspected at least monthly (in addition to after each use)?
Is a record kept of the inspection of "emergency use" respiratory protective equipment?

Repair
Are replacement parts used in repair those of the manufacturer of the respirator?
Are repairs made by knowledgeable individuals?
Are repairs of SCBA made only by certified personnel or by a manufacturer's representative?

Training
Are users trained in proper respirator usage?
Are users trained in the basis for selection of respirators?

A.12.2 Respirator Training Outline

Instructor(s): (List names in appendix or refer to log)
Time: (60 minutes)
1. Background information

Respirator used to protect from air contaminants
Use when engineering controls not feasible or inadequate, alone, or during their installation
Use when airborne concentration is known to exceed PEG or where degree of exposure is unknown and suspect problem may exist but has not yet been evaluated by air sampling

2. Reason for training
To get the expected protection must know how to use respirator properly
To use properly must understand:

1. How to wear, maintain, etc.
2. Proper application, i.e., appropriate, cartridge, etc., and its limitations
3. Choice of respirator

Air purifying

Choosing proper filter for hazard
Cartridges are color coded to match specific hazard
Cartridges must match!
Concentration limitation
Toxicity limitation for dust, mist, fumes
Must have adequate O_2 (19–25%)
Negative-pressure respirators have potential for leakage

CHAPTER 18

How to Apply the Microcomputer to Occupational Safety and Health

BRAD T. GARBER
University of New Haven, New Haven, CT 06516

18.1 INTRODUCTION
18.2 SAFETY STATISTICS
18.3 DATABASES
18.4 SPREADSHEETS
18.5 LITERATURE SEARCHING SERVICES
18.6 REPETITIVE OR COMPLICATED CALCULATIONS
18.7 THE INTERNET
18.8 CD-ROMs
18.9 HARDWARE CONSIDERATIONS
18.10 SUMMARY
ADDITIONAL READING

18.1 INTRODUCTION

Efficient systems for information handling and retrieval are of substantial value to occupational safety and health professionals. As a result, microcomputers have been widely adopted as a tool for managing information. The current generation of hardware and software is sufficiently powerful and priced low enough to make it suitable for use in most safety and health offices. This chapter will provide an overview of the many ways in which microcomputers can be used to assist in performing common safety-related tasks.

18.2 SAFETY STATISTICS

Measuring performance is a key element of a safety and health program. A thorough analysis of the trends and patterns in injury and illness experience allows one to identify problem areas and allocate resources where they will do the most good. Unfortunately,

Handbook of Occupational Safety and Health, Second Edition, Edited by Louis J. DiBerardinis, ISBN 0-471-16017-2

collecting, analyzing, and reporting safety statistics can be very time consuming. The use of microcomputers can ease the burden of performing these tasks while enhancing the accuracy of reported information and improving the quality of presentation graphics.

There are two major approaches to using microcomputers for managing safety statistics. One is to purchase a commercially available software package, and the other is to develop customized safety software "in-house." The biggest advantage of using commercial packages is lower initial costs. However, these packages tend to lack the flexibility that is needed to tailor them to the special needs of specific operations and organizations. This is an important issue in organizations where management tends to be very particular about how safety and health data are analyzed and presented.

In organizations with sophisticated safety and health programs, methods of analyzing data have usually been developed using a process of continual refinement over a period of many years. As a result, procedures used to handle safety statistics are often unique and tailored to the types of operations and organizational structures in place. Consequently, when computers are used to manage safety statistics, custom software must be developed. This is best accomplished using either a programming language (such as BASIC, COBOL, or FORTRAN), spreadsheet software, or a database management program. The fastest and easiest approach is using a spreadsheet, but it is also the most limiting. The most flexible, but most time-consuming, method utilizes a programming language. It is also possible to use a combination of these approaches wherein different functions such as data entry, report generation, and preparation of presentation graphics are modularized and each is handled using the method that is best suited to the function. Links can then be created to transfer data automatically between the various modules (current generation application software packages contain sophisticated systems for making these links). For example, it is possible to use a database management program for data entry, a programming language for calculations, and a presentation graphics program to produce reports.

If you choose to develop custom software to set up a complex system for managing safety and health statistics, it is usually necessary to employ the services of professional programmers. When a relatively simple system is all that is desired, it is practical for a safety and health professional, with an aptitude with computers, to prepare one on their own. Developing the skills needed to use spreadsheet software is not particularly demanding and it also is not difficult to employ an easy to master programming language such as BASIC.

An example of a safety statistics system that uses a program written in BASIC and a commercial presentation graphics package follows:

A company has six plants and a corporate headquarters. At the end of every month each of the seven reporting units sends data on its employee hours and OSHA-recordable injuries and illnesses to the corporate safety department. The injuries and illnesses are categorized as follows: fatalities, lost workday cases involving days away from work, lost workday cases involving restricted activity, and nonlost workday cases requiring medical treatment. Total recordable injuries and illnesses are computed by summing the numbers in the categories. Incidence rates are calculated as follows:

$$\text{Incidence rate} = \frac{\text{number of injuries or illnesses} \times 200,000 \text{ hours}}{\text{employee hours of exposure}}$$

A monthly report on safety performance is prepared by the safety department. The

BASIC program is used to perform the following functions: (1) record the incoming data, (2) compile statistics and compute incidence rates, and (3) generate tables. A commercial software package is used to prepare tables and charts.

The BASIC program (Fig. 18.1) was designed to be easy to use. It is menu driven, which allows the operator to receive instructions in a convenient and easy to understand manner. The starting menu is shown in Fig. 18.2.

To enter data, function 1 is chosen from the starting menu (Fig. 18.3). Function 2 permits the viewing of data (Fig. 18.4), and function 3 allows corrections to be made. Tables are generated by function 4 (Fig. 18.5). At the beginning of the year it is necessary to run function 5 to initialize the database. Graphs and charts may be generated using a commercially available program such as Microsoft Powerpoint (Microsoft Corporation, Redmond, Washington). Typically, bar, pie and line charts as well as scatter diagrams are used to present safety data (Figs. 18.6–18.9).

If you desire to work in a completely GUI (graphical user interface) environment, programming may be performed using a graphical version of BASIC instead of the text-based version employed in this example. This would make software development slightly more difficult but would make for a more visually appealing system and facilitate linking data to other applications such as spreadsheets, database managers, and presentation graphics programs.

When designing safety statistics programs, it is important to consider the manner in which the generated reports will be used. Organizations that are serious about employee safety and health generally give safety statistics reports a high degree of visibility. Many organizations use competition between reporting units to promote safety awareness and to increase the incentive for good safety performance. Others use safety records as a factor in determining the level of merit salary increases. It is therefore obvious that the numbers in the reports must be accurate. Equally important is ensuring that the statistics chosen for emphasis in safety reports be as reflective of the quality of safety programs as possible.

The following should be decided before developing a safety statistics software system as they will have a significant effect on the way it is designed (see Chapter 3 for additional information):

What safety statistics will be included in reports: Many organizations generate custom statistics, not required by regulatory agencies, to enhance their ability to analyze safety performance. For example, variants of the old ANSI [Z16.1, Z16.2, American National Standards Institute, New York, NY (212) 642-4900] system for tracing severity are often used to supplement the statistics mandated by governmental agencies.

What safety statistics will be emphasized: Some organizations use total OSHA recordables as their primary indicator of safety performance, while others use lost workday cases involving days away from work.

How and when to correct errors in historical data: Where safety performance is linked to substantial awards or where competition between reporting units is intense, the time frame during which corrections of errors in previously recorded data are permitted can have great significance. Many organizations "close their books" shortly after a final yearly tally and prohibit subsequent changes so as to prevent disruptive "second guessing" that can go on for extended periods of time.

How to handle minor organizational changes: If charts, graphs, or tables are used to identify trends in safety performance over time (for example, a graph of 12 month moving average incidence rates over a multiyear period), the way in which changes in organizational structure are handled can affect the perception of the quality of a divi-

```
100 DIM D(7,12,5),M$(12)
110 FOR N = 1 TO 7
120 READ L$(N)
130 NEXT N
140 DATA "HEADQUARTERS","ATLANTA","DETROIT","CLEVELAND","NEWARK","PHILADELPHIA","WILMINGTON"
190 ON ERROR GOTO 7000
200 OPEN "SDATA.TXT" FOR INPUT AS # 1
205 ON ERROR GOTO 0
210 FOR N1 = 1 TO 7
220 FOR N2 = 1 TO 12
230 FOR N3 = 1 TO 5
250 INPUT # 1,D(N1,N2,N3)
260 NEXT N3
270 NEXT N2
280 NEXT N1
290 CLOSE
402 FOR N = 1 TO 12 : READ M$(N) : NEXT N
403 DATA JANUARY,FEBRUARY,MARCH,APRIL,MAY,JUNE,JULY,AUGUST,SEPTEMBER,OCTOBER,NOVEMBER,DECEMBER
500 CLS
510 PRINT "                    ABC CORPORATION"
520 PRINT "                 SAFETY STATISTICS PROGRAM"
530 PRINT:PRINT:PRINT:PRINT:PRINT:PRINT
540 PRINT "          ENTER <1> TO ENTER DATA FOR MONTH"
550 PRINT "          ENTER <2> TO VIEW DATA"
560 PRINT "          ENTER <3> TO CORRECT DATA"
570 PRINT "          ENTER <4> TO GENERATE TABLES"
580 PRINT "          ENTER <5> TO SET UP DATABASE FOR NEW YEAR"
590 PRINT "          ENTER <6> TO END PROGRAM"
595 PRINT
600 INPUT CHOICE
605 IF CHOICE > 6 THEN GOTO 500
610 CLS
620 ON CHOICE GOTO 1000,2000,3000,4000,5000,6000
1000 INPUT "ENTER NUMBER OF MONTH (EX. 7 FOR JULY)  ";MONTH
1010 CLS
1020 FOR N = 1 TO 7
1030 PRINT "                ENTER DATA FOR ";L$(N)
1035 PRINT :PRINT
1040 INPUT "ENTER NUMBER OF FATALITIES ";D(N,MONTH,1)
1050 INPUT "ENTER NUMBER OF LOST WORKDAY AWAY FROM WORK CASES ";D(N,MONTH,2)
1060 INPUT "ENTER NUMBER OF LOST WORKDAY RESTRICTED CASES ";D(N,MONTH,3)
1070 INPUT "ENTER NUMBER NON LOST WORKDAY MEDICAL CASES ";D(N,MONTH,4)
1075 INPUT "ENTER NUMBER OF HOURS FOR THE MONTH ";D(N,MONTH,5)
1080 CLS
1100 NEXT N
1110 OPEN "SDATA.TXT" FOR OUTPUT AS # 1
1120 FOR N1 = 1 TO 7
1130 FOR N2 = 1 TO 12
1140 FOR N3 = 1 TO 5
1165 PRINT # 1,D(N1,N2,N3)
1170 NEXT N3
1180 NEXT N2
1190 NEXT N1
1200 CLOSE
1210 GOTO 500
2000 CLS
2010 INPUT "ENTER MONTH NUMBER ";M1
2015 PRINT
2020 PRINT
2030 PRINT"LOCATION        FATAL   LWCAFW   LWCR   NLWC   HOURS"
2035 PRINT
2040 FOR N = 1 TO 7
2050 PRINT L$(N) TAB(20) D(N,M1,1) TAB(30) D(N,M1,2) TAB(40) D(N,M1,3) TAB(50)  D(N,M1,4) TAB(60) D(N,M1,5)
2065 PRINT
2070 NEXT N
2075 PRINT
2080 INPUT "PRESS RETURN TO CONTINUE ";A$
2090 GOTO 500
3000 CLS
3010 INPUT "ENTER MONTH OF INCORRECT DATA ";M1
3015 PRINT "ENTER NUMBER OF PLANT "
3020 PRINT : FOR N = 1 TO 7 : PRINT L$(N) ; " = " ; N : NEXT N
```

Fig. 18.1 A safety statistics program written in BASIC.

```
3025 PRINT
3030 INPUT NUM
3035 PRINT
3040 INPUT "ENTER NUMBER OF FATALITIES ";D(NUM,M1,1)
3050 INPUT "ENTER NUMBER OF LOST WORKDAY AWAY FROM WORK CASES ";D(NUM,M1,2)
3060 INPUT "ENTER NUMBER OF LOST WORKDAY RESTRICTED CASES ";D(NUM,M1,3)
3070 INPUT "ENTER NUMBER NON LOST WORKDAY MEDICAL CASES ";D(NUM,M1,4)
3075 INPUT "ENTER NUMBER OF HOURS FOR MONTH ";D(NUM,M1,5)
3080 OPEN "SDATA.TXT" FOR OUTPUT AS # 1
3090 FOR N1 = 1 TO 7
3100 FOR N2 = 1 TO 12
3110 FOR N3 = 1 TO 5
3120 PRINT # 1,D(N1,N2,N3)
3130 NEXT N3
3140 NEXT N2
3150 NEXT N1
3160 CLOSE
3170 GOTO 500
4000 CLS
4010 INPUT "GENERATE TABLES FOR WHICH MONTH NUMBER ";M2
4025 IF D(1,M2,5) <> 99 THEN GOTO 4035
4026 PRINT
4027 PRINT "NO DATA EXISTS FOR THE MONTH OF ";M$(M2)
4028 PRINT
4029 INPUT "PRESS RETURN TO CONTINUE"; A$
4031 GOTO 500
4035 CLS
4036 PRINT "                        TABLE I"
4040 PRINT "                      ABC CORPORATION"
4050 PRINT "                    SAFETY REPORT FOR ";M$(M2)
4060 PRINT
4070 PRINT "                   YEAR TO DATE INCIDENCE RATES"
4073 TRC = 0 : HRC = 0 : FC = 0 : LAC = 0 : LRC = 0 : MC = 0
4074 FOR N = 1 TO 8
4075 F(N) = 0 : LA(N) = 0 : LR(N) = 0 : MC(N) = 0 : HR(N) = 0
4076 NEXT N
4080 FOR P = 1 TO 7
4090 FOR M4 = 1 TO M2
4100 F(P)=F(P) + D(P,M4,1)
4101 LA(P)=LA(P) +D(P,M4,2)
4102 LR(P)=LR(P) +D(P,M4,3)
4103 MC(P)=MC(P) +D(P,M4,4)
4105 HR(P)=HR(P) +D(P,M4,5)
4110 NEXT M4
4115 TR(P) = F(P) + LA(P) + LR(P) + MC(P)
4120 NEXT P
4130 PRINT:PRINT:PRINT
4140 FOR N = 1 TO 7
4150 FC = FC + F(N)
4160 LAC = LAC + LA(N)
4170 LRC = LRC + LR(N)
4180 MC = MC + MC(N)
4190 HRC = HRC + HR(N)
4195 TRC = TRC + TR(N)
4200 NEXT N
4300 PRINT "LOCATION         TOTAL RECORDABLES INCIDENCE    AWAY FROM WORK INCIDENCE"
4305 PRINT "                       RATE                   RATE"
4306 PRINT
4310 FOR N = 1 TO 7
4320 PRINT L$(N) TAB(30);
4321 PRINT USING "####.##";TR(N)*200000!/HR(N);
4322 PRINT TAB(60);
4323 PRINT USING "####.##";LA(N)*200000!/HR(N)
4325 NEXT N
4330 PRINT : PRINT
4340 PRINT "CORPORATION" TAB(30)
4341 PRINT USING "####.##" ;TRC*200000!/HRC;
4342 PRINT TAB(60);
4343 PRINT USING "####.##" ; LAC*200000!/HRC
4345 PRINT
4350 INPUT "PRESS RETURN TO CONTINUE"; A$
4360 CLS
```

Fig. 18.1 *(Continued)*

```
4436 PRINT "                    TABLE II"
4440 PRINT "                  ABC CORPORATION"
4450 PRINT "               SAFETY REPORT FOR ";M$(M2)
4460 PRINT
4470 PRINT "              NUMBERS OF INJURIES BY PLANT"
4471 PRINT "                  YEAR TO DATE"
4472 PRINT : PRINT
4475 PRINT "LOCATION      FATAL    LOST-AWAY    LOST REST   MEDICAL   TOTAL"
4476 PRINT : PRINT
4490 FOR N = 1 TO 7
4500 PRINT L$(N) TAB(20) F(N) TAB(33) LA(N) TAB(48) LR(N) TAB(60) MC(N) TAB(70)TR(N)
4510 NEXT N
4520 PRINT : PRINT
4530 PRINT "CORPORATION" TAB(20) FC TAB(33) LAC TAB(48) LRC TAB(60) MC TAB(70) TRC
4531 PRINT
4550 INPUT "PRESS RETURN TO CONTINUE"; A$
4551 CLS
4636 PRINT "                    TABLE III"
4640 PRINT "                  ABC CORPORATION"
4650 PRINT "               SAFETY REPORT FOR ";M$(M2)
4660 PRINT
4670 PRINT "          YEAR TO DATE INCIDENCE RATES BY MONTH"
4671 PRINT "                 TOTAL CORPORATION"
4672 PRINT : PRINT
4680 PRINT "MONTH        TOTAL RECORDABLES INCIDENCE    AWAY FROM WORK INCIDENCE"
4681 PRINT "             RATE                  RATE"
4682 PRINT
4872 FOR Q = 1 TO M2
4873 TRC = 0 : HRC = 0 : FC = 0 : LAC = 0 : LAR = 0 : MC = 0
4874 FOR N = 1 TO 8
4875 F(N) = 0 : LA(N) = 0 : LR(N) = 0 : MC(N) = 0 : HR(N) = 0
4876 NEXT N
4880 FOR P = 1 TO 7
4890 FOR M4 = 1 TO Q
4900 F(P)=F(P) + D(P,M4,1)
4901 LA(P)=LA(P) +D(P,M4,2)
4902 LR(P)=LR(P) +D(P,M4,3)
4903 MC(P)=MC(P) +D(P,M4,4)
4905 HR(P)=HR(P) +D(P,M4,5)
4910 NEXT M4
4915 TR(P) = F(P) + LA(P) + LR(P) + MC(P)
4920 NEXT P
4940 FOR N = 1 TO 7
4950 FC = FC + F(N)
4960 LAC = LAC + LA(N)
4970 LRC = LRC + LR(N)
4980 MC = MC + MC(N)
4990 HRC = HRC + HR(N)
4991 TRC = TRC + TR(N)
4992 NEXT N
4993 PRINT M$(Q) TAB(30);: PRINT USING "####.##"; TRC*200000!/HRC;
4994 PRINT TAB(60);:PRINT USING "####.##"; LAC*200000!/HRC
4996 NEXT Q
4997 PRINT
4998 INPUT "PRESS RETURN TO CONTINUE"; A$
4999 GOTO 500
5000 OPEN "SDATA.TXT" FOR OUTPUT AS # 1
5010 FOR N1 = 1 TO 7
5020 FOR N2 = 1 TO 12
5030 FOR N3 = 1 TO 5
5040 D(N1,N2,N3) = 99
5045 PRINT # 1,D(N1,N2,N3)
5050 NEXT N3
5060 NEXT N2
5070 NEXT N1
5080 GOTO 500
6000 CLOSE
6010 END
7000 CLS
7010 PRINT "NO DATA FILE EXISTS, RUN FUNCTION 5 "
7015 INPUT "PRESS RETURN TO CONTINUE ";A$
7020 GOTO 500
```

Fig. 18.1 *(Continued)*

```
ABC CORPORATION
SAFETY STATISTICS PROGRAM

ENTER <1> TO ENTER DATA FOR MONTH
ENTER <2> TO VIEW DATA
ENTER <3> TO CORRECT DATA
ENTER <4> TO GENERATE TABLES
ENTER <5> TO SET UP DATABASE FOR NEW YEAR
ENTER <6> TO END PROGRAM
```

Fig. 18.2 Starting menu generated by safety statistics program.

sion's safety performance. If, as happens not too infrequently in many organizations, a reporting unit is transferred from one division to another, does it bring its past performance to the new division? Many organizations choose not to have reporting units carry along "baggage." Instead, they have incidents and employee hours that accrued in the past remain with the former division. Provisions also must be made for handling new acquisitions and facilities that are sold or closed.

Who will mediate disputes regarding the recordability or classification of accidents: This is typically handled by an organization's safety director.

Choosing what statistics to record and report should be done with great care. A major consideration is the expense of collecting, entering, analyzing, and disseminating safety data. Even more important is ensuring that the information presented to managers and employees is sufficiently concise and understandable so as to have a maximum positive effect. Obviously, all regulatory requirements must be met. Since each organization has its own characteristics and needs, there is no single answer to the question of what information should be collected and how it should be analyzed. Based on individual requirements, carefully formulated decisions must be made about what data are recorded for each reporting unit and what information is collected for each specific incident. Possibilities are shown below:

For reporting units:

Employee hours
Number of fatalities
Number of lost workday cases involving days away from work
Number of lost workdays
Number of lost workday cases involving restricted duty
Number of cases requiring medical treatment
Number of first aid cases

```
ENTER DATA FOR HEADQUARTERS

ENTER NUMBER OF FATALITIES ? 0
ENTER NUMBER OF LOST WORKDAY AWAY FROM WORK CASES ? 0
ENTER NUMBER OF LOST WORKDAY RESTRICTED CASES ? 0
ENTER NUMBER NON LOST WORKDAY MEDICAL CASES ? 1
ENTER NUMBER OF HOURS FOR THE MONTH ? 293845
```

Fig. 18.3 Data entry format for safety statistics program.

```
ENTER MONTH NUMBER ? 2
```

LOCATION	FATAL	LWCAFW	LWCR	NLWC	HOURS
HEADQUARTERS	0	0	0	1	123001
ATLANTA	0	0	0	1	434298
DETROIT	0	0	0	1	333299
CLEVELAND	0	0	0	0	666752
NEWARK	0	1	0	0	126472
PHILADELPHIA	0	0	0	1	126599
WILMINGTON	0	0	0	0	663230

Fig. 18.4 Table generated by view function.

Number of near misses
Number of property damage cases
Cost of property damage cases

For individual incidents:

Classification (medical case, first aid, etc.)
Narrative of what happened
Reporting unit
Physical location
Part of body injured
Type of injury (laceration, burn, etc.)
Remedial action

For each category of information, it is necessary to develop a classification scheme that specifies how entries are to be made. A combination of OSHA guidelines, ANSI standards, and medical department practices can be used to assist in this task (see Chapter 26). For example, OSHA guidelines can be used to help decide how to classify and define injuries while medical department practices can be used as the basis for a classification scheme for body part and type of injury. Since there is no general consensus in the safety community, "home-grown" systems for recording information about incidents are common. This is not necessarily bad, since the needs and characteristics of different organizations vary greatly.

18.3 DATABASES

A database is a collection of records, each of which contain pieces of information referred to as *fields*. Programs used to administer these records are called *database managers*. Utilizing this type of software can ease the burden of performing a wide variety of safety-related record keeping tasks. In addition to safety statistics, typical applications include managing material safety data sheets, industrial hygiene sampling records,

TABLE I
ABC CORPORATION
SAFETY REPORT FOR FEBRUARY

YEAR TO DATE INCIDENCE RATES

LOCATION	TOTAL RECORDABLES INCIDENCE RATE	AWAY FROM WORK INCIDENCE RATE
HEADQUARTERS	3.24	0.81
ATLANTA	0.46	0.00
DETROIT	0.30	0.00
CLEVELAND	0.15	0.00
NEWARK	0.80	0.80
PHILADELPHIA	1.58	0.00
WILMINGTON	0.15	0.15
CORPORATION	0.49	0.12

TABLE II
ABC CORPORATION
SAFETY REPORT FOR FEBRUARY

NUMBERS OF INJURIES BY PLANT
YEAR TO DATE

LOCATION	FATAL	LOST-AWAY	LOST REST	MEDICAL	TOTAL
HEADQUARTERS	0	1	0	3	4
ATLANTA	0	0	0	2	2
DETROIT	0	0	0	1	1
CLEVELAND	0	0	1	0	1
NEWARK	0	1	0	0	1
PHILADELPHIA	0	0	0	2	2
WILMINGTON	0	1	0	0	1
CORPORATION	0	3	1	8	12

TABLE III
ABC CORPORATION
SAFETY REPORT FOR FEBRUARY

YEAR TO DATE INCIDENCE RATES BY MONTH
TOTAL CORPORATION

MONTH	TOTAL RECORDABLES INCIDENCE RATE	AWAY FROM WORK INCIDENCE RATE
JANUARY	0.57	0.16
FEBRUARY	0.49	0.12

Fig. 18.5 Tables generated by safety statistics program.

training records, and contact lists. While learning to use database management software can usually be done fairly easily, formulating a record structure for a specific database is often much more demanding. It is necessary not only to determine what information should be included in each record, but also to develop effective coding schemes that facilitate entry and retrieval of information. Many of the types of data collected by safety professionals pose significant problems. For example, keeping records that incorporate the identities of chemicals is difficult because compounds often have several commonly used names. Furthermore, entry of chemical names is subject to error,

ABC CORPORATION
1997 YEAR TO DATE
TOTAL RECORDABLE INJURY AND ILLNESS
INCIDENCE RATES

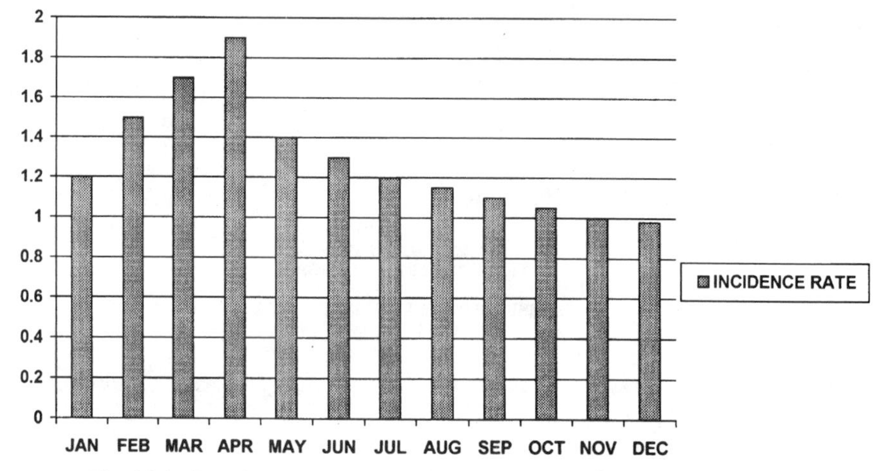

Fig. 18.6 Bar chart generated by commercial presentation graphics software.

ABC CORPORATION
TYPES OF INJURIES
1997

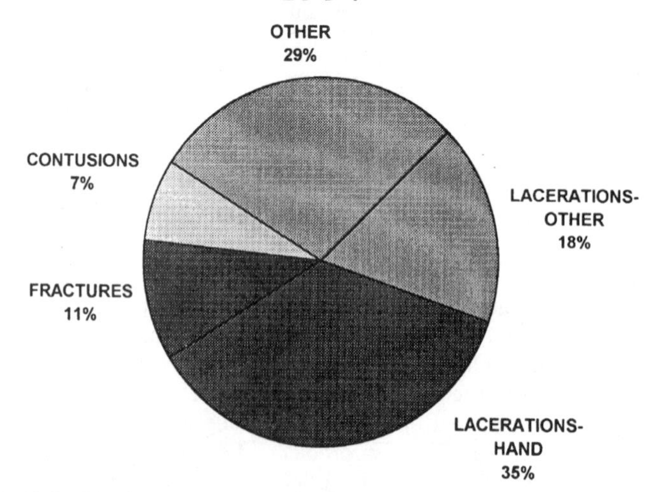

Fig. 18.7 Pie chart generated by commercial presentation graphics software.

ABC CORPORATION
TYPES OF INJURIES
1997

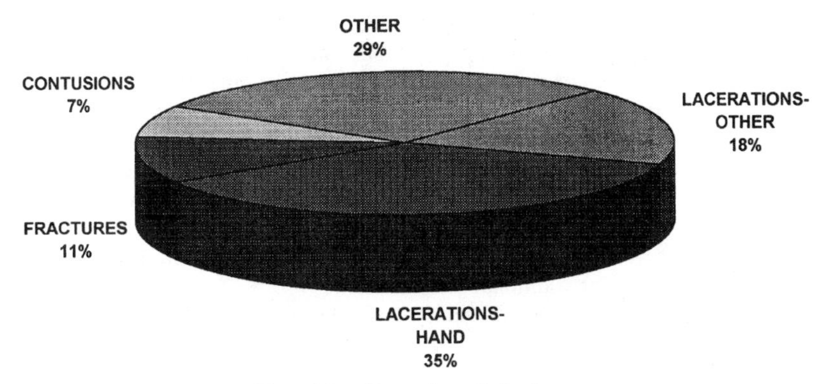

Fig. 18.8 3D version of pie chart.

TOTAL RECORDABLE
INJURY AND ILLNESS
INCIDENCE RATES
BY YEAR

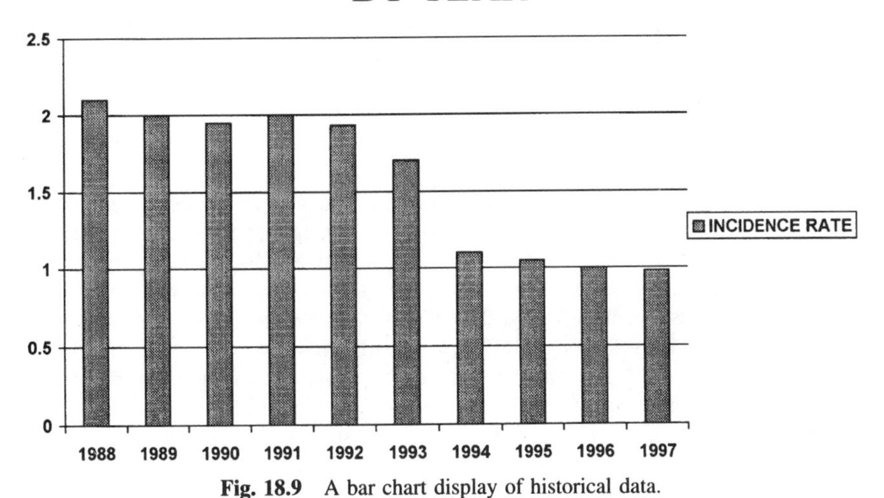

Fig. 18.9 A bar chart display of historical data.

especially if it is done by persons not proficient in chemical nomenclature. As a result, for databases whose records include chemical names, numbers are often assigned to each unique substance for purposes of identifying them. This speeds data entry and decreases the likelihood of error. Deciding on a numbering system, however, poses its own problems. Does one use a widely adopted scheme such as the CAS (Chemical Abstracts Service, Columbus, Ohio 614-447-3600) system or develop a custom approach? One of the advantages of using the CAS system is that it makes use of methodologies that have been developed and fine tuned by experts who have put many hours of effort into producing an efficient and effective way of recording the identities of specific chemicals. On the other hand, the numbering system may be too cumbersome for applications that track only a small number of compounds. It also has limited capabilities to handle substances of variable composition such as total dust.

Problems such as those described for coding chemical names also exist for many of the other fields that might be found in safety database records. One way to avoid the complication of developing coding schemes is to buy commercial software that includes prepackaged record structures and coding schemes. Information on currently available safety software is periodically presented in magazines such as Occupational Health and Safety (see Chapter 26 for additional information).

The following guidelines should be followed in choosing a commercial package designed specifically for managing safety records:

Be sure it utilizes widely accepted, general-purpose database management software and that the source code for compiled programs is provided: This allows the user to customize the system to meet specific needs. It also serves as insurance that the system can be maintained if the vendor goes out of business.

Carefully examine the record structures and coding schemes that are used: In many cases data entry is the most expensive part of managing a database. Efficient record structures and coding schemes can minimize costs and reduce the number of errors in the database.

Purchase a system whose software is optimized for speed: When databases get large, some routine tasks can become very time consuming. The quality of the database engine can have a major impact on the speed with which operations are performed.

Be sure that your hardware is capable of running the software with a reasonable level of performance: An underpowered microcomputer can make using database management software much more tedious and time consuming than is necessary.

18.4 SPREADSHEETS

Spreadsheet programs generate grids that contain cells into which information can be entered. A major advantage of using spreadsheets for applications such as managing safety and health statistics is that simple systems can usually be set up much more rapidly than is possible using programming languages or database management software. The major disadvantage is that they tend to lack the flexibility needed to handle complex or unusual tasks efficiently. Most spreadsheet software contains moderately capable graphics and database management components that can be used where needs are not too demanding. Spreadsheets are ideal for managing information that consists mainly of numbers and for which "what if" analyses are common. Budgets, for example, are often prepared using spreadsheets because the data involved are mostly numerical and managers frequently want to know "what" would the effect be on the bottom-line

A	A	B	C	D	E	F	G	H
1				ABC CORPORATION				
2				INCIDENCE RATES				
3								
4								
5	LOCATION	FATALITIES	LOST	LOST	NON LOST	HOURS	TOTAL	TOTAL
6			WORKDAY	WORKDAY	WORKDAY		LOST	RECORDABLES
7			AWAY	RESTRICTED			WORKDAY	I.R.
8							I.R.	
9								
10								
11	HEADQUARTERS	0	1	0	2	123543	1.62	4.86
12	ATLANTA	0	0	0	1	434312	0.00	0.46
13	DETROIT	0	0	0	0	333214	0.00	0.00
14	CLEVELAND	0	0	1	0	665432	0.30	0.30
15	NEWARK	0	0	0	0	123298	0.00	0.00
16	PHILADELPHIA	0	0	0	1	126543	0.00	1.58
17	WILMINGTON	0	1	0	0	654879	0.31	0.31
18								
19	CORPORATION	0	2	1	4	2461221	0.24	0.57
20								

Fig. 18.10 A spreadsheet for computing safety statistics. Columns G and H are calculated automatically from raw data in columns B through F.

costs "if" the amount budgeted to various items is changed. A large number of possibilities can be considered in a short period of time since changes to cell entries result in an almost instantaneous retabulation of the entire spreadsheet.

An example of a spreadsheet used to prepare safety statistics is shown in Figs. 18.10 and 18.11. Cells in the spreadsheet contain numbers, labels, or formulas for doing calculations. If a cell containing raw data is modified, results are automatically and quickly recalculated. A "what if" analysis, such as determining the effects of increases or decreases in the number of incidents on overall rates, can easily be performed. For example, you can determine "what" happens to the incidence rate "if" the number of recordable injuries increases by a specific number such as two.

ABC CORPORATION
INCIDENCE RATES

LOCATION	FATALITIES	LOST WORKDAY AWAY	LOST WORKDAY RESTRICTED	NON LOST WORKDAY	HOURS	TOTAL LOST WORKDAY I.R.	TOTAL RECORDABLES I.R.
HEADQUARTERS	0	1	0	2	123543	1.62	4.86
ATLANTA	0	0	0	1	434312	0.00	0.46
DETROIT	0	0	0	0	333214	0.00	0.00
CLEVELAND	0	0	1	0	665432	0.30	0.30
NEWARK	0	0	0	0	123298	0.00	0.00
PHILADELPHIA	0	0	0	1	126543	0.00	1.58
WILMINGTON	0	1	0	0	654879	0.31	0.31
CORPORATION	0	2	1	4	2461221	0.24	0.57

Fig. 18.11 Spreadsheet formatted for inclusion in a report.

18.5 LITERATURE SEARCHING SERVICES

A large number of literature searching services of interest to safety professionals are in existence (see Chapter 3). They allow users to search through databases that contain indexes of articles that appear in journals that cover topics of relevance to occupational safety and health. If an article of interest to the searcher is located, it is possible to obtain information such as the name and date of the source publication, view an abstract or, in some cases, retrieve a paper in its entirety.

Typically, literature searching services are accessed over telephone lines using a microcomputer and a modem. Recently, the Internet has also become a popular means of access. This has, for many people, dramatically reduced the cost of using these services since going online can involve making an expensive long distance telephone call (access to the Internet can be procured for a relatively low monthly flat rate for unlimited usage in most areas of the United States and many places throughout the world).

A widely used literature searching service is run by the National Library of Medicine (Bethesda, Maryland, 800-338-7657). It is of most use for locating articles dealing with medical or toxicological subjects. For instance, if you want to find out what is known about factors that influence the gastrointestinal absorption of lead, a search using the key words *gastrointestinal* and *lead* could be performed (Fig. 18.12). When searching using only one or two key words, finding relevant articles is relatively straightforward. If, however, three or more conditions are imposed upon a search, the situation is more complex. For example, if you want to find information on the effects of lead on both the kidney and liver, telling the computer to find articles dealing with lead and kidney or liver might produce unintended results. You might end up with hits for "lead and kidney" as well as all the articles dealing with the liver even if they have nothing to do with the effects of lead. In order to avoid this type of situation, each literature search engine has a carefully defined method for preparing queries that avoid this and other types of ambiguities. Although virtually all search languages are based on Boolean logic (a branch of mathematics which is well suited to defining and analyzing search conditions), the specific details of how search statements are constructed vary considerably. Unless you are proficient at doing searches, it is advisable to obtain outside help from persons who regularly perform this task. Typically, reference librarians have received extensive training on literature searching.

18.6 REPETITIVE OR COMPLICATED CALCULATIONS

The microcomputer is an excellent tool for doing repetitive or complicated calculations. It provides a number of benefits over using electronic calculators or doing arithmetic by hand. In addition to the obvious speed advantages, the microcomputer allows the user to format output into an easily understandable form. This allows you to minimize the likelihood of mistakes since, when using a computer, the only major source of error is inputting raw data incorrectly.

An example of a simple program written in BASIC, which calculates incidence rates from numbers of injuries and illnesses and employee hours of exposure, is shown in Fig. 18.13. The output (Fig. 18.14) is formatted so that both the raw data and the results can be viewed. Since the major source of error is typing in numbers incorrectly, the program was designed to display raw data as well as results. This facilitates spotting data entry mistakes.

```
SS 5 /C?
USER:
PRINT FULL 1 INDENTED SKIP 17
PROG:

18
AUTHOR                  GARBER BT
AUTHOR                  WEI E
TITLE                   Influence of dietary factors of the
                        gastrointestinal absorption of lead.
SECONDARY SOURCE ID     HEEP/74/09469
SOURCE                  TOXICOL APPL PHARMACOL; 27 (3). 1974 685-691
ABSTRACT                HEEP COPYRIGHT: BIOL ABS. Gastrointestinal
                        absorption of Pb was investigated in mice
                        after oral administration of lead acetate
                        labeled with 210-Pb. When doses of 0.2, 2
                        and 20 mg of Pb/kg were given, the magnitude
                        of the dose did not appear to affect
                        significantly the percent absorbed. The
                        presence of food in the gastrointestinal
                        tract reduced Pb absorption when a tracer
                        dose was administered but did not affect
                        absorption after 2 mg of Pb/kg orally. The
                        chelators nitrilotriacetic acid and sodium
                        citrate increased absorption of Pb, as did
                        orange juice, a source of citric acid. Milk
                        and the chelating agents, EDTA and
                        diethylenetriaminepentaacetic acid, did not
                        affect significantly Pb absorption.
```

Fig. 18.12 Part of the output from a National Library of Medicine Literature Search on factors influencing the gastrointestinal absorption of lead.

```
10 DIM P$(500),H(500),L(500),R(500),LIR(500),RIR(500),H$(500)
20 C = C + 1
30 INPUT "ENTER PLANT NAME ";P$(C)
40 INPUT "ENTER EMPLOYEE HOURS ";H$(C)
45 H(C) = VAL(H$(C))
50 INPUT "ENTER NUMBER OF LOST WORKDAY CASES ";L(C)
60 INPUT "ENTER TOTAL NUMBER OF RECORDABLES ";R(C)
70 INPUT "IS THIS THE LAST PLANT ";B$
90 LIR(C) = L(C) * 200000! / H(C)
100 RIR(C) = R(C) * 200000! / H(C)
110 IF LEFT$(B$,1) <> "Y" THEN GOTO 20
130  PRINT "                INCIDENCE RATE CALCULATIONS"
140 PRINT : PRINT
150 PRINT"          RAW DATA                      CALCULATED DATA"
160 PRINT"-------------------------------------  -------------------"
170 PRINT"PLANT         EMPLOYEE  LOST   TOTAL      LOST      TOTAL    "
180 PRINT"NAME          HOURS   WORKDAY RECORDABLE WORKDAY    RECORDABLE "
190 PRINT"                CASES   CASES     I.R.      I.R.    "
200 PRINT"_____________ __________ _____ _____   _________ _________"
205 PRINT
210 FOR N = 1 TO C
220 PRINT LEFT$(P$(N),14) TAB(20) H$(N) TAB(31) L(N) TAB(40) R(N) TAB(53);
230 PRINT USING "####.##"; LIR(N);
235 PRINT TAB(66) USING "####.##"; RIR(N)
250 NEXT N
260 END
```

Fig. 18.13 BASIC program that computes incidence rates.

```
ENTER PLANT NAME ? WILMINGTON
ENTER EMPLOYEE HOURS ? 654879
ENTER NUMBER OF LOST WORKDAY CASES ? 1
ENTER TOTAL NUMBER OF RECORDABLES ? 1
IS THIS THE LAST PLANT ? Y
                    INCIDENCE RATE CALCULATIONS
```

RAW DATA				CALCULATED DATA	
PLANT NAME	EMPLOYEE HOURS	LOST WORKDAY CASES	TOTAL RECORDABLE CASES	LOST WORKDAY I.R.	TOTAL RECORDABLE I.R.
HEADQUARTERS	123543	1	3	1.62	4.86
ATLANTA	434312	0	1	0.00	0.46
DETROIT	333214	0	0	0.00	0.00
CLEVELAND	665432	1	1	0.30	0.30
NEWARK	123298	0	0	0.00	0.00
PHILADELPHIA	126543	0	1	0.00	1.58
WILMINGTON	654879	1	1	0.31	0.31

Fig. 18.14 Output from incidence rate calculating program.

18.7 THE INTERNET

The Internet is a network of millions of linked computers spread throughout the world (and some have even found their way into space). Typically these computers are either servers that house Internet applications or they are client machines that act as sophisticated terminals for net users. The Internet can be productively employed by safety and health professionals in a number of ways. The most useful services are the World Wide Web, electronic mail, discussion groups, and remote operation of distant computers. Other applications include live (real-time) conversation groups, video and audio transmissions, and transferring files.

The amazing growth of the Internet is due mostly to its most popular application, the World Wide Web. The Web consists of information pages, most of which contain colorful graphical presentations. By using a mouse pointer to click on displayed objects (usually pictures or highlighted words) you are automatically transferred to a new Web page that is related to that object. The location of the computer on which a page resides is transparent to the user. It is possible to be viewing a page on a computer in California, for example, and with the click of a button be transferred to a page in Australia. A number of Web servers that deal specifically with occupational safety and health are currently in existence. Because of the rapid evolution of the World Wide Web, no attempt will be made here to provide a comprehensive list of safety and health resources as it would, no doubt, be obsolete by the time this book is printed. Two excellent servers that are likely to have considerable longevity can be accessed at the following URLs (uniform resource locators): http://www.osha.gov (Fig. 18.15) and http://hazard.com. These provide a wide variety of services and contain lists of many current safety-related Web resources (see Chapter 3 for more information).

One of the beauties of the World Wide Web is how easy it is to use. Persons familiar with operating microcomputers having graphical user interfaces can become proficient at using the Web in a short period of time. One of the features of the Web that makes it so very useful is search engines, which allow users to locate pages that deal with topics of interest. If, for example, you enter words *occupational safety* into the Alta Vista search engine (http://www.altavista.com) and follow what appear to be the most relevant of

OSHA Strategic Plan FY1997-FY2002

- The Assistant Secretary
- Information About OSHA
- What's New
- Media Releases
- Publications
- Programs & Services
- OSHA Software/Advisors
- Office Directory
- Statistics and Data
- Compliance Assistance
- Ergonomics
- Federal Registry Notices
- Frequently Asked Questions
- Standards
- Other OSHA Documents
- Technical Information
- Vanguard & Customer Service
- US Government Internet Sites
- Safety & Health Internet Sites

[Webmaster I OSHA Home Page I OSHA-OCIS I US DOL Web Site I Disclaimer]

Fig. 18.15 Items available on the OSHA World Wide Web Page.

the links that are listed, you can rapidly get to the most important safety-related Web pages. If you want to locate other search engines, just enter the words *search engines* into Alta Vista and follow the links. By trying all the many popular search engines, you can choose the one that best fits your needs and preferences.

The software packages used to access the Web are referred to as *browsers*. One of the most important features of browsers is their ability to save the location of previously visited sites. This is accomplished using a feature known as *bookmarks*. You simply click on the portion of the screen that activates the "add bookmarks" feature and all the information needed to go back to a site is stored on your computer. This removes the necessity of typing in URLs, whose obscure format makes them cumbersome to type. When a page is added to your bookmarks, you can subsequently revisit it by pulling down an automatically generated list of sites and clicking on the name of the one you wish to go to. Since bookmarks are listed using narrative page titles, rather than often hard-to-decipher URLs, finding a specific page is relatively easy. It is only when you have collected a large number of bookmarks that finding a site becomes difficult. Many browsers have the ability to organize bookmarks, which eases the burden of navigating the Web when you regularly visit many different sites.

Electronic mail (e-mail), file transfers, and news groups are carried using the Internet protocols smtp (simple mail transfer protocol), ftp (file transfer protocol), and nntp (network news transfer protocol), respectively. Remote operation of computers uses the telnet protocol. Software packages, with highly sophisticated user interfaces, are available for each of these functions. World Wide Web browsers often also contain smtp, ftp, nntp, and telnet capabilities but sometimes lack the advanced features found in dedicated programs. It is possible with some browsers, however, to have them automatically activate external programs of the user's choice when functions such as remote operation of computers using the telnet protocol are required. This allows users to select the software that best meets their individual needs for each of the major Internet functions, while using a browser as a single master control program for all commonly performed tasks.

Electronic mail as a means of communication offers many advantages over other methods of sending documents. It is less expensive than using the postal service and minimizes the use of paper. In addition, messages reach their destination very rapidly (in most cases in a matter of seconds). Maintaining mailing lists and address books is also facilitated, as most e-mail software packages have these functions built in.

While most news groups are dedicated to recreational topics, many focus on serious technical topics of interest to safety and health professionals. News is transmitted in the form of postings, which usually contain questions, answers to questions, or announcements. Some groups deal specifically with occupational safety and health issues. Since new groups are frequently created and old ones sometimes go out of existence, it would not be worthwhile to list those safety-specific groups that currently exist as the information would rapidly become obsolete. Most software packages have the capability to locate, among the thousands news groups active at any time, those that deal with specific topics. This is usually accomplished by doing a search using keywords such as *safety*. The names of news groups that focus on specific topics can also frequently be gotten from Web pages that deal with the same topics.

As an alternative to news groups, discussions of specific topics can be accomplished using mailing lists. With this technique a sponsor sets up a server that receives and stores postings that are subsequently distributed via e-mail. Individuals who are interested in receiving posted messages must subscribe by sending a request to a type of Internet computer called a *list server*. Some saftey-related mailing lists have become quite popular and contain lively discussions of many interesting issues. As with new groups, locations of safety-related mailing lists can best be found by examining descriptions of available resources which appear on many safety-oriented Web pages (visit the Vermont SIRI Web Site, http://hazard.com/#safety for current information on active lists).

Live (real-time) conversations can take place on the Internet using Internet Relay Chats (IRCs). Those discussion groups dealing with safety are usually sponsored by businesses, organizations, or individuals. When using IRCs, typed comments from individuals are transmitted in real time to all the people who are currently logged on. In order to keep track of who is saying what, the name of the poster is displayed at the beginning of each line. Moderators are often present to facilitate orderly interchange of information. Sometimes well-known experts are invited to lead discussions on technical subjects. This tends to increase participation. Scheduled sessions are frequently announced on safety-oriented Web pages. Unscheduled chats can also take place since IRC servers tend to be active all the time. Thus discussions can take place whenever two or more individuals happen to be logged on to a particular chat group. Scheduling simply enhances the probability that a large number of people will be on at the same time.

Transferring files to or from other computers on the Internet is usually accomplished using software packages that utilize the ftp protocol. Transfer can also be performed using e-mail packages that support attachments. The types of files most commonly transmitted are those containing text, programs, graphics, or word processing documents. The speed at which transmission occurs is dependent on the quality of the path between the sender and receiver. For all but the largest of files, it usually takes just a few minutes to send them anywhere in the world.

Many of the early Internet services like archie (for locating files), gopher (a menuing system), and WAIS (a document search system) have been subsumed by, incorporated into, or made obsolete by the World Wide Web and its associated browsers. Virtually all the major Internet functions can be performed using browsers or software packages that use browsers as a frontend. As the Web has developed, new and interesting features, such as real-time audio and video, two-way voice communications, and animations, have been implemented.

18.8 CD-ROMs

CD-ROMs (compact disk read-only memory) are removable disks with very high storage capacities (about 650 megabytes, which is enough to store tens of thousands of pages of text). Once they are pressed they cannot be altered or erased. Due to the low expense of mass producing CD-ROMs, they have become very popular for distributing large amounts of data. Many of the databases, which at one time were available only online, can now be obtained on CD-ROM and thus be accessed locally. Examples include NIOSHTIC (an occupational safety and health literature searching database) and the databases from the National Library of Medicine. Since the biggest expense incurred in using online literature searching services tends to come from the high price of making long-distance phone calls and the high hourly fees charged for using a remote system, CD-ROMs can be used to reduce greatly the cost of locating and retrieving information.

In addition to literature searching, CD-ROMs can also be used for many other safety-related tasks. Disseminating large collections of material safety data sheets, providing interactive multimedia training and distributing toxicology databases are three applications that have been implemented with great success. In the near future CD-ROMs will be displaced with equally inexpensive devices with far greater storage capacity. When this happens, the sophistication and usefulness of the types of applications currently distributed on CD-ROMs will be significantly enhanced.

18.9 HARDWARE CONSIDERATIONS

It is impractical to relate specific hardware requirements for using safety-related software as changes in technology are occurring too rapidly. It has been estimated that major advances in microcomputer technology or a significant lowering of the price of hardware needed to achieve a given level of performance occurs at least once every 90 days. What is considered a high-end system at the time of purchase will likely be an entry-level system in about a year, severely behind the times in 2 years and obsolete by year 3. This trend will probably continue for many years into the future with no end in sight. Nevertheless, it is possible to make some generalizations about hardware purchases. Since state-of-the-art safety software will require state-of-the-art microcomputer hardware, it is important to be sure that components are up-to-date. The most important considerations are listed below:

Hardware component	Major considerations
CPU (microprocessor)	Speed, cache size
Memory (RAM)	Size, speed
Hard disk drive	Capacity, speed
Monitor	Size, resolution, ergonomics (glare, flicker, etc.)
Video card	Speed, memory size
Backup device	Speed, capacity, software quality
Network card	Speed, compatibility
CD-ROM	Speed
Modems	Speed, compatibility, line-noise handling capabilities
Audio/video electronics	Varies according to intended use

In addition, reliability is a major issue for all components.

18.10 SUMMARY

The microcomputer is a tool that can be used to great advantage by the safety and health professional. Technological advances continue to lower the cost of hardware and improve the quality of software. The number of safety-specific software packages that have been produced is large, and the pace of development is increasing. The most dramatic changes are likely to occur, however, in the area of communications and information dissemination. The growth and development of the Internet has added a whole new dimension to microcomputer-based safety and health applications. The future is likely to hold an era in which anyone with a microcomputer and a connection to the Internet will be able to retrieve virtually anything that is available in the technical literature. High-quality, real-time transmissions of audio and video over high-speed networks will allow people to see or hear programs at a time of their own choosing and not be at the mercy of the scheduling whims of providers. Discussion groups and communications between individuals will flourish, and important applications we have not even thought of yet will arise. These developments will benefit safety and health professionals and others for whom effectively communicating is critical to their mission.

ADDITIONAL READING

ACGIH, *Microcomputer Applications in Occupational Safety and Health*, Lewis Publishers, Chelsea, MI, 1987.

Garber, Brad T., "Industrial Hygiene Applications of Microcomputers I. An Air Sampling Data Base Management System," *Am. Ind. Hyg. Assoc. J.*, August 1981.

Garber, Brad T., "A Microcomputer Approach to Safety Recordkeeping," *Occ. Hlth. and Safety*, January, 1983.

Rash, Jr., Wayne, "Choices in Health and Safety Software," *Occ. Hlth. and Safety*, October, 1995.

Rawls, Gregorie (Ed.), *Computers in Health and Safety*, American Industrial Hygiene Association, Akron, OH, 1990.

Ross, Charles W., *Computer Systems for Occupational Safety and Health Management*, Marcel Dekker, New York, NY, 1984.

CHAPTER 19

Occupational Dermatoses

DAVID E. COHEN

Dermatology Department, NYU Medical Center, 560 First Ave., New York, NY 10016

19.1 INTRODUCTION
- 19.1.1 Historical
- 19.1.2 Incidence
- 19.1.3 Structure and Function of the Skin

19.2 PERCUTANEOUS ABSORPTION
- 19.2.1 Metabolism

19.3 CAUSAL FACTORS
- 19.3.1 Indirect Factors

19.4 DIRECT CAUSES OF OCCUPATIONAL SKIN DISEASE
- 19.4.1 Chemicals
- 19.4.2 Chemical Burns
- 19.4.3 Allergic Skin Disease
- 19.4.4 Plants and Woods
- 19.4.5 Photosensitivity
- 19.4.6 Photoallergy
- 19.4.7 Phototesting
- 19.4.8 Mechanical
- 19.4.9 Physical

19.5 PHYSICAL FINDINGS OF OCCUPATIONAL SKIN DISEASE
- 19.5.1 Acute Eczematous Contact Dermatitis
- 19.5.2 Chronic Eczematous Contact Dermatitis
- 19.5.3 Folliculitis, Acne, and Chloracne
- 19.5.4 Sweat-Induced Reactions
- 19.5.5 Pigmentary Abnormalities
- 19.5.6 Neoplasms
- 19.5.7 Ulcerations
- 19.5.8 Granulomas
- 19.5.9 Other Clinical Patterns

19.6 SOME SIGNS OF SYSTEMIC INTOXICATION FOLLOWING PERCUTANEOUS ABSORPTION

19.7 DIAGNOSIS
- 19.7.1 Patient History
- 19.7.2 Appearance of the Lesions
- 19.7.3 Sites Affected

Adapted from "Occupational Dermatoses" by Donald J. Birmingham, MD, FACP, Chapter 10 in Clayton and Clayton, *Patty's Industrial Hygiene and Toxicology, 4th ed., Vol. IA*. New York: John Wiley and Sons, Inc., 1991., pp. 253–287.

Handbook of Occupational Safety and Health, Second Edition, Edited by Louis J. DiBerardinis, ISBN 0-471-16017-2

19.7.4 Diagnostic Tests
19.7.5 The Patch Test
19.8 TREATMENT
19.8.1 Prolonged and Recurrent Dermatoses
19.9 PREVENTION
19.9.1 Direct Measures
19.9.2 Indirect Measures
19.9.3 Control Measures
GLOSSARY
REFERENCES

19.1 INTRODUCTION

The skin represents the largest organ in the body, encompassing 1.5–2 m^2 of surface area. In its role as a primary defender against external insult, the skin is particularly vulnerable to damage by physical and chemical assaults in the workplace. Barrier function represents only a fraction of the duties performed by the entire integumentary system, which participates directly in thermal, electrolyte, hormone, and immune regulation without which life is not possible. The metabolic potential of the skin is impressive, and rather than merely repelling chemical or physical assaults, the skin may compensate by metabolizing and biotransforming agents to less harmful ones. Hence, the skin is far from a passive coat of armor, but rather an interactive organ that is in constant flux with its environment.

The skin's precarious location has rendered it the most commonly injured organ from chemical agents and physical conditions of the workplace. Pathologic responses of the skin can vary from excessive dryness and mild redness to more generalized exfoliative dermatitides that are life threatening. Neoplasms of the skin may occur as the result of primary skin exposures or through systemic absorption via the skin or other route of entry. Benign or malignant, such events can have catastrophic consequences to the host. Historically, occupational skin disease has been morphologically documented and often have very descriptive nomenclature that easily identifies the purported causative agent. Among the work force, a number of more descriptive titles associated with cause are commonly used, for example, asbestos wart, cement burn, chrome holes, fiber glass itch, hog itch, oil acne, rubber rash, and tar smarts. In view of the variety of skin lesions known to result from contactants within the workplace, the term *occupational dermatoses* is preferred because it includes any abnormality of the skin resulting directly from or aggravated by the work environment (Schwartz, 1957).

19.1.1 Historical

Just when and how occupational affections of the skin first occurred is a matter of conjecture; however, if we apply our past and present knowledge of diseases associated with work, it can be reasonably suspected that skin disorders in one or another form were expressed soon after humans began to perform various types of work. Archaeology has shown that ancient inhabitants invented and used primitive tools and weapons made of stone, flint, bone, and wood (History of Technology, 1978). It is quite likely that abrasions, blisters, bruises, lacerations, punctures, and probably more serious traumas were incurred as part of daily living associated with hunting for food, building shelter, making clothing, and gaining protection. Poisonous plants and biological agents, including various parasites, probably took their toll; however, such harmful effects remain suspect rather than documented in medical history. Perhaps the earliest references occurred in the writings of Celsus about 100 A.D. (White, 1934) when he described ulcers of the skin caused by corrosive metals. During later centuries, several authors enhanced the knowledge of certain occupational diseases, but cutaneous ulcerations seem to have been the major occupational skin disease of record. An explanation may reside in the fact that ulcerations of the skin were easily recognized, especially among those handling metal salts in mining, smelting, tool and weapon making, creating objects of art, glass making, gold and silver coinage, casting, and similar metallics. It would be strange indeed if none of these tradesmen incurred skin problems caused by a substance or condition met with at work. Nonetheless, little was recorded about occupational skin disease until Ramazzini's historic treatise on diseases of tradesmen in 1700 (Ramazzini, 1864). In this tome he described skin disorders experienced by bath attendants, bakers, gilders, midwives, millers, and miners, among other tradespeople. Seventy-five years later Sir Percival Pott published the first account of occupational skin cancer when he described scrotal cancer among chimney sweeps (Pott, 1775).

The Industrial Revolution of the eighteenth century changed an agricultural and guild economy to one dominated by machines and industrial expansion. As cities and industries grew, so did the study of science and the eventual discovery and use of new materials such as chromium, mercury, and petroleum, among many others. The chemical age brought enormous numbers of materials, natural and synthetic, into industrial and household use (History of Technology, 1978). As a result, physicians began to recognize occupational dermatoses and publish their observations in England, Germany, Italy, and France. Similarly, industrialization within the United States led to the recognition of old and new causes of occupational skin disease. Numerous dermatologists, industrial physicians, practitioners, and allied scientists have added to the information bank dealing with clinical investigations, clinical manifestations, causal factors, diagnostic procedures, and treatment and prevention of these disorders. A number of updated texts and related publications are available (Suskind, 1959; Samitz, 1985; Maltwn, 1964; Adams, 1983; Maibach, 1987; Fisher, 1995; Cronin, 1980; Gellin, 1972; Rycroft, 1986; Birmingham, 1978). Today's technology has brought about entirely new exposure patterns that directly affect the health of the skin. The astonishing expansion of medical technology coupled with the emergence of increasingly dangerous infectious agents has resulted in tens of thousands of workers potentially exposed to agents never before encountered by such large numbers of people. Today, health care workers not only face peril from the infectious waste they may handle, but may develop skin disease from the personal protective equipment they use as well as the chemicals used to neutralize biologic hazards.

19.1.2 Incidence

The National Institute of Occupational Safety and Health (NIOSH) has considered the skin a vitally important organ with respect to occupational diseases. In fact, NIOSH has characterized skin disease as one of the most pervasive problems facing workers in the United States. Since 1982, skin disease is listed in the top ten work-related diseases based on potential for prevention, incidence, and severity (NIOSH, 1996). In the 1978 edition of Patty's *Industrial Hygiene and Toxicology* text, dermatologic diseases were shown to account for about 40 percent of all occupational diseases reported to the U.S. Department of Labor (OSHA, 1978). In the 1984 Bureau of Labor Statistics, dermatologic disease had decreased to 34 percent of all reported occupational disease. For disease rates, a Bureau of Labor Statistics survey show a declining trend for occupational skin disease from a high in 1972 of 16.2 events per 10,000 full-time workers to a low in 1986 of 6.9 events per 10,000 workers. Recently, however, there has been an upward trend in incidence, with a rate of 7.9/10,000 workers in 1990 or about 61,000 new cases per year. Skin disease resulting from exposures in the agriculture and manufacturing industries were responsible for the greatest number of cases with incidence rates of 86 and 41/10,000 workers, respectively (NIOSH, 1996). Since skin disease is often not life threatening, many believe that the rate at which it is reported to government agencies is underrepresented 10 to 50-fold. Although the health care field has a relatively low rate of disease, the large number of workers in this industry results in almost 3,900 cases of illness per year. As indicated, while skin disease most often does not involve a life-threatening hazard, it is often assumed that the cost of occupational skin disease to society is diminutive. It has been estimated that up to 20 to 25% of persons with occupational skin disease lose an average of 11 days of work annually. This translates to an economic loss of 222 million to 1 billion dollars annually.

Occupational skin *injuries* such as thermal and chemical burns, lacerations, and blunt skin trauma are extremely common. NIOSH estimates that there are approximately 1.07 to 1.65 million skin injuries per year, accounting for a rate of 1.4 to 2.2 cases per 100 workers. These potentially disabling injuries are probably the most preventable illnesses and should not be overlooked. The economic costs of injuries to the skin have not been calculated.

19.1.3 Structure and Function of the Skin

Any type of acute or chronic exposure to workplace chemical or physical agents can result in a skin disease. In the overwhelming majority of cases, the skin is able to compensate adequately for such assaults and thus disease is averted. This ability to protect through an impressive reservoir of compensatory mechanisms stems from the multifunctional capabilities of the entire skin system. As will be described later, the skin is not composed of a monomorphous group of cells, but rather is a complex dynamic system of differentiating cells that closely interact with almost every other organ system in the body.

Injury to the host from occupational stressors may result from toxicity to the cells of the skin or through interference of normal homeostatic functions that the skin performs. Only short periods of malfunction of thermoregulatory and electrolyte homeostasis of the skin are compatible with life. Hence, widespread destruction of the skin may result in immediate breakdown of these mechanisms and can result in death. Such insults include chemical and thermal burns, overexposure to heat and humidity, or prolonged occlusion of the skin.

Anatomically, the skin is composed of two main levels, epidermis and dermis. They are separated from each other by a basement membrane, which forms a wavy interface between the layers. The skin appendages, which include the hair follicle unit and sebaceous, eccrine, and apocrine sweat glands, have their respective ductal structure crossing the epidermis to the surface. The concentration of these appendageal structures varies greatly by location. For most, hair follicles abound on the scalp, sebaceous glands are concentrated on the face, and eccrine glands are on the palms, soles, and axillae. Apocrine glands, which have only an incompletely characterized function, are found in the axillae, areolae, and groin.

The outer or epidermal layer varies in thickness, being most protective on the palms and soles. Because it is contiguous with the dermis, it also acts as the outer cover of the cushion of connective and elastic tissue that guards the blood and lymph vessels, nerves, secretory glands, hair shafts, and muscles. Epidermal resiliency provides protection within limits against blunt trauma and its flexibility accounts for the return of stretched skin to its normal location (Schwartz, 1957; History of Technology, 1978; White, 1934; Ramazzini, 1864; Pott, 1775; Suskind, 1959; Samitz, 1985; Maltwn, 1964; Adams, 1983; Maibach, 1987; Fisher, 1995; Cronin, 1980; Gellin, 1972; Rycroft, 1986; Birmingham, 1978; Blank, 1979).

The epidermis has many layers; however, for the purpose of this review, it can be functionally divided into the noncornified layer and the cornified layer or stratum corneum. The principle cell of the epidermis is called the *keratinocyte* and is arranged as a stratified squamous epithelium. Keratinocytes begin as basal cells abutting the basement membrane and progress upward through the epidermis via a terminal differentiation pathway that lasts approximately 14 days (Fig. 19.1). During this period there are substantial changes in cell surface markers and a concomitant accumulation of keratin proteins. By the end of the 2-week cycle, the keratinocyte has lost its nucleus and intracellular organelles and has flattened. It is at this point that the nonviable keratinocytes enter the stratum corneum and are known as *corneocytes*. The cell structure of the keratinocyte will be largely replaced by a tightly linked layer of fibrous proteins. It will take another 14 days for the original corneocyte to reach the surface of the skin, where it will be sloughed (Blank, 1979; Kligman, 1964). The stratum corneum is highly important in resisting the mass entrance and exit of water and electrolytes. Stresses such as friction, pressure, and natural and artificial ultraviolet light can induce compensatory thickening of the stratum corneum in the form of a callous. Besides acting as a water barrier and a physical shield, the stratum corneum provides modest protection against acids and acidic substances. In contrast, it is quite vulnerable to the action of organic and inorganic alkaline materials. Such chemical substances attack the stratum corneum by denaturing the keratin proteins, thus altering the cohesiveness and the capacity to retain water, which is essential in the maintenance of the barrier layer. In short, any physical or chemical force such as lowered temperature and humidity or repetitive action of soaps, detergents, and organic solvents generally leads to impairment of the barrier efficiency because of water loss and dryness (Blank, 1979).

Located also in the basal layer are *melanocytes*, which are neuroendocrine-derived cells that are responsible for the production of melanin pigment. Melanin is packaged into pigment granules called melanosomes and imparts the natural pigmentation to the skin. Differences in skin pigmentation occur mostly from differences in melanosome structures than from gross quantities of melanin. The pigment granules that arise from complex enzymatic reactions within the melanocytes are picked up by the epidermal cells and eventually are shed by way of the keratin exfoliation (Kligman, 1964; Fitzpatrick, 1993). Melanin acts as the principal defense against ultraviolet light, since it acts as a broad-spectrum chromophore or light absorber. Besides the natural production of melanin, certain agents such as coal, tar, pitch, selected aromatic chlorinated hydrocarbons, petroleum products, and trauma can cause excess melanin production, leading to hyperpigmentation (Birmingham, 1978; OSHA, 1978). In contrast, members of the quinone family and selected phenolics can inhibit pigment formation following

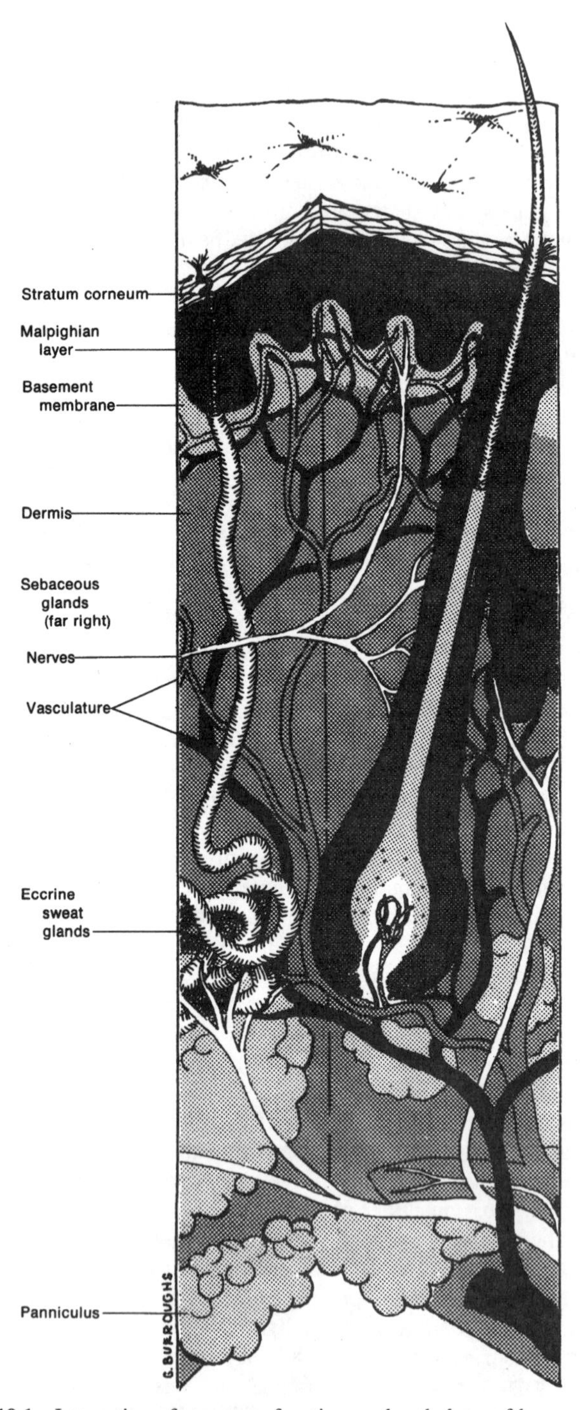

Fig. 19.1 Integration of structure, function, and pathology of human skin.

percutaneous absorption by direct action upon the melanin enzymatic system (Schwartz, 1957; Suskind, 1959; Adams, 1983; Gellin, 1972; Kahn, 1979; Gellin, 1970; Malten, 1971). This could result in depigmentation, impacting a marked lightening of the skin or ivory-white appearance.

Immediately below the epidermal region lies the *dermis*, which is much thicker than the epidermis. It is composed of connective tissue made up of collagen, elastic tissue, and ground substance that constitutes an encasement for the sweat glands and ducts, the hair follicles, sebaceous glands that secrete natural sebum (fatty substance), the blood and lymph vessels, and the nerve endings (Suskind, 1959; Adams, 1983; Birmingham, 1978; Blank, 1979, Fitzpatrick, 1993).

Thermoregulation is modulated by the excretion of eccrine sweat and changes in the superficial circulation of blood, all of which is controlled by the central nervous system. Thus, core body temperature and circulating blood are physiologically stabilized at a constant temperature despite climatic variations. Eccrine sweat is composed primarily of water and electrolytes and participates in temperature control through heat loss via evaporation from the surface. Simultaneously radiant heat loss is facilitated by the dilation of the cutaneous blood vessels. The opposite occurs when a decreased core and/or surface temperature causes blood vessels of the skin to constrict and shunt warm blood away from exposed surfaces thus preserving heat (Suskind, 1959; Adams, 1983; Birmingham, 1978; NIOSH, 1996; Blank, 1979).

Secretory functions within the skin are relegated to sweat gland and sebaceous gland activity. Sweat gland function and sweat delivery is essential for physiological normalcy. Too much or too little sweat delivery can have deleterious effects on the whole physiological behavior. Sebaceous glands reside within the dermis as part of the hair follicle unit. Their functional product, sebum, is excreted through the hair follicle and the orifice on the skin surface. Overfunction of this gland is a frequent problem of the adolescent, but the sebaceous glands are also target sites for occupational acne resulting from working with coal tar, heavy oils, greases, and certain aromatic chlorinated hydrocarbons (Suskind, 1959; Samitz, 1985; Adams, 1983; Birmingham, 1978; Blank, 1979).

Special receptors within the skin are part of a network of nerve endings and fibers that receive and conduct various stimuli, later recognized as heat, cold, pain, and other perceptions such as wet, dry, sharp, dull, smooth, and rough (Suskind, 1959; Adams, 1983; Birmingham, 1978; OSHA, 1978).

19.2 PERCUTANEOUS ABSORPTION

Historically, the skin was believed to be an impervious barrier to external chemicals. More recently, however, it is understood that percutaneous absorption occurs frequently and depends on a variety of properties of the agent. The skin maintains a relatively hydrophobic character, which is imparted from the sphingolipids present in the epidermis (Elias, 1992). The water content, which makes up about 20 percent of the weight of the epidermis, is concentrated around intracellular proteins within keratinocytes. Prolonged submersion can cause a several-fold increase in skin hydration and loss of barrier function.

Percutaneous absorption tends to follow the physics of typical membrane kinetics. Hence increasing the concentration gradient across the epidermis will result in increasing percutaneous absorption. However, two other significant factors must be considered in this complex biological membrane barrier. Molecular weight and hydrophobicity are important in determining or predicting percutaneous absorption. Low-molecular-weight molecules pass through the epidermis with greater alacrity than do larger weight compounds, and those with high octanol/water partition coefficients (more hydrophobic) will pass more easily. Hence low-molecular-weight hydrophobic compounds may pass readily through an intact epidermis and cause toxicity. Since percutaneous absorption is becoming increasingly important, pharmacokinetic models have been developed to predict percutaneous absorption of substances (Potts, 1992).

Absorption may be accomplished via two routes: initially through the skin appendages such as hair follicles and sweat glands, then later through diffusion through the epidermis. Occlusion may serve to enhance penetration by hydrating the epidermis and by preventing evaporation or mechanical removal of a chemical from the skin surface. From a practical perspective, percutaneous absorption has been a particular problem for only certain chemicals such as pesticides, cyanides, aromatic hydrocarbons, mercury, lead, and others. Organophosphate pesticides like parathion are sufficiently absorbed through the skin to be fatal after moderate skin exposure. Until recently organic leads were thought to be the primary form of the heavy metal capable of percutaneous absorption. More recent work has indicated that inorganic lead in dust may be result in substantial dermal levels of lead and can cause increases in total body lead burden (Stauber, 1994).

Body site has long been described as an important variable in percutaneous absorption. The scrotum has been identified as an area where absorption is greatest, with the face as a moderately penetrable site, and the abdomen the least (Scheuplein, 1971).

19.2.1 Metabolism

The barrier features of the skin do not solely lie in the physical obstacle of the epidermal lipids and crosslinked proteins of the stratum corneum. The keratinocytes of the lower epidermis are capable of a legion of biotransformation reactions that may act to render toxic xenobiotics innocuous by the time they reach the superficial vascular plexes that will carry them inward. In fact, the skin possesses most of the phase 1 and phase 2 metabolic enzymes necessary for xenobiotic metabolism and biotransformation. Overall, the skin has 2 percent of the metabolic capacity of the liver. Specific enzyme pathways may also be inducible and result in dramatic rises in enzyme activity. This is particularly important for biotransformation reactions when carcinogenic intermediates are formed during detoxifying metabolism of certain polycyclic hydrocarbons (Rice, 1996). Animal studies have demonstrated that when skin is exposed to polychlorinated biphenyls or benzo-a-pyrene, the activity of the phase 1 enzyme aryl hydrocarbon hydroxylase can be increased as to account for 20 percent of the total body activity of the enzyme. Hence, under conditions of repeated or significant dermal exposure, the skin may be responsible for a marked amount of detoxifying and metabolism of the chemical. Exposure to these procarcinogens on the skin, however, may result in biotransformation to active carcinogens locally, without the aide of hepatic enzymes (McNulty, 1985; Mukthar, 1981).

19.3 CAUSAL FACTORS

To detect the cause of an occupational dermatosis it is fundamental to consider a spectrum of factors that can have indirect relationship to the disease.

19.3.1 Indirect Factors

In determining the cause of an occupationally acquired disease, it is important to recognize host and environmental factors that may predispose or aggravate an already existing dermatosis. These include host factors such as genetic predisposition, preexisting dermatologic disease, age, and environmental factors such as temperature, humidity, and season (Schwartz, 1957; White, 1934; Suskind, 1959; Samitz, 1985; Adams, 1983; Gellin, 1972; Birmingham, 1978; OSHA, 1978).

19.3.1.1 Genetic Predisposition This particular subheading represents a wide variety of potentially confounding factors with regard to occupational skin disease. They can include genetic predisposition to skin diseases as described below or extraordinary sensitivity to routinely encountered occupational and environmental stressors. Sensitivity to chemical irritants and ability to acquire allergic contact sensitivity are both likely to be genetically determined. Extreme examples are metabolic genetic diseases that predispose individuals to disease after banal exposures. An example would be a patient with xeroderma pigmentosa (inability to repair DNA after routine sun exposure) who works outdoors and would have an enormously high risk of developing skin cancer compared to even the fairest worker without the disease.

Another genetically determined characteristic potentially impacting on the development of occupational skin disease is skin type. This nomenclature refers to the sun reactivity of the skin and can be loosely correlated to the degree of skin pigmentation. It can be particularly important for workers exposed to ultraviolet light either from sunlight or from artificial sources. The susceptibility of some skin cancers may be linked to those having lower skin types. Sunburn or ultraviolet-induced skin damage will reduce the barrier function of the skin as well as increase sensitivity to irritant chemicals. Clearly, the sunburned worker will also be less productive than a comfortable one. The use of sunscreen should be encouraged for all workers exposed to ultraviolet light on a regular basis for the prevention of skin cancer. Table 19.1 illustrates their characteristics (Fitzpatrick, 1993).

19.3.1.2 Presence of Other Skin Diseases New and old employees with preexisting skin disease are prime candidates for a supervening occupational dermatitis or an aggravation of a preexisting disease of the skin. Hirees with adolescent acne may be at higher risk of flaring their disease if exposed to insoluble oils, tar products, greases, and certain polychlorinated aromatic hydrocarbons known to cause chloracne (Schwartz, 1957; White, 1934; Suskind, 1959; Samitz, 1985; Adams, 1983; Gellin, 1972; Birmingham, 1978).

TABLE 19.1 Characteristics of Skin Types

Skin type	Characteristics
1	Always burns, never tans
2	Usually burns, tans less than average (with difficulty)
3	Sometimes mild burn, tan about average
4	Rarely burns, tan more than average (easily)
5	Rarely burn, tan profusely, brown skin
6	Never burn, tan profusely, black skin

Patients with dermatitis (inflammation of the skin) are at greater risk when exposed to any number of irritant chemical agents. The capricious skin behavior of these people does not tolerate exposure to dusty or oil-laden jobs (Adams, 1983; Gellin, 1972; Birmingham, 1978; OSHA, 1978; Schmunes, 1986).

Other skin diseases known to be worsened by physical or chemical trauma, even though mild, are psoriasis, lichen planus, chronic recurrent eczema of the hands, and those skin conditions to which light exposure can be detrimental. The need of careful skin evaluation and job placement is clear (Suskind, 1959; Samitz, 1985; Adams, 1983; Birmingham, 1978; Schmunes, 1986).

19.3.1.3 Age Young workers, particularly those in the adolescent group, often incur acute contact dermatitis. More often, young people are placed in service jobs, for example, fast food, janitorial, or car wash, where wet work prevails and protection is difficult. At times, disregard for safety and hygiene measures may be the reason.

Older workers usually are more careful through experience, but aging skin is often dry, and when work is largely outdoors, sunlight can cause skin cancer (Schwartz, 1957; White, 1934; Suskind, 1959; Adams, 1983; Birmingham, 1978; OSHA, 1978).

19.3.1.4 Environmental Factors: Temperature, Humidity, Season Normal seasonal variations in temperature, humidity, wind, and incident ultraviolet light can affect the normal physiologic defense mechanisms of the body. Increased temperature coupled with high humidity can overload the normal thermoregulatory mechanisms. This may result in abnormally excessive perspiration to more extreme temperature-related diseases such as heat exhaustion and heat stroke. (See Chapter 21 for more information on heat stress.)

Excessive delivery of sweat can cause maceration of the skin in the groin, the armpits, and other sites were skin surfaces are opposed to each other. This can foster the trapping of dusts and suspended particles allowing percutaneous absorption in this hydrated and occluded environment.

Sweat can also partially solubilize nickel, cobalt, and chromium in small amounts, a situation troublesome to those individuals with cutaneous allergy to these metals. Dry chemical agents can be put into aqueous solution, which can be irritating, destructive, or harmful if absorbed. Beneficially, sweat can act as lavage to keep the skin relatively free of irritant contact (Schwartz, 1957; White, 1934; Suskind, 1959; Adams, 1983; Fisher, 1995; Cronin, 1980; Birmingham, 1978; OSHA, 1978). High temperature and humidity may alone produce skin disease if pores are occluded. Miliaria or occlusive eccrine sweat gland disease may produce annoying symptoms such as itching and stinging. Widespread involvement can interfere with temperature regulation secondary to altered sweat pattern. Mild forms of the disease have been termed *prickly heat* in the vernacular. Diseases caused by temperature are elaborated on later in the chapter. Hot weather may discourage the use of protective clothing gear, thus allowing more unprotected skin for exposure to environmental contactants.

Cold weather is associated with dry skin because of lowered temperatures and humidity. Further, during cold weather, some workers simply do not like to take a shower and then go into the cold after working, allowing prolonged skin contact with potentially dangerous chemicals. Workers with inherently dry skin (xerosis and ichthyosis) usually have a worsening of their condition under the low-humidity conditions of the winter. They are at increased risk or irritation when exposed to alkaline agents, acids, detergents, and most solvents (Schwartz, 1957; White, 1934; Suskind, 1959;

Samitz, 1985; Adams, 1983; Gellin, 1972; Birmingham, 1978; Blank, 1979; Schmunes, 1986).

19.3.1.5 Personal Hygiene Poor washing habits breed prolonged occupational contact with agents that harm the skin. Personal cleanliness is a sound preventive measure, but it depends upon the presence of readily accessible washing facilities, quality hand cleansers, and the recognition by the workers of the need to use them (Schwartz, 1957; White, 1934; Suskind, 1959; Samitz, 1985; Adams, 1983; Birmingham, 1978; OSHA, 1978).

19.4 DIRECT CAUSES OF OCCUPATIONAL SKIN DISEASE

Chemical agents are unquestionably the major cutaneous hazards; however, there are multiple additional agents that are categorized as mechanical, physical, and biological causalities.

19.4.1 Chemicals

Chemical agents have always been and will most likely continue to be a major cause of work-incurred skin disease. Organic and inorganic chemicals are used throughout modern industrial processes and increasingly on farms. They act as primary skin irritants, as allergic sensitizers, or as photosensitizers to induce acute and chronic contact eczematous dermatitis, which accounts for the majority of cases of occupational skin disease, probably no less than 75 to 80 percent (Schwartz, 1957; White, 1934; Suskind, 1959; Samitz, 1985; Maltwn, 1964; Adams, 1983; Maibach, 1987; Fisher, 1995; Cronin, 1980; Gellin, 1972; Rycroft, 1986; Birmingham, 1978; OSHA, 1978).

19.4.1.1 Primary Irritants Most diseases of the skin caused by work result from contact with primary irritant chemicals. As contact dermatitis overwhelmingly represents the most common occupational skin disease, and approximately 80 percent of cases of contact dermatitis are of the irritant variety, irritant contact dermatitis is singularly the most common occupational disease involving the skin. (American Academy of Dermatology). These materials cause an irritant dermatitis by direct action on normal skin. The eruption will occur at the site of contact if sufficient quantity, concentration, and time is allowed. In other words, any normal skin can be injured by a primary irritant. This form of dermatitis represents a nonallergic variety and is dependent on a number of factors relating to the exposure (Table 19.2). Certain irritants, such as sulfuric, nitric, or hydrofluoric acid, can be exceedingly powerful in damaging the skin within moments. Similarly, sodium hydroxide, chloride of lime, or ethylene oxide gas can produce rapid damage. These are absolute or strong irritants, and they can produce necrosis and ulceration resulting in severe scarring. More commonly encountered are the low-grade or marginal irritants that through repetitious contact produce a slowly evolving contact dermatitis. Marginal irritation is often associated with contact with soluble metalworking fluids, soap and water, and solvents such as acetone, ketone, and alcohol. Wet work in general is associated with repetitive contact with marginal irritants (Schwartz, 1957; White, 1934; Suskind, 1959; Samitz, 1985; Maltwn, 1964; Adams, 1983; Maibach, 1987; Fisher, 1995; Cronin, 1980; Gellin, 1972; Rycroft, 1986; Birmingham, 1978; OSHA, 1978).

TABLE 19.2 Factors involved in irritant contact dermatitis

Exposure duration
Repeated exposures
Concentration
pH
Occlusion
Body site

19.4.1.2 Primary Irritant Action on Skin Clinical manifestations produced by contact with primary irritant agents are readily recognized but not well understood. General behavior of many chemicals in the laboratory or in industrial processes is fairly well known, and at times their chemical action can be applied theoretically, and in some instances actually, to chemical action on human skin. For example, we know that organic and inorganic alkalis damage keratin; that organic solvents dissolve surface lipids and remove lipid components from keratin cells; that heavy metal salts, notably arsenic and chromium, precipitate protein and cause it to denature; that salicylic acid, oxalic acid, and urea, among other substances, can chemically and physically reduce keratin; and that arsenic, tar, methylcholanthrene, and other known carcinogens stimulate abnormal growth patterns. Regardless of the specific agent eliciting an irritant contact dermatitis, multiple chemotactic cytokines are released from both the ailing keratinocytes and neighboring vascular endothelial cells. This results in infiltration of the epidermis and upper dermis with inflammatory cells and concomitant edema resulting in the hallmark dermatitis (Schwartz, 1957; White, 1934; Suskind, 1959; Samitz, 1985; Adams, 1983; Fisher, 1995; Cronin, 1980; Gellin, 1972; Rycroft, 1986; Birmingham, 1978; OSHA, 1978; Rice, 1996).

Irritant chemicals are commonly present in agriculture, manufacturing, and service pursuits. Hundreds of these agents classified as acids, alkalis, gases, organic materials, metal salts, solvents, resins, and soaps, including synthetic detergents, can cause absolute or marginal irritation (Table 19.3) (Birmingham, 1978).

19.4.2 Chemical Burns

Under irritant dermatitis conditions, epidermal necrosis of scattered keratinocytes is typical (Lever, 1990). If, however, the damage to relatively large areas of epidermis is overwhelming to the defenses of the skin, necrosis of the epidermis may occur as a result of a chemical exposure. The mechanisms of such chemical burns are similar to those of irritant dermatitis, but on a more destructive scale. For many irritants, extreme exposures can result in significant chemical burns.

Table 19.4 lists chemicals with characteristics that make them particularly corrosive to skin. Many of their mechanisms lie in their ability to denature epidermal proteins rapidly or cause damage by their ability to cause extreme temperature changes. A more exhaustive list can be found in de Groot's text on untoward effects of chemicals on the skin.

19.4.3 Allergic Skin Disease

Allergic contact dermatitis represents the quintessential delayed-type hypersensitivity (type IV) allergic reaction to an external chemical. Over 2,800 allergens have thus far

TABLE 19.3 Typical Primary Irritants

Acids	
Inorganic	*Organic*
Arsenious	Acetic
Chromic	Acrylic
Hydrobromic	Carbolic
Hydrochloric	Chloracetic
Hydrofluoric	Cresylic
Nitric	Formic
Phosphoric	Lactic
	Oxalic
	Salicylic
Alkalis	
Inorganic	*Organic*
Ammonium	Butylamines
Carbonate	Ethylamines
Hydroxide	Ethanolamines
Calcium	Methylamines
Carbonate	Propylamines
Cyanamide	Triethanolamine
Hydroxide	
Oxide	
Potassium	
Carbonate	
Hydroxide	
Sodium	
Carbonate (soda ash)	
Hydroxide (caustic soda)	
Silicate	
Trisodium phosphate	
Cement	
Soaps	
Detergents	
Surfactants	
Metal salts	
Antimony trioxide	
Arsenic trioxide	
Chromium and alkaline chromates	
Cobalt sulfate and chloride	
Nickel sulfate	
Mercuric chloride	
Silver nitrate	
Zinc chloride	

TABLE 19.3 (*Continued*)

Solvents	
Alcohols	*Ketones*
Allyl	Acetone
Amyl	Methyl ethyl
Butyl	Methyl cyclohexanone
Ethyl	
Methyl	
Propyl	
Chlorinated	*Petroleum*
Carbon tetrachloride	Benzene
Chloroform	Ether
Dichloroethylene	Gasoline
Epichlorohydrin	Kerosene
Ethylene chlorohydrin	Varsol
Perchloroethylene	White spirit
Trichloroethylene	
Coal tar	*Turpentine*
Benzene	Pure oil
Naphtha	Turpentine
Toluene	Terpineol
Xylene	Rosin spirit

been identified, and allergic contact dermatitis represents approximately 20 percent of cases of contact dermatitis (de Groot, 1994). Agents capable of causing contact dermatitis are generally small-molecular-weight chemicals that act as haptens. Haptens are not intrinsically allergenic but become so when they bind to an endogenous protein, usually in the epidermis or dermis, and mark the formation of a complete allergen. As with all truly allergic reactions, initial sensitization must occur. Sensitization occurs when an

TABLE 19.4 Chemicals capable of severe skin burns

Ammonia
Calcium oxide
Chlorine
Ethylene oxide
Hydrochloric acid
Hydrofluoric acid
Hydrogen peroxide
Methyl bromide
Nitrogen oxide
Phosphorous
Phenol
Sodium hydroxide
Toluene diisocyanate

allergen is incorporated in cutaneous immune cells called Langerhans cells. These cells possess surface proteins (human leukocyte antigens, HLA, class II), which allow them to directly present the digested allergen to a helper T lymphocyte. This lymphocyte will become activated and migrate to regional lymph nodes, where a clone of similarly sensitized cells will proliferate under the stimulation of cytokines like interleukin 1 and 2. These cells may now migrate back to the area of skin that had the original contact with the allergen and cause a typical dermatitis to form. This initial sensitization may take several days to complete and last a lifetime. Now sensitized, subsequent challenges from the same allergen will result in a rapid elaboration of lymphocytes into the dermis and result in a dermatitis within 48–96 hours (Rietschel, 1995).

The ability to become sensitized appears to be strongly mediated by genetics. Hence, one is probably born with all the potential allergies that an individual will possess for a lifetime. Recent work has linked specific HLA proteins to allergy during a lifetime will depend on whether exposure is sufficient to trigger the immune cascade necessary for sensitization.

Clinically the dermatitis caused by allergic mechanisms and irritant mechanisms may be indistinguishable. In the field, to distinguish between a primary irritant and allergic contact dermatitis, it is necessary to recognize that (a) allergic reactions usually require a longer induction period than occurs with primary irritation effects; (b) cutaneous sensitizers generally do not affect large numbers of workers except when dealing with very potent sensitizers such as epoxy resin systems, phenol–formaldehyde plastics, poison ivy, and poison oak. Some other well-known sensitizers associated with occupation exposures are potassium dichromate by itself or contained in cement, nickel sulfate, hexamethylenetetramine, mercaptobenzothiazole, and tetramethylthiuram disulfide, among several other agents (Schwartz, 1957; Suskind, 1959; Samitz, 1985; Adams, 1983; Maibach, 1987; Fisher, 1995; Cronin, 1980; Gellin, 1972; Birmingham, 1978). Table 19.5 lists common allergens and their likely sources of exposure (Rice, 1996).

19.4.4 Plants and Woods

Many plants and woods cause injury to the skin through direct irritation or allergic sensitization by their chemical nature. Additionally, irritation can result from contact with sharp edges of leaves, spines, thorns, and so on, which are appendages of the plants. Photosensitivity may also be a factor.

Although the chemical identity of many plant toxins remains undetermined, it is well known that the irritant or allergic principal can be present in the leaves, stems, roots, flowers, and bark (Lampke, 1968; Barber, 1977).

High-risk jobs include agricultural workers, construction workers, electric and telephone linemen, florists, gardeners, lumberjacks, pipeline installers, road builders, and others who work outdoors (Gellin, 1971).

Poison ivy and poison oak are major offenders. In California, several thousand cases of poison oak occupational dermatitis are reported each year. Poison ivy, oak, and sumac are members of the Anacardiaceae, which also includes a number of chemically related allergens as cashew nut shell oil, Indian marking nut oil, and mango. The chemical toxicant common to this family is a phenolic (catechol), and sensitization to one family member generally confers sensitivity or cross-reactivity to the others (Schwartz, 1957; Adams, 1983; Fisher, 1995; Cronin, 1980; Gellin, 1972; Rycroft, 1986; Lampke, 1968; Gellin, 1971).

Plants known to cause dermatitis are carrots, castor beans, celery, chrysanthemum,

TABLE 19.5 Common Contact Allergens

Source	Common allergens	
Topical medications/ hygiene products	*Antibiotics*	*Therapeutics*
	Bacitracin	Benzocaine
	Neomycin	Corticosteroids
	Polymyxin	α-Tocopherol (vit E)
	Aminoglycosides	
	Sulfonamides	
	Preservatives	*Others*
	Benzalkonium chloride	Cinnamic aldehyde
	Formaldehyde	Ethylenediamine
	Formaldehyde releasers	Lanolin
	Quarternium 15	*p*-Phenylenediamine
	Imidazolidinyl	Propylene glycol
	Diazolidinyl urea	Benzophenones
	DMDM hydantoin	Fragrances
	Methylchloroisothiazolene	Thioglycolates
Plants and trees	Abietic acid	Pentadecylcatechols
	Balsam of Peru	Sesquiterpene lactone
	Rosin (colophony)	Tuliposide A
Antiseptics	Glutaraldehyde	Hexachlorophene
	Chlorhexidine	Mercurials
	Chloroxylenol	Thimerosal
		Phenylmercuric acetate
Rubber products	Diphenylguanidine	Resorcinol monobenzoate
	Hydroquinone	Benzothiazolesulfenamides
	Mercaptobenzothizole	Dithiocarbamates
	p-Phenylenediamine	Thiurams
		Thioureas
Leather	Formaldehyde	Potassium dichromate
	Glutaraldehyde	
Paper products	Abietic acid	Rosin (colophony)
	Formaldehyde	Dyes
Glues and bonding agents	Bisphenol A	Epoxy resins
	Epichlorohydrin	*p*-(*t*-butyl)formaldehyde resin
	Formaldehyde	Toulene sulfonamide resins
	Acrylic monomers	Urea formaldehyde resins
	Cyanocrylates	
Metals	Chromium	Mercury
	Cobalt	Nickel
	Gold	Palladium

hyacinth, tulip bulbs, oleander, primrose, ragweed, and wild parsnip. Other plants including vegetables have been reported as causal in contact dermatitis (Adams, 1983; Fisher, 1995; Cronin, 1980; Gellin, 1972; Rycroft, 1986; Lampke, 1968; Gellin, 1971).

A number of woods are known to provoke skin disease. Woods do not cause as many cases as are reported from plants, but carpenters, cabinetmakers, furniture builders, lumberjacks, lumberyard workers, and model makers (patterns) can incur primary irritant, allergic dermatitis or traumatic effects from the wood being handled. Colophony or rosin is a particularly common cause of allergic contact dermatitis and is derived from pine wood. Its color and physical properties make it a useful ingredient in many nonwood products. It may be found as a colorant in yellow soaps, and its tackiness is exploited in its use for baseball players, preparations for violin bows, as well as adhesives, tapes, paints, and polishes (Rietschel, 1995). Sawdust, wood spicules, and chemical impregnants in the wood may cause irritation, whereas most cases of allergic dermatitis are caused by oleoresins, the natural oil, or chemical additives. Woods best known for their dermatitis-producing potential are acacia, ash, beech, birch, cedar, mahogany, maple, pine, and spruce. Other agents capable of causing cutaneous injury are the chemicals used for wood preservation purposes, such as arsenicals, chlorophenols, creosote, and copper compounds (Adams, 1983; Birmingham, 1978; Barber, 1977; Can. Dept. Forestry, 1996).

19.4.5 Photosensitivity

19.4.5.1 Phototoxicity Dermatitis resulting from photoreactivity is an untoward cutaneous reaction usually to the ultraviolet (UV) band of the electromagnetic spectrum. This band of light spans from 200 to 400 nm. It is further stratified into three sub-bands of light; UV-A, UV-B, UV-C. UV-C, spanning 200–280 nm, does not penetrate the upper atmosphere and has no clinical significance from natural light sources. In artificial settings, UV-C can cause marked sunburn within hours of an exposure, faster than other bands of UV light. It has been exploited for its antimicrobial potential in sterilizing air handling systems. UV-B, 280–320 nm, has potent effects on keratinocytes and is capable of causing significant acute damage to the epidermis. It can penetrate the epidermis but cannot reach the upper portions of the dermis. It is the sub-band responsible for sunburn or UV-induced erythema. UV-A, 320–400 nm, reaches the earth's surface in greater quantity but has less potency in causing acute epidermal damage than UV-B. It is a potent stimulator of melanin production, is capable of penetrating the skin to the upper dermis, and can cause DNA damage to keratinocytes. These combined effects result in damage to elastic tissue and dermal supporting structure, which causes wrinkling and are likely responsible for carcinogenesis in the skin (Lim, 1993).

The effect of exposure to UV light may be phototoxic, which is similar to primary irritation, or it may be allergic. While sunburn represents a phototoxic reaction, phototoxicity classically represents the interaction of the skin with a combination of a chemical and ultraviolet light. Phototoxic reactions occur when a specific chemical under the influence of ultraviolet light produces free radicals capable of inducing cell death. In that regard, the epidermal damage is similar to irritant dermatitis. Thousands of outdoor workers in construction, road building, fishing, forestry, gardening, farming, and electric and phone line erection are potentially exposed to sunlight and photosensitizing chemicals. Additionally, exposure to artificial ultraviolet light is experienced by electric furnace and foundry operators, glassblowers, photoengravers, steelworkers, welders, and printers in contact with photocure inks. Phototoxic reactions due to certain plants, a number of medications, and some fragrances have been well documented (Schwartz, 1957; Adams, 1983; Fisher, 1995; Cronin, 1980; Gellin, 1972; Birmingham, 1978; Epstein, 1971; Birmingham, 1968; Malten, 1976; Emmett, 1977; DeLeo, 1995). In the coal and tar industry, distillation can offer exposure to anthracene, phenanthrene, and acridine, all of which are well-known phototoxic chemical agents. Related products such as creosote, pitch roof paint, road tar, and pipeline coatings have caused hyperpigmentation from the interaction of tar vapors or dusts with sunlight (see Table 19.12) (Schwartz, 1957; White, 1934; Adams, 1983; Birmingham, 1978; Epstein, 1971; Birmingham, 1968; DeLeo, 1995).

Occupational photosensitivity is complicated by a number of topically applied and ingested drugs that can interact with specific wavelengths of light to produce a phototoxic or photoallergic reaction. Among such agents known to produce these effects are drugs related to sulfonamides, certain antibiotics, tranquilizers of the phenothiazine group, and a number of phototoxic oils that are used in fragrances (Fisher, 1995; Cronin, 1980; Gellin, 1972; Birmingham, 1978; Epstein, 1971; Birmingham, 1968; DeLeo, 1995).

Among the plants known to cause photosensitivity reaction are members of the Umbellifera. They include celery that has been infected with pink rot fungus, cow parsnip, dill, fennel, carrot, and wild parsnip (Birmingham, 1978; Epstein, 1971; Birmingham, 1968; Emmett, 1977). The development of classic signs of allergic contact dermatitis resulting from the combined interaction of a purported plant photoallergen and ultraviolet light is termed *phytophotodermatitis*. The photoactive chemicals in these plants are psoralens or furocoumarins and have been used for decades as therapeutic agents in the treatment of photoresponsive dermatoses like psoriasis and eczematous dermatitis.

19.4.6 Photoallergy

Allergic contact dermatitis may occur in the setting of ultraviolet light exposure. In photocontact allergic dermatitis prior sensitization to the allergen is required, in contrast to phototoxic reaction, which may occur on initial introduction to a chemical. The photoallergens, in the absence of ultraviolet light, are not inherently allergenic and are incapable of causing a dermatitis in the sensitized individual. The introduction of ultraviolet light to the chemical can cause changes in the molecule that render it allergenic and hence capable of inducing a rash.

19.4.7 Phototesting

It is often necessary to test for sensitivity to potentially photoactive chemicals since the cause of a dermatitis may not be obvious. In such a setting photopatch testing may be performed where suspected photoallergens are placed on a patient's back in duplicate. After 24 hours one test set is exposed to ultraviolet A light. (See patch test section for greater details.) Reactions on the irradiated side with negative reactions on the nonirradiated side confirm the diagnosis of photocontact allergic dermatitis. A majority of the chemicals listed in Table 19.6 are capable of causing photoallergic contact dermatitis and may be tested by photopatch testing methods. Those that typically cause phototoxic reaction such as tar derivative are generally not tested in clinical settings since history and physical exam are generally revealing. Table 19.6 outlines potentially photosensitizing chemicals and those used for photopatch testing at New York University Medical Center as of January, 1997 (Occupational and Environmental Dermatology Unit, Department of Dermatology, New York University Medical Center, 1997). Table 19.7 lists potentially photosensitizing plants.

TABLE 19.6 Photosensitizing Chemicals

Acridine[a]	Bithionol	Octyl methoxycinnamate
Anthracene[a]	Chlorhexidene	Promethazine
Certain chlorinated hydrocarbons[a]	Chlorpromazine	Sandalwood oil
Coal tar[a]	Cinoxate	Selected plant and pesticides
Creosote[a]	Dichlorophen	Sulfanilamide
Phenanthrene[a]	Diphenhydramine	Thiourea
Tar pitch[a]	Fentichlor	Tribromosalicylanilide
1-(4-Isopropylphenyl)-3-phenyl-1,3 proandione	Hexachlorophene	Trichlorocarbanilide
3-(4-Methylbenzylidene) camphor	Methyl anthranilate	Triclosan
6-Methylcoumarin	Musk ambrette	
Benzophenone-4	Octyl dimethyl PABA	

[a]Not tested under normal phototesting conditions.

TABLE 19.7 Photosensitizing Plants

Moracae
- Ficus carica (fig)

Rutaceae
- Citrus aurantifolia (lime)
- Dictamnus (gas plant)
- Ruta graveolens (rue)
- Citrus aurantium (bitter orange)
- Citrus limon (lemon)
- Citrus bergamia (bergamot)

Umbelliferae
- Anthriscus sylvestris (cow parsley)
- Apium graveolens (celery, pink rot)
- Daucus carota var, savita (carrot)
- Pastinaca sativa (garden parsnip)
- Foeniculum vulgare (fennel)
- Anethum graveolens (dill)
- Peucedanum ostruthium (masterwort)
- Heracleum spp. (cow parsnip)

Compositae
- Anthemis cotula (stinking mayweed)

Ranunculaceae
- Ranunculus (buttercup)

Cruciferae
- Brassica spp. (mustard)

Hypericaceae
- Psoralea corylifolia (scurfy pea, bavchi)
- Hypericum perforatum (St. John's wort)

Fitzpatrick, 1993.

19.4.8 Mechanical

Work-incurred cutaneous injury may be mild, moderate, or severe. The injuries include cuts, lacerations, punctures, abrasions, and burns, which account for about 35 percent of occupational injuries for which worker's compensation claims are filed (National Electric Injury Surveillance System, 1983). This translates to almost 1.5 million injuries to workers annually (American Academy of Dermatology, 1994).

Contact with spicules of fiber glass, copra, hemp, and so on induce irritation and stimulate itching and scratching. Skin can react to repetitive friction by forming a blister or a callus; to pressure by changing color or becoming thickened or hyperkeratotic; and to shearing by sharp force by denudation or a puncture wound. Any break in the skin may become the site of a secondary infection (Schwartz, 1957; White, 1934; Suskind, 1959; Adams, 1983; Gellin, 1972; Rycroft, 1986; Birmingham, 1978).

Thousands of workmen use air-powered and electric tools that operate at variable frequencies. Exposure to vibration in a certain frequency range can produce painful fingers, a Raynaud-like disorder resulting from spasm of the blood vessels in the tool-holding hand. Slower-frequency tools such as jackhammers can cause bony, muscular, and tendon injury (Williams, 1975; Suvorov, 1983).

19.4.9 Physical

Heat, cold, electricity, ultraviolet light (natural and artificial), and various radiation sources can induce cutaneous injury and sometimes systemic effects. Chemical irritants and ultraviolet light are covered in the aforementioned text.

19.4.9.1 Heat Thermal burns are common among welders, lead burners, metal cutters, roofers, molten metalworkers, and glass blowers.

Miliaria (prickly heat) often follows overexposure to increased temperatures and humidities. Increase in sweating causes waterlogging of the keratin layer with blockading of the sweat ducts.

Excessive failure of thermoregulation under hot, humid climates may result in heat exhaustion. Symptoms include muscle cramping, nausea, vomiting, and fainting. Treatment by moving to a cooler environment and rehydrating with an electrolyte solution is necessary. Untreated heat exhaustion can progress to heat stroke. Heat stroke is characterized by elevated core temperature, neurologic symptoms, and lack of sweating. This disease has a high fatality rate if untreated. Aggressive resuscitation efforts including fluid and electrolyte replacement as well as core temperature cooling are required (De Galan, 1995; Schwartz, 1957; Suskind, 1959; Samitz, 1985; Adams, 1983; Meso, 1997; Dukes-Dobos, 1977).

19.4.9.2 Cold Frostbite is a common injury caused by intracellular crystallization of water. As ice crystals form, their sharp points are capable of puncturing membranes and fatally disrupting the cell's homeostatic mechanisms. Fingers, toes, ears, and nose are the usual sites of injury (policemen, firemen, postal workers, farmers, construction workers, military personnel, and frozen food storage employees are at risk). The hallmark of therapy rests in the preservation of viable tissue through rapid rewarming. Adjuvant approaches such as medication, surgery, and hemoperfusion continue to be investigated (Schwartz, 1957; Suskind, 1959; Samitz, 1985; Adams, 1983; Gellin, 1972; Dukes-Dobos, 1977; Foray, 1992).

19.4.9.3 Electricity Severe cutaneous burns of local or widespread proportions can result from electrical injury. Cutaneous signs of high-intensity electrical injury such as lightning strikes produce a pathognomonic sign consisting of ramifying fern-shaped red lesions emulating the track of the electricity. Less dramatic electrical burns can cause local necrosis of skin, similar in nature to a chemical burn (Schwartz, 1957; Adams, 1983; Gellin, 1972; Braun-Falco, 1991).

19.4.9.4 Microwaves Thermal burn is the major hazardous potential in contact with this radiation source (Meso, 1977).

19.4.9.5 Lasers Laser radiation has enjoyed exponentially expanding utility in medicine, but particularly in dermatology. Dozens of lasers are now available for therapeutic use in the treatment of cutaneous disease. They work by selectively destroying pathological tissue while sparing normal surrounding tissue. Lasers with light outputs in the visible spectrum are capable of destroying tissue containing complementary colors. For example, a laser producing yellow light will cause destruction of red tissue. This may be useful in eliminating disfiguring vascular birthmarks without the necessity of surgery and the resultant scar. Normal tissue is spared because the tissue without overt redness will not absorb the energy of the light. This protective phenomenon is augmented by an extremely short pulse time that allows normal surrounding tissues to cool before heat damage occurs. This is termed *thermal relaxation time*. Other lasers like the carbon dioxide laser produce nonvisible light that causes destruction to any tissue through the production of intense heat. Adverse effects of inadvertent exposure of the skin to lasers primarily rest in the thermal damage cause by absorption of the light (Wheeland, 1994).

19.4.9.6 Ionizing Radiation Modern industry and technology have many applications of this radiation type. It is important in the production and use of fissionable materials, radioisotopes, X-ray diffraction machines, electron beam operations, industrial X-rays for detecting metal flaws, and various uses in diagnostic and therapeutic radiology. Accidental exposures may result in severe acute cutaneous and systemic injury depending upon the level of radiation received (Schwartz, 1957; Gellin, 1972; Meso, 1977). Lower-level exposures to ionizing radiation may often produce skin changes that manifest years after exposure. Skin thinning, scarring, and ulcerations occurring in sites of previous radiation exposure is termed *chronic radiation dermatitis* (Fitzpatrick, 1993).

19.4.9.7 Biological Primary or secondary infection can happen in any occupation following exposure to bacteria, viruses, fungi, or parasites. Simple lacerations or embedment of a thorn or a wood splinter or metal slug can lead to infection. Certain occupations are associated with greater risk of bacterial infection, for example, anthrax among sheepherders, hide processors, and wool handlers; erysipeloid infection among meat, fish, and fowl dressers; and folliculitis among machinists, garage workers, candymakers, sanitation and sewage employees, and those exposed to coal tar (Schwartz, 1957; Suskind, 1959; Samitz, 1985; Adams, 1983; Birmingham, 1978; Wilkinson, 1982).

Fungi can produce localized cutaneous disease. Yeast infections (*Candida*) can occur among those employees engaged in wet work, for example, bartenders, cannery workers, fruit processors, or anyone who works in a wet environment. Sporotrichosis is seen among garden and landscape workers, florists, farmers, and miners. Ringworm infection of animals can be transmitted to farmers, veterinarians, laboratory personnel, and anyone in frequent contact with infected animals (Schwartz, 1957; Adams, 1983; Gellin, 1972; Birmingham, 1978; Wilkinson, 1982).

Certain parasitic mites inhabit cheese, grain, and other foods and will attack bakers, grain harvesters, grocers, and longshoremen. Mites that live on animals and fowl similarly are known to attack humans. In the southeastern states, animal hookworm larvae from dogs and cats are deposited in sandy soils and lead to infection among construction workers, farmers, plumbers, and, of course, anyone who works in the infected soils. Ticks, fleas, and insects can produce troublesome skin reactions and, in certain instances, systemic disease such as Rocky Mountain Spotted Fever, Lyme disease, yellow fever, and malaria, among other vector-borne diseases (Schwartz, 1957; Gellin, 1972; Wilkinson, 1982).

Several occupational diseases are associated with virus infections, for example, Q fever, Newcastle disease, and ornithosis. In fact, viral diseases are becoming the most important class of biological agents to cause severe illness. Well-known causes of occupational dermatoses caused by virus infection are often contracted from infected sheep, milker's nodules from infected cows, chicken pox from infected children, and herpes infections from infected patients. Herpetic infections are particularly problematic for dentists, nurses, physicians, and others whose work occasions contact with open lesions (Schwartz, 1957; Birmingham, 1978; OSHA, 1978).

Animals such as snakes, sharks, dogs, and cats can result in aggressive necrotizing infections of the skin. Outdoor workers are particularly at risk for injury and also are placed at risk for rabies via similar circumstances. Insects such as spiders (particularly brown recluse spiders) can cause very painful and life-threatening necrotic bites in victims. Hornet, wasp, and bee stings can cause painful local reactions and are life threatening if individuals are sensitized to their venoms. Such severe allergic reactions can result in compromised breathing and blood pressure with potential shock or death (Schwartz, 1957; Gellin, 1972; Birmingham, 1978; OSHA, 1978).

19.5 PHYSICAL FINDINGS OF OCCUPATIONAL SKIN DISEASE

The hazardous potential of the work environment is unlimited. Chemical and physical agents can produce a wide variety of clinical displays that differ in appearance and in histopathological pattern. The nature of the lesions and the sites of involvement may provide a clue as to a certain class of materials involved, but only in rare instances does clinical appearance indicate the precise cause. Except for a few strange and unusual effects, the majority of occupational dermatoses can be placed in one of the following reaction patterns. Several materials known to be causal for each clinical type are included (Schwartz, 1957; White, 1934; Suskind, 1959; Samitz, 1985; Adams, 1983; Gellin, 1972, Birmingham, 1978; OSHA, 1978).

19.5.1 Acute Eczematous Contact Dermatitis

Most of the occupational dermatoses can be classified as acute eczematous contact dermatitis. Heat, redness, swelling, vesiculation, and oozing are the clinical signs; itch, burning, and general discomfort are the major symptoms experienced. The backs of the hands, the inner wrists, and the forearms are the usual sites of attack, but acute contact dermatitis can occur anywhere on the skin. When forehead, eyelids, ears, face, and neck

TABLE 19.8 Classes of chemicals causing allergic contact dermatitis

Acids, dilute	Herbicides	Resin systems
Alkalies, dilute	Insecticides	Rubber accelerators
Anhydrides	Liquid fuels	Rubber antioxidants
Detergents	Metal salts	Soluble emulsions
Germicides	Plants and woods	Solvents

are involved, airborne agents such as dust and vapors are suspected. More subtle clues such as upper lip, posterior ear, and mid-neck sparing may point to a photosensitivity disease since these areas are often less exposed to ultraviolet light. Generalized contact dermatitis may occur from massive exposure, the wearing of contaminated clothing, autosensitization from a preexisting dermatitis, or from systemic exposure.

Usually a contact dermatitis is recognizable as such, but whether the eruption has resulted from contact with a primary irritant or a cutaneous sensitizer can be ascertained only through a detailed history, a working knowledge of the materials being handled, their behavior on the skin, and a proper application and evaluation of diagnostic tests. Severe blistering or destruction of tissue generally indicates the action of an absolute or strong irritant; however, the history is what reveals the precise agent.

Acute contact eczematous dermatitis can be caused by hundreds of irritant and sensitizing chemicals, plants, and photoreactive agents. Some examples are listed in Table 19.8 (Schwartz, 1957; White, 1934; Suskind, 1959; Samitz, 1985; Adams, 1983; Fisher, 1995; Cronin, 1980; Gellin, 1972; Rycroft, 1986; Birmingham, 1978; Klauder, 1946, 1951). More specific agents are listed by category and type of exposure in Table 19.5.

19.5.2 Chronic Eczematous Contact Dermatitis

Hands, fingers, wrists, and forearms are the favored sites affected by chronic eczematous lesions. The skin is dry, thickened, and scaly with cracking and fissuring of the affected areas. Concurrent with skin thickening is the accentuation of normal skin lines and topical features. This is known as *lichenification.* Chronic nail dystrophy is a common accompaniment. Periodically, acute weeping lesions appear because of re-exposure, imprudent treatment, or secondary infection. Chronic contact dermatitis occurs when exposure is perpetuated to irritant chemicals. Less often, low-level exposure to allergens can produce similar findings. In the latter case, signs and symptoms of acute contact dermatitis persist for long periods before chronic changes occur. A large number of materials (Table 19.9) have the potential to sustain the marked dryness that accompanies this chronic recurrent skin problem (Schwartz, 1957; White, 1934, Suskind, 1959; Samitz, 1985; Adams, 1983; Fisher, 1995; Cronin, 1980; Gellin, 1972; Rycroft, 1986; Birmingham, 1978; Key, 1966).

TABLE 19.9 Chemicals capable of perpetuating chronic contact dermatitis

Abrasive dusts (pumice, sand, fiber glass)	Chronic fungal infections
Alkalis	Oils
Cement	Resin systems
Cleansers (industrial)	Solvents
Cutting fluids (soluble)	Wet work

TABLE 9.10 Chemicals capable of inducing acne or folliculitis

Asphalt
Creosote
Crude oil
Greases
Insoluble cutting oil
Lubricating oil
General-purpose petroleum oils
Heavy plant oils
Pitch
Tar

19.5.3 Folliculitis, Acne, and Chloracne

Hair follicles on the face, neck, forearms, backs of hands, fingers, lower abdomen, buttocks, and thighs can be affected in any kind of work entailing heavy soilage. Comedones (blackheads/whiteheads) and follicular infection are common among garage mechanics, certain machine tool operators, oil drillers, tar workers, roofers, and tradesmen engaged in generally dusty and dirty work.

Acne caused by industrial agents usually is seen on the face, arms, upper back, and chest; however, when exposure is severe, lesions may be seen on the abdominal wall, buttocks, and thighs. Machinists, mechanics, oil field and oil refinery workers, road builders, and roofers exposed to tar are at risk. Such effects are far less prevalent than was noted in the past.

The term *folliculitis* implies inflammation of the follicles caused either by an irritant chemical or infection. *Acne* is a term reserved for a skin eruptions affecting follicles on the skin. The hallmark signs include comedones, papules (bumps less than 1 cm), cysts, pustules, and subsequent scars (Plewig, 1993). Often folliculitis and acne are coexistent in the same person, since chemicals known to cause one can cause the other. Site of exposure will also determine the type of follicular disease that is manifested such that the face will usually manifest with acne, and the arms and legs with folliculitis. Table 19.10 lists some known causes of folliculitis and acne.

Chloracne is a disease with similarities to acne that are limited to follicular involvement and similar-sounding names. Chloracne is not merely severe acne but differs from regular acne entirely. Clinically, yellow, straw-colored cysts first appear on the sides of the forehead, around the lateral aspects of the eyelids, and behind the ears. Comedones, pustules, pigmentary disturbances, and subsequent scarring abound. Typically the head and neck, chest, back, groin and buttocks are involved. Atypical areas may be involved if these areas are heavily or chronically exposed (Urabe, 1985).

Chloracne differs from acne in several ways. First, the histopathology of chloracne is distinct from acne. In chloracne sebaceous glands are conspicuously destroyed rather than being large and overactive (Plewig, 1993). Second, the exposure patterns of chloracne are distinct and for the most part include exposure to halogenated aromatic hydrocarbons (Table 19.11). Third, the natural history of chloracne and acne are divergent. Chemically induced acne and folliculitis will resolve without treatment within a week or two, and sooner if treatment is instituted. Chloracne is classically resistant to treatment, including systemic retinoids, and may last for decades after exposure.

Polychlorinated biphenyls (PCB) and dioxins are epidemiologically the most fre-

TABLE 9.11 Cloracnegens

Hexachlorodibenzo-*p*-dioxin
Polybrominated dibenzofurans
Polybrominated biphenyls
Polychlorinated biphenyls
Polychlorinated dibenzofurans
Polychloronaphthalenes
Tetrachloroazobenzene
Tetrachloroazoxybenzene
Tetrachlorodibenzo-*p*-dioxin

quent cause of chloracne. During the Vietnam war, chloracne occurred from exposure to a 2,3,7,8-tetrachlorodibenzo-*p*-dioxin–contaminated chlorophenoxyacetic acid herbicide called Agent Orange. These herbicides are capable of causing a variety of neurologic and irritant dermatological findings. The chloracne was likely produced by the dioxin contaminate, a potent chloracnegen.

Domestically, chloracne most commonly occurs in the occupational setting from polychlorinated biphenyls. These dielectric compounds were commonly utilized in electrical transformers. While production of PCBs has been discontinued since the 1970s, hundreds of millions of pounds have escaped into the environment or are still present in some transformers in relatively high concentration. PCB exposure has been reported to cause illness in practically every organ system and has been described as a potential carcinogen. Despite this, rigorous epidemiologic studies in humans have failed to link PCB exposure with any specific illness other than chloracne. Hence chloracne remains the only reliable indicator of PCB exposure in humans (James, 1993). Further studies are necessary before conclusions can be drawn about other disease linkages.

19.5.4 Sweat-Induced Reactions

Miliaria is discussed above. Intertrigo occurs at sites where the skin opposes skin and allows sweat and warmth to macerate the tissue. Favored locations are the armpits, the groin, between the buttocks, and under the breasts (Schwartz, 1957; Suskind, 1959; Samitz, 1985; Adams, 1983; Gellin, 1972; Rycroft, 1986; Birmingham, 1978).

19.5.5 Pigmentary Abnormalities

Color changes in skin can result from percutaneous absorption, inhalation, or a combination of both entry routes. The color change may represent chemical fixation of a dye to keratin or an increase or decrease in epidermal pigment through stimulation or destruction of melanocytes, respectively (melanin).

Hyperpigmentation from excessive melanin production may follow an inflammatory dermatosis, exposure to sunlight alone, or the combined action of sunlight plus a number of photoactive chemicals or plants. The opposite (loss of pigment or leukoderma) results from direct injury to the epidermis and melanin-producing cells by burns, chronic dermatitis, trauma, or chemical interference with the enzyme system that produces melanin. Antioxidant chemicals used in adhesives, cutting fluids, sanitizing agents, and rubber have caused complete loss of pigment (leukoderma).

Inhalation or percutaneous absorption of certain toxicants as aniline or other aromatic

TABLE 9.12 Chemicals Involved in Pigmentary Disturbances

Discoloration	Hyperpigmentation	Hypopigmentation
Arsenic	Chloracnegens	Antioxidants
Certain organic amines	Coal tar products	Hydroquinone
Carbon	Petroleum oils	Monobenzyl ether of hydroquinone
Dyes	Photoactive chemicals	Tertiary amyl phenol
Mercury	Photoactive plants	Tertiary butyl catechol
Picric acid	Radiation (sunlight)	Tertiary butyl phenol
Silver	Radiation (ionizing)	Burns
Trinitrotoluene		Chronic dermatitis
		Trauma

nitro and amino compounds causes methemoglobinemia. Jaundice can result from hepatic injury by carbon tetrachloride or trinitrotoluene, among other hepatotoxins.

Pigmentary abnormalities are caused chemical agents are listed in Table 19.12 (Schwartz, 1957; Suskind, 1959; Samitz, 1985; Adams, 1983; Maibach, 1987; Gellin, 1972; Birmingham, 1978; Kahn, 1979; Gellin, 1970; Malten, 1971).

19.5.6 Neoplasms

Mutagens are abundant in the workplace and are easily able to contact the skin under many circumstances. Such exposures may result in the development of benign or malignant neoplasms in the skin. These tumors, while often occurring at sites of contact with a suspected carcinogen, may also occur at distant sites, or from systemic exposures. Asbestos warts and tar papillomas associated with petroleum and tar exposures are examples of benign neoplasms associated with repeated contact. Basal and squamous cell carcinomas (malignant neoplasms) are clearly associated with recurrent ultraviolet light exposure (Fitzpatrick, 1993; Lim, 1993). Controversy exists whether malignant melanoma is associated with specific occupational exposures, but epidemiologic evidence has mounted to implicate ultraviolet light (particularly exposure during childhood) as a causal factor. Melanomas may also arise in locations classically shielded from ultraviolet light exposure (Birmingham, 1978; Fitzpatrick, 1993; Lim, 1993; Combs, 1954).

Several chemical and physical agents are classified as industrial carcinogens, but only a few frequently cause skin cancer (Table 19.13). Admittedly, more cancers appear on the skin than at any other site; however, the number of these that are of occupational origin is not known. Sunlight is probably the major cause of occupational skin cancer, particularly among those engaged in agriculture, construction, fishing, forestry, gardening and landscaping, oil drilling, road building, roofing, and telephone and electric line installations (Schwartz, 1957; White, 1934; Birmingham, 1978; Epstein, 1971; Combs, 1954; Emmett, 1975; Buchanan, 1962; Berenblum, 1943; Bingham, 1965; Rothman, 1988).

In European countries, mule spinners exposed to shale oil and pressmen exposed to paraffin experienced a high frequency of carcinomatous lesions of the scrotum and lower extremities. Similar experiences with paraffin happened in the United States, but improved industrial practices and hygienic controls have all but eliminated the problem. In 1984, the International Agency for Research on Cancer determined that mineral oils containing various additives and impurities used in mule spinning, metal machining, and jute processing were carcinogenic to humans. Oils formerly in use that were responsible

TABLE 19.13 Chemicals and physical agents implicated as cutaneous carcinogens

Anthracene	Mineral oils containing various additives and impurities
Arsenic	Crude oils
Burns	Radium and roentgen rays
Coal tar	Shale oil
Coal tar pitch	Soot
Creosote oil	Sunlight
	Ultraviolet light

for cutaneous cancers including those affecting the scrotum were not as well refined as the lubricating oils being used today (Rothman, 1988).

Systemic exposures to carcinogens is clearly problematic when discussing internal malignancies such as hepatic and pulmonary cancers. For the skin there are relatively few systemic toxins capable of producing malignant neoplasms. Arsenic, however, is well known to cause a variety of benign and premalignant growths as well as basal and squamous cell carcinomas. The clinical and histopathologic appearance of these growths are often so distinct that they have been termed arsenical keratoses and arsenical carcinomas (Lever, 1990).

It is important to understand that skin cancers noted on workers in contact with carcinogenic agents are not necessarily of occupational origin. A certain number of people develop skin cancer irrespective of their jobs. For instance, residents in the southwestern United States or in Australia are associated with a high frequency of skin cancer because of the exposure to sunlight. Ascertaining whether a skin cancer is truly of occupational origin in individuals residing in sun belt regions can be controversial (Schwartz, 1957; White, 1934; Gellin, 1972; Birmingham, 1978; Epstein, 1971; Combs, 1954; Emmett, 1975; Eckhard, 1959; Buchanan, 1862; Berenblum, 1943; Bingham, 1965; Rothman, 1988).

19.5.7 Ulcerations

Cutaneous ulcers were the earliest documented skin changes observed among miners and allied craftsmen. In 1827, Cumin (1827) reported on skin ulcers produced by chromium. Today the chrome ulcer (hole) caused by chromic acid or concentrated alkaline dichromate is a familiar lesion among chrome palters and chrome reduction plant operators. Perforation of the nasal septum also occurs among these employees, though in smaller numbers than occurred 20 years ago because many of the operations are now well enclosed (Cohen, 1974). Punched-out ulcers on the skin can result from contact with arsenic trioxide, calcium arsenate, calcium nitrate, and slaked lime (Birmingham, 1964). Nonchemical ulcerations may be associated with trauma and ulcers of the lower extremities in diabetics, pyogenic infections, vascular insufficiency, and sickle cell anemia (Schwartz, 1957; Birmingham, 1978; OSHA, 1978; Samitz, 1978).

19.5.8 Granulomas

Cutaneous granulomas are caused by many agents of animate and inanimate nature. Such lesions are characterized by chronic, indolent inflammatory reactions that can be localized or systemic and result in severe scar formation. Granulomatous lesions can be the result of bacterial, fungal, viral, or parasitic elements such as atypical mycobacterium, sporotrichosis, milker's nodules, and tick bite, respectively. Additionally, minerals such as silica, zirconium, and beryllium and substances such as bone, chitin, coral, thorns, and grease have produced chronic granulomatous change in the skin (Schwartz, 1957; Birmingham, 1978; OSHA, 1978; Pinkus, 1976).

19.5.9 Other Clinical Patterns

The clinical patterns described above represent well-known forms of occupational skin diseases. However, there are a number of other disorders affecting skin, hair, and nails that do not fit into these patterns. Some examples follow.

19.5.9.1 Contact Urticaria Contact urticaria (hives) result secondary to contact with an allergen or nonallergenic chemical urticariant. Allergic contact urticaria represents immediate-type hypersensitivity and requires sensitization to occur. Occupational contact urticaria has become an extremely important issue, particularly in the health care setting. In the mid 1980s during the emergence and recognition of the human immunodeficiency virus and the escalating incidence of hepatitis, standards of personal protection developed. Ultimately the term *universal precautions* arose and charged health care personnel to handle all blood and body fluids as if they were infected with a serious transmissible agent. As such, the use of protective gloves made of latex became widespread. This resulted in a tremendous increase in the number of gloves being used by over 5 million health care workers (Centers for Disease Control, 1987; Turjanmaa, 1987).

A recently recognized syndrome principally manifesting as contact urticaria with other symptoms such as rhinitis, conjunctivitis, asthma, and less commonly anaphylaxis and death has been associated with latex proteins found in rubber products. For health care workers, the main exposure is through rubber gloves and rubber material devices. The exact source of initial sensitization, however, is extremely difficult to ascertain since latex is so ubiquitous in society. The combination of extremely itchy skin and respiratory symptoms not only makes it uncomfortable for an employee to work with latex gloves, but may make it life threatening. Glove powders are known to bind latex protein allergen and become airborne when gloves are removed. Inhaled latex protein can trigger severe allergic reactions, causing compromised breathing and occasionally shock and death. Those at high risk include people who frequently use disposable latex gloves, patients with spina bifida, those with a history of multiple surgical procedures, and those with a history of hand dermatitis (Jaeger, 1992; Tomazic, 1994; Charous, 1994; Asa, 1994). Since dermatitis of the hands is common in many occupational settings, the frequent use of disposable latex gloves should alert health care workers to the sentinel signals of latex allergy. Health care institutions are now recognizing this problem and are developing strategies to detect latex allergy in employees and cope with them. At New York University Medical Center an algorithm has been developed to screen potentially allergic employees (Table 19.14).

Here, employees are first screened with a latex radioallergosorbent assay (RAST) test to detect latex-specific immunoglobulin E. If negative, a "use" test utilizing a latex glove under supervised setting is performed, first with one finger, than an entire hand. If these tests are negative, eluted latex protein in solution is used for prick or scratch testing. The presence of a hive at the test site indicates latex hypersensitivity. If found to be positive, employees use alternative gloves, and the latex in the immediate worksite

TABLE 19.14 New York University Medical Center Latex Energy Evaluation

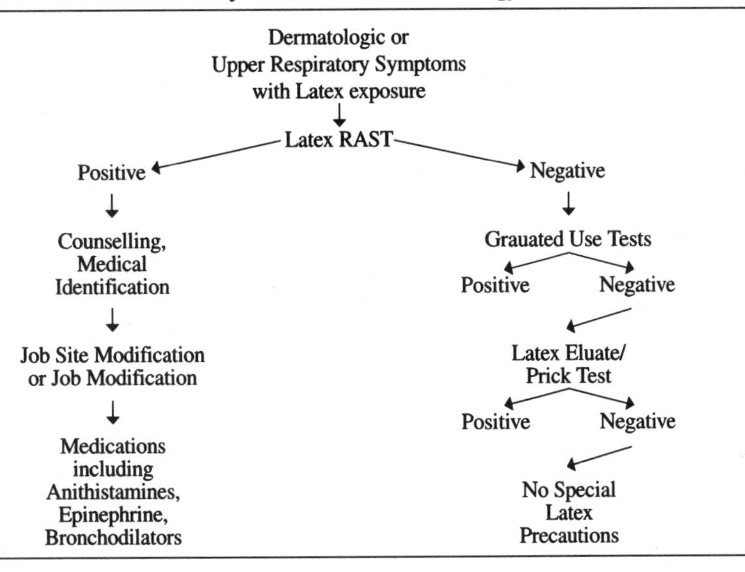

is minimized. To date all new latex-sensitive employees at NYU have successfully had job site or job modification protocols enacted.

Occupational contact urticaria may also occur from a variety of other allergenic sources such as foods like apple, carrot, egg, fish, beef, chicken, lamb, pork, and turkey. Other sources have arisen from animal viscera and products handled by veterinarians and food dressers; from contact with formaldehyde, rat tail, guinea pig, and streptomycin. It is likely that many of these cases go unreported and unrecognized as such (Maibach, 1987; Fisher, 1995, Chapter 34; Odom, 1977).

19.5.9.2 Nail Discoloration and Dystrophy Chemicals such as alkaline bichromate induce an ochre in nails; tetryl and trinitrotoluene induce yellow coloring; dyes of various colors may change the nail color; carpenters may have wood stains on their nails. Dystrophy can follow chronic contact from acids and corrosive salts, alkaline agents, moisture exposures, sugars, trauma, and infectious agents such as bacteria and fungi. Long-standing contact dermatitis of the finger tips may disrupt the nail matrix and bed, causing dystrophic nail plates to form (Schwartz, 1957; Adams, 1983; OSHA, 1978; Ronchese, 1948, 1965; Fisher, 1995, Chapter 6, pp. 66–92).

19.5.9.3 Facial Flush This peculiar phenomenon has been reported from the combination of certain chemicals such as tetramethylthiuram disulfide, trichloroethylene, or butyraldoxime following the ingestion of alcohol. Because trichloroethylene is known to cause liver damage, the facial flush is related to the intolerance associated with alcohol ingestion (Schwartz, 1957; Birmingham, 1978; Lewis, 1956; Stewart, 1974).

19.5.9.4 Acroosteolysis Several years ago, a number of workers involved in cleaning vinyl chloride polymerization reaction tanks incurred a peculiar vascular and bony abnormality involving the digits, hands, and forearms. Bone resorption of the dig-

TABLE 9.15 Percutaneous absorption and target organ systems

Aniline → red blood cell → methemoglobinemia
Benzidine → urinary bladder → carcinoma
Carbon disulfide → central nervous system → psychopathological symptoms and signs as peripheral neuritis + cardiac disease
Carbon tetrachloride → liver, kidney, and CNS → liver and kidney damage, depression
Chlorinated naphthalenes, biphenyls, and dioxins → skin → chloracne, hepatitis, peripheral neuritis
Ethylene glycol ethers → CNS → lungs, liver and kidney damage
Methyl butyl ketone → CNS and peripheral nerves → polyneuritis, depression
Organophosphate pesticides → inhibition of cholinesterase → cardiovascular, gastrointestinal, neuromuscular, and pulmonary disturbances
Tetrachlorethane → CNS → depression and liver damage
Toluene → CNS and liver → confusion, dizziness, headache, paresthesias

ital tufts was accompanied by Raynaud's syndrome and scleroderma-like changes of the hands and forearms. Removal from the tank cleaning duties led to vascular bone improvement, but the skin changes did not always improve (Wilson, 1967; Harris, 1967). Acroosteolysis has become a rare disease in the United States since the inception of strict exposure standards for vinyl chloride monomer. Polymerized vinyl chloride does not cause this illness.

19.6 SOME SIGNS OF SYSTEMIC INTOXICATION FOLLOWING PERCUTANEOUS ABSORPTION

A number of chemicals with or without direct toxic effect on the skin can cause systemic intoxication following percutaneous entry. Transepidermal passage may occur with or without metabolism by keratinocytes. Once into the upper dermis, chemicals have direct access to a vascular plexus and the systemic circulation, where target organs can be reached prior to excretion or metabolism. Table 19.15 lists examples of chemicals capable of significant percutaneous absorption with target organs and demonstrable effects (OSHA, 1978; Tabershaw, 1977; Proctor, 1978). More details may be found in the Wang et al. text on percutaneous absorption.

Over a period of years, industrial and contract laboratories have developed and used tests for predicting the toxicologic behavior of a new product or one in use whose toxic action becomes questioned. Such tests on animals and humans are designed to demonstrate signs of systemic and cutaneous toxic effects, for example, primary irritation, allergic hypersensitivity, phototoxicity, photoallergenicity, interference with pigment formation, sweat and sebaceous gland activity, dermal absorption routes, metabolic markers, and cellular aberrations indicating mutation or frank carcinoma. Conducting these tests leads to prediction of toxicologic potential and diminishment thereby of the number of untoward reactions that might appear if such tests had not been done (OSHA, 1978; Hood, 1977).

19.7 DIAGNOSIS

It is a common assumption among employees that any skin disease they incur has something to do with their work. At times the supposition is correct, but often there is no true relationship to the work situation. Arriving at the correct diagnosis may be quite easy, but such is not a routine occurrence. The industrial physician has a distinct advantage in being familiar with agents within the work environment and the conditions associated with contacting them. The dermatologist may find the diagnosis difficult if unfamiliar with contact agents in the work environment. The practitioner with little or no interest in dermatologic problems associated with work and also lacking dermatologic skill will find it most difficult to make a correct diagnosis. At any rate, the attending physician, specialist or otherwise, should satisfy certain basic tenets in establishing a diagnosis of occupational skin disease. The health and safety professional should provide the medical provider as much information as possible about the materials used in the workplace. An understanding of the work process is important because the offending agent may be a byproduct or waste and not appear on the "list" of materials used. Likewise, workplace conditions such as temperature, humidity, and physical activities may also be important.

19.7.1 Patient History

Only through detailed questioning can the proper relationship between cause and effect be established. Taking a thorough history is time consuming, because it should cover the past and present health and work status of the employee. Areas requiring thorough coverage include family history, allergies, past medical illnesses, job titles, nature of work performed, materials handled, frequency of potential exposures, history and distribution of the skin findings, and any treatments used. It is important to ascertain whether the employee has had any prior dermatologic illnesses, and if any systemic or topical drugs are used for any medical or surgical illness. Finally, materials used for home hobbies such as gardening, woodworking, painting and model building may cause skin disease indistinguishable from occupationally acquired exposures (Schwartz, 1957; Suskind, 1959; Samitz, 1985; Adams, 1983; Maibach, 1987; Gellin, 1972; Birmingham, 1978; OSHA, 1978).

19.7.2 Appearance of the Lesions

The eruption should fit into one of the clinical types in its appearance. Although the majority of the occupational dermatoses are either acute or chronic eczematous contact dermatitis, other clinical types of disease such as follicular, acneiform, pigmentary, neoplastic, ulcerative, and granulomatous can occur. Further, one must be on the lookout for the oddities that show up in unpredictable fashion, for example, Raynaud's disease and contact urticaria (Schwartz, 1957; Suskind, 1959; Samitz, 1985; Adams, 1983; Gellin, 1972; Birmingham, 1978).

19.7.3 Sites Affected

Most cases of occupational skin disease affect the hands, digits, wrists, and forearms, for the upper extremities are truly the instruments for work. However, the forehead, face, "V" of the neck, and ears may also display active lesions, particularly when the employee is exposed to dusts, vapors, or fumes. Legs may be favored when splashing of material on the floor occurs or when suspected contactant are at ground level such as in farming. Although most of the work dermatoses are usually seen in the above sites, generalization can occur from massive exposure, contaminated work clothing, and also from autosensitization (spread) of an already existent rash (Schwartz, 1957; Suskind, 1959; Samitz, 1985; Adams, 1983; Gellin, 1972; Birmingham, 1978; OSHA, 1978).

19.7.4 Diagnostic Tests

Laboratory tests should be employed when necessary for the detection of bacteria, fungi, and parasites. Such tests include direct microscopic examination of surface specimens, culture of bacterial or fungal elements, and biopsy of one or more lesions for histopathological definition. When allergic reactions are suspect, diagnostic patch tests can be used to ascertain occupational, as well as nonoccupational allergies, including photosensitization. At times, useful information can be obtained through the use of analytical chemical examination of blood, urine, or tissue (skin, hair, nails) (Schwartz, 1957; Suskind, 1959; Samitz, 1985; Adams, 1983; Maibach, 1987; Fisher, 1995; Cronin, 1980; Gellin, 1972; Rycroft, 1986; Birmingham, 1978; OSHA, 1978).

19.7.5 The Patch Test

Diagnostic patch testing, properly performed and interpreted, is a highly useful procedure. The test is based on the theory that when an acute or chronic eczematous dermatitis is caused by a given sensitizing agent, application of the suspected material to an area of unaffected skin for 48 hours will cause an inflammatory reaction at the site of application. A positive test usually indicates that the individual has an allergic sensitivity to the test material. When the employee is working with primary irritants and fellow employees also are affected with dermatitis, the cause is self-evident and patch testing is neither necessary nor indicated. Exception to this rule occurs when the employees are working with irritant agents that can also sensitize (epoxy and acrylic systems, resin hardeners of the amine group, formaldehyde, or chromates) or when an agent is highly sensitizing and many employees have become allergic. There are approximately 65,000 known irritant chemicals and 2,800 described allergic chemicals, and so the patch test is invaluable in identifying a causal agent or distinguishing an irritant from allergic contact dermatitis (Schwartz, 1957; Suskind, 1959; Adams, 1983; Fisher, 1995; Cronin, 1980; Rycroft, 1986; Birmingham, 1978; Rietschel, 1995; Sulzberger, 1931).

If the test is to have relevance and reliability, it must be performed by one with a clear understanding of the difference between a primary irritant and a sensitizer.

When patch tests are conducted with strong or even marginal irritants, a skin reaction is inevitable and usually not relevant or interpretable. However, this does not mean that a patch test cannot be performed with a diluted primary irritant. There is an abundance of published material pointing out proper patch test concentrations and appropriate vehicles considered safe for skin tests (Adams, 1983; Maibach, 1987; Fisher, 1995; Cronin, 1980). The goal of testing is to use concentrations of chemicals known to cause reactions in sensitized individuals and be negative in those not allergic. Hence, test concentrations are ideally below the threshold necessary to cause an irritant reaction, but above that necessary to cause an allergic reaction. For the most part, these concentrations do not overlap for most chemicals. Further, the one performing the tests should have a working knowledge of environmental contactants, particularly those well known as potential cutaneous allergens (Schwartz, 1957; Suskind, 1959; Adams, 1983; Fisher, 1995; Cronin, 1980; Rycroft, 1986; Birmingham, 1978; Sulzberger, 1931; Maibach, 1974).

The technique of the test is simple. Liquids, powders, or solids are applied under occlusive conditions under stainless steel discs or in a hydrogel suspension to the back.

TABLE 9.16 Interpretive criteria for patch testing

(?) =	doubtful; faint, macular redness only
(+) =	weak (nonvesicular) positive reaction; redness, infiltration, possibly papules
(++) =	strong (vesicular) positive reaction; redness, infiltration, papules, vesicles
(+++) =	extreme positive reaction; bullous reaction
(−) =	negative reaction
(IR) =	irritant reaction

The test panel should include relevant chemicals based on the history and distribution of the rash. A standard allergen panel is not useful since exposure patterns across various occupations is so divergent (Cohen, in press).

The North American Contact Dermatitis Group and the International Contact Dermatitis Group advocate standardized test concentrations applied in vertical rows on the back and covered by hypoallergenic tape. Contact with the test material is maintained on the skin for 48 h, and readings are made 30 min or later after removal, and an additional time at the 72nd to 168th hour.

Reading the tests and interpreting them for the degree of reaction requires experience. The levels of reaction currently in use are noted in Table 19.16.

True allergic reactions tend toward increased intensity for 24 to 48 h after test removal, whereas irritant reactions usually subside within 24 to 48 h after removal (Adams, 1983; Fisher, 1995; Birmingham, 1978; Sulzberger, 1931; Fregert, 1974; Malten, 1976; Maibach, 1974).

Interpreting the significance of the test reactions is of paramount importance. A positive test can result from exposure to an irritant or a sensitizer. When specific sensitization is the case, it means that the patient is reactive to the allergens at the time of the test. When the positive test coincides with a positive history of contact, it is considered strong evidence of an allergic etiology. Conversely, the examiner must be aware that clinically irrelevant positive tests can occur if the patient is tested (1) during an active dermatitic phase leading to one or several nonspecific reactions; (2) with a marginal irritant; or (3) with a sensitizer to which the patient had developed an early sensitization, for example, nickel, but which is not relevant to the present occupational dermatitis. The patch test is incapable of testing the irritancy potential of a suspect chemical. Irritant reactions on a patch test should never be correlated with any workplace exposures. *No conclusions regarding the cause of a dermatitis can be drawn from an irritant reaction on a patch test.*

A negative test indicates the absence of an irritant or an allergic reaction. However, a negative reaction can also mean (1) testing omitted an important allergen; (2) insufficient strength and quantity of the test allergen; (3) poor test condition; or (4) hyporeactivity by the patient at the time of the test.

Performing the patch test with unknown substances the employee has brought to the physician's office can be most misleading and potentially hazardous. The material could be a caustic and thus produce a strongly positive chemical burn or perhaps no reaction will occur. In either case, the test can be misused and provide no help in diagnosing the cause. Useful information concerning unknown materials can be obtained by contacting the plant manager, physician, nurse, industrial hygienist, or safety supervisor (Adams, 1983; Fisher, 1995; Cronin, 1980; Fregert, 1974; Malten, 1976; Maibach, 1974).

A patch test being employed more frequently has to do with suspected photoallergy. This is discussed in the photosensitivity section of the chapter. Most of the photodermatoses incurred in the work area are phototoxic and thus do not require photo patch tests for diagnosis. Some dermatologists perform photo patch tests in their offices; however, many of these patients suffering with suspected dermatoses are referred to a center where photo patch tests are performed routinely (Adams, 1983; Fisher, 1995; Cronin, 1980; Williams, 1975; Fregert, 1974; Malten, 1976).

19.8 TREATMENT

Immediate treatment of an occupational dermatosis does not differ essentially from that used for a similar eruption of nonoccupational nature. In either case, treatment should be directed toward providing fairly rapid relief of symptoms. The choice of treatment agents depends upon the nature and severity of the dermatitis. Most of the cases are either an acute or a chronic eczematous dermatitis, and most of these can be managed with ambulatory care. However, hospitalization is indicated when the severity of the eruption warrants in-patient care.

Acute eczematous dermatoses caused by a contactant generally respond promptly to wet dressings and topical steroid preparations, but systemic therapy with corticosteroids should be used when deemed necessary. Corticosteroids have definitely lessened the morbidity in the acute and chronic eczematous dermatoses caused by work. Once the dermatitis is under good control, clinical management must be directed toward:

1. Ascertaining the cause;
2. Returning the patient to the job when the skin condition warrants, but not before;
3. Instructing the patient in the means necessary to minimize or prevent contact at work with the offending material.

In any contact dermatitis it is essential to establish the causal agents or situations that contributed to the induction of the disease. Follicular or acneform skin lesions, notably chloracne, are notoriously slow in responding to treatment. Pigmentary change similarly may resist the run of therapeutic agents and remain active for months. New growths can be removed by an appropriate method and studied histopathologically. Ulcerations inevitably lead to the formation of scar tissue. Similarly, granulomatous lesions generally scar.

Almost all cases of occupational skin disease respond to appropriate therapy within 2 to 8 weeks; however, when chloracne, pigmentary changes, or allergic contact dermatitis due to chrome or nickel are the problem, therapeutic response may take months or years. It cannot be overemphasized that contact with the causative agents must be minimized, if not eliminated; otherwise return to work is accompanied by the return of the rash (Schwartz, 1957; Suskind, 1959; Samitz, 1985; Maltwn, 1964; Adams, 1983; Gellin, 1972; Birmingham, 1978).

19.8.1 Prolonged and Recurrent Dermatoses

As a rule, an occupational dermatosis can be expected to disappear or to be considerably improved within a period of 2 to 8 weeks after initiating treatment. Yet there are cases that are recalcitrant to appropriate treatment and continue to plague the patient with chronic recurrent episodes. This situation is commonly noted when the dermatosis was caused by chromium, nickel, mercury, or certain plastics and glues. However, all cases of recurrent disease are not necessarily associated with the above materials. The following situations may be operable in prolonged and recurrent disease (Morris, 1952; Birmingham, 1986):

1. Incorrect clinical diagnosis;
2. Failure to establish cause;
3. Failure to eliminate the cause even when direct cause has been established;
4. Improper treatment, often self-directed;
5. Poor hygiene habits at work;
6. Supervening secondary infections;
7. Cross-reactions with related chemicals;
8. Self-perpetuation for gain.

19.9 PREVENTION

The key to preventing occupational skin disease is to eliminate or at least minimize skin contact with potential irritants and sensitizers present in the workplace. To do so requires:

1. Recognition of the hazardous exposure potentials (see Chapters 3 and 20);
2. Assessment of the workplace exposures (see Chapters 4, 6, 7, 8, and 25);
3. Establishment of necessary controls (see Chapters 9, 10, 13, 14, 15, and 16).

Achieving these steps is more likely to occur in large industrial establishments with trained personnel responsible for the maintenance of health and safety practices. In contrast, many small plants or workplaces that employ the largest percentage of the work force have neither the money nor personnel to initiate and monitor effective preventive programs. Nonetheless, any work establishment has the responsibility of providing those preventive measures that at least minimize, if not entirely eliminate, contact with hazardous exposures.

19.9.1 Direct Measures

Time-tested control measures known to prevent occupational diseases are classified as primary (immediate) and indirect. The primary categories are:

1. Substitution;
2. Process change;
3. Isolation/enclosure;
4. Ventilation;
5. Good housekeeping;
6. Personal protection.

19.9.2 Indirect Measures

The indirect measures include:

1. Education and training of management, supervisory force, and employees;
2. Medical programs;
3. Environmental monitoring;

Although small plants generally lack in-house medical and industrial hygiene services, they do have access to such services through state health departments or through private consultants knowledgeable in health and safety measures.

19.9.2.1 Substitution and Process Change When a particular agent or process is recognized as a trouble source, substituting a less hazardous agent or process can minimize or eliminate the problem. This has been done with a number of toxic agents, for example, substituting toluene for benzene, tetrachloroethylene, or carbon tetrachloride. Substitution has potential value in allergen replacement of known offenders such as with chromium, nickel, certain antioxidants and accelerators in rubber manufacture, and certain biocides in metalworking fluids. The addition of iron salts to cement will change hexavalent chromium, a highly allergic species to trivalent chrome, a low-grade sensitizer without affecting the quality of the cement (Turk, 1993). When feasible, substituting a nonallergen for a hazardous agent is a recommended procedure.

19.9.2.2 Isolation and Enclosure Isolation of an agent or a process can be used to minimize hours of exposure or the number of people exposed. Isolation can mean creation of a barrier, or distance, or time, as the means of isolation to lessen exposure. Enclosure of processes provides a high level of safety when hazardous agents are involved. Local enclosures against oil spray and splash from metalworking fluid lessen the amount of exposure to the machine operators.

Radiation exposures can be shielded with proper barriers and remote control systems. Bagging operations can be enclosed to lessen, it not entirely eliminate, exposure to the operators involved.

19.9.2.3 Ventilation Movement of air can mean general dilution and or local exhaust ventilation used to reduce exposures to harmful airborne agents. Local exhaust ventilation is effective in controlling vapors of degreasing tanks and in mixing, layup, curing, and tooling of epoxy, polyester, phenol–formaldehyde resin systems.

19.9.2.4 Good Housekeeping A clean shop or plant is essential in controlling exposure to hazardous materials. This means keeping the workplace ceilings, windows, walls, floors, workbenches, and tools clean. Adequate storage space, properly placed warning signs, and sanitary facilities adequate in number should be available. Expeditious cleanup of spills and emergency showers for use after accidental heavy exposure to harmful chemicals are essentials of good industrial hygiene.

19.9.2.5 Personal Protection

19.9.2.5.1 Clothing. It is not necessary for all workers to wear protective clothing, but for those jobs for which it is required, good-quality clothing should be issued as a plant responsibility. Protective clothing against cold, heat, and biological and chemical injury to the skin is advisable. Depending upon the need, equipment such as hairnets, caps, helmets, shirts, trousers, coveralls, aprons, gloves, boots, safety glasses, and face shields should be available. Similarly, clothing and chemical screens to protect against ultraviolet light and ionizing, microwave, and laser radiation should also be readily available.

Once protective clothing has been issued, its laundering and maintenance should be the responsibility of the plant. When work clothing is laundered at home, it becomes a ready means of contaminating family members' apparel with chemicals, fiber glass, or harmful dusts.

Specific information concerned with protective equipment can be obtained from the National Institute for Occupational Safety and Health and from any of the manufacturers listed in the safety and hygiene journals. For Internet users, OSHA can be accessed via the world wide web at address http://www.osha.gov/. NIOSH may be reached at http://www.cdc.gov/NIOSH/homepage.html.

19.9.2.5.2 Gloves. Gloves are an important part of protective gear because the hands are most frequently exposed to chemicals work. Leather gloves, though expensive, offer fairly good protection against mechanical trauma (friction, abrasion, etc.). Cotton gloves suffice for light work, but they wear out sometimes in a matter of hours. Neoprene, butadiene-nitrile, and vinyl-dipped cotton gloves are useful in protecting against mechanical trauma, chemicals, solvents, and dusts. Unlined rubber gloves and plastic gloves can cause maceration because of occlusion and contact dermatitis from the chemical accelerators or antioxidant in the material. Much time and money can be saved by reviewing the catalogs and tables provided by the manufacturers.

19.9.2.5.3 Hand Cleansers. Of all the measures advocated for preventing occupational skin disease, personal cleanliness is paramount. Although ventilating systems and monitoring are important in controlling the workplace exposures, there remains no substitute for washing the hands, forearms, and face and keeping clean. To do this the plant must provide conveniently located wash stations with hot and cold running water, good-quality cleansers and clean towels.

Several varieties of acceptable cleansers are available on the market, and these include conventional soaps of liquid, cake, or powdered variety. Conventional soaps are used each day by millions of people and are generally considered safe. Liquid varieties including "cream" soaps are satisfactory for light soil removal. Powdered soaps are designed for light frictional removal of soil and may contain pumice, wood fiber, or corn meal.

Waterless cleansers are popular among those who contact heavy tenacious soilage such as tar, grease, and paint. They should not serve as a substitute for conventional removal of soilage. Daily use of waterless cleansers leads to dryness of the skin and, at times, eczematous dermatitis from the solvent action of the cleanser.

In choosing an industrial cleanser the following dicta are suggested:

1. It should have good cleansing quality;
2. It should not dry out the skin through normal usage;
3. It should not harmfully abrade the skin;
4. It should not contain known sensitizers;

5. It should flow readily through dispensers;
6. It should resist insect invasion;
7. It should resist easy spoilage and rancidity;
8. It should not clog the plumbing.

19.9.2.5.4 Protective Creams. Covering the skin with a barrier cream, lotion, or ointment is a common practice in and out of industry. Easy application and removal plus the psychological aspect of protection account for the popularity of these materials. Obviously, a thin layer of barrier cream is not the same as good environmental control or an appropriate protective sleeve or glove. There are currently no protective creams available that can provide adequate defense against irritants and allergens outside of laboratory testing conditions. Contact dermatitis may only be controlled with avoidance of offending allergens and good hygiene. The use of sunscreens offer incontrovertible protection against sunburn from ultraviolet light exposure. Their use is encouraged for all workers routinely exposed to sunlight or artificial ultraviolet light sources. DEET and citronella oil containing lotions can offer protection for outdoor worker against arthropod assault; however, they may on occasion cause allergic contact dermatitis.

The development of protective creams has clearly been an effort to reduce the necessity for wearing gloves that may reduce manual dexterity. Current glove technology has sufficiently advanced to allow excellent dexterity for workers performing intricate duties. For decades now neuro-, cardiac, and microsurgeons have successfully performed surgery requiring the highest degree of adroitness while having their hands completely enveloped in latex. Few if any occupations require higher degrees of deftness that make it impossible to wear protective gloves. However, workers exposed to strong solvent chemicals may find it difficult to locate gloves that are resistant to breakdown or provide adequate barrier protection under these harsh circumstances (Schwartz, 1957; Suskind, 1959; Samitz, 1985; Adams, 1983; Gellin, 1972; Fregert, 1974; OSHA, 1978; Calnan, 1970; Birmingham, 1975; NIOSH, 1988).

19.9.3 Control Measures

19.9.3.1 Education An effective prevention and control program against occupational disease in general, including diseases of the skin, must begin with education. A joint commitment by management, supervisory personnel, workers, and worker representatives is required. The purpose is to acquaint managerial personnel and the workers with the hazards inherent in the workplace and the measures available to control the hazards. The training should be in the hands of well-qualified instructors capable of instructing the involved people with:

1. Identification of the agents involved in the plant;
2. Potential risks;
3. Symptoms and signs of unwanted effects;
4. Results of environmental and biological monitoring in the plant;
5. Management plans for hazard control;
6. Instructions for emergencies;
7. Safe job procedures.

Worker education cannot be static. It must be periodic through the medical and hygiene personnel, during job training, and periodically thereafter through health and safety meetings.

Special training courses are available at several universities, with departments specializing in occupational and environmental health and hygiene.

19.9.3.2 Environmental Monitoring Periodic sampling of the work environment detects the nature and extent of potential difficulties and also the effectiveness of the control measures being used. Monitoring for skin hazards can include wipe samples from the skin as well as the work sites and use of a black light for detecting the presence of tar product fluorescence on the skin before and after washing. Monitoring is particularly required when new compounds are introduced into plant processes. (See Chapters 8 and 9.)

19.9.3.3 Medical Controls Sound medical programs contribute greatly to preventing illness and injury among the plant employees. Large establishments have used in-house medical and hygiene personnel quite effectively. Daily surveillance of this type is not generally available to small plants; however, small plants do have access to well-trained occupational health and hygiene specialists through contractual agreements. At any rate, medical programs are designed to prevent occupational illness and injury, and this begins with a thorough placement physical examination, including the condition of the skin. (See Chapters 10 and 11.) When the preplacement examination detects the presence of or personal history of chronic eczema (atopic), psoriasis, hyperhidrosis, acne vulgaris, discoid lupus erythematosus, chronic fungal disease, dry skin, or other skin diseases, extreme care must be used in placement to avoid worsening a preexisting disease.

Plant medical personnel, full time or otherwise, should make periodic inspections of the plant operations to note the presence of skin disease, the use or misuse of protective gear, and hygiene breaches that predispose to skin injury.

When toxic agents are being handled, periodic biological monitoring of urine and blood for specific indicators or metabolites should be regularly performed.

Plant medical and industrial hygiene personnel should have constant surveillance over the introduction of new materials into the operations within the plant. Failure to do so can lead to the unwitting use of toxic agents capable of producing serious problems.

Of great importance in the medical control program is the maintenance of good medical records, indicating occupational and nonoccupational conditions affecting the skin, as well as other organ systems. Medical records are vital in compensation cases, particularly those of litigious character (Schwartz, 1957; Suskind, 1959; Samitz, 1985; Adams, 1983; Gellin, 1972; Birmingham, 1978; NIOSH, 1988).

A well-detailed coverage of prevention of occupational skin diseases is present in the publications, "Proposed National Strategies for the Prevention of Leading Work-Related Diseases and Injuries. Part 2" (107) and the "Report of the Advisory Committee on Cutaneous Hazards to the Assistant Secretary of Labor, OSHA, 1978." Current information regarding occupational and environmental skin disease may be further studied in the texts listed in the reference section or via specialized dermatology journals such as the *American Journal of Contact Dermatitis*, the *Journal of the American Academy of Dermatology*, and *Contact Dermatitis*.

Glossary

Carter, R. L. (1992). *A Dictionary of Dermatologic Terms.* Fourth Edition. Williams and Wilkins, Baltimore.

Appendages: In dermatology, they are applied to internal structures of the skin like the nail unit, the hair unit, and sweat glands.

Apocrine gland: A gland that secretes a fatty substance with the rest of the discharge secretory product.

Bulla: A blister of the skin.

Bullous: Relating to many blisters.

Comedones: Acne lesions such as whiteheads and blackheads.

Chromophore: A chemical capable of absorbing various wavelengths of light.

Cyst: A circumscribed nodule filled with fluid or solid matter usually located directly under the skin.

Cytokines: Chemicals capable of specific cellular action or recruitment.

Depigmentation: Complete loss of pigment or color.

Eccrine gland: A term used to designate a sweat gland.

Erythema: Relating to redness of skin.

Exfoliative dermatitis: A red, scaling skin rash involving the majority of skin from head to toe. Causes may relate to external exposure or internal disease.

Granuloma: An inflammatory response of specific cells intended to wall off infectious or inanimate foreign objects.

HLA proteins (human leukocyte antigens): Proteins on the surface of cells that identify them to the immune system as belonging to “self.”

Hyperpigmentation: Refers to excessive coloration of the skin.

Hypopigmentation: Decreased pigmentation of the skin.

Ichthyosis: A genetic disease of marked dryness and scaling of the skin.

Integument: A Latin term referring to a covering, which in medical use refers to the skin.

Keratinocyte: A principal cell making up the epidermis.

Langerhans cells: A dendritic cell found in the skin that is capable of processing antigenic material.

Necrosis: Death of a cell.

Papule: A solid, raised lesion above the skin.

Raynaud’s disease/syndrome/phenomenon: A disease of blanching of the digits upon exposure to ordinary tolerable cold. It may also be accompanied by pallor, a purple color, and pain in the affected areas.

Stratified squamous: Relating to the stacked nature of epidermal cells.

T-lymphocyte: A specific subtype of white blood cell that regulates immune responses.

Vesicle: A small blister.

Xerosis: Relating to dry skin.

REFERENCES

Adams, R. M. (1983). *Occupational Dermatology*, Grune & Stratton, New York.

American Academy of Dermatology (1994). *Proceedings of the National Conference on Environmental Hazards to the Skin*, Schaumburg, IL, pp. 61–79.

Asa, R. (1994). "Allergens Spur Hospitals to Offer Latex-Free Care," *Mater. Manag. Health Care*, 28–34.

Barber, T., and Husting, E. (1977). "Plant and Wood Hazards," in *Occupational Diseases—A Guide to Their Recognition*, rev. ed., M. M. Key et al., Eds., U.S. Department of Health, Education, and Welfare, PHS, CDC, NIOSH, DHEW-NIOSH Publications 181, U.S. Government Printing Office, Washington, DC.

Berenblum and Schoental, R. (1943). "Carcinogenic Compounds of Shale Oil," *Br. J. Exp. Pathol.* **24,** 232–39.

Bingham, L., Horton, A. V., and Tye, R. (1965). "The Carcinogenic Potential of Certain Oils," *Arch. Environ. Health* **10,** 449–51.

Birmingham, D. J. (1968). "Photosensitizing Drugs, Plants, and Chemicals," *Michigan* **67,** 39–43.

Birmingham, D. J. (1975). *The Prevention of Occupational Skin Disease*, Soap and Detergent Association, New York.

Birmingham, D. J. (1978). "Occupational Dermatoses," in *Patty's Industrial Hygiene and Toxicology*, 3rd ed., Vol. I, G. D. Clayton and F. E. Clayton, Eds., John Wiley, New York.

Birmingham, D. J. (1986). *Prolonged and Recurrent Occupational Dermatitis. Some Why and Wherefores, Occupational Medicine—State of Art Reviews*, Vol. 1, No. 2, Hanley & Belfus, Philadelphia, April–June.

Birmingham, D. J., Key, M. M., Holaday, D. A., and Perone, V. B. (1964). "An Outbreak of Arsenical Dermatoses in a Mining Community," *Arch. Dermatol.* **91,** 457–65.

Blank, I. H. (1979). "The Skin as an Organ of Protection against the External Environment," in *Dermatology in General Medicine*, 2nd ed., T. B. Fitzpatrick et al., Eds., McGraw–Hill, New York.

Braun-Falco, O., Plewig, G., Wolff, H. H., and Winkelman, R. K. (1991). *Dermatology.* Springer-Verlag, New York, p. 376.

Buchanan, V. D. (1962). *Toxicity of Arsenic Compounds*, Elsevier, Amsterdam.

Calnan, C. D. (1970). Studies in Contact Dermatitis XXIII Allergen Replacement, *Trans. St. John's Hosp. Dermatol. Soc.* **56,** 131–38.

Canadian Dept. of Forestry (1966). *Wood Preservation around the Home and Farm.* Forest Products Laboratory Publication No. 1117, Ottawa.

Centers for Disease Control, (1987). "Recommendations for Prevention of HIV Transmission in Health-Care Settings," *MMWR* **36**(2), 1S–18S.

Charous, B. L., Hamilton, R. G., and Yungunger, J. W. (1994). "Occupational Latex Exposure: Characteristics of Contact and Systemic Reactions in 47 Workers," *J. Allergy Clin. Immunol.* **94,** 12–18.

Cohen, D. E., Brancaccio, R., Andersen, D., and Belsito, D. V. (in press). "Utility of the Standard

Allergen Series Alone in the Evacuation of Allergic Contact Dermatitis: A Retrospective Study of 732 Patients," *J. Amer. Acad. Dermatol.*

Cohen, D., Davis, D., and Kozamkowski, R. (1974). "Clinical Manifestations of Chromic Acid Toxicity Nasal Lesions in Electroplate Workers," *Cutis* **13,** 558–68.

Combs (1954). *Coal Tar and Cutaneous Carcinogenesis in Industry*, Charles C. Thomas, Springfield, IL.

Cronin, E. (1980). *Contact Dermatitis*, Churchhill Livingston, London.

Cumin, N. (1827). "Remarks on the Medicinal Properties of Madar and on the Effects of Bichromate of Potassium on the Human Body," *Edinburgh Med. & Surg. J.*, **28,** 295–302.

De Galan, B. E., and Hoekstra, J. B. (1995). "Extremely Elevated Body Temperature: A Case Report and Review of Classical Heat Stroke," *Netherlands Journal of Medicine* **47**(6), 281–87.

de Groot, A. C., and Nater, J. P. (1994). *Unwanted Effects of Cosmetics and Drugs Used in Dermatology*. 3rd edition, Elsevier, Amsterdam.

De Leo, V., and Harber, L. C. (1995). "Contact Photodermatitis," Chapter 23, in *Contact Dermatitis*, Fisher, A. A., Ed., Williams and Wilkins, Baltimore.

Dukes-Dobos, F. N., and Badger, D. W. (1977). Physical Hazards—Atmospheric Variance, in *Occupational Disease—A Guide to Their Recognition*, rev. ed., U.S. Depart HEW, NIOSH, U.S. Government Printing Office, Washington, DC.

Eckhard, L. E. (1959). *Industrial Carcinogens*, Grune and Stratton, New York.

Elias, P. M. (1992). Role of Lipids in Barrier Function of the Skin, in Muktar H. (Ed.). *Pharmacology of the Skin.* CRC Press, Boca Raton, FL, pp. 389–416.

Emmett, E. (1975). "Occupational Skin Cancer—A Review," *J. Occup. Med.* **17,** 44–49.

Emmett, E., and Kaminski, J. R. (1977). "Allergic Contact Dermatitis from Acrylates Violet Cured Inks," *J. Occup. Med.* **19,** 113.

Emtestam, L., Zetterquist, H., and Olerup, O. (1993). HLA-DR, -DQ, and -DP alleles in nickel, chromium, and/or cobalt sensitive individuals: Genomic analysis based on restriction fragment length polymorphisms." *Journal of Invest Dermatol.* **100,** 271–74.

Epstein, J. (1971). "Adverse Cutaneous Reactions to the Sun," in *Year Book of Dermatology*, F. D. Malkinson and R. W. Pearson, Eds., Year Book Medical Publishers, Chicago.

Fisher, A. A. (1995). *Contact Dermatitis*, 4th ed., Williams and Wilkins, Baltimore.

Fitzpatrick, T. B. (1993). "Biology of the Melanin Pigmentary System," in *Dermatology in General Medicine*, T. B. Fitzpatrick et al., Eds., 4th Edition, McGraw–Hill, New York.

Fitzpatrick, T. B., Eisen, A. Z., Wolff, K., Freedberg, I. M., and Austen, K. F., Eds. (1993). *Dermatology in General Medicine*, 4th ed., McGraw–Hill, New York.

Foray, J. (1992). "Mountain Frostbite. Current Trends in Prognosis and Treatment (from Results Concerning 1261 Cases)," *International Journal of Sports Medicine* **13**(1), S193–6.

Fregert, S. (1974). *Manual of Contact Dermatitis*, Munksgaard, Copenhagen.

Gellin, G. A. (1972). *Occupational Dermatoses*, Department of Environmental, Public, and Occupational Health, American Medical Association, 1972.

Gellin, G. A., Possick, P. A., and Perone, V. B. "Depigmentation from 4-Tertiary Butyl Catechol-An Experimental Study," *J. Invest. Dermatol.*, **55,** 1970. pp. 190–197.

Gellin, G. A., Wolf, C. R., and Milby, T. H. (1971). "Poison Ivy, Poison Oak, and Sumac—Common Causes of Occupational Dermatitis," *Arch. Environ. Health.*

Harris, D. K., and Adams, W. G. (1967). "Acroostereolysis Occurring in Men Engaged in the Polymerization of Vinyl Chloride," *Br. Med. J.* **3,** 712–14.

"History of Technology" (1978), in *The New Encyclopedia Brittanica Macropedia*, 15th ed., Vol. 18, 1978.

Hood, D. (1977). "Practical and Theoretical Considerations in Evaluating Dermal Safety," in *Cutaneous Toxicity*, Drill, V., and Lazar, P., Eds., Academic Press, New York.

Hunt, L. W., Fransway, A. F., Reed, C. E., et al. (1995). "An Epidemic of Occupational Allergy to Latex Involving Health Care Workers." *J. Occup. Environ. Med.* **37,** 1204–9.

Jaeger, D., Kleinhaus, D., Czuppon, A. B., and Baur, X. (1992). "Latex-Specific Proteins Causing Immediate-Type Cutaneous, Nasal, Bronchial, and Systemic Reactions," *J. Allergy Clin. Immunol.* **89,** 759–68.

James, R. C., Busch, H., Tamburro, C. H., et al. (1993). Polychlorinated Biphenyl Exposure and Human Disease, *J. Occupational Med.* **35,** 136–48.

Kahn, G. (1979). "Depigmentation Caused by Phenolic Detergent Germicides," *Arch. Dermatol.* **102,** 177–87.

Key, K. M., Ritter, E. J., and Arndt, K. A. (1966). "Cutting and Grinding Fluids and Their Effects on Skin," *Am. Ind. Hyg. Assoc. J.* **27,** 423–27.

Klauder, J. V., and Gross, B. A. (1951). "Actual Causes of Certain Occupational Dermatitis—A Further Study with Special Reference to Effect of Alkali on the Skin, Effect on pH of Skin, Moderate Cutaneous Detergents," *Arch. Dermatol. Syphilol.*

Klauder, J. V., and Hardy, M. K. (1946). "Actual Causes of Certain Occupational Dermatitis—Further Study of 532 Cases with Special Reference to Dermatitis Caused by Petroleum Solvents," *Occup. Med.* **1,** 168–81.

Kligman, A. M. (1964). "The Biology of the Stratum Corneum," in *The Epidermis*, W. Montagna and W. C. Lobitz, Jr., Eds., Academic Press, New York, pp. 387–433.

Lampke, K. F., and Fagerstrom, R. (1968). *Plant Toxicity and Dermatitis* (A Manual for Physicians), Williams and Wilkins, Baltimore.

Lever, W. F., and Schaumburg-Lever, G. (1990). *Histopathology of the Skin.* Lippincott, New York, pp. 243–38.

Lewis, W., and Schwartz, L. (1956). "An Occupational Agent (N-Butyraldoxime) Causing Reaction to Alcohol," *Med. Ann.* **25,** 485–490.

Lim, H., and Soter, N., Eds. (1993). *Clinical Photomedicine*, Marcel Dekker, New York.

Maibach, H. I. (1974). "Patch Testing—An Objective Tool," *Cutis.* **13,** 4.

Maibach, H. I. (1987). *Occupational and Industrial Dermatology*, 2nd ed., Year Book Medical Publishers, Chicago.

Malten, K. E., and Beude, W. J. M. (1976). "2-Hydroxy-Alkyl Methacrylate and Di and Ethylene Glycol D-Methacrylate, Contact Photo-Sensitizers in Photo Polymer Plate Procedure," *Contact Dermatitis* **5,** 214.

Malten, K. E., Nater, J. P., and Von Ketel, W. G. (1976). *Patch Test Guidelines*, Dekker Van de Vegt, Nigmegan.

Malten, K., Seutter, E., and Hara, I. (1971). "Occupational Vitiligo due to *p*-Tertiary Butyl Phenol and Homologues," *Trans. St. John Hosp. Dermatol. Soc.* **57,** 115–34.

Maltwn, K. E., and Zielhius, R. L. (1964). "Industrial Toxicology and Dermatology," in *Production and Processing of Plastics*, Elsevier, New York.

Mathias, C. G. T. (1986). "Contact Dermatitis from Use and Misuse of Soaps, Detergents, and Cleansers in the Workplace," *Occupational Medicine—State of Art Reviews*, Vol. 1, Hanley and Belfus, Philadelphia, 205–18.

McNulty, W. P. (1985). "Toxic and Fetotoxicity of TCDD, TCDF and PCB Isomers in Rhesus Macaques," *Environ. Health Perspectives* **60,** 77–88.

Meso, E., Murray, W., Parr, W., and Conover, J. (1977). "Physical Hazards: Radiation," *Occupational Diseases—A Guide to Their Recognition*, rev. ed., U.S. Department of Health, Education and Welfare, NIOSH, U.S. Government Printing Office, Washington, DC.

Morris, G. E. (1952). "Why Doesn't the Worker's Skin Clear Up? An Analysis of Facts Complicating Industrial Dermatoses," *Arch. Ind. Hyg. Occup. Med.* **10,** 43–49.

Mukthar, H., and Bickers, D. R. (1981). "Comparative Activity of the Mixed Function Oxidases, Epoxide Hydratase, and Glutathione-S-transferase in Liver and Skin of the Neonatal Rat," *Drug Metab. Dispos.* **9,** 311–14.

National Electric Injury Surveillance System, (1984). U.S. Consumer Product Safety Commission, 1984; U.S. Bureau of Labor Statistics Supplementary Data Systems [SDS], 1983 data.

National Institute for Occupational Safety and Health (NIOSH) (1996). National Occupational Research Agenda. Cincinnati, U.S. Department of Health and Human Services, DHHS, Publication (NIOSH) 96-115.

NIOSH (1988). Prevention Planning, Implementation, Evaluation and Recommendations in Proposed National Strategy for the Prevention of Dermatological Conditions in Proposed National Strategies for the Prevention of Leading Work-Related Diseases and Injuries, Part 2, The Association of Schools of Public Health under a Cooperative Agreement with NIOSH.

Occupational and Environmental Dermatology Unit, Department of Dermatology, (1977). New York University Medical Center, New York, NY.

Odom, R. B., and Maibach, H. I. (1977). "Contact Urticaria: A Different Contact Dermatitis, in Dermatotoxicology and Pharmacology," in *Advances in Modern Toxicology*, Vol. 4, F. N. Marzulli and H. I. Maibach, Eds., Hemisphere Publishing Corp., Washington, D.C.

OSHA (1978). Report of Advisory Committee on Cutaneous Hazards to Assistant Secretary of Labor, U.S. Department of Labor.

Pinkus, H., and Mehregan, A. H. (1976). "Granulomatous Inflammation and Proliferation," Section IV, in *A Guide to Dermatopathology*, 2nd ed., Pinkus, H., and Mehregan, A. H., Eds., Appleton–Century–Crofts, New York.

Plewig, G., and Kligman, K. M. (1993). *Acne and Rosacea*, 2nd ed., Springer-Verlag, New York.

Pott, P. (1775). *Cancer Scroti, Chiruigical Works*, London, p. 734; 1790 ed., pp. 257–61.

Potts, R. O., and Guy, R. H. (1992). "Predicting Skin Permeability," *Pharmacol. Res.* **9,** 663–69.

Proctor, N. H., and Hughes, J. P. (1978). *Chemical Hazards in the Workplace*, J. B. Lippincott, Philadelphia.

Ramazzini, B. (1864). *Diseases of Workers* (translated from the Latin text, *De Morbis Artificum*, 1713, by W. C. Wright), Hafner, New York–London.

Rice, R. H., and Cohen, D. E. (1996). "Toxic Responses of the Skin," Casarett and Doull's Toxicology. *The Basic Science of Poisons.* 5th ed., Pergamon Press, Jan.

Rietschel, Robert L. and Joseph F. Fowler, Eds. (1995). *Contact Dermatitis*, 4th. ed., Williams and Wilkins, Baltimore.

Ronchese, F. (1948). *Occupational Marks and Other Physical Signs*, Grune and Stratton, New York.

Ronchese, F. (1965). "Occupational Nails," *Cutis* **5,** 164.

Rothman, A., and Emmett, E. A. (1988). "The Carcinogenic Potential of Selected Petroleum Derived Products," Chapter 7, in *Occupational Medicine—State of the Art Review*, Hanley and Belfus, Philadelphia, July–Sept., pp. 475–94.

Rycroft, R. J. G. (1986). "Occupational Dermatoses," Chapter 16 in *Textbook of Dermatology*, 4th ed., Vol. I, A. Rook, D. S. Wilkinson, F. J. G. Evling, and J. L. Burton, Eds., Blackwell Scientific Publications, Oxford.

Samitz, M. H., and Cohen, S. R. (1985). "Occupational Skin Diseases," in *Dermatology*, Vol. II, S. L. Moschella and H. J. Hurley, Eds., W. B. Saunders, Philadelphia.

Samitz, M. H., and Dana, A. S. *Cutaneous Lesions of the Lower Extremities*, J. B. Lippincott, Philadelphia.

Scheuplein, R. J., and Blank, I. H. (1971). "Permeability of the Skin," *Physiol. Rev.* **51,** 702–47.

Schmunes, E. (1986). "The Role of Atopy in Occupational Skin Disease," in *Occup. Medicine-State of Art Review*, Vol. I, Haney & Belfus, Philadelphia, p. 28.

Schwartz, L., Tulipan, L., and Birmingham, D. J. (1957). *Occupational Diseases of the Skin*, 3rd ed., Lea & Febiger, Philadelphia.

Stauber, J. L., Florence, T. M., Gulson, B. L., and Dale, L. S. (1994). "Percutaneous Absorption of Inorganic Lead Compounds," *Science of the Total Environment* **145**(2), 55–70.

Stewart, R. D., Hake, C. L., and Peterson, I. E. (1974). "Degreaser's Flush, Dermal Response to Trichloroethylene and Ethanol," *Arch. Environ. Health* **29,** 1–5.

Sulzberger, M. B., and Wise, F. (1931). "The Patch Test in Contact Dermatitis," *Arch. Dermatol. Syphilol.* **23,** 519.

Suskind, R. R. (1959). "Occupational Skin Problems. I. Mechanisms of Dermatologic Response. II. Methods of Evaluation for Cutaneous Hazards, III. Case Study and Diagnostic Appraisal," *J. Occup. Med.* **1.**

Suvorov, G. A., and Razumov, I. K. (1983). "Vibration," in *Encyclopedia of Occupation and Health*, 3rd ed., Vol. II, International Labor Office, Geneva.

Tabershaw, I. R., Utudjian, H. M. J., and Kawahara, B. L. (1977). "Chemical Hazards," Section VII in *Occupational Diseases—A Guide to Their Recognition*, rev. ed., U.S. Department of HEW-NIOSH Publication No. 77-181.

Tomazic, V. J., Shampaine, E. L., Lamanna, A., Withrow, T. J., Adkinson, N. F., Jr., and Hamilton, R. G. (1994). "Cornstarch Powder on Latex Products Is an Allergen Carrier," *J. Allergy Clin. Immunol.* **93,** 751–58.

Turjanmaa, K. "Incidence of Immediate Allergy to Latex Gloves in Hospital Personnel," *Contact Dermatitis*, **17,** 270–75.

Turk, K., and Rietschel, R. L. (1993). "Effect of Processing Cement to Concrete on Hexavalent Chromium Levels," *Contact Dermatitis* **28**(4), 209–11.

Urabe, H., and Asahi, M. (1985). "Past and Current Dermatological Status of Yusho Patients," *Environmental Health Perspectives* **59,** 11–15.

Wang, R. G., Knaack, J. B., and Maibach, H. I. (1993). *Health Risk Assessment and Dermal and Inhalation Exposure and Absorption of Toxicants.* CRC Press, Boca Raton.

Wheeland, R. G. (1994). *Cutaneous Surgery*, W. B. Saunders Co., Philadelphia.

White, R. P. (1934). *The Dermatoses or Occupational Affections of the Skin*, 4th ed., H. K. Lewis & Company, London.

Wilkinson, D. S. (1982). "Biological Causes of Occupational Dermatoses," in *Occup. and Industrial Dermatology*. Maibach, H. I., and Gellin, G. A., Eds., Year Book Medical Publishers, Chicago.

Williams, N. (1975). "Biological Effects of Segmental Vibration," *J. Occup. Med.* **17.**

Wilson, R. H., McCormick, W. I. G., Tatum, C. F., and Creech, J. L. (1967). "Occupational Acroosteolysis: Report of 31 Cases," *J. Am. Med. Assoc.* **201,** 577–81.

CHAPTER 20

Indoor Air Quality in Nonindustrial Occupational Environments

PHILIP R. MOREY

AQS Services, Inc., 2235 Baltimore Pike, Gettysburg, PA 17325

20.1 INTRODUCTION
- 20.1.1 Historical
- 20.1.2 Sick Building Syndrome
- 20.1.3 Building-Related Illness

20.2 INDOOR AIR POLLUTANTS
- 20.2.1 Microbials
- 20.2.2 Volatile Organic Compounds
- 20.2.3 Other Indoor Pollutants

20.3 PROBLEM BUILDING STUDIES
- 20.3.1 Approaches to Studying SBS Complaints in Buildings
- 20.3.2 Thermal Environmental Conditions
- 20.3.3 Building-Related Illness
- 20.3.4 Economic Cost of Poor IAQ

20.4 VENTILATION SYSTEMS IN COMMERCIAL BUILDINGS
- 20.4.1 Description of a Typical HVAC System
- 20.4.2 Humidification and Dehumidification
- 20.4.3 CO_2 Concentration
- 20.4.4 ASHRAE Standard 62-1989 and Its Revision
- 20.4.5 Importance of HVAC Systems in IAQ

20.5 IAQ EVALUATION PROTOCOL AND CHECKLISTS
- 20.5.1 Qualitative Evaluation
- 20.5.2 Checklist for the Qualitative Evaluation
- 20.5.3 Quantitative Evaluation

20.6 CONTROL MEASURES
- 20.6.1 Microbials
- 20.6.2 Volatile Organic Compounds
- 20.6.3 Other Indoor Pollutants
- 20.6.4 Thermal Environmental Conditions
- 20.6.5 Communications
- 20.6.6 Conclusions

REFERENCES

Adapted from "Indoor Air Quality in Nonindustrial Occupational Environments" by Morey and Jaswant Singh, Ph.D. and CIH, Chapter 17 in Clayton & Clayton, *Patty's Industrial Hygiene and Toxicology, 4th ed., Vol. IA.* New York: John Wiley and Sons, Inc., 1991, pp. 531–594.

Handbook of Occupational Safety and Health, Second Edition, Edited by Louis J. DiBerardinis, ISBN 0-471-16017-2

20.1 INTRODUCTION

20.1.1 Historical

Concern over the quality of air in indoor environments has historically included viewpoints that outdoor ventilation air is required both to prevent *adverse health effects* and to provide for *comfort* of occupants. Thus, over 2 centuries ago, Benjamin Franklin wrote that "... I am persuaded that no common air from without is so unwholesome as the air within a closed room that has been often breathed and not changed" (Morey and Woods, 1987). He further stated that outdoor air and "cool air does good to persons in the smallpox and other fevers. It is hoped that in another century or two we may find out that it is not bad even for people in health." Accordingly, ventilation codes by the late nineteenth century recommended the provision of large amounts of outdoor air to lower the risk of disease from certain infective agents such as that causing tuberculosis.

20.1.2 Sick Building Syndrome

Studies in the 1930s (Yaglou et al., 1936) showed that in order to prevent malodor complaints in the majority of buildings occupants, a minimum of 10 to 30 cubic feet per minute (cfm) of ventilation by outdoor air was required. An important objective of the current American Society of Heating, Refrigerating, and Air-Conditioning Engineers (ASHRAE) Standards 55 and 62 (ASHRAE, 1989, 1992) is to provide for comfortable environmental conditions for a majority (80 percent or more) of occupants in indoor environments. This means that some occupants (up to about 20%) may be uncomfortable even if these guidelines are met.

In the mid-1980s the term sick building syndrome (SBS) was first applied to describe occupant complaints associated with nonspecific symptoms that went away when people left the building. Typical symptoms reported by occupants included irritation of the mucous membranes and eyes, fatigue, headache, difficulty in concentrating or performing mental tasks, odor annoyance, and skin irritation. A finding of many studies is that SBS incidence is greater in air-conditioned than in naturally ventilated buildings (Sundell, 1994). Consistent associations between SBS and concentrations of airborne contaminants have not been made. The causes of SBS are thought to be multifactorial, including associations with low concentrations of many air contaminants, with thermal environmental factors, with building design variables, and with design, operation, and maintenance of heating, ventilation, and air-conditioning (HVAC) systems. In general, when higher-quality ventilation air is delivered to the occupant breathing zone, the incidence of SBS is reduced.

A number of factors in the 1990s collectively make indoor air quality (IAQ) and SBS complaints more important issues compared to 40 or 50 years ago. Construction practices today are vastly different from those employed half a century ago. Windows in modern buildings generally do not open, whereas in the 1940s natural ventilation (opened windows) was common. In modern buildings, reliance is placed on the HVAC system to transport outdoor air to the breathing zone. Insufficient outdoor air is often transported to the occupant breathing zone as a result of energy conservation measures that reduce HVAC system operational costs during the cooling and heating season. Poor maintenance and cleaning of today's complex building systems are other reasons cited for increased attention to IAQ issues.

Construction materials have markedly changed over the past 50 years. Stone, wood,

TABLE 20.1 Representative Kinds of Indoor Air Pollutants and Their Sources

Pollutant	Source(s)
4-Phenylcyclohexene	New carpet with styrene butadiene backing
2-Butoxyethanol	Glass cleaners
C_{15} to C_{18} aliphatic hydrocarbons	Hydraulic elevator fluid
Chlorpyrifos	Pesticide used to control fleas
Nicotine	Environmental tobacco smoke
Sulfur dioxide	Diesel fuel combustion
Manmade fibers	Some ceiling tiles
Fungi	Moist or damp finishing materials

and other "natural" construction/finishing materials have largely been replaced by synthetics. The ceiling tiles, wall and floor coverings, and even desks and chairs in modern buildings are primarily synthetic in nature. Volatile organic compounds (VOCs) are emitted into the indoor environment from modern finishing and construction materials. Volatile agents from cleaning and graphics materials, pesticides, hydraulic elevator fluid, and personal care products are additional sources of indoor air pollution. Table 20.1 provides a list of some representative kinds of air contaminants that have been associated with SBS symptoms.

20.1.3 Building-Related Illness

Building-related illness (BRI) is used to distinguish those instances where occupant health problems are clearly recognizable as disease upon medical examination and are associated with indoor environmental exposure. Perhaps the most well-known case of BRI was the outbreak of pneumonia (later known as Legionnaire's disease) that occurred in a Philadelphia hotel in 1976 as a result of indoor exposure of guests to a bacterium called *Legionella pneumophila*. This case of BRI is still remembered by the public because of the 29 fatalities that occurred during the disease outbreak. Examples of BRI and other air contaminants that are involved in disease etiology include:

- Cancer; caused by gaseous and particulate components of environmental tobacco smoke (ETS), some VOCs, and radon;
- Dermatitis; caused by fibers from manmade insulation and some irritant molds;
- Hypersensitivity pneumonitis, humidifier fever, allergic rhinitis, asthma, and non-allergic respiratory disease; caused by microorganisms or their toxins.

It is important to realize that the diagnosis of a BRI can be based on medical examination of one or more occupants confirming the presence of a specific disease plus an environmental assessment showing the likelihood of a contaminant exposure that would cause the disease. Thus the occurrence of antibodies to *Legionella pneumophila* following pneumonia plus the finding of the bacterium in a building water system is the basis for a diagnosis of building-related Legionnaire's disease. The diagnosis of SBS, on the other hand, is a group diagnosis based on review of the nonspecific discomfort symptoms (for example, headache, fatigue, eye irritation) of a group of building occupants.

A complaint rate of more than 20% during a formal epidemiologic survey is generally required to qualify as an SBS. Since SBS symptoms remit upon exiting the building, medical examination of individuals at the physician's office is usually of little use in making SBS diagnosis. The same environmental agents may cause both SBS and BRI. For example, exposure to VOCs, ETS, or combustion products may cause SBS. Chronic exposure to the same air contaminants may also elicit BRI.

20.2 INDOOR AIR POLLUTANTS

In the following discussion the pollutants that are important in IAQ evaluations are reviewed. The reader is also referred to the proceedings of international conferences on indoor air quality and climate (Seppänen, 1993; Yoshizawa, 1996), where numerous papers are presented on each type of pollutant covered in this section.

20.2.1 Microbials

Microbial contaminants in indoor environments include viruses, bacteria, fungi, protozoa, and cellular or toxic components such as endotoxins and mycotoxins. It should be understood that surfaces of construction and finishing materials, as well as the indoor air, are *not* sterile. Thus background concentrations of various kinds of microorganisms are normally present in indoor environments. Microbial contaminants and the diseases they cause (Table 20.2) are associated with moisture incursion or lack of maintenance of building water systems and this is emphasized in the discussion that follows.

20.2.1.1 Sources and Illnesses Most of the fungal or mold spores commonly found in indoor environments in nonproblem buildings originate from outdoor sources. The air outdoors is dominated by common fungal spores such as *Cladosporium* or *Alternaria* species, which grow on the leaves of plants; so called phylloplane (leaf-derived) spores. Other fungi such as *Penicillium* and *Aspergillus* species occur in topsoil. When air enters a building such as through the HVAC system, outdoor air inlet, through an open door, or by infiltration through cracks in the building envelope, phylloplane spores are carried into the building. In nonproblem buildings, the mix of fungal species found indoors is similar to that which occurs outdoors.

TABLE 20.2 Diseases Caused by Microbial Contaminants in Buildings

Disease	Symptoms	Agent and sources
Legionnaire's disease	pneumonia	hot water service system; cooling towers (HSE, 1994)
Humidifier fever	influenzalike	endotoxin in humidifier water (Rylander et al., 1978)
Hypersensitivity pneumonitis	acute fever and cough, fibrosis of lung	fungi and bacteria growing in the HVAC system (Arnow et al., 1978)
Asthma	constriction of the airways	fungi, bacteria, house dust mites (NAS, 1993)
Pulmonary hemosiderosis	bleeding in lungs	Stachybotrys and other fungi (MMWR, 1994; Sorenson et al., 1996)

In buildings with highly efficient HVAC system filters, the total concentration of airborne fungi indoors is generally much lower (exception, during winter snow cover when mold growth does not occur outdoors) than that found outdoors. Tracking of soil indoors often leads to indoor accumulation of *Penicillium* and *Aspergillus* species, especially in poorly maintained carpet.

While a diversity of fungi are normally present in the air in nonproblem buildings, the occurrence of only one or two dominating kinds of fungi indoors (same kind not present outdoors) is a sure indication of a moisture problem in the building infrastructure or the HVAC system. The kinds of fungi that grow in a building are determined by the amount of moisture available in the microscopic pores of construction and finishing materials. Damp materials subject to biodeterioration with a surface relative humidity (RH) in the 65 to 85% range can support the growth of xerophilic (dry-loving) fungi such as *Aspergillus versicolor* or *Wallemia sebi*. Porous materials whose capillaries are nearly water saturated (surface RH about 95%) can support the growth of hydrophilic (water-loving) fungi such as *Stachybotrys chartarum* and *Chaetomium* species. Liquid water, for example in a HVAC system drain pan, supports the growth of other hydrophilic fungi such as yeasts and *Fusarium* species.

There are many areas in buildings where moisture problems can lead to fungal growth. These include:

- The building envelope. In cold climates, the growth of mold may occur on the inside surface of poorly insulated envelope walls because moisture in room air condenses on the cold wall surface. In air-conditioned buildings in hot, humid climates, molds may grow on the inner surface of the envelope wall because moisture in infiltrating humid air condenses on cool (air-conditioned) surfaces.
- Porous materials in damp locations. Fungal growth may occur on damp HVAC system insulation especially in locations downstream of cooling coils when the RH is >90%. Insulation that is dirty (poor HVAC system filtration) is most susceptible to fungal growth because dirt is hydrophilic (attracts moisture). In like manner, carpet that is placed in building locations that are damp or chronically flooded is susceptible to fungal growth.
- Gypsum wall board. Signature flood molds such as *Stachybotrys* and *Chaetomium* grow on cellulose (paper) covered gypsum wall board that is chronically wet. Other molds such as *Aspergillus versicolor* grow on wallboard that is damp but not wet.

Fungi can cause various respiratory and nonrespiratory diseases. In homes there is a clear causal connection between dampness, fungal growth, and the occurrence of respiratory symptoms. Rhinitis and asthma have been associated with exposure to fungi such as species of *Penicillium*, *Aspergillus*, and *Cladosporium*. Exposure to toxigenic fungi (fungi that produce mycotoxins) such as *Aspergillus flavus*, *Aspergillus versicolor*, and *Stachybotrys chartarum* is considered dangerous in agricultural settings. Mycotoxins produced by toxigenic fungi can cause adverse cellular responses such as inhibition of the activities of pulmonary macrophages, inhibition of protein synthesis, and carcinogenesis. Exposure to *Stachybotrys chartarum* has been associated with mortality (pulmonary hemosiderosis) in infants. Hypersensitivity pneumonitis, a form of respiratory allergy in which affected people manifest symptoms such as malaise, fever, chills, shortness of breath, and cough can be caused by exposure to elevated concentrations of fungi as well as other microorganisms.

Lukewarm water (30 to 40°C) in cooling towers, evaporative condensers, and potable water service systems all provide niches for the growth of *Legionella*, which are gram-negative bacteria. *Legionella pneumophila* is the species that caused the outbreak of Legionnaire's disease in Philadelphia in 1976. Legionnaire's disease is an infection (the lung of a susceptible person is the target organ) that results in pneumonia, which may be fatal. *Legionella* may cause a milder form of illness known as Pontiac Fever, which is nonpneumonic and similar to a severe flu. Both Legionnaire's disease and Pontiac Fever result from exposure to water droplets containing *Legionella*. Sources of water droplets containing *Legionella* include drift (fog) from cooling towers, showers, splash from faucets, and aerosols from jacuzzis and spas.

Gram-negative bacteria such as species of *Pseudomonas*, *Flavobacterium*, and *Blastobacter* grow in stagnant water (e.g., humidifier sumps) or on wet surfaces of HVAC system cooling coils or drain pans. Humidifier fever, a disease characterized by influenzalike symptoms that remit within a day, has in some cases been associated with endotoxin (a toxic cell wall component characteristic of gram-negative bacteria) present in water droplets emitted from humidifiers in office buildings.

Endotoxins have been causally associated with disease (for example, byssinosis from cotton dust; organic dust toxicity syndrome in silos and barns) in industrial and agricultural settings. The endotoxin in cotton dust is derived from gram-negative bacteria that grew on the cotton plant prior to harvest. Gram-negative bacteria growing on moist silage and hay are sources of endotoxin in some agricultural settings. Metalworking fluids and the water in humidifiers and air washers are sources of endotoxin in industrial environments. In general, airborne endotoxin concentrations that are two or three orders of magnitude greater than background (outdoor) levels are considered significant exposures (ACGIH, 1989; Milton, 1995). Recent work in Danish buildings has shown that the concentration of endotoxin is settled dust was correlated with prevalence of SBS symptoms (Gyntelberg et al., 1994).

Other kinds of microorganisms may be found in buildings. These include gram-positive bacteria such as *Staphylococcus* and *Micrococcus* species present on human skin scales and *Streptococcus* species emitted as aerosols from the nasal/pharynx when a person is talking. Other than being an indicator of the occurrence of people in the indoor environment, adverse health effects have not been directly ascribed to these gram-positive bacteria in nonhospital settings.

Outbreaks of infective illness in the indoor environment may be caused by airborne exposure to specific human-shed microorganisms such as the influenza virus and the viruses and bacteria that cause the common cold. Provision of adequate amounts of outdoor air or highly filtered recirculated air to occupants in buildings is thought to reduce the risk of infection due to the common cold. The concept of provision of dilution ventilation to reduce the risk of "contagion" has been recognized in ventilation codes since the late nineteenth century.

20.2.1.2 Sampling and Interpretation The principles of sampling and analysis of microorganisms and microbial products (e.g., endotoxins) are reviewed elsewhere (AIHA, 1996; Flannigan, 1995). An enormous diversity of microbial contaminants may occur in buildings, including thousands of different kinds of culturable fungi and bacteria, nonculturable (nonviable) microorganisms, and microbial toxins (e.g., mycotoxins, endotoxins, B 1-3 glucans). Procedures used to collect microbial contaminants are also varied, including air, surface, and source (bulk) sampling methods (AIHA, 1996). Because moisture problems and fungal growth frequently occur in buildings, the following discussion on sampling is restricted primarily to the collection of culturable fungi. The reader is referred to the AIHA (1996) field guide for background information on the collection and analysis of other kinds of microbial contaminants.

A plan for data interpretation must be in place prior to the start of sample collection. The objective of the sampling plan as well as appropriate controls must be clearly defined. Thus surface or bulk samples should be collected from moisture problem areas as well as from background (reference control) areas not affected by moisture or dampness. Air sampling should include outdoor controls, generally obtained from high-quality air on the roof of the building. Regardless of the kind of sampling methodology used, enough replicates must be collected to allow for data interpretation in spite of the natural order-of-magnitude variation in fungal populations normally encountered in air or surface samples. At least three kinds of culture media (see AIHA, 1996, for formulation of culture media), as follows, should be considered for the collection and enumeration of the varied kinds of fungi that may occur as a result of moisture problems in buildings:

- DG18 or malt extract agar plus 20 or 40% sucrose for xerophilic fungi;
- Malt extract for hydrophilic fungi;
- Cellulose agar for *Stachybotrys* that may be present on wet cellulosic surfaces (e.g., wallboard).

Sampling data for culturable fungi may be interpreted as follows: (a) Building surfaces are not sterile. However, the occurrence of sustained visible fungal growth in a building is unacceptable. Visible mold should be removed in such a manner that spores are not dispersed into occupied or clean areas; (b) in nonproblem buildings the diversity of airborne fungi indoors and outdoors should be similar; (c) the dominating presence of one to two fungal species indoors and the absence of the same species outdoors indicates a moisture problem and degraded air quality in the building; and (d) the presence of *Stachybotrys chartarum*, *Aspergillus versicolor*, *Aspergillus fumigatus*, *Aspergillus flavus*, or *Fusarium moniliforme* indoors (over and beyond background concentrations) is atypical and requires risk management decisions to be made. The reader is referred to the AIHA (1996) field guide and ISIAQ (1996) for further guidance on interpretation of microbial sampling data.

20.2.2 Volatile Organic Compounds

The VOCs are commonly associated with indoor air pollution and SBS. VOCs are characterized by boiling points ranging from about 50 to 260°C and include alcohols, aldehydes, alkanes, aromatic hydrocarbons, halogenated hydrocarbons, terpenes, ketones, esters, and cycloalkanes. Those VOCs with boiling points less than 50 to 100°C are referred to as *very* volatile organic compounds (VVOC). Those with boiling points above 240–260°C are called *semivolatile* organic compounds (SVOC). SVOCs include pesticides, polynuclear aromatic compounds, and certain plasticizers such as phthalate esters (Lewis and Wallace, 1988).

Even in those geographic areas where outdoor air pollution from sources such as heavy vehicle traffic and petroleum refineries is significant, indoor total VOC (TVOC) concentrations have been shown to be 2 to 10 times higher than those outdoors (Hartwell et al., 1987). In new office buildings, the TVOC concentration at the time of initial occupancy is often 50 to 100 times that present in outdoor air. The variety of VOCs found in indoor air is almost always greater than that in outdoor air, largely because of the very large number of possible sources in indoor environments.

TVOC cannot be directly correlated with incidence of SBS. However, the presence of elevated TVOC (for example, greater than 3,000 micrograms [μg]/m^3) or the dominating presence of specific classes of VOCs indoors (for example, terpenes such as limonene; aromatic hydrocarbons, such as benzene; halocarbons such as methylene chloride) indicates strong contaminant sources in the building. Elevated TVOC or the dominating presence of a single class of VOC indoors is not characteristic of well-operated and maintained buildings (Seifert, 1990; Molhave and Clausen, 1996).

20.2.2.1 Sources and Health Effects Major VOC sources indoors include the building and its finishing products, activities performed by people including cleaning and photocopying, and occupants themselves.

Alkanes and aromatic hydrocarbons from building finishes are commonly present in indoor air in most buildings. Alkanes such as *n*-decane and *n*-undecane may be present indoors in new buildings (and during renovation of portions of older buildings) at concentrations of 100 to 1000 times that present outdoors. Toluene, xylene, and ethylbenzene are almost always found in indoor air because of the extensive use of these aromatic hydrocarbons in interior finishes (for example, linoleum and paints). 4-Phenylcyclohexene from styrene–butadiene carpet backing has been associated with allergic responses in some buildings.

Activities such as cleaning and photocopying may be associated with strong VOC emissions. Halogenated hydrocarbons (e.g., methylene chloride and trichloroethane) may be released into indoor air from solvents used to clean workstation panels and to strip paint from finishes. Liquid process photocopies and photocopied paper itself are strong sources of branched C_{10} and C_{11} alkanes (Shields et al., 1996). Alcohols and alkanes from cleaning solvents and soaps can become predominant VOCs in buildings where original interior finishes are no longer significant emission sources. In the example in Table 20.3, isopropanol and various alkanes present in cleaning agents accounted for almost 90% of the TVOC found in indoor air.

The occupants themselves are major sources of indoor VOCs. Benzene is present in ETS and higher concentrations of this aromatic hydrocarbon are expected in designated smoking areas. Limonene (a terpene) and various siloxane compounds (e.g., decamethylcyclopentasiloxane) may be present in many personal care products including antiperspirants and deodorants. Tetrachloroethylene is emitted from clothing that has

TABLE 20.3 Concentrations of Alcohols and Alkanes in a 5-Year-Old Building with and without Cleaning Activities[a]

	Concentration (μg/m^3)	
Cleaning status	Alcohols	Alkanes
No cleaning[a]	3	250
Routine cleaning	1,030[b]	1,300[c]

[a]Zero cleaning was performed during a six-month period.
[b]Mostly isopropanol.
[c]Mostly C_9 to C_{11} alkanes.

been recently dry cleaned. Hand and body lotions, moisturizing soaps, and cosmetics can be strong sources of C_{12} to C_{16} alkanes (Shields et al., 1996).

Among the VOCs, formaldehyde is most closely associated by the public with indoor air pollution. Although formaldehyde may be emitted from a number of sources such as from gas stoves and from smoking, its major source indoors is from construction materials such as particle board, fiberboard, and plywood. Concentrations of formaldehyde in residential buildings are higher than in office buildings because of the relatively large ratio of pressed wood products to air volume in the former as compared to the latter type of buildings (Girman, 1989).

Adverse health responses potentially caused by VOCs in nonindustrial indoor environments fall into three categories, namely, (a) irritant effects including the perception of unpleasant odors and mucous membrane irritation, (b) systemic effects such as fatigue and difficulty concentrating, and (c) toxic effects such as carcinogenicity (Girman, 1989).

Many of the VOCs emitted from new furnishings and buildings materials are mucous membrane irritants. Molhave et al. (1986) exposed 62 people with a history of difficulty with poor air quality in a chamber setting to a mixture of 22 irritant VOCs (TVOC 5 or 25 mg/m^3). Subjective tests showed that the perception of poor air quality and odor intensity increased with increase in the total concentration of VOCs. Similarly, the perception of mucous membrane irritation was elevated at both 5 and 25 mg/m^3 concentrations.

It has been hypothesized that IAQ complaints of eye, nose, and throat irritation so common in SBS may be due to exposure to VOC mixtures that are often found in new or renovated buildings (Kjaergaard et al., 1987). Occupant complaints are almost always encountered when the TVOC is about 3 mg/m^3 or higher (Molhave and Clausen, 1996). At TVOC levels in the range of 0.2 to 3.0 mg/m^3 occupant discomfort and irritation complaints are manifested if other kinds of exposure occur simultaneously (Molhave and Clausen, 1996). At TVOC levels below 0.2 mg/m^3 discomfort and irritation complaints due to VOCs should be minimal.

The dose–response relationship of TVOC to adverse health responses of occupants is negated when highly reactive VOCs (e.g., aldehydes or amines) or highly odorous VOCs are present in indoor air. In other words, even low concentrations of these VOCs can elicit complaints. Highly odorous aldehydes, esters, and alkoxy alcohols are known to degrade the human perception (in sniffing experiments) of air quality even at minute concentrations. Thus the quality of the air perceived by building occupants may be governed not by TVOC but by specific VOCs having high odor indices (Wolkoff et al. 1996).

An additional complication to the association of VOCs and adverse health effects occurs in damp buildings, where growing fungi and bacteria produce microbial VOCs (MVOCs). MVOCs may include compounds such as ethanol and acetone, which are commonly found in nonmicrobial sources in most buildings as well as "signature" volatiles such as geosmin, 3-methylfuran, and 1-octen-3-ol rarely encountered in dry buildings (Wessen et al., 1995). Although some MVOCs are characteristically odorous, their potential adverse health effects are at present unknown.

20.2.2.2 Sampling and Interpretation Existing methods of sampling used for VOCs in industrial workplaces are not readily adaptable to nonindustrial indoor studies. Methods developed for industrial workplaces are often bulky, noisy, and validated for concentrations about an order of magnitude below the applicable threshold limit value

(TLV). Concentrations of VOCs found in nonindustrial indoor air are usually two or more orders of magnitude lower than industrial TLVs (Lewis and Wallace, 1988). More sensitive sampling and analytical methods are therefore required for characterization of VOCs in indoor air.

Direct-reading photoionization, flame ionization, and infrared detectors may be used for qualitative sampling of VOCs. The utility of these instruments is restricted to walk-through field survey work where emphasis is placed on instantaneous detection of strong VOC sources and where analytical sensitivity of measurement to the microgram per cubic meter ($\mu g/m^3$) level for specific VOCs is not required.

Most IAQ sampling evaluations for VOCs utilize in-field trapping of pollutants in tubes containing various sorbents and subsequent laboratory analysis of specific VOCs. A variety of sorbents (often two or more sorbents per tube), including Tenax, graphatized carbon black, or charcoal, are often used to trap VOCs. After desorption in the laboratory, gas chromatography and mass spectrometry (GC/MS) are used to provide the sensitivity (generally 1 $\mu g/m^3$ for specific VOCs) required for IAQ studies. The use of passive sampling devices based on the principle of molecular diffusion onto charcoal combined with GC/MS provides another method of measuring VOCs in indoor air at levels of 1 $\mu g/m^3$ or less (Shields et al., 1996).

Interpreting VOC sampling results is difficult because of the absence of dose–response data, the complexity and changing nature of VOC mixtures, and the variation caused by changes in ventilation and occupant activities indoors. In general, sampling for VOCs should be performed only when a hypothesis for data interpretation is available. Thus sampling is appropriate when activities in one zone (for example, interior renovation, paint spraying, or dry cleaning) may be degrading air quality elsewhere in the building. Sampling should be performed in close proximity to the suspected source of the VOCs, in zones where air quality may be degraded by ingress of VOCs, and in control locations such as in the outdoor air on the roof in an area distant from building vents and exhausts. Analytical methods used in the laboratory (usually GC/MS) must be able to distinguish individual compounds representative of each VOC class (Shields et al., 1996; Seifert, 1990). Indoor/outdoor TVOC ratios or VOC class ratios consistently exceeding 10 or 20 suggest the presence of strong indoor sources of VOCs, which are not expected in well-operated and maintained buildings.

To determine whether or not a mixture of VOCs is typical or atypical, it is useful to establish both the TVOC as well as the rank-order concentration of specific VOCs both indoors and outdoors. For example, sampling performed shortly after the opening of Building 1 (Table 20.4) showed that the TVOC indoors was almost 30 times higher than that in the outdoor air. The presence of elevated concentrations of four specific VOCs indoors (Table 20.4) suggested the presence of strong emission sources such as from cleaning chemicals and from unreacted solvents present in new construction materials.

The equivalent TVOC indoors and outdoors for Building 2 (Table 20.4) might at first suggest an absence of unusual VOC sources. However, examination of the concentration data for three specific VOCs indicates that methylene chloride from the outdoor air is being entrained in the building. In addition, there is indication of indoor emission sources for alkanes and aromatic hydrocarbons. Inspection of Building 2 showed that methylene chloride in the outdoor air was originating from a roofing operation near the HVAC air intake where sampling occurred. The presence of alkanes and aromatic hydrocarbons indoors was due to use of solvents on a laboratory bench without proper local exhaust ventilation. Sampling in Building 2 thus showed that quantification of

TABLE 20.4 Volatile Organic Compounds in Two Office Buildings[a]

Compound	VOC concentrations in Outdoor air ($\mu g/m^3$)	VOC concentrations in Indoor air ($\mu g/m^3$)
	Building 1	
Total VOCs	62	2200
Tetrachloroethylene	6	700
Cyclohexanes	1	140
Dodecane	4	200
Chloroform	1	40
	Building 2	
Total VOCs	542	448
Methylene chloride	400	40
Decane	6	200
Ethylbenzene	2	80

[a]Sample collection on Tenax sorbent; analysis by gas chromatography/mass spectrometry.

concentrations of specific VOCs is required to reach a correct interpretation. The determination of TVOC alone was inadequate for data interpretation.

Variables such as building age and outdoor air ventilation must also be considered when interpreting VOC sampling results. In new office buildings or in portions of older buildings undergoing renovation, the indoor/outdoor concentration ratio of TVOC may be 50 or 100 to 1 (Sheldon et al., 1988a,b). With adequate outdoor air ventilation, these ratios fall to less than 5 to 1 after 4 or 5 months of aging. In older buildings, an indoor/outdoor concentration ratio of TVOC may vary from nearly 1 when maximum amounts of outdoor air are being used in HVAC systems to greater than 10 during winter and summer months when minimum amounts of outdoor air are being used (Morey and Jenkins, 1989).

20.2.3 Other Indoor Pollutants

20.2.3.1 Pesticides Pesticides including insecticides, termiticides, and fungicides are often used in interior spaces to control a wide variety or organisms including wood-boring insects, moths, and fungi. Although pesticides are by definition poisons, their toxicity varies toward different types of organisms. Several million pounds of naphthalene and paradichlorobenzene are used annually in U.S. homes as moth repellants (Reinert, 1984). Similar quantities of pentachlorophenol are used annually in U.S. homes to protect wood and paints from degradation by insects and fungi. Pesticides such as chlordane, heptachlor, aldrin, dieldrin, and chlorpyrifos have been used or are still used in the soil or under the foundations of buildings (Dingle and Tapsell, 1996).

In 1985, the U.S. Environmental Protection Agency (EPA) initiated studies on pesticide exposures in the general population. The Non-Occupational Pesticide Exposure Study (NOPES) has facilitated the development of improved methodologies for sampling for pesticides in indoor environments (Lewis and Bond, 1987; EPA, 1987). Using techniques described in these studies, it is now possible to sample and analyze for over 50 specific organochlorine, organophosphate, organonitrogen, and pyrethroid pesticides using polyurethane foam sorbent with sensitivities as low as 0.01 $\mu g/m^3$. For example, among 50 residences monitored in the Jacksonville, Florida, area, 46 contained detectable levels of the pesticide chlorpyrifos (mean concentration 0.47 $\mu g/m^3$). Chlorpyrifos was detectable in outdoor air around only 9 of the 50 residences, with a mean concentration of 0.059 $\mu g/m^3$. Concentrations of chlorpyrifos as high as 37 $\mu g/m^3$ have been found in other residences (Lewis and Wallace, 1988). After application, pesticides may be present not only in indoor air, but also in settled dusts (which may become airborne) and as secondary contaminants in interior finishes such as flooring and door frames (Krooss and Stolz, 1996). Thus, following application of the pesticide lindane to ethnological objects in museum store rooms, it could be found in the air (0.2 to 3.7 $\mu g/m^3$), in dust (up to 128 mg/kg), and on room finishes (4 to 32 mg/kg) (Krooss and Stolz, 1996). The pyrethroid insecticides permethrin and deltamethrin used for cockroach control has been shown to be very persistent (for periods longer than 70 weeks) in settled dust (Berger-Preiss et al., 1996) on interior surfaces.

A combination of air and dust (surface) sampling is probably best for characterization of pesticide residues that may be present in indoor environments. If sampling is performed, the analytical methods used must be sufficiently sensitive (for example, 0.01 $\mu g/m^3$ for air samples) so as to detect background concentrations of specific pesticides that may be present in the outdoor air. The objective of sampling is often to determine if pesticide concentrations in areas of suspect contamination exceed background outdoor levels and average concentrations that have been previously reported in the literature (Lewis and Bond, 1987).

Health effect data available for pesticides are based primarily on oral or dermal exposures from animal studies. Little information is available on adverse health effects on inhalation exposures. It is significant, however, that most pesticide-related injuries (including those from organophosphate insecticides) occur in the home (Reinert, 1984). Of the accidents that were not related to ingestion, inhalation and dermal exposures were found to be equally important.

Because of the potential chronic health effects from exposure to termiticides, the National Research Council has recommended that airborne concentrations in indoor air be limited as follows: chlorpyrifos (10 $\mu g/m^3$), chlordane (5 $\mu g/m^3$), heptachlor (2 $\mu g/m^3$), and dieldrin and aldrin (1 $\mu g/m^3$) (NRC, 1982). It should be noted that these exposure levels are more than an order of magnitude lower than concentrations considered acceptable in industrial workplaces.

20.2.3.2 Combustion products When combustion occurs in air, carbon dioxide (CO_2) and water vapor are emission products. However, carbon monoxide (CO), nitrogen dioxide (NO_2), sulfur dioxide (SO_2), and respirable particles are the byproducts of combustion that can most often cause indoor air pollution. The type and amount of combustion byproducts in indoor air of commercial buildings depend upon the type of fuel consumed (for example, diesel oil used by trucks at a building loading dock contains significant amounts of sulfur, and SO_2 is an important combustion byproduct) and the location of outdoor air inlets and emission sources. Combustion processes that are oxygen-starved and characterized by yellow-colored flames are characterized by elevated CO emissions. Combustion under oxygen-rich conditions results in higher flame temperatures, thus emitting greater amounts of oxides of nitrogen. Considerable literature is available describing characteristics of combustion devices found in indoor environments (primarily residences) and factors affecting the emission rates of various combustion byproducts (DOE, 1985; Woodring et al., 1985).

Indoor air pollutants such as CO and NO_2 originate from multiple sources, and concentrations indoors depend on parameters such as source emission rates, the volume of air indoors, outdoor air ventilation rates, and HVAC system characteristics. Pollutants such as CO and NO_2 are emitted intermittently and are usually concentrated only in certain areas of the building. Outdoor conditions, such as ambient concentrations of NO_2, have a strong influence on indoor pollutant concentrations (Parkhurst et al., 1988).

CO combines with hemoglobin to form carboxyhemoglobin, thereby resulting in a decline in the oxygen-carrying capacity of the blood. Carboxyhemoglobin levels higher than 4 to 5 percent are known to exacerbate symptoms of individuals with preexisting cardiovascular disease. Limiting average CO exposures to 9 ppm (maximum) for 8 h or to 35 ppm to 1 h, as specified in the National Ambient Air Quality Standards, is intended to provide a margin of safety with regard to carboxyhemoglobin buildup in individuals with cardiovascular disease.

As a general rule the emission or transport of combustion byproducts such as CO in or into a building is unacceptable. Complaints of exhaust odors coupled with occupants' symptoms such as headache, nausea, fatigue, dizziness, rapid breathing, and confusion are indicators of the presence of combustion products such as CO. The source of combustion byproducts should be found and eliminated or exhausted out of the building. The measurement of relatively low CO concentrations in the range of 3 to 5 ppm (assume that ambient levels are lower) still indicates the occurrence of an indoor source that should be eliminated.

Establishing possible pathways for entry of combustion products into buildings is relatively simple. Thus pathways for entry of combustion products are often obvious in buildings with attached garages and loading docks or in buildings near heavily traveled roads.

HVAC outdoor air intakes near the dock or garage are obvious transport conduits for combustion products. Proving that combustion products can enter a building and cause occupant complaints, however, can be made difficult because of the intermittent nature of pollutant generation and HVAC and building operational variables. In one large building where a portion of an office floor had been vacated because of suspected exposure to combustion products, round-the-clock sampling for NO_2 for 6 days was necessary before it could be demonstrated that combustion products from a loading dock were being entrained in the HVAC system outdoor air inlet serving the affected office (Morey and Jenkins, 1989). The concentrations of NO_2 at the outdoor air inlet, in the vacated office, and in the outdoor air on the roof far removed from emission sources were 2.0, 0.7, and 0.08 ppm, respectively, only at a time when a garbage truck was unloading a dumpster at the loading dock near the HVAC inlet.

In another building with a tuck-under loading dock, combustion products only entered the buildings interior when the lower floors were under negative pressure relative to ambient air (stack effect) during the winter. Concentrations of CO in corridors and elevator lobby areas (15 to 50 ppm) connecting with the dock were equivalent to those in the tuck-under dock (25 to 50 ppm). While tracer gas techniques can be used to demonstrate pathways of entry of combustion products into a building, smoke pencils (air current tubes), if used with care, can provide useful information on points of entry during an initial walk-through evaluation.

The National Ambient Air Quality Standards (see Table 20.5) have been used in ASHRAE Standard 62-1989 for determining if concentrations of certain combustion products and other contaminants present in outdoor air must be reduced by cleaning prior to introduction of air into HVAC systems. Thus air cleaning is needed if the NO_2

TABLE 20.5 Air Quality Standards for Certain Outdoor Contaminants

Contaminant	Concentration[a] ($\mu g/m^3$) [ppm]
Sulfur dioxide	80 [0.030]
Nitrogen dioxide	100 [0.055]
Particles (PM10)	50

[a]Yearly averaging.

concentration outdoors exceeds 0.055 ppm and the SO_2 concentration exceeds 0.030 ppm. While these are not intended to indicate "safe" levels indoors, it is often argued that the quality of indoor air should at least be as good or acceptable as that outdoors. Thus the occurrence of NO_2 levels of 0.60 ppm in an indoor ice skating rink where ice resurfacing machines with gasoline combustion engines periodically operate (Jantunen, 1993) would be considered unsatisfactory for acceptable indoor air quality.

20.2.3.3 Environmental Tobacco Smoke Environmental tobacco smoke (ETS) is composed of a complex mixture of chemicals, including combustion gases and respirable particles. Because tobacco leaf does not burn completely, ETS contains more than 4700 chemical compounds (EPA, 1989a), including nicotine, tars (containing a variety of polynuclear aromatic hydrocarbons), vinyl chloride, formaldehyde, benzene, styrene, ammonia, NO_2, SO_2, CO, hydrogen cyanide, and arsenic. Many carcinogenic compounds are found in ETS.

ETS is derived from mainstream smoke that is drawn through the cigarette by the smoker and from sidestream smoke that is emitted from the cigarette itself directly into room air between puffs. Individual chemical components of ETS may be found in both the gaseous and the particulate phases. During smoking, particulate ETS may adsorb on surfaces in the occupied space and ventilation system. Gaseous components can then be reemitted into the indoor air from adsorbed particulate. Smoking characteristically results in a significant rise in the concentration of respirable particulate with a mass median diameter of about 0.2 to 0.4 μm (Miesner et al., 1988). Although many chemical components of ETS can also be attributed to other sources, nicotine is considered a specific marker for this indoor pollutant.

Adverse effects of ETS can be divided into annoyance and discomfort complaints and chronic disease including increased frequency of respiratory illness, loss of lung function, and increased risk of lung cancer. Eye, nose, and throat irritation are frequent complaints in people exposed to ETS. Odor annoyance due to ETS is perceived at concentrations much lower than those that cause irritation (Clausen, 1988); consequently much greater amounts of outdoor air are required to control odor annoyance as opposed to the acute irritation effects of ETS (Cain et al., 1983). Both odor annoyance and irritational effects are thought to be caused by the gaseous phase of ETS. In studies in a series of Danish buildings, authors estimated that about 25 percent of the perception of stale, unacceptable air is due to smoking (Fanger et al., 1988).

The major chronic health effect from ETS is lung cancer. In 1985, major study panels were convened by the U.S. Public Health Service, the National Research Council, and the Federal Interagency Task Force on Environmental Cancer, Heart and Lung Disease, to consider the risk associated with breathing ETS in indoor air (passive smoking). All groups arrived at the same conclusion that passive smoking significantly increased the risk of lung cancer in adults (EPA, 1989a; USPHS, 1987).

The presence of ETS in indoor air is associated with increased concentrations of respirable particulate, nicotine, and a variety of other contaminants. One smoker consuming a pack of cigarettes daily in a residence contributes approximately 20 $\mu g/m^3$ of respirable particulate to the (24-h) particle concentration (Samet et al., 1987). Concentrations of respirable particulates less than 2.5 μm in aerodynamic diameter in indoor environments where smoking is permitted may rise up to 500 $\mu g/m^3$ (Miesner et al., 1988).

Nicotine, which is the most specific marker for ETS, can be measured at concentrations less than 1 $\mu g/m^3$ with personal or area monitors (Phillips et al., 1996). Miesner et al. (1988) measured nicotine concentrations in a number of office environments with varying smoking policies and found an increase in nicotine levels as respirable particulate concentrations increased. Nicotine and respirable particulate levels in a designated smoking room in one office were 26.5 and 520.8 $\mu g/m^3$. In a nonsmoking office located directly above a floor where smoking was permitted, the concentration of nicotine was 2.0 $\mu g/m^3$. Nicotine concentrations in nightclubs, taverns, and bars range from about 10 to 100 $\mu g/m^3$ (Collett et al., 1992).

ASHRAE Standard 62-1989 recommends that a minimum of 20 cfm of outdoor air per occupant is needed to provide acceptable indoor air quality in indoor environments with a moderate amount of smoking. In smoking lounges, a minimum ventilation rate of 60 cfm/occupant (including transfer air) is required for acceptable IAQ.

Because there is no acceptable level of ETS relative to its carcinogenicity, a proposed revision to ASHRAE Standard 62-1989 makes the fundamental change from the current Standard by stating that acceptable IAQ can only be achieved in environments that are smoke free. For indoor environments with smoking allowed, acceptable "perceived" air quality is achieved by following procedures, where adequate volumes of dilution air (based on the number of cigarettes smoked) are provided for adapted or nonadapted (nonsmoking visitors) persons.

In commercial buildings in the United States, ETS is becoming less of an issue because of the increasing number of smoke-free facilities. For those buildings where smoking lounges are provided, ETS-IAQ problems can be minimized by complete exhaustion of smoke outdoors. The measurement of airborne concentrations of nicotine provides a sensitive analytical measure of ETS in cases where smoke may not be completely exhausted from the building. Because ETS is carcinogenic and because nicotine is a specific marker for ETS, the presence of airborne nicotine in a building is an indication of degraded IAQ.

20.2.3.4 Particles A great variety of particles, most solid (for example, dusts, smoke, and microorganisms) and a few liquid (for example, mist from humidifiers) are present in indoor environments. Particles present in indoor air may originate in the outdoor air, in the HVAC system, or in occupied spaces. Particles in the outdoor air, for example, from heavy vehicular traffic, may enter HVAC air intakes, and, especially if filtration is poor, be transported to the breathing zone in ventilation supply air. Particulates from soil are tracked into buildings and can accumulate in porous furnishing, especially carpet. The HVAC system becomes a major source of particles when fibers are eroded from damaged or improperly installed porous insulation or when fungal spores growing on damp surfaces are released.

A major source of particles in most buildings is occupants (skin scales and fiber from

TABLE 20.6 Quick Reference Guide for Major Categories of Indoor Air Pollutants and Appropriate Sampling Methodology

Pollutant	Guideline considerations	Source(s)
Molds	The kinds of molds found indoors should be similar to those found outdoors (ISIAQ, 1996)	AIHA (1996)
VOCs	Indoor/outdoor TVOC ratios >10 indicate strong indoor sources; a single class of VOC should not dominante indoor air (Seifert, 1990)	EC (1995)
Pesticides	Following post-treatment ventilation, pesticides should not be detectable in air above background outdoor levels	Polyurethane foam (EPA, 1987)
Combustion products	Avoid entrainment from external sources; remove internal sources by local exhaust ventilation	EPA (1987)
Particulate	PM10 seldom exceeds 50 $\mu g/m^3$; avoid dust raising activities	Direct reading instruments
Nicotine	Presence indicates degraded IAQ	ASTM (1995)

clothing) and their activities (settled dust aerosolized by foot traffic, particulates from copiers). Dusts from interior renovations, if not properly contained, may be dispersed throughout the entire building. Activities such as vacuuming carpets and upholstered surfaces may disperse settled dust into the indoor air. Maintenance activities such as moving ceiling tiles or replacing dirty filters may inadvertently lead to dispersal of particles if not carefully performed.

In general, airborne particles of a size from about 0.1 to 10 μm are of concern for human health. Particles less than 0.1 μm are exhaled, and those greater than 10 μm do not enter the lower regions of the lung. ASHRAE Standard 62-1989 provides guidance on conditioning of outdoor air for use in HVAC systems in terms of acceptable maximum concentrations of airborne particles. When the yearly average concentration of particles less than 10 μm (PM_{10}) exceeds 50 $\mu g/m^3$, filtration is required.

As a general rule, particle concentrations in indoor air in nonproblem buildings seldom exceed 25 $\mu g/m^3$. SBS complaints can occur when smoking is allowed or when dusts from interior renovation activities are dispersed, and often the indoor particulate concentrations in these cases exceed 50 or even 100 $\mu g/m^3$. Studies in some buildings have associated the manmade mineral fibers (Hedge et al., 1993) and macromolecular organic or microbial components (Gravesen et al., 1993) in dusts with SBS complaints.

Table 20.6 provides a quick reference quide on data interpretation for major categories of air pollutants as well as references useful for sampling and analysis methodology.

20.3 PROBLEM BUILDING STUDIES

The different approaches that can be used during evaluations reflect the varied kinds of problems that can occur in buildings. Complaints may be related epidemiologically to ventilation or to defects in building performance. Thermal environmental parameters can interact strongly with indoor air quality problems in buildings. In some buildings where environmental conditions are greatly deteriorated, BRI occurs in addition to SBS. Finally, problem buildings can be viewed in terms of the economic losses associated with lost productivity of occupants.

20.3.1 Approaches to Studying SBS Complaints in Buildings

SBS is characterized by a number of nonspecifics symptoms including mucous membrane irritation, eye irritation, headache, odor annoyance, sinus congestion, and fatigue. Ideally, in order to understand the etiology of SBS, the prevalence of symptoms in occupants should be studied, preferably in problem and nonproblem buildings. Objectives of such case-control studies should include a determination if the percentage of dissatisfied occupants exceeds some unacceptable level, and if dissatisfaction can be related to building, air contaminant, or work-practice variables.

20.3.1.1 NIOSH Studies By 1990, the U.S. National Institute for Occupational Safety and Health (NIOSH) had investigated over 500 problem buildings (Crandall and Sieber, 1996). In each building NIOSH determined the single, most important factor likely relating to occupant complaints. Factors associated with complaints in descending order of frequency were: inadequate ventilation (53%), indoor pollutant sources (15%), entrainment of outdoor contaminants (10%), microbial problems (5%), building fabric contamination (4%), and unknown causation (13%).

In 1993, NIOSH investigated 104 buildings with complaints and identified major problems of the following type: (a) 63% had facility defects such as water damage, water intrusion, or poor housekeeping; (b) 60% had defective HVAC operation such as insufficient outdoor air ventilation, poor air distribution to the breathing zone, and flooded drain pans; (c) 58% had defective HVAC maintenance such as systems that were dirty and in disrepair or suffered from an absence of written operation and maintenance plans; (d) 51% had defective HVAC design problems such as outdoor air inlets that needed to be moved, inefficient or poorly fitting filters, or absence of minimum stops on VAV terminals; (e) 30% had thermal environmental problems; (f) 24% had indoor contaminant sources such as VOCs, cooking odors, renovation dusts, and chemicals in mechanical equipment rooms; (g) 22% had combustion gas and restroom entrainment problems; and (h) 12% had ergonomic (workstation design) or physical agent (lighting/glare) problems (Crandall and Sieber, 1996). While NIOSH studies do not provide information on the cause of SBS complaints, they do provide practical information on the frequency of building defects that give rise to IAQ problems.

20.3.1.2 British Epidemiologic Studies Several systematic studies on complaints related to building characteristics have been carried out in British office buildings (Burge et al., 1987; Harrison et al., 1987). In the largest study, complainant symptoms were recorded using a common protocol in 47 different groups of occupants in 42 buildings. The questionnaire used in these studies elicited information on whether the following 10 symptoms occurred during the past year and whether symptoms disappeared during periods when occupants were away from the building: dryness of eyes, itching of eyes, stuffy noise, runny noise, dry throat, lethargy, headache, fever, breathing difficulty, and chest tightness.

Ventilation systems serving the 47 different occupant groups were categorized as *natural* (open windows, no forced air), *mechanical* (forced air system without cooling or humidification), *local induction units*, *central induction/fan coil unit*, and *variable-* or *constant-volume system*. The latter three ventilation categories were characterized by cooling of air, and in some cases, also by humidification. Questionnaire results showed that the lowest prevalence of work-related symptoms was found in the naturally or mechanically ventilated categories. Although there were considerable variations between buildings of each ventilation type, the highest symptom prevalence rates were found in ventilation systems with induction or fan coil units. A somewhat intermediate symptom prevalence occurred in variable/constant-air systems.

These British studies show that naturally ventilated and non-air-conditioned, mechanically ventilated buildings are the healthiest workplaces. Because the other three ventilation types examined are characterized by air-conditioning, and, in some cases, also by humidification, it has been hypothesized that microbiological air contaminants may be responsible for some of the higher prevalence rates of work-related symptoms found in these studies (Burge et al., 1987).

20.3.1.3 Danish Epidemiologic Studies Systematic questionnaire studies of complaints were carried out in 14 town halls and 14 affiliated buildings in the Copenhagen area (Skov et al., 1987). The buildings chosen in this study were not previously categorized as being sick or healthy (Valbjorn and Skov, 1987). Environmental measurements such as the concentration of TVOCs and CO_2, the microbial content in floor dusts, and thermal/air moisture parameters were made in one representative office in each building. Two unique parameters, the "fleece" and "shelf" factors, were measured in each study office. The fleece factor is a measure of the surface area of all porous room furnishings (e.g., carpets, drapes, upholstery) divided by the room volume. The shelf factor is the length of open shelves in the study room divided by total volume of the room studied (Skov et al., 1987).

Work-related symptoms reported most frequently in the Danish Town Hall Studies included eye, nose, and throat irritation, fatigue, and headache. A great variation in the prevalence of work-related symptoms was found between buildings. Workplace environmental factors such as total amount of floor dusts and the fleece and shelf factors of studied offices were related to symptom prevalence. Certain indoor environmental factors such as CO_2 concentrations, however, could not be related to the prevalence of work-related symptoms.

The Danish Town Hall Study is important in IAQ research because it indicates that there is considerable variation in the prevalence rate of SBS symptoms between buildings. In addition, factors such as the amount of adsorptive and absorptive (fleece and shelf factors) area for VOCs, combustion products, and microorganisms have been postulated to explain SBS causation.

20.3.1.4 Perceived Air Quality A prominent complaint in buildings is the perception of odors. Indeed, ventilation rates for large commercial buildings are based primarily on the provision of adequate amounts of outdoor air in order to dilute human bioeffluents. An approach toward understanding SBS (Fanger et al., 1988) uses bioeffluents from humans as a Standard for quantifying the perception of air quality in a building.

The bioeffluents emitted from a Standard person (age 18 to 30 years; skin area 1.8 m^2, daily change of underwear; 0.7 baths per day; sedentary, metabolic rate of 58 W/m^2) is defined as one Standard *olf*. The unacceptability of the indoor air caused by other pollutant sources such as building furnishings or the HVAC system itself is defined in terms of the number of Standard olfs needed to cause the same degree of dissatisfaction. Thus the moist dirt in a HVAC system might be the source of enough MVOCs to cause the same equivalent dissatisfaction as 14 Standard persons or 14 olfs. The measurement of olf values for different sources requires that a panel of judges determine the acceptability of the air in a space ventilated by a given flow rate of unpolluted or outdoor air.

The *decipol* is used to quantify the perception (by the nose) of air pollutant concentrations. One decipol is equivalent to the perceived air pollution that would cause the same dissatisfaction as the bioeffluents from a Standard person (one olf) diluted by 20 cfm or 10 L/sec of unpolluted or outdoor air. In the panel studies carried out by Fanger and colleagues, 1 decipol is the perceived air pollution that results in causing about 15 percent of judges to find the air unacceptable when entering the occupied space. Three decipols is the perceived air pollution that causes slightly more than 30 percent of judges to be dissatisfied with the air quality.

The olf and decipol concepts (widely used in research studies in the European community) were used to quantify air pollution sources in 15 office buildings and five assembly halls in the Copenhagen area (Fanger et al., 1988). Panels of judges visited each building and immediately assessed air quality during periods when the building was unoccupied and unventilated, unoccupied and ventilated, and occupied and ventilated. Although the ventilation rate in offices was on average 25 L/sec (50 cfm) per occupant, the percentage of dissatisfied judges was in excess of 30 percent. For each olf associated with human occupancy (bioeffluents), 6 or 7 olfs were estimated to be emitted from other sources in the buildings. The perception of poor air quality in these studies was assigned to the following pollutant sources: the HVAC system (42 percent), ETS (25 percent), furnishings and construction materials (20 percent), and occupants themselves (13 percent). The Copenhagen study showed that people themselves (body odors) were not the primary source of indoor air pollutants. Contaminants from the HVAC system and interior furnishings plus ETS were major contributors to the perception of poor air quality. In addition, the Copenhagen study showed that excessive amounts of outdoor air ventilation (25 L/sec) were insufficient to dilute strong indoor contaminant sources that affect perceived air quality.

20.3.2 Thermal Environmental Conditions Human acceptance of a thermal environment is related to a number of variables such as metabolic heat production and the transfer of heat between the occupant and the environment. Heat transfer is influenced by variables including the temperature, relative humidity, and velocity of the air around the occupant. The body is at thermal equilibrium when the net heat gain or loss is zero.

ASHRAE Standard 55-1992 provides performance criteria for environmental conditions that 80% of more occupants in indoor environments will find acceptable. Among the more important parameters recommended by this Standard are the following (a) a temperature range of 68 to 75°F in the winter and 73 to 79°F during the summer, (b) the vertical temperature gradient measured 4 and 67 in. above the floor should not exceed 5°F, (c) air velocity in the vicinity of the occupant should be about 30 fpm during the winter and about 50 fpm during the summer, and (d) the floor temperature for people wearing typical footwear should range from 65 to 84°F.

The RH in indoor air strongly affects thermal environmental acceptability. When the RH is too low (<10 to 20%), complaints of dry mucous membranes and dry skin may occur. An addendum to ASHRAE Standard 55-1992 recently eliminated the upper acceptable RH limit of 60%. Consequently, according to Standard 55-1992, it is acceptable in terms of thermal comfort for the RH to exceed 70 and even 75%. However, it has recently been shown that elevated RH (for example, 70%) and temperature (for example, 82°F) cause occupants to perceive air quality as being unacceptable (Fanger, 1996). It is best, therefore, that the RH in indoor environments not exceed 60%.

Thermal complaints in buildings are often associated with temperature problems such as those caused by installation of additional heat-producing equipment or with solar loads and improperly zoned thermostats. Poorly insulated envelope walls and floors (for example, above a tuck-under garage) often cause "too cold" complaints during the winter.

Lack of adequate airflow in the occupied space (often indicated by the presence of portable fans) is commonly associated with thermal environmental complaints. In these cases, air flow from diffusers may be blocked by newly added modular partitions or walls, VAV terminals may be closed, or the airflow to the zone has become unbalanced.

20.3.3 Building-Related Illness

Because of the long latency period and relatively small populations involved, some BRI, such as cancer, that may be caused by exposure indoors to carcinogens such as asbestos, ETS, radon, and some VOCs are estimated primarily through risk analysis (Kreiss, 1989). Other BRIs attributable to specific agents such as organophosphate or CO poisoning can be successfully evaluated as a case study in a building. In the section that follows the etiology of some BRIs of microbial origin are reviewed.

20.3.3.1 Hypersensitivity Pneumonitis Hypersensitivity pneumonitis (HP) is an immunologic lung disease that occurs in some individuals after inhalation of organic dusts. Hypersensitivity pneumonitis is suspected when symptoms such as fever, cough, and chest tightness occur several hours after exposure. The diagnosis of the disease is made on the basis of review of patient symptomology plus a battery of tests (Kreiss, 1989), including restrictive pulmonary function measurements, the formation of antibodies to extracts of microbial agents collected in the building, and, in rare instances, experimental inhalation of suspect antigens in a clinical setting by the patient.

Although cases of HP are rare, they are usually associated with HVAC components such as water spray systems that are heavily contaminated with microorganisms (Arnow et al., 1978) or with heavy moisture intrusion into the building. The specific microorganism(s) causing HP remain unidentified even in well-studied cases probably because of the numerous antigens present in different growth sites in the building and its HVAC system.

20.3.3.2 Humidifier Fever Episodes of fever, muscle aches, and malaise with only minor pulmonary function changes have been associated with inhalation of aerosols from humidifiers or water sumps contaminated with gram-negative bacteria, bacterial endotoxins, and protozoa (MRC, 1977). Symptoms of this flulike illness generally subside within a day after exposure without any long-term adverse effects.

An example of this type of BRI occurred in an office where three of the seven occupants reported attacks of fever and chills that started late during the workday and lasted well into the night (Rylander et al., 1978). Illness was associated with the use of a humidifier on occasions when the air was considered too dry for comfort. Analysis of water from the humidifier reservoir showed that *Flavobacterium* (a gram-negative bacterium) was present at a concentration of about 8×10^4/ml. The concentration of airborne *Flavobacterium* increased from nondetectable when the humidifier was not running to about 3000/m^3 within 15 min of operation.

20.3.3.3 Legionellosis Sources of *Legionella* in indoor air include entrainment of aerosols from cooling towers and evaporative condensers into HVAC system outdoor air intakes and the generation of aerosols containing *Legionella* from indoor sources such as shower heads, humidifiers, whirlpools, and saunas. A hotel-associated outbreak of Legionnaire's disease in Wisconsin (Band et al., 1981) illustrates the unexpected pathways that may lead to *Legionella* exposure in the indoor air. A cooling tower on the roof of the hotel was shown to be the reservoir where the bacterium was amplifying. Aerosol from the tower was disseminated into a meeting room via a nearby open chimney that connected to a room fireplace. The operation at the same time of a large exhaust fan in the meeting room kept this indoor space under negative pressure relative to the ambient air. The proximity of an open building vent to a poorly maintained cooling tower and the operation of an exhaust fan collectively were associated with the high attack rate of Legionnaire's disease in the meeting room.

20.3.4 Economic Cost of Poor IAQ The exact number of occupants affected by SBS in nonresidential, nonindustrial buildings is unknown. However, estimates suggest that about 20 to 30 percent of the nonindustrial building stock may be classified as "problem buildings"; that is, those where SBS and impaired productivity occur because of poor IAQ (Woods, 1989). Since the total number of nonindustrial commercial buildings in the United States is about 4,000,000, this means that about 1,000,000 buildings have occupants with SBS symptoms.

Some estimates suggest that as many as one-third of problem buildings are also characterized by BRI (Woods, 1989). Although this estimate may be too high, it is clear that the costs associated with each instance of BRI can be significant. Several instances where BRI has led to the evacuation of a building or portion of a building and the renovation/reconstruction of the affected structure are found in the literature. The building studied by Arnow et al. (1978) originally housed about 1000 employees and was vacated for about 3 years during renovation. Renovations in the buildings studied by Hodgson et al. (1985, 1987) required 1 to 3 years. Costs associated with renovations of these buildings varied from $150,000 to $10,000,000. Direct health care and litigation costs in these types of problem buildings were also significant. Costs of renovating two newly constructed Florida courthouses that had to be vacated because of fungal contamination have exceeded $20,000,000 in one case and $40,000,000 in the second case.

It has been estimated that the indirect cost to U.S. employers for 3 sick days per year *or* a loss of 6 minutes of concentration ability per day is about $10 billion annually (Woods, 1989). Costs for improvement in operation and maintenance of existing buildings and better design and construction for new buildings to improve IAQ should be balanced against potential gains in occupant productivity. Various costs associated with the construction and operation of nonindustrial buildings are estimated as follows (Woods, 1989; EPA, 1989b):

- Salary cost: \$100 to \$300/(ft^2) (year)
- Lease cost: \$15 to \$50/(ft^2) (year)
- Building construction cost: \$50 to \$125/gross ft^2
- Capital assets cost (furnishings, equipment): \$20 to \$100/ft^2 of floor area
- Maintenance and operation costs: \$2 to \$4/(ft^2) (year)
- Environmental control costs: \$2 to \$10/(ft^2) (year)
- Utility cost: \$2 to \$4/(ft^2) (year)
- Costs to improve operation and maintenance programs: \$0.25 to \$1.0/(ft^2) (year)

The costs associated with improved building performance are potentially more than offset through improved productivity of building occupants. It thus makes good economic as well as environmental sense to transform a problem building into healthy building, and also to prevent a healthy building from degrading into a problem building.

20.4 VENTILATION SYSTEMS IN COMMERCIAL BUILDINGS

A HVAC system should provide the occupied space with acceptable IAQ and acceptable thermal conditions. Thus the HVAC system should provide conditioned air so that the range of air temperatures and moisture levels in occupied space conforms to the guidelines of ASHRAE Standard 55-1992 described in Section 20.3.2.

Ventilation should also provide the occupied zone with an appropriate amount of clean outdoor air and remove (dilute) air contaminants to the extent that a majority of people do not express dissatisfaction. While dilution ventilation offers the only practical method to control odorous bioeffluents from occupants, other building contaminants such as formaldehyde, VOCs, and ETS are optimally removed at their source.

20.4.1 Description of a Typical HVAC System

(See also Morey and Shattuck, 1989.) The HVAC system of a large mechanically ventilated building contains one or more air handling unit (AHU). In the mixed air plenum of an AHU, outdoor air is mixed with a portion of the HVAC system's return air. The amount of outdoor air that enters the mixing plenum must be sufficient to deliver a minimum of 15 cfm (7.5 L/s) per person to the occupied zone (ASHRAE, 1989). The mixture of outdoor and return air enters a filter bank that contains filters of varying efficiency as measured by the dust spot efficiency method (ASHRAE, 1996a). Highly efficient bag filters found in some HVAC systems remove 60 to 90 percent of the fine airborne dusts that would otherwise visually soil interior surfaces in occupied spaces. Unfortunately, in many buildings only low-efficiency (20 percent or less) filters are utilized to remove particulates from the ventilation air.

Special filters that are seldom found in commercial building HVAC systems are needed to remove gases and VOCs. ASHRAE Standard 62-1989, however, requires that if contaminant concentrations exceed limits set by the EPA for ambient air (see Table 20.5), outdoor air entering HVAC systems must be appropriately filtered. Thus HVAC filter systems of buildings located in urban areas where ozone and NO_2 frequently exceed ambient air quality limits should (but seldom are) equipped to remove these gases.

The air mixture after filtration enters the heat exchanger section, where heat is either added to or removed from the airstream as required to maintain the thermal comfort of occupants in the building.

During the summer air-conditioning season, moisture is removed from the airstream as it passes over the cooling coils. This moisture collects in drain pans beneath the heat exchanger and should exit the AHU through drain lines with deep sealed traps. Water should not be allowed to stagnate in drain pans or in other portions of the AHU.

AHUs must be inspected and maintained in order to provide acceptable performance. Easy access into the mixed air, filter, heat exchanger, and fan plenums is required for routine maintenance.

After passing through the heat exchanger and the supply fan, conditioned air is distributed to the occupied spaces. The main air supply ductwork is usually constructed of

TABLE 20.7 Supply Air Flow Rates from Variable Air Volume Terminal Serving Office Zone with Seven Occupants (Approximately 1000 sq. ft. floor space)

Supply air measurement condition	Total cfm from supply diffusers	Outdoor air per occupant[c]
Design maximum	750	54
Actual maximum[a]	800	57
Design minimum	200	14[d]
Actual minimum[b]	0	0[d]

[a]Maximum flow rate measured when VAV terminal thermostat set to minimum temperature (55°F) setting, calling for maximum cool air.
[b]Minimum flow rate measured when VAV terminal thermostat set to maximum temperature (85°F) setting, calling for minimum cool air.
[c]Outdoor air comprises approximately 50% of supply air.
[d]Does not comply with ASHRAE Standard 62-1989.

sheet metal, and it, as well as the plenums housing the fan and heat exchanger, may be internally insulated with a porous liner for both thermal and acoustic control. Internal fiber glass liner in these plenums and ducts must be undamaged and have a structural integrity that does not allow for loose fibers to be entrained in the airstream.

Air from main supply ducts enters rigid branch ducts which in modern buildings (designed since the 1970s) often contain variable air volume (VAV) terminals. In a VAV system, the volume of air delivered to the occupant zone is varied to maintain the interior temperature, and this is accomplished by a control system including a thermostat in the occupied space. In a constant-air-volume system that is often found in older buildings, the interior space temperature is maintained by varying the temperature of the conditioned air.

According to the requirements of zone thermostats, the VAV terminals modulate the flow of conditioned air to each zone. VAV terminals should always provide some continuous airflow including a minimum of 15 cfm of outdoor air per occupant (ASHRAE, 1989) to the occupied zone even when the thermostat is satisfied. Defective hardware or control systems for VAV terminals in buildings often result in HVAC system performance that does not meet design specifications.

Table 20.7 illustrates the kinds of problems that may be encountered with VAV systems with defective hardware. In an office of 1000 square feet with 7 occupants, the measured supply air flow (approximately 50% outdoor air) was 850 cfm or 57 cfm outdoor air per person. Under minimum design flow rate conditions (see Table 20.7) 14 cfm of outdoor air per person would be provided, which is somewhat less that the minimum recommended by ASHRAE Standard 62 for offices. Under actual minimum conditions (thermostat thermally satisfied) no outdoor air was being provided to office occupants by the defective VAV terminal. ASHRAE Standard 62 requires that when the total supply of conditioned air to an office space is reduced a minimum amount of outdoor air (20 cfm per person) must still be provided for acceptable IAQ.

Air from rigid branch ducts usually enters a flexible duct and then passes into the occupied space through a ceiling-mounted diffuser. A portion of room air is then entrained into the supply airstream being discharged into the occupied space. If air supply inlets and return air outlets are properly sized and selected, outdoor air can be effectively distributed to the occupied zone.

The fraction of the outdoor air delivered to the occupied space (via the diffuser) that actually reaches the occupied zone (generally 3 to 72 in. above the floor) is defined as ventilation effectiveness. The minimum outdoor air ventilation rates recommended in ASHRAE Standard 62-1989 assume a ventilation effectiveness that approaches 100 percent. Actual measurements of ventilation efficiency indicate that most HVAC systems perform at less than 100 percent efficiency in delivering outdoor air to the occupied zone (Persily, 1986). One study reported ventilation effectiveness measurements for different office configurations in the range of 0.57 to 0.76 (Offerman and Int-Hout, 1987). It is implicit in ASHRAE Standard 62-1989 that more than the minimum amount of outdoor air as recommended by the ventilation rate procedure is required to compensate for imperfect mixing of outdoor air in the occupied space.

In many modern buildings, the above-ceiling cavity is used for the passage of return air from the occupied space back to the AHU. The use of a ceiling plenum instead of return ducts is associated with a number of problems such as entrainment of fibrous fireproofing into return air and blockage of airflow by walls that extend to the underside of the floor above.

Some return air is usually discharged from the HVAC system through a relief louver. The amount of outdoor air brought into the HVAC system should be slightly more than the combined total of relief air plus air exhausted from the building by toilet and local exhaust fans. The building as a whole is thus maintained slightly positive or neutral with respect to the atmosphere so as to prevent the infiltration of unfiltered and unconditioned air through loose construction in the envelope and through other building openings.

In many buildings, fan coil units (FCUs) and induction units (IUs) located along exterior walls are used to condition the air in perimeter zones. FCUs contain small fans, low-efficiency filters, and small heat exchangers. These units condition and recirculate supply air (usually without any outdoor air supply) in peripheral zones. IUs are generally supplied with conditioned outdoor air from a central AHU. Conditioned air from the unit passes through nozzles and mixed with a portion of the room air. Because a large building may contain several hundred FCUs or IUs, maintenance of this part of the HVAC system is often neglected.

20.4.2 Humidification and Dehumidification

In some types of building zones, such as hospital critical care areas, computer rooms, and in animal research facilities, moisture is often added to ventilation air to maintain humidity levels usually between 40 and 50 percent. Moisture is preferably added to the ventilation air in commercial buildings by humidifiers that inject steam or water vapor into the AHU (often in fan plenum) or in the main supply air ductwork. Injection nozzles should not be located near porous interior insulation that can become wet and offer a niche for the amplification of microorganisms. HVAC system humidifiers should preferably use steam as a moisture source. The supply of steam introduced into the HVAC system should be free of volatile amine corrosion inhibitors, especially morpholine (NRC, 1983).

Cold water humidifiers that contain reservoirs of water can become contaminated with microorganisms that may cause BRIs such as HP, humidifier fever, and asthma. The use of biocides in these types of humidifiers may be ineffective or may in itself cause asthmatic reactions. Preventive maintenance programs for cold-water humidifiers used in HVAC systems includes the following essential aspects: (a) thorough draining and cleaning of the unit so as to prevent development of biofilm (slime), (b) physical removal of biofilm and disinfection of wet surfaces, (c) removal of disinfectants prior to recommissioning of unit so as to avoid aerosolization of chemicals into the air stream, (d) bleedoff of sump water at a constant rate so as to minimize the buildup of solids and impurities, and (e) installation of highly efficient upstream HVAC filters to minimize deposit of lint, fly, and particulate in the open water system.

During the air-conditioning season, as air passes around cooling coils of FCUs and AHUs, the dry bulb temperature approaches the dew point temperature. Consequently, air downstream of cooling coils has a humidity close to 100 percent. Organic dusts and debris not removed during filtration can pass through the heat exchanger section and become entrained in moist, porous, internal insulation in downstream plenums and air supply ductwork. Amplification of fungi and bacteria may thus occur on the nutrients trapped in these HVAC system locations.

During the air-conditioning season, the cooling coil's chilled water temperature may be raised in an effort to save energy costs. This results in an elevated RH in occupied spaces. As the RH in occupied spaces rises above about 65 to 70 percent, the equilibrium moisture content in organic dusts and in the capillary structure of porous material such as wall board and carpet approaches a level sufficient to support fungal spore germination and proliferation (ISIAQ, 1996). As a general principle for controlling mold growth (in all buildings), it is important to prevent the surface relative humidity in building finishing materials from consistently exceeding the 65 to 70% level.

20.4.3 CO_2 Concentration

Carbon dioxide concentrations above 1000 ppm have been associated with an increase in occupant complaints in buildings. It must be noted, however, that the CO_2 concentration in buildings is merely a predictor of the adequacy of outdoor air ventilation to remove low levels of contaminants (e.g., human body odors). CO_2 itself (at the levels found in buildings) is neither toxic nor a contaminant. The normal concentration of CO_2 in the outdoor air is about 325 ppm. In the indoor environment the CO_2 concentration is always somewhat elevated relative to the outdoor level because a CO_2 concentration of about 38,000 ppm is present in the air exhaled from the lung.

ASHRAE Standard 62-1989 recommends that steady-state levels of CO_2 in indoor air not exceed 1000 ppm (a level approximately 675 ppm higher than that in outdoor air), which is equivalent to the provision of about 15 cfm of outdoor air per occupant (Janssen, 1989). Panel studies (Berg-Munch et al., 1986) have shown the approximately 15 cfm of outdoor air per occupant is needed to dilute human body odor adequately to a concentration where 20 percent or less of visitors are dissatisfied (80 percent acceptability) with the IAQ. The maintenance of CO_2 at 1000 ppm or less when the building is fully occupied is a convenient surrogate for assuring that outdoor air ventilation rates do not fall below 15 cfm per occupant.

Although CO_2 concentration remains a good surrogate for human-generated contaminants such as body odor and a good predictor of overoccupancy conditions, epidemiologic studies show that prevalence of SBS is not related to CO_2 levels (Kreiss, 1989). Air contaminants derived from nonhuman sources such as HVAC systems and interior construction and finishing materials are major sources of perceived dissatisfaction with indoor air (see Section 20.3.1.4). The concentration of CO_2 in indoor air would, of course, have no relationship with these emission sources.

20.4.4 ASHRAE Standard 62-1989 and Its Revision

ASHRAE Standard 62-1989 is probably the most important document in the IAQ literature. A key feature of Standard 62-1989 and its *ventilation rate procedure* is the increase in the minimum outdoor ventilation rate compared to its predecessor standard from 5 to 15 cfm per person. Outdoor air requirements recommended by the ventilation rate procedure make no distinction between "smoking-allowed" and "smoking-prohibited" areas. A minimum of 15 cfm of outdoor air per person as specified in the ventilation rate procedure is recommended because it was believed that this is the *minimum* amount of outdoor air needed to dilute body and tobacco smoke odors to acceptable levels (Janssen, 1989). The outdoor air requirements specified by ASHRAE Standard 62-1989 must be delivered to the occupant breathing zone.

Standard 62-1989 also requires that the design documentation for a HVAC system state clearly which assumptions are used in design. This allows others to estimate the limits of the HVAC system in removing air contaminants prior to commissioning and prior to the introduction of new contaminant sources into the occupied space.

A key provision in Standard 62-1989 requires that when the supply of air to the occupied zone is reduced (for example, in VAV systems), provision be made to maintain minimum flow rate of outdoor air throughout the occupied zone.

Building maintenance is recognized as an important factor in providing acceptable IAQ. Thus AHUs and FCUs should be easily accessible for both inspection and maintenance. Specific mention is made of avoiding stagnant water in HVAC system. Caution is urged in the use of recirculated water spray systems that are prone to microbial contamination. Special care is urged to prevent entrainment of moisture drift from cooling towers into outdoor air inlets. However, a minimum distance between makeup air inlets and external contaminant sources such as cooling towers, sanitary vents, and loading docks is not specified.

The outdoor air used in HVAC systems must meet ambient air quality standards for priority pollutants before it is introduced into the occupied zone. According to this criterion, the average long-term concentration of NO_2 and SO_2, for example, in outdoor air introduced into HVAC systems, shall not exceed 0.055 and 0.03 ppm, respectively. Ozone shall not exceed a short-term concentration of 0.12 ppm. An addendum to Standard 62-1989 recommends that in accordance with the 1987 revision of the National Ambient Air Quality Standards, the long-term respirable particulate concentration in outdoor air used in HVAC systems shall not exceed 50 $\mu g/m^3$. When outdoor air is unacceptable, contaminant levels must be reduced by cleaning to acceptable limits.

An *indoor air quality* procedure for achieving acceptable IAQ is present in Standard 62-1989. In this procedure it is required that the concentration of air contaminants in the occupied zone be held below acceptable limits. The outdoor air ventilation rate is left unspecified. Guidance in Standard 62-1989 as to what concentrations of indoor air contaminants are acceptable is furnished for outdoor contaminants covered in the ambient air quality standards and for four additional air contaminants of indoor origin, namely, CO_2 (1000 ppm), chlordane (5 $\mu g/m^3$), ozone (0.05 ppm), and radon (0.027 WL). Because acceptable limits and health effects data are currently unavailable for the vast majority of other indoor air contaminants, most designers do not use the air quality procedure of the Standard. The Standard does, however, make provision for the subjective evaluation of indoor air contaminants much like that carried out by Fanger et al. (1988) as one means of implementing the IAQ procedure.

Some experimental work has been initiated with sensors having a capacity to detect thermal parameters as well as air contaminant concentrations (Klein et al., 1987). In theory, these devices would replace thermostats in the occupied space and control HVAC system operation and IAQ much like that envisioned in the IAQ procedure of Standard 62-1989. Major difficulties with this approach include the limitations in the sensitivity of detector systems for concentrations of interest in indoor air and the absence of acceptable concentration guidelines for most indoor air pollutants.

Although at the time of writing, the revision to ASHRAE Standard 62-1989 (1996b) has entered the public review process and the final outcome of the document is uncertain, several important proposed changes are noteworthy. Standard 62-1989R is formatted in code (shall) and advisory (should) language so as to make it easier for portions of the Standard to be incorporated into building codes. Separate sections of ASHRAE Standard 62-1989R cover general requirements for buildings and HVAC systems, design ventilation rates, construction and system startup, and operating and maintenance procedures. Because ETS poses a significant health risk, acceptable IAQ, according to ASHRAE 62-1989R, can only be achieved in the absence of ETS.

Several pathways of varying complexity will be available to choose design ventilation rates. In calculating design ventilation rates for achieving acceptable IAQ emphasis will be placed on estimating the ventilation needed to dilute sensory irritants emitted from building materials as well as bioeffluents from occupants. Significant emphasis is placed on contaminant source control and designing HVAC system components for ease of maintenance (e.g., access doors required both upstream and downstream of AHU cooling coils) and for performance of actual maintenance activities in operating systems. The current and proposed revision of ASHRAE Standard 62 should be reviewed for more details, and building codes should be reviewed to determine other requirements that may be in effect.

20.4.5 Importance of HVAC Systems in IAQ

In mechanically ventilated buildings, both the quantity and the quality of air supplied to occupants is highly dependent upon the HVAC system. In most studies where the quantity of outdoor air supplied to occupants has been measured, the risk of SBS increased as the ventilation decreased below 20 cfm/person (Seppänen, 1996). The occurrence of SBS symptoms has also been associated with air-conditioning (Mendell and Smith, 1990). In studies where the perception of conditioned or supply air from a building's HVAC system has been compared to that of the outdoor air, the former was considered worse or poorest quality. The perception of poor quality of supply air was thought to be caused by pollutants originating from the HVAC system itself.

HVAC systems can become major sources of pollutants that lead to degraded air quality for the following reasons:

- External, often malodorous air contaminants such as water droplets from cooling towers (may contain *Legionella* and water treatment chemicals) and combustion products from garages or loading docks may enter the building through the HVAC system outdoor air intakes.
- Dirt on airstream surfaces in HVAC systems is a reservoir for various kinds of pollutants (e.g., VOCs, ETS, fungal spores, pollen), which can be subsequently emitted into supply air. The accumulation of dirt in HVAC systems is especially a problem where filter deck efficiency is poor or where porous materials line airstream surfaces.

- Microbial contaminants including MVOCs originate in moist niches in HVAC system mechanical equipment. Fungi grow on dirty surfaces such as in moist filters and damp insulation. Standing water in drain pans and humidifiers and water droplets emitted from these sources as well as from the surfaces of dehumidifying cooling coils contain yeasts, gram-negative bacteria, and endotoxins.
- Disinfectants and biocides used in humidifiers and drain pans may be emitted into supply air.
- VOCs from paints, solvents, and other chemicals often stored in mechanical equipment rooms can enter the HVAC system through openings in negatively pressurized plenums (e.g., fan plenum of AHU).

20.5 IAQ EVALUATION PROTOCOL AND CHECKLISTS

Approaches to solving SBS problems using traditional industrial hygiene techniques are generally inadequate. The traditional industrial hygiene approach to solving occupational health problems involves recognition of the hazard (because input and output materials are known, the identity of air pollutants is generally preestablished), measuring the contaminant(s) by NIOSH or Occupational Safety and Health Administration (OSHA)-approved procedures, interpretation of analytical data in terms of current health standards, and suggestion of corrective action such as containment of the contaminant at its source. A common result of this approach when applied to IAQ evaluations is that compliance with industrial workplace standards is demonstrated but building-associated complaints from occupants persist (Woods et al., 1989).

For IAQ evaluations, more sensitive sampling methods and evaluation protocols are needed because comfort, occupant well-being, and general population susceptibilities are parameters that are not addressed by industrial workplace health standards. In addition, a clear understanding of how a building and its HVAC system function is required to evaluate and resolve IAQ problems. The evaluation approach described here is divided into a qualitative and a quantitative phase.

20.5.1 Qualitative Evaluation

The qualitative IAQ evaluation includes an extensive telephone interview and a site visit. During this phase of the evaluation, objectives and scope are defined and a preliminary hypothesis as to the likely reason for occupant complaints is formulated.

At the building site, the nature of the complaints is reviewed, environmental factors that may be responsible for complaints are visually evaluated, and an engineering assessment of the HVAC system is conducted. A primary objective of the qualitative evaluation is to provide recommendations for remedial actions in a manner that avoids the necessity of costly sampling of air contaminants or measurement of HVAC system performance parameters. Instrumentation used during the qualitative evaluation is limited, usually consisting of monitors to measure temperature, RH, CO_2, and respirable particulate concentrations.

A key aspect of the qualitative evaluation is the analysis of a building's HVAC system. HVAC system mechanical components are visually examined for deficiencies with regard to original and current design and for operation and maintenance parameters. The control strategies that govern HVAC system operation must be understood. Essential items for the qualitative evaluation of the HVAC system include a smoke pencil for visualization of airflow patterns, a complete set of mechanical plans and specifications, and the on-site availability of the facility engineer of the building being investigated.

When health complaints are reviewed, it is essential to determine if the perceived problem is one of occupant discomfort and annoyance (SBS, thermal discomfort) or if one or more occupants have a BRI. When a BRI is apparent, medical attention for the affected occupants is required along with the initiation of appropriate remedial actions.

Interviews are performed with office managers and occupants or their representatives in order to form a conclusion as to the presence of absence of SBS. The goal of the interview is to determine if SBS occurs in certain occupied zones. Information is elicited on whether or not the prevalence of SBS symptoms follows a daily, weekly, or seasonal pattern.

The interview process should also determine if occupant complaints have a thermal environmental basis. Is there a seasonal or a zonal aspect (for example, thermal complaints in interior zones only) to thermal problems? Finally, determine if complaints in the building are multifactorial, suggesting that SBS, BRI, and thermal environmental problems may coexist.

20.5.2 Checklist for the Qualitative Evaluation

Observations on possible contaminant sources (see checklist that follows) made during the qualitative evaluation are important both in verifying the hypothesis of complaint etiology and in formulating recommendations for corrective actions. Because a building is dynamic in operation, observations should be made with regard to present and past environmental conditions. It is important during the qualitative evaluation to examine not only the complaint area but also noncomplaint areas and the HVAC system serving both areas.

20.5.2.1 Microbial Contaminants (see Section 20.2.1; see Table 20.8)

______ Is there evidence of current of past water damage? How extensive or localized is the water damage and what are the likely sources of the water? (Table 20.8, #1)

______ Is visible mold present on interior finishes and construction materials? How extensive is the surface area covered by mold? (Table 20.8, #2)

______ Are moisture problems evident in the building envelope? Is there evidence of biodeterioration (condensation, mold growth) on the room (occupied) side of the envelope? Are vapor diffusion barriers correctly positioned for the building's climatic location? (Table 20.8, #3; Fig. 20.1)

______ Is there evidence of hidden microbial growth such as musty odors (MVOCs)? (Fig. 20.2)

______ Do records indicate that RH consistently exceeds 60%? Measure the temperature and RH of room air plus the surface temperature and RH of finishes and construction material that may be moist or show evidence of mold growth.

______ Are porous materials such as carpet and insulation present in damp niches in the building? (Table 20.8, #4; Figs. 20.3 and 20.4)

______ Are dead botanical materials such as bark chips used in moist locations such as in an atrium?

TABLE 20.8 Corrective Actions for Microbial Contamination

Event	Corrective action
1. Floor and leaks	Remove moisture from building infrastructure so as to prevent mold growth (Morey, 1996)
2. Visually moldy materials	Remove so as to prevent dispersion of fungal spores
3. Condensation in building envelope	Prevent condensation by proper use of air barriers and vapor diffusion retarders; eliminate locally cool surfaces; See ISIAQ (1996)
4. Materials susceptible to biodeterioration	Avoid their use in wet niches in occupied spaces or the HVAC system (ISIAQ, 1996)
5. Water stagnation	Design drain pans for self-drainage (ASHRAE, 1996b); continuous bleedoff required for sump humidifiers (Health Canada, 1993)
6. Biofilm in HVAC	Physical removal required (ASHRAE, 1996b; ISIAQ, 1996)
7. Biocides and disinfectants	Must not be aerosolized in functioning HVAC systems (ISIAQ, 1996; ASHRAE, 1996b).

_____ Are the HVAC outdoor air inlet and other building openings positioned such that bioaerosols from cooling towers are likely to be entrained? Consider horizontal and vertical separation distances as well as air flow patterns around the building and prevailing wind direction.

_____ Is there standing water in the HVAC (e.g., drain pans, humidifier sumps)? Are biofilms (slime) present on wet surfaces? (Table 20.8, #5 and 6; Fig. 20.5)

_____ Are airstream surfaces in the outdoor air intake, mixed air, filter, fan, and supply air (and supply ducts) plenums damp and dirty? Is visible mold present? How much of the airstream surface is lined with porous insulation? (Table 20.8, #4)

_____ Are HVAC filters wet? Check both visually and by touching. Discard wet filters.

_____ Does moisture carry over from cooling coils to downstream HVAC surfaces? (Fig. 20.6)

_____ What protocols are used to clean HVAC components? Is cleaning done during off-hour periods? Are biocides used during cleaning? Are biocides and odor-maskers (fragrances) used in active HVAC systems? (Table 20.8, #7)

_____ What cleaning protocols are used to clean and control moisture in unit ventilators, FCUs, and IUs that may be present?

20.5.2.2 *Volatile Organic Compounds and Odors* (see Section 20.2.2)

_____ Are malodors present? The presence of body odors indicates inadequate ventilation. Chemical odors might be associated with cleaning compounds and restroom deodorants. Control is achieved by reducing the amount of cleaning chemicals and deodorants.

_____ Has renovation recently occurred? If yes, were VOCs exhausted directly out of the building or recirculated by the HVAC system?

_____ Are specialized processes such as printing, graphics production, and darkroom work occurring in occupied spaces? Are chemicals from these processes exhausted directly out of the building? If not, install local exhaust ventilation.

_____ Have pressed wood products that contain formaldehyde been used during reno-

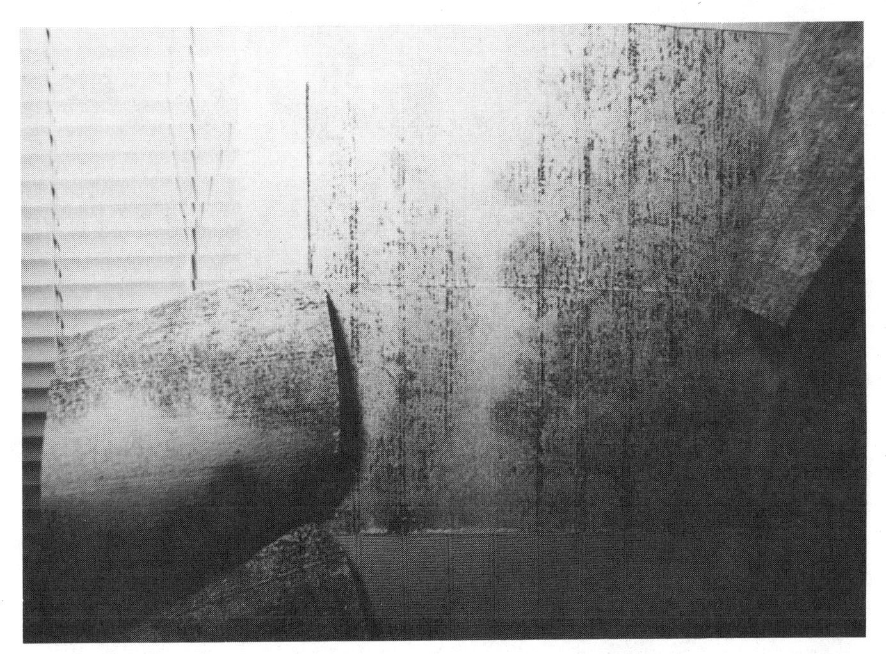

Fig. 20.1 Mold growth occurs beneath vinyl wall covering on perimeter wall because of condensation. Moldy porous materials must be removed in a manner that prevents dispersion of spores.

Fig. 20.2 Mold growth occurred on gypsum board in wall cavity because of moisture incursion. Note rusty electrical socket.

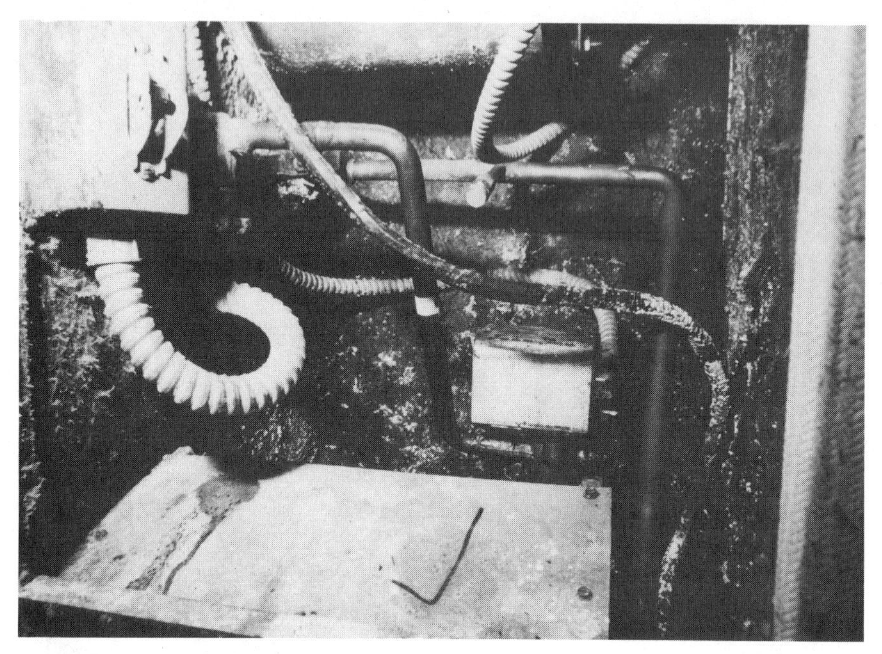

Fig. 20.3 Mold growth occurred on/in porous insulation on airstream surface in unit ventilator.

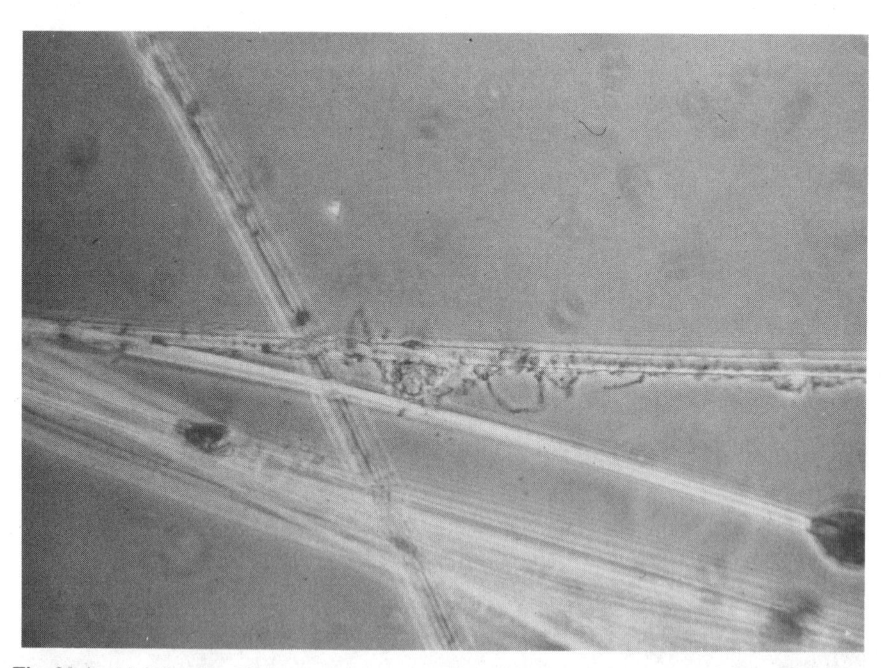

Fig. 20.4 Mold hyphae (filaments) grow around fiber glass in moldy insulation. Insulation must be discarded.

Fig. 20.5 Biofilm occurs in drain pan of AHU. Biofilm must be physically removed and maintenance program must prevent its reoccurrence.

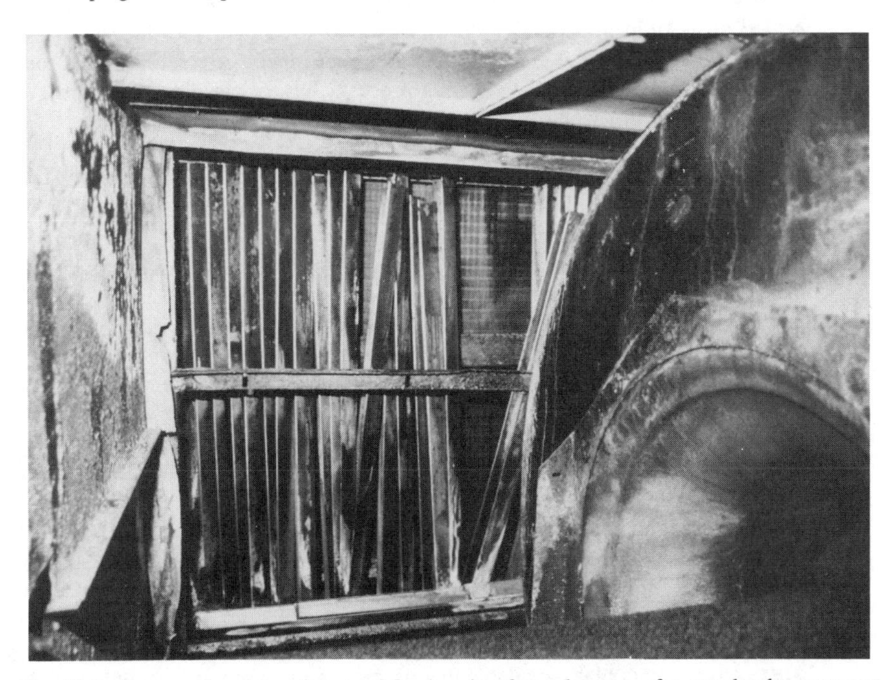

Fig. 20.6 Rust on de-mister plates and fan housing forms because of water droplet carryover. Prevent carryover of water droplets from coils and spray humidifiers.

vation? If chemical odors are present increase outdoor air ventilation. Use low VOC (including formaldehyde) emitting wood products for next renovation.

______ Determine who is responsible for monitoring the use of cleaning chemicals and solvents by occupants and cleaning personnel. Are protocols in place to restrict usage of toxic, irritative, and odorous chemicals in the building and its HVAC system?

20.5.2.3 Other Indoor Pollutants (see Section 20.2.3)

______ Are HVAC outdoor air inlets located within three stories of the ground (or at grade or below-grade levels) where they may be contaminated by combustion products from garages, loading docks, or vehicular traffic? Does the time of peak occupant complaint correlate with vehicular activity? Move the air inlet or reduce vehicular activity if combustion product entrainment occurs.

______ Are intermittently operated sources of combustion products near HVAC outdoor air intakes (e.g., gas fired heaters in rooftop AHUs or diesel powered emergency generators)?

______ Does stack effect induce infiltration of combustion gases during winter months?

______ Are pesticides applied in the building, and what protocol is followed in their application?

______ Can ETS be transported to nonsmoking area? This is important in multiple-tenant buildings where ETS can be transported from a smoking floor to a nonsmoking floor. ETS can be transported by negative pressurization and stack effect into smoke-free buildings from smokers standing just outside the lobby entrance.

______ What kinds of practices are used by facilities and housekeeping staff to remove settled dusts from floors and walls and from airstream surfaces in the HVAC system? When are these cleaning activities performed? Are settled dusts dispersed into the air during the cleaning process? Initiate cleaning with a HEPA filter vacuum if dust control is a problem.

20.5.2.4 HVAC System Design

______ Are HVAC outdoor air inlets and other building openings inappropriately located near external contaminant sources?

______ Are original HVAC plans available? Do available plans incorporate remodeling that has occurred or changes in occupant and/or thermal loads?

______ Is porous liner located on airstream surfaces in moist niches in the HVAC system? Are liners moldy? Remove moldy liners.

______ Was the building designed to meet the minimum outdoor air ventilation rates recommended by ASHRAE 62-1989?

______ What is the design efficiency of the HVAC system filter deck? Can filtration efficiency be increased from 60% for 3 μm particles to 60% for 1 μm particles?

______ Do VAV terminal boxes have minimum set points that provide at least 15 to 20 cfm of outdoor air per occupant when the zone thermostat is satisfied? If not, replace or fix the equipment and/or controls.

20.5.2.5 HVAC System Operation

______ Are control system plans available that describe how minimum amounts of outdoor air ventilation (20 cfm/occupant for offices) are provided during the heating and cooling season and during transitional climatic periods?

______ Do outdoor air inlet dampers open and close (check for minimum stops) according to design criteria?

______ Is the occupied space (especially complaint area) stuffy and warm? Do VAV terminal boxes close when the thermostat is thermally satisfied? Does supply air short circuit to return air vents? Are return air vents present? Do CO_2 concentration measurements (generally 500 to 600 ppm above ambient levels) suggest inadequate outdoor air ventilation?

______ Are thermostats properly positioned to control heating and cooling requirements in perimeter (envelope) and core (interior) areas? Are temperature recommendations of ASHRAE Standard 55-1992 being met for all occupied zones?

______ Are exhaust systems properly removing combustion products from garage and dock areas and keeping these areas negatively pressurized relative to occupied spaces?

20.5.2.6 HVAC System Maintenance

______ Are written preventive maintenance protocols available for the HVAC system including FCUs and IUs? Are the protocols being followed?

______ Is a written preventive maintenance plan available for the cooling tower including bacterial, scale, and corrosion control?

______ Are HVAC facilities staff trained with regard to recognition and control of potential IAQ problems?

______ Are mechanical equipment rooms and HVAC plenums used for storage of solvents, paints, fertilizers, pesticides, and other indoor pollutants?

______ Are airstream surfaces in the HVAC system clean and dry? This includes dampers and pressure sensor stations, the filter plenum, plenums downstream of the heat exchanger, and air supply ductwork.

20.5.3 Quantitative Evaluation

In some building investigations, measurement of contaminant concentrations and HVAC system operational parameters are required to justify recommendations or for litigation purposes. Many variables can affect sampling strategy. Note that the sampling strategy used in IAQ evaluations can be very different from that discussed in Chapter 7 (Personal Exposure Measurement). Consider first if the contaminant is from the building itself (e.g., VOC emissions from finishes), from occupant activities (e.g., VOCs from photocopied paper), or externally sourced (e.g., VOCs and combustion gases from vehicular emission). The sampling strategy used should always include contaminant collection in a worst-case location. In the case of VOCs from building finishes, the worst-case samples should be collected early in the workday before ventilation has diluted contaminant concentrations. If VOCs are likely being emitted from photocopied paper, sampling should occur in the copy room or where printed paper is stored. In the case of entrainment of vehicular emissions, sampling at the outdoor air inlet and along the ventilation air stream is required. In addition to sampling in worst-case location, quantitative assessment of contaminants should occur in best-case situations (e.g., in locations where minimal pollutants are anticipated), as well as in the zone of highest-quality outdoor air (e.g., facing into the wind on the roof at a site remote from external contaminants). Data interpretation often involves a comparison of worst case with indoor and outdoor control locations.

Sampling is performed differently depending upon the objective of the quantitative evaluation. If the objective is to determine occupant exposure, the samples are collected in breathing zone locations under normal building operation conditions (HVAC system is on, routine activities occur in the occupied space). However, if the objective is to demonstrate that a contaminant source is present, then the air sample, for example, for VOCs may be collected adjacent to the finishing material (or determine emission characteristics of finish directly in the laboratory). In the case of fungi, source sampling may include dislodgement of spores from a suspect surface (aggressive sampling), sticky tape surface sampling, or simply collection and analysis of a piece of the bulk surface material (Morey, 1995a).

Quantitative measurements are highly affected by the operation of the HVAC system. Contaminants sourced in the building will likely build up rapidly in occupied zones if a VAV terminal completely closes. Since less outdoor air is used in HVAC systems of many buildings during periods when it is very warm or very cold outdoors, the season of the year is important with regard to dilution of building sourced contaminants like VOCs. Season of the year is an important consideration when air samples of culturable fungi are obtained. Different concentrations and viability of fungi are expected in the cooling season when the building and its HVAC system are humid and moist as compared to the heating season when materials are dry.

20.6 CONTROL MEASURES

A major objective of both the qualitative and quantitative IAQ evaluation is to provide recommendations to remediate building problems, including the causes of occupant complaints that may lead to SBS, BRI, or thermal discomfort. Two very different approaches can be taken when performing remedial actions to upgrade building IAQ. Remedial actions can be directed at *source control*, which involves removal of the pollutant (Morey, 1995b). Alternatively, remedial actions may be directed at *exposure* control, where the concentrations of pollutants that may be present in indoor air are diluted by provision of relatively clean outdoor air or by cleaning (filtration) of indoor air.

Source control is usually pollutant specific and the more effective remediation strategy because occupant exposure can be eliminated or greatly reduced. For example, if occupant complaints are being caused from emission of 4-phenylcyclohexene from newly installed carpet, replacement of the carpet with a brand that does not contain this chemical (replacement carpet should also have low emission rate of other VOCs) is an effective remediation strategy. Alternatively, the exposure control approach using the same example would be to increase the ventilation rate to the complainant space to dilute 4-phenylcyclohexene to levels that hopefully would not be perceived by occupants.

In general, recommendations for both source and exposure control in a building are often made simultaneously. For example, the minimum outdoor air ventilation rates recommended by ASHRAE Standard 62-1989R are adequate for dilution of VOCs emitted from interior finishes with low source strengths. For intense sources of VOCs (e.g., liquid process photocopier) the appropriate recommendation is removal of the source (the photocopier) or effective removal of the contaminant by local exhaust ventilation.

IAQ evaluations may be reactive or proactive. The reactive evaluation is usually accompanied by recommendations to lessen the concentration of pollutants in occupied spaces (exposure control) rather than to eliminate exposure (source control). For example, in an existing building with a tuck-under loading dock, recommendations such as negative pressurization of the dock area, installation of double doors and air curtains, and limiting vehicle idling time are made to limit entry of combustion products into occupied spaces (exposure control). Moving the loading dock (source control) would eliminate exposure, but the cost associated with this form of remediation often inhibits action. Proactive approaches toward control of indoor air pollutants through source control are almost always easiest to implement during design of new buildings or during major retrofit of existing buildings.

20.6.1 Microbials

Sustained growth of microorganisms in buildings is caused by moisture and dampness in building infrastructure. The primary approach toward control is achieved by eliminating sources of moisture that support growth. Since dirt, dust, and soiling of building materials provide nutrient for microbial growth when moisture is nonlimiting, the proper cleaning of the building and its equipment is also essential for microbial control. While biocides may have some role in controlling microbial growth, the fundamental aspect of control should always by exercised through elimination of excess moisture and nutrients (ISIAQ, 1996).

The following actions should be taken to prevent the growth of microbials in buildings:

- Eliminate sites of water accumulation and dampness;
- Keep the surface RH of interior construction and finishing materials from consistently exceeding the 65–70% range;
- Avoid the use of fleecy, extended surface finishing materials such as porous insulation and carpet in portions of the building or building equipment where conditions of dampness occur;
- Dry building infrastructure as rapidly as possible following a leak or a flood so as to prevent microbial growth on susceptible materials. Porous finishes such as gypsum wall board, ceiling tiles, and carpet that remain wet for periods of 24 hours or more are likely to be heavily colonized by gram-negative bacteria, yeasts, and molds;
- Develop a plan of action for dealing with moisture problems such as floods and water spills that may occur in buildings; Removing moisture hidden within building infrastructure (e.g., water in wall cavities) is an important aspect of the plan;
- Incorporate a highly efficient cleaning program into facilities maintenance so that dust and dirt are physically removed from surfaces, especially those niches that may be affected by dampness.

Condensation or near-condensation (surface RH 65 to 100%) conditions should be prevented in building envelopes in all climates. In air-conditioned buildings in hot humid climates, the following actions are intended to reduce the likelihood of moisture and microbial problems:

- The building should be provided with a vapor diffusion retarder and an air barrier system, typically located toward the exterior of the envelope. Avoid the use of low-permeance materials on interior surface of external walls.

- Maintain a net positive pressurization in the building so that infiltration of humid air through the envelope is prevented.
- Prevent rain from entering the envelope.
- Avoid cooling the interior space below the average monthly outdoor dew-point temperature.
- Outdoor air that enters the HVAC system even during unoccupied periods must be dehumidified.

In cold climates, the following actions are intended to reduce the likelihood of moisture and microbial problems in the building envelope:

- Ensure that the envelope has an adequate thermal resistance to prevent interior surfaces from becoming too cold.
- Prevent room moisture from entering the envelope. This may be accomplished by placing the vapor diffusion retarder toward the warm (interior) side of the envelope. The movement of moist room air into the envelope can also be reduced by provision of an air barrier system typically toward the warm (interior) side.
- Lower the RH in indoor air by source control (e.g., locally exhaust water vapor from simmering or boiling of foods) or by ventilation with outdoor air.

Moisture and dirt in HVAC systems result in microbial growth, and the mechanical equipment then becomes a source of microbial pollutants for the rest of the building. The following actions are useful for controlling moisture and microbial problems in HVAC systems:

- Locate outdoor air inlets sufficiently away from (preferably above-grade) standing water, leaves, soil, dead vegetation, and bird droppings. The outdoor air inlet should be designed so as to prevent the intake of rain, snow, and in coastal areas, fog.
- Keep the floors, walls, and ceilings of HVAC plenums clean and dry. Air stream surfaces in wet or damp niches in HVAC systems should not accumulate or retain moisture, should be smooth or easily cleanable, and should be resistant to biodeterioration. Avoid use of uncovered fiberglass insulation in the HVAC system, since it cannot be effectively cleaned and cannot be decontaminated once it is colonized.
- Upgrade the efficiency of HVAC filtration to 65% for particles in the 1 to 3 μm size range (approximately 85% of dust spot efficiency). This will remove mold spores from the airstream and reduce dust accumulation in portions of the system that are difficult to clean.
- Prevent filters from becoming wetted by rain, snow, fog, or water droplets from dehumidifying cooling coils or the humidification system.
- Conditions of water stagnation (accumulation) must not occur in operating HVAC systems. Prevent the buildup of biofilm on wet surfaces of drain pans and dehumidifying cooling coils by frequent thorough cleaning. Biocides must be physically removed after cleaning of pans and coils. Microbiocidal chemicals and disinfectants used in cleaning must not be aerosolized into occupied spaces.
- Humidifiers and dehumidifying cooling coils in HVAC systems should not wet downstream components. When humidifiers are used, the types that emit water vapor rather than droplets are desirable. Control of microbial growth in recircu-

lating water humidifiers is achieved by keeping the unit clean, using high-quality water, and continuous blowdown or replacement of some of the water from the sump.

- Avoid use of unit ventilators, FCUs, and IUs unless units can be kept clean and dry through high-quality maintenance.

Controlling the growth of *Legionella* involves design, operation, and maintenance of cooling towers and potable water systems where this organism can amplify. Some general guidelines follow:

- Design hot water tanks and piping so that the water system can be disinfected by superheating throughout to 60°C in the event that *Legionella* contamination occurs.
- Avoid conditions in hot water tanks and piping that are conducive to *Legionella* growth such as poor mixing, lukewarm water, and water stagnation.
- Locate HVAC outdoor air inlets and other building openings, at a minimum horizontal separation distance of 25 feet (50 feet better) from cooling towers and evaporative condensers. Air flow patterns around buildings and prevailing wind direction are important considerations with regard to protecting inlets and openings from cooling tower drift.
- Cooling tower water systems should be treated to control the growth of microbial contaminants. Microbiocidal control involves the use of biocides and/or physical methods to kill bacteria. A testing program to verify the effectiveness of microbiocidal treatment may be considered (ASTM, 1996).

Consensus has been reached that the occurrence of sustained visual growth of microorganisms on building surfaces is unacceptable (NYC, 1993; Health Canada, 1995). When more than about 3 m^2 of interior surface is covered by visible mold such as *Stachybotrys*, the contaminated materials should be physically removed according to containment procedures similar to those used during removal of hazardous materials. General principles to be followed during removal of visible mold include the following actions:

- Fix the moisture problem in building infrastructure that led to microbial growth.
- Remove visually moldy materials under negative pressure containment in such a manner that dusts and spores are not dispersed into adjacent clean or occupied areas. Biocide treatment or encapsulation of the moldy surface does not substitute for physical removal of the contaminant.
- Remove all fine dusts (particulate) from the formerly moldy area through damp wiping and/or HEPA vacuuming prior to installation of new finishes in the occupied space.

20.6.2 Volatile Organic Compounds

VOC emission from finishing and construction materials used in new buildings or major renovations in existing buildings are best controlled by source reduction or elimination. Voluntary emission rate guidelines for VOCs have been established for carpet as follows (Black et al., 1993; Morey, 1995b):

- TVOC <0.6 mg/m^2 h
- Formaldehyde <0.05 mg/m^2 h
- 4-Phenylcyclohexene <0.1 mg/m^2 h

According to the guidelines established by the State of Washington, the finishes installed in new buildings should not increase certain VOC concentrations above the following levels (Black, 1993):

- TVOC 0.5 mg/m^3
- Formaldehyde 0.05 ppm
- 4-Phenylcyclohexene 1 ppb

The concentration guidelines for VOCs (above) should be achievable for low-source-strength finishes provided that at least 30 days of continuous preoccupancy ventilation at a rate of 20 cfm of outdoor air per 140 square feet of floor space. ASHRAE Standard 62-1989R (1996b) requires that prior to occupany, no less than the design rate of outdoor air ventilation be provided continuously for 48 h to zones following completion of major construction.

VOCs may be controlled during new construction and major renovation by following general principles such as:

- Avoid use of finishes and construction materials that contain carcinogens, toxins, or teratogens.
- Avoid the use of materials containing highly odorous VOCs that are known to degrade perceived air quality at minute concentration levels (Wolkoff et al., 1996; see Section 20.2.2)
- Avoid the use of finishing and construction materials that increase VOC concentrations above the guidelines described by Black et al. (1993).
- Minimize the use of solvents, sealants, caulks, paints, and other products that have high emission rates for VOCs (Levin, 1989) or that are highly odorous.
- Provide continuous preoccupancy outdoor air ventilation at the design rates recommended by ASHRAE preferably for 30 days, to flush out VOCs emitted from new finishes. Increased outdoor air ventilation will probably be needed during the first 3 months of occupancy to dilute VOCs from weak sources (Health Canada, 1993).
- Isolate areas of existing buildings undergoing major renovation by construction of critical barriers (temporary walls, plastic sheeting, or other vapor retarding layer). The construction area should be maintained at a negative pressure relative to adjacent occupied area. Air from the construction area should be locally exhausted outdoors in order to prevent VOC adsorption into finishes elsewhere in the building.

Many diverse sources of VOCs potentially occur in existing buildings not undergoing major renovation. Sources include cleaning chemicals and products, minor touchup and painting, equipment operated by occupants, and occupants' personal care products. Principles useful in controlling VOCs from diverse sources in occupied buildings include the following:

- Monitor and reduce to the lowest feasible level the usage of volatile chemicals used in cleaning agents and work processes in the building.
- Use local exhaust ventilation to remove VOCs and other air contaminants from printing machines, graphics operations, darkrooms, closets or rooms where paints, solvents, and cleaning agents are stored, or from any other strong VOC sources.
- Prohibit the storage of cleaning chemicals, solvents, paints, etc. in HVAC mechanical equipment rooms or in HVAC plenums.
- Carpets, modular partitions, or other finishes brought into the building during minor touchup should be previously off-gassed in a clean, well-ventilated warehouse.
- Determine, based on emission data provided by manufacturers, that new computers, printers, copiers, etc., are not significant sources of VOCs.

20.6.3 Other Indoor Pollutants

The following principles should be considered for control of air pollutants including pesticides, combustion products, ETS, and particles:

- *Pesticides.* Limit pesticide usage to building areas where its application is required. Avoid environmental conditions (e.g., food in offices) that cause pest problems. Apply pesticides only during unoccupied periods and accurately follow manufacturer's directions. Ventilate the space (follow manufacturer's directions) prior to occupancy. Never use or store pesticides in active HVAC equipment or plenums (e.g., in FCUs or mechanical equipment rooms).
- *Combustion products.* Avoid building designs where sources of combustion products such as garages or loading docks are located beneath or within buildings or where combustion sources are located near HVAC outdoor air inlets or other building openings.
- Keep garages and loading docks that may be present within or adjacent to buildings negatively pressurized relative to office (occupied) area. Block paths (air curtains, doors) through which air contaminants may travel from the garage or dock into occupied spaces. An alarm should be triggered at the building central security desk when CO sensors in the garage or dock exceed high limits so that prompt action can be taken to control combustion product emissions.
- Eliminate stack effect as a means through which combustion products (and other pollutants) may enter tall buildings (Morey, 1995b; Tamblyn, 1993). Prevent idling of vehicles in dock or garage areas.
- *ETS.* Contaminants from smoking lounges must be totally exhausted outdoors. Acceptable IAQ is not compatible with the recirculation of ETS from any source (ASHRAE, 1996b).
- *Particulates.* Avoid activities that generate particulate contaminants. These activities may include dry mopping that disturbs settled dust and use of inefficient vacuum cleaners that allow fine particles to pass through the instrument and back into the indoor environment.
- Prevent dusts generated during major renovation from contaminating occupied spaces by use of containment barriers and exhausting particulate laden air outdoors through HEPA filters.

- Avoid the use of extended surface, fleecy finishes in building areas where dust control is critical or where soiling is difficult to prevent.
- Provide occupied spaces with highly filtered conditioned air at a sufficient air exchange rate so as to dilute particles that may be aerosolized by occupant activities. Air cleaning is almost always more effective when performed in the central HVAC system rather than by portable air cleaners.

20.6.4 Thermal Environmental Conditions

Thermal environmental conditions strongly affect occupant perception of acceptable IAQ (Fanger, 1996). Control of thermal comfort is achieved by following the recommendations present in ASHRAE Standard 55-1992 especially those dealing with temperature, temperature gradients, and air motion in occupied spaces. Provision of RH in the 30 to 50% range is best for maintaining acceptable perceived air quality (Fanger, 1996).

Control of thermal environmental problems involves an evaluation of the HVAC system's capacity to adequately heat or cool the occupied spaces. Changes in thermal loads (e.g., more occupants or heat generating equipment) necessitates addition to capacity perhaps by upgrading fan capacity or addition of new AHUs. Other considerations for controlling thermal environmental problems include the following (Health Canada, 1993):

- Rebalance the HVAC system to reflect current loads. Verify that conditioned air being supplied to and returned from zones is in balance.
- Verify that HVAC controls are operating correctly. For example, cooling may be provided to core (interior) zones while heating is occurring in envelope area.
- Insulate envelope surfaces to control thermal loads and to reduce temperature gradients in perimeter zones.

20.6.5 Communications

Open communications between building management and occupants are necessary for the resolution of IAQ complaints (Health Canada, 1993). Successful resolution of IAQ complaints requires communication and participation of occupants and/or their representatives, housekeeping and HVAC facility staff, health and safety committee representatives, and the building manager (or owner). Cooperation and early action to investigate and solve problems usually results in a successful resolution of complaints. Denial of the existence of a problem in the face of continued occupant complaints often results in a situation where problem solving is difficult and delayed.

When IAQ complaints occur in a building, the following communication actions should occur: (a) openly discuss the complaints (or problem) with all concerned parties; solicit occupant views on reasons for the complaints; establish a policy through which all parties will be appraised of progress during complaint investigation and resolution; (b) develop a written procedure for recording the time, location, and nature of complaints as they occur; housekeeping and facilities staff should record building operational parameters that may coincide with the complaint (e.g., occurrence of renovation or cleaning activities); a team approach is often needed to resolve the problem; for example, the building manager may be unaware that a renovation or cleaning activity is degrading IAQ, but occupants may readily recognize the source of the problem; (c) all parties, including occupants, should be provided with information on compliance with remedial recommendations; highly technical reports dealing with problem resolution should be accurately summarized in plain English so as to be understandable by the general public.

20.6.6 Conclusions

Control and prevention of IAQ problems is facilitated by a written plan dealing with all aspects of building performance. Components of the written IAQ plan should include a description of building design, construction documents, location of mechanical equipment, commissioning documents, a description of HVAC automatic temperature control strategy, and a description of the HVAC operation and maintenance controls, including plans to meet outdoor air ventilation requirements during the various seasons of the year. The written IAQ plan should be sufficiently detailed and documented such that a knowledgable facilities engineer or professional can determine if current building performance meets plan expectation. A very important aspect of the written plan is a description of training of facilities and housekeeping staff (for example, HVAC maintenance personnel should receive instruction on the importance of their activities with regard to providing acceptable IAQ) in IAQ matters, including potential causes of SBS and BRI. The reader is referred to Sections 5, 7, and 8 of ASHRAE 62-1989R (1996b) and to pages 16035–38 of the OSHA-proposed rule on IAQ (OSHA, 1994) for guidance on content of a written IAQ program.

REFERENCES

American Conference on Governmental Industrial Hygienists (1989). *Guidelines for the Assessment of Bioaerosols in the Indoor Environment.* ACGIH, Cincinnati.

American Industrial Hygiene Association (1996). *Field Guide for the Determination of Biological Contaminants in Environmental Samples.* AIHA, Fairfax, Virginia.

American Society of Heating, Refrigerating and Air-Conditioning Engineers (1989). *Ventilation for Acceptable Indoor Air quality*, Standard 62-1989, ASHRAE, Atlanta.

American Society of Heating, Refrigerating and Air-Conditioning Engineers (1992). *Thermal Environmental Conditions for Human Occupancy*, Standard 55-1992, ASHRAE, Atlanta.

American Society of Heating, Refrigerating and Air-Conditioning Engineers (1996a). Air Cleaners for Particulate Contaminants, Chapter 24, in *Systems and Equipment Handbook*, 24.1–24.12, ASHRAE, Atlanta.

American Society of Heating, Refrigerating and Air-Conditioning Engineers (1996b). *Ventilation for Acceptable Indoor Air Quality.* Public Review Draft Standard 62-1989R, ASHRAE, Atlanta.

American Society for Testing & Materials (1995). Standard Test Method for Nicotine in Indoor Air, D5075-90a, *Annual Book of ASTM Standards.* 11.03, 418–24.

American Society for Testing & Materials (1996). *Guide for Inspecting Water Systems for Legionella and Investigating Possible Outbreaks of Legionellosis* (Legionnaires' Disease and Pontiac Fever), ASTM.

Arnow, P., J. Fink, D. Schlueter, J. Barboriak, G. Mallison, S. Said, S. Martin, G. Unger, G. Scanlon, and V. Kurup (1978). "Early Detection of Hypersensitivity Pneumonitis in Office Workers," *Am. J. Med.* **64,** 236–42.

Band, J., M. LaVenture, J. Davis, G. Mallison, P. Skaliy, P. Hayes, W. Schell, H. Weiss, D. Greenberg, and D. Fraser (1981). "Epidemic Legionnaires' Disease. Airborne Transmission Down a Chimney," *J. Am. Med. Assoc.* **245,** 2404–7.

Berg-Munch, B., G. Clausen, and O. Fanger (1986). "Ventilation Requirements for the Control of Body Odor in Spaces Occupied by Women." *Environmental International* **12,** 195–99.

Berger-Preiss, E., A. Preiss, and K. Levsen (1996). "Indoor Exposure to Pyrethroid Insecticides," *Proc. 7th Intern. Conf. on Indoor Air Quality and Climate* **1,** 507–12.

Black, M., L. Work, A. Worthan and W. Pearson (1993). "Measuring the TVOC Contributions of Carpet, Using Environmental Chambers," in *Proc. 6th Intern. Conf. on Indoor Air Quality and Climate* **2,** 401–5.

Burge, S., A. Hedge, S. Wilson, J. Bass, and A. Robertson (1987). "Sick Building Syndrome; A Study of 4373 Office Workers." *Ann. Occup. Hyg.* **31,** 493–504.

Cain, W., B. Leaderer, R. Isseroff, L. Berglund, R. Hury, E. Lipsitt and D. Perlman (1983). "Ventilation requirements in buildings—I. Control of Occupancy Odor and Tobacco Smoke Odor," *Atmos. Environ.* **17,** 1183–97.

Clausen, G. (1988). "Comfort and Environmental Tobacco Smoke," in *IAQ '88 Engineering Solutions to Indoor Air Problems*, ASHRAE, Atlanta, 267–74.

Collett, C., J. Ross, and K. Levine (1992). "Nicotine, RSP and CO_2 Levels in Bars and Nightclubs," *Environ. Intern.* **18,** 347–52.

Crandall, M., and W. Sieber (1996). The National Institute for Occupational Safety and Health Indoor Environmental Evaluation Experience. Part One: Building Environmental Evaluations. *Appl. Occup. Environ. Hyg.* **11,** 533–39.

Department of Energy (1985). *Indoor Air Quality Environmental Information Handbook: Combustion Sources*, DOE/EV/10450-1.

Dingle, P., and P. Tapsell (1996). "Pesticide Contamination Resulting from the Over-Use of Pesticides in Perth (Australia) Homes," *Proc. 7th International Conference on Indoor Air Quality and Climate*, **1,** 595–98.

Environmental Protection Agency (1987). *Indoor Air Quality Implementation Plan, Appendix A, Preliminary Indoor Pollution Information Assessment*, Washington, D.C. EPA/600/8-87/014.

Environmental Protection Agency (1989a). "Environmental Tobacco Smoke," *Indoor Air Facts*, No. 5, ANR-445.

Environmental Protection Agency (1989b). *Report to Congress on Indoor Air Quality; Volume II: Assessment and Control of Indoor Air Pollution*, EPA/400/1-89/001C.

European Commission (1995). *Sampling Strategies for Volatile Organic Compounds (VOCs) in Indoor Air.* Report No. 14, EUR 16051.

Fanger, O. (1996). "The Philosophy behind Ventilation; Past Present and Future," *Proc. 7th Intern. Conf. on Indoor Air Quality and Climate* **4,** 3–12.

Fanger, O., Lauridsen, Bluyssen, P., and G. Clausen (1988). "Air Pollution Sources in Offices and Assembly Halls, Quantified by the Olf Unit," *Energy and Buildings* **12,** 7–19.

Flannigan, B. (1995). "Guidelines for Evaluation of Airborne Microbial Contamination of Buildings," in Johanning, E., and Yang, C., Eds., *Fungi and Bacteria in Indoor Environments*, pp. 123–30, Eastern New York Occupational Health Program, Latham.

Girman, J. (1989). "Volatile Organic Compounds and Building Bake-Out," in J. Cone and M. Hodgson, Eds., *Occupational Medicine: State of the Art Reviews* **4,** 695–712.

Gravesen, S., H. Ipsen, and P. Skov (1993). "Partial Characterization of the Components in the Macromolecular Organic Dust (MOD) Fraction and Their Possible Role in the Sick Building Syndrome (SBS)," *Proc. 6th Intern. Conf. on Indoor Air Quality and Climate* **4,** 33–35.

Gyntelberg, F., P. Suadicani, J. Nielsen, P. Skov, O. Valbjorn, P. Nielsen, T. Schneider, O. Jorgensen, P. Wolkoff, C. Wilkins, S. Gravesen, and S. Norm (1994). "Dust and the Sick Building Syndrome," *Indoor Air* **4,** 223–28.

Harrison, J., A. Pickering, M. Finnegan, and P. Austwick (1987). "The Sick Building Syndrome. Further Prevalence Studies and Investigation of Possible Causes," *Proc. 4th Intern. Conf. on Indoor Air Quality and Climate* **2,** 487–91.

Hartwell, T., E. Pellizzari, R. Perritt, R. Whitmore, H. Zelon, and L. Wallace (1987). "Comparison of Volatile Organic Levels between Sites and Seasons for the Total Exposure Assessment Methodology Study," *Proc. 4th Intern. Conf. on Indoor Air Quality and Climate* **1,** 112–16.

Health Canada (1993). *Indoor Air Quality in Office Buildings: A Technical Guide*, Federal-Provincial Advisory Committee on Environmental and Occupational Health, H43-3/3-1993E, Ottawa.

Health Canada (1995). *Fungal Contamination in Public Buildings: A Guide to Recognition and Management.* Environmental Health Directorate, Ottawa.

Health and Safety Executive (1994). *The Control of Legionellosis Including Legionnaires' Disease*, Health and Safety Series Booklet HS (G) 70, Sudbury, UK.

Hedge, A., W. Erickson, and G. Rubin (1993). "Effects of Man-Made Mineral Fibers in Settled Dust on Sick Building Syndrome in Air-Conditioned Offices," *Proc. 6th Intern. Conf. on Indoor Air Quality and Climate*, **1,** 291–96.

Hodgson, M., P. Morey, M. Attfield, W. Sorensen, J. Fink, W. Rhodes, and G. Visvesvara (1985). "Pulmonary Disease Associated with Cafeteria Flooding," *Anch. Env. Health* **40,** 96–101.

Hodgson, M., P. Morey, J. Simon, J. Waters, and J. Fink (1987). "An Outbreak of Recurrent Acute and Chronic Hypersensitivity Pneumonitis in Office Workers," *Am. J. Epidemiology* **125,** 631–638.

International Society of Indoor Air Quality (1996). *Control of Moisture Problems Affecting Biological Indoor Air Quality.* Task Force 1, ISIAQ, Ottawa.

Janssen, J. (1989). "Ventilation for acceptable indoor air quality," *ASHRAE Journal*, 40–48, Oct.

Jantunen, M. (1993). "Inorganic Combustion Products, Acid Aerosols, and Oxidants," in *Indoor Air '93 Summary Report*, 125–32.

Kjaergaard, S., L. Molhave, and O. Pedersen (1987). "Human Reactions to Indoor Air Pollution: N-Decane," *Proc. 4th Intern. Conf. on Indoor Air Quality and Climate* **1,** 97–101.

Klein, C., S. Relwani, and D. Moschandreas (1987). "Evaluation of Proprietary Detector as an Indoor Air Quality Sensor," in *Proc. 4th Intern. Conf. on Indoor Air Quality and Climate* **3,** 249–52.

Kreiss, K. (1989). "The Epidemiology of Building-Related Complaints and Illness," in J. Cone and M. Hodgson, Eds., *Occupational Medicine: State of the Art Reviews* **4,** 575–92.

Krooss, J., and P. Stolz (1996). "Indoor Air Pollution in Museum Store Rooms due to Biocide Substances," *Proc. 7th Intern. Conf. on Indoor Air Quality and Climate* **2,** 681–85.

Levin, H. (1989). "Building Materials and Indoor Air Quality," in J. Cone and M. Hodgson, Eds., *Occupational Medicine: State of the Art Review* **4,** 667–93.

Lewis, R., and A. Bond (1987). "Non-Occupational Exposure to Household Pesticides," *Proc. 4th Intern. Conf. on Indoor Air Quality and Climate* **1,** 195–99.

Lewis, R., and L. Wallace (1988). "Toxic Organic Vapors in Indoor Air," *ASTM Standardization News* **16,** 40–44.

Medical Research Council (1977). "Humidifier Fever," *Thorax* **32,** 653–63.

Mendell, M., and A. Smith (1990). "Consistent Pattern of Elevated Symptoms in Air-Conditioned Office Buildings: A Reanalysis of Epidemiologic Studies," *Am. J. Public Health*, **80,** 1193–99.

Miesner, E., S. Rudnich, F-C. Hu, J. Spengler, L. Preller, H. Ozkaynak, and W. Nelson (1988). "Aerosol and ETS Sampling in Public Facilities and Offices," Paper 88-76·4, *Air Pollution Control Association.*

Milton, D. (1995). "Bacterial Endotoxins: A Review of Health Effects and Potential Impact in the

Indoor Environment," in R. Gammage and B. Berven, Eds., *Indoor Air and Human Health*, pp. 179–95, Lewis Publishers, Boca Raton.

Molhave, L., B. Bach, and O. Pedersen (1986). "Human Reactions to Low Concentrations of Volatile Organic Compounds," *Environ. Intern.* **12,** 167–75.

Molhave, L., and G. Clausen (1996). "The Use of TVOC as an Indicator in IAQ Investigations." *Proc. 7th Intern. Conf. On Indoor Air Quality and Climate* **2,** 37–46.

Morbidity and Mortality Weekly Report (1994). "Acute Pulmonary Hemorrhage/Hemosiderosis among Infants," Centers for Disease Control, Atlanta. Vol. 43, 881–83.

Morey, P. (1995a). "Studies on Fungi in Air-Conditioned Buildings in a Humid Climate," in E. Johanning and C. Yang, Eds., *Fungi and Bacteria in Indoor Air Environments*, Eastern NY Occupational Health Program, Latham, 79–92.

Morey, P. (1995b). "Control of Indoor Air Pollution," in P. Harber, M. Schenker, and J. Balmes, Eds., *Occupational and Environmental Respiratory Disease*, Mosby, St. Louis, 981–1003.

Morey, P. (1996). "Mold Growth in Buildings: Removal and Prevention," in *Proc. 7th Intern. Conf. on Indoor Air Quality and Climate* **2,** 27–36.

Morey, P., and B. Jenkins (1989). "What Are Typical Concentrations of Fungi, Total Volatile Organic Compounds, and Nitrogen Dioxide in an Office Environment," in *Proceedings of IAQ '89. The Human Equation: Health and Comfort*, ASHRAE, Atlanta, 67–71.

Morey, P., and D. Shattuck (1989). "Role of Ventilation in the Causation of Buildings-Associated Illnesses," in J. Cone and M. Hodgson, Eds., *Occupational Medicine: State of the Art Reviews* **4,** 625–42.

Morey, P. R., and J. E. Woods (1987). "Indoor Air Quality in Health Care Facilities," *Occup. Med. State of Art Reviews* **2,** 547–63.

National Academy of Sciences (1993). *Indoor Allergens*. National Academy Press, Washington D.C.

National Research Council (1982). *An Assessment of the Health Risks of Seven Pesticides Used for Termite Control*, Committee on Toxicology, Washington, D.C.

National Research Council (1983). *An Assessment of the Health Risks of Morpholine and Diethylaminoethanol*, National Academy Press, Washington, D.C.

New York City, Department of Health, NYC Human Resources Administration, and Mount Sinai—Irving J. Selikoff Occupational Health Clinical Center (1993). *Guidelines on Assessment and Remediation of Stachybotys atra in Indoor Environments*, NYC.

Occupational Safety and Health Administration (1994). *Indoor Air Quality, Proposed Rule*, 29 CFR Parts 1910, 1915, 1926, and 1928, Federal Register 59, No. 65, 15967-16039.

Offermann, F., and D. Int-Hout (1987). Ventilation Effectiveness and ADPI Measurements of Three Supply/Return Air Configurations. In *Proc. 4th Intern. Conf. on Indoor Air Quality and Climate* **3,** 313–18.

Parkhurst, W., J. Harper, J. Spengler, L. Fraumeni, A. Majahad, and J. Cropp (1988). "Indoor Nitrogen Dioxide in Five Chattanooga, Tennessee, Public Housing Developments. Paper 88-109.4, *Air Pollution Control Association.*

Persily, A. (1986). "Ventilation Effectiveness Measurements in an Office Building," in *Proceedings of IAQ '86, Managing Indoor Air For Health and Energy Conservation*, ASHRAE, Atlanta, 548–58.

Phillips, K., D. Howard, M. Bentley, and G. Alván (1996). "Assessment of Air Quality in Europe by Personal Monitoring of Nonsmokers in Homes/Workplaces for RSP, ETS and VOCs Inside and Outside Homes," *Proc. 7th Intern. Conf. on Indoor Air Quality and Climate* **1,** 495–500.

Reinert, J. (1984). "Pesticides in the Indoor Environment," *Proc. 3rd Inter. Conf. on Indoor Air Quality and Climate* **1,** 233–38.

Rylander, R., P. Haglind, M. Lundholm, I. Mattsby, and K. Stengvist (1978). "Humidifier Fever and Endotoxin Exposure," *Clin. Allergy* **8,** 511–16.

Samet, J., M. Marbury, and J. Spengler (1987). "Health Effects and Sources of Indoor Air Pollution," Part 1, *Am. Rev. Respir. Dis.* **136,** 1486–508.

Seifert, B. (1990). "Regulating Indoor Air," *Proc. 5th Intern. Conf. on Indoor Air Quality and Climate* **5,** 35–49.

Seppänen, O. (1993). *Proceedings of the 6th International Conference on Indoor Air Quality and Climate*, Helsinki, Finland (6 volumes).

Seppänen, O. (1996). "Ventilation and Air Quality," in *Proc. 7th Intern. Conf. on Indoor Air Quality and Climate* **3,** 15–32.

Sheldon, L., R. Handy, T. Hartwell, R. Whitmore, H. Zelon, and E. Pellizzari (1988a). *Indoor Air Quality in Public Buildings: Volume 1*, Washington D.C., EPA/600/S6-88/009a.

Sheldon, L., H. Zelong, J. Sickles, C. Eaton, T. Hartwell, and L. Wallace (1988b). *Indoor Air Quality in Public Buildings: Volume 2*, Washington, D.C., EPA/600/S6-88/009b.

Shields, H., D. Fleischer, and C. Weschler (1996). "Comparisons among VOCs Measured in Three Types of U.S. Commercial Buildings with Different Occupant Densities," *Indoor Air* **6,** 2–17.

Skov, P., O. Valbjorn, and DISG (1987). "The Sick Building Syndrome in the Office Environment: The Danish Town Hall Study," *Environ. Intern.* **13,** 339–49.

Sorenson, W., G. Kullman, and P. Hintz (1996). *NIOSH Health Hazard Evaluation Report No. 95-0160*, Centers for Disease Control, Atlanta.

Sundell, J. (1994). "On the Association between Building Ventilation Characteristics, Some Indoor Environmental Exposures, Some Allergic Manifestations and Subjective Symptom Reports," *Indoor Air*, Supplement 2/94, 9–49.

Tamblyn, R. (1993). HVAC System Effects for Tall Buildings, *ASHRAE Transactions* Paper DE-93-10-1.

U.S. Public Health Service (1987). *Surgeon General's Report: The Health Consequences of Involuntary Smoking*, USDHHS, CDC, 87-8398.

Valbjorn, O., and P. Skov (1987). "Influence of Indoor Climate on the Sick Building Syndrome Prevalence," *Proc. 4th Intern. Conf. on Indoor Air Quality and Climate* **2,** 593–97.

Wessen, B., G. Strom, and K.-O. Schoeps (1995). "MVOC-Profiles—A Tool for Indoor Air Quality Assessment," in L. Morawska, N. Bofinger, and M. Maroni, Eds., *Indoor Air an Integrated Approach*, 67–70.

Wolkoff, P., B. Jensen, U. Kjaer, and P. Nielsen (1996). "Characterization of Emissions from Building Products: Short-Term Chemical and Comfort Evaluation after Two Days," in *Proc. 7th Intern. Conf. on Indoor Air Quality and Climate*, **4,** 331–36.

Woodring, J., T. Duffy, J. Davis, and R. Bechtold (1985). "Measurements of Combustion Product Emission Factors of Unvented Kerosene Heaters," *Am. Ind. Hyg. Assoc. J.* **46,** 350–56.

Woods, J. (1989). "Cost Avoidance and Productivity in Owning and Operating Buildings," in J. Cone and M. Hodgson, Eds., *Occupational Medicine State of the Art Reviews* **4,** 753–70.

Woods, J., P. Morey, and D. Rask (1989). "Indoor Air Quality Diagnostics: Qualitative and Quantitative Procedures to Improve Environmental Conditions," in N. Nagda and J. Harper, Eds., *Design and Protocol for Monitoring Indoor Air Quality*, ASTM STP 1002, 80–98.

Yaglou, C., E. Riley, and D. Coggins (1936). "Ventilation Requirements," *ASHVE Trans.* **42:** 133–63.

Yoshizawa, S. (1996). *Proceedings of the 7th International Conference on Indoor Air Quality and Climate.* Nagoya, Japan (4 volumes).

CHAPTER 21

Heat Stress

ANNE M. VENETTA RICHARD

Lucent Technologies, 1600 Osgood St., North Andover, MA 01845

RALPH COLLIPI, JR.

AT&T, 40 Elwood Rd., Londonderry, NH 03053

21.1 Significance of Heat Stress in Industry
21.2 Physiology of Heat Stress
21.3 Heat Exchange and Heat Balance
  21.3.1 Heat Balance Equation
21.4 Factors Affecting Heat Tolerance
  21.4.1 Acclimatization
  21.4.2 Age
  21.4.3 Gender
  21.4.4 Obesity
  21.4.5 Physical Fitness
  21.4.6 Wellness Programs
  21.4.7 Water and Electrolyte Balance
  21.4.8 Alcohol and Drugs
21.5 Heat Stress Disorders
  21.5.1 Heat Rash
  21.5.2 Heat Syncope
  21.5.3 Heat Cramps
  21.5.4 Heat Exhaustion
  21.5.5 Heat Stroke
21.6 Measurement of the Thermal Environment
  21.6.1 Indexes of Heat Stress
  21.6.2 Dry-Bulb and Wet-Bulb Temperatures
  21.6.3 Effective Temperature
  21.6.4 Equivalent Effective Temperature Corrected for Radiation
  21.6.5 Predicted 4-Hour Sweat Rate
  21.6.6 Heat Stress Index
  21.6.7 Job Ranking

Adapted from "Heat Stress: Its Effects, Measurement, and Control" by John E. Mutchler, CIH, Chapter 21 in Clayton & Clayton, *Patty's Industrial Hygiene and Toxicology, 4th ed., Vol. IA*. New York: John Wiley & Sons, Inc., 1991, pp. 763–837.

Handbook of Occupational Safety and Health, Second Edition, Edited by Louis J. DiBerardinis, ISBN 0-471-16017-2

21.6.8 Wet-Bulb Globe Temperature
21.6.9 Wet-Globe Temperature
21.7 Instrumentation
21.7.1 Thermometers
21.7.2 Humidity
21.7.3 Air Velocity
21.7.4 Radiant Heat
21.8 Control Measures
21.8.1 Control at the Source
21.8.2 Local Exhaust Ventilation
21.8.3 Localized Cooling at Work Stations
21.8.4 General Ventilation
21.8.5 Moisture Control
21.9 Management of Employee Heat Exposure
21.9.1 Education and Training
21.9.2 Medical Supervision
21.9.3 Effect of Clothing and Heat Exchange
21.9.4 Protective Clothing
21.9.5 Work–Rest Regimen
21.10 Summary
References
Appendix A—List of Abbreviations Used

21.1 SIGNIFICANCE OF HEAT STRESS IN INDUSTRY

As industry has developed, through the Industrial Revolution to our present highly technological society, on-the-job potential for injury and illness from acute exposure to heat has increased far beyond that known earlier to home-centered craftsmen. Among the more dangerous original industrial vocations were those using molten materials, such as glass and metals. In these first "hot industries" the ever-present danger of burns, explosions, and spills of molten material were well known and accepted, as were potential illness and death from very hard physical work in excessively hot environments.[1]

The traditional hot work industries (i.e., foundries, smelting, firefighting, military, mining, construction, utilities, glass working, tire and rubber, and textile industry workers) are being augmented by industries at even greater risk. Hazardous chemical handlers, lead and asbestos abatement workers, emergency responders, nuclear and radiation containment specialists, and laboratory and hospital personnel who must wear nonpermeable personal protective equipment, along with warm or hot ambient conditions are new sources that need special attention to protect them from potential heat stress.[2] Knowledge of the process involved in the exchange of heat between the worker and the environment and the effects of heat can reduce the potential adverse effects of heat exposure. More details on different work environments and the health hazards associated with them can be found in Chapter 3.

21.2 PHYSIOLOGY OF HEAT STRESS

Basic to any understanding of heat stress is comprehension of the interaction that occurs between our bodies and our environment. Humans must maintain an internal body temperature within a narrow range, near 37°C (98.6°F), to remain healthy and efficient. If the "core temperature" of the body falls below 35°C (95°F), hypothermia results, and death is likely at core temperatures below 27°C (80.6°F) and above 42°C (107.6°F).[3] Internal body temperature is maintained at the regulated level when heat loss is in balance with heat production, which depends on the controlled exchange of heat between the body and the environment.

The heat that must be exchanged is a function of (1) the total heat produced by the body (*metabolic heat*) and (2) the heat gained, if any, from the environment. The rate of heat exchange with the environment depends essentially on conditions such as air temperature and velocity, the amount of humidity or vapor pressure of water in the air, and radiant temperature. The body's inability to accommodate these contributions can result in heat stress. Body conditions that influence the heat equilibrium are skin temperature, amount of evaporated sweat produced, and the type, amount, and characteristics of the clothing worn. Respiratory heat loss is generally of minor consequence except during vigorous activity in very dry environments.

21.3 HEAT EXCHANGE AND HEAT BALANCE

There are several concepts that must be considered when performing hazard assessments for heat stress. These concepts are the basis for education and training programs. *Conduction* is the transfer of heat between two objects when they are in direct physical contact with each other. A person immersed in a cool bath will lose heat from the body via conduction as the body heat is transferred to the object of lower temperature, the water. Conduction of heat to or from solid objects usually can be ignored since workers are generally not in direct contact with surfaces hotter than normal body temperatures for any sustained length of time. However, heat loss by conduction into air occurs when the air in contact with the skin is below body temperature (negative heat load). Conversely, heat gain by conduction from the air occurs when air temperature exceeds body temperature (positive heat load).

Convective heat exchange between the skin and air immediately around the skin is a function of the difference in temperature between the skin and the air and rate of air movement over the skin. *Convection* can produce a negative or positive heat load. For example, when the air temperature is higher than the skin temperature, the contribution to heat load by convection will be positive.

Another aspect of heat exchange is *radiation*, or the transfer of heat from one object to another via electromagnetic waves through a vacuum or air. The heat is not perceived until it is absorbed by the object that the electromagnetic waves strike. A body gains heat by radiation when the temperature of the surrounding surfaces is above body surface temperature (i.e., the sun, steam pipes, and blast furnaces), and conversely a body loses heat when the surrounding objects have surface temperatures lower than the temperature of the body surface. Radiant heat can produce a negative or positive heat load and is independent of air motion.

Evaporation is generally the mechanism most in use by the body for the dissipation of large amounts of heat generated by working muscles. Evaporation is determined by air speed and the difference between the vapor pressure of perspiration on the skin and the partial pressure of water in the air. In industries with high levels of ambient water vapor from wet processes or escaping steam (mining, laundries), high relative humidity reduces evaporative cooling capacity. In a hot, dry environment the maximum sweat production, maintained over 8 hours, is about 1 liter per hour (L/h). Heat is also lost through respiration in negligible amounts.

21.3.1 Heat Balance Equation

The total heat load is a combination of the metabolic heat load and the environmental factors. This condition is often expressed by the "heat balance equation":

$$H = M + R + C + D - E$$

where H is the heat load, which should be 0; M, the metabolic heat gain; R, the radiant heat gain or loss; C, the convective heat gain or loss; D, the conductive heat gain or loss; and E, the evaporative loss.

If ambient conditions rise along with the metabolic load, the body has an increasingly difficult task of cooling itself and "balancing" the equation.[4] The heat stress disorders resulting from the heat load in excess of "0" are explained in Section 11.5. Complicated measurements of metabolic heat production, air temperature, air water vapor pressure, wind velocity, and mean radiant temperatures are required to compute this equation.

21.4 FACTORS AFFECTING HEAT TOLERANCE

21.4.1 Acclimatization

Acclimatization refers to a set of adaptive physiological and psychological adjustments that occur when an individual accustomed to working in a temperate environment undertakes work in a hot environment. These progressive adjustments occur over periods of increasing duration and reduce the strain experienced on initial exposure to heat. This enhanced tolerance allows a person to work effectively under conditions that they might have been unable to endure before acclimatization.[5] Acclimatization is marked by a lower heart rate, lower body temperature, an increase in sweat rate, which increases evaporative cooling, reduced circulatory load, and enhanced heat conduction through the skin. Not only does the sweat rate increase, but the composition of the sweat changes. Sweat has a lower osmotic pressure than water, has one-third of the electrolyte concentration of extracellular fluids, and is even more dilute in persons not acclimated to the environment.[6] Failure to replace the water lost in sweat will retard or even prevent the development of the physiological adaptations characteristic of acclimatization. Thus acclimated people have less strain on their bodies while working in heat. Full acclimatization is usually achieved within 7 days of beginning work in a hot environment. Subjective discomfort of working in the hot environment dissipates by 4 to 7 days, with a decrease in heart rate and body temperature and increased sweat rate. Unacclimated workers should work at 50% of required work on the first day and increase production at a rate of 10% each day thereafter until full acclimatization is achieved.[7]

After heat exposure on several successive days, the individuals perform the same work with a much lower core temperature and heart rate and higher sweat rate (reduced thermoregulatory strain), and with none of the distressing symptoms that may be experienced initially. Heat acclimatization represents a dynamic state of conditioning rather than a long-term change in innate physiology. The level of acclimatization is relative to the initial level of physical fitness and the total heat stress experienced by an individual.

Thus a worker who does only light work indoors in a hot climate will not achieve the level of acclimatization needed to work outdoors (with the additional heat load from the sun) or to do harder physical work in the same hot environment indoors.

21.4.2 Age

Older workers (40 to 65 years of age) are at some disadvantage when working in hot environments. As the body ages, its maximal oxygen intake decreases and the body is more likely to reach the stage where its metabolism becomes *anaerobic* rather than *aerobic*. *Anaerobic* is defined as able to live and grow where there is no air or free oxygen, as certain bacteria, and *aerobic* is the ability to live, grow, or take place only where free oxygen is present. Older adults may have up to 30% less cardiovascular reserve than younger adults. Also, older workers have less efficient sweat glands, causing a delay in the onset of sweating and a lower sweat rate, and thus are less able to compensate for heat load by evaporative cooling. Underlying degenerative diseases of the heart and lungs may further compromise the older worker's circulatory capacity to move heat away from the body core and to the surface. Given these limitations, the older worker can work effectively in a hot environment if allowed to work at an independent pace. It has been suggested that with the current emphasis on fitness in today's society, that physiological responses to heat stress, rather than age, should be a major consideration for working in a hot environment.[8]

21.4.3 Gender

Men tolerate an imposed heat stress slightly better than women in hot dry climates, but not as well as women in humid conditions, according to a study in 1980.[9] Actual body size is more predicative of risk. Smaller persons (less than 50 kg) of either sex are at a disadvantage with regard to heat tolerance due to an increased surface area to mass (SA:M) ratio and a lower aerobic capacity.

Recent retrospective epidemiological studies have associated hyperthermia during the first trimester of pregnancy with birth defects, especially in the CNS development (e.g., anencephaly). It is prudent to monitor the body temperature of a pregnant woman exposed to total heat loads above the recommended exposure limit every hour to assure that the body temperature does not exceed 39 to 39.5°C (102 to 103°F) during the first trimester.

Heat exposure has been associated with temporary infertility in both males and females, with the effects more pronounced in males. Workers who report problems with fertility should be evaluated to determine whether a temporary transfer is needed or whether the amount of work exposure in hot environments should be limited.

21.4.4 Obesity

It is well established that obesity predisposes individuals to heat disorders. The acquisition of fat means that additional weight must be carried; therefore, a greater expenditure of energy is required to perform a given task, and a greater proportion of aerobic capacity is used. In addition, obese individuals have less surface area for their weight, thus limiting the rate of heat exchange with the environment. Obesity also alters sweating and sweat gland distribution, and the increased layer of subcutaneous fat provides an insulating barrier between the skin and the deep-lying tissues. The fat layer theoretically reduces the direct transfer of heat from the muscles to the skin. The increased metabolic load due to excess weight and decreased surface area for cooling puts the obese worker at risk when performing in a hot environment.

21.4.5 Physical Fitness

Levels of fitness in workers are important criteria for determining their ability to work in a heat stress environment. Fitness increases heat tolerance by improving cardiovascular capacity via increasing the capillary bed and vascular tone. Also, improved cardiac output decreases heart rate during hot work with less need to accelerate the heart. Therefore, a physically conditioned person, by virtue of having a higher maximum ventilatory capacity, has a wider margin of safety in coping with the added circulatory strain of working under heat stress.

21.4.6 Wellness Programs

There has been a thrust to implement "wellness programs" in some business sectors. The general concept is that business needs a healthy workforce to keep pace with the daily demands of the job. It is felt that employees who are "well" have more energy, resilience, and stamina and are more likely to be part of a high-performing team and will feel respected and valued.

Healthy employees also help business to manage health care costs. There is strong empirical evidence that employees at lower levels of health risk use fewer health plan dollars.

Wellness programs can include such things as employee assistance programs, health and well-being survey data, smoking cessation sessions, health fitness centers or membership assistance programs, lifestyle change counseling, and lifting and back education sessions.

Employee educational and informational programs that are part of wellness programs may also increase employee awareness about potential risk factors associated with their occupational duties. These programs can address issues like ergonomics awareness associated with computer use or in the case of heat stress, how acclimatization and proper hydration and work breaks will help to reduce the potential for illness.

21.4.7 Water and Electrolyte Balance

Effective work performance in heat requires a replacement of body water and electrolytes lost through sweating. If this water is not replaced by drinking, continued sweating will draw on water reserves from both tissues and body cells, leading to dehydration. Water loss, through sweat, of up to 1.4% body weight can be tolerated without serious effects.[10] At water loss of 3% to 6% of body weight, work performance is impaired; continued work under such conditions leads to heat exhaustion. Acclimated workers are better able to maintain appropriate water balance, although sweating workers should be supplied with cool drinking water and encouraged to drink small amounts every 15 to 20 minutes to assure adequate fluid replacement. One to three gallons per day per worker is typically supplied in hot environments.

The typical American diet provides enough salt for proper salt balance in most acclimatized workers. Unacclimated workers may be given salt supplements in the form of salt tablets, preferably impregnated to avoid gastric irritation, if ample water is avail-

TABLE 21.1 Relationship of Common Medications and Heat Stress

1. Amphetamines—increase the metabolic needs and put additional stress on the person
2. Anticholinergic medications—inhibit perspiration
3. Antihistamines—inhibit perspiration
4. Atropine—inhibits perspiration
5. Beta blocking agents—may enhance dehydration by promoting an inappropriate increase in sweat production
6. Central nervous system inhibitors—affect the hypothalamus' ability to regulate heat properly
7. Diuretics—inhibit necessary expansion of bodily fluid volume; lower circulating fluid volume, inhibit cutaneous vasodilation
8. Muscle relaxants—may cause postural hypotension
9. Tranquilizers and sedatives—phenothiazines, tricyclic antidepressants, monoamine oxidase inhibitors and glutethamide (Doriden)—implicated in lower heat tolerance
10. Vasodilators—cause a relaxing action on smooth muscles; decreased peripheral resistance; reflex increase in heart rate

Source: AAOHN J. **39**(8), 372, Aug. 1991.

able. A preferable practice is to use salted water (one tablespoon per gallon) or ingest a rehydration beverage that contains sodium, specifically any electrolyte replenisher. Salt supplements should be decreased or discontinued after acclimatization so as not to suppress normal hormonal mechanisms of balancing salt and water.

21.4.8 Alcohol and Drugs

Alcohol is a risk factor for heat strain by physiologically or behaviorally altering thermoregulatory functions. Alcohol interferes with central and peripheral nervous function and is associated with dehydration by suppressing alcohol dehydrogenase (ADH) enzyme production, leading to hypohydration. The ingestion of alcohol before or during work in the heat should not be permitted. Therapeutic drugs such as diuretics, antihypertensives, anticholinergic drugs, antihistamines, CNS inhibitors, muscle relaxants, atropine, tranquilizers, sedatives, beta blockers, and amphetamines can interfere with thermoregulation and could potentially affect heat tolerance. Over-the-counter medications should not be overlooked, as indicated in Table 21.1. Any worker subject to heat stress taking therapeutic medications who is exposed even intermittently or occasionally to a hot environment should be under the supervision of a physician who understands the potential ramifications of medications on heat tolerance. It is difficult to separate the heat-disorder implications of drugs used therapeutically from those that are used socially. Nevertheless, there are many drugs other than alcohol that are used on social occasions. Some of these have been implicated in cases of heat disorder, sometimes leading to death.[11]

21.5 HEAT STRESS DISORDERS

Environmental heat and the inability to remove metabolic heat can lead to well-known reactions in humans, including increased cardiovascular activity, sweating, and increased body core temperature. A variety of heat stress disorders resulting from the body's reaction to excessive heat, in order of increasing severity, are heat rash, heat syncope, heat cramps, heat exhaustion, and heat stroke. In addition to the physiological effects of heat, overexposure to heat and accompany dehydration may precipitate psychological effects, including irritability, an increase in the frequency of errors, a higher frequency of accidents, and a reduction in efficiency in the performance of skilled physical tasks.

21.5.1 Heat Rash

Heat rash is a profuse, red vesicular rash accompanied by prickly sensations of areas affected by heat. Heat rash is caused by plugged sweat glands, retention of sweat, and an accompanying inflammatory reaction. This is precipitated by hot, humid conditions in which sweat cannot adequately evaporate. The most common heat rash is prickly heat (miliaria rubra), which appears as red papules, usually in areas where the clothing is restrictive, and gives rise to a prickly sensation, particularly as sweating increases. The papules may become infected unless they are treated. Another skin disorder (miliaria crystallina) appears with the onset of sweating in skin previously injured at the surface, commonly in sunburned areas, although it has been reported to occur without clear evidence of previous skin injury. Heat rashes disappear when the individuals are returned to cool environments. Heat rash may be treated by keeping the skin clean and with light applications of mild drying lotions or powders.

21.5.2 Heat Syncope

Heat syncope is fainting while standing immobile in a hot environment, and is caused by the pooling of blood in the lower extremities. Syncope is due to poor or inadequate acclimatization, and recovery is typically prompt and complete after moving the worker to a cooler environment. Syncope can be prevented by intermittent activity to assist the return of venous blood return to the heart.[12]

21.5.3 Heat Cramps

Heat cramps are painful spasms of the voluntary muscles (i.e., arm, leg, abdomen, or back) used during work. The onset may be during work or after several hours of heavy work or while showering. Heat cramps are caused by a reduction of the concentration of sodium chloride (salt) in blood below a certain critical level. Heat cramps are caused by a continued loss of salt in the sweat and accompanying dehydration, drinking large volumes of water without appropriate replacement of salt, depletion of body electrolytes, and lack of acclimatization. The dilution of tissue fluid and resultant transfer into muscle tissue precipitates the cramps. Heat cramps can be readily alleviated by replacing the water and salt in body fluids that has been lost and resting the muscle.

21.5.4 Heat Exhaustion

Heat exhaustion is a milder form of heat disorder linked to depletion of body fluids and electrolytes. This is a state of collapse due to insufficient blood supply to the brain. It is precipitated by dehydration, electrolyte loss, low arterial blood pressure, widespread vasodilation, or a compromised cariovascular system (i.e., competing demands for blood, poor physical fitness, alcohol consumption). Heavy exertion, heat, and poor acclimatization also are causative factors. The symptoms of heat exhaustion include fatigue, fainting, weak pulse, low blood pressure, headache, nausea, vertigo, weakness, giddiness, skin clammy and moist, complexion pale, muddy, or with hectic flush. Oral temperature is normal or low, but rectal temperature is usually elevated (37.5 to 38.5°C, or 99.54 to 101.34°F). The use of occlusive clothing limits skin evaporation, creating a rise in skin temperature. As skin temperature approaches core temperature, the blood has less capacity to transport heat from the core to the skin. Under these conditions, heat exhaustion can occur at core body temperatures less than 43°C (109.44°F) and with heart rates of 120 to 130 beats per minute.

Treatment includes rest in a cool place and liquids high in electrolytes. Administer fluids by mouth or give intravenous infusions of normal saline (0.9%) if patient is unconscious or vomiting. Keep at rest until urine volume and content indicate water and electrolyte balances have been restored.

Heat exhaustion can be prevented by proper acclimatization using a breaking-in schedule for 5 to 7 days. Ample drinking water must be available at all times during the work shift, and saline liquids or dietary salt supplements can be taken during acclimatization only.

21.5.5 Heat Stroke

Heat Stroke is a serious disorder that is linked to failure of the body's thermoregulatory mechanisms and can be life threatening. Sweat production, the normal body coping mechanism, is blocked, and, consequently, the opportunity for evaporative cooling is reduced. The body temperature rises and heart rate increases. The classic symptoms of heat stroke include (1) a major disruption of central nervous function (unconsciousness, convulsions, or coma), (2) lack of sweating, causing hot, dry, flushed, or cyanotic skin, and (3) a rectal temperature in excess of 41°C (105.8°F) continuing to rise if the condition goes untreated. Irreversible changes occur in the body organs beyond 40°C (104°F).

Heat stroke is a medical emergency, and immediate medical action is required. Any procedure from the onset that will cool the patient will improve the prognosis. Cooling ensures that the vasoconstriction of superficial blood vessels is avoided, as this hinders the body's cooling ability by diverting blood away from the skin. Cooling by placing the person in a shady area, removing the outer clothing, and tepid sponging and increasing air movement by fanning are all necessary and appropriate until professional methods of cooling and assessing the degree of the disorder are available. Excessive cooling with ice water should be avoided. Cooling attempts should be discontinued once the core temperature reaches 37.7°C (99.9°F). Treat for shock and anticipate possible cardiac arrest or cardiovascular collapse due to electrolyte imbalance. It is important that transfer to a hospital is quickly arranged so that further close monitoring can be undertaken to identify possible cardiovascular and renal problems and the extent of organ and tissue involvement. Deaths do occur, the majority within the first 12 hours, but the rest within 2 weeks.

Heat stroke typically occurs in young men undertaking moderate to hard physical exercise, and it is more likely to occur in those who are unacclimatized, obese, dehydrated, and inappropriately dressed and those who have consumed alcohol. Prevention involves medical screening of workers, and placement must be based on health and physical fitness. Workers must be acclimatized for 5 to 7 days by graded work and heat exposure and monitored during sustained work in severe heat.

21.6 MEASUREMENT OF THE THERMAL ENVIRONMENT

Heat stress refers to the total heat load on the body that is the contribution by both environmental and physical factors. The assessment of heat stress includes the use of environmental variables to describe the thermal environment and may also include variables to reflect how the environment exchanges heat with the worker. The degree of discomfort and the level of stress caused by environmental heat depend on air temperature, relative humidity, velocity of air movement, and the contribution of radiant heat sources.

21.6.1 Indexes of Heat Stress

There are several indexes that can be used for assessing heat stress. They range from simple dry-bulb and wet-bulb temperature measurements to algebraic combinations of multiple environmental variables, which may include correction factors and modifications. Several indexes of thermal stress are commonly used in the assessment of industrial heat stress. These include:

- The effective temperature (ET) index;
- The equivalent effective temperature corrected for radiation (ETCR) index, a modification of ET that helps to determine the contribution made by radiant heat;
- The predicted 4-hour sweat rate (P4SR);
- The heat stress index (HSI);
- The wet-bulb globe temperature (WBGT) index.

The development of heat stress indexes has been based on use of the thermometric scale, sweat rates, and calculations of heat loads and the evaporative capacities of the environment.

21.6.2 Dry-Bulb and Wet-Bulb Temperatures

The dry-bulb air temperature (TA) is the temperature of the ambient air as measured with a thermometer or equivalent instrument (see Section 21.7.1). It is the simplest climatic factor to measure.

The natural wet-bulb temperature (NWB) is the temperature measured by a thermometer covered by a wetted cotton wick and exposed only to the naturally prevailing air movement. Accurate measurement of NWB requires use of a clean wick, distilled water, and shielding to prevent radiant heat gain.

The psychrometric wet-bulb temperature (WB) is determined by forcing air over the wetted wick that covers the sensor. The WB is generally measured with a psychrometer, which consists of two mercury-in-glass thermometers mounted alongside each other on the frame of the psychrometer. One thermometer is used to measure the dry-bulb temperature (TA), and the second is used to measure the wet-bulb temperature. The air movement can be manual, as with a sling psychrometer, or mechanical, as with a motor-driven psychrometer.

These simple temperature parameters are easy to measure; however, they do not provide sufficient information to help determine the thermal exchange between the worker and the occupational environment.

21.6.3 Effective Temperature

Effective temperature is an index that combines dry-bulb temperature, wet-bulb temperature, and air velocity to estimate a thermal sensation to that of a given temperature of still, saturated air. When taking measurements, ensure that the dry-bulb thermometer is shielded from radiation. Determination of effective temperature also requires an anemometer to measure air velocity in feet per minute (fpm). Refer to Section 21.7.3 for more information on taking air velocity measurements. Nomograms were developed that characterized these equivalent environments, as shown in Fig. 21.1.[13]

To use the nomogram, connect the dry-bulb thermometer temperature to the wet-bulb temperature with a straight line. The effective temperature (ET) can be read where this line intersects the measured corresponding air velocity. Studies have shown that the risk of fatal heatstroke begins at 28°C ET, and this risk increased sharply above 33°C ET. There is little risk of heat stroke at less than 26°C ET.

21.6.4 Equivalent Effective Temperature Corrected for Radiation

The use of a black-globe temperature in place of a dry-bulb temperature will allow a correction for the contribution by surrounding radiation sources. This is known as the equivalent effective temperature corrected for radiation (ETCR). The nomogram in Fig. 21.1 is also used to determine the ETCR; however, the dry-bulb temperature reading is substituted by the black-globe temperature reading, and the ETCR is then determined in the same manner as the ET. This particular ET nomogram scale pertains to workers wearing lightweight summer clothing. There is another scale for seminude men called the "basic" scale.[13]

The limitations of ET and ETCR as heat stress indices are because of their basis on sedentary workers, with no consideration given towards the type of clothing worn and its inherent absorption/desorption properties (refer to Section 21.9.3).

Example: Given a dry-bulb temperature = 76°F, a wet-bulb temperature = 55°F, and air speed = 100 ft/min, find ET. Using the nomogram shown in Fig. 20.1, use a rule to join the dry-bulb and wet-bulb temperatures. At the point where the rule intersects with the air velocity line, read the effective temperature on the grid lines. In this case the ET = 67°F. That is, the environment in which these readings were taken would provide the same thermal sensation as one with dry-bulb and wet-bulb temperatures of 67°F and "still" air (approximately 25 ft/min). The ETCR can be determined by using the black-globe temperature reading in place of the dry-bulb temperature and following the same procedure.[13]

21.6.5 Predicted 4-Hour Sweat Rate

The predicted 4-hour sweat rate (P4SR) is an index based on observations of sweat rates under various environmental conditions. This index is expressed in liters and is a representation of the amount of sweat generated by a fit, well-acclimatized worker. It is believed that most workers will be unable to tolerate 4 hours of exposure as the P4SR rises above 4.5 liters. It has been recommended that the safe limit for exposure of unacclimatized workers is 2.5 to 3 liters.

The P4SR index requires globe temperature, wet-bulb temperature, air velocity, and metabolic rate of the workers. Although the index works well for temperatures that

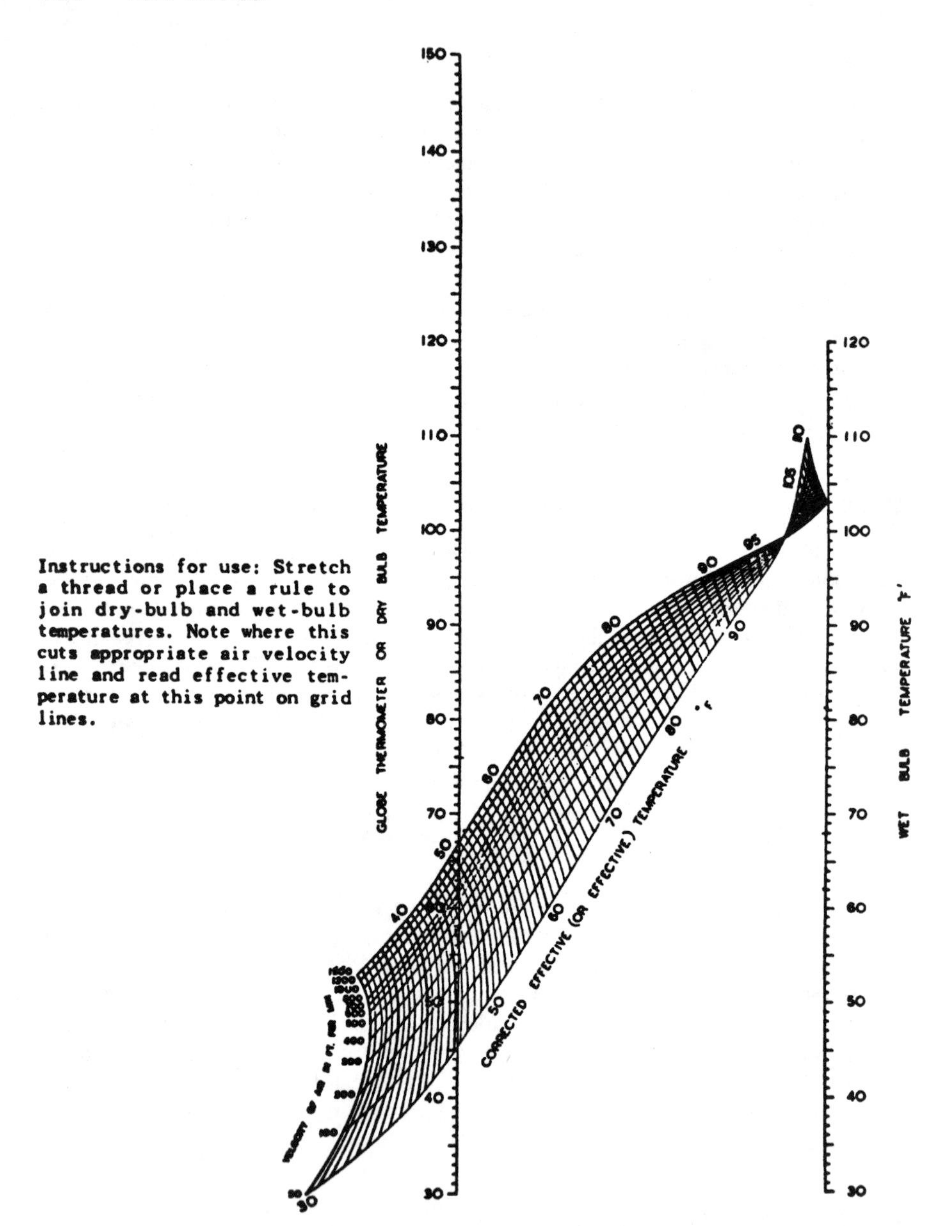

Fig. 21.1. Normal scale of corrected effective (or effective) temperature from *The Industrial Environment—Its Evaluation and Control*, 2nd ed., C. H. Powell and A. D. Hosey, Eds., Public Health Service Publication No. 614, Government Printing Office, Washington, D.C., 1965.

cause moderate sweating, it does not account for worker acclimatization. This index is probably of more value to physiologists than industrial hygienists.

1. *Calculation of the P4SR.* Calculation of the P4SR requires the use of a three-part nomogram, as displayed in Fig. 21.2.[14] The scale on the left represents the globe temperature. The five straight lines on the right constitute the psychometric wet-bulb scales, which are dependent upon air velocity. The third part of the nomogram consists of a group of curves running downward from right to left, which represent the basic 4-hour sweat rate (B4SR). The appropriate B4SR should be used depending on the air velocity of the environment.

a. *Modified Wet-Bulb.* The wet-bulb value requires modification in the following circumstances:

1. If the globe thermometer temperature differs from the dry-bulb temperature, the wet-bulb temperature is corrected by the addition of an amount equal to 0.4 (globe temperature minus dry-bulb temperature). Monitors similar to the one shown in Figure 21.4 which measure dry-bulb, wet-bulb and globe temperature are available.

2. If the energy expenditure exceeds 54 kcal/m^2/h (the metabolic rate of men sitting in chairs), an amount that is read off from the small inset chart is added to the wet-bulb temperature. The metabolic rate should be converted to an hourly rate and then divided by a factor of 1.8 to express it in kcal/m^2/h.

3. If the men are wearing clothing, imposing a greater stress than that imposed by shorts, an appropriate amount, dependent upon the amount of clothing worn, is added to the wet-bulb temperature. In the case of men wearing overalls over shorts, add 1.8°F.

b. *B4SR.* Draw a line between the globe temperature on the scale on the left and the modified wet-bulb temperature on the appropriate wet-bulb scale on the right. The B4SR is the point where this line intersects the appropriate curve for the given air speed on the B4SR scale.

c. *P4SR*

- Men sitting in shorts: P4SR = B4SR
- Men working in shorts: P4SR = B4SR + 0.014 (M – 54) where M is the metabolic rate in kcal/m^2 h
- Men sitting in overalls worn over shorts: P4SR = B4SR + 0.25
- Men working in overalls over shorts: P4SR = B4SR + 0.25 + 0.02 (M – 54)

2. *Exposure limits.* As P4SR rises above 4.5 liters, an increasing number of men will be unable to tolerate 4 hours of exposure. Lower P4SR values have never been very clearly related to physiological strain. It has been suggested that the safe limit for exposure of unacclimatized men may be 2.5 to 3 liters. Use the nomograph in Fig. 21.2 for the following example.

Example: This sample solution demonstrates how to calculate the P4SR.[13]

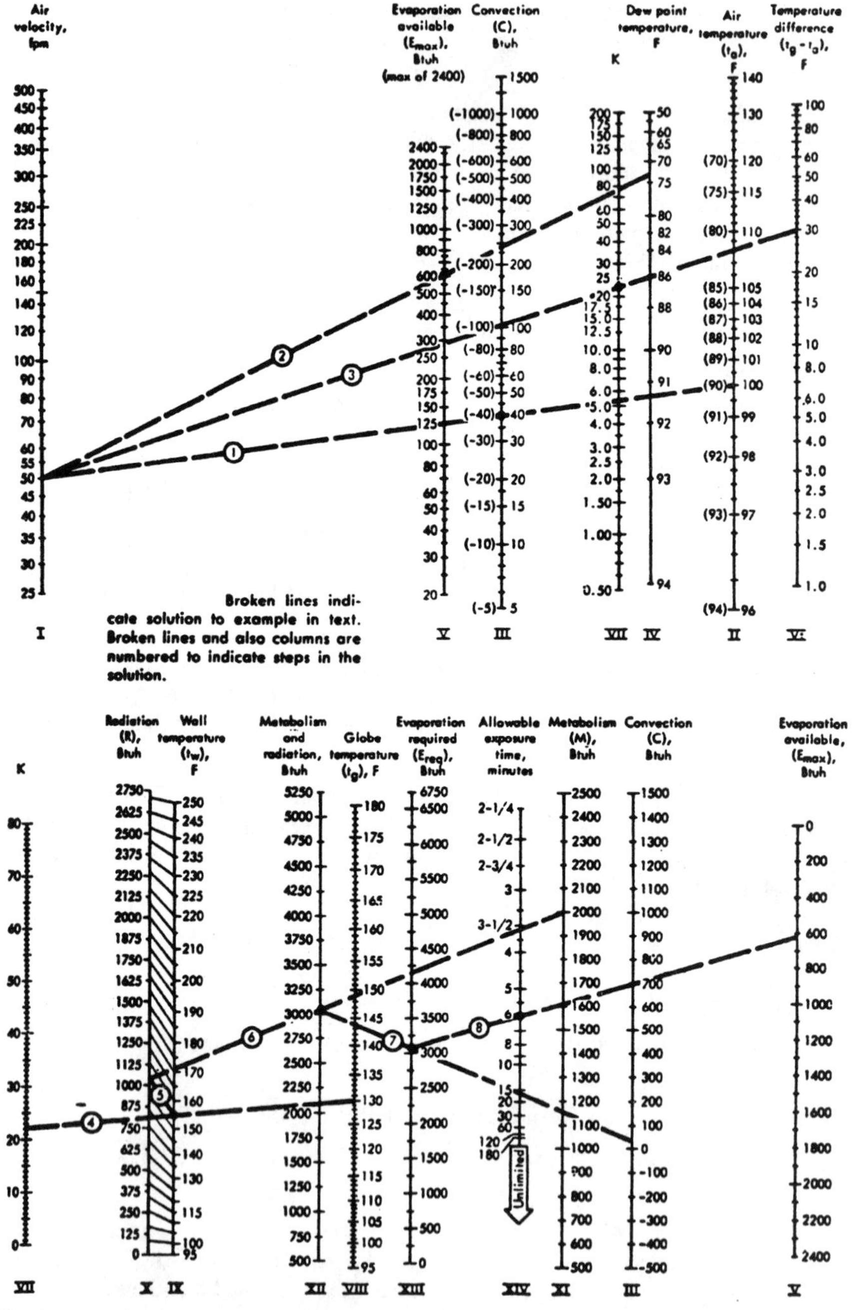

Fig. 21.2. Nomogram for the determination of heat stress index (HSI) and allowable exposure time (AET). Reprinted from *Heating/Piping/Air Conditioning*, 1976, Reinhold, Penton/IPC.

$$\begin{aligned} \text{Given:} \quad \text{GT} &= 105 \\ \text{WB} &= 80 \\ V &= 70\,\text{ft/min} \\ M &= 100\,\text{kcal/m}^2\,\text{h}. \end{aligned}$$

Step 1. From the small graph, find 4°F to be added to the wet-bulb temperature to compensate for M above resting.

Step 2. At the right side of the P4SR nomogram, enter T_{wb}(WB) = 84(80 + 4). Follow the 84 T_{wb} line to the intersection with the 70 ft/min line.

Step 3. Using a ruler, connect this point to T_g(GT) = 105.

Step 4. Read P4SR where this transverse line intersects with air velocity, $V = 70$. P4SR = 1.2

This value represents conditions that pose little physiologic strain, as the upper tolerance limit for fit young men dressed in shorts is P4SR = 4.5. This index is less accurate in predicting strain as the upper level of tolerance is approached.

21.6.6 Heat Stress Index

Another metabolic rate method to measure heat stress is the heat stress index (HSI). The HSI is generally used by heating, ventilating, and air conditioning engineers as a good diagnostic tool but does not provide an accurate assessment of heat stress in the work environment. This method was developed by Belding and Hatch at the University of Pittsburgh in the mid-1950s[15] and combines the heat exchange of radiation (R) and convection (C) components with metabolic heat (M) in terms of the required sweat evaporation (E_{req}). Stated algebraically:

$$E_{\text{req}} = M \pm R \pm C$$

The HSI is determined by 100 times the ratio of (E_{req}) to the maximum evaporative capacity of the environment (E_{max}).

$$\text{HSI} = \frac{E_{\text{req}}}{E_{\text{max}}} \times 100$$

Body heating occurs when the HSI exceeds 100, and body cooling occurs when the HSI is less than 100. Table 21.2 shows the physiological and hygienic implications of 8-h exposures to various levels of the HSI.[16]

Values of E_{req} and E_{max} may be computed by means of appropriate equations or by use of the nomogram method developed by McKarns and Brief[17] and shown in Fig. 21.2.

To use the nomogram, follow these steps:

1. Determine the convective heat load by extending a line from the air velocity scale

TABLE 21.2 Evaluation of Heat Stress Index

Index of Heat Stress	Physiological and Hygienic Implications of 8-hr Exposure to Various Heat Stresses
−20 −10	Mild cold strain. This condition frequently exists in areas where individuals recover from exposure to heat
0	No thermal strain.
+10 +20 +30	Mild to moderate heat strain. Where a job involves higher intellectual functions, dexterity, or alertness, subtle to substantial decrements in performance may be expected. In performance of heavy physical work, little decrement expected unless ability of individuals to perform such work under no thermal stress is marginal.
+40 +50 +60	Severe heat strain, involving a threat to health unless individuals are physically fit. Break-in period required for workers not previously acclimatized. Some decrement in performance of physical work is to be expected. Medical selection of personnel desirable because these conditions are unsuitable for those with cardiovascular or respiratory impairment or with chronic dermatitis. These working conditions are also unsuitable for activities requiring sustained mental effort.
+70	Very severe heat strain. Only a small percentage of the population may be expected to qualify for this work. Personnel should be selected (a) by medical examination and (b) by trial on the job (after acclimation). Special measures are needed to assure adequate water and salt intake. Amelioration of working conditions by any feasible means is highly desirable, and may be expected to decrease the health hazard while increasing efficiency on the job. Slight "indisposition," which in most jobs would be insufficient to affect performance may render workers unfit for this exposure.
+100	The maximum strain tolerated daily by fit, acclimatized young workers.

Source: Adapted from H. S. Belding and T. F. Hatch, "Index for Evaluating Heat stress in Terms of Resulting Physiological Strains," *Heat. Piping Air Cond.* **27:**129 (1955).

(column 1) to the air temperature scale (column II). The convective load is indicated at the intersection of this line with column III. Note whether the heat load is positive or negative, as indicated by an air temperature above or below 35°C (95°F), respectively.

2. Obtain the maximum available evaporative cooling (E_{max}) from column V by extending a line from the air velocity scale (column I) to the dew point temperature scale (column IV). The dew point temperature is obtained from a psychometric chart at the intersection of the lines for wet-bulb and dry-bulb air temperatures.

3. Determine K in column VII by extending a line from the air velocity scale to the temperature difference scale (column VI). Transfer the value of K to column VII in the second set of alignment charts.

4. Extend the line from the K scale to the globe-temperature scale (column VIII) and read the radiant wall temperature in column IX.

5. Project this value to the radiation scale (column X) by extending a line parallel to the given slanting lines.

6. Estimate the appropriate metabolic rate (see Section 21.6.7) and locate this value

of M on the metabolism scale (column XI). Connect columns X and XI and determine the sum of metabolism and radiation in column IX.

7. Transfer the convective load determined in step 1 to column III, noting whether it is negative or positive. Connect this point with the sum in column XII and read the required rate of evaporation (E_{req}) in column XIII.

8. Locate the available evaporative cooling in column V. Extend a line from this point to column XIII. The approximate allowable continuous exposure time (AET) is indicated at the intersection of this line with column XIV.

The heat stress index provides an analysis tool for occupational heat exposure that also serves as a predictor of the "allowable exposure time" (AET).[18] For an average man, AET is given by the equation:

$$\text{AET} = \frac{250 - 60}{E_{req} - E_{max}}$$

The HSI loses some applicability at very high heat stress conditions. It also does not identify correctly the heat stress differences resulting from hot, dry and hot, and humid conditions. The strain resulting from metabolic versus environmental heat may not be differentiated because E_{req}/E_{max} is a ratio that may disguise the absolute values of the two factors.[19]

The HSI not only provides an excellent starting point for specifying corrective measures for heat stress, but also offers information that helps to determine feasible engineering control measures.

Example: This sample solution demonstrates how to use the Belding and Hatch nomogram to find E_{req}, HSI, and AET.[13]

$$\begin{aligned} \text{Given}: \text{GT} &= 130°\text{F} \\ \text{TA} &= 100°\text{F} \\ \text{WB} &= 80°\text{F} \\ V &= 50\,\text{ft/min} \\ M &= 2000\,\text{Btu/h} \end{aligned}$$

Step 1. Determine C (convection). Connect 50 fpm (column I) with TA = 100°F (column II). Read $C = 40$ Btu/h (column III).

Step 2. Determine E_{max} (maximum evaporative heat loss). From a psychometric chart using the dry-bulb and wet-bulb temperatures, read the dew point of 73°F. Connect 50 fpm (column I) to the dew point = 73°F (column IV). Read $E_{max} = 620$ (column V).

Step 3. Determine the constant, K. Connect $V = 50$ fpm (column I) with $T_g - T_a$ (GT − TA) or (130 − 100) = 30 (column VI). Read $K = 22$ (column VII).

Step 4. Determine T_w (mean radiant temperature). Enter $K = 22$ in column VII. Connect this to $T_g = 130$, column VIII. Read $T_w = 155$ (column IX).

Step 5. Determine R (radiant heat exchange). Follow the slanting line to column X; read $R = 1050$ Btu/h.

Step 6. Determine $R + M$ (radiation and metabolism). Connect $R = 1050$ with $M = 2000$ (column XI). Read $R + M = 3050$ on column XII.

Step 7. Determine E_{req} (requirement for evaporation of sweat). Enter $C = 30$ in column III of the lower figure. Connect with $R + M = 3050$ on column XII. Read $E_{req} = 3090$ Btu/h on column XIII.

Step 8. Determine AET (allowable exposure time). Enter E_{max} in column V of the lower diagram. Connect with $E_{req} = 3090$ Btu/h (column XIII). Read AET = 6 min.

Compute the HSI:

$$\text{HSI} = \frac{3090}{620} \times 100 = 500$$

An HSI value in excess of 100 is viewed as the maximum tolerated strain tolerated daily by fit, acclimatized young men.

21.6.7 Job Ranking

Job ranking involves the establishment of categories for each job into light, medium, and heavy based on the type of operation. The permissible heat exposure limit for that job can be determined by use of the ACGIH table in Fig. 21.3.[20] Some examples of job ranking are as follows:

1. Light work (up to 200 kcal/h or 800 BTU/h); e.g., sitting or standing to control machines, performing light hand or arm work.
2. Moderate work (200-350 kcal/h) or 800-1400 BTU/h); e.g., walking about with moderate lifting and pushing.
3. Heavy work (350-500 kcal/h or 1400-2000 BTU/h); e.g., pick and shovel work.

Table 21.3 shows some recommended work/rest regimen strategies to help stay within the permissible heat exposure limits.

Table 21.4 also lists energy expenditures for various types of tasks.

21.6.8 Wet-Bulb Globe Temperature

The WBGT index has proved very successful in monitoring heat stress and minimizing heat casualties in the United States and has been widely adopted. The WBGT index provides a fast and convenient method to quickly assess conditions that pose threats of thermal strain. It has been adopted as the principal index for the TLV for heat stress established by the ACGIH.[21] NIOSH endorses WBGT as the preferred measure of severity of occupational exposures to heat stress.[22,19] The main criteria used by NIOSH for the selection of a suitable index were that (1) the measurements and calculations must be simple, and (2) index values must be predictive of the physiological strain of heat exposure.

The WBGT index is an algebraic approximation of the effective temperature concept. Air velocity does not have to be measured directly to calculate the intensity of WBGT, as allowances are made for this factor by the use of the naturally convected wet-bulb sensor.

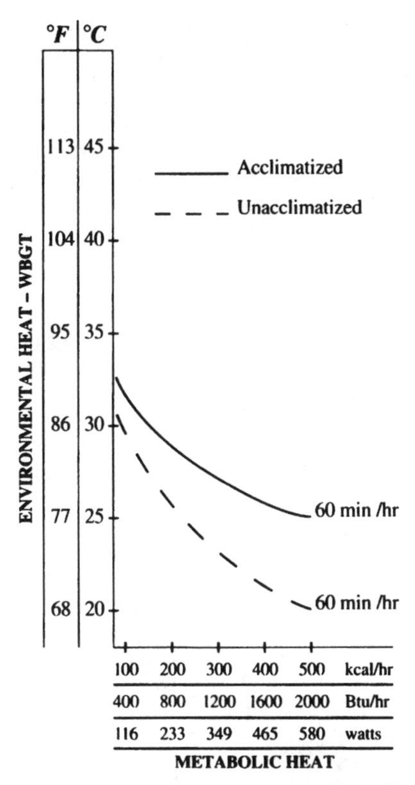

Fig. 21.3. ACGIH permissible heat exposure threshold limit values.

For outdoor use with solar load, the index is derived from the formula:

$$\text{WBGT} = 0.7\text{NWB} + 0.2\text{GT} + 0.1\text{TA}$$

where NWB is the natural wet-bulb temperature; GT, the globe temperature; and TA, the dry-bulb (air) temperature. For indoor use, the weighted expression becomes:

$$\text{WBGT} = 0.7\text{NWB} + 0.3\text{GT}$$

TABLE 21.3 Permissible Heat Exposure Threshold Limit Values [WGBT-°F (°C)]

Work load (BTU/h)	Continuous work	75% Work, 25% rest each hour	50% Work, 50% rest each hour	25% Work, 75% rest each hour
Light (800)	86 (30.0)	87 (30.6)	89 (31.4)	90 (32.2)
Medium (1400)	80 (26.7)	82 (28)	85 (29.4)	88 (31.1)
Heavy (2000)	77 (25)	79 (25.9)	82 (28)	86 (30)

TABLE 21.4 Energy Expenditures, *M*, for Various Activities

Activity	*M* (BTU/h)	Activity	*M* (BTU/h)
Typing, electrical[a]	227–330	Hoeing	1045
Typing, mechanical[a]	300–375	Mixing cement	1115
Lying at ease	334–360	Walking on the job	1165–1610
Sitting or standing at ease	405–450	Pushing a	1190–1660
Draftsman, drilling machine, light assembly line	430	Shoveling	1285–2495
Armature winding, printer	525	Digging trenches	1425–2090
Light machine work, machine wood sawing	570	Gardening, digging Brush clearing	1450
Medium assembly work	650	Sawing wood	1500–1780
Driving a car	670	Forging	1520–1595
Sheet metal worker	715	Furnace tending	1595–3850
Casual walking	715–925	Scrubbing, hand drilling wood	1665
Machinist	740	Cross cutting with bucksaw	1780–2500
Drilling rock, drilling coal	900–2255	Climbing stairs or ladder	1830–3140
Weeding	905–1855	Planing wood	1925–2160
Bricklayer	950	Tree felling	1950–3020
Timbering	975–2140	Trimming felled trees	2070–2760
Machine fitting, tractor plowing, grass cutting	1000–1200	Slag removal	2520–3000

[a]Women.

The WBGT has limitations as a heat stress index, especially at high levels of severity.[23] Studies have shown clearly that environmental combinations yielding the same WBGT levels result in different physiological strains in individuals working at a moderate level.[23] This problem is compounded when impermeable clothing is worn, because evaporative cooling (wet-bulb temperature) will be limited, and in this instance, the WGBT values are irrelevant. Correction factors for clothing are found in Table 21.6 (see p. 818). Nevertheless, WBGT has become the index most commonly used and recommended throughout the world.

When taking WGBT measurements, the dry- and wet-bulb thermometers should be shielded from the sun and other radiant surfaces without restricting airflow around the bulb. The globe thermometer is a 6-in.-diameter hollow copper sphere painted on the outside with a black matte finish.

21.6.9 Wet-Globe Temperature

The wet-globe thermometer includes a hollow, 3-in. copper sphere covered by a black cloth that is kept at 100-percent wetness from a water reservoir. The wet sphere exchanges heat with the environment by the same mechanism that a person with a totally wetted skin would use in the same environment. In this regard, heat exchange by convection, radiation, and evaporation is integrated into a single instrument reading.[24]

During the past several years, the wet-globe temperature (WGT) has been used in many laboratory studies and field situations, where it has been compared with the WGBT.[25–29] In general, the correlation between the WGT and WBGT is high. Nevertheless, the relationship between the two is not constant for all combinations of environmental variables. A simple approximation of the relationship is WBGT = WGT + 2°C for conditions of moderate radiant heat and humidity and is applicable for general monitoring in industry.[19]

21.7 INSTRUMENTATION

The basic parameters measured to help determine the level of heat stress are temperature, humidity, air velocity, and radiant heat from the sun or infrared sources. Technological advances have rendered instrumentation more compact and bundled for ease of handling and use in the field. There are many instrument manufacturers who have heat stress monitoring devices available. The type of instruments needed for assessment of heat stress depends on the heat stress index to be used for the assessment and the parameters needed to use the chosen index.

21.7.1 Thermometers

A wide variety of thermometers is available for use in measuring temperature. The liquid-in-glass thermometer is probably the most commonly used, is relatively inexpensive, and comes in many temperature ranges. Bimetallic thermometers are generally of the type used in dial thermometer configurations. Thermocouples are made of a wires of two different metals and operate on the principle of electromotive force variation at the junction of the wires. A thermistor is a semiconductor that will show a significant change in resistance of the metal wire with even small temperature changes.

Air temperature measurements can easily be affected by the presence of surrounding surfaces that vary significantly from the ambient air temperature. Thermocouples and thermistors are less sensitive to this problem than are liquid-in-glass thermometers because of their small size. It is possible to shield the thermometers from these sources by the use of heavy aluminum foil and/or by increasing air flow over the sensor.

21.7.2 Humidity

Humidity is a representation of water vapor presence in the atmosphere. For heat stress assessment, the relative humidity can be measured by a psychrometer. A psychrometer is an instrument with both a dry-bulb and a wet-bulb temperature sensor. The wet-bulb sensor is covered by a wetted wick, which is cooled by evaporation that is the result of air movement of at least 900 feet per minute (fpm).[30] This air movement can be achieved by a small fan or squeeze bulb that is part of the instrument or by the use of a sling psychrometer. The sling psychrometer is hand spun by the operator to achieve this effect. The WGBT index uses a natural wet-bulb (NWB) temperature, which is not exposed to forced air movement, making this monitoring parameter always less than or equal to the psychometric wet-bulb temperature.

Hygrometers are direct reading instruments that measure relative humidity or dew point that are generally found in humidity recorders and control instruments. As a result, they have little use in field heat stress assessments and are better suited for obtaining data in laboratory or other controlled settings.

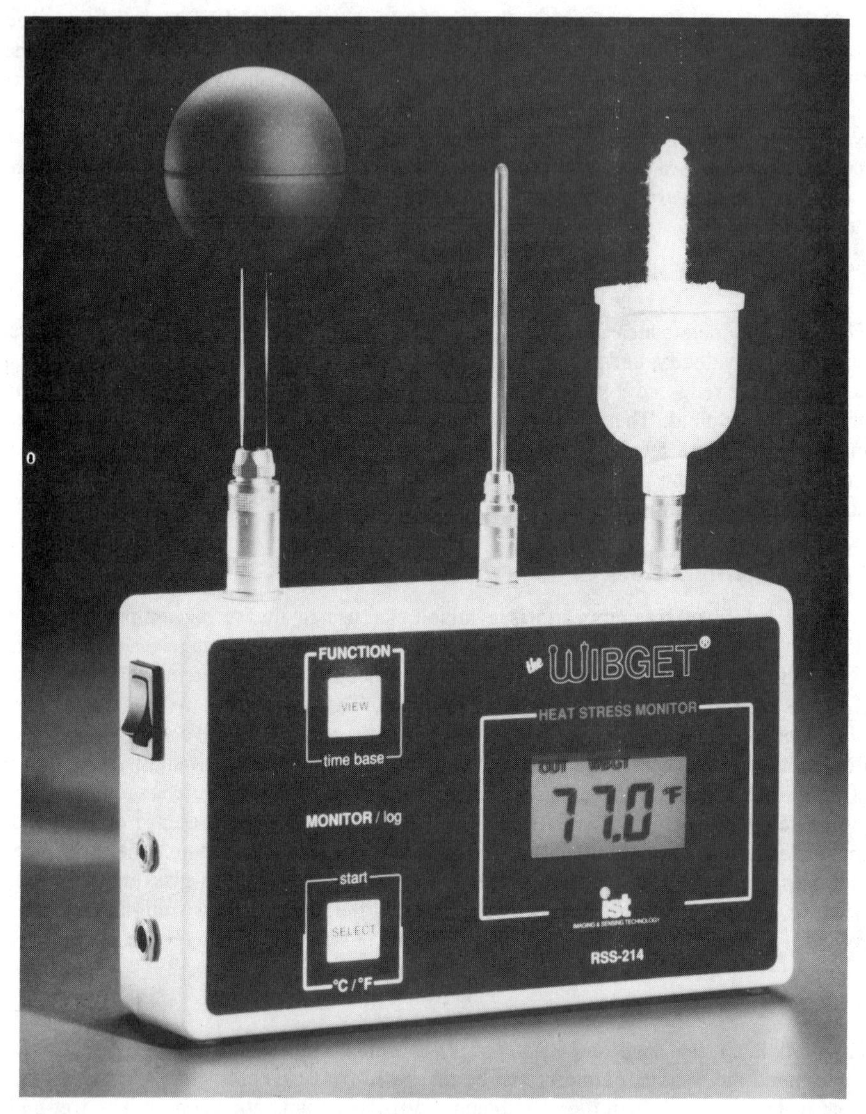

Fig. 21.4. Heat stress monitor that measures dry-bulb, wet-bulb, and globe temperature.

21.7.3 Air Velocity

Thermal anemometers are the best choice for measuring nondirectional air movement at relatively low air velocities. This instrument passes air over a heated sensor, which responds when it is cooled by the air mass flow. Heated thermocouple anemometers operate by sensing the temperature difference between a heated and unheated thermocouple that are exposed to the air flow. Hot-wire anemometers use a fine wire that is heated and measure the variation of temperature and electrical resistance of the air flow to which it is exposed. Vane anemometers are better suited for high air velocities and may be of the rotating type, which measure revolutions per elapsed time, or the swinging vane type, which measure deflection of the vane.

21.7.4 Radiant Heat

The black-globe thermometer is probably the best choice for measuring thermal radiation. The radiant heat is absorbed by the black surface of the sensor, causing the globe temperature (GT) to be greater than the air temperature. As the globe cools by convection, the temperature stabilizes when the heat gained by radiation equals the rate of heat lost by convection. This effect may take as long as 15 minutes to complete. The exchange of radiant heat with a worker is best represented by the mean radiant temperature (MRT). This can be calculated with the globe temperature, air velocity, and air temperature with the following equation:[19]

$$\mathrm{MRT} = \mathrm{GT} + 1.8\mathrm{V} \times 0.5(\mathrm{GT} - \mathrm{TA})$$

where MRT is the mean radiant temperature in °C, GT is black-globe temperature in °C, V is the air velocity in m/sec, and TA is the dry-bulb temperature in °C.

21.8 CONTROL MEASURES

A number of elements should be considered when attempting to control heat stress. These include bodily heat production, number and duration of exposures, heat exchange components as affected by environmental factors, thermal conditions of the rest area, and worker clothing and protective equipment.[13] Before selecting control measures, it is crucial to identify the components of heat stress to which workers are exposed in new operations. Using the basic heat balance equation, $S = (M - W) + C + R - E$, heat stress can be reduced by modifying metabolic heat production $(M - W)$ or environmental heat production. Environmental heat load (C, R, and E) responds to engineering controls (e.g., ventilation, air conditioning, screening, insulation, and modification of process or operation), as well as protective clothing and equipment. Metabolic heat production $(M - W)$ can be modified by work practices and application of labor-reducing devices.

21.8.1 Control at the Source

The most fundamental approach to engineering control of heat in the workplace is eliminating heat at its point of generation. One feasible alternative is to change the operation or substitute a process component of lower temperature for one of higher temperature. One example is the use of induction heating rather than direct-fired furnaces for certain forging operations. Heat is controlled most effectively if it is regulated at the source and the options for control of heat at the source are isolation, reduction in emissivity, insulation, radiation shielding, and local exhaust ventilation.

The most practical method for limiting heat exposure from hot processing operations that are difficult to control or for operations that are extremely hot is to isolate the heat source. Such operations might be partitioned and separated from the rest of the facility, located in a separate building, or relocated outdoors with minimal shelter (e.g., an industrial boiler, segregated from other operations in the same facility).

The rate at which heat is radiated from the surface of a hot source can often be lowered if the emissivity of the source is reduced through surface treatment. The emissivity of a hot source can be lowered if the emissivity of the source is reduced through surface treatment. The emissivity of a hot source can be lowered by painting with aluminum paint or covering the source with sheet aluminum. When an oven, boiler, or other hot surface is covered with aluminum paint or sheet aluminum, less heat is radiated to workers nearby and heat is conserved inside the unit, representing a substantial savings in energy cost.

Insulation is not mutually exclusive of "isolation" in the context of engineering control of heat stress. Insulation also prevents the escape of sensible and radiant heat into the work environment. An example of insulation that also has implications for energy conservation is that of pipe-covering insulation on steam lines. In addition to reducing radiant heat exchange, insulation reduces the convective heat transfer from hot equipment to the work environment by minimizing local convective currents that form when air that contacts very hot surfaces is heated.

When pipe leakage occurs in urban heating systems, it is often difficult to close the circuit in order to perform the necessary repairs. In the event of an escape of liquid and steam in environments that may be confined spaces, workers may be exposed to temperatures of 80 to 85°C (176–185°F). Although these exposures are infrequent and may be short term, workers must be protected from the high heat and burn potential posed by the steam. In these instances, lockout/tagout procedures must be implemented to control hazardous energy. In the event that this is not feasible, measures should be taken to insulate or shield workers from the hazards.

Workers who operate high-temperature furnaces throughout the day may face significant radiant heat exposures. One control strategy for this type of exposure is the use of heat-reflective curtains or shields between the furnace and the worker. Workers would stand behind this barrier whenever they do not have to tend the furnace to insert or remove work. Frequent breaks away from the high heat area and electrolyte replacement are also an important part of worker protection in this type of environment.

Radiation shielding represents an extremely important control measure. Radiant heat passes through air without heating the air; it heats only the objects in its path that are capable of absorbing it. Shielding of radiant heat sources means putting a barrier between the worker and the source to protect the worker from being a receptor of the radiant energy. Radiation shielding can be classified into reflecting, absorbing, transparent, and flexible shields.

Reflective shields are constructed from sheets of aluminum, stainless steel, or other bright-surface metallic materials. The advantage of using aluminum is 85 to 95 percent reflectivity, and the successfulness of using aluminum as shielding depends on the following:

1. There must be an aluminum-to-air surface; the shield cannot be embedded in other materials.

2. The shield should not be painted or enameled.

3. The shield should be kept free of oil, grease, and dirt to maximize reflectivity.

4. The shield should be separated from the hot source by several inches when used as an enclosure.

5. Corrugated sheeting should be arranged so that the corrugations run vertically to help maintain a surface free of foreign matter.

21.8.2 Local Exhaust Ventilation

Canopy hoods with natural draft or mechanical exhaust ventilation are used commonly over furnaces and hot equipment. Because heat has a tendency to rise, it must be remembered that local ventilation removes only convective heat. Also, as the temperature of air increases, the volume increases as a result of expansion; therefore, the ventilation system must be designed with this in mind. Radiant energy losses, whose magnitude often overrides convective losses, are not controlled by local exhaust hooding. Radiation shielding must be used as well to control what is likely to be the larger fraction of the total heat escaping from the hot process.[14] More information regarding design of ventilation systems can be found in Chapter 15.

21.8.3 Localized Cooling at Work Stations

Relief can be provided at localized areas by the introduction of cool air in sufficient quantities to surround the worker with an "independent" atmospheric environment. This local relief or "spot cooling" serves two functions, depending on the relative magnitudes of the radiant and convective components in the total heat load. If the overall thermal load is primarily convective in the form of hot air surrounding the worker, the local relief system displaces the hot air immediately around the individual with cooler air having a higher velocity. If such air is available at a suitable temperature and is introduced without mixing with the hot ambient air, no further cooling of the worker is necessary. The introduction of cooling air must not interfere with local exhaust ventilation systems used for contaminant control.

When there is a significant radiation load, the local relief system must provide actual cooling to offset the radiant energy that penetrates the mass of air surrounding the worker. The temperature of the supplied air must be low enough to make the convection component (C) negative in the heat exchange model to offset the radiation component (R).

In extreme heat, workers should be stationed inside an insulated, locally cooled observation booth or relief room to which they can return after brief periods of high heat exposure. Air-conditioned crane cabs represent one application of this concept now in common use.

21.8.4 General Ventilation

A common method for heat removal in the hot industries is general ventilation by utilizing wall openings for the entrance of cool outside air and roof openings for the discharge of heated air. For an ideal system of general ventilation, combined with radiation shielding, the outside air must be cooled either by an evaporative cooler (i.e., water sprays or wetted filters) or a chilled coil system before it is distributed throughout the plant. The inlet air should enter near floor level, be directed toward the workers, and flow toward the hot equipment; thus the coolest air available is received by the workers before its temperature is increased by mixing with warm building air or circulation over hot processes. This air then flows toward the hot equipment and, as its temperature increases, rises and escapes through vent openings in the roof. Provisions for proper distribution of the air supply should receive the same consideration that is given to the selection of exhaust equipment. The basic strategy when general ventilation is used to remove heated air is to position the exhaust openings, either natural draft or mechanically operated, above the sources of heat and as close to them as practical.[15] The key point with air movement is that any air movement can promote cooling by evaporation even if the air temperature is above that of the skin.

21.8.5 Moisture Control

Moisture control includes both prevention of increased humidity and the use of dehumidification procedures. Effective controls such as enclosing hot water tanks, covering drains carrying hot water, and repairing leaking joints and valves in steam pipes offer direct measures to alleviate heat stress in warm-moist industries. Dehumidification can be accomplished by refrigeration, absorption, or adsorption. In the context of occupational heat exposures, refrigeration is the most widely used technique to condition the air in relief areas, operating booths, or other local or regional portions of an industrial facility.

21.9 MANAGEMENT OF EMPLOYEE HEAT EXPOSURE

21.9.1 Education and Training

The development of an education and training program is an important aspect of controlling heat stress. The program should include workers and supervisors and should be conducted prior to assignment on jobs where heat exposure is a concern and periodically as determined by the level and frequency of exposure. Workers should be educated to the dangers of work in hot conditions and the early signs and symptoms of heat fatigue, heat exhaustion, and heat stress in themselves and in their colleagues. Supervisors and selected personnel should be trained to recognize the signs and symptoms of heat disorder and emergency first aid measures. They must be trained in the safe use of any personal protective clothing with which they are supplied. Training must also explain the need to take frequent breaks and not to accumulate them with the aim of leaving work early at the end of the day. The need to report accidents, illnesses, and ill health as early as possible needs to be emphasized and that failure to report illness promptly may lead to a worsening of conditions with the potential for serious physical consequences.

The importance of acclimatization should also be stressed. The body will adapt physiologically over a few days of heat exposure so that it becomes more efficient at dealing with raised environmental temperatures. It does not, however, allow the body to tolerate a raised core temperature. Sweat output increases, and the pulse rate and deep body temperature decrease. This adaptation, however, may be lost in as little as 3 days away from work; so workers should be aware of the increased demands they will place on their bodies after a holiday, long weekend, or a period of illness.

During hot work regular replacement of fluids is needed, the amount and composition of which will depend on the physical effort involved and the ambient temperatures. Physical training should be encouraged, as it aids the body's ability to cope with the increased demands that heat places on the body. Instruction on the possible combined effects of heat and alcoholic beverages, prescription and nonprescription drugs, and other physical agents should also be provided.

21.9.2 Medical Supervision

A health monitoring program should include a preplacement physical examination and history, with concentration on the cardiovascular, metabolic, renal, skin, and pulmonary

TABLE 21.5 Heat Stress Evaluation Form (Sample)[2]

Name ____________________ SS#__________ DOB ______________

Date ___________ Department/Job Description ________________________

Yes	No	History of heat tolerance (e.g., heat stroke, heat exhaustion)
Yes	No	History of nonacclimatability
Yes	No	More than 40 years old, no history in hot environments
Yes	No	Obese (body fat greater than 15%)
Yes	No	Weight less than 50 kg
Yes	No	Skin disease symptomatology over large areas
Yes	No	History of alcohol or substance abuse
Yes	No	Hypertension (greater than 160 mm/95 mm)
Yes	No	Diuretics, anticholinergic, vasodilators, antihistamines, CNS inhibitors, muscle relaxants, atropine, MAOs, sedatives, beta blockers, amphetamines
Yes	No	Organic disease of the heart/vascular system
Yes	No	COPD, active lung disease, asthma
Yes	No	Liver, renal, endocrine, metabolic, digestive disease
Yes	No	Pregnancy
Yes	No	Infertility problems
Total # of positive risks_____		Recommendations __________________

Signature ___________________________ Date ___________

systems. Previous intolerance of hot environments or pregnancy are reasons for special consideration. It is recommended that occupational health professionals refer to a checklist (Table 21.5) at the time of employment and review annually those clients at risk. Each facility should determine the degree of severity and the number of positive risk factors that should be considered to preclude employment in high heat environments. After employment, workers should receive periodic examinations and annual examinations after the age of 45.[4] Ongoing attention should be placed on nutritional status, weight gain, and accident and injury records. Management and health care professionals should anticipate and prepare for unseasonably hot weather, summer temperature changes, and heavy production requirements.

21.9.3 Effect of Clothing on Heat Exchange

Clothing can have a profound effect on the heat exchange process. It is the insulating effects of clothing that reduce heat loss to the environment. When it is cold, of course, reduced heat loss is beneficial, but when it is hot, clothing interferes with heat loss and can be harmful.

The insulation value of most materials is a direct linear function of its thickness. The material itself (whether it is wool, cotton, or nylon) plays only a minor role. It is the amount of trapped air within the weave and fibers that provides the insulation. If the

TABLE 21.6 clo Insulation Values for Individual Items of Clothing

Men's clothing	clo	Women's clothing	clo
T-shirt	0.09	Bra & panties	0.05
Underpants	0.05	Half-slip	0.13
Light-weight short sleeve shirt	0.14	Light-weight blouse	0.20
Light-weight long sleeve shirt	0.22	Light-weight dress	0.22
Light-weight trousers	0.26	Light-weight slacks	0.26
Light-weight sweater	0.27	Light-weight sweater	0.17
Heavy sweater	0.37	Heavy sweater	0.37
Light jacket	0.22	Light jacket	0.17
Heavy jacket	0.49	Heavy jacket	0.37
Socks	0.04	Stockings	0.01
Shoes (oxfords)	0.04	Shoes (pumps)	0.04

material is compacted or gets water-soaked, it loses much of its insulating properties because of the loss of trapped air.

The unit of measure for the insulating properties of clothing is the *clo*. The *clo unit* is a measure of the thermal insulation necessary to maintain in comfort a sitting, resting subject in a normally ventilated room at 70°F (21°C) and 50 percent relative humidity. Because the typical individual in the nude is comfortable at about 86°F (30°C), one clo unit has roughly the amount of insulation required to compensate for a drop of about 16°F (9°C). The typical value of clothing insulation is about 4 clo per inch of thickness (1.57 clo/cm). clo values for some typical articles of men's and women's clothing are given in Table 21.6. The formula for computing overall clo values for a clothing ensemble is:

$$\text{Total clo units} = 0.8 \times (\text{sum of individual items}) + 0.8$$

Another feature of clothing that affects heat transfer is the permeability of the material to moisture. It is the permeability that permits evaporative heat transfer through the fabric. In general, the greater the clo value of the fabric, the lower is its permeability. The index of permeability (im) is a dimensionless unit that ranges from 0.0 for total impermeability to 1.0 if all moisture that could be evaporated into the air could pass through the fabric. Typical im values of most clothing materials in still air is less than 0.5. Water-repellant treatments, very tight weaves, and chemical protective impregnation can reduce the im values significantly. This is an issue when trying to protect workers wearing impermeable clothing, such as that used at hazardous waste sites or for emergency response activities. It is imperative that these workers are closely monitored and that a strict work-rest regimen is planned and enforced. In hot environments, evaporation of sweat is vital to maintain thermal equilibrium, and materials that interfere with this process can result in heat stress. In a cold environment, if evaporation of sweat is impeded, a garment can become soaked with perspiration, thus reducing its insulating capacity.

TABLE 21.7 TLV WBGT Correction Factors in °C for Clothing

Clothing type	clo[a] value	WBGT correction
Summer work uniform	0.6	0
Cotton coveralls	1.0	−2
Winter work uniform	1.4	−4
Water barrier, permeable	1.2	−6

[a]clo: insulation value of clothing. One clo unit = 5.55 kcal/m^2/h of heat exchange by radiation and convection for each °C of temperature difference between the skin and adjusted dry-bulb temperature (the average of the ambient air dry temperature and the mean radiant temperature[20].

21.9.4 Protective Clothing

Water-cooled garments include (a) a hood that provides cooling to the head, (b) a vest that provides cooling to the heat and torso, (c) a short undergarment that provides cooling to the torso, arms, and legs, and (d) a long undergarment that provides cooling to the head, torso, arms and legs. None of these water-cooled garments provides cooling to the hands and feet. Water-cooled garments and headgear require a battery-driven circulating pump and container where the circulating fluid is cooled by the ice. The amount of ice available determines the effective time of the water-cooled garment.

Air-cooled suits and/or hoods that distribute cooling air next to the skin are available. The total heat exchange from sweat-wetted skin when cooling air is supplied to the air-cooled suit is a function of cooling-air temperature and cooling-airflow rate. Both the total heat exchanges and the cooling power increase with cooling airflow rate and decrease with increasing cooling air inlet temperature. Attaching a vortex tube to the worker with a constant source of compressed air supplied through an air hose is a method of body cooling in many hot industrial situations. However, the vortex tube is noisy, and the hose limits the area in which the worker can operate.

Cooling with an ice packet vest will vary with time and with its contact pressure with the body surface, plus any heating effect of the clothing and hot environment. Because the ice packet vest does not provide continuous and regulated cooling over an indefinite period of time, exposure to a hot environment would require replacement of the frozen vests every 3 to 4 hours. Ice packet vests can add an additional 10 to 15 pounds of weight, which can increase the metabolic load and negate much of the benefits. However, ice packet vests allow increased mobility of the worker unlike other control measures. The greatest potential for use of an ice packet vest is short-duration tasks and emergency repairs, and it is relatively less expensive than other cooling approaches.

Metallized reflecting fabrics can provide protection against radiant heat, but if their surfaces are damaged or become dirty, then their effectiveness is reduced. Workers need to be trained and aware of the limitations of clothing, which may also need to have other characteristics such as flame resistance. The majority of workers in hot environments will not be wearing protective clothing and will don the minimum of garments so as to reduce heat retention. Wetted cottons tend to be cooler and more comfortable to wear than mammade fabrics.

21.9.5 Work–Rest Regimen

Shortening the duration of exposure and/or increasing the frequency and length of rest periods, allowing workers to self-limit exposure, are administrative measures that may be taken to control heat stress. The length of the work/rest is dependent on the nature of the work and the environmental conditions. Workers should rest before becoming fatigued and should remain at rest until the heart rate drops below 100 beats per minute. Air-conditioned (about 24°C), low-humidity rest areas speed the rate and degree of recovery. Work load can be modified by the use of mechanized tasks and shared work loads.

According to NIOSH, there are several ways to control the daily length of time and temperature to which a worker is exposed to heat stress conditions:

- Schedule hot jobs for a cooler part of the day (early morning, late afternoon, or evening), if possible
- Schedule routine maintenance and repair work in hot areas for the cooler seasons of the year
- Alter the work-rest regimen to permit more rest time
- Provide shaded or air-conditioned for rest and recovery
- Use extra personnel to reduce the exposure time of each member of the work team
- Employ a buddy system for high-risk tasks
- Permit freedom to interrupt work when a worker feels extreme heat discomfort

21.10 SUMMARY

The success of any heat stress program begins with hazard assessment. Processes or operations where heat stress may be a concern should be reviewed prior to installation for engineering controls needed to minimize exposure. For operations already in place, a field assessment using one of the heat stress indices described earlier should be performed to determine if additional engineering controls are needed. The use of personal protective equipment and work–rest regimens can be effective measures to reduce exposures until engineering controls are in place or when they are not feasible. Fitness, acclimatization, and hydration are key elements to any heat stress program; however worker education and awareness are probably the most important aspects of keeping at-risk workers healthy in heat stress situations. It is important to realize that heat stress can be an insidious occupational hazard when it is not well understood by workers and management.

REFERENCES

1. *Patty's Industrial Hygiene and Toxicology*, Fourth Edition, Volume 1, Part A, p. 763.
2. Heat Stress Disorders, Old Problems New Implications," *AAOHN Journal*, August 1991, M. V. Barrett, p. 269.
3. *Patty's*, p. 773.
4. "Heat Stress—Its Effect and Control," *AAOHN Journal*, 1993, **41**(6):268–74.
5. NIOSH (1993) Chapters 30 and 31. The Industrial Environment.

6. Smith, N. J., and Slarinski, C. L. (1987). *Sports Medicine*, W. B. Saunders, Philadelphia, pp. 161–63.

7. NIOSH, 1972.

8. Dukes-Dubos, F. N., and Henschel, A. (1980). *Proceedings of a NIOSH Workshop on Recommended Heat Stress Standards*, Cincinnati, Ohio, USDEH, Public Health Service, CDC, NIOSH.

9. Shapiro, Y., Hubbard, R. W., Kimbrough, C. M., and Pandolf, K. B. (1981). "Physiological and hematological responses to summer and winter dry-heat acclimation," *Journal of Applied Physiology: Respiratory, Environmental and Exercise Physiology* **50,** 792–98.

10. *Patty's*.

11. *Patty's*.

12. *Patty's*, p. 786.

13. *The Industrial Environment—Its Evaluation and Control*, 2nd ed., C. H. Powell and A. D. Hosey, Eds., Public Health Service Publication No. 614, Government Printing Office, Washington, D.C., 1965.

14. *Heating/Piping/Air Conditioning*, 1976, Reinhold, Inc., Penton/IPC.

15. H. S. Belding and T. F. Hatch, *Heat. Pip. Air Cond.* **27,** 129–35 (1955).

16. B. A. Hertig, "Thermal Standards and Measurement Techniques," in *The Industrial Environment—Its Evaluation and Control*, U.S. Department of Health, Education and Welfare.

17. J. S. McKarns and R. S. Brief, *Heat. Pip. Air Cond.* **38,** 113 (1966).

18. R. S. Brief and R. G. Confer, *Am. Ind. Hyg. Assoc. J.* **32,** 11–16 (1971).

19. Criteria for a Recommended Standard ... Occupational Exposure to Hot Environments, Revised Criteria, 1986, U.S. Department of Health and Human Services, NIOSH, April 1986.

20. "TLV's—Threshold Limit Values for Chemical Substances & Physical Agents in the Workroom Environment with Intended Changes for 1981," ACGIH, Cincinnati, Ohio (1981).

21. Occupational Safety and Health Act PL91-596, 91st Congress, S.2193, U.S. Department of Labor, Washington, DC, 1970.

22. Criteria for a Recommended Standard ... Occupational Exposure to Hot Environments, U.S. Department of Health, Education and Welfare, NIOSH, HSM-72-10269, 1972.

23. N. L. Ramanathan and H. S. Belding, *Am. Ind. Hyg. Assoc. J.* **34,** 375–83 (1973).

24. J. H. Botsford, *Am. Ind. Hyg. Assoc. J.* **32,** 1–10 (1971).

25. V. M. Ciricello and S. H. Snook, *Am. Ind. Hyg. Assoc. J.* **38,** 264–71 (1971).

26. A. T. Johnson and G. D. Kirk, *Am. Ind. Hyg. Assoc. J.* **41,** 361–66 (1980).

27. M. Y. Beshir, J. D. Ramsey, and C. L. Burford, *Ergonomics* **25,** 247–54 (1982).

28. M. Y. Beshir, *Am. Ind. Hyg. Assoc. J.* **42,** 81–87 (1981).

29. R. D. Parker and F. D. Pierce, *Am. Ind. Hyg. Assoc. J.* **45,** 405–15 (1984).

30. World Meteorological Organization, *Guide to Meteorological Instruments and Observing Practices*, WMO, Geneva, 1971.

APPENDIX A
LIST OF ABBREVIATIONS USED

AET	Allowable exposure time as calculated from elements of the heat stress index (HSI)
C	Rate of heat exchange (net) by convection between an individual and the environment
C	Ceiling limits (WBGT) recommended by NIOSH for all workers in hot jobs
E	Rate of heat loss by evaporation of water from the skin
EDZ	Environment-drive zone—the range of heat load $(M + R + C)$ beyond the prescriptive zone in which physiological response is affected drastically by the thermal environment
E_{max}	Maximum evaporative heat loss by water vapor uptake in the air at prevailing meteorologic conditions
E_{req}	Heat loss required solely by evaporation of sweat to maintain body heat balance
ET	Effective temperature—an index used to estimate the effect of of temperature, humidity, and air movement on the subjective sensation of warmth
ETCR	Effective temperature corrected for radiation—an index for estimating the effect of temperature, humidity, and air movement on the subjective sensation of warmth using globe temperature rather than air temperature
GT	The temperature inside a blackened, hollow, thin copper globe measured by a thermometer whose sensing element is at the center of the sphere
HSI	An index of heat stress derived from the ratio of E_{req} to E_{max}
M	Rate of transformation of chemical energy into energy used for performing work and producing heat
MRT	The mean radiant (surface) temperature of the material and objects totally surrounding an individual
NWB	The wet-bulb temperature measured under conditions of the prevailing (natural) air movement
PZ	Prescriptive zone—that range of environmental heat load $(M + R + C)$ at which the physiologic strain (heart rate and core body temperature) is independent of the thermal environment

R	Rate of heat exchange by radiation between two radiant surfaces of different temperatures
RAL	Recommended alert limits (WBGT) specified by NIOSH for unacclimated, healthy workers
REL	Recommended exposure limits (WBGT) specified by NIOSH for acclimated, healthy workers
RH	Relative humidity—the ratio of the water vapor pressure in the ambient air (VPA) to the water vapor pressure in saturated air at the same temperature
DS	Change in heat content of the body
TA	The temperature of the air surrounding a body (also dry-bulb temperature)
TLV	Threshold limit values specified by the American Conference of Governmental Industrial Hygienists
TS	The mean of skin temperatures taken at several locations and weighted for skin area
ULPZ	The level of heat stress at the interface of the PZ (upper limit) and the EDZ (lower limit)
V	Air velocity
VPA	The partial pressure exerted by water vapor in the air
VPS	Vapor pressure exerted by water on the skin
WBGT	Wet-bulb globe temperature—an empirical index of heat stress obtained by weighting NWB, GT, and TA (outdoors with solar load)
WGT	Wet-globe temperature—an empirical index of heat stress as measured within a 3-in. copper sphere covered by a black cloth kept at 100 percent wetness

CHAPTER 22

Occupational Noise Exposure and Hearing Conservation

CHARLES P. LICHTENWALNER

Lucent Technologies, Bell Laboratories, Murray Hill, NJ 07974

KEVIN MICHAEL

Michael & Associates, 246 Woodland Drive, State College, PA 16803

22.1 INTRODUCTION
22.2 PHYSICS OF SOUND
22.2.1 Frequency
22.2.2 Wavelength
22.2.3 Amplitude
22.2.4 Abbreviations and Letter Symbols
22.3 THE EAR
22.3.1 Outer Ear
22.3.2 Eardrum
22.3.3 Middle Ear
22.3.4 Inner Ear
22.3.5 Conductive and Sensorineural Hearing Loss
22.3.6 How Noise Damages Hearing
22.3.7 Occupational Noise-Induced Hearing Loss
22.3.8 Nonoccupational Hearing Loss
22.3.9 Nonauditory Physiologic Effects of Noise
22.4 HEARING MEASUREMENT
22.4.1 Audiometers
22.4.2 Test Rooms
22.4.3 Hearing Threshold Measurement
22.4.4 Records
22.4.5 Audiometric Database Analysis
22.5 NOISE MEASUREMENT
22.5.1 The Sound Level Meter
22.5.2 Sound Exposures and Long-Term Average Sound Levels

Adapted from "Industrial Noise Exposure and Conservation of Hearing" by Paul L. Michael, Ph.D., Chapter 23 in Clayton & Clayton, *Patty's Industrial Hygiene and Toxicology, 4th ed., Vol. IA*. New York: John Wiley & Sons, Inc., 1991, pp. 937–1039.

Handbook of Occupational Safety and Health, Second Edition, Edited by Louis J. DiBerardinis, ISBN 0-471-16017-2

22.5.3 Frequency Analyzers
22.5.4 Tape Recording of Noise
22.5.5 Level Recording (Data Logging)
22.5.6 Instrument Calibration
22.5.7 Making Occupational Noise Exposure Measurements
22.5.8 Making Environmental Noise Measurements
22.6 OCCUPATIONAL NOISE EXPOSURE CRITERIA
22.6.1 Background—United States Federal Regulations
22.6.2 Noise Regulations
22.6.3 Noise Measurements
22.6.4 Exposure Calculations
22.6.5 Other Noise Exposure Criteria
22.6.6 Steady-State and Impulsive Noise
22.7 HEARING CONSERVATION PROGRAMS
22.7.1 General
22.7.2 Requirements
22.7.3 Setting up a Hearing Conservation Program
22.8 HEARING PROTECTORS
22.8.1 General
22.8.2 Performance Limitations
22.8.3 Types of Hearing Protectors
22.8.4 Selection of Hearing Protectors
22.8.5 Communication without Hearing Protectors
22.8.6 Communication with Hearing Protectors
22.8.7 Integration of Communication Requirements and the Use of Field Monitoring Systems
22.9 NOISE CONTROL PROCEDURES
22.9.1 Introduction
22.9.2 Source Control
22.9.3 Noise in Rooms
22.10 COMMUNITY NOISE
22.10.1 Community Noise Regulations
ACKNOWLEDGMENT
TERMINOLOGY
REFERENCES

22.1 INTRODUCTION

Few locations or circumstances in nature produce noise loud enough or for long enough duration to damage hearing. It remained for man, himself, to produce noises capable of causing injury. The knowledge that loud noises can produce these injuries has been known for several hundred years.

The most obvious and best quantified injury from noise is deterioration of hearing ability. However, except for extremely loud sounds, noise-induced hearing loss is a slowly progressive debility that usually goes unnoticed by those affected until the loss is significant. There are no visible effects and usually no pain.

Hearing loss occurring naturally from aging is called *presbycusis*, although some have argued that much of the presbycusis seen in industrialized societies may actually be caused by ambient (non-occupational) noise exposures.[1] Hearing losses can also be pathological from medical abnormalities, ranging from simple, easily correctable conditions such as impacted ear wax and middle ear infections, to severe problems such as deafness from rubella during gestation. As in many other areas of industrial hygiene, it is difficult to determine the contribution of these various elements: (1) occupational noise exposure, (2) non-occupational noise exposure, (3) aging, (4) medical pathology, and (5) normal human variability, to one individual's hearing loss.

In addition to hearing loss, noise causes a number of other problems: (1) communication interference, (2) annoyance, (3) performance degradation, and (4) physiological effects. High noise levels make it difficult to communicate. At moderate noise levels, telephone conversation becomes difficult. At higher levels people must raise their voices or shout to be understood. Communication problems lead to safety problems when the noise masks speech or alerting signals. Disturbing or distracting noises can lead to accidents.

In the United States, legislation has been used, first to provide compensation to those injured,[2] then later to limit noise exposures. In 1970, the Occupational Safety and Health Administration (OSHA) used noise regulations established by the Department of Labor under the Walsh–Healy Public Contracts Act. The regulations established a permissible exposure limit (PEL) based on an 8-hour time-weighted average noise exposure. In 1981, OSHA amended the noise standard—adopted in final form in 1983[3]—establishing requirements for a hearing conservation program. From 1990 through 1995, 8106 inspections were conducted by Federal OSHA and state regulators for noise, resulting in 14,138 citations and assessment of $7,703,027 in penalties.[4]

22.2 PHYSICS OF SOUND

Technically, the sensation of sound results from oscillations in pressure, stress, particle displacement, and particle velocity in any elastic medium that connects the sound source with the ear. When sound is transmitted through air, it is usually described in terms of changes in pressure that alternate above and below atmospheric pressure. These pressure changes are produced when vibrating objects (sound sources) cause regions of high and low pressure that propagate from the sound source. The characteristics of a particular sound depend on the rate at which the sound source vibrates, the amplitude of the vibration, and the characteristics of the conducting medium. A sound may have a single rate of pressure/vacuum alternation or frequency (f), but most sounds have many frequency components. Each of these frequency components, or bands of sound, may have a different amplitude.

22.2.1 Frequency

Frequency is defined as the rate at which a sound is emitted from a source. Physically it is measured by the number of times per second the pressure oscillates between levels above and below atmospheric pressure (Fig. 22.1). Frequency is denoted by the symbol f and is measured in Hertz (1 Hz = 1 cycle per second). Frequency is inversely related to the period (T), which is the time a sound wave requires to complete one cycle. The frequency range over which normal young adults are capable of hearing sound at moderate levels is 20–20,000 Hz. Pitch or tone is the sensation associated with frequency.

A sound may consist of a single frequency (i.e., a pure tone); however, most common sounds contain many frequency components. It is generally infeasible to report the characteristics of all the frequencies emitted by noise sources; so measurements are made including the sound energy from a broad range of frequencies. As will be discussed in the measurement section, weighting networks have been standardized for single-number

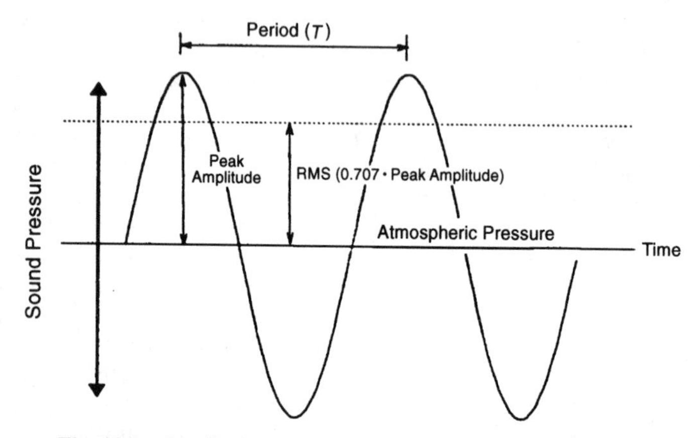

Fig. 22.1 Amplitude and frequency of a pure tone sound wave.

assessments of sounds having properties similar to the response of the human ear. In addition to these broad-band weightings, the frequency range in acoustics is frequently divided into smaller ranges. The most common range of frequencies is the octave band, where the upper edge of the band is twice the frequency of the lower band edge. One-third octave (three bands are used to measure one octave) and narrower bands are also used.

22.2.2 Wavelength

The distance a sound wave travels during one sound pressure cycle is called the wavelength (λ). The wavelength is related to the speed of sound by:

$$\lambda = \frac{c}{f} \tag{22.1}$$

where c is the speed of sound in meters per second (m/s), f is the frequency in Hz, and λ is the wavelength in meters (m).

For most work, it is sufficient to note that sound travels in air at approximately 344 m/sec at normal atmospheric pressures and room temperatures.

Example 1: Calculate the wavelength of 20, 1,000, and 20,000 Hz waves.
at 20 Hz:

$$\lambda = \frac{c}{f} = \frac{344 \text{ m/sec}}{20/\text{sec}} = 17.2 \text{ m}$$

at 1,000 Hz

$$\lambda = \frac{c}{f} = \frac{344\,\text{m/sec}}{1000/\text{sec}} = 0.344\ \text{m}$$

at 20,000 Hz

$$\lambda = \frac{c}{f} = \frac{344\,\text{m/sec}}{20,000/\text{sec}} = 0.017\ \text{m}$$

When considering the behavior of sound, one important characteristic is the ratio of the length of the item affected by the noise to the wavelength of the sound. For instance, sound will not efficiently couple to vibrations of the entire eardrum when the $\frac{1}{4}$ wavelength of sound is less than the dimensions of the eardrum. It is not surprising that frequencies greater than 20,000 Hz cannot be heard since $\frac{1}{4}$ wavelength is less than the 5-mm-diameter dimension of a typical eardrum. As will be shown later, the effectiveness of a barrier in shielding one side from a noise source on the other depends on the path differences between going around the barrier and the straight line distance in terms of the number of wavelengths.

22.2.3 Amplitude

22.2.3.1 Sound Pressure and Sound Pressure Level The amplitude of the sound pressure is caused by the vibrations of the sound source. Loudness is the sensation from sound pressure. Pressure is measured in pascals [abbreviated Pa; 1 Pa = 1 newton/meter^2 = 1 kilogram/(meter second^2)]. One atmosphere = 1.013×10^5 Pa at sea level. Sound pressures are very small variations around atmospheric pressure. Normal speech at 1 meter distance from a talker averages about 0.1 Pa—one millionth of an atmosphere.

The range of sound pressures commonly measured is very wide. Sound pressures well above the pain threshold, about 20 Pa, are found in some work areas, whereas sound pressures down to the threshold of hearing at about 20 μPa (micropascals) are used for hearing measurements. The range of common sound exposures exceeds 10^6 Pa. This range cannot be scaled linearly with a practical instrument while maintaining the desired accuracy at low and high levels. To be able to see "just noticeable differences" in hearing at sound pressures in noisy environments would require a scale many miles long. To cover this wide range of sound pressure with a reasonable number of scale divisions and to provide a scale that responds more closely to the response of the human ear, the logarithmic decibel (dB) scale is used. The abbreviation dB is used with upper and lower case—"d" for deci, and "B" for Bell—in honor of Alexander Graham Bell. The measurement unit was invented at Bell Laboratories to facilitate calculations involving loss of signals in long lengths of telephone lines.

By definition, the decibel is a unit without dimensions; it is the logarithm to the base 10 of the *ratio* of a *measured quantity* to a *reference quantity when the quantities are proportional to power.*[5] The decibel is sometimes difficult to use and to understand because it is often used with different reference quantities. Acoustic intensity, acoustic power, hearing thresholds, electric voltage, electric current, electric power, and sound pressure level may all be expressed in decibels, each having a different reference. Obviously the decibel has no meaning unless a specific reference quantity is specified, or understood. Any time a level is referred to in acoustics, decibel notation is implied.

Most sound-measuring instruments are calibrated to provide a reading (called root mean square or rms) of sound pressures on a logarithmic scale in decibels. The decibel reading taken from such an instrument is called the sound pressure level (L_p). The term *level* is used because the measured pressure is at a particular level above a given pressure reference. For sound measurements in air 0.00002 Pa, or 20 μPa, commonly serves as the reference sound pressure. This reference is an arbitrary pressure chosen many years ago because it is approximately the normal threshold of human hearing at 1000 Hz. Since the eardrum responds to the intensity of the sound wave, and since intensity is proportional to pressure squared, sound pressure levels are calculated from the square of sound pressures. Mathematically, L_p is written as follows:

$$L_p = 10 \log \left(\frac{p}{p_r} \right)^2 \tag{22.2}$$

where p is the measured sound pressure, p_r is the reference sound pressure [generally in n/m^2 or pascals (Pa)], and the logarithm is to the base 10.

An equivalent form of the preceding equation is frequently found in acoustics textbooks.

$$L_p = 20 \log \left(\frac{p}{p_r} \right) \tag{22.3}$$

Specifying sound pressure levels in this form masks the fact that levels are ratios of quantities equivalent to power. For technical purposes, L_p should always be written in terms of decibels relative to the recommended reference pressure level of 20 μPa. Reference quantities for acoustic levels are specified in ANSI S1.8.[6] The reference quantity should be stated at least once in every document.

Figure 22.2 shows the relationship between sound *pressure* (in pascals) and sound *pressure level* (in dB re 20 μPa). This figure illustrates the advantage of using the decibel scale rather than the wide range of direct pressure measurements. It is of interest to note that any doubling of a pressure is equivalent to a 6-dB change in level. For example, a range of 20 to 40 μPa, which might be found in hearing measurements, or a range of 1 to 2 Pa, which might be found in hearing conservation programs, are both ranges of 6 dB. Measuring in decibels allows reasonable accuracy for both low- and high-sound pressure levels.

22.2.3.2 Sound Intensity and Sound Intensity Level Sound intensity at any specified location may be defined as the average acoustic energy per unit time passing through a unit area that is normal to the direction of propagation. For a spherical or free-progressive sound wave, the intensity may be expressed by:

$$I = \frac{p^2}{\rho c} \tag{22.4}$$

where p is the rms sound pressure, ρ is the density of the medium, and c is the speed of sound in the medium. Sound intensity units, like sound pressure units, cover a wide

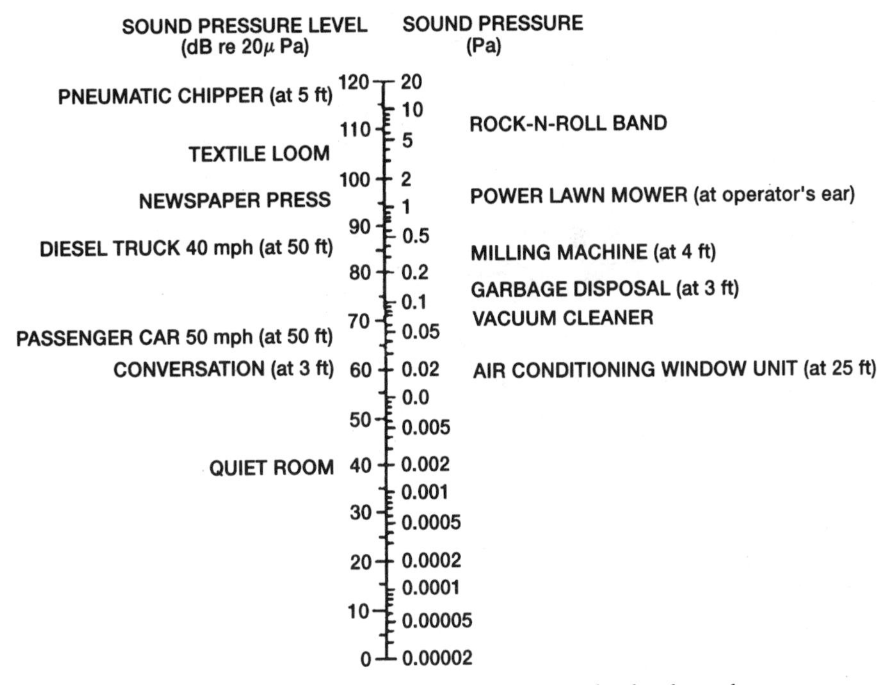

Fig. 22.2 Relation between A-weighted sound pressure level and sound pressure.

range, and it is often desirable to use decibel levels to compress the measuring scale. To be consistent, intensity level is defined as

$$L_I = 10 \log \left(\frac{I}{I_r} \right) \tag{22.5}$$

where I is the measured intensity at some given distance from the source and I_r is a reference intensity, 10^{-12} W/m^2. In air, this reference closely corresponds to the reference pressure 20 μPa used for sound pressure levels.

$$L_I = L_p \tag{22.6}$$

22.2.3.3 Sound Power and Sound Power Level Consider a vibrating object suspended in the air (Fig. 22.3). The vibrations will create sound pressure waves that travel away from the source—decreasing by 6 dB as the distance from the source doubles. The sound *power* of this source is independent of its environment, but sound *pressure* around the source is not.

Sound power (represented by W) is used to describe the sound source in terms of the amount of acoustic energy that is produced per unit time (watts). Sound power is related to the average sound intensity produced in free-field conditions at a distance r from a point source by:

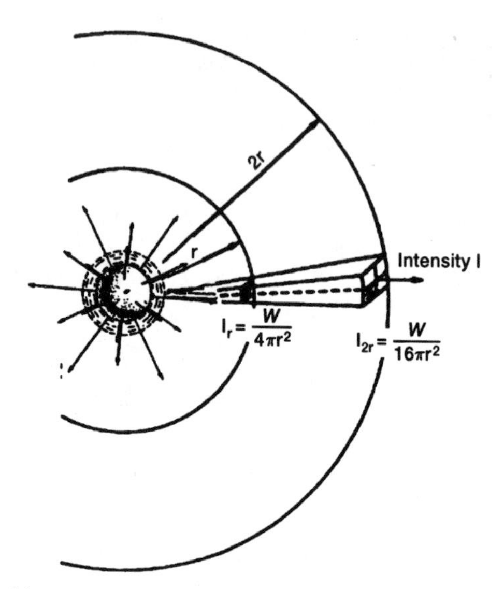

Fig. 22.3 Pulsating object suspended in air (free field). Sound Intensity is inversely proportional to the square of the distance from the source.

$$W = I_{\text{avg}} 4\pi r^2 \tag{22.7}$$

where I_{avg} is the average intensity at a distance r from a sound source whose acoustic power is W. The quantity $4\pi r^2$ is the area of a sphere surrounding the source over which the intensity is averaged. The intensity decreases with the square of the distance from the source, hence the well-known inverse square law.

Units for power are also usually given in terms of decibel levels because of the wide range of powers covered in practical applications. Choosing a reference surface S_r of 1 m^2 defines power as:

$$\frac{W}{W_r} = \frac{I_{\text{avg}} 4\pi r^2}{I_r S_r} = \frac{I_{\text{avg}}}{I_r} \frac{4\pi r^2}{S_r}$$

$$10 \log\left(\frac{W}{W_r}\right) = 10 \log\left(\frac{I_{\text{avg}}}{I_r}\right) + 10 \log\left(\frac{4\pi r^2}{S_r}\right)$$

$$L_p = L_1 + 10 \log\left(\frac{4\pi r^2}{S_r}\right), \text{ and since } L_1 = L_p\text{:}$$

$$L_{\text{power}} = L_{\text{pressure}} + 10 \log\left(\frac{4\pi r^2}{S_r}\right) \tag{22.8}$$

where W is the power of the source in watts (1 W = 1 N m/sec) and W_r is the reference power (10^{-12} watts since $W_r = I_r S_r$ and $I_r = 10^{-12}$ W/m^2 and $S_r = 1$ m^2).

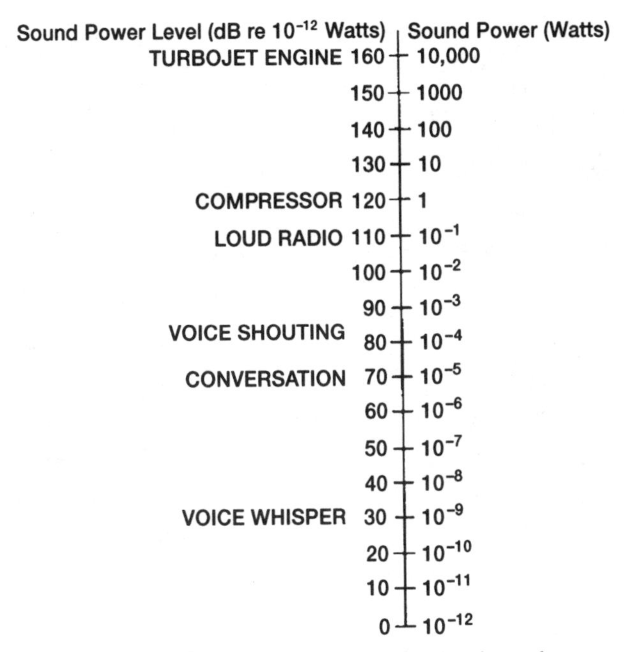

Fig. 22.4 Relation between sound power level and sound power.

Figure 22.4 shows the relation between sound power in watts and sound power level in dB re 10^{-12} W. Note that the distance must be specified or inferred to determine sound pressure level from the sound power. Equation (22.8) is used to predict sound pressure levels if the sound power level of a source is known and the acoustic environment is known or can be estimated.

In a free field (no surfaces to reflect sound waves), sound waves spread out from the source, losing power as the square of the distance. Normally, however, reflecting surfaces are present or the sound source is not omnidirectional. A reflecting surface will increase the sound intensity since the volume in which the sound radiates is reduced. The *directivity factor* (Q) is a dimensionless quantity used to describe the ratio of volume to which a sound is emitted relative to the volume of a sphere with the same radius (Fig. 22.5).

22.2.3.4 Sound Power versus Sound Intensity versus Sound Pressure

Noise control problems require a practical knowledge of the relationship between pressure, intensity, and power. For example, consider the prediction of sound pressure levels that would be produced around a proposed machine location from the sound power level provided by the machine.

Example 2: The manufacturer of a machine states that this machine has an acoustic power output of 1 W. Predict the sound pressure level at a location 10 m from the machine.

Answer: From Eqs. (22.4) and (22.7) in a free field and for an omnidirectional source the sound pressure is:

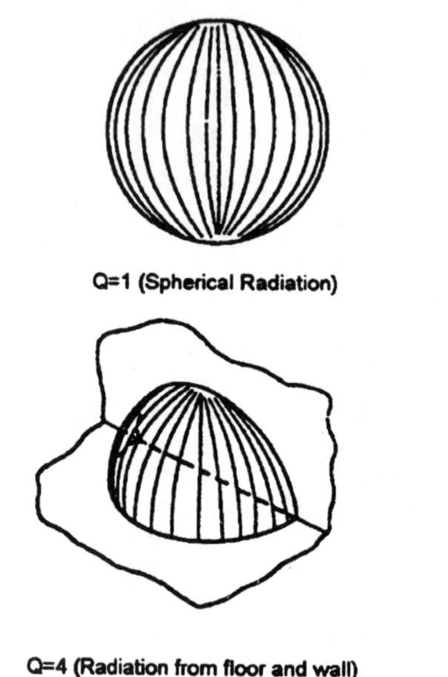

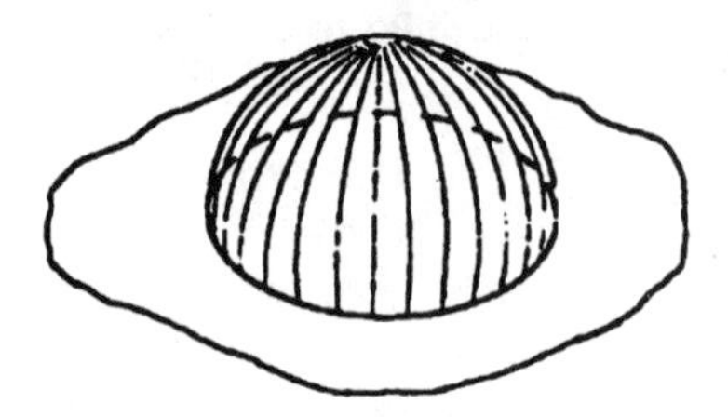

Q=2 (1/2 Spherical Radiation)

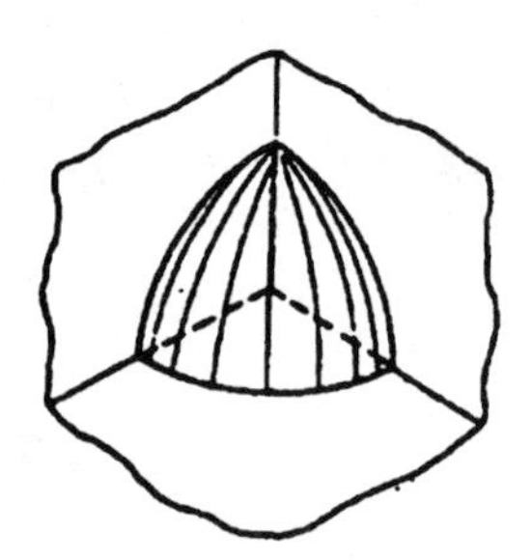

Q=8 (Radiation into 1/8 Sphere)

Fig. 22.5 Directivity factor (Q).

$$I = \frac{p^2}{\rho c} \quad \text{and} \quad W = 4\pi r^2 \quad \text{so} \quad p = (I\rho c)^{1/2} = \left(\frac{W\rho c}{4\pi r^2}\right)^{1/2}$$

$$p = \left(\frac{(1N\ \text{m/sec})(1.18\ \text{kg/m}^3)(344\ \text{m/sec})}{4\pi(10\ \text{m})^2}\right)^{1/2} = 0.32\ \text{Pa}$$

but since the source will be sitting on the ground, reflections will double the sound pressure level, so:

$$L_p = 10 \log\left(\frac{0.64\ \text{Pa}}{0.00002\ \text{Pa}}\right)^2 = 90\ \text{dB}(\text{re}\ 20\ \mu\text{Pa})$$

Admittedly, there are few truly free-field situations, and few omnidirectional sources; so the above calculation can only give a rough estimation of the absolute value of the sound pressure level. However, comparison or rank ordering of different machines can be made, and at least a rough estimate of the sound pressure is available. Noise levels in locations that are reverberant, or where there are many reflecting surfaces, can be expected to be higher than that predicted because noise is reflected back to the point of measurement.

TABLE 22.1 Reference Formulae and Quantities[6]

Name	Definition	SI reference quantity	English reference quantity
Sound pressure level (dB)	$L_p = 20\log_{10}(p/p_0)$	$p_0 = 20\mu\text{Pa}$ $= 2\times10^{-5}\ \text{N/m}^2$	$p_0 = 2.9\times10^{-9}\text{lb/in.}^2$
Sound power (dB)	$L_W = 10\log_{10}(W/W_0)$	$W_0 = 1\ \text{pW}$	$W_0 = 5\times10^{-10}\ \text{in. lb/s}$
Sound intensity level (dB)	$L_I = 10\log_{10}(I/I_0)$	$I_0 = 1\ \text{pW/m}^2$	$I_0 = 5.71\times10^{-15}\ \text{lb/in. s}$
Sound exposure level (dB)	$L_E = 10\log_{10}(E/E_0)$	$E_0(20\mu\text{Pa})^2\ \text{s}$ $= (2\times10^{-5}\ \text{Pa})^2\ \text{s}$	$E_0 = 1\times10^{-18}\ \text{lb}^2/\text{in.}^4\ \text{s}$

22.2.4 Abbreviations and Letter Symbols

Terminology abbreviations and letter symbols for reporting noise measurements are defined by consensus standards[57] and should be used. By convention, the exponential-time-weighted, frequency-weighted sound pressure levels measured in dB re 20 μPa are known as *sound levels* (Table 22.1).

Sound levels should be reported using either the letter or symbol abbreviations. Fast, slow, and impulse exponential-time-weighting carry the letters F, S, and I, respectively, A and C frequency weightings carry their respective letters. Therefore, measurements taken using the OSHA requirement for measuring with A frequency weighting and slow exponential-time weighting should be reported using either the letter symbol L_{AS} or the abbreviation SAL. Similarly, C frequency-weighted measurements taken with fast exponential-time weighting would be indicated with the letter symbol L_{CF} or abbreviated as FCL. When no frequency weighting is specified, the A-frequency weighting is assumed.

It is good practice, however, to specify both the frequency weighting and the duration of the measurement. For instance, a 1-hour measurement should be reported as $L_{\text{Aeq, 1h}}$. Since L_{eq} is understood for time-averaged measurements, the "eq" is sometimes omitted and a 1-hour measurement could be reported using the letter symbol $L_{\text{A, 1H}}$ or abbreviated as 1HL. Table 22.2 describes this convention.

TABLE 22.2 Letter Symbols and Abbreviations Used in Acoustics[6]

Description	Letter/symbol	Meaning
Slow	S	1.0 second exponential-time weighting
Fast	F	0.125 second exponential-time weighting
Impulse	I	0.035 sec rise time, 1.5 sec decay time
Peak	Pk	Greatest level with no exponential-time weighting except as provided by frequency weighting
A	A	A frequency weighting
C	C	C frequency weighting
	EC, L_{eq}	Equivalent continuous sound pressure level
	Mx	Greatest level for specific exponential-time averaging during a specified period
	SEL, L_E	Sound exposure level, measured in dB re (20 μPa)^2 sec
	$E_{A,T}$	Sound exposure (A-frequency weighted) over a specified time T, measured in Pa^2 sec

22.2.4.1 Adding Sound Levels When Noise Sources Are Independent

Most industrial noises have random frequencies and phases, and can be combined as described later. In the few cases when noises have significant pure tone components, these calculations are not accurate, and phase relationships must be considered. In areas where significant tones are present, standing waves often can be recognized by the presence of rapidly varying sound pressure levels over short distances. It is not practical to try to predict levels in areas where standing waves are present.

When the sound pressure levels of two pure tone sources are added, the resultant sound pressure level L_p may be greater than, equal to, or less than the level of a single source (Fig. 22.6). In most cases, however, the resultant L_p is greater than either single source.

At zero phase difference, the resultant of two identical pure tone sources is 6 dB greater than either single level. At a phase difference of 90°, the resultant is 3 dB greater

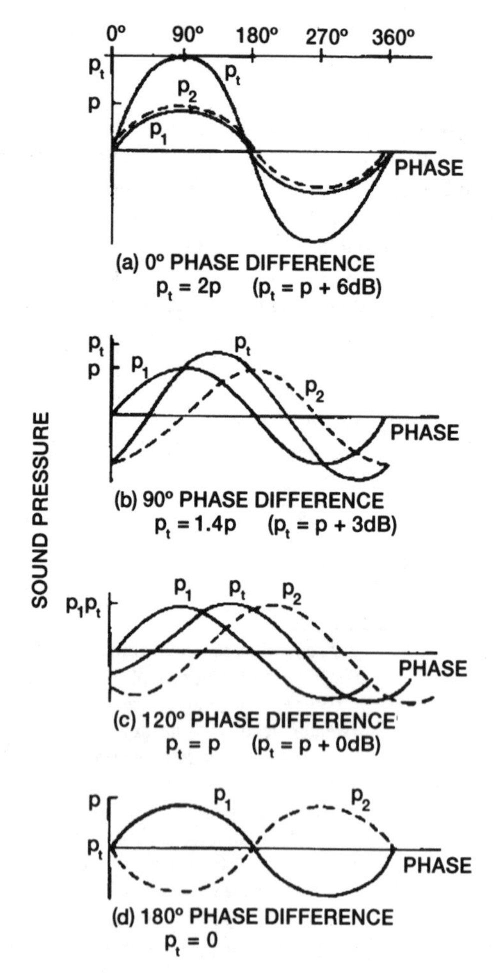

Fig. 22.6 Combination of two pure tones of the same frequency (p_1 and p_2) with various phase differences.

than either level. Between 90° and 0°, the resultant is somewhere between 3 and 6 dB greater than either level. At a phase difference of 120°, the resultant is equal to the individual levels; and between 120° and 90°, the resultant is between 0 and 3 dB greater than either level. At 180° there is complete cancellation of sound. The resultant $L_p(R)$ is greater than the individual levels for all phase differences from 0° and 120°, but less than individual levels for phase differences from 120° and 180°. Also, most tonal noises are not single tones but combinations of frequencies.

Thus, at almost all points in the noise field the pressure levels exceed the individual levels, and overall there is a 3 decibel average increase in sound pressures by adding two randomly phased sources. This allows industrial hygienists to estimate resultant sound pressure levels when noise sources are added or removed from industrial environments by combining sound levels as shown below.

22.2.4.2 Combining Sound Levels It is often necessary to combine sound levels—for example, to combine frequency band levels to obtain the overall or total sound pressure level from band levels within the overall noise. Another example is the estimation of total sound pressure level resulting from adding a machine of known noise spectrum to a noise environment of known characteristics. Simple addition of individual sound pressure levels, which are logarithmic quantities, constitutes multiplication of pressure ratios; therefore, the sound pressure corresponding to each sound pressure level must be determined and added.

For the most part, industrial noise is made up of sources having many frequencies (broad band), with nearly random phase relationships. Sound pressure levels of different random noise sources can be added by (1) converting the levels of each to pressure units, (2) converting to intensity units that may be added arithmetically, (3) then converting the resultant intensity to pressure, and finally, (4) converting the resultant pressure value to sound pressure levels in decibels.

$$L_{\text{total}} = 10\log\left(\sum_{i=1}^{N} 10^{L_i/10}\right) \quad (22.9)$$

Example 3: If the octave band sound pressure levels measured for a random noise source are as shown in the table, what is the overall sound pressure level?

Freq. (Hz)	31.5	63	125	250	500	1000	2000	4000	8000
L_i (dB)	85	88	91	94	95	100	97	90	88

$$L_{\text{total}} = 10\log\,(10^{85/10} + 10^{88/10} + 10^{91/10} + \cdots)$$

$$L_{\text{total}} = 10\log\,(10^{8.5} + 10^{8.8} + 10^{9.1} + \cdots) = 10\log\,(24.5 \times 10^{10}) = 103.9\text{ dB}$$

Acceptable accuracy is found in most cases, however, by using Table 22.3. Begin by adding the highest levels so that calculations may be discontinued when there is no significant difference when lower values are added to the total.

TABLE 22.3 Simplified Table for Combining Decibel Levels of Noise with Random Frequency Characteristics

Numerical differences between levels	Amount to be added to the higher level
0–1	3
2–4	2
5–9	1
>10	0

Example 4: Using Example 3 measurements, determine the overall sound pressure level using Table 22.3.

The difference between 100 and 97 dB is 3 dB; therefore, opposite the range 2–4 in the left-hand column of the table, read a value of 2 in the right-hand column and add this value to the higher of the two levels: 100 + 2 = 102 dB. This resultant is now added to the next highest level by repeating the process: 102 − 95 = 7; from the table read an amount to be added of 1. Thus 102 + 1 = 103 dB. The next highest level is 94, which is 9 dB lower than the total; so another 1 dB is added, giving 103 + 1 = 104 dB. All other readings are more than 10 dB lower than the total; so the procedure is halted, giving a final L_{total} = 104 dB—acceptably close to the precise calculation.

22.3 THE EAR

The normal human ear responds to a remarkable frequency range, ranging from about 20 to 20,000 Hz at common loudness levels.[8] The characteristics of any individual ear over this wide frequency range are extremely complex, and understanding the ear's capabilities is made even more difficult because of large differences among individuals. An ear's response characteristics may change as a result of physical or mental conditions, sound level, medications, environmental stresses, diseases, and other factors.

A normal healthy human ear also effectively transduces a remarkable range of sound pressures. It is sensitive to very low sound pressures that produce a displacement of the ear drum no greater than the diameter of a hydrogen molecule. At the other extreme, it can transduce sounds with sound pressures that are more than 10^6 times greater than the ear's lower threshold value; however, exposure to high-level sounds may cause temporary or permanent damage to the ear.

The ear is divided into three sections (Fig. 22.7), the outer ear, the middle ear, and the inner ear.[8a] Sound incident upon the ear travels through the ear canal to the eardrum, which separates the outer and middle ear sections. The combined alternating sound pressures that are incident upon the eardrum cause it to vibrate with the same relative characteristics as the sound source(s). The mechanical vibration of the eardrum is then coupled through the three bones of the middle ear to the oval window of the inner ear. The vibration of the oval window is then coupled to the fluid contained in the inner ear.

22.3.1 Outer Ear

The auricle, sometimes called the pinna, plays a significant role in the hearing process only at very high frequencies, where its size is large compared with that of a wavelength.

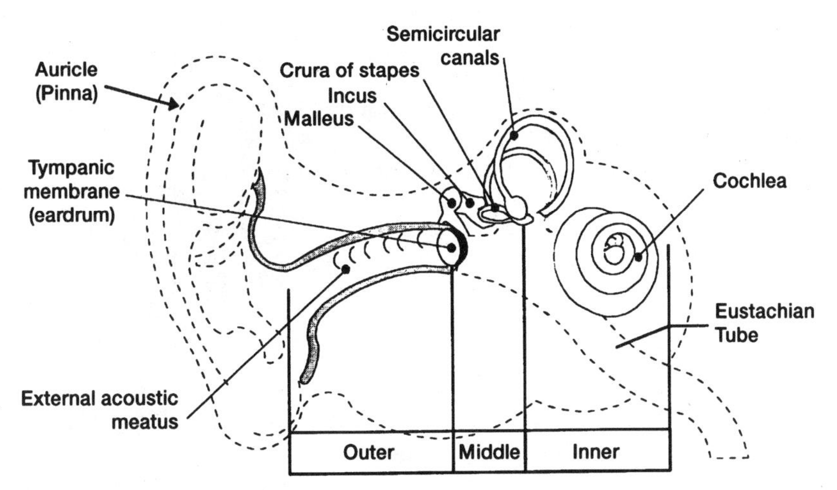

Fig. 22.7 Cross section of the ear showing the outer, middle, and inner ear configurations.

The auricle helps direct these high-frequency sounds into the ear canal, and it assists the auditory system in localizing the sound source.

Ear canals have many sizes and shapes. They are seldom as straight as indicated in the figures, and the shape and size of ear canals differ significantly among individuals and can even differ between ears of the same individual. The average length of the ear canal is about 2.5 cm. When closed at one end by the eardrum, it has a quarter-wavelength resonance of about 3000 Hz. This resonance increases the response of the ear by about 10 dB at 3000 Hz—aproximately the frequency at which the ear is most sensitive, and also approximately the frequency at which the greatest noise-induced hearing loss occurs (see Fig. 22.8).

The hairs at the outer end of the ear canal help to keep out dust and dirt; further into the canal are the wax-secreting glands. Normally, ear wax flows toward the entrance of the ear canal, carrying with it the dust and dirt that accumulate in the canal. The flow of wax is normal, and may be interrupted by changes in body chemistry that can cause the wax to become hard and to build up within the ear. Too much cleaning or the prolonged use of earplugs may cause increased production of wax or dryness and irritation in the ear canal. At times, the wax may build up to the point of occluding the canal, and a conductive loss of hearing will result. Any buildup of wax deep within the ear canal should be removed very carefully, by a trained person, to prevent damage to the eardrum and middle ear structures.

During welding or grinding operations a spark may enter the ear canal and burn the canal or a portion of the eardrum. Although very effective surgical procedures have been developed to repair or replace the eardrum, this painful and costly accident can be prevented by wearing hearing protectors.

The surface of the external ear canal is extremely delicate and easily irritated. Cleaning or scratching with match sticks, nails, hairpins, and other objects can break the skin and cause a very painful and persistent infection. Infections can cause swelling of the canal walls and, occasionally, a loss of hearing if the canal swells shut. An infected ear should be given prompt attention by a physician.

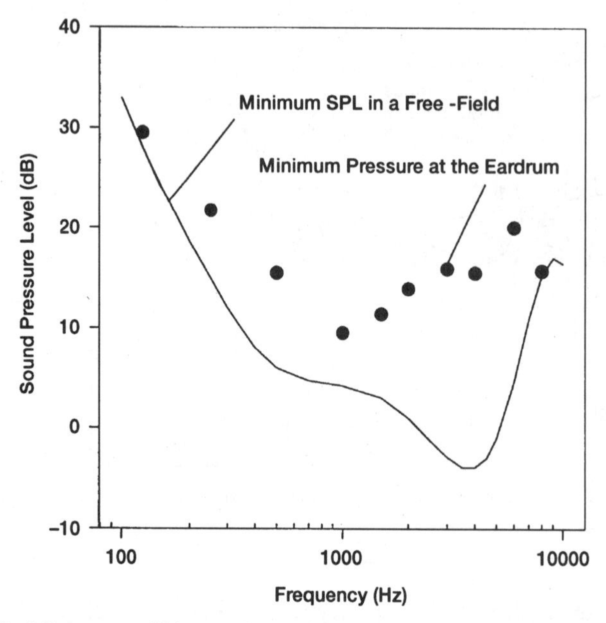

Fig. 22.8 Minimum audible sound pressure levels for young adults with normal hearing.[9]

22.3.2 Eardrum

The eardrum is a thin and delicate membrane that responds to the very small sound pressure changes at the lower threshold of normal hearing; yet it is seldom damaged by common continuous high-level noises. Although an eardrum may be damaged by an explosion or a rapid change in ambient pressure, the often repeated statement "the noise was so loud it almost burst my eardrums" is rarely true for common steady-state noise exposures.

When an eardrum is ruptured, the attached middle-ear bones (or ossicles) may be dislocated; thus the eardrum should be carefully examined immediately after the injury occurs to determine whether realignment of the ossicles is necessary. In a high percentage of cases, surgical procedures are successful in realigning dislocated ossicles, so that little or no significant loss in hearing acuity results from this injury.

22.3.3 Middle Ear

The air-filled space between the eardrum and the inner ear is called the *middle ear* (Fig. 22.9). The middle ear contains three small bones—the malleus (hammer), the incus (anvil), and the stapes (stirrup)—that mechanically connect the eardrum to the oval window of the inner ear.

The eardrum and middle ear act as a transformer to convert sound pressure waves in air to motion in the cochlear fluids.[10] The eardrum has an area about 20 times that of the oval window, thereby providing a mechanical advantage of about 20 : 1. The ossicles provide an additional mechanical advantage.

The transmission through the middle ear is frequency dependent. Low-frequency

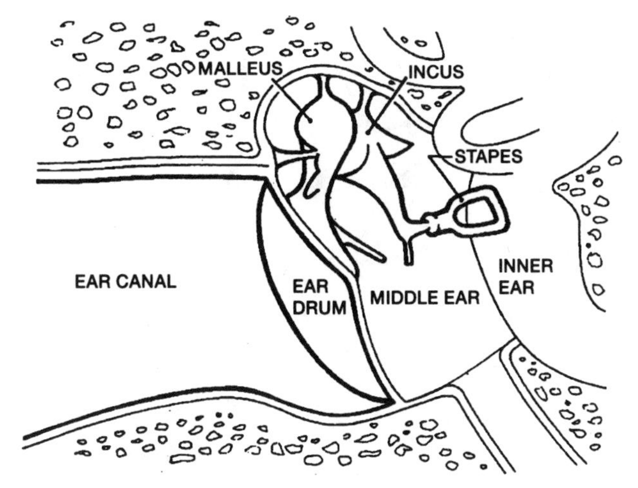

Fig. 22.9 The middle ear.

waves are reduced by stiffness of the middle ear structures and compression of air in the middle ear cavity. The mass of the ossicles, less efficient modes of vibration of the eardrum, and inability of high-frequency sound waves to couple to the eardrum also reduce transmission at high frequencies. This complex system also acts as a hearing protector, since the involuntary relaxation of coupling efficiency between the ossicles can reduce the transmission of pressure from the eardrum to the oval window. The reaction time for this relaxation in the middle-ear system is approximately 10 milliseconds.

The most common problem encountered in the middle ear is infection. This warm and humid air-filled space is completely enclosed except for the small eustachian tube that connects the middle ear to the back of the throat; thus it is susceptible to infection, particularly in children. If the eustachian tube is closed as a result of an infection or an allergy, the pressure inside the middle ear cannot be equalized with that of the surrounding atmosphere. In such an event, a significant change in atmospheric pressure, such as that encountered in an airplane or when driving in mountainous territory, may produce a loss of hearing sensitivity and extreme discomfort as a result of the inward displacement (or retraction) of the eardrum. Even a healthy ear may suffer a temporary loss of hearing sensitivity if the eustachian tube becomes blocked, but this loss of hearing can often be restored simply by swallowing or moving the jaw to open the eustachian tube momentarily.

22.3.4 Inner Ear

The inner ear is completely surrounded by bone (Fig. 22.10). One end of the space inside the bony shell of the inner ear is shaped like a snail shell; it contains the cochlea. The other end of the inner ear has the shape of three semicircular loops. The fluid-filled cochlea serves to detect and analyze incoming sound signals and to translate them into nerve impulses that are transmitted to the brain. The semicircular canals contain sensors for balance and orientation.

In operation, sound energy is coupled into the inner ear by the stapes, whose base is coupled into the oval window of the cochlea. The oval window and the round window

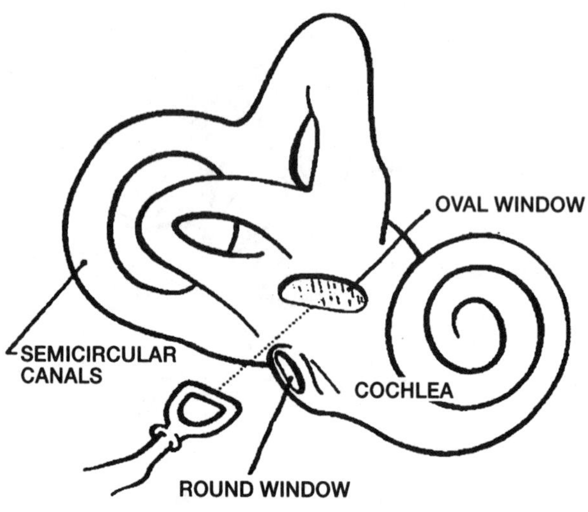

Fig. 22.10 The inner ear.

located below it are covered by thin, elastic membranes to contain the fluid within the cochlea. As the stapes forces the oval window in and out with the dynamic characteristics of the incident sound, the fluid of the cochlea is moved with the same characteristic motions. Thousands of hair cells located along the two and one-half turns of the cochlea detect and analyze these motions and translate them into nerve impulses. The nerve impulses, in turn, are transmitted to the brain via the eighth nerve for analysis and interpretation.

The hair cells within the cochlea may be damaged by aging, disease, medication, blows to the head, and exposure to high levels of noise.[11] Unfortunately the characteristics of the hearing losses resulting from these various causes are often very similar, and therefore it may be difficult or impossible to determine the etiology of a particular case from an audiogram.

22.3.5 Conductive and Sensorineural Hearing Loss

Hearing loss can be either conductive or sensorineural. Abnormalities in the outer or middle ear create conductive losses, such as ossification of the middle ear bones, buildup of wax in the outer ear, and fluid in the middle ear. Conductive hearing loss is a result of any occurrence that changes the transfer of vibrations from the external ear to the oval window. Conductive losses frequently occur at or below 500 Hz. In many cases the loss is amenable to treatment by antibiotics or in severe cases, by surgical intervention.

Another middle-ear problem may result from an abnormal bone growth (otosclerosis) around the ossicles, restricting their normal movement. The cause of otosclerosis is not totally understood, but heredity is considered to be an important factor. The conductive type of hearing loss from otosclerosis is generally observed first at low frequencies, it then extends to higher frequencies, and eventually may result in a severe loss in hearing sensitivity over a wide frequency range. Hearing aids may often restore hearing sensitivity lost as a result of otosclerosis, and effective surgical procedures have been refined to such a point that they are often recommended. An important side benefit of an effective hearing conservation program is the early detection of such hearing impairments as otosclerosis.

Hearing loss in the cochlea or auditory nerve is known as *sensorineural hearing loss.* Tumors are frequently the cause of hearing losses arising in the auditory nerve. Cochlear hearing losses can be caused by drugs, infections, or may be congenital. In older age, senile changes produce a progressive incurable impairment[12] known as *presbycusis*, but the major cause of sensorineural loss is from acoustic trauma or noise exposure. The sensitive hair cells of the cochlea are usually destroyed, no treatment for replacement is known, and hearing aids are frequently of limited efficacy.

22.3.6 How Noise Damages Hearing

Noise-induced hearing loss may be temporary or permanent depending on the level and frequency characteristics of the noise, the duration of exposures, and the susceptibility of the individual. A temporary loss of hearing sensitivity is usually restored within about 16 h[11]; however, temporary losses may last for weeks in some cases. When temporary losses do not recover completely before other significant exposures, permanent losses may be produced. Permanent losses (noise-induced permanent threshold shifts, or NIPTS) are irreversible and cannot be corrected by conventional surgical or therapeutic procedures.

Noise-induced damage generally occurs in hair cells located within the cochlea. For common broadband noise exposures, hearing acuity is generally affected first in the frequency range from 3000 to 6000 Hz, with most affected persons showing a loss or "dip" at 4000 Hz. For noise exposures having significant components concentrated in narrow frequency bands below 4000 Hz, impairments usually are found about one-half to one octave above the predominant exposure frequencies. If high-level exposures are continued, the loss of hearing generally increases around 4000 Hz and spreads to higher and lower frequencies[13] (see Fig. 22.11).

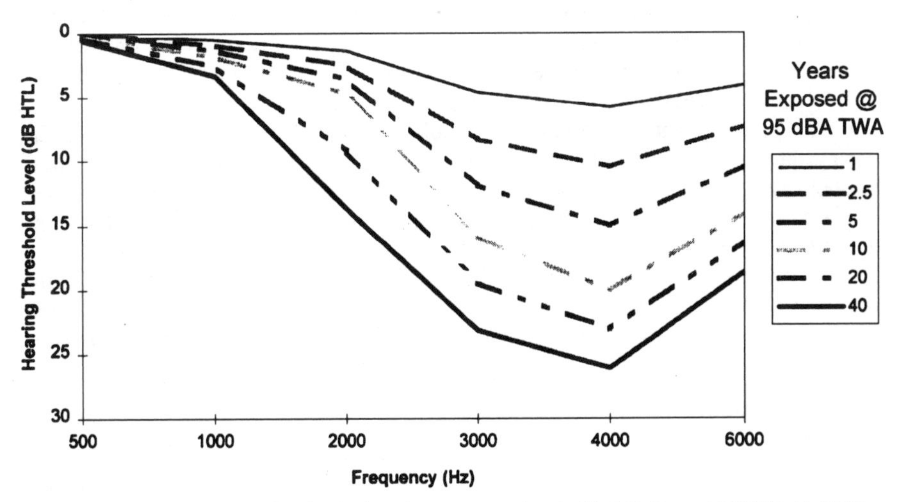

Fig. 22.11 Progression of noise-induced permanent threshold shift from ANSI S3.44–1996.

Noise-induced hearing loss is an insidious problem because a person does not necessarily experience pain before learning that severe hearing damage has taken place. The damage may occur instantaneously, or over a long period of time, depending upon the noise characteristics. Generally, impulsive or impact noises are most likely to produce significant losses with short exposure periods, and steady-state continuous noises are responsible for impairments that develop over a long period of time.

Even after a significant amount of damage, a person with noise-induced hearing loss is able to hear common, low-frequency (vowel) sounds, but the high frequencies (consonants) in speech are not heard as well. Perceived loudness levels may be nearly normal, but intelligibility will be poor in some situations. A noise-induced hearing loss becomes noticeable when speech communication is attempted in noisy, reverberant areas. Speech is masked most effectively by background noises having major frequency components in the speech frequencies, such as those found in areas where many people are talking in the background, sometimes called the "cocktail party" effect.

22.3.7 Occupational Noise-Induced Hearing Loss

Because of differences in the susceptibility of individuals to noise-induced hearing impairment, studies leading to the development of comprehensive damage–risk criteria require complex statistical studies of large groups over long perods of time. It is therefore impossible to set a single exposure level as a dividing line between safe and unsafe conditions that would apply for all individuals. Damage–risk criteria are also compromised by practical limits. Usually, exposure level limits are established as a compromise between[14] the amount of hearing impairment that may result from a specified exposure dose and[15] the economic or other impact that may result from noise control expenditures.

The best estimates of the number of persons who have significant hearing impairment as a result of overexposure to noise are based on a comparison of the number of those with hearing impairments found in high-noise work areas with members of the general population, who have relatively low noise exposures.[16] These studies show that significant hearing impairments for industrial populations are 10 to 30 percent greater for all ages than for general nonexposed populations. At age 55, for example, 22 percent of a group that has had low noise exposures may show significant hearing impairment, whereas in an industrial high-noise exposure group, the figure is 46 percent. In this study, significant hearing loss was defined as greater than 25-dB hearing level.

Hearing impairment for an exposed population can be determined by adding the hearing threshold level associated with age (HTLA) to the noise-induced permanent threshold shift (NIPTS) to calculate what is called the hearing threshold level associated with age and noise (HTLAN). Figure 22.11 shows the median noise-induced hearing loss expected for a group of workers exposed to $L_{\text{Aeq}} = 95$ dB for a number of years. In addition to their noise-induced threshold shifts, there will also be threshold shifts from presbycusis. Figure 22.12 shows the expected median threshold shift (and 10% and 90% ranges) for 60-year-old male workers exposed to $L_{\text{Aeq}} = 95$ dB for 40 years.

22.3.8 Nonoccupational Hearing Loss

Most persons in our society are exposed to potentially hazardous noise away from work. These "off-the-job" exposures must be considered when studying noise-induced hearing loss. The magnitude of noise exposures vary significantly from one location to another and

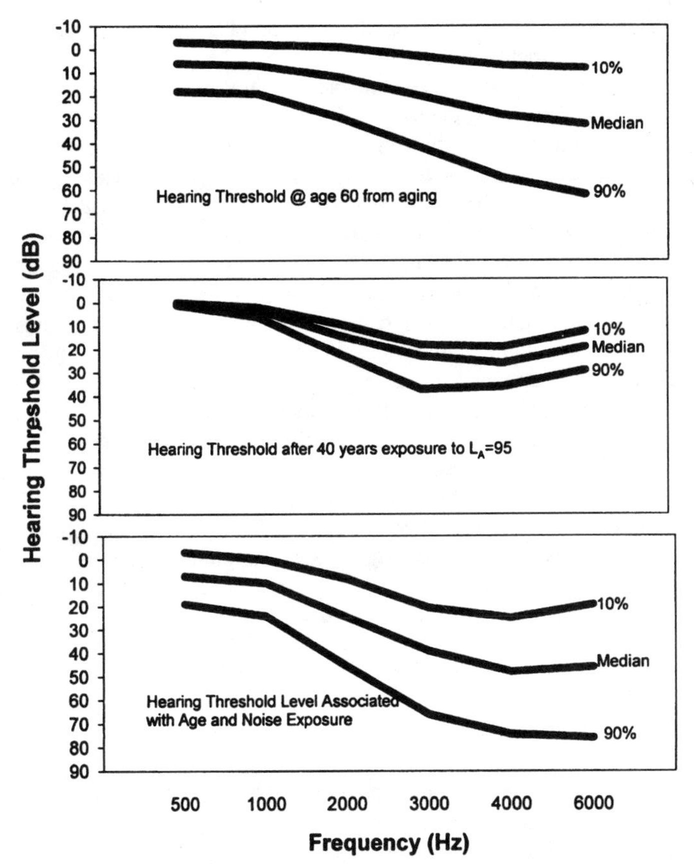

Fig. 22.12 Hearing threshold shifts (median, 10%ile, and 90%ile) for male workers exposed to L_{eq} = 95 for 40 years.

from one time to another. Potentially hazardous exposures away from work may result from noises from guns, chainsaws, airplanes, lawn mowers, motorboats, motorcycles, automobile or motorcycle races, farm equipment, loud music, and shop tools. Even riding in a car at legal speeds with the window down or riding in a subway may be harmful to some noise-sensitive individuals. Again, these variances must be considered in hearing loss analyses. There is a great potential for inaccuracy in studies that consider only on-the-job exposures at one work site with an assumed exposure duration.

A hearing conservation program cannot be effective unless each individual observes the rules limiting noise exposure both at work and away from work. All hearing conservation programs must emphasize the need for constant awareness of noise hazards through a continuing program of education and enforcement.

22.3.9 Nonauditory Physiologic Effects of Noise

Clearly, the most documented effect of noise on man is hearing loss. Other physiologic changes occur with exposure to high levels of noise in the cardiovascular, neurological, and metabolic systems. Effects have also been reported in the reproductive system and in vision and physical orientation.[17]

The vegetative nervous system is responsible for many uncontrollable reactions to both sudden and continuous high-level noise. Cardiovascular reactions include constriction of the blood vessels, an increase in diastolic blood pressure, and an increase in adrenalin level in the bloodstream. Typically, the constriction of the blood vessels diminishes over a period of about 25 minutes if the noise is discontinued or maintained at a relatively constant level. Pulse rate may increase or decrease with exposure to noise. Increased hormonal activity in humans has also been documented, with potential negative long-term effects, including tendencies toward degenerative arterial and cyocardial tissues.

Neural impulses are generated in response to noise, and these impulses spread throughout the gray and white matter of the brain. It is likely that these impulses cause some of the unrelated reactions to noise exposure by affecting areas of the brain other than areas related to pure auditory function. Sudden noise causes voluntary and involuntary muscles to contract. The eyes blink, pupil size changes, facial muscles contract, and the knees bend. Familiarization with sound may or may not have an effect on the startle responses, depending on the individual.

Many of the effects associated with noise have been documented in humans from exposures to other stressors, such as bright light. Continued exposure to any stressor can lead to maladies known as diseases of adaption. These are highly dependent on the individual, and may include gastrointestinal disorders, high blood pressure, and even arthritis.

22.4 HEARING MEASUREMENT

The only way to monitor the overall effectiveness of a hearing conservation program is to periodically check the hearing of all persons exposed to potentially hazardous noises. Air conduction hearing thresholds must be checked at 500, 1000, 2000, 3000, 4000, and 6000 Hz. It is recommended that testing also be performed at 8000 Hz, since data at this frequency will help discriminate between noise-induced and presbycusic hearing loss.

To assure the accuracy of air conduction thresholds, audiometers must be calibrated periodically, and a quiet test environment must be maintained.[18,19] A well-trained audiometric technician or hearing conservationist must be used to perform hearing threshold measurements.[20] Requirements for audiometer performance and for the background noise limits in test rooms have been specified in standards published by the American National Standards Institute (ANSI).[21–23] Guidelines for training hearing conservationists have been established by the Council for Accreditation in Occupational Hearing Conservation[20] (CAOHC).

22.4.1 Audiometers

The audiometer is used for measuring pure tone, air conduction hearing thresholds. The audiometer may be designed for manual, self-recording, or automatic operation.

The ANSI Standard S3.6-1996, Specifications for Audiometers, provides the requirements that the instruments must meet to provide accurate information. Instrumentation inaccuracies are seldom obvious, and there is a strong tendency to accept readings as being accurate. Audiometers may lose their specified accuracy very quickly if they are handled roughly. Earphones are particularly susceptible to damage from rough handling; a janitor may drop them on the floor while cleaning and not inform the person in charge of testing. Rough handling during shipment could cause unseen damage so that even a new instrument may be out of calibration when received.

Dust or dirt inside the audiometer can cause switches to produce electrical noise or to wear excessively. Poor electrical contacts may produce intermittent operation or poor accuracy. Dust covers should always be used to protect the instrument when not in use, and the exterior of the instrument should be cleaned periodically to prevent dust and dirt from getting inside the case.

High humidity, salt air, and acid fumes may corrode electrical contacts in switches within an audiometer. The increased resistance that results from corroded contacts may cause electrical noise or affect the accuracy of the instrument.

If the operating characteristics of the audiometer change significantly, the instrument must be serviced. Many changes in the instrument occur slowly, however, and may not be noticed; so instruments must be calibrated periodically. OSHA requires that biological calibrations be performed before each day's use, and that acoustic calibrations be made at least annually.[24] Accuracy checks should also be made any time there are any reasons to suspect a problem.

Biologic calibrations are audiograms taken on persons with known, stable hearing thresholds. These tests are typically performed at the beginning of the day before any evaluations are performed. Because hearing thresholds may increase as much as 20 dB temporarily because of allergies, colds, or other causes; it is strongly recommended that at least two normal-hearing persons be made available as biological test subjects. Relatively inexpensive artificial test heads are also available for daily biologic testing.

Audiometers that have not been calibrated for several years may not meet the ANSI standard specifications. Furthermore, differences of several decibels may be found in threshold measurement data taken with two audiometers that meet opposite extremes of the allowed calibration limits. Thus the need for dependable calibration services that will produce accurate adjustments and correction data is emphasized. The pertinent performance specifications may not be well understood by some of the laboratories offering calibration services. Therefore, a statement that the audiometer meets specifications should not be accepted without an explanation of the specific calibration procedures used and a copy of the calibration data.

22.4.2 Test Rooms

The sound pressure level of the background noise in rooms used for measuring hearing thresholds must be limited to prevent masking effects that cause misleading, elevated threshold values. The maximum allowable sound pressure levels for industrial audiometric test rooms specified by the Occupational Safety and Health Administration (OSHA)[25] may mask some thresholds that are better (lower hearing levels) than above 10-dB hearing level; so quieter test areas are recommended.

More specifically, the maximum allowable sound pressure levels in ANSI S3.1-1991 are recommended in order to measure 0 dB hearing levels with no more than 1-dB threshold elevation for one-ear listening. These background noise levels are difficult to obtain in noisy areas, and may be below the practical limit for room noise in many industrial locations. Thus it has been the accepted practice for some industries to use the 10-dB hearing level as the lowest hearing threshold measurement level.[26] If the

TABLE 22.4 Maximum Permissible Ambient Sound Pressure Level for Audiometric Testing with Ears Covered[23]

Octave band center frequency (Hz)	125	250	500	1000	2000	4000	8000
L_{max} (dB SPL re 20 μPa)	34	22.5	19.5	26.5	28	34.5	43.5

lowest hearing level to be measured is 10-dB HL, the background noise limits shown in Table 22.4 can be adjusted upward by 10 dB.

In addition to the noise level requirements, subjective tests should be made inside the closed test booth on location to determine that no noises (e.g., talking and heel clicking) are audible. Any audible noise may distract the subject and interfere with the hearing threshold measurements. These noises must be eliminated, or the tests must be delayed until no noise can be heard. Short impulse-type noises, in particular, may be heard even though the measured sound pressure levels are below the limits in the table.

In areas where ambient noise in audiometric test rooms is of particular concern, the hearing conservationist may wish to consider continuous monitoring of background noise level. Devices are available that continuously sample the ambient noise and compare it to preset standard levels. A visual indication is provided if the noise sample exceeds the preset levels. If background noise level is measured once per day, for example, the activation of a machine or process that produces excessive noise in the booth may not be recognized. Continuous monitoring of background noise alerts the technician of this problem.

Factors to consider for audiometric test booths include:

1. What are the size and appearance?
2. Will opening and closing the door result in wearing of contact material (seals), necessitating frequent replacement?
3. Are the interior surfaces durable and easily cleaned?
4. Is the door easily opened from the inside, so that subjects will not feel "locked in"?
5. The observation window and seating arrangement must provide an easy view of the subject, but the subject must not be able to see the technician or audiologist.
6. If the booth is equipped with a ventilation system, the noise should be below the limits specified for the room.
7. Portability of the room is seldom an important factor, because test rooms are rarely moved.

22.4.3 Hearing Threshold Measurements

Normally, the purpose of hearing testing in industrial hearing conservation programs is to monitor the effectiveness of hearing conservation procedures. If this is the case, there is no need for diagnostic information that would require the use of sophisticated audiometric techniques. Pure tone, air conduction hearing thresholds are usually adequate for industrial monitoring purposes.[27]

Properly calibrated manual, self-recording, or automatic audiometers may be used to monitor hearing thresholds in industry. Self-recording audiometers print or store a range of hearing threshold levels between the barely perceptible sound pressure level and not perceptible sound pressure level. The range is frequently 10 dB wide, and the audiometric technician usually chooses the midpoint of the range as the subject's hearing threshold.

Microprocessor-based automatic audiometers are commonly in use in industry and in clinical environments. These units measure hearing threshold by presenting the patient with one or a series of tones and adjusting the level of the tones according to the patient responses. A common method of defining a hearing threshold with an automatic audiometer (and also via manual audiometry) is referred to as the modified Hughson–Westlake method. In this procedure, the stimulus is presented to the patient at an audible level and then decreased in 10 or 15 dB steps until it is no longer audible. After the tone is inaudible, the level is increased in 5-dB steps until the patient responds. After a positive response, the tone is lowered by 10 dB and testing resumes, with the level again raised in 5-dB steps until the patient gives a positive response. This is referred to as an ascending audiometric technique. The threshold is then defined as the level at which the patient responds to 50% of the stimuli with a minimum of two responses at a single level.

Hearing the thresholds measured by a manual or automatic audiometer are typically 2–3 dB higher than those measured with self-recording audiometers because of the differences in data analysis. For precision, and to assure consistency, it is important that the method of threshold determination remain constant throughout the hearing conservation program.

22.4.4 Records

The hearing conservationist's records must be complete and accurate if they are to have medico-legal significance. Records should be kept in ink, without erasures. Employee signatures should be required.

If a mistake is made in recording, a line should be drawn through the erroneous entry, and the initials of the person making the recording should be placed above the line along with the date. The entry must not be erased.

The model, serial number, and calibration dates of the instruments should be recorded on each audiogram. Records should also be kept of periodic noise level measurements in the test space.

In addition to the threshold levels, an audiogram should have space provided for the recording of pertinent medical and noise exposure information. Some typical questions are listed on Table 22.5.

22.4.5 Audiometric Database Analysis

The occupational noise exposure regulations require that annual audiograms of all persons included in the hearing conservation program (Section 22.7) are performed and that appropriate records are kept. A convenient measure of the effectiveness of a hearing conservation program is to compare these records to those of a non-noise-exposed population. The number of standard threshold shifts (STSs) in the industrial population are counted and compared against the number of STSs that may be recorded in the control population. An STS is defined in 29 CFR 1910.95 as a change in hearing threshold by an average of 10 dB or more at 2000, 3000, and 4000 Hz in either ear. This calculation is made after making the correction for presbycusis. Obviously, if the number

TABLE 22.5 Noise Exposure and Medical Information for Audiograms

Name______________________ Date: _______ Recorded by: _______________
Employee Signature _____________________________

History	Yes	No	Comments
Have you had a previous hearing test? (When)			
Have you EVER had trouble hearing?			
Do you NOW have trouble hearing?			
Have you ever worked in a noisy industry?			
Do you think you can heard better in your right ear left ear			
Have you ever heard noises in your ears?			
Have you ever had dizziness?			
Have you ever had a head injury?			
Has anyone in your family lost hearing before age 50?			
Have you every had measles, mumps, or scarlet fever?			
Do you have any allergies?			
Are you now taking or ever regularly taken drugs, antibiotics, or medication?			
Have you ever had an earache?			
Have your ears ever run? Right ear Left ear			
Have you ever been in the military service? Describe—especially experience with firearms.			
Have you ever been exposed to any sort of gunfire? Describe.			
Do you have a second job?			
What are your hobbies?			

of STSs in industry compares favorably with the control population, then the hearing conservation program is successful. This would indicate that either (1) the noise exposure levels are not too high, (2) the engineering noise control activities are successful, or (3) the use of personal hearing protection is successful, or most likely, (4) a combination of all three. ANSI S12.20, Method for Evaluating the Effectiveness of a Hearing Conservation Program, includes additional information on this topic.

22.5 NOISE MEASUREMENT

Since the human ear is a pressure-sensitive device, measurements of sound pressure levels are usually sufficient to determine the hazard potential of the noise. Two broad categories of measurements are:

- Measuring noise from a specific source;
- Measuring noise to characterize a certain environment.

Personal noise exposure measurements for compliance or hearing conservation purposes would fall into the latter category. Simple sound level instruments can be used to estimate the potential risk of high noise levels. More sophisticated measurements are taken when the objective is to obtain data on which engineering changes are to be made. ANSI[28] and ISO standards specify the accuracy of various sound measuring equipment. Qualities range from Laboratory Grade (type 0), Precision (type 1) to General Purpose (type 2), and Special Purpose (type S). OSHA has specified the instrument used for industrial noise measurements to be Type 2 or better.

22.5.1 The Sound Level Meter

The basic sound level meter consists of a microphone that converts the pressure variations into an electrical signal, an amplifier with frequency weighting, an exponential-time-averaging device, a device to determine the logarithm of the signal, and either a display or other means of storing the result in dB re 20 μPa (see Fig. 22.13).

Since sound pressures are rapidly fluctuating pressure levels above and below atmospheric pressure, the average pressure is simply atmospheric pressure. Sound-measuring instruments measure the pressure levels as the exponential time-averaged square root pressure (voltage) signal multiplied by itself (voltage^2). Modern sound measuring instruments cover a dynamic range of more than 40 to 140 dB. More expensive equipment permits measurements in octave bands from about 0 dB to well above 160 dB, depending on the microphone selected.

Tolerance limits for Type 2 sound level meters are less than a half-decibel change in reading for: (1) 10% change in static pressure, (2) a 20°C change in ambient temperature over -10 to 50°C, (3) for a change in relative humidity between 30% and 90%, or (4) any change within 1 hour of operation. Modern sound level meters are capable of repeatable, accurate, and precise measurements.

22.5.1.1 Microphones and Directional Characteristics Most noises encountered in industry are produced from many different noise sources combined with their reflected energies. However, microphones may also be used outdoors, where conditions are closer to free field, and may be used in the near field close to the noise source.

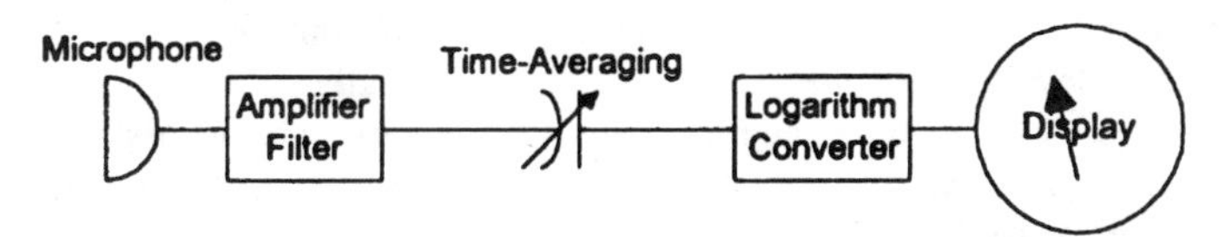

Fig. 22.13 Components of a basic sound level meter.

Depending on the design and purpose of a microphone, it may be calibrated for grazing, perpendicular, or random incidence sounds, or it may be calibrated for use in couplers (pressure calibration). Care must be taken to use the microphone in the manner specified by the manufacturer.

A microphone calibrated with randomly incident sound should be pointed at an angle to the major noise source that is specified by the manufacturer. An angle of about 70° from the axis of the microphone is often used to produce characteristics similar to randomly incident waves, but the angle for each microphone should be supplied by the manufacturer.

A free-field microphone is calibrated to measure sounds perpendicularly incident to the microphone diaphragm; thus it should be pointed directly at the source to be measured. A pressure-type microphone is designed for use in a coupler such as those used for calibrating audiometers; however, this microphone can be used to measure noise over most of the audible spectrum if the noise propagation is at grazing incidence (90° to the diaphragm).

Microphones used with sound measuring equipment are often said to be nearly omnidirectional, but, strictly speaking, this is generally true only for the lower frequencies. Above 1,000 Hz, 1-in.-diameter microphones (above 5,000 Hz for 1/2-in.-diameter microphones) may respond differently depending on the orientation of the microphone to the sound. When measurements are to be made of high-frequency noise produced by a directional noise source (i.e., where a high percentage of the noise energy is coming from one direction), the orientation of these microphones becomes important, or smaller-diameter microphones should be used. The tradeoff is that smaller-diameter microphones are not as sensitive as larger microphones and thus are not able to measure low sound pressure levels. For measuring levels of concern for OSHA compliance or hearing conservation, 1/2-in., 1/4-in., and even 1/8-in.-diameter microphones are capable of handling sound pressure level measurements above 70 dB.

When measuring sound pressure it is best to position the microphone several wavelengths away from any surface that may reflect sound. In particular, this includes the body of the person conducting the measurement. For most accurate measurements, the microphone or sound level meter is mounted on a tripod with the person making the measurements standing several feet away. When measuring exposure of a machine operator, accurate measurements may be made by asking the operator to step aside, and placing the microphone where the center of the operator's head is while standing in the normal operating position. However, measurements of adequate precision can usually be made by placing the microphone within 0.5–1 meter of the machine operator's ear.

22.5.1.2 Frequency-Weighting Networks General-purpose sound level meters are normally equipped with frequency-weighting networks that are used to characterize the frequency distribution of noise over the audible spectrum.[29] Frequency weightings, given in Table 22.6, were chosen because (1) they approximate the response characteristics of the ear at various sound levels, and (2) they can be easily produced with a few common electronic components. A linear, flat, or overall response, also included on many sound level meters, weights all frequencies equally.

The A-frequency weighting approximates the response characteristics of the ear for low-level sounds (below $L_A = 55$ dB). C-frequency weighting corresponds to the response of the ear for levels above $L_A = 85$ dB and may be used to estimate loudness when significant low-frequency sounds are present. B-frequency weighting, approximat-

TABLE 22.6 Random Incidence Relative Response as a Function of Frequency[28,30]

Band number	Nominal frequency (Hz)	A-weighting (dB)	C-weighting (dB)	D-weighting (dB)
14	25	−44.7	−4.4	−18.7
15	31.5	−39.4	−3.0	−16.7
16	40	−34.6	−2.0	−14.7
17	50	−30.2	−1.3	−12.8
18	63	−26.2	−0.8	−10.9
19	80	−22.5	−0.5	−9.0
20	100	−19.1	−0.3	−7.2
22	125	−16.1	−0.2	−5.5
22	160	−13.4	−0.1	−4.0
23	200	−10.9	0	−2.65
24	250	−8.6	0	−1.6
25	315	−6.6	0	−0.8
26	400	−4.8	0	−0.4
27	500	−3.2	0	−0.3
28	630	−1.9	0	−0.5
29	800	−0.8	0	−0.6
30	1000	0	0	0
31	1250	+0.6	0	+2.0
32	1600	+1.0	−0.1	+4.9
33	2000	+1.2	−0.2	+7.9
34	2500	+1.3	−0.3	+10.4
35	3150	+1.2	−0.5	+11.6
36	4000	+1.0	−0.8	+11.1
37	5000	+0.5	−1.3	+9.6
38	6300	−0.1	−2.0	+7.6
39	8000	−1.1	−3.0	+5.5
40	10000	−2.5	−4.4	+3.4

ing the response of the ear for levels of $55 < L_A < 85$ dB is still listed as part of ANSI S1.4, but is not commonly used today. For measuring annoyance of aircraft and other environmental noise, D-weighting[30] has been developed.

The frequency response at any frequency is specified relative to the meter's response at 1,000 Hz. Positive and negative values indicate that the specific frequency contributes more or less to the overall level than a 1000 Hz tone. The values for octave and third octave frequencies are given numerically in Table 22.6 and graphically in Fig. 22.14.

In use, the frequency distribution of noise energy can be approximated by comparing the levels measured with each of the frequency weightings. For example, if the noise levels measured using the A and C networks are approximately equal, it can be reasoned that most of the noise energy is above 1000 Hz, because this is the only portion of the spectrum in which the weightings are similar. On the other hand, a large difference between these readings indicates that most of the energy will be found below 1000 Hz.

For determining compliance with OSHA standards and to estimate the risk of noise-induced hearing loss, measurements using A-weightings and slow response are required. A-weighting has become the default for determining noise exposures for continuous, intermittent, or varying noises and is used for most work except impulsive sounds and some annoyance measurements.

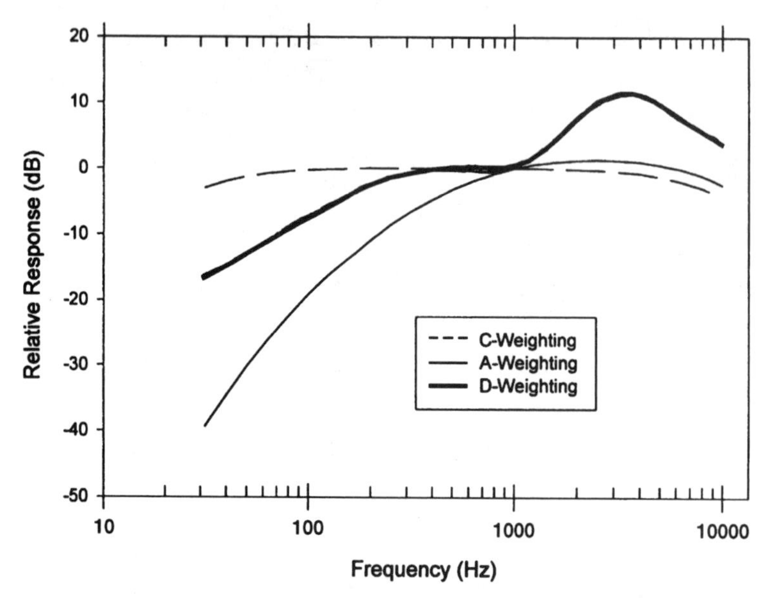

Fig. 22.14 Relative response of A, C, and D frequency weighting filters.

22.5.1.3 Exponential-Time Averaging Before the advent of digital displays, sound measuring instruments relied on galvanometers for displaying sound levels. These indicating meters had ballistic characteristics that were not constant over their entire dynamic range, often resulting in different readings, depending on the attenuator setting and the portion of the meter scale used, requiring the reader to adjust attenuators so the meter reads in a particular portion of the scale. More modern meters with electronic displays do not suffer from this limitation.

"Slow," "Fast," and "Impulse" time constants are specified for sound level meters. The *slow* response (1 second time constant) is intended to provide an averaging effect that will make widely fluctuating sound levels easier to read; however, this setting will not provide accurate readings if the sound levels change significantly in less than 0.5 second. The *fast* response (0.125 second time constant) enables the meter to reach within 1 dB of its calibrated reading for a 0.2 second pulse of 1000 Hz; thus it can be used to measure, with reasonable accuracy, noise levels that do not change substantially in periods less than 0.2 second.

Impulse time averaging introduces a peak detector into the circuitry. For sound pressures that increase with time, the time constant is 0.035 second. When the sound pressure decreases, the peak detector introduces a 1.5 second decay time. Impulse exponential-time averaging was introduced on earlier generations of sound level meters, so the maximum sound pressures of varying sounds could be measured. However, because of the asymmetric nature of the rise and decay times, it is not possible to calculate integrated exposures from recordings of impulse measurements. Today, many sound level meters are built with the capability of measuring the instantaneous *peak* sound level. They generally are measured with the A-frequency weighting circuitry giving the equivalent of a 30 millisecond exponential-time weighting.

Figure 22.15 shows fluctuations in sound pressures for a speech signal with slow, fast, and impulse time-averaged sound pressure levels. Notice how the sound pressure level changes by more than 15 decibels for a fast time constant measurements. Without time averaging the fluctuations in sound pressure levels would be too fast to measure accurately. Table 22.7 shows differences between fast, impulse, and peak time-averaged levels for some commonly encountered sounds.

22.5.2 Sound Exposures and Long-Term Average Sound Levels

Because of the large variations of typical sound levels as noted above, and because of the desire to characterize exposures with a single number, it is necessary to develop a long-term (typically 8-h) sound level descriptor. The equivalent-continuous sound pressure *level* of a time-varying sound is equal to the level of an equivalent steady sound *pressure* that has the same acoustic energy as the time-varying sound actually measured. The symbol is L_{eq} or L_{Aeq}, and the abbreviation is TEQ. Mathematically:

$$L_{\mathrm{eq}} = 10 \log \left(\frac{1}{(t_2 - t_1)} \int_{t_1}^{t_2} \frac{p_{\mathrm{A}}^2(t)\, dt}{p_{\mathrm{r}}^2} \right) \tag{22.10}$$

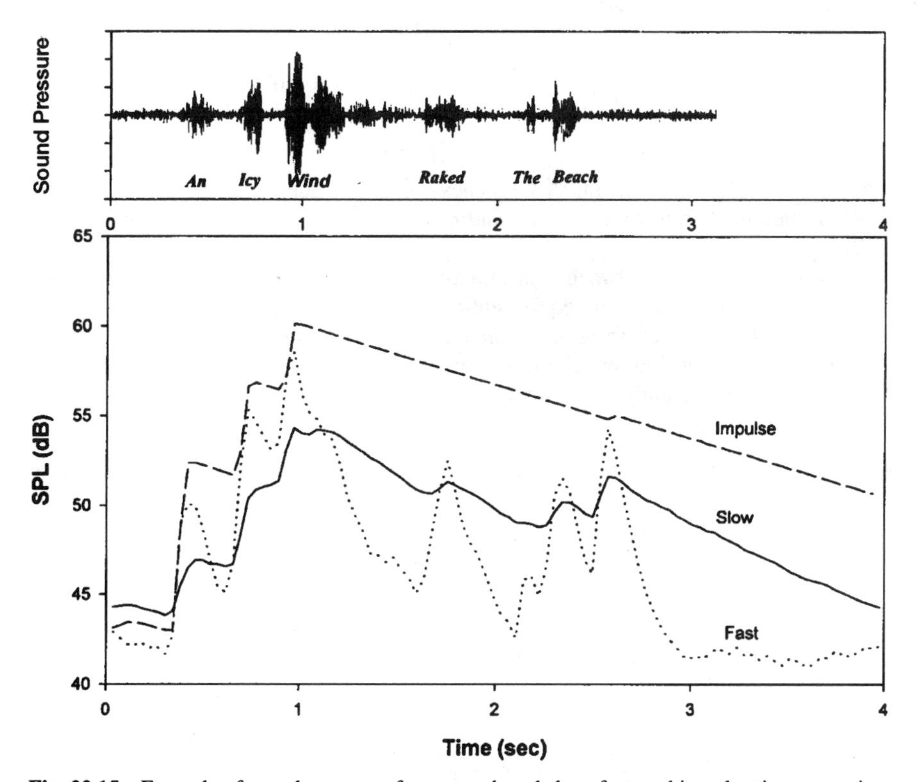

Fig. 22.15 Example of sound pressures from speech and slow, fast, and impulse time averaging.

TABLE 22.7 Examples of Typical Sound Levels Taken with Different Exponential-Time Weightings[31]

Sound source	L_{AF}	L_{AI}	L_{APk}
Sinusoidal pure tone @ 1000 Hz	94	94	97
Highway traffic 15 m distance	80	80	89
Train 18 m distance	85	87	94
Noise in car 80 km/h	75	75	86
Lawn mower 1 m distance	97	99	116
Diesel motor in electricity generating plant	100	101	113
Pneumatic nailing machine near operator's head	116	120	148
Air compressor room	92	92	104
Large machine shop	81	82	98
Large punch press near operator's head	93	98	122
Small automatic punch press	100	103	118
Small high-speed drill	98	101	109
Machine saw near operator's head	102	102	113
Vacuum cleaner 1.2 m distance	85	88	105
Bottling machinery in brewery	98	99	122
Pistol, 9 mm, 5 m distance from side	111	114	146
Shotgun, 5 m distance from side	106	110	139

where $p_A^2(t)$ is the square of the A frequency-weighted sound pressure, $(t_2 - t_1)$ is the duration measured from time t_1 to t_2, and p_r is the standard reference sound pressure of 20 μPa.

Note the difference between this and the response time of sound level meters, i.e. exponential-time weighting-equivalent-continuous sound levels have uniform time weighting. All sounds occurring *at any time* during the measurement period are included with equal weighting. Exponential-time-weighting yields sound levels primarily influenced by the most recently occurring sounds.

Example 5: What is L_{eq} for an 8-h workday consisting of four exposures as shown in the table?

L_{AS} (dB)	Duration (h)
82	6.0
88	1.4
97	0.5
105	0.1

Converting levels to sound pressure.

$$L = 10\log\left(\frac{p}{p_r}\right)^2$$

$$\frac{L}{10} = \log\left(\frac{p}{p_r}\right)^2$$

$$10^{L/10} = \left(\frac{p}{p_r}\right)^2$$

and hours to seconds gives the following table:

L_{AS} (dB)	$(p/p_r)^2$	Duration (h)	Duration (s)	$(p_A/p_r)^2 dt$
82	1.58×10^8	6.0	21600	3.42×10^{12}
88	6.31×10^8	1.4	5040	3.18×10^{12}
97	5.01×10^9	0.5	1800	9.02×10^{12}
105	3.16×10^{10}	0.1	360	1.14×10^{13}
		Total	28800	2.70×10^{13}

$$L_{\text{Aeq, 8 h}} = 10\log\left(\frac{270\times 10^{13}}{2.88\times 10^{4}}\right) = 90 \text{ dB}$$

Integrating sound level meters calculate L_{eq} automatically; so the industrial hygienist is generally spared the necessity of making these calculations.

22.5.3 Frequency Analyzers

For noise control purposes, the rough estimate of frequency–response characteristics provided by the weighting networks of a sound level meter is not always adequate. In such cases frequency analyzers are used. The most common analyzers are octave band, one-third octave band, and narrow-band analyzers using filters or fast-Fourier-transform digital calculations.[32,33]

22.5.3.1 Octave Band Analyzers The octave band (OB) analyzer is the most common type of noise filter used for many noise analyses because the octave band normally provides adequate spectral information with a minimum number of measurements. An octave is defined as any bandwidth having an upper band-edge frequency equal to twice the lower band-edge frequency.

Filters are used to distinguish the characteristics of a noise source. Octave bands "filter out" frequencies below and above a one-octave bandwidth. Therefore, the upper edge of the band of frequencies f_u is twice the frequency of the lower band edge f_l.

$$f_u = 2f_l \tag{22.11}$$

When more detail is required, one-third octave bands are used. For one-third octaves, the upper band edge is the cube root of 2 times the lower band edge.

$$f_u = \sqrt[3]{2}\, f_l \tag{22.12}$$

Note that not all sounds outside the upper and lower frequencies are excluded. The band-edge frequencies represent the point at which the filter response is 3 dB down from the center frequency response. Specifications for octave and fractional octave filters can be found in ANSI S1.11.

When describing industrial noise, frequencies less than 20 Hz and greater than 20,000 Hz are generally not used. The bands and frequencies commonly used are shown in Table 22.6. Frequencies shown in bold are the octave band preferred frequencies—chosen to include 1000 Hz.

22.5.3.2 Narrow-Band Analyzers The A- or C-weighting networks are often used for hearing conservation purposes. For noise control purposes, octave band and third octave band analyzers are adequate and satisfy the requirements listed in national and international consensus standards. In rare cases, finer resolution of frequency information is needed. Fast Fourier transform (FFT) analyzers are the most common technique used for these measurements.

FFT analyzers convert time-domain signals (amplitude versus time) to the frequency domain (amplitude versus frequency). Input signals need to be sampled at least twice the minimum frequency of interest. Thus for noise control work up to 20,000 Hz, sound pressure levels must be sampled at least 40,000 times per second. The frequency resolution of each spectral line is equal to the sampling rate divided by the number of lines chosen. Therefore, the resolution of a 1024-line FFT analysis of a signal sampled 40,000 times per second would be about 40 Hz.

$$\frac{\dfrac{40000 \text{ samples}}{\text{second}}}{\dfrac{1 \text{ sample}}{1024 \text{ lines}}} = \frac{39 \text{ lines}}{\text{second}} \tag{22.13}$$

22.5.4 Tape Recording of Noise

It is sometimes convenient to record a noise so that it can be analyzed at a later date. This is particularly helpful when lengthy narrow-band analyses are to be made, or when very short, transient-type noises are to be analyzed. Unfortunately, most recorders are meant for audio measurements and incorporate automatic gain control (AGC) circuitry. These circuits constantly monitor the incoming signal, increasing the gain when the levels are low, and decreasing the gain at higher sound pressure levels. Consequently, all information of the absolute level of the signal is lost. Professional or specialized recorders are needed for industrial noise measurements.

It is important to calibrate the combination of tape recorder and sound level meter at a known level, preferably throughout the frequency range. Before each series of measurements, a pressure level calibration should be made by recording the overall sound pressure level reading by stating the levels orally, along with the tape recorder dial settings. It is also good practice to state orally on each recording the type and serial numbers of the microphone and sound and surroundings, and other pertinent information. This practice of noting information orally on the tape often prevents information from being lost or confused with other tapes.

22.5.5 Level Recording (Data Logging)

Previously, graphic level recorders were connected to the output of a sound level meter or analyzer to provide a continuous paper and ink record of the output level. Currently, most recording of levels is performed electronically by sound level meters or noise dosimeters with data logging (recording) capabilities. Data are stored inside the instruments and later printed or transferred to personal computers.

Sound pressure levels are sampled typically several times per second, and the distribution of sound pressure levels is saved in electronic memory. Depending on the mode of operation chosen by the user, either the distribution of levels is integrated over the measurement period or, periodically, some measures are recorded from the distribution, then a new distribution is started for the next sampling period. After sampling, the user has a record of either the distribution of levels over the entire sampling period or indicators of the distribution for each period.

Distribution results are normally plotted as amount of time (or fraction of total time) sound pressure levels fall into 1, 2, or 5 decibel bins. The data provide a discrete-valued function of sound pressure levels.

22.5.6 Instrument Calibration

If valid data are to be obtained, it is essential that all equipment for the measurement and analysis of sound is calibrated. When equipment is purchased from the manufacturer, it should have been calibrated to the pertinent ANSI or IEC specifications. However, it is the responsibility of the equipment user to keep the instrument in calibration by periodic checks.

Acoustic calibrators are available for checking the overall acoustical and electrical performance at one or more frequencies. These calibrations should be made according to the manufacturer's instructions at the beginning and at the end of each day's measurements. A battery check should also be made at these times. These calibration procedures cannot be considered to be of high absolute accuracy, nor will they allow the operator to detect changes in performance at frequencies other than that used for calibration. They do serve as a warning of most common instrument failures, thus avoiding many invalid measurements.

Periodically sound measuring instruments should be sent back to the manufacturer or to a competent acoustic laboratory for calibration at several frequencies throughout the instrument range. These calibrations require technical competence and the use of expensive equipment. How frequently these complete calibrations should be made depends on the purpose of the measurements and how roughly the instruments are handled. ANSI and IEC standards require a complete calibration performed at least once a year, and at any time a calibration shift greater than 1 dB is found.

22.5.7 Making Occupational Noise Exposure Measurements

Although noise measuring equipment may be simple to operate and read, the results will not be meaningful unless the proper measurements are taken and recorded using appropriate scales and units in a manner that can be interpreted by others.

The following guide can be used before making any acoustical noise measurements.

1. Classify the problem: Generally noise can be classified as:

A. Generation or transmission of noise from one or more sources
   The purpose is to measure a quantity—sound pressure level or frequency spectrum—at a certain point or time relative to the source.
B. Measurement of noise at a receiver
   Measure a quantity—integrated sound pressure level or energy—related to the effects of noise on the exposed individual.

2. Consider the type of noise
   A. Frequency spectrum
      Continuous spectrum
         Measure overall level or octave band levels
      Spectrum with audible tones
         Third-octave band or narrow-band analysis is required.
   B. Variation of level with time
      Steady noise
         For noise with small (generally less than 3 dB) fluctuations in level, any measurement technique may be used.
      Nonsteady noise
         Measure the levels of noise and duration for each identifiable noise level. Alternatively, measure equivalent continuous sound pressure level or sound exposure.
      Impulsive
         Measure impulse sound levels and frequency of impulses or number per day, as well as background ambient noise level. Measure sound exposure with meter set to fast or peak exponential-time averaging.
3. Consider the sound field (path)
   A. Free-field, reverberant field, or semireverberant
      Document location of measurement relative to the source. Consider how sound pressure level decays with distance from source.
   B. Personal exposure monitoring
      Best accuracy for short-duration monitoring is made with microphone positioned at a location that would be the center of the employee's head while operating the machine. For noise dosimetry, place the microphone on the employees shoulder.
4. Consider the quality[34] of measurement desired
   A. Precision—Generally not used for hearing conservation purposes. Requires a thorough description of the noise source and environment.
   B. Engineering—Useful for collecting data of existing equipment or environments for noise control purposes. Sound pressure levels are supplemented by band pressure levels. Usually need to characterize noise with sources operating and background noise.
   C. Survey—Primarily measurements of sound pressure level, OSHA noise dose, or equivalent continuous sound level. Measure the noise "as is," without modifying sources or path. Used for exposure monitoring—generally does not supply enough detail for noise control purposes.
5. Consider the purpose of the measurements
   A. Compliance to a standard
      Make sure the equipment is capable of measuring full range of sound levels, frequency weighting, and time averaging specified by the standard.

Integrating sound level meters and dosimeters must measure according to the specified exchange rate.

B. For evaluating engineering controls or source characterization:
The equipment should be capable of measuring at least octave band levels. Frequently one-third octave band or an FFT analyzer is used. Recorders—either digital or tape—permit later analysis.

A checklist of equipment usually required for acoustic evaluations would include:

Sound level meter or dosimeter(s)
Acoustical calibrator for all meters
Spare batteries for meters and calibrators
Windscreen(s) for protecting the microphones
Watch for recording measurement start and stop times
Forms for recording measurement conditions
Camera for photographic documentation
Hearing protection (and any other necessary personal protective equipment)

22.5.7.1 Measurement Procedures Procedures for performing occupational noise exposure measurements can be found in ANSI S12.19[35] "American National Standard Measurement of Occupational Noise Exposure" or ISO-9612[36] "Acoustics—Guidelines for the Measurement and Assessment of Exposure to Noise in the Working Environment."

Sound level meters shall meet the requirement of Type 2 meters per ANSI S1.4 (or IEC 651). Noise dosimeters are specified in ANSI S1.25 (IEC 942), and calibrators in ANSI S1.40 (IEC 942). All instruments should be calibrated annually by a qualified test laboratory. The battery in each meter should be checked and the calibration of each meter checked before and after each measurement period. If there is a change of more than 1.0 dB from the initial to the final calibrations, the measurements must be rejected.

As far as possible, nothing should be done to disturb the acoustic environment at the measurement site, and measurements should be arranged so there is minimal disturbance of normal work patterns. The worker being monitored (and their supervisor) should be informed of the purpose of the measurement. At the conclusion of the measurement, input from the employee and their management should be sought to determine if there were any unusual events or work conditions that would make the measurement not representative of normal operations.

22.5.7.1.1 Placement of Microphone of a Sound Level Meter Measurements are taken in the employee's "hearing zone." This is a location chosen to be representative of a worker's noise exposure. When a sound level meter is used, the preferred location is in a position at the center of where the head of a worker would be while performing the task being measured. This is not always feasible; so measurements are frequently taken with the microphone placed approximately 0.1 m from the entrance to the ear canal of the ear receiving the higher sound level. For areas where there is one dominant source, care must be taken that a direct line of sight is maintained between the microphone and the source; that is, it should not be blocked by the body of the employee or the industrial hygienist taking the measurement. Usual practice is to hold the sound level meter out at arm's length—away from any reflections or absorption created by the individual taking the measurement. The sound level meter's instruction manual must be consulted to determine proper orientation of the microphone relative to the source.

22.5.7.1.2 Dosimeter Placement The hearing zone for a dosimeter microphone is specified as on the top of the shoulder, midway between the neck and the outside edge of the shoulder. The shoulder facing the source should be used if the worker is consistently exposed to noise from one side. The microphone diaphragm should be oriented parallel to the plane of the shoulder. The microphone cable should be routed so it does not interfere with the worker's performance or safety. The body of the dosimeter can be placed wherever convenient. Typically, the microphone cord lies diagonally across the worker's back, leading to the electronics package, which is clipped to a belt at the employee's hip. The cable is often clipped or taped to the worker's shirt in the back to keep it from swinging loose. It may be necessary to reset the dosimeter after the initial calibration, and the dosimeter results should be recorded before performing the final calibration to avoid adding calibration signals to the workers dose.

22.5.7.1.3 Measurements Measurements should be of sufficient duration to get a representative noise exposure. If the level does not vary by more than ±3 dB, it is considered steady, and an average reading from a sound level meter may be taken. Frequently the level varies by more than this; so an integrating meter should be used. If there is some periodicity to the sound level, the readings must be taken over at least one full period. If there are several distinct noise levels, each level should be measured as well as its duration and the number of times it occurs each work shift.

When dosimeters are used, all workers should be under visual observation to assure they do not perform any activities which will invalidate the measurement.

22.5.7.2 Recordkeeping Measurements, in particular those collected for determining exposure, should be recorded in a manner that assures future occupational assessments will be possible. It is important that adequate data are collected using precise and uniform definitions. The American Conference of Governmental Industrial Hygienists and the American Industrial Hygiene Association have issued guidelines and recommendations[37] for occupational exposure databases. Also, whenever industrial hygiene measurements are collected, some form of sampling data sheet is recommended. The sampling data sheet acts as a reminder to ensure all relevant information is collected. A sampling data sheet formed from the recommendations of the joint committee is in Table 22.8. Letter–number pairs follow from the "data groups" specified in the guidelines. Starred items are required data fields.

22.5.8 Making Environmental Noise Measurements

Measurements taken for community noise assessment[38] use all the above principles. However, the position at which outdoor measurements should be taken is not always as straightforward as it is for occupational noise measurements; measurements frequently need to be taken for longer duration, and much more attention should be paid to microphone placement. A windscreen is virtually a requirement for environmental noise measurements, and a microphone cover for protection from precipitation is usually needed.

Generally the positioning of the microphone depends on the purpose of the measurements. At times the industrial hygienist will be called upon to make property line

TABLE 22.8 Noise Monitoring Recording Form

NOISE MONITORING FORM	
Facility/Site Information	
A-1* Company/Organization:	
A-2* Facility Name:	A-3* Facility Address
A-4* SIC Code	A-5* Industrial Category (descr.)
A-7* # employees at facility	
A-6* (Contractor name, type, SIC)	
Survey Tracking Information	
B-1* Survey (reference) #	B-2* Survey Date
B-4 Report #	
B-3* Person Performing Survey	
B-5 Is Followup Required? [ ] Yes, followup required [ ] No, followup is not required	
B-6 Followup Summary	
B-7 Person Responsible for Followup	B-8 Date Followup completed
B-9* Quality Control Reviewer Name, position, SSN	B-10* Date Reviewed
B-11 General Survey Comments	

TABLE 22.8 (*Continued*)

Work Area Information	
C-1* Building/Zone(s)	C-2* Room/Area
C-3* Department	
C-4* Type of Work Area: [ ] Open Air [ ] Enclosed Indoor Space [ ] Confined Space (descr.) [ ] Equipment Cab [ ] Other	
C-5 Location Comments:	
C-6 Climatic Conditions:	
Employee Information	
D-1* Employee Name	D-2* Employee Id:
D-3* Administrative Job Title	D-4* Occupational Title
D-5* Work or Task Description	
D-6 Similar Exposure Groups (SEG)	
D-7* Shift __:__ start __:__ end	D-8 Union
D-9 Job Safety Training: [ ] Yes [ ] No	
D-10 Comments:	
Process & Operation Information	
E-1* Process	
E-2* Task	
E-3* Frequency of Process: [ ] Continuous [ ] Frequency: __times/____ (hr., day, etc.)	

TABLE 22.8 (*Continued*)

Process & Operation Information (*Continued*)
E-4 Comments on Process:
E-5* Source(s): [ ] Single source directly associated with employee [ ] Single source associated with employee AND additional sources [ ] Single source distant from employee [ ] Multiple sources distant from employee [ ] Other (descr.)
Exposure Modifier Information
G-1* Exposure Representative? [ ] Yes (identify SEG) [ ] No [ ] Unknown
G-2* Exposure Representativeness Comments
G-3* Exposure Conditions: [ ] Typical [ ] Higher than Normal [ ] Lower than Normal [ ] Unknown
G-4 Basis for Estimate of Conditions:
G-5* Exposure Pattern on Day of Sample [ ] Continuous exposure throughout day [ ] Continuous throughout part of day (specify) [ ] Intermittent (specify frequency & duration) [ ] Other (descr.)
G-6* Exposure Frequency over Extended Time Period: [ ] Daily [ ] Regular Frequency (specify) [ ] Occasional (Estimate frequency) [ ] Other (descr.)
G-9* Exposure Modifier Comments:
Sample Information
H-1* Sample Collected? [ ] Yes H-3* Sample # ______________ [ ] NO
H-2* Reason No Sample Collected

TABLE 22.8 (*Continued*)

<table>
<tr><th colspan="2">Sample Information (Continued)</th></tr>
<tr><td>H-5* Sample Duration:</td><td>H-4* Sample Date:</td></tr>
<tr><td colspan="2">H-6* Reason for Sample: [ ] Baseline [ ] Scheduled [ ] Complaint [ ] Compliance [ ] Diagnostic [ ] Emergency (Response) [ ] Unusual Activity (descr.) [ ] Other (descr.)</td></tr>
<tr><td colspan="2">H-7* Type of Sample – Duration: [ ] Single Sample for full-shift TWA [ ] Multiple partial periods for TWA [ ] Task [ ] Peak Sample [ ] Other (descr.)</td></tr>
<tr><td colspan="2">H-8* Type of Sample – Location: [ ] Personal (outside Hearing Protection) [ ] Personal (inside Hearing Protection) [ ] Area [ ] Source (specify distance) [ ] Other (descr.)</td></tr>
<tr><td colspan="2">H-9 Sample Information Comments:</td></tr>
<tr><th colspan="2">Sampling Device Information</th></tr>
<tr><td colspan="2">I-1* Sampling Device Type: [ ] Sound Level Meter [ ] Dosimeter [ ] Impact Noise [ ] Other (descr.)</td></tr>
<tr><td colspan="2">I-2* Sampling Device Identification (name, manufacturer, model #):</td></tr>
<tr><td colspan="2">I-3* Calibration Documentation:</td></tr>
<tr><td colspan="2">I-6 Comments:</td></tr>
<tr><th colspan="2">Administrative/Engineering Controls</th></tr>
<tr><td colspan="2">J-1* Administrative Controls: [ ] Yes (descr.) [ ] No</td></tr>
</table>

TABLE 22.8 (*Continued*)

Administrative/Engineering Controls (*Continued*)
J-3* Type of Acoustic Engineering Controls: [ ] Enclosure [ ] Vibration Isolation [ ] Dampening [ ] Noise Absorption [ ] Noise Cancellation [ ] None [ ] Other
J-4 Specific Engineering Controls:
J-5 Estimated Effectiveness of Engineering Controls: [ ] Effective [ ] NOT Effective (check all that apply and descr.) [ ] Improper Choice [ ] Improper Design [ ] Improper Installation [ ] Poor Condition [ ] Improperly Modified [ ] Not working according to design specifications [ ] Other [ ] Not Evaluated
J-6 Comments about Effectiveness of Engineering Controls:
Personal Protective Equipment Information
K-23* Hearing Protection Worn: [ ] Worn [ ] Not Worn
K-24* Hearing Protection Requirements: [ ] Required [ ] Not Required [ ] Determination not made
K-25* Hearing Protection Type: [ ] Plugs/Inserts [ ] Circum-aural [ ] Other (descr.)
K-26 Hearing Protection Specific (manufacturer, model, NRR):
K-27 Estimated Effectiveness of Hearing Protection: [ ] Effective [ ] NOT Effective (check all that apply and descr.) [ ] Improper Choice [ ] Poor Condition [ ] Improper Use [ ] Not Worn when Required [ ] Other [ ] Not Evaluated
K-28 Comments about Effectiveness of Hearing Protection:

TABLE 22.8 (*Continued*)

Noise Exposure Results	
M-1* Noise Exposure Dose: _____ Criterion: 8 hours at ____ dB = 100% [ ] A-Weighting [ ] C-Weighting [ ] Slow Response [ ] Fast Response	M-2* Exchange Rate: [ ] 3 dB [ ] 5 dB
M-3* L_{eq}: _____	M-4 L_{max} _____
M-5* Sound Level Measurements: L = _____ dB [ ] A-Weighting [ ] C/Flat Weighting [ ] D-Weighting [ ] Slow Response [ ] Fast Response	
M-6 Impact Noise Measurements: _____ (dB)	
M-7 Comments	

noise measurements to determine compliance with community noise criteria. The microphone should be oriented so it is most sensitive to noise from the source. It is normally placed 1.2–1.5 m above the ground and should be more than 3.5 m from any reflecting surfaces. Measurements taken near a building should be taken 1–2 m from the facade and 3 dB should be subtracted from any readings.

Measurements can be taken over a range of meteorological conditions. Choose conditions such that:

1. Wind direction is within ±45° of a line connecting the source to the microphone, blowing from the source to the microphone.
2. Wind speed is 1–5 m/s at 3–11 m above the ground.
3. There is no strong temperature inversion near the ground.
4. There is no heavy precipitation.

Consideration must be paid to the duration of measurements. To cover typical human activities reference intervals may be days and nights. It may be necessary to include intervals for evenings, weekends, and holidays. Long-term measurements are frequently performed on the order of months. In this case, seasonal variations may be important. In any case the intervals should be chosen so all significant variations are covered.

Since the noise is generally not steady, sufficient samples are needed to estimate long-term levels that characterize the noise. Frequently this is reported using percentile levels, for example, L_1, L_5, L_{10}, L_{50}, etc.—the sound pressure levels exceeded 1%, 5%, 10%, and 50% of the time. Many dosimeters and integrating sound level meters can measure, accumulate, and report these values. Another common measurement required for community noise is brief disturbances such as aircraft flyover. Sound exposure levels (SEL) are the criteria commonly used to assess sources of this type.

Note if any tones are present. A source is considered tonal if the sound level in any one-third octave band exceeds adjacent bands by more than 5 dB. If the tonal components are clearly audible, a 5–6 dB "penalty" will added to the sound pressure level. 2–3 dB are added if the tonal components are just detectable.

When it is necessary to characterize an area or a line (such as the impact of traffic noise from a highway), measurements will be taken at a number of locations on a grid. The grid spacing should be adjusted so that there is less than 5 dB difference between adjacent grid points. See Section 22.10 for more information on environmental noise.

22.6 OCCUPATIONAL NOISE EXPOSURE CRITERIA

22.6.1 Background—United States Federal Regulations

The Occupational Safety and Health Administration (OSHA) is tasked with enforcing the Department of Labor Occupational Noise Exposure Standard, 29 CFR 1910.95. In 1983, the occupational noise exposure hearing conservation amendment, final rule, was put into effect. The regulation stipulates aspects of the hearing conservation program such as (1) noise exposure measurement, (2) audiometric testing programs, (3) the identification of employees that must be included in the program, and (4) the record-keeping activities that must be performed. In addition, the role of hearing protectors and training issues are addressed.

OSHA is the enforcement branch of the Federal government's industrial health and safety program. The National Institute for Occupational Safety and Health (NIOSH) is the research and training division. NIOSH is responsible for performing pertinent research in the areas of industrial health and hygiene, providing training to industrial hygienists and safety inspectors, and developing recommended regulations. OSHA does not have enforcement power in the mining industry; instead the Mine Safety and Health Administration (MSHA) has this responsibility. OSHA also does not have enforcement power over the military or the railroad industry.

22.6.2 OSHA Noise Regulations

22.6.2.1 Noise Exposure OSHA regulations state that no employee shall be exposed to greater than 90 dBA 8-hour time-weighted average (TWA) noise exposure. Any employee exposed to greater than 90 dBA TWA shall be provided protection from the effects of noise exposure through feasible engineering controls, administrative controls, or with hearing protective devices.

22.6.2.2 Noise Exposure Measurements for Hearing Conservation A limit of 85 dBA for 8 hours was added as an action level in the 1983 amendment. The 1983 OSHA amendment states that the "employer shall administer a continuing, effective hearing conservation program, ... whenever employee noise exposures equal or exceed an 8-hour TWA of 85 dB measured on the A-scale (slow response), or equivalently, a dose of fifty percent." Monitoring must be conducted to determine if any employees exceed these limits. All those exceeding the limits must be placed in a hearing conservation program consisting of:

- Noise monitoring
- Audiometric evaluations at least annually
- Hearing protective devices
- Annual training to employees
- Maintenance of proper records

22.6.2.3 Practical Considerations The limits specified in the OSHA noise exposure regulations were the most restrictive limits deemed feasible with due consideration given to other important factors, such as economic impact. These limits were not intended to provide complete protection for all persons. It is estimated that 85 percent of persons exposed to the OSHA limits of $L_{A,8h} = 90$ dB for 8 hour/day, 5 days a week, for about 10 years, will not develop a significant hearing impairment. The other 15 percent of persons exposed to these limits would probably have various levels of hearing impairment depending upon their susceptibility. Exposures outside the workplace may also contribute significantly to the hearing impairment for some persons.

OSHA exposure limits are intended only as *minimum* action levels. Wherever feasible, hearing conservation measures should be extended to reduce exposure levels *below the limits specified* both at and away from the workplace. Companies may choose a more conservative approach and require the use of hearing protectors at 85 dBA.

22.6.3 Noise Measurements

Most occupational noise measurements are taken using meters set to A-weighted slow response as specified by OSHA and international standards. Usually, $L_{A,S} = 90$ dB is the *criterion sound level* (LC), that is, the sound pressure level that, if present for a full 8-hour *criterion duration* (TC) workday would give a *dose* (D) of 1.0 or 100%. An *exchange rate* (Q) of 5 dB is used, which is the difference in sound level that would give the same dose if the exposure time were doubled or halved. The above terminology is that used in the ANSI standards.

To establish compliance to the hearing conservation amendment, A-weighted sound levels must be measured from 80 to 130 dB. The level $L_{A,S} = 80$ dB is called the *threshold level* (TL), the lowest sound pressure level used to compute the dose. If the dose exceeds 50%, the employee must be placed in a hearing conservation program. The OSHA hearing conservation noise exposure dose criteria are given in Table 22.9.

Compliance to the noise standard is established with much the same criteria except that the threshold level is set at 90 dB ($TL_{AS} = 90$ dB). A dose exceeding 100% would be considered exceeding the noise standard. The dual nature of the standard arose from the fact that the hearing conservation program was amended to the noise regulation, and subsequently interpreted by the U.S. courts.

TABLE 22.9 Allowable Time and 8-Hour Noise Dose per OSHA 29 CFR 1910.95, Table G-16a.

$L_{A,S}$ (dB)	Time allowed (h)	Dose (%) if exposed for 8 h	$L_{A,S}$ (dB)	Time allowed (h)	Dose (%) if exposed for 8 h
80	32	25	85	16	50
90	8	100	95	4	200
100	2	400	105	1	800
110	0.5	1600	115	0.25	3200

22.6.4 Exposure Calculations

When the daily noise exposure is composed of two or more periods of exposure at different levels, their combined effect is determined by adding the individual contribution as follows:

$$D_{\text{OSHA}} = \sum_{i=1}^{n} \frac{C_i}{T_i} \tag{22.14}$$

This method is called the *time-weighted average noise dose*. The values C_1 to C_n indicate the times of exposure to specified levels of noise, and the corresponding values of T_i indicate the total time of exposure permitted at each of these levels (see Table 22.9). If the sum of the individual contributions exceeds 1.0, the mixed exposures are considered to exceed the overall limit value.

For $L_{A,S} > 80$ dB, the time allowed for any sound pressure level can be calculated using the formula:

$$T_{\text{OSHA}} = \frac{8\text{h}}{2^{(L_{A,S} - 90)/5}} \tag{22.15}$$

In general the values of T_i can be determined for any criterion[39] by

$$T_i = \frac{\text{TC}}{2^{((L_i - \text{LC})/Q)}} \tag{22.16}$$

where TC is the criterion duration (h), and LC is the criterion sound pressure level (dB). Most organizations specify a criterion duration TC of 8 h, corresponding to a normal work day.

Example 6: If a person were exposed to 90 dBA for 5 h, 100 dBA for 1 h and 75 dBA for 3 h during an 8-h working day, the times of exposure are $C_1 = 5$ h; $C_2 = 1$ h; and $C_3 = 3$ h. The corresponding OSHA time limits for these exposures are $T_1 = 8$ h, $T_2 = 2$ h, and T_3 = infinity. Therefore, because 3 divided by infinity is zero, there is no contribution from the 75-dBA exposure

$$D = \frac{5}{8} + \frac{1}{2} + \frac{3}{\infty} = 1.125 \text{ or } 112\%$$

Hence the time-weighted average noise dose for this person slightly exceeds the specified limit of 1.0 (100%).

Note that an 8-h exposure to a continuous sound level of 90 dB would result in a dose of 100%. Similarly continuous $L_A = 95$ dB for 8 h would lead to a 200% dose, and $L_A = 85$ dB would give $D = 50\%$.

Therefore, a noise dose calculated according to the OSHA regulations can be converted to a time-weighted average (TWA) sound pressure level. Expressing dose D as a fraction.

$$\mathrm{TWA} = 16.61 \log_{10}(D) + 90 \tag{22.17}$$

***Example** 7:* Determine the time-weighted average sound pressure level for the 112% OSHA dose calculated above.

$$L_A = 16.61 \log(1.12) + 90 = 90.8\mathrm{dB}$$

22.6.5 Other Noise Exposure Criteria

For determining occupational noise exposure and estimating hearing impairment, the International Standards Organization[40] and the ACGIH[41] have recommended use of $L_{eq,8h}$ (equivalent continuous A-weighted sound pressure level normalized to an 8-hour working day). ANSI S3.44-1996 recommends the same but notes that other exchange rates (Q) (notably 5 dB) can be used. In this case the exposure is deemed "equivalent effective level" or EEL[13]. ANSI 3.44 and ISO 1999 and some experts[45] also use *sound exposure* $E_{A,T}$ (measured in Pa^2 sec) to determine the risk of hearing loss from noise exposures.

$$E_{A,T} = \int_{t}^{t_2} p_A^2(t)\, dt \tag{22.18}$$

The above calculation would be done by an integrating sound level meter. When the exposure consists of only a few levels and durations, the discrete form of the above integral is used to calculate the sound exposure.

TABLE 22.10 Criterion Sound Level, Thresholds, and Exchange Rates[42]

Organization	LC[a]	TL[b]	Q[c]
OSHA 29 CFR 1910.95 for noise exposures	90	90	5
OSHA for hearing conservation	85	80	5
ACGIH 1996 TLVs and BEIs	85	80	3

[a] LC is the criterion sound pressure level (dB).
[b] TL is the threshold level (dB).
[c] Q is the exchange rate (dB).

$$E_{\mathrm{A,T}} = \sum_{i=1}^{n} (p_r^2 \, 10^{L_{\mathrm{A}i}/10} \, T_i) \tag{22.19}$$

Example 8: Using the same data for the previous example calculating L_{eq}: What is the sound exposure $E_{\mathrm{A,8\,h}}$ for an 8-hour workday consisting of four exposures as shown in the table?

L_{AS} (dB)	Duration (h)
82	6.0
88	1.4
97	0.5
105	0.1

L_{AS} (dB)	p_{A}^2 (Pa^2)	Duration (h)	Duration (sec)	Exposure (Pa^2 s)
82	0.063	6.0	21600	1.37×10^3
88	0.252	1.4	5040	1.27×10^3
97	2.00	0.5	1800	3.61×10^3
105	12.64	0.1	360	4.55×10^3
			Total	1.08×10^4

$$E_{\mathrm{A,8\,h}} = 1.08 \times 10^4 \text{ Pa}^2 \text{ s}$$

Sound exposure levels, normalized to an 8 hour working day, are calculated from sound exposures using the following.

$$L_{\mathrm{Aeq,8\,h}} = 10 \log \left(\frac{E_{\mathrm{A,T}}}{E_r} \right) \tag{22.20}$$

where the reference exposure $E_r = p_r^2 T_r = (0.00002 \text{ Pa})^2 \; 8 \text{ h} \; \frac{3600 \text{ sec}}{\text{h}} = 1.15 \times 10^{-5} \text{ Pa}^2$ sec.

Example 9: Show that the sound exposure $E_{\mathrm{A,8\,h}} = 1.08 \times 10^4$ is equivalent to the equivalent continuous sound level $L_{\mathrm{Aeq}} = 90$ calculated previously.

$$L_{\mathrm{Aeq,8\,h}} = 10 \log \left(\frac{1.08 \times 10^4}{1.15 \times 10^{-5}} \right) \cong 90 \text{ dB}$$

When a noise *dose* is given, the equivalent time-weighted average *level* (TWA) can be determined using the generalized formula

$$\mathrm{TWA} = \mathrm{LC} + 10\left(\frac{Q}{10\log_{10}(2)}\right)(D) \tag{22.21}$$

where LC is the criterion sound level, Q is the exchange rate, and D is the dose (expressed as a fraction). For the OSHA noise standard, this formula becomes that given in Eq. (22.17).

22.6.6 Steady-State and Impulsive Noise

The integration of continuous and impulsive noise is done automatically if exposure levels are measured using a dosimeter. The current ANSI standard S1.25-1991 requires noise dosimeters have an operating range of at least 50 dB. For measurement of impulse noises, instrumentation and suggested presentation of results is available and defined by ANSI S12.7.[43] There is evidence[44] that, above a critical intensity, the damage creating hearing loss involves a mechanical mechanism rather than the metabolic mechanism that is responsible for damage from steady noises of moderate intensity. A damage risk criterion that considers steady-state and impulsive sounds above a critical level and periods of quiet[45] may be helpful in predicting hearing loss.

22.7 HEARING CONSERVATION PROGRAMS

22.7.1 General

The objective of hearing conservation programs is to prevent noise-induced hearing loss. This obvious fact is often forgotten, or ignored, when pressures are applied for compliance with local, state, or Federal rules and regulations on noise exposures. Because compliance with rules and regulations will not always prevent noise-induced hearing impairment in susceptible individuals, every effort should be made to reduce noise exposures wherever possible.

The lowest feasible noise exposure levels are obviously desirable for the health, safety, and well-being of workers. In addition, these lower limits are of significant value to employers because morale and work productivity of workers should be maximized,[46] and the number of compensation claims for noise-induced hearing impairment should be minimized.

22.7.2 Requirements

An effective hearing conservation program should provide for:

1. The identification of noise hazard areas, and the performance of noise exposure measurements.

2. The reduction of the noise exposure to safe levels preferably using engineering controls, or if infeasible, through administrative controls or hearing protective devices.

3. The measurement of exposed worker's hearing thresholds to monitor the effectiveness of the program.

4. The education and motivation of employees[47] and management about the need for hearing conservation, and the instruction of employees in the use and care of personal hearing protectors.

5. The maintenance of accurate and reliable records of hearing and noise exposure measurements.

6. The referral of employees who have abnormal hearing thresholds for examination and diagnosis.

22.7.2.1 *Identification of Noise Hazard Areas* Noise hazard areas having continuous noise characteristics can be identified with sound level meters or with dosimeters. When significant impulse-type noises are present, hazard areas may be established with noise dosimeters, integrating sound level meters, or impulse measuring devices.

Action levels for exposure must be established that are at least as low as those specified by the OSHA Rules and Regulations. In order to select the best hearing protectors for particular noise spectra, octave band sound pressure levels should be measured.

22.7.2.2 *Reduction of Noise Exposure Levels* As soon as a noise hazard area has been identified, hearing protective devices should be provided to reduce the exposures to safe levels. Engineering control means should then be employed where they are feasible. This will often require the use of experts in noise control to work with those at the plant, and/or others, who understand the machines and their operation. If it is not feasible to reduce exposure levels to the limits selected by engineering control means, the use of hearing protectors must be continued.

Continued monitoring of the effectiveness of hearing protector devices (hearing threshold measurements) must be maintained until engineering control procedures have reduced the noise exposures to safe levels. It is strongly advisable to continue monitoring the effectiveness of the hearing conservation program even after engineering control measures have been successfully installed if there is any reason to suspect that noise exposures at, or away from, work are significant.

22.7.2.3 *Hearing Measurement and Recordkeeping* Periodic hearing threshold measurements are necessary to monitor the effectiveness of a hearing conservation program. High-level noise exposures away from work may be partially responsible for any STS (standard threshold shift); hence, it is important to make a significant effort to determine the cause of any threshold shift. Affected persons should be interviewed and their audiological historical data sheets should be updated periodically.

Hearing conservation measures away from work can be encouraged or assisted by lending or giving hearing protectors to employees for use with noisy activities. The cost of supplying hearing protectors for this use is far less than hearing loss compensation costs, and it often directs more attention to the hearing conservation program at work.

If a standard threshold shift (see below) is found in an annual audiogram, a letter must be sent to the employee informing him or her of the results of the audiogram. Even though follow-up letters to employees are not required after the baseline audiogram, letters should be used to advise new employees of hearing impairments greater than would be expected for their age. These employees may be more susceptible to noise-induced hearing impairment; so this early warning should result in lower exposures at and away from work during the following year.

Records should be retained after an employee is no longer employed, as required by Federal or state medical record-keeping requirements. These records could be valuable if there is a question of legal responsibility for an employee's hearing impairment, or if the employee is to return to this job. Also, *termination audiograms* are not required by OSHA, but they should be taken because of potential liability for hearing impairments after termination.

22.7.2.4 Baseline Audiograms A baseline audiogram must be established within 6 months of an employee's first exposure at or above an 8-h time-weighted average (TWA) of 85 dBA or, equivalently, a dose of fifty (50) percent. All subsequent audiograms are to be compared to the baseline audiogram to identify changes in hearing thresholds. If mobile audiometric testing services are used, the new baseline may be established within one year of first exposure. Hearing protectors must be worn during the second six months of this one-year period until the baseline is established.

New baselines may be established if the audiologist or physician in charge of the hearing conservation program determines that old audiograms are not valid, if a standard threshold shift (STS) is established, or if the annual audiogram indicates significant improvement over the baseline audiogram.

Note: Not all audiologists or physicians are qualified industrial hearing conservationists. Employers should carefully investigate the professional qualifications and experience of any person employed for industrial hearing conservation. Generally, expertise only in clinical practice or certification in specialty fields other than hearing conservation is not sufficient.

22.7.2.5 Annual Audiograms and Standard Threshold Shifts Annual audiograms must be taken on all employees who are exposed to an 8-h time-weighted average of 85 dBA or greater. If an STS, "a change in hearing threshold relative to the baseline audiogram of an average of 10 dB or more at 2000, 3000, and 4000 Hz in either ear," is established by an annual audiogram, it must be reported to the affected employee within 21 days.

In practice, a retest is justified in most cases when an STS is found because there are many common reasons why a 10-dB STS may be temporary. All retests should be completed within 30 days. Retests performed within 21 days may show the STS was temporary and eliminate the requirement of preparing a notification letter for the employee.

Old audiograms must not be discarded when new baselines are established for an employee. The letter used to inform an employee of a STS should also be used to educate and motivate the employee to take better care of his or her hearing both at and away from work. A sample letter is presented here:

Sample Letter to Persons Showing a Standard Threshold Shift on an Audiogram

Dear ______:

Your hearing thresholds measured on ______ indicate a change of more than ______ dB from (worse than) your baseline audiogram. This change may be temporary, or it may be permanent. Temporary changes may be the result of colds, allergies, medications, recent exposures to high-level noises, blows to the head, etc. Depending upon the severity of these and other factors, permanent impairments may also result.

While your hearing impairment may be temporary, we recommend strongly that you have an audiological examination at the ______ Speech and Hearing Clinic (tel. xxx-xxx-xxxx) We have arranged for payment for this examination. Please continue to obey our hearing conservation rules at and away from work.

Sincerely,

22.7.3 Setting up a Hearing Conservation Program

Setting up a hearing conservation program entails the development and maintenance of all the factors mentioned above. Ideally, a team made up of management, industrial hygiene, occupational medicine, safety, and production personnel will work together towards building an effective hearing conservation program. The first objective for this group should be to become aware of their problems and to consider possible ways to limit noise exposure levels as much as possible, at least to levels set by OSHA.

In practice, these committees may not work well except as advisors on general items such as company policy. Many successful hearing conservation programs have been highly dependent upon the competence and dedication of one person *who is given the time and support required to develop and maintain the program.* This person must be motivated and knowledgeable, and respected by both management and workers. Extensive formal training in any specific field is not usually necessary, but he or she must have a good practical knowledge of noise-induced hearing impairment, hearing threshold measurement, and personal hearing protectors.

Outside specialty firms may be useful for specific jobs such as audiometry, or noise control projects, but it is usually to the company's benefit to have a knowledgeable employee work with the specialist. Information on individuals and detailed information on machine operation (access areas needed for operation and production flow, etc.) are often needed for the best results. Otherwise, costly mistakes can be made that may require repeating expensive projects.

Personnel and work conditions may vary widely from one plant, or work area, to another. Monitoring safety procedures is much easier where workers are concentrated in areas where they are easily visible than in situations where they are widely scattered. Significant differences are found in management/employee relationships, and in motivation, education, and communication skills of hearing conservationists from one location to another, etc. As a result, a program that works well in one place may fail in others. It is generally necessary, therefore, to customize the program for each situation.

Obviously, top management must support the program, and others in middle management should be enthusiastic and knowledgeable. Perhaps even more important are the floor supervisors and the plant nurses. These key persons often have the respect of workers, and they may be the only persons at the plant with whom some workers will communicate freely. The floor supervisor must be genuinely concerned about the health and safety of the employee and be able to answer questions about the effects of noise exposures, the use of safety equipment, and the overall importance of the program.

22.8 HEARING PROTECTORS

22.8.1 General

Personal hearing protectors can provide adequate protection against noise-induced hearing impairment in a high percentage of industrial work areas if the protectors are properly selected, fitted, and worn. In many cases, however, protectors are not worn effectively, if at all.

Even though the use of hearing protectors is one of the most important parts of a hearing conservation program, only a fraction of the total budget is spent for this purpose. Ironically, a much larger percentage of hearing conservation effort and money is often spent over long time periods on hearing threshold measurements that essentially monitor the effectiveness of poor or nonexistent hearing conservation programs.

Hearing threshold measurements are important when action is taken based on the results. In fact, the only practical means for evaluating the effectiveness of personal protectors is to monitor thresholds periodically. If no hearing losses are observed—other than those due to the aging process—the program may be considered to be successful. Noise-induced hearing impairments usually develop slowly, however; so it may take years for the results from a hearing monitoring program to become meaningful. The careful selection of protectors and a continuing hearing conservation program, including close supervision by floor supervisors, are therefore very important.

An effective hearing conservation program seldom develops automatically simply by making hearing threshold or noise measurements. Noise does not have to be painful to be potentially harmful; so many employees do not understand the need for wearing protectors. A significant effort must be made to develop and maintain an effective program.

22.8.2 Performance Limitations

A primary limitation of protection afforded by a hearing protector is the way it is fitted and worn. Other important limitations of a hearing protector depend upon its construction and on the physiological and anatomical characteristics of the wearer. Sound energy may reach the inner ears of persons wearing protectors by four different pathways: (1) by passing through bone and tissue around the protector; (2) by causing vibration of the protector, which in turn generates sound into the external ear canal; (3) by passing through leaks in the protector; and (4) by passing through leaks around the protector. These pathways are illustrated in Figure 22.16.

Even if there are no acoustic leaks through or around a hearing protector, some noise reaches the inner ear by bone and tissue conduction or protector vibration if noise levels are sufficiently high.[48] The practical limits set by the bone- and tissue-conduction threshold vary significantly among individuals, and among protector types, generally from about 40 to 55 dB. Limits set by protector vibration also vary widely, generally from about 25 to 40 dB, depending upon the protector type and design and on the materials used. Contact surface area and compliance of materials are major contributing factors. The results of studies on these limitations are influenced significantly by

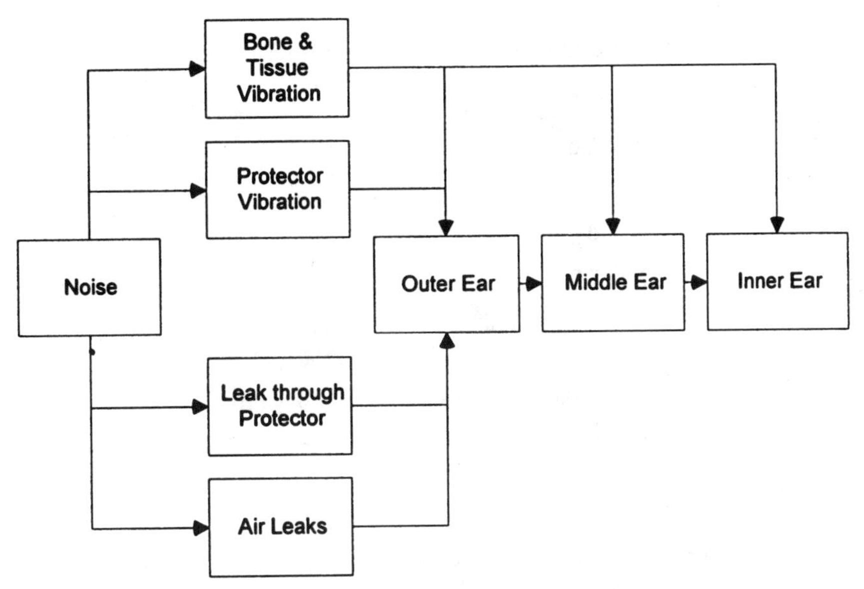

Fig. 22.16 Noise pathways to the inner ear.

procedures and techniques used and by the choice of subjects. These wide ranges of performance are therefore of limited value for individuals.[49,50] If hearing protectors are to provide noise-reduction values approaching practical limits, acoustic leaks through and around the protectors must be minimized by proper fitting and wearing.

Perhaps the major reason why hearing protective devices fail to protect employees from noise exposure is because the protection is not always worn. Workers fail to wear hearing protection for comfort, because they misplaced or forgot their protection, or because they want to communicate or listen to the noise without the frequency shaping caused by the device. Whatever the reason, failure to wear HPDs for even a small fraction of a working shift will greatly reduce the effective protection. For instance, an HPD with a Noise Reduction Rating (NRR) of 25, not worn for 15 min out of an 8-h workday, will lose about 10 dB of protection (Fig. 22.17).

22.8.3 Types of Hearing Protectors

Hearing protectors usually take the form of either insert types that seal against the ear canal walls or muff types that seal against the head around the ear. There are also concha-seated devices that provide an acoustic seal at the entrance of the external ear canal. There is no "best" type for all situations. However, some types are better than others for use in specific noise exposures, for some work activities, or for some environmental conditions. The 1994 publication "The NIOSH Compendium of Hearing Protection Devices" contains descriptions and laboratory attenuation data for all the HPDs currently available in the USA.[51]

22.8.3.1 Insert Types Ear canals differ widely in size, shape, and position among individuals and even within the same individual. Several different types of earplugs may

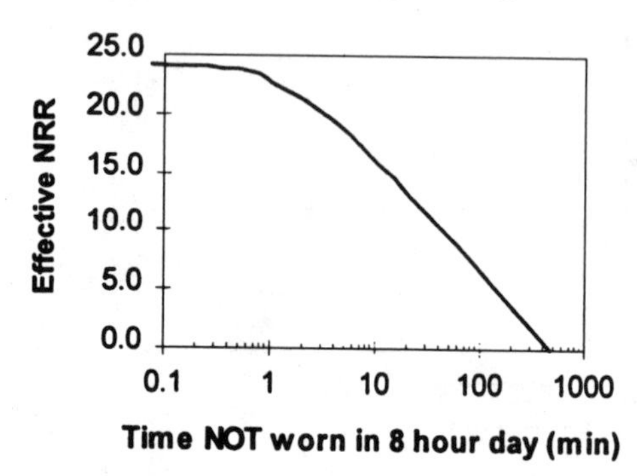

Fig. 22.17 Effective NRR vs. reported NRR when an HPD with an NRR of 25 dB is not worn full time.

be required to fit the wide differences in ear canal configurations. No single insert-type protector is best for all individuals, nor for all situations.

Ear canals vary in cross-sectional diameter from about 3 to 14 mm, but most are between 5 and 11 mm. Most ear canals are elliptically shaped, but some are round, and many have only a small slit-like opening that may open into a large diameter. Some canals are directed in a straight line toward the center of the head, but most are directed toward the front of the head, and they may bend in various ways.

HPDs must fit snugly and have an air-tight seal to be effective. Most ear canals can be opened and straightened by pulling the pinna back, or directly away from the head, making it possible to seat many earplugs securely. For comfort and plug retention, ear canals must return to their approximate normal configuration once the protector is seated.

22.8.3.1.1 Single- and Multiple-Sized Molded Ear Plugs Molded ear plugs (for instance, Fig. 22.18) are usually made of a silicon-type material and are available in a variety of styles. Often, these plugs are of a "flanged" design, with one or more flanges. The single-sized molded earplugs may not provide consistently high levels of protection for a large range of ear canal sizes and shapes. Hence, multiple-sized molded plugs are available from many manufacturers. The number of sizes available usually ranges from two to five. The best molded earplugs are made of soft and flexible materials that will conform readily to the shape of the ear canal for a snug, airtight, and comfortable fit. Earplugs must be nontoxic and have smooth surfaces that are easily cleaned with soap

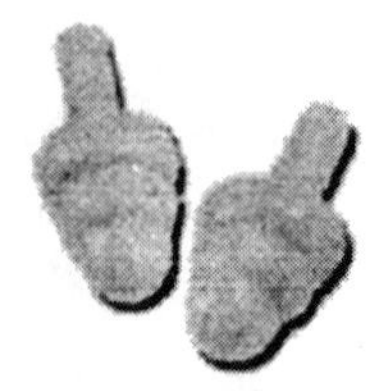

Fig. 22.18 Ear plugs.

and water. Earplugs are usually made of materials that will retain their size and flexibility over long periods of time.

The earplug size distribution for a large group of males is approximately as follows: 5 percent extra small, 15 percent small, 30 percent medium, 30 percent large, 15 percent extra large, and 5 percent larger than those supplied by many earplug manufacturers. The equal percentage of wearers for medium and large sizes indicates that many persons are fitted with earplugs that are too small.

If individuals are permitted to fit themselves, they often select a size on the basis of comfort rather than on the amount of protection provided. Some ear canals increase in size slightly with the regular use of earplugs, and earplugs may shrink in size after extended periods. Thus if a given ear size falls between two plug sizes, it is advisable to choose the larger rather than the smaller of the two. It follows that the fit of earplugs should be checked periodically. Proper storage of molded plugs is required if long usage periods are expected. These plugs often come in zip-type plastic bags, or, preferably, hard plastic cases.

22.8.3.1.2 Malleable Earplugs Malleable earplugs are made of materials such as silicon (swimmer's plugs), cotton, wax, glass wool, and mixtures of these materials. Typically, a small cone or cylinder of the material is hand-formed and inserted into the ear with sufficient force so that the material conforms to the shape of the canal and holds itself in position. Manufacturer's instructions should be followed regarding the depth of insertion of the plug into the canal. In any case, care should always be taken to avoid deep insertions that may cause the material to touch the eardrum.

Malleable earplugs should be formed and inserted with clean hands, because dirt or foreign objects placed in the ear canal may cause irritation or infection. Malleable plugs should be carefully inserted at the beginning of a work shift, and, if removed, they should not be reinserted until the hands are cleaned. For this reason, malleable plugs (and to a somewhat lesser extent, all earplugs) may be a poor choice if the work area is dirty or if the worker is subjected to intermittent high-level noises, where it may be desirable to remove and reinsert protective devices during the work period.

On the other hand, malleable plugs have the obvious advantage of fitting almost any ear canal, eliminating the need to keep a stock of various sizes, as is necessary for most molded earplugs. The cost of any disposable protector may be relatively high, depending upon how they are used.

22.8.3.1.3 Foam-Type Plugs Earplugs made of cylindrical or tapered foam are very popular (for instance Fig. 22.19). Generally, these plugs must be inserted about

Fig. 22.19 Foam-type ear plugs.

three-fourths of their length into the canal and held while they expand to fill the canal in order to provide the rated protection level. If these plugs are not inserted properly, their performance may drop significantly. It may not be possible to insert foam plugs properly in some small, rapid-bending, or slit-shaped canals.

Foam plugs are typically either composed of either vinyl or urethane-based material. The vinyl plugs are usually "stamped" out of sheets of foam material into cylinders, or less commonly, into hexagon-shaped plugs. The urethane plugs are usually tapered and manufactured in a molding process. Both types of plugs provide excellent noise attenuation *when they are fitted and worn properly*. Both types of plugs are generally comfortable to wear, although there are sometimes complaints of vinyl plugs being more "abrasive" to the ear canal. Urethane plugs are typically somewhat softer than the vinyl plugs but may be more expensive due to the more complicated manufacturing process. Both types are reusable on a limited basis, although the urethane plugs may tend to swell when wet.

22.8.3.2 Concha-Seated Protectors Protectors that provide an acoustic seal in the concha and/or at the entrance to the ear canal can be referred to as concha-seated, semi-aural or semi-insert devices (for instance, Fig. 22.20). Hearing aid-type molds used as hearing protectors are considered in this category, although some of these molds extend far into the canal. Another protector design in this class makes use of various plug shapes attached to a lightweight headband that holds the plugs in the entrance of the external ear canal. The performance and comfort of this kind of protector varies significantly among different models.

Because the materials used for hearing aid molds do not usually expand or conform after insertion, the molds must be made to fit very well in order to get the tight fit required for good attenuation. Hence significant differences in attenuation and comfort ratings may be found among these protectors made by different manufacturers.

If individually molded protectors are well made, protection levels are typically very good. Also, when made properly, these protectors can be expected to provide among the most consistent protection levels in daily use of all protector types. The initial cost of individually molded protectors is relatively high, but most should last a long period of time.

22.8.3.3 Earmuffs Manufacturers of muff-type hearing protectors (for instance, Fig. 22.21) usually offer a choice of two or more models having different physical characteristics.

Fig. 22.20 Concha-seated hearing protectors.

Fig. 22.21 Earmuffs.

The performance levels of muff-type protectors depend upon many individual factors and how well these factors blend together. Generally, larger and heavier protectors provide greater protection, but the wearing comfort may not be as good for long wearing periods. Hence workers often accept lightweight protectors more readily than the larger ones. Other important factors include:

1. The suspension must distribute the force uniformly around the seal, with slightly more force at the bottom to prevent leaks in the hollow behind the ear.
2. The suspension mounting should be such that the proper seal is made against the head automatically without adjustments by the wearer.
3. The earcups must be formed of a rigid, dense, imperforate material to prevent leaks and significant resonances.
4. The size of the enclosed volume within the muff shell (particularly for low frequencies) and the mass of the protectors are directly related to the attenuation provided.
5. The inside of each earcup should be partially filled with an open-cell material to absorb high-frequency resonant noises and to damp movement of the shell. The material placed inside the cup should not contact the external ear; otherwise discomfort to the wearer and soiling of the lining may result.

Ear seals should have a small circumference so that the acoustic seal takes place over the smallest possible irregularities in head contour, and leaks caused by jaw and neck movements are minimized. However, a compromise between the small seal circumference and the number of persons that can be fitted properly must be made.

Earmuff cushions are generally made of a smooth, plastic envelope filled with a foam or fluid material. Because skin oil and perspiration may have adverse effects on cushion materials, the soft and pliant cushions may tend to become stiff and to shrink after extended use. Fluid-filled cushions can provide superior performance, but they occasionally have a leakage problem. Foam-filled seals should have small holes to allow air to escape within the muff when mounted. Seals filled with trapped air may vibrate, causing a loss of attenuation in low frequencies. Most earmuffs are equipped with easily replaceable seals.

Earmuffs normally provide maximum protection when placed on flat and smooth surfaces; thus less protection should be expected when muffs are worn over long hair, glasses, or other uneven surfaces. Glasses with plastic temples may cause losses in attenuation of from 1 to 8 dB. In some cases, this loss of protection can be reduced substantially if small, close-fitting, wire or elastic-band temples are used. Acoustic seal covers provided to absorb perspiration may also reduce the amount of attenuation by several decibels, because noise may leak through porous materials.

Because the loss of protection is directly proportional to the size of the uneven obstructions under the seal, every effort should be made to minimize these obstructions. If long coarse hair or other significant obstructions cannot be avoided, it may be advisable to use earplugs.

The force applied by muff suspensions is often directly related to the level of noise attenuation provided. On the other hand, the wearing comfort of a muff-type protector is generally inversely related to the suspension forces; so a compromise must be made between performance and comfort. Muff suspensions should never be deliberately sprung to reduce the applied force. Not only may a loss of attenuation be expected, but the distribution of force around the seal may be changed, which may cause an additional reduction in performance. To assure the expected performance, the applied force should be measured periodically, and the muff should be visually inspected.

Muff-type protectors are sometimes chosen because their use can be monitored from greater distances than insert-type protectors. They are also easier to fit than insert types because one size fits most persons. Comfort may also be better for muff-type protectors for some persons in some work areas. However, muff-type protectors may be very uncomfortable to wear in hot work areas, particularly when the work involves a vigorous activity. Muff-type protectors are also a poor choice when work is performed in areas having limited head space. Contact with vibrating machinery is likely to introduce high sound pressure levels inside the muff.

22.8.3.4 Electronic Earmuffs Several types of electronic earmuffs are currently available. These are usually referred to as "active" muffs, and they include muffs with electronic clipping circuits and with noise-canceling circuits.

The clipping circuits typically have a microphone on the outside shell of one or both muff cups that senses the sound outside the protector and provides input to an amplifier. The amp is used with a speaker inside the cup to let the wearer hear the ambient sounds at a safe level, usually limited to about $L_A = 80$ dB. In some of these devices, equalization is performed on the signal to enhance the speech frequencies.

Noise-canceling hearing protectors have recently made a modest impact in the hearing protector market. These devices sample the noise that has penetrated the muff cup, reverse the phase of the sampled signal, and reintroduce the out-of-phase signal into the cup through a speaker. The magnitude of the attenuation provided by the noise-canceling circuits is greatest at low frequencies, and there is usually no attenuation provided at frequencies above about 1000 Hz. In fact, some of these devices unintentionally provide a small amount of amplification in a narrow frequency range around 1000 Hz. Noise canceling is an evolving technology, and it is likely the future will bring better performance and lower costs.

Electronic muffs are typically heavier and more costly than passive protectors. Electronic clipping-type protectors are particularly useful in intermittent noise exposures where communication is required. These muffs do not require removal for effective communication in this type of noise environment. Noise-canceling muffs are typically used in very high-noise environments where traditional muff-type protectors do not provide sufficient low-frequency attenuation.

When used in passive mode (i.e., with the electronics deactivated), electronic muffs typically perform somewhat poorer than the muff would if there were no electronics present. The electronics circuit and speakers occupy space in the muff cup and therefore reduce enclosed volume of the cup. The reduced cup volume leads to poorer performance, especially in the low frequencies. Care must be taken when designing these types of protectors that all boards in the cup are solidly mounted, preventing vibration and resonance.

22.8.3.5 *Moderate or Flat Attenuation Hearing Protectors* Employees and hearing conservationists are often attracted to protectors advertised as providing a filter that allows speech to be heard but blocks harmful noise. These filter-type devices generally provide less protection than conventional protectors, particularly at frequencies below 1000 Hz. These protectors may afford adequate protection in modest noise exposure levels (see Section 7.6), and they may provide better communication than conventional protectors. In most cases, filter-type protectors are more likely to be acceptable for use in noises having major components above 1000 Hz. The simplest of these protectors are simply vented with small holes through the protectors, permiting low frequencies (below about 1000 Hz) to pass without significant attenuation.

More recently, "flat attenuation" hearing protectors have been introduced, in both muff and insert styles. The insert-types are sometimes referred to as "musician's plugs," as they are intended to provide a moderate amount of attenuation that is approximately equal at all frequencies. The "even" or flat attenuation has the benefit of not distorting the speech or music signal by attenuating some frequencies more than others. If flat attenuation protectors are not fitted properly, it is likely that the low-frequency attenuation will decrease, and the overall attenuation will not be as flat as intended.

There is a genuine need for moderate-protection hearing protectors in industry. These protectors will provide adequate protection in most work areas and allow better communication for many individuals. The hearing conservationist should not fall into the trap of thinking that "more is better" in terms of NRRs. Instead, protectors should be selected that provide adequate protection, that maximize communication, and that are worn consistently throughout the duration of the noise exposure.

22.8.3.6 *Double Protection—Protection Provided by Using Both Insert and Muff Types* The combined attenuation from wearing both a muff-type and an insert-type protector cannot be predicted accurately because of complex coupling factors. If the attenuation values of the muff and insert protectors are about the same in a frequency band, the combined attenuation should be 3 to 6 dB greater than the higher of the two individual values. If one of the two protectors has an attenuation value that is significantly higher in a band, the increased attenuation by wearing both will be only slightly greater than the higher of the two.

22.8.3.7 *Summary of Advantages and Disadvantages of Protector Types* Both insert- and muff-type hearing protectors have distinct advantages and disadvantages. Some *advantages of insert-type protectors* are:

1. They are small and easily carried.
2. They can be worn conveniently and effectively without interference from glasses, headgear, earrings, or hair.
3. They are normally comfortable to wear in hot environments.
4. They do not restrict head movement in close quarters.
5. The cost of sized earplugs (except for some throwaway types and molded protectors) is significantly less than muffs.

Some *disadvantages of insert-type protectors* are:

1. Almost all insert protectors require more time and effort for proper fitting than do muffs.
2. The amount of protection provided by plugs is more variable between wearers than that of muff-type protectors.
3. Dirt may be introduced into the ear canal if earplugs are inserted with dirty hands.
4. The wearing of earplugs is difficult to monitor because they cannot be seen at a significant distance.
5. Earplugs should be worn only in healthy ear canals, and even then, acceptance may take time for some individuals.

Some *advantages of muff-type protectors* are:

1. A good muff-type protector generally provides more consistent attenuation among wearers than good earplugs.
2. One size fits most heads.
3. Muffs are more easily seen at a distance, making program enforcement easier.
4. At the beginning of a hearing conservation program, muffs are usually accepted more readily than are earplugs.
5. Muffs can be worn despite minor ear infections.
6. Muffs are less easily misplaced or lost.

Some *disadvantages of muff-type protectors* are:

1. They may be uncomfortable in hot environments.
2. They are not easily carried or stored.
3. They are not convenient to wear without interference from glasses, headgear, earrings, or hair.
4. Usage or deliberate bending of suspension band may reduce protection significantly.
5. They may restrict head movement in close quarters.
6. They are more expensive than most insert-type protectors.

22.8.3.8 Summary The proper use of a good hearing protector can provide adequate protection in most work environments where it is not feasible to use engineering control measures. Special care should be taken to obtain the best protectors for a given purpose to assure adequate protection and communication. A continuing effort must also be made to ensure that protectors are used properly and to monitor hearing thresholds regularly to be sure that workers are not being exposed to harmful noises at or away from work, or to ascertain that they may have other problems with their hearing.

An employer may be held responsible for any high-frequency hearing impairment regardless of the cause, because it is often impossible to apportion responsibility for a hearing impairment. If an STS is found on an annual audiogram (see Section 22.6), the cause must be investigated. If it is determined that noise exposures may be the problem, it may be necessary either for this person to use different protectors or a combination of insert- and muff-type protectors, or to limit the duration of exposure by administrative means.

22.8.4 Selection of Hearing Protectors

Reasons for ineffective hearing protector programs often may be pinpointed by complaints from persons wearing protectors. The most common complaints are related to comfort and communication. Ideally, a hearing protector should be selected with the following objectives in mind:

1. The ear must be protected with an adequate margin of safety, but without unnecessarily reducing important communication (hearing warning signals, machines, speech, etc.). The protector's attenuation characteristics therefore should be selected to match those of the noise exposure spectra as closely as possible.
2. Environmental conditions and the kind of activity required by the job may significantly influence the comfort and general acceptability of a hearing protector. For example, insert-type protectors may be a better choice than muff-types for use in high temperatures, vigorous activities, or close quarters. Muff-types may be preferred to insert-types in very dirty areas, for very sensitive ear canals, or for ease of monitoring.
3. Some wearers cannot be fitted properly by certain types of protectors. Performance, comfort, and/or effectiveness may be significantly enhanced by the proper choice of protector for some individuals. Wearing time is often an important consideration in making this decision.
4. It is difficult to predict or understand the acceptance of protector types at times. Important factors may include (a) the wearer's appearance or hair style while wearing the protector, and (b) something said about particular models, or (c) just a desire to be different from others.

The effectiveness of a hearing conservation program, job performance, health and safety considerations, and morale may be adversely affected if attention is not given to all these factors. In particular, attention should be given to requests for changes in protector types. Obviously, there is no single protector, or protector type, that is best for all individuals or all situations. A choice of several protectors should be offered. This action also reinforces the importance of the program.

The acceptance of protectors can be helped, in some cases, by issuing selected protectors to management a few days before they are issued to workers. This procedure may cause some employees to react with demands that they be given these protectors, too, thereby creating a more positive attitude when protectors are issued. In all cases supervisory personnel and visitors must be required to obey all safety and health rules if safety equipment is to be used effectively. Exceptions cannot be tolerated.

22.8.4.1 Hearing Protector Ratings Hearing protectors are often selected with consideration being given only to the magnitude of a single-number performance rating. Generally, this rating is the Environmental Protection Agency (EPA) Noise Reduction Rating (NRR).[52] Although a single-number rating is useful in initiating a program when more precise octave or one-third octave noise measurement data are not available, its use requires large safety factors because of the lack of precision.

OSHA regulations specify that a safety factor of 7 dB is to be subtracted from the NRR when A-weighted exposure levels are used. The calculation of the NRR includes additional safety factors of two times the standard deviation plus 3 dB for spectral uncertainty. Therefore, the total safety factor applied to laboratory data because of the poor accuracy of the NRR can easily exceed 15 dB.

Even with the large safety factors now being used with the NRR, there are situations where some individuals may not be adequately protected. For example, the NRR misleads the user of muff-type hearing protectors in typical steel industry noise exposure spectra by more than ±8 dB as compared to calculations based on the more accurate NIOSH Method 1.[53]

Steps for calculating the amount of protection provided by an HPD using the NIOSH Method 1[54,55] ("long method")

1. Measure and record the octave band exposure levels.
2. Adjust the octave band values with the A-frequency weightings (Table 22.6). The overall A-frequency-weighted sound exposure level (L_A) for unprotected ears is equal to the logarithmic sum (see Section 22.2.4.2).
3. Record the mean hearing protector attenuation values supplied by the protector manufacturer and two times the corresponding standard deviation value.
4. Subtract the mean attenuation values from the corresponding A-frequency-weighted levels, then add the two standard deviations to the result to obtain the A-frequency-weighted exposure levels under the protector.
5. Combine the A-frequency-weighted exposure levels to obtain an estimate ($L_{A(\text{protector})}$) of the highest sound pressure level to which a population of wearers of this protector would be exposed 97.5 percent of the time *if the protectors are fitted and worn properly.*

Three examples of noise exposure calculation using the NIOSH Method 1 are presented. In Example 10, the sound pressure levels in each octave band are equal resulting in a "flat" exposure. This example demonstrates how the NIOSH Method 1 used with a flat noise exposure is similar to the NRR calculation. In Example 11, the noise exposure is "falling," with lower sound pressure levels in the higher-frequency octave bands than in the lower-frequency bands. In this case, the exposure to the ear is significantly greater than indicated by calculations using the NRR because the NRR tends to overestimate the protection afforded in primarily low-frequency exposures. Example 12 demonstrates the opposite situation with a "rising" noise exposure. The NRR tends to underestimate the protection afforded a HPD wearer in this type of noise exposure, and therefore the exposure to the ear is less than is expected if calculations using the NRR are used.

The NRR is calculated assuming an exposure to pink noise (equal energy in each octave band) from 125 to 8000 Hz.

$$\text{NRR} = L_{C(\text{pink noise})} - L_{A(\text{protector})} - 3 \tag{22.22}$$

where $L_{C(\text{pink noise})}$ is the C-frequency-weighted sound pressure level for the pink noise exposure, $L_{A(\text{protector})}$ is the A-frequency-weighted sound pressure level under the protector and 3 (dB) is a safety factor to correct for spectral differences between the assumed pink noise spectrum and actual noise exposures.

Example 10: The manufacturer of an HPD reports an NRR of 23 with the following octave band attenuation values and standard deviations:

Frequency (Hz)	125	250	500	1000	2000	4000	8000
Attenuation (dB)	18	20	30	39	34	32[a]	36[b]
Standard deviation	3.5	2.5	3.5	3	3	3[a]	3.5[b]

[a]Average of 3150 and 4000 Hz.
[b]Average of 6300 and 8000 Hz.

Verify the NRR.

The C-frequency-weighted sound pressure levels are determined from a spectrum with constant octave band levels. 100 dB is arbitrarily chosen.

Frequency (Hz)	125	250	500	1000	2000	4000	8000
Constant octave band level	100	100	100	100	100	100	100
C-frequency-weighting (from Table 22.6)	−0.2	0.0	0.0	0.0	−0.2	−0.8	−3.0
Calculate L_C	99.8	100	100	100	99.8	99.2	97

$L_C = 10 \log (\sum 10^{9.98} + 10^{10.0} + 10^{10.10} + \cdots) = 108.0$

Using NIOSH Method 1 and the pink noise "flat" spectrum, calculate the exposure to the ear and the attenuation provided by the hearing protector.

Step	Octave band center frequency (Hz)						
	125	250	500	1000	2000	4000	8000
1. Octave band exposures (from above)	100.0	100.0	100.0	100.0	100.0	100.0	100.0
2a. A-frequency weighting (from Table 22.6)	−16.1	−8.6	3.2	0.0	+1.2	+1.0	−1.1
2b. L_A (100 dB + step 2)	83.9	91.4	96.8	100.0	101.2	101.0	98.9
2c. Optional (combine octave band levels)	L_A = 107.0 dB						
3a. Mean attenuation (from manufacturer)	18	20	30	39	34	32	36
3b. Standard deviation times 2 (from manufacturer)	7	5	7	6	6	6	7
4. A-frequency-weighted levels in protected ear (step 2b-step 3a + step 3b)	72.9	76.4	73.8	67.0	73.2	75.0	69.9
5. Combine octave band levels	$L_{A(\text{protector})}$ = 81.9 dB						

$\text{NRR} = L_C - L_{A(\text{Protector})} - 3 = 108.0 - 81.9 - 3 = 23.1$, which is rounded to 23 dB.

Example 11: What is the protection calculated according to NIOSH Method 1, for the same hearing protector for a noise source with the following "falling" octave band levels?

Frequency (Hz)	125	250	500	1000	2000	4000	8000
SPL (dB)	85	96	86	80	75	68	67

Using the procedure listed above the following table is constructed.

	Octave band center frequency (Hz)						
Step	125	250	500	1000	2000	4000	8000
1. Octave band exposures (measured)	85	96	86	80	75	68	67
2a. A-frequency weighting (from Table 22.6)	−16.1	−8.6	−3.2	0.0	+1.2	+1.0	−1.1
2b. $L_{protector}$ (step 1 + step 2)	68.9	87.4	82.8	80	76.2	69	65.9
2c. Optional (combine octave band levels)	L_A = 89.5 dB						
3a. Mean attenuation (from manufacturer)	18	20	30	39	34	32	36
3b. Standard deviation times 2 (from manufacturer)	7	5	7	6	6	6	7
4. A-frequency-weighted levels in protected ear (step 2b-step 3a + step 3b)	57.9	72.4	59.8	47	48.2	43	36.9
5. Combine octave band levels	$L_{A(protector)}$ = 72.8 dB						

Therefore, the protection provided by the HPD in the specified "falling" noise spectrum would be 89.5 - 72.8 = 16.7 dB, which is 6.3 dB less than the NRR.

Example 12: Again, using the same hearing protector, what would be the protection calculated according to NIOSH Method 1, for a noise source with the following "rising" octave band levels?

Frequency (Hz)	125	250	500	1000	2000	4000	8000
SPL (dB)	85	96	100	109	111	110	104

Using the procedure listed above the following table is constructed:

Step	Octave band center frequency (Hz)						
	125	250	500	1000	2000	4000	8000
1. Octave band exposures (measured)	85	96	100	109	111	110	104
2a. A-frequency weighting (from Table 22.6)	−16.1	−8.6	−3.2	0.0	+1.2	+1.0	−1.1
2b. $L_{\text{protector}}$ (step 1 + step 2)	68.9	87.4	96.8	109	112.2	111	102.9
2c. Optional (combine octave band levels)	L_A = 116.0 dB						
3a. Mean attenuation (from manufacturer)	18	20	30	39	34	32	36
3b. Standard deviation times 2 (from manufacturer)	7	5	7	6	6	6	7
4. 4-frequency-weighted levels in protected ear (step 2b-step 3a + step 3b)	57.9	72.4	73.8	76	84.2	85.0	73.9
5. Combine octave band levels	$L_{A(\text{protector})}$ = 88.4 dB						

Therefore, the protection provided by the HPD in the specified "rising" noise spectrum would be 116 − 88.4 = 27.6 dB, which is 4.6 dB greater than the NRR.

The following examples demonstrate exposure calculations to be followed if octave band noise exposure data are not known and single number estimates of exposure, such as the A-weighted or C-weighted sound pressure level, are used.

Steps for estimating worker's noise exposure when C-frequency-weighting sound pressure level L_C is known

Exposure is estimated by subtracting the NRR from the C-frequency weighting.

$$L_{A(\text{protector})} = L_C - \text{NRR} \tag{22.23}$$

where $L_{A(\text{protector})}$ is the *A-weighted* sound level exposure estimated under the protector (dB), and L_C is the long-term *C-weighted* sound pressure level noise exposure *outside* the protector.

Example 13: Calculate the estimated worker exposure for the above three frequency spectra using the C-frequency-weighted level.

From the calculations in Example 10, L_C = 108 dB

$$L_{A(\text{protector})} = 108 - 23 = 85 \text{ dB}$$

L_C from Example 11 is:

Frequency (Hz)	125	250	500	1000	2000	4000	8000
SPL (dB)	85	96	86	80	75	68	67
C-frequency weighting (from Table 22.6)	−0.2	0.0	0.0	0.0	−0.2	−0.8	−3.0
L_C	84.8	96	86	80	74.8	67.2	64

$$L_C = 10\log(10^{8.48} + 10^{9.6} + 10^{8.6} + \ldots) = 10\log(4.8 \times 10^{9}) = 96.8 \text{ dB}$$

$$L_{A(protector)} = 96.8 - 23 \cong 74 \text{ dB}$$

L_C from Example 12 is:

Frequency (Hz)	125	250	500	1000	2000	4000	8000
SPL (dB)	85	96	100	109	111	110	104
C-frequency weighting (from Table 22.6)	−0.2	0.0	0.0	0.0	−0.2	−0.8	−3.0
L_C	84.8	96	100	109	110.8	109.2	101

$$L_C = 10\log(10^{8.48} + 10^{9.6} + 10^{10.0} + \ldots) = 10\log(3.10 \times 10^{11}) = 114.9 \text{ dB}$$

$$L_{A(protector)} = 114.9 - 23 \cong 92 \text{ dB}$$

Steps for estimating worker's noise exposure when A-frequency-weighting sound pressure level L_A or L_{Aeq} is known

Accuracy is lost in calculating hearing protection when only A-weighted sound pressure levels are known since the NRR is based on a flat spectrum. Because of this OSHA specifies that a 7-dB safety factor must be used to compensate for the $L_c - L_A$ differences found in "typical" industrial noise. The exposure under the protector is estimated by the following.

$$L_{A(protector)} = L_A - (NRR - 7) \tag{22.24}$$

where $L_{A(protector)}$ is the A-weighted sound level exposure estimate under the protector (dB), and L_A is the long-term A-weighted noise exposure *outside* the protector (dB).

Example 14: Calculate the estimated worker exposure for the above three frequency spectra using the A-frequency-weighted levels.

In Example 10, L_A = 107.0 dB. Therefore, the estimated exposure according to the NIOSH formula would be:

$$L_{A(protector)} = 107.0 - (23 - 7) = 91$$

Similarly in Example 11, L_A = 89.4 dB; so the estimate exposure is:

$$L_{A(protector)} = 89.5 - (23 - 7) \cong 74$$

From the calculations in Example 12, $L_A = 116$ dB; so the estimated exposure is:

$$L_{A(\text{protector})} = 116 - (23 - 7) = 100$$

	Estimate exposure $L_{A(\text{protector})}$		
Spectrum	NIOSH Method 1	Using L_C	Using L_A
Falling	73	74	74
Flat (pink noise)	82	85	91
Rising	88	92	100

The table demonstrates the potential differences between the various methods of exposure calculation. The NIOSH Method 1 calculation is considered to be the most accurate of the three because it uses all available measurement data and does not rely on any single-number estimate. In these examples, the L_C exposure calculation tends to be greater than the NIOSH Method 1, since it includes the 3-dB safety factor from the NRR calculation. The L_A exposure calculations vary from the L_C calculations because of the additional 7-dB safety factor imposed by OSHA and the significant deemphasis of low-frequency noise in the A-weighting.

The large safety factors used in the single-number NRR rating may result in many workers being significantly overprotected, causing a variety of problems, ranging from injuries when warning signals are not heard to reduced work efficiency when important machine sounds are inaudible. In addition, unnecessarily inhibited communication may cause significant annoyance and stress-related effects that can in turn encourage wearers to deliberately to disable their protectors to decrease attenuation.

Another weakness of single-number ratings, such as the NRR, is that they are often based on just one or two of the nine third-octave test signals used in laboratory measurements. Generally, the controlling test signals are below 1000 Hz and performance levels for other test signals may have little or no effect on the final NRR. For example, two protectors may have the same EPA NRR because of their limiting attenuation values at test signals centered at 250 or 500 Hz, but one of these protectors may provide more than 15 dB greater protection than the other for higher frequencies, above 1000 Hz, without this information being indicated. The NRR values in this example would be a reasonably accurate assessment of protector performance if the highest exposures were centered at 250 or 500 Hz, but if the noise exposures contain prominent high-frequency components, the NRR values may be very misleading.

The more precise NIOSH Method 1 for calculating hearing protector performance ratings can be used to obtain a more accurate estimate of protection, while at the same time affording a more accurate means of maximizing communication. None of the safety factors discussed above are required when octave band sound pressure levels are used to determine exposure levels under hearing protectors.

In Europe, the most common single-number rating of hearing protector performance is the single number rating (SNR).[56] The SNR has the same pitfalls as the NRR, although it is somewhat less sensitive to outlier data during the laboratory testing. The European market also utilizes a three-number rating system, called the high–middle–low, or HML, system. The three numbers in the HML rating estimate the attenuation provided in the high test frequencies, middle test frequencies, and low test frequencies, respectively. The HML values are used in conjunction with measurements of A- and C-frequency-weighted sound pressure levels to obtain the HPD attenuation that can be expected in specific noise environments. The European tests are performed according to International Standards Organization (ISO) 4869-1,[57] whereas the U.S. testing is performed according to ANSI S3.19-1974.

22.8.4.2 Derating the NRR The literature contains many references to the differences between laboratory ratings of hearing protectors and the amount of attenuation measured in the field. This discrepancy has led to derating procedures that are applied to laboratory data, which is labeled on the hearing protector packaging. Some industrial hygienists derate laboratory data by 50% before calculating exposure under the protector, and other hearing conservation programs use a straight 10 dB derating scheme. These procedures can lead to significant overprotection, which may unnecessarily inhibit communication. None of these inaccurate derating methods are necessary if the hearing conservation program utilizes (1) the laboratory ratings for guidance only, (2) a field monitoring system as a training tool and to verify protector fit (see Section 22.8.7.1), and (3) an HPD selection procedure that ensures that devices are selected that provide sufficient protection while avoiding overprotection. The derating schemes do not help solve the problem of poor protector usage and may exacerbate the problem of communication in noise.

22.8.5 Communication without Hearing Protectors

Performance and safety aspects of a job often depend on the workers' ability to hear warning signals, machine sounds, and speech in the presence of high noise levels. The effect of noise on communication depends to a large extent on the spectrum of the noise, the hearing characteristics of the worker, and the attenuation characteristics of hearing protectors, if they are used.

For a normal-hearing person, speech communication is affected most when the noise has high-level components in the speech frequency range from about 400 to 3000 Hz. Speech interference studies[58] show that conversational speech begins to be difficult for a speaker and a normal-hearing listener, separated by about 2 ft when broad-band noise levels approach about 88 dBA. Hearing-impaired persons have much more difficulty communicating in noise than do persons having normal hearing; the degree of difficulty depends upon the amount and type of impairment.

22.8.6 Communication with Hearing Protectors

Few hearing protectors are selected and purchased with any thought being given to communication of any kind. Overprotection may be considered as acceptable, or even desirable, against many health and safety hazards, but overprotection against noise exposure may cause significant communication problems. Maximizing communication while wearing hearing protectors often improves safety and work efficiency conditions, and can prolong the working lifetime of skilled workers having high-frequency hearing impairment. Overprotection from noise may also lead to the deliberate misuse or rejection of hearing protectors.

Hearing protectors interfere with speech communication in quiet environments for most persons, regardless of their hearing characteristics. Normal-hearing persons can often raise their voice levels to provide satisfactory communication in moderate levels of noise.

When wearing hearing protectors in noise levels between 88 and 97 dB, normal-hearing persons sometimes complain that the protectors prevent communication, although the HPDs usually do not affect the overall speech-to-noise ratio and probably do not impede communication. Above about 97 dBA background noise levels, normal-hearing persons are often able to communicate about as well with as without wearing protectors, albeit very poorly.[59–61] In fact, protectors may improve speech communication for some normal-hearing persons when background noise levels are higher than 97 dBA because speech-to-noise ratios are held relatively constant, and distortion is reduced. Optimal communication is usually provided when the protector's attenuation characteristics are matched to those of the noise spectra. Individuals with hearing loss almost always have difficulty communicating in noisy areas.

22.8.7 Integration of Communication Requirements and the Use of Field Monitoring Systems

Among the problems associated with the use of hearing protective devices in industry are poor training and motivation on proper wearing techniques, differences between labeled noise reduction ratings and achievable field attenuation and impaired communication with the use of HPDs. The use of hearing protector field monitoring systems and the careful selection of hearing protectors that maximize communication provides an opportunity for the hearing conservationist to address these issues in a methodical and well-accepted manner.

The careful selection of HPDs and verification of protector fit on the individual wearer will result in reduced incidence of noise-induced hearing loss and fewer injuries caused by a lack of communication. The comprehensive documentation of HPD selection and usage verifies effective program administration and apportions responsibility to the hearing protector wearer.

22.8.7.1 Field Monitoring Systems Field monitoring systems are designed to allow the hearing conservationist to measure the attenuation provided by hearing protectors *on the individual wearer*. This capability eliminates the need for single-number ratings, such as the NRR, and misleading safety factors. The NRR is derived from laboratory test methods that are designed to estimate attenuation afforded to a *population*, not to an *individual*. Current laboratory performance ratings represent a "best-fit" situation using trained and motivated subjects under closely supervised conditions. It is likely that the attenuation achieved in the field will be significantly less than the laboratory values *unless* the wearers are properly fitted and motivated.

Field measurement systems perform several functions for the industrial hearing conservation program administrator, including: (1) training of wearers in correct fitting procedures, (2) random field sampling of protector effectiveness, (3) documentation that training was provided and that proper protection was provided to the employee, and (4) identification of failing or deteriorating protectors and changes in ear physiology.

Experience has shown that individual measurement systems are, in general, well received by the HPD wearers. The employees are typically interested in how the protectors function, and they appreciate the attention to their individual needs. Individual

HPD attenuation measurement is particularly valuable as a training tool during the initial selection of insert-type HPDs. The hearing conservationist can assist the wearer during the initial fitting and measure the attenuation that is provided. If the attenuation is sufficient, the employee should then re-fit the HPD and the measurement procedure should be repeated. If the measured attenuation is sufficient after the wearer has fitted the device, documentation is provided that adequate training and protection was provided to the employee.

Portable field monitoring systems are currently available for both muffs and insert-type hearing protectors.[62,63]

22.8.7.1.1 Field Monitoring of Insert-Type Protectors The amount of noise attenuation provided to wearers of insert protectors varies widely across the general population. The attenuation provided to poorly trained plug wearers or to individuals with narrow, sharply bending, or slit-shaped ear canals may be much lower than the labeled laboratory values. Persons with large ear canals may be able to insert protectors deep into the canal without achieving a satisfactory seal, resulting in poor attenuation, especially at the lower frequencies. Visual inspection of HPD fit is not always sufficient.

For insert-type protectors, field monitoring systems are available that essentially replicate the laboratory tests as defined in ANSI S3.19-1974, except that the stimuli are presented via headphones. This test involves measuring the hearing thresholds of the HPD wearer at selected test frequencies with and without the HPD in place. The difference in hearing threshold at each test frequency is equal to the amount of noise attenuation provided by the hearing protector.

22.8.7.1.2 Field Monitoring and Muff-Type Protectors For muff-type protectors, existing field measurement systems include a handheld microprocessor-based unit that utilizes two microphones, one located at the entrance to the ear canal and the second located outside the cup of the muff. A digital readout displays the difference in sound pressure level between the two microphones. The unit is designed to be used at the employees workplace; therefore, the attenuation measured will be accurate while the wearer is in *that particular noise exposure*. Use of the protector in noises with differing spectral characteristics will affect the amount of noise attenuation provided.

22.8.7.2 Selection of Hearing Protection Devices to Maximize Communication There is currently a European guidance document, EN 458, entitled "Hearing Protectors—Recommendations for Selection, Use, Care and Maintenance," that addresses the issue of selecting hearing protectors that maximize the ability to communicate. The document includes several methods of HPD selection, including the octave band method, the HML method, the HML Check and the SNR method. All the methods are dependent on the level of the exposure at the ear (i.e., under the protectors).

A summary of the octave band method follows:

1. Using octave band noise exposure data, octave band hearing protector attenuation characteristics and the NIOSH Method 1, calculate the A-weighted exposure level under the protector.

2. The allowable criteria are based on 8-h exposures and may be selected based on the legal exposure limits (i.e., 90 dBA) or a more conservative level (i.e., 85 dBA). Compare the A-weighted exposure under the protector to the criteria level in Table 22.11.

TABLE 22.11 Recommendations for Hearing Protector Attenuation

Relationship between exposure under the protector (in dBA) and allowable exposure criteria (in dBA)[a]	HPD selection
Exposure > criteria	Insufficient protection
Criteria > exposure > (criteria − 5 dB)	Acceptable
(Criteria − 5 dB) > exposure > (criteria − 10 dB)	Good
(Criteria − 10 dB) > exposure > (criteria − 15 dB)	Acceptable
(Criteria − 15 dB) > exposure	Overprotection

[a]Exposure (dBA) = A-weighted exposure level under the protector.

For communication purposes, the careful selection of HPDs is more critical for individuals with hearing loss than for normal hearing individuals. The range of "acceptable" or "good" HPD selections is greater for normal-hearing individuals than the range specified in EN 458. The intelligibility of noise-degraded speech does not decrease from a maximum for normal-hearing individuals until the presentation level is about 50–60 dB SPL, whereas the EN 458 definition of "overprotection" starts at 70 dBA for a criteria level of 85 dBA. As the degree of hearing loss increases, the number of "good" or "acceptable" HPD selections is reduced because of the increased possibility of overprotection.

The algorithm may be used for HPD selection for large groups of wearers. In this case, a conservative selection criteria (i.e., selection assuming the wearer has at least a moderate hearing loss) should be employed, since it is likely that at least some of the individuals in the group will be hearing impaired.

The algorithm is dependent on laboratory HPD attenuation data. The attenuation afforded to an individual HPD wearer, of course, is not necessarily equal to the lab data. The best way to implement these criteria is to use a field monitoring system to verify HPD performance and the HPD selection algorithm to determine the range of acceptable protectors. This approach documents that the employee has been fitted with protectors that provide sufficient protection while optimizing the ability to communicate.

Proper HPD selection according to these guidelines does not guarantee good communication. Obviously even the best HPD selection will not significantly improve a highly adverse communication situation. To predict the ability to communicate in noise, ANSI S3.79, American National Standard for the Calculation of the Speech Intelligibility Index, should be referenced.

Properly fitted individuals are more likely to wear their protectors consistently since they will not be unnecessarily "isolated" from other workers and they will be less likely to intentionally disable protectors to decrease attenuation. Fitting HPDs using these criteria will dissuade hearing conservationists from the "more attenuation is better" attitude and create a greater demand for HPD manufacturers to develop and market comfortable HPDs with a range of attenuation characteristics. The use of a field measurement system and an HPD selection procedure is educational for both the hearing conservationist and HPD wearer. The individual wearer becomes involved in the fitting process and is required to assume responsibility for the correct wearing of the protector.

22.8.7.3 Documentation Continued periodic use of a field measurement system and careful selection of HPDs that provide sufficient protection without needlessly impairing communication provides the best defense possible against litigation based on hearing loss or lack of communication ability. To date, the most common type of litigation against industry in this area has been hearing loss and related physiological problems such as tinnitus. It is likely, however, that the number of cases will increase where industry is held responsible for "overprotecting" employees.

Measurement of the attenuation provided by hearing protectors, especially insert-type, on the individual wearer is extremely valuable. This type of measurement should be required when the individual is initially fitted with the HPD and at least annually thereafter. Individual measurement of hearing protector performance documents that adequate and proper protection was provided to the worker and *shifts responsibility of effective HPD use to the employee*. The time and cost required for these measurements is more than justified by the reduction in liability, and more important, the protection of the individual. These measurements, used in conjunction with HPD selection criteria that optimize the ability to communicate while wearing HPDs, result in a truly comprehensive hearing protector management program.

22.9 NOISE CONTROL PROCEDURES

22.9.1 Introduction

Noise control efforts should be approached using the paradigm:

source → path → receiver

The noise from most equipment is waste energy. For this and efficiency reasons, the best way to reduce noise is to tackle the problem at the source. Generally, reducing the noise at the source also offers the most options. Changes to the path generally involve adding barriers or enclosing the equipment, but may involve adding sound-absorbent materials. Reductions of more than a few decibels are difficult to achieve by these modifications. At the other end of the path is the receiver or affected employee. Reduction of noise exposures here are achieved by either removing the employees from the sound field, limiting time in the area, or through the use of hearing protection. While the latter is really a modification of the noise path, personal protective equipment is usually considered noise control at the receiver, and as with other industrial hygiene uses of personal protective equipment, should be the last step taken for controlling exposures—only used when other measures have proven ineffective.

22.9.2 Noise Control

The first step in providing quiet equipment is to make a strong effort to have purchase orders include noise limits. The desired quiet equipment will not always be available, but at least these specifications will provide an incentive for the design of quiet products. Modifications to equipment to reduce noise includes closer tolerances, better assembly, balancing of rotating machinery, redesign of components, and other quality control measures. Usually these changes must be left to the manufacturer of the equipment. Users of equipment can specify the noise level that will be tolerated in new equipment purchases. Prior to the promulgation of OSHA noise regulations, few manufacturers were concerned with noise emissions. Currently, many manufacturers attempt to distinguish their products by emphasizing their lowered noise emissions. Purchasers of equipment, however, may not be aware of the cost advantages if quiet equipment is purchased—avoiding the need for hearing conservation programs and incurring less hearing loss by employees. The health and safety professional should assure that management and purchasing departments are aware of the need to purchase quiet equipment. Proper maintenance of equipment must also be stressed since virtually all mechanical equipment becomes noisier as components wear.

If the need is to reduce the noise from existing machinery, consider both generation and radiation of sound for possible noise control measures. Once generated, the noise will be from (1) the direct sound field, (2) reverberant sound field, or (3) a structure-borne path. Finally, the only true measure for reducing operator exposure at the receiver is to reduce the amount of time the operator spends in the sound field. Providing a noise enclosure for the operator can also be considered as a receiver noise control measure.

The following systematic approach could be used for controlling noise. Each will be discussed later in somewhat greater detail.

Source: Generation

1. Reduce impact noise.
2. Reduce or eliminate aerodynamically generated noise.
3. Reduce or eliminate any resonance effects.
4. Modify or replace gears or bearings.
5. Reduce unbalance in rotating systems.

Source: Radiation

6. Move the machinery to a new location—distant from exposed personnel.
7. Provide vibration isolation to reduce the radiation of noise from the surface on which the machinery is mounted.
8. For large heavy machinery, use an inertia block.
9. Insert flexible connectors between the machine and any ductwork, conduit, or cables.
10. Reduce or modify the surfaces that radiate noise.
11. Apply vibration isolation to machine housing.
12. Use active noise cancellation or active vibration cancellation.

Path: Direct

13. Provide a partial or full enclosure around the machine.
14. Use sound-absorptive materials.
15. Use an acoustical barrier to shield, deflect, or absorb energy.
16. Reduce the leakage paths permitting noise to leak from enclosures.

Path: Reverberant

17. Apply sound absorptive materials to walls, ceiling, or floor.
18. Reduce reflections by moving equipment away from corners or walls.

Path: Structure-Borne

19. Use ducts lined with absorptive materials.
20. Use wrapping or lagging on pipes to increase their sound insulation.
21. Reduce turbulence from liquid or gaseous flows.
22. Add mufflers.

Receiver

23. Use enclosure or control room to house operators.
24. Provide hearing protection to operators.
25. Reduce amount of time operators are allowed to work in high-noise areas.

22.9.2.1 Reduce Generated Noise

22.9.2.1.1 Reduce Impact Noise Mechanical and materials handling devices commonly produce noise from impact. Noise can be reduced by:

1. Reducing the dropping height of goods collected in boxes or bins (Figs. 22.22 and 22.23).
2. Using soft rubber or plastic to receive hard impacts.
3. Increasing the rigidity of containers receiving impact goods and add damping materials—especially to large surfaces.
4. Using belt conveyors, which are generally quieter than roller conveyors.
5. Regulating the speed or cycle time of conveyors to prevent collisions and excess noise.

22.9.2.1.2 Reduce or Eliminate Aerodynamically Generated Noise A wide variety of methods can be used for noise control from aerodynamically generated noise.

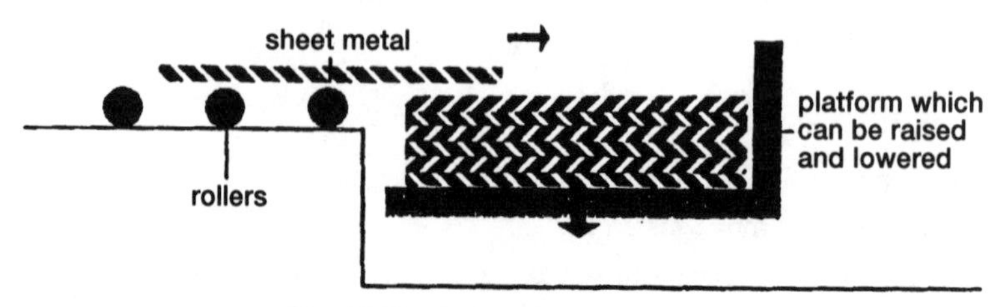

Fig. 22.22 Reduce height of drops.

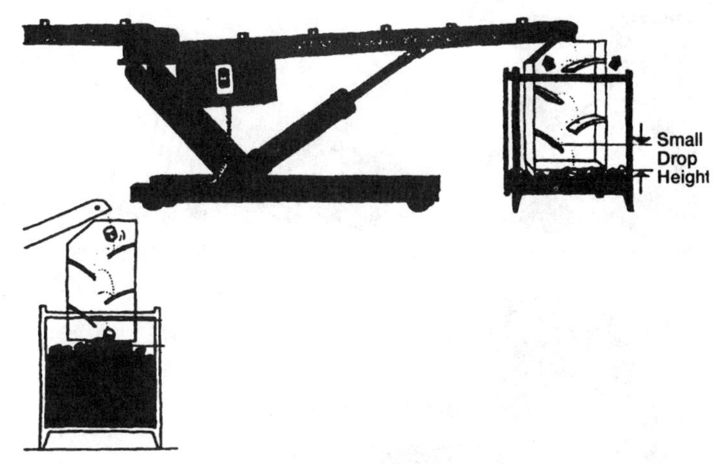

Fig. 22.23 Reduce drop height and use rubber flaps to slow fall speed.

22.9.2.1.2.1 CHANGE THE CHARACTER OF THE NOISE Where great distances are involved, such as outdoors, air reduces high-frequency sounds more than lower frequencies. Atmospheric attenuation for pure tones at 70% relative humidity and 10°C is given by:

$$A_{atm,f} = \frac{r}{1000\ \mathrm{m}}\left[0.6 + 1.6\left(\frac{f}{1000\ \mathrm{Hz}}\right) + 1.4\left(\frac{f}{1000\ \mathrm{Hz}}\right)^2\right] \qquad (22.25)$$

where r is the distance between the source and receiver (m) and f is the frequency (Hz).

For environmental noise control, replacing a source with one of higher frequency may reduce the sound level at typical property line distances (Fig. 22.24). Note, however, that shifting to higher frequencies will likely bring the source into a frequency range where the ear is more sensitive. Therefore, careful consideration of perceived loudness must be made before applying this solution.

22.9.2.1.2.2 REDUCE THE AREA OF THE SOURCE Large surfaces when vibrating will produce high sound levels. Consider replacing solid plates, when possible, with expanded metal, wire mesh, or perforated metal (Fig. 22.25).

22.9.2.1.2.3 CHANGE THE SOURCE DIMENSIONS SO NOISE IS CANCELLED AT THE EDGES At the edges of large vibrating plates, the pressure and rarefaction waves tend to cancel. The same principle can be applied using long narrow surfaces instead of square surfaces (Fig. 22.26).

22.9.2.1.2.4 REDUCE OR REMOVE INTERRUPTED WIND NOISE CAUSING TONES Wind instruments use the principle of air blowing across an edge to produce standing-wave pressure vibrations. When tonal noise is produced by machinery, it may be possible to eliminate the wind and thus the noise (Fig. 22.27).

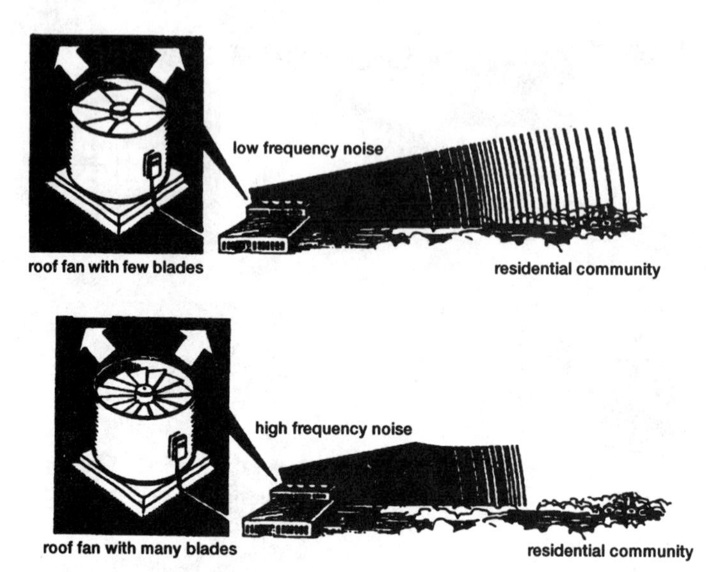

Fig. 22.24 Convert fan noise to higher frequencies, which are more attenuated by atmosphere.

22.9.2.1.2.5 REDUCE TURBULENCE IN FLUIDS Smooth laminar flow of fluids through ducts and pipes does not generate noise. The more turbulent the flow, the greater the noise. Vapor bubbles can be created by abrupt changes in the flow of fluids. Providing gradual transitional changes in cross-sectional area reduces the chance of forming these bubbles (Fig. 22.28).

Implosion of vapor bubbles or cavitation can be severely damaging to plumbing as well as causing severe noise problems. Cavitation occurs when static pressure downstream of a valve is greater than the vapor pressure, and at some point within the valve the static pressure is less than the liquid vapor pressure. Reducing the pressure in several smaller steps prevents cavitation and reduces noise (Fig. 22.29).

Turbulence at the walls of ducts and pipes is always present. To reduce noise, interior walls should be smooth, free of protrusions at joints, and sharp bends at tees and wyes should be avoided. Turning vanes can be placed inside ductwork when construction methods utilize sharp bends. Straightening vanes can be used to smooth the flow downstream of any changes in direction, diameter, or system branches (Fig. 22.30).

22.9.2.1.2.6 METHODS FOR REDUCING FAN NOISE Fans, in particular, produce large amounts of turbulence, which produces noise. Since the sound power generated by a fan varies as the fifth power of rotational speed, the most cost-effective noise control is to reduce the speed when possible. When purchasing new fans, consider also quieter designs. For instance, backward curved blades on squirrel cage fans are quieter than straight blades or forward curved blades.

Fans mounted inside ductwork create significant noise, especially when mounted in regions where a great deal of turbulence is present. In-line duct fans should be mounted in low-turbulence regions of ductwork (Fig. 22.31).

Pneumatic equipment produces noise from turbulence when the high-speed gas stream mixes with the ambient air at the tool outlet. The simplest control measure is to

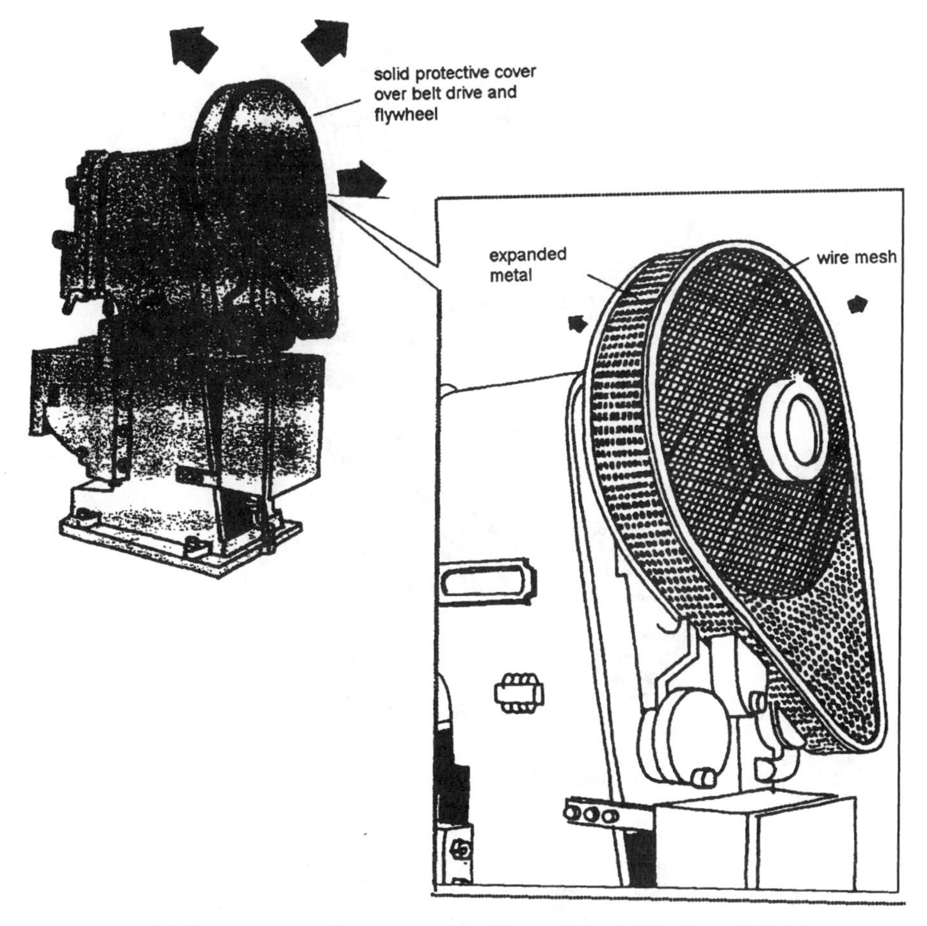

Fig. 22.25 Lower sound power generated by perforated plate than solid plates.

reduce the exit speed of the gas. Reducing the speed by a factor of two can decrease the noise level by 15 dB. Silencers can also be installed.

22.9.2.1.2.7 SILENCERS Silencers or mufflers can be classified into two fundamental groups—absorptive or reactive. They are made to reduce noise while permitting flow of the air or gas. Absorptive silencers contain porous or fibrous materials and use absorption to reduce noise. The basic mechanisms for reactive silencers is expansion or reflection of sound waves, leading to noise cancellation.

22.9.2.1.2.7.1 Absorptive Silencer The simplest form of absorptive silencer is a lined duct. Generally long sections of ducts are lined with absorptive material, but lining is particularly effective along duct bends. Typically 2 to 5 cm acoustical grade fibrous glass is used. Where dust is present or humid conditions could create microbiological growth, the absorptive material is covered with thin plastic or mylar, etc. In the past, HVAC

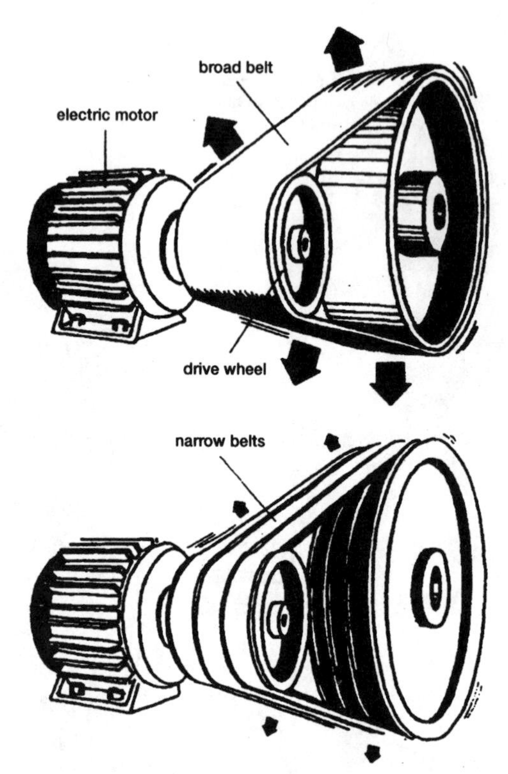

Fig. 22.26 Less sound power produced by several narrower belts than a single broad belt.

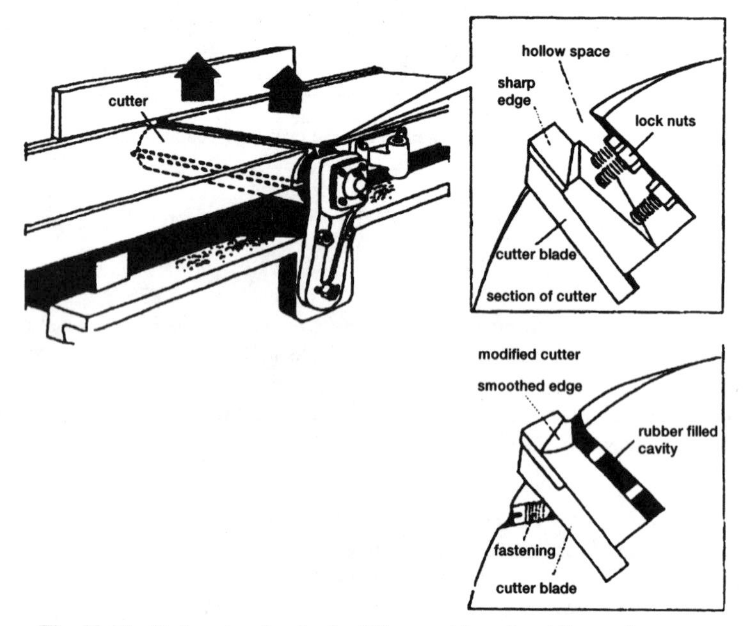

Fig. 22.27 Reduce tonal noise by filling cavities of rapidly moving parts.

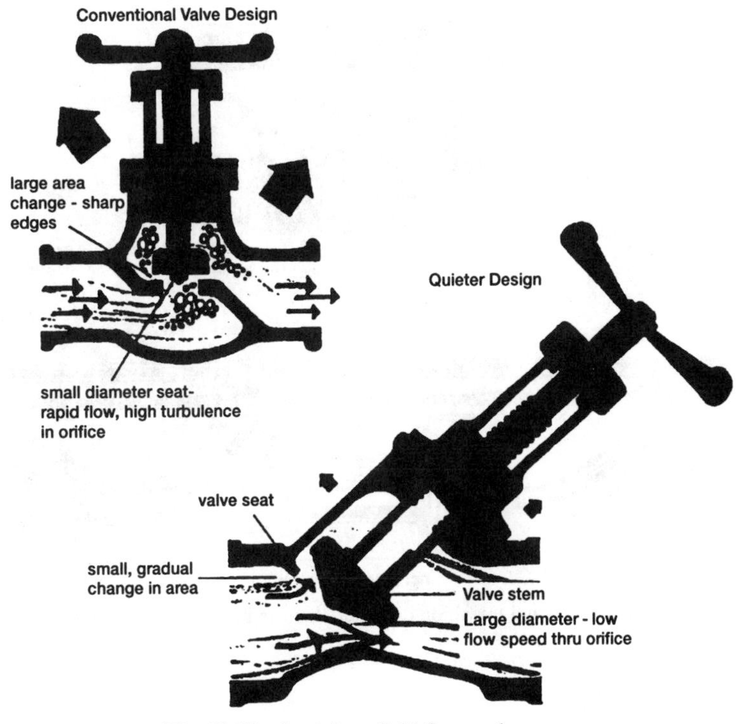

Fig. 22.28 Straighten fluid flow pathways.

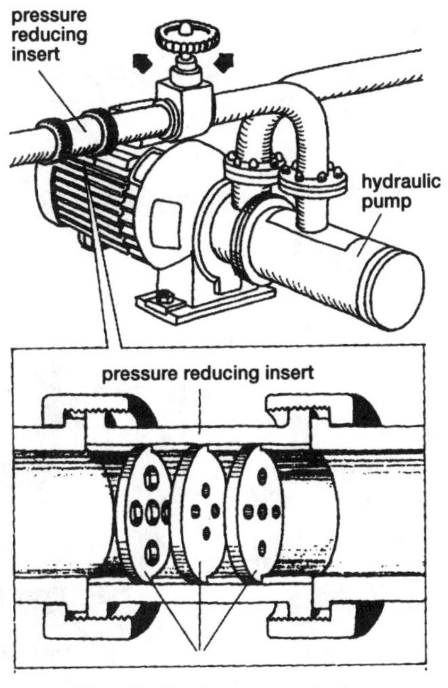

Fig. 22.29 Prevent cavitation.

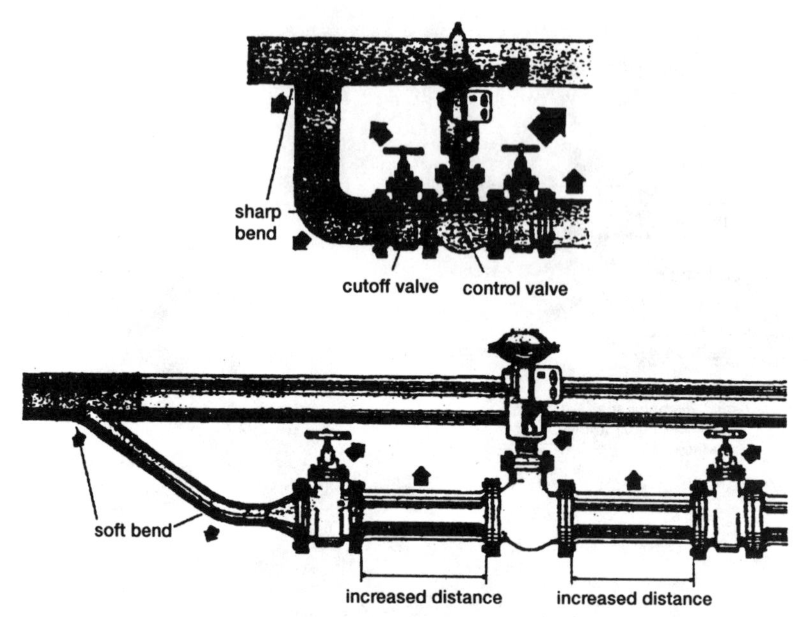

Fig. 22.30 Straighten bends and transition diameter changes in fluid flow to reduce turbulence.

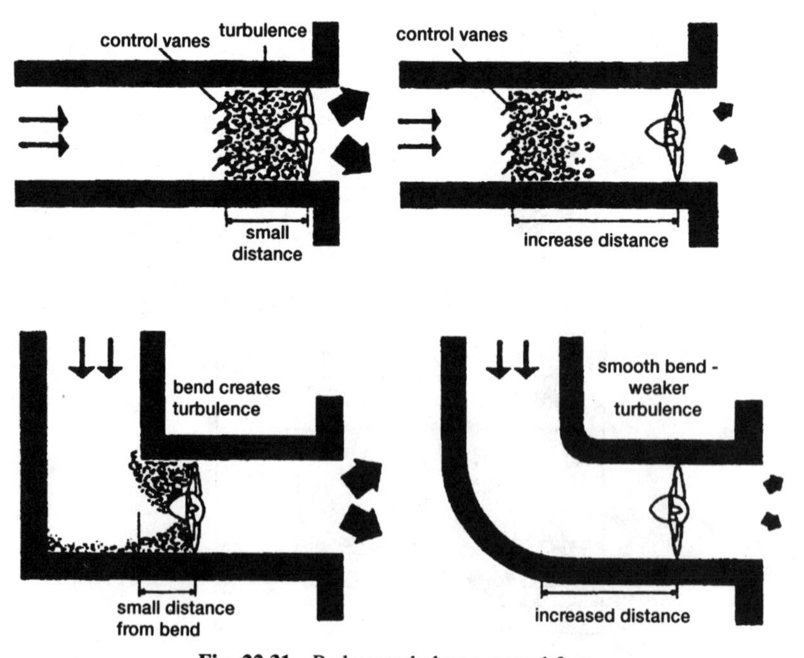

Fig. 22.31 Reduce turbulence around fans.

ducts were lined with absorptive materials downstream of fans in office buildings to reduce fan noise. Several of these installations have had severe microbiological growth, causing other (nonacoustic) industrial hygiene problems (see Chapter 19).

Another form of absorptive silencer is parallel baffles. Good design includes aerodynamically streamlined entrance and exit ends with perforated spaces filled with highly absorbent acoustical materials. The first few feet of length are highly absorbent; so the attenuation is not linear with length. Thick absorbent material with wide spaces between absorbers are effective for low frequencies, while thin absorbers and narrow spaces are effective for higher frequencies. They should be considered for cooling and exhaust air whenever sources are to be enclosed (Fig. 22.32.).

22.9.2.1.2.7.2 Reactive Silencer The simplest form of a reactive silencer is a single expansion chamber. As the air enters and leaves the chamber, the expansion and contraction in pressure cause reflection of sound waves. The reflected wave added to the incoming sound wave results in destructive interference, leading to noise reduction. This reduction only occurs when:

$$l = \frac{n\lambda}{4}, \qquad n = 1, 3, 5, \ldots \tag{22.6}$$

where l is the length of the muffler, λ is the wavelength of the tone, and n is an integer. This equation can be used to calculate the length needed for a reactive silencer.

22.9.2.1.3 Reduce Vibrations In many cases, machinery noise is created by a vibrating source coupling to a large radiating surface. When possible, large radiating surfaces should be detached from vibrating sources (Fig. 22.33).

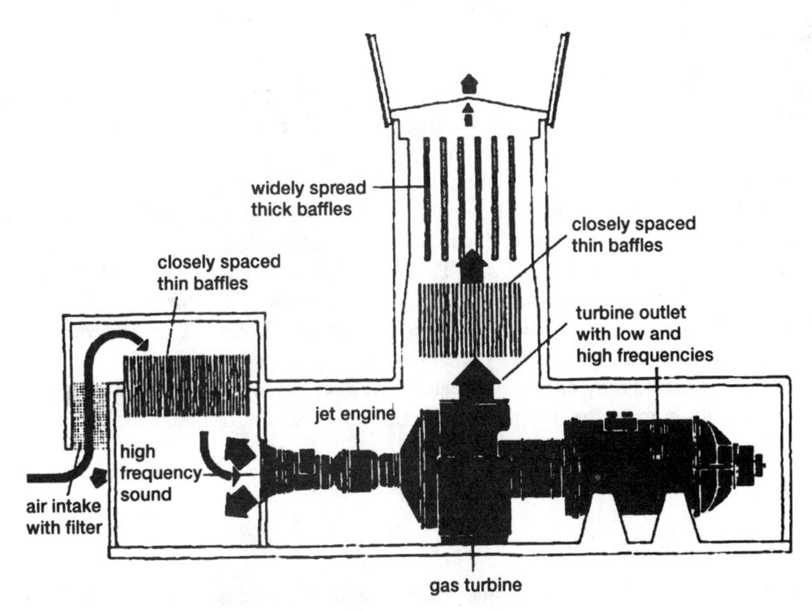

Fig. 22.32 Absorption mufflers sized for dominant frequencies.

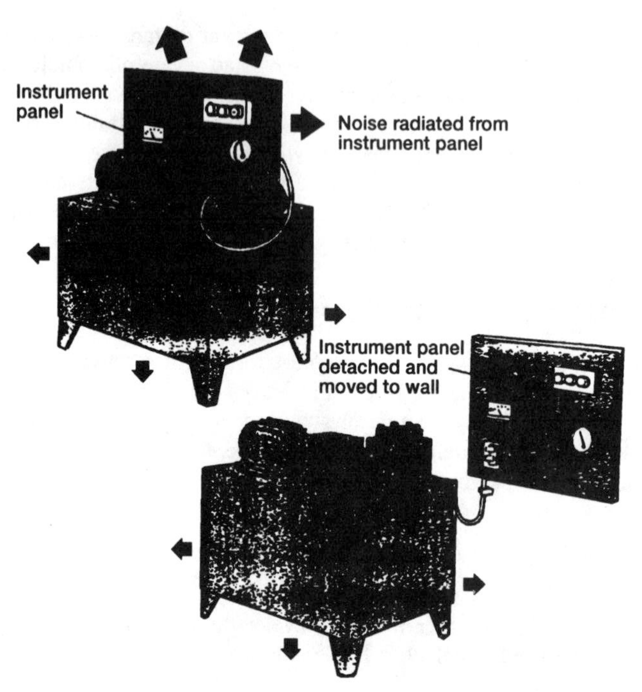

Fig. 22.33 Remove panels from vibrating equipment.

The vibrating surface may be stiffened to limit the motion, or may be covered with a damping material (Fig. 22.34). Alternatively, the source might be detached from the building structure and mounted on the floor.

If possible, the equipment should be placed on a concrete base plate resting directly on the ground. More effective isolation is achieved when the base plate is not part of the building structure, and is mounted directly on the ground (Fig. 22.35).

When equipment is attached to the building structure, care must be taken to provide vibration isolation or the noise may be transmitted throughout the building by vibrating the building structural members. The equipment may be mounted on concrete inertia blocks or directly to steel frames. Regardless of the mounting, some form of vibration isolators are usually used (Fig. 22.36). The degree of isolation achieved with vibration isolators depends on the frequency of vibrations relative to the natural frequency of the system and the amount of damping built into the isolator. Formulae for determining the isolation achievable are provided by manufacturers of vibration isolators.

Vibration isolation may not be completely effective when noise is transferred through piping or conduits from the equipment. Flexible connectors to mount the tubing to the building must also be considered (Fig. 22.37).

22.9.3 Noise in Rooms

The sound *power* of a source is independent of its environment, but sound *pressure* around the source is not (Fig. 22.38). Considering the equipment as a point source, if it

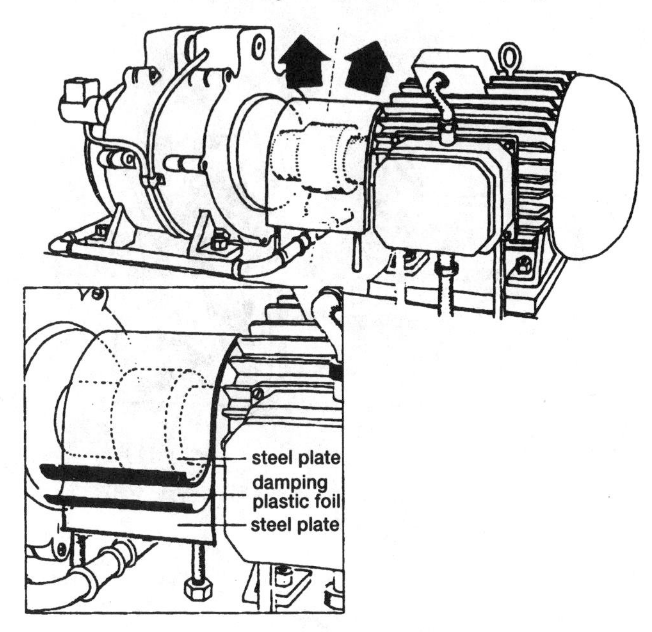

Fig. 22.34 Provide vibration damping on flexible panels.

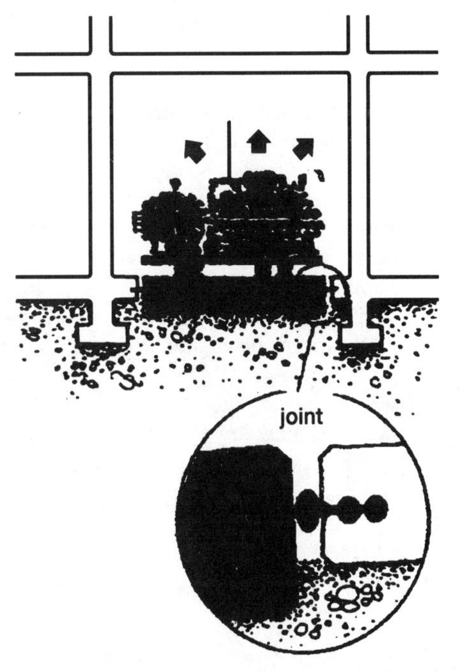

Fig. 22.35 Heavy vibrating equipment should be isolated from building.

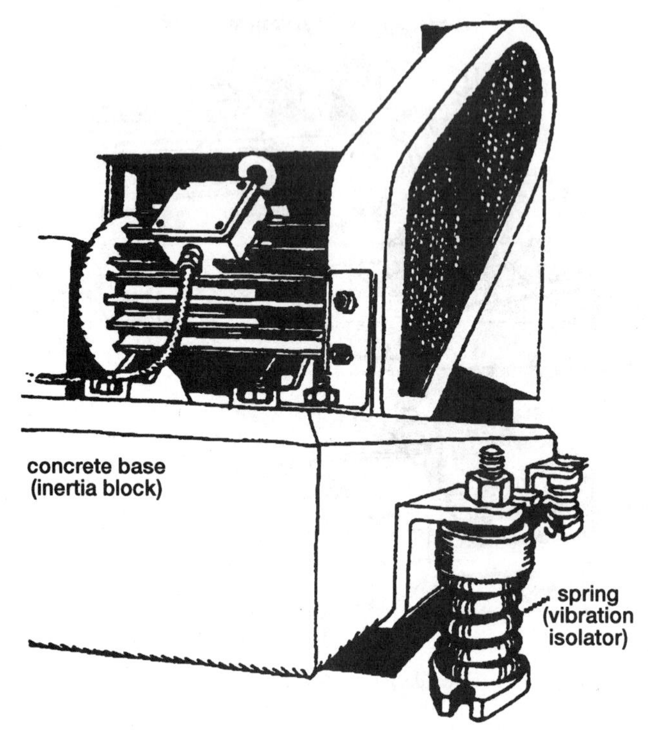

Fig. 22.36 Heavy vibrating equipment can be placed on inertial blocks with vibration isolators and dampers.

could be suspended in midair away from any reflecting surfaces, including the ground, the equipment would be in a free field, and sound pressure levels would decrease by 6 dB as the distance from the source was doubled.

In practice, machinery is neither a point source nor in free-field conditions. Noise is usually emitted from all parts of equipment, and not all parts vibrate in phase with each other. Consequently, some parts will be moving in while other parts move out, leading to a partial cancellation of sound pressure. At other parts of the equipment, parts may be moving in phase, reinforcing the sound pressure levels. Close to a machine, in the near-field, sound pressure levels may be higher or lower than predicted from sound power and distance.

Outdoors, at distances about two times the longest dimension of the machine, these effects disappear, and the inverse square law or 6 dB decrease per doubling of distance, becomes valid. Indoors, reverberations from walls, floors, and ceilings will result in less of a decrease in sound pressure level as the distance increases than predicted in a free-field environment.

Noise sources should be kept away from walls. The worst placement is in the corners, where the reflections are greatest (Fig. 22.39).

22.9.3.1 Treating Rooms with Absorbent Materials Absorbing materials are used when it is desired to reduce the noise within a particular environment. Sound

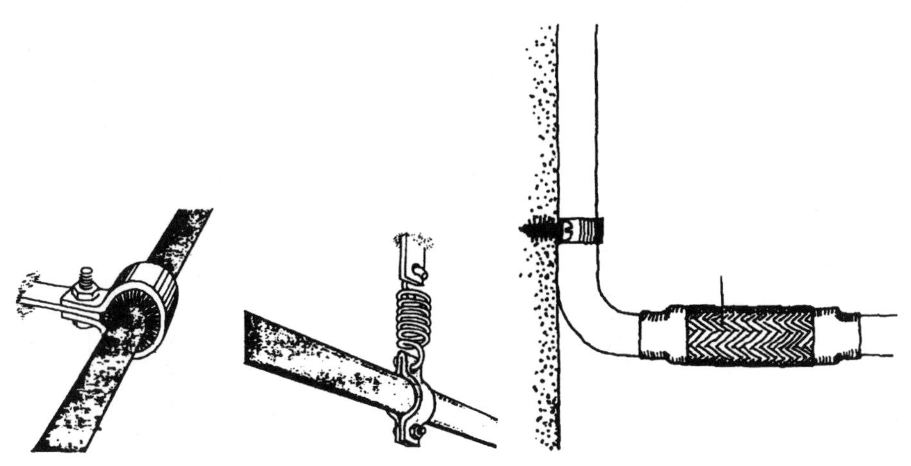

Fig. 22.37 Flexible couplings prevent transmission of vibration to building structure.

reaching the ear consists of two components: sound transmitted directly from the source (direct sound), and sound reflected from all the room's surfaces (reverberant sound). The level of direct sound depends only on distance from the source and is reduced by 6 dB for each doubling of distance from the source (inverse square law). In contrast, reverberant sound depends on the room size, shape, and absorption, and does not depend on distance from the source.

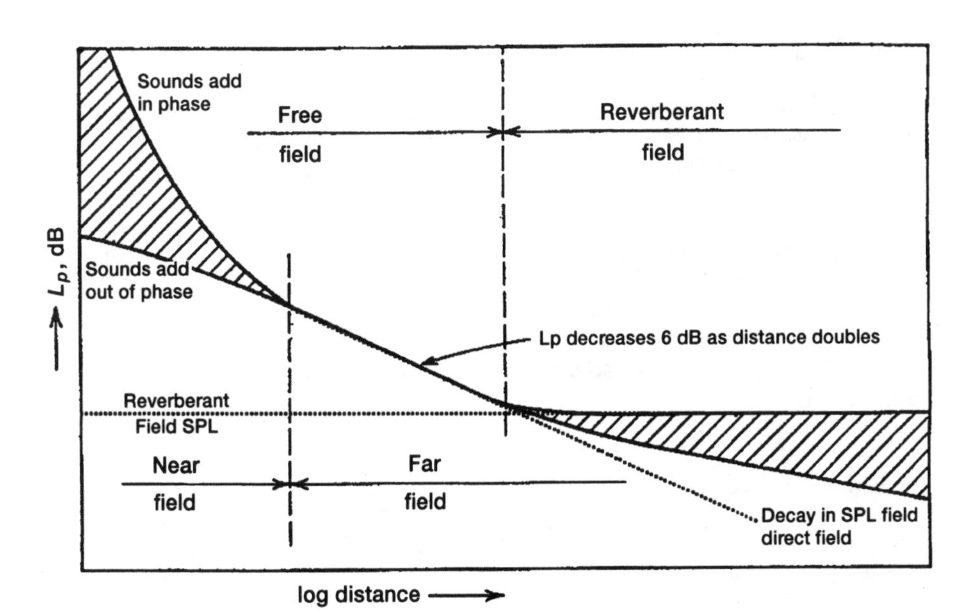

Fig. 22.38 Sound pressure levels in the near, free, and far fields.

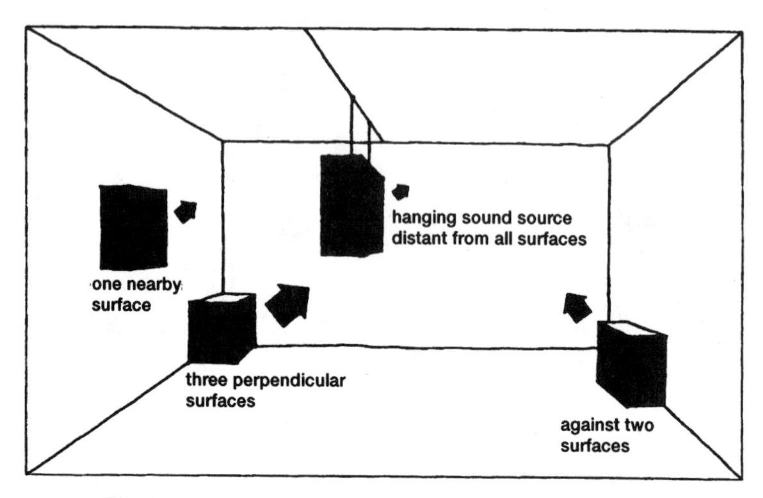

Fig. 22.39 Avoid placing equipment near walls and corners.

The unit for sound absorption in a room is the room constant R, measured in m^2.

The relative sound pressure level in rooms can also be calculated using the Eyring theory, which for a point source is calculated[81] as:

$$L_p = L_{p1} - L_{p2} = 10\log\left(\frac{Q}{4\pi r^2} + \frac{4}{R}\right) \tag{22.27}$$

where L_p is the relative sound pressure level (dB); Q the directivity factor (taken to be 1); r, the source–receiver distance (m); and R, the room constant (m^2).

From Fig. 22.40 we see that close to the source the sound pressure level is determined primarily by distance, whereas far from the source it is strongly influenced by the amount of absorption. Thus increasing the room acoustic absorption is not effective in reducing the sound pressure level close to the source.

In practice, the room constant can be difficult to calculate or requires some relatively sophisticated equipment to measure. For a rough estimation of the room constant, determine the fraction of the total room surface covered with absorption material (i.e., carpeting, drapes, acoustic ceiling tiles, absorbent wall panels, etc.). Table 22.12 can be used to estimate the acoustic characteristics of the room.

Use the room acoustic characteristic and room volume to estimate the room constant (Fig. 22.41).

Example 15: A computer room 50 feet by 50 feet with 10 foot ceiling with acoustic tile ceiling and some sound absorbtive panels on the walls has an "average" room constant ($R = 80\ m^2$). Installing carpeting was suggested to reduce the noise from a supplemental HVAC unit located about 15′ (4.5 m) from several employee's work stations. Estimate the reduction in noise.

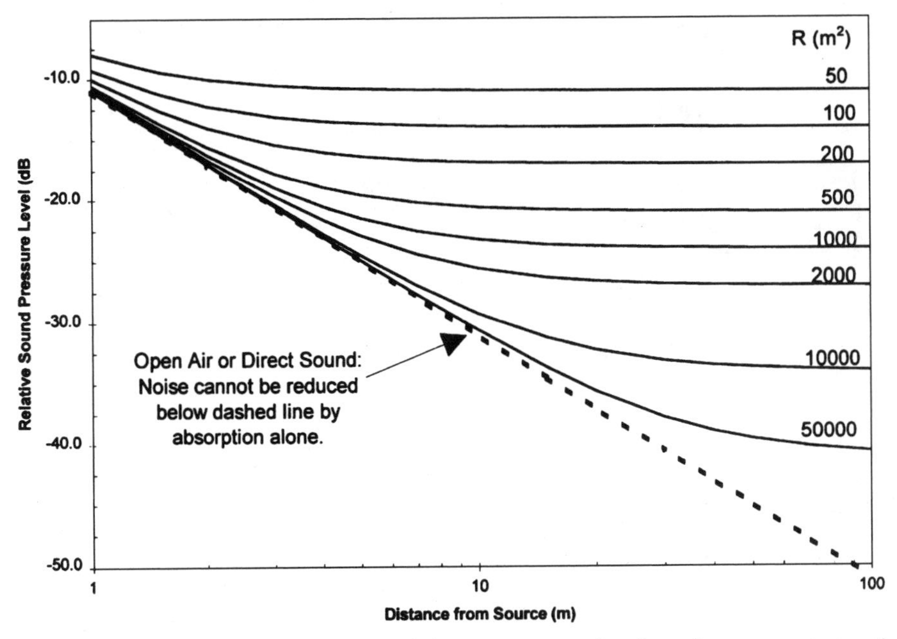

Fig. 22.40 Relative sound pressure level in a room as a function of room constant and source–receiver distance.

Heavy carpeting on the floor would increase the fraction of room surface covered by 0.36 [(floor area)/(room surface area) = (50′ × 50′)/(2 × 50′ × 50′ + 4 × 50′ × 10′) = 0.36]—changing the room reverberation characteristic from "average" to "dead" and increasing the room constant R to 300 (see Fig. 22.41). Using Eq. 22.27 or Fig. 22.40, the relative sound pressure level change is 5 dB.

Note the above is an extremely crude estimate, and it relies on a number of simplifying assumptions. A more accurate estimation should be made measuring octave band sound pressure levels from the source and calculating the change in room constant at each octave band level from the absorption coefficients. In particular, for this example,

TABLE 22.12 Estimation of Room Acoustic Character from Fraction of Surface Area covered with Absorptive Material

Fraction of total room surface area covered with absorption material	Room acoustic characteristics
0	"Live room"
0.1	"Medium-live room"
0.15–0.2	"Average room"
0.3–0.35	"Medium-dead room"
0.5–0.6	"Dead room"

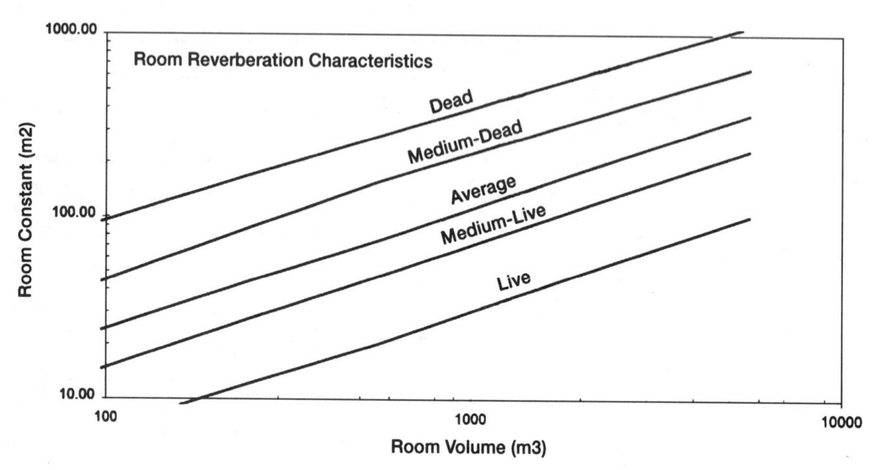

Fig. 22.41 Approximate relationship between room constant and room volume for spaces with varying acoustic characteristics.

the noise from the HVAC unit is likely to be predominantly in the low frequencies, where carpeting is not particularly absorbent. Most machinery is used inside buildings, and even outdoor equipment is seldom suspended above the ground. Consequently, free-field conditions seldom exist.

22.9.3.2 Sound Absorption Coefficients Machines that contain cams, gears, reciprocating components, and metal stops are often located in large, acoustically reverberant areas that reflect and build up noise levels in the room. Frequently the noise levels in adjoining areas can be reduced significantly by using sound-absorbing materials on walls and ceilings. However, the amount of reduction close to the machines may be slight because most of the noise exposure energy is coming directly from the machines and not from the reflecting surfaces. The type, amount, configuration, and placement of absorption materials depend on the specific application; however, the choice of absorbing materials can be guided by the absorption coefficients listed in Table 22.13. Absorption coefficients of 1.0 will absorb all sound randomly impinging on the surface; coefficients close to 0 mean the material absorbs very little acoustic energy.

22.9.3.3 Noise Barriers and Enclosures The amount of noise reduction that can be attained with barriers depends on the characteristics of the noise source, the configuration and materials used for the barrier, and the acoustic environment on each side of the barrier. It is necessary to consider all these complex factors to determine the overall benefit of a barrier.

The noise reduction of barriers or enclosures vary significantly. Single-wall barriers with no openings may provide as little as 2 to 5 dB reduction in the low frequencies and a 10 to 15 dB reduction in the high frequencies. Higher reduction values are possible with heavier barriers with greater surface areas. Higher values may also be expected when the source and/or the persons exposed are close to the barrier. The effects of two- and three-sided barriers are difficult to predict on a general basis. However, well-

TABLE 22.13 Sound Absorption Coefficients of Surface Materials[89]

Material	Frequency (Hz)					
	125	250	500	1000	2000	4000
Brick: Glazed	0.01	0.01	0.01	0.01	0.02	0.02
Unglazed	0.03	0.03	0.03	0.04	0.05	0.06
Unglazed, painted	0.01	0.01	0.02	0.02	0.02	0.03
Carpet: Heavy (on concrete)	0.02	0.06	0.14	0.37	0.60	0.65
On 40 oz. hairfelt or foam rubber (carpet has coarse backing)	0.08	0.24	0.57	0.69	0.71	0.73
With impermeable latex backing on 40 oz. hairfelt or foam rubber	0.08	0.27	0.39	0.34	0.48	0.63
Concrete block: Coarse	0.36	0.44	0.31	0.29	0.39	0.25
Painted	0.10	0.05	0.06	0.07	0.09	0.08
Poured	0.01	0.01	0.02	0.02	0.02	0.03
Fabrics: Light velour: 10 oz/yard^2, hung straight, in contact with wall	0.03	0.04	0.11	0.17	0.24	0.35
Medium velour: 14 oz/yard^2, draped to half-area	0.07	0.31	0.49	0.75	0.70	0.60
Heavy velour: 18 oz/yard^2, draped to half-area	0.14	0.35	0.55	0.72	0.70	0.65
Floors: Concrete or terrazzo	0.01	0.01	0.015	0.02	0.02	0.02
Linoleum, asphalt, rubber, or cork tile on concrete	0.02	0.03	0.03	0.03	0.03	0.02
Wood	0.15	0.11	0.10	0.07	0.06	0.07
Wood parquet in asphalt on concrete	0.04	0.04	0.07	0.06	0.06	0.07
Glass: Ordinary window glass	0.35	0.25	0.18	0.12	0.07	0.04
Large panes of heavy plate glass	0.18	0.06	0.04	0.03	0.02	0.02
Glass fiber: Mounted with impervious backing, 3 lb./ft^3, 1 in. thick	0.14	0.55	0.67	0.97	0.90	0.85
Mounted with impervious backing, 3 lb. /ft^2, 2 in. thick	0.39	0.78	0.94	0.96	0.85	0.84
Mounted with impervious backing, 3 lb./ft^3, 3 in. thick	0.43	0.91	0.99	0.98	0.95	0.93
Gypsum board: 1/2 in. thick nailed to 2″ × 4″s, 16″ on center	0.29	0.1	0.05	0.04	0.07	0.09
Marble	0.01	0.01	0.01	0.01	0.02	0.02
Plaster: Gypsum or lime, smooth finish on tile or brick	0.013	0.015	0.02	0.03	0.04	0.05
Gypsum or lime, rough finish on lath	0.14	0.10	0.06	0.05	0.04	0.05
With smooth finish	0.14	0.10	0.06	0.04	0.04	0.03
Plywood paneling, 3/8 in. thick	0.28	0.22	0.17	0.09	0.10	0.11
Sand: Dry, 4 in. thick	0.15	0.35	0.40	0.50	0.55	0.80
Dry, 12 in. thick	0.20	0.30	0.40	0.50	0.60	0.75
14 lb. H_2O/ft^3, 4 in. thick	0.05	0.05	0.05	0.05	0.05	0.15
Water	0.01	0.01	0.01	0.01	0.02	0.02
Air, per 10 m^3 at 50% RH (for larger spaces, air attenuates sound—particularly the higher frequencies)				0.32	0.81	2.56

designed partial enclosures may provide noise reduction values of more than twice as much as single wall barriers. Complete enclosures from simple practical designs may provide noise reduction values in excess of 10 to 15 dB in the low frequencies and in excess of 30 dB in the high frequencies.

The best noise isolation is achieved with a complete enclosure constructed around the noise source. The enclosure should be lined with absorbent material or the reverberation inside the enclosure will increase the sound pressure level negating some of the transmission loss achieved by the enclosure walls. In addition, care should be taken to limit the openings in the enclosure. Generally openings are necessary for ventilation for cooling, piping, conduits, or material handling. However, as shown below, even small openings will reduce the transmission loss. For instance, an enclosure capable of reducing noise by 40 decibels will provide less than 20 decibels of attenuation if 1% of the total surface area is left open.

Partial enclosures (Figs. 22.42 and 22.43) must be lined with absorptive material to obtain maximum effectiveness. If the workers are not in the direct line of the opening, a shadow effect of 3 to 15 dB for high-frequency sounds may be achievable.[83] The shadow effect is limited to high-frequency sounds where the dimension of the enclosure is several times the wavelength of the noise.

22.9.3.3.1 Calculating Barrier Noise Reduction

If a sound source is in a room with a large amount of absorption (i.e., few reflections), blocking the direct path with a partial barrier may provide adequate noise control. (This technique is more commonly used outdoors, since even a modest amount of reverberation will destroy the effectiveness of a shield.) Based on theoretical considerations, the decrease in sound pressure level L_p in the shadow of a barrier is as given in Fig. 22.44, where N is a geometric factor; $A + B$, the shortest distance over the barrier between

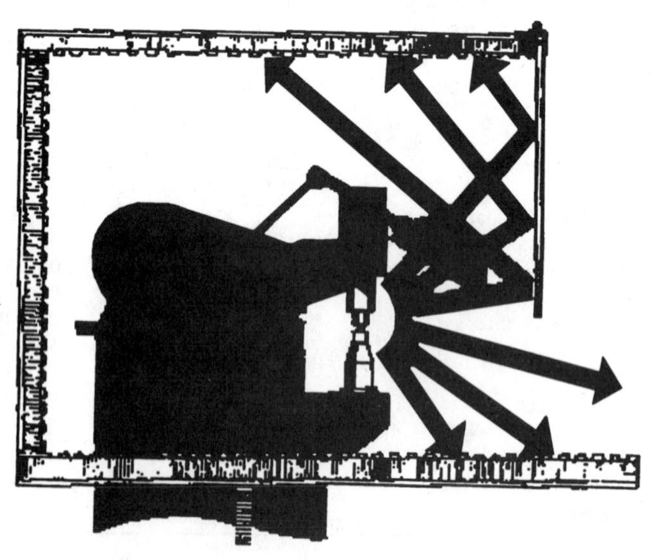

Fig. 22.42 Reducing high-frequency noise with a barrier and partial enclosure lined with absorbent material.

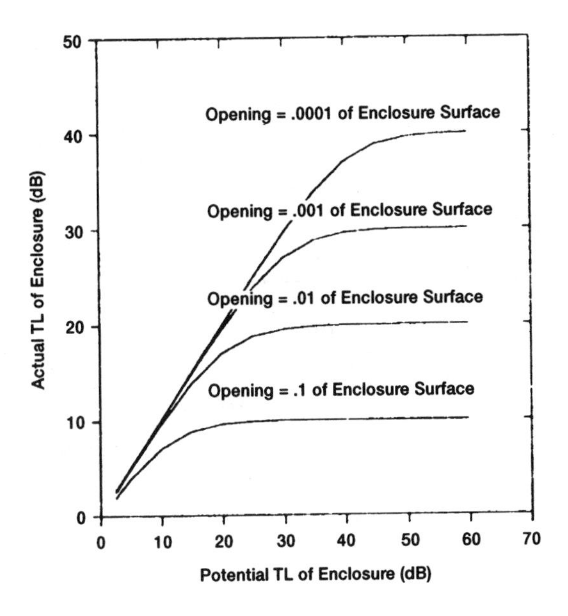

Fig. 22.43 Reduction in transmission loss if an enclosure has openings.

source and receiver; d, the straight line distance between source and receiver; λ, the wavelength of sound; and L_p, the decrease in sound pressure level.

Example 16: What is the expected decrease in sound pressure levels at typical speech frequencies (500, 1000, 2000 Hz) by increasing the height of a barrier between 2 reservation agents from 1.2 to 1.5 m.

The distance between source and receiver of seated telephone agents is taken to be 1.05 m. Assume agents are seated 1.2 m apart. Therefore, d = 1.2 m, and for a 4 ft

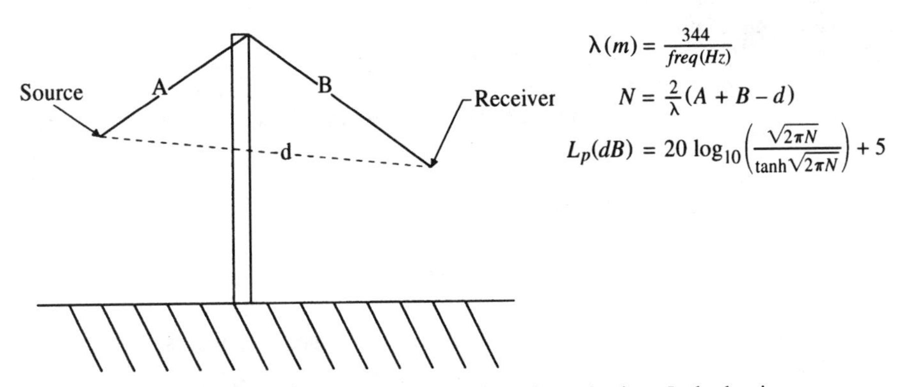

Fig. 22.44 Geometrical considerations for noise reductions L_p by barriers.

(1.22 m) barrier:

$$A = B = \sqrt{(0.6)^2 + (1.22 - 1.05)^2} = 0.624 \text{ m}$$

similarly for a 5 ft (1.52 m) barrier, $A = B = 0.762$ m.

Using the above formulas or graphs:

Frequency (Hz)	S	N (@4′)	N (@5′)	L_p (4′)	L_p (5′)	Difference (dB)
500	0.7	0.1	1.0	7.0	12.9	5.9
1000	0.3	0.3	1.9	8.6	15.8	7.2
2000	0.2	0.5	3.8	10.7	18.8	8.1

Under ideal conditions, a decrease in sound from the adjacent reservation agent could be reduced by 6 to 8 dB. In practice, there will be reflections from nearby surfaces (in particular from the ceiling and floor), and end effects from the barrier will be important.

Recommendations for use of indoor barriers:[84]

1. The barrier should be as close to the source or receiver as possible. However, the source should not touch the barrier to prevent transferring vibrations.
2. Barrier should extend beyond the line of sight of the source plus $\frac{1}{4}$ wavelength of the lowest frequency of interest for attenuating the noise.
3. Select a solid material without any holes or openings. Barriers made of material with more than 4 pounds per square foot are usually adequate. The TL of the barrier should be at least 10 dB greater than the required attenuation.

22.10 COMMUNITY NOISE

Most community noise exposures, by themselves, do not cause noise-induced hearing impairment. Noise problems from communities commonly entail complaints of communication and stress-related problems such as annoyance.

The effects of noise on communication can be measured with a reasonable degree of accuracy[85,86]; however, other psychological and physiological responses to noise are extremely complicated and difficult to measure in a meaningful way. Attempts to correlate noise exposure levels and annoyance, or the general well-being of humans, are complicated by many factors such as:

- The attitude of the listener toward the noise source
- The history of individual noise exposures
- The activities of the listeners and stresses on the listeners during the noise exposures
- The hearing sensitivities of the listeners and other differences in individual responses

- Whether the noise source contains tonal or impulsive components
- Ambient noise in the environment
- Season of the year

Corrections may be made to measured sound levels to attempt to account for the some of these factors (Table 22.14).

22.10.1 Community Noise Regulations

For the most part, old rules and regulations were based on vaguely defined nuisance factors. Nuisance-type regulations often take the form, "There shall be no unnecessary

TABLE 22.14 Factors Affecting Community Response to Noise and Suggested Corrections[87]

Type of correction	Description	Correction added to measured L_{dn} (dB)
Seasonal	Summer (or year-round operations)	0
	Winter only (or windows always closed)	−5
Outdoor residual noise level	Quiet suburban or rural community (away from large cities, industrial activities, and trucking)	+10
	Normal suburban community (away from industrial activity)	+5
	Urban residential community (not near heavy traffic or industry)	0
	Noisy urban residential community (near relatively busy roads or industrial areas)	−5
	Very noisy urban residential community	−10
Previous exposure or community attitudes	No prior experience with intruding noise	+5
	Community has had some exposure to intruding noise; little effort is being made to control noise. This correction may also be applied to a community that has not been previously exposed to noise, but the people are aware a bona fide effort is being made to control it.	0
	Community has had considerable exposure to intruding noise; noise maker's relations with the community is good	−5
	Community is aware that operations causing noise is necessary but will not continue indefinitely. This correction may be applied on a limited basis and under emergency conditions.	−10
Pure tone or impulse	No pure tone or impulsive character	0
	Pure tone or impulsive character present	+5

nor disturbing noise. . . ." An obvious weakness of this form of regulation is the failure to specify the conditions of how, when, and to whom noise is unnecessary or disturbing. Innumerable arguments may result in the interpretation of these laws when an attempt is made to enforce them. On the other hand, this kind of law may be useful in some cases where it is not possible to make reasonable decisions based on exposure (sound pressure) levels.

Most recent zoning codes have specified maximum noise limits for various zones. These so-called performance zoning codes are more objective and easier to enforce than the nuisance laws. Unfortunately, many problems must be considered individually on a nuisance basis because of the many possible noise exposure situations. For example, a noise limit of L_A = 55 dB may be generally acceptable for a given neighborhood. However, it is reasonable to assume that lawn mowers should be permitted, and they may produce a noise level of L_A = 88 dB at the property line. The lawn mower noise should be acceptable at certain hours when other persons' activities are not unreasonably affected. On the other hand, this same noise may not be acceptable if neighbors have guests on the adjoining lawn, if they are trying to sleep, and so on.

Obviously performance laws may be used to establish guidelines for reasonable noise exposure levels, but nuisance laws may also be needed in some instances. No law can replace the fact that all individuals must show consideration for their neighbors.

Most performance-type noise ordinances establish limiting levels that are well below the level where physiological damage may occur. Usually the limits are based on the number and level of adverse responses in the community rather than trying to measure individual annoyance reactions to noise. This decision is made in most cases because the group statistics involved in describing "community responses" to noise avoid many of the variables that are extremely difficult to account for individual annoyance reactions.

Most performance-type noise codes specify limits of sound pressure level that are based on:

1. A selected frequency weighting or noise analysis procedure,
2. The pattern of exposure times for various noise levels,
3. The ambient noise levels that would be expected in that particular kind of community without the offending noise source(s), and
4. A land-use zoning of the area.

Regulations based on specific exposures—such as those from aircraft or from ground transportation—often work well at other locations having the same kinds of noise (see below). Caution must be used, however, in applying these results to other kinds of noise exposure.

The Noise Guidebook document[88] "provides guidance on noise policy, attenuation of outdoor noise, and noise assessment. This document does not constitute a standard, specification, or regulation, but it does present available technical knowledge that may be used in a uniform and practical way by communities of various sizes to tailor ordinances to their specific conditions and goals.

The most common time-averaged descriptor used for community noise problems is the day–night-time average sound level, abbreviated DNL or L_{dn} (Eq. 22.28). This is the 24-h equivalent sound level obtained after the addition of a 10 dB penalty for sound levels that occur at night between 10 PM and 7 AM. The penalty is based on the fact that more people are disturbed by noise at night than any other time. Background noise is usually much lower at night, and sleep disturbance is one of the prime motivators of community reactions to noise.

$$L_{dn} = 10 \log \frac{1}{24} \left(\frac{\int_{07:00}^{22:00} p_A^2(t)\, dt}{p_{ref}^2} + \frac{\int_{22:00}^{07:00} 10 p_A^2(t)\, dt}{p_{ref}^2} \right) \quad (22.28)$$

Sound level meters equipped with clocks and appropriate computing power can directly measure L_{dn}. It is also common practice to measure L_{eq}, add 10 dB for nighttime readings, then combine the sound levels using one of the techniques shown in Section 22.2.4.2.

22.10.1.1 Transportation Noise A number of social surveys of community annoyance have been analyzed used as criteria for evaluating the environmental impact of aircraft noise and recommended for predicting the effects of general transportation noise.[89] An updated analysis[90] recommended a minor change to the calculation (Fig. 22.45).

Slight differences were noted in community response to various kinds of transportation noise (rail, road traffic, and aircraft). Therefore, caution should be used equating community noise created by industrial sources to transportation noise, but the above equation and modifiers such as those suggested in Table 22.14 can provide an initial estimation for planning purposes.

22.10.1.2 A Guide for Community Noise Criteria Acceptable noise levels vary from one community to another depending on the history of noise exposures and other

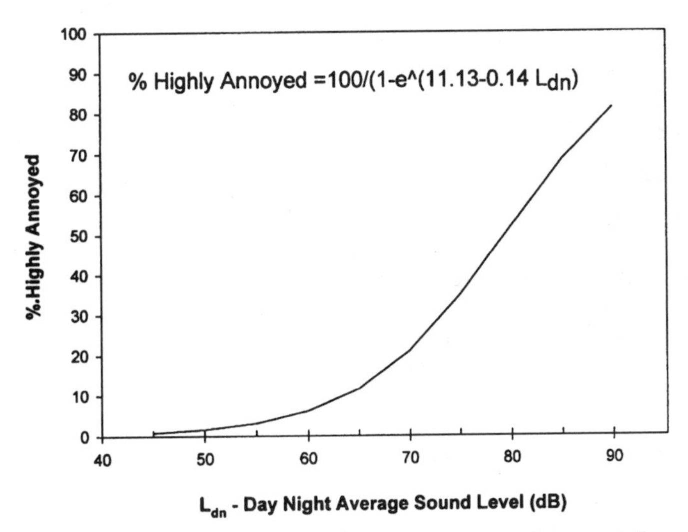

Fig. 22.45 Community annoyance in response to general transportation noise.

TABLE 22.15 Community Noise Exposure Guidelines

Authority	Specified sound levels	Criterion
EPA levels document[92]	$L_{dn} \leq 55$ dB (outdoors)	Protection of public health and welfare with adequate margin of safety
	$L_{dn} \leq 45$ dB (indoors)	
WHO document (1995)[93]	$L_{eq} \leq 50/55$ (outdoors: day)	Recommended guideline values (Task Force Consensus)
	$L_{eq} \leq 45$ dB (outdoors: night)	
	$L_{eq} \leq 30$ dB (bedroom)	
	$L_{max} \leq 45$ dB (bedroom)	
U.S. Interagency Committee (FICON)[94]	$L_{dn} \leq 65$ dB	Considered generally compatible with residential development
	$65 \leq L_{dn} \leq 70$ dB (indoors)	Residential use discouraged
Various European road traffic regulations[95]	$L_{dn} \leq 45$ dB (indoors)	Remedial measures required

variables. In most cases, noise codes are tailored to the character and requirements of the particular community.

Annoyance, sleep interference, and speech interference are the effects used as a basis for establishing positions of acceptable community noise. Community noise exposures guidelines have been summarized[91] as shown in Table 22.15.

ACKNOWLEDGMENT

Both authors need to acknowledge Paul L. Michael for the original text[96] that formed the basis for this chapter, and for his education of the industrial hygiene community in principles of hearing conservation (and, in the case of one of the authors, education in everything else). The authors are also indebted to David Byrne for his careful review of the chapter.

Terminology

Absorption coefficient: The sound absorption coefficient of a given surface is the ratio of the sound energy absorbed by the surface to the sound energy incident upon the surface.

Acoustic intensity (I): The average rate at which sound energy is transmitted through a unit area normal to the direction of propagation. The units used for sound intensity are joules per square meter (J/m^2 sec). Sound intensity is also expressed in terms of a sound intensity level (L_I) in decibels referred to 10^{-12} W/m^2.

Acoustic power: See Sound power.

Acoustic pressure: See Sound pressure.

Ambient noise: The overall composite of sounds in an environment.

Amplitude: The quantity of sound produced at a given location away from the source, or the overall ability of the source to emit sound. The amount of sound at a location away from the source is generally described by the sound pressure or sound intensity, whereas the ability of the source to produce noise is described by the sound power of the source.

Anechoic room: A room that has essentially no boundaries to reflect sound energy generated therein. Thus any sound generated within is referred to as being in a free field (see definition).

Atmospheric pressure (*P*): is measured in pascals (Pa) $P = 1.013 \times 10^5$ Pa = 1.013×10^5 N/m^2.

Audiogram: A recording of hearing levels referenced to a statistically normal sound pressure as a function of frequency.

Audiometer: An instrument for measuring hearing thresholds.

Continuous spectrum: A spectrum of sound that has components distributed over a known frequency range.

Cycle: A cycle of a periodic function, such as a single-frequency sound, is the complete sequence of values that occur in a period.

Cycles per second: See Frequency.

Criterion duration (TC): Duration used a basis for noise exposure measurements—typically 8 h.

Criterion level (LC): The sound pressure level that would give 100% of the allowable noise exposure is continued for the criterion duration.

Decibel (dB): A convenient means for describing the logarithmic level of sound intensity, sound power, or sound pressure above arbitrarily chosen reference values (see text).

Density of air: At 1 atmosphere and 22°C, $\rho = 1.18$ kg/m^3.[97]

Diffuse sound field: A sound field that has sound pressure levels that are essentially the same throughout, with the directional incidence of energy flux randomly distributed.

Directivity factor (*Q*): Measure of how much the sound is concentrated in a given direction.

Effective sound pressure: The sound pressure at a given location, derived by calculating the root mean square (rms) value of the instantaneous sound pressures measured over a period of time at that location.

Exchange rate (*Q*): The number of decibels change in sound level that gives an equivalent exposure when the duration is halved or doubled. Frequently called the *doubling rate*.

Free field: A field that exists in a homogeneous isotropic medium, free from boundaries. In a free field, sound radiated from a source can be measured accurately without influence from the test space. True free-field conditions are rarely found, except in anechoic test chambers. However, approximate free-field conditions may be found in any homogeneous space where the distance from the reflecting surfaces to the measuring location is much greater than the wavelengths of the sound being measured.

Frequency: The rate at which complete cycles of high- and low-pressure regions are produced by the sound source. The unit of frequency is hertz (Hz). 1 Hz = 1 cycle per second.

Hertz (Hz): Unit for measuring frequency 1 Hz = 1 cycle per second.

Infrasonic frequency: Sounds having major frequency components well below audible range (below 20 Hz).

Level: The level of any quantity, when described in decibels, is the logarithm of the ratio of that quantity to a reference value in the same units as the specific quantity.

Loudness: An observer's impression of the sound's amplitude, which includes the response characteristics of the ear.

Noise: The terms *noise* and *sound* are often used interchangeably, but generally, sound describes useful communication or pleasant sounds, such as music, whereas noise is dissonance or unwanted sound.

Noise dose: Noise exposure expressed as a percentage.

Noise reduction coefficient (NRC): The arithmetic average of the sound absorption coefficients of a material at 250, 500, 1000, and 2000 Hz.

Octave band: A frequency bandwidth that has an upper band-edge frequency equal to twice its lower band-edge frequency.

Peak sound pressure level: The maximum instantaneous level that occurs over any specified period of time.

Period (T): The time (in seconds) required for one cycle of pressure change to take place; hence it is the reciprocal of the frequency.

Pitch: A measure of auditory sensation that depends primarily on frequency but also on the pressure and wave form of the sound stimulus.

Pure tone: A sound wave with only one sinusoidal change of pressure with time.

Random incidence sound field: See Diffuse sound field.

Random noise: A noise made up of many frequency components whose instantaneous amplitudes occur randomly as a function of time.

Resonance: A system is in resonance when any change in the frequency of forced oscillation causes a decrease in the response of the system.

Reverberation: Reverberation occurs when sound persists after direct reception of the sound has stopped. The reverberation of a space is specified by the reverberation time, which is the time required, after the source has stopped radiating sound, for the rms pressure to decrease 60 dB from its steady-state level.

Root mean square sound pressure: The root-mean-square (rms) value of a changing quantity, such as sound pressure, is the square root of the mean of the squares of the instantaneous values of the quantity.

Sound intensity (I): The average rate at which sound energy is transmitted through a unit area normal to the direction of propagation. The units used for sound intensity are joules per square meter sec (J/m^2 sec). Sound intensity is also expressed in terms of a sound intensity level (L_I) in decibels referred to 10^{-12} W/m^2.

Sound power (P): The total sound energy radiated by the source per unit time. Sound power is normally expressed in terms of a sound power level (L_P) decibels references to 10^{-12} W.

Sound pressure (p): The rms value of the pressure above and below atmospheric pressure of steady-state noise. Short-term or impulse-type noises are described by peak pressure values. The units used to describe sound pressures are pascals (Pa), newtons per square meter (N/m^2), dynes per square centimeter (dyn/cm^2), or microbars (μbar).

Sound pressure level (L_p): Sound pressure is also described in terms of sound pressure level in decibels referenced to 20 μPa.

Standard threshold shift (STS): Defined in 29 CFR 1910.95 as "a change in the hearing threshold relative to the baseline audiogram of an average of 10 dB or more at 2000, 3000, and 4000 Hz in either ear.

Standing waves: Periodic waves that have a fixed distribution in the propagation medium.

Threshold level: The lowest sound level used for measuring noise exposure.

Transmission loss (TL): Ten times the logarithm (to the base 10) of the ratio of the incident acoustic energy to the acoustic energy transmitted through a sound barrier.

Ultrasonic: The frequency of ultrasonic sound is higher than that of audible sound—usually considered as frequencies above 20 kHz.

Velocity: The speed at which the regions of sound-producing pressure changes move away from the sound source is called the velocity of propagation. Sound velocity (c) varies directly with the square root of the density and inversely with the compressibility of the transmitting medium; however, the velocity of sound is usually considered constant under normal conditions. For example, the velocity of sound is approximately 344 m/sec (1130 ft/sec) in air, 1433 m/sec (4700 ft/sec) in water, 3962 m/sec (13,000 ft/sec) in wood, and 5029 m/sec (16,500 ft/sec) in steel.

Wavelength (λ): The distance required to complete one pressure cycle. The wavelength, a very useful tool in noise control, is calculated from known values of frequency (f) and velocity (c): $\lambda = c/f$.

White noise: White noise has an essentially random spectrum with equal energy per unit frequency bandwidth over a specified frequency band.

REFERENCES

[1]See, e.g., Kryter, K. D. *The Handbook of Hearing and the Effects of Noise*, Academic Press, San Diego, CA, 1994.

[2]Newby, H. A. (1964) *Audiology*, 2nd Ed., Appleton-Century-Crofts, New York, N.Y.

[3]Suter, A. H., *Hearing Conservation*, in Berger, E. H., W. E. Ward, J. C. Morrill, and L. H. Royster, *Noise and Hearing Conservation Manual*, American Industrial Hygiene Association, Fairfax, Virginia, 1986.

[4]Results of Freedom of Information request for inspection reports citing 29 CFR 1910.0095 and 1926.0052.

[5]ANSI S1.1, "Acoustical Terminology," American National Standards Institute, New York.

[6]ANSI S1.8, "Reference Quantities for Acoustical Levels (ASA 84)," American National Standards Institute, New York.

[7]ASME Y10.11-1984, "Letter Symbols and Abbreviations for Quantities used in Acoustics" (Revision of ASA Y10.11-1953 (R1959).

[8]J. R. Anticaglia, "Physiology of Hearing," in *The Industrial Environment—Its Evaluation and Control*, U.S. DHEW, PHS, CDC, National Institute for Occupational Safety and Health, 1973. A. Feldmen and C. Grimes, Hearing Conservation in Industry, Williams and Wilkins, Baltimore, MD, 1985, pp. 77–88. A. Glorig, *Noise and Your Ear*, Grune & Stratton, New York, 1958. *Industrial Noise Manual*, 2nd, 3rd, and 4th eds., American Industrial Hygiene Association, Akron, OH, 1966, 1975, 1987. American National Standards Institute, American National Standard Specifications for Audiometers, ANSI S3.6-1989, New York, 1989.

[8a]E. H. Berger, W. D. Ward, J. C. Morrill, and L. H. Royster. *Noise and Hearing Conservation Manual*, 4th Edn., AIHA, 1986, Chapter 5.

[9]Killion, Mead C., "Revised Estimate of Minimum Audible Pressure: Where Is the "Missing 6 dB?," *J. Acoust. Soc. Am.* **63**(5), May 1978, p. 1501–8.

[10]Pickles, James O., *An Introduction to the Physiology of Hearing*, 2nd Edition, Academic Press, 1988, p. 5.

[11]ANSI S3.28-1986, "Methods for the Evaluation of the Potential Effect on Human Hearing of Sounds with Peak A-Weighted Sound Pressure Levels above 120 decibels and Peak C-Weighted Sound Pressure Levels below 140 Decibels (Draft ASA 66)," American National Standards Institute, New York.

[12]Pickles, p. 297.

[13]ANSI S3.44, "Determination of Occupational Noise Exposure and Estimation of Noise-Induced Hearing Impairment," American National Standards Institute, New York.

[14]NIOSH, Criteria for a Recommended Standard: Occupational Exposure to Noise, NIOSH 73-11001.

[15]Safety and Health Standards for Federal Supply Contracts (Walsh–Healy Public Contracts Act), U.S. Department of Labor, Fed. Regist., 34, 7948 (1969).

[16]A. Lawther and D. W. Robinson, "Further Investigation of Tests for Susceptibility to Noise-Induced Hearing Loss," ISVR Technical Report 149, 1987.

[17]Miller, J. D. (1974). Effects of Noise on People. *Journ. Acoust. Society of Amer.* **56,** 3, p. 729.

[18]Occupational Safety and Health Standards (Williams–Steiger Occupational Safety and Health Act of 1970), U.S. Department of Labor, Fed. Regist., 36, 10518 (1971).

[19] ANSI S3.21 "Manual Pure-Tone Threshold Audiometry, Thresholds for (ASA 19) (R 1992)," American National Standards Institute, New York.

[20]Council for Accreditation in Occupational Hearing Conservation Manual, Fischler's Printing, Cherry Hill, NJ, 1978.

[21]Occupational Noise Exposure: Hearing Conservation Ammendment, Occupational Safety and Health Administration, 29 CFR Part 1910, Fed. Regist., 48(42) (March 3, 1983).

[22]ANSI S3.6, "Specification for Audiometers," American National Standards Institute, New York.

[23]ANSI S3.1, "Maximum Permissible Ambient Noise Levels for Audiometric Test Rooms (ASA 99)," American National Standards Institute, New York.

[24]Occupational Safety and Health Standards (Williams–Steiger Occupational Safety and Health Act of 1970), U.S. Department of Labor, Fed. Regist., 36, 10518 (1971).

[25]Occupational Noise Exposure; Hearing Conservation Amendment, Occupational Safety and Health Administration, 29 CFR Part 1910, Fed. Regist., 48(42) (March 3, 1983).

[26]Occupational Noise Exposure; Hearing Conservation Amendment, Occupational Safety and Health Administration, 29 CFR Part 1910, Fed. Regist., 48(42) (March 3, 1983).

[27]Council for Accreditation in Occupational Hearing Conservation Manual, Fischler's Printing, Cherry Hill, NJ, 1978.

[28]ANSI S1.4, "Specification for Sound Level Meters," American National Standards Institute, New York.

[29]P. L. Michael, "Noise Measurements and Personal Protection," Industrial Hygiene Foundation Transactions of the 20th Meeting held in Pittsburgh, PA, 1955, pp. 1–10.

[30]ANSI S1.42, "Design Response of Weighting Networks (ASA 64)," American National Standards Institute, New York.

[31]Bruel, P. V. Do We Measure Damaging Noise Correctly?, Noise Control Engineering, March–April 1977.

[32]ANSI S1.6, "Preferred Frequencies, Frequency Levels, and Band Numbers for Acoustical Measurements (ASA 53)," American National Standards Institute, New York.

[33]ANSI S1.11, "Specifications for Octave-Band and Fractional-Octave-Band Digital Filters (ASA 65)," American National Standards Institute, New York.

[34]ISO 2204, "Acoustics—Guide to International Standards on the Measurement of Airborne Acoustical Noise and Evaluation of its Effects on Human Beings," International Organization for Standardization, Geneve 20, Switzerland.

[35]ANSI S12.19, "Measurement of Occupational Noise Exposure," American National Standards Institute, New York.

[36]ISO 9612-1991, "Acoustics—Guidelines for the Measurement and Assessment of Exposure to Noise in the Working Environment," International Organization for Standardization (ISO), Geneve 20, Switzerland.

[37]Joint ACGIH-AIHA Task Group on Occupational Exposure Databases, "Data Elements for Occupational Exposure Databases: Guidelines and Recommendations for Airborne Hazards and Noise," *Appl. Occup. Environ. Hyg.* **11**(11), November 1996, pp. 1294–311.

[38]ISO 1996, "Acoustics—Description and Measurement of Environmental Noise: Part 1: Basic Quantities and Procedures," "Acoustics—Description and Measurement of Environmental Noise: Part 2: Acquisition of Data Pertinent to Land Use," "Acoustics—Description and Measurement of Environmental Noise: Part 3: Application to Noise Limits," International Organization for Standardization, Geneve 20, Switzerland.

[39]ANSI 12.19, "Measurement of Occupational Noise Exposure," American National Standards Institute, New York.

[40]ISO 1999, "Acoustics—Determination of Occupational Noise Exposure and Estimation of Noise-Induced Hearing Impairment," International Organization for Standardization, Geneve 20, Switzerland.

[41]ACGIH, "1996 TLVs and BEIs Threshold Limit Values for Chemical Substances and Physical Agents Biological Exposure Indices," Cincinnati, OH, 1996.

[42]ACGIH, "1996 TLVs and BEIs Threshold Limit Values for Chemical Substances and Physical Agents Biological Exposure Indices," ACGIH, Cincinnati, OH, 1996, pp. 110–12.

[43]ANSI S12.7, "Method for Measurement of Impulse Noise," American National Standards Institute, New York.

[44]Lataye, R., and P. Campo, "Applicability of the L_{eq} as a damage-risk criterion: An animal experiment," *J. Acoust. Soc. Am.* **99**(3), March 1996, pp. 1621–32.

[45]Kryter, K. D., *The Handbook of Hearing and the Effects of Noise*, Academic Press, San Diego, CA, 1994, pp. 275–77.

[46]Cohen, A., *Nat. Saf. News* **108**(2), 93–99 (1973). Cohen, A., *Nat. Saf. News* **108**(3), 68–76 (1973). Cohen, A. "Effects of Noise on Performance," *Proceedings of the International Congress of Occupational Health*, Vienna, Austria, 1966, pp. 157–60. Cohen, A. "Noise and Psychologic State," *Proceedings of National Conference on Noise as a Public Health Hazard*, American Speech and Hearing Association, 1969, pp. 89–98. Cohen, A. *Trans. N.Y. Acad.*

Sci. **30,** 910–18 (1968). Grether, W. F. "Noise and Human Performance," Report AMRL-TR-70-29, AD 729 213, Aerospace Medical Research Laboratory, Wright–Patterson Air Force Base, OH, 1971. Michael, P. L., and G. R. Bienvenue, "Industrial Noise and Man," in *Physiology and Productivity at Work—The Physical Environment*, John Wiley, Sussex, England, 1983.

[47]Berger, E. H., "EARLog 7 Motivating Employees to Wear Hearing Protective Devices," Cabot Safety Corp., Indianapolis, IN, 1981.

[48]Berger, E. H., EARLog 13 "Attenuation of Earplugs Worn in Combination with Earmuffs," Cabot Safety Corporation, Indianapolis, IN, 1984.

[49]Nixon, C. W., and W. C. Knoblach, "Hearing Protection of Ear Muffs Worn Over Eyeglasses," AMRL-TR-74-61, Aerospace Medical Research Laboratory, Wright–Patterson AFB, OH, June 1974.

[50]Nixon, C. W., and H. E. von Gierke, "Experiments in Bone Conduction Threshold in a Free Sound Field," *J. Acoust. Soc. Am.* **31,** 1121–25 (1969).

[51]Franks, J., C. Themann and C. Sherris, "The NIOSH Compendium of Hearing Protection Devices," NIOSH Publication 95-105, U.S. Dept. of Health and Human Services, Public Health Service, Centers for Disease Control and Prevention, National Institute of Occupational Safety and Health.

[52]Environmental Protection Agency, Noise Labeling Requirements for Hearing Protectors, Fed. Regist., 40 CFR Part 211, Subpart B, 56139-56147, No. 190, 1979.

[53]American Iron and Steel Institute, "Steel Industry Hearing Protector Study," Vol. 1, *Applications Handbook*, Vol. 2, *References*, AISI, Washington, DC, 1983.

[54]NIOSH, "A Report on the Performance of Personal Noise Dosimeters," 1982.

[55]Lempert, B., *Sound Vib.* **18,** 26–39 (1984).

[56]"Hearing Protectors—Recommendations for selection, use, care and maintenance—Guidance Document," European Committee for Standardization, CENprEN 458 : 1993.

[57]ISO 4869-1, "Acoustics—Hearing protectors—Part 1: Subjective Method for the Measurement of Sound Attenuation," International Organization for Standardization, Geneve 20, Switzerland.

[58]ANSI S3.14, "Rating Noise with Respect to Speech Interference (ASA 21)," American National Standards Institute, New York.

[59]Michael, P. L., *Arch. Environ. Health* **10,** 612–18 (1965).

[60]Bienvenue, G. R., and P. L. Michael, "Digital Processing Techniques in Speech Discrimination Testing," in *Rehabilitation Strategies for Sensorineural Hearing Loss*, Grune & Stratton, New York, 1979.

[61]Michael, P. L., et al., "Intelligibility Test on a Family of Electroacoustic Devices (FEADS)—Final Report," Contract No N00953-70-2620, U.S. Navy, 1970.

[62]HPD Evaluation Meter, Liberty Technology, Hopkinton, MA.

[63]FitCheck Model 700, Michael & Associates, State College, PA.

[64]Witt, M., Ed., "Noise Control, A Guide for Workers and Employers," U.S. Department of Labor Occupational Safety and Health Administration Office of Information OSHA 3048, (1980), p. 106.

[65]Witt, M., Ed., "Noise Control, A Guide for Workers and Employers," U.S. Department of Labor Occupational Safety and Health Administration Office of Information OSHA 3048, (1980), p. 37.

[66]Witt, M., Ed., "Noise Control, A Guide for Workers and Employers," U.S. Department of Labor Occupational Safety and Health Administration Office of Information OSHA 3048, (1980), p. 25.

[67]Witt, M., Ed., "Noise Control, A Guide for Workers and Employers," U.S. Department of Labor

Occupational Safety and Health Administration Office of Information OSHA 3048, (1980), p. 31.

[68]Witt, M., Ed., "Noise Control, A Guide for Workers and Employers," U.S. Department of Labor Occupational Safety and Health Administration Office of Information OSHA 3048, (1980), p. 33.

[69]Witt, M., Ed., "Noise Control, A Guide for Workers and Employers," U.S. Department of Labor Occupational Safety and Health Administration Office of Information OSHA 3048, (1980), p. 47.

[70]Witt, M., Ed., "Noise Control, A Guide for Workers and Employers," U.S. Department of Labor Occupational Safety and Health Administration Office of Information OSHA 3048, (1980), p. 59.

[71]Witt, M., Ed., "Noise Control, A Guide for Workers and Employers," U.S. Department of Labor Occupational Safety and Health Administration Office of Information OSHA 3048, (1980), p. 61.

[72]Witt, M., Ed., "Noise Control, A Guide for Workers and Employers," U.S. Department of Labor Occupational Safety and Health Administration Office of Information OSHA 3048, (1980), p. 49.

[73]Witt, M., Ed., "Noise Control, A Guide for Workers and Employers," U.S. Department of Labor Occupational Safety and Health Administration Office of Information OSHA 3048, (1980), p. 57.

[74]Witt, M., Ed., "Noise Control, A Guide for Workers and Employers," U.S. Department of Labor Occupational Safety and Health Administration Office of Information OSHA 3048, (1980), p. 79.

[75]Witt, M., Ed., "Noise Control, A Guide for Workers and Employers," U.S. Department of Labor Occupational Safety and Health Administration Office of Information OSHA 3048, (1980), p. 29.

[76]Witt, M., Ed., "Noise Control, A Guide for Workers and Employers," U.S. Department of Labor Occupational Safety and Health Administration Office of Information OSHA 3048, (1980), p. 39.

[77]Witt, M., Ed., "Noise Control, A Guide for Workers and Employers," U.S. Department of Labor Occupational Safety and Health Administration Office of Information OSHA 3048, (1980), p. 83.

[78]Witt, M., Ed., "Noise Control, A Guide for Workers and Employers," U.S. Department of Labor Occupational Safety and Health Administration Office of Information OSHA 3048, (1980), p. 87.

[79]Witt, M., Ed., "Noise Control, A Guide for Workers and Employers," U.S. Department of Labor Occupational Safety and Health Administration Office of Information OSHA 3048, (1980), p. 15, 92.

[80]Witt, M., Ed., "Noise Control, A Guide for Workers and Employers," U.S. Department of Labor Occupational Safety and Health Administration Office of Information OSHA 3048, (1980), p. 62.

[81]Hodgson, M., and A. C. C. Warnock, "Noise in Rooms," in Beranek, L. L., and I. L. Ver, *Noise and Vibration Control Engineering Principles and Applications*, John Wiley & Sons, New York, 1992.

[82]Witt, M., Ed., "Noise Control, A Guide for Workers and Employers," U.S. Department of Labor Occupational Safety and Health Administration Office of Information OSHA 3048, (1980), p. 21.

[83]Bruce, R. D., and E. H. Toothman, "Engineering Controls," in *Noise and Hearing Conservation Manual*, AIHA, Farfax, VA, 1986.

[84]Gordon, Colin G., and Robert S. Jones, Control of Machinery Noise, in Harris, Cyril M.,

Handbook of Acoustical Measurements and Noise Control, Third Edition, McGraw–Hill, p. 40.8.

[85]von Gierke, H. E., and K. M. Eldred, "Effects of Noise on People," *Noise/News International*, Vol. 1, No. 2, June 1993, pp. 75–76.

[86]Beranek, Leo L., "Criteria for Face-to-Face Speech Communication," in Beranek, Leo L., and I. L. Vér, *Noise and Vibration Control Engineering: Principles and Practice*, John Wiley, New York, 1992, pp. 618–21.

[87]von Gierke, H. E., and K. M. Eldred, "Effects of Noise on People," *Noise/News International*, Vol. 1, No. 2, June 1993, p. 79.

[88]U.S. Dept. of Housing and Urban Development, *The Noise Guidebook*, U.S. Government Printing Office, HUD-953-CPD, March 1985.

[89]*Federal Agency Review of Selected Airport Noise Analysis Issues* (Federal Interagency Committee on Noise [FICON], Environmental Protection Agency, Washington, DC, 1992).

[90]Finegold, L. S., C. S. Harris, and H. E. von Gierke, "Community Annoyance and Sleep Disturbance: Updated Criteria for Assessing the Impacts of General Transportation Noise on People," *Noise Control Eng. J.* **42**(1), 1994, Jan-Feb. pp. 25–30.

[91]Shaw, Edgar A. G., "Noise Environments Outdoors and the Effects of Community Noise Exposure," *Noise Control Eng. J.* **44**(3), 1996, May–Jun. p. 115.

[92]*Information on Levels of Environmental Noise Requisite to Protect the Public Health and Welfare with an Adequate Margin of Safety*, (Document EPA 550/9-74-004, U.S. Environmental Protection Agency, Washington, D.C. 1974.

[93]Berglund, B., and T. Lindvall, Eds., Community Noise, document prepared for the World Health Organization (Center for Sensory Research, Stockholm, Sweden, 1995).

[94]*Guidelines for Considering Noise in Land Use Planning and Control*, Federal Interagency Committee on Urban Noise (Document 1981-338-006/8071, U.S. Government Printing Office, Washington, D.C., 1980).

[95]Gottlob, D., "Regulations for Community Noise," *Noise/News International* **3**(4), 223–26 (1995).

[96]Michel, Paul L., "Industrial Noise and Conservation of Hearing," in *Patty's Industrial Hygiene and Toxicology*, Fourth Edition, Vol. I, Part A, *General Principles*, Edited by Clayton, G. D., and Clayton, F. E., John Wiley, New York, 1991, pp. 937–1039.

[97]Beranek, Leo L., and I. L. Ver, *Noise and Vibration Control Engineering Principles and Applications*, John Wiley, New York, p. 32.

CHAPTER 23

Ergonomics: Achieving System Balance through Ergonomic Analysis and Control

GRACIELA PEREZ

Department of Work Environment, College of Engineering, University of Massachusetts at Lowell, Lowell, MA 01854

23.1 INTRODUCTION
23.2 JOB ANALYSIS
23.3 WORK FACTORS TO CONSIDER WHEN PERFORMING JOB ANALYSES
- 23.3.1 Forces Required to Perform the Task
- 23.3.2 Contact Stresses
- 23.3.3 Postures Assumed during the Task
- 23.3.4 Frequency of Muscle Contraction
- 23.3.5 Work Duration
- 23.3.6 Vibration
- 23.3.7 Cold Temperatures
- 23.3.8 Poorly Fitted Gloves
- 23.3.9 Obstructions
- 23.3.10 Standing Surfaces
- 23.3.11 Prolonged High Visual Demands
- 23.3.12 Physical Energy Demands

23.4 WHEN TO CONDUCT JOB ANALYSES
- 23.4.1 The Reactive Approach

23.5 JOB ANALYSIS STEPS
- 23.5.1 Who Does the Analysis?
- 23.5.2 Basic Equipment Needed for Conducting Most Job Analyses
- 23.5.3 Guidelines for Videotaping for Ergonomic Analysis

23.6 JOB ANALYSIS METHODS
- 23.6.1 Document the Job
- 23.6.2 Identify and Evaluate Exposure to Work Factors
- 23.6.3 Summarize Findings
- 23.6.4 Compare Findings to Normative Values Where Available

23.7 ASSESSING MANUAL HANDLING TASKS
- 23.7.1 Strength Data

Handbook of Occupational Safety and Health, Second Edition, Edited by Louis J. DiBerardinis, ISBN 0-471-16017-2

23.7.2 Design Criteria: Anthropometry
23.7.3 Physiological Data
23.7.4 Psychophysical Data
23.8 IDENTIFY POSSIBLE SOLUTIONS
23.9 SELECTION AND EVALUATION OF SOLUTIONS
23.9.1 Analysis
23.9.2 Controls
23.9.3 Medical Management Program
23.9.4 Training Programs
23.10 AN EXAMPLE OF JOB ANALYSIS FORMS
23.11 SUMMARY
APPENDIX A: NIOSH LIFT GUIDELINES
APPENDIX B: ASSESSMENT OF AND SOLUTIONS TO WORKSITE RISK FACTORS
APPENDIX C: TABLES OF ERGONOMIC STUDIES
APPENDIX REFERENCES
TEXT REFERENCES
RESOURCES

23.1 INTRODUCTION

Healthy work systems require a balance between the task, the technology, the organization, the environment, and the individual (Fig. 23.1). When any one of these basic connections is not functioning optimally, the work system is impaired (Hosey, 1973). For example, such a system would recognize where humans excel and where machines excel and the work system would then be balanced accordingly (Table 23.1). When humans work in environments that are not compatible with their strengths, the workers may become injured or ill. Thus a healthy work system is designed to optimize the human–machine interface and keep workers safe and healthy.

In essence, this is the science called *ergonomics*. The word ergonomics derives its meaning from the greek roots of *ergon*, meaning work, and *nomos*, meaning law, or the *work law*. In plain English, ergonomics has been defined as applying the laws of work to design the work and work environment to fit the capabilities of the people who perform the work. The benefits of a work system, which is ergonomically designed, include

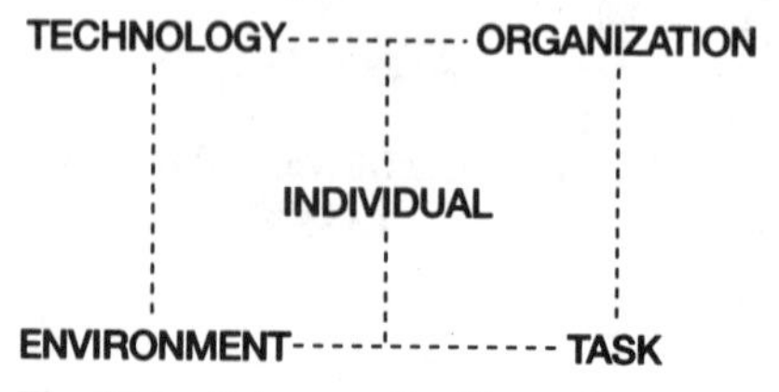

Fig. 23.1 Balance of health work systems.

TABLE 23.1 Basic Strengths of Humans and Machines

Human strengths over machines:

- Sensitive to a wide variety of stimuli
- Ability to react to unexpected, low-probability events
- Ability to exercise judgment where events cannot be completely defined
- Perception of patterns and ability to make generalizations about them

Machine strength over humans:

- Perform routine, repetitive, or very precise operations
- Exert great force, smoothly and with precision
- Operate in environments that are toxic or hazardous to humans or beyond human tolerance
- Can be designed to be insensitive to extraneous factors
- Can perform many different functions at the same time

a healthy, productive, and efficient work environment. Conversely, the results of system imbalance may result in poor production, poor product quality, as well as additional hidden costs due to preventable musculoskeletal and cardiovascular disorders (MSDs). Perhaps that is why international quality standards, such as ISO 9000, require similar system balances.

When people are involved in the work system, and musculoskeletal disorders occur, or other indicators of system imbalance are present (i.e., low quality, decreases in productivity, absenteeism, etc.), an analytical procedure needs to be implemented to identify the cause of the system imbalance. Only after analysis can procedures be put into place to bring the system back into balance. If an analysis is not performed, the repair of the system is reduced to guesswork, at best. "Guesswork" will not be able to justify the capital investment in workstation and work redesign to a skeptical CEO or thrifty purchasing manager.

When dealing with musculoskeletal disorders, the job improvement should be a process rather than a "bandaid." Job analysis is part of the job improvement cycle.

Recently, the National Institute of Occupational Safety and Health (NIOSH) has suggested the following seven elements of an effective program that comprise a "pathway" for addressing system balance (NIOSH, 1997).

Step 1. Look for signs of potential MSDs in the system, such as frequent worker reports of aches and pains, or job elements that require repetitive, forceful exertions.

Step 2. Show management commitment in addressing possible problems and encourage worker involvement in achieving system balance.

Step 3. Offer training to expand management and worker ability to evaluate potential system imbalances associated with MSDs.

Step 4. Gather data to identify jobs or work conditions that are most problematic. Sources such as injury and illness logs, symptom surveys, medical records, first reports of injuries, worker compensation data, and job analysis may be used.

Step 5. Identify effective controls for tasks that pose a risk of MSDs. Evaluate these approaches once they have been implemented to see if they have reduced or eliminated the problem.

Step 6. Establish a solid health care management process that integrates early detection and conservative treatment to prevent impairment and disability. This should include return-to-work programs and job accommodation.

Step 7. Minimize risk factors for MSDs during the design phase of new work processes, products, tools, and operations. Thus the design team should be part of the ergonomic process to ensure system balance.

The purpose of this chapter is to provide the reader with some basic tools and references so that the above steps can be carried out in a systematic fashion. In particular, methods of ergonomic job analysis will be presented so that the professional can develop a basic understanding of how to work towards system balance.

The etiology of musculoskeletal disorders is beyond the scope of this chapter. However, it should be noted that medical management, including conservative return to work and early detection of MSDs, are paramount in an ergonomic program. In some cases, solutions to system imbalances will be intuitive, while others may require the services of a qualified ergonomist who may assist with the identification of risk factors, controls, and the evaluation of the controls. The references at the end of this chapter are offered so that the safety and health professional may build a library of resources for training, education, job analysis, and solutions.

23.2 JOB ANALYSIS

Ergonomic job analysis has been described as the systematic investigation of jobs to describe individual and combined work factors and work activities for the purpose of designing task demands to match the human capabilities of those who perform the task. Job analysis has also been used to identify reasons (root causes) for increased risk of MSDs. It should be stated up front that this type of analysis is not possible without resource and time commitment from top management. Resource allocation, in the form of a budget, should be secured prior to the commencement of analysis. Too often, analyses are performed only to have the resulting recommendations collect dust. This type of scenario only breeds mistrust among employees and compromises the reality of a balanced system. In the same breath, it is imperative that workers be involved in the job analysis process. After all, they are the true experts of their jobs. Workers are also a vital component for the ultimate success of the implementation of controls.

Important areas for consideration when performing job analyses include:

- Identify the population who does the job;
- Determine why the job is done (consider its relation to the whole work process and organizational design);
- Break the job into tasks, and then break the tasks into elements; determine the physical demands and mental information demands associated with each element of the tasks;
- Identify equipment, tools, and information that are needed to accomplish each task (technology);
- Locate where the tasks are done in relation to the person who performs the tasks;
- Identify the environmental conditions under which the tasks are performed (consider seasonal and product changes).

Under all circumstances, the frequency, duration, and level of exposure to work factors can be determined. Where available, these data can be compared with normative data (quantitative or qualitative) to assess the risk of MSDs. This in turn can contribute to finding integrated solutions that work towards a balanced system. Job analysis procedures are then used to evaluate the effectiveness of implemented control measures.

The following sections will present work factors to consider such as when to conduct a job analysis, as well as the steps and methods involved in such a job analysis. In addition, the selection and evaluation of solutions will be explained.

23.3 WORK FACTORS TO CONSIDER WHEN PERFORMING JOB ANALYSES

There are many work factors that should be considered during the ergonomic job analysis process. Some of these are more obvious and easier to identify than others.

23.3.1 Forces Required to Perform the Task

Force requirements have a direct impact on the muscular effort that must be expended by the worker to perform an action.

The amount of force exerted by the worker depends upon:

- Work postures (awkward postures decrease the ability to exert force);
- Speed of movement;
- The weight handled;
- The friction characteristics (ability to grip) of the objects handled.

23.3.2 Contact Stresses

Contact stress occurs when soft tissue of the human body comes in contact with a hard object. High contact forces (due to hard or sharp objects and work surfaces) may create pressure over one area of the body and inhibit nerve function and blood flow.

23.3.3 Postures Assumed during the Task

The position of the body has a direct effect on the moment (torque) about a joint. The moment is the rotation produced when a force is applied The smaller the moment, the less force required by the muscles that support movement about the joint. Neutral postures tend to optimize the ability of the body to exert maximum force.

- Greater force is required when awkward postures are used because muscles lose their "mechanical advantage" and cannot perform efficiently.
- Prolonged postures (e.g., holding the arm out straight to hold a fixture) produces a moment (torque), around a joint. This is referred to as "static posture." Static postures may decrease the time to muscle fatigue, especially when force is required by the working muscles. This fatigue occurs as muscles continue to work with impaired blood flow caused by the static posture.

- The farther away from the body a load is carried, the greater the compressive forces on the spine.
- Shear and compressive forces on the spine increase when lifting, lowering, or handling objects with the back bent or twisted and when objects are bulky and/or slippery.
- Uneven work surfaces or surfaces that are slippery or difficult to walk on may also affect forces required to perform the job.

23.3.4 Frequency of Muscle Contraction

Frequent repetition of the same work activities can exacerbate the effects of awkward work postures and forceful exertions. Tendons and muscles can often recover from the effects of stretching or forceful exertion if sufficient time is allotted between exertions. However, if movements involving the same muscles are repeated frequently, without rest, fatigue and strain can accumulate, producing tissue damage. It is important to note that different tasks may require repetitive use of the same muscle groups even though the tasks are not similar. This should be taken into consideration when workers are rotated among different tasks.

23.3.5 Duration of Muscle Contraction

The amount of time a muscle is contracted can have a substantial effect on both localized and general fatigue. In general, the longer the period of continuous work (muscle contraction), the longer the recovery or rest time will be required. An example of prolonged muscle contraction is standing for long periods without a foot rest, chair, or sit/stand stool. Another example would include constantly gripping a tool.

23.3.6 Vibration

Hand–arm (segmental) vibration from handheld power tools increases force requirements and has been associated with the development of hand–arm vibration syndrome (vibration white finger or Raynaud's Disease) and carpal tunnel syndrome. Guidelines for measuring continuous, intermittent, impulsive, and impact segmental vibration have been provided in the ISO standard ISO5349 (1986), entitled "Guide for the Measurement and the Assessment of Human Exposure to Hand Transmitted Vibration," and the ANSI standard ANSI S3.34-1986 entitled "Guide for the Measurement and Evaluation of Human Exposure to Vibration Transmitted to the Hand."

Whole body vibration exposures have been observed in truck drivers, large metal stamping operations, and with the use of large vibrating tools, such as jackhammers. These exposures have been associated with back and neck disorders, including microfractures of vertebral endplates, and urinary and digestive discomfort. Currently, whole body vibration standards are under consideration by the ISO and ANSI.

23.3.7 Cold Temperatures

Cold temperatures can reduce the dexterity and sensory perception of the hand due to vasoconstriction (a shunting of the blood away from the extremities to the organs to maintain a core temperature of over 96.8°F). This may cause workers to apply more grip force to tool handles and objects than would be necessary under warmer conditions, and may increase risk for accidents due to the loss of hand dexterity. There is also evidence that cold tends to exacerbate the effects of segmental vibration. The following guidelines have been provided by the American Conference of Governmental Industrial Hygienists:

- If fine work is performed with the bare hands for more than 10 min in an environment below 60.8°F (16°C), special provisions should be established for keeping the workers' hands warm.
- Metal handles on hand tools and control bars should be covered by thermal insulating materials at temperatures below 30.2°F (−1°C).
- If the air temperature falls below 60.8°F (16°C) for sedentary work, 39.2°F (4°C) for light work, 19.4°F (−7°C) for moderate work, and fine manual dexterity is not required, then gloves should be used by the workers. If gloves are not appropriate (see below), then warm air jets, radiant heaters (fuel burner or electric radiator), or contact warm plates may be utilized.
- Avoid cold exhaust exposure to hands when using hand tools.
- Avoid cold exposures to the cooling effect associated with handling evaporative liquids such as gasoline, alcohol, or cleaning fluids.

At temperatures below 39.2°F (4°C), an evaporative cooling effect may occur and exacerbate other musculoskeletal risk factors listed in the previous sections.

23.3.8 Poorly Fitted Gloves

Poorly fitted gloves can reduce sensory feedback from the fingertips and result in increased grip force. Force is also increased by working against tight-fitting gloves or trying to get a tight grip with big bulky gloves. In addition, loose gloves may present a safety hazard around moving parts and tools, such as chain drives and drills.

23.3.9 Obstructions

Obstructions such as rails, jigs, and fixtures, which hamper smooth free lifting and reaching motions. Obstructions should be evaluated to analyze their effect on awkward postures and the forces required to accomplish the job. An example of obstructions is seen in warehouses, where the size of the first slot may not allow the worker to use safe lifting practices while they select cases. Workers may not be able to stand up in the slot when the height of the second slot is at 50 in. Thus they have to lift and carry in a stooped or bent position due to the obstruction created by the second slot.

23.3.10 Standing Surfaces

Prolonged standing on hard surfaces increases back and leg fatigue. In addition, standing on inclines or different levels may put stress on the spine and legs. The use of a foot pedal while standing may also increase the stress on the back, hip, and ankle. Standing on slippery surfaces may increase the risk of sudden force exertion while trying to prevent slips and falls.

To reduce the risk factors associated with prolonged standing, employers can provide foot rests or rest bars, as long as they do not pose a safety hazard. Chairs may also reduce the stress associated with standing only if the chairs are well suited for the job and the workers understand how to easily adjust them. Sit/stand stools have also been used in environments where seated postures increased other risk factors such as excessive reaching. It is imperative to analyze the effect of a chair on other risks factors before purchasing the chair. Workers are usually able to determine how a chair will affect their work environment, and they should always be included in the purchase decision process.

23.3.11 Prolonged High Visual Demands

Prolonged visual demands (vigilance), processing information quickly over time, making inspection decisions rapidly, visually scanning complicated displays for slight evidence of malfunction may increase fatigue, blood pressure, and muscle tension. Inadequate lighting or poorly designed displays may result in a decreased time to fatigue. In addition to increased musculoskeletal load, these health effects are accompanied by increases in errors and decreases in quality.

23.3.12 Physical Energy Demands

Fatigue is believed to increase a worker's risk for musculoskeletal injury and illness. Oxygen consumption has been accepted as one of the best measures of energy demands of the job. Once the energy demands are known, the appropriate work–rest cycles can be established. Heart rate and ratings of perceived exertion have also been used with success. It is recommended that a professional with experience in the measurement of energy demand and the design of work cycles be employed to address fatigue in the work environment.

23.4 WHEN TO CONDUCT JOB ANALYSES

Job analyses can be conducted on jobs in a reactive or proactive manner. In the reactive approach, the health and safety professional responds to illnesses or injuries that are associated with occupational hazards. Conversely, the proactive approach commands that risk factors are actively sought through job analyses, even in the absence of injuries and illnesses. Job analyses can be as simple as checklists or as complex as electromyography to measure muscle force. Since this chapter is written for the safety and health professional who may have little or no experience with ergonomic design, the simpler methods will be presented with suggestions of how to obtain resources for more complex analyses.

23.4.1 The Reactive Approach

In the reactive approach, records review, single incidents of MSDs, first aid cases, reports of near misses, reports of pain, and workers compensation cases can prompt a response from the health and safety professional in the form of a job analysis. This is analogous to the job hazard analysis and similar reviews performed in Chapter 5. The analyst may investigate the root cause of a single case or review OSHA 200 logs for trends and incidents of many cases that are occurring in a facility. If a facility is very large, records review may provide a better picture of where the problems exist and will allow for the identification of high-, medium-, and low-risk areas (see also Chapter 25).

For example, the statistics of incidence rate and severity rates may be used. The *incidence rate* is usually calculated for the period of one calendar year and represents the number of new cases per 100 workers that occurs for a specific industry (SIC code), a specific company, a specific department, a specific job, and/or a specific type of injury. Incidence rates are calculated as follows:

$$\frac{(\text{number of new cases} \times 200,000)}{\begin{array}{c}\text{total number of hours worked by the}\\ \text{exposed population (SIC, company, or dept., etc.)}\end{array}} \times 100$$

The 200,000 represents 2000 full-time hours of 40 hours per week for 50 weeks in a year, multiplied by 100 workers. This number normalizes the data so that companies, departments, and jobs may be compared despite the number of employees they have. For comparisons with fewer than 20 workers, it may be better to use the absolute number of injuries and illnesses instead of the incidence rate. Thus the incidence rate for 1995 for the nursing home industry could be described as "16.8 cases per 100 workers in 1995." This number could be compared with other industries to determine if specific nursing homes are high risk, despite their size, because the data have been normalized per 100 workers.

Another statistic that is frequently used is the *severity rate*. The severity rate is usually calculated for a period of 1 year and represents a severity measure (the amount of medical costs, the number of lost work days, the number of restricted work days, or the dollars of workers compensation, etc.) per 100 workers for a given calendar year. The severity is calculated as follows:

$$\frac{\begin{array}{c}\text{severity measure} \times 200,000\\ \text{(lost work days, or medical costs, etc.)}\end{array}}{\begin{array}{c}\text{total number of hours worked by the}\\ \text{exposed workers}\end{array}} \times 100$$

Since there is usually an initial increase in the reporting of MSDs when a company increases awareness among its employees, severity can be used as the real indicator of a program's success. Since workers are encouraged to report earlier, the severity of the injuries should be decreased, even though reporting (i.e. the number of cases) may increase. Workers will report injuries when they first notice symptoms instead of waiting until they are impaired (when a more costly treatment may then be needed). Thus, as the incidence rate rises, initially, the severity rate should decrease accordingly. The combination of these statistics allows a safety and health professional to prioritize ergonomic intervention in the work environments when there are too many workers to address every job in a short time frame. In addition, these statistics may be used to explain the long-term benefits associated with system balance.

23.5 JOB ANALYSIS STEPS

Before the job analysis is initiated, it is important to secure management commitment. This commitment includes a budget, a commitment to fund the cost of controls, a commitment to a continuous improvement process that includes a followup, and the inclusion of responsibility for the ergonomic program in the job performance reviews of managers. Employees should also be included as a vital part of the job analysis team.

This inclusion will necessitate that they be trained and educated in ergonomic principles related to their work and work environment.

23.5.1 Who Does the Analysis?

The job analysis can be done by an individual or team who has knowledge of ergonomics, how the body works, and who is knowledgeable about potential control measures for the identified exposures. It is advisable that the analyst also be familiar with the type of job that is analyzed and the types of controls that are available to abate the hazards. If a consultant is used, it is preferable that they have had previous experience in the industry in which they are working. For example, if a consultant is hired to evaluate an office environment, he/she should provide references from other office environments.

Management, the affected employees, health care professionals, engineering, and maintenance are all vital contributors to the process. In talking with employees from many companies, their biggest impression was formed early in the hiring process from plant and office managers who were actively involved in the job analysis and abatement procedures. At The American Saw Company, in East Longmeadow, Massachusetts, the plant manager is the person who walks each new employee through the facility to introduce the importance of various safety and health aspects in the work environment. In interviewing employees of that company, the employees stated that it is the dedication from top management that formed their safety and health "culture" from their first day on the job. Subsequently, the company has consistently maintained injury rates below their industry average.

In addition to management commitment, informal discussions with workers and supervisors can provide useful information for the job analysis. Workers who perform a job on a daily basis are often the best source of information about the specific elements that may pose an increased risk for MSDs. The use of symptoms surveys or postural discomfort surveys (Fig. 23.2) may also be used to identify specific tasks that are causing discomfort or pain and can be quite useful in the analysis. These surveys can also be used for before/after comparisons after measures are implemented.

Supervisors can provide a macro view of their operations, as well as insight into the feasibility of proposed changes. Interviews also provide workers with the opportunity to participate and to provide input to the job analysis process. Soliciting worker input early in the process may also lead to the identification of more feasible and accepted control strategies. Kennebec Nursing Home in Maine found that when their employees were given the opportunities to choose controls to reduce injuries associated with resident handling, injuries were reduced. The workers were receptive to the interventions they had chosen to reduce risk factors, and unlike other nursing homes who purchased lift devices without employee input, the nursing aides actively followed the risk reduction strategies.

23.5.2 Basic Equipment Needed for Conducting Most Job Analyses

Equipment for recording occupational exposures can be quite sophisticated, such as 3D static biomechanical models, dynamic measures of torsion, compressive and shear force models, and the kinematic analysis of a person in motion using light sensors. These tools include electrogoniometers, lumbar motion monitors, and other electrical equipment (pocket protectors and slide rules are optional). However, in simple, non-research-oriented cases, the following basic instrumentation is sufficient.

Symptoms Survey: *Ergonomics Program*

Date ____/____/____/

Plant ____ Dept # ____ Job Name ____

Shift ____ Hours worked/week ____ ____ years ____ months Time on THIS Job

Other jobs you have done in the last year (for more than 2 weeks)

Plant ____ Dept # ____ Job Name ____ ____ months ____ weeks Time on THIS Job

Plant ____ Dept # ____ Job Name ____ ____ months ____ weeks Time on THIS Job

(If more than 2 jobs, include those you worked on the most)

Have you had any pain or discomfort during the last year?

☐ Yes ☐ No (If NO, stop here)

If YES, carefully shade in area of the drawing which bothers you the MOST.

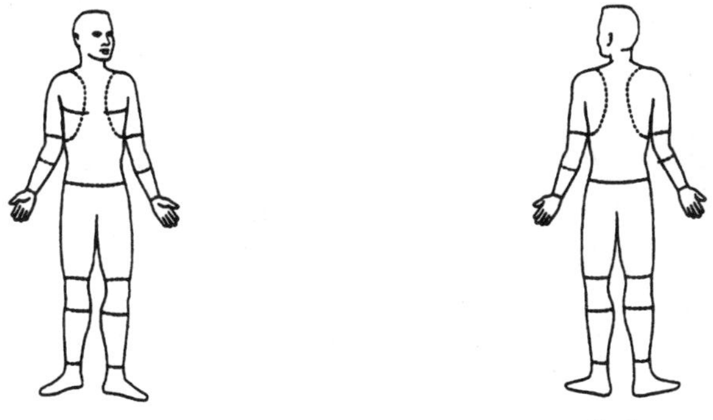

Front

Back

(Continued)

Figure 23.2 Symptoms survey form.

1. Cameras and film for recording workers' postures and motions during job activities. This equipment is also very useful for recording "before and after" improvements and for developing training materials (see notes on filming below).

2. Tape measures and rulers for measuring workstation dimensions, tool dimensions, and reach distances. In addition, a goniometer may be used to measure angles of the body, materials handled, viewing angles, and workstation attributes. This information may be used as inputs to computer programs.

Symptoms Survey Form (Continued)

(Complete a separate page for each area that bothers you)

Check Area: ☐ Neck ☐ Shoulder ☐ Elbow/Forearm ☐ Hand/Wrist ☐ Fingers
☐ Upper Back ☐ Low Back ☐ Thigh/Knee ☐ Low Leg ☐ Ankle/Foot

1. Please put a check by the words(s) that best describe your problem

☐ Aching	☐ Numbness (asleep)	☐ Tingling
☐ Burning	☐ Pain	☐ Weakness
☐ Cramping	☐ Swelling	☐ Other
☐ Loss of Color	☐ Stiffness	

2. When did you first notice the problem? ________(month) ________(year)

3. How long does each episode last? (Mark an X along the line)

________/________/________/________/________/
1 hour 1 day 1 week 1 month 6 months

4. How many separate episodes have you had in the last year? ________________

5. What do you think caused the problem? ________________________________

____

6. Have you had this problem in the last 7 days? ☐ Yes ☐ No

7. How would you rate this problem? (mark an X on the line)
NOW

____
None Unbearable

When it is the WORST

____
None Unbearable

8. Have you had medical treatment for this problem? ☐ Yes ☐ No

8a. If NO, why not? ________________________________

8b. If YES, where did you receive treatment?

☐ 1. Company Medical Times in past year ____________

☐ 2. Personal doctor Times in past year ____________

☐ 3. Other Times in past year ____________

Did treatment help? ☐ Yes ☐ No ____________

9. How much time have you lost in the last year because of this problem?_____ days

10. How many days in the last year were you on restricted or light duty because of this problem?
_____ days

11. Please comment on what you think would improve your symptoms

____

____

____

____

Figure 23.2 *(Continued)*

3. Force gauges or spring scales (fish scales) for measuring the force of exertions (e.g., the force needed to push or pull a hand cart) and the weight of tools or objects in manual handling tasks.

4. Timers (stopwatches) for measuring the duration of work activities, breaks, etc. (care should be taken to assure employees that the stopwatch is not being used to establish production standards). A real-time simulation program works quite well in the job analysis process (Keyserling, 1986). Real-time analysis can measure the percentage and absolute time spent in various postures. ROPEM is a technique that allows the analyst to record postures verbally into a microcasette for transfer to a computer at another time, and works well in environments where the use of a computer would be difficult (Perez-Balke, 1993).

5. A checklist that is designed for the task at hand and provides feedback as to what needs to be done in response to the answers. For example, a checklist for the analysis of a computer workstation would be different from one used to evaluate jobs on a construction site and might stipulate that all negative responses of "no" signify the need for further analysis. An example of a checklist from Lifshitz and Armstrong is included in Fig. 23.3 and examples of checklists that have been proposed by the Occupational Safety and Health Administration (OSHA) are found in Appendix A.

6. Borg scales (Fig. 23.4) provide the analysis with a tool to subjectively measure the force requirements of the job as well as the effect of controls on the effort needed to perform the job. These scales will be discussed in further detail below.

23.5.3 Guidelines for Videotaping for Ergonomic Analysis

Guidelines for recording work activities on videotape are available from the National Institute of Occupational Safety and Health (USA) at 1-800-35NIOSH, and are summarized as:

- If the video camera has the ability to record the time and date on the videotape, use these features to document when each job was observed and filmed. Recording the time on videotape can be especially helpful if a detailed motion study will be performed at a later date (time should be recorded in seconds). Make sure the time and date are set properly before videotaping begins.
- If the video camera cannot record time directly on the film, it may be useful to position a clock or a stopwatch in the field of view.
- At the beginning of each recording session, announce the name and location of the job being filmed so that it is recorded on the film's audio track. Restrict subsequent commentary to facts about the job or workstation.
- For best accuracy, try to remain unobtrusive; that is, disturb the work process as little as possible while filming. Workers should not alter their work methods because of the videotaping process.
- If the job is repetitive or cyclic in nature, film at least 10–15 cycles of the primary job task. If several workers perform the same job, film at least 2–3 different workers performing the job to capture differences in work method.
- If necessary, film the worker from several angles or positions to capture all relevant postures and the activity of both hands. Initially, the worker's whole body posture

Risk factors

1. Physical stress
   a. Can the job be done without hand/wrist contact on sharp edges?
   b. Is the tool operating without vibration?
   c. Are the worker's hands exposed to >21 degrees C?
   d. Can the job be done without using gloves?
2. Force
   a. Does the job require less than 4.5 kg of force?
   b. Can the job be done without using a finger pinch grip?
3. Posture
   a. Can the job be done without wrist flexion or extension?
   b. Can the tool be used without wrist flexion or extension?
   c. Can the job be done without deviating the wrist from side to side?
   d. Can the tool be used without deviating the wrist from side to side?
   e. Can the worker be seated while performing the job?
   f. Can the job be done without a "clothes wringing" motion?
4. Workstation hardware
   a. Can the orientation of the work surface be adjusted?
   b. Can the height of the work surface be adjusted?
   c. Can the location of the tool be adjusted?
5. Repetitiveness
   a. Is the cycle time longer than 30 seconds
6. Tool design
   a. Are the thumb ane finger slightly overlapped in a closed grip?
   b. Is the span of the tool's handle between 5 and 7 cm?
   c. Is the hand of the tool made from material other than metal?
   d. Is the weight of the handle below 4 kg?
   e. Is the tool suspended?

Note: "No" responses are indicative of conditions associated with the risk of cumulative trauma disorders.

Figure 23.3 Checklist for work related musculoskeletal disorders of the upper extremities (Lifshitz, Y., and Armstrong, T. J., 1986, A design checklist for control and prediction of cumulative trauma disorders in hand intensive manual jobs. *Proceedings of the 30th Annual Meeting of the Human Factors Society*, 837–41).

should be recorded (as well as the work surface or chair on which the worker is standing or sitting). Later, closeup shots of the hands should also be recorded if the work is manually intensive or extremely repetitive.

- If possible, film jobs in the order in which they appear in the process. For example, if several jobs on an assembly line are being evaluated, begin by recording the first job on the line, followed by the second, third, etc.
- Avoid making jerky or fast movements with the camera while recording. Mounting the camera on a tripod may be useful for filming work activities at a fixed workstation where the worker does not move around much.

Ratings of Perceived Scales

Rating of Peceived Exertion: Whole Body Effort. Borg, 1962.	Category Scale for Rating of Perceived Exertion: Large-Muscle-Group Activity. Borg, 1980.
20	*Maximal
19 Very, very hard	10 Very, very strong (almost max)
18	9
17 Very hard	8
16	7 Very strong
15 Hard	6
14	5 Strong (heavy)
13 Somewhat hard	4 Somewhat strong
12	3 Moderate
11 Fairly	2 Weak (light)
10	1 Very weak
9 8 Very light	0.5 Very, very weak (just noticeable)
7 Very, very light	0 Nothing at all
6	

[Borg, G.A.V. (1962). *Physical Performance of Perceived Exertion*. Lund, Sweden: Gleerups. Borg, G.A.V. (1980). A Category Scale with Ratio Properties for Intermodal and Interindividual Comparisons. Paper presented at the International Congress of Psychology, Leibig, West Germany, 1980.]

Figure 23.4 Borg scales.

Videotapes can then be reviewed using slow-motion or real-time playback more accurately to measure task durations or detect subtle or rapid movements. As mentioned, more advanced techniques such as "lumbar motion monitors," "peak performance models," and "joint motion analysis," to name a few methods, are available. These types of analysis are often used where the cost of a less than perfect abatement plan could cost many more times the amount of money spent on ergonomic controls. It is sometimes worthwhile to quantify the root cause so that design criteria can be implemented early in the setup stage of manufacturing processes.

The videotape analysis can also be invaluable as a tool for documenting visible risk factors for MSDs in the workplace. It is imperative to gain worker consent before videotaping or analyzing a work task. If the worker is not involved, the analyst may not get an accurate "picture" of the task. Perhaps an equal mistake is the use of surrogate workers for ergonomic analyses. Since people are different sizes and use different techniques to perform the same job, workers who perform the task on a regular basis, and who will be affected by the worksite changes, should be an integral part of the analytical process.

23.6 JOB ANALYSIS METHODS

The primary objective of job analyses is to collect sufficient information to allow the analyst to completely describe tasks as they are currently being performed (i.e., what the worker is doing and how it is being performed over time) as well as the system attributes that interact with the worker. This information allows for the assessment of risk for MSDs and describes the system balance. Generally, this requires the analyst to observe the job during a "typical" work period under "normal" operating conditions. It may be that there is only one task ("cut thigh from chicken" or "enter tax form data into computer"). However, in general, a job consists of a group of tasks that can be defined by related task elements (Niebel, 1989) Fig. 23.7 (see Section 23.10) depicts examples of work elements.

A job may consist of several regular and several irregular tasks (tasks not done with any regularity). It is important to identify the frequency and duration of both regular and irregular tasks as well as the frequency and duration of any "recovery" periods. The more complex the job tasks (including rotation between tasks or jobs), the more complex the analysis strategy. With the advent of cell manufacturing, there are new challenges presented for job analysis, especially when rotation is "informal" (Armstrong et al., 1986; Rodgers, 1992).

Workers who perform the same job may use different methods due to differences in training, stature, or strength. Therefore, several workers who perform the same job should be observed. The analyst should also attempt to observe the same task at different times during the work shift to determine if fatigue affects workers' performance or if the workload changes during the workday. However, sometimes production records may provide this information. The common steps used in ergonomic job analysis are listed below.

23.6.1 Document the Job

1. Describe the job briefly, the goal of the job, what is to be accomplished.

2. Describe the workers: How many workers are employed in each job? What are the characteristics of the workforce (e.g., gender, age, education level, seniority, union)?

3. Define the work schedule. What is the work–rest cycle? How frequently are their breaks? How many hours do employees work per week? Is work organized into shifts? How much overtime is worked per week?

4. Determine production information (work pace). Is there an established work rate? If so, how is the work rate determined (e.g., machine paced, piece rate, time standards, etc.)?

5. Take measurements (see discussion above on tools). A sketch or drawing of the workstation or worksite is useful for identifying the location of fixtures, equipment items, etc. Sketches should be labeled with dimensions indicative of work surface heights, reach distances, clearance, walking distances, etc. (Keyserling et al., 1991; Putz-Anderson, 1988). If workers handle tools or objects, the location, size, shape, and

weight of these items should be recorded. Force measurements (e.g., using spring scales or force gauges) should be recorded for push, pull, and lift activities.

6. Measurements may also be taken to apply to the NIOSH lifting equation (Appendix B, Fig. 23.8, p. 969) to evaluate exposures associated with manual material handling.

7. Describe how the work is organized. How much control do workers have over the way they perform their job tasks, and what is the level of job demand? For example, is there an opportunity for workers to communicate, receive feedback quickly, provide input regarding problems in production quality or health and safety? Do workers perform the same tasks over and over throughout the workshift, or do workers perform a large number of different tasks? Is there an opportunity for workers to rotate to other jobs? What type of pay system is utilized (e.g., hourly wage, piece rate)?

23.6.2 Identify and Evaluate Exposure to Work Factors

Record work factors, such as those mentioned above, associated with specific tasks or subtasks. Describe the reason that the work factor(s) are present (root cause). Perhaps the biggest error seen in job analysis is the failure to perform root cause analysis. For example, a safety and health professional might list the cause of a hazard of bent wrists with a pinch grip as "poor worker training." The resulting recommendation would then be "train the worker," which would not address the hazard.

A root cause analysis would question the underlying exposures. For example, is it the orientation of the part that causes the worker to need to bend the wrist? If so, then a jig with a clamp might be recommended. A common error seen in warehouse and distribution centers is to retrain workers in "safe lifting procedures" or mandate that their workers "wear a back belt" when they injure their backs. Safe lifting procedures would tell workers to bend their knees even though they cannot possibly lift safely given the constraints of their warehouse design. For example, the last boxes on a pallet in the first tier, or slot, of a warehouse may be located 48 in. inside an area only 44 in. high. There would be no conceivable way for the worker to select the box "using safe lifting techniques" unless crawling was acceptable (which it is not), even if a back belt was used. Instead, a root cause analysis would show that the physical constraints of the environment result in poor work postures and lifting methods. This finding might result in the recommendation that pallets be rotated by forklift operators when the front of the pallet has been picked or that slots be reconfigured to provide a safe lifting environment.

23.6.3 Summarize Findings

For each task (subtask) list each body part and work factor present, including the frequency and duration of the exposure as quantitatively as possible. List the root cause for the work factor present. Fig. 23.6 presents an example of a form that may be used for a summary of findings.

23.6.4 Compare Findings to Normative Values Where Available

In some cases, it will be obvious that task demands exceed the capabilities of the workers performing the job. For example, most analysts will recognize a job as being too repetitive if the worker has difficulty keeping up with the required pace or if the body is in constant motion. However, in other instances it may not be as clear that force or postural demands are appropriate for the population of workers performing a task. This is especially true if a new job is in the design stages. However, it is possible to estimate the percentage of workers for whom a job may be difficult. Information that can be used for this purpose is available from the following sources. In addition, the reader is encouraged to peruse the list of resources at the end of this chapter for more information.

1. For lifting analysis, use the NIOSH Work Practices Guide (1981; and Waters et al., 1993).

2. For push/pull/carry problems, use Snook tables (1978) or computerized biomechanical models (Chaffin, 1992). ANSI standard B-11-1994 also has this information as well as recommendations on human machine interface.

3. For workstation and tool design, use anthropometry tables such as those in Figure 23.5.

4. For frequency/duration analysis, use work–rest cycle data based on energy expenditure models that use oxygen consumption or heart rate data. Tables for these data can be found in textbooks (Kodak, 1986; Astrand, 1986).

5. For visual demand/information processing and displays analysis, use signal detection theory and guidelines such as those published by the American National Standards Institute (ANSI-HFES 100).

6. For vibration exposures, use NIOSH, ACGIH, or ISO guidelines on segmental and whole body vibration. Guidelines for measuring continuous, intermittent, impulsive, and impact segmental vibration have been provided in the ISO standard ISO5349 (1886), entitled "Guide for the Measurement and the Assessment of Human Exposure to Hand Transmitted Vibration" and the ANSI standard ANSI S3.34-1986, entitled "Guide for the Measurement and Evaluation of Human Exposure to Vibration Transmitted to the Hand."

Standing Body Dimensions

	5th Percentile				95th Percentile			
	Ground Troops		Females		Ground Troops		Females	
Standing Body Dimensions	Cm	In.	Cm	In.	Cm	In.	Cm	In.
1. Stature	162.8	64.1	152.4	60.0	185.6	73.1	174.1	68.5
2. Eye height (standing)	151.1	59.5	140.9	55.5	173.3	68.2	162.2	63.9
3. Shoulder (acromale) height	133.8	52.6	123.0	48.4	154.2	60.7	143.7	56.6
4. Chest (nipple) height[a]	117.9	46.4	109.3	43.0	136.5	53.7	127.8	50.3
5. Elbow (radial) height	101.0	39.8	94.9	37.4	117.8	46.4	110.7	43.6
6. Waist height	96.6	38.0	93.1	36.6	115.2	45.3	110.3	43.4
7. Crotch height	76.3	30.0	68.1	26.8	91.8	36.1	83.9	33.0
8. Gluteal furrow height	73.3	28.8	66.4	26.2	87.7	34.5	81.0	31.9
9. Kneecap height	47.5	18.7	43.8	17.2	58.6	23.1	52.5	20.7
10. Calf height	31.1	12.2	29.0	11.4	40.6	16.0	36.8	14.4
11. Functional reach	72.6	28.6	64.0	25.2	90.9	35.8	80.4	31.7
12. Functional reach extended	84.2	33.2	73.5	28.9	101.2	39.8	92.7	36.5

Source: Adapted from DOD, 1981.
[a]Bustpoint Height for Women

Figure 23.5

Seated Body Dimensions

Seated Body Dimensions	5th Percentile				95th Percentile			
	Ground Troops		Females		Ground Troops		Females	
	Cm	In.	Cm	In.	Cm	In.	Cm	In.
13. Vertical arm reach sitting	128.6	50.6	117.4	46.2	147.8	58.2	139.4	54.9
14. Sitting height erect	83.5	32.9	79.0	31.1	96.9	38.2	90.9	35.8
15. Sitting height relaxed	81.5	32.1	77.5	30.5	94.8	37.3	89.7	35.3
16. Eye height sitting erect	72.0	28.3	67.7	26.6	84.6	33.3	79.1	31.2
17. Eye height sitting relaxed	70.0	27.6	96.2	26.1	82.5	32.5	77.9	30.7
18. Midshoulder height	56.6	22.3	53.7	21.2	67.7	26.7	62.5	24.6
19. Shoulder height sitting	54.2	21.3	48.9	19.6	65.4	25.7	60.3	23.7
20. Shoulder-elbow length	33.3	13.1	30.6	12.1	40.2	15.8	36.6	14.4
21. Elbow-grip length	31.7	12.5	29.6	11.6	36.3	15.1	35.4	14.0
22. Elbow-fingertip length	43.8	17.3	40.0	15.7	52.0	20.5	47.5	18.7
23. Elbow rest height	17.5	6.9	16.1	6.4	26.0	11.0	26.9	10.6
24. Thigh clearance height	-	-	10.4	4.1	-	-	17.5	6.9
25. Knee height sitting	49.7	19.6	46.9	18.5	60.2	23.7	55.5	21.8
26. Popliteal height	39.7	15.6	36.0	15.0	50.0	19.7	45.7	18.0
27. Buttock-knee length	54.9	21.6	53.1	20.9	65.8	25.9	63.2	24.9
28. Buttock-popliteal length	45.5	17.9	43.4	17.1	54.5	21.5	52.6	20.7
29. Functional leg length	110.6	43.5	96.6	38.2	127.7	50.3	118.6	46.7

Source: Adapted from DOD, 1981.

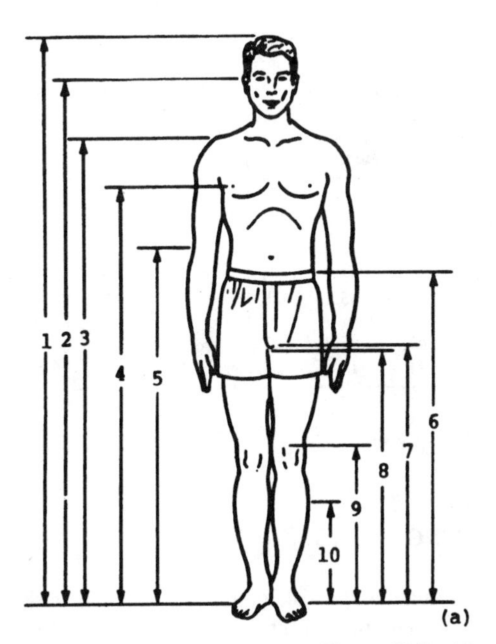
1
2
3
4
5
6
7
8
9
10
(a)

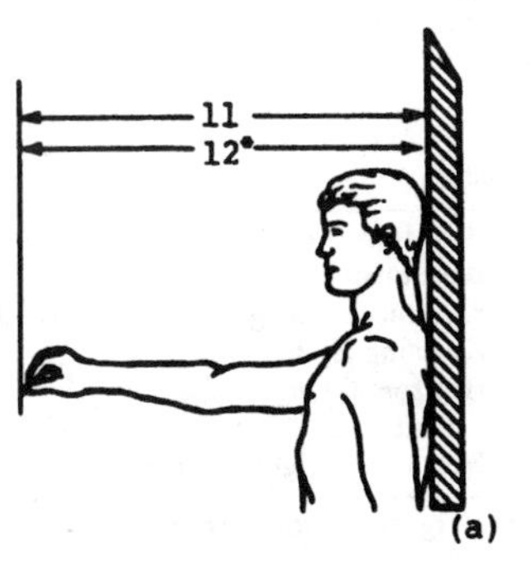
11
12*
(a)
*SAME AS 11; HOWEVER, RIGHT SHOULDER IS EXTENDED AS FAR FORWARD AS POSSIBLE WHILE KEEPING THE BACK OF THE LEFT SHOULDER FIRMLY AGAINST THE BACK WALL.

Figure 23.5 *(Continued)*

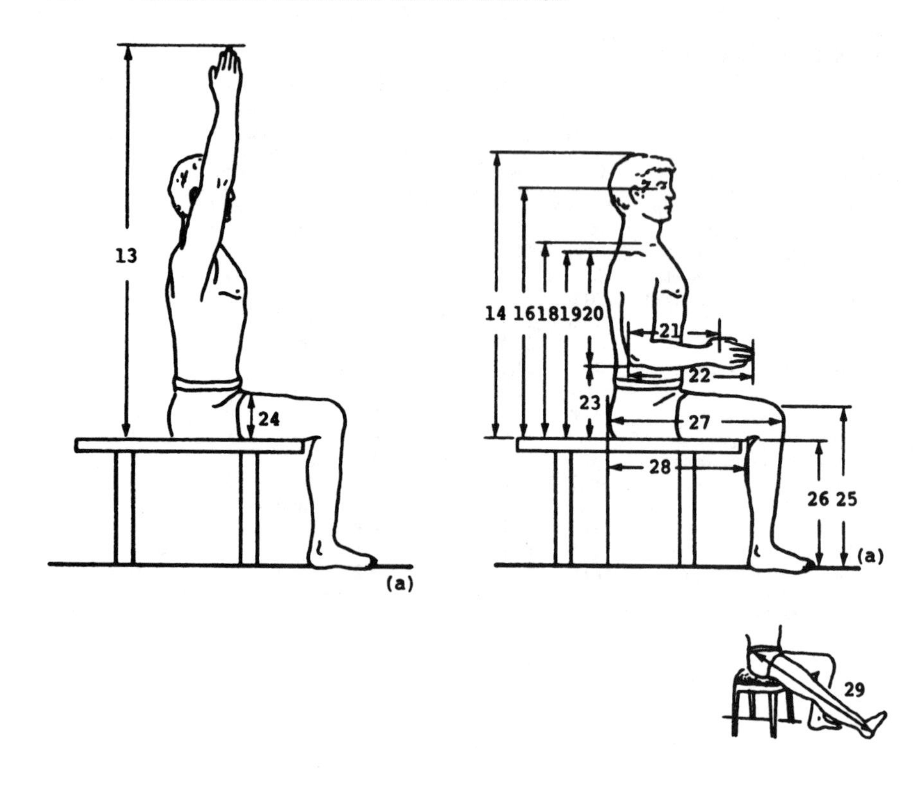

Anthropmetric Data for Working Positions[a]

Position	5th Percentile Men Cm	In.	Women Cm	In.	95th Percentile Men Cm	In.	Women Cm	In.
1. Bent torso breadth	40.9	16.1	36.8	14.5	48.3	19.0	43.5	17.1
2. Bent torso height	125.6	49.4	112.7	44.4	149.8	59.0	138.6	54.6
3. Kneeling leg length	63.9	25.2	50.2	23.3	75.5	29.7	70.5	27.8
4. Kneeling leg height	121.9	48.0	114.5	45.1	136.9	53.9	130.3	51.3
5. Horizontal length, knees bent	150.8	59.4	140.3	55.2	173.0	66.1	163.8	64.5
6. Bent knee height, supine	44.7	17.6	41.3	16.3	53.5	21.1	49.6	19.5

[a]See Figure 3.4 for illustration.

Figure 23.5 *(Continued)*

23.7 ASSESSING MANUAL HANDLING TASKS

Factors that need to be considered in evaluating the risk of manual handling tasks include the relationship between the task, the environment, the object handled, and the person (people) performing the task (Imada, 1993; Waters, 1993).

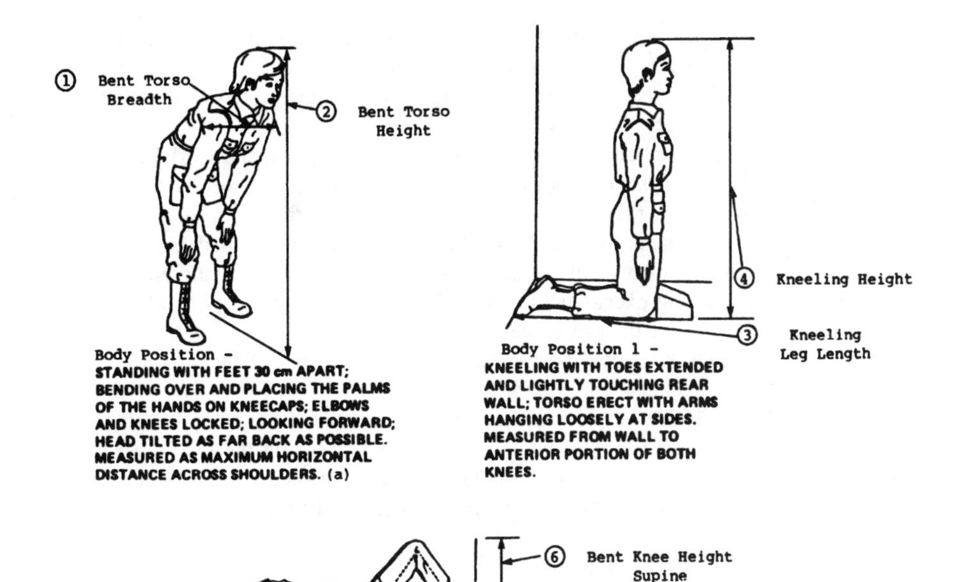

Figure 23.5 (*Continued*)

Considerations	Factors to Consider
The tasks	Location/position of the object; frequency and duration of handling; precision, type of handling; velocity and acceleration; time constraints, pacing, incentives; work/rest cycle, shift work, job rotation; availability of assistance
Environment	Size and layout of workplace (obstructions); terrain; lighting/visibility/humidity/temperature; motion (vibration/transport)
Object handled	Nature (animate/inanimate); resistance to movement; size and shape; center of gravity; physical/chemical hazards; mechanical status; handling interface; information/instructions
Worker	Gender, age, strength, health status; physical status; motivation; skill/knowledge, perception; size and shape; handedness; protective equipment

Evaluation techniques that integrate biomechanical, physiological, and psychophysical considerations to assess the appropriateness of job tasks are available. Probably the best known method for evaluating the demands of material handling tasks is the NIOSH lifting equation. The NIOSH lifting equation was first published in 1981 to assist safety and health practitioners evaluating sagittal plane lifting tasks (NIOSH, 1981; Habes,

1985). The equation has recently been revised to reflect new research findings and provide methods for evaluating asymmetrical lifting tasks, lifts of objects with less than optimal hand–container couplings, and jobs with a larger range of work durations and lifting frequencies (Waters, 1993).

Using criteria from the fields of biomechanics, psychophysics, and work physiology, the equation defines a recommended weight limit (RWL) based on specific task parameters (e.g., the location of the load relative to the body and the floor, the distance the load is moved, the frequency of the lift). The RWL represents a load that nearly all healthy workers can lift over a substantial period of time without placing an excessive load on the back, causing excess fatigue, or otherwise increasing the risk of low back pain. The actual weight of lift can be compared to the RWL for a given task to derive an estimate of the risk presented by the task, and to determine if measures to reduce the risk of injury to workers are needed.

NIOSH (Waters, 1993) has recommended no more than 51 pounds be lifted in a "perfect" lift. The perfect lift includes using two hands, holding the load close to the body, the load having good handles, the load is located between knuckle and shoulder height, and no awkward postures or obstructions. This load limit is reduced:

- The further the object is from the body
- The distance the object has to be lifted (distance)
- The lower or higher the lift is from knuckle to shoulder
- The more frequent the lift
- The more difficult it is to handle (coupling)
- The more twisted the torso is during the lift

An example of the NIOSH lifting equation and its application is presented in Appendix B.

23.7.1 Strength Data

Strength data for different populations and muscle groups have been published in a number of sources (Ulin, 1992; Kamon, 1978; Mathiowetz, 1985). If the force requirements of a task are known, it may be possible to compare these requirements against existing strength data to estimate the percentage of the worker population for which the job may be difficult. Successfully applying strength data requires knowledge of the subject population upon which the data are based, the posture in which the measurements were made, whether the measurement was static or dynamic, and how long the effort was sustained (Keyserling, 1991).

Computerized biomechanical models have been developed to predict the percentage of males and females capable of exerting static forces in certain postures. The advantage of these models is that most recognize that a worker's capacity for force exertion is rarely dependent on the strength of a single muscle group. Rather, the capacity for exerting force is dependent on the moment created at each joint by the external load and the muscle strength at that joint. The models compute the moment created at each joint by an exertion, and compare the moments against static strength data to estimate the percent of the population capable of performing a specific exertion for each joint and muscle function (Ulin, 1990). Currently, both two- and three-dimensional models are available for these analyses (Chaffin, 1992).

23.7.2 Design Criteria: Anthropometry

Anthropometric data on body size and range of joint motion can be used to assess the appropriateness of workplace, equipment, and product designs relative to the capacities for reach, grasp, and clearance of the work force. Compilations of anthropometric data for different populations are available from numerous sources (NASA, 1978; Kodak, 1983). An example of these data can be seen in Fig. 23.5.

23.7.3 Physiological Data

In activities such as repetitive lifting and load carrying, large muscle groups perform submaximal, dynamic contractions. During these activities, a worker's endurance is primarily limited by the capacity of the oxygen transporting and utilization systems (maximum aerobic power) (Astrand, 1986). Data on the maximum aerobic capacities of working populations and the energy demands of common industrial tasks have been compiled by a number of sources (Kodak, 1986; Astrand, 1986; NIOSH, 1981). In general, maximum aerobic capacity generally declines with age, increases with physical fitness level, and is 13 to 30 percent lower for women compared with men (Durnin, 1967). Several researchers suggest that the maximum energy expenditure rate for an 8-hour workday should not exceed 33 percent of maximum aerobic power (Kodak, 1986). Limits of 5.2 kcal/min for average healthy young males, or 3.5 kcal/min for populations containing women and older workers, have been proposed based on this recommendation (Bink, 1962; Bonjer, 1962; Waters, 1993).

Because table values provide only a rough approximation of the metabolic costs of a given job, models to predict metabolic energy expenditure for simple tasks based on a combination of personal and task variables have also been developed (Givoni, 1971; Vanderwalt, 1973). It has been demonstrated that the energy expenditure rate of complex jobs can be predicted if the energy expenditure rates of the simple tasks that comprise the job and the time duration of the job are known. Comparison of measured and model-predicted rates for 48 tasks indicated that models can account for up to 90.8 percent of the variation in measured metabolic rates (Garg, 1978).

23.7.4 Psychophysical Data

It is difficult to apply strength data to dynamic tasks involving more than one muscle group. In addition, motivational factors play an important role in determining an individual's capacity for physical work. This has resulted in a "Gestalt" method of analysis of human capability, called "psychophysics," to develop guidelines for the evaluation and modification of repetitive work tasks (Snook, 1969; Gamberale, 1990; Putz-Anderson, 1993). Psychophysical limits are generally based on data derived from laboratory simulations of a specific task in which the participants are allowed to adjust their workload to a level subjectively defined as the maximum acceptable. Limiting workload in this manner should allow workers to perform work tasks without overexertion or excessive fatigue (Snook, 1978).

Although there is little data to indicate how psychophysically derived limits relate to the risk of injury during work, many researchers believe that use of these limits may be the most accurate method of determining if a given task is acceptable (Chaffin, 1991). Psychophysical limits for various lifting, pushing, and pulling tasks and other manual operations have been developed and are widely available for application (Snook, 1991, 1996a,b).

An example of two psychophysical scales that have been used to rate the effort required to do a job are presented in Fig. 23.4 (p. 945). This is the Borg scale. The scale on the left has been used to estimate whole body exertion, since it has been validated against heart rate. The scale has been used to represent a rating of "7" as the baseline heart rate of approximately 70 beats per minute and a rating of "20" has been proposed as the maximum heart rate of 200 beats per minute. The Borg scale on the right has been used to estimate maximum voluntary contraction (MVC) of large muscle groups. A rating of 1 corresponds to approximately 10 percent of MVC. In general, subjects tend to be more accurate in their ability to approximate their heart rate and MVC at the tail ends of the scale. However, these scales are excellent tools for looking at before and after results of the implementation of controls.

23.8 IDENTIFY POSSIBLE SOLUTIONS

Ultimately, the results of a job analysis should provide facts that allow the ergonomics team to recommend controls to eliminate or reduce identified work factors. This may be achieved through the modification of equipment and tools, workstations, or work methods that contribute to excessive work demands. In all cases, the best ergonomic solutions are those in which safe work is a natural result of the job design and are independent of specific worker capabilities or work techniques. However, in situations where design changes are infeasible, it may be possible to limit exposure to work factors by the identification of administrative controls, such as rest breaks and worker rotation. In some situations, reducing work factors may require a combination of engineering or administrative controls.

23.9 SELECTION AND EVALUATION OF SOLUTIONS

When possible, the proposed solution should originate from a team that includes the workers who will be affected by the solution and by management who will be responsible for providing funds to purchase new equipment (if needed) or to approve of changes to work practices. All solutions should be tested among a small group of workers, to allow adjustments to be made before widespread changes are implemented. A common flaw seen in industry is that large capital investments dare often made, usually due to artifacts such as fiscal year budget deadlines and purchase order minimums, before the solutions have been tested. Once changes are implemented, followup job analyses should be performed to ensure that the solutions have effectively reduced the work factors without imposing new demands on the worker.

Beyond the specific controls, a multifaceted abatement program is often necessary to properly control hazards leading to musculoskeletal disorders and to proactively protect employees. Some elements of an ergonomic solution process to bring the work system into balance are summarized below and were introduced early in this chapter as NIOSH recommendations.

23.9.1 Analysis

If in-house expertise does not exist, a consultant should assist an in-house team to perform a systematic evaluation with regard to existing and new work practices and workstation design. The consultant should work with the ergonomic team to recommend engineering and administrative controls to reduce or eliminate ergonomic stressors. The consultant can also assist in the implementation of the recommendations, the evaluation of the effectiveness of the controls implemented, and can make new recommendations if necessary.

23.9.2 Controls

Hazards associated with the development of MSDs can be controlled through proper engineering design of the job, workstation, and equipment so the work can be performed independent of specific worker characteristics and techniques. *This requires the job to be designed to fit the worker and not vice versa.* Engineering controls attempt to reduce extreme postures, excessive forces, and repetitive motions. To be effective, employee input is necessary, since improperly designed workstations and controls will not be used if employees believe they interfere with their work. Also, after installation, the effectiveness of the controls must be evaluated and modified if necessary to ensure their effectiveness. Appendix C contains a list of examples and references from NIOSH that demonstrate the effectiveness of engineering controls for reducing exposures to ergonomic risk factors.

Administrative controls can include but are not limited to: training of new employees in safe work techniques including lifting, working with minimum strain on the body, and minimizing the application of forces with the fingers; job rotation and job enhancement; adequate mandatory rest breaks; implementation of an exercise program.

23.9.3 Medical Management Program

A medical management program is necessary to monitor employees and prevent early symptoms from progressing to injuries and illnesses. This program should include:

1. Determining the extent of injuries and illnesses; determining if injuries and illnesses are caused or aggravated by work;
2. Educating all employees and supervisors on early signs of injuries and disorders and encourage early reporting;
3. Instituting a formal documented tracking and surveillance program to monitor injury trends in the plant;
4. Providing adequate treatment of ergonomic-related cases (including not reassigning employees to a job until it has been modified to minimize the hazards that resulted in the injury);
5. Allowing adequate time off for recovery after surgery or other aggressive intervention;
6. Preventive measures, including early physical evaluation of employees with musculoskeletal symptoms;
7. Allowing adequate time off after a cumulative trauma disorder is diagnosed; and
8. Providing access to trained medical personnel for development and implementation of conservative treatment measures upon detection of cumulative trauma disorder symptoms.

23.9.4 Training Programs

A training program is necessary to alert employees on the hazards of cumulative trauma disorders and controls and work practices that can be used to minimize the hazards. This includes designing and implementing a written training program for managers, supervisors, engineers, union representatives, health professionals, and employees on the nature, range, and causes and means of prevention of ergonomic-related disorders. The training program for new and reassigned workers should allow the following:

1. Demonstrations of safe and effective methods of performing their job;
2. Familiarization of employees with applicable safety procedures and equipment;
3. The new or reassigned employee to work with a skilled employee and/or provide on the job training for specific jobs; and
4. New or reassigned employees to condition their muscle/tendon groups prior to working at full capacity rate, which has been determined to be safe and will not cause adverse effects.

Workers should be instructed in the basics of body biomechanics and work practices to minimize the ergonomic hazards associated with their jobs. This should include, but not be limited to:

Avoid postures where:

- The elbow is above midtorso
- The hand is above the shoulder
- The arms must reach behind the torso

Avoid wrist postures where there is:

- Inward or outward rotation with bent wrist
- Excessive palmar flexion or extension
- Ulnar or radial deviation
- Pinching or high finger forces with above postures

Avoid mechanical stress concentrations on elbows, base of palm, and backs of fingers.

General lifting guidelines include:

- Keep the load close to the body
- Use the most comfortable posture
- Lift slowly and evenly (do not jerk)
- Do not twist the back
- Securely grip the load
- Use a lifting aid or get help

See Chapter 7 for more advice on training techniques.

23.10 AN EXAMPLE OF JOB ANALYSIS FORMS

The forms in Figs. 23.6 and 23.7 provide an outline for how the methods and information discussed above can be used to identify work factors present in jobs in the work environment so that system balance can be restored. The implementation plan is especially useful for planning which solutions will be implemented and when. Appendix

Implementation Plan	Date Initiated
Job Name Department Tracking #	
Process	
Number of Employees 1st Shift 2nd Shift 3d Shift	
Musculoskeletal Disorders in Last Year Neck___ Wrist___ Back___ Ankle___ Shoulder___ Hand___ Hip Elbow Fingers Knee Other	

Action	**Responsibility**	**Date Assigned**	**Date Completed**
Checklist Score Upper Lower			
Discomfort Survey			
Employee Input No Yes (attach)			
Job Analysis			
Videotape			
Employee Input			

Risk Factor	**Cause**	**Solution (Controls)**	**Status**	**Responsible**	**Finish Date**
A					
B					
C					
D					
E					
F					
G					
H					
I					

Implementation Plan		Date Initiated	
Followup	**Responsibility**	**Date Assigned**	**Date Completed**
Checklist Score Upper Lower			
Discomfort Survey			
Employee Input No Yes (attach)			
Postchange Videotape			
Vendor Notification of Problems			
Maintenance Requirements			
Describe			
Remaining Problems			

Risk Factor	**Cause**	**Solution (Controls)**	**Status**	**Responsible**	**Finish Date**
A					
B					
C					
D					
E					
F					
G					
H					
I					

Attachments
Risk factor checklists
Body part discomfort surveys
Comments from the employee interview
Job analyses
Specifications
Work order numbers
Vendor costs
Vendor notified of problems

Figure 23.6 Sample forms for job analysis summary.

Example of Job, Tasks, Work Cycle, and Elements

Job Name: Sew uppers
Task: Sew elastic and liner together for shoe upper
Cycle Time: 20 seconds
Task Elements: The task consists of the following six elements.

1. Get unit from the stack (left hand, wrist flexed).
2. Place unit over sewing mount (both hands, pinch grip 1 in 10 requires more than 2 pounds).
3. Push through sewing machine (both hands, finger press, low force).
4. Turn knob to lift needle (left hand, pinch grip less than 2 pounds, wrist extended, forearm rotation).
5. Cut thread with scissors (right hand, wrist flexed, fingers contact hand surface).
6. Discard unit into finished bin (left hand, shoulder extended).

Example of Job, Tasks, Work Cycle, and Elements

Job Name: Order picker
Task: Fill order
Cycle time: 15 seconds
Task Elements: The task consists of the following six elements.

1. Remove backing strip from label (right hand).
2. Retrieve stock unit from shelf (either hand).
3. Place stock unit on cart shelf (either hand).
4. Place label on stock unit (right hand).
5. Push stock unit into cart (left hand).
6. Push cart to next pick (both hands).

Fig. 23.7 Examples of job tasks and elements.

C provides a list of selected studies from NIOSH on the various control strategies for reducing MSDs and discomfort.

23.11 SUMMARY

The information in this chapter has provided a basic framework, with examples, for ergonomic analysis and the control of poorly designed work environments. Since ergonomics is a dynamic science, the reader is encouraged to consult the resource list below to maintain a working library of information specific to the exposures of interest. In addition, the references provided by NIOSH in Appendix C provide an excellent source of networking opportunities and examples of companies that have successfully worked towards healthy work systems. Perhaps the most important "take-home message" that should be gained from this chapter is that management and employees need to work together in a concerted effort to control workplace exposures through a continuous process. It is hoped that the information provided in this chapter will facilitate that process.

APPENDIX A

NIOSH Lift Guidelines

Example: Loading Punch Press Stock

- Requires control at the destination of lift.
- Duration < 1 hour
- Weight of the supply reel = 44 lbs

Origin	Destination
H = 23″	23″
V = 15″	64″
D = 49″	49″
A = 0	0
C = Fair	Fair
F = <0.2	<0.2

Origin

$$
\begin{aligned}
HM &= (10/23) = 0.43 \\
VM &= (1 - 0.0075|15 - -30|) = 0.89 \\
DM &= (0.82 + 1.8/49) = 0.86 \\
AM &= (1 - 0.0032 \times 0) = 1 \\
CM &= 0.95 \\
FM &= 1 \\
RWL &= 51 \times HM \times FM \times DM \times AM \times CM \times FM \\
&= 51 \times 0.43 \times 0.89 \times 0.86 \times 1 \times 0.95 \times 1 \\
&= 15.9 \text{ lbs}
\end{aligned}
$$

Destination

$$
\begin{aligned}
HM &= (10/23) = 0.43 \\
VM &= (1 - 0.0075|64 - -30|) = 0.75 \\
DM &= (0.82 + 1.8/49) = 0.86 \\
AM &= (1 - 0.0032 \times 0) = 1
\end{aligned}
$$

$$\begin{aligned}
\text{CM} &= 1 \\
\text{FM} &= 1 \\
\text{RWL} &= 51 \times \text{HM} \times \text{VM} \times \text{DM} \times \text{AM} \times \text{CM} \times \text{FM} \\
&= 51 \times 0.43 \times 0.75 \times 0.86 \times 1 \times 1 \times 1 \\
&= 14.1 \text{ lbs}
\end{aligned}$$

$$\begin{aligned}
\text{RWL} &= \text{lower of 15.9 and 14.1 lbs} \\
&= 14.1 \text{ lbs} \\
\text{LI} &= \frac{44}{14.1} \\
&= 3.12
\end{aligned}$$

1981, AL = 15.8 lbs based on H = 21”

Recommendations

- Engineering controls: Use a mechanical device to load supply reel (hoist, small crane, etc.)
- Administrative controls: Use two workers and load the reel from the side
- Education and training of workers
- Enforcement

Calculation for Recommended Weight Limit

$$\text{RWL} = LC \times HM \times VM \times DM \times AM \times FM \times CM$$

Recommended weight limit

Component		*Metric*	*U.S. customary*
LC = load constant	=	23 kg	51 lbs
HM = horizontal multiplier	=	$(25/H)$	$(10/H)$
VM = vertical multiplier	=	$(1 - (0.003\lvert V - 75\rvert))$	$(1 - (0.0075\lvert V - 30\rvert))$
DM = distance multiplier	=	$(0.82 + (4.5/D))$	$(0.82 + (1.8/D))$
AM = asymmetric multiplier	=	$(1 - (0.0032A))$	$(1 - (0.0032A))$
FM = frequency multiplier (from table 7)			
CM = coupling multiplier (from table 6)			

where:

H = horizontal distance of hands from midpoint between the anles. Measure at the origin and the destination of the lift (cm or in.).

V = vertical distance of the hands from the floor. Measure at the origin and destination of the lift (cm or in.).

D = vertical travel distance between the origin and the destination of the lift (cm or in.).

A = angle of asymmetry—angular displacement of the load from the sagittal plane. Measure at the origin and destination of the lift (degrees).

F = average frequency rate of lifting measured in lifts/min. Duration is defined to be: ≤ 1 h; ≤ 2 h; or ≤ 8 h assuming appropriate recovery allowances.

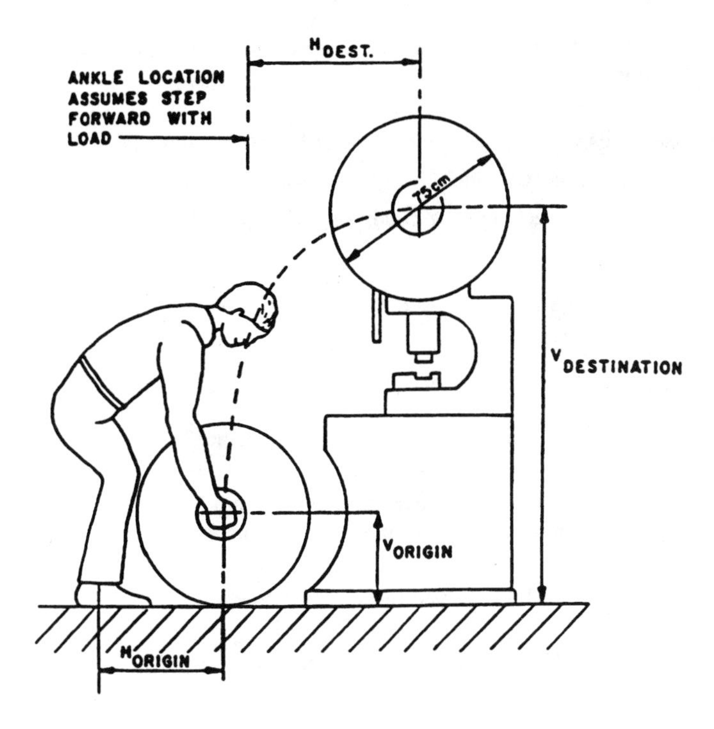

APPENDIX B

ASSESSMENT OF AND SOLUTIONS TO WORKSITE RISK FACTORS (OSHA, 1995)

I. WORKSITE ASSESSMENT CHECKLISTS

The following are sample checklists that you may wish to use as a guide in developing your own worksite assessment checklist. These checklists cover:

- Manual handling
- Task/work methods
- Video display unit and keyboard issues
- Workstation layout
- Hand tool use

Note: You should consider modifying these checklists to include the specific concerns of your site. Other areas to assess may include:

- Information displays and controls (such as in control rooms)
- Environmental concerns (such as lighting, noise, or heat)
- Facility concerns (such as floor condition, ventilation, or vibration)

MANUAL HANDLING CHECKLIST

Yes No

☐ ○ 1. Is material moved over a minimum distance?

☐ ○ 2. Is the horizontal distance between the middle knuckle and the body less than 4 in. during manual material handling?

☐ ○ 3. Are obstacles removed to minimize reaches?

☐ ○ 4. Are walking surfaces level, slip resistant, and well lit?

☐ ○ 5. Are objects:

☐ ○ • Able to be grasped by good handholds?
☐ ○ • Stable?
☐ ○ • Able to be held without slipping?

☐ ○ 6. When required, do gloves improve the grasp without bunching up or resisting movement of the hands?

☐ ○ 7. Is there enough room to access and move objects?

☐ ○ 8. Are mechanical aids easily available and used whenever possible?

☐ ○ 9. Are working surfaces adjustable to the best handling heights?

☐ ○ 10. Does material handling avoid:

☐ ○ • Movements below knuckle height and above shoulder height?
☐ ○ • Static awkward postures?
☐ ○ • Sudden movements during handling?
☐ ○ • Twisting of the trunk?
☐ ○ • Excessive reaching while holding/moving the load?

☐ ○ 11. Is help available for heavy or awkward lifts?

Yes No

☐ ○ 12. Are high rates of repetition avoided by:

☐ ○ • Job rotation?
☐ ○ • Self-pacing?
☐ ○ • Sufficient rest pauses?

☐ ○ 13. Are pushing/pulling forces reduced/eliminated by:

☐ ○ • Casters that are sized correctly and roll freely?
☐ ○ • Handles for pushing/pulling?
☐ ○ • Availability of mechanical assists?

☐ ○ 14. Are objects rarely carried more than 10 ft?

☐ ○ 15. Is there a prevention maintenance program for manual handling equipment?

TASK/WORK METHODS CHECKLIST

Yes No

☐ ○ 1. Does the design of the task reduce or eliminate:

☐ ○ • Bending or twisting of the trunk?
☐ ○ • Squatting or kneeling?
☐ ○ • Elbows above midtorso?
☐ ○ • Extending the arms?
☐ ○ • Bending the wrist?
☐ ○ • Static muscle loading?
☐ ○ • Forceful pinch grips?

☐ ○ 2. Are mechanical devices used to lift or move objects that are heavy or require repetitive lifting?

☐ ○ 3. Can the task be performed with either hand?

☐ ○ 4. Are the materials:

☐ ○ • Able to be held without a forceful grip?
☐ ○ • Easy to grasp?
☐ ○ • Free from sharp edges?

☐ ○ 5. Do containers have good handholds?

☐ ○ 6. Are fixtures used to reduce or eliminate hard grasping forces (e.g., part held by the fixture, not by the hands)?

Yes No

☐ ○ 7. If gloves are needed, do they fit properly?

☐ ○ 8. Does the task avoid contact with sharp edges or corners?

☐ ○ 9. Is exposure to repetitive motions reduced by:

☐ ○ • Job rotation?
☐ ○ • Self-pacing?
☐ ○ • Sufficient rest pauses?

VIDEO DISPLAY UNIT (VDU) AND KEYBOARD ISSUES CHECKLIST

Yes No

☐ ○ 1. Can the workstation be adjusted to ensure proper posture by:

☐ ○ • Adjusting knee and hip angles to achieve comfort and variability?
☐ ○ • Supporting heels and toes on the floor or a footrest?
☐ ○ • Placing arms comfortably at the side and hands parallel to the floor (plus or minus 2 in.)?
☐ ○ • Holding wrists nearly straight and resting them on a padded surface? (Note: Wrists should not rest on the padded surface while keying).

☐ ○ 2. Does the chair:

☐ ○ • Adjust easily from the seated position?
☐ ○ • Have a padded seat pan (soft but compresses about 1 in.)?
☐ ○ • Have a seat that accommodates the worker?
☐ ○ • Have a back rest that provides lumbar support and can be used while working?
☐ ○ • Have a stable base with casters that are suited to the type of flooring?

☐ ○ 3. Does the chair manufacturer offer different seat pan lengths that have a waterfall design?

☐ ○ 4. Does the seat pan adjust for both height and angle?

☐ ○ 5. Is there adequate clearance for the feet, knees, and legs relative to the edge of the work surface?

☐ ○ 6. Is there sufficient space for the thighs between the work surface and the seat?

☐ ○ 7. Are the keyboard height from the floor and the slope of the keyboard surface adjustable?

Yes	No	
☐	○	8. Is the keyboard prevented from slipping when in use?
☐	○	9. Is the keyboard detachable?
☐	○	10. Does the keyboard meet ANSI/HFS 100-1988 (or ISO 9241) standards?
☐	○	11. Is the mouse, pointing device, or calculator at the same level as the keyboard?
☐	○	12. Are the head and neck held in a neutral posture?
☐	○	13. Are arm rests provided for intensive or long-duration keying jobs?
☐	○	14. Is the screen clean and free from flickering?
☐	○	15. Is the top of the screen slightly below eye level? (For non bi-focal users.)
☐	○	16. Can the screen swivel horizontally and tilt or elevate vertically?
☐	○	17. Does the monitor have brightness and contrast controls?
☐	○	18. Is the monitor between 18 and 30 in. from the worker? (Some workers may need glasses specifically for computer use.)
☐	○	19. Is there sufficient lighting without glare on the screen from lights, windows, and surfaces?
☐	○	20. Are headsets used when frequent telephone work is combined with hand tasks such as typing, use of a calculator, or writing?
☐	○	21. Is the job organized so that workers can change postures frequently?
☐	○	22. Does the worker leave the workstation for at least 10 min after every hour of intensive keying and for a least 15 min after every 2 h of intermittent keying?
☐	○	23. Is intensive keying avoided by:
☐	○	• Job rotation?
☐	○	• Self-pacing?
☐	○	• Job enlargement?
☐	○	• Adequate recovery breaks?
☐	○	24. Is there the possibility of alternating tasks during the shift (e.g., intensive keying or mouse work, filing, copying telephone calls, intermittent keying)?

Yes No

□ ○ 25. Are employees trained in:

□ ○ • Healthy work postures?
□ ○ • Safe and healthy work methods?
□ ○ • How to make adjustments to the workstation?
□ ○ • Awareness of risk factors for musculoskeletal disorders?
□ ○ • How to seek assistance with concerns?

□ ○ 26. Are workers able to set their own pace, without electronic monitoring or incentive pay?

WORKSTATION LAYOUT CHECKLIST

Yes No

□ ○ 1. Is the workstation designed to reduce or eliminate:

□ ○ • Bending or twisting of the trunk?
□ ○ • Squatting or kneeling?
□ ○ • Elbows above midtorso?
□ ○ • Extending the arms?
□ ○ • Bending the wrist?
□ ○ • Hands behind the body?

□ ○ 2. Are mechanical aids and equipment available?

□ ○ 3. Can the work be performed without repetitive (or static) bending or reaching?

□ ○ 4. Are awkward postures reduced by:

□ ○ • Providing adjustable work surfaces and supports (such as chairs or fixtures)?
□ ○ • Tilting the surface?

□ ○ 5. Are all job requirements visible without awkward postures?

□ ○ 6. Is an arm rest provided for precision work?

□ ○ 7. Is a foot rest provided for those who need it?

□ ○ 8. Are cushioned floor mats and foot rests provided for employees who are required to stand for long periods?

□ ○ 9. Have jobs been reviewed to determine if they are best suited for sit/stand, seated, or standing work?

Yes No

☐ ○ 10. Are seated workers provided chairs that:

☐ ○ • Adjust easily from the seated position?
☐ ○ • Have a padded seat pan (soft but that compresses no more than 1 in.)?
☐ ○ • Have a seat that is wide enough to accommodate the worker?
☐ ○ • Have a back rest that provides lumbar support and can be used while working?
☐ ○ • Have a stable base with casters that are suited to the type of flooring?

☐ ○ 11. Do chair manufacturers offer different seat pan lengths that have a waterfall design?

☐ ○ 12. Does the seat pan adjust for both height and angle?

☐ ○ 13. Are body parts (e.g., hands, arms, legs) free from contact stress from sharp edges on work surfaces?

☐ ○ 14. Is there a preventive maintenance program for mechanical aids, tools, and other equipment?

HAND TOOL USE CHECKLIST

Yes No

☐ ○ 1. Are tools selected to:

☐ ○ • Minimize exposure to localized vibration?
☐ ○ • Reduce hand force?
☐ ○ • Reduce/eliminate bending or awkward postures of the wrist?
☐ ○ • Avoid forceful pinch grips?
☐ ○ • Avoid the use of the hand as a hammer?

☐ ○ 2. Are tools powered where necessary (e.g., to reduce forces, repetitive motions)?

☐ ○ 3. Are tools evenly balanced?

☐ ○ 4. Are heavy tools suspended or counterbalanced?

☐ ○ 5. Does the tool allow adequate view of the work?

☐ ○ 6. Does the tool grip/handle prevent slipping during use?

Yes No

☐ ○ 7. Are tools equipped with handles that:

☐ ○ • Do not press into the palm area?
☐ ○ • Are made of textured, nonconductive material?
☐ ○ • Have a grip diameter suitable for most workers, or are different size handles available? (women tend to have smaller hands than men)

☐ ○ 8. Can the tool be used safely with gloves?

☐ ○ 9. Can the tool be used with either hand?

☐ ○ 10. Is there a preventive maintenance program to keep tools operating as designed?

☐ ○ 11. Can triggers be operated by more than one finger to avoid static contractions?

☐ ○ 12. Does the tool design or workstation minimize the twist or shock to the hand (in particular, observe the reaction of power tools after the torque limit is reached)?

Revised NIOSH Guide for Manual Lifting
Analysis of Individual Task

Analyst:		Object Weight: Average:	20 lbs.
JOB:	Loader in Receiving Area	Maximum:	44 lbs.
Task:	Product B		

	ORIGIN	DESTINATION
Horizontal Multiplier, HM	0.83	
Vertical Multiplier, VM	0.78	
Distance Multiplier, DM	1.00	
Asymmetric Multiplier, AM	1.00	
Frequency Multiplier, FM	0.65	
Coupling Multiplier, CM	0.95	

RWL	20.4 lbs.
Lifting Index, LI	0.98
Rest Allowance	0 Minutes

Estimated percent capable population excluding compressive force.

Male = More than 99% Capable

Female = 76% Capable

Help | Print Screen | Print | Edit Task | Edit Job | Calculate Job

NOTE: RWL and LI are based on the Average Weight, Maximum Weight was ignored.

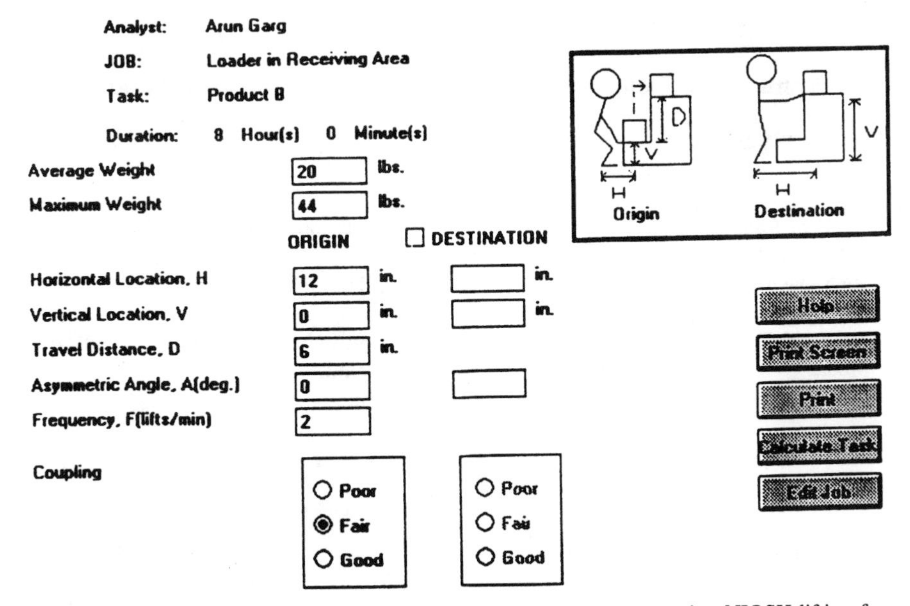

Individual Task Data

Analyst: Arun Garg

JOB: Loader in Receiving Area

Task: Product B

Duration: 8 Hour(s) 0 Minute(s)

Average Weight 20 lbs.

Maximum Weight 44 lbs.

	ORIGIN	DESTINATION
Horizontal Location, H	12 in.	in.
Vertical Location, V	0 in.	in.
Travel Distance, D	6 in.	
Asymmetric Angle, A(deg.)	0	
Frequency, F(lifts/min)	2	
Coupling	○ Poor ◉ Fair ○ Good	○ Poor ○ Fair ○ Good

Fig. 23.8 Example of a computer model to analyze lifting exposures using NIOSH lifting formula (Courtesy of Dr. Arun Garg).

Revised NIOSH Guide for Manual Lifting
Analysis of Entire Job

Analyst: Arun Garg

JOB: Loader in Receiving Area

Frequency Weighted RWL (FWRWL)	42.5	lbs.
FW Average Weight (FWW)	13.9	lbs.
Composite Recommended Weight Limit (CRWL)	7.6	lbs.
Composite Lifting Index (CLI)	1.83	
Recommended Rest Allowances	0 Minutes	

Help
Print Screen
Print
Edit Job

Estimated percent capable population excluding compressive force.
Male = 86% Capable
Female = 28% Capable

Task		Frequency Independent (Origin)	Using Task Frequency (Origin)	Frequency Independent (Dest.)	Using Task Frequency (Dest.)
Product A	RWL	22.3	16.7		
	LI	1.48	0.96		
Product B	RWL	31.4	20.4		
	LI	1.40	0.98		
Product C	RWL	51.0	17.9		
	LI	0.43	0.61		

System Requirements

- IBM 80286 or Better Compatible Computer
- 5.25" or 3.5" Floppy Disk Drive
- 1 MB of Ram
- VGA Color Monitor
- High Quality Dot-Matrix or Laser Printer
- Microsoft Windows 3.x
- Mouse

Four Different Windows

- Analyst, Job, Duration, and Task List Windows
- Individual Task Data Window
- Simple Task Analysis Window
- Multiple Task (Entire Job) Analysis Window

Fig. 23.8 *(Continued)*

APPENDIX C
TABLES OF ERGONOMIC STUDIES

Select Studies Demonstrating Effectiveness of Engineering Controls for Reducing Exposure to Ergonomic Risk Factors

Study	Target Population	Problem/Risk Factor	Control Measure	Effect
Miller et al. (1971)	Surgeons (use of bayonet forceps)	Muscle fatigue during forceps use, frequent errors in passing instruments	Redesigned forceps (increased surface area of handle)	Reduced muscle tension (determined by EMG) and number of passing errors
Armstrong et al. (1982)	Poultry cutters (knives)	Excessive muscle force during poultry cutting tasks	Redesigned knife (reoriented blade, enlarged handle, provided strap for hand)	Reduced force grip during use, reduced forearm muscle fatigue
Knowlton and Gilbert (1983)	Carpenters (hammers)	Muscle fatigue, wrist deviation during hammering	Bent handle of hammer and its diameter	Smaller decrement in strength, reduced ulnar wrist deviation
Habes (1984)	Auto workers	Back fatigue during embossing tasks	Designed cutout in die to reduce reach distance	Reduced back muscle fatigue as determined by EMG
Goel and Rim (1987)	Miners (pneumatic chippers)	Hand–arm vibration	Provided padded gloves	Reduced vibration by 23.5% to 45.5%

Wick (1987)	Machine operators in a sandal plant	Pinch grips, wrist deviation, high repetition rates, static loading of legs and back	Provided adjustable chairs and bench-mounted armrests; angled press, furnished parts bins	Reduced wrist deviation, compressive force on lumbar–sacral discs from 85 to 13 lb
Little (1987)	Film notchers	Wrist deviation, high repetition rates, pressure in the palm of the hand imposed by notching tool	Redesigned notching tool (extended, widened and bent handles, reduced squeezing force)	Reduced squeezing force from 15 to 10 lb, eliminated wrist deviation, productivity increased by 15%
Johnson (1988)	Power handtool users	Muscle fatigue, excessive grip force	Added vinyl sleeve and brace to handle	Reduced grip force as determined by EMG
Fellows and Freivalds (1989)	Gardeners (rakes)	Blisters, muscle fatigue	Provided foam cover for handle	Reduced muscle tension and fatigue buildup as determined by EMG
Andersson (1990)	Power handtool users	Hand–arm vibration	Provided vibration damping handle	Reduced hand-transmitted vibration by 61% to 85%

Reducing Exposure to Ergonomic Risk Factors

Study	Target Population	Problem/Risk Factor	Control Measure	Effect
Radwin and Oh (1991)	Trigger-operated power hand tool users	Excessive hand exertion and muscle fatigue	Extended trigger	Reduced finger and palmar force during tool operation by 7%
Freudenthal et al. (1991)	Office workers	Static loading of back and shoulders during seated tasks	Provided desk with 10 degree incline, adjustable chair and tables	Reduced moment of force on lower spinal column by 29%; by 21% on upper part
Powers et al. (1992)	Office workers	Wrist deviation during typing tasks	Provided forearm supports and a negative slope keyboard support system	Reduced wrist extension
Erisman and Wick (1992)	Assembly workers	Pinch grips, wrist deviation	Provided new assembly fixtures	Eliminated pinch grips, reduced wrist deviations by 65%, reduced cycle time by 50%
Luttmann and Jager (1992)	Weavers	Forearm muscle fatigue	Redesigned workstation (numerous changes)	Reduced fatigue as measured by EMG, improved quality of product
Fogleman et al. (1993)	Poultry workers (knives)	Excessive hand force, wrist deviation	Altered blade angle and handle diameter	Wrist deviation reduced with altered blade angle
Lindberg et al. (1993)	Seaming operators	Awkward, fixed (static) neck and shoulder postures, monotonous work movements, high work pace	Automated seaming task	Freer head postures during automated seaming; loads on neck/shoulder muscles reduced as indicated by EMG; perceived exertion was reduced
Nevala-Puranen et al. (1993)	Dairy farmers	Whole-body fatigue, bent and twisted back postures, static arm postures	Installed rail system for carrying milking equipment	Heart rate decreased; bent and twisted back/trunk postures decreased by 64%; above-shoulder arm postures cut in half; mean milking time per cow decreased by 24%

Select Studies of Various Control Strategies for Reducing Musculoskeletal Injuries and Discomfort

Study	Industry	Study Group	Intervention Method	Summary of Results	Additional Comments
Itani et al. (1979)	Film manufacturing	124 film rolers in two groups	Reduced work time, increased number of rest breaks	Reduction in neck and shoulder disorders and low back complaints, improved worker health	Productivity after the intervention was found to be 86% of the pre-intervention level
Luopajarvi et al. (1982)	Food production	200 packers	Redesigned packing machine	Decreased neck, elbow, and wrist pain	Not all recommended job changes implemented; workers still complained
Drury and Wick (1984)	Shoe manufacturing	Workers at 6 factory sites	Workstation redesign	Reduced postural stress, increased productivity	Trunk and upper limbs most affecte by changes
Westgaard and Aaras (1984, 1985)	Cable forms production	100 workers	Introduced adjustable workstations and fixtures, counter-balanced tools	Turnover decreased, musculoskeletal sick leave reduced by 67% over 8-year period, productivity increased	Reductions in shoulder, upper back muscle load verified by EMB
McKenzie et al. (1985)	Telecommunication equipment manufacturing	660 employees	Redesigned handles on power screwdrivers and wire wrapping guns, instituted plant-wide ergonomics program	Incidence rate of repetitive trauma disorders decreased from 2.2 to 0.53 cases/200,000 work hours; lost days reduced from 1001 to 129 in 3 years	
Echard et al. (1987)	Automobile manufacturing		Redesigned tools, fixtures, and work organization in assembly operations	Reduced long-term upper extremity and back disabilities; CTS surgeries reduced by 50%	
LaBar (1992)	Household products manufacturing	800 workers	Introduced adjustable workstations, improved the grips on handtools, improved parts organization and work flow	Reduced injuries (particularly back) by 50%	Company also had a labor management safety committee to investigate ergonomics-related complaints

Orgel et al. (1992)	Grocery store	23 employees	Redesigned checkout counter to reduce distances, installed a height-adjustable keyboard, and trained workers to adopt preferred work practices	Decreased self-reported neck, upper back, and shoulder discomfort; no change in arm, forearm, and wrist discomfort	Study lacked a reference group not subject to the same interventions for making suitable comparisons
Rigdon (1992)	Bakery	630 employees	Formed union management committee to study cumulative trauma problems which led to workstation changes, tool modifications; improved work practices	Cumulative trauma cases dropped from 34 to 13 in 4 years; lost days reduced from 731 to 8 over the same period	Union advocated more equipment to reduce manual material handling
Garg and Owen (1992)	Nursing home	57 nursing assistants	Implemented patient transferring devices	IR of back injuries decreased from 83 to 43 per 200,000 work hours following the invervention; no lost or restricted work days during the 4 months following the intervention	
Halpern and Davis (1993)	Office	90 office workers	Adjusted workstations according to the workers' anthropometric dimensions	Body part discomfort decreased; perceived efficiency and usability of the equipment increased	
Lutz and Hansford (1987)	Medical products manufacturing	More than 1000 workers	Introduced adjustable workstations and fixtures, mechanical aids to reduce repetitive motions, job rotation	Medical visits reduced from 76 to 28 per month	Exployees also expressed enthusiasm for exercise program introduced with other interventions

(*Continued*)

Study	Industry	Study Group	Intervention Method	Summary of Results	Additional Comments
Jonsson (1988)	Telephone assembly, printed circuit card manufacturing, glass blowing, mining	25 workers	Job rotation	Job rotation in light-duty tasks not as effective as in dynamic heavy-duty tasks	Measured static load on shoulder upper back muscles with EMG
Geras et al. (1988)	Rubber and plastic parts manufacturing	87 plants within one company	Ergonomics training and intervention program introduced, added material handling equipment, workstation modifications to eliminate postural stresses	Lost time prevalence rates at two plants reduced from 4.9 and 9.7/200,000 hours to 0.9 and 2.6, respectively, within 1 year and maintained over a 4-year period	Success attributed to increased training, awareness of hazards, and improved communication between management and workers
Tadano (1990)	Office	500 VDT operators	Provided training, redesigned workstations, and incorporated additional breaks and exercises into the work schedule	Cumulative traum disorder cases reduced from 49 in the 6 months preceding the intervention to 24 in the 6 months following the intervention	
Hopsu and Fouhevaara (1991)	Office	8 female cleaners	Provide training and greater flexibility in their work and eliminated strictly proportioned work areas and time schedules	Average sick leave decreased from 20 days/year before the intervention to 10 days/year 2 years after intervention	Mean maximum VO_2 rate increased, mean heart rate decreased after intervention
Aaras (1994)	Telephone exchange manufacturing, office	96 workers (divided into 4 groups)	Provided adjustable workstations and additional work space; tools were suspended and counterbalanced	Significant reduction in intensity and duration of neck pain reported after intervention	Reductions in static loading on the neck and shoulder muscles after intervention were confirmed via EMG

Moore (1994)	Automotive engine and transmission manufacturing	5 workers	Eliminated manual flywheel truing operation by implementing a mechanical press	29% decrease in musculoskeletal disorders, 78% decrease in upper extremity CTDs, 82% reduction in restricted or lost work time	Used participatory (team) approach to select intervention method
NIOSH (1994)	Red meatpacking	3 beef/pork processing companies	Implemented participatory (labor management) ergonomics program	Results varied: only two teams able to introduce changes to address identified problems; some evidence that incidence and severity of injury was reduced following introduction of an ergonomics program	Additional followup needed to evaluate intervention effectiveness
Narayan and Rudolph (1993)	Medical device assembly plant	316 employees	Redesigned workstation to reduce reach distances, provided adjustable chairs and footrests, provided fixtures and pneumatic gripper to eliminate pinch grips	Plant-wide CTD incidence rate reduced from 13.7 to 11.3 per 200,000 worker hours after intervention, plant-wide severity rate reduced from 154.9 lost-time days to 67.8 lost-time days per 200,000 worker hours	Not all jobs in plant affected by changes
Parenmark et al. (1993)	Chain saw assembly plant	279 workers	Increased number of workers and tasks, provided training, reduced work pace, adopted new wage system and flexible working hours	Sick leave dropped from 17 to 13.7 days per worker per year, labor turnover dropped from 35% to 10%; assembly errors cut by 3% to 6%; total production cost reduced by 10%; productivity not affected	Difficult to pinpoint which factor had biggest impact

(Continued)

Study	Industry	Study Group	Intervention Method	Summary of Results	Additional Comments
Shi (1993)	County government (various occupations represented)	205 workers	Education, back safety training, physical fitness activities, equipment/facility improvements (e.g., additional material handling equipment)	Back pain prevalence declined modestly; significant improvement in satisfaction and a reduction in risky lifting behaviors were reported; a savings of $161,108 was realized, giving a 179% return in the investment	
Reynolds et al. (1994)	Apparel manufacturing	18 operators	Introduced height- and tilt-adjustable work stands, additional jigs, antifatigue mats, and automatic thread cutters	Body part discomfort reduced in shoulders, arms, hands, and wrists; no injury costs incurred in 5 months following intervention	Used worker participation approach; productivity significantly increased after intervention

APPENDIX REFERENCES

Aaras, A. (1994). Relationship between trapezius load and the incidence of musculoskeletal illness in the shoulder. *Int J Ind Ergonomics* **14**(4):341–348.

Andersson, E. R. (1990). Design and testing of a vibration attenuating handle. *Int J Ind Ergonomics* **6**(2):119–126.

Degani, A., Asfour, S. S., Waly, S. M., Koshy, J. G. (1993). A comparative study of two shovel designs. *Appl Ergonomics* **24**(5):306–12.

Drury, C. G., Wick, J. (1984). Ergonomic applications in the shoe industry. In: *Proceedings of the International Conference on Occupational Ergonomics*, pp. 489–93.

Echard, M., Smolenski, S., Zamiska, M. (1987). Ergonomic considerations: Engineering controls at Volkswagen of America. In: *Ergonomic interventions to prevent musculoskeletal injuries in industry*. Cincinnati, OH: American Conference of Governmental Industrial Hygienists, Industrial Hygiene Science Series, pp. 117–31.

Erisman, J., Wick, J. (1992). Ergonomic and productivity improvements in an assembly clamping fixture. In: Kumar, S., Ed., *Advances in industrial ergonomics and safety IV*. Philadelphia, PA, Taylor & Francis, pp. 463–68.

Fellows, G. L., Freivalds, A. (1989). The use of force sensing resistors in ergonomic tool design. In: *Proceedings of the Human Factors Society*, 33rd Annual Meeting, pp. 713-17.

Fogleman, M. T., Freivalds, A., Goldberg, J. H. (1993). An ergonomic evaluation of knives for two poultry cutting tasks. *Int J Ind Ergonomics* **11**(3):257–65.

Freudenthal, A., van Riel, M. P. J. M., Molenbroek, J. F. M., Snijders, C. J. (1991). The effect on sitting posture of a desk with a ten-degree inclination using an adjustable chair and table. *Appl Ergonomics* **22**(5):329–36.

Gallimore, J. J., Brown, M. E. (1993). Effectiveness of the C-sharp: Reducing ergonomics problems at VDTs. *Appl Ergonomics* **24**(5):327–36.

Garg, A., Owen, B. (1992). Reducing back stress to nursing personnel: An ergonomic intervention in a nursing home. *Ergonomics* **35**(11):1353–75.

Geras, D. T., Pepper, C. D., Rodgers, S. H. (1988). An integrated ergonomics program at the Goodyear Tire and Rubber Company. In: Mital, A., Ed., *Advances in industrial ergonomics and safety*. Bristol, PA, Taylor & Francis, pp. 21–28.

Goel, V. K., Rim, K. (1987). Role of gloves in reducing vibration: An analysis for pneumatic chipping hammer. *Am Ind Hyg Assoc J* **48**(1):9–14.

Habes, D. J., (1984). Use of EMG in a kinesiological study in industry. *Appl Ergonomics* 15(4):297–301.

Halpern, C. A., Davis, P. J. (1993). An evaluation of workstation adjustment and musculoskeletal discomfort. In: *Proceedings of the 37th Annual Meeting of the Human Factors Society*, Seattle, WA, pp. 817–21.

Hopsu, L., Louhevaara, V. (1991). The influence of educational training and ergonomic job redesign intervention on the cleaner's work: A follow-up study. In: *Designing for everyone: Proceedings of the Eleventh Congress of the International Ergonomics Association*, Paris, Vol. 1, pp. 534–36.

Itani, T., Onishi, K. Sakai, K., Shindo, H. (1979). Occupational hazard of female film rolling workers and effects of improved working conditions. *Arh Hig Rada Toksikol* **30**(Suppl.):1243–51.

Johnson, S. L. (1988). Evaluation of powered screwdriver design characteristics. *Hum Factors* **30**(1):61–69.

Jonsson, B. (1988). Electromyographic studies of job rotation. *Scand J Work Environ Health* **14**(1):108–9.

Knowlton, R. G., Gilbert, J. C. (1983). Ulnar deviation and short-term strength reductions as affected by a curve handled ripping hammer and a conventional claw hammer. *Ergonomics* **26**(2):173–79.

LaBar, G. (1992). A battle plan for back injury prevention. *Occup Hazards* **54:**29–33.

Little, R. M. (1987). Redesign of a hand tool: A case study. *Semin Occup Med* **2**(1):71–72.

Lindberg, M., Frisk-Kempe, K., Linderhed, J. (1993). Musculoskeletal disorders, posture and EMG temporal pattern in fabric seaming tasks. *Inter J Ind Ergonomics* **11**(3):267–76.

Luopajarvi, T., Kuorinka, I., Kukkonen, R. (1982). The effects of ergonomic measures on the health of the neck and upper extremities of assembly-line packers—a four year follow-up study. In: Noro, K., Ed., *Proceedings of the 8th Congress of the International Ergonomics Association*, Tokyo, Japan, pp. 160–61.

Luttmann, A., Jager, M. (1992). Reduction in muscular strain by work design: Electromyographical field studies in a weaving mill. In: Kumar, S., Ed., *Advances in industrial etrgonomics and safety IV*, Philadelphia, PA: Taylor & Francis, pp. 553–60.

Lutz, G., Hansford, T. (1987). Cumulative trauma disorder controls: The ergonomics program at Ethicon, Inc. *J Hand Surg* **12A**(5, Part 2):863–66.

McKenzie, F., Storment, J., Van Hook, P., Armstrong, T. J. (1985). A program for control of repetitive trauma disorders associated with hand tool operations in a telecommunications manufacturing facility. *Am Ind Hyg Assoc J* **46**(11):674–78.

Miller, M., Ransohoff, J., Tichauer, E. R. (1971). Ergonomic evaluation of a redesigned surgical instrument. *Appl Ergonomics* **2**(4):194–97.

Moore, J. S. (1994). Flywheel tuning: A case study of an ergonomic intervention. *Am Ind Hyg J* **55**(8):236–44.

Narayan, M., Rudolph, R. (1993). Ergonomic improvements in a medical device assembly plant: A field study. In: *Proceedings of the 37th Annual Meeting of the Human Factors Society*, Seattle, WA, pp. 812–16.

NIOSH (1994). Participatory ergonomics interventions in meatpacking plants. Cincinnati, OH: U.S. Department of Health and Human Services, Public Health Service, Centers for Disease Control and Prevention, National Institute for Occupational Safety and Health, DHHS [NIOSH] Pub. No. 94-124.

Nevala-Puranen, N., Taattola, K., Venalainen, J. M. (1993). Rail system decreased physical strain in milking. *Int J Ind Ergonomics* **12**(4):311–16.

Orgel, D. L., Milliron, M. J., Frederick, L. J. (1992). Musculoskeletal discomfort in grocery express checkstand workers: An ergonomic intervention study. *J Occup Med* **34**(8):815–18.

Parenmark, G., Malmvisk, A. K., Ortengren, R. (1993). Ergonomic moves in an engineering industry: Effects on sick leave frequency, labor turnover and productivity. *Int J Ind Ergonomics* **11**(4):291–300.

Peng, S. L. (1994). Characteristics and ergonomic design modifications for percussive rivet tools. *Int J Ind Ergonomics* **913**(3):171–87.

Powers, J. R., Hedge, A., Martin, M. G. (1992). Effects of full motion forearm supports and a negative slope keyboard system on hand–wrist posture while keyboarding. *Proceedings of the Human Factors Society 36th Annual Meeting*, Atlanta, GA, pp. 796–800.

Radwin, R. G., Oh, S. (1991). Handle and trigger size effects on power tool operation. In: *Proceedings of the Human Factors Society 35th Annual Meeting*, pp. 843–47.

Reynolds, J. L., Dury, C. G., Broderick, R. L. (1994). A field methodology for the control of musculoskeletal injuries. *Applied Ergonomics* **25**(1):3–16.

Rigdon, J. E. (1992). The wrist watch: How a plant handles occupational hazard with common sense. *The Wall Street Journal* 9128192.

Snook, S. H., Campanelli, R. A., Hart, J. W. (1978). A study of three preventive approaches to low back injury. *J Occup Med* **20:**478–81.

Tadano, P. (1990). A safety/prevention program for VDT operators: One company's approach. *J Hand Therapy* **3**(2):64–71.

Westgaard, R. H., Aaras, A. (1984). Postural muscle strain as a causal factor in the development of musculoskeletal illnesses. *Appl Ergonomics* **15**(3):162–74.

Westgaard, R. H., Aaras, A. (1985). The effect of improved workplace design on the development of work-related musculoskeletal illnesses. *Appl Ergonomics* **16**(2):91–97.

Wick, J. L. (1987). Workplace design changes to reduce repetitive motion injuries in an assembly task: A case study. *Semin Occup Med* **2**(1):75–78.

Wick, J. L., Deweese, R. (1993). Validation of ergonomic improvements to a shipping workstation. In: *Proceedings of the 37th Annual Meeting of the Human Factors Society*, Seattle, WA, pp. 808–11.

TEXT REFERENCES

ANSI. 1986. Guide for the Measurement and Evaluation of Human Exposure to Vibration Transmitted to the Hand. ANSI 53 34. New York: ANSI.

ANSI. 1993. Ergonomic Guidelines for the Design, Installation, and Use of Machine Tools in the Work Environment. ANSI B11 TR1. New York: ANSI.

ANSI. 1995. Control of Work-Related Cumulative Trauma Disorders (working draft). ANSI Z 365. New York: ANSI, 17 April.

Armstrong, T. J., R. G. Radwin, D. J. Hansen, and K. W. Kennedy (1986). Repetitive trauma disorder: Job evaluation and design. *Human Factors* **28**(3), 325–36.

Astrand, P. O., and K. Rodahl (1986). *Textbook of Work Physiology. Physiological Bases of Exercise*, 3rd ed., McGraw–Hill, New York.

Bink, B. (1962). The physical working capacity in relation to working time and age. *Ergonomics* **5**(1), 25–28.

Bonjer, F. H. (1962). Actual energy expenditure in relation to the physical working capacity. *Ergonomics* **5**(1), 29–31.

Chaffin, D. B. (1992). Biomechanical modeling for simulation of 3D static human exertions. *Computer Applications in Ergonomics, Occupational Safety and Health*, M. Mattila and W. Karwowski, Eds., Elsevier Science Publishers, Netherlands, pp. 1–11.

Chaffin, D. B., and G. B. J. Andersson (1991). *Occupational Biomechanics*, John Wiley & Sons, New York.

Durnin, J. V. G. A., and R. Passmore (1967). *Energy, Work and Leisure*, Heineman Educational Books, London.

Eastman Kodak Co. (1983). *Ergonomic Design for People at Work*, Vol. 1, Van Nostrand Reinhold, New York.

Eastman Kodak Co. (1996). *Ergonomic Design for People at Work*, Vol. 2, Van Nostrand Reinhold, New York.

F. Gamberale (1990). Perception of effort in manual materials handling. *Scand. J. Work Environ. Health* **16**(Suppl. 1), 59–66.

Garg, A., D. B. Chaffin, and G. D. Herrin, Prediction of metabolic rates for manual materials handling jobs. *Am. Ind. Hyg. Assoc. J.* **39,** 661–74.

Givoni, B., and R. F. Goldman (1971). Predicting metabolic energy cost. *J. Appl. Physiology* **30,** 429–33.

Habes, D. J., and V. Putz-Anderson (1985). The NIOSH program for evaluating biomechanical hazards in the workplace. *J. Safety Research* **16,** 49–60.

Hosey, A. D. (1973). General principles in evaluating the occupational environment. *The Industrial*

Environment—Its Evaluation and Control, (G. D. Clayton, Ed., National Institute for Occupational Safety and Health, Cincinnati, Ohio, p. 95.

Imada, A. S. (1993). Macroergonomic approaches for improving safety and health in flexible, self-organizing systems. *The Ergonomics of Manual Work*, W. S. Marras, W. Karwowski, J. L. Smith, and L. Pacholski, Eds., Taylor & Francis, Philadelphia, pp. 477–80.

Kamon, E., and A. Goldfuss (1978). In-plant evaluation of muscle strength of workers. *Am. Ind. Hyg. Assoc. J.* **39,** 801–7.

Keyserling, W. M. (1986). Postural analysis of the trunk and shoulders in simulated real time. *Ergonomics* **29**(4), 569–83.

Keyserling, W. M., T. J. Armstrong, and L. Punnett (1991). Ergonomic job analysis: A structured approach for identifying risk factors associated with overexertion injuries and disorders, *Appl. Occ. Environ. Hyg.* **6,** 353–63.

Mathiowetz, V., N. Kashman, G. Volland, K. Weber, M. Dowe, and S. Rogers (1985). Grip and pinch strength: Normative data for adults. *Arch. Phys. Med. Rehab.* **66,** 69–72.

NASA (1978). *Anthropometric Source Book, Volumes I, II, and III* (Reference Publication 1024), NASA Scientific and Technical Information Office, Yellow Springs, Ohio.

Niebel, B. W. (1989). *Motion and Time Study*, 8th ed. Irwin, Homewood, IL.

NIOSH (1981). *Work Practices Guide for Manual Lifting*, NIOSH Technical Report No. 81-122, U.S. Department of Health and Human Services, National Institute for Occupational Safety and Health, Cincinnati, OH.

OSHA (1990). *Ergonomics Program Management Guidelines for Meatpacking Plants*, OSHA 3123, U.S. Department of Labor, Occupational Safety and Health Administration, Washington, D.C.

Pacholski, Eds., Taylor & Francis, Philadelphia, pp. 477–80.

Perez-Balke, G. M. (1993). The recording of observed physical exposures (ROPEM): A simple and low cost method of analyzing upper extremity postures (ABSTRACT). Proceeding of Am. Ind. Hyg. Assoc. Conf., New Orleans, LA.

Putz-Anderson, V. (1988). Cumulative trauma disorders: A manual for musculoskeletal diseases of the upper limbs. Taylor & Francis.

Putz-Anderson, V., and T. L. Galinsky (1993). Psychophysically determined work durations for limiting shoulder girdle fatigue from elevated manual work. *Int. J. of Ind. Ergon.* **11**(1), 19–28.

Rodgers, S.H. (1992). Functional job analysis technique. *Occupational Medicine: State of the Art Reviews*, J. S. Moore and A. Garg, Eds., Hanley & Belfus, Philadelphia, p. 680.

Snook, S. H. (1978). The design of manual handling tasks. *Ergonomics* **21**(12), 963–85.

Snook, S. H., and C. H. Irvine, Psychophysical studies of physiological fatigue criteria. *Human Factors* **11**(3), 291–300.

Snook, S. H., and V. M. Ciriello (1991). The design of manual handling tasks: Revised tables of maximum acceptable weights and forces. *Ergonomics* **21,** 1197–213.

Snook, S. H., D. R. Vaillancourt, V. M. Ciriello, and B. S. Webster (1996a). Psychophysical studies of repetitive wrist motion: Part I: Two day per week exposure. *Ergonomics* (1996a).

Snook, S. H., D. R. Vaillancourt, V. M. Ciriello, and B. S. Webster (1996b). Psychophysical studies of repetitive wrist motion: Part II: Five day per week exposure. *Ergonomics* (1996b).

Ulin, S. S., and T. J. Armstrong (1992). A strategy for evaluating occupational risk factors of musculoskeletal disorders. *J. Occup. Rehab.* **2**(1), 35–50.

Ulin, S. S., T. J. Armstrong, and R. G. Radwin (1990). Use of computer aided drafting for analysis and control of posture in manual work. *Appl. Ergon.* **21**(2), 143–51.

VanderWalt, W. H., and C. H. Wyndham (1973). An equation for prediction of energy expenditure of walking and running. *J. Appl. Physiology* **34,** 559–63.

Waters, T. R., V. Putz-Anderson, A. Garg, and L. Fine (1993). Revised NIOSH equation for the design and evaluation of manual lifting tasks. *Ergonomics* **36**(7), 749–76.

RESOURCES
(Publications available from area OSHA offices, except as noted)

Ergonomics Program Management Guidelines for Meatpacking Plants (1991) (reprinted). OSHA 3123 (free).

Working Safely with Video Display Terminals (1997) (reprinted). OSHA 3092*, Order number 029-016-00127-1.

Ergonomics: The Study of Work (1991). OSHA 3125*, Order number 029-016-00124-7.

"Back Injuries—Nation's Number One Workplace Safety Problem." This is one of a series of fact sheets highlighting USDOL programs. Fact Sheet Number OSHA 89-09.

ErgoFacts. OSHA's 1-page fact sheet on ergonomics hazards in the workplace, their solutions and benefits. Available from the OSHA publications office.

OSHA Handbook for Small Businesses (1996) (revised). OSHA 2209*.

*These publications cost one to six dollars. They are available from your local OSHA area office or U.S. Government bookstore or from the Office of Information and Consumer Affairs at:

OSHA Publications Office
200 Constitution Avenue N.W.
Room N-3647
Washington, DC 20210

Job Safety & Health Quarterly. The agency's quarterly magazine contains various articles on health and safety, including ergonomic issues. Available for $5.50/year from the GPO.

The Federal Register published the Advanced Notice of Proposed Rulemaking on August 3, 1992, and will publish the proposed standard. Order forms for the Federal Register can be obtained from the GPO.

Superintendant of Documents
Government Printing Office
Washington, DC 20402-9325
202-275-0019 (to fax orders and inquiries)
202-783-3238 8 a.m. to 4 p.m. Mon.–Fri. EST for charge orders

OSHA FAX on demand provides a variety of information available, much of which is less expensive to access via the internet.
900-555-3400
$1.50 per minute

Internet addresses to get you started:

http://www.ergoweb.com/
http://www.OSHA.gov
http://www.OSHA-slc.gov/SLTC/ergonomics/index.html

For direct dial to the Labor News Bulletin Board, which has a gateway to the Salt Lake City Technical Lab and OSHA Computerized Information System (OCIS) dial 202-219-4784 (this is a toll call outside Washington, DC).

ADDITIONAL RESOURCES

- National Institute of Occupational Safety and Health (NIOSH)
  Public Dissemination, DSDTT
  4676 Columbia Parkway
  Cincinnati, OH 45226
  USA
  513-533-8573 (Fax)
  1-800-35-NIOSH

NIOSH has a free help line (8 A.M. to 4:00 P.M. EST) to answer technical questions and disseminate publications on ergonomics and other occupational safety and health questions such as blood borne pathogens and indoor air quality. NIOSH and DHHS publications noted in the reference section of this chapter may be available through the help line.

- Human Factors and Ergonomics Society
  PO Box 1369
  Santa Monica, CA 90406-1369
  USA
  310-394-2410 (Fax)
  310-394-1811

The Human Factors and Ergonomic Society (HFES) is the primary society that represents human factor engineers, engineering psychologists, and ergonomists in the United States. HFES publishes an annual list of consultants and a list of ergonomic and human factors programs and maintains an active employment service for ergonomists and employers who seek ergonomists. The society also publishes a newsletter and a journal entitled *Human Factors*.

- The Ergonomics Society
  Department of Human Sciences
  University of Technology
  Loughborough, Leicestershire LE11 3TU
  United Kingdom

Like HFES, the Ergonomics Society is a professional society that assists in the publication of an international journal. The journal is entitled *Applied Ergonomics*. The journal reports on applied studies of people's relationships with equipment, environments, and work systems.

- Massachusetts Coalition on New Office Technology
  (CNOT)
  1 Summer Street
  Somerville, MA 02142
  617-776-2777

CNOT provides information on risk factors in the office environment and a checklist for work analysis for VDU workstations. They have ongoing support groups for workers who suffer from WMSDs and provide speakers for related topics.

- Center for Workplace Health Information
  CTDNEWS
  410 Lancaster Avenue, Suite 15
  PO Box 239
  Haverford, PA 19041
  610-896-2762 (Fax)
  610-896-2770 or 1-800-554-4283

CTDNEWS is a monthly newsletter that provides up-to-date information on legal cases, policy, journal articles, case studies of ergonomic intervention, dates of ergonomic conferences and seminars, and ergonomic resources. *CTDNEWS* will send a sample of their newsletter upon request. Their web site can be found at http://www.ctdnews.com on the internet.

- Ergoweb http://www.ergoweb.com

Ergoweb is a place to meet and discuss ergonomics with ergonomists and those with ergonomic issues. Other useful resources on Ergoweb include a full copy of the OSHA draft proposed ergonomic protection standard (in the reference section of Ergoweb), a copy of the RULA method of analysis for the upper extremities and neck (as described in the methods section of this chapter), six different programs to analyze hazards associated with lifting, and the UTAH Process Diagnostician (an "expert" system for evaluating and designing systems to minimize ergonomic risk factors).

- Board Certification in Professional Ergonomics (BCPE)
  PO Box 2811
  Bellingham, WA 98227-2811
  360-671-7681 (Fax)
  360-671-7601

BCPE is a nonprofit organization responsible for the certification of human factor and ergonomic professionals. Certification requires a master's degree in ergonomics or human factors, or equivalent, four years of full time practice as an ergonomist practitioner with emphasis on design involvement, documentation of education, employment history, and ergonomic project involvement, a passing score on an eight hour written examination, and payment of fees for the certification process. BCPE publishes a newsletter and a directory of board-certified ergonomists and human factors practitioners.

CHAPTER 24

Radiation: Nonionizing and Ionizing Sources

DONALD L. HAES, JR. and MITCHELL S. GALANEK

Massachusetts Institute of Technology, 77 Massachusetts Ave., Cambridge, MA 02139

The workplace is potentially the site of a varied number of radiation sources. From lasers, heat sealers, and communications antennae in the nonionizing spectrum to density gauges, analytical instrumentation, and x-ray machines in the ionizing region, there is potential for worker exposures to these sources and thus the need for a program of recognition, evaluation, and control to ensure a safe working environment. This chapter will deal with the various sources found in each area and their significance with respect to control of exposures. Control is accomplished via administrative procedures, effective worker training, engineering and environmental controls, and ongoing surveillance and radiation safety audits.

24.1 NONIONIZING RADIATION
- 24.1.1 Introduction
- 24.1.2 Nonionizing Radiation in Industry
- 24.1.3 Summary

24.2 IONIZING RADIATION
- 24.2.1 Specific Licensing
- 24.2.2 General Licensing
- 24.2.3 Source Material
- 24.2.4 Machine Radiation Sources
- 24.2.5 Radiation Safety Officer
- 24.2.6 Radiation Safety Liaison
- 24.2.7 Administrative Procedures
- 24.2.8 Licensed Radiation Sources or Machines
- 24.2.9 Radiation Worker Training
- 24.2.10 Safe Handling and Dose Reduction Techniques
- 24.2.11 Radiation Detection and Measurement
- 24.2.12 Survey Instruments
- 24.2.13 Analytical Instruments
- 24.2.14 Maximum Permissible Exposure Limits
- 24.2.15 ALARA Concept

Handbook of Occupational Safety and Health, Second Edition, Edited by Louis J. DiBerardinis, ISBN 0-471-16017-2

24.2.16 Biological Effect and Risks from Occupational Exposures
24.2.17 Worker Exposure Monitoring
24.2.18 Radiation and Contamination Surveys
24.2.19 Engineering and Environmental Controls
24.2.20 Transportation of Radiation Sources
24.2.21 Low-Level Radioactive Waste Disposal and Source Disposition
24.2.22 Laboratory Surveillance and Management Audits
24.2.23 Emergency Procedures
24.3 CONCLUSIONS
REFERENCES
ADDITIONAL REFERENCES
APPENDIX A

24.1 NONIONIZING RADIATION

24.1.1 Introduction

While the use of nonionizing radiation (NIR) in the workplace is not new, acknowledgement for its potential hazards is still unfolding. Standards for NIR safety are numerous, but are often incomprehensible and hard to find. It is imperative that NIR sources in industry be identified, evaluated, and their potential for hazards controlled. While sources of nonionizing radiation can be found in the workshop, they can also be found in administrative and physical plant activities. These sources include, but are not limited to, the following:

- Static (DC) magnets
- Industrial alternating current (AC) line voltage
- Radiofrequency (RF) induction ovens
- RF heaters/sealers
- Hand-held two-way communications
- Microwave ovens
- Cellular telephones
- Infrared (IR)
- Visible light (VL)
- Ultraviolet (UV) radiation
- Lasers (visible and invisible)

Comprehending the basic physics of exposures will be central to this struggle.

24.1.2 Nonionizing Radiation in Industry

24.1.2.1 Physics of Nonionizing Radiation *Nonionizing radiation* is a practical term for the frequency band of the electromagnetic spectrum that lacks the energy to break chemical and/or molecular bonds, hence being unable to cause *ionization* (ionization is the energetic removal or addition of orbital electrons from the outer shells of atoms, leading to ion production).

There are several different classes of NIR, differentiated according to wavelength, and hence by frequency (see Fig. 24.1). NIR consists of electromagnetic waves composed of electric and magnetic fields, mutually orthogonal to the direction of propagation (see Fig. 24.2). The waves move through a vacuum with the velocity of light. There is a mathematically inverse relationship between the frequency and wavelength:

$$\text{frequency}[f] = (\text{velocity of light}[C] \div \text{wavelength}[\lambda])(\text{in free space})$$

	Frequency (Hertz)	Wavelength (Meters)	Function Device(s)	Band
	10^0 - 1Hz	3×10^8 -		
	10^1 -	3×10^7 -		
	10^2 -	3×10^6 - 3Mm	Power (50/60Hz)	
	10^3 - 1kHz	3×10^5 -		
	10^4 -	3×10^4 -	CRT/TV circuitry (17-31kHz)	
	10^5 -	3×10^3 - 3km		RADIO FREQUENCY (RF)
	10^6 - 1MHz	3×10^2 -	AM Radio (535-1705kHz)	
	10^7 -	3×10^1 -	VHF Television (54-88MHz) FM Radio (88-108MHz)	
	10^8 -	3×10^0 - 3m	VHF Television (174-216MHz) UHF Television (470-890MHz)	
	10^9 - 1GHz	3×10^{-1} -	Microwave ovens (2.45GHz) AT&T Telephone (3.7-4.2GHz)	MICROWAVE (MW)
NON-IONIZING	10^{10} -	3×10^{-2} - 3cm	Radar	
	10^{11} -	3×10^{-3} - 3mm		INFRARED (IR)
	10^{12} - 1THz	3×10^{-4} -		
	10^{13} -	3×10^{-5} -		
	10^{14} -	3×10^{-6} - 3μm		
			Visible Light	
	10^{15} - 1PHz	3×10^{-7} -		ULTRAVIOLET (UV)
	10^{16} -	3×10^{-8} -		
IONIZING	10^{17} -	3×10^{-9} - 3nm		X RAYS
	10^{18} - 1EHz	3×10^{-10} -		
	10^{19} -	3×10^{-11} -		GAMMA RAYS
	10^{20} -	3×10^{-12} - 3pm		
	10^{21} -	3×10^{-13} -		
	10^{22} -	3×10^{-14} -		
	10^{22} -	3×10^{-15} - 3fm		
	10^{23} -	3×10^{-16} -		COSMIC RAYS
	10^{24} -	3×10^{-17} -		
	10^{25} -	3×10^{-18} - 3am		
	10^{26} -	3×10^{-19} -		
	10^{27} -	3×10^{-20} -		
	10^{28} -	3×10^{-21} -		

Fig. 24.1 Electromagnetic spectrum.

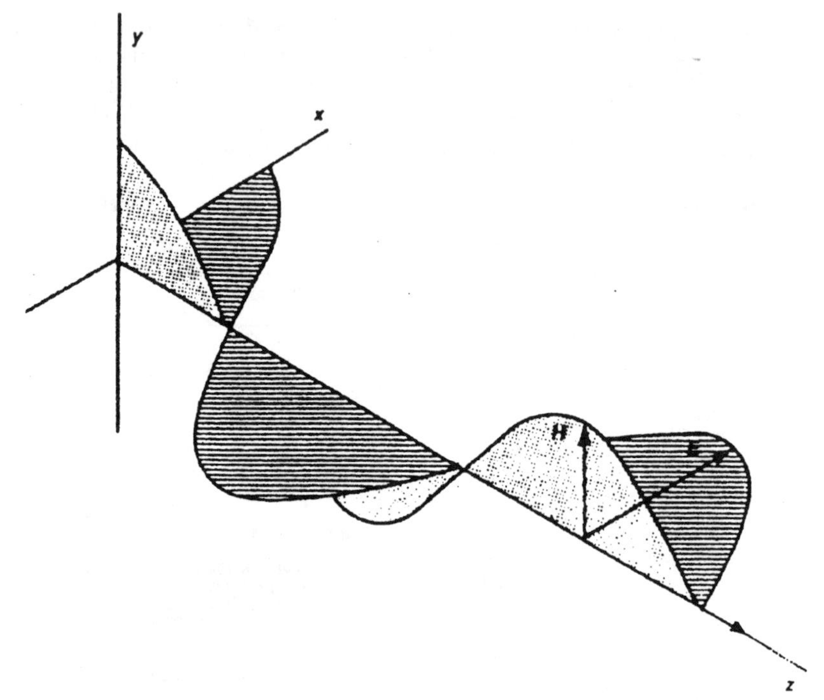

Fig. 24.2 The electromagnetic wave. The magnetic (*H*) field component is vertical, the electric field (*E*) component is horizontal, with the direction of propagation along the *z* axis.

The frequency is usually expressed in units of hertz (Hz) with one hertz defined as one cycle per second. Figure 24.1 shows the electromagnetic spectrum from the static fields to the very high frequencies found in cosmic rays. To help simplify this chapter, the elements of the electromagnetic spectrum included in this discussion are as follows: static field (also known as DC, direct current), extremely low frequency (ELF), very low frequency (VLF), radiofrequency (RF) (including microwaves), infrared (IR), visible light (VL), ultraviolet (UV), and lasers (which may use IR, VL, and/or UV). Table 24.1 gives the frequency ranges and acronyms associated with these frequency bands.

A number of devices emit a portion of the electromagnetic spectrum as either their primary product, or as consequences of their use. Lasers (light amplification by stimulated emission of radiation) are designed to emit light and may be dangerous because of their primary function. Video display screens (e.g., televisions and video display terminals) emit visible light as a design function. In addition, these same products, and a legion of other manmade equipment, emit other frequencies of the electromagnetic spectrum as a consequence. The generated electromagnetic field is an unavoidable technological consequence.

All devices that employ the use of NIR can be considered to contain three basic parts: the source, a transmission pathway, and a receiver. For the confined classification, leakage may occur along the transmission pathway between the source and the receiver. For open classifications, the source itself may be broadcasting large amounts of energy through open space, the pathway. The receivers may be chambers for energy absorption,

TABLE 24.1 Frequency Bands of the Electromagnetic Spectrum

Frequency Range	Designate/Acronym
0–30 Hz	Sub-extremely low frequency/SELF[a]
30–300 Hz	Extremely low frequency/ELF
0.3–3 kHz	Voice frequency/VF
3–30 kHz	Very low frequency/VLF
30–300 kHz	Low frequency/LF
0.3–3 MHz	Medium frequency/MF
3–30 MHz	High frequency/HF
30–300 MHz	Very high frequency/VHF
0.3–3 GHz	Ultra-high frequency/UHF
3–30 GHz	Super-high frequency/SHF
30–300 GHz	Extremely high frequency/EHF
0.3–3 THz	Super-extremely-high frequency/SEHF[a]

[a]No "official" acronym.

or mobile devices designed to absorb the signal for processing. In either arrangement, the surveyor must be certain of all source strengths, pathways, and intended receivers.

24.1.2.2 Static Fields Static fields do not vary in amplitude over time. Static *electric* fields abound (walk across a carpet in a dry room and touch something metallic), but are usually harmless. Very large sources of static electric fields are not usually found in the work environment. Static *magnetic* fields, however, can be found in the workplace in sufficient strength to present a potential for harm.

In industry, the typical sources are large magnets used for metallic coupling and research purposes. While very strong magnetic fields are present in close proximity to the magnets, the intensity of the field reduces dramatically with distance. These weaker fields are also known as *fringe fields*.

The two main sources of very strong magnetic fields found in the industrial environment are magnets used to compile metallic objects and nuclear magnetic resonance (NMR) imaging devices used for research. Near (within about 1 meter) NMRs, workers can be exposed to static magnetic fields up to 0.1 tesla in intensity, which is about 2000 times that of the background magnetic field of the earth. For many of these devices, the magnetic field strengths can be roughly determined by maps provided by the manufacturer. These maps have contour lines that relate field strength with distance from the magnetic's center (bore). The demarcation of the 0.5 mT line (often referred to as the "5-gauss line") is essential for pacemaker warning. The placement of this posting should be verified with measurements by a competent health physicist or other similarly qualified individual.

Before we discuss measurements, it should be noted that varying the orientation of the measuring device within the lines of flux will greatly vary the results. This is because most measuring devices are highly directional, while the sources are too. Thus orientation becomes critical for peak field evaluation. The root-mean square (RMS) value can be determined by acquiring the peak field intensities in three orthogonal directions, and calculating the square root of the sum of the squares. Many measuring devices already accomplish this through design.

Electric fields can be measured by inserting a displacement sensor (a pair of flat conductive plates) into the field and measuring the electric potential between the plates.

The electric field lines land on one plate and create voltage that drives a current through a meter to the other plate, where the field lines continue. The electric field strength is expressed typically in units of volts per meter (V/m).

Measuring magnetic field strength is often accomplished by loops of conducting wire. The lines of magnetic field passing through the loop induce current flow. The field can be calculated by measuring the amperes of induced current and divided by the circumference of that loop, with units of field strength expressed in amperes per meter (A/m). The relative intensity of magnetic fields can also be determined by measuring the magnetic flux density using the Hall effect. Here, an object in a magnetic field will develop a voltage in a direction perpendicular to the magnetic field, which can be measured. Hall effect sensors are less sensitive than loops, and are often used for determining strong DC fields. Typical units are the tesla (T) or gauss (G), where 10,000 gauss = 1.0 tesla.

Standards for exposure to static magnetic fields can be found published by the American Conference of Governmental Industrial Hygienists (ACGIH).[1] These published exposure standards are as follows:

- Whole body: 60 mT routine occupational time weighted daily average
- Limbs: 600 mT
- Ceiling value: 2 T
- Pacemakers, etc.: 0.5 mT

Additional standards for exposure to static magnetic fields can be found published by the International Council on Non-Ionizing Radiation Protection (ICNIRP).[2] These published exposure standards are as follows:

- Whole body: 200 mT average over a working day
- Limbs: 5 T
- Ceiling value: 2 T
- Public: 40 mT continuous exposure

Very strong electric fields greater than 5 kV/m could cause irritating sparks, while fields in excess of 15 kV/m could cause painful sparks. Very strong magnetic fields have been postulated to increase blood pressure and rotate sickle cells, but this is yet to be verified scientifically. The main concern in strong magnetic fields above about one millitesla (mT, 10^{-3}T) is the attraction of loose ferromagnetic objects (for example, iron gas cylinders, steel tools, etc.), which may be propelled toward the magnet at considerable velocity. The object will adhere to the magnet, as will anything that remained in the way. Additionally, surgical implants may be torqued, dislodged, or rotated, resulting in serious injury or death. Field strengths roughly above 0.5 mT may also interfere with unshielded electronic equipment, such as implanted cardiac pacemakers.[3] Even though field strengths may be "in compliance" with established pacemaker safety standards, patients must be warned that the strength of these fields may be enhanced by ferromagnetic objects held in close contact to the wearer.[4]

Control of exposure to static electric and magnetic fields can best be accomplished by the use of distance. While shielding is purported to be available for magnetic fields, the manufacturers' claims are dubious at best. As a final note, magnetic media, such as the magnetic stripes on the backs of credit cards and diskettes, can be erased by fields above 1 mT. Digital watches can also be damaged by intense magnetic fields.

24.1.2.3 Extremely Low Frequency and Very Low Frequency

For time-varying fields, the "wave" consists of an electric field component and a magnetic field component, each perpendicular to the direction of motion and to each other (see Fig. 24.2). This is referred to as being "mutually orthogonal." The "polarity" of the field is determined by the direction of the electric field relative to the direction of propagation. In Fig. 24.2, the electric field is horizontal relative to the direction of motion; thus that wave is considered to be horizontally polarized. The polarization of the incident wave plays an important part in determining the amount of energy absorbed.

Extremely low frequency (ELF) and very low frequency (VLF) vary in intensity over time, as shown in Fig. 24.1. While ELF fields are often referred to alternating current fields, it is technically incorrect to refer to them as "radiation." When the wavelength is extremely long, as in ELF, exposed objects, including people, will be enveloped within a wavelength of the source. The physical characteristics of such "confined" fields dictate that radiant energy effects will be trivial.[5] What the exposed object encounters is an "induction field," in which charges from the nearby electromagnetic source induce currents and charges on and within the exposed body. Understanding this technical difference will help avoid confusion between the very low imparted energies and the physiological changes that occur in cells exposed to induction fields.

The generation, transport, and use of common industrial and household electricity is associated with ELF fields. The ubiquitous potential source of exposure is the power line, a frequently visible (unless buried) source of ELF. Although the electric and magnetic field components are necessarily generated together, the magnetic field has been implicated to health outcomes.

As in DC electric fields, ELF electric fields can be measured by inserting a displacement sensor into the field and measuring the electric potential between the plates. The electric field strength is typically expressed in units of volts per meter (V/m), or even thousands of volts per meter (kV/m). The electric field lines are severely perturbed by the presence of anything in the field, including the survey equipment and even the surveyor. Some equipment has been developed to aid this problem, including dielectric extension arms and remote readouts coupled to the meter via fiber optic cables.

Again as in DC magnetic fields, measuring ELF magnetic field strength is often accomplished by loops. The current can be augmented by increasing the number of turns in the loops, or by putting a core of permeable material in the loop. While the appropriate units are A/m, field strengths are often expressed in microtesla (μT, 10^{-6}T), or milligauss (mG). Meter readings are often converted to mG assuming the permeability of free space = $4\pi \times 10^{-7}$ henry/meter; 1 mG $\approx$ 80 mA/m. Measuring ELF magnetic fields are not as problematic as electric fields, and can be accomplished accurately with little training and practice.

Standards for exposure to ELF fields can be found published by the ACGIH. These published exposure standards are as follows:

- <100 Hz electric fields: 25 kV/m
- 1–30 kHz magnetic fields: $60{,}000/f\,\mu$T where f is the frequency of the source in Hz. As is the case with DC fields, AC fields may interfere with the proper operation of cardiac pacemakers and defibrillators. ACGIH has set limits of exposure to 1

kV/m electric fields, and 0.01 mT (1 G) magnetic fields. These values are also mentioned by the manufacturer's notes to wearers.[6]

Despite persistent methodologic problems of dose and exposure, residence near power lines and work with or around electrical power has been associated with the development of leukemia in several epidemiologic studies of children and adults, respectively. Wertheimer and Leeper were the first to attempt to show a positive association between a higher "dose" magnetic fields (albeit by a surrogate exposure assessment) and leukemia.[7] Evidence for an association between adult occupational leukemia and ELF exposure is not compelling so far. Published data show odds ratios at or below unity as well as above. Almost all 95% confidence intervals of cohort and case control studies included unity. Recent studies have attempted to overcome some of the deficiencies of earlier studies by measuring ELF field exposures at the workplace and taking work duration into consideration.

In apparent contrast to the epidemiology results, short-term measurements with magnetic field meters that were also used in some of the studies provided no evidence for an association between exposure to 50/60 Hz magnetic fields and the risk of any form of cancer in children. These studies prompted the National Academy of Science Committee to suggest that, confounded by some unknown risk factor for children, leukemia, which is associated with residency in the vicinity of power lines, might be the explanation.[8]

Taken together, the epidemiology results from residential magnetic field exposure studies and childhood cancer is not strong enough in the absence of support from experimental research to form the basis for exposure standards. For this reason, published standards list allowable exposures several orders of magnitude above those levels specified in epidemiology studies to produce a positive effect.

Computers area associated with NIR exposures. Most computers use video display terminals (VDT), which employ cathode ray tubes. Both have a source of electrons (cathode) at one end, and a phosphorous coating on the inside of the viewing screen (anode) at the other. The electrons, accelerated by high voltage, are focused onto the screen. They display information when the rapidly moving electron beam strikes the screen to produce visible light (the only type of NIR emitted by design!). The resultant NIR fields are in both the ELF and VLF frequency range. Extensive measurements at MIT yielded fields at 50 cm from the screen ranging from 1.9 to 87.5 mA/m.[9] Larger displays allow the operator to sit farther from the screen, with correpondingly less exposure. Thus seating positions are crucial in assigning any "dose" restrospectively. In fact, it is questionable if any information from measurements at fixed locations from the VDT's can be useful assigning VDT "dose."

Epidemiologic studies of reproductive outcomes of VDT users bear the generic problem familiar to all NIR field studies. Exposures have never been quantified. Instead, reproductive outcomes have been compared retrospectively to survey reports of VDP "use" time at the keyboard. This approach introduces two important limitations. First, both the outcome and the exposure are reconstructed retrospectively, with an implied assumption that one will not bias the other. Second, any reproductive effect detected, if real, would be associated with VDT use in general and not necessarily with electromagnetic emissions specifically.

Goldhaber et al. were the first to report a statistically significant positive association of VDT use with spontaneous abortions among members attending three clinics of a large, prepaid group health care plan.[10] Previous, similarly designed studies had concluded that no reproductive hazard is associated with VDT use during pregnancy. Windham and colleagues reviewed earlier data and reported the entire positive effect was limited to those respondents in the initial studies who were contacted by telephone after not responding to questionnaires.[11] The latest major study was conducted by the National Institute of Occupational Safety and Health (NIOSH), with the conclusion that: "The use of VDTs and exposure to the accompanying electromagnetic fields were not associated with an increased risk of spontaneous abortion in this study."[12] The body of accessible data do not imply a clear association between VDT use and adverse reproductive outcomes. Based on a large number of epidemiological studies, there do not appear to be adverse effects on reproductive outcomes as a result of exposure to low-frequency fields from VDTs, or other sources.

Very few standards apply to VDT exposure. A number of countries and agencies have created exposure limits that include the NIR emission from a VDT. Table 24.2 shows the wide variations in frequencies covered and exposure intensities permitted.

As with static fields, the control of exposure to ELF and VLF electric and magnetic fields is best done through distance. However, shielding is also available. Electric field shielding can be sheets of grounded conductors, allowing induced charges to flow between the conductor and the earth. The material can be solid, but a mesh of at most $\frac{1}{4}$ wavelength will also do. Magnetic fields can be controlled using a permeable alloy that confines the magnetic flux lines and diverts them. This material is usually made of high nickel alloys called "mu metal," or soft iron. However, neither shielding material is easy to form into the necessary complex shapes. As another source of magnetic field control, field cancellation can be used. Here, fields of similar magnitude, but opposition orientation, can be used to cancel out of the first field, since the vector sum is near zero.

24.1.2.4 Radio Frequency

These bands are generally defined to include the frequencies 10 kHz to 300 GHz. Although a wide spectrum of frequencies is included,

TABLE 24.2 Various Magnetic Field Exposure/Emission Standards in the Frequency Range ≤300 kHz

Organization or Country/Year	Frequency Range	Magnetic Field (μT)[a]	Exposure Duration and/or Comments
ANSI C95.1-1992	3–100 kHz	205	"Uncontrolled environment"
ACGIH/1996-97	4–30 kHz	2.01	"Threshold values"
105 CMR[b] 122.00/1986	10–3000 kHz	1.99	Occupational
MPR-III/1994[c]	5–2000 Hz	0.250	At 0.5 m; 3 planes:
	2–400 kHz	0.025	Centerline, 0.25 m above and below

[a]Most exposure limits list magnetic field intensity (H) in units of (A m^{-1}). These values have been converted to magnetic flux density (B) in units of (μT, 1×10^{-6} Tesla), $B = \mu H$, by assuming the permeability (μ) of free space ($4\pi \times 10^{-7}$ henry m^{-1}) is the same for tissue.

[b]Commonwealth of Massachusetts Department of Public Health (105 CMR 122.000: Massachusetts Department of Public Health, Fixed facilities which generate electromagnetic fields in the frequency range of 300 kHz to 100 GHz & microwave ovens).

[c]This standard is a "performance" standard. It is based on VDT emissions achievable with current technology, and not on any "safe" level of EMF emissions.

exposures are commonly grouped together under the collective term "RF exposures." RF is used in industry in varying frequency ranges and applications. The major sources of RF exposure found in industry are RF induction ovens, RF heater/sealers, two-way communication devices, microwave ovens, and cellular telephones. Most all these sources operate at a frequency assigned by the Federal Communications Commission (FCC) in the ISM (Industrial, Scientific, and Medical) band of the electromagnetic spectrum. Typical frequencies are 13.56, 27.12, 40.68, and 2450 MHz. RF exposures can be from sources, leakage along the transmission pathway, or receivers. The areas affected in the body can be whole or partial body, or even from contact with a reflective object within an intense low-frequency (from 3 kHz to 100 MHz) RF field.

RF induction ovens operate as convection ovens, but use RF as their initial heat source. These usually operate in the kHz range (typically 300–450 kHz) and produce intense electric and magnetic fields, which are very difficult to measure. Luckily, the thermal component of the operation of these ovens often precludes worker RF exposure above safety standards due to the impossibility to remain near the sources.

In the hospital setting, diathermy, first used in 1907, uses RF to produce a therapeutic effect. Common diathermy units use 27.12 MHz RF to stimulate circulation and promote healing. Diathermy units operating at 2450 MHz have also been used, but are not as common. A specialized form of diathermy is pulsed electromagnetic stimulation (PES). PES is a noninvasive procedure used for the palliative treatment of postoperative pain and edema in superficial soft tissue. RF ablation is a nonsurgical procedure to treat some types of rapid heart beating. In these procedures, a physician guides a catheter with an electrode at its tip to the affected area of the heart muscle. RF is transmitted to a very small area to kill the muscle cells conducting the extra impulses that cause rapid heart beats. RF exposures in hospital settings are usually confined to the patients receiving electrosurgery or diathermy treatments. However, hospital staff may be exposed to RF from the unintentional use of the sources or the near-field region (within a few wavelengths) of portable transmitters, including portable transceivers and cellular telephones.

Microwave ovens are ubiquitous in industry and households. These devices use RF at 2450 MHz to heat nonmetallic materials. While microwave ovens may be ordinary in the home, they have recently been used in both laboratory and industrial settings as a quick method to heat. Since microwave ovens are designed to keep all the microwaves within the RF chamber, emission standards, vice exposure standards, apply. The present emission standard for microwave ovens, 21 CFR 1930.10 from the Food and Drug Administration (FDA), allows new ovens to leak no more than 1 mW/cm^2 when measured no closer than 5 cm from the device, and no more than 5 mW/cm^2 once the ovens leave the store.[13]

Cellular telephones utilize wireless communication technology (operating between 800 and 900 MHz in the UHF portion of the RF band of the electromagnetic spectrum) to interlink their information to an existing wire-line telephone network. The communication network consists of many systems. The call can be initiated by any of the following sources: a portable hand-held phone operating at very low power (usually less than 1 watt); a mobile unit operating at a higher power (several watts); or a land-line telephone. The call is transferred to the nearest cell. The latest entry into the wireless communications industry is PCS, personal wireless systems. PCS phones operate at much higher frequencies, typically around 1900 MHz, and send and receive digital signals. While older cellular technology used analog signals, the carriers are slowly converting to all-digital networks.

Below about 300 MHz, the field intensity cannot be suitably quantified by measurements of the electric nor magnetic fields alone; thus, in the absence of scientific evidence to the contrary, individual readings of both the electric fields and magnetic fields are necessary to determine regulatory compliance. Measurements of both fields are always necessary below 30 MHz. It is beyond the scope of this chapter to attempt to train personnel to make RF measurements. It should be mentioned, however, that the probe should be held no closer than 20 cm from the source of leakage. For broadcast sources, calculations of the safe distances for both survey equipment and personnel must be performed prior to actually making measurements. For more details on the procedures to make RF measurements, refer to the National Council of Radiation Protection and Measurements (NCRP) Report number 119[14] and the Institute of Electrical and Electronics Engineers/American National Standards Institute (IEEE/ANSI) C95.3-1992.[15]

Electric fields at frequencies above 100 kHz are measured using small dipole antennae. The electric field induces a surge of current in the dipole, which is connected to a diode. An amplified signal is applied to a meter, with results displayed in field strength units (FSUs) such as V/m, or V^2/m^2, or even power density (a measure of the power going through an area of space) (S) in watts per square meter (W/m^2) or milliwatts per square centimeter (mW/cm^2). *Note:* Power density (S in mW/cm^2) is derived from the following relationship between the electric (E in V/m) and magnetic (H in A/m) field strengths:

$$E^2/3770 = S = H^2 \times 37.7$$

Dipoles and diodes can be used at lower frequencies (below about 100 kHz), but displacement sensors are also used. As before, the resultant RMS field intensity should be determined. Most field survey instruments gang three sensors together so they are mutually orthogonal to provide a nearly isotropic response.

Magnetic fields at frequencies below 300 MHz are measured with single loops, and the field orientation is critical to determine peak field strengths. Three mutually orthogonal loops have been used to allow nearly isotropic responses. As before, the magnetic field induces a current that drives a meter, with results displayed in FSUs such as A/m, or A^2/m^2, or even power density milliwatts per square centimeter (mW/cm^2). The resultant RMS field intensity should be determined. Most field survey instruments gang three sensors together so they are mutually orthogonal to provide a nearly isotropic response.

Recently, equipment to measure the "grasping" or "contract" currents have been made commercially available. The grasping or contact current measuring device uses electrodes to measure electrically charged objects within a relatively intense RF field. Similarly, "induced current" meters have been made available. These devices measure the current induced through either foot, or both feet, to determine if current flowing through a constriction in the body, such as the ankle, exceeds the maximum permissible.

Measurements of RF fields to determine regulatory compliance should only be undertaken by a competent health physicist or other suitably trained individual. The appropriate equipment used is not only very expensive to purchase and calibrate, but is often difficult to use.

RF exposure standards, including contact/grasping currents and induced current exposure limits, have been published by many organizations around the world. ANSI,[16] NCRP,[17] and ACGIH are but a few. The RF standard most cited is that published by ANSI. Refer to Table 24.3a for RF exposure limits for "uncontrolled areas" (those areas

TABLE 24.3a IEEE/ANSI Maximum Permissible Exposures for Uncontrolled Environments

Frequency Range (MHz)	Electric Field Strength (E) (V/m)	Magnetic Field Strength (H) (A/m)	Power Density (S) E Field; H Field (mW/cm^2)	Averaging Time $\lvert E\rvert^2$, S, or $\lvert H\rvert^2$ (min)	
0.003–0.1	614	163	(100; 1,000,000)[a]	6	6
0.1–1.34	614	$16.3/f$	(100; $10{,}000/f^2$)[a]	6	6
1.34–3.0	$823.8/f$	$16.3/f$	($180/f^2$; $10{,}000/f^2$)[a]	$f^2/0.3$	6
3–30	$823.8/f$	$16.3/f$	($180/f^2$; $10{,}000/f^2$)[a]	30	6
30–100	27.5	$158/f^{1.668}$	(0.2; $940{,}000/f^{3.336}$)[a]	30	$0.0636/f^{1.337}$
100–300	27.5	0.0729	0.2	30	30
300–3,000			$f/1{,}500$	30	
3,000–15,000			$f/1{,}500$	$90{,}000/f$	
15,000–300,000			10	$616{,}000/f^{1.22}$	

Induced and Contact RF Currents

Frequency range (MHz)	Maximum current (mA)		
	Through both feet	Through each foot	Contact
0.003–0.1	$900f$	$450f$	$450f$
0.1–100	90	45	45

[a]Values of plane wave equivalent power densities given for comparison.

not restricted for reasons of RF exposure, where personnel are exposed to RF as a consequence of their employment). Table 24.3b contains RF exposure limits for "controlled areas" (those areas restricted for reasons of RF exposure, where personnel are exposed to RF as a consequence of their employment).

Exposure to electric and magnetic fields that vary with time results in internal body currents and energy absorption in tissues, depending on the coupling mechanisms and the incident frequency. Table 24.4 lists the dosimetric quantities to be considered and corresponding units.

Cellular telephones operating in the frequency range of 800–900 MHz were implicated in the causation and/or promotion of brain cancer through the evening syndicated "talk show" media circuit. The show reported that a lawsuit was filed due to a man's wife succumbing to brain cancer, which was detected shortly after her continual use of a cellular phone; the tumor was reported to be in the shape of an antenna. This particular show generated a flurry of controversy, continuing until the present day. Sadly, the dismissal of the lawsuit without finding was not as widely publicized. Nevertheless, numerous studies have been published attempting to determine possible carcinogenic effects of exposure to RF with frequencies in the range used by communication systems. A concise summary of laboratory research findings was recently published by the International Commission on Non-Ionizing Radiation Protection (ICNIRP)[20]. There is substantial evidence that RF is not mutagenic, and exposure to these fields is therefore unlikely to initiate carcinogenesis. A recent study did indicate that use of cellular telephones may be detrimental; it was reported that use of cellular telephones in a moving automobile increases the risk of accidents to the level of driving drunk.[21]

Recent concern has been expressed about the possible interference of cellular tele-

TABLE 24.3b IEEE/ANSI Maximum Permissible Exposures for Controlled Environments

Frequency Range (MHz)	Electric Field Strength (E) (V/m)	Magnetic Field Strength (H) (A/m)	Power Density (S) E Field; H Field (mW/cm^2)	Averaging Time $\|E\|^2$, S, or $\|H\|^2$ (min)
0.003–0.1	614	163	(100; 1,000,000)[a]	6
0.1–3.0	614	$16.3/f$	(100; $10{,}000/f^2$)[a]	6
3–30	$1842/f$	$16.3/f$	($900/f^2$; $10{,}000/f^2$)[a]	6
30–100	61.4	$16.3/f$	(1.0; $10{,}000/f^2$)[a]	6
100–300	61.4	0.163	1.0	6
300–3,000			$f/300$	6
3,000–15,000			10	6
15,000–300,000			10	$616{,}000/f^{1.2}$

Induced and Contact RF Currents

Frequency range (MHz)	Maximum current (mA)		
	Through both feet	Through each foot	Contact
0.003–0.1	$2{,}000f$	$1{,}000f$	$1{,}000f$
0.1–100	200	100	100

[a] Values of plane wave equivalent power densities given for comparison.

phone usage with the proper operation of cardiac pacemakers. The FDA has researched this concern, and concluded: "Based on these preliminary findings, cellular phones do not seem to pose a significant health problem for pacemaker wearers."[22]

A significant difference between RF and ELF exposures is that RF exposures, especially RF exposures in the microwave (MW band, may have thermal effects. Exposure to low-frequency electric and magnetic fields in air results in a negligible amount of energy absorbed and no measurable temperature rise in the body. Exposures of greater intensity than 1 mW/cm^2 at frequencies above about 10 MHz can lead to signficant energy absorption and temperature increases. Exposures of greater intensity than 10 mW/cm^2 may cause increases in whole body temperature. Heating from RF is no different in producing thermal effects than other sources of exogenous sources of heat except that the degree and distribution of heating are less predictable. The initial physiologic response in laboratory animals is cutaneous vasodilation, which reverses rapidly if the source

TABLE 24.4 Dosimetric Quantity for RF Exposure Assessment

Frequency Range	Dosimetric Quantity	Units
1 Hz–10 MHz	Current density	amperes per meter (A/m^2)
1 Hz–100 MHz	Contact currents	amps (A)
100 kHz–10 GHz	Specific absorption rate (SAR)	watts per kilogram (W/kg)
300 MHz–10 GHz	Specific absorption (SA)	joules per kilogram (J/kg)
10–300 GHz	Power density (S)	watts per square meter

of RF is removed. Surface evaporation, metabolism, cell growth, and cell division, as well as immune response, can be altered by increasing thermal RF exposures. Heated humans have been noted to suffer psychological symptoms, intermittent hypertension, and burns, while heated laboratory rats and monkeys suffered decreased task performance at specific absorption rates (SAR) (a measure of energy absorbed in the body) in the range of 1–3 W/kg. At levels of body temperature elevation above about 1–2°C, a large number of biological effects have been observed, including increased blood–brain barrier permeability, ocular impairment, stress-associated effects on the immune system, decreased sperm production, teratogenic effects, and alterations in neural functions.

Thermally vulnerable organs such as the eye may sustain damage from high-energy exposures. Microwave exposure to the eye may cause injury, including cataracts (at frequencies less than 2 GHz), corneal damage (at frequencies above 10 GHz), and retinal lesions. Microwave cataracts have been induced in experimental animals and reported, rarely, in man. The present consensus is that very high-intensity (greater than 1.5 kW/m^2) exposures are required to produce detectable eye damage. Frequent reports of an auditory sensation from powerful RF fields are due to rapid thermoelastic expansion and secondary pressure waves, which can be detected by the cochlea. This is known as a *microthermal effect*, where a high local rate of tissue absorption and expansion occurs in the absence of any measurable change in temperature.

In the frequency range between 100 kHz and 110 MHz, shocks and burns can result either from touching an ungrounded metal object that has acquired a charge in the field or by contact between a charged body and a grounded metal object. Threshold currents that result in biological effects that range from perception to pain have been measured in controlled human experiments. Generally, when the current between the point of contact of the human body and the conductor exceeds 50 mA, there is a risk of severe burns and irreversible tissue damage.

Control of RF exposures by shielding can be relatively inexpensive, but may be an arduous task for the novice. Metal-conducting wire mesh, particularly copper, is very useful, as long as the openings are less than $\frac{1}{4}$ wavelength, and all sections are securely grounded. Foam can be used to gradually diminish the fields through insulators of varying impedance. Distance is perhaps the best tool for control of stray RF fields. However, it must be mentioned that the proper maintenance of equipment to prevent RF leakage is of the utmost importance. Refer to Fig. 24.3 for the internationally recognized RF caution sign published by ANSI.[23]

24.1.2.5 *Infrared, Visible Light, Ultraviolet* *Note:* In the part of the electromagnetic spectrum above about 1 teraherz (THz, 10^{12} Hz), it is traditional to refer to wave-

Fig. 24.3 The radio frequency (RF) hazard warning symbol. Courtesy ANSI C95.2-1982.

length vice frequency. All objects with temperatures above absolute zero emit infrared radiation, which extends in a band between 760 nanometers (nm, 10^{-9}) and 1 millimeter (mm, 10^{-3}). The human eye detects visible light in the narrow band between 760 nm at the red end, and 380 nm at the violet end, with peak sensitivity at 555 nm. UV light extends in a narrow band between 400 and 40 nm.

The major industrial sources of IR are heating and drying devices and lasers (to be discussed later in this chapter). Large sources of visible light exist in industry, mainly for the expected application: artificial lighting. Ultraviolet sources in industry are very diverse. They may be in the form of intentionally harmful UV lights, such as those used in biological safety cabinets to kill entering or exiting bacteria, or unnecessarily harmful, such as UV lasers. Welding is also a major source of harmful UV.

The units used to express IR and UV exposures are typically mW or μW per cm^2. Standards for IR, VL, and UV can be found published by the ACGIH for nonlaser sources. Standards for IR exposure vary with the coinciding presence of VL, and ambient temperatures. UV standards are used to calculate "safe stay times" by using tables and/or formulae to calculate the time necessary to exceed to TLV. For UV exposures incident on the unprotected eyes or skin, between 180 and 400 nm, use the following formula to calculate the appropriate "safe" stay time, in seconds, for every 8-hour period: Divide (0.003 J/cm^2) by E_{eff} in W/cm, where $E_{eff} = \Sigma E_\lambda S_\lambda \Delta\lambda$[$E_\lambda$ = spectral irradiance in $W/cm^2/nm$; S_λ = relative spectral effectiveness (unitless); and $\Delta\lambda$ = bandwidth in nm)].

The major hazard from IR exposure is the coinciding rise in the temperature of the absorbing tissue. IR is absorbed by tissue water and therefore nonpenetrating; energy transfer is directly to outer surfaces. Pain is the best warning property in normal skin for thermal injury due to IR. Conversely, the eyelid and eye do not have ample thermal warning properties. Eyelid blisters are common following heating exposures, and the eye is at risk for injury from IR exposures. IR overexposure produces corneal burns and choroid and retinal damage.

There is laboratory evidence that chronic exposure to excessively bright light may lead to premature degeneration of the cones responsible for color vision. The retina is sensitive to excess light.

The harmful effects of UV are due primarily to the range of 315–280 nm, also known as "UV-B." Acute effects upon the eye include photokeratitis and photoconjunctivitis. Human eyesight perceives UV poorly (or not at all) because of absorption by ocular media. Acute damage may occur before there is awareness of exposure. Keratitis and conjunctivitis are usually reversible within several days. Burns may occur from lateral UV-B exposure to the cornea, and not merely through the pupil and lens. Exposures to UV around 297 nm induce cataract formation with a clear dose–response relationship.

Keratitis has been observed in biotechnology environments, where UV is used for sterility and for DNA sequencing operations. Irradiance levels of germicidal UV lamps vary markedly with distance from the lamp or nearby reflecting surfaces. Some UV-B exposure is necessary for vitamin D_3 production where nutritional supplies are inadequate. However, there is ample evidence that UV exposure causes substantial skin pathology. There has been an alarming epidemic of skin cancers in the parts of the world where increases in outdoor recreation and admiration for tanning have occurred.

Control of IR, VL, and UV sources can be accomplished through specially designed eyewear and, to some extent, protective clothing. Only qualified personnel should make a determination of the effectiveness of any particular form of personal protection.

24.1.2.6 Lasers "Laser" is an acronym for light amplification by stimulated emission of radiation. An energized laser emits intense monochromatic (that is, one wavelength color) visible or invisible coherent (that is, all photons of light are in "phase" with one another) radiation continuously or in pulses.

IR lasers include the neodymium yttrium garnet (Nd-YAG) and carbon dioxide (CO_2) lasers. VL lasers include dye lasers, argon, helium–neon lasers, and krypton lasers. UV lasers include "excimer" lasers, and Nd-YAG lasers tripled in frequency into the UV spectrum.

Measurements of the optical radiation from lasers is usually accomplished with either thermal or quantum detectors. Thermal detectors measure the rise in temperature in crystals, although these are best used for IR. UV, IR, and VL measurements can be obtained with quantum detectors. These detectors emit electrons in response to being struck by radiation.

The focusing ability of the eye increases the retinal hazard of any visible laser beam. Laser classification and associated hazards are described in Table 24.5. Eye protection issues are outlines in ANSI Z136.1-1993 for the Safe Use of Lasers.[24] Skin protection is also important for noninterlocked operations of class 4 lasers. CO_2–N_2 lasers emit invisible energy; so burns may potentially be severe before the hazard is appreciated. For the proper posting of laser warning signs (see Fig. 24.4), refer to ANSI Z136.1-1993.

The most consistent hazard is not related to the radiation, but to the possibility of electric shock. Fire, cryogenic, and x-rays hazards are also associated with laser use. For these reasons, only appropriately qualified personnel should be involved with the design, installation, personnel, procedures, and any changes in conditions of laser use. Additionally, there is a significant acute hazard to the eye and skin from certain types of laser equipment. These depend upon the wavelength, intensity, and duration of exposure. Lasers may emit infrared, visible, and ultraviolet light.

Personal protective equipment has serious limitations when used as the only protective measure for class 4 lasers. Eyes and skin are best kept out of the beam by interlock devices, which turn off the laser electronically or else provide physical barriers between

TABLE 24.5 Laser Classification

Class of Laser[a]	Potential danger
1	Essentially harmless
2; 2a	Do not stare into the beam
3; 3a; 3b	Hazardous
	Direct viewing must not occur
	Specular reflections are also dangerous
	Skin exposure is harmful
4	Extremely hazardous
	Direct viewing must not occur
	Specular reflections are extremely dangerous
	Diffuse reflections dangerous
	Skin exposure extremely hazardous

[a]When Class 3 and 4 lasers are fully embedded into an enclosed system that has full safety interlocks, the system may be classified as a Class 1 system [per instructions of the Laser Safety Officer (LSO)].

Fig. 24.4 The laser warning symbol. Courtesy Laser Institute of America.

operators and bystanders and the beam during operations. Not all research laser activities are amenable to interlock protection. Personal protective devices are then used. Eye shielding must be selected in accordance with the wavelength employed, and should include side shielding. Absorptive filters are generally preferred to reflective types, as absorptive filters are reliable regardless of the incident angle of the beam. A problem with optical density markings on laser-protective eyewear has been uncertainty with the reliability of manufacturers' markings concerning transmittance. Independent checks are not truly achievable with available equipment; so it is important to pick the most reputable supplier. Another problem with eye protection is that complete protection renders otherwise visible beams invisible.

Laser surgery or any other tissue use of class 4 lasers will inevitably generate heat and pressure, creating smoke and fume. The smoke is malodorous. Designed ventilation is the appropriate industrial hygiene measure, but this fails to address adequately that the smoke plume may carry infectious organisms.

Calculations such as "optical density" (OD) (the protection factor necessary to reduce a given laser hazard to the level below which permanent injury is prevented) and "nominal hazard zone" (NHZ) (the distance necessary to reduce a given laser hazard to the level below which permanent injury is avoided) are necessary in determining the safe use of any class 3 or 4 laser systems. Confirmatory measurements should be made with radiometry instrumentation to ensure that appropriate eye (and skin) protection is chosen.

Medical surveillance of laser workers and support staff is highly recommended to ascertain eye injury in the event of accidental exposure. Baseline eye exams before working with the laser is essential in making this determination. Further eye examinations will be required postincident, and postemployment.

24.1.3 Summary

The three elements that must be considered in any effective safety program designed to control actual or potential hazards are as follows:

- Recognition
- Evaluation
- Control

This chapter was designed to help the reader to *recognize* NIR hazards that may be found in the industrial setting. Upon recognition of each potential or actual hazard, a competent individual should perform a safety *evaluation*. Based on the outcome of the evaluation, *control* may or may not be necessary.

There is an immense potential for serious injury considering all the NIR sources in use by industry. It is imperative that an NIR safety program be initiated and maintained in each facility that uses NIR sources. The development of an NIR safety program, as well as the evaluation of any NIR actual or potential hazard, must be carried out by an individual properly trained and experienced in this complex field of occupational safety. Consultants are available for such work. Qualified individuals may be found by contacting the Health Physics Society. (1313 Dolly Madison Blvd, Suite 402, McLean, VA, 22101).

24.2 IONIZING RADIATION

These sources can come in the form of sealed radioactive material used for its radiation emissions, unsealed radioactive material used in a research or manufacturing process, or machine-produced radiation in the form of an x-ray machine or the byproduct of a high-voltage supply. It is important to be able to recognize potential sources of ionizing radiation and have an effective evaluation program to limit worker exposures. It is the responsibility of management to provide the necessary training, facilities, equipment, and personnel to maintain levels of radiation exposure to its employees, the general public, and the environment to as low as reasonably achievable (ALARA). The following sections outline administrative, process, and engineering controls that can be used to help achieve this goal.

24.2.1 Specific Licensing

Users of radioactive sources typically must obtain a specific license from an appropriate Federal or state agency depending on the state where the facility will be located. States fall into two categories: agreement and nonagreement states. Agreement states are those that have entered into an agreement with the Nuclear Regulatory Commission (NRC) to license, regulate, and inspect the use of radioactive material within their boundaries. Nonagreement states are directly licensed, regulated, and inspected by NRC. See Appendix A for the current list of agreement states.

The specific license application outlines to the agency the types and quantities of radioisotopes that will be possessed and used, a description of the proposed use of the radioisotopes, and a comprehensive radiation safety program to ensure that radioisotopes will be possessed, stored, used, and disposed in a responsible and safe manner. Administrative control in the safe use of radioisotopes begins with the license application process.

24.2.2 General Licensing

Many radiation sources used as part of a mechanical system or process do not require a specific license to possess and use them. These sources are possessed and used under a general license agreement. Any person or company in the United States has the right to possess and use generally licensed quantities of radioactive material. Because a license is not required to buy these sources, it is often difficult to track these sources within the work environment. Furthermore, it is difficult to involve workers in a comprehensive radiation safety program if it is unknown that radioactive sources are possessed. Some examples of generally licensed radioactive sources are static eliminator bars, ionizing-type smoke detectors, gas chromatography analyzers that use a radioactive source, and density gauges. Although these sources are generally licensed, it does not exempt the employer from performing certain checks and balances on the sources during their use. These requirements are typically outlined in instructions from the device manufacturer. They typically require some form of monitoring or radiation level measurements at specified intervals.

24.2.3 Source Material

As with the generally licensed sources, certain quantities of naturally occurring radioactive material (source material) are exempt from licensing. Uranium- and thorium-containing materials are used widely in industry from shielding and counterweights to coating processes on high polished optical surfaces. Although these materials can be possessed and used without a license (and many times without radiation safety considerations), the potential for worker exposure, especially internal exposures from some manufacturing processes, can be significant.

24.2.4 Machine Radiation Sources

X-ray machines such as diagnostic machines as part of a company health clinic to analytical machines such as diffraction or fluorescence equipment used to study manufactured goods may present an exposure potential to users and surrounding environments. The emissions from these machines or the area where the machines are used must be controlled to ensure worker safety. Most state radiation control programs require a formal registration of radiation-producing machines. The registration process requires that users comply with the agency rules and regulations for the safe use of these types of machines.

24.2.5 Radiation Safety Officer

A qualified person should be identified as the Radiation Safety Officer (RSO) for the facility. The responsibilities of the RSO may include:

1. Ensuring facility compliance with all applicable Federal, state, and local regulations and ordinances.
2. Updating license parameters, as necessary, via amendment request to the granting agency.
3. Ensuring compliance with conditions of generally licensed sources, exempt sources, and radiation-producing machine registrations.
4. Reviewing all proposed uses of radiation, especially new procedures that require substantially increased amounts of radiation.
5. Provide radiation safety training to all radiation workers, ancillary personnel, or other employees whose duties require them to be in areas where radiation sources or machines are used and stored.
6. Approve all purchases of radiation sources or machines to ensure compliance with the license conditions or registrations.

7. Perform radiation safety audits of all radiation source and machine use.

8. Be available to assist in any emergencies, special decontamination efforts, or worker exposure evaluations.

9. Maintain all appropriate records as required by specific license conditions, general license requirements, and machine registration regulations.

10. Properly dispose of radiation sources, radioactive materials, or radiation-producing machines when they are no longer of any use to the facility.

11. Approve of all transportation of radioactive sources.

The RSO duties need not translate into a full-time position and typically will be only a small percent of a person's responsibilities. This will depend on the size of the program and the number of radiation sources and workers. Typically, certified health physics consultants are used to assist RSOs of small programs. Consultants can evaluate the adequacy of radiation shielding, perform required leak testing of radioactive sources, provide radiation worker training seminars, and assist the RSO in worker exposure evaluations. However, it is very important that ample time be scheduled to allow the RSO to audit the radiation safety program on an ongoing basis.

24.2.6 Radiation Safety Liaison

A person at the user level should be identified to assist the RSO in the administrative functions necessary to ensure compliance with license conditions and regulations. The liaison's responsibilities may include:

1. Maintaining a current inventory of radiation sources stored in the lab. Perform physical inventories of sealed sources as required.

2. Exchange of personnel monitoring devices on a timely basis.

3. Perform and maintain records of radiation surveys or source leak tests.

4. Inform the RSO when new persons join the user group to ensure the person receives proper training.

5. Be responsible for radioactive source security and use logs.

6. Inform the RSO whenever new generally licensed or exempt natural sources are purchased.

7. Act as the liaison between the user group and the RSO.

Small facilities with only a few radiation sources and a small number of radiation workers may not need a radiation safety liaison. The RSO would be directly responsible for the above-listed duties. However, if radiation sources or machines are used in the field or by various different work groups within the facility, the need for this position becomes evident.

24.2.7 Administrative Procedures

Licensed facilities using radiation sources or machines that produce radiation should have a set of standard operating procedures addressing the following functions:

1. Radiation worker training.
2. Ancillary personnel training.
3. Assignment of personnel monitoring devices and record of radiation exposures.
4. Radiation source ordering, receipt, and inventory.
5. Routine bioassay or *in vivo* measurements of radiation workers, as necessary.
6. Emergency medical surveillance after accidents or known exposure above allowable limits.
7. Policy for pregnant radiation workers.
8. Environmental monitoring of experiments that potentially release airborne containments.
9. Radiation survey meter calibration and records of calibration certificates.
10. Procedures for and record of safe disposal of low-level radioactive waste, or final disposition of radioactive sources or machines.
11. Emergency procedures.

These standard operating procedures will be used as part of an effective radiation safety program. Administrative control will include maintenance of all records required by regulations, registrations, and license conditions. These records will be scrutinized during licensing agency inspections. Records of personnel exposure histories and radioactive waste disposal will be maintained indefinitely.

24.2.8 Licensed Radiation Sources or Machines

A thorough understanding of the radioisotopes used in the sealed sources or the type of radiation-producing machine to be used is essential to determine the control methods necessary for safe handling of the sources or machines. The appropriate instrumentation and calibration needed to detect the material, the potential pathways of exposure, handling tools or shielding necessary to reduce external exposures, the need for personnel monitoring devices, personal protective equipment, emergency procedures, and waste disposal requirements are safety areas that need to be addressed.

24.2.9 Radiation Worker Training

The most important aspect of radiation worker safety is providing an effective training program to ensure the worker understands the potential hazards involved while working with the radioactive sources or radiation-producing machines. Radiation worker training is a requirement of the NRC or state licensing agency. Radiation safety training is mandatory for all radiation workers. The following subjects should be covered in a radiation worker training seminar:

- Units of radioactivity and radiation dose
- Concept of radioactive decay and half-life
- Radiation detection and measurement
- Type and amounts of radioactive material licensed
- Safe handling techniques and dose reduction techniques
- Standard operating procedures for the source or machine being used

- Personal protective equipment
- Maximum permissible exposure limits
- ALARA concept
- Biological effects and risks from occupational exposures
- Bioassay and *in vivo* measurement of ingested radioisotopes
- Radiation and contamination survey techniques
- Leak testing of sealed sources
- Emergency procedures and decontamination methods
- Machine calibrations and maintenance
- Transportation of radiation sources for field work

The radiation safety training seminar should afford the worker an opportunity to ask questions concerning the safe use of the radiation sources. Attendance at training seminars will be documented.

24.2.10 Safe Handling and Dose Reduction Techniques

Once information is known about the types and amounts of radiation sources to be handled, the worker can develop safe handling and dose reduction techniques. Whenever radioactive sources are handled, the worker must wear protective clothing to reduce the possibility of personnel contamination. Lab coats, gloves, and safety glasses are required during all handling. Double gloving is appropriate. The outise glove would be changed if contamination is detected. A dry run of all new procedures, especially when significant increases in activity are required, is important to preclude unforeseen handling problems. Benches should be covered with plastic-backed absorbent bench paper whenever liquid sources are handled. Continual contamination monitoring with a portable survey meter during handling will reduce the potential for widespread contamination.

Dose reduction can be accomplished by utilizing the concepts of time, distance, and shielding or a combination of the three. Radiation dose is measured as a rate, usually in millirads per hour. If one can reduce the amount of time they are exposed to a source, their cumulative radiation dose equivalent will be lowered. Of the three concepts, time may be the most difficult to change due to experimental procedures.

Distance from a source of radiation can significantly reduce the worker's exposure. The dose rate from a point source of radiation, for example a 10-millicurie sealed source of ^{137}Cs found in density gauges, is reduced by the square of the distance as one moves away from the source (inverse square law). Thus the dose rate at 1 cm is 100 times greater than the dose at 10 cm. Dose reduction to the hands and fingers can be significantly reduced with use of tongs to handle sealed sources.

Appropriate shielding, depending on the type and energy of the emitted radiation, can also significantly reduce worker exposures. For x-ray and gamma-ray emitters, the recommended shielding is a high-atomic-number material such as lead. The thickness of the shielding will be determined by the energy of the incident radiation.

24.2.11 Radiation Detection and Measurement

To work safely with sealed or unsealed sources of radioactivity or radiation-producing machines, workers must be fully trained in the use of radiation detection equipment to measure potential radiation exposure rates and contamination levels. In addition, radioactive sealed sources are required to be leak tested at a certain intervals (usually 6 months). These leak-test samples must be analyzed in analytical equipment to quantify any radioactive material leakage.

24.2.12 Survey Instruments

Lightweight, portable radiation survey instruments must be available in radiation laboratories to enable workers to monitor themselves and their work areas while handling sources. The most versatile radiation detector used is the Geiger Mueller (GM) detector. It is capable of detecting beta, x-ray, and gamma-ray radiations emitted by the most commonly used radioisotopes used in sealed sources. In addition to the GM detector, there are portable specialty detectors such as scintillation detectors or ionization chambers whose use becomes necessary depending on the source of radiation and the measurements required. However, the GM detector serves most purposes fromm a radiation worker exposure evaluation potential.

All survey instruments must be routinely calibrated against certified radioactive standards. In addition, small radioactive check sources should be attached to each instrument to allow the worker to check operability each time the instrument is used.

24.2.13 Analytical Instruments

These instruments are used to quantify amounts of radioisotopes. For most small users or users of sealed sources that require leak testing, these instruments are prohibitively expensive. Most small users contract with laboratory facilities for radioactivity analysis on such things as leak tests or contamination smears. The types of analyzer typically used in these analysis laboratories are liquid scintillation counters, gas flow proportional counters, or gamma-ray spectroscopy analyzers.

24.2.14 Maximum Permissible Exposure Limits

Regulations outline the maximum permissible exposure limits allowed radiation workers, pregnant radiation workers, and the general public. The limits are reported in units of millrems. The following are the current limits of exposure:

Area exposed	Annual limit (mrems)
1. Total effective dose equivalent (internal and external whole body)	5000
2. Lens of the eye	15000
3. Skin and extremities	50000
4. Pregnant radiation worker	500
5. General public	100

Of the commonly used radiation sources, beta emitters are classified as nonpenetrating radiation and will give the worker a skin exposure. X-ray and gamma-ray emitters, whether from radioactive sources or machine produced, are classified as penetrating radiation and will give the worker a whole body exposure.

24.2.15 ALARA Concept

Although there are limits for exposures to workers as outlined above, it is the responsibility of the radiation safety officer to operate the radiation safety program such that workers utilize safe handling and dose reduction techniques to keep radiation exposures as low as reasonably achievable.

24.2.16 Biological Effect and Risks from Occupational Exposures

Although there is some risk associated with all exposures to radiation, workers who keep their exposures within the maximum permissible exposure limits maintain the associated risk at acceptable levels.

Exposures from uses of radiation sources or machine-produced radiation should be kept with 10% of the maximum permissible limits, and any exposures above 10% should be investigated. Typically, workers are kept within 1–5% of the maxima. The effects associated with small radiation exposures are practically unmeasurable due to the varying levels of natural background radiation received by everyone and exposures to other potentially harmful materials. Radiation workers should be encouraged to read NRC Regulatory Guide 8.29 “Risks from Occupational Radiation Exposure.” This document reviews potential health effects from radiation exposures and enables the worker to compare the risk associated with occupational exposures to the risk associated with other occupations and the risk associated with normal daily life activities.

24.2.17 Worker Exposure Monitoring

Workers handling radioactive sources or working with radiation-producing machines that may result in potential exposure to radiation in excess of 10% of the maximum permissible exposure limits should be monitored. Typically monitoring is accomplished by use of film badge or thermoluminescent dosimeters capable of measuring exposures for x-rays, gamma rays, and beta particles. The dosimeter should be worn at all times when radiation exposures are possible. The film badge or thermoluminescent dosimeter is used to measure whole body and skin exposures. If workers are handling large radioactive sources, extremity monitors should also be provided. The dosimeter should be exchanged on a monthly or quarterly basis. More frequent exchange of dosimeters is recommended for workers who consistently receive significant radiation exposures (greater than 25% of the MPEs). Radiation dosimetry reports should be made available to the workers, and all results greater than 10% of the MPEs should be investigated to establish if additional dose reduction controls are needed. Radiation exposure histories must be maintained indefinitely.

Monitoring of individual workers may not be feasible or necessary for certain uses of radioactive sources or radiation-producing machines. Area monitors, either passive systems such as the film badge monitor or an active system such as a GM detector–based radiation monitor, can be used to estimate radiation doses to all workers who are present in the area.

24.2.18 Radiation and Contamination Surveys

Radiation workers must be trained to perform radiation and contamination surveys whenever radioactive sources or radiation-producing machines are used. Radiation surveys are simply measuring the exposure rate in millirems per hour in the areas where radiation sources are used or stored. The results of these surveys will determine the approximate amount of exposure the worker can anticipate receiving during the time spent in the area. These results can determine whether shielding may be necessary to reduce potential exposures. Radiation surveys should be made during all handling phases of the sources. The GM detector–equipped survey meter is an appropriate instrument for these measurements. An ion chamber survey instrument is a good choice for measuring potential exposures from machine-produced radiation (typically x-rays).

Contamination surveys are performed to determine if any unsealed radioactivity has been spilled or transferred to areas where it is not welcome. Such areas are bench tops, gloves, lab coats, survey instruments, hood aprons, and equipment. If detected, these areas must be decontaminated to license condition or regulatory limits for removable contamination. The GM detector–equipped survey meter is the most versatile instrument for these types of surveys.

In addition to the everyday surveys required while working with radioisotopes, routine surveys including swipe tests should be performed in all radiation laboratories. Swipe tests are performed by rubbing a piece of filter paper or cotton swab over the area to be tested. The typical area surveyed is 100 square centimeters. The swipe tests are then analyzed in an appropriate analytical instrument such as a liquid scintillation counter or a gas flow proportional counter.

Sealed radioactive sources must be tested for radioactivity leakage at certain time intervals, typically every 6 months. Swipe testing the source surface or the closest accessible surface followed by analytical analysis is appropriate. The results of the test must be less than ≤ 0.005 microcuries of activity. If the results exceed this value, the source must be removed from service, and further testing must be done on the source.

The frequency of routine surveys can be weekly, monthly, quarterly, semiannually, or annually depending on the types and amounts of radioactivity or the radiation producing equipment handled in the facility. The RSO or RSO liaison is usually responsible for the routine surveys.

24.2.19 Engineering and Environmental Controls

For the use of unsealed radioactive material, engineering control is in the form of carefully planned work space with surfaces that are easily cleaned. Benches and floors should have nonporous surfaces. Sinks used for radioisotope disposal should be stainless steel. The physical layout of the facility is very important. Workers' desks or lockers should not be in the middle of the work area. Since smoking, eating, and drinking are not permitted in radiation work areas, an area should be made available for these activities.

Work with volatile radioactive material or processes that generate airborne radioactivity such as grinding, polishing, or coating must be done in a fume hood or in a work space with localized exhaust. The exhaust from this hood/area should be filtered through an appropriate filter (HEPA/activated carbon filter) to trap any potential releases to the environment. The exhaust air from any hood/area where volatile or airborne radioactive materials are used must be sampled to prove compliance with the EPA NESHAPS regulatory limits for release of airborne pollutants. In addition, the room air should be sampled via a breathing zone sampler to alert workers to potential airborne radioactivity levels and possible internal radiation exposures.

24.2.20 Transportation of Radiation Sources

The transportation of radioactive sealed sources or devices containing such sources must be done in compliance with Department of Transportation rules and regulations found in Title 49 of the Code of Federal Regulations. Most devices are sold with a carrying case that is acceptable for transportation. However, there is specific documentation that must accompany each shipment. A record of all shipments should be maintained for inspections by licensing authorities. There are basic training requirements for shippers of hazardous materials that will have to be met for shipment of radioactive sources.

If radioactive sources or devices are to be used outside the jurisdiction of the licensing authority, the user may have to apply for reciprocity to use these devices in different geographical locations. Consult your state radiation control department or the radiation control department in the state where the work will be done for reciprocity details.

24.2.21 Low-Level Radioactive Waste Disposal and Source Disposition

The management and disposal of low-level radioactive waste and final source disposition is an integral part of any radiation safety program. There are several licensed low-level waste brokers within the United States who can assist licensees in the disposal of any wastes or sources. Current waste disposal fees are very expensive. Users of unsealed radioactive materials must incorporate a waste minimization plan as part of an effective radiation safety program. Users of sealed radioactive sources should negotiate an agreement with the source manufacturer to accept return of the source when it is no longer of use to the project. Currently, there are only three sites within the United States accepting low-level waste. These sites are located in Richmond, Washington; Clive, Utah; and Barnwell, South Carolina. Most states have appointed boards to manage the low-level wastes generated within their boundaries. These boards as well as state radiation control agencies can be a good resource for up-to-date information on waste disposal requirements.

24.2.22 Laboratory Surveillance and Management Audits

Routine radiation laboratory inspections should include surveys for removable contamination, radiation exposure level measurements, and observations for compliance with the licensee radiation safety program (i.e., evidence of eating or drinking in the lab). Development of a compliance checklist serves as a useful tool during radiation audits. Radiation workers should be observed handling radioactive sources to ensure good technique as well as regulatory compliance. Deficiencies should be noted and reports sent to the laboratory supervisors along with suggested corrective actions for improvement. Group retraining seminars are an effective avenue to discuss problems with the entire laboratory and to work on effective solutions. It is important to include management in periodic audits as well as all deficiency reports sent to groups. This will keep management informed of potential regulatory compliance problems.

24.2.23 Emergency Procedures

Whenever radioactive sources are used, the potential exists for accidents or spills that result in personnel or facility contamination. Facilities should establish emergency procedures to enable efficient and effective handling of accident situations. There are three general categories of accidents involving radioactive material: (1) radioactive contamination of personnel and facilities including personnel injury, (2) radioactive contamination of personnel and facilities, and (3) radioactive contamination of facilities only.

The following are general procedures to follow in the event of a radiation accident.

24.2.23.1 Contamination and Personnel Injury

1. Immediately notify appropriate Emergency Response Personnel (police, fire, etc.).
2. Remove person from contaminated area, if possible, depending on the extent of the injury.
3. Remove contaminated clothing from person and wait for emergency medical help to arrive.
4. Isolate the contaminated area of the facility.
5. Inform emergency respondents that patient has potential radioactive contamination. Give information on radioisotope, amount, etc. if known.
6. Accompany patient to treatment area. Bring your radiation survey meter to help identify contaminated areas.
7. Notify the Radiation Safety Officer.

Whenever a person sustains an injury in an accident involving radioactive material, first aid comes first. After appropriate medical attention is received, decontamination of the facilities can commence.

24.2.23.2 Contamination of Personnel and Facilities (No Injury)

1. Isolate the contaminated persons in a noncontaminated area.
2. Isolate the contaminated area or equipment with physical barriers.
3. Survey persons and remove any contaminated clothing articles.
4. Decontaminate personnel to minimize total radiation exposure.
5. Notify the Radiation Safety Officer.
6. Begin decontamination of facilities.

The facilities must be decontaminated to levels acceptable to license conditions. Perform a radiation survey to determine the physical extent of the contamination. Working from a clean or noncontaminated area, begin decontamination of facility. Clean a small area and wipe test to check for remove contamination. Repeat as necessary until removable contamination levels are below license limits. Perform a thorough radiation contamination survey to ensure all areas are clean. Continue cleaning until all areas have been successfully decontaminated and the facility can be cleared for further use. Keep appropriate records in an accident or incident file. Radiation laboratories should be equipped with radioactive spill kits to aid in cleanup efforts when an accident occurs.

24.2.23.3 Loss of Radioactive Source or Device An additional emergency scenario may be the loss of a radioactive sealed source or a device containing radioactivity. Such loss of control usually requires notification of competent authorities such as the state radiation control department or the Nuclear regulatory Commission. It is advisable to procure the services of a competent health physics consultant to assist facility personnel in emergency situations.

24.3 CONCLUSIONS

There are many sources of radiation commonly found in the workplace environment. These sources span the energy spectrum from nonionizing to ionizing radiation. Each source carries its own potential for worker occupational exposure, especially if not handled or managed properly. Many facilities have a combination of sources that are used as part of an overall process but are only a small part of that process. It is sometimes difficult for a facilities or plant manager to have a handle on all the different regulatory and safety compliance considerations that must be followed to assure safe use of these sources and a safe work environment for all workers. All too often an accident or incident is the catalyst for the development of a solid radiation safety program to deal effectively with these sources. Many times facilities lack the technical resources in house to deal with the potential problems.

Due to the wide variety and uses of these radiation sources, retaining the services of a health physics consultant certified by the American Board of Health Physics and with applicable experience in the type of radiation source being handled is a viable option for the facility.

REFERENCES

1. ACGIH, *Threshold Limit Values for Chemical Substances and Physical Agents and Biological Indices.* Cincinnati, Ohio, American Conference of Governmental Industrial Hygienists, 1997, 123–25.
2. International Commission on Non-Ionizing Radiation Protection (ICNIRP). Guidelines on limits of exposure to static magnetic fields, *Health Physics* **66**:100–6, 1994.
3. Tenforde, T. S. Biological Effects of Stationary Magnetic Fields, in: *Biological Effects and Dosimetry of Static and ELF Electromagnetic Fields.* Grandolfo, M., Michaelson, S., Rindi, A. (Eds.), Plenum Press, New York, 1985, pp. 93–128.
4. Hansen, D. J., Baum, J. W., Weilandics, C., Barnhardt, J. F., Meninger, M. H. Enhancement of Low-Level Magnetic Fields by Ferro-Magnetic Objects. *Appl. Occup. Env. Hyg.* **5**(4):236–41, 1990.
5. Polk, C., Postow, E. P. (Eds.). *Handbooks of Biological Effects of Electromagnetic Fields,* CRC Press, Boca Raton, Florida, 1996.
6. Medtronics Pacing Services.
7. Wertheimer, N., Leeper, E. Electrical wiring configurations and childhood cancer. *Am. J. Epidemiol.* **109**:273–84, 1979.
8. NAS National Academy of Science/National Research Council. *Possible Health Effects of Exposure to Residential Electric and Magnetic Fields,* 1996.
9. Haes, D. L., Jr., Fitzgerald, M. F. VDT VLF Measurements: The Need for Protocols in Assessing VDT User "Dose." *Health Physics* **68**(4):572–78, 1995.
10. Goldhaber, M. K., Polen, M. G., Hiatt, P. A. The risk of miscarriage and birth defects among women who use visual display terminals during pregnancy. *Am. J. Industr. Med.* **13**:695–706, 1988.

11. Windham, G. C., Fenster, L., Swan, S., Neutra, R. R. Use of video display terminals during pregnancy and the risk of spontaneous abortion, low birthweight, or intrauterine growth retardation. *Am. J. Industr. Med.* **18**:675–88, 1990.
12. Schnorr, T. M., Grajewski, B. A., Hornung, R. W., Thun, M. J., Egeland, G. M., Murray, W. E., Conover, D. L., Halperin, W. E. Video Display Terminals and The Risk of Spontaneous Abortion. *N. Engl. J. Med.* **324**:727–33, 1991.
13. 21 CFR 1930.10; Food and Drug Administration.
14. NCRP 119: National Council on Radiation Protection and Measurements, 1993; *A Practical Guide to the Determination of Human Exposure to Radiofrequency Fields.*
15. IEEE/ANSI C95.3-1992: American National Standard, *Recommended Practice for the Measurement of Potential Electromagnetic Fields—RF and Microwave*. The Institute of Electrical and Electronics Engineers, Inc., 345 East 47th Street, New York, NY, 10017.
16. IEEE/ANSI C95.1-1992. American National Standard, Safety levels with respect to human exposure to radio frequency electromagnetic fields, from 3 kHz to 300 GHz. The Institute of Electrical and Electronics Engineers, Inc. 345 East 47th Street, New York, NY, 10017.
17. National Council on Radiation Protection and Measurements (NCRP). Biological Effects and Exposure Criteria for Radiofrequency Electromagnetic Fields, NCRP Report 86, 1986.
18. Federal Register, Federal Communications Commission Rules; *Radiofrequency radiation; environmental effects evaluation guidelines*, Vol. 1, No. 153, 41006–199, August 7, 1996 [47 CRF Part 1; Federal Communications Commission].
19. Telecommunications Act of 1996, 47 U.S.C.; Second Session of the 104th Congress of the United States of America, January 3, 1996.
20. ICNIRP. Health issues related to the use of hand-held radiotelephones and base transmitters, *Health Physics* **70**(4):587–93, 1996.
21. Redelmeier, D. A., Tibshirani, R. J. Association between cellular-telephone calls and motor vehicle collisions. *N. Engl. J. Med.* **336**:453–58, 1997.
22. Food and Drug Administration Department of Health & Human Services, *Update on Cellular Phone Interference with Cardiac Pacemakers*, May 3, 1995.
23. IEEE/ANSI C95.2-1982 (Reaffirmed 1996). American National Standard, *RF Safety Warning Symbol*. The Insitute of Electrical and Electronics Engineers, Inc., 345 East 47th Street, New York, NY, 10017.
24. ANSI Z136.1-1993. American National Standards Institute for the Safe Use of Lasers. Laser Institute of America, 12424 Research Parkway, Suite 130, Orlando, FL, 32826.

ADDITIONAL REFERENCES

The following are some suggested references that would be useful to a person responsible for ionizing radiation safety:

1. United States Codes of Federal Regulations, Title 10, Chapter 1, Nuclear Regulatory Commission Rules and Regulations.
2. U. S. Nuclear Regulatory Commission, Regulatory Guide 8.13, "Instruction Concerning Prenatal Radiation Exposure," Revision 2; 12/87.
3. U. S. Nuclear Regulatory Commission, Regulatory Guide 8.29, "Instruction Concerning Risks from Occupational Radiation Exposure," 2/96.
4. Cember, H., *Introduction to Health Physics*, 2nd Edition, Pergamon Press, 1985.
5. Turner, J., *Atoms, Radiation, and Radiation Protection*, Pergamon Press, 1986.
6. National Research Council, Committee on the Biological Effects of Ionizing Radiations, "The

Effects on Populations of Exposure to Low Levels of Ionizing Radiation: 1980," National Academy Press, BEIR IV.

7. National Research Council, Committee on the Biological Effects of Ionizing Radiations, "Health Effects of Exposure to Low Levels of Ionizing Radiation," BEIR V, National Academy Press, 1990.
8. U.S. Code of Federal Regulations, Title 40, Protection of the Environment.
9. U.S. Code of Federal Regulations, Title 49, Department of Transportation.
10. U.S. Nuclear Regulatory Commission, Regulatory Guide 8.10, "Operating Philosophy for Maintaining Occupational Radiation Exposures As low As Is Reasonably Achievable," 5/77.
11. Radiological Health Handbook, U.S. Department of Health, Education, and Welfare, Public Health Service, Revised Edition, January 1970.
12. Nuclear Lectern Associates, *The Health Physics and Radiological Health Handbook*, Second Printing, June 1984.

APPENDIX A

Currently, the following 30 states have been granted Agreement State status by the Nuclear Regulatory Commission:

Alabama	Kentucky	New York
Arizona	Louisiana	North Carolina
Arkansas	Maine	North Dakota
California	Maryland	Oregon
Colorado	Massachusetts	Rhode Island
Florida	Mississippi	South Carolina
Georgia	Nebraska	Tennessee
Illinois	Nevada	Texas
Iowa	New Hampshire	Utah
Kansas	New Mexico	Washington

Information concerning a state's status with regard to the use of radioactive material can be obtained by calling the State Radiation Control office.

CHAPTER 25

How to Deal With The Troubled Employee

MARGIE MAGOWAN

25.1 INTRODUCTION
25.2 THE DIFFICULT EMPLOYEE
 25.2.1 Mediocre Employee
 25.2.2 Marginal Employee
 25.2.3 Inconsistent Employee
 25.2.4 Intolerable Employee
25.3 WARNING SIGNS
25.4 CORRECTIVE ACTION
25.5 DISCIPLINARY ACTION
25.6 TERMINATION
25.7 GUIDELINES FOR TERMINATION
25.8 INFORMING THE EMPLOYEE
25.9 INFORMING THE CO-WORKERS
25.10 LIMITATIONS ON THE EMPLOYER'S RIGHT TO DISCHARGE
25.11 VOLUNTARY RESIGNATION
25.12 DRUGS AND ALCOHOL IN THE WORKPLACE
25.13 DRUG AND ALCOHOL ABUSE: WRITTEN POLICIES
 25.13.1 Prescription Medication
25.14 RECOMMENDATIONS FOR DEALING WITH AN INTOXICATED EMPLOYEE
25.15 COMPANY LAYOFFS, DOWNSIZING, RE-ENGINEERING
25.16 THE EMPLOYEES LEFT BEHIND
25.17 STRESS IN THE WORKPLACE

25.1 INTRODUCTION

Managers and supervisors are required to address issues pertaining to performance and behavioral problems associated with members of their staff. It is imperative to deter-

Handbook of Occupational Safety and Health, Second Edition, Edited by Louis J. DiBerardinis, ISBN 0-471-16017-2

mine the cause of the problem and work to improve or correct the cause rather than to personalize the problem. In most cases, this approach will avoid a defensive response from the employee. The methods addressed in this chapter represent constructive ways to improve employee performance to achieve the result that all individuals involved will be winners.

The problem employee, in many cases, does not just involve the manager and the employee. Additional staff may, and in many cases are affected by the substandard performance levels of others. Clearly, it is not the responsibility of support staff members to correct the problem. It is a management concern to take the necessary steps and to initiate corrective action to improve unacceptable performance to a standard that meets company expectations.

A troubled employee may be so preoccupied with their problem that they neglect to pay attention to safety and health regulations. It is imperative that close attention be paid by management in these situations to avoid the occurrence of any unnecessary accidents.

Employees are covered by comprehensive laws intended to protect their safety and general health. The Federal Occupational and Safety Act of 1970 (OSHA) was passed by the United States Congress in 1970.

Under OSHA, employers are given broad duties to provide a place of employment free from obvious hazards likely to cause injury or death. In addition to such general responsibilities, employers also must comply with many very specific standards depending on the particular industry. An example: In Massachusetts, OSHA and the Massachusetts law apply to every employer in both production and office settings, including offices with only one employee.

All employees have certain rights and responsibilities under OSHA.

- Employees are obliged to comply with all standards and regulations. Sole responsibility for compliance rests with the employer.
- Employees may not be fined or cited for noncompliance. Employers are expected to establish disciplinary procedures for those employees who violate standards, rules and regulations provided by OSHA, which may include proper use of safety devices, care for wearing of protective equipment, and the requirement to report all injuries to the appropriate supervisor.
- Employees may participate in the development, revision, and revocation of standards.
- Employees may be notified of an employer's application for variances and may take part in hearings held in connection with it.
- Employees must be informed of exposure to toxic materials and be given medical examinations, paid for by the employer, when required by a health and safety standard.
- Employees may request inspection by the Department of Labor and in all cases are protected from discharge, retaliation, or discriminatory treatment for using any of their rights granted by the act.
- Specifically, any employee who refuses to work (in good faith) because of likely exposure to workplace hazards is protected from discrimination or future retaliation for filing a complaint.

25.2 THE DIFFICULT EMPLOYEE

A difficult employee is an individual who does not completely meet the performance standards as required by his or her job description.

The substandard performance may or may not be intentional. In some cases, the requirements of the position may not have been clearly articulated to the employee. After a thorough discussion regarding performance standards between the manager and the employee; if performance continues to remain below established standards, a performance problem exists and management must initiate corrective action at once.

When dealing with a difficult employee, it is important to remember that most employees want to and try to do a good job. Many situations can be corrected by spending some time with the employee to clarify any confusion regarding the expectations of the position. In addition, it is important to continue to provide the employee with constructive feedback on a regular basis addressing what the employee does well and discussing existing performance deficiencies.

If it is noted that some employees are experiencing accidents on a regular basis, an evaluation of their equipment and tools should be performed. In addition, their training should be evaluated to determine if it was adequate.

There is a variety of categories of difficult employees. Some examples include:

25.2.1 Mediocre Employee

This is an individual who meets the basic requirements of the job; performance is at the minimum level expected.

25.2.2 Marginal Employee

This represents someone who does not meet the required standards of the position, a nonproductive employee, one who wastes time, and gets by via the efforts of others in the department.

25.2.3 Inconsistent Employee

This individual is an employee whose performance reflects good performance to poor performance, which is often associated with varying employee moods.

25.2.4 Intolerable Employee

This employee's work performance is at a level considered to be poor, usually associated with a high absentee rate; individual may be annoying and disruptive to others in the department.

In most situations if managers work closely with difficult employees, problems can be worked out effectively to the satisfaction of all concerned. Although managers may view time spent with a difficult employee as excessive and costly, it is still cheaper for organizations to take this route than to proceed with termination. Only as a last resort should termination be the action of choice.

25.3 WARNING SIGNS

Since employees are unable to read the manager's mind, it is necessary for problems or potential problems to be addressed quickly and directly. Under no circumstance should the manager remain silent and hope the problem takes care of itself or goes away. Employees are entitled to feedback, and if not informed otherwise, they have the right to assume their performance is at an acceptable level.

Warning signs of developing problems include:

- Quality and quantity of work decrease
- Missed deadlines
- Lack of initiative
- Tardiness and/or increased absences
- Lack of cooperation
- Defensive attitude
- Irritability
- Decrease in interaction with other employees
- Increased accidents

Reasons why substandard performance exists:

- Employees are not aware of what is expected of them
- Employees view their performance as acceptable
- A lack of necessary skills exists to prioritize their work
- There is no consequence for performing poorly
- Their assignment represents a job they do not want to do

When any of these signs are noticed, the manager should take a calm, controlled approach and discuss the situation with the employee. This represents a constructive approach and should result in a noticed improvement.

If a manager takes a threatening approach, the result will be an angry and defensive employee. This will also close off communication channels, create an atmosphere of mistrust, and will not be helpful in solving the problem. Such a punitive approach will also magnify the problem into being far greater than initially identified.

25.4 CORRECTIVE ACTION

Managers should clearly identify the problem to the employee, not belabor it, and move on to suggestions of corrective action. Employees prefer to be told what they should be doing, rather than being told what they should not be doing.

Several attempts to counsel, coach, and provide the employee with feedback are required before considering disciplinary action. The goal is to work together to improve the employee's performance. When this action is implemented, employee performance usually improves. However, the employee may opt to seek employment at another organization if he/she feels unable to meet the required standards of the position. If performance does not improve after a reasonable amount of time and coaching and the employee does not elect to voluntarily resign, termination may be required. If dismissal is necessary, the reason must be made explicitly clear to the employee.

Participation and support from the Human Resources Department is imperative at this time. Terminations must be for just cause only except in very extreme situations.

25.5 DISCIPLINARY ACTION

If an improvement in employee performance is not noticed after initial discussions, disciplinary action may be necessary. This step represents a formal statement of actions that will be taken by the organization to cease a performance problem from going further. The statement also specifically states the improvement necessary for employment to continue. This action is most effective if the initial efforts to counsel and coach have not resulted in a noticed improvement. During the counseling sessions, the manager is responsible for informing the employee that disciplinary action will be taken if an improvement does not occur. Therefore, when disciplinary action does occur, it should not be a surprise to the employee.

Disciplinary action is also needed when the employee demonstrates gross misconduct. Immediate dismissal may be necessary depending on the severity of the employee's conduct.

25.6 TERMINATION

At this stage, involvement from the Human Resources Department and/or legal counsel is required. Actions must be legal, appropriate, and within the guidelines established in the policies and procedures of the organization. The proper handling of these situations is of utmost importance.

The most common lawsuit in the past decade against employers involves wrongful termination. Many employees win these cases due to errors employers made in handling problem performance issues. Lack of adequate documentation, failure to inform the employee of how to improve his/her performance, and employer actions that are viewed as unfair represent common employer errors during termination proceedings.

The termination of an employee due to performance difficulties represents a stressful and difficult time for all individuals involved. This action should always be a last resort except in cases when employee behavior is severe in nature.

25.7 GUIDELINES FOR TERMINATION

The decision to terminate an employee should never be made solely by the supervisor or manager. In an effort to assure the termination is a fair and equitable solution to the problem, a human resource professional should be involved in the process.

All written documentation should be in order before going forward with the termination process. Paperwork should be reviewed thoroughly to avoid hasty actions that may in time come back to haunt the manager or the organization.

25.8 INFORMING THE EMPLOYEE

When informing the employee of the decision to terminate, the manager must be calm in demeanor and not demonstrate signs of anger or frustration. The employee should always be provided an opportunity to speak and the manager should listen carefully and attentively. The manager must demonstrate a caring and understanding attitude to assure the employee of an awareness of the difficulty he/she is experiencing with the termination process. The manager must be certain that all comments specifically address the performance and not the individual. If a difficult encounter with the employee is anticipated, it is advisable to have a neutral third party present, ideally a member of the Human Resources Department. Privacy must always be provided for the meeting. In addition, managers must be sensitive of the concerns that will develop among staff members as a result of the termination of a co-worker. It is critical to keep in mind that a human being is attached to the decision of termination. The manager must be supportive and caring during the process.

The terminated employee should be permitted a departure from the building with the smallest possible audience. Immediate arrangements should be made to collect the employee's personal belongings and to have these items shipped via mail to the employee's home.

25.9 INFORMING THE CO-WORKERS

Communications with other employees regarding the dismissal should be limited. Only information necessary to proceed with business should be shared with others in the department and organization. Managers must never make an "example" out of the terminated employee.

25.10 LIMITATIONS ON THE EMPLOYER'S RIGHT TO DISCHARGE

During the 1980s employer/employee relations took a dramatic turn. Many employers were sued in court by terminated employees who believed their dismissal was unjust or wrongful. The term frequently used in such cases is *wrongful discharge*.

There are Federal and state statutes prohibiting the termination of employees for certain reasons. The following list represents a few of the Federal laws.

- Employee Retirement Income Security Act
- Occupational Safety and Health Act
- Americans with Disabilities Act
- Age Discrimination in Employment Act
- 1964 Civil Rights Act (Title VII)

State laws frequently prohibit terminating an employee for service in the military, for service on a jury, or for filing a worker's compensation claim.

On the spot discharges represent a great risk for litigation for employers. Such situations should be limited to extreme cases only. Employers should seek legal advise

TABLE 25.1 Examples of Federal Laws That Create "Protected Classes"

Act	Protected Class
Family and Medical Leave Act	Employees requesting leave
Age Discrimination in Employment Act	People over 40 years of age
COBRA (Consolidated Omnibus Budget Reconciliation Act)	Employees who want to convert health insurance
National Labor Relations Act of 1935	People with union activity affiliations
Fair Labor Standards Act of 1938	Employees who work more than 40 hours per week (overtime)
Title VII of the Civil Rights Act of 1964	Race, color, sex, religion, or national origin
Occupational Safety and Health Act of 1970	Employees who report OSHA violations
Rehabilitation Act of 1973	People with disabilities
ERISA (Employment Retirement Income Security Act of 1973)	Employees with pension/benefit plans
Pregnancy Discrimination Act of 1978	Pregnancy
Americans with Disabilities Act of 1990	People with disabilities

from a labor attorney whenever a discharge is being considered. If the employee is anticipating discharge, it is highly possible he/she has received legal advice.

The importance of documentation and explicit recordkeeping is critical before, during, and following the discharge process. Employers and managers today must be aware of the employees who are eligible members of "protected classes." A lack of familiarity with these categories can prove to be damaging to managers and employers. (Refer to Table 25.1 for specifics of each category.)

25.11 VOLUNTARY RESIGNATION

In an effort to avoid the termination process, many employers and/or managers encourage the employee to resign voluntarily. Managers often believe a voluntary resignation will avoid any recourse by the affected employee. This belief is not always correct.

In situations where the employee is not adequately performing the duties of his/her position and the manager is concerned about the job performance, the employer should not encourage a voluntary resignation. This avoids the potential for the action to be viewed as an involuntary resignation. It is the manager's responsibility and duty to address and deal with the performance issue and if necessary termination.

To avoid cases where the employee could argue he/she was forced into resigning, the following considerations should be adhered to. If possible, the employee should be given the opportunity to transfer to another department in a similar position. The manager should allow the employee adequate time to make a decision regarding a transfer on their own.

If the particular employee has been with the company for many years, the company should review the possibility of offering a severance package. The employee's decision to accept or refuse the severance package should be voluntary. The manager must advise the employee of the option to remain in the present position if the employee elects to refuse the severance package. If the employee's decision is resignation, this should be put in writing with a statement acknowledging the alternatives given to the employee, including the fact that the employee's choice was a voluntary resignation.

25.12 DRUGS AND ALCOHOL IN THE WORKPLACE

The Drug Enforcement Administrator (DEA) has estimated that drug abuse costs U.S. industry between $60 and $100 billion annually. Drug abusers are absent from work three times as much as nonabusers and have four times as many accidents. Approximately 50 train wrecks in the past 10 years have been directly related to substance abuse.

Per OSHA regulations, an employer has a duty to provide a safe workplace for employees. The employer's expectation is for staff members to perform their work-related responsibilities in a safe and efficient manner.

Drug and alcohol use seriously impair the mental and physical capabilities of an employee to perform work. They may also increase adverse effects from exposure to some chemicals. Many employers are enforcing drug and alcohol abuse policies in an effort to control this growing problem. The Drug-Free Workplace Act of 1988 requires employers who are Federal contractors to implement such policies. Policies must be properly enforced or the employer may be subject to litigation. Drug testing may make a company more efficient, but the exposure of the employer to tremendous litigation must also be a consideration. Employers who consider drug and alcohol testing should have a valid reason for testing existing employees and/or applicants. Employees and applicants must be informed of the need for the drug testing prior to instituting a drug testing policy. The policy must be consistently applied to all employees. Employers must obtain written consent prior to testing. All test results, whether positive or negative, must be kept in strict confidence. The assurance of confidentiality must be communicated to employees. A reputable laboratory must be used for testing, results must be evaluated with care, and positive tests should be confirmed with a second testing.

Costs and benefits of a drug testing program should be carefully evaluated in advance of implementation. Employers who elect to Institute a Drug Testing Program should work closely with the Human Resource Department to develop an Employee Assistance Program to deal effectively with drug-related problems.

25.13 DRUG AND ALCOHOL ABUSE: WRITTEN POLICIES

It is advisable for organizations to have a written drug and alcohol abuse policy whether the company tests or does not test. The availability of a written policy will eliminate any confusion among the employees. The written policy should include work rules against drug and alcohol abuse and a procedure for the manager to deal with policy violations.

The written policy should prohibit any employee from being at work if under the influence of alcohol, drugs, or controlled substances. In addition, the policy should prohibit the possession, sale, transfer, or use of such substances while at work, on duty, or on the company premises. Included are not only illegal drugs, but legal prescription and over-the-counter drugs.

25.13.1 Prescription Medication

If an employee is taking prescription medication that may affect his physical or mental abilities, the employer should be informed. The employer is required to take appropriate action to assure safety and productivity levels when legal or illegal drug use is present. The underlying problem is the same for both legal and illegal drug use. However, the action taken for the use of legal drugs will be vastly different from that for illegal drug use.

The policy should inform employees that medical testing may be required if drug and alcohol use is suspected. The policy should state that drug use may result in discharge for the first offense. This clearly informs employees that the organization will not tolerate drug and alcohol abuse in the work place. The policy should also state that the employer may search the employee, his/her work area, his/her locker, and his/her personal property, including a vehicle parked on the premises of the company. In addition, the policy should state if the employee refuses to a search, the company will consider this a voluntary resignation. Companies should avoid spur of the moment decisions to subject an employee to drug or alcohol testing. Such actions may have serious legal implications for the employer.

As with all company policies, the Drug and Alcohol Abuse policy should be clearly communicated to all employees via meetings, memos, handbook distribution, or a letter to the employee's home. Employees must be made aware of the company's expectations so they may comply as required.

Managers should be given clear, easy to follow guidelines so they will know their responsibilities when dealing with an employee who is under the influence of drugs or alcohol. The procedure for managers should include approval from senior-level management to assure fairness is exercised in the enforcement of the drug and alcohol abuse policies. Companies should not wait until an incident occurs to develop a Drug and Alcohol Abuse policy.

25.13.2 Example: Sample Policy on Alcohol/Drugs in the Workplace

"Company name" recognizes that alcohol and drug abuse in the workplace has become a major concern. Management believes that by reducing drug and alcohol use, there will be an improvement in the safety, health, and overall productivity of employees. The objective of the company's alcohol and drug policy is to provide a safe and healthy workplace for all employees, to comply with Federal and state health and safety regulations, and to prevent accidents.

The use, possession, sale, transfer, purchase, or one's being under the influence of alcoholic beverages, illegal drugs, or other intoxicants by employees at any time on company premises or while on company business is strictly prohibited. Employees must not report for duty or be on company property while under the influence of, or have in their possession while on company property, any alcoholic beverage, marijuana, or illegally obtained drug, narcotic, or illegal substance.

Note: Employers dealing with issues concerning employment, labor, or personnel law should consult an attorney. The previous information presented should serve as guidance and be used only for reference purposes.

25.14 RECOMMENDATIONS FOR DEALING WITH AN INTOXICATED EMPLOYEE

The manager must be certain his/her actions are reasonable and fair when dealing with an intoxicated employee. It is important that the well-being of the intoxicated individual, as well as the well-being of other employees, be of paramount concern. Immediate action must be taken to avoid the possibility of an accident occurring. In an effort to avoid potential litigation, it is important for the manager to respond to the situation with professionalism, empathy, and common sense.

The intoxicated employee should not be confronted in the presence of his/her co-workers. The manager should approach the employee in a calm manner, and ask the employee to accompany him/her to a private office or area. If the employee does not agree to go with the manager, an inquiry in front of others may be unavoidable. The manager must not accuse the employee of being intoxicated. The employee should be asked if he/she has been drinking or taking drugs. Due to lowered inhibitions, the employee may admit to intoxication. However, in many cases, the employee may deny that he/she has consumed alcohol. In this case, the manager should state the symptoms the employee is demonstrating and request an explanation. The employee at this point must be informed by the manager that he/she cannot work the remainder of the day. The manager should request a meeting with the employee on the following day to discuss the matter in detail.

The manager must take the necessary steps to prevent the intoxicated employee from driving home. An offer to call a friend or relative or to pay for a taxi to transport the employee home is appropriate. If the employee insists on driving, the manager must advise him/her of the company's concern for the safety and well-being of the employee and advise the employee that the police will be contacted if the employee attempts to drive.

25.15 COMPANY LAYOFFS, DOWNSIZING, RE-ENGINEERING

The trauma caused to employees by layoffs, downsizing, or re-engineering are profound. It is important for managers to be understanding and compassionate about the significance one's job has in a person's life. An attitude of indifference is not acceptable or appropriate during layoff proceedings. Many employees strongly identify who they are by what they do. With a significant number of layoffs taking place in the 1990s, the prospects for lower skilled workers, middle managers, and older employees may be somewhat glum. A severance package and outplacement services will provide laidoff employees with a temporary cushion while seeking new employment. However, such a package in no way makes up for the loss the laidoff employee endured. It is imperative for managers to recognize this fact. During the layoff proceedings, managers should insist on participation by a professional human resources staff member. It is of utmost importance for the laidoff employee to have a clear understanding of health insurance, pension plans, severance payments, life insurance, early retirement packages if applicable, and unemployment compensation. The newly laidoff employee in most cases will be more comfortable discussing these issues with a member of Human Resources rather than his/her former manager.

Companies must pay close attention to adhering to the legislature concerning "protected classes" during layoffs to avoid any potential litigation. The laws are very explicit concerning "protected classes." Some organizations have been required to pay large sums of money to laidoff employees as a result of the downsizing. The affected employees were able to prove in the judicial system that only "over age 40" employees were selected for the layoff.

Prior to initiating any layoff proceedings, it is advisable for organizations to seek legal counsel. This can avoid a costly and unnecessary aftermath.

25.16 THE EMPLOYEES LEFT BEHIND

During the 1990s, the terms *downsizing*, *right-sizing*, and *re-engineering* were heard throughout industries, including those that had previously been somewhat recession proof. An example of such an industry is health care. The employees left behind have different issues to deal with than the laidoff employee. A sense of closure exists for the employee who was laidoff, while the employees left behind continue to deal with the day-to-day work-related issues.

It is of utmost importance for managers to pay special attention to those employees "left behind" after a layoff has occurred. Many of the individuals who experienced being laid off represented longer-term employees. The loss of such individuals in an organization has a profound impact on the remaining employees. On the day following a layoff, the remaining employees work alongside an empty office or empty desk where their co-worker, boss, and/or friend previously worked. In some cases remaining employees compare the sudden, unexpected departure to that of a death. Many of the same feelings are present: denial, anger, betrayal, depression, and finally acceptance of the loss.

Managers must offer support, compassion, and understanding to those left behind, while setting clear goals to move forward. The loss of these employees must be addressed and spoken about with those remaining. Employees will have to be given an opportunity to deal with the loss that may be in a variety of ways. Additional resignations may occur after a layoff from employees who no longer feel a sense of "trust or commitment" to the organization. Managers must accept such resignations graciously and be supportive of the employee's decision. The entire workforce has experienced difficult times during this decade. Change is never easy and is always challenging. These rapid changes have increased tension and anxiety to significant levels. The work environment has become emotionally charged and difficult even for the most sophisticated managers. The need to keep communication channels open has never been so critical. A caring and supportive environment will be extremely beneficial to help the remaining employees meet and achieve the goals of the company.

25.17 STRESS IN THE WORKPLACE

Without stress, there would be no life. Stress in many ways can be the best part of life, when managed appropriately. The problem occurs when stress takes over, and the individual feels overwhelmed, out of control, or "stressed out."

A certain degree of stress belongs in the workplace. The problem of stress overload occurs when the employee is unable to release the daily stresses of life, and the result can be damaging physical and emotional symptoms. Samples of stress-related symptoms include: sleep disorders, headaches, and muscular pains. In addition, fatigue may also lead to accidents. Stress is such a major factor in everyday life, it is considered by some health professionals to be at epidemic proportions. The key to managing stress is to find an outlet that works for the individual. Exercise, meditation, reading, walking, and jogging work wonders for many. In addition, it is important to manage one's work schedule. It is advisable to schedule time off for vacation and relaxation in an effort to avoid burnout. Stress in the workplace occurs in situations when an employee feels threatened either by change, a deadline, or a new situation. When the tension is not released, the employee may experience a variety of symptoms including: trembling, pounding heart, dizziness, and/or fatigue. Employees who are experiencing problems in the workplace (as previously discussed in this chapter) will no doubt have some stress-related symptoms. It is not advisable to attempt to eliminate all stress in life, but to develop coping mechanisms to deal with the stressful aspects of our daily routines.

When one's physical or emotional resistance is at a low point, stress can be overwhelming. If a series of stressful situations occur, one may feel overwhelmed or out of control.

Stress management programs are an ideal perk for organizations to offer employees. These programs benefit both the employee and the employer. Some of the methods of stress reduction are fairly simple in nature. Employees who experience significant stress should consult their physician or health care provider for advice and guidance to bring their stress level under control. It is also important for the manager to take time to discuss the causes of the employee's stress to determine if some of the stressors could easily be eliminated, particularly if the stress is work related.

Managers can also strive to create a harmonious and pleasant work environment for employees where ideas and thoughts are openly shared and communication channels remain open at all times.

CHAPTER 26

Managing Workers' Compensation

JAMES J. PAUGH III

Lynch, Ryan & Paugh, Inc., 120 Front St., Worcester, MA 06108-1408

26.1 AN OVERVIEW OF WORKERS' COMPENSATION
- 26.1.1 Why the No-Fault System Has Trouble Functioning
- 26.1.2 Current Status of Workers' Compensation
- 26.1.3 United States: 50 Different Laws
- 26.1.4 International Workers' Compensation
- 26.1.5 What Is Workers' Compensation?
- 26.1.6 General Benefits
- 26.1.7 Profile of the Occupational Injury Litigant

26.2 DETERMINING THE EMPLOYER'S SITUATION
- 26.2.1 Finding and Using Data
- 26.2.2 Incidence Rates
- 26.2.3 Injury Cause–Cost Analysis—A Vital Few Analysis
- 26.2.4 Costs per Employee or Percent of Payroll
- 26.2.5 The Vital Few by Exposure Hours
- 26.2.6 Weekly Claims Control
- 26.2.7 The Lynch Ryan Shift
- 26.2.8 Other Measures

26.3 A BASIC UNDERSTANDING OF WORKERS' COMPENSATION INSURANCE PROGRAMS
- 26.3.1 Premium Calculation
- 26.3.2 Types of Workers' Compensation Insurance Policies
- 26.3.3 Terminology

26.4 STRATEGIES TO DEVELOP A POSTINJURY RESPONSE SYSTEM
- 26.4.1 Accept Responsibility for the System
- 26.4.2 Incorporate Injury Management Controls into Safety Programming
- 26.4.3 Retain Accountability for Injury Management at Each Worksite
- 26.4.4 Establish Dedicated Medical Provider Relationships at the Worksite
- 26.4.5 Set Performance Objectives
- 26.4.6 Develop Early Return to Work Opportunities
- 26.4.7 Effective Claims Management through Communications
- 26.4.8 The Hearing and Settlement Processes
- 26.4.9 Concentrate on Managing Today's Injuries
- 26.4.10 Involve the Claims Service Provider in the Program

Handbook of Occupational Safety and Health, Second Edition, Edited by Louis J. DiBerardinis,
ISBN 0-471-16017-2

26.5 MAKING THE RIGHT DECISION
26.5.1 Cost–Benefit Analysis
26.5.2 Steps Performed in Cost–Benefit and Cost-Effectiveness Analysis
26.5.3 The Risk Score Method
26.5.4 A Graphical Calculation of Risk Score and Effectiveness
26.6 QUICK HELP
26.6.1 The Americans with Disabilities Act and Workers' Compensation
26.6.2 What Do I Do When ...
26.6.3 Glossary of Terms
26.7 MAINTAINING A POSITIVE COMMITMENT FOR THE LONG TERM

26.1 AN OVERVIEW OF WORKERS' COMPENSATION

The workers' compensation system is the oldest form of "no fault" insurance. Workers' compensation was first developed in Germany in 1856 and adopted soon after by England and most of Western Europe. In 1912, Wisconsin adopted the first workers' compensation statute, and by 1947 all states had mandated workers' compensation coverage in America.

Workers' compensation is a "no fault" insurance system in the sense that benefits are paid regardless of who or what caused the injury or illness that "arises out of and in the course of employment."

Before workers' compensation was established, an employer could be sued for negligence, and could only defend against such lawsuits by proving that the employee was at least partially at fault, in one of the following ways:

- Assumption of risk (the employee knew the risk of injury was high)
- Fellow-servant rule (a co-worker caused the injury to happen)
- Negligence (the employee was careless)

It was usually a cumbersome and unfair system for the employee.

In exchange for this "no-fault" system on benefits, employees gave up the right to sue their employers. This is now referred to as the "sole remedy." Though nearly a hundred years old, this exchange is constantly being challenged in the court systems and through constant legislative reforms.

At the same time, the insurance industry devised a means of rating an owner/employer's loss experience, which was then considered nothing more than a manifestation of the owner/employer's attitude toward his employees. According to this rating method, the severity of an injury drove the length of time away from the job, which, in turn, drove the cost of lost wages.

> This cornerstone premise is still the key concept to managing workers' compensation:
> SEVERITY (DAYS LOST) DICTATES COSTS (LOST WAGES).

26.1.1 Why the No-Fault System Has Trouble Functioning

However, many employers are all too quick to find fault in the failings of this no-fault system with any number of groups: the legal and medical communities, the legislature, the labor union or the actual injured employee being the cause. The insurance industry has failed to show employers how its own experience rating method should be used, as it was designed, to modify the behavior of errant business owners. Employers who did not understand insurance practices or who were pleased to pass both their risks and their responsibilities onto the insurance carrier suffered high costs. Under this scenario, length of time away from the job continued to drive the cost of lost wages, but became increasingly less related to the medical severity of the injury than to the employer's attitude toward the injured employee.

We could continue, as we have for years now, to blame each other for the shortcomings in the workers' compensation system. All of us are inclined to find fault in the other. But the most vociferous party is the employer, who also pays all the bills. Therefore, the employer has the most to gain from understanding the importance—historical and current—of attitudes toward injured employees.

26.1.2 Current Status of Workers' Compensation

Workers' compensation is one of the most costly types of business insurance and one of the least understood. Perhaps the most common misunderstanding among employers is that their costs come ultimately from the rates charged to them by the insurance industry. In reality, the workers' compensation rate schedule is simply a temporary compromise between the state regulators and the insurance industry.

According to *The John Liner Letter* of December 1987 (Volume 25, No. 1, Standard Publishing Corporation, Boston), there is a never-ending conflict between the insurance industry and state regulators over rates ... "a constant seesaw between state legislators who hike compensation benefits and insurance underwriters who seek rate hikes to offset the cost of those higher benefits. In addition, the base of compensable claims is expanding to include asbestosis, stress, etc. In an effort to gain some cost control over rising and expanding liabilities, insurers and their clients are often foiled by politically motivated regulators. And so, the rate–benefit seesaw seems destined to go on forever."

This "cycle" of give and take closely models a particular state's economic cycle. In periods of high unemployment, there is a rise in workers' compensation claims and its resultant insurance premiums.

26.1.3 United States: 50 Different State Laws

Since the beginning of the twentieth century, American employers have come under 50 state laws providing a "no-fault" medical and financial safety net for those who are injured. Workers' compensation laws are designed to ensure that employers can, will, and shall provide equitable financial and medical support to injured workers. These laws fix the employer's obligation, ensure the employer is solvent to pay bills, and provide a path for dispute resolution without resort to lawsuits. (Note: Workers' compensation laws typically do not address accident prevention, job design, training, health risk screening, wellness, or other internal management practices.)

Recently many states have revised their laws for workers' compensation. Changes in statutes have resulted in:

- Stronger incentives to avoid employer–employee disputes.
- Stronger incentives for all parties to resolve disputes without delay.

- More state resources devoted to regulation.
- Encouragement of return to work and rehabilitation.

Reforms are continually being developed. Therefore, for up-to-date information, check with your insurer or state workers' compensation board or commission or order, free of charge, a booklet, published each January, entitled "State Workers' Compensation Laws," from the U.S. Department of Labor, Employment Standards Administration, Office of Workers Compensation Programs, Washington, DC. 20036.

26.1.4 International Workers' Compensation

While most countries have a social welfare system for disabilities (on or off the job), some have utilized U.S. laws as a basis for a separate workers' compensation act. Canada, for instance, closely resembles typical U.S. workers' compensation laws. The European Union is considering a form of workers' compensation that would remove the costs from the public sector and place it squarely on the private employers. However, regardless of the country and the employer, the straightforward techniques explained later in the chapter will ensure the lowest costs possible.

26.1.5 What Is Workers' Compensation?

Workers' compensation is an insurance system that is mutually beneficial to both employees and their employers. It serves two basic purposes:

1. To promptly provide employees who have suffered a work-related injury or illness with medical and indemnity (time away from work) coverage.
2. To protect employers from costly litigation over claims of work-related injuries and illness.

26.1.6 General Benefits

Employee benefits for workers' compensation are the following:

A. Indemnity (wage loss replacement)—Typically, there is a waiting period for benefits to begin. This varies among states.
   - Temporary—usually for a defined time period (3 to 5 years).
   - Partial—A make-whole provision (commonly used when an employee is paid at a lower wage rate due to the effects of the injury). Also, for a set time period (3 to 7 years).
   - Permanent—As implied, there is no time period and benefits continue indefinitely.

B. Medical

- All medical treatment—Includes all hospital visits and surgeries/treatments. In most states, chiropractors and osteopathic physicians are included.
- Rehabilitation—Both physical (exercise, work-hardening, massage, etc.) and vocational (retraining for suitable jobs).

C. Loss of function (scarring/disfigurement)—A predetermined amount paid to the injured employee for the permanent residuals of an injury.

For a comprehensive and up-to-date digest of the provisions of each U.S. state's and Canadian province's law, consider purchasing the latest edition of the U.S. Chamber of Commerce's Analysis of Workers' Compensation Laws (Pub. No. 6983) (800-638-6582).

All states require employers be covered by workers' compensation rates, which is available privately in most states. Prominent exceptions are Ohio, Nevada, North Dakota, Washington, West Virginia, and Wyoming, where coverage is provided through a monopolistic state fund.

26.1.7 Profile of the Occupational Injury Litigant

Data gathered from Lynch, Ryan & Paugh, Inc.'s workers' compensation projects indicate that employees who litigate claims for workers' compensation benefits most frequently are older, less educated, and lower-wage earners than the general work force. Most are in blue-collar or service occupations. Three of every four are married. Slightly more than half have worked less than 2 years for the same employer, and there is even a chance the litigating employee is a union member. Three out of four have no knowledge of workers' compensation prior to their injury. Back injuries account for every third litigated claim. The vast majority of litigants perceived their injury as serious.

When confronted with a traumatic injury and lacking information and affirmative support from the employer, the injured employee will turn elsewhere. Without recognizing it, each of these uninformed, injured employees is looking for the mythical pamphlet, entitled "What Every Employee Wants to Know about Workers' Compensation, But Whose Employer Doesn't Seem to Know Either." Uncertainty, instigated by the injury event and promulgated by the employer, lays fertile ground for litigation.

The typical litigant does not contact an attorney until at least 2 weeks after the injury and after contact with the employer, the physician, and the insurer, none of whom helps reduce the anxiety. Handicapped by the level of education and sometimes the command of the English language the employee finally receives the first wage-replacement check (less than the usual weekly pay check) along with a confusing, legalistic from that indicates the right to consult an attorney. Even unions were not able to allay the injured employee's concerns, except to refer the employee to an attorney. Nevertheless, the prime source of attorney referral is the employee's peers—spouse, friends, relatives, co-workers—who collectively account for nearly twice as many referrals as labor officials.

Fully a third of the represented litigants do not recognize the adversary relationship; presumably they see the application for benefits as a necessary first step rather than the start of litigation. However, among those who recognize the adversary relationship most think they're suing the insurer, usually due to the insurer's investigative and uninformative treatment. Less than 10% realize they are suing the employer.

Researchers find that employers can do much to reduce the time and expense of litigation through (1) keeping all responsible and interested parties well informed; (2) enhancing the quantity and quality of timely communications during benefit delivery; and (3) training claims coordinators and other intermediaries between the injured employee and the system.

26.2 DETERMINING THE EMPLOYER'S SITUATION

Most employers sense that their workers'compensation costs may be higher than they should be. There are several measures available to employers for evaluating whether they manage workers' compensation fairly and cost effectively and are most meaningful when used together. Some are classic measures; others are new.

Traditional, descriptive measures may take a year or more to compile and fail to provide management with sufficient understanding of any opportunity to reverse unfavorable trends. All these measures to be described have merit, but as they are typically used, they share the disadvantage of not being timely enough to tell an employer how well they are managing today. They are, however, the easiest and most available data for an employer.

Comprehensive, prescriptive measures can be effective, within 30-day measurement cycles, for tracking progress to aid well-defined and realistic targets, based on data specific to the injury, the worksite, the industry, and the state. These measures help establish standards or guidelines. Their value lies in that they allow (1) for evaluating loss control work in progress and (2) for evaluating the combined impact of safety (preinjury) and postinjury response. Prescriptive standards help lead employers to early identification of controllable causes of preinjury and postinjury costs.

26.2.1 Finding and Using Data

There are two primary locations to find data on injuries. The first and most useful is required by law with few exceptions, to be kept by every employer . . . the OSHA 200 log (Fig. 26.1). Simply, it is a listing of all injuries and illnesses beyond first aid treatment. Data kept include: date of injury, name, department, type of injury, length of restricted work, length lost time. It is useful to use 3–5 years of OSHA data to conduct an analysis.

The second source, called a loss run, is available from the insurance carrier (Fig. 26.2). It is a report, usually on a quarterly basis, of every claim, showing the amounts paid and expected to be paid (reserves). Most insurance companies will provide detail reports by department, by injury type and cost.

26.2.2 Incidence Rates

Incidence rates are classic measures, used regularly by those in the safety profession and government, to assess the frequency of injuries and illnesses, and their severity, calculated in terms of the number of lost workdays. Both rates are figured on the basis of units per 200,000 hours worked (including overtime) or 100 full-time equivalents (FTEs). (The 200,000 is derived from 100 workers times 50 weeks time 40 hours.)

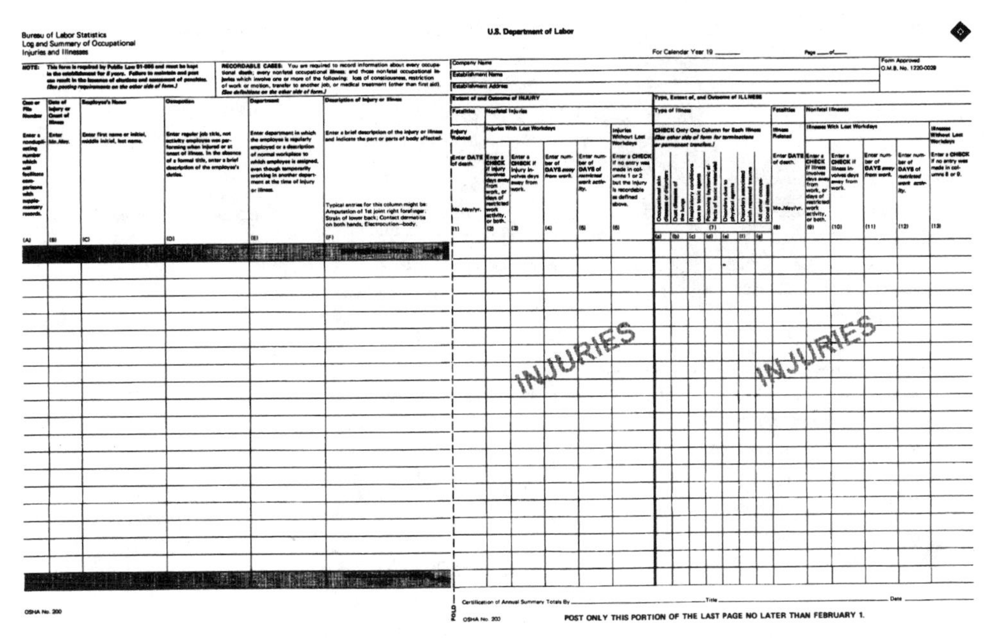
Bureau of Labor Statistics
Log and Summary of Occupational
Injuries and Illnesses

U.S. Department of Labor

For Calendar Year 19 ____ Page __ of __

Form Approved
O.M.B. No. 1220-0029

NOTE: This form is required by Public Law 91-596 and must be kept in the establishment for 5 years. Failure to maintain and post can result in the issuance of citations and assessment of penalties. (See posting requirements on the other side of form.)

RECORDABLE CASES: You are required to record information about every occupational death; every nonfatal occupational illness; and those nonfatal occupational injuries which involve one or more of the following: loss of consciousness, restriction of work or motion, transfer to another job, or medical treatment (other than first aid). (See definitions on the other side of form.)

Company Name
Establishment Name
Establishment Address

Case or File Number — Enter a nonduplicating number which will facilitate comparisons with supplementary records. (A)
Date of Injury or Onset of Illness — Enter Mo./day. (B)
Employee's Name — Enter first name or initial, middle initial, last name. (C)
Occupation — Enter regular job title, not activity employee was performing when injured or at onset of illness. In the absence of a formal title, enter a brief description of the employee's duties. (D)
Department — Enter department in which the employee is regularly employed or a description of normal workplace to which employee is assigned, even though temporarily working in another department at the time of injury or illness. (E)
Description of Injury or Illness — Enter a brief description of the injury or illness and indicate the part or parts of body affected. Typical entries for this column might be: Amputation of 1st joint right forefinger; Strain of lower back; Contact dermatitis on both hands; Electrocution—body. (F)

Extent of and Outcome of INJURY
Fatalities — Injury Related — Enter DATE of death. Mo./day/yr. (1)
Nonfatal Injuries — Injuries With Lost Workdays — Enter a CHECK if injury involves days away from work, or days of restricted work activity, or both. (2)
Enter a CHECK if injury involves days away from work. (3)
Enter number of DAYS away from work. (4)
Enter number of DAYS of restricted work activity. (5)
Injuries Without Lost Workdays — Enter a CHECK if no entry was made in columns 1 or 2 but the injury is recordable as defined above. (6)

Type, Extent of, and Outcome of ILLNESS
Type of Illness — CHECK Only One Column for Each Illness (See other side of form for terminations or permanent transfers.) (7) (a) (b) (c) (d) (e) (f) (g)
Fatalities — Illness Related — Enter DATE of death. Mo./day/yr. (8)
Nonfatal Illnesses — Illnesses With Lost Workdays — Enter a CHECK if illness involves days away from work, or days of restricted work activity, or both. (9)
Enter a CHECK if illness involves days away from work. (10)
Enter number of DAYS away from work. (11)
Enter number of DAYS of restricted work activity. (12)
Illnesses Without Lost Workdays — Enter a CHECK if no entry was made in columns 8 or 9. (13)

INJURIES
INJURIES

Certification of Annual Summary Totals By ____ Title ____ Date ____

OSHA No. 200
FOLD
OSHA No. 200 POST ONLY THIS PORTION OF THE LAST PAGE NO LATER THAN FEBRUARY 1.

Fig. 26.1 Sample of an OSHA 200 log.

Each year the Bureau of Labor Statistics (BLS) and the National Safety Council (NSC) publish average incidence rates by type of industry (SIC code). You should know your SIC code. If you do not, contact your regional BLS office for assistance.

The frequency rate functions primarily as a measure of safety (preinjury) controls and in the insurance arena, as a driver of the experience modification factor. It measures the number of incidents for a given time period.

$$\text{frequency rate} = [(\text{number of injuries and illnesses})(200{,}000)/\\ \text{number of hours worked by the employee}]$$

The severity rate functions primarily as a measure of injury management controls and in insurance, as a driver of reserves. It measures how long injured employees are out of work.

$$\text{severity rate} = [(\text{number of days lost})(200{,}000)/\\ \text{number of hours worked by the employee}]$$

Employers are encouraged to compare their incidence rates with the averages published by the Bureau of Labor Statistics in its annual "Occupational Injuries and Illnesses in the United States by Industry" (published each May) and by NSC in the annual "Accident Facts" (published in October). Sometimes trade associations or other employer groups will also have data on incidence rates for comparison (see Fig. 26.3).

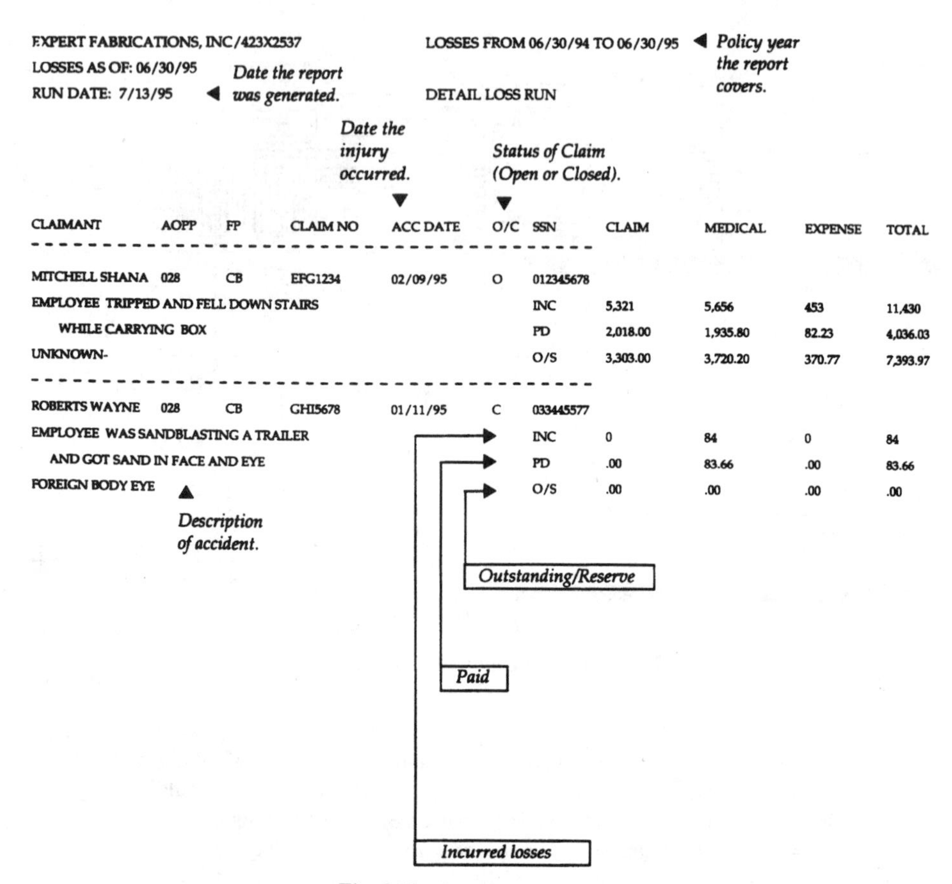

Fig. 26.2 Loss run example.

Industry Division (Incidence Rate)	Total Cases	Lost Workday Cases	Cases without Lost Workdays
Private Sector	8.5	3.8	4.8
Agriculture	11.2	5.0	6.2
Mining	6.8	3.9	2.9
Construction	12.2	5.5	6.7
Transportation	9.5	5.4	4.1
Wholesale/Rentail	8.1	3.4	4.7
Finance/Insuranace	2.9	1.2	1.7
Services	6.7	2.8	3.9
Manufacturing	12.1	5.3	6.8

Fig. 26.3 BLS estimates of nonfatal occupational injuries and illness incidence rates, 1993.

Company	Industry Wide Sheet Metal Shops WCC #3066	
% of Injuries	% of Injuries	Accident types
23%	37%	Handling containers, boxes
11%	8%	Struck by falling objects
4%	7%	Caught-in machines
1%	4%	Slipping & Falling

Fig. 26.4 Injury cause-cost profile.

26.2.3 Injury Cause–Cost Analysis—A Vital Few Analysis

In the 1890s an Italian economist, Vitredo Pareto, observed from analysis of monetary patterns that the significant items in a group will normally constitute a relatively small portion of the total.

Application of this approach as an effective strategy for preventing injuries and reducing their costs come from a thorough understanding of how a given employer's injury profile compares in incidence and cost to like employers. The National Council on Compensation Insurance (NCCI) and the insurer can provide data to help make comparisons on the basis of specific injury types and their costs by workers' compensation classification code (WCCC) [see Section 26.3.1]. Employers seriously committed to injury prevention and management should consider a thorough injury cause–cost analysis for the purpose of effectively isolating and identifying opportunities to save and quickly taking advantage of those opportunities. Without an injury cause–cost analysis, employers run the risk of placing too much emphasis on areas with little or slow return and ignoring less apparent opportunities.

An injury cause–cost profile for a sheet metal shop may look like Fig. 26.4: It is always useful to compare employer data to the industry data.

By computing average claim costs, number and costs of claims under $100, and number and costs of claims over $2000, additional industry comparisons can be made. These comparisons would lead to strategies for focusing on those injury types for which intervention will be the most effective. Usually, one-third of the injuries account for 94% of all injury cost.

26.2.4 Costs per Employee or Percent of Payroll

The average annual costs per covered employee as seen in Fig. 26.5. should only serve as performance guidelines. To calculate your costs, simply divide workers' compensation costs (either losses or premiums paid) by the number of full-time equivalent employees.

26.2.5 The Vital Few by Exposure Hours

Another tool to help identify problem areas is a comparison of percent of cost of injuries (or of days lost) to percent of hours worked for a particular department or plant location. The data become apparent as they make sense that a location with a small number of hours worked should have a proportional number of claims or losses.

Industry Type	Cost per FTE	Percent of Payroll
Total, all industries	378	1.1
Total, all manufacturing	591	1.4
Food, Beverage, and tobacco	899	2.5
Textile and apparel	583	2.3
Pulp, paper, lumber, and furniture	749	2.1
Printing and publishing	320	0.9
Chemicals and allied industries	324	0.7
Petroleum	986	1.9
Rubber, leather, plastic	500	1.2
Stone, clay and glass	707	2.0
Primary metals	853	2.5
Fabricated metals products	607	1.7
Machinery	581	1.4
Electrical machinery	261	0.6
Transportation equipment	690	1.4
Instrumentals and misc.	615	1.4
Total, all non-manufacturing	345	1.1
Public utilities	393	0.8
Department stores	335	1.8
Other trade	474	1.6
Banks, finance, and trust co.	173	0.5
Insurance companies	182	0.4
Hospitals	393	1.2
Miscellaneous non-manufacturing	421	1.3

(Source: 1997 Employee Benefits Report—US Chamber of Commerce, Washington, D.C.)

Fig. 26.5 Cost per FTE.

In Fig. 26.6, the HIP Isothermal Department worked 22% of the hours, yet had no claims, while the Inspection Department worked 16% of the hours and experienced 12% of the claims and 27% of the cost.

The Vital Few departments in this example are both the Machine Shop and the Inspection Departments, as collectively they account for two-thirds of both total cost and total claims, which indicates both a frequency and severity issue. These data can be gathered from insurer-supplied loss runs and internal company payroll records.

26.2.6 Weekly Claims Control

Weekly Claims Control (Fig. 26.7) is a quick measure simply to gauge how a particular firm or department(s) is performing on a weekly basis regarding lost time claims.

This measure works for higher-frequency claims. The calculation is the number of out-of-work employees divided by the entire number of employees. There are no national or industry data; so it is best to measure against an internal index. This measure can also be used to redefine out of work as daily absence or disability (non-work related). These conditions can also be compared internally by department or by quarter to quarter and year to year.

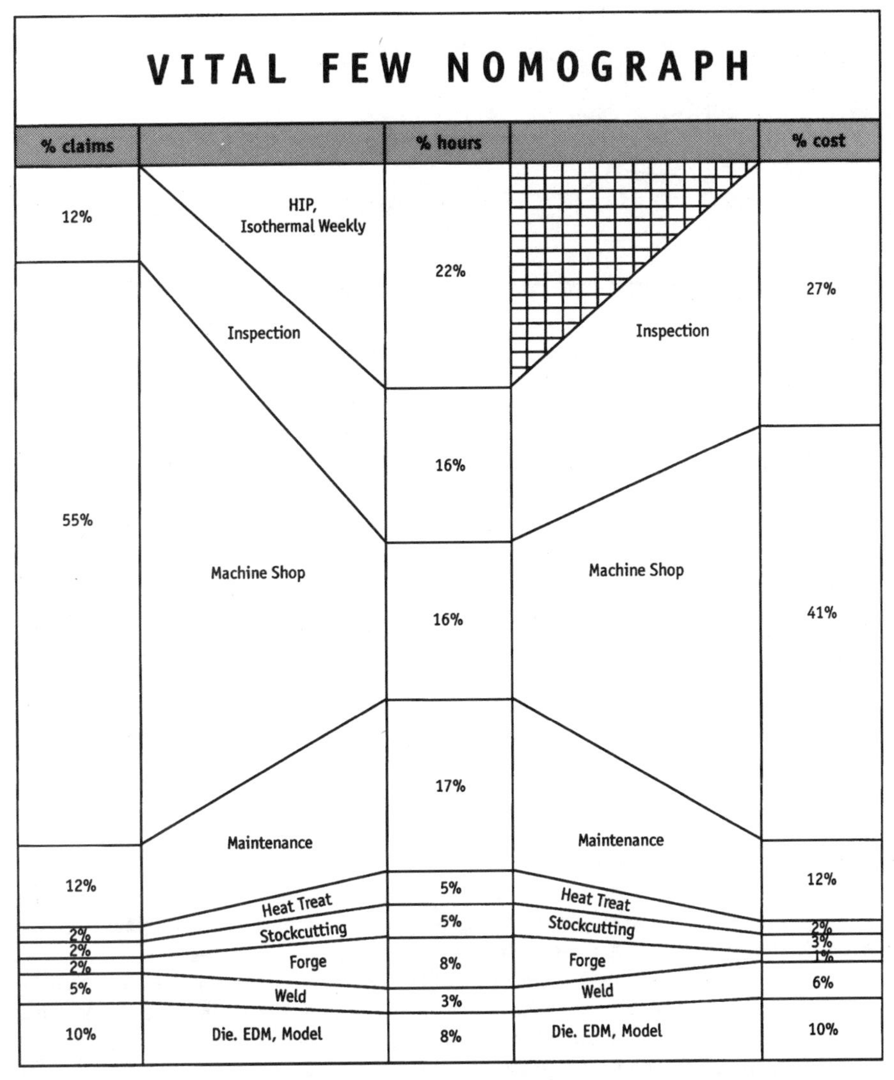

Fig. 26.6 Vital few nomograph.

26.2.7 The Lynch Ryan Shift

The "Lynch Ryan Shift" is a straightforward measure to understand how quickly injured employees return to work. By segmenting data from a year or two of OSHA logs, the employer can obtain a clear picture of which types of claims to focus upon.

An employer with unnecessarily high workers' compensation costs will commonly show 55% of lost-time cases under 5 days, 10% between 6 and 10 days, 30% under 6 months, and 5% above 6 months. Such a profile (Fig. 26.8) will reveal that employees are out too long and that time out is not directly related to the medical nature of the injury.

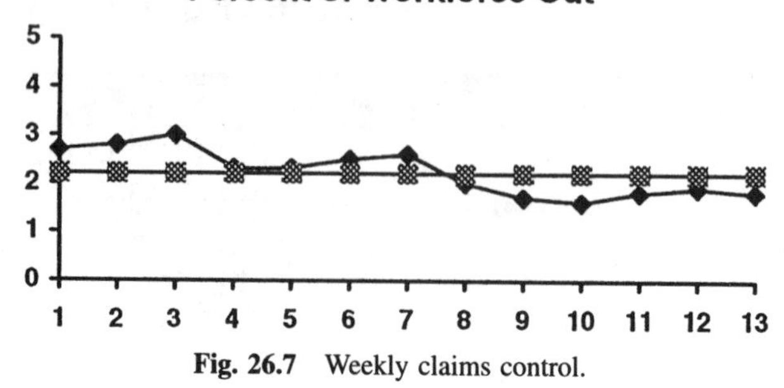

Fig. 26.7 Weekly claims control.

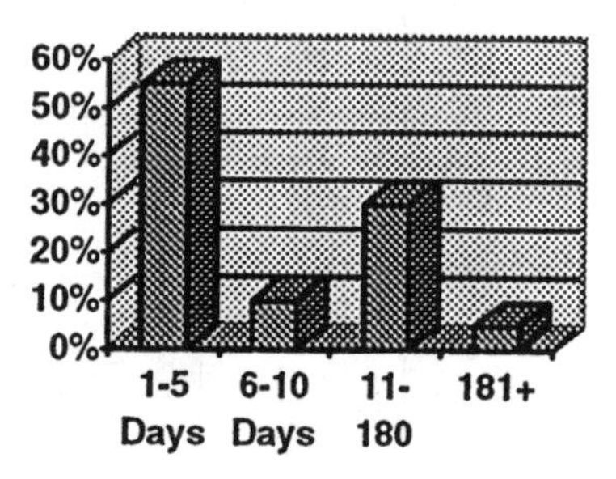

Fig. 26.8 Typical employer.

The high number of cases in the 11–180 days range may indicate that dialogue has broken down between the employer, the employee, and the insurance company. To zero in on the reasons, break the data into smaller groups, such as 11–21, 22–42, and so on. High percentages in the 11–21 area points more to a need to increase the return-to-work strategies with medical networks, while the 60+ categories indicate deeper case management problems with the either the insurer or the "system."

A healtheir profile, seen in Fig. 26.9 is 80% under 5 days, 15% under 10 days, 3% under 6 months, and 2% over 6 months.

The "Lynch Ryan Shift" represents the movement from the 60–30–5–5 curve to the 80–15–3–2 curve. This indicates a 50% reduction in incurred losses and should take no longer than 3 to 6 months to effect.

26.2.8 Other Measures

Two other types of measures that may prove useful are to gauge the impact of injuries on new or transferred workers as well as extended use of overtime. Theory and practice show that new employees are significantly high risk for injuries due to lack of training, unfamiliarity with equipment and process, etc.

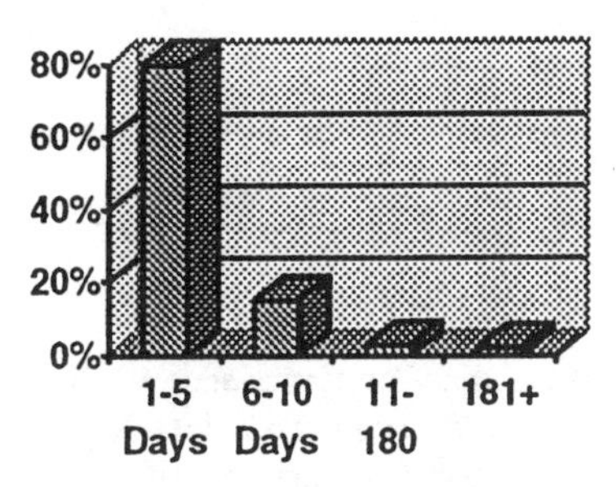

Fig. 26.9 Model employer.

Finally, it is always useful to conduct an analysis of employees and dates of injuries. At times, significant patterns may highlight problem employees or possible fraudulent claim activity. It is always best to use 3–5 years of data. Any significant findings should be shared with the appropriate company personnel.

These simple and straightforward measures will highlight the cost saving opportunities areas in an operation.

26.3 A BASIC UNDERSTANDING OF WORKERS' COMPENSATION INSURANCE PROGRAMS

Workers' compensation insurance programs are simple in concept, but can quickly become complex based on the type of policy offered or negotiated. In 70–80% of the cases, they all use a basic rate structure to develop a premium. They may be underwritten by private insurers or state funds. In the case of private insurers it is advisable to check the A.M. Best Rating Service for financial strength.

26.3.1 Premium Calculation

In its basic form, premiums are paid to an insurer to cover the direct costs of work-related accidents. The premium is determined by multiplying rates and the employer's payroll for particular types of jobs.

Rates. Calculated on the basis of $100 of payroll for each of the separate worker classifications in your employment. Set by state regulators, rates differ for varying degrees of hazard in workers' jobs.

Workers' Compensation Classification Code (WCCC). Workers' compensation insurance rates are set by each state for each types of jobs (workers' compensation classification codes or WCCCs). They are specific to an industry and an employer's payroll. Each class code is rated by the loss history of that class of business across a particular state. It does make sense that a clerical employee, exposed to fewer hazards than a sheet metal installer, would have a lower rate.

Here is an example of developing a basic premium for a sheet metal shop:

WCCC	Description	Rate	Payroll/100	Premium
3066	Sheet Metal Work-Shop	5.56	600	$ 3,336
5538	Sheet Metal Work-Outside	15.44	3000	46,320
8742	Salespersons-Outside	0.63	500	315
8810	Clerical-Inside	0.30	1250	375
			Total	$50,346

Experience Modification Factor. The Experience Modification Factor is a multiplier of premium designed to reflect an employer's loss experience. The "mod" is more responsive to injury frequency than to injury severity and is specific to an individual owner/employer. It tracks and holds the employer accountable for safety performance (in the broad sense).

A mod factor of 1.0 means the rating bureau in your state has determined that as an employer either (a) your loss experience is at average for your type of industry in your state or (b) your loss experience cannot be established because you are a new owner. The underlying assumption here is that differences in mod factors reflect differences in "owner-employer" attitude, not differences in employees, unions, or locales. In other words, insurance rating procedures do, in fact, appropriately hold the employer accountable for their own safety and injury management practices.

In our example, the standard premium is multiplied by 1.0, and the employer receives no penalty or incentive.

Accordingly, a mod factor of 1.5 shows the employers loss experience is 50 percent higher than other similar employers in the state. A mod factor of 0.5 indicates the employers injury and loss record is than those of its competitors in the state. Again, using the above example, the employer's premium of $50,346 is multiplied by 1.5 and becomes the standard modified premium of $75,519.

The mod factor is used to multiply the basic premium to obtain the standard or modified premium. In policy year 1998, the mod would be determined by the losses in policy years 1994, 1995, and 1996, as shown below. It is always based on a three-year time line.

Many employers will find their insurance carriers are very anxious to close any open cases older than four years, otherwise any additional costs incurred by those cases will have to be borne wholly by the carrier.

The mod is calculated by the state rating bureau or the National Council of Compensation Insurance (NCCI). The first $5,000 of every claim is valued at 100%, called

Experience Modification Time Line

Claims from these 3 years are used for the current year calculation

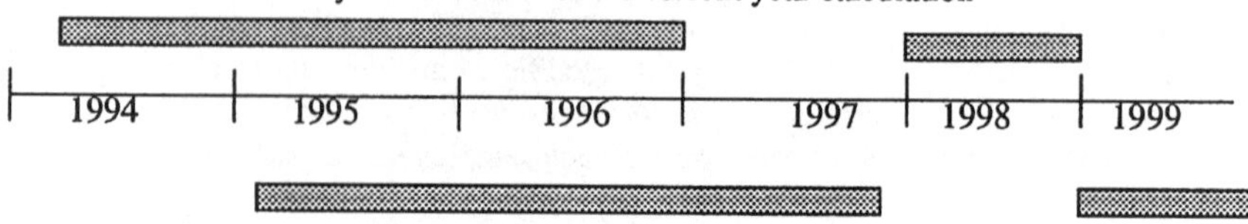

Next time period, the oldest year drops off and picks up a new year.

the primary losses. The remainder of the loss is weighted and discounted. Finally, any losses for a single claim above state rating point (ex. $175,000) are not included in the calculation. The state rating point varies for each state. An example of an Experience Rating Calculation can be see in Fig. 26.10.

The figures and calculations that go into determining the experience modification factor may appear complex and confusing. They are! To gain a better understanding of this rating procedure, contact the state's rating bureau for a copy of the latest rating data, called the Experience Rating Form or the Unit Statistical Report. Consult the agent, broker, or carrier regarding the origin, significance, and accuracy of the various figures and calculations. Most states set a premium threshold of $5,500 before the experience rating is used.

The formula ensures that one catastrophic injury does not ruin a company's experience rating over the 3-year period.

> Keep in mind that a lot of little injuries will impact the "mod" negatively, since the first $5,000 counts the most in the formula.

THE WORKERS' COMPENSATION RATING AND INSPECTION BUREAU

EXPERIENCE RATING CALCULATION

BUREAU FILE NO. PAGE 1 1 06

PART II - EXHIBIT OF EXPECTED LOSSES

(1) CLASS CODE	(2) YR	(3) PAYROLL	(2) YR	(3) PAYROLL	(2) YR	(3) PAYROLL	(2) YR	(3) PAYROLL	(4) EXPECTED LOSS RATE PAYROLL LIMITATIONS UNLIMITED	$300	(5) EXPECTED LOSSES (3) X (4) ÷ 100	(6) "D" RATIO	(7) PRIMARY EXPECTED LOSSES (5) X (6)
7380	4	10190	3	21722					3.81		1216	.24	292
8742	4	154448	3	59353	2	160001			.24		897	.21	188
8868	4	4791475	3	4348903	2	3948449			.37		48429	.27	13076
9101	4	1217640	3	1265398	2	1248525			1.91		71292	.29	20675
TOTAL	4	6173753	3	5696376	2	5356975							

EXPECTED EXCESS (f) = (d)—(e)	(d) TOTAL EXPECTED LOSSES	(e) TOTAL PRIMARY EXPECTED LOSSES
87603	121834	34231

PART I — EXHIBIT OF ACTUAL LOSSES

(8) YR	(9) CLAIM NUMBER	(10) ACTUAL INCURRED LOSSES	(11) PRIMARY ACTUAL LOSSES	(8) YR	(9) CLAIM NUMBER	(10) ACTUAL INCURRED LOSSES	(11) PRIMARY ACTUAL LOSSES
2	06526	25739	5000				
2	08375	132436	5000				
2	09064	171006	5000				
3	03329	25963	5000				
3	03485	36388	5000				
3	05728	12345	5000				
4	08871	11109	5000				
4	08976	24440	5000				

TOTAL BY POLICY YEAR OF ALL CASES [illegible] OR UNDER $5000	YR	ACTUAL INCURRED LOSSES	PRIMARY ACTUAL LOSSES	TOTAL BY POLICY YEAR OF ALL CASES	YR	ACTUAL INCURRED LOSSES	PRIMARY ACTUAL LOSSES
	2	2559	2559		2	331740	17559
	3	4697	4697		3	79393	19697
	4	1641	1641		4	37190	11641

PCL 7 21644 10/25/1995

ACTUAL EXCESS (c) = (a)—(b)	(a) ACTUAL INCURRED LOSSES	(b) PRIMARY ACTUAL LOSSES
399426	440323	48897

TO – WORCESTER INS CO
2 WORCESTER

BUREAU FILE NO.
NAME OF RISK

WESTBOROUGH MA

ARAP SURCHARGE = 1.48

PART III — RATING PROCEDURE

1. PRIMARY ACTUAL AND TOTAL EXPECTED LOSSES	(b) 48897	(d) 121834
2. "B" VALUE	28000	28000
3. RATABLE EXCESS W = .12	(w) X (c) 47931	
4. (1.00—W) = .88	(1 — w) X (f) 77091	
5. TOTALS	(g) 201919	(h) 149834
EXPERIENCE MODIFICATION (g)/(h)		1.35

CODE NO.	MANUAL CLASSIFICATION	AUTHORIZED RATE	DATE EFFECTIVE
8868	SCHOOL-PROFESSIONAL EMP & CLERICAL		[illegible] 1996
9101	SCHOOL-A/O EMPLOYEES		DATE ISSUED
8742	SALESPERSON,COLLECTOR,MESSENGER-OUTSIDE		
7380	DRIVERS AND HELPERS-NOC-COMMERCIAL		AUG. [illegible], 1996

Fig. 26.10 Sample of an experience rating calculation.

The information and understanding gained from this exercise, even though it may be time consuming, will significantly enhance the ability to manage the relationship with the insurance providers.

Cost-saving hint: With the insurance company's agreement, many employers self-pay small, incidental medical expenses, usually under $500. The claim is recorded by the insurer as zero since they did not pay it. Remember to report the claim to the insurer as this substantiates the case and covers the employer should medical complications arise.

The All Risk Adjustment Program (ARAP) is the final "incentive" used by the insurance industry to motivate employers. It is akin to a "bad driver" penalty for employers with higher-than-average losses. The ARAP has no limit on individual claims and carriers a maximum surcharge of 49%. Typically, the ARAP applied to employers in the assigned risk pool and with a "mod" over 1.0. In the example with the sheet metal shop, the modified premium of $75,519 with an ARAP, for example, of 1.31 would become $98,929 ($50,346 × 1.5 × 1.31). Due to poor loss experience, the sheet metal firm's premium has nearly doubled.

26.3.2 Types of Workers' Compensation Insurance Policies

Workers' compensation may be underwritten in many forms depending on the employer characteristics, for example, size, risk, workforce, historical loss experience, as well as the insurer's financial strength, appetite for risk, and the general economic climate. Here are the three most common forms of workers' compensation insurance.

Guaranteed Cost. This plan guarantees a fixed premium for the policy year, regardless of the firm's injury record. Even with no injuries, the employer pays the full premium. This is costly protection, but it does cap the amount of insurance paid. Employers in the assigned risk pool (sometimes referred to as involuntary market) are covered by this type of plan—the least loss-sensitive and least risky. A common variation on this plan includes a dividend to the employer for good to superior loss experience. Dividends must be declared by the insurer's board of directors. Since they are not guaranteed, a consistent history of paying dividends is recommended. The higher the A.M. Best Rating, the more likely indicator of dividend payment. Premium sizes for this plan range from $10,000, to $100,000.

Retrospective Rating. These plans come in six or more varieties, based on the degree of risk assumed by the employer and the preference for cash-flow alternatives. Net premium costs are determined by the total incurred or paid losses within a negotiated minimum or maximum premium range. An employer with an effective internal injury management system will find some type of "retro" program a fiscally responsible choice ... a form of limited self-insurance program. The premium range for a retro and its variations is $75,000 and up.

Self-Insurance. This is a best buy for the employer who has effective internal cost-containment procedures in place, provided state-mandated qualifications for this type of coverage are also met. Self-insurance is also available in most other states to employers who meet strict financial requirements. Usually self-insurance is appropriate for an employer whose annual workers' compensation premiums are above $500,000.

25.3.3 Terminology

A.M. Best's Rating. A Rating Report from the A.M. Best Company represents an independent opinion of an insurance company's financial strength and ability to meet its obligations to policyholders. The rankings range from A++ and A+ (superior) to F (in liquidation).

Incurred Losses. The total of actual paid losses plus reserves. The employer has greatest and quickest control over these. It is this figure that is used in the experience modification calculation.

Outstanding Reserves. Dollars set aside to cover anticipated losses for open cases. Reserves are driven more by injury than severity. The reserve is usually a "most-likely" scenario, presuming that full recovery is needed prior to return to work. Initial reserves are set within 30 days and are updated periodically. They are always updated prior to the submission of the unit statistical report. Reserve estimating is the most contentious of the insurance premium, as it may be based upon an individual insurer history or that of a particular adjuster's interpretation of the case.

Paid Losses. Dollars that have actually been paid on a particular case. This amount usually includes medical, indemnity benefits as well as expenses incurred.

Status of claim (open or closed). Claims are considered closed when there is no foreseeable activity on the case. Usually, a case will close after an employee has returned to work from a lost time injury for a period ranging from 90 to 180 days.

26.4 STRATEGIES TO DEVELOP A POSTINJURY RESPONSE SYSTEM

A systematic response to an injury will significantly reduce lost time and uncertainty as to the outcome. The employer can guide the employee through a designed path of medical assessment and treatment, restricted duty, and return to normal occupational duties. In a very few cases, the outcome will be long-term rehabilitation and/or separation from the work force.

Example: 6 years ago a loyal carpenter, John, sustained a back injury on a worksite managed by a firm we will call Wiley Sheet Metal Contractors. John had been working with Wiley Sheet Metal for 2 years, was well liked, and considered dependable by all. Accordingly, he submitted a Workers' Compensation claim to Wiley Sheet Metal's insurer, XYZ Mutual. By the time John's case was eventually settled, 3 years after the accident, for $80,000 by the insurer, Wiley Sheet Metal's management had become both embittered and suspicious of all injured workers' claims. Management felt abused by and vulnerable to a system of laws and benefits over which it had no control and through which it was destined to remain a victim.

Further, Wiley Sheet Metal began to sense some loss of competitiveness in bids it made on new jobs because of the skyrocketing costs projected for Workers' Compensation premium. Depending on the circumstances of each new claim, Wiley Sheet Metal management would find some target(s) on which to fix the blame: the insurance carrier, state fund, the doctors, the union, the lawyers, the state regulators and legislators, or the poor work ethic of its employees.

In fact, not until 2 years ago did Wiley Sheet Metal consider blaming its own internal management policies and practices for any of the costs it experienced in Workers' Compensation. At a loss-control seminar, Wiley Sheet Metal learned that its costs were higher than state and national averages for similar construction and property management firms. Further, they discovered that, like most other "victimized" employers, over 50% of its prevailing Workers' Compensation costs were directly attributable to how Wiley Sheet Metal responded to injured employees and their claims. Shortly after applying specific loss control strategies into its management practices, Wiley Sheet Metal experienced another back injury claim, similar to the one above. This time the injured worker was back at his job within 29 days at a total claims cost of $6800.

This experience and others like it have convinced many employers, like Wiley Sheet Metal Contractors, that much of the responsibility for controlling Workers Compensation costs rests with the employer. This assumption of responsibility is not simply a financial one (i.e., risk financing). But purely those of risk control. Nevertheless, these strategies are applicable regardless of the mode of risk financing utilized: self-insurance, state fund, or traditional guaranteed insurance.

What follows are ten loss control strategies that are built into the operations of those employers whose costs are lowest because they have accepted direct responsibility for controlling their Workers' Compensation experience.

26.4.1 Accept Responsibility for the System

The best reason to adopt this strategy is to save unproductive and frustrating effort. This law is the law, and there is no evidence, in all 50 states and the provinces of Canada, that a change in the law and benefits structure guaranteed employers any cost savings.

> A firm should accept responsibility internally and channel its energy productively by reducing lost-time cases, its modification factor, and incurred losses. These steps will, in turn, reduce the cost per covered worker.

There is much avoidance among managers and business owners in accepting responsibility for controlling their Workers' Compensation experience. Some say they have tried everything. Others say these strategies will not work in their state or in their line of work. Basically, they are saying they will not accept any responsibility on their own and are more comfortable blaming their state system, their work force, or their type of industry. Employers who are seriously committed to taking constructive action to do something about Workers' Compensation costs do not hide behind these other excuses. These strategies work in all states and work environments. It is acknowledged that certain environments are easier than others. For example, most employers (and insurers) would tend to agree that Maine is one of the most costly and adversarial states for Workers' Compensation cost control, while the Federal U.S. Longshore and Harbor Workers Act provides the most difficult environment.

Here are other examples of objections, which in reality are without substance. "Unions in old smokestack industries will not allow their members to be controlled by company doctors." Another is "a plant closing will precipitate a host of fraudulent injury claims."

One example involved a national manufacturer who had just announced to the union that the plant would be closed in 4 months. Workers' Compensation losses were already high for this employer. Nevertheless, strategies were applied as vigorously and intensively as time would permit, and within 12 weeks the employer had reduced the rate of lost time by 60% and prevailing per-worker losses were cut from $1600 to $400 per year. The union asked why had these strategies not been applied sooner.

In situation after situation, it points out that the responsibility rests squarely with the employer for controlling their own costs.

26.4.2 Incorporate Injury Management Controls into Safety Programming

Most firms already have a safety program that may include descriptive accident prevention materials distributed to employees. The firm may have a safety committee, accident investigation reports, and OSHA and state injury records. Some employers may also have incentive programs designed to sustain employee awareness of safety and to reward those whose records improve. The vast majority of safety programs focus specifically on the number of injury incidents, especially the frequency of lost-time cases. Further, and even more specifically, they focus on how to avoid the acts and conditions leading to those incidents. Independent safety consultants and loss control specialists from insurance companies strongly support this focus on frequency. (See Chapter 5 for more information on this topic.)

Yet there are two problems with this singular focus. In the first instance, frequency is not impacting costs as much as severity—the amount of direct time (and Workers' Compensation dollars) lost due to injuries. (Let the reader understand: A frequency analysis, as part of an overall effective risk control/risk financing analysis, is crucial.) Second, a focus solely on frequency can lead to a "prevention myopia," potentially obscuring an employer's willingness or ability to see costly failings in postinjury response. Such an employer commonly assumes that after an accident the safety program has collapsed and the claims program now takes over. The employer has essentially given up and abrogated their responsibility to others, usually outside of the firm (the third-party administrator or an insurer's claims department).

To illustrate why the concentration should be on total lost time and its management, consider this case study: two moderately high-risk employers, both in the same metropolitan area. They were both drawing from the same labor market and paying claims under the same state Workers' Compensation law. Both employ over 1000 people. In the latest year, both employers had an identical frequency rate of lost-time injuries; this rate was 40% higher than the national average for their type of work force. Finally, both firms had identical retrospectively rated Workers' Compensation plans, administered by the same insurer. However, only one of these two employers was experiencing high Workers' Compensation costs. Why? One was managing lost time; the other was not.

Here are the results for the most recent calendar year:

	Employer "A"	Employer "B"
Frequency rate (per 100 employees)	5.3	5.3
Losses/covered worker	$450	$120
Severity rate (#) of days) (per 100 employees)	127.0	50.0

Employer "A" had 5.3 lost-time claims for every 100 workers, and those claims resulted in 127 days lost for every 100 workers. Employer "B" had the same number (and type) of cases but managed to lose no more than 50 workdays for every 100 workers.

By comparison, employer "B" experienced costs per covered worker of almost one-fourth of "A's" experience. In absolute terms, "A" was paying over $200,000 more in Workers' Compensation losses per year than "B." In a competitive environment, "B" enjoyed opportunities "A" had lost unnecessarily.

The concept of injury management starts with proper planning. Each employer must have a specific procedure for lost-time management before an employee starts to lose any time.

> What happens (or doesn't happen) within the first 8–24 hours after an accident occurs is the primary determinant of how much time will be lost.

In fact, unnecessary lost time from one injury may be a lot more costly than the total cost of ten well-managed injuries.

The first step is to establish an action-oriented policy for safety and injury management. This should set the tone for the program and communicate its crucial link in the management of the business.

A written policy on workers' compensation and injury management should meet these four objectives:

1. State the employers official position and general guidelines regarding workers' compensation management;
2. Define program goals in terms of types of losses the employer wants to avoid or reduce, in terms of insurance coverage, and in terms of cost or incidence targets;
3. Describe the functions of the department in charge of workers' compensation management and its relationship to other operating departments; and
4. State the employer's philosophy towards worker injuries, injured workers, and the means of meeting those liabilities.

Recordkeeping and forms. The proliferation of forms, mandated by government and preferred by the employer, can themselves become an impediment to efficiency and early resolution of claims. Depending on the state, the type of industry, and available computer access, some means exist to simplify forms and forms processing to meet policy performance objectives rather than bureaucratic objectives. Most insurers now report the so-called, First Report of Injury to the state on behalf of the employer. This saves time, ensures accuracy of the data, and avoids penalties that can arise due to late reporting of the claim.

> The purpose for keeping data is not for government or insurance regulations, but to manage the process. Therefore, the better the data collected, the better the results of the action taken.

In addition to mandated first-report-of-injury forms and OSHA logs, an employer should consider developing a key form to help manage injuries:

A combined accident report and medical treatment form (Fig. 26.11), which includes a short description of the injury, a medical records release signed by the employee, a medical report of diagnosis and specific physical restrictions or impairments, and instructions to the employee to return to the supervisor after seeing the doctor. The second part of the form can be utilized for a supervisory accident analysis (Fig. 26.12).

In addition, reevaluate your compliance with mandated recordkeeping requirements. Occupational safety and health records must be kept in accordance with guidelines of OSHA and various state agencies. OSHA supplies two instructional publications: "Recordkeeping Guidelines for Occupational Injuries" and the shorter, "A Brief Guide to Recordkeeping Requirements for Occupational Injuries and Illnesses." These are available from the U.S. Bureau of Labor Statistics. (*Note:* At the time of this writing, these OSHA Recordkeeping regulations are being changed. Be sure to consult OSHA/BLS for the latest requirements.)

ACCIDENT REPORTING AND TREATMENT (ART) FORM

ACCIDENT DATA

Name of Organization ____________ Department ____________ Time ______
Date __/__/__ Contact ____________ Phone ____________
Employee Name ____________ Birthdate ____________
Home Address ____________ Home Phone # ____________
Job Title ____________ Date of Injury ____________
Describe what happened including exact location or machine # ____________
____
____
Supervisor ____________ Witness ____________

I hereby authorize (Company Name) ____________ (or any of its representatives including the insurance company) to be furnished any information and facts regarding this injury, including reports and records, results of diagnosis, treatment and prognosis, estimates of disability, and recommendations for further treatment. This information is to be used for the purpose of evaluating and handling my claim for injury as a result of an incident occurring on or about the above-noted date of injury and for no other purpose, now or in the future.

Employee Signature ____________ Date ____________

FIRST AID

Nature of Injury:
☐ New Injury ☐ No injury/illness found ☐ Recurrence/aggravation of existing condition ☐ Employee refused treatment
Type of injury ____________ Body part injured ____________
Treatment ____________
Action taken: ☐ Return to work ☐ Needs additional treatment

MEDICAL PROVIDER

Where was the employee taken for initial treatment? ____________ Arrival Time ____________
Nature of Injury: ☐ New Injury ☐ No injury/illness found ☐ Recurrence/aggravation of existing condition
Type of injury ____________ Body part injured ____________ Dominant Hand ___ R ___ L

RECOMMENDATIONS FOR WORK:
☐ Regular Work
☐ Modified Duty
☐ No Work

LIFTING LIMITED TO:
☐ 10 lbs.
☐ 20 lbs.
☐ 30 lbs.
☐ 40 lbs.
☐ 50 lbs.
☐ No Lifting
☐ Lifting Limited to _____ times per hour.

CARRYING LIMITED TO:
☐ 10 lbs.
☐ 20 lbs.
☐ 30 lbs.
☐ 40 lbs.
☐ 50 lbs.
☐ No Carrying
☐ Carrying Limited to _____ times per hour.

PUSHING/PULLING LIMITED TO:
☐ 10 lbs.
☐ 20 lbs.
☐ 30 lbs.
☐ 40 lbs.
☐ 50 lbs.
☐ No Pushing/Pulling
☐ Pushing/Pulling Limited to _____ times per hour.

POSITION LIMITATION:
☐ No Repetitive Body Motion (list body part) ____________
☐ No Reaching Above Shoulders
☐ No Reaching Below Knees
☐ No Exposure to Substances (specify): ____________
☐ Sitting/Standing Limited to _____ hrs/day
☐ No Climbing
☐ Not to Drive Vehicles
☐ No Exposure to Vibrating Tools

Treatment and Comments: ____________
____
____

Patient disposition:
☐ Return to Supervisor; no modifications needed.
☐ Return to Supervisor with modifications (above) for _____ days.
☐ Return to Supervisor; send home (employee can return to work on _______ (date).
Follow-up appointment on ____________ with ____________

Medical Provider Signature ____________ Date ____________

RTW

To be completed by the employee and his/her supervisor
The above restrictions (if applicable) have been reviewed and the employee:
☐ Has been placed in an appropriate modified-duty position
☐ Cannot be accommodated at this time
☐ Was sent home per medical instructions
☐ Returned to full duty, no modifications
☐ Employee refused modified duty
☐ Employee did not return to work

Supervisor ____________ Date ____________
Note: To facilitate the best care for your employee, the supervisor must also adhere to the above restrictions.

WHITE — Human Resources/Injury Coordinator
YELLOW — Medical Provider

Fig. 26.11 Sample of an accident report and treatment (ART) form.

ACCIDENT ANALYSIS (To be completed by the Supervisor)

In detail, describe the job the employee was doing at the time of the accident (include accident location and exact activity).

____

____

____

____

☐ Additional sheet attached for more comments.

Was it the employee's regular work assignment? ☐ Yes ☐ No   If no, was this person trained for this assignment? ☐ Yes ☐ No

Is there training required for the employee to complete this procedure? ☐ Yes ☐ No

	CAUSAL FACTORS	Yes	No	COMMENTS	CORRECTIVE ACTION
Employee	1.1 Is there a written job procedure for this job?	☐	☐		
	1.2 Did the employee understand the procedure?	☐	☐		
	1.3 Was the employee able to follow the procedure? If no—comment.	☐	☐		
	1.4 Was the employee the only staff person available at the time of the incident?	☐	☐		
	1.5 Was the employee working overtime?	☐	☐		
	1.6 Is safety equipment specified for this job? List all.	☐	☐		
	1.7 Was this safety equipment being used?	☐	☐		
	1.8 Was this safety equipment being used correctly?	☐	☐		
Environment	2.1 Did the work area contribute to the incident? (For example, clutter, noise, weather conditions, wet floor or pavement, lighting.)	☐	☐		
	2.2 Did the injury occur during behavioral intervention with a client?	☐	☐		
	2.3 Was there any change in client behavior prior to the incident?	☐	☐		
Equipment	3.1 Did any defect(s) in equipment/vehicle contribute to hazardous conditions?	☐	☐		
	3.2 Is there regular maintenance provided on the equipment?	☐	☐		
	3.3 Are there maintenance logs?	☐	☐		
	If accident involves a motor vehicle accident:				
	3.4 Was it reported to the police?	☐	☐		
	3.5 Was the vehicle towed? By whom?	☐	☐		
	3.6 Describe the incident. (Include the name, address, telephone number of the party or parties; year, make, model of car; ins. co. name and policy no.; description of damage; operator(s) license no. and license plate no.)				
Management	4.1 Were the employee behaviors which caused the accident observed before?	☐	☐		
	4.2 Are safety practices/procedures enforced? How? By whom?	☐	☐		
	4.3 Are safety practices/procedures supported by all?	☐	☐		
	4.4 Are there specific safety issues or concerns that require management intervention? If yes, please specify.	☐	☐		
	4.5 Was the employee's performance reviewed recently? Explain.	☐	☐		
	4.6 Was safety discussed during the review? Explain.	☐	☐		

SUMMARY OF CORRECTIVE ACTIONS TO BE REVIEWED WITH EMPLOYEE

(Example: Review or change of procedures, specific training, eliminating the conditions, submitted work request, etc.)

Causal Factor (Refer to Number Above)	Assigned To	Target Date	Completion Date (Safety Committee/I.C.)

Employee's Signature: ______________________ Date ________

Supervisor's Signature: ______________________ Date ________ Telephone Extension ________

(Forward completed form to Injury Coordinator within 24 hours of accident.)

Fig. 26.12 Sample of the accident analysis form.

26.4.3 Retain Accountability for Injury Management at the Worksite

Managers, foremen, and supervisors can and should be responsible for initiating injury management at the scene of an accident. Top management will need to provide support and management mechanism for starting and monitoring this process. The timeliness of their intervention is critical in treating the injured worker in the most humane and the least costly manner.

Company policy should explicitly state that on-site managers ensure that all injured workers receive immediate and appropriate medical attention, regardless of how minor the injury may be. Managers on site should spare no effort in assuring the injured worker of the company's desire to have him/her return to work. Indeed, the supervisor/foreman is the logical candidate for being the key liaison between the worker and the company in communicating concern for the injured worker and his/her family and to allay their anxieties over health, income, and the earliest possible return to gainful employment. Some training on understanding Workers' Compensation (procedures, etc.) may be necessary and well worth the additional time and dollars. If a supervisor is not available, then another candidate must be selected to perform this vital task.

If, for some reason, a company is not prepared to adopt such a worker-sensitive policy on injuries, then many of the following strategies will be ineffectual. The simple (and commonplace) strategy of having supervisors/foremen simply complete accident investigation reports is little more than a paper exercise. Rarely does this practice alone impact on reducing lost time.

Why is this supervisor liaison so important? Experience from Lynch Ryan projects shows that many worker injury claims, especially those involving back injuries, are indirectly affected by supervisors' performance rating of the injured workers. Specifically, supervisor behavior was a contributing factor in reported back injuries. This fact underscores the importance of the supervisor's role and the value of training to modify supervisory behavior before and after worker injuries. If, for example, a supervisor views worker injuries merely as statistics or as evidence of unproductive workers, then his unsympathetic or negative attitude will be communicated to the workers. They, in turn, when injured, feel uncared for and may seek ways of recovery. Employers, like Wiley Sheet Metal Contractors, have learned to accept responsibility for injury management and to show their on-site supervisors how to share that responsibility.

The environment in which the responsibility of injury management is clearly and fairly shared among the main office, the supervisor, and—yes—the employee is the environment in which the least amount of time is unnecessarily lost to injuries.

26.4.4 Establish Dedicated Medical Provider Relationships at Each Worksite

No groundbreaking at a new job site should occur until management has arranged, in advance (by contract or simple letter of agreement), that a specific medical provider will be available to immediately treat injured employees. The objective is not to control the provider but rather to offer workers the benefit of the earliest possible professional care of their injuries. Neither employer nor employee wants time lost due to an unresponsive medical provider or to the employee's frustration and delay in trying to locate a physician after the injury. Should the employee refuse immediate medical attention by the designated provider, an appointment for him should be immediately at the provider of his choice. If, on the other hand, he is treated by the firm's designated provider but wants a second opinion from the family physician, encourage him to do so. Make the appointment for him, and have the records of the first visit sent to the physician.

Medicine or occupational health professionals not only have more experience with worksite injuries, but they also understand the value of defining an injured employee's physical restrictions. Through a clear understanding of the job functions at the worksite, the trained medical provider will prefer to cite an injured employee's restrictions rather than to arbitrarily "prescribe" a week or two off. The referral patterns of trained occupational health practitioners are also designed to avoid unnecessary lost time between doctors' appointments.

The primary objective is to keep medical providers who treat the workers informed of the firm's genuine concern. Further, the physician must be advised of the nature of work required and the circumstances of the accident. Otherwise, any opportunity to aid the doctor in treatment and ready the employee for work without any unnecessary lost time is lessened.

> When employees are left to do their own medical case management and physicians are left uninformed about actual physical job functions, days and weeks are unnecessarily lost ... at the employer's expense.

The more that is known about local medical providers and specialists to whom they refer and, conversely, the more these physicians know about the employer, the more likely they will respond in the mutual interest of both employer and employee. Nevertheless, the firm must bear the responsibility in developing and maintaining these relationships. If the firm relies exclusively on the insurance carrier and/or claims administrator for these relationships, then there has been a forfeiture of dollars to outsiders without holding them accountable. The dollars lost to unnecessary lost time are the largest contribution to rising Workers' Compensation costs. Neither the medical providers nor the claims adjusters have any incentive to keep those dollars under control.

Shepherding injured workers along a predesigned path way of immediate and appropriate medical assessment can be done without compromise to the employee's health and employability, as long as both the employer and the employee agree that the primary objective is to make the injured employee healthy again (or as nearly so as medically possible) and to do so without delay in returning to gainful employment.

Labor unions and state workers' compensation systems will embrace this objective as the primary goal.

The following table provides guidance on finding and selecting medical providers. It is helpful to solicit input from employee groups (labor) as to preferred providers. Employees prefer a "selection" of medical providers to choose from as this eliminates the "company doc" issue.

Finding the Right Medical Provider
• Staffed by a Board Certified Occupational Health Physician.
• Utilizes a Sports Medicine or Active Treatment Approach.
Sports Medicine consists of exercises, stretching, as well as ultrasound and massage
• Refers to specialists within two weeks.
• Schedules visits every 5–7 days for employees on light duty.
• Visits the worksite to familiarize the practice with your work environment.
• Has extended hours and back-up coverage for off-hours.
• Follow up for off-hour coverage is initiated by the medical provider.
• Designation of single point of contact/access to the physician.
• On-site X-ray, MRI, and lab services.
• Completes company-required forms and claims information.
• Remains the treating physician or continues to follow the employee through other specialists.

26.4.5 Set Performance Objectives

All successful firms have quantifiable business objectives. By the same token, the firm should have quantifiable injury management objectives. They should reach all levels of the organization. A good starting point is to refer to Sections 26.2.2–26.2.7 and establish baseline and reasonable goals.

Here are a few specific measurements:

- Evaluate the total net cost for Workers' Compensation per year. If the cost exceeds more than $500–1000 per year per full-time employee (total hours average 2000 h/yr/employee), it is too much. Set targets for reducing that prevailing cost 1 year from now, 2 years out, and 3 years out. Manage this budget item as a controllable expense.
- Spend time focusing on incurred losses and techniques for reducing your experience modification factor.
- Develop departmental/division chargeback for losses.
- Most important, document how many lost-time cases in the past year ot two have been closed within 10 working days of the date of injury. If the percentage of cases resolved within 10 days from among all lost-time cases is less than 90%, then the Workers' Compensation costs are higher than they should be for an employer prepared to assume some responsibility for injury management.
- Begin all company meetings (at all levels) with a discussion of recent injuries and techniques used to prevent them.
- Have returning employees meet with senior management to discuss the incident and corrective actions that have taken place.
- Educate all managers and supervisors on claims management practices.

For each successive year, goals should be adjusted to reflect a 10–20% improvement. Ideally, each manager should be as familiar with the company injury management goals as they are for production and profit goals.

26.4.6 Developing Early Return to Work Opportunities

Productivity and earnings depend on how a firm manages and motivates its workforce. Building a sense of teamwork is essential; an injured player still should be considered a member of the team. Creativity in designing light-duty or modified-duty assignments for injured workers on the mend is recommended, where possible. Not only does it maintain the team spirit, but it also gets workers back on their jobs sooner. Light-duty assignments should be kept to within a specified time limit; such as 2–6 weeks. Although some industries have few light-duty opportunities, too many employers resign themselves to paying Workers' Compensation benefits without really evaluating such options.

> For every 500 employees working in a no-light-duty environment, the employer is experiencing unnecessary losses of at least $100,000 per year as a result of that no-light-duty policy.

Additionally, a firm should require that medical providers assist in this area by citing physical restrictions or impairments for injured workers and by specifying date-certain endpoints for full recovery. Qualified occupational health providers can do this easily and willingly. Avoid doctors who regularly prescribe days off for any worker complaint without explaining what lost function prevents the workers' return to work and when that return is medically appropriate. Again, recognize that there are doctors who will obstinately refuse to cooperate. Nevertheless, a planned management strategy to engage the physician's help and avoid, where possible, the uncooperative ones will pay handsome dividends in the long run.

Current medical research shows that extended bed rest (beyond a day or two) is one of the worst possible treatments for most back sprains. In fact, the majority of work-related back injuries can be diagnosed and their medical endpoints established within 3 weeks of the injury. Also, do not let union contract language muddle the strategy. When sensibly and sensitively approached, no union will oppose an injury management program designed as a benefit to the employee. The greater barrier to well-designed light-duty programs is management, not labor.

26.4.6.1 Tips for Working with a Union When trying to implement new programs aimed at controlling workers' compensation costs, management should look to the union for their approval. Most individuals, whether represented or not, are interested in programs designed to return an injured employee to the work force. In fact, union officials agree that by providing timely, high-quality medical care and modified duty, an injured worker will return to the job quickly. About 23% of the U.S. workforce participates in unions. In companies with collective bargaining, a unique set of rules and procedures require some adjustments in your postinjury management program.

- Collective bargaining agreements negotiated between unions and management address a number of areas that can impact the design of an injury management system: Seniority can be a major factor in work assignments and thus in modified duty.
- The collective bargaining contract may require that "lighter"-duty jobs be offered not on the basis of injury or functional restrictions, but seniority. The union may also seek to convert a temporary modified-duty opportunity into a permanent work assignment.
- Transitional duty can impact the definition of work and thus reduce the wages paid to an injured worker (resulting in payments under partial disability).
- The transitional duty job may reflect the work of a lower pay class, resulting in reduction of hourly wage; in this situation, the injured employee is eligible for "partial" disability, which pays a portion of lost wages.
- Unions may have their own medical providers and may discourage members from going to the company designated provider. The union medical provider may be generous both in interpretation of "work-related" injury and in the length of time prescribed away from work.
- Unions are extremely sensitive to the loss of established benefits; the program must not be perceived in this manner.

A good injury management program must be presented to unions as a benefit for all employees. No rights are compromises; there is no loss of seniority or change in the bargaining position (unless unions resist modified-duty assignments due to seniority considerations, in which case there is a bargaining issue).

Unions respond well to the concept that "sports medicine" is good for workers and that returning to work as soon after an injury as possible is better for the worker than staying out of work until totally cured and able to return to full-time, regular work. They often recognize that an active worker is more satisfied with his/her situation. On the other hand, unions may feel threatened by the employer controlling or attempting to control the medical path; in this case it is crucial that the medical services be timely and of the highest quality.

> SAMPLE CONTRACT CLAUSES
>
> Many union contracts include clauses protecting people with disabilities. Examples follow:
>
> Nondiscrimination—The employer shall hire, train, transfer, and promote employees without regard to age, sex, race, creed, color, or disability.
>
> Employment Rights—The employer agrees that no employee shall be denied any aspect of his/her employment rights solely on the basis of disability, and shall work towards that end.
>
> Reasonable Accommodations—Employers shall make reasonable accommodation to the physical and mental limitations of an employee or applicant.
>
> Job Transfer—Any employee who, as result of an accident on or off the job, or chronic disease or condition, is unable to perform his/her duties, shall be transferred to an other position if work is available for which he/she is qualified or can be retrained within a reasonable period of time. He/she shall retain full seniority rights and wages.

If an employer is sincerely committed to fair treatment of injured employees and to significant reduction of workers' compensation costs, some form of light-duty jobs has to be created, and they require creative, open-minded thinking. Coming up with a list of light-duty jobs based on function is a difficult and challenging task. The payback, however, within the first year would be between $5,000 and $10,000 per hour of effort put into designing and implementing an effective light-duty program.

26.4.6.2 The Window of Suggestibility Why does one person return to work following a disability while another becomes captive to what is sometimes called the "worker's disability syndrome"? The determinants include the policies and procedures of employer and insurance programs as well as characteristics of the disabled individuals themselves. Various studies by the Menninger Foundation strongly suggest that early intervention is a variable that can make a major difference in outcomes. The sooner return to work can be established as an individual's objective, the more likely it is that the goal will be met.

Early studies point out that the industrially injured worker commonly enters a highly suggestible state just following a disabling incident . . . the conclusion is that this "window of suggestibility" is the key to the rehabilitation success. The injured worker's immediate response to the injury is an important one. The longer the "nonwork" situation exists, the easier it becomes for the injured worker to respond negatively to it and the chances of successfully returning to a productive meaningful life are significantly decreased.

> In general, the employer stands the best chance to return the employee to work within 60 days of the injury.

Other studies support the contention that successful rehabilitation outcomes depend on prompt initiation of the process. Studies using national samples have found that personality characteristics—especially those relating to independence—begin to change within 60 days after injury. To illustrate, 80% of a group of cardiac patients who were given counseling within 2 weeks after their heart attacks returned to work faster than those who were not given counseling. As these studies show, it is important to have "early warning systems" or "prompt contact programs" for instant reporting of injuries or signs of illness. These programs offer early support and guidance and have proven their value in dealing with the "invalid psychology" that ensnares some disabled workers for life.

26.4.6.3 Quick Return to Work What qualities motivate certain employees to return to work more quickly than others (after a disabling injury or illness)? A recent study, *Development of a Rehabilitation Strategy for Fortis Benefits Insurance Company's Group Long Term Disability Claimants*, provides some answers (Fortis, 1994).

Overall, the study shows that employees who return to work quickly tend to be more resilient, proactive, and conscientious than those who do not. "[Quick returners] have a dependable work ethic and style because they view work as a normal—even a good—part of life that bestows a sense of self-worth and self-expression," explains the study. "A disabling impairment is a disruption of the person's normal work routine, but it [does not signify] the end of the person's self-concept." In the study, researchers conducted focus groups and interviewed 275 Fortis disability claimants in order to probe the personal ethics and values of workers.

Among the findings:

- 70 percent of quick returners say they rarely think about their impairment, compared to 34 percent of slow returners;
- 76 percent of quick returners say they refuse to feel victimized about their condition, compared to 43 percent of slow returners;
- 90 percent of quick returners say they have a good relationship with their supervisors, compared to 83 percent of slow returners;
- Quick returners are more likely to take an active interest in their therapy, even to the point of reading medical journals and conversing with doctors. In addition, quick returners are more likely to have an intense work ethic, the study shows.

24.4.6.4 A Practical Approach Providing suitable early return to work opportunities for an injured employee in the face of tight production scheduling can often seem like a burdensome situation. Initially, the supervisor may believe that he/she is being held solely responsible for finding and implementing solutions to difficult problems that he/she has no control over.

While the supervisor is the leader in guiding the employee through the return to work process, supportive tools are available:

1. Customize the employee's current work tasks by reducing pace or assigned work hours, or by other means.

2. Identify a plant-wide access to work within physical restrictions and prepare a restricted-duty position.

3. Using input from all sources in the plant, develop a job bank of tasks. Depending on the injury, a person could work at another job without restriction.

4. Include ergonomic focus to engineer out and eliminate inappropriate or unsafe work tasks and processes.

5. Provide education for employees and supervisors on adaptive work techniques and behavioral change.

6. Ensure medical treatment and short-term therapy to increase the injured employee's strength and endurance to safely complete the assigned work duties.

Application of these measures is provided throughout the text.

26.4.7 Effective Claims Management through Communications

Once an accident has been responded to, it is important to make sure all activity (medical, legal, benefits) is being managed professionally and the employee is being returned to work as early as possible. This requires frequent contact with the injured employee, the insurer, and the medical provider. The employee's supervisor and if available, the plant nurse should call the injured employee at least once a week to check on his/her condition and rehabilitation progress. During the call to employees, the following should be covered:

- Express concern for the employee's health;
- Help the employee understand benefit policies and regulations;
- Ensure the employee is getting all the benefits to which he or she is entitled;
- Describe return to work assistance available;
- Convey to the employee that the convalescence/recovery period will be monitored;
- Assess the employee's motivation and resources, the personal and interpersonal factors that will support return to work;
- Remind the employee of obligations during disability (e.g., furnishing doctor's statements, timely completion of paperwork, regular communication with the supervisor or personal departments);
- Create a "return to work" expectation in the employee's mind,
- Assure the employee that a job is waiting—that light-duty work will be made available if needed.

One of the biggest obstacles to supervisors or managers making phone calls is the lack of knowledge about the process. In those situations, it is most helpful to provide a script for the supervisor to use in the phone call.

Weekly calls to employees should try to determine how much monitoring and how much support will be necessary to get the employee back to work.

- How is the problem affecting the employee's support group?
- How supportive is the support group?
- Have there been other episodes of disability? How had they coped before? What worked? What didn't?
- Is the employee optimistic or apprehensive? How will the employee maintain motivation?
- What does the employee really want?

The depth of employee interviews depends on the circumstance of each case. By bringing up important, difficult topics, the employee will appreciate the willingness to talk about real concerns. Modelling a willingness to confront problems, and demonstrating that the employee does not have to cope with disability alone, underscores the fact that it is a shared responsibility. An internal checklist for each lost-time case will provide a roadmap on determining appropriate action:

Claim management checklist:

1. On what date is the employee expected to achieve maximum medical improvement? Will the medical treatment currently provided result in meeting that target date?
2. What is the expected date that the employee will return to full time at the regular job? When will the employee be able to begin alternative duty and work hardening? Has preparation begun for return to work?
3. If after reaching maximum medical improvement, the employee cannot be returned to his/her regular job, can reasonable accommodations be made to allow the employee to return to the regular job with the remaining impairments? If not, are other jobs available? If not, is vocational rehabilitation being considered and implemented?
4. When did the employee's supervisor call the employee? The plant nurse? A call should be made once a week. What is the information from the last phone call? What is the employee's state of mind, attitude, and family situation? Is the employee eager to come back to work?
5. When did management last discuss this case with the insurance company? Is the claim being actively managed?
6. Are there any psychosocial problems that would prevent the employee from coming back to work?
7. Any problems with doctors not cooperating with treatment plan to get employee back to work?
8. Has the employee hired a lawyer? If so, why did the employee think a lawyer was necessary? Has the lawyer created an adversarial climate that will delay a return to work? If so, what is being done to improve the situation?
9. Is a medical case manager assigned to the claim?
10. Is there any suspected fraud involved with the claim? If so, is a vigorous investigation ongoing and when will it be completed?
11. Have the conditions that caused the injury been eliminated or reduced to prevent future injuries?
12. Are any and all controversies regarding this claim being resolved to speed return to work and minimizing litigation?

If employers miss the early window of opportunity to establish a positive, constructive attitude toward convalescence and return to work they will be reacting to what others—doctors, lawyers, employees—do. Offering early support and guidance has proven its value in short-circuiting the psychology of invalidism that can snare disabled workers.

Make sure the disabled worker knows a job is waiting. It need not be the same job—but it should be equivalent pay.

It is tough to sort out disability from performance or attendance issues. If the employee was an unsatisfactory performer before the illness or injury, the problem should have been documented and progressive discipline applied. There is no question that employees sometimes use disability as a haven from progressive discipline. They do it because it works. To avoid legal confrontation, manage the disability to resolution, return the employee to work, then reinitiate, rebuild, and document the case for future corrective action.

A plan of action is the bridge over which you and your employee progress from injury to successful return to work. It is the mutually understood blueprint for successful reentry to the work site. The point of disability management is to avoid leaving things to chance. A written statement of understanding can specify the actions to be taken by each party, spell out respective responsibilities and obligations, help ensure clear communication, enlist commitment, and motivate positive actions by the employee. Such an agreement, or series of agreements, can head off "yes, but's" later on.

Control the outcome of disability by controlling the process.

Motivation is key. What does the disabled employee really want? Does he (or she) have a realistic appreciation of alternatives? Does he have the information necessary to decide? Does he know what he wants? Is he willing to say? Going back to work promptly may be the best idea in the world—for the employee and for you. But it may still have to be packaged and sold in a way that will be attractive to the employee. Recovering from disability can be hard work—harder than working. The process takes resolve and committment.

Motivation is essential in the disabilities management process.

26.4.7.1 Indicators of workers' compensation fraud

These indicators of potential fraud should help identify claims that merit closer scrutiny and further investigation. The presence of several indicators, while significant, does not mean that a fraud has been committed; however, it does suggest that fraud may be a factor and should be given careful consideration. In general, fraud constitutes less than 5% of all field claims. Indicators of fraud include (Source: *Issues in Workers' Compensation: A Collection in White Papers*, 1994):

1. The injured worker is a disgruntled employee or one on the verge of being fired.
2. The injured worker has received poor job performance evaluations.
3. The injured worker is a new employee.
4. The injured worker is never home or his/her spouse or relative always answers the phone call and indicates that the injured worker will call back in a few minutes. (The injured worker may actually be working.)
5. The injured worker appears to be taking more time off from work than that injury suggests.
6. The injured worker has prior unexplained absences from work.
7. Details of the accident are not reported promptly.
8. There is a history of prior claims.
9. There are no witnesses to the accident.
10. There are discrepancies in the injured worker's story.
11. The injured worker has a history of moving from job to job.
12. The injured worker is having financial difficulties.

13. The injured worker does not show up for an I.M.E. (independent medical examination) or cancels an appointment.
14. A "tip" is received that the injured worker is employed elsewhere or is engaging in an activity inconsistent with the injury.
15. Disability is not substantiated by objective medical findings.
16. The injured worker inquires about a settlement early in the life of the claim.
17. The injured worker recently purchased disability policies.
18. The injured worker treats with multiple physicians.
19. The injured worker calls from a pay phone.
20. The injured worker has a side business or attends school.
21. The accident occurs in an area where an employee would normally not be working.
22. The accident is not the type that an employee should be involved with (Example: An office worker would not be lifting heavy equipment).
23. The claim is reported on a Monday. (This may indicate that the injury is the result of an incident that occurred over the weekend.)
24. The injured worker participates in contact sports or physically demanding hobbies.

It is extremely important to communicate with the claim provider when a number of these indicators are present. It is the claims provider's responsibility to investigate any allegations. Employers should not undertake surveillance on their own and certainly without advice of legal counsel.

26.4.8 The Hearing and Settlement Processes

In each state, cases progress through an adjudication process that can typically take 6 months to a few years. It becomes extremely complicated and prejudicial. When this happens, employers should prepare to document each and every contact with the injured employee as though it was a high-profile civil case.

Employees have a right to hire an attorney to represent them for Workers' Compensation benefits. However, this should be totally unnecessary if employees are educated in the workers' compensation system and injuries are properly responded to with care, compassion, and quality medical care. Employers must ensure that the insurer pays all the benefits necessary and on a timely basis. Employees should understand that the retention of an attorney may result in added fees, which could reduce a workers' compensation settlement by 20%.

Employers do not have to be passive bystanders once a disputed compensation claim enters litigation. There are litigation management techniques that can be used to monitor, manage, and improve the responsiveness of outside counsel. The selection of a qualified attorney to handle disputed workers' compensation cases and the maintenance of service standards for those attorneys are the most important steps in managing the litigation process.

In most routine cases an attorney will be selected by the insurance company. Do not assume the assigned attorney is representing your interests. Inquire frequently about what the attorney is doing to prepare for hearings and trials. Be actively involved in the process. Ask questions. Make sure the assigned attorney has goals, strategies, and an action plan for each claim.

Be certain that hearing dates are communicated to all parties with enough advance notice. Speak with the assigned attorney and ask what is needed to do to prepare for the case and what he/she is doing to prepare for the case.

- Form an effective defense strategy with the insurer.
- Discuss cases that are scheduled for hearings/trial with the insurer in a timely fashion.
- Implement a defense counsel evaluation program.
- Manage the activities of defense counsel.
- Maintain contact with defense counsel during hearing/trial and consider attending trial, if feasable.

Finally, with most longer-term cases (3–5 years), the issue of a lump sum settlement is typically discussed. The lump sum settlement is an agreement between the parties to end workers' compensation benefits in exchange for one-time cash settlement. There is no science to determining an amount for a particular case, as it becomes a negotiation between the parties. However, to help keep the value in the employer's favor, substantiate as best as possible the rationale for the settlement. For example, a certain amount for retraining and lost wages during that period, plus an estimate of ongoing medical payments. Workers' compensation does not recognize the value of "pain and suffering," only lost wages and medical costs. Above all, remember that hearing officers will view the intent of the law as true workers' compensation, not employers compensation. You must prove your case.

26.4.9 Concentrate on Managing Today's Injuries

Today's injuries become next year's major settlements. If the major effort is focused on claims that exceed 1 year in age, it becomes pointless, even counterproductive. Statistics show that seven out of eight long-term claimants never return to work. Therefore, these claims are beyond most efforts of the claims administrator, insurer, or employer. According to this strategy, one must significantly reduce the rate at which new cases enter the "one-year-plus" club. In other words, the firm should not allow the older, higher-valued claims to consume its attention while smaller, more recent ones are maturing or festering. Rather, the greatest energy should be put into resolving an increasing percentage of cases in the first 2 weeks.

> The majority of techniques used to control workers' compensation claims quickly disappear as lost time grows.

26.4.10 Involve the Claims Service Provider in the Program

It is highly recommended to establish (or maintain) a high-profile, good working relationship with your claims service provider. The more the claims staff know about a firm's loss control philosophy and concern for its employees, the greater likelihood its claims will receive better attention. Third party administrators and especially in-house claims departments are much more receptive to their client's desires. Yet, despite which claims organization is involved, a working, communicative relationship is important, since it is an integral part in the claims management process.

Employers rely on their workers' compensation insurer to provide high-quality service, yet most employers cannot define an appropriate level of service. Ask the insurer in writing to:

- Submit loss runs at least quarterly, if not monthly;
- Code loss-run claims by department or cost center;
- Provide a copy of the Unit Statistical Report for review prior to submission to the rating bureau or NCCI;
- Explain available premium discounts and how they are computed;
- Give an analysis of potential savings at various levels of loss ratios;
- Describe the frequency and effectiveness of loss-prevention services;
- Describe training materials or courses available to you;
- Update all reserves monthly;
- Confer immediately regarding any claim valued at more than $5000;
- Confer prior to any claim denials or lump-sum settlements;
- Hold claim review meetings at least quarterly;
- Consult in selecting defense counsel and;
- Name an account manager who will respond to requests, become familiar with the employer's operation, and cooperate with employer selected outside medical and safety services, and who has authority to access insurer's data and decision makers.

26.5 MAKING THE RIGHT DECISION

One would think that making the right decision for a cost–benefit analysis is easy. When the discussion centers on pure economic or financial variables, it generally is. The investment decision to commit the funds must equal or surpass the firm's internal cost of capital. This is truly a dollars and cents approach. It works well until the human factor is considered as a variable.

Some have attempted to put a value on human life and limb. Workers' compensation laws have specific benefits that pay for such losses. For instance, the loss of a life in some states pays $100,000, and the loss of a hand ranges from $12,000 in Puerto Rico to $142,000 in Iowa. Life insurance policies have a dollar amount assigned, though that amount is determined by the buyer. Regulators must make similar decisions when contemplating raising the interstate highway speed limit from 55 miles per hour to 65 mph. The 10 mph change certainly increases the likelihood of both an accident as well an increase in the injury severity. Yet public opinion overwhelmingly supports the tradeoff between a faster trip home versus the increased risk.

Certainly, there are risk/hazard situations that should not be put to economic scrutiny, and in doing so may introduce unnecessary problems and confrontation.

Consider these two business examples:

> The president of a company who said, when the economics of safety were being discussed, "We're in business to make a profit, but we make it from our product, not our people."
>
> The plant manager, who while being given cost alternatives to control a crane runway repair hazard, interrupted the presentation and said, "What's with all these options? Let's assume our sons are to be working on the crane runways, which alternative would we choose?"

Since there is no perfect world, plant management view their desire to prevent loss of life and human suffering in terms of reducing exposure to the risk at question.

26.5.1 Cost–Benefit Analysis

A cost–benefit analysis provides the framework for comparison of costs and benefits of a single program to determine if the benefits of the program exceed the costs. Where the costs outweigh the benefit values, an inference could be made that the monetary values evaluate not only the economics of applying a control but also whether the hazard deserves to be controlled.

Performing economic valuations that seem to produce hard numbers may cause neglect of other significant decision factors. Factors other than economics (social, political, legal) may be the most significant decision determinants. The economics of a choice must never be overlooked, nor should it be the sole or major determinant, either in business safety practice or in national policy and program issues. True, the economic bottom line must be the immediate and continuing objective in the survival of a business. But, generally, it is not looked upon as the only purpose of the business.

Appropriate uses of cost–benefit analysis and cost-effectiveness analysis are:

> Cost-Effectiveness Analysis: Use where identical outcomes exist to help select among alternatives for achieving that outcome.
> Cost–Benefit Analysis: Use where there are projects with differing objectives or outcomes to help choose among the alternatives, or use for an actual monetary cost versus monetary benefit evaluation of a single program/project.

When used properly, both processes can be of assistance to: (1) Make choices among alternative courses of action; (2) establish a framework for measuring program performance after implementation.

Since the steps in making CBA and CEA are identical except for benefit evaluation, this section will concentrate primarily on making cost–benefit analyses. Making cost–benefit analyses includes:

- Objectives of the program.
- Financial requirements.
- Benefits expected to result.
- Uncertainties involving both cost and benefit estimates.

- The time when the costs are expected to be incurred and the benefits expected to result.

The scope or range of analysis and the basic question to be resolved may vary. However, there are two methods to consider. One is a fixed budget approach, which assumes the availability of a given level of funds. The question involved here is: With a fixed level of resources available, what course of action should be taken to achieve maximum benefits? The objective is to weigh alternatives in terms of costs, benefits, risks, and timing to select the measures that would do most towards achieving goals, while still remaining within the constraints of available funds.

The other method, the fixed effectiveness approach, assumes that some specific objective is to be accomplished. The question here is: Considering that some specific objectives must be accomplished, what courses of action should be taken to minimize the cost of achieving these objectives?

26.5.2 Steps Performed in Cost–Benefit and Cost-Effectiveness Analysis

From a plant manager's perspective, there are basic steps that should be used:

1. Will the program pay for itself? In what length of time?
2. Is control of the hazard affordable; that is, how much will correction cost versus available funds?
3. Which hazard produces the most serious exposure?
4. Which corrective measure will be most efficient in reducing the exposure?
5. Which corrective measure will produce the most exposure reduction?
6. Which corrective measure under consideration costs the least? the most?
7. Which exposure problem is most urgent?
8. Which corrective measure will be most acceptable to all the audiences concerned?

Questions such as these are not always subject to resolution via cost analyses. Whether or not cost–benefit analysis is applicable will depend on the degree of uncertainty, the level of expenditures involved, and the amount of time needed to make the evaluation. For the general run of occupational health and safety situations, where a single objective is to be achieved, many of the factors (e.g., societal benefits, cost reduction or increase, fewer production interruptions) may be common to all alternatives and factor out.

Thus the economic evaluation may boil down, simply, to a comparison of implementation costs. In spite of its humanitarian objective, safety must be subjected more and more to cost evaluations before and after implementation. The efficient use of money for the support of safety programming (as contrasted to its use for competing non-health-and-safety activities), the efficient selection among competing safety ventures, and the selection of the most efficient way to achieve a given safety result demand such evaluation.

26.5.3 The Risk Score Method

A method of mathematically deciding priorities has been developed to weigh the controlling factors and calculate the risk of a hazardous situation, giving a "risk score" to indicate the urgency of remedial action. An excerpt from *Selected Readings in Safety* by W. Fine, entitled, "Mathematical Evaluations for Controlling Hazards" (Fine, 1973) explains this process in detail.

> "Normal industrial safety routines such as inspections and investigations usually produce or reveal numerous hazardous situations which, due to limitations of time, maintenance facilities and money, cannot be corrected at once. The safety director must then decide which problems he should attack first. A great aid in making this decision would be a method to establish priorities for all hazardous situations, based on the relative risk caused by each hazard. By means of such a priority system, safety personnel can allocate their own time and effort and request expenditure of funds to correct situations in proportion to the actual degrees of risk involved in the various situations. Such a priority system is created by the use of a simple formula to "calculate the risk" in each hazardous situation and thereby arrive at a risk score which indicates the urgency for remedial attention.

Another closely related problem deals with economics. When the safety department comes up with a proposed remedy for a hazard, it may be necessary to convince management that the cost of the corrective action is justified. Since most budgets are limited, the safety effort must compete with other organizations for funds for safety projects. Unfortunately in many cases, the decision to undertake a costly project depends to a great extent on the salesmanship of safety personnel. As a result, due to a poor selling job, an important safety project may not be approved; or due to excellent selling by safety, a highly expensive project may get approval when the risk may not actually be great. This difficulty is solved by use of an addition to the risk score formula that weighs the estimated cost and effectiveness of contemplated corrective action against the risk and gives a determination as to whether the cost is justified.

26.5.3.1 Risk Score Formula The seriousness of the risk due to a recognized hazard is calculated by use of the "risk score formula." A numerical evaluation is determined by considering three factors: the consequences of a possible accident due to the hazard, the exposure to the basic cause, and the probability that the complete accident sequence and consequences will occur:

$$\text{risk score} = \text{consequences} \times \text{exposure} \times \text{probability}$$

In using the formula, the numerical ratings or weights assigned to each factor are based upon the judgment and experience of the investigator making the calculation. A detailed review of the elements of this formula follows.

The first element, consequences, is defined as the most probable results of an accident due to the hazard that is under consideration, including both injuries and property damage.

Numerical ratings are assigned for the most likely consequences of the accident, from 100 points for a catastrophe down through various degrees of severity to 1 point for a minor cut or bruise.

Degree of Severity of Consequences	Rating
Catastrophe: numerous fatalities; extensive damage (over $1,000,000); major disruption	100
Several fatalities; damage $500,000 to $1,000,000	50
Fatality; damage $100,000 to $500,000	25
Extremely serious injury (amputation, permanent disability); damage $1,000 to $100,000	15
Disabling injuries; damage up to $1,000	5
Minor cuts, bruises, bumps; minor damage	1

The next factor, exposure, is defined as the frequency of occurrence of the hazard event being the first undesired event that could start the accident sequence. The frequency at which the hazard event occurs is rated from continuously with 10 points through various lesser degrees down to 0.5 for extremely remote.

Hazard-Event Occurrence	Rating
Continously (or many times daily)	10
Frequently (approximately once daily)	6
Occasionally (from once per week to once per month)	3
Usually (from once per month to once per year)	2
Rarely (it has been known to occur)	1
Very rarely (not known to have occurred, but considered remotely possible)	0.5

The third factor, probability, is defined as the likelihood that, once the hazard event occurs, the complete accident sequence of events will follow with the timing and coincidence to result in the accident and consequences. The ratings go from 10 points if the complete accident sequence is most likely and expected, down to 0.1 for the "one in a million" or practically impossible chance.

Probability of the Accident-Sequence, Including the Consequences:	Rating
Is the most likely and expected result if the hazard event takes place	10
Is quite possible, would not be unusual, has an even 50/50 chance	6
Would be an unusual sequence or coincidence	3
Would be a remotely possible coincidence; it has been known to have happened	1
Extremely remote but conceivably possible; has never happened after many years of exposure	0.5
Practically impossible sequence of coincidence; a "one in a million" possibility; has never happened in spite of exposure over many years	0.1

Example. A building of an explosive-processing laboratory contains a number of ovens that are used for environmental testing (heating) of explosive material with up to five pounds of high-explosive material in each oven. This type of oven has been known to heat excessively due to faulty heat controls and thereby cause the explosives in the oven to detonate. People walk past the outside of this building. The potential hazard considered here is the endangering of persons who occasionally walk past the building.

The first step in calculating the risk is to study the situation and list the most probable sequence of events for an accident. These are as follows:

1. Several ovens are in use, each containing explosives.
2. Persons are present outside the building.
3. The thermostat of one oven fails and the oven temperature rises above the proper operating range. (This is the hazard-event.)
4. The secondary emergency shutoff control also fails to function.
5. The oven overheats.
6. The explosive detonates.
7. A passerby near the building is fatally injured by flying debris.

Factors are considered and evaluated for use in the formula:

$$\text{risk score} = \text{consequences} \times \text{exposure} \times \text{probability}$$

1. Consequences. Considering that a fatality was most likely, the rating for consequence = 25.

2. Exposure. The hazard event is the failure of the thermostat. Experience shows that this has happened before, but very "rarely." Therefore, Exposure = 1.

3. Probability. Based on judgment and experience, it must be decided on what the probability is that the complete accident sequence will follow the hazard event, considering each step in the sequence.

Considerations include the facts that all ovens have been equipped with secondary emergency shutoff controls, and that thorough maintenance procedures ensure the proper functioning of both the thermostatic controls and emergency shutoff controls. Failure of either set of controls is quite unlikely. For both sets to fail at the same time and on the same oven would be a very remote possible coincidence. Therefore, the probability rating is remotely possible, and probability = 1.

4. Substituting in the formula:

$$\text{risk score} = 25 \times 1 \times 1 = 25$$

The significance of this risk score will be seen when risk scores for other hazards are calculated, using the same criteria and judgment, and then there will be a basis for comparison of risks.

In the same manner as demonstrated in the above example, the risk scores for many other hazardous situations have been calculated, using the same criteria and judgment. These cases are now listed in order of their risk scores, or order of the relative seriousness of their risks on one sheet that is called the risk score summary and action sheet (Fig. 26.13).

The listing of hazardous situations in the order of the seriousness of their risks, the higher risks first, becomes an actual priority list. On the right side of the chart, horizontal brackets have been drawn. These are the critical dividing lines, which signify the

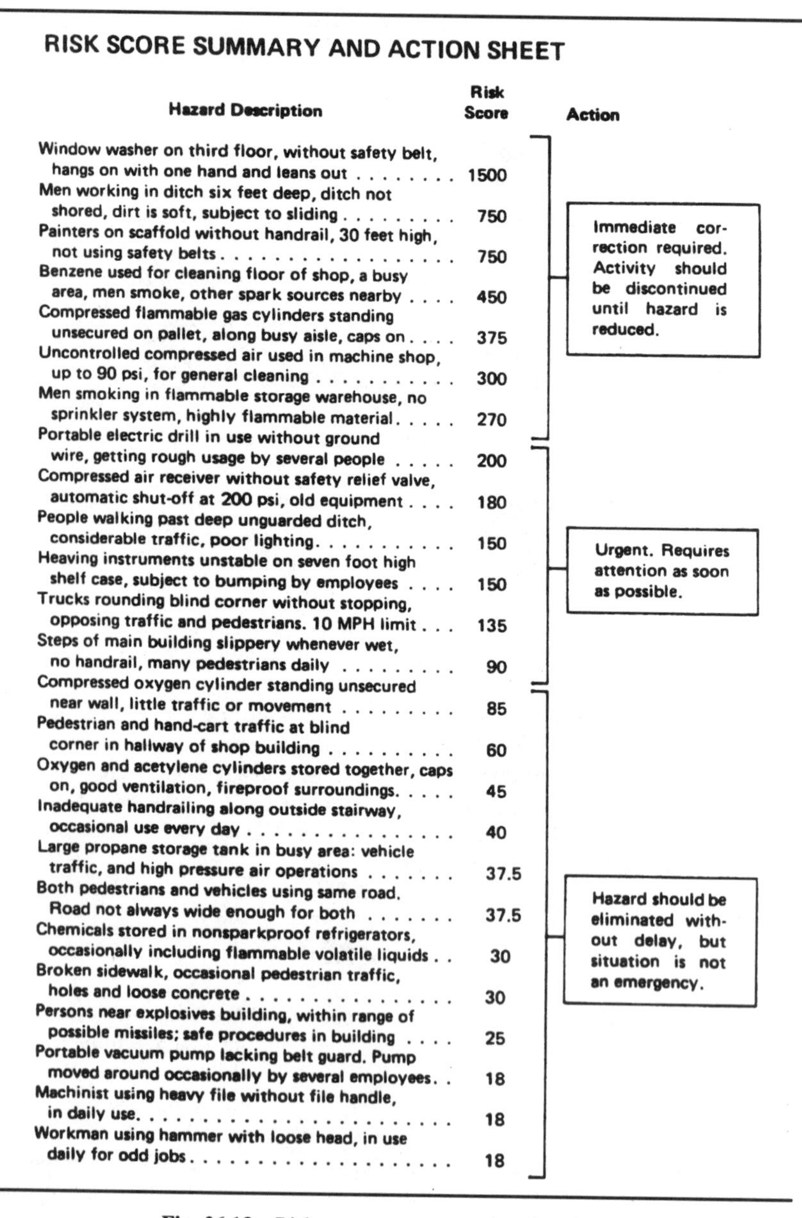

RISK SCORE SUMMARY AND ACTION SHEET

Hazard Description	Risk Score	Action
Window washer on third floor, without safety belt, hangs on with one hand and leans out	1500	Immediate correction required. Activity should be discontinued until hazard is reduced.
Men working in ditch six feet deep, ditch not shored, dirt is soft, subject to sliding	750	
Painters on scaffold without handrail, 30 feet high, not using safety belts	750	
Benzene used for cleaning floor of shop, a busy area, men smoke, other spark sources nearby	450	
Compressed flammable gas cylinders standing unsecured on pallet, along busy aisle, caps on	375	
Uncontrolled compressed air used in machine shop, up to 90 psi, for general cleaning	300	
Men smoking in flammable storage warehouse, no sprinkler system, highly flammable material	270	
Portable electric drill in use without ground wire, getting rough usage by several people	200	Urgent. Requires attention as soon as possible.
Compressed air receiver without safety relief valve, automatic shut-off at 200 psi, old equipment	180	
People walking past deep unguarded ditch, considerable traffic, poor lighting	150	
Heaving instruments unstable on seven foot high shelf case, subject to bumping by employees	150	
Trucks rounding blind corner without stopping, opposing traffic and pedestrians. 10 MPH limit	135	
Steps of main building slippery whenever wet, no handrail, many pedestrians daily	90	
Compressed oxygen cylinder standing unsecured near wall, little traffic or movement	85	Hazard should be eliminated without delay, but situation is not an emergency.
Pedestrian and hand-cart traffic at blind corner in hallway of shop building	60	
Oxygen and acetylene cylinders stored together, caps on, good ventilation, fireproof surroundings	45	
Inadequate handrailing along outside stairway, occasional use every day	40	
Large propane storage tank in busy area: vehicle traffic, and high pressure air operations	37.5	
Both pedestrians and vehicles using same road. Road not always wide enough for both	37.5	
Chemicals stored in nonsparkproof refrigerators, occasionally including flammable volatile liquids	30	
Broken sidewalk, occasional pedestrian traffic, holes and loose concrete	30	
Persons near explosives building, within range of possible missiles; safe procedures in building	25	
Portable vacuum pump lacking belt guard. Pump moved around occasionally by several employees	18	
Machinist using heavy file without file handle, in daily use	18	
Workman using hammer with loose head, in use daily for odd jobs	18	

Fig. 26.13 Risk score summary and action sheet.

different zones based on the degrees of risk, and indicate the required corrective action commensurate with the degrees of risk. For the hazards with the higher risk scores (in the high-risk zone) the action column calls for immediate corrective action. In these cases, or for any other hazardous situation whose risk score is calculated to be in the high-risk zone, any operation should be stopped until something is done to lower the risk, and get the score into a less urgent category.

The medium-risk hazards are in the second bracketed zone, and as the action column states, these cases are "urgent" and require corrective action as soon as possible. But for these degrees of risk or urgencies we do not say "Stop the job!"

The hazardous situations in the lowest zone of the chart are lesser ordinary hazards that, as stated in the action column, should be acted upon without undue delay, but not as emergency situations.

The risk score summary and action sheet can be a very useful device:

It establishes priorities for attention by both safety and management, since all the hazards are now listed in order of their importance. The position of any item can be lowered by corrective measures that will decrease any one of the factors: consequences, exposure, or probability. For example, the consequences can be reduced by providing protective clothing or equipment. Better machine guarding or improved procedures could decrease both exposure and probability.

For a newly discovered hazard, the list provides guidance to indicate its urgency. Once its risk score is calculated, its urgency will be indicated by the action area in which its risk score places it.

It can be used to evaluate a safety program, or to compare safety programs of various plants, a more realistic method than accident statistics. At any given time the complete chart for a plant represents the actual status of safety; that is, let us say the chart shows seven "immediate actions" for emergency items; six items in the "urgent" category; and twelve "minor" hazards. Accomplishment of the safety program over a period of time will then be demonstrated by reducing risk scores and moving items downward on the chart, from the high-risk categories into lower-risk areas. For example, it would show progress if the number of "emergency action" items were reduced from seven to two, and "urgent" items from six to four; or if the overall average risk score is reduced from 140 to 115; etc. If it is desired to compare the actual safety status of each of a number of industrial plants, it could be done simply by comparing the average of the risk scores of all the principal hazardous situations at each plant. For example, the plant with an average risk score of 90 would be a safer place to work than one with an average risk score of 120.

26.5.3.2 The Justification Formula To determine whether proposed corrective action to alleviate a hazardous situation is justified, the estimated cost of the corrective measures is balanced or weighed against the degree of risk. This is done by integrating two additional factors into the risk score formula. The justification formula is as follows:

$$\text{justification} = \frac{\text{consequences} \times \text{exposure} \times \text{probability}}{\text{cost factor} \times \text{degree of correction}}$$

Note that the numerator of this fraction is actually the risk source. A denominator has been added, made up of two additional elements: cost factor and degree of correction.

The cost factor is a measure of the estimated dollar cost of the proposed corrective action. Ratings are as follows:

Cost	Rating
Over $50,000	10
$25,000–50,000	6
$10,000–25,000	4
$1,000–10,000	3
$100–1,000	2
$25–100	1
Under $25	0.5

The degree of correction is an estimate of the degree to which the proposed corrective action will eliminate or alleviate the hazard. Its ratings are as follows:

Description	Rating
Hazard positively eliminated, 100%	1
Hazard reduced at least 75%, but not completely	2
Hazard reduced by 50–75%	3
Hazard reduced by 25–50%	4
Slight effect on hazard (less than 25%)	6

To use the formula and make a determination as to whether a proposed expenditure is justified, values are substituted and a numerical value for justification is computed. The critical justification rating has been arbitrarily set at 10. For any rating over 10, the expenditure will be considered justified. For a score less than 10, the cost of the contemplated corrective action is not justified.

To demonstrate the use of the justification formula, the same examples used in the demonstration of the risk score will be used. Consider the example of the hazard to persons near a building in which explosives are processed. The corrective action that was proposed was the construction of a barricade along the outside of the building to protect passersby in the event of an explosion within. The estimated cost was $5,000. Using the "J" formula:

1. The consequences, exposure, and probability as already discussed were evaluated at 25, 1, and 1, respectively.

2. Cost factor. The estimated cost is $5,000. Therefore, based on the rating chart, the cost factor = 3.

3. Degree of correction. The effectiveness of the barricade to protect passersby is considered to be over 75%. Therefore, the degree of correction = 2.

4. Computation:

$$\frac{25 \times 1 \times 1 = 25}{3 \times 2 = 6} = 4.20$$

5. Conclusion. The expenditure of $5,000 to construct a barricade to protect passersby is well below 10, and therefore is not justified.

6. Further consideration. Since the risk score is 25, this situation still requires attention.

Review of this problem revealed that other steps could be taken to lower the risk. The probability of the complete accident sequence occurring was considered to be remote, but it could be made much more remote (and the risk score halved) by administrative controls such as portable barriers and warning signs, to reduce or even eliminate the presence of passersby in the danger zone.

The preceding examples demonstrate how the formula can save money. In addition, it can be a highly valuable management tool. For example, immediately after a very serious accident occurs, let us say an explosion with a fatality, there is usually a tendency for managers as well as safety directors to over-react, to go to extremes in favor of safety. Judgment may become somewhat biased in favor of excessive safety measures. Such action actually hurts management's image in the long run, because when situations cool down and people again become rational and reasonable, the poor judgment of such projects is apparent. Therefore, under excitable circumstances when costly projects are being considered and may be too hastily approved, the justification formula can show whether or not the measures are justified, logically and simply. This formula is a simple and positive management tool to help management make a proper decision.

The hazard evaluation systems presented herein can be used effectively by anyone who has sound judgment and experience in safety, with nominal training and guidance.

For convenience, all the factors used in this hazard evaluation system are given as a single chart (see Fig. 26.14).

26.5.4 A Graphical Calculation of Risk Score and Effectiveness

An even easier method of the mathematical risk score model was developed in nomograph form (Figs. 26.15 and 26.16). This allows quick assessment and decision making as to the worthiness of a particular project or repair. The steps in using this nomographs are:

1. Determine the likelihood of an injury and mark it on the chart.
2. Determine the exposure rating and mark it on the chart.
3. Connect the marks and intersect with the tie-line.
4. Assign the possible consequences of the injury and mark it on the chart.
5. Connect the tie-line mark through the possible consequences and intersect with the risk score.
6. Transfer the risk score onto the second chart.
7. Estimate the risk reduction and mark it.
8. Intersect the tie-line.
9. Estimate the costs for correction and mark it.
10. Connect the tie-line mark through the costs for correction and intersect with cost effectiveness.

The ease of this technique allows safety committee members, plant operations, and other concerned parties to complete their own analysis and collectively, determine the most effective solution.

Factor	Classification	Rating
1. **Consequences**	a. Catastrophe; numerous fatalities; damage over $1,000,000; major disruption of activities	100
	b. Multiple fatalities; damage $500,000 to $1,000,000	50
Most probable result of the potential accident.	c. Fatality, damage $100,000 to $500,000	25
	d. Extremely serious injury (amputation, permanent disability); damage $1000 to $100,000	15
	e. Disabling injury; damage up to $1000	5
	f. Minor cuts, bruises, bumps; minor damage	1
2. **Exposure**	*Hazard-event occurs:*	
	a. Continuously (or many times daily)	10
	b. Frequently (approximately once daily)	6
The frequency of occurrence of the hazard event.	c. Occasionally (from once per week to once per month)	3
	d. Usually (from once per month to once per year)	2
	e. Rarely (it has been known to occur)	1
	f. Remotely possible (not known to have occurred)	0.5
3. **Probability**	*Complete accident sequence:*	
	a. Is the *most likely* and expected result if the hazard-event takes place	10
Likelihood that accident sequence will follow to completion.	b. Is *quite possible*, not unusual, has an even 50/50 chance	6
	c. Would be an *unusual* sequence or coincidence	3
	d. Would be a *remotely possible* coincidence	1
	e. *Has never happened* after many years of exposure, but is conceivably possible	0.5
	f. *Practically impossible* sequence (has never happened)	0.1
4. **Cost Factor**	a. Over $50,000	10
	b. $25,000 to $50,000	6
Estimated dollar cost of proposed corrective aciton.	c. $10,000 to $25,000	4
	d. $1,000 to $10,000	3
	e. $100 to $1,000	2
	f. $25 to $100	1
	g. Under $25	0.5
5. **Degree of Correction**	a. Hazard positively eliminated, 100%	1
	b. Hazard reduced at least 75%	2
Degree to which hazard will be reduced.	c. Hazard reduced 50% to 75%	3
	d. Hazard reduced by 25% to 50%	4
	e. Slight effect on hazard (less than 25%)	6

Fig. 26.14 Justification formula rating summary sheet.

26.6 QUICK HELP

26.6.1 The Americans with Disabilities Act and Workers' Compensation

The purpose of the Americans with Disabilities Act (ADA) is to remove barriers, artificial and real, to independent living for persons with disabilities. The ADA focuses on what an individual can do as opposed to what he cannot do. The ADA prohibits discrimination based on physical or mental disabilities in:

- Public and private places of employment
- Public accommodations

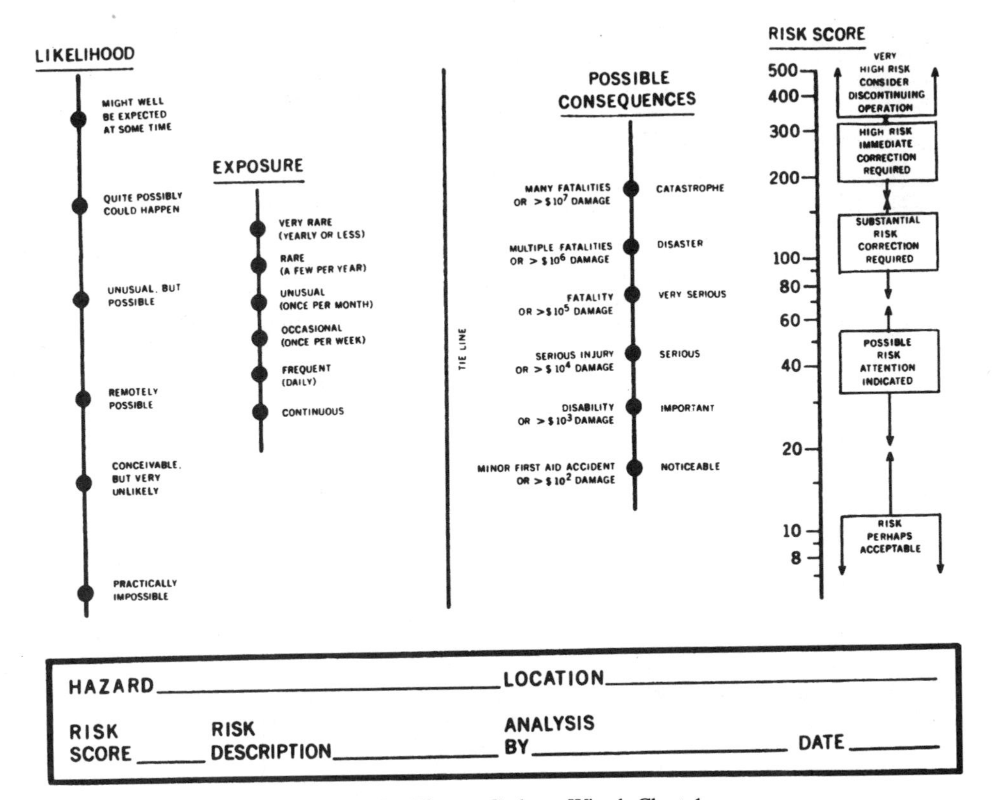

Fig. 26.15 Kinney–Graham–Wiruth Chart 1.

- Transportation systems
- Communication systems

The ADA prohibits disqualification of disabled applicants or employees because of their inability to perform marginal or nonessential job functions. The ADA also requires that employers make reasonable accommodations to help disabled applicants or employees perform the essential functions of a given job.

Who is covered? To be covered by the ADA, an individual must have a disability that amounts to a physical or mental impairment that substantially limits one or more major life activity, or a record of such an impairment or be regarded as having such an impairment. To be protected by the ADA, an individual must meet the definition of disabled as well as have the requisite skill, experience, education, and other job-related requirements of the position, and who, with or without reasonable accommodation, can perform the essential functions of the position.

When does coverage begin? Employers of more than 25 people were required to follow the ADA and its regulations beginning July 26, 1992. For employers of 15 or more, the implementation date was July 26, 1994.

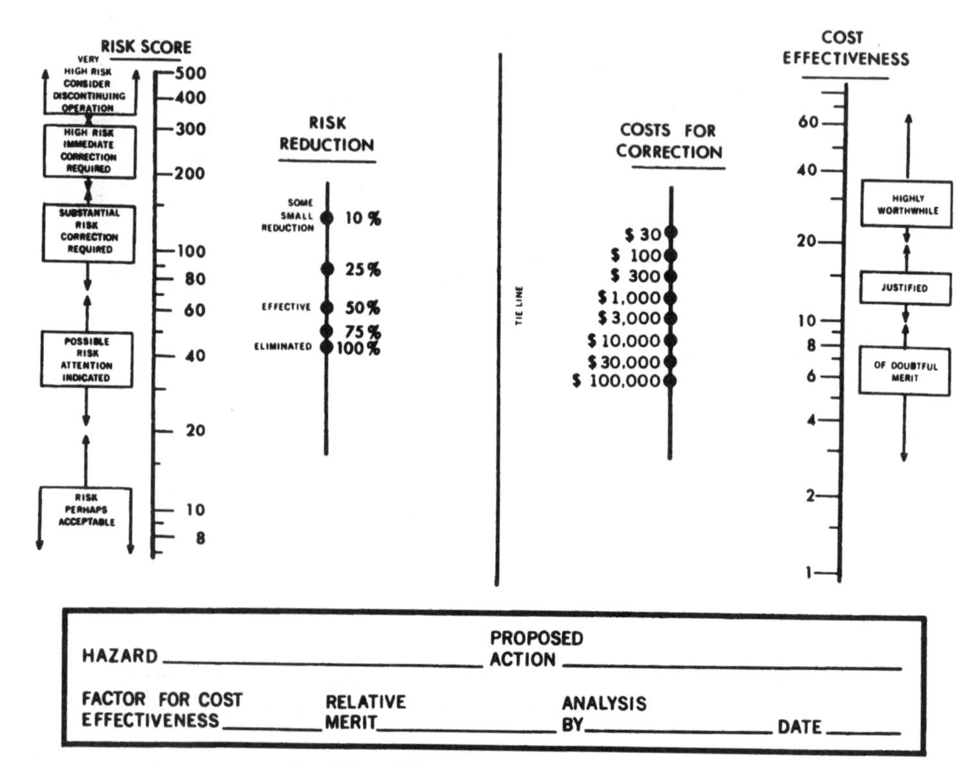

Fig. 26.16 Kinney–Graham–Wiruth Chart 2.

ADA accommodation and transitional duty. Transitional duty is an essential tool to enhance productivity and speed the recovery of injured workers. It is a tool that is fair to both the employer and the employee. It is important to keep in mind the key differences between short-term accommodation under early retrun to work and long term under the ADA. Here is a quick study of the differences.

Under the ADA	Under Early Return to Work
The disability is permanent.	The disability is temporary.
The job changes are intended to be permanent.	The job changes are intended to be temporary.
The need to accommodate is driven by law.	The employer seeks to speed recovery and reduce costs by making a temporary change in the work assignment.
The accommodations must enable the individual to perform the essential functions of the job.	The accommodations need not relate to the original job; any function may be assigned.

26.6.2 What Do I Do When ...

1. What do I do when an employee is disrupting the early return to work program; he/she is boasting about the lighter duties?

First, explain to this individual your concerns about their attitude toward the job and the effect it is having on co-workers. If this does not correct the situation, give the person a less attractive job. If all else fails, you may find it necessary to remove the employee from the work site all together. Remember, if you do this you are making a decision to pay the individual's workers' compensation.

2. What do I do with a supervisor/manager who refuses to take someone back on early return to work?

Supervisors/managers are responsible for their employees when they are healthy and also when they have suffered a work-related injury. You may have to discuss the roles and responsibilities that supervisors/managers have in your organization with respect to safety and injury management. The individual's accountability to the pertinent budget may also be an issue of discussion. If necessary, you may find yourself having to discuss the situation with the supervisor/manager's superior.

3. What do I do if the employee complains of physical inability to do the job?

Remember to respond in a caring manner. The employee should be interviewed to find out exactly what physical problems he/she is having with the individual tasks of the current job. You may find that the employee is performing more duties or heavier duties than the restrictions allow. If this does not correct the problem, have the employee medically re-evaluated for a new set of restrictions. After medical reevaluation, assign the employee a new job that fits the new restrictions. You should discuss these changes with the supervisor/manager. The employee should also be observed frequently by you and the supervisor/manager.

4. What do I do when I have too many employees on transitional duty?

If the transitional-duty placements are concentrated in a specific department, look for interdepartmental job placements. As a last resort, you may have to make a human and financial cost decision to put employees on workers' compensation.

5. What do I do when there is no transitional duty in a particular department?

Help the supervisor/manager search the department to see if there is anything that the employee can do. If nothing can be found in the employee's department, look interdepartmentally for proper job placement. In addition, you may have some extra work that must be completed.

6. What do I do if someone refuses an initial offer of transitional duty?

If the employee has been cleared for work with specific restrictions, you have the job that is in line with the restrictions, and the employee refuses the job, try to find out why the employee rejected the initial offer. Based on this information, reoffer the job by certified mail, requesting an answer within 5 working days. If you do not hear from the employee or if he/she refuses the second job offer, contact the insurer and verify procedures under the state law.

7. What do I do if someone refuses to change jobs because of an upgrade in physical restrictions?

Explain that the transitional-duty program is progressive and is constantly changed and customized to meet the physical restrictions cited through medical re-evaluation. In addition, transitional duty is a work-hardening procedure. Job upgrades help the physical recuperation.

8. What do I do if someone is not fully recuperated at the end of a specified time frame for transitional duty?

The employee should be medically re-evaluated to determine the specific restrictions and the duration the employee must follow their restrictions. Extensions of transitional duty should only be approved based upon medical criteria. The employee should then be re-evaluated on a weekly basis.

9. What do I do if someone is taking advantage of a transitional-duty job (i.e., rooting in)?

The employee should be medically re-evaluated for new restrictions. Whether the restrictions change or not, you should place the employee in a new job. This takes the employee out of the "too comfortable" job while still adhering to the restrictions cited in the medical re-evaluation.

10. What do I do if a supervisor or manager will not adhere to the physician's restrictions?

Explain the dangers of their actions. The employee could be reinjured or complicate the current injury. This will also jeopardize your relationship with the physician and with the employee. There could be financial liabilities. If this continues, the supervisor/manager should be disciplined. Never leave the employee in a department with pressure to perform more than his/her capabilities.

11. What do I do if an employee wishes to perform more than the restrictions allow?

You should meet with the employee immediately to discuss the problems associated with their overzealousness. Let him/her know that he/she is a valued employee that you cannot afford to lose and that no one will look down on him/her because of the restrictions. In addition, explain that the physician has cited specific restrictions that must be followed to prevent injury, and you will be more than happy to discuss the concerns about their restrictions with the physician. You should also discuss this with the supervisor/manager and observe the employee on a regular basis.

26.6.3 Glossary of Terms

The following is a brief summary of key definitions and terms that may be encountered in the workers' compensation arena.

Accident–An unplanned event that is invariably preceded by an unsafe act or unsafe condition or a combination of both.

Average Weekly Wage–The base amount used to calculate an injured employee's compensation rate of pay. It averages all wages, including overtime shift differential for that particular employee.

Benefits–An injured employee's reimbursable medical expenses and indemnity (in lieu of wages) payments.

Carelessness–Not paying sufficient attention to avoid an accident. Not thinking of the safety aspects of a behavior before acting. "Carelessness" should not be given as the cause of an accident.

Carpal Tunnel–Carpal means wrist; the carpal tunnel is a narrow, bony cavity, pathway, or tunnel in the center of the wrist, which contains major nerves and tendons that extend to the hand.

Carpal Tunnel Syndrome–A cumulative trauma or repetitive motion injury that can be caused by inflamed tendons, which compress the nerve.

Compensation Rate–The weekly amount paid to a workers' compensation claimant.

Compensable Injury–A work-related injury that prevents an employee from returning to work within the waiting period.

Cumulative Trauma/Repetitive Motion Injury–Excessive strain caused by repetition in task.

Direct Cost–The sum of compensation payments and medical expenses for an injury.

Disability–A legal term meaning the inability to perform normal occupational duties at full capacity. It may be partial or total, temporary or permanent. It may be deemed permanent after it has continued for a specified period.

Ergonomics–The science of utilizing adaptive engineering to modify the workplace to reduce stress and fatigue.

Experience Modification–An employer's specific loss history over a 3-year period, with other employers performing similar work as the point of comparison.

Frequency Rate–The number of lost-time injuries requiring more than first aid per 100 full-time workers per year. This formula is used by OSHA to compare injury rates within and between industries.

Hazard–Any dangerous condition that can cause an interruption or interference with the expected orderly progress of an activity.

Impairment–A medical term referring to the inability or lessened ability to perform specified functions.

Independent Medical Exam (IME)–A third-party medical opinion given in order to resolve an insurance claim.

Injury–Physical harm or damage to a person.

Job Analysis/Functional Job Requirements–The breaking down of any job into several components to determine the requirements necessary to perform the job. This analysis is needed from the manager when medical restrictions are specified by a physician.

Loss–A payment on behalf of an injured employee, or money reserved by the insurance carrier for future payments. "Incurred losses" are the combinations of paid and reserved losses.

Loss Control–A program of injury response and safety-related activities including injury prevention and case management.

Lost-Time Accident–A work injury that results in disability and prohibits the employee from reporting to the next scheduled shift.

Occupational Illness–An abnormal physical condition caused by exposure to environmental factors associated with employment.

OSHA–Occupational Safety and Health Administration. A federal regulatory agency charged with upholding and enforcing safety and health codes.

Partial Disability–An inability to perform a part of one's duties due to occupational injury.

Permanent Disability–Disability that is expected to continue for the lifetime of the injured person. It may be deemed permanent after it has continued for a specific period.

Preferred Provider Network (PPN)–An association of medical providers who have agreed to follow certain practices and procedures to effectively provide treatment of work-related injuries.

Reserves–The insurer's estimate of the ultimate cost of a claim.

Risk–Chance of injury that can be reduced through loss prevention programs.

Risk Factors–Job task demands that increase an employee's vulnerability to injury loss.

Safety–In the context of Workers' Compensation, the ability to perform a job in a risk-free manner and in a risk-free environment.

Sports Medicine–An aggressive, proactive approach to recovery that emphasizes staying active, as opposed to prolonged periods of unnecessary and debilitating bed rest.

Transitional Duty–Modification of a (particular) job for a prescribed period of time for a particular employee. Also called modified work or light duty.

Unit Statistical Report–The actual record of an employer's losses (paid to date and incurred) used to calculate the experience modification. Submitted about 6 months after the end of the policy year.

Waiting Period–In Workers' Compensation, the number of days after the occurrence of an accident for which the injured employee does not receive indemnity (lost wage) compensation.

Workers' Compensation–A "no-fault" system of insurance to protect injured workers.

26.7 MAINTAINING A POSITIVE COMMITMENT FOR THE LONG TERM

Taking responsibility for controlling Workers' Compensation is challenging. Yet, for most employers, it does not require any additional staff. It does require senior management commitment to the objectives and responsibilities ties of sound injury management practices over the long term. By redirection of the negative energy toward outsiders and by reorganization of internal practices, significant progress can be made. Further, support and enthusiasm from the employees are a likely, and welcome, side benefit.

By weaving these strategies systematically into the culture of an organization, the costs of its Workers' Compensation program will drop typically by as much as $100,000 a year for every 250 workers.

ACKNOWLEDGMENTS

The author is indebted to the many workers' compensation projects undertaken by Lynch, Ryan & Paugh, Inc. These projects helped form the basis for the key strategies offered in this chapter. In particular, special thanks to Thomas C. Lynch, and Jonathan Coppelman, for their contributions in this chapter.

BIBLIOGRAPHY

Accident Facts (1995). 1995 Edition, National Safety Council, Itasca, IL.

Analysis of Workers' Compensation Laws (1995). U.S. Chamber of Commerce, Washington, D.C.

Best's Rating Report (1996). A. M. Best Company, Oldwick, NJ.

Development of a Rehabilitation Strategy for Fortis Benefits Insurance Company's Group Long Term Disability Claimants (1994). Fortis Benefits Insurance Co., Kansas City, MO, March.

Employee Benefits, 1997 Edition 50th Anniversary, U.S. Chamber of Commerce, Washington, D.C.

Fine, W. (1973). *Selected Readings in Safety*, Academy Press, Macon.

Hester, E. J., Decelles, P. G. (1987). *Decision Making in Referral for Rehabilitation Services*, The Menninger Foundation, Kansas.

Hester, E. J., Decelles, P. G., and Gaddis, E. L. (1989). *Analysis of Alberta Workers' Compensation Claimants with the Menninger Return to Work Scale*, The Menninger Foundation, Kansas.

Hester, E. J., Decelles, P. G., and Gaddis, E. L. (1986). *Predicting Which Disabled Employees Will Return to Work: The Menninger RTW Scale*, The Menninger Foundation, Kansas.

Kinney, G. F. and Wiruth, A. D. (1976). *Practical Risk Analysis for Safety Management*, Naval Weapons Center, China Lake, CA, June 1976.

Partnership for Best Practices (1994). Lynch, Ryan & Associates, Inc., Westborough, MA.

Petersen, Dan (1980). *Analyzing Safety Performance*, Garland Publishing, New York, NY 10016.

Russo, Carmen. *Issues in Workers' Compensation: A Collection of White Papers*, Attorney General's Task Force to Reduce Waste, Fraud and Abuse in the Workers' Compensation System, Commonwealth of Massachusetts, July, 1994.

State Workers' Compensation Laws (1996). U.S. Department of Labor, Employment Standards Administration, Office of Workers' Compensation Programs, Washington, D.C., January.

The Employer's Guide to the American with Disabilities Act (ADA) (1994). Lynch, Ryan & Associates, Inc., Westborough, MA.

The John Liner Letter **25,** Standard Publishing Corp., Boston, 1987.

The Vital Few (1971). Employers Insurance of Wausau, Wausau, Wisconsin.

Williams, C. A. and Heins, R. M. (1981). *Risk Management and Insurance*, McGraw–Hill, New York.

CHAPTER 27

How to Evaluate Your Occupational Safety and Health Program

ROSEMARY COLVIN and RAY COLVIN

Safety Training Dynamics, Inc. 27497 AL Hwy 91, Hanceville, AL 35077

27.1 AN INTRODUCTION TO EVALUATION

27.2 ENCOURAGING AND SELLING MANAGEMENT ON THE NEED TO EVALUATE PROGRAMS

27.3 UNDERSTANDING THE DIFFERENT TYPES OF EVALUATION METHODS
- 27.3.1 Overview of Inspection Procedures
- 27.3.2 Overview of Audit Procedures

27.4 AN OVERVIEW OF HOW TO DEVELOP AND ORGANIZE A COMPANY SAFETY AND HEALTH EVALUATION PROGRAM

27.5 MANAGEMENT'S ROLES AND RESPONSIBILITIES IN EVALUATING

27.6 HOW MANAGEMENT CAN KEEP KNOWLEDGEABLE OF ONGOING SAFETY AND HEALTH ACTIVITIES TO PREPARE FOR EVALUATING PROGRAMS

27.7 THE IMPORTANCE OF THE EVALUATION PROCESS

27.8 THE POTENTIAL LEGAL DANGERS OF EVALUATING SAFETY AND HEALTH PROGRAMS

27.9 THE OSHA PROGRAM EVALUATION PROFILE: AN EXAMPLE FOR ALL ORGANIZATIONS

27.10 PRE-EVALUATION PREPARATION—SCORING GUIDELINES FOR AUDITORS

27.11 AN EXAMPLE OF RECORDKEEPING AND DOCUMENTATION THAT WOULD BE REQUIRED DURING A COMPANY PEP-TYPE AUDIT

27.12 OSHA PEP EVALUATION FORMAT

27.13 AN EXAMPLE OF EVALUATING SAFETY AND HEALTH ACTIVITIES

27.14 COMPILING THE FINAL PEP EVALUATION REPORT

27.15 WHEN THE AUDIT EVALUATION IS COMPLETED

27.16 HOW EVALUATIONS IMPROVE THE SAFETY AND HEALTH PROGRAM

REFERENCES

APPENDIX A OVERVIEW OF OSHA PROGRAM EVALUATION PROFILE

Handbook of Occupational Safety and Health, Second Edition, Edited by Louis J. DiBerardinis, ISBN 0-471-16017-2

27.1 AN INTRODUCTION TO EVALUATION

There are many reasons management should want to evaluate their safety and health programs. The first and foremost is "because it is good business to evaluate the important business activities in an organization." If some activity or program is beneficial and has value to the company, then that activity or program should be measured for its effectiveness. If a program or a business activity becomes ineffective, it should be corrected, eliminated, or redesigned.

If an organization's senior management understood the potential impact accidents can have upon the company, it would welcome a formal review (audit) of the safety and health programs so it could be made factually aware of the potential losses.

A major question that many company managers face is "how far do we go, and how much time and money do we spend to control injuries, illnesses, and accidents to the facility & process?" It would be wise if management would start with a factual, formal assessment of what is currently being done and measure that against what needs to be done to comply with industry practices and government and legal regulations. Then an action plan should be developed to accomplish it within a "reasonable" amount of time and expense.

A formal evaluation that factually measures what a company is actually doing to prevent employee injuries and illnesses, and possible damaging accidents to the facility (fire, explosion, etc.) and production interruption, offers senior management a powerful business tool to help them evaluate potential future losses and expenses.

27.2 ENCOURAGING AND SELLING MANAGEMENT ON THE NEED TO EVALUATE PROGRAMS

Selling management on the need to perform a formal evaluation of the safety and health programs is a major first step in gaining support for actually conducting an evaluation. Management must be motivated to want to have a formal review of the safety and health programs.

To help in this process, the following are some descriptions of potential results accidents could have on the financial impact on an organization's profits:

1. The total time lost by an injured employee beyond the actual time at home away from the job, for example, after he returns to work while still being treated, examinations, filling out questionnaires, talking about injury with supervisor or accident investigation team, short-term light duty.

2. Long-term physical restrictions (light duty) on employee's work activities from causes such as carpal tunnel or repetitive work motion problems, back injuries, strains, etc.

3. Lost and/or reduced production until the injured employee *fully* returns to work.

4. Reduced production until a replacement employee comes up to speed with the operations, if employee's duties are being done by someone else.

5. Overtime for other employees who must perform injured employee's duties, if a replacement cannot be obtained.

6. Increased insurance costs. Insurance companies add a surcharge onto the premium for the money they pay out over and above the basic insurance contract.

7. OSHA fines as a result of violations involved in the employee accident. [There is a maximum of $70,000 for each violation—egregious (flagrant violations).]

8. Local or state government fines for violations of their codes and requirements.

9. Decreased production after a serious accident when workers may become fearful for their safety.

10. Damage to machinery, equipment, product, facility during the accident (forklift truck, materials handling, process, etc.).

11. Production interruption. Accidents may cause production to stop or decrease. (Sometimes machines cannot run until government officials investigate accident, or health officials approve use of machine, i.e., blood on food processing machine.)

12. Delayed filling of production orders or contractual deadlines to complete work.

13. The cleanup of blood in machine or about the general area (requires specially trained persons and equipment according to OSHA laws).

14. Negative publicity by media can cause damage to company's reputation. (Good employees do not want to work in a company where employees are injured. Larger companies are afraid to do business with suppliers who might not be able to fill orders due to accidents. It can also produce negative neighborhood and community attitudes against the company. (Since the serious accident to the employee made the TV and/or newspapers, community leaders might ask: What about your company's environmental hazards?

15. Major companies that purchase materials, supplies, and/or services for other smaller companies are now requiring the companies they do business with to have an effective safety and health program. (Many OSHA laws require the employer and the contractor to share their programs and to ensure their employees are properly trained to recognize and avoid the hazards they may produce. Therefore, the larger companies are requiring written programs and formal training by the contractors with whom they will do business.) Smaller companies having formal written safety and health evaluations of their programs can use them as a method to obtain contracts from larger organizations.

Senior management wants to know how to control the elements that allow accidents to occur that would impact their profits. A formal review (audit) process could identify the areas and activities that need to be corrected, eliminated, or changed and that would improve and establish a more effective safety and health program.

The foregoing observations may appear to exaggerate the significance of the costs of a poor safety record. Very few companies, only those with outstanding safety and health programs, capture the costs described. NIOSH[1] studies show that companies with outstanding safety and health programs also have excellent profits. One company's public image may not be as critical as another company's. Nevertheless, a poor safety record does affect the company in the marketplace. While appropriate safety programming generates a positive return on investment, it also attacks the indirect costs with positive results.

27.3 UNDERSTANDING THE DIFFERENT TYPES OF EVALUATION METHODS

There is a major difference between the methodology and reasoning between an inspection and an audit.

27.3.1 Overview of Inspection Procedures

When making an inspection, there is a need to identify the purpose of exactly what needs to be done. For example, if there is a concern for electrical safety and for the physical condition of the electrical equipment, then a plan on what to inspect for on the electrical equipment must be developed (i.e., ground fault circuit interrupters, the polarity check, physical condition of the outlet, cords, plugs, a resistance check, resistance/faults to ground tripping checks, etc.). Before actually starting the inspection, an inventory of all electrically related equipment should be compiled for every department and where all the equipment is located and who owns or is responsible for maintaining the equipment in the event it needs repairs and or replacement.

If there is a concern for fire safety, a plan should be developed on what would be inspected, for example, housekeeping, flammable liquids, or combustible materials, and where they are located in the plant or on the property. The heating and air conditioning systems, vents, kitchen or cooking areas, etc., should also be inspected. The fire protection system could be a separate inspection procedure, due to the complexity of the system. If chemical safety concerns arise, and a chemical safety inspection becomes necessary, specific areas and/or processes should be identified and a plan developed on what specifically is going to be inspected for chemical safety before starting the inspection.

An inspection plan should identify what exactly is going to be inspected what will be looked for during the inspection, the location and who owns or is responsible for its condition, maintenance and/or repairs or replacement. It would also be advisable to meet with the owner of the equipment when developing an inspection plan.

For more details on inspections and risk assessments, see Chapters 4 and 5.

27.3.2 Overview of Audit Procedures

An audit is entirely different from a physical inspection. Its purpose is to measure management's role and its success and failure in controlling safety and health activities in the organization. A combination physical inspection and an administrative procedure identify the "root causes" of safety and health problems and also the failure in the management systems that allow accidents due to unsafe conditions and employee human error to occur in the workplace.

The process in doing an inspection and an audit is entirely different. The audit is more of an administrative process to identify broader managerial issues, whereas the inspection is more of a physical inspection process to find unsafe conditions. After a physical inspection, especially if a number of unsafe conditions are identified, management might ask, what has failed in our management system that allows these unsafe conditions to exist? Isn't our preventive maintenance safety program adequate? Now, management should want to perform an audit to identify the weaknesses in their management system that are responsible for the unsafe conditions found during the inspection.

27.4 AN OVERVIEW OF HOW TO DEVELOP AND ORGANIZE A COMPANY SAFETY AND HEALTH EVALUATION PROGRAM

If a company is going to go through the process of conducting an in-house audit, it should get some value out of it and use it as an opportunity to effect positive change in the safety and health program.

1. Start with getting top management's support and involvement in developing a quantitative review of the present program and its approval to conduct the self-audit. Make top management aware that as part of its approval, it will have to follow up and address all issues that might be uncovered during the audit, such as a documentation of lack of support for the safety and health program by midmanagement.

2. After getting its approval, make a presentation to the senior staff of the entire process that will be used over the next 4–5 months, spelling out what support and involvement will be needed from them. (Ideally, they will become part of the audit team, which would go out onto the floor to obtain the data needed, or at least give support in their areas of responsibility.)

3. Emphasize that the audit is based upon a generic evaluation process that *could be* used by OSHA after a serious incident as part of their "wall-to-wall" investigation process. *Note:* You might want to work with corporate legal council to propose the audit to top management and have it involved in the process. You probably would get better support for the audit and the followup measures.

4. Working through company safety committees, organize a "game plan" to implement the audit process for your organization.

5. When organizing an inspection team, management should allow selected employees, supervisors, and managers to participate in the evaluation. Other special employees should be drawn from the appropriate disciplines needed to intelligently inspect the areas or equipment, such as lab personnel to inspect labs, fork truck "experts" when inspecting fork trucks, safety, engineering (facilities, industrial, process), manufacturing people, and so on. There should be a mixed group of expertise, as everyone can contribute to a good inspection. Nontechnical experts might ask "dumb" questions, but will open the door to detailed discussions about safety. Sometimes an outsider can spot basic safety problems where the "experts" do not see them.

27.5 MANAGEMENT'S ROLES AND RESPONSIBILITIES IN EVALUATING

In order for management to establish credibility for the total safety effort in the company, the importance of their participation in conducting safety audits and inspections is a major responsibility. Typically most companies conduct an occasional physical inspection of the operations by management, but they are not the most effective way of identifying the "root causes" of potential accidents and/or hazards. Frequent, extensive physical inspections must be conducted by each department or operating unit to identify and correct unsafe equipment, conditions, processes, and work practices. This is a necessary activity of every good safety and health program. It is management's role to evaluate the results of these inspections to determine the effectiveness of the inspections and the people conducting them.

The supervisor/manager who initiates the inspections is responsible for ensuring that essential safety devices and protective equipment are provided and used on each job requiring them. In addition, it is the supervisor's/manager's responsibility to make sure that each employee understands that willful violations of established safety rules will be subject to discipline. It is also important to remember that the supervisor/manager is accountable for the safety of the machinery, process, people, and all related equipment in the areas directly under his/her control, and is answerable to senior management for any deficiencies. Their safety responsibilities should carry equal value with all their other performance responsibilities and be reflected in their annual performance evaluation for pay raises and/or rewards. It is important that management provide adequate training for the supervisors and managers to perform their safety, health, and environmental responsibilities before they are held accountable in their performance evaluations. Once trained, every supervisor should have a department safety manual that documents and records all their safety activities, such as employee training, accident reports and investigations, hazards and job safety analysis, MSDSs, inspections findings, meetings, and safety-related prolems.

Periodically, a review of the supervisor's safety manual should be made by the supervisor's manager or the safety person to evaluate the supervisor's safety, health, and environmental activities and progress.

Company management can show their visible support for the safety and health program by accompanying department inspection teams on their periodic safety inspections. Management participation in safety and health inspections has high visibility for employees, as they are seen as "caring" about the safety and health of employees.

More important, evaluations serve as an effective tool for identifying safety rule violations, hazardous conditions, and failures in the management system.

Safety evaluation findings should be reviewed and noted by senior management and the information distributed to the areas and departments inspected. Since positive feedback is often more effective than negative feedback, it would be wise to provide praise to those people and work areas in compliance with the safety and health rules and company procedures.

Observations made during safety evaluations must be followed up within an appropriate time frame and in some visible way. This practice is one characteristic of record-holding plants, as noted in a NIOSH[1] government study of excellent company safety programs.

"In four plants (of the five studied), management and workers were cooperatively involved in plant inspection and hazard identification programs. In all cases, hazard identification was quickly followed up by hazard correction and reinspection. Line management was typically responsible for hazard correction within a period of time designated by the inspection team."[1]

27.6 HOW MANAGEMENT CAN KEEP KNOWLEDGEABLE OF ONGOING SAFETY AND HEALTH ACTIVITIES TO PREPARE FOR EVALUATING PROGRAMS

Safety programs cannot be successful without a high level of "safety awareness" being maintained throughout the organization. Positive safety and health communications must be constant in order to be effective. Management cannot simply give its approval for the distribution of information. They must actively review data and provide feedback. This means that the organization's safety and health record and all its activities should be integrated into the company's regular reviews of business operations as a continuing agenda item. (Other forums, such as routine safety meetings, are more appropriate for analyzing the incidence of injury in terms of implications and recommendations.)

Routine reviews of the organization's accident record underscores management's special accountability for the welfare of those whom it leads. While employees themselves must be responsible for following established safety rules and procedures and for mature behavior guided by simple common sense, management should monitor employee's safety and health actions, through reviewing incidents that adversely affect the company and/or employees. In this way, operational decisions may be made to prevent accidents and injuries before they occur.

Senior management will tell you they are "concerned" with the safety of the employees, their product, and the environment. But ... is safety really a priority issue of management?

Does management view safety as equally important as production? Can senior management quickly tell you the cost of safety for their organization per hour worked? There is no question that it would surely know the cost of making the product.

Many company executives of successful safety, health, and environmental programs know exactly the number of accidents and the cost of controlling these accidents and how they are impacting their operations and profits. When senior management places safety alongside production and quality, accidents to the employees, products, production, and the environment tend to be *controlled.*

In order to obtain the necessary funds for implementing a safety program, you must construct a sound business proposal. Selling safety as "nice to do" or "taxing" budgets to defend against the force of government regulations will earn only minimum funding and token acceptance from management. Gather your data, prepare your argument, and be open, honest, and realistic in your bid to obtain management's involvement and support.

Would large corporations with outstanding safety records spend time and money on safety activities if it were not profitable? You will find that many of the larger companies cost justify their programs, and that feasibility studies are mandatory in such companies before sizable amounts of money are spent for safety programs.

27.7 THE IMPORTANCE OF THE EVALUATION PROCESS

A significant principle of safety and health programs that is often overlooked is that the company management may delegate some of its responsibility for safety and health activities, but none of its *accountability* for safety. This means that, ultimately, management is legally responsible for the injured employee and to the federal regulatory agency, or even to the company board of directors. Since management cannot pass the blame, it must create some means of determining whether its commitment to safety, as established by its policy statement, is carried out by everyone else in the organization.

Conducting safety and health auditing by the company management is essentially a business review of the effectiveness of the overall safety and health effort, and also an assessment of the safety and health program payback, in terms of reduced employee injuries and facilities/process damages.

Corporatewide or division/site auditing can help improve the effectiveness of the overall safety and health program by developing peer pressure and fostering a sense of positive competition between department/divisions for excellence in safety and health performance.

Historically, in order to quantify how well a safety program is performing, management usually requires an annual report from the safety and health departments that details how many inspections, meeting, drills, and other activities were conducted during the year. Additionally, an accident/injury summary report is submitted, showing the trend for the year in accident/injury occurrence or frequency. This is a very poor indicator of the success of a safety program. Measuring accidents and injuries is measuring negatives, things that went wrong. Management does not measure productivity by counting defective parts. They evaluate progress toward quotas, goals, or objectives aimed at producing the least number of defective parts. (This does not mean that we should ignore good measurable feedback obtained by analyzing accident information and facts.) In the same way, we should measure progress toward accident prevention according to specific goals and objectives. The plan to meet these goals and objectives is the safety program.

A well-developed audit program forces us to define our objectives and then measure our efforts against them. Remember, all safety work is measurable, We can assign a numerical value to each safety activity and use this value to grade the success of that particular activity. A safety audit is not an evaluation of the safety person, but of management's effectiveness in making sure that safety activities are being carried out according to agreed-on plans. It is management's report card.

Safety audit objectives should be defined before the auditing process begins, so that measurable objectives can be written. The purpose of an audit may be to determine one or more of the following:

- Manager's safety and health participation and their activities
- Employee's knowledge of safety and health policies and programs
- Employee's actions as they relate to safety and health programs
- Supervisor's participation in safety activities
- Financial losses to the company
- Machinery and process safety equipment and safeguards
- Safety and health management and administrative procedures
- Housekeeping and physical conditions of the facilities and processes
- The actual performance against stated policies
- Strengths and weaknesses of the safety and health programs
- Safety attitudes of management and workforce

As there are different purposes for audits, so are there different auditing methods. These must be appropriate to the purpose. There may be an audit by a written questionnaire to employees, supervisors, or managers, physical observations by a team on site, or by reviewing written materials only, such as job safety analyses, accident investigation reports, and other reports. There may be division/department heads conducting self-audits. Each method has its own advantages and requirements. For example, developing a questionnaire takes one kind of administrative effort, and selecting and training team observers to conduct physical audits, takes another. Whichever method is chosen, make sure it will meet the goals and objectives established for the purpose of the audit. Know specifically what it is that is trying to be quantified and/or documented.

In our present-day legal and adverse business culture with OSHA fines reaching the tens of millions of dollars and public opinion being critical of irresponsible management, legally, company managers do not have a choice of whether they will or will not support the company safety and health programs and activities.

Company managers must formally become involved with the company safety and health programs and make safety and health an equal partner with the product or the service they supply or generate. They must develop a management line item budget for safety, health, and environmental activities.

When employers find themselves in a courtroom, and are asked by the prosecuting attorney how they participate in making the safety and health program work and if they have a budget to make the program work, the court, and possibly a jury, will decide on the sincerity of the employer if an approved budget is not part of the program.

27.8 THE POTENTIAL LEGAL DANGERS OF EVALUATING SAFETY AND HEALTH PROGRAMS

A formal audit will create a "legal paper trail" of any safety and health problems. Company management and especially the legal department or company lawyer should be supportive of conducting a company audit. If they have concerns for potential legal problems, the corporate legal counsel should formally sponsor the audit by outside safety and health consultants. The audit results may then become privileged information and not subject to outside legal investigators.

The question must be asked, why do you want to measure how well the safety and health programs and their activities are doing?

- Do you want the results to influence top management on the need to improve the programs?
- Do you want to legally measure how well the programs are doing to comply with industry practices and/or OSHA standards?
- Do you want to use the audit as a legal defense foundation, to challenge questionable employee, contractor, or public claims against the company?

If your company has an effective safety and health program and management is open and forthright about its day-to-day dealings, it will not be legally threatened by an audit, and will welcome the opportunity to document how well the programs are measured against general industry practices and standards.

27.9 THE OSHA PROGRAM EVALUATION PROFILE: AN EXAMPLE FOR ALL ORGANIZATIONS

OSHA has published auditing guidelines in the Program Evaluation Profile (PEP)[2] document based upon their *1989 Safety & Health Program Guidelines*[3] for employers to use to assess their safety and health programs. This suggested process quantitatively evaluates an employer's safety and health programs.

The following is an outline of the OSHA-recommended program for their compliance officers when assessing an employers safety and health programs. Its format can be used by companies that need to administratively measure the effectiveness of their safety and health programs against effective industry safety and health programs.

There are six major subjects to be scored in the PEP evaluation:

1. Management leadership and employee participation
2. Workplace analysis
3. Accident and record analysis
4. Hazard prevention and control

5. Emergency response
6. Safety and health training

Each of the subject is subdivided into smaller factors, which will be scored individually, and the composite of each subject's totals will be averaged as a total score for the evaluation.

The subdivisions of each subject are as follows, with relevant chapters in this text that provide more detail noted:

1. Management leadership and employee participation
   - Management leadership (Chapters 1, 2)
   - Employee participation
   - Implementation
   - Contractor safety

2. Workplace analysis
   - Survey and hazards analysis (Chapters 3, 5–9, 24)
   - Inspection (Chapter 8)
   - Reporting

3. Accident and record analysis
   - Investigation of accidents and near-miss incidents (Chapter 5)
   - Data analysis (Chapter 7)

4. Hazard prevention and control
   - Hazard control (Chapters 12–17, 19–23)
   - Maintenance
   - Medical program (Chapters 10 and 11)

5. Emergency response
   - Emergence preparedness
   - First aid

6. Safety and health training
   - Safety and health training (Chapters 13, 14)

27.10 PRE-EVALUATION PREPARATION—SCORING GUIDELINES FOR AUDITORS

OSHA takes the following into account in assessing specific subject factors (subject subdivisions):

1. **Written programs.** All programs should be in writing in order to be effectively implemented and communicated.

2. **Employee participation.** Employee involvement in an establishment's safety and health program is essential to the overall effectiveness of the programs.

3. **Comprehensives.** An effective safety and health program shall address all known and potential sources of workplace injuries and illnesses, whether or not they are covered by a specific OSHA standard, for example, lifting hazards and workplace violence problems if they pertain to specific conditions in the employer's workplace.

4. **Consistency with violations found.** The PEP evaluation and scores assigned to the individual subjects and factors should be consistent with the types and numbers of violations or hazards found during the inspection of the facility and/or the recordkeeping.

As a general rule, high scores will be consistent with low incidence rates of injuries, accidents, or near misses. High scores would be inconsistent with numerous or grave violations or a high injury/illness rate.

27.11 AN EXAMPLE OF RECORDKEEPING AND DOCUMENTATION THAT WOULD BE REQUIRED DURING A COMPANY PEP-TYPE AUDIT

In order to properly evaluate safety activities, extensive records must be maintained throughout the organization. Without this documentation, it would be difficult to determine exactly what is being done to control accidents. For example, every area of safety work must be recorded and then evaluated on how well the work was done (i.e., in training employees, we need to know who was trained, when they were trained, and did they successfully pass the training). Therefore, a simple form should be developed to capture this information (see Chapters 6, 14, 18)

Accidents and injuries must be recorded for local, state, and Federal government requirements, as well as for legal and insurance purposes. OSHA has set the basic standards for injury/accident reporting. Any self-designed report must at least contain the information on the OSHA reports. In addition to the OSHA report forms, an organization should have incident reports to record close calls and noninjury accidents.

Other recordkeeping should include safety meeting times, dates, attendees, and subjects discussed; safety inspection times, dates, findings and recommendations; cost of personal injury accidents; and/or near-miss property damage accidents.

Where special hazards exist and preplanning must be done to avoid accident or injury, this activity should be documented (i.e., confined space entry, lockoff/tagoff procedures, open flame permits, sprinkler shutdowns, hazardous work permits).

Another area of documentation is the medical history of employees, including pre- and postwork employment exams, special hazard evaluations, such as confined space entry, respirator use and special chemical exposures, and emergency team members. Also, first aid training and special medical emergency training for emergency team personnel require documentation. Additionally, records of emergency drills and training activities should be kept.

Recordkeeping in any organization is a massive task. But with today's information technology equipment, the challenge of maintaining records is within reach of any resourceful safety person. With a basic computer system, all recordkeeping can be captured and effectively used to set goals and make midcourse corrections of safety activities.

At present, many computer software packages are available through the National

Safety Council (Chicago, Illinois) and local safety councils and associations as well as other sources that can be used for accident investigation, hazards communications, OSHA recordkeeping, MSDS maintenance, etc.

The primary reason for maintaining records of safety, health, and environmental activities is for legal requirements, and the secondary reason is for management auditing of the programs and their effectiveness. It is very important that a competent person maintain all these records that become part of an organization's legal documentation of their compliance activities.

Safety work can be made more exacting and effective with better control and understanding of information. Senior management is influenced by detailed and exact numbers. Senior management will respond to safety requests supported by facts and figures.

27.12 OSHA PEP EVALUATION FORMAT

In 1996 OSHA developed their Program Evaluation Profile. It provides a concise picture of an entire safety and health program for all types of companies and organizations, including all the basic elements and activities involved in the programs. Additionally, the profile details the important activities of each of the elements expected of management, supervisors, and employees.

Companies can use this OSHA profile to develop their own customized evaluation format for their safety and health programs. It is suggested that when developing a customized evaluation format that it not be made too detailed, as it will become too long to be useful. Keep it simple!

OSHA's (PEP) assessment program of safety and health conditions in the workplace provides a clear understanding of the programs and the *management systems* that an employer is using for his/her safety and health compliance. The evaluation is designed to *identify weaknesses in the management system* that allow unsafe conditions and/or human errors to occur in the workplace.

Weaknesses and/or faults in the management system are the "root cause" of unsafe conditions and human error in the workplace. By identifying these "root causes," management could better control and/or eliminate unsafe conditions or human error that cause accidents and the resulting financial losses to the company.

The OSHA (PEP) evaluation is an excellent tool to measure a company's safety and health program's effectiveness. Appendix A is an overview of the OSHA Program Evaluation Profile used in assessing employer safety and health programs in general industry workplaces.

27.13 AN EXAMPLE OF EVALUATING SAFETY AND HEALTH ACTIVITIES

Before starting a training audit, an administrative review of what the employer has done to prepare for an effective training program should be conducted.

For example, employers should have identified their workplace "recognized hazards," and after they have completed them with a written hazards analysis (job safety analysis), they should have developed a training plan. This plan should be based upon the written hazards analysis, which identifies how the employee will avoid the hazards of the workplace. These then become the *training and/or learning objectives*, after which a *method of communicating the learning objectives* should be selected (i.e., lecture, video, workshop, hands on, interactive, brochures, booklets, etc., or a combination). Finally, a *method of evaluating the training* effectiveness must be developed (i.e., written test, skills evaluation, verbal test, workshop activity, etc.).

All training activities should be documented, for example:

1. The learning objectives, based upon the hazards analysis,
2. The methods used to communicate and obtain the learning objectives,
3. The evaluation and/or testing process to determine if objectives were met,
4. Standard recordkeeping, such as employee's name, date of training, etc.

Effective and above average safety programs include educating and training supervisors, managers, and senior officers about their safety and health roles and responsibilities, as well as the contents of the subjects being taught to the employees. *OSHA has stated this in its PEP outline.* Ideally, supervisors and managers should be directly involved in the employee education and training, by their active support and involvement in the training process.

Additionally, the employer has the responsibility to ensure that *all* employees and contractor's employees are adequately trained to recognize and avoid the hazards of the workplace.

27.14 COMPILING THE FINAL PEP EVALUATION REPORT

Some key evaluators of the OSHA PEP that weigh heavily on final audit results include:

- Management leadership and its direct involvement in the safety and health program is mandatory and more important that any other single activity.
- How management "motivates and leads" the employees to work safely.
- How well the program is documented.
- How employees "feel" about the safety and health program, for example, is it working? Does management "really" care about employee safety? Do employees feel management is doing all it should for safety? Do employees really know the hazards of their jobs and how to avoid them?

OSHA suggests specific scoring guidelines. All employer safety and health programs *should be in writing in order to be effective.* (If an employer does not have written programs, this should raise a "red" flag according to OSHA). But it should not be the only evaluation criteria of an effective program, especially for smaller companies. Compliance officers are referred to the OSHA Compliance Officer's Guidelines, "Citation Policy for Paperwork and Written Program Requirement Violation." The principle of "performance counts more that paperwork" should be considered when evaluating management's actual effectiveness in conducting the safety and health program, not just the documentation and paperwork. *The actual employee knowledge and management's participation is the bottom line of an effective safety and health program. The employer's final grade or score that is obtained on the PEP audit is only secondary to the value achieved by having management and employees go through the PEP audit process:*

- The value of having management and employees formally "working" through the internal audit process together.
- The value of formally reviewing the basic elements of a safety and health program as outlined in the PEP.
- The value of getting management involved in the audit process.
- The value of bringing an awareness to management, supervisors, and employees of the totality and complexity of a formal safety and health program, and their roles and responsibilities in the success of the company safety and health program.

27.15 WHEN THE AUDIT EVALUATION IS COMPLETED

Share the results with top management and your legal department. Remember the process of doing an internal audit creates a paper trail of your safety program's weak points. This could be used against your company, and possibly against yourself if unsafe conditions, processes, or activities are not corrected within a reasonable time and a serious accident occurs. Reminder: To avoid creating a potential legal paper trail of the audit, corporate legal counsel could request an audit, which may then become "privileged information" and not be subject to investigators. Many times corporate attorneys hire outside auditors to conduct a company safety and health program evaluation. Internal company safety and health personnel who conduct their own audit subject the audit results to possible evaluation by outside legal sources (see Section 27.8.) The purpose of an audit is to formally identify weaknesses or faults in the *management system* (the "root causes" of accidents) that allow unsafe work practices and unsafe conditions that could cause employee injury or diminished health, production interruption, or product/property damage.

27.16 HOW EVALUATIONS IMPROVE THE SAFETY AND HEALTH PROGRAM

A well-planned and -executed evaluation program can greatly improve a good safety and health program. It can:

1. Get managers, supervisors, and employees involved in a common effort.
2. Build morale.
3. Teach everyone in the organization about the safety and health programs activities.
4. Provide management with factual data on company activities.
5. Help meet government and legal obligations.
6. Build competition between departments.
7. Formally document safety and health activities.
8. Be the beginning of working toward membership in OSHA's VPP program.
9. Develop a formal followup process to improve and/or correct unsafe deficiencies.
10. Professionalize the company safety and health program.
11. Give top management an opportunity to participate in the safety and health program.

It can be clearly seen from the foregoing that there are many benefits of auditing and evaluating the safety and health program. It is one of the single most important administrative management techniques a safety and health professional can use to bring senior management a detailed review of the company safety and health program and all its related activities. It shows how the company safety and health activities compare with internal company policies, industry practices, and government regulations and guidelines.

REFERENCES

Raymond J. Colvin, *The Guidebook to Successful Safety Programming*, Lewis Publishers, Inc.

OSHA Program Evaluation Profile (PEP) Program, OSHA Notice CPL 2, August 1, 1996, Form OSHA-195 and OSHA Instruction CPL 2.103, 9/26/94 Field Inspection Reference Manual (FIRM).

Rosemary Colvin, *OSHA Safety & Health Program Guidelines Manual*, Safety Training Dynamics, Inc.

APPENDIX A

OVERVIEW OF OSHA PROGRAM EVALUATION PROFILE

Tables A.1–A.4 cover management leadership and employee participation. Visible management leadership provides the motivating force for an effective safety and health program (Table A.1).

TABLE A.1 Management Leadership

1	Management demonstrates no policy, goals, objectives, or interest in safety and health issues at this worksite.
2	Management sets and communicates safety and health policy and goals, but remains detached from all other safety and health efforts.
3	Management follows all safety and health rules, and gives visible support to the safety and health efforts of others.
4	Management participates in significant aspects of the site's safety and health program, such as site inspections, incident reviews, and program reviews. Incentive programs that discourage reporting of accidents, symptoms, injuries, or hazards are absent. Other incentive programs may be present.
5	Site safety and health issues are regularly included on agendas of management operations meetings. Management clearly demonstrates—by involvement, support, and example—the primary importance of safety and health for everyone on the worksite. Performance is consistent and sustained or has improved over time.

Source: 1989 Voluntary Safety and Health Program Management Guidelines, (b)(1) and (c)(1).

Employee participation provides the means through which workers identify hazards, recommend and monitor abatement, and otherwise participate in their own protection (Table A.2).

TABLE A.2 Employee Participation

1	Worker participation in workplace safety and health concerns is not encouraged. Incentive programs are present that have the effect of discouraging reporting of incidents, injuries, potential hazards, or symptoms. Employees/employee representatives are not involved in the safety and health program.
2	Workers and their representatives can participate freely in safety and health communication between employer and workers on safety and health matters. Worker rights under the Occupational Safety and Health Act to refuse or stop work that they reasonably believe involves imminent danger are understood by workers and honored by management. Workers are paid while performing safety activities.
3	Workers and their representatives are involved in the safety and health program, involved in inspection of work area, and permitted to observe monitoring and receive results. Workers' and representatives' right of access to information is understood by workers and recognized by management. A documented procedure is in place for raising complaints of hazards or discrimination and receiving timely employer responses.
4	Workers and their representatives participate in workplace analysis, inspections and investigations, and development of control strategies throughout facility, and have necessary training and education to participate in such activities. Workers and their representatives have access to all pertinent health and safety information, including safety reports and audits. Workers are informed of their right to refuse job assignments that pose serious hazards to themselves pending management response.
5	Workers and their representatives participate fully in development of the safety and health program and conduct of training and education. Workers participate in audits, program reviews conducted by management or third parties, and collection of samples for monitoring purposes, and have necessary training and education to participate in such activities. Employer encourages and authorizes employees to stop activities that present potentially serious safety and health hazards.

Source: Guidelines, (b)(1) and (c)(1).

Implementation means tools, provided by management, that include (Table A.3):

- Budget
- Information
- Personnel
- Assigned responsibility
- Adequate expertise and authority
- Means to hold responsible persons accountable (line accountability)
- Program review procedures

TABLE A.3 Implementation

1	Tools to implement a safety and health program are inadequate or missing.
2	Some tools to implement a safety and health program are adequate and effectively used; others are ineffective or inadequate. Management assigns responsibility for implementing a site safety and health program to identified person(s). Management's designated representative has authority to direct abatement of hazards that can be corrected without major capital expenditure.
3	Tools to implement a safety and health program are adequate, but are not all effectively used. Management representative has some expertise in hazard recognition and applicable OSHA requirements. Management keeps or has access to applicable OSHA standards at the facility, and seeks appropriate guidance information for interpretation of OSHA standards. Management representative has authority to order/purchase safety and health equipment.
4	All tools to implement a safety and health program are more than adequate and effectively used. Written safety procedures, policies, and interpretations are updated based on reviews of the safety and health program. Safety and health expenditures, including training costs and personnel, are identified in the facility budget. Hazard abatement is an element in management performance evaluation.
5	All tools necessary to implement a good safety and health program are more than adequate and effectively used. Management safety and health representative has expertise appropriate to facility size and process, and has access to professional advice when needed. Safety and health budgets and funding procedures are reviewed periodically for adequacy.

Source: Guidelines, (b)(1) and (c)(1).

Contractor safety (Table A.4) is an effective safety and health program protects all personnel on the worksite, including the employees of contractors and subcontractors. It is the responsibility of management to address contractor safety.

TABLE A.4 Contractor Safety

1	Management makes no provision to include contractors within the scope of the worksite's safety and health program.
2	Management policy requires contractor to conform to OSHA regulations and other legal requirements.
3	Management designates a representative to monitor contractor safety and health practices, and that individual has authority to stop contractor practices that expose host or contractor employees to hazards. Management informs contractor and employees of hazards present at the facility.
4	Management investigates a contractor's safety and health record as one of the bidding criteria.
5	The site's safety and health program ensures protection of everyone employed at the worksite, i.e., regular full-time employees, contractors, temporary and part-time employees.

Source: Guidelines, (b)(1) and (c)(1).

Tables A.5–A.7 concern workplace analysis. Survey and hazard analysis. Survey and hazard analysis (Table A.5) is an effective, proactive safety and health program will seek to identify and analyze all hazards. In large or complex workplaces, components of such analysis are the comprehensive survey and analyses of job hazards and changes in conditions.

TABLE A.5 Survey and Hazard Analysis

1	No system or requirement exists for hazard review of planned/changed/new operations. There is no evidence of a comprehensive survey for safety or health hazards or for routine job hazard analysis.
2	Surveys for violations of standards are conducted by knowledgeable person(s), but only in response to accidents or complaints. The employer has identified principal OSHA standards that apply to the worksite.
3	Process, task, and environmental surveys are conducted by knowledgeable person(s) and updated as needed and as required by applicable standards. Current hazard analyses are written (where appropriate) for all high-hazard jobs and processes; analyses are communicated to and understood by affected employees. Hazard analyses are conducted for jobs/tasks/workstations where injury or illnesses have been recorded.
4	Methodical surveys are conducted periodically and drive appropriate corrective action. Initial surveys are conducted by a qualified professional. Current hazard analyses are documented for all work areas and are communicated and available to all the workforce; knowledgeable persons review all planned/changed/new facilities, processes, materials, or equipment.

TABLE A.5 (*Continued*)

5	Regular surveys, including documented comprehensive workplace hazard evaluations, are conducted by certified safety and health professional or professional engineer, etc. Corrective action is documented, and hazard inventories are updated. Hazard analysis is integrated into the design, development, implementation, and changing of all processes and work practices.

Source: Guidelines, (c)(2)(1).

The definition of inspection is to identify new or previously missed hazards and failures in hazard controls. An effective safety and health program will include regular site inspections (Table A.6).

TABLE A.6 Inspection

1	No routine physical inspection of the workplace and equipment is conducted.
2	Supervisors dedicate time to observing work practices and other safety and health conditions in work areas where they have responsibility.
3	Competent personnel conduct inspections with appropriate involvement of employees. Items in need of correction are documented. Inspections include compliance with relevant OSHA standards. Time periods for correction are set.
4	Inspections are conducted by specifically trained employees, and all items are corrected promptly and appropriately. Workplace inspections are planned, with key observations or check points defined and results documented. Persons conducting inspections have specific training in hazard identification applicable to the facility. Corrections are documented through followup inspections. Results are available to workers.
5	Inspections are planned and overseen by certified safety or health professionals. Statistically valid random audits of compliance with all elements of the safety and health program are conducted. Observations are analyzed to evaluate progress.

Source: Guidelines, (c)(2)(ii).

A reliable hazard reporting system (Table A.7) enables employees, without fear of reprisal, to notify management of conditions that appear hazardous and to receive timely and appropriate responses.

TABLE A.7 Hazard Reporting

1	No formal hazard reporting system exists, or employees are reluctant to report hazards.
2	Employees are instructed to report hazards to management. Supervisors are instructed and are aware of a procedure for evaluating and responding to such reports. Employees use the system with no risk of reprisals.
3	A formal system for hazard reporting exists. Employee reports of hazards are documented, corrective action is scheduled, and records maintained.
4	Employees are periodically instructed in hazard identification and reporting procedures. Management conducts surveys of employee observations of hazards to ensure that the system is working. Results are documented.
5	Management responds to reports of hazards in writing within specified time frames. The workforce readily identifies and self-corrects hazards; they are supported by management when they do so.

Source: Guidelines, (c)(2)(iii).

Tables A.8 and A.9 concern accident and record analysis. An effective program will provide for investigation of accidents (Table A.8) and "near-miss" incidents, so that their causes, and the means for their prevention, are identified.

TABLE A.8 Accident Investigation

1	No investigation of accidents, injuries, near misses, or other incidents is conducted.
2	Some investigation of incidents takes place, but root cause may not be identified, and correction may be inconsistent. Supervisors prepare injury reports for lost time cases.
3	OSHA-101 is completed for all recordable incidents. Reports are generally prepared with cause identification and corrective measures prescribed.
4	OSHA-recordable incidents are always investigated, and effective prevention is implemented. Reports and recommendations are available to employees. Quality and completeness of investigations are systematically reviewed by trained safety personnel.
5	All loss-producing accidents and "near misses" are investigated for root causes by teams or individuals that include trained safety personnel and employees.

Source: Guidelines, (c)(2)(iv).

An effective program will analyze injury and illness records (Table A.9) for indications of sources and locations of hazards, and jobs that experience higher umbers of injuries. By analyzing injury and illness trends over time, patterns with common causes can be identified and prevented.

TABLE A.9 Data Analysis

1	Little or no analysis of injury/illness records; records (OSHA 200/101, exposure monitoring) are kept or conducted.
2	Data are collected and analyzed, but not widely used for prevention. OSHA-101 is completed for all recordable cases. Exposure records and analyses are organized and are available to safety personnel.
3	Injury/illness logs and exposure records are kept correctly, are audited by facility personnel, and are essentially accurate and complete. Rates are calculated so as to identify high-risk areas and jobs. Workers compensation claim records are analyzed and the results used in the program. Significant analytical findings are used for prevention.
4	Employer can identify the frequent and most severe problem areas, the high-risk areas and job classifications, and any exposures responsible for OSHA recordable cases. Data are fully analyzed and effectively communicated to employees. Illness/injury data are audited and certified by a responsible person.
5	All levels of management and the workforce are aware of results of data analyses and resulting preventive activity. External audits of accuracy of injury and illness data, including review of all available data sources are conducted. Scientific analysis of health information, including nonoccupational databases, is included where appropriate in the program.

Source: Guidelines, (c)(2)(v).

Tables A.10–A.12 concern hazard prevention and control. Workforce exposure to all current and potential hazards should be prevented or controlled by using engineerng controls wherever feasible and appropriate work practices and administrative controls and personal protective equipment (PPE) (Table A.10).

TABLE A.10 Hazard Control

1	Hazard control is seriously lacking or absent from the facility.
2	Hazard controls are generally in place, but effectiveness and completeness vary. Serious hazards may still exist. Employer has achieved general compliance with applicable OSHA standards regarding hazards with a significant probability of causing serious physical harm. Hazards that have caused past injuries in the facility have been corrected.
3	Appropriate controls (engineering, work practice, and administrative controls, and PPE) are in place for significant hazards. Some serious hazards may exist. Employer is generally in compliance with voluntary standards, industry practices, and manufacturers' and suppliers' safety recommendations. Documented reviews of needs for machine guarding, energy lockout, ergonomics, materials handling, bloodborne pathogens, confined space, hazard communication, and other generally applicable standards have been conducted. The overall program tolerates occasional deviations.

TABLE A.10 (*Continued*)

4	Hazard controls are fully in place, and are known and supported by the workforce. Few serious hazards exist. The employer requires strict and complete compliance with all OSHA, consensus, and industry standards and recommendations. All deviations are identified and causes determined.
5	Hazard controls are fully in place and continually improved upon based on workplace experience and general knowledge. Documented reviews of needs are conducted by certified health and safety professionals or professional engineers, etc.

Source: Guidelines, (c)(3)(1).

An effective safety and health program will provide for facility and equipment maintenance, so that hazardous breakdowns are presented (Table A.11).

TABLE A.11 Maintenance

1	No preventive maintenance program is in place; breakdown maintenance is the rule.
2	There is a preventive maintenance schedule, but it does not cover everything and may be allowed to slide or performance is not documented. Safety devices on machinery and equipment are generally checked before each production shift.
3	A preventive maintenance schedule is implemented for areas where it is most needed; it is followed under normal circumstances. Manufacturers' and industry recommendations and consensus standards for maintenance frequency are complied with. Breakdown repairs for safety related items are expedited. Safety device checks are documented. Ventilation system function is observed periodically.
4	The employer has effectively implemented a preventive maintenance schedule that applies to all equipment. Facility experience is used to improve safety-related preventative maintenance scheduling.
5	There is a comprehensive safety and preventive maintenance program that maximizes equipment reliability.

Souce: Guidelines, (c)(3)(ii).

An effective safety and health program will include a suitable medical program where it is appropriate for the size and nature of the workplace and its hazards. (Table A.12, OSHA Notice CPL 2, August 1, 1996, Directorate of Compliance Programs).

TABLE A.12 Medical Program

1	Employer is unaware of, or unresponsive to medical needs. Required medical surveillance, monitoring, and reporting are absent or inadequate.
2	Required medical surveillance, monitoring, removal, and reporting responsibilities for applicable standards are assigned and carried out, but results may be incomplete or inadequate.

TABLE A.12 (*Continued*)

3	Medical surveillance, removal, monitoring, and reporting comply with applicable standards. Employees report early signs/symptoms of job-related injury or illness and receive appropriate treatment.
4	Health care providers provide followup on employee treatment protocols and are involved in hazard identification and control in the workplace. Medical surveillance addresses conditions not covered by specific standards. Employee concerns about medical treatment are documented and responded to.
5	Health care providers are on site for all production shifts and are involved in hazard identification and training. Health care providers periodically observe the work areas and activities and are fully involved in hazard identification and training.

Source: Guidelines, (c)(3)(iv).

Tables A.13 and A.14 concern emergency response. Table A.13 lists levels of emergency preparedness: There should be appropriate planning, training/drills; and equipment for response to emergencies. *Note:* In some facilities the employer plan is to evacuate and call the fire department. In such cases, only applicable items listed should be considered (OSHA Notice CPL 2, August 1, 1996, Directorate of Compliance Programs).

TABLE A.13 Emergency Preparedness

1	Little or no effective effort to prepare for emergencies.
2	Emergency response plans for fire, chemical, and weather emergencies as required by 29 CFR 1910.38, 1910.120, or 1926.35 are present. Training is conducted as required by the applicable standard. Some deficiencies may exist.
3	Emergency response plans have been prepared by persons with specific training. Appropriate alarm systems are present. Employees are trained in emergency procedures. The emergency response extends to spills and incidents in routine production. Adequate supply of spill control and PPE appropriate to hazards on site is available.
4	Evacuation drills are conducted no less than annually. The plan is reviewed by a qualified safety and health professional.
5	Designated emergency response team with adequate training is on site. All potential emergencies have been identified. Plan is reviewed by the local fire department. Plan and performance are reevaluated at least annually and after each significant incident. Procedures for terminating an emergency response condition are clearly defined.

Source: Guidelines, (c)(3)(iii) and (iv).

First aid/emergency care should be readily available to minimize harm if an injury or illness occurs (Table A.14).

TABLE A.14 First Aid

1	Neither on-site nor nearby community aid (e.g., emergency room) can be ensured.
2	Either on-site or nearby community aid is available on every shift.
3	Personnel with appropriate first aid skills commensurate with likely hazards in the workplace and as required by OSHA standards (e.g., 1910.151, 1926.23) are available. Management documents and evaluates response time on a continuing basis.
4	Personnel with *certified* first aid skills are always available on site; their level of training is appropriate to the hazards of the work being done. Adequacy of first aid is formally reviewed after significant incidents.
5	Personnel trained in advanced first aid and/or emergency medical care are always available on site. In larger facilities a health care provider is on site for each production shift.

Source: Guidelines, (c)(3)(iii) and (iv).

Safety and health training should cover the safety and health responsibilities of all personnel who work at the site or affect its operations. It is most effective when incorporated into other training about performance requirements and job practices. It should include all subjects and areas necessary to address the hazards at the site (Table A.15, OSHA Notice CPL 2, August 1, 1996, Directorate of Compliance Programs). Figure 27.1 is an OSHA form for PEP.

TABLE A.15 Safety and Health Training

1	Facility depends on experience and peer training to meet needs. Managers/supervisors demonstrate little or no involvement in safety and health training responsibilities.
2	Some orientation training is given to new hires. Some safety training materials (e.g., pamphlets, posters, videotapes) are available or are used periodically at safety meetings, but there is little or no documentation of training or assessment of worker knowledge in this area. Managers generally demonstrate awareness of safety and health responsibilities, but have limited training themselves or involvement in the site's training program.
3	Training includes OSHA rights and access to information. Training required by applicable standards is provided to all site employees. Supervisors and managers attend training in all subjects provided to employees under their direction. Employees can generally demonstrate the skills/knowledge necessary to perform their jobs safely. Records of training are kept and training is evaluated to ensure that it is effective.

TABLE A.15 (*Continued*)

4	Knowledgeable persons conduct safety and health training that is scheduled, assessed, and documented, and addresses all necessary technical topics. Employees are trained to recognize hazards, violations of OSHA standards, and facility practices. Employees are trained to report violations to management. All site employees—including supervisors and managers—can generally demonstrate prepardness for participation in the overall safety and health program. There are easily retrievable scheduling and recordkeeping systems.
5	Knowledgeable persons conduct safety and health training that is scheduled, assessed, and documented. Training covers all necessary topics and situations, and includes all persons working at the site (hourly employees, supervisors, managers, contractors, part-time and temporary employees). Employees participate in creating site-specific training methods and materials. Employees are trained to recognize inadequate responses to reported program violations. Retrievable recordkeeping system provides for appropriate retraining, makeup training, and modifications to training as the result of evaluations.

Source: Guidelines, (b)(4) and (c)(4).

PEP Program Evaluation Profile Employer: Inspection No.: Date: CSHO ID:		Management Leadership and Employee Participation				Workplace Analysis			Accident and Record Analysis		Hazard Prevention and Control			Emergency Response		Safety and Health Training	
		Management Leadership	*Employee Participation*	*Implementation*	*Contractor Safety*	*Survey and Hazard Analysis*	*Inspection*	*Reporting*	*Accident Investigation*	*Data Analysis*	*Hazard Control*	*Maintenance*	*Medical Program*	*Emergency Preparedness*	*First Aid*	*Training*	
Outstanding	5																5
Superior	4																4
Basic	3																3
Developmental	2																2
Absent or Ineffective	1																1
Score for element																	
Overall Score																	

OSHA-195 (3/96)

Fig. 27.1 OSHA PEP form.

CHAPTER 28

Occupational Health and Safety Regulatory Affairs

ROY DEITCHMAN and JAMES DEGEN

28.1 INTRODUCTION
28.2 SAFETY REGULATORY HISTORY
28.3 CREATING FEDERAL LEGISLATION
28.4 ADDITIONAL FEDERAL LAWS IMPACTING HEALTH AND SAFETY PROGRAMS
- 28.4.1 Toxic Substance Control Act
- 28.4.2 Clean Air Act
- 28.4.3 Clean Water Act
- 28.4.4 Safe Drinking Water Act
- 28.4.5 Federal Insecticide, Fungicide and Rodenticide Act
- 28.4.6 Resource Conservation and Recovery Act
- 28.4.7 Comprehensive Environmental Response, Compensation and Liability Act
- 28.4.8 Hazardous Materials Transportation Act
- 28.4.9 Consumer Product Safety Act

28.5 DEVELOPMENT OF STANDARDS
28.6 STRATEGIES TO PROVIDE INPUT TO STANDARDS DEVELOPMENT
- 28.6.1 National Consensus Standards
- 28.6.2 Trade and Professional Associations
- 28.6.3 Stakeholder and Town Meetings
- 28.6.4 Prepublication Drafts
- 28.6.5 Advanced Notice of Proposed Rulemaking
- 28.6.6 Notice of Proposed Rulemaking
- 28.6.7 Legal Challenge

28.7 OSHA INSPECTION TARGETING
28.8 MANAGING OSHA INSPECTIONS
- 28.8.1 Types of Inspections
- 28.8.2 Employee Complaints
- 28.8.3 The Beginning of an Inspection
- 28.8.4 The Opening Conference
- 28.8.5 Records Review
- 28.8.6 The Walkaround

Adapted from "Job Safety and Health Law" and "Compliance and Projection" by Martha Hartle Munsch, J.D. and Robert L. Potter, J.D., Chapters 20 & 21 in Cralley and Cralley, *Patty's Industrial Hygiene and Toxicology, 1st ed., Vol. III*. New York: John Wiley & Sons, Inc., 1979, pp. 681–736.

Handbook of Occupational Safety and Health, Second Edition, Edited by Louis J. DiBerardinis, ISBN 0-471-16017-2

28.8.7 The Closing Conference
28.8.8 Employer Preparedness
28.9 AFTER THE OSHA INSPECTION
28.9.1 Types of Violations
28.9.2 Posting Requirements
28.9.3 Employer Options
28.9.4 Informal Conference
28.9.5 The Contest Process
28.9.6 Petition for Modification of Abatement
28.9.7 Employee Rights and Courses of Action
28.10 COMPLIANCE STRATEGIES
28.11 THE REVIEW COMMISSION
28.11.1 Preemption
28.11.2 Defenses
28.11.3 Case Review
28.12 VOLUNTARY PROTECTION PROGRAMS
28.13 ISO GUIDELINES
28.14 CLOSING
REFERENCES
GLOSSARY
APPENDIX A

28.1 INTRODUCTION

Whenever laws and regulations require individuals or corporations to act in ways contrary to their perceived short-term interests, it may be necessary to provide authority and resources to the government for monitoring and enforcement. There must be a method to monitor actions subject to the law and to create a "level playing field." Credible threats of punishment must also be available. These common-sense requirements apply to laws intended to discourage dangerous drug use, local housing codes, environmental regulations issued by state agencies or U.S. Environmental Protection Agency (EPA), and regulations promulgated by the Occupational Health and Safety Administration (OSHA).

In order to have an effective compliance system, the needs include:

- A simple means of identifying and describing requirements for compliance
- Compliance monitoring methods
- Inspection targeting techniques
- Authority to impose administrative penalties and techniques for calculating penalties that can withstand challenges (Russell, 1986)

The purpose of this chapter is to describe the elements of federal occupational health and safety regulations, strategies for compliance, and a method of analysis for regulatory decision making. There is the need to develop regulatory affairs management procedures to successfully conduct an occupational health and safety program. Those responsible for management of safety and health, including industrial hygienists, safety managers, and plant personnel must understand the nature and scope of applicable regulations and how to work within the legal process framework. A glossary of legal and regulatory terms used in occupational health and safety is provided at the end of this chapter.

28.2 SAFETY REGULATORY HISTORY

Legislative and regulatory interest in industrial safety began at the state level. In 1877, Massachusetts passed a statute requiring the guarding of hazardous parts of machinery, such as shafts and gears. By 1890, 21 states had passed occupational safety and health laws, and by 1920, nearly every state had an industrial safety law. These laws were usually not substantive.

Despite the lack of preventive legislation in the early 1900s, states made the first efforts to relieve the hardship of industrial accidents from individual workers and their families. Many states enacted employer liability laws that modified the common law and made it easier for injured employees to recover from their employers. Before these laws, the doctrines of contributory negligence and assumption of risk often prevented employees from recovering for their injuries. By 1921, 46 states had some form of worker compensation law. These laws also set limits on recoveries by workers (Rothstein, 1970).

At the Federal level, legislation to create a Federal Mining Bureau was proposed in 1865, but it was not until 1891 that the first coal mine legislation was enacted. In 1893, safety equipment was specified for railroad cars and engines. In 1902, the Public Health Service was established, and Congress passed a safety act that regulated the sale and control of viruses, serums, and toxins. The Bureau of Mines was established in the Department of the Interior in 1910. In 1912 the Esch Act was passed, which placed a prohibitive tax on phosphorous to curtail its use in match factories. This used economic pressure to eliminate the occupational illness of "phossy-jaw," a disabling injury causing necrosis of the lower jaw from exposure to phosphorus.

Although Federal safety legislation was limited in this period, other Congressional enactments in the field of labor law had indirect effects on workplace safety and health. In 1916, Congress enacted a child labor law, although this act was later declared unconstitutional. The Fair Labor Standards Act of 1938 contained the first child labor regulations to be upheld by the U.S. Supreme Court.

In 1936, Congress passed the Walsh–Healey Public Contracts Act, which limited working hours and set some limited standards for working conditions in factories. The Act required that contracts entered into by any agency of the United States for the manufacture or furnishing of materials in any amount exceeding $10,000 must contain a requirement that the working conditions of the contractor's employees must not be unsanitary, hazardous, or dangerous to health and safety. The Walsh–Healey Act, however, had limited coverage and failed to provide and enforce strict industrial health and safety standards.

The Labor Management Relations Act (Taft–Hartley Act), passed in 1947 over President Truman's veto, contained a provision (502) permitting employees to walk off a job if it was "abnormally dangerous." In 1948 President Truman attempted to remedy industrial accidents by organizing the first Presidential Conference on Industrial Safety.

During the mid and late 1960s Congress continued to enact specialized or limited safety statutes. In 1965 the McNamara–O'Hara Public Service Contract Act was passed to provide labor standards for the protection of employees of contractors who performed maintenance service for federal agencies. Also in 1965, the National Foundation on the Arts and Humanities Act was passed, conditioning receipt of Federal grants on the maintenance of safe and healthful working conditions for performers, laborers, and mechanics.

Congress took the first significant federal step in job safety and health when it passed the Metal and Nonmetallic Mine Safety Act of 1966. In January, 1968, President Johnson proposed the nation's first comprehensive occupational safety and health program.

By 1970, interest in job safety and health regulations had peaked. Many members of Congress were ready to give serious consideration to more comprehensive Federal regulations in industrial safety and health. The result was the passage of the Occupational Safety and Health Act of 1970 (OSHAct).

The legislative history indicates that OSHA was the result of numerous legislative compromises. This is clear in two important ways. First, the Act is not well drafted: Various sections of the Act are vague, redundant and paradoxical. Second, in construing the Act, the legislative history is seldom conclusive because the members of Congress never considered the Act in its present form until after the House–Senate Conference Joint Report. Thus judicial use of the legislative history as an aid in interpreting the Act has not been helpful.

28.3 CREATING FEDERAL LEGISLATION

It is important to understand how Federal legislation is created so that a party can make its view understood. Most legislative proposals begin as bills, although much groundwork may have been done in earlier legislative sessions. They may be introduced in either house of Congress. Bills introduced in the Senate are designated with an "S" and numbered in sequence of introduction. Bills introduced in the house are designated with an "H.R." and also numbered in sequence of introduction. Bills may be sponsored by one or more senators or representatives. Table 28.1 indicates the chronology of the Federal legislative process.

In addition to the letter(s) and number indicating the house of origin, type of legislative measure, and unique identity of the measure, the front page of a newly introduced bill or resolution includes the Congress and session in which it was introduced; the house in which it originated; date of introduction; sponsor(s); committee(s) to which it was referred; the notation "A Bill," "Joint Resolution," "Concurrent Resolution," or "Resolution;" the title; and the enacting clause. The full text of the bill or resolution follows, with each line numbered.

Senate and House committees hold hearings when considering legislation and performing oversight duties. Occupational health and safety matters are usually referred

TABLE 28.1 The Federal Legislative Process

Process
A. Introduction and referral
B. Committee action
Hearing
Subcommittee or committee markup
C. Scheduling and Floor Action
Scheduling
Floor action
D. Conference action
E. Final passage
F. Presidential action

to Human Resources, Labor, or Commerce-related committees. Witnesses present oral and written testimony at these hearings and are questioned by committee members, and occasionally by committee staff. Statistical data, reports, copies of correspondence and other information also are submitted by committee members and witnesses for inclusion in the hearing record.

The hearings are recorded and transcripts are made. At a later date, most transcripts are printed and simply called "hearings." There is no numbering system for identifying hearings as there is with bills or committee reports. Research can be done by date, subject matter, or reference to the hearings in other documents comprising a bill's legislative history.

Hearings are held not only by committees with legislative jurisdiction over a subject but also by other committees with an interest in the subject. The Senate Appropriations, Budget, Finance, and Governmental Affairs committees and the House Appropriations, Budget, Energy and Commerce, and Government Operations committees all have a significant interest in many matters and may use hearings to obtain information even if no legislation is planned. Other standing committees of each house may have an ancillary interest in legislation pending in another committee. Select, special, and ad hoc committees of each house and the joint committees, any of which may or may not have authority to report legislation, may use hearings as their primary means of influencing congressional action.

A committee or subcommittee "marks up" a bill after the conclusion of the hearings. The markup is where committee members redraft portions of the bill, attempt to insert new provisions and delete others, bargain over final language, and generally determine the final committee product. The markup is a key stage for a piece of legislation; if there is no markup, there probably will not be floor action on the bill. Most markups are held in open session.

Following committee markup, committees have several options when they vote to report a bill out of committee. They may report the bill without any changes or with various amendments. A committee that has extensively amended a bill may instruct the chairman to incorporate the modifications in a new measure, known as a "clean bill." This bill will be reintroduced, assigned a new bill number, referred back to the committee, and reported by the committee.

A committee report is a document prepared by a committee on a bill and sets forth the committee's recommendation to its house on that bill. Most reports favor passage. When a committee does not consider a bill fully or reports a bill unfavorably to its house, the committee normally does not prepare a written report on the measure. The phrase "report the bill" has become synonymous with the terminology "report favorably."

Generally, a report contains information on the purpose and scope of a bill and sets forth the committee's arguments for its passage. The report also contains information on the budgetary impact of the bill and the committee's work on the subject of the bill. The report must contain certain information prepared by the Congressional Budget Office and the text of laws amended or repealed by the bill, showing the changes in existing law. Each house requires other specific information to be included in committee reports on bills.

When a committee votes to report a bill, a member of the committee may have views that do not agree with the committee majority's position. Any member or group of members of the committee may file a statement containing these views. These statements are called dissenting, minority, supplemental, or additional views, and they must be

TABLE 28.2 Intervention Points in the Regulatory Process

Contacts with legislators prior to bill introduction
Testimony at Congressional hearings
Regular contacts with the agency
Mediation rulemaking
Review of regulatory calendar
Advanced notice of proposed rulemaking (ANPR)
Notice of proposed rulemaking (NPRM)
Regulatory hearing
Challenge of new proposed rule
Contest of an administrative citation

included in a report. These statements can be arguments against passage of the bill or in favor of a floor amendment.

There is no easy way to find out when a piece of legislation will be considered on the floor of the House or Senate. In the House, a rule spells out the environment for floor action because it sets the time for debate, and the type and number of amendments allowed. A rule is always in the form of a simple resolution. The Senate does not use rules to govern floor action of legislation. Instead, it uses unanimous consent agreements, which are verbal agreements between the leadership and bill managers about when and how the legislation will be considered on the floor.

If a bill passes one chamber of Congress and is sent to the other, the legislation technically becomes an "act." The words, "An Act," will appear after the enacting clause of the legislation. After both houses pass an identical version of a bill or joint resolution and it is enrolled and sent to the President, the President has four choices for acting on the bill. He may sign the bill within 10 days (Sundays excluded); allow the bill to become law without his signature by not acting on it within the 10-day period; veto the bill within 10 days and return it with his objections to the house in which it originated; or pocket-veto the bill if Congress has adjourned.

In Table 28.2, intervention points in the regulatory process are listed. The first two points involve congressional matters. This type of intervention can be most useful for general policy concerns. If following proposed legislation, it is useful to contact the Senate or House Committee staff to receive information on hearings and status. Also, maintaining contact with your Senator's or Representative's office can be helpful. Commercial services and newspapers such as the *Washington Post* and *Washington Times* publish daily schedules of Congressional hearings and conferences. The other intervention points involve interactions with OSHA or administrative law or court actions which will be described later in this chapter.

28.4 ADDITIONAL FEDERAL LAWS IMPACTING OCCUPATIONAL HEALTH AND SAFETY PROGRAMS

Additional federal laws impacting occupational health and safety programs are discussed in the following (Worobec and Hogue, 1992) (full copies of the laws and implementing regulations can be obtained from the sponsoring federal agency or commercial services).

28.4.1 Toxic Substance Control Act, Public Law Number 94-469, Enacted September 28, 1976

Regulations found at 40 Code of Federal Regulations (CFR) 700–799; responsible Federal agency: Environmental Protection Agency (EPA). The Toxic Substance Control Act (TSCA) provides the regulatory framework for controlling exposure and use of raw industrial chemicals. This Act requires chemicals be evaluated before use to determine that they pose no unnecessary risk to health or the environment.

Under TSCA, chemical manufacture, use, import, or disposal may be banned, controlled, or restricted. The Act provides for listing all chemicals that must be evaluated before manufacture or use in the United States. Existing chemicals are ranked by their hazard potential and are subjected to toxicity testing when necessary to evaluate that benefits outweigh risks. The Asbestos Hazard Emergency Response Act of 1986 (Pub. L. No. 99-519) was added at Title II of TSCA, requiring asbestos inspections and control measures for school systems. TSCA was also amended in 1988 (Pub. L. No. 100-551) to add a radon abatement program and authorize grants and technical assistance to states for indoor radon control.

28.4.2 Clean Air Act, Public Law Number 101-549, Enacted December 31, 1970

Regulations found at 40 CFR 50–80; responsible Federal agency: EPA. The Clean Air Act (CAA) provides for the prevention and control of discharges and emissions into the air of substances that may harm public health or natural resources. Included are both stationary sources of pollutants (such as factories) and mobile sources (such as motor vehicles).

Under the Act, the EPA sets national ambient air quality standards for specific air pollutants which all areas in the United States must meet. Asbestos demolition and air emissions are also regulated under this Act.

Substances entering the air from any source are regulated through a system of national emission limits and permits for individual discharges. Both new and existing sources are required to comply with standards under the Act. The discharge limits are based on the type and age of the plant and the type of substance being released. Various types of pollution control equipment may be needed to bring a pollution source into compliance.

Most enforcement and issuing of permits for pollution sources is carried out by states that adopt EPA-approved State Implementation Plans (SIPs) for controlling air pollution.

The Act was amended in 1990, making it the most comprehensive environmental legislation ever enacted. The new law has a goal to reduce emissions of air pollutants by 56 billion tons per year through incorporation of market-based strategies that allow trading of air pollution credits.

The primary focuses of the act are to clean up the air in urban areas, to regulate air toxics more stringently and to tackle the problem of acid rain. Instead of mandating a command-and-control form of regulation, the Act sets goals and will allow industries to find cost-effective alternative control measures.

28.4.3 Clean Water Act, CWA Public Law Number 92-500, Enacted October 18, 1972

Regulations found at 40 CFR 100–140, 400–470; responsible Federal agency: EPA. The Clean Water Act (CWA) regulates the discharge of nontoxic and toxic pollutants into surface waters by municipal sources, industrial sources, and other specific and nonspecific sources. The Act's ultimate goal is to eliminate discharges of pollutants into surface waters. Its interim goal is to make all waters in the United States usable for fishing and swimming.

Under the Act, EPA sets effluent guidelines for various types of industries and municipal sewage treatment plants. These guidelines are minimum, technology-based levels of required pollution reduction. Using these guidelines, states issue an individual facility (municipal or industrial) a permit to discharge wastes into surface waters. The individual permit is a National Pollutant Discharge Elimination System (NPDES) permit and specifies the types of control equipment and discharge limits for the specific facility. It is written with the quality of the receiving waterway as criteria.

Amendments in 1987 phased out the subsidy program for local sewage treatment plants, replacing it with state revolving loan funds. The amendments also set up programs to reduce polluted runoff from "nonpoint" sources, such as city streets, farm land, and mining sites. Also, it established a new program to control the discharge of toxic pollutants.

28.4.4 Safe Drinking Water Act, SDWA Public Law Number 93-523, Enacted December 16, 1974

Regulations found at 40 CFR 140–149; responsible Federal agency: EPA. The SDWA mandates establishment of uniform Federal standards for drinking water quality and sets up a system to regulate underground injection of wastes and other substances that could contaminate underground water sources. (Surface water is protected under the CWA.)

Under the law, EPA sets two types of drinking water standards. Primary standards apply to substances that may have an adverse effect on health. These are enforced by the states, and compliance is mandatory. Secondary standards provide guidelines on substances or conditions that affect color, taste, smell, and other physical characteristics of drinking water. These standards are advisory, not mandatory. Drinking water obtained from underground sources must be tested to evaluate that it meets primary standards. Some states also require that water meet the secondary standards.

The law also bans underground injection of certain materials in or near an underground water source, and requires permits, monitoring, and recordkeeping for underground injection that is allowable.

Amendments created a Federal groundwater protection program, under which EPA gives grants to states to prevent contamination of wellfields that provide public drinking water. Although regulation of groundwater is retained by the states, EPA can require that minimum criteria be met before the grants are awarded.

28.4.5 Federal Insecticide, Fungicide and Rodenticide Act, Public Law Number 92-516, Enacted October 21, 1972

Regulations found at 40 CFR 162–180; responsible Federal agency: EPA. FIFRA provides the regulatory framework for the registration and use of pesticides and other products intended to kill or control insects, rodents, weeds, and other living organisms. Key to the definition of "pesticides" is the concept of intended use, which allows a broad range of regulatory authority over chemicals and devices that function as pest control agents, regardless of their original purpose of manufacture. Under the law, if a product is represented in such a way as to result in use as a pesticide, the product is considered to be a pesticide under section 2 of FIFRA.

A manufacturer offering a new pesticide in the marketplace must register with EPA. This procedure includes submission of test data, proposed uses, and suggested labeling. If the product is to be used on agricultural crops, EPA must establish tolerances for residues of the substance before use of the crop for food is permitted. The manufacturer requests EPA to set a recommended tolerance, but the agency makes the final decision.

Under FIFRA, pesticide manufacture, use, import, or disposal may be banned, controlled, or restricted. Regulations under the Act allow EPA to establish safety standards for pesticide products and to remove from the market, restrict use of or refuse registration for products that do not meet those standards.

A June 1991 Supreme Court decision upheld the authority of cities and towns to control and even ban the use of pesticides by ruling that FIFRA does not preempt local regulations dealing with pesticide use (*Wisconsin Public Intervenor v. Mortier*).

28.4.6 Resource Conservation and Recovery Act, Public Law Number 94-580, Enacted October 21, 1976

Regulations found at 40 CFR 240–271; responsible Federal agency: EPA. Although the Act was passed to controll all varieties of solid waste disposal and to encourage recycling and alternative energy sources, its major emphasis is control of hazardous waste disposition. The law establishes a system to identify wastes and track their generation, transport, and ultimate disposal. Standards for disposal sites and state hazardous waste programs also are part of RCRA.

Under the law, EPA lists substances that are considered hazardous if land disposed. Anyone who generates listed wastes above a certain quantity must register with the EPA and comply with requirements applying to generators of waste. Transporters of hazardous wastes and disposal sites also must be registered with the agency, and a permit must be obtained for disposal sites to receive hazardous wastes. A multipart manifest must accompany each shipment of wastes from generator to ultimate disposal site so that waste may be tracked and that each site has a record of its constituents. Records on wastes generated, shipped, and disposed of must be available to regulatory authorities and submitted to them periodically.

Major amendments made in 1984 expand waste management requirements, ban land disposal of bulk liquid hazardous wastes, require regulation of underground storage tanks, and bring generators of smaller amounts of hazardous wastes under the law's regulatory requirements.

28.4.7 Comprehensive Environmental Response, Compensation and Liability Act, Public Law Number 96-510, Enacted December 11, 1980

Regulations found at 40 CFR 300; responsible Federal agency: EPA. CERCLA, known as Superfund, requires cleanup of releases of hazardous substances in the environment: air, water, groundwater and land. Both new spills and leaking or abandoned sites are covered. Releases of reportable quantities of a substance listed as hazardous must be immediately reported to the National Response Center at (800)424-8802. CERCLA also establishes a trust fund to pay for cleaning up hazardous substances in the environment and gives EPA authority to collect the cost of cleanup from the parties responsible for the contamination.

Funds for the various types of cleanups authorized under the law comes from fines and other penalties collected by the government, from a tax imposed on chemicals and petrochemical feedstocks and from the U.S. Treasury. A separate fund established under the law is authorized to collect taxes imposed on active hazardous waste disposal sites, to finance monitoring of sites after they close.

The Superfund Amendments and Reauthorization Act (SARA) passed in 1986 added strict cleanup standards strongly favoring permanent remedies, stronger EPA control over the process of reaching settlement with responsible parties, a mandatory schedule for initiation of cleanup work and studies, individual assessments of the potential threat to human health posed by each waste site, and increased state and public involvement in the cleanup decision-making process.

A separate Title III of SARA provides a framework for emergency planning and preparedness and requires facilities to provide community groups with information on their inventories of hazardous chemicals and for manufacturers to report releases of chemicals to the environment.

28.4.8 Hazardous Materials Transportation Act, Public Law Number 93-633, Enacted January 3, 1975

Regulations found at 49 CFR 106,107,171–179; responsible Federal agency: Department of Transportation (DOT). This Act requires the Secretary of the DOT to regulate hazardous materials transportation in intrastate, interstate, and international commerce. Hazardous materials that are shipped by road, rail, air, or water must be correctly packaged, labeled, marked, and placarded. In addition, hazardous materials loads must be handled according to instructions promulgated by DOT's Research and Special Programs Administration (RSPA).

Under HMTA, RSPA shares regulatory authority with the Federal Highway Administration, the Federal Railroad Administration, the Federal Aviation Administration, and the U.S. Coast Guard, all DOT agencies. Transport of single-mode shipments are enforced by the corresponding DOT modal agency (for example, air shipments are enforced by the Federal Aviation Administration). RSPA enforces hazardous materials shipments transported by more than one mode (for example, by air and highway). RSPA also promulgates and enforces nonbulk packaging requirements and shares jurisdiction with other modal agencies.

Various types of accidents and releases of reportable quantities of a substance listed as hazardous must be immediately reported to the National Response Center at (800)424-8802. Information necessary for immediate handling of spills (such as how to identify contents, how to contact shipment owner) is available to registered members toll free from CHEMTREC, a service of the Chemical Manufacturers Association (CMA), at (800)424-9300.

In 1990, HMTA was reauthorized by Congress, which made significant changes to the statute. These changes sought to achieve greater uniformity of hazardous materials transportation rules at the Federal, state, and local levels. Training of hazardous materials employees is required to prevent releases of such materials during loading, unloading, and transport. Training of emergency response officials is being enhanced to improve incident coordination and a national registration system is sought for shippers, carriers, and package manufacturers.

28.4.9 Consumer Product Safety Act, Public Law Number 92-573, Enacted October 27, 1972

Regulations found at 16 CFR 1000–1406; responsible Federal agency: Consumer Product Safety Commission (CPSC). The goals of the CPSC are to assist consumers in evaluating the safety of consumer products, to protect the public against unreasonable risks associated with consumer products, to develop uniform safety standards for consumer products, and to research and prevent product-related deaths, illnesses, and injuries. Under the Act, the CPSC maintains product safety information, promulgates safety standards, bans unsafe products, and requires recalls or corrective action for unsafe products. Standards promulgated under the act may require specific labeling, design, packaging, or composition of products intended for sale to the public.

Chemical regulations include rules on consumer uses of asbestos and formaldehyde and on labeling of hazardous substances such as paints and cleaning compounds. The Act also requires manufacturers, importers, distributors, and retailers to notify the commission immediately if they receive information that one of their products fails to comply with a product safety rule or may contain a defect that could create a substantial product hazard. The CPSC was reauthorized in 1990 and the legislation established a new requirement for any firm to file a report with the agency if its products has been the subject of three product liability suits within a 2-year period.

28.5 DEVELOPMENT OF STANDARDS

The passage of the Occupational Safety and Health Act of 1970 set the framework of how occupational safety and health issues would be regulated, but did not include the actual specification of what employers would be required to do to make their places of employment free of hazards. The Act authorized procedures by which the Secretary of Labor could develop and promulgate occupational safety and health standards.

> Section 6(a) provided a means to quickly establish a basic set of standards: "Without regard to chapter 5 of title 5, United States Code, or to the other subsections of this section, the Secretary shall, as soon as practicable during the period beginning with the effective date of this Act and ending two years after such date, by rule promulgate as an occupational safety or health standard any national consensus standard, and any established Federal standard, unless he determines that the promulgation of such a standard would not result in improved safety or health for specifically designated employees. In the event of conflict among any such standards, the Secretary shall promulgate the standard which assures the greatest protection of the safety or health of the affected employees."

The 2-year "window" (4/28/71–4/28/73) created by Section 6(a) allowed the newly formed Occupational Safety and Health Administration to quickly adopt a vast array of existing national consensus standards (see Table 28.3) and some Federal standards (primarily in the areas of construction and maritime) into a comprehensive set of mandatory safety and health standards. In some instances OSHA utilized only portions of the consensus standards and reprinted the adopted text in the Code of Federal Regulations (CFR) (see Appendix A, Section A.1), while in other cases OSHA adopted the entire consensus standard by referencing it in the CFR. The most notable example of this was the adoption by reference of the entire 1971 edition of the National Electrical Code. While the ability to adopt existing consensus standards provided a means to quickly establish a set of mandatory safety and health rules, some problems came with the process:

- It became necessary for those affected by the standards to purchase a large collection of consensus standards to supplement the CFR.

TABLE 28.3 Sources of 6(b) Standards

American Conference of Governmental Industrial Hygienists
American National Standards Institute
American Petroleum Institute
American Society of Heating, Refrigeration, and Air Conditioning Engineers
American Society of Mechanical Engineers
American Society for Testing & Materials
American Welding Society
Compressed Gas Association
Crane Manufacturers Association of America
Factory Mutual Engineering Corporation
National Fire Protection Association
Rubber Manufacturers Association
Society of Automotive Engineers
Underwriter's Laboratories, Incorporated
United States of America Standards Institute

- Although generally widely accepted and used, consensus standards are voluntary, not mandatory.
- Consensus standards are regularly revised to reflect latest technology and best practice, but the revised requirements were never intended to be applied retroactively to existing equipment or facilities, as they were when adopted by OSHA.
- Revised consensus standards could not be automatically adopted by OSHA after 4/28/73.
- Many provisions of the 6(a) standards were found to be advisory and not enforceable (see Appendix A, Section A.2).

Some of the original set of OSHA Standards, commonly referred to as the 6(a) standards, are still in effect in their original versions. Some have been updated, and a few have been either partially or completely revoked.

Following the initial 2-year period, the Secretary could no longer adopt consensus standards or other Federal standards, but instead had to follow the formal rulemaking procedures of Section 6(b) to promulgate, modify, or revoke any occupational safety or health standard. These procedures require that the proposed standard or modification be published in the *Federal Register* as a Notice of Proposed Rulemaking (NPRM) and that all interested parties may submit written comments (e.g., objections, endorsements, suggested changes, data not considered by OSHA, etc.) to be considered in the final rulemaking process (see Appendix A, Section A.3). Interested parties may also request that public hearings be held as another means to provide input to the process. These hearings are held before an Administrative Law Judge, and all testimony as well as documentary evidence presented is entered into the official record. Public hearings are generally followed by a short period to allow participants to file post-hearing comments as well. Following the public participation period, the record is closed to further input, and the proposed rulemaking action is finalized for publication.

The 6(b) rulemaking procedure, as prescribed in the Act, is relatively simple and straightforward. Application of the procedure, however, has proven to be a slow and often difficult process to bring to closure. For example, the rulemaking for a relatively noncontroversial standard may take 2 years to complete, while controversial standards may take 10 or more years to complete (the rulemaking process for the General Industry Permit Required Confined Spaces Standard lasted for 17 years). Early attempts to promulgate new standards were met with court challenges that often resulted in the new standard being withdrawn or sent back to the Administration for partial revision. As a result of experience, each step of the rulemaking has evolved into a more complex process than the Act itself would indicate, beginning with a thorough analysis of the benefits to be gained by the new standard and the cost or burden that it will impose on the impacted business sectors. Hard choices need to be made to balance incremental benefit with incremental costs when establishing the performance endpoint of a standard in order to avoid a legal challenge (employers arguing that the standard goes too far, while employee representatives claiming it does not provide sufficient protection).

One of the factors that has contributed to the complexity of the rulemaking process is the changing nature of the standards themselves. The early standards, particularly those adopted under the 6(a) procedure, could be characterized as point-by-point specification standards (i.e., railing heights, aisle widths, machine guarding specifications). These standards are easily understood by employers, employees, and enforcement personnel; compliance is not a question of interpretation or judgment. The later additions to the OSHA Standards, however, can be characterized as performance-oriented program standards. The first major program standard was Hazard Communication (29 CFR 1910.1200), followed by Hazardous Waste Operations and Emergency Response (29 CFR 1910.120) and Process Safety Management of Highly Hazardous Chemicals (29 CFR 1910.119). These program standards require the employer to develop, implement, and continuously maintain a comprehensive program to achieve compliance.

OSHA has taken several initiatives to improve the rulemaking process by obtaining better and earlier public input. The first of these initiatives was the use of the Advanced Notice of Proposed Rulemaking (ANPR), published in the *Federal Register.* By means of an ANPR, OSHA can communicate its intent to develop a standard focused on a particular hazard or industry process and solicit input from all interested parties before actually proposing a standard. The ANPR will typically explain the issues that OSHA plans to address, estimates of the seriousness of the hazard(s), benefits that a standard may achieve, alternative methods of obtaining those benefits, estimated costs to implement, and other topics relevant to the issue. The ANPR will usually include a list of specific questions and topics that OSHA would like respondents to address in their comments. The use of ANPRs has enabled OSHA to obtain more information early in the rulemaking process to better identify those issues that will be difficult to resolve during the formal process after the standard has actually been proposed.

In addition to identifying contentious issues early in the process, OSHA has also introduced several procedures to attempt to resolve these issues prior to promulgating a final rule. These include circulating prepublication drafts of standards or portions of standards to stakeholder groups for comments and suggestions, holding "town meetings" to explain issues and the direction that OSHA plans to take, and conducting "stakeholder meetings" facilitated by consultants skilled in issue resolution. In a few rulemaking issues that only impact a small, well-defined industry segment, OSHA has tried, with some degree of success, negotiated rulemaking with representatives of the impacted employers and employees.

28.6 STRATEGIES TO PROVIDE INPUT TO STANDARDS DEVELOPMENT

The OSHA rulemaking procedure explained in the previous section provides multiple opportunities for interested parties to become part of the process by providing data, industry specific information, and suggestions for controlling hazards that the OSHA project management staff can utilize in the development of the standard. In order to be effective in determining the final outcome of a standard, it is imperative to get involved early in the process. Waiting for the official public comment period following the *Federal Register* publication of the Notice of Proposed Rulemaking is generally too late to begin making a meaningful contribution to the process. Requesting a significant change to a proposed rule at that stage of the process is an adversarial undertaking, as OSHA has already published their complete rationale and justification for the proposal.

It can be far more effective to participate in the development process "upstream" of the proposed rule stage when the OSHA project staff are gathering information regarding the nature of the hazard to be addressed, successful techniques for controlling the hazard, unique conditions specific to particular industries, etc. Simply put, the earlier and more frequent the input, the better. The following paragraphs describe the various opportunities to provide input to the standards development process in the approximate sequence that they would occur. Naturally, all of the informal stages will not necessarily be part of every rulemaking endeavor.

28.6.1 National Consensus Standards

Even though the 2-year period when OSHA could adopt national consensus standards has long passed, the consensus standards still play in important role in OSHA standard development, both formally and informally (see Table 28.3). Section 6(b)(8) of the Act requires that OSHA consider national consensus standards, specifically: "Whenever a rule promulgated by the Secretary differs substantially from an existing national consensus standard, the Secretary shall, at the same time, publish in the *Federal Register* a statement of the reasons why the rule as adopted will better effectuate the purposes of this Act than the national consensus standard." Beyond the statutorial mandate for OSHA to justify any deviation from an existing consensus standard, the consensus standards are a natural starting point for a standards development project. Utilizing consensus standards as a means of providing direction to OSHA is a long-term commitment. Made up of volunteers representing various interest groups, the committees that develop and periodically revise the consensus standards meet on an infrequent basis. The cycle time to develop or significantly revise a major consensus standard frequently spans a 2- or 3-year period.

In addition to the formal requirement to consider the content of the consensus standards, participation on the committees frequently provides an opportunity to interface directly with OSHA staff members, who frequently participate in the consensus standard process. This interface on the committee level encourages mutual understanding of the problems and conditions faced by both the regulators and the parties subject to the regulations long before a standard is formally proposed.

28.6.2 Trade and Professional Associations

In addition to providing a conduit to be represented on consensus standards committees, trade and professional associates provide another avenue to interact directly with OSHA, both through the formal comment and testimony stages of rulemaking, and also earlier in the process. Trade and professional associations provide a means for OSHA to gather vital information regarding industry practices and procedures necessary to the rulemaking process. In cases where a standard under consideration or actual development could have a major impact on an industry or industry segment, the applicable trade association will often establish a task force of experts from member companies to provide input to the OSHA staff. In some situations it has been possible to invite OSHA staff on field trips or plant tours to experience firsthand unique conditions or problem solutions.

28.6.3 Stakeholder and Town Meetings

OSHA frequently elects to utilize both town meetings and stakeholder meetings as a means of gathering information from the public and also providing information as to its views and intended position on a particular issue. Notice of both town meetings and stakeholder meetings is published in the *Federal Register*, and frequently reprinted in trade publications and safety and health news service bulletins. The two forums are similar; however, there are some definite differences between the two. Town meetings are less structured and generally open to anyone with a desired to attend. Persons desiring to make presentations are requested to provide notice of their intent to speak and a description of the points that they intend to cover. Observers are admitted subject to the capacity of the facility. In most cases, written comments are also solicited from parties who either cannot attend or who choose not to make an oral presentation.

In contrast to the almost complete openness of the town meeting, stakeholder meetings are more focused to specific issues and the meeting is structured, frequently moderated by a third-party facilitator hired for the purpose. Attendance at stakeholder meetings is open to all parties with a genuine stake or interest in the issues; however, attendance at the sessions is scheduled in advance in order to have a balance of interests at each session. Advance registration usually includes responses to a questionnaire or other solicited information in order to establish the agenda for the meeting.

28.6.4 Prepublication Drafts

On some occasions, particularly when developing standards that are highly controversial, OSHA has made draft materials available for review and comment as a means of "testing the water" prior to a formal rulemaking proposal. These draft materials are generally provided to the major industry, labor, and professional organizations known to have interest in the subject. Individuals with an interest in reviewing the draft material can generally obtain a copy directly from OSHA, if unable to obtain the material from other sources. In many cases the drafts are reprinted in the various occupational safety and health publications that follow OSHA proceedings. It should be kept in mind, however, that these drafts are unofficial and that OSHA is under no obligation to consider all comments received.

28.6.5 Advance Notice of Proposed Rulemaking

The Advance Notice of Proposed Rulemaking, or ANPR, is the first opportunity for interested parties to officially enter information into the rulemaking record. The ANPR published in the *Federal Register* does not include a draft of the rule that will ultimately be proposed. Rather, it describes OSHA's intentions to propose a rule or standard in considerable detail concerning the need for the standard, the industry segments expected to be impacted, alternative hazard control strategies being considered, and any other information deemed to be relevant to the issue. Most important, the ANPR solicits public input on the intended rulemaking. Comments are accepted on any aspect of the rulemaking; however, most ANPRs invite specific comments on a detailed list of issues and questions. Comprehensive responses to many of these questions may require generation of company- or industry-specific data regarding feasibility of control strategies, quantification of employees at risk, implementation costs, etc. Although the ANPR may require significant effort to prepare a comprehensive response, it is an opportunity to impact the rulemaking process that should not be missed. In some cases, OSHA has scheduled Town Meetings in conjunction with the ANPR.

28.6.6 Notice of Proposed Rulemaking

The Notice of Proposed Rulemaking, or NPRM, is the official publication in the *Federal Register* of the proposed rule and all of the supporting justification (e.g., expected benefit, industry-specific cost, alternatives considered, etc.). Comments from the public are generally solicited on a number of specific issues concerning the proposed rule, as well as on the text of the rule itself. Even though this stage of the rulemaking process is somewhat late to effect a significant change, unless OSHA has committed a gross omission of factual detail; it is nonetheless a very important stage: The official record for the rulemaking closes at the end of the public comment and hearing period. Input to the NPRM can be delivered in two ways: written comments and also oral testimony at public hearings. Public hearings are scheduled if there is sufficient interest from affected parties expressed to OSHA. Major rulemakings almost always include multiple hearings to accommodate the number of people requesting to appear.

Comments submitted in response to the NPRM will not be effective unless they are substantive, well constructed, and credible. When objecting to a standard or part of a standard, the comment must clearly explain the rationale for the objection and offer a realistic alternative to achieve the desired result. Most comments on proposed standards focus only on those aspects of the standard to which the commentor objects. It is equally important, however, to provide positive, supportive comments on those parts of the proposal that the commentor likes, keeping in mind that others may object to these parts. All comments, positive and negative, are considered in promulgating the final rule.

Public hearings are an extension of the comment process and provide another opportunity to restate important points and input to the official record. All testimony, as well as supporting documentation, presented at the public hearing is entered into the record and transcripts of the hearing are made available to the public. The hearings are conducted before an administrative law judge, although strict courtroom proceedings are not followed. Cross-examination is not allowed; however, persons presenting testimony at public hearings should expect to be questioned by members of the OSHA staff, a representative of the Solicitor of Labor's office, and possibly by other participants at the hearing. Observers at the hearing who are not presenting testimony are not allowed to question presenters, nor can they submit written comments at that time.

28.6.7 Legal Challenge

Following the promulgation of a final standard, Section 6(f) of the Act provides for legal challenge: "Any person who may be adversely affected by a standard issued under this section may at any time prior to the sixtieth day after such standard is promulgated file a petition challenging the validity of such standard with the United States court of appeals for the circuit wherein such person resides or has his principal place of business, for a judicial review of such standard. A copy of the petition shall be forthwith transmitted by the clerk of the court to the Secretary. The filing of such petition shall not, unless otherwise ordered by the court, operate as a stay of the standard. The determinations of the Secretary shall be conclusive if supported by substantial evidence in the record considered as a whole."

The importance of making comprehensive input to the official record is underscored by the last sentence of Section 6(f). The resolution of a legal challenge to a final standard is made solely on the official record. In order to be considered during a legal challenge, the facts must have been introduced into the record during the rulemaking process. Legal challenges to final standards have resulted in significant amendments to some standards where the courts have determined that OSHA did not adequately consider factual information presented by written and oral comment.

28.7 OSHA INSPECTION TARGETING

Occupational health and safety enforcement agencies use various methods to target compliance inspections. The methods depend on agency objectives, available data, public concerns, and past enforcement actions. Targeting can focus on a single type of work operation or type of industry or may be general in nature (Lofgen, 1996).

Safety targeting enforcement systems may differ from occupational health systems since they focus on preventing different adverse outcomes. Safety inspections generally focus on preventing deaths or injuries from acute incidents. The relatively extensive reporting of deaths and acute trauma assists with the targeting of safety inspections. Safety inspections may also be directed to industries that have long been recognized as high hazard (e.g., construction).

Health inspection targeting includes chronic illness prevention, and as a result presents more challenges. Many occupational illnesses suffer from lack of recognition and reporting, complicated by the latency period of the disease. Specific targeting may be successful for those serious illnesses better recognized and reported. An occupational health compliance program may choose to use several types of targeting systems. The types of systems an agency selects would presumably be based on a well-conceived and tested plan aimed at preventing the most significant and damaging illnesses.

The Federal OSHA general occupational health targeting system defines high-hazard industries as those with the greatest rate of serious health violations. OSHA has used the serious health violation rate for over 10 years to select employers in high-hazard industries. The violation rate continues to be the basis for the general targeting program. The list is initiated by conducting a computer search of health inspection files covering a 5-year period. Employers are placed into industry groups by their four-digit SIC code assigned to the file at the time of inspection. The serious, willful, and/or repeat (S/W/R) health violations issued within each four-digit SIC code are counted.

By policy, occupational health violations include violations of standards for ventilation, respirators, comprehensive carcinogen or toxic chemical standards, hazard communication, and noise. They also include violations for other standards addressing compressed gases, eye and face protection, medical services and first aid, commercial diving, and sanitation.

Federal OSHA also uses a local emphasis program (LEP). OSHA regional and area offices may propose and, with approval, initiate a targeting system separate from the primary national program. A local targeting scheme may be developed by an area office if they find certain industries, hazards, or other workplace characteristics deserving of an inspection. Industries in this program have included waterborne seafood processors, pulp and paper mills, and radiator shops. Some area offices have also utilized state disease registries for local and specific targeting. Local emphasis health target inspections accounted for about 20 percent of the total target or planned occupational health inspections in 1994.

The more recently developed Maine 200 targeting system selects employers with the greatest number of compensation claims in the state of Maine with emphasis on ergonomic-related hazards. Thus, Maine 200 system is a general safety and health targeting system for a specific region. Federal OSHA is applying variations of the Maine 200 system to other areas.

Federal OSHA's top ten of the 200 high-hazard industries for 1994 are shown in Tables 28.4a and b. The 1994 list was used to provide assignments to occupational health inspectors during 1994. The list was created from inspection data from the previous 5 years, including most inspections conducted from 1989 to 1993. Rank was determined by the S/W/R violation rate for the industry. The list represents the types of industries the OSHA targeting program finds most hazardous.

Automotive stamping outranked all others, followed by nonferrous die casting (excluding aluminum), kidney dialysis centers, and copper foundries. The top ten list is dominated by manufacturing, as is the case for the rest of the 200 list, in part reflecting a policy to exclude agriculture and construction from general health targeting. Table 28.4 also contains the top 10 manufacturing classifications investigated by OSHA in 1996.

TABLE 28.4a Top Ten Targeted Industries for 1994, Federal OSHA Program

Rank	Industry Group	SIC code
1	Manufacture of automotive stamping	3465
2	Manufacture of nonferrous die casting, except aluminum	3364
3	Kidney dialysis centers	8092
4	Copper foundries	3366
5	Manufacture of railroad equipment	3743
6	Primary smelting and refining of nonferrous metals, except copper and aluminum	3339
7	Manufacture of custom compound purchased plastic resins	3087
8	Manufacture of small arms	3484
9	Manufacture of plastics plumbing fixtures	3088
10	Industrial supplies, wholesale trade	5085

Source: U.S. Department of Labor, OSHA Report, "FY 1994 Health High Hazard Industries Ranked by Serious Health Violations by Inspection."

TABLE 28.4b Top Ten Manufacturing Classifications to be Investigated by OSHA in 1996

Rank	Standard Industrial Code (SIC)	Industry
1	3089	plastic products
2	3599	industrial machinery
3	3272	concrete products
4	2033	canned fruits and vegetables
5	3441	fabricated structural metal
6	3444	sheet metalwork
7	3273	ready-mixed concrete
8	2434	wood kitchen cabinets
9	2499	wood products
10	3451	screw machine products

Source: Compliance Magazine, "Top-10 OSHA Targets Published On-line," IHS Publishing Group, November/December, 1996.

28.8 MANAGING OSHA INSPECTIONS

Authorized by Section 8 of the OSHAct, the primary responsibility of the Compliance Safety and Health Officer (CSHO) is to carry out the mandate given to the Secretary of Labor, namely, "to assure so far as possible every working man and woman in the Nation safe and healthful working conditions . . ." To accomplish this mandate the Occupational Safety and Health Administration employs a wide variety of programs and initiatives, one of which is enforcement of standards through the conduct of effective inspections to determine whether employers are:

1. Furnishing places of employment free from recognized hazards that are causing or are likely to cause death or serious physical harm to their employees, and
2. Complying with safety and health standards and regulations.

This section will explain the different types of OSHA inspections, the inspection procedures, the rights of both employers and employees during and after the inspection, and things that employers can do to be better able to manage the OSHA inspection when the CSHO knocks on the door.

28.8.1 Types of Inspections

There are two general types of OSHA inspections, programmed and unprogrammed. Programmed inspections of worksites are scheduled based upon objective selection criteria and targeting systems discussed in the preceding section. Unprogrammed inspections are conducted in response to alleged hazardous working conditions that have been identified at a specific worksite. This type of inspection responds to reports of imminent dangers, fatalities/catastrophes (employers are required to notify OSHA of all incidents that result in a fatality or require the hospitalization of three or more employees), employee complaints, and referrals from other agencies. Reports of imminent danger and fatalities/catastrophes almost always result in a prompt inspection.

All inspections, either programmed or unprogrammed, fall into one of two categories depending on the scope of the inspection: comprehensive or partial. A comprehensive inspection is a substantially complete inspection of the potentially high-hazard areas of the establishment. In large establishments with multiple operations and processes, a comprehensive inspection may last for several weeks or even months. A partial inspection is limited in scope to only certain potentially hazardous areas or operations. A partial inspection may be expanded based on information gathered by the CSHO during the inspection process, such as observations of apparent hazards in areas not programmed for inspection, information obtained from on-site employee interviews, or preliminary findings that major programs (e.g., lockout/tagout, respiratory protection, etc.) have not been implemented.

28.8.2 Employee Complaints

Employee complaints received by OSHA are prioritized for scheduling of inspections or other investigation depending on the seriousness of the complaint and also on how it is transmitted to OSHA. Previously known as formal or nonformal, complaints are classified as those that result in on-site inspections and those that result in investigations using telephone, telefax, and similar means. A signed complaint (form OSHA-7 or letter) alleging an imminent danger or the existence of a violation threatening physical harm, submitted by a current employee, a representative of employees, or any other individual knowledgeable of the alleged hazardous condition usually results in an inspection. A complaint of imminent danger will be scheduled for inspection with the highest priority if there is a reasonable basis for the allegation even if unsigned. Other complaints will be given an inspection priority based upon the classification and gravity of the alleged hazards.

Signed complaints concerning workplace conditions that have no direct relationship to safety or health and do not threaten physical harm, as well as oral and unsigned complaints of a nonimminent danger nature, will be investigated without a workplace inspection. An "investigation" also differs from an "inspection" in that in an investigation, OSHA advises the employer of the alleged hazards by telephone and telefax, or by letter if necessary. The employer is required to provide a written response. OSHA will provide copies of the response(s) to the complaint. Citations are generally not issued based solely on the results of an investigation.

28.8.3 The Beginning of an Inspection

Inspections will be made with no advance notice, generally during regular working hours of the establishment except when special circumstances indicate otherwise. At the beginning of the inspection the CSHO will present credentials and ask for the owner, representative, manager, or operator in charge at the workplace. The manager in charge should look at the credentials offered. If there is any question as to their authenticity, the OSHA Area Office can be contacted for verification. If an appropriate management official is not present at the time of the inspection, the CSHO will wait for a reasonable amount of time for a management official to be present before starting the inspection. This delay should normally not exceed 1 hour.

Section 8 of the Act provides that CSHOs may enter without delay and at reasonable times any establishment covered under the Act for the purpose of conducting an inspection. An employer has the right, however, to require that the CSHO seek an inspection warrant prior to entering an establishment and may refuse entry without such a warrant.

28.8.4 The Opening Conference

The CSHO will begin the inspection with an explanation of why the establishment is being inspected. An inquiry will be made as to whether or not the employees are represented by a union or other organization. If the employees are represented, a joint management/labor opening conference will be held. (If either management or labor requests it, separate conferences will be held.) During the opening conference, the CSHO will:

- Explain the scope of the inspection to be conducted
- Inform the management official and the labor representative whether or not he intends to conduct private employee interviews
- Indicate the records that will be inspected
- Explain that as a result of conditions observed during the inspection there could be referrals to other agencies (e.g., Environmental Protection Agency, Department of Transportation, etc.)
- Review that portion of the Act that prohibits discrimination against employees for cooperating with OSHA
- Review the employer's safety and health programs
- Calculate the establishment's lost work day injury and illness rate based on the OSHA Form 200 Logs (see Chapter 26)

28.8.5 Records Review

The next step in the inspection process generally consists of a review of the records and written programs that are required to be kept either by the Act itself, or by specific standards. The CSHO will not ask to see records that clearly have no application to the establishment being inspected. The records review may encompass the following:

- OSHA Form 200 Logs for the past 5 years
- Written Hazard Communication Program
- Employee exposure and medical records
- Hazard assessments relative to use of personal protective equipment
- Respiratory protection program
- Hazardous energy control program (lock out/tag out)
- Exposure control program (bloodborne pathogens)
- Confined space entry program
- Chemical hygiene plan (laboratories not following hazard communication standard)

Generally, the OSHA Form 200 Logs and the Written Hazard Communication Program will be reviewed on all inspections, while the other records will be reviewed as applicable to the particular establishment being inspected. The appropriate sections of this Handbook should be consulted for specific details about each topic noted above.

28.8.6 The Walkaround

The walkaround is the name given to the actual physical inspection of the workplace. The CSHO will ask that a representative of the employer and the employees be given the opportunity to accompany the CSHO during the inspection. Where there is no authorized employee representative, the CSHO may consult with a reasonable number of employees concerning matters of health and safety in the workplace. During a comprehensive inspection, the walkaround may include the entire facility, whereas a partial inspection may be limited to a specific process or area of the facility. During a partial inspection the CSHO may expand the scope of the inspection if in his professional judgment there is reason to do so based on information gathered during the records review and the walkaround.

In addition to making observations of operations, processes, and potentially exposed employees, the CSHO may chose to collect evidence in the form of photographs, video tapes, measurements, sketches, etc. Where employees are potentially exposed to chemicals or harmful physical agents, the inspection may also include exposure monitoring, both area and personal. In addition, the CSHO may make specific measurements of ventilation systems where the standards require minimum ventilation flow rates. Whenever the CSHO or his assistants collect evidence or perform exposure monitoring, employers are well advised to collect duplicate samples, photos, etc. in the event that they may wish to contest any citations that might result from the inspection.

28.8.7 The Closing Conference

At the completion of the inspection, the CSHO will hold a closing conference with the employer and the employee representative. As with the opening conference, separate closing conferences will be held if either party so requests. During the closing conference the CSHO will discuss all apparent violations that were observed at any stage of the inspection, and possibly recommend specific abatements. In addition to discussing apparent violations, the CSHO will explain the rights and responsibilities of both employers and employees with regard to the findings and outcome of the inspection.

Beyond discussing specific conditions and apparent violations observed during the inspection, the CSHO may offer an evaluation of the strengths and weaknesses of the employers program. The CSHO may also indicate the he will recommend enforcement referrals to other agencies for observed or suspected conditions not within the jurisdiction of OSHA.

The manager representing the employer at the closing conference should take detailed notes of everything that the CSHO says regarding possible violations and citations. If something is unclear, the CSHO should be asked to explain why something is a violation. It is not inappropriate to challenge alleged violations at the closing conference if there is a difference of opinion regarding the alleged violation or suggested abatement. Citations issued as a result of the inspection may or may not be as stated at the closing conference. It is not uncommon for the CSHO to re-evaluate his findings with regard to specific standards to cite for an observed condition, or the likelihood that a citation will stand up to challenge. In any event, the OSHA Area Director makes the final determination with regard to issuing citations.

28.8.8 Employer Preparedness

An employer's preparations to effectively manage an OSHA inspection cannot begin when the compliance officer arrives at the door. Well-meaning, but uninformed, employees could easily increase the severity of an inspection or create unwanted adversarial relationships through inappropriate behavior. To prevent this from occurring, employers should prepare written procedures to be followed in the event of an OSHA inspection or request for information. All managers should be made aware of the procedures, and they should be included in whatever format the employer keeps standard operating procedures or contingency plans readily available (e.g., administration handbooks, site binders, safety manuals, etc.).

The instructions for managing OSHA inspections do not need to be voluminous in order to be effective, but should cover the basic questions and decision making common to all inspections. The fundamental issues to be addressed include:

- Whom to call immediately, with alternates in the event that the primary contact is not available. Many employers want the Safety Director or member of the Safety Staff to be a part of any inspection. Some employers want the legal department involved.
- The company policy regarding search warrants.
- Identification of the employee representative who will be present and the company policy regarding joint or separate conferences (opening and closing).
- The location of records that are likely to be requested during the inspection.
- The importance of taking accurate notes throughout the inspection process.
- How to act during the inspection (i.e., answer questions honestly, stick to the facts, do not speculate, ask questions of the CSHO if unsure of things he/she says).
- The importance of taking dual samples and photographs.
- Taking the CSHO directly to and from the requested sites only.
- What to do with OSHA correspondence (i.e., requests for information, responses to informal complaints, citations) received at locations other than company headquarters or local administrative centers.

28.9 AFTER THE OSHA INSPECTION

Following an inspection, the Compliance Safety and Health Officer will report the findings and recommendations for citations to the Area Director for evaluation and concurrence. If a violation(s) exist, OSHA will issue a Citation and Notification of Penalty detailing the exact nature of the violation(s) and any associated penalties. In addition to describing the violation(s) and penalties, the citation will set a proposed time period within which to correct the violation(s). The OSH Act requires employers to perform specific duties within a prescribed period of time when a citation is issued. The Act also provides specific rights that must be exercised within prescribed time limits to both employers and employees following the receipt of a citation.

28.9.1 Types of Violations

The Act provides for four categories of violations, with minimum and maximum penalties for each category (the penalty structure was amended by public law in 1990 to provide for a tenfold increase in the maximum panelties).

- **Willful:** A willful violation is defined as a violation in which the employer knew that a hazardous condition existed but made no reasonable effort to eliminate it and in which the hazardous condition violated a standard, regulation, or the OSH Act. Penalties range from $5,000 to $70,000 per willful violation, with a minimum penalty of $25,000 for a willful serious violation. Examples include: continued use of machine with nonfunctional interlock, failure to protect employees working in trenches from collapse of trench walls.

- **Serious:** A serious violation exists when the workplace hazard could cause an accident or illness that would most likely result in death or serious physical harm, unless the employer did not know or could not have known of the violation. A mandatory penalty of up to $7,000 may be proposed for each violation. Examples include: failure to maintain adjustment of machine guard, lack of guard rails where there is a potential for a fall, ungrounded electrical equipment.
- **Repeated:** An employer may be cited for a repeated violation if that employer has been cited previously for a substantially similar condition and the citation has become a final order of the Occupational Safety and Health Review Commission. A citation is viewed as a repeated violation if it occurs within 3 years either from the date that the earlier citation becomes a final order or from the final abatement date, whichever is later. Repeated violations can bring a fine of up to $70,000 for each such violation. For employers with multiple facilities, a violation can be cited as repeated if the employer has been cited for the same violation anywhere in the nation within the past 3 years.
- **Other:** A violation that has a direct relationship to job safety and health, but is not serious in nature, is classified as "other." Examples include: lack of exit sign at an obvious exit, omission from Hazard Communication Program of a list of chemicals at worksite, failure to have copy of Hearing Conservation Program Standard available in the workplace

28.9.2 Posting Requirements

Citations will be sent to the employer by certified mail, unless it has been determined that hand delivery would be more effective. A copy of the citation will also be mailed to the appropriate employee representative no later than 1 day after the citation is sent to the employer. Citations may also be mailed to any employee upon request.

Upon receiving a Citation and Notification of Penalty, the employer must post the citation (or a copy of it) at or near the place where each violation occurred to make employees aware of the hazards to which they may be exposed. The citation must remain posted for 3 working days or until the violation is corrected, whichever is longer (Saturdays, Sundays, and Federal holidays are not counted as working days). These posting requirements must be followed even if the citation is contested.

28.9.3 Employer Options

An employer who has received a citation for one or more violations may take either of the following courses of action:

- Agree to the Citation and Notice of Penalty; correct the condition by the date set in the citation and pay the penalty; or
- Within 15 working days of the date the citation was received, contest in writing

any or all of the following: the citation, the proposed penalty, and/or the abatement date.

28.9.4 Informal Conference

Before deciding whether to agree to a citation or file a Notice of Intent to Contest, employers may request an Informal Conference with the OSHA Area Director to discuss the Citation and Notification of Penalty. This opportunity may be used to do any of the following:

- Obtain a better explanation of the violations cited,
- Obtain a more complete understanding of the specific standards that apply,
- Resolve disputed citations and penalties,
- Discuss ways to correct violations,
- Discuss problems concerning the abatement dates,
- Negotiate and enter into an Informal Settlement Agreement,
- Discuss problems concerning employee safety practices, and
- Obtain answers to any other questions.

Employers are encouraged by OSHA to take advantage of the Informal Conference if they foresee any difficulties in complying with any part of the citation. *Employers are cautioned, however, that an Informal Conference will neither extend the 15-working-day period to formally contest a citation nor take the place of actually filing the written notice of contest.* If a contest is not filed within the 15-day period, the citation becomes a final order. After this occurs, the OSHA Area Director may continue to provide information and assistance on how to abate the cited hazards, but may not amend or change any citation or penalty that has become a final order. Also OSHA has begun to consider and use precitation consent decrees with certain employers and proposed large fines.

28.9.5 The Contest Process

A formal contest to a citation is initiated by filing a Notice of Intent to Contest with the OSHA Area Director within 15 working days of receiving the citation. The notice must clearly state what is being contested—the citation, the penalty, the abatement date, or any combination of these factors. In addition, the notice must state whether all the violations on the citation, or just specific violations, are being contested.

A proper contest of any item suspends the employer's legal obligation to abate and pay until the item contested has been administratively resolved. If only the penalty is being contested, all violations must still be corrected by the dates indicated on the citation. Likewise, if only some items on the citation are contested, the other items must be corrected by the abatement date and the corresponding penalties paid within 15 days of notification. A contest must be made in good faith. A contest filed solely to avoid responsibilities for abatement or payment of penalties will not be considered a good-faith contest.

If the written Notice of Intent to Contest has been properly filed and is within the 15 working day period, the OSHA Area Director will forward the case to the Occupational Safety and Health Review Commission (see Section 28.11). At this point the case is officially in litigation, and any settlements must be negotiated according to the rules of procedure of the Commission. The Commission will assign the case to an administrative law judge, who will usually schedule a hearing in a public place close to the employer's workplace. Both employers and employees have the right to participate in this hearing, which contains all the elements of a trial, including examination and cross-examination of witnesses. The administrative law judge may affirm, modify, or eliminate any contested items of the citation or penalty.

As with any other legal procedure, there is an appeals process. Once the administrative law judge has ruled, any party to the case may request a further review by the full Review Commission. In addition, any of the three commissioners may, on his or her own motion, bring the case before the entire Commission for review. The Commission's ruling, in turn, may be appealed to the U.S. Court of Appeals for the circuit in which the case arose or for the circuit where the employer has his or her principal office.

28.9.6 Petition for Modification of Abatement

Abatement dates are assigned on the basis of the best information available at the time the citation is issued. Employers unable to meet an abatement date because of uncontrollable events or other circumstances after the expiration of the 15-working-day contest period may file a Petition for Modification of Abatement (PMA) with the OSHA Area Director. The petition must be in writing and submitted no later than 1 working day after the abatement date. The PMA must include all of the following information before it can be considered:

- Steps taken in an effort to achieve compliance, and the dates they were taken
- Additional time needed to comply
- Why additional time is needed
- Interim steps being taken to safeguard employees against the cited hazard(s) until the abatement is completed
- A certification that the petition has been posted, the date of posting and, when appropriate, a statement that the petition has been furnished to an authorized representative of the affected employees. The petition must remain posted for 10 working days, during which employees may file an objection.

A PMA may be granted or opposed by the OSHA Area Director. If it is opposed, it automatically becomes a contested case before the Review Commission. If a PMA is granted, a monitoring inspection may be conducted to ensure that conditions are as they have been described and that adequate progress toward abatement has been made.

28.9.7 Employee Rights and Courses of Action

Employees or their authorized representatives may contest any or all of the abatement dates set for violations if they believe them to be unreasonable. A written Notice of Intent to Contest must be filed with the OSHA Area Director within 15 working days after the employer receives the citation. The filing on an employee contest does not suspend the employer's obligation to abate. Employees also have the right to object to a PMA. Such objections must be in writing and must be sent to the Area Office within

10 days of service or posting. A decision regarding the PMA will not be made until the issue is resolved by the Review Commission.

In addition to specific rights regarding abatement, employees and their representatives have a right to participate in any Informal Conference or negotiations between the Regional Administrator or Area Director and the employer, and also any hearings held as a result of a formal contest. The Act does not, however, give employees the right to contest the classification of a citation or the dollar amount of any penalty assessed.

28.10 COMPLIANCE STRATEGIES

In developing an occupational health and safety program there are several conceptual models for compliance with government regulations. These models include:

1. Full compliance—absolute compliance with the literal words and actions required under OSHA
2. Substantial compliance—a full understanding of the goals and objectives of OSHA and a full attempt to comply with the spirit of the regulations
3. "Pick and choose" compliance—allowing institutional judgment on complying with so-called "important" standards only
4. Noncompliance—intentional or negligent disregard of government standards applicable to a company's operations

In terms of OSHA compliance reviews the full compliance mode should prevent citations. The substantial compliance model could lead to citations that might pit an institution's professional health and safety judgment against an OSHA interpretation. "Pick and choose" will be likely found in noncompliance at an inspection. Besides ethical concerns, noncompliance will result in citations.

28.11 THE REVIEW COMMISSION

The Occupational Safety and Health Review Commission (OSHRC) was created by the Occupational Safety and Health Act (OSHAct) to serve as a "court" to adjudicate contested matters arising out of the enforcement activity by the Secretary of Labor. An employer may contest citations, penalties, or both within 15 days or affected employees may contest the abatement date within the same period, and both are entitled to a hearing before an Administrative Law Judge (ALJ). ALJs are appointed for life by the Chairman of the Commission under Section 12(e) of the Act.

The Occupational Safety and Health Administration notifies the Commission that an enforcement action is being contested. The case is given a docket number and is assigned to an ALJ. The hearings are adversary proceedings and are conducted according to Federal procedure and evidence rules. The burden of proof lies with the Secretary. Following the hearing, the Judge must issue an order, based on findings of fact and conclusions of law, affirming, modifying or vacating the Secretary's citation or proposed penalty, or directing other appropriate relief. The order will become final within 30 days unless a Commission member directs that the decision shall be reviewed by the Commission itself. If the Commission does review the case, it will reconsider the record and make another decision. The Commission will make a final order. If the employer or employees adversely affected wish to appeal, they may do so by filing a petition for review with the United States Court of Appeals within 60 days.

The Commission itself, pursuant to Section 12(a) of the Act, is composed of three members appointed by the Present, "by and with the advice and consent of the Senate." The term of service is 6 years. The President appoints one of the members to be Chairman. The President can remove one of the members only for "inefficiency, neglect of duty, or malfeasance in office" [S12(b)]. Section 12(a) says that members "shall be appointed . . . from among persons who by reason of training, education, or experience are qualified to carry out the function of the Commission under this Act."

The practice of industrial hygiene and safety has been profoundly affected by the Occupational Safety and Health Act of 1970. Under the Act, the Occupational Safety and Health Review Commission was created as an independent adjudicatory body to review safety and health cases, examine evidence in the record, and issue an impartial decision.

The trends and major viewpoints of the Commission towards industrial hygiene and safety issues are important. One way to review the decisions is to analyze the factors used by the Commissions for recognition, evaluation, and control issues. Recognition parameters in the decisions include the degree of experience and expertise needed by an investigating professional and standards for proof. Evaluation concerns include types of methods and instrumentation to be used and application of Occupational Safety and Health Administration Standards. Control factors involve tests of technical and economic feasibility and proper engineering controls to be implemented.

Although decisions are reviewable on appeal to the Federal Courts, analysis of industrial hygiene and safety practice aspects usually ends with the Commission's decision. Review of the decisions provides a good indicator of the "legal" aspects of industrial hygiene and safety practice such as requirements for evidence.

28.11.1 Preemption

The OSHA Act preempted, at least temporarily, state job safety and health legislation. The constitutional doctrine of Federal supremacy precludes states from enacting or enforcing any conflicting law. Under section 18(a) of the Act, if the Secretary of Labor determines that a state has standards comparable to OSHA's and has an enforcement plan meeting the criteria of 18(c), jurisdiction may be given back to the state.

The Act leaves the choice of submitting a plan to the individual state. If a state does not submit a state plan, it is precluded from enforcing state laws, regulations, or standards relating to issues covered by the Act. This preclusion, however, does not extend to a state's enforcement of a law or standard directed to an issue upon which there is no effective OSHA standard. An "issue" is defined as "an industrial or hazard grouping contained in any of the subparts to the general industry standards." Boilers and elevators are two issues over which OSHA has not promulgated standards and, therefore, over which state enforcement is not preempted.

States without an approved plan also retain jurisdiction in three other areas. States may enforce standards, such as state and local fire regulations, which are designed to protect a wider class of persons than employees. Also, states may conduct consultation, training, and safety information activities. They may enforce standards to protect state and local government employees.

Current OSHA State Plan States as of 1997 are:

Alaska	Michigan	South Carolina
Arizona	Minnesota	Tennessee
California	Nevada	Utah
Connecticut*	New Mexico	Vermont
Hawaii	New York*	(Virgin Islands)
Indiana	North Carolina	Virginia
Iowa	Oregon	Washington
Kentucky	(Puerto Rico)	Wyoming

*Public sector only.

28.11.2 Defenses

Although OSHA standards have been developed for literal compliance, these standards do not impose strict liability; that is, the liability is not absolute and there can be factors used to limit liability. There are procedural defenses concerned with the validity of the enforcement procedures used by OSHA. These include challenges to the issuance of citations and inspection procedures.

Substantive defenses concern the validity and application of a specific standard to the facts of a case, the nature of the employer's conduct, and the impact on the safety of an employee. These defenses must be proved by a preponderance of the evidence.

Examples of substantive defenses include challenging a citation by proving no exposure, no knowledge, no control of the environment; applying the wrong standard to the situation or employer; or vagueness or improper promulgation of an OSHA standard. Other possible defenses include isolated misconduct by an employee or the refusal of an employee to comply. More difficult to prove defenses include impossibilities of compliance, technical infeasibility, economic infeasibility, or compliance creating a greater hazard.

28.11.3 Case Review

Decisions of the Occupational Safety and Health Review Commission reflect the range of issues that must be resolved in interpreting the OSHAct. Thus, recent issues addressed by OSHRC decisions include:

- **Validity of compliance sampling methods.** While OSHA laboratory analysis indicated that air in the breathing zone of grinder operator contained respirable crystalline quartz silica exceeding permissible exposure limit, the employer successfully showed that sampling results were not reliable, because the grinding wheel was composed of 40 percent zirconium oxide, which can interfere with silica analysis, and the lab was not informed of this fact when it conducted its analysis; accordingly, citation items alleging serious violations of 29 CFR 1910.1000(c) (PEL) and 29 CFR 1910.1000(e) (administrative, engineering controls) were vacated. [*Secretary of Labor v. EBAA Iron, Inc. (OSHRC, 2/7/95) 1051*]
- **Quality of compliance officer evidence.** Evidence was insufficient to sustain compliance officer's allegation that an employee painting a bridge 75 feet above

water was not protected by safety net or other means of fall protection, as his testimony was equivocal and inconsistent, and he confused times and places; accordingly, citation alleging serious-repeat violation of 29 CFR 1926.105(a), which requires safety nets for workplaces more than 25 feet above ground or water, is vacated. [*Secretary of Labor v. Dynamic Painting Corp. (OSHRCJ, 1/27/95) 1086*]

- **Application and validity of a standard.** Employers argument that confined space standards were inapplicable to its underground sump because it turned out that sump did not contain any hazards that its evaluators thought were possible is rejected because employer never reclassified sump from permit-only confined space to nonpermit space, and evidence showed that atmospheric hazards were possible in sump that had limited ventilation. [*Secretary of Labor v. NI Industries, Riverbank Army Ammunitions Plant (OSHRCJ, 6/17/95) 1399*]
- **Validity of an industry practice.** Sixteen-foot-long planks used by workers to gain access to their work areas across steel girders qualified as "runways," and the fact that the bridge building industry may not have guarded such planks does not excuse employer's failure to guard in violation of 29 CFR 1926.500(d)(2). [*Secretary of Labor v. Armstrong Steel Erectors, Inc. (OSHRC, 9/20/95) 1385*]
- **Citation under the general duty clause.** Objectives and policies of Occupational Safety and Health Act, and the plain language of OSH Act's general duty clause at Section 5(a)(1), do not support Secretary of Labor's view that he can cite separate Section 5(a)(1) violation for each employee exposed to same hazard; accordingly, unit of prosecution under general duty clause is condition or conditions constituting recognized hazard, rather than number of exposed employees, and case is remanded to allow Secretary opportunity to amend complaint, if appropriate. [*Secretary of Labor v. Arcadian Corp. (OSHRC, 9/15/95) 1345*]
- **Explanation of a control/abatement method.** Employer failed to equip extension cord with ground-fault circuit interrupter, and employer's argument that plugging extension cord into permanent wiring of building made it part of permanent wiring is rejected, as is employer's argument that violation was de minimis (insignificant); accordingly, nonserious violation of 29 CFR 1926.404(b)(1)(i), which requires GFCI or assured equipment grounding conductor program to protect employees, is affirmed. [*Secretary of Labor v. Otis Elevator Co. (OSHRC, 4/4/95) 1167*]
- **Knowledge of a hazard.** Employer knew safety mechanism on machine used to dry towels at car wash was inoperative, and knew employees could open machine while drum was spinning, but allowed machine to operate and exercised little supervision over teenage employees who said they lacked safety training; accordingly, violation of machine guarding standard at 29 CFR 1910.212(a)(4), following incident in which minor employee's arm was pulled off when he inserted his hand in spinning machine, is affirmed as willful [*Secretary of Labor v. Valdak Corp. (OSHRC, 3/29/95) 1135*]
- **Deference: *Secretary of Labor v. OSHRC*.** The U.S. Supreme Court, in *Secretary of Labor v. OSHRC*, ruled that a reviewing court should defer to the Secretary of Labor's reasonable interpretation of an ambiguous OSHA standard rather than the OSHRC reasonable, but conflicting, interpretation. The Court found that the power to render authoritative interpretations of the OSHAct regulations is a "necessary adjunct" of the OSHA powers to promulgate and enforce national health and safety

standards. This gives much more weight to OSHA interpretations over OSHRC interpretations of standards.

This case involved an effort to enforce compliance with the OSHA coke-oven emissions against CF&I Steel Corporation. The standard establish maximum permissible emissions levels and required the use of respirators in certain circumstances in 29 CFR 1910.1029. An investigation by OSHA found that CF&I Steel Corporation had equipped 28 of its employees with respirators that failed an "atmospheric test" designed to determine whether a respirator provides a sufficiently tight fit to protect its wearer from carcinogenic emissions. As a result of being equipped with these loose-fitting respirators, some employees were exposed to coke-oven emissions exceeding the regulatory limit. Based on these findings, the compliance officer issued a citation to CF&I and assessed it a $10,000 penalty for violating 29 CFR 1910.1029(g)(3), which requires an employer to "institute a respiratory protection program in accordance with (29 CFR) 1910.134." CF&I contested the citation.

The ALJ sided with the Secretary, but the full Commission subsequently granted review and vacated the citation. In the Commission's view, the "respiratory protection program" referred to in 1910.1029(g)(3) expressly requires only that an employer train employees in the proper use of respirators; the obligation to assure proper fit of an individual employee's respirator, the Commission noted, was expressly stated in another regulation, namely, 1910.1029(g)(4)(i). Reasoning that the Secretary's interpretation of 1910.1029(g)(3) would render 1910.1029(g)(4) superfluous, the Commission concluded that the facts alleged in the citation and found by the ALJ did not establish a violation of 1910.1029(g)(3). Because 1910.1029(g)(3) was only asserted basis for liability, the Commission vacated the citation.

The question before the Supreme Court in this case was to determine which administrative actor—the Secretary or the Commission—did Congress delegate "interpretive" lawmaking power under the OSHAct. To put this question in perspective, it was necessary to take account of the unusual regulatory structure established by the Act. Under most regulatory schemes, rulemaking, enforcement, and adjudicative powers are combined in a single administrative authority. Under the OSHAct, however, Congress separated enforcement and rulemaking powers from adjudicative powers, assigning these respective functions to two *independent* administrative authorities. The purpose of this "split enforcement" structure was to achieve a greater separation of functions than exists within the traditional "unitary" agency.

Although the Act does not expressly address the issue, the Supreme Court inferred from the structure and history of the statute that the power to render authoritative interpretations of OSH Act regulations is a "necessary adjunct" of the Secretary's powers to promulgate and to enforce national health and safety standards. The Secretary was found to enjoy a readily identifiable structural advantages over the Commission in rendering authoritative interpretations of OSHAct regulations. Because the Secretary promulgates these standards, the Secretary is in a better position than is the Commission to reconstruct the purpose of the regulations in question. Moreover, by virtue of the Secretary's statutory role as enforcer, the Secretary comes into contact with a much greater number of regulatory problems than does the Commission, which encounters only those regulatory episodes resulting in contested citations. Consequently, the Secretary is more likely to develop the expertise relevant to assessing the effect of a particular regulatory interpretation. Because historical familiarity and policymaking expertise account for the presumption that Congress delegates interpretive lawmaking power to the agency rather than to the reviewing court, the Supreme Court found here that Congress intended to invest interpretive power in the administrative actor in the best position to develop these attributes.

It is important properly to review and interpret a decision of the Occupational Health and Safety Review Commission and its administrative law judges. These cases can be accessed through services such as the Bureau of National Affairs (BNA) Occupational Health and Safety Reporters. The case of *Secretary of Labor v. J.L.P., Inc.* is used as an example. The important aspects of this case include, by footnote:

(1) The name of the case, its docket number and date
(2) Name of the attorney for the government
(3) Name of the defense attorney
(4) Name of the administrative law judge (ALJ)
(5) The report by the ALJ summarizing the facts and findings in making a decision
(6) Reference to the specific section of the Code of Federal Regulations OSHA standard that is alleged to be violated
(7) ALJ's findings and conclusion for first citation
(8) Description of other alleged violation
(9) ALJ's findings and conclusion dismissing items

Secretary of Labor v. J.L.P., Inc.
Occupational Safety and Health Review Commission Judge
SECRETARY OF LABOR, Complainant v. J.L.P., Inc., Respondent, OSHRC Docket No. 92-0406, Dec. 28, 1992. **(1)**
H. P. Baker, U.S. Department of Labor, Philadelphia, Pa., for complainant. **(2)**

R. Leonard Vance, Richmond, Va., for respondent. **(3)**

Administrative Law Judge Michael H. Schoenfeld. **(4)**
Digest of Judge's Report **(5)**

Following a fatal accident, the Occupational Safety and Health Administration inspected a work site at the Norfolk Naval Shipyard in Portsmouth, VA. J.L.P. Inc. was the asbestos removal subcontractor at the site.

As a result of the inspection, the company was issued a serious citation that alleged a violation of 29 CFR 1926.500(b)(1), for failure to guard floor openings by a standard railing and toeboard or cover. **(6)**

J.L.P. was removing panels containing asbestos from the building. One of its employees was on the roof when he fell through an opening in the roof. The employee died from his injuries.

The removal of large exhaust fans from their roof mountings left openings in the roof large enough for a person to fall through. Earlier, when J.L.P.'s employees were working on the roof in anticipation of their contracted work, the company had placed covers over the holes. After they were finished with their work, the employees removed the covers. At this time, the company did not anticipate that it would need to perform any other work on the roof.

However, the company received a change order that required its workers to return to the roof to perform additional work. It was at this time the fatal accident occurred.

J.L.P.'s sole defense was the multiemployer work site defense. The company argued that the roof openings were created by another subcontractor when it removed the fans. J.L.P. also argued that another contractor had control over the entire site. Additionally, J.L.P. argued that its hiring of its own safety consultant constituted reasonable steps to protect its employees.

The defense is rejected. Prior to the accident, the company installed temporary covers over the openings. When it returned its employees to the roof, J.L.P. created the violative condition. The fact that it previously installed the covers established that the company had the expertise and personnel necessary to abate the hazard.

Accordingly, the serious citation is affirmed. A penalty of $2,500 is assessed. **(7)**

J.L.P. was also issued an other-than-serious citation that alleged violations of 29 CFR 1926.404(a)(2), for attaching a grounded conductor to a terminal or lead so as to reverse the designated polarity; 29 CFR 1926.404(f)(6), for failure to ensure that the paths to ground from circuits, equipment, and enclosures are permanent and continuous; and 29 CFR 1926.405(g)(2)(iii), for failure to use flexible cords only in continuous lengths without a splice or tap. **(8)**

J.L.P. did not contradict the compliance officer's testimony that a ground-fault circuit interrupter used by employees had reversed polarity, that two extension cords had missing ground prongs, and that the pendant cord of an impact wrench had been spliced.

The record showed, however, that none of these violative conditions had a direct or immediate effect on employee safety. In each instance, the compliance officer testified that serious injury or death would not result from the violative condition. But he did not state what other consequences that were.

Even though the GFCI had reversed polarity, it was still operating. Therefore, no employees were exposed to a shock hazard from using the GFCI.
Given that there was no direct or immediate relationship to employee safety and health, the violations are classified as de minimis. No penalty is assessed. **(9)**

28.12 VOLUNTARY PROTECTION PROGRAMS

The voluntary protection programs (VPP) represent an effort by OSHA to extend worker protection beyond the minimum required by OSHA standards. The programs are designed to recognize and promote effective safety and health management by:

- Recognizing outstanding achievement of those who have successfully incorporated comprehensive safety and health programs into their total management system;
- Motivating others to achieve excellent safety and health results in the same outstanding way; and
- Establishing a relationship between employers, employees, and OSHA that is based on cooperation rather than coercion.

The VPP concept recognizes that compliance enforcement alone can never achieve the objectives of the Occupational Safety and Health Act. Good safety management programs that go beyond OSHA standards can protect workers more effectively than simple compliance. The participants are a select group of employers/facilities whose implementation of health and safety programs qualify them for one of two VPP levels:

- The Star Program is the most demanding and the most prestigious, requiring that all program elements be fully met, and that the facility injury rates be below the national average for the industry. Specific requirements for the program include: management commitment and employee participation; a high-quality worksite analysis program; hazard prevention and control programs; and comprehensive safety and health training for all employees.
- The Merit Program is primarily an intermediate step to Star Program participation. Participants must have a basic safety and health program build around the Star requirements and demonstrate the potential and willingness to implement the necessary steps to meet all Star requirements fully.

The VPP application process is designed to be rigorous, to assure that only the best programs qualify. The benefits of participation, however, are commensurate with the rigor of the program:

- Improved employee motivation to work safely, leading to better quality and productivity
- Reduced workers' compensation costs
- Recognition in the community
- Improvement of programs that are already good, through the internal and external review that is part of the application process
- Exemption from programmed OSHA inspections (OSHA may still investigate major accidents, formal employee complaints, and chemical spills)

28.13 ISO GUIDELINES

The International Organization for Standardization (ISO) was organized just after World War II. ISO is a nongovernmental, international organization based in Geneva, with over 100 member bodies, or countries. It is not affiliated with the United Nations.

Countries are represented in ISO by designated authorities within those countries. For instance, the United States is represented by the American National Standards Institute (ANSI) a private-sector organization.

Traditionally, ISO focused almost exclusively on product and safety standards. These technical standards have been of great value and have helped international commerce, product uniformity and interconnectivity. All the standards that ISO develops are voluntary, consensus, private-sector standards. Since ISO is nongovernment, it has no authority to impose its standards on any country or organization. In addition, technical experts from ISO's member bodies develop ISO Standards through a process of extensive discussion, negotiation, and international consensus. Although the standards are developed for the private sector and are meant to be voluntary, government bodies may elect to convert an ISO standard to a required or legal standard. Such standards may also become conditions of doing business in commercial transactions so that parties may no longer view them as strictly voluntary (Cascio, 1996).

During the 1980s, ISO developed a quality management series, ISO 9000. With the success of the ISO 9000 series, ISO became confident it could develop other organizational standards. With the global concern for the environment, ISO 14000 has been developed to address environmental management systems.

There is growing interest in the United States about using ISO 14001 for regulatory compliance and enforcement programs. While the U.S. Environmental Protection Agency (EPA) and the U.S. Department of Justice (DOJ) have not taken official positions on its use, there is some interest from both government bodies. Official positions from these authorities are not expected before the standards are finalized and judged to be successful. To a significant extent, that success will depend on the integrity and reliability of the third-party conformity assessment system. Government authorities will want some evidence or justification for placing their reliance on ISO 14001 registration. Such evidence must cover the accreditation and registration processes, including the rigor of third-party assessment, the independence of auditors, and the use of appropriate professional safeguards similar to those used in financial audits.

It has been proposed to either include occupational health and safety management as part of the ISO 14000 series or a new series of standards. Concerns exist about the difference of environmental management versus occupational health and safety management such as the impact of labor–management relationships, efficacy of an international management standard, and third-party certifications.

28.14 CLOSING

Regulations have clearly served as a driver for occupational health and safety programs. The passage of the OSHA Act in 1970 encouraged and required employers to implement workplace health and safety initiatives and led to the growth in safety and industrial hygiene professionals. Although regulatory policies may under- or over-target workplace issues, there is the opportunity to influence the process.

Appendix A, Section A.4, indicates possible measurements that could be used to measure regulatory success. The best measure of success would certainly be no workplace accidents or illnesses. However, for management control systems, these measurement alternatives offer a method to sustain an occupational health and safety program.

Occupational health and safety managers must understand the regulatory process and have a strategy to interact at key points along the rulemaking pathway. Input to the process can help develop more focused, cost-effective regulations that put adequate resources in place to prevent workplace accidents and illnesses.

REFERENCES

Cascio, J., et al. (1996). *ISO 14000 Guide: The New International Environmental Management Standards*, McGraw–Hill, New York, NY.

Lofgren, D. J. (1996). "Targeting Industrial and Work Sites for OSHA Health Inspections," *Applied Occupational and Environmental Hygiene Journal* **11**(5), May.

Rothstein, M. A. (1970). *Occupational Safety and Health Law*, West Publishing Company, St. Paul, MN.

Russell, C. (1986). *Enforcing Pollution Control Laws*, Winston Harrington and William Vaughan, Resources for the Future, Washington, D.C.

Worobec, M. D., and C. Hogue (1992). *Toxic Substances Controls Guide: Federal Regulations of Chemicals in the Environment*, The Bureau of National Affairs, Inc., Washington, D.C.

GLOSSARY

Legal Terms in Safety and Industrial Hygiene*

Access: Freedom of approach or communication. The right of access to public records includes not only a legal right of access, but a reasonable opportunity to access. In relation to OSHA's recordkeeping requirements, authorized state or Federal representatives, employees, former employees, or their representatives have access to the log and summary of all recordable occupational injuries and illnesses.

Administrative Law Judge (ALJ): "Judge" means a hearing examiner appointed by the Chairman of the Occupational Safety and Health Review Commission. A hearing examiner appointed by the Commission shall hear, and make a determination upon, any proceeding instituted before the Commission and any motion assigned to such hearing examiner by the Chairman of the Commission, and shall make a report of any such determination which constitutes his final disposition of the proceeding. The report of the hearing examiner shall become the final order of the Commission within 30 days after such report by the hearing examiner, unless within such period any Commission member has directed that such report shall be reviewed by the Commission.

Citation: If, upon inspection or investigation, the Secretary of Labor or his authorized representative believes that an employer has violated a requirement of the OSHAct of 1970, of any standard, rule, or order promulgated pursuant to this Act, or of any regulations prescribed pursuant to this Act, he shall with reasonable promptness issue a citation to the employer. Each citation shall be in writing and shall describe with particularity the nature of the violation, including a reference to the provision of the Act, standard, rule, regulation, or order alleged to have been violated. In addition, the citation shall fix a reasonable time for the abatement of the violation. The Secretary may prescribe procedures for the issuance of a notice in lieu of a citation with respect to de minimis violations that have no direct or immediate relationship to safety or health.

Consensus Standard: [OSHA Act Section 3(9)] The term *national consensus standard* means any occupational safety and health standard or modification that (1) has been adopted and promulgated by a nationally recognized standards—producing organization under procedures whereby it can be determined by the Secretary that persons interested and affected by the scope or provisions of the standard have reached substantial agreement on its adoption, (2) was formulated in a manner which afforded an opportunity for diverse views to be considered, and (3) has been designated as such a standard by the Secretary after consultation with other appropriate Federal agencies.

General Duty Clause: Each employer has a specific duty to comply with occupational safety and health standards promulgated under the OSHAct. In all cases not covered by specific standards, the employer has a general duty "to furnish to each of his employees employment and a place of employment that are free from recognized hazards that are causing or are likely to cause death and serious physical harm to his employees." The general duty clause has been the basis of many citations issued against employers as well as the subject of legal analysis.

*From "Glossary of Legal Terms for the Industrial Hygienist," prepared by the American Industrial Hygiene Association Law Committee, May 1984.

Judicial Review: Any person adversely affected or aggrieved by an order of the Occupational Safety and Health Review Commission issued under the OSHAct may obtain a review of such order in any United States Court of Appeals for the circuit in which the violation is alleged to have occurred or where the employer has its principal office, by filing in such court within 60 days following the issuance of such a order a written petition praying that the order be modified or set aside. Upon filing, the court shall have jurisdiction of the proceeding and of the question determined, and shall have power to grant such temporary relief or restraining order as it deems just and proper, and to make and enter upon the pleadings, testimony, and proceedings set forth in such record a decree affirming, modifying, or setting aside in whole or part, the order of the Commission and enforcing the same to the extent that such order is affirmed or modified. Upon the filing of the record with the court, the jurisdiction of the court shall be exclusive and its judgment and decree shall be final, except that the same shall be subject to review by the Supreme Court of the United States.

Multiemployer Worksite: A multiemployer worksite exists where several employers are working on a single worksite. For example, on a construction site, several employers are involved with varying degrees of responsibility for completion of the work. The general contractor has overall responsibility for all work, and the various subcontractors have responsibility for various portions of the work (e.g., electrical, steel erection, plumbing, concrete). The OSHRC has fashioned a rule that where an employer is in control of an area and responsible for its maintenance, the Secretary can sustain a citation against such employer by showing that a standard has been violated and that employees—those of the cited employer or those of other employers engaged in a common undertaking—had access to the zone of danger. Stated more simply, the rule is that the employer who creates a hazard may be held to answer for it under OSHA whether his, or employees of others, are imperiled.

Occupational Safety and Health Review Commission: The Commission is composed of three members who are appointed by the President, by and with the advice and consent of the Senate, from among persons who by reason of training, education, or experience are qualified to carry out the functions of the Commission under this act. The President shall designate one of the members of the Commission to serve as Chairman. A member of the Commission may be removed by the President for inefficiency, neglect of duty, or malfeasance in office.

The Commission's function is to carry out adjudicatory functions under the Act. It is empowered to consider and decide enforcement actions initiated by the Secretary of Labor under the Act that are contested by employers, employees, or representatives of employees.

OSHA: The Occupational Safety and Health Administration, a branch of the Department of Labor. Established by the Occupational Safety and Health Act of 1970 (OSHAct), this law is also known as P1 91-596 or as the Williams–Steiger Act.

The purpose of OSHA is found in Section II of the OSHAct. The key points are:

1. "by encouraging employers and employees in their efforts to reduce the number of occupational safety and health hazards at their places of employment, and to stimulate employers and employees to institute new and to perfect existing programs for providing safe and healthful working conditions."

2. "by authorizing the Secretary of Labor to set mandatory occupational safety and health standards applicable to business effecting interstate commerce, and by creating an Occupation Health and Safety Review Commission for carrying out adjudicatory functions under the Act."

3. "by providing for the development and promulgation of occupational safety and health standards,"

4. "by providing an effective and enforcement program which shall include a prohibition against giving advance notion of any inspection and sanctions or any individual violating this prohibition."

OSHA Record Keeping: The Occupational Safety and Health Act of 1970 requires employers to prepare and maintain records of occupational illness and injury. Ordinarily, records must be maintained at each establishment, although in some instances records can be located at a central place or where employees report to work each morning. Only two forms must be maintained (1) the Log and Summary, form #200, and (2) the Supplementary record, form #101. The Log is a convenient way of classifying injury and illness cases and for noting the extent of and outcome of each. The Supplementary record is designed to record additional information about the occurrence. All records must remain in the establishment for 5 years after the year to which they relate. Federal or state government representatives have the authority inspect or copy these records. Small employers, which employ no more than 100 full- or part-time employees at any one time, are exempt form the record-keeping requirements.

Recognized Hazard: An actual or publicly known risk or peril.

Right of Entry: The right of taking or resuming possession of land by entering on it in a peaceable manner.

Right to Know: State and Federal regulation requiring manufactures, importers, and employers to assess the hazards of materials which they produce, import, or have in their work place and provide information to their employees concerning the hazards of these materials. The purpose of the OSHA Hazard Communication Standard is that the hazards of all chemicals produced or imported by chemical manufacturers or importers are evaluated and that information concerning their hazards is transmitted to affected employers and employees. This transmittal of information is to be accomplished by means of comprehensive hazard communication programs, which are to include container labeling and other forms of warning, material safety data sheets, and employee training.

Settlement: A settlement can be defined as an act or process of adjusting or determining; an adjustment between persons concerning their dealings or difficulties; an agreement by which parties having disputed matters between them reach or ascertain what is coming from one to the other. A settlement could also be a compromise between the adverse parties in a civil suit before the final judgment, thus eliminating the necessity of a judicial resolution.

State Plan: Provision in OSHAct that allows the State to assert jurisdiction over occupational safety and health matters. There are a number of provisions that must be met before the Secretary may approve the State plan. These provisions are found in Section

18(c) of the OSHAct and include establishing of a State Agency to administer the plan, establish and enforce safety and health standards that must be at least effective in providing safe and healthful employment as Federal standards, and provide for means of inspection and enforcement of standards.

Statutes of Repose: A legislative act defining or delineating a time limit upon which action ceases. A statute of limitations is an example of a statute of repose.

Statue of Limitations: A legislative act prescribing a certain time to the right of action on certain described causes of action or criminal prosecutions. Describes the time limits within which certain actions may be brought.

Statutory Law: That law that is created by an act of a legislature declaring, commanding, or prohibiting something; a particular law enacted and established by the will of the legislative department of government; the written will of the legislature, solemnly expressed according to the forms necessary to constitute it the law of the state.

Stay: The act of interrupting or postponing the legal effectiveness of an action usually by order of a court. As an example, when the validity of the OSHA Emergency Temporary Standard for Asbestos was challenged, it was temporarily stayed by the court with jurisdiction of the lawsuit pending a final decision on the validity of the standard.

Unsafe working conditions: One that violates an OSHA standard or the general duty clause. In regard to the OSHAct of 1970, each employer has a responsibility to comply with the occupational safety and health standards promulgated under the act.

Variance: Any employer or class of employers may apply for a variance from an occupational safety and health standard and the variance sought may be temporary or permanent. Variance applications to the Federal Government must be submitted in accordance with OSHA regulations set forth in 29 CFR part 1905. Applications to OSHA for a permanent variance, for example, must include the name and address of the applicant, the addresses of any places of employment involved, a description of the work process proposed, a statement that the proposed work process would provide a level of safety equal to what would exist if the standard were met, certification that affected employees were informed of the application, any request for a hearing, and a description of how the employees were informed of the application and their right to petition for a hearing. Application for a variance also may include a request for an interim order to be effective until action is taken on the application. A request for an interim order may include statements of fact and arguments as to why the order should be granted.

Walkaround Rights: The right to accompany an OSHA inspector on the inspection tour is guaranteed under Section 9(e) of the Act for both the employer and a representative of the employees:

> "Subject to the regulations issued by the Secretary, a representative of the employer and a representative authorized by his employees shall be given an opportunity to accompany the Secretary of his authorized representative during the physical inspection of any workplace under subsection (a) for the purpose of aiding such inspection. Where there is no authorized employee representative, the Secretary or this authorized representative shall consult with a reasonable number of employees concerning matters of health and safety in the workplace."

Warrant: A writ or precept from a competent authority in pursuance of law, directing the doing of an act, and addressed to an officer or person competent to do the act, and affording him protection from damage, if he does it. Particularly, a writ or precept issued by a magistrate, justice, or other competent authority, addressed to a sheriff, constable, or other officer, requiring him to arrest the person named, and bring him before the magistrate or court, to answer, or to be examined, touching some offense which he is charged with having committed.

APPENDIX A

A.1 Code of Federal Regulations

The Code of Federal Regulations is a systematic method of codifying, organizing, and publishing the various regulations promulgated by the agencies of the Federal Government. The CFRs are organized by governmental department, e.g., 29 CFR pertains to regulations of the Department of Labor, while 49 CFR is the Department of Transportation. Copies of the CFRs may be purchased from the U.S. Government Printing Office, and in some cases obtained from the Regional Offices of the respective departments. Many of the CFRs are now available free of charge on the Internet. All the OSHA Standards, Regulations, Compliance Directives, and Letters of Interpretation may be found via the OSHA Web Site (http://www.osha.gov/).

A.2 "Should" vs. "Shall" Standards

Many of the standards adopted under Section 6(a) were American National Standards Institute (ANSI) or National Fire Protection Association (NFPA) consensus standards that were incorporated by reference and contained advisory provisions (e.g., use the word "should" rather than "shall").

In the past, OSHA maintained that all standards, regardless of whether the term "should" or "shall" is used, created mandatory compliance responsibilities. Employers consistently challenged this position on the basis that Section 6(a) of the Act only gave OSHA the authority to adopt ANSI standards verbatim. In ANSI standards, use of the term "should" means that the provision is only advisory. Therefore, employers maintained that ANSI "should" standards could only be advisory when adopted or incorporated by reference by OSHA under Section 6(a). Enforcement of "should" standards has been denied by the Occupational Safety and Health Review Commission, and by most of the appellate courts in which contested cases have been heard.

Although the "should" standards have not been enforceable in and of themselves, OSHA has employed them to demonstrate the existence of "recognized hazards" under the general duty clause [Section 5(a)(1)] of the Act. However, the Review Commission has ruled that, as long as the "should" provision remains in effect as an OSHA standard, OSHA may not issue a general duty clause citation for the hazard it addresses.

In 1984, OSHA conducted a rulemaking for 29 CFR part 1910 (General Industry Standards) to clarify that only the mandatory provisions of standards incorporated by reference are adopted as OSHA standards and removed 153 advisory provisions.

A.3 Notice of Proposed Rulemaking

When proposing a new standard, or modification to an existing standard, the official notice to the public is a Notice of Proposed Rulemaking (NPRM) published in the *Federal Register*. The NPRM follows a standardized format: Summary, Supplemental Information, and Text of Proposed Standard.

The Summary contains a very brief synopsis of what is being proposed, the basis for the proposal, and the response date and address for public comment. The Proposed Text section contains the exact text and structure of the proposed standard, including illustrations, forms, or appendices that will accompany the standard.

While the general format of the Supplemental Information section is the same for all ANPRs, the number and content of the subsections will vary from standard to standard, depending on the complexity of the issue. A typical Supplemental Information section will contain the following major subsections:

Background—review of existing standards, applicable consensus standards, petitions, and requests for a standard, basis for proposal

Technical information—review of the equipment, substance, or procedure that is being considered for a new standard

Hazard analysis—review of the hazards associated with the subject of the proposal

Accident, illness, and injury data—detailed, technical analysis of the prevalence and severity of the hazards by Standard Industrial Classification

Summary and explanation of the proposed rule—description of the various requirements of the proposed rules, including alternatives considered and rationale for the proposal

Specific issues for comment—in some NPRMs, OSHA solicits comments on specific issues and alternatives in addition to general comments on the proposal

Regulatory impact assessment—expected effectiveness of the proposal in terms of industry specific reduction in injuries, fatalities, etc.

Economic and feasibility analysis—review of the technical feasibility of the proposal and the cost impact to affected industries

Public participation—invitation for public comments with closing deadline and address for submission, dates and locations for public hearings if scheduled

In addition to the major subsections, the Supplemental Information section also addresses issues that are necessary to satisfy statuary requirements: Environmental Impact, Compliance with Paperwork Reduction Act, Regulatory Flexibility Analysis, Impact on State Plan States, and Legal Authority and Signature.

A.4 Measurements of Regulatory Success

- Successful integration of occupational health/safety position into laws/regulation
- Defeat or passage of specific occupational health/safety legislation/rulemaking
- Lack of OSHA administration citations
- Successful challenges of administrative OSHA citations
- Sustainable national interest—protection of working women and men from illness/injuries

CHAPTER 29

Occupational Safety and Health Legal Issues

THEA DUNMIRE

ENLAR Compliance Services, Inc., PO Box 3520, Clearwater, FL 33767

29.1 UNDERSTANDING THE LAW
- 29.1.1 Statutory Law
- 29.1.2 Common Law
- 29.1.3 Evolution of the Law

29.2 OVERVIEW OF COMMONLY OCCURRING LEGAL CLAIMS
- 29.2.1 Tort Claims
- 29.2.2 Contract Claims
- 29.2.3 Claims Based on Statutory Law

29.3 HOW A LEGAL CASE PROCEEDS
- 29.3.1 Initial Pleadings
- 29.3.2 Discovery
- 29.3.3 The Trial and Decision

29.4 RECORDKEEPING—LEGAL ISSUES
- 29.4.1 Good Records Management Helps Prevent Legal Actions
- 29.4.2 Dealing with Privileged and Confidential Documents
- 29.4.3 Record Retention Policies

29.5 WHEN TO CONSIDER SEEKING LEGAL ADVICE
- 29.5.1 To Determine Legal Obligations
- 29.5.2 When Negotiating Contracts
- 29.5.3 When Determining Legal Compliance
- 29.5.4 When Served with Legal Documents

REFERENCES

29.1 UNDERSTANDING THE LAW

Most Safety and Health Professionals are familiar with the Occupational Safety and Health Administration (OSHA) regulatory requirements. Many are also familiar with other regulations such as those promulgated by the Environmental Protection Agency,

Handbook of Occupational Safety and Health, Second Edition, Edited by Louis J. DiBerardinis,
ISBN 0-471-16017-2

the Department of Transportation, and various state agencies.[1] Regulations are not, however, the only legal issues that may impact safety and health professionals. In addition to regulations, there are the laws passed by Congress and state legislatures as statutes and an entire body of law that is simply referred to as "common law."

29.1.1 Statutory Law

Statutes are usually written down, or codified, in the United States Code (for Federal laws) or state statute books (for state laws).[2] Although statutory laws are written down, interpreting their meaning is not simply a matter of reading the words. The meaning of a particular law can be significantly altered because of subsequent court decisions. In addition, a statute or regulation may be unenforceable because it is unconstitutional or because it is pre-empted[3] by another law.

An example of the pre-emptive effect of a Federal law is provided by the case of *Gade v. National Solid Wastes Management Association.*[4] In 1988, Illinois enacted the Hazardous Waste Crane and Hoisting Equipment Operators Licensing Act. Under the Act, persons working at hazardous waste cleanup sites were required to obtain a state license. In order to get the license, persons had to pass a written examination prescribed by the Illinois Environmental Protection Act. The Act authorized employers to be fined if they permitted employees to work without having the required state licenses. The Supreme Court invalidated the state law by finding that this state regulation of workers at hazardous waste sites had been pre-empted by the Occupational Safety and Health Act and the regulations OSHA had promulgated to protect workers at hazardous waste sites.[5]

29.1.2 Common Law

The common law is a set of legal principles determined by analyzing how and why judges have decided individual cases. As such, the common law is not something one can simply look up in a book. Instead, common law evolves over time and may vary from state to state depending on the decisions made by individual judges in various courts.

Because case decisions are based on specific disputes between individual parties, they are case or fact specific. Therefore, similar disputes may have very different outcomes either because of differences in the underlying facts or because a court's view of what the law requires is different. These differences in the interpretation of the law are, in some instances, resolved by the highest state court or the U.S. Supreme Court.

An example of such a circumstance involves the admissibility of scientific testimony in Federal court. Following the enactment of the Federal Rules of Evidence in 1975, several Federal district courts came to different conclusions about when scientific principles or discoveries were sufficiently credible to be admissible as evidence or, in the alternative, were inadmissible as "junk science." Different courts used different criteria until the Supreme Court resolved the issue in 1991 in a case called *Daubert v. Merrell Dow Pharmaceuticals.*[6]

In *Daubert*, a suit was filed alleging that Bendectin, an anti-morning-sickness drug manufactured by Merrell Dow, caused limb deformities. In support of their case, the plaintiffs proposed offering a reanalysis of previously published epidemiological studies that had been developed solely for the purposes of the trial.[7] This reanalysis was unpublished and had not been peer reviewed. Based on the lack of peer review, the district court determined that the plaintiff's evidence was insufficient and ruled in favor of the defendant. The circuit court upheld the decision. The Supreme Court overturned the lower courts' decisions and, in its opinion, set out the guidelines courts should follow in determining the admissibility of scientific expert testimony under the Federal Rules of Evidence.

29.1.3 Evolution of the Law

Under both statutory and common law, legal requirements may also change over time. This change may be the result of one or more of the following factors: Society's views about certain issues may have changed; there may be changes in technology; new information may be available about an issue or the courts may interpret the law differently. An example of an issue that has been treated differently at different times is protection of wetlands. Prior to the 1960s, wetlands were believed to be useless, nonproductive swamps that bred snakes and mosquitoes. Public policy and laws encouraged projects to drain and fill wetlands to convert them to "better" uses. This policy was reversed in the 1970s and Federal laws were enacted to protect wetlands and to prevent them from being filled in. Then in the 1980s, substantial controversy arose over wetland delineation, the science of determining exactly when property should be considered a wetland subject to environmental protection.

Given the fact-specific nature of the common law, the evolution of the law over time, and the differences in legal requirements geographically from jurisdiction to jurisdiction, no attempt is made in this chapter to define the specific legal obligations of safety and health professionals. Determining the potential legal ramifications of particular activities can be exceedingly complex. Instead, the goals of this chapter are to highlight some commonly occurring legal claims, provide an overview of the legal process, discuss specific legal concerns related to record keeping, and highlight specific circumstances when safety and health professionals may want to consider seeking legal advice.

29.2 OVERVIEW OF COMMONLY OCCURRING LEGAL CLAIMS

Cases may be brought under a variety of legal theories. Some are based on the rights and obligations created as the result of Federal or state statutes. Others are brought based on common law claims. Two categories of common law claims are tort claims and contract claims.

29.2.1 Tort Claims

A tort is "a private or civil wrong or injury, other than a breach of contract, for which the court will provide a remedy in the form of an action for damages."[8] The tort claims that most commonly arise in the safety and health field are negligence, misrepresentation, and strict products liability.

Negligence. The most frequently alleged tort claim is negligence. In order to establish negligence, the plaintiff must prove that the defendant owed him or her a duty, the defendant breached that duty, and the breach proximately caused injury or damage to the plaintiff.[9] Whether a duty exists on the part of the defendant is dependant on both the relationship between the parties and the existence of laws or regulations imposing some sort of legal obligation. The types of relationships that may create legal duties include landlord–tenant, consultant–client, and employer–employee.

An example of a negligence action brought because of a landlord–tenant relationship is *Wright v. McDonald's Corp.*[10] In this case, Mr. Wright alleged he suffered permanent neurological injury caused by carbon monoxide exposure resulting from a defectively installed water heater in the McDonald's restaurant he operated under a franchise and lease agreement. In a decision on a summary judgment motion,[11] the court stated that a jury could find McDonald's liable if they concluded that McDonald's "designed and constructed a ventilation system with a latent dangerous defect of which it should have been aware and which Mr. Wright could not uncover by reasonable inspection."

In certain circumstances, OSHA or environmental regulations may be used to establish the existence of the legal duty giving rise to a negligence claim. An example of such a circumstance is the case of *Arnett v. Environmental Science & Engineering, Inc.*[12] In this case, Mr. Arnett alleged he was injured by the fumes generated as the result of his job duties involving the pouring of chemicals into 55-gallon drums during an asbestos abatement project. The suit was filed against Environmental Science and Engineering (ESE) and an individual, Charles Jenkins, in their capacities as the asbestos project managers for the project.[13] ESE was not Mr. Arnett's employer. In the case, ESE argued that it did not have any legal duty to protect Mr. Arnett from exposure to fumes because that was his employer's responsibility. The court disagreed. In reaching its decision, the court referred to the provisions of the Illinois Asbestos Abatement Act, which imposed certain duties and obligations on the project manager. Based on those provisions, the court stated that the project manager was not free to ignore the conditions existing at the abatement work site and, further, that it was the duty of the project manager to require the contractor to take appropriate safety precautions.

Misrepresentation. Another tort claim a plaintiff might allege is misrepresentation—either fraudulent or negligent. The elements of fraudulent misrepresentation are as follows: a misrepresentation of fact which the defendant knew was false, an intent on the part of the defendant that the plaintiff rely on the misrepresentation for some action, and damage or injury to the plaintiff caused by that person's justifiable reliance on the misrepresentation. The kind of acts that could be considered fraudulent misrepresentations include removing manufacturers' warning labels on toxic substance containers, knowingly providing inadequte safety equipment, and misrepresenting the danger or extent of the toxicity of materials and the need for proper safety equipment.[14]

An example of a case alleging fraudulent misrepresentation is *AltairStrickland Inc. v. Chevron U.S.A. Products Co.*[15] On March 6, 1995, AltairStrickland Inc. (ASI) began work as the prime contractor for maintenance and construction work to be done during a refinery shutdown at Chevron's El Paso, Texas, refinery. Almost immediately ASI's workers started complaining of rashes and respiratory problems. On March 17, ASI refused to continue work, claiming that its workers were being exposed to high levels of sulfur dioxide. On March 18, Chevron locked ASI out of the refinery and replaced it as the contactor for the project. ASI sued Chevron, charging it has committed fraud "by intentionally misrepresenting that work conditions in and around the crude towers at the refinery were safe, when they were not." The jury awarded ASI $5 million in compensatory damages and $38.5 million in punitive damages.

A defendant would engage in negligent misrepresentation, as set out in the Restatement of Torts (a summary of the common law often referred to by the courts), if "in the course of his or her business, profession or employment, he or she supplied false information for the guidance of others in their business transaction and failed to exercise reasonable care or competence in obtaining or communicating the information.[11]

In both fraudulent and negligent misrepresentation the information transmitted and relied on was false. The difference lies in whether the defendant knew the information being transmitted was false (fraudulent misrepresentation) or the defendant was careless in obtaining and communicating the information (negligent misrepresentation). A number of cases have been brought against consultants alleging negligent misrepresentation. These cases have involved misrepresentation of laboratory data,[16] misrepresentations concerning the condition of property in environmental assessment reports,[17] and misrepresentation of the regulatory compliance status of facilities in audit reports.[18]

Strict Products Liability. Strict liability is a legal concept often applied by the courts in product liability cases pursuant to which the seller of "any product in a defective condition unreasonably dangerous to the user" is liable for any damages caused by the defective product. The difference between negligence and strict liability is that strict liability is liability without fault. Lawsuits alleging strict liability focus on the product that caused the injury rather than the actions of the defendants.

A product may be held to be defective because it does not carry an adequate warning to prevent it from being inherently dangerous. "Failure to warn" is one of the primary claims made in the lawsuits brought alleging asbestos-related injuries of workers. One of the earliest cases holding that manufacturers had a duty to warn workers of the dangers associated with the use of their products was *Borel v. Fibreboard Paper Products Corp.*[19] In this case, the court stated that a manufacturer "must keep abreast of scientific knowledge, discoveries and advances and is presumed to know what is imparted thereby. ... A product must not be made available to the public without disclosure of those dangers that the application of reasonable foresight would reveal."

Damages. The one element common to all tort cases is that the plaintiff has suffered some damage or injury. Therefore, one of the most effective ways of avoiding tort liability is to prevent injury to persons or property. Safety and health professionals can help to prevent tort claims being brought against their companies by developing programs to identify and then eliminate safety and health hazards that cause injuries and illnesses.

29.2.2 Contract Claims

A contrast is simply an agreement between two or more parties to engage in certain acts in the future. Except in certain circumstances, an agreement does not have to be in writing to be valid. It can consist of the statements made during a series of telephone calls. It is also possible for a "contract" to be the agreement reached as the result of a series of separate documents (e.g., a proposal, response, purchase order, and invoice for a particular project). Common contract claims include breach of contract, breach of warranty, or indemnification claims.

Breach of Contract. A breach of contract is a failure, without legal excuse, to comply with a provision, or obligation, of the contract. Typical contract obligations include promises to perform certain tasks within certain time periods, obligations to meet certain conditions (e.g., licensing requirements), and promises that the work will meet certain specifications. Other contract provisions include warranties, indemnification, and dispute resolution mechanisms. Contract problems often arise because of vague contract language or because the contract is incomplete.

An example of a breach of contract case is *Power Engineering Co. v. Envirochem Service, L.C.*[20] In this case the court held that an environmental services broker that agreed to arrange for the disposal of a generator's hazardous waste materially breached its contract when it failed to properly dispose of the waste.

Breach of Warranty. A warranty is a contractual promise that a product or service will attain a particular level of quality. If it does not, then there is a potential "breach of warranty" claim. An example of a warranty would read:

> Consultant warrants to Company that all services supplied by Consultant in the performance of this Agreement shall be supplied by personnel who are licensed or certified, as required by applicable law, and experienced and skilled in their respective professions.

If, in fact, personnel were then used on the job who were not licensed, certified, or experienced, the Company would have a breach of warranty claim agains the consultant.

Indemnification Claims. An indemnification provision in a contract is an agreement on the part of one party that it will compensate the other party for any loss or damage arising from some specified act or acts. A typical indemnification provision reads:

> "[Party A] will indemnify and hold harmless [Party B] from and against any claims, damages, losses and expenses arising out of or resulting from the performance of the Work [as defined in the agreement].

Such an indemnification provision might be limited by adding certain conditions to the indemnification obligation, such as requiring that the party providing the indemnification be negligent before compensation would be required.

Indemnification provisions are similar to insurance policies in that they shift the risk of loss to other parties. Just like insurance policies, indemnification provisions will not provide compensation if the party providing the indemnification has no assets to cover the potential loss.

In addition, statutory provisions can affect the extent to which parties can shift the risk to others. An example of the effect of an indemnification provision in a contract is provided by the case of *Bradford v. Kupper Associates.*[21] In this case, a municipal utility contracted with a company, Agate Construction Company, to replace and rehabilitate portions of its sewer lines. Prior to the work commencing, the municipal utility retained an engineering firm, Kupper Associates, to provide engineering documents and engineering review of the work as it progressed. In the course of the work, two employees of Agate Construction Company entered into the sewer system. Bradford died as a result of "asphyxia due to exposure to toxic fumes and stagnant air syndrome." The executrix of Mr. Bradford's estate sued the municipal utility and Kupper Associates for negligence. In its decision, the court held that, although the state's workers compensation statute would have barred a suit by the municipality and Kupper Associates against Agate Construction Company, since the parties had negotiated a contract with a valid indemnification provision, Agate Construction Company was required to indemnify those parties for the attorney's fees they incurred in defending the action brought by Mr. Bradford's estate.

29.2.3 Claims Based on Statutory Law

In a number of areas, the common law has been either supplemented or replaced by statutory law. Examples of such laws would include the Comprehensive Environmental Response, Compensation and Liability Act (CERCLA), workers compensation, and the

Americans with Disabilities Act (ADA). Safety and health professionals also need to be aware that cases may be brought against them or their companies based on violations of Federal or state criminal statutes.

CERCLA. The Comprehensive Environmental Response, Compensation and Liability Act, commonly referred to as "CERCLA" or "Superfund" was enacted in 1980. One of the principal effects of CERCLA is that it imposes strict liability on parties who owned or operated sites where cleanup actions are initiated or who generated or transported hazardous substances that wound up at such sites. There are numerous court decisions interpreting the CERCLA liability provisions.

One of the principal results of CERCLA's liability provisions is that buyers and sellers of real estate began conducting environmental assessments in conjunction with property transfers. The content and format of these environmental assessments has evolved over time and, in some cases, now include consideration of occupational safety and health issues. Safety and health professionals may become involved in compiling information for use in environmental assessments.

Workers Compensation. All states have workers compensation laws that have, to some extent, replaced common law claims on the part of injured workers. Under workers compensation, employees are compensated for injuries of illnesses arising out of or in the course of their employment without regard to the fault of the worker or the employer. The award to the injured employee is limited to lost wages, medical expenses, and set compensation levels based on the state's workers compensation schedule. Intentional misconduct on the part of the employee or the employer can, however, eliminate the exclusive nature of workers compensation.

Americans with Disabilities Act. The Americans with Disabilities Act of 1990 was enacted to establish a national program to eliminate discrimination against individuals with disabilities. Pursuant to this law the Equal Employment Opportunity Commission (EEOC) has developed regulatory standards dealing with equal employment opportunities for disabled individuals. No employee may discriminate on the basis of disability with regard to any terms, conditions, or privileges of employment.

The ADA was written to ensure that qualified individuals are not discriminated against in hiring and employment. An individual is qualified if he or she can perform the *essential functions* of a job. In addition, employers must make reasonable accommodation for the known physical or mental limitations of an otherwise qualified individual with a disability. As a result, it is important for employers to define the essential functions of each job position. For many jobs, written job descriptions will include safety and health issues such as the requirement that certain personal protective equipment be worn or that periodic medical testing is required (e.g., audiometric testing).

Criminal Statutes. Both Federal and state prosecutors may seek criminal indictments when they believe there has been sufficiently egregious misconduct or violations of safety, health, or environmental laws. Almost all Federal environmental statutes, as well as the Occupational Safety and Health Act, have criminal penalty provisions. In addition, Title 18 of the United States Code sets out criminal offenses that are generally applicable to a wide variety of activities (e.g., making false statements to government officials). In addition, criminal charges may be brought under general criminal statutes such as those prohibiting involuntary manslaughter or criminally negligent homicide.

In particular, safety and health professionals should be aware that many environmental states contain provisions imposing substantial penalties for making false statements and for destroying, altering, or concealing records.

29.3 HOW A LEGAL CASE PROCEEDS

Just as there are differences between regulatory requirements and statutory or common law obligations, there are also significant differences between a regulatory enforcement action and a court case.[22]

The conduct of a case is governed by the procedural rules of the court in which it is filed. In the Federal courts, the Federal Rules of Civil Procedure govern all civil cases. These rules govern how documents are prepared, how many pages a document filed with the court may have, and the time periods the parties are required to meet in answering pleadings and motions. Once a case is filed, the court has substantial power over the parties. The court can order that certain documents be released, that parties appear in front of the court at specific times, and that certain case-preparation deadlines be met. If the court's orders are not compiled with, the court can impose penalties on the parties.

All legal cases start out in a trial court—a municipal court, state court, or Federal district court. If one of the parties believes that the judge has committed some error in the case, the case may be appealed to a higher court. In the Federal court system, cases are appealed to the Court of Appeals and, finally, the U.S. Supreme Court. Appellate courts do not, however, retry cases. If they decide some mistake has been made, they send the case back to the trial court for further proceedings.

29.3.1 Initial Pleadings

In most cases, the first formal notice of a lawsuit is the receipt of a complaint. The purpose of the complaint is to allege some legal cause of action and put the opposing party on notice concerning the allegations the plaintiff believes supports his or her claim for damages. Unless some sort of motion to dismiss the complaint is filed, the defendant files a response called an answer. In the answer, the defendant typically responds to each of the allegations in the complaint by admitting it, denying it, or stating that there is insufficient information to either admit or deny the allegation. The answer will often also contain what are called affirmative defenses and counterclaims. Counterclaims are claims for damages that the defendant makes against the plaintiff. It is not uncommon for other paties to be brought into the action at this point by the defendant, the plaintiff, or both. Typically there will also be a variety of motions filed with the court that will need to be resolved before the case can proceed.

29.3.2 Discovery

Once the initial pleadings are filed and the preliminary motions are resolved, the parties enter into an exchange of information concerning the issues in the case. This is typically called discovery. In the course of discovery each party tries to obtain detailed information concerning the other party's claims. The Federal Rules of Civil Procedure set forth a number of different discovery mechanisms, including depositions, interrogatories, requests for production of documents, requests for entry upon land, and requests for admission. Certain types of records are protected from discovery (see Section 29.4.2).

Depositions. Lawyers for the parties can subpoena individuals associated with the case to give oral testimony under oath. The witness is sworn to tell the truth, and a transcript of the testimony is prepared. This testimony is used by the parties to help them prepare their cases. The deposition transcript can be used during the trial to impeach a witness if his or her testimony conflicts with the statements made during the deposition.

Interrogations and Requests for Production of Documents. Interrogatories are written questions sent to the other side to be answered. Typically, a request for production of documents is sent with the interrogatories. This request typically asks for copies of any documents referenced in preparing answers to the interrogatories and copies of all records related to the case.

29.3.3 The Trial and Decision

Once discovery is completed and all the pretrial motions are resolved, the case is set for trial. Depending on the case, the case may either be decided by a jury or by the judge. In some instances, decisions made in the lower court may be appealed to a higher court for review.

29.4 RECORDKEEPING—LEGAL ISSUES

Developing and maintaining a safety and health program typically involves a substantial amount of document creation and record management. For example, safety and health professionals are required to develop various written policies and procedures, maintain information on toxic materials (e.g., material safety data sheets), maintain training records, and keep records concerning employee injuries. As a result, many safety and health professionals soon find they are dealing with a vast accumulation of paper. This "ocean of paper" eventually necessitates the development of a records management program. There are several legal considerations that need to be kept in mind in developing a records management program.

29.4.1 Good Records Management Helps Prevent Legal Actions

A good records management program can help avoid legal actions. First, by having all legally mandated records maintained so they are easily retrieved, a legal action can be avoided because records are available when they were needed (e.g., when requested by a government inspector or when an emergency happens). Second, it is much easier to identify gaps in safety and health program documentation when documents are well organized. Once such gaps are identified, they can be corrected.

29.4.2 Dealing with Privileged and Confidential Documents

A safety and health records management program needs to adequately protect confidential documents. Written procedures should be established to prevent the dissemination of confidential documents. In particular, confidential documents should *never* be filed with documents that are required to be provided to governmental agencies upon their request (e.g., permits, OSHA logs, emergency procedures). Documents with confidentiality concerns would include privileged documents (those subject to attorney–client or attorney–work product privileges), documents containing confidential business information (e.g., trade secrets), and documents containing personal information (e.g., employee medical information).

Attorney–Client and Attorney–Work Product Privileges. Certain documents are protected from disclosure in a legal proceeding. These are typically referred to as privileged documents. There are two well-recognized privileges—attorney–client correspondence and attorney–work product.[23] It should be noted that simply marking a document as "confidential" does not establish it as a document protected from disclosure.[24] Specific steps must be taken in order to establish these privileges. In addition, if privileged documents are released to other parties, their privileged status will be lost.

The purpose of the attorney–client correspondence privilege is to promote the freedom of discussion between a lawyer and client without the fear that disclosure can later be compelled from the layer without the client's consent. In order to establish the privilege, the correspondence must be prepared by the attorney, acting in a legal role, at the request of the client. The correspondence must include a legal analysis (not just a discussion of a factual investigation) and should include a specific reference to the likelihood of litigation. It should be noted that the privilege does not apply to all documents addressed or sent to a lawyer, nor can it be used to protect documents prepared prior to the lawyer being consulted.[25]

When an attorney directs specific information gathering "in anticipation of litigation," the documents produced during that information gathering process may be protected as attorney work product. Again, the important point in establishing this privilege is that the investigation is directed by an attorney. The privilege cannot be applied to documents prepared without an attorney's involvement, nor does it apply to documents prepared in the regular course of business.

This privilege may protect experts' reports (for example, an accident investigation report)[26]; however, the protection may not be absolute. For example, disclosure may be compelled if the report contains factual information that cannot be obtained in any other way by the opposing party. Therefore, experts' reports should typically be divided into two parts—data compilation and opinions. The data compilation may be required to be disclosed, whereas the opinion of the expert will typically remain privileged.

Self-Evaluative Privilege and Self-Audit Privilege. Several states have passed so-called "audit privilege" laws. In addition, the U.S. EPA has issued its own policy concerning the handling of environmental audit results.[27] It should be noted, however, that although these laws and the EPA policy are often referred to as creating an "audit privilege," most *require* disclosure and therefore have completely the opposite effect. Rather than creating a disclosure privilege, they typically create a limited immunity if proper disclosure is made. Although there may be occasions where taking the steps to qualify for a limited immunity would be appropriate, it should not be considered to offer the same protection against disclosure as the traditional attorney–client and attorney–work product privileges.

There is another privilege that has been developed in some contexts, the self-evaluative privilege. Originally developed in the context of protecting the self-evaluative dialogue among hospital staff, the privilege has been extended to additional types of institutional self-evaluations: employee personnel files, university faculty reviews, and documents submitted to the government pursuant to Title VII of the Civil Rights Act of 1964.[28] However, this privilege may not be effective in resisting a discovery request made by the government.[29] In addition, it is not clear that judges will allow this privilege to be applied to audit reports except in relatively narrow circumstances.

Confidential Business Information. Companies may have certain information that is considered "trade secret." A trade secret is defined as "any confidential formula, pattern, process, device, information or compilation that is used in one's business and which gives the business an opportunity to obtain an advantage over competitors who do not know or use it."[30] There are specific legal protections available for trade secrets under both common law and certain safety, health, and environmental regulations. In a court action, a trade secret may be protected under a protective order.

Employee Privacy Issues. In the performance of their job functions, safety and health professionals may have access to private information concerning employees (e.g., medical test results). There are state and Federal laws relevant to the protection of employee's personal privacy. In general, no personal information about employees should be disclosed to others without an evaluation of the privacy issues raised.

29.4.3 Record Retention Policies

An important component of a records management program is a records retention policy. A records retention program improves an organization's ability to handle valuable information. By getting rid of paper, valuable information is more easily retrieved and less likely to get lost, misfiled, or accidentally destroyed. Without such a program, it can be extremely difficult to locate critical information when it is needed or establish that, if information is unavailable, it has been properly destroyed. A written records retention program also helps to ensure that records that are required to be kept are not improperly destroyed.

29.5 WHEN TO CONSIDER SEEKING LEGAL ADVICE

Safety and health professionals should consider seeking legal advice in the following circumstances: when they are uncertain of their legal obligations, when entering into contracts that raise significant legal issues, when determining legal compliance, and when served with a complaint or other legal summons.

29.5.1 To Determine Legal Obligations

As discussed in Section 29.1, it is sometimes difficult to determine your legal obligations. Yet "ignorance of the law" is not considered a defense in a legal action. Since several environmental statutes can impose individual criminal liability, it may be prudent to seek legal advice to help clarify what you are legally required to do.

29.5.2 When Negotiating Contracts

There are certain legal protections, such as the worker's compensation liability limitation, which may be impacted by the terms of contracts you execute. In addition, certain contractual provisions, such as limitations of liability and indemnifications, may greatly increase your potential legal liability. If the terms of a contract are unclear to you, consult with an attorney *before* you execute the contract.

29.5.3 When Determining Legal Compliance

Although there are several benefits associated with evaluating whether a company is complying with safety, health, and environmental requirements, the documentation created as a result of such an evaluation, or audit, can end up being the "smoking gun" evidence used against a company in a subsequent legal proceeding. Consulting with an attorney *before* conducting compliance audits can help in structuring the audit to minimize the potential for increasing the company's legal liability.

29.5.4 When Served with Legal Documents

As discussed in Section 29.3, there are documents that will be sent, or served on, parties as the result of a legal action. Typically, these must be responded to within certain time periods or there can be serious legal consequences. It is always advisable to consult with an attorney if you are served with such a document.

REFERENCES

1. Refer to Chapter 28 for a detailed discussion of regulatory requirements.
2. It should be noted, however, that not all laws are codified, and new laws are continually being promulgated that modify the laws already set out in the U.S. Code or state statute books.
3. Pre-emption is a doctrine adopted by the U.S. Supreme Court that holds that certain matters are of such a national character that states are prohibited from passing a law inconsistent with the Federal law.
4. 60 U.S.L.W. 4587.
5. 29 CFR 1910.120.
6. 113 S Ct 2786 (1993).
7. The person bringing a lawsuit is referred to as the *plaintiff*; the person defending the lawsuit is referred to as the *defendant*.
8. H. C. Black *Black's Law Dictionary*, Fifth Edition, West, 1979.
9. *Proximate cause* can be a difficult legal issue. The proximate cause has been alternatively defined by courts as the primary cause, dominant cause, immediate cause, or the cause that played a substantial part in bringing about or actually causing the injury.
10. 1993 U.S. Dist. LEXIS 2635.
11. In a summary judgment motion, a party requests that the court dismiss the case because there are no facts that would allow the other party to win the case.
12. 275 Ill. App. 3d 938; 657 N.E.2d 668, 1995.
13. Prior to this case being filed, Mr. Arnett had settled a workers' compensation claim against his employer.
14. *Cunningham v. Anchor Hocking Corporation*, 558 So. 2d 93, 1990 Fla. App. (Court found that plaintiffs alleging these actions sufficiently pleaded that company had intentionally misrepresented the danger to employees so the case was not barred by workers' compensation statute).
15. b152339 (Dist. Ct. Jefferson Co., Texas). Information based on case summary reported in *National Law Journal*, Monday, January 27, 1993, A13.
16. One case alleging the misrepresentation of laboratory test data is *Metal Finishing Technologies Inc. V. Fuss and O'Neill, Inc.* [1991 WL 172833 (Conn. Super 1991)]. In this case, Fuss and O'Neill, an environmental consulting firm, was hired to help the plaintiff comply with an environmental consent decree and Federal and state environmental laws. In a decision on a summary judgment motion, the court stated that the plaintiff could bring a claim of negligent misrepresentation based upon the allegations that Fuss and O'Neill knew or should have known that sludge test results provided by its subcontractor were false, the plaintiff would rely on the results, and the plaintiff did, in fact, reasonably rely on the firm's representations.

17. A case including such allegations is *Nicholas J. Murlas Living Trust v. Mobil Oil Corporation and Ground Water Technology, Inc.* In this case the plaintiffs alleged that Ground Water Technology had engaged in fraudulent and/or negligent misrepresentation when it issued a report that was misleading because other information concerning environmental contamination was not disclosed. The court stated that "An omission coupled with an affirmative statement or act is actionable when the omission causes the affirmative statement or act to be misleading."
18. Such a case arose based on a report prepared by a consultant for a prospective purchaser that stated that a facility was in compliance with RCRA. When Ohio EPA subsequently conducted a compliance inspection at the facility, it found numerous prepurchase violations and prohibited the facility from operating. A jury awarded the purchaser over $2 million in damages for negligent misrepresentation. *Titanium Industries v. SEA, Inc.* (Mahoney County Common Pleas), as reported in the *GibbyMoon Report*, May/June 1996.
19. 493 F.2d 1076 (5th Cir 1973).
20. No. 95-CV-1941 (Colo. Dist. Ct, Nov. 1, 1996) reported in *ELR Update* **26**(34), December 9, 1996.
21. 283 N.J. Super. 556; 662 A.2d 1004; 1995 N.J. Super.
22. Issues related to regulatory enforcement actions, such as inspection issues, are discussed in more detail in Chapter 28. It should be noted that although most regulatory enforcement actions are resolved either directly with the agency or in a quasijudicial forum, such as before the Occupational Safety and Health Review Commission, they may eventually end up in court if there is an appeal.
23. There are two other privileges that are frequently mentioned—the self-evaluative privilege and the self-audit privilege. Neither of these has the same "protective" nature as the attorney–client or attorney–work product privileges. The self-evaluative privilege is limited in application, and the so-called self-audit privilege typically requires disclosure of information rather than protecting information from disclosure.
24. There is a strong legal bias that favors the candid and complete discovery of all information relevant to a case. There is a presumption that exemptions to the liberal discovery rules are unusual and to be discouraged. Consequently, courts treat privilege rules cautiously and often narrowly construe the claimed privilege.
25. In *United States v. Chevron*, No. 88-6681 (E.D. Pa. Act 16, 1989), the court ruled that the attorney–client privilege did not apply to a "Corporate Environmental Compliance Review Status Report" because the attorney did not act as an attorney but was merely present as one of the parties to whom the communication was made.
26. In *West Management, Inc. v. Fla. Power & Light Co.*, No. 90-02280 (Fla. Dist. Ct. App. Oct. 26, 1990), the court ruled that work product protection applied to materials prepared, at the direction of Waste Management's counsel, after the work-related death of a man employed by a Waste Management subsidiary because of the foreseeability of a legal claim.
27. Incentives for self-policing: discovery, disclosure, correction, and prevention of violations, 60 Fed. Reg. 66, 706 (1995).
28. *The Self-Critical Analysis Privilege and Environmental Audit Report*, Gish, Northwestern School of Law of Lewis and Clark College, Winter, 1995.
29. In *United States v. Dexter Corp.*, 132 F.R.D 8 (D. Conn. 1990), an enforcement action brought under the Clean Water Act, the court rejected Dexter's arguments that certain self-evaluative materials were exempt from discovery because the privilege would impede EPA's ability to enforce the Clean Water Act.
30. Restatement of Torts, Section 757.

CHAPTER 30

How to Provide Liability Insurance Coverage

DONALD M. WEEKES, JR.

Abacus Environmental, Inc., 123 Pinney St., Ellington, CT 06029

30.1 INTRODUCTION
30.2 EXPOSURES
30.2.1 Employees
30.2.2 Customers
30.2.3 Third Parties
30.3 COVERAGES
30.3.1 General Requirements
30.3.2 Uniqueness of Each Coverage
30.4 MECHANICS
30.4.1 The Broker/Agent
30.4.2 Insurance Companies
30.4.3 Underwriters
30.5 CONCLUSION
REFERENCES

30.1 INTRODUCTION

For the occupational health and safety practitioner, insurance can be one of the great unknowns. The concepts and theories that govern insurance are often alien to the safety and health professional. Even the jargon and terminology common to insurance can be incomprehensible. Yet the system of the protection of a company's assets can have a profound effect upon the job of those individuals responsible for safety and health.

This chapter will examine liability insurance and will explain as thoroughly as possible the relationship between this form of insurance and the responsibilities of the safety and health professional. A safety and health practitioner following the guidance provided in this entire handbook should be assisted in reducing the potential liability of

Handbook of Occupational Safety and Health, Second Edition, Edited by Louis J. DiBerardinis, ISBN 0-471-16017-2

their company. This will be done by exploring three areas of the insurance field: exposures, coverages, and mechanics.

Each area will be further divided into categories, which will be examined in light of the effects they may have on the job of the safety and health practitioner.

30.2 EXPOSURES

The safety and health practitioner will be familiar with *exposures*. It is the area of insurance upon which the practitioner can have the most notable effect by reducing and controlling the amount of exposures that may result in a claim against the appropriate insurance policy. In safety, examining exposures is comparable to safety audits or surveys; that is, it is looking for potential hazards such as a tripping hazard, or the failure to use proper safety controls on a potentially dangerous machine. In health, exposures are similar to air sampling for a potentially hazardous chemical at an operation where the chemical is used.

By requiring the delineation of the exposures at a risk, an underwriter is attempting to find the possible ways that a loss could occur under the policy to be placed for the potential insured. The underwriter will use the previous loss experience of the potential insured and the underwriter's knowledge of the potential insured's operations to determine if the risk is worth the premium to be generated.

Thus the first step in any successful underwriter's job is recognition of the exposures. How the practitioner can assist in the recognition process will be shown in each of the exposure categories.

Exposures can be divided into three major categories: employees, customers, and third parties.

30.2.1 Employees

The *employees* category is most familiar to safety and health professionals. In fact, many of the current safety and health efforts undertaken by industry, insurance companies, and various associations and groups were begun shortly after the introduction of Workers' Compensation insurance in the early 1900s. Many state legislatures required such insurance before a firm could operate within the state. It was in response to the high costs of insurance that many companies instituted safety departments. Due to the high number of claims from Workers' Compensation insurance, and the perceived likelihood that additional claims would be filed, many insurance companies started safety departments to assist their clients in proper safety and health procedures and techniques.

So insurance and occupational safety and health have worked in concert for many years. However, each has developed its own methods of addressing the problem of employee injuries. For example, safety and health professionals examine losses in order to recommend measures to prevent future accidents (see Chapter 5). While underwriters are in favor of that practice, they look at the losses from the viewpoint of a premium/loss ratio (see Chapter 26). The methods of underwriting (use of deductibles, exclusions, etc.) need to be adjusted to reduce the underwriting losses anticipated next year.

Both are looking at the same data but they are seeing different problems, due to their differing functions. However, the work of the practitioner can be very useful to the underwriter in determining future trends in losses. Let us look at employee exposure categories to illustrate this.

30.2.1.1 Bodily Injury by Accident Data concerning the accidental losses at an insured's facility are often kept by the safety and health practitioner, who is responsible for keeping and updating the company accident records. These data are used to report the company's safety record to the federal and state occupational safety and health departments. In addition, accident investigations are conducted by the practitioner to determine the cause of the accident and to recommend appropriate actions to reduce or eliminate the cause.

For the underwriter, these statistics and trends form the basis for an opinion of the risk. By reviewing these factors, the underwriter determines whether the account is an insurable risk, at what premium, and at what deductible.

30.2.1.2 Occupational Disease The industrial hygienist and occupational health practitioner (nurse or doctor) are mainly concerned with this type of employee exposure. Unlike accidents, which tend to injure the employee immediately, these injuries may not develop until many years after the initial exposure. Both the occupational health practitioner and the underwriter must think about the past exposures at the facility that may result in present occupational disease. In addition, the occupational health practitioner must deal with the current exposure in order to help prevent future disease.

Statistics on this exposure tend to be nebulous at best, due to the lack of a verifiable connection between the disease and the cause. Occupational disease potential is evaluated much as a detective evaluates evidence for clues to a murder.

The occupational health practitioner can examine the evidence biologically, in the results of blood, urine, and other tests on the employees (see Chapter 10), and atmospherically, by the results of the air sampling throughout the plant (see Chapter 8). Measurable changes in these two criteria may indicate that problems exist.

The underwriter is often not well versed in the technical aspects of occupational disease. Often the underwriter will react to the toxicity of the materials at a facility and not necessarily to the exposure of the employees. It is the responsibility of the occupational health specialist to interpret the sampling data for the underwriter and to put it into understandable terms.

30.2.1.3 Vehicular Accidents As it is a specialty among safety practitioners (i.e., traffic safety manager), so it is in insurance (commercial automobile liability underwriter). Statistically, this is a well-documented risk; careful record keeping is a must for all involved with the evaluation of vehicular accidents.

The basis units are similar to the bodily-injury-by-accident statistics, which are the number of accidents per million man-hours worked. For vehicular accidents, statistics are kept as to the number of accidents per million man-hours on the road. Investigations of accidents tend to be the same, in that the investigator is trying to determine, in both cases, the "how, who, when, where, and why" of the accident. The main difference is that, with vehicular accidents, there are consistencies and trends that may not exist for employee work-related accidents.

The commercial auto underwriter looks at the statistics and the accident reports from the viewpoint of the potential for additional accidents. Often the actions taken by the company to correct problem areas and help prevent reoccurrences are most important. The safety practitioner can be most helpful by reporting what was done.

30.2.1.4 Off-Site Accidents Off-site accidents are placed statistically into other categories, such as Workers' Compensation or Automobile Liability. However, often the cause of an off-site accident is quite different from the cause of one occurring on site or involving a vehicle. For example, sales personnel make visits to facilities at which they may not be given the appropriate safety equipment. All the safety precautions at the insured's plant and the extra care taken while the employee is driving will not help at the nonowned facility.

Also, many service firms employ maintenance and service personnel who must work at nonowned sites as contract workers. This is particularly true with construction companies employing craft workers. Often the individual worker may be completely alone while performing difficult and potentially dangerous tasks.

The underwriter looks at these types of companies as having a higher potential for loss; so, in general, such risks are avoided. However, the safety/health practitioner can mitigate the situation by providing a well-organized, well-written safety plan for those employees who work away from the owned facility. This may include recommendations for safety apparatus, personal protective equipment, procedures for emergencies, and so on.

In general, inclusion of such a written plan would help to convince an underwriter that the company is taking action to reduce employee accidents off site. If implementing the plan results in a reduced number of off-site accidents, the firm has been successful in two ways: (1) improved employee safety; and (2) reduced insurance costs.

30.2.2 Customers

In general, in the past, interaction between the customers of a firm and the company's safety and health personnel was minimal. This was due primarily to the traditional role of the safety and health professionals, which was to evaluate the workplace and make recommendations to control hazards.

Today, however, many of the activities of the safety and health practitioner directly or indirectly affect the customer. This is due largely to the reaction of underwriters to large losses from customers. The underwriter requires stricter controls and better engineering to help prevent future losses of the same nature. The goals of the underwriter and the safety and health professional are the same: reduced exposure for the customer and reduced liability for the company and its insurance carrier.

30.2.2.1 Product Liability *Product hazards* mean bodily injury or property damage arising from the insured's products or reliance upon representation or warranty made at any time with respect to the product, provided that the loss occurs away from the insured's premises and after physical possession of the products has been relinquished to others. The insured's products means goods or products manufactured, sold, handled, or distributed by the insured or by others trading under the insured's name, including its container.

In recent years, product liability has received considerable publicity due to court rulings that have favored the plaintiffs. For the general liability underwriter, this has resulted in a very cautious attitude towards any risk with the possibility of loss. In general, the insurance industry has been attempting to get all firms with product exposure to undertake quality control measures to reduce the number of opportunities for loss. In addition, the claims departments have been requesting that companies have preplanned responses to emergency situations, such as product tampering or sabotage. Product recall systems that address the public information problem are also requested by insurance representatives of their larger clients with potential product-loss situations.

In these areas, the safety and health practitioner can be very helpful. Many can apply the principles of their fields to the product liability question, by recognizing the risk, evaluating the products and their manufacturing process, and recommending quality-control measures to reduce the risk. In plants, the occupational safety personnel often are also responsible for product safety. In both cases, the objective is the same: the reduction of probable loss by means of control methods, in particular, engineering controls.

The key item from an underwriting viewpoint is that the implementation and evaluation of the controls are documented in writing. This allows the underwriter to feel comfortable with the risk, because the written documentation can be utilized in a court case to indicate the quality control measures taken by the insured. This may not exonerate the company totally if there is a defective product, but it may help to reduce the amount of monetary damages, particularly punitive damages to the plaintiff. The written material may show that the insured took all reasonable precautions to ensure that a defective product did not get on the market.

The safety and health professional may also be helpful in the planning and the implementation of a product safety material and instructions for product use to be included in the packaging. This may include the Material Safety Data Sheet (MSDS), which outlines the safe and healthful use of the product. Thorough and easy-to-understand safety guidelines may help the user avoid accidents, and, again, the instructions may be helpful in a lawsuit. Failure to follow instructions may indicate negligence on the part of the user.

30.2.2.2 Completed Operations *Completed operations* are that portion of the operations performed by or on behalf of an insured under a contract at the site of the operations that have been completed. Additionally, it can be defined as that portion of the work that has been put to its intended use by any person or organization other than another contractor or subcontractor engaged in performing operations for a principal part of the same project. Finally, operations that may require further service or maintenance work, repair, or replacement because of any defect or deficiency, but that are otherwise complete, shall also be deemed completed.

Although it has not received as much publicity as product liability, this also is a growing area of litigation, due primarily to the growth of the service industry, which has overtaken the manufacturing industry as the primary employer in the United States. More and more employees are involved in providing a service (e.g., repair and maintenance of equipment) than in manufacture.

This type of operation presents problems both to the underwriter and the company's safety and health professionals due to the lack of supervision over the employees. Many of the losses as the result of completed operations occur because the employees do not exercise the care necessary to ensure that the job is safe.

The safety and health professional, in a traditional arrangement, would not be responsible for the problems of a completed operations. However, many professionals are now part of loss prevention, or risk management, departments. This risk and the potential for loss associated with it fall under the jurisdiction of such departments.

For the underwriter, the risk management of the completed operations hazards is important as an indication that the company is attempting to control losses. The underwriter looks for a written completed-operations-loss-prevention manual of all contractors accepted for coverage. In addition, the underwriter requires that all subcontractors submit certificates of insurance, prior to the job, showing that they have coverage if there is a problem.

In general, completed operations represents a new area for safety and health professional. To go beyond the Workers' Compensation implications requires that the professional think as a risk manager. In that regard, the exposures to the customer at a construction site or during a service visit can be equally important as the exposures to the worker and, therefore, should be evaluated. The evaluation should be followed by careful and thorough recommendations to control the potential for damage to the customers' property and any bodily injury that may result.

30.2.2.3 On-Site Visitor (or Contractor) Accidents In many ways, these accidents are similar to employee accidents that occur on the premises. Many of the same hazards that exist for the company's employees will cause problems for the visitor or contractor. Unfortunately, personal protective equipment for visitors is often inadequate or lacking, and visitors have not been trained to handle the potentially hazardous conditions within the facility. This is changing somewhat due to a number of fatalities of contracted workers at explosions that took place at petrochemical facilities in the late 1980s and early 1990s. Inadequate training and personal protective equipment can result in serious bodily injury losses that can and should be easily prevented.

The key for the safety and health practitioner is involvement: The professional should be involved in setting up safety procedures for all visitors and contractors. For the underwriter, a written procedure is essential to an understanding and evaluation of the facility. Both are looking for written steps that will help to ensure the safety of visitors and contractors while they are in the plant.

Although the visitory may not be exposed to the hazards of the operations to the same extent as the workers, it is always best to provide the visitor with the same protection as the workers. Although this may be overprotection, it does provide the visitor with a margin of safety to compensate for a possible lack of knowledge of the dangers associated with the operation. At the same time, it will tell workers that all personnel, not just themselves, are required to wear the personal protective equipment, helping to reinforce the need for the employees to wear the protection.

Contracted workers are often exposed to the same hazards of the operations based on the nature of the services they are contracted to provide. Therefore, these workers should be provided with the same protection as the workers at these operations.

Educational and training materials on the potential harm should always be given to visitors at the beginning of the visit, and to contractors at the beginning of the contract (see Chapter 13). The greater the danger, the more training should be done. Training should also be given on the correct use of all protective equipment (see Chapters 16 and 17).

The safety and health professional should be responsible for the technical matters associated with personal protective equipment and training materials, and should ensure that the correct equipment be used and maintained. Also, educational materials should be reviewed for technical correctness. All instruction, procedures, and manuals should be carefully written and their use documented.

The underwriter will review any losses that may have occurred, but the documented loss prevention techniques will also be evaluated. A well-organized and well-written program may help to sway an underwriter concerning the commitment of the company to visitor safety.

30.2.3 Third Parties

The third major class of risk exposure is the largest group in terms of potential claimants and the one that receives the least amount of attention by the safety and health professional: That is the *third party* situation. A third party, by definition, is anyone who is not an employee (first party) or a customer (second party). This large nebulous group of individuals does not have direct contact with the company but may be affected indirectly by the actions of the company.

In general, the claims resulting from third parties are the result of long-term conditions or policies of the insured. Two exceptions occur frequently: sudden and accidental pollution and third party involvement in incidents with first or second parties.

The underwriters for general liability, auto liability, and environmental impairment liability policies are interested in third party liability. In each case, the underwriter is looking for the methods used by the insured to limit its liability by preventive methods and effective loss management techniques. It is in these areas that the safety and health professional, acting as a risk manager, can provide expertise that may make a difference.

For a risk manager, it is important to recognize each of the following areas as potential loss situations:

Environmental impairment
Third party involvement (in employee or customer incidents)
Class action suits
Stockholders' suits
Professional liability

In this chapter, only the first three of these areas will be addressed because those are the areas that most pertain to the work of a safety and health professional.

30.2.3.1 Environmental Impairment Environmental matters and their relationship to insurance is a relatively new topic. Until 1973, the insurance industry did not address this in its policies and coverage was provided in effect, although it was not the subject of many claims. However, since 1973, the number of environmentally related claims and the problems associated with them have been growing very quickly.

In 1973, the Insurance Service Office (ISO) promulgated a general liability policy that for the first time addressed the pollution question.[1] That policy, which is used as the standard throughout the industry, excluded all pollution on and into the land, air, or water, except for those incidents that were "sudden and accidental." Those words were undefined in the policy.

Until that time, the general liability policy was silent on pollution, and it was considered likely that any claims for bodily injury and property damage would be handled by this policy. With the 1973 revision, that was supposed to be the case no longer, but the courts saw it differently.

At the same time, certain insurance companies began to fill the gap in coverage that the exclusion created by offering a pollution liability policy, which was designed to provide gradual pollution coverage. By these two developments, the insurance industry had split environmental pollution incidents into two, sudden and accidental situations and gradual problems. The question became, where does one (sudden and accidental) end and the other (gradual) begin?

In 1985, ISO issued a revision to the general liability policy that included an "absolute" pollution exclusion.[2] The intention of this exclusion was to eliminate any and all pollution coverage from the general liability policy, including "sudden and accidental" coverage.

In addition, the pollution liability underwriter has also become more aware of the difficulties associated with certain classes of business and pollution. Many of these policies have resulted in large claims, which have greatly affected the profitability of this type of insurance. Many of the insurance companies previously offering this policy have stopped writing pollution liability due to the high incidence of losses and the exceptionally high claims that are paid. The remaining underwriters have become very cautious about the type of account that is acceptable.

The basis for the underwriting of a risk with potential pollution problems is the *risk assessment report*. The report is written either by an outside consultant or by an internal insurance representative. However, in either case, the company's safety and health professional staff can have a major impact on the type of information in the report and how it is presented. In many companies, due to downsizing and cutbacks, the professional has taken on the additional responsibility of environmental protection. The professional is then in a good position to provide the underwriter with details about the pollution potential at the company.

The key question for the risk assessor is: How does the company control its wastes and discharges to reduce exposure to the outside? The safety and health professional can have input into the company's action plan for wastes and discharge. The toxicological data on the effects of chemicals can be gathered and presented by the professional. The emergency response plan in case of an incident would also benefit from the work of the safety and health staff.

At any rate, it is the responsibility of the risk manager to ensure that the full and complete data are presented to the risk assessor for the report. In this way, the underwriter can have the information necessary to make the correct decision about premiums, exclusions, and deductibles for the account.

30.2.3.2 Third Party Involvement The key item here for both the underwriter and the safety and health professional is proper loss control and claims management. Beyond what was described earlier regarding controls over employer or customer exposures, further control of third party exposures that result due to actions of the employee or customer is difficult. When an incident occurs despite the best efforts to control the exposure, it is important that the third party involved, the innocent bystander, be handled initially by the employee or customer, and by the Claims Department of the insured or the insurance company when the claim is made.

By initiating and implementing steps that result in better handling, the safety and health professional may help to reduce the third party's pain and anguish over the loss. For example, a concerned and courteous truck driver involved in an accident with a motor vehicle may help to dispel the anger of the car driver. In addition, if someone is hurt, a first-aid trained driver may be able to help with the injury until professional help arrives.

In catastrophic occurrences such as an explosion, a tank car overturning and spilling a toxic chemical, or contamination of a product by one of the company's chemicals, the safety and health specialist can be helpful in two ways: providing professional advice on the plan or action to reduce and control the exposure, and serving as a technical spokesperson on the incident to the public. First, the professional can consult on ways to clean up and remove the substance to minimize the exposure to the public. Second, the professional can explain in reasoned and dispassionate fashion what is taking place at the site and what is being done to alleviate the damage.

The underwriter looks for printed material indicating that the company has plans for effective loss management. The underwriter wants to see that each employee or customer has been given adequate information in order to handle any situation that comes up involving a third party. For employees, this may invoke a training session that discusses what to do in an emergency situation. The preparedness of the company and its employees for potential disasters is important.

For customers, this may require quick and easy access to necessary information from the company when an accident occurs. Toll-free customer information numbers or hot lines have been effective.

In addition, the company should prepare a method of handling the public information detail. Knowledgeable, confident spokespersons who can speak intelligently to the technical issues should be ready to step in when a catastrophe occurs. In many cases, the safety and health professional fits the bill.

30.2.3.3 Class Action Suits Initially, it may seem farfetched that the safety and health professional can affect the losses involved with the class action type of third party suit. However, a review of past class actions indicates that the specialist may be helpful. For example, two of the largest class action suits ever initiated involved environmental matters: the asbestos litigation against the asbestos manufacturers, and the Vietnam veterans' suit against the manufacturers of Agent Orange. In both cases, epidemiological and toxicological data gathered by safety and health experts indicated that potential problems were developing. In each case, the information was not utilized fully by the management of those companies.

A class action suit can be filed by groups on behalf of people, things, or places that cannot file suit themselves, for example, the environmental class action suits against polluters on behalf of rivers being contaminated with PCBs, on behalf of wild birds whose eggs were softened due to the bird's exposure to DDT. In many of these cases, the filing of the suit requires that the company think about its potential liability. This, in turn, resulted in a review of the policies that caused the situation to develop. Eventually, the safety and health personnel may be asked for their opinion of the situation and advice on what the company could do now to alleviate the situation.

For the health and safety professional, the ideal method of handling these cases is not to let them develop in the first place. This requires that the professional have input into the managerial decision-making process. Whenever a decision is to be made that may affect safety and health, whether that of employees, customers, or third parties, the safety and health professionals should be asked for an opinion on the matter. Thus decisions concerning policies can be made with all the necessary input placed before the managers.

In addition, when class action suits involving safety and health issues are filed, the staff should be involved as soon as possible. A full professional evaluation of the issues involved may help the company decide what action to take. The safety and health professionals can also serve as public spokespersons to answer technical questions about the suit or can provide technical data to those who will be answering the questions.

For the underwriter, the key items are the company's previous suits, its reaction to them, and the company's plan to avoid future suits by involving safety and health personnel in decisions. Lawsuits that may affect the profitability of the company by a minimum of 10% of its gross income must be listed in a Securities and Exchange Commission (SEC) 10-K report and in the sompany's annual report. The information should be complete enough so that when the underwriter reads it, additional questions are kept to a minimum. However, if additional explanation is needed to explore further the technical aspects of the problem, safety and health personnel should be involved in writing the report.

All input of the safety and health professionals should be documented and available to the underwriter for review, if requested. Companies with a good system for communication on safety and health matters are more likely to be treated as good accounts.

30.3 COVERAGES

The exposures listed in the first section indicate the variety of problems that may affect the company from a safety and health perspective. This section will address the available insurance that provides liability coverage for these exposures. Not every type of insurance policy will be covered, but only those most likely to involve input from the health and safety professional.

First, a review will be made of the key sections in all insurance policies, including those areas that are common in all policies and the aspects of each. Next, each major coverage provided today by the insurance industry will be reviewed. The key differences between policies and the basis for rating each type of policy will be examined.

For the safety and health practitioner, knowledge about different policies is critical when discussing exposures and their relationship to the profitability of a company. The comptroller or insurance buyer of the firm will want to know if changes in the engineering controls will affect the insurability of the company and the size of the premium. Each should work closely with the other in order to provide support for each one's positions with regard to such matters as boiler safety, workers' compensation insurance costs, and claims settling for third party involvement in traffic accidents.

30.3.1 General Requirements

When purchasing a policy, one of the key questions is whether it is an "ISO Standard" policy or "manuscript" form. The Insurance Services Office, or ISO, disseminates sample policies produced by their committees, which consist of representatives of a variety of insurance companies. Many companies will place their own logo on the policy and issue it as their own. The advantage of doing this is that ISO has received approval from the state insurance boards for the policy to be issued within the state. In addition, the policy is acceptable to a majority of the insurance field, thereby reducing friction concerning exclusions. Third, a track record with regard to claims and the courts' interpretation of the policy will be developed relatively quickly due to the large number of insureds who have received that policy.

A "manuscript" policy is one that has been written (1) by the insurance company for its own clients, or (2) jointly with the insured to handle the uniqueness of the insurance client's liability situation. The insurance company issues its own policy when there is no ISO Standard policy for that type of coverage, or when the insurance company wants to change the ISO form to make the coverage broader or more restrictive.

Sometimes the manuscript form addresses the uniqueness of the liability associated with the business. For example, a crop-dusting outfit may need specific coverage to address errant sprays that may affect the neighbors' crops instead of the designated site. A modification in the contractual liability section may be necessary.

In general, all policies, whether they are ISO Standard or manuscripted, have the following five features:

1. Insuring agreements
2. Limits and deductibles

3. Exclusions
4. Definitions
5. Claims procedures

30.3.1.1 Insuring Agreements As defined in the Fire, Casualty, and Surety Bulletins issued by the National Underwriter Company, Insuring Agreements are defined as:

> (The Insuring Agreement) expresses the insurer's promise to pay on behalf of the insured for damages and the insurer's duty to settle or defend claims against the insured alleging bodily injury or property damage covered under the policy, even if the claims are groundless, false, or fraudulent.

This is the basis for the coverage offered under the policy. Any and all coverage should be laid out in this section so that it is clear to all who read the policy what circumstances would allow a claim against the policy. This is the heart of the policy. If it is not covered here, it is not covered.

This is why the insuring agreement should be reviewed carefully *before* the policy is bought. The language should be straightforward, with no unexplained qualifications or clauses that limit the coverage significantly.

Additionally, the insuring agreement should be evaluated to determine if it is an "occurrence" form or a "claims-made" form. This is a critical difference, which has become more important due to the emergence of more and more claims-made forms for all lines of coverage.

The occurrence form provides coverage (e.g., bodily injury, BI, and property damage (PD) for occurrences during the policy period. This seems fine until you realize that no coverage is provided if the BI and PD occur before (or after) the policy period, regardless of whether the resulting claim is made during the policy period. In effect, that means that not all claims made now are covered by the present policy. This becomes particularly important if the previous policy provided inadequate limits to cover the resultant claim.

At the same time, remember that coverage is provided for a claim made after the policy period ends, so long as the occurrence took place during the policy period. The limits available on each of the policies may be applicable to these cases all the way through the last date of employment.

Claims-made policies pay for claims brought against the insured during the policy period, regardless of when the BI or PD occurs. But once the policy expires, no coverage is available for any claim made after the expiration date, regardless of when the occurrence took place. The claims-made form "cuts the tail" of claims, which means that the insurance company no longer has to worry about a policy years after it was issued.

As you see, each type of policy has its advantages and disadvantages. Thus it is important to know what type of form your company has and important that your company is aware of the restrictions and advantages of the form.

Another item to be aware of in the insuring agreements is the "retroactive date," which is important for the claims-made form. This is the data before which no coverage is afforded. For the insurance company, this again cuts the other end of the claims by limiting the retroactive coverage available under this policy. This date should be differentiated from the "inception date," which is the date the policy begins coverage. The retroactive date is usually a date set chronologically before the inception date, and coverage is provided for the time period between those dates. For a claim-made policy, this time period should be sufficient to provide your company with coverage for most occurrences that might generate claims during the policy period.

The clauses outlining the defense-costs coverage should also be reviewed carefully. Some policies offer to defend all claims and include the cost in the policy limits. In effect, the limits of liability available to pay a claim are being deleted by the costs to define the claim. Defense costs (or claims expenses) may be offered as a sublimit of the policy, which means that only the amount noted in the sublimit is available for these expenses. Others offer to pay defense costs in addition to the limit of liability, which means the full limits to pay a claim remain intact. The applicability of the deductible to the defense costs should also be checked. In general, the best option, if available and cost effective, is defense costs in addition to the limits of liability with no deductible applied to it. The insuring agreements should make that clear.

30.3.1.2 Limits and Deductibles This section will delineate how much money is available to pay for the damages that are found to be covered by the insuring agreements. It also shows, by a deductible provision, how much the insured is responsible for before the insurance company will begin to pay.

The limits of liability can be divided into two categories (1) per "occurrence," "incident," or "claim" limits, or (2) "aggregate" limits.

Occurrence limits are used for occurrence policies, which means that only the sum listed at this limit is available to pay for all damages that result from any one occurrence. This works well when there is one occurrence and one claimant, because the policy cannot pay to that one claimant any more than is on the policy limit line. However, this is complicated when there are multiple occurrences involving multiple claimants.

The insurance industry is addressing the problem of multiple occurrences and multiple claimants in two ways: (1) by the "claims-made" policy, where there is a per incident limit, or (2) by placing a "per claim" limit on the occurrence policy.

The *per incident* limit theoretically reduces the insurance company's liability to one set limit for each situation regardless of the number of claimants involved. Each claimant in a catastrophic situation would not receive the full limits available for the damages suffered during the incident.

The *per claim* limit of liability is the latest attempt by the insurance industry to decrease the potential for loss. All claims for damages sustained by any person or organization as the result of any one incident will be deemed to have been made when the first of those claims is made. In other words, all claims resulting from one occurrence will be treated as one claim, and the maximum possible loss will be the per claim limit.

The second limit listed on a policy is the aggregate limit. The *aggregate limit* is the most that the insurer will pay during the policy year regardless of the number of occurrences.

There are two types of aggregate limits: (1) "per site" aggregate, and (2) "per policy" aggregate. The *per site aggregate*, important when the insured has multiple locations, indicates the total money available to pay all claims for incidents that occurred at that one site. For example, if you purchase a policy with a "per site" aggregate of \$1 million and your company owns four sites, that means that each site is protected under that policy for up to $1 million. The disadvantage of this type of policy is that the premiums can be quite high.

The *per policy aggregate* limits the insurance protection to the aggregate limit, regardless of the umber of sites. This policy is less expensive but does not offer the protection of the per site aggregate.

Regardless of the policy, it is important to clarify what types of limits are listed. Since the type of limit can have major effect on premiums, the claims services available, the extent of loss control available from the insurance company, and a variety of other factors, the safety and health practitioner should always be aware of the types of limits listed on the company policies.

30.3.1.3 Exclusions Just as the insuring agreement indicates what is covered, the exclusions section states what is not covered. This area of the policy is often twice as long as the insuring agreement section. Insurance companies have learned, through litigation, that when the policy is silent on an area, the courts will find that coverage exists. Therefore, the policy will list every possible area that the insurance company will not cover.

Many different types of exclusions are applicable to the circumstances of the insured and the type of policy to be written. However, there are some general characteristics of exclusions and some common exclusions that appear in many policies, including:

1. Liabilities that are covered by other policies.
2. Liabilities that took place in the past or at past locations.
3. Contractual liability.
4. Liabilities that take place outside the coverage territory.
5. Liabilities resulting from damages that are intentional or expected by the insured.
6. Liabilities that result from war.

It should be noted that some of the exposures that are excluded in the policies cannot be covered, because the insurance industry as a whole will not provide insurance for that exposure. For example, war is considered a unique situation where considerable damage and bodily injury, completely out of the control of the insured, can take place.

Most policies contain specific time periods of coverage, usually running from the inception date or retroactive date to the expiration date. Many policies include a list of designated locations that are covered under the terms of the policy, and all other sites, whether owned by the insured or not, are not covered. This would also apply to sites that are no longer the property of the insured, whether they have been sold, abandoned, or given away. Even with policies that do not designate sites, an "alienated permises" exclusion is usually included in the policy.

For the majority of insured, a coverage territory of the United States, its territories and possessions, and Canada is sufficient. Most policies issued by U.S. insurance companies will restrict coverage to these areas. The defense costs and all liabilities associated with law suits filed overseas are excluded under the policy.

One of the most important exclusions is the *contractual liability exclusion*. The comprehensive general liability policy for example, does not include coverage for contracts other than incidental contracts. The insured under a contract can become liable in three ways: (1) negligence, where the insured bears responsibility for the incident; (2) statute, such as workers' compensation laws; and (3) by assuming the liability of others. A separate contractual liability policy must be written to provide coverage when the insured has assumed liability under other than an incidental contract.

30.3.1.4 Definitions This section allows the insurance company to define what the terms in the other sections of the policy mean. It is particularly important to review this, because the expected coverage may be restricted by what is defined. Conversely, this is also the area where additional coverage may be hidden among the clauses used to define the terms of the policy. It is also important to note what is not defined, because when a policy is "silent" on something, often more coverage than expected is in the policy.

For example, the definition of insured site may include any and all sites currently occupied by the insured. This would be considered "blanket" coverage and would be part of a very broad policy. In some policies, any new sites acquired during the policy year would be covered automatically.

In other policies, only the specific locations as designated in the policy are covered. This restricts coverage and is called a "site-specific" policy. It is imperative that the accompanying list of sites be reviewed carefully, to ensure that the necessary locations are covered. In addition, any new locations acquired during the term of the policy must be added by endorsement.

The definition of *property damage* may include a clause, "including consequential loss or use thereof." This means that part of the losses paid can include compensation for the time that the insured is unable to use the property while it is being remediated.

One subject often not addressed in policies is the issue of punitive damages. Many courts have ruled that "damages" covered in the policy include punitive damages. This is not true if the term *compensatory damages* is used, as this term is interpreted to exclude punitive damages. However, it should be noted that many states do not allow the exclusion of punitive damages.

At any rate, it is worthwhile to review this section because these situations may prove to be important when a claim arises.

30.3.1.5 Claims Procedures In order to file a claim under the terms of the policy, it is necessary to know the correct procedure of notification. In addition, we will note what the insured must do to facilitate the claim, and also what the insured cannot do.

In general, written notice of the claim must be sent to the insurance company, outlining the circumstances of the claim. This includes time, place, and nature of the incident, injury, or damage sustained by any persons or organizations (including their names and addresses) and names and addresses of available witnesses. All summons, notices, demands, or other processes must also be forwarded.

The insured must also cooperate fully with the insurance company's representatives with regard to settlements, conducting suits or proceedings, or enforcing the right of contribution or indemnity against anyone else who may be liable for the damages. The insured must attend hearings and trials and assist in securing and giving evidence and obtaining the attendance of witnesses.

The one thing that the insured *cannot* do is make any payment or incur any expense unless it is approved by the insurance company. This should be made clear to all that it may affect, so that no disagreement will arise later as to the extent of the liability that the insurance company will assume under the policy.

All in all, this is straightforward, but it is important to be sure that the correct procedures are completed and no adverse actions are taken.

30.3.2 Uniqueness of Each Coverage

This section will discuss the various available insurance coverages that can be purchased from many property and casualty (P&C) insurance companies. Keep in mind that the previous section discussed the general format of all policies and what they all have in common. What will be discussed here is the way each type of insurance differs from the others.

Each coverage requires different data for the underwriter to evaluate the risk. Those data will be outlined under each coverage type. Much of the data relate directly to the exposures of the risks, and the practitioner will be the employee most aware of the situation of that exposure. The various coverages to be examined are:

1. General liability
2. Workers' compensation
3. Commercial auto

30.3.2.1 General Liability General liability is the most universal of all policies and also, in many ways, the most comprehensive. In fact, it is often called comprehensive general liability (CGL). Currently, the rules and rates of the insurance are written in the General Liability Division of the *Commercial Lines Manual.*

It was named *comprehensive* for two reasons: (1) It contained one comprehensive insuring agreement covering all hazards within the scope of the insuring agreement that were not otherwise excluded, and (2) it provided automatic coverage for new locations and business activities of the named insured that arose after the policy inception.

In general, the CGL provides coverage for bodily injury and property damage of third parties that are the responsibility of the insured. It covers the hazards of premises and operations, independent contractors, and products and completed operations.

The "premises and operations" hazard encompasses liability for accidental bodily injury or property damage that results either from a condition on the insured's premises. Many of the exclusions in the CGL are applicable primarily to premises and operations coverage, including contractual liability, automobiles, liquor liability, pollution, employee injuries, damage to property in the insured's care, and others.

The "independent contractors" hazard exists whenever the insured hires an independent contractor to do work on behalf of the insured. The insured has potential liability for injury to others as a result of the work performed by the contractor.

The "products liability" hazard exists for any insured that manufactures, sells, handles, or distributes goods or products. Any bodily injury or property damage that arises from the goods or products is the insured's liability. It should be noted that the CGL form restricts by exclusion the extent of coverage available under this policy.

The "completed operations" hazard is the insured's potential liability for bodily injury or property damage arising out of the insured's completed work, for example, the bodily injury and property damage that may result from the collapse of a building after its completion. Again, like the "products liability" hazard, the coverage is restricted by exclusion.

A CGL policy consists of three documents: (1) a declarations page, (2) the policy jacket, which contains the definitions and provisions common to various standard coverage parts, and (3) the CGL coverage part, which contains the insuring agreement, exclusions, and other provisions specific to CGL coverage.

The declarations page outlines the data on the risk itself, such as name, address, inception, expiration, and retroactive dates, policy limits and deductibles, broker's name and address, commissions, and other such information. This is a legal document, and a record that the insured does have insurance; so it is imperative that all data are correct.

The policy jacket is the standard terms and conditions of the policy. Section 30.3.1 covered these items in depth. Suffice it to say that all sections of the policy jacket should be reviewed carefully to determine what is and what is not covered by the policy.

The coverage section is where the coverage becomes specific for the insured in question. In the CGL policy, this is where the other coverage parts (contractual liability, broad from property damage, etc.) are added. In contrast, various exclusions that restrict coverage can also be added.

These additional coverages and exclusions must be studied carefully to determine what is actually covered by the policy.

Though this all sounds complicated, which it is, one rule must be kept in mind when reviewing a policy: "If it ain't excluded, it's covered!" Once you look at the policy language with that in mind, the exclusions can be read one at a time and their effect on the basic coverage noted. In the same way, all coverage parts can be read for content with regard to the additional coverage extended by the endorsement.

At any rate, it is helpful for the safety and health professional to have a good working knowledge of what a CGL policy does and does not cover. There are two reasons for this: (1) The professional then can determine if the exposures that exist at the company are adequately covered by insurance under the terms and conditions of the policy, and (2) the professional can assist in gathering the information needed to present clearly and concisely to the underwriter what is taking place at the account.

The information given to the underwriter is known as the "submission." For comprehensive general liability coverage, the following are necessary as part of the submission:

1. Application
2. Financial data on the firm
3. Loss data and analysis
4. Requests for unique coverage features
5. The firm's loss control program information

Depending upon the company, the application can be quite extensive or very short. For example, the company's annual gross sales figure would always be requested, as well as the main business activity, number of employees, size of facility(ies), number of vehicles, location of facility(ies), and types and numbers of products. Additionally, the application may request data on previous business activity(ies), previous business location(s), plans for expansion, growth figures in sales, personnel, and products, and other business-related matters.

The applications are usually designed to elicit basic facts about the company, but not necessarily the in-depth data needed to determine ultimately the acceptability of a risk or what premium should be requested.

The application may request two items where the professional can be helpful: (1) loss data and (2) loss control program information.

Loss data are any and all information concerning situations that either resulted or could have resulted in a claim against the policy. This would include the cause of the loss, actual dollar amounts paid, the anticipated future payments (called *reserves*), the anticipated future claims related to incidents that took place during the policy period (called *incurred but not reported*, IBNR, losses), and any other pertinent data. As will be seen later, this information is the basis for setting premiums for the risk.

Remember that the current CGL carrier (insurance company) does not automatically provide these data to any new carrier who has been approached to replace it. The company interested in getting premium "quotations" from a new carrier must supply the loss data to them. The "old" carrier, therefore, has an advantage over the others in that they have the loss experience data, perhaps for a number of years. Of course, if that record is poor, it would be anticipated that, at renewal, the data would be used against the insured. The insurance company often has computer printouts on current losses including all the pertinent details. It would behoove all insureds to request these runs on a regular basis for evaluation purposes. Also, many insurance brokers can provide these data through their computer connections to the insurance companies' data files.

The occupational safety and health professionals on the insured's staff should be part of the evaluation process. The professionals can evaluate accident trends and note in a report what has been done to alleviate the exposure. In addition, for large losses, a full inspection report and analysis can be completed by the professional staff. This report should be part of the submission made to the underwriter, who may be swayed by the actions listed in the report to control the number and severity of the accidents. Substantial differences in premium may be involved.

Also, a section of the submission should be devoted to enumerating and describing the elements of the loss control program of the insured. The focus should be on those parts of the program that relate to the coverage being requested. A systematic review of the procedures for quality control of products would be important if products liability is included in the request for insurance.

The safety and health personnel should have direct input to the submission in these areas. The professional should be able to complete the information in a technical fashion and add to the completeness and thoroughness of the submission.

The financial data for the firm are considered the jurisdiction of the comptroller or financial officer of the insured. This usually involves the annual report of the company and certain SEC reports, such as the 10-K. The main function of the Safety and Health professional with regards to these data will be to make sure that the department's budget figures are correct. Also, under the Legal Proceedings section of the 10-K, the company may list various claims that may affect the financial capabilities of the firm. The professionals should make sure that any loss information is correct and that the company has listed all relevant losses.

The Loss Control department should also be part of decision making about unique coverage features. Situations that have a large potential for loss may not be covered under the provisions of the standard CGL policy. For example, a large part of a construction company's business is conducted by contract, and, therefore, coverage for contractual liability is important. Such a situation would be apparent to the broker and the insurance buyer for the insured. However, other circumstances may be more visible to the professional that others may not see. For example, the professional may be able to provide insight into the effectiveness of the construction safety and health program on the number of accidents that occur on other company's property.

Once the underwriter has all the information in the submission and any additional data that are requested, a decision is made as to the acceptability of the risk. In general, an underwriter is given guidelines as to the types of business that the insurance company considers to be good risks and poor risks. These rules are usually contained in internal underwriting manuals that are periodically updated to reflect changes in the judicial and public environment with regard to the exposures associated with a particular type of account.

Often there is a grading system, which indicates what current thinking is on the potential for loss at a class of business. These classifications come into play early in the submission process. Once the class of business is known, the underwriter will review the underwriting guidelines to determine whether or not the risk is in an acceptable class.

Another source of classification of risk is *Best's Underwriting Manual*.[3] This book, also updated yearly, classifies many different types of businesses with regard to the standard lines of coverage (GL, Workers' Compensation, Commercial Auto, etc.) and assigns a rating from 1 (excellent) to 10 (poor) to each type of exposure for each business class. This rating is often used to establish parameters for premium rates.

Also keep in mind that the insurance company often establishes rates different from the standard rates sent to the State Insurance board for approval by ISO or other insurance groups. This change in rate, called a *deviation*, may be lower or higher than the established rates depending upon the current market for this product line. The underwriter is thus allowed certain leeway with regard to setting premiums, which often depends on the quality of the data included in the submission. Also the underwriter will often compare the risk to others previously evaluated to determine where this one fits in the scale.

The rating device that underlies CGL coverage is the insurer's right to conduct an audit at policy expiration and assess additional premium if necessary. Ordinarily, at the beginning of each policy period the CGL insurer surveys the insured's existing or anticipated exposures and charges an appropriate premium called the *deposit premium*. At the end of the policy period, the insurer conducts an audit to determine actual exposures and, using this information, either charges an additional premium or refunds part of the deposit premium.

The safety and health department of the insured should be prepared to provide any additional data concerning new operations and their engineering controls requested by the insurer. This will enable the underwriters to get a clear picture of what is taking place and should assist the underwriter in setting a fair and equitable additional premium.

All in all, the safety and health professional can be very helpful in providing data to the Corporate Insurance department on the existing exposures and evaluating the potential for loss. This information will help the underwriter in fully evaluating the risk. The safety and health professional should be aware of this and should press the Insurance department about the quality of the information going into the submission. In the long run, that input can result in better data going to the insurance company, which may have a salutary effect on the premium to be charged.

30.3.2.2 Workers' Compensation This coverage is the one most familiar to the safety and health field. Because of the cost of this insurance, Safety and Health departments of manufacturing companies were created, as were the Engineering departments in insurance companies. However, in many ways the coverage still remains somewhat mysterious to the practitioner, and the principles that govern its use and pricing are often foreign. This section will attempt to dispel some of that strangeness so that the professional will have a good working knowledge of the product.

Officially called the Workers' Compensation and Employers Liability Policy, the new form is the result of a compromise between employers and employees on this critical issue. The employees give up the right to sue their employers for employment-related injuries, and the employers agree to pay for the damages through insurance, under a statute mechanism providing specifically scheduled benefits. Insurance has been, and most likely will continue to be, the most effective method for an employer to compensate employees and their families for work-related injuries or diseases. The Employee Liability portion of the policy protects employers against suits filed to collect damages that are not compensable under the WC coverage (see Chapter 26 for more technical information).

The revision of the policy, which also utilized "easy read," or simplified, language, was done by the National Council on Compensation Insurance (NCCI). It should be noted that all states and territories have WC statutes. But six states–Nevada, North Dakota, Ohio, Washington, West Virginia, and Wyoming—administer their own programs and are called "monopolistic" states. In effect, all employees in these states are covered under the state insurance plan, not by private company insurance.

Finally, it should be noted that certain entities will self-insure their exposures, which in effect means that the entity has taken on to itself the liability of any employees' injuries or illnesses. Usually, the practice is that the entity purchases excess workers' compensation insurance for claims that exceed a certain self-insured amount, such as $1 million, which the self-insured entity will pay. This chapter will not explore this option because it is only available to a select few very large entities.

The workers' compensation policy contains three distinct areas of coverage: (1) workers' compensation, (2) employers' liability, and (3) other states' insurance.

Under Part 1, the workers' compensation (WC) section, the insurer agrees to assume liability imposed upon the insured by the workers' compensation laws of the states listed on the declarations page. The language is simple and straightforward: "We will pay promptly when due the benefits required of you by the Workers' Compensation Law." *Workers' Compensation* is defined in the policy as it is defined in the laws of each state listed on the declarations page. Coverage applies to bodily injury by accident and bodily injury by disease (including resulting death).

There are two restrictions: (1) bodily injury by accident must occur during the policy periods; and (2) bodily injury by disease must be caused or aggravated by conditions of employment, with the employee's last day of exposure falling during the policy period. This section also covers reasonable expenses incurred at the insurer's request, premiums for appeal bonds, litigation costs taxed against the insured, and interest on a judgment as required by law. The insurer also has a right and duty to defend the insured in any proceeding for benefits payable and policy.

The employer's liability portion of the policy, Part 2, is similar to other liability insuring agreements. This section covers the insured for liabilities arising out of the injury of an employee is the course of employment not covered by workers' compensation. The insured damages include: (1) "third-party-over" suits, which involve the insured's liability for damages claimed against a third party by one of the insured's employees; (2) damages assessed for care and loss of services; and (3) consequential bodily injury to a spouse or other close relative of the injured employee.

Part 3 of the policy provides other states' coverage. This is necessary because Part 1 applies only to obligations imposed on the insured in the states listed on the declarations page.

Part 4 is "Your (the insured's) duties if injury occurs." Part 5 is "premium," and

Part 6 is "conditions." Both of the latter parts (5 and 6) are subject to exclusions listed under Section F, "Payments you must make." These limitations include any payments in excess of the benefits regularly required by Workers' Compensation statutes due to: (1) serious and willful misconduct by the insured; (2) the knowing employment of a person in violation of the law; (3) failure to comply with health and safety laws or regulations; or (4) the discharge, coercement, or other discrimination against employees in violation of the WC law.

There is no coverage for liability of another assumed under a contract. Punitive damages are excluded, as is coverage for bodily injury to an employee while employed in violation of the law with the insured's "actual knowledge." For the Employee's Liability section, there are additional exclusions relating to no coverage for damages imposed by a Workers' Compensation, Occuptional Disease, Unemployment Compensation, or Disability Benefits Law. In addition, there is no coverage for bodily injury intentionally caused or aggravated by the insured, or for damages arising out of the discharge of, coercion of, or discrimination against any employee in violation of the liability.

Although most companies consider WC to be the exclusive remedy for work-related disabilities, there are several reasons why Employees' Liability coverage is desirable. Some states do not make WC insurance compulsory or do not require coverage unless there are three or more employees. In some cases, an on-the-job injury or disease is not considered to be work related and therefore is not compensable. The employee may still sue if there is reason to believe that the employer should be accountable.

Certain state laws have been interpreted to permit suits against employers by spouses and dependents of disabled workers to collect damages for themselves beyond what the employee has gotten. The basis of such suits may be loss of companionship, comfort, and affection.

Finally, employers are being confronted with claims and suits when an injured employee sues a negligent third party for compensation beyond WC coverage. The third party may in turn sue the employer for contributory negligence.

From this review of the WC policy, it should be clear how the occupational safety and health professional can be of help regarding the coverage. All the covered items should fall within the jurisdiction of the professional. Perhaps the most important clause in the policy from the viewpoint of the professional is the exclusion related to the failure to comply with health and safety laws or regulations. The practitioner should utilize this possibility to press for the necessary controls to eliminate such violations.

In addition, the safety and health staff can be helpful in the submission process. Many of the same features required of the general liability submission are required for WC coverage, for instance, and application, and loss data and analysis. The safety and health personnel can help to prepare the needed data and review it to ensure its high technical quality.

This would be particularly true with the loss data and analysis. The same information used to justify expenditure for new engineering control equipment can also be geared towards any underwriting submission. Again, insurance company computer loss runs can be helpful as data sources. The interpretation of the information to indicate trends, to report on large losses, and to show the engineering changes that have been made to address the problems should be done by the health and safety staff.

In addition to these data, the submission should also include a written safety and health program and a safety training program. These would be good indications of the insured's attitude towards safety and health and would demonstrate that the insured has addressed the major potential loss areas. For the underwriter, any deviations from the "book" premium will be based upon the evaluation of the additional material by the underwriter. *Often the underwriter will be influenced favorably by a well-written and thorough program that provides good guidelines to the insured's employees about safety and health matters.*

The underwriter will request an insurance loss control representative to visit the facility to investigate what is actually taking place. This usually occurs after the underwriter has indicated an interest in the risk and the broker has been given an approximate premium amount, called a "preliminary premium indication." This is before the risk is "bound," which means that the coverage is in effect.

This visit of the loss control representative is the most important responsibility of the safety and health professional regarding securing Workers' Compensation Insurance. The report filed by the loss control representative can overwhelm all the previous work, whether it is favorable or not. The safety and health staff must make sure that the loss control representative understands the prospective insured's programs and can see the future plans to address any problems that may arise.

First and foremost, the safety and health professional should accompany the insurance company representative on the inspection and answer all questions as completely as possible. Although the loss control representative may not be an expert on the particular operations at your facility, remember that the representative is a professional, who will see hundreds of facilities in a year. The questions asked often reflect this experience as well as the information being requested by the underwriter.

Second, the professional should make sure that the supervisors of the operations examined are informed in advance of the visit and asked to cooperate with the inspector. Selected managers of main operations should be available for interviews by the representative, if requested, in order to provide a description of the operation.

Last, a wrapup session should be conducted with the loss control representative. If the representative has recommendations for the operations, these should be noted and, if necessary, discussed in detail. Any misconceptions about the operations of the loss control programs should be cleared up at this time. The session will enable the safety and health professional to report back to management about what they may expect to find in the loss control representative's report.

Once the underwriter has the loss control report, the finalization of the premium begins. It should be noted that the rating basis for all WC premiums is the total remuneration earned during the policy period by employees and officers of the insured.

The National Council on Compensation Insurance is a statistical agency that accumulates and prepares data for the various state boards for consideration in setting "manual" rates. The National Council takes the latest available data for a 2-year period as a basis for computing future manual rates. These rates are promulgated on a regular basis and are the rates used initially by all insurance companies.

Within a given business class, there will be a wide variety of companies that may have had different loss experience. To accommodate this possibility, the underwriter will use "experience rating." A *experience rate* is a manual rate modified on the basis of the deviation of the loss experience of an individual risk from that of the average established by the classification. If the accident history shows a loss cost below normal for the class, a credit is allowed, whereas unfavorable experience results in a debit. This credit or debit is applied to the manual rate to find the adjusted, or experience, rate.

The larger the risk, the more likely it is that it will have credible experience data, whereas small risks with few employees and fewer hours worked per year may not have good data. A good rule of thumb, therefore, is that the larger the risk, the larger will be its credit for good experience and the larger its debit for poor experience.

Experience rating thus enhances the importance of a good safety and health program aimed at reducing the frequency and severity of accidents.

One additional rating system may be affected favorably by the safety and health professional by the loss control. *Retrospective rating* is a plan or method that permits adjustment of the final premium on the basis of the employer's own loss experience, subject to maximum and minimum limits. Such plans can be valuable to insureds with modest premium who feel that their experience will be better than average and would like to benefit from the expected lower loss ratios. For larger risks, the chief advantage will be providing legal liability to satisfy government regulation while still gearing the premium to the actual losses of the particular insured.

Whatever method is used to calculate the premium, one general principle stands out: The lower the losses, the lower the premium. There is a "bottom-line" effect that the professional can point to when negotiating for additional personnel, equipment, or a larger say in the engineering controls that are used. Safety and health can then be seen as a direct money-saving device that is absolutely vital in reducing costs, particularly insurance costs. Of course, the key is to make sure that the occupational safety and health programs capable of reducing the number of accidents and controlling the severity of those losses that do occur. But once that program is in place, the case can be made about the relationship between a strong program and good loss experience.

30.3.2.3 Business Auto For many companies, the transportation of raw materials, products, and employees is an important component of the firm's business. For others, that moving of materials and people is the business. In either case, the accidents that can result on the road are multitudinous and the Business Auto policy is the means used to protect the company against the resultant losses. The type of losses covered by this policy are:

1. Physical damage
2. Automobile liability

Automobile physical damage insurance for commercial and public vehicles involves four basic coverages: comprehensive, specified perils, collision, and towing. The insurer agrees in the insuring agreement to pay for loss to a covered auto or its equipment under any one of the first three coverages listed above.

Comprehensive coverage means that the insurer will pay for loss from any cause except collision or overturn. No list of perils that specifies, one way or another, what is or is not covered, is attached.

On the other hand, *specific perils* coverage does list what is covered. These perils are: (1) fire or explosion; (2) theft; (3) windstorm, hail, or earthquake; (4) flood; (5) mischief or vandalism; (6) sinking, burning, collision, or derailment of any conveyance transporting the covered auto. These are offered as a package because experience shows that most insureds who do not buy comprehensive coverage select all of the above.

Collision coverage involves the damage caused by a covered auto's collision with another object or by its overturn.

The key item for a business auto policy is the selection of the covered autos. There are five physical damage insurance categories:

1. Owned (company's) autos only.
2. Owned private passenger autos only.
3. Owned autos other than private passenger autos only.
4. Specifically described autos (basic auto).
5. Hired (rented) autos only.

By picking one of these categories, the insured narrows the coverage that has been purchased. It is further limited by the "Persons Insured" section. This defines the three categories of insured individuals as follows.

1. The persons or organization as named on the policy for any covered auto.
2. Those other than the named insured who have been given permission to use the covered auto, such as employees, executive officers, those who service or park the covered auto, and those who move property off or from the covered auto.
3. Anyone who is liable for the conduct of an insured described above is an insured but only to the extent of that liability.

The exclusions for physical damage are as follows:

1. Wear and tear, freezing, mechanical or electrical breakdown, unless these are caused by other loss covered by the policy.
2. Blowouts, punctures, and other road damage to tires, unless the damage is caused by other loss covered by the policy.
3. Loss caused by declared or undeclared war.
4. Loss caused by nuclear weapon.
5. Nonpermanent installed tape decks or CD players, loss of tapes or CDs, or loss of CBs, mobile radios, or telephone equipment.

Automobile liability provides coverage for bodily injury and property damage to third parties caused by the insured's covered autos and the persons insured. In many ways, this is similar to commercial general liability in the scope of coverage. However, all coverages for this policy is restricted to the covered autos and the persons insured.

Nine Automobile Liability classifications from which the insured must choose:

1. Any auto (comprehensive auto liability coverage).
2. Owned (company's) autos only.
3. Owned private passenger autos only.
4. Owned (company's) autos other than private passenger autos.
5. Owned autos subjected to no fault (applicable in certain states).
6. Owned autos subject to compulsory uninsured motorist law (applicable in certain states).
7. Specifically described autos (basic auto).
8. Hired (rented) autos only.
9. Nonowned autos only (applicable to private autos used by employees for work).

The persons-insured section of the physical damage section is applicable to automobile liability. For purposes of contrast, your personal auto liability policy would most likely be a "Basic Auto" policy, whereas your company, if it has a fleet, most likely has a "Comprehensive Auto" policy.

The exclusions are familiar, including exclusions of contractor liability and pollution, and a broad form nuclear energy liability exclusion. Also, no coverage is afforded for injuries covered under Workers' Compensation, Disability Benefits, or similar laws. No bodily injuries to employees or to fellow employees is included for coverage. In addition, there is an exclusion for damage to property owned or transported by the insured.

As part of the overall loss control visit to a facility, the insurance company representative may review the auto losses, review the safety and health procedures, and inspect such fleet areas as garages, depots, and repair facilities. The purpose of this will be the same as the visits to the manufacturing areas for workers' compensation, that is, to clarify for the underwriter in a report what is happening at the facility. The company's traffic safety manager or other individual responsible for the safety and health of the drivers should meet with the loss control representative during that visit to ensure that correct and complete data are being gathered.

The traffic safety manager should be prepared to show the representative a written safety and health program that outlines the major potential exposures and the procedures used by the company to control those exposures. For example, the daily inspection of the vehicles should be described and any forms used as verification of the review should be included in the written program material. Also, the procedures to follow in an accident should be carefully delineated and the training methods used to present the procedures to the drivers should be shown.

The loss control representative's report will form a major part of the material that the underwriter will want in order to evaluate and rate a business auto risk fairly. The data needed for a submission should include:

1. Application (includes complete data on all vehicles).
2. Loss data and analysis.
3. Motor vehicle registration for all covered autos.
4. Drivers' licenses and records for all person insured.
5. Safety and health programs.
6. Training program.

With these materials and the loss control representative's written report, the underwriter will have a good picture of what is going on at the facility and can then decide whether or not it is an acceptable risk. If it is considered acceptable, the account would be rated to determine the premiums.

The rating of these coverages is based upon the rates set forth in ISO's *Commercial Lines Manual*. The classification tables in this book list all industries and their class code for each line of business, including auto. For business auto, the main factors that are used to rate a risk are those for trucks, and those for autos. For risks with trucks, the factors are:

1. Size of trucks (by weight)
2. Business class (in classification tables)

3. Radius class (distances traveled)
4. Zones (geographic location of main depots)
5. Collision versus noncollision coverage

For risks with private passenger autos, the factors are:

1. Original cost
2. Age group (up to 6 years)
3. Deductibles per type of coverage
4. Comprehensive versus collision only versus combined coverages
5. Geographic location

Each and every truck and private passenger auto is rated separately and included in the cost of the premium. In addition, the underwriter will request motor vehicle registrations (MVRs) on all vehicles to be sure that the autors and trucks are properly registered and, in many states, inspected. Data on all drivers will also be requested, including previous driving records of violation, revocations, and the like. Copies of current drivers licenses often provide this type of data.

The safety and health professional can play an important part in obtaining insurance for commercial auto exposures. The data accumulated by the professional on auto losses can be vital in determining the acceptability of the risk, and then in deciding the preliminary premium to be offered. Interaction between the insurance loss control representative and the safety and health professional focuses the representative's report to underwriting on the aspects of the loss control program that help prevent and reduce the frequency and severity of auto accidents. Often the extent to which the loss control program is seen as being effective sways an underwriter about an account.

This section has provided a picture of three liability policies as they now exist. Be aware that this is strictly an overview and the actual underwriting and policy wording are likely to be far more complicated. It is suggested that those interested in more details on a particular policy should contact the ISO for information.[4]

Now that we have reviewed the exposures that may exist where we work, and have taken a look at the various policies that will provide insurance coverage for these exposures, what remains is the mechanics of the actual purchase of the policies. The next section addresses this question.

30.4 MECHANICS

For the safety and health professional with risk management responsibilities, the purchasing of insurance can be a harrowing experience. As can be seen from the previous section, insurance has its own jargon, which is often difficult for an outsider to understand. In addition, many of the provisions of the policies are quite legalistic and subject to interpretation not only by nonlegal personnel but by lawyers as well.

Although insurance can be a tangled web, it is extremely helpful when losses arise. Therefore, the necessity of having the correct policies with the adequate limits of liability far outweighs the hassle it may be to deal with any problems that may arise. The main rule when dealing with insurance is: Use common sense. If the terms and conditions do not make sense to you, have your broker or the underwriter explain them. If the premium seems too high, find out the basis for the premium. If an exclusion seems particularly tricky and endlessly confusing to you, have it clarified by a lawyer.

Your main contact with the insurance community will be either your insurance agent or broker, or, in the case of a direct insurance writer, the underwriter. When purchasing a policy, there are three main components that must be taken into account:

1. The broker/agent.
2. The insurance company.
3. The specific underwriter assigned to your account.

It is important to understand the relationship that you and your company should establish with each of these, what information you should have about them, and what you can expect them to do for you. We will examine only the mechanics of placing an insurance program using an agent/broker. Although there are some differences in the mechanics when dealing with a direct writer, the placement of insurance with a direct writing insurance carrier requires the acquisition of much of the same underwriting information, which most likely will be the safety and health professional's role in the placement of insurance coverage.

30.4.1 The Broker/Agent

To begin, there is a difference between a broker and an agent. A broker usually works for a nationwide firm with offices everywhere and can work with all the major insurance companies. These large firms are called "alphabet houses" because they often are called by their initials, such as M&M for Marsh & McLennan and A&A for Alexander & Alexander. An agent usually works for a much smaller firm that may work with a number of insurance companies or with only one.

The chief advantage of a broker is the size of the firm, which allows it to do many things that a smaller agent cannot do for you, such as loss control, claim services, and computer loss runs. The chief disadvantage may also be size, because your company's insurance program may not receive as much attention as a larger company's insurance program. This reflects the fact that the basis for profit for the broker is commission, and the larger the insurance program, the larger the commission.

For an agent, on the other hand, your insurance program may be the largest piece of business the firm handles. Therefore, the amount of time and attention spent on it may be much greater. However, the disadvantage of using an agent is getting access to the insurance companies and products you may want.

An intemediate solution is what is known as the "second line" agencies, which are large regional agencies. These firms have access to the majority of insurance companies, can often match the services provided by the brokerage houses, and are still small enough to spend the time on a moderate-sized account, with premium dollars ranging from $50,000 to $100,000.

Whatever route you go, there are three items that you should check on before choosing a broker/agent:

1. The services that can be provided.
2. The experience the firm has.
3. The market access of the firm.

Many agent/brokers have additional services they can offer you besides placing your business with an insurance company. Many firms have loss control personnel on staff or as a separate entity, which can be called upon by the client to provide inspections and evaluations. The accessibility of these representatives and their costs should be discussed and, if possible, determined in advance.

Claim services can be extremely valuable for lines of insurance in which you can expect losses. Some brokers/agents have claims specialists who can investigate a loss and make recommendations concerning settlements. This can help to resolve many of the disputes between parties before they become lawsuits.

Brokers and agents may also have access through computer terminals and modems to loss data generated by the insurance company in terms of loss reserves (money set aside to pay a claim at a future date when settled), losses actually paid, information on the incidents that generated the losses, and what is called "incurred but not reported" (IBNR) losses. The last figure is a statistically derived estimate of the amount of losses that have occurred during the policy period but will not be reported until the policy period is over. Such information is often available only on computer terminals to an agent who has a very close relationship with a particular insurance company.

The risk manager should ask about all services even if there is no immediate intention of utilizing any particular service. Those services that the risk manager wants upon policy inception should be delineated and priced in advance by the broker/agent.

The risk manager should have a good picture of the coverages needed before choosing the broker/agent. In that way, the risk manager can inquire about the experience of the broker/agent with placing such coverages. This is particularly true if the risk manager is planning to purchase a policy that is out of the ordinary, such as professional liability or umbrellas at high limits of liability. Also, the broker/agent should be familiar with your company's type of exposure and your company's financial status, product lines, and history. This information should be presented to the broker/agent, and the risk manager should attempt to gauge the broker/agent's ability to understand what is taking place at the company and what exposures need to be insured.

A professional broker/agent can assist the risk manager by evaluating the company's risk and determining the best method of safeguarding the company against loss. A thorough broker/agent will ask many questions of the risk manager and others to discover any hidden exposures and to clarify those exposures that may not be readily apparent. The best agent/brokers are thinking "exposure first, commission second."

Once the risk manager has been satisfied about the services and the experience of the agent/broker, the next item is to find out to what markets the agent/broker has access. It may not be necessary to have a broker/agent who is capable of working with every insurance company if the agent/broker can work with all the markets with the coverages you need. This can easily be discovered by asking the agent/broker which companies it represents. If you do not hear the name of an insurance company, you are particularly interested in, ask the broker/agent about the company.

Access is important because in some lines of coverage, there are only a few markets available that will write that line of business. If that coverage is important to you and that particular agent/broker cannot get it, that may not be the broker/agent for you. Of course, you always have the option of splitting the lines of coverage between two brokers, or going direct to the insurance company that provides the type of insurance coverage you need. The disadvantage to this is that certain companies will not offer the special lines of coverages unless you also buy the standard (CGL, Auto, WC) lines from the company. It is therefore often best to make sure your agent/broker can get quotes on all the lines you need.

30.4.2 Insurance Companies

The key decision on insurance companies is whether to approach one company for all the policies you need or to approach different carriers for different policies. Certain key common factors must be evaluated when choosing the exact insurance company for your firm.

The advantage of putting all policies at one insurance company is the possibility of discounted pricing for the whole package based upon the sheer volume of premium. Insurance companies are aware that many incidents affect a variety of policies because of their interlocking nature. It is expected that there is a better chance to get a handle on a loss if the group of policies is all in one company.

In addition, a large premium will enable the insurance carrier to have a pool of cash available if an incident that affects any one line takes place. That money can be earning revenue until such time as a loss is paid, thus enabling the insurance company to have another source of income.

From the viewpoint of the insured, the advantage of having one carrier is the ease of claims handling. All losses would be reported in one fashion to one company, and there would be no need to determine the applicability of coverage since all policies are in one place.

Conversely, the advantage of placing different lines of coverage with different carriers is that each company may provide you with the exact coverage you need, particularly with regard to "exotic" coverages that not everyone needs. A *manuscripted*, or designed, policy may provide coverage that an ISO manual form does not. Depending on the needs of your company, this flexibility in coverage may be important.

Also, with each line of coverage, competition between carriers may be more intense than between carriers who can offer package deals. The premium one carrier may offer on WC may be substantially lower than the price offered on WC as part of a package. That can vary consideraly from year to year, but if your company is price sensitive with regard to insurance, this consideration may be important.

Another important fact in choosing insurance companies should be the financial stability of the carrier. Each year, Best's Insurance Guides, Inc.,[5] rates each insurance company on the basis of its financial conditions. A+ is the highest rating. Best's takes into account the gross writing premium, the net profit, the loss reserves set aside, investment income, and other fiscal consideration in making the rating. As a rule of thumb, it would be best to work only with companies that have Best's ratings of B+ or better.

In addition, annual reports of the insurance companies will give you a better picture of the company as a whole or as part of a larger entity. Many insurance companies today are part of a large financial service company or are a subsidiary of a conglomerate. A review of the annual reports of the larger companies will indicate what priority the insurance company has within the larger firm.

Also, the annual report may indicate the direction of the insurance company regarding certain lines of coverage. Also, new products are often discussed in the report, and these may also fulfill a need of your company. A fuller spreadsheet on the financial state of the insurance company is also included in the report.

Insurance companies have many services that are often available to the insureds for little or no charge. Brochures and booklets concerning these services should be obtained and reviewed. Some are duplicates of the ones that the broker/agent can provide (loss control, claims, loss runs) and may be used in lieu of the broker's services. Also, carriers can provide training and educational materials to supplement what you have and enhance the insurance-related program.

One of the primary services, loss control, can be quite extensive in terms of the range of specialties. Often, the cost of their services can be negotiated as part of the premium paid up front. A service visit program can be established, and the various specialists' visit can be scheduled in advance. To do this, it will be necessary to know about the capabilities of the loss control staff in advance. Your broker or insurance company representative should have up-to-date information on the insurance company's services, which you should request for review.

In today's market, it is often just as important to know the reinsurers that back each product line you will be buying as it is to know the insurance carrier. Reinsurance is the insurance company's insurance company. The reinsurer will pay the losses on a certain portion of the limits of liability in return for a piece of the premium. Therefore, the type of reinsurer used is important to the insured because that reinsurer is at a risk for any loss that occurs. The reinsurer must also be financially stable and capable of absorbing the loss if it occurs. Best's also rates reinsurers and the same rule of thumb "B+ or better" applies to them. The key point is that the risk manager should know what reinsurers back each line of coverage that has been purchased, so that any news concerning a reinsurer can be viewed in light of that knowledge. The financial soundness of the companies backing your policies is of the utmost importance.

All the data must be weighed before any final decisions are made on which insurance companies and their policies should be the ones you as the risk manager should choose. Fortunately, most of the information is easily obtainable for review. If the insurance terminology becomes too tangled, your insurance agent or insurance company representative should be able to straighten it out for you.

30.4.3 Underwriters

Ultimately, the individual who will have the most impact upon the premium and the details of your insurance program is the underwriter. The underwriter will evaluate the material sent in as part of the submission and will decide whether your company is a viable risk for the insurance company to insure against certain types of damages. Their need for information and documentation must be met in order that you can put together the best insurance package possible for your company.

In many insurance transactions, the insured does not meet with the underwriter. The correspondence is handled through the broker/agent or the internal insurance company sales representative, and, in most cases, this system works well. However, it should be noted that if you are having a problem understanding what is needed by the underwriter, you should meet directly with the underwriter to discuss that problem.

Because of the sheer volume of paperwork that an underwriter will see in a typical year, the initial submission must make a good impression so that the underwriter can distinguish your submission from others. To accomplish this, the underwriter will measure the data on three points:

1. Adequacy of the information for the coverage desired.
2. Accuracy of the information.
3. No loose ends left in the data.

The adequacy of the information relates, first and foremost, to the items essential to a submission for each line of coverage reviewed in the previous section. This is where the two rules of underwriting come into play, particularly the second, "underwrite to sell" and "underwrite to cover your assets." The first rule indicates that pricing has to be within the current marketplace pricing, otherwise there will be no sale. The second rule state that the file's data must justify the pricing that the underwriter has determined. The underwriting file must show how the premium was derived and all scheduled credits and adjustments must be documented.

That is one reason why the underwriter may request additional followup data after the initial submission. The way to limit that request is to make sure that the facts and figures most likely to be wanted by the underwriter are in the submission in the first place. A well-organized and thorough submission may include an index, clear, concise text, and uncomplicated, precise figures and charts.

The accuracy of the data should be documented with collaborating outside sources, if possible. A typical collaborating outside source would be a report from a consultant focussing on the safety aspects of a particular process. Another would be an industrial hygiene report on air samples taken on employees. When outside collaboration is not possible, statements concerning an item should indicate the basis for the remarks. There should be no inaccuracies in the submission.

The bane of the underwriter's existence is loose ends. The underwriter must question anything for which the data are inadequate or inaccurate so that the evaluation of the risk and the judgments made by the underwriter cannot be questioned later. For example, an underwriter will want the latest loss run with data on accidents and illnesses, and a previous year's data will not be satisfactory. Also, an underwriter will not be satisfied with a submission that is not complete, with additional information to be sent after the policy is written. This will make the underwriter wary and think that the prospective insured has something to hide. Any unclear statement will be picked out, and if the evidence for such a statement is not presented, the underwriter will request such evidence. When inquiries come from the underwriter, the reply should also provide complete and accurate answers to all questions. In addition, all answers should be documented so that no more doubts will remain in the underwriter's mind about this matter.

When meetings with the underwriter are required in the initial stages of contact, a good first impression is important. You should be prepared for the meeting with all necessary data available for review. It is always best to include the broker/agent or insurance company internal sales representative in all meetings, so that the broker/agent, as a third party, can break an impasse and steer the conversations to a productive conclusion. Also, if a specific area is in question, the person within your company with the best information on that situation should be present at the meeting.

The underwriter's product is the terms and conditions of the quotation on your account, including the premium. Once you have this, it is up to you to react to the quotation. Generally, you are given a period of time, usually 30 days, to get back to the underwriter with your answer. If you are dissatisfied with the quotation, you should ask the basis for the pricing and why certain terms (particularly exclusions and deductibles) were included. The answers to these inquiries may give you some leeway with regard to negotiations.

The risk manager should be suspicious of a very low price as well. It is possible that the coverage from the company for this line is very restrictive. Alternatively, the underwriter may be inexperienced with such coverage and an audit down the line may force cancellation of the policy. The search for adequate insurance coverage would then begin all over again. A third reason may be that the insurance company is underpricing the rest of the market in order to write a large volume in a short period of time, and then withdraw from the market. This again would leave you high and dry, with the possibility of no coverage.

A reasonable price and good terms and conditions with a solid basis in the facts and figures of the account should be the goal of the risk manager. Once that is accomplished to everyone's satisfaction, the account can be "bound," or insurance purchased. A binder should be signed by the insurance company outlining all the terms, conditions, and prices in writing. The risk manager should make sure that the binder is complete and accurate before accepting it. Generally, a binder indicates coverage for a period of time running from 30 to 90 days. Within that time, a policy as described earlier will be assembled by the underwriter and other members of the insurance company. A careful reading of all endorsements and exclusions, the declarations page, and certificates of insurance should be done by the risk manager or the staff. Any discrepancies or inaccuracies should be immediately reported to the broker/agent or the insurance company sales representative, and the underwriter, for correction.

30.5 CONCLUSION

The purpose of this chapter was to provide the safety and health practitioner with practical knowledge about liability insurance and to enable the practitioner to use that knowledge to adequately protect the assets of the company through liability insurance policies. Three major aspects of insurance—exposures, coverages, and mechanics—were explored in depth. Examples were give to illustrate many of the points to be made about these subjects.

By illuminating the subject in this fashion, it is hoped that safety and health professionals will have a better understanding of insurances in their future dealings. If insurance is, as they say, a "necessary evil," it is best to know the evil as thoroughly as possible in order to conquer it.

One final point: Real-life situations involving insurance are almost never exactly as they are written in a book. It is only by experiencing the actual placement of a book of insurance business that an individual can say that he or she knows what insurance is about. The material in this chapter can serve only as a beginning when delving into this subject. But even a beginning is better than attempting to dive into insurance without any knowledge at all.

Some additional sources of information about insurance in general, check with the Insurance Services Office (ISO) in New York, and the American Insurance Association (AIA), which is also based in New York.

REFERENCES

1. *Analysis of Workers' Compensation Laws*, Washington D.C.: U.S. Chamber of Commerce, 1983.
2. *Best's Insurance Report—Property-Casualty*, Oldwick, NJ: A. M. Best, 1993.
3. *Best's Loss Control Engineering Manual*, Oldwick, NJ: A. M. Best, 1994.
4. Castle, G., R. F. Cushman, and P. R. Kensicki, *The Business Insurance Handbook*, Homewood, IL: Dow Jones–Irwin, 1981.

5. *Fire, Casualty and Surety Bulletins*, 3rd Printing, Cincinnati: The National Underwriter Company, 1980.
6. Hinley, J. W., "Problems in Relation to Contractual Liability Insurance," in *Proceedings of the Casualty Actuarial Society*, Volume XXV, Number 51, published by the Society, 1935.
7. Insurance Service Office, *Commercial Lines Manual*, 3rd ed., New York: Insurance Services Office, 1984.
8. *Rimco's Manual of Rules, Classifications and Interpretations for Workers' Compensation Insurance*, Dallas, TX: International Risk Management Institute, 1984.
9. Williams, C. A., Jr., G. L. Head, R. C. Horn, and G. W. Glendenning, *Principles of Risk Management and Insurance*, Vol. II, 2nd ed., Malvern, PA: American Institute for Property and Liability Underwriters, 1981.

INDEX

In this index, page numbers followed by the letter "t" designate tables; *See also* cross-references designate related topics or more detailed topic breakdowns.

Abbreviations, 288–289. *See also* Terminology
Abrasive blasting, 54–56
Abrasive blasting hoods, 581t
Abrasives, for grinding, polishing, and buffing, 68–70, 69t
Absorbent tube sampling, 257–258
Absorption
  percutaneous, 703–704
  of sound, 910–918, 915t
  of toxicants, 382
Absorption coefficient, 922
AC (alternating current) electromagnetic fields, 993. *See also* Radiation: nonionizing
Acceptance testing, 582–586
Access, defined, 1142
Accident(s). *See also* Liability insurance
  bodily injury by, 1165
  as compared with injuries, 16–18
  defined, 15
  direct versus indirect costs of, 14–15
  financial impact of, 14–18
  liability insurance aspects of. *See* Liability insurance
  mortality and morbidity from, 104–105, 105t
  off-site, 1165–1166
  response time as cost factor, 1048
  root causes of, 22–23
  severity rate, 939
  SIC code, 938–939
  vehicular, 1165
Accident, defined, 1077
Accident investigation, 103–126, 602–603
  benefits of, 104–105
  management systems in, 121–124
  need for, 107–108
  overview of, 109–113
  participants in, 108–109
  phases of
    step 1: fact determination, 113–115
    step 2: immediate causes (acts and conditions), 115–117
    step 3: underlying factors, 117–119
    step 4: organizational prevention, 119–121
  Program Evaluation Profile (PEP), 1100t
  reporting requirements, 105–107
  summary of, 124–125
Accident reports, in hazard identification, 436
Accident triangle, 105, 111
Acclimatization, to heat, 794–795
Accountability, 1048-1051
Acetone, biological exposure index (BEI), 392t
2-Acetylaminofluorene, 388t
ACGIH (American Conference of Governmental Industrial Hygienists)
  Air Sampling Instruments for Evaluation of Atmospheric Contaminants, 269
  biological exposure indices, 391–400
  Chemical Substances TLVs, 265
  cold temperature work standards, 936–937
  heat exposure job ranking, 808
  Industrial Ventilation Manual, 578
  IR and UV light exposure standards, 1001
  magnetic field exposure standards, 992, 993–994, 995t
  noise exposure criteria, 872–874
  Ventilation Manual, 57, 67, 70
  vibration standards, 948
Acid cleaning, of metals, 56–57
Acids, primary skin irritant, 709t
Acne, 720t, 720–721, 721t
  skin disease risk and, 705
ACOEM (American College of Occupational and Environmental Medicine), 420, 422
Acoustic intensity, 922
Acoustic pressure, 922
Acoustics, 827–838. *See also* Hearing; Noise; Sound
Acquired immunodeficiency, 362
Acquired immunodeficiency syndrome (AIDS). *See* HIV/AIDS
Acroosteolysis, 725–726
Acrylonitrile, 388t

Activated charcoal, as sorbent, 258–259
Active immunity, 352
Acute eczematous contact dermatitis, 718–719, 719t
Acute versus chronic disorders, 5
ADA (Americans with Disabilities Act), 1072–1074, 1155. *See also* Workers' compensation
Adaption, diseases of, 846
Administrative law judge, defined, 1142
Adsorbents, for passive badges, 258–259
Advance Notice of Proposed Rulemaking (ANPR), 1119, 1121–1122
Aerobic capacity, 953. *See also* Ergonomic job analysis
Aerobic metabolism, 795
Aerosol(s). *See also* Airborne contaminants
  collection and evaluation of, 264–271
    by direct reading instruments, 269–271, 270t
    by filter sampling, 265–268, 266t
    by impactors, 268
    by impingers, 269
  defined, 664
Age
  hearing loss and (presbycusis), 826
  hearing threshold level associated with (HTLA), 845
  heat tolerance and, 795
  skin diseases and, 706
Agents, insurance, 1188-1190
Aggregate liability, 1174–1175
Agriculture, infectious diseases related to, 316–317
AIDS (acquired immunodeficiency syndrome). *See* HIV/AIDS
AIHA (American Industrial Hygiene Association), 2, 7
  guidance document of, 11–12
  Proficiency Analytical Testing program, 273
  sampling size recommendations, 240
Air, density of, 923
Airborne contaminants. *See also* Dusts; Gases; Respirators; Vapors
  in abrasive blasting, 55–56
  chemical
    aerosols, 264–271
    gases and vapors, 257–264
  Clean Air Act (P.L. 101–549), 1113
  control equipment and processes, 562–587
    equipment
      ancillary, 564–567
      major processing, 567–569
    facility layout, 572–577
      general considerations, 573–574
      material transport, 572–573
      segregation of operation, 576–577
      service and maintenance, 575–576
      worker isolation, 577
      workstations, 574–575
    processes, 569–572
    testing and maintenance, 586–587
    ventilation, 577–586. *See also* Ventilation
  control strategies for, 544–562
    dusts, 559–561
    impurities in production chemicals, 561–562
    toxic materials, 544–559
      elimination, 545
      replacement with alternate, 545–559, 548t, 551t, 553t, 554t, 556t–558t
  critical zones for, 540
  in electroplating, 65–67
  equipment and processes
    diagnostic air sampling, 562–564
    equipment, 564–569
  in gas welding, 96–97
  general control guidelines for, 541–544
  indoor air quality (IAQ) and, 743–789
    building-related illness, 745–746, 762–764
      economic cost of, 763–764
      humidifier fever, 762
      hypersensitivity pneumonitis, 746t, 762
      legionellosis, 762–763
    combustion products, 754–756
    control methods, 778–785
      combustion products, 783
      communications, 784–785
      environmental tobacco smoke, 783
      microbials, 779–781
      particulates, 783–784
      pesticides, 783
      thermal environmental conditions, 784
      volatile organic compounds (VOCs), 781–783
    environmental tobacco smoke, 756–757
    evaluation and checklists, 770–778
      qualitative evaluation, 770–777, 772t
      quantitative evaluation, 777–778
    historical background, 744
    microbials, 746t, 746–749
    particles, 757–758
    pesticides, 753–754
    problem building studies, 758t, 759–764
      British epidemiologic, 759–760
      Danish epidemiologic, 760
      NIOSH, 759
      of perceived air quality (odors), 760–761
      of thermal environment, 761–762
    representative kinds and sources, 745t
    sick building syndrome, 744–745
    ventilation systems (HVACs) and standards, 764–770
      ASHRAE 62-1989 and revision, 768–770

CO_2 concentration, 767
humidification and dehumidification, 766–767
system description, 764–766, 765t
volatile organic compounds (VOCs), 749–753, 750t, 753t
in metal cutting, 73–74
in metal descaling, 57
OSHA hazard communication regulations, 514
risk assessment specialists for new construction, 543t
in thermal spraying of metals, 80
from welding, 91, 97t
Airborne precautions, 351t
Airborne transmission, 328
Air conditioners, diseases spread by. *See* Indoor air quality (IAQ)
Air-cooled clothing, 819
Air coolers, safety design of, 472
Airflow testing, 583–584
Airline respirators, 664
Air nozzles, 565
Air-purifying respirators, 645–652, 664, 671
particulate filter, 647–648
regulations and standards, 645–647, 647t
Air sampling, 562–564, 563t. *See also* Airborne contaminants
Air Sampling Instruments for Evaluation of Atmospheric Contaminants (ACGIH), 269
Air velocity measurement, 812
Alcohol(s)
chemical metabolism and, 384–385
heat tolerance and, 797
in indoor air, 750t, 750–751
labeling requirements for, 482
primary skin irritant, 710t
Alcohol abuse, 1024–1025
Alkaline cleaning
electrolytic, 58
of metals, 57–58
immersion, 57–58
salt baths, 58
Alkalis, primary skin irritant, 709t
Alkanes, in indoor air, 750t, 750–751
Allergy (hypersensitization), 708–714
latex, 632
photoallergy, 714, 715t
phototesting for, 714
phototoxicity, 713–714
to plants and woods, 711–713, 712t
Alloys, for cutting tools, 73
All Risk Adjustment Program (ARAP), 1044
AltairStrickland, Inc. v. Chevron U.S.A. Products Co., 1152
A.M. Best rating reports, 1044–1045
Ambient monitoring, 378
Ambient noise, 922
Amebiasis, 308t, 345t
Amenometers, 812
American Board of Preventive Medicine, 420
American College of Occupational and Environmental Medicine (ACOEM), 420, 422
American Conference of Governmental Industrial Hygienists (ACGIH). *See* ACGIH
American Industrial Hygiene Association (AIHA). *See* AIHA
American Iron and Steel Institute, 579
American National Standards Institute (ANSI). *See* ANSI
American Society for Testing and Materials (ASTM). *See* ASTM
American Society of Heating, Refrigeration, and Air-Conditioning Engineers (ASHRAE). *See* ASHRAE
American Society of Safety Engineers (ASSE). *See* ASSE
Americans with Disabilities Act (ADA), 1072–1074, 1155. *See also* Workers' compensation
4-Aminodiphenyl, 388t
Amplitude (of sound), 829–834, 922–923
abbreviations and letter symbols for, 834–835, 835t
pressure, intensity, and power relationship and, 833–834
sound intensity/sound intensity level, 830–831
sound power/sound power level, 831–833
sound pressure/sound pressure level, 829–830
summing of independent sources, 835–837
summing of levels and frequencies, 837–838
Anaerobic metabolism, 795
Anechoic room, 923
Anergy, 342
Aniline, BEI for, 392t, 418t
Animal-associated occupations, infectious diseases related to, 316–317
Animal contact, infectious diseases transmitted by, 292t–293t
Annealing, 71–72
Anodizing, 64–65. *See also* Electroplating
ANPR (Advance Notice of Proposed Rulemaking), 1119, 1121–1122
ANSI (American National Standards Institute), 2
audiometer/audiometry standards, 846
eye and face protection standards, 625–626, 626t
footwear standards of, 632–633
hard hat types, 622
hazardous materials labeling standard, 507
instrument certification by, 275

ANSI (*cont'd*)
laser standards, 1002
magnetic field exposure standards, 995t
noise exposure criteria, 861–862, 872–874
personal protective equipment standards, 608–610
respirator standards, 641, 646, 656
vibration standards, 936, 948
Anthrax, 303t
immunization for, 353t
incidence, 345t
Anthropometrics, 950–954. *See also* Ergonomic job analysis
Anthropometric working position data, 950t
Antibiotic resistance, 313–314
Antibiotics interfering with biological monitoring, 383t–384t
Antibodies, 330
Antibody tests, 337, 340–341
Antibody titer, 340
Antigens
HLA, 736
in skin irritation, 711
Antigen tests, 341
Antiseptics, as allergens, 712t
Apocrine glands, 736
Appendages, skin, 736
Aqueous cleaning processes, 554t, 554–555
Arc welding
gas metal, 93–95
gas tungsten, 92–93
shielded metal, 90–92
Argon, in thermal spraying of metals, 79–80
Arnett v. Environmental Science and Engineering, Inc., 1152
Aromatic hydrocarbons. *See also specific compounds*
in indoor air, 750–751
ARPA (All Risk Adjustment Program), 1044
Arsenic
biological exposure index (BEI), 392t
biological monitoring case study, 404–405
primary health effects, 388t
Arthropod-borne viral infections, 301t
Asbestos, 245–247, 546
primary health effects, 388t
ASHRAE 110-1995, 585
ASHRAE (American Society of Heating, Refrigeration, and Air-conditioning Engineers), 744
environmental tobacco smoke standards, 757
standard 62-1989 and revision, 768–770
Aspergillosis, 309t
Aspergillus, 746–748
ASSE (American Society of Safety Engineers), 2, 7
Assigned protection factor (APF), 664
ASTM personal protective equipment standards, 608
Asymptomatic infection, 329
Atmosphere supplying respirators, 671–672
Atmospheric pressure, 923
Attitudes
of employees, 18–19, 27–28
of management, 18–19
Attorney-client/attorney-work product privilege, 1158
Attorneys
employer's representation by, 1060–1061
instances for consulting, 1159–1160
in workers' compensation, 1033–1034
Audiogram, defined, 923
Audiometers, 846, 923
Audiometry, 846–850. *See also* Hearing measurement
baseline, 876
maximum permissible ambient sound levels for, 849
Audit privilege, 1158
Audits
hazard communication, 508–512
program evaluation, 1081–1105. *See also* Program evaluation
safety, 460–465
Auricle, of ear, 839
Automation, 570–571
Automobile insurance, 1184–1187. *See also* Liability insurance
Automobiles, recyclable, 548
Average weekly wage, 1077

Back belts, 636
Bacteria
defined, 323
indoor airborne, 746t, 746–748. *See also* Microbials
Bacterial dysentery (shigellosis), 300t
Bacterial infections
general characteristics of, 299t–302t, 303t–307t, 310t
modes of transmission of, 291t, 295t
skin, 717–718
Badges, passive sampling, 258–259
Bags, sampling, 260–261
Barrier creams, 734
Baseline specimen, 340
Batteries, mercury, 545
Battery limit valves, 473
Behavior-based programs, 6
Belts
body, 636
seat, 635
Benzene, 388t
biological exposure index (BEI), 392t, 418t
PEL for, 562
as solvent in tire manufacture, 562

Benzidine, 388t
(A.M.) Best rating reports, 1044–1045
Best's Safety Directory, 606
Biological evaluation, of infectious disease exposure, 336–339, 338t. *See also under* Infectious disease(s)
Biological exposure indices (BEIs), 416, 418t
Biological monitoring, 373–413. *See also* Toxicity; Toxicology
  ACGIH exposure indices, 383t–396t, 391–397
    implementation of, 391, 397, 398t–399t
    interpretation of, 397
    timing of sampling, 397, 400
    variations affecting sample determinants, 401t–403t
  ambient, 378
  biological exposure indices, 391–400, 392t–395t, 396t–397t, 398t–399t
  case studies
    arsenic toxicity, 404–405
    cholinesterase-inhibiting (organophosphate) pesticide toxicity, 405–406
    tetrachloroethylene toxicity, 408–410
    toluene toxicity, 406–408
  clinical application of biomarkers, 386–391
    definitions used in, 386–387
    in illness evaluation, 390
    occupational physician and, 387–390
    OSHA medical screening standards, 388t–389t
  as compared with diagnosis, 378
  definition and background, 374–375
  ethical issues in, 391
  of indoor air. *See* Indoor air quality (IAQ)
  limitations of, 421
  practical considerations in, 377–379
    communication of results, 379
    personnel, 379
    sampling strategies, 377–379
  protocols for, 423–424, 423–425
  scope and purpose of, 375–377
  toxicologic and pharmacokinetic considerations in, 379–386
    distribution, 382
    dose, 380t, 380–381
    duration, 381–382
    excretion, 385–386
    metabolism, 382–385
      drug effects, 383t–384t
    routes of entry, 380
  types of, 375–377
  variables affecting, 401t–403t
Biological monitoring case studies, 404–411
Biomarker, defined, 386–387
Biomarkers, defined, 416
Biomechanical strength data, 952–953. *See also* Ergonomic job analysis
Biosafety levels (BSLs), 350
Blasting
  abrasive, 54–56
  ventilation for, 581t
Blastobacter, 748
Blastomycosis, 309t
Blast-resistant construction, 474–475
Blood, as index of exposure, 377–379
Bloodborne pathogens standard, OSHA, 110, 313
Bloodborne pathogen standard, OSHA, 334t
Bloodborne transmission, 326
Blood samples, preemployment, 336–337
BLS incidence and frequency rates, 1035
Blue M* ovens, 581t
Body belts, 636
Body dimensions
  seated, 949t
  standing, 948t
Body harnesses, 636
Body (metabolic) heat, 793. *See also* Heat stress
Body position, 935–936. *See also* Ergonomic job analysis
Bone disease, acroosteolysis, 725–726
Bordetella sp., 340
Borel v. Fibreboard Paper Products Corp., 1153
Borg scale of maximum voluntary muscle contraction, 954
Borrelia burgdorferi. See Lyme disease
Botulinum toxin, 328–329
Bovine spongiform encephalopathy (mad-cow disease), 323
Bradford v. Kupper Associates, 1154
Brazing, 88–89. *See also* Soldering
  health hazards of, 88–89
Breach of contract, 1153–1154
Breach of warranty, 1154
Breathing area measurements, 245
Bright dip processes, 56–57
BRIs (building-related illnesses). *See* Indoor air quality (IAQ)
British epidemiologic problem building studies, 759–760
British system, 12
Brokers, insurance, 1188-1190
Brucella sp., 340
Brucellosis, 303t–304t, 345t
BS 8800, 12
BSLs (biosafety levels), 350
Buffing, 68–70, 69t
Building-related illness, 745–746, 762–764. *See also* Indoor air quality (IAQ)
  humidifier fever, 762
  hypersensitivity pneumonitis, 746t, 762
  legionellosis, 762–763
Bulk samples, 348
Bullae, defined, 736

Bullous, defined, 736
Bump caps, 623–624
Bureau of Labor Statistics (BLS), 1035
Burn-in testers, 580t
Burns, 603
  chemical, 708, 710t
  radiation, 1000

Cadmium
  BEI for, 418t
  biological exposure index (BEI), 392t
Calibration
  of acoustic testing equipment, 859
  of instruments, 274–275
Canada
  Office of Biosafety Materials Safety Data Sheets (MSDSs), 350
  workers' compensation in, 1032
Cancer. *See also* Carcinogenicity; Neoplasms
  cellular telephones and risk of, 998
  magnetic fields and risk of, 994
  skin, 722–723, 723t
Carbon dioxide(CO_2), 94
Carbon dioxide (CO_2), 754–756
  concentration, 767
  indoor concentration of, 767
Carbon disulfide, biological exposure index (BEI), 392t, 418t
Carbon fire extinguishing systems, 447–448
Carbon monoxide (CO), 754–756
  biological exposure index (BEI), 392t
Carboxyhemoglobin, as biological marker, 390
Carburizing, 71
Carcinogen(s)
  defined, 484
  skin, 704
  substitution for, 551
Carcinogenicity. *See also* Cancer; Carcinogens; Neoplasms
  of magnetic radiation, 996, 998
  MDSD requirements, 485–486
  of metal cutting materials, 73–74
  potential of environmental tobacco smoke, 756–757
Cardiopulmonary resuscitation (CPR), 110
Carelessness, legal definition of, 1077
Carpal tunnel syndrome, 1077
Cartridges, respirator, 649
Case definitions, 344
Case procedure, legal, 1156–1157
Cataracts, 1001
Cat-scratch disease, 304t
CDC (Centers for Disease Control and Prevention)
  hazardous materials classification, 350
  infection control guide for healthcare workers, 313
  infectious disease regulations of, 333
  Morbidity and Mortality Weekly Report, 343
  notifiable disease regulations, 343–346, 344t, 345t
  tuberculosis prevention guidelines, 313
CD-ROMs, 695
CEFIC occupational exposure analysis chart, 242
Cellular telephones, 996, 998
Cellulose ester filters, 266t
CE mark, 6
Centers for Disease Control and Prevention. *See* CDC
Centrifugal pumps, 565–567
Certification
  of instruments, 275
  of vaccination, 319–320
CGL (comprehensive general liability) policy, 1177–1180
Chain of custody, 273–274
Checklists, safety, 133–137
Check, Plan, Do, Act, 9, 10, 104
Chemical burns, 708, 710t
Chemical processing, fired heater safety design, 469–470
Chemical protective clothing. *See also* Personal protective equipment (PPE)
  decontamination of, 615, 622
  ensembles, 615, 616t–618t
  selection considerations, 612
    chemical resistance rating, 612–615
      degradation, 613
      permeation, 612–613
    durability, 615
    duration of use, 615
    ease of decontamination, 615
    flexibility, 615
    heat transfer, 613–615
    vision, 615
Chemicals
  exposure evaluation
    air-borne contaminants, 257–284
      aerosols, 264–271
      gases and vapors, 257–264
    general considerations in, 271–277
      analytical lab services and chain of custody, 273–274
      documentation, 277–278
      instrumentation, 274–277
      sample collection planning, 271–273
    sampling strategy, 240–257
      duration, 247–249
      environmental variability and, 243–246
      frequency, 249–251
      general objectives and principles, 240–242
      location, 246–247

measurement method, 251–257
measurement purpose, 243
surface-borne contaminants, 277–278
hazard communication program elements, 483–486. *See also* Hazard communication
highly toxic defined, 484
irritant, 484
PMN (premanufacturing notification), 585–586
sensitizer, 484
skin irritants, 707–708, 708t, 709t–710t
toxic defined, 484
Chest films, protocols for, 423
Chickenpox (varicella), 296t, 315t, 354t, 363t
Child care
infectious diseases related to, 314–316, 315t
regulations affecting, 332
Chloracne, 705, 720t, 720–721, 721t
Chlorinated degreasers, 61
Chlorinated skin irritants, 710t
Chlorobenzene, biological exposure index (BEI), 392t, 418t
Chlorofluorocarbon solvents (CFCs), 553t, 554t
bis-Chloromethyl ether, 388t
Cholinesterase-inhibiting pesticides (organophosphates)
biological exposure index (BEI), 394t, 418t
biological monitoring case study, 405–406
Chromatography, filters for, 277–278
Chromium, biological exposure index (BEI), 392t
Chromium plating, 66. *See also* Electroplating
Chromophores, defined, 736
Chronically infected workers, 338–339
Chronic disorders, 5
Chronic eczematous contact dermatitis, 719, 720t
Citation, defined, 1142
Cladosporium, 746–748
Claims, workers' compensation, 1045–1062. *See also* Workers' compensation: postinjury response strategy
Claim status ratings, 1045
Class action suits, 1171–1172. *See also* Legal issues; Liability insurance
Clean Air Act (P.L. 101–549), 1113
Cleaning, of respirators, 661–662, 670
Cleansers, hand, 733–734
Clean Water Act (P.L. 92–500), 1113–1114
Clinical disease, defined, 329
Clostridium botulinum, 328–329
Clothing. *See also* Personal protective equipment (PPE)
insulation value of, 817–818, 819t
105 CMR, magnetic field exposure standards, 995t
CO_2. *See* Carbon dioxide (CO_2)
Coal tar, as skin irritant, 710t
Coatings
powder, 555, 556t
radiation-cured, 556t
supercritical fluid spray, 556t
Cobalt, biological exposure index (BEI), 392t
CO (carbon monoxide), 754–756
Coccidioidomycosis, 309t
Cochlea (of ear), 841–842
Codes of ethics, 7
Coil backflow protection, 470
Coke ovens, 388t, 577
Cold degreasing, 59
Cold injury, 706–707, 936–937
Cold temperatures, musculoskeletal injury and, 936–937
Collective bargaining, return to work and, 1054–1055
Collision insurance, 1184
Color, identification by, 631, 647t
Color-coding, of respirators, 647t
Colorimetric sampling, of gases, 259–260
Combustion products, 754–756
indoor, 783
Comedones, defined, 736
Committees, congressional, 1111–1112
Common law, 1150–1151
Common-vehicle transmission, 328
Communicable disease, defined, 321–322
Communicable period, 330
Communication
with hearing protectors, 894–898
respirators and voice, 664
in workers' compensation claims management, 1057–1059
Community noise, 862–869, 921
Company safety and health policy statement, 26
Compensable injuries, 1077
Completed operations, 1167–1168
Compliance Safety and Health Officer (CSHO), 1125
Comprehensive Environmental Response, Compensation, and Liability Act (CERCLA, Superfund) (P.L. 96-510), 482, 1115–1116, 1154–1155
Comprehensive general liability (CGL) policy, 1177–1180
Comprehensive vehicular coverage, 1184
Compression injury, 603
Compressors, safety design of, 472
Computers/computer systems. *See also* Computer simulation
in engineering control, 677–696
CD-ROMs, 695
databases, 684–688
hardware considerations, 695, 696t

Computers/computer systems (*cont'd*)
internet, 692–695
literature search, 690
repetitive or complicated calculations, 690–692
spreadsheets, 688–689
statistical analysis, 677–684
radiation exposure and VDTs, 994–995
for ventilation testing, 586–587
Computer simulation, 179–237
applications of
effectiveness of safety organization, 225–230
maintenance manual evaluation, 230–231
troubleshooting model, 231–235
cost of, 184–185
fundamentals of
computation of replications, 201–202
computer simulation languages, 219–223
cumulative probability distribution function, 194–196
distribution sampling by computer, 198–201
games and gaming, 223–224
Monte Carlo sampling, 193
simulation algorithms for, 216–219
from standard distributions, 202–214
random number generation, 193–194
sample values, 196–198
spreadsheet functions useful in, 215–216
of infectious disease transmission, 330–332
management's role in, 182
modeling and analysis for, 185–193
documentation and user manuals, 192
experimentation in, 191–192
input modeling, 187–188
model design and implementation, 185–187
model testing, verification, and validation, 188-190
output analysis, 190–191
problem definition and simulation objectives, 185
results implementation, 192–193
as problem-solving methodology, 181–185
project team for, 182–183
of strength tasks, 952
system specification for, 183–184
Conduction, thermal, 793
Conductive footwear, 633
Confirmatory tests, 341
Conjunctivitis
bacterial, 299t
viral, 297t
Consensus standard, defined, 1142
Consequence analysis, 433–434
Construction details, 439
Construction safety checklist, 139
Consumer Product Safety Act (P.L. 92-573), 1116–1117
Contact dermatitis. *See* Skin diseases
Contact/grasping RF currents, 996–997
Contact lenses, safety of, 626–628
Contact precautions, 351t–352t
Contact stress, 935. *See also* Ergonomic job analysis
Contact transmission, 326
Contact urticaria (hives), 724–725
Containment, 356
Contamination, radioactive, 1012–1014
Continuous-flow respirator, 664
Continuous spectrum, of sound, 923
Contract, breach of, 1153–1154
Contract claims, 1153–1154
Contractors
accidents to or involving, 1168
hazard communication and, 502–503
safety Program Evaluation Profile (PEP), 1098t
Control buildings, 474–475
Control practices, 429–675. *See also individual subtopics*
air contaminants. *See also* Air contaminants
equipment and processes for, 562–587
general guidelines for, 541–544
NIOSH control technology reports, 591–599
strategies for, 544–562
computer applications in, 677–696
CD-ROMs, 695
databases, 684–688
hardware considerations, 695, 696t
internet, 692–695
literature search, 690
repetitive or complicated calculations, 690–692
spreadsheets, 688–689
statistical analysis, 677–684
in ergonomics, 955
hazard communication, 479–515
implementation tools, 503–507
OSHA regulations related to, 513–515
program management, 492–503
program requirements, 481–492
program unit, 507–513
health and safety instruction, 517–537
evaluation of instruction, 536–537
feedback, 537
goal setting, 520–530
learning audience, 534
learning objectives, 532–533
needs analysis, 518–520
presentation, 534–536
task analysis, 530–531
loss control and fire prevention, 429–478
audits and self-inspection, 460–465
basic program components, 430–431

emergency planning, 457–460
fire protection elements, 440–457
emergency action plan, 454–456
evacuation drills, 456–457
facilities and equipment, 440–449
fire prevention plan, 451–454
fireproofing, 449–451
life safety principles, 457
hazard impact, 433–434
loss control program elements, 434–440
control, 439–440
evaluation, 438–439
identification of hazards, 436–438
organization, 434–436
policy statement, 434
plant site selection, 465–469
process equipment and facilities design, 469–475
risk quantification, 2–433
training, 475–476
noise control, 898–922. *See also* Noise control
Program Evaluation Profile (PEP), 1101t–1102t
protective equipment, personal, 601–638. *See also* Personal protective equipment (PPE) *and specific types*
for radiation exposure, 1011
for skin disease, 734–735
workers' compensation and, 1047–1049
Convalescent specimen, 340
Convection, thermal, 793
Copper, in wiring boards, 548
Core temperature, 793
Corneocytes, 701. *See also* Skin; Skin diseases
Corrective action, in employee relations, 1019–1021
Corrosion Proof Fittings v. EPA, 546
Corrosive, defined, 484
Cost-benefit analysis, in workers' compensation/disability, 1063–1072. *See also under* Workers' compensation
Costs
of building-related illness, 763–764
of health and safety training, 526–530
of respirator programs, 657–658
Cotton dust, 388t
CPSC (Consumer Product Safety Commission), 1116–1117
Cramps, heat, 798
Creams, barrier, 734
Creatinine excretion, medicines affecting, 384t
Criterion duration (TC), of noise, 923
Criterion level (LC), of noise, 923
Cryptonecrosis, 309t
Cryptosporidiosis, 299t
Culture, organizational, 2, 7–8
Culture isolation, 340
Cumulative probability distribution function, 194–196
Cumulative trauma, 5
Cutaneous larva migrans, 308t
Cutting fluids, 73, 74t
Cyaniding, for surface hardening, 71
Cyclone separators, 267–268
Cytokines, defined, 736
Cytomegalovirus, 297t, 315t, 363t

Damages, legal, 1153
Damper controls, remote, 470
Dangerous to life and health
defined, 664–665
respirator use in, 672–673
Danish epidemiologic problem building studies, 760
Data
as management tool, 1048
workers' compensation, 1034
Data analysis, Program Evaluation Profile (PEP), 1101t
Databases, 684–688
for Material Safety Data Sheets (MDSD), 503–504
Data loggers, 277
Daubert v. Merrell Dow Pharmaceuticals, 1150–1151
Day-night-time-averaged sound level (DNL), 920–921
dB (decibels), 830, 923. *See also* Noise; Sound
DC (direct current) electromagnetic fields, 993. *See also* Radiation: nonionizing
Deaths. *See* Mortality
Decibels (dB), 830, 923. *See also* Noise; Sound
Definitions. *See also* Terminology
in insurance policy, 1174
Degradation, of protective clothing, 613
Degreasing, 58–64
health hazards of
cold, 59
vapor-phase, 59–64, 62t.567
ventilation for, 580t
Dehumidification, 766–767
Demand respirator, 665
Deming's Plan, Do, Check, Act, 9, 10, 104
Denmark, step model of toxic use reduction, 548t, 548–549
Density of air, 923
Depigmentation, defined, 736
Depressuring valves, 470–471
Dermal absorption, 382
Dermal exposure assessment, chemicals, 278
Dermatitis. *see also* Skin disorders
defined, 706

Dermatitis (*cont'd*)
exfoliative, 736
from metal cutting, 73
from paints, 84–85
radiation, 717
Dermatophytoses, 299t. *see also* Skin disorders
Descaling, of metals, 56–57
Design aspects, of hazard elimination, 440
Design for the Environment Program (EPA), 548
Detectors, heat and smoke, 448
Diagnosis, as compared with biological monitoring, 378
Diarrhea, childcare-related risks of, 315t
Diathermy units, 996. *See also* Radio frequency radiation exposure
1,2-Dibromo-3-chloropropane, 388t
3,3-Dichlorobenzidine, 388t
Difficult employees, 1019–1023. *See also* Employee relations
Diffusion furnaces, 580t
Dilution ventilation, 357
N,N-Dimethylacetamide, biological exposure index (BEI), 393t
4-Dimethylaminoazobenzene, 388t
N,N-Dimethylformanide (DMF), 393t, 395t, 418t
Diphtheria, immunization for, 353t
Direct-contact transmission, 326
Directivity factor, 923
Disability. *See also* Workers' compensation
defined, 1077
permanent, 1078
Disability syndrome, 1055–1056
Discharge, employer's right to, 1022–1023, 1023t
Disciplinary action, 1021
Discovery (legal), 1156–1157
Disease nomenclature, 325–326
Diseases
of adaption, 846
clinical, 329–330
communicable defined, 321–322
infectious, 287–364. *See also* Infectious diseases *and specific diseases*
notifiable, 343–346, 344t, 345t
Disinfectants, regulations for, 335–336
Disinfection, 356
of respirators, 661, 670
Disposable respirators, 665
Distribution, bodily of toxicants, 382
Distribution curves, 245
Distribution sampling by computer, 198–201
DNA, as biological marker, 390
DNL (day-night-time-averaged sound level), 920–921
Documentation
in ergonomic job analysis, 946–947
in hazard communication, 500–503
Documents, privileged and confidential, 1157–1159
Dose
biological monitoring of, 380–381
effective (infectious), 329
Dose reduction, for radiation, 1008
Dose-response curves, 381
Dosimeters, 262
DOT (Department of Transportation), 1116
Downsizing, 6, 7, 1026–1027
Drainage, fire considerations in, 473–474
Droplet precautions, 351t
Droplet transmission, 327
Drug abuse written policies, 1024–1025
Drug-Free Workplace Act of 1988, 1024
Drugs, use and abuse of, 1024–1026. *See also* Medications
Drying ovens, 581t
Durability, of protective clothing, 615
Dusts. *See also* Aerosols; Respirators
in abrasive blasting, 54–56
control of, 559–560
by dustiness testing, 560
by moisture content, 559–560, 560t
by particle size, 560
by slurry methods, 560–561
cotton, 748
in indoor nonindustrial environments, 757–758
Duty, transitional, 1078. *See also* Return to work
Dysentery, bacterial (shigellosis), 300t
Dystrophy, nail, 725

Ear(s), 838–846. *See also* Hearing loss; Noise; Sound
anatomy of, 838–842
eardrum (tympanic membrane), 840
external, 839
inner ear, 841–842
middle ear, 840–841
in hearing loss, 842–846
Earthquakes, 468
Eccrine gland, defined, 736
Echinococcus (hydatid disease), 302t
ECM (electrochemical machining), 75
Eczematous contact dermatitis, 718–719
acute, 718–719, 719t
chronic, 719, 720t
EDM (electrical discharge machining), 75–76
Education. *See also* Training
about skin hazards, 734–735
defined, 517
in environmental medicine, 420
in infection prophylaxis, 357

Effect, biological markers of, 387
Effective (infectious) dose, 329
Effective sound pressure, 923
Effective temperature (ET) index, 800, 801
Electrical classification, of facilities, 474
Electrical discharge machining (EDM), 75–76
Electrical hazard footwear, 633–634
Electrical protection, 636
Electricity, skin disease related to, 717
Electrochemical machining (ECM), 75
Electrochemical sensors, 263t
Electrolytic alkaline cleaning, 58
Electroplating, health hazards of, 64–67, 68t
  air contamination, 65–66
  technique-related, 64–65
ELF (extremely low frequency) radiation, 990, 991t, 993–995
Emergencies
  heat stroke as, 799
  medical. *See* Medical emergencies
Emergency action plan, for fires, 454–456
Emergency planning, 457–460
Emergency Planning and Community Right-to-Know Act (EPCRA), 481
Emergency preparedness, Program Evaluation Profile (PEP), 1103t–1104t
Emergency procedures, for radiation exposure, 1012–1014
Emergency shut-down valves, 473
Employee involvement, 18-19
Employee participation, Program Evaluation Profile (PEP), 1096t
Employee privacy, 1159
Employee relations, 1017–1028
  corrective action, 1019–1021
  difficult employee types, 1019
  disciplinary action, 1021
  drugs and alcohol, 1024–1026
  employer responsibilities, 1017–1018
  layoffs, downsizing, and re-engineering, 1026–1027
  stress, 1027–1028
  termination, 1021–1023
    employer's right to discharge, 1022–1023, 1023t
    guidelines for, 1021–1022
    informing co-workers, 1022
    informing employee, 1022
  voluntary resignation, 1023–1024
  warning signs, 1019
Employee right to know, 1144
Employees. *See also* Personnel; Staff; Workers
  attitudes of, 27–28
  as insurance category, 1164–1166
  intoxicated, 1026
  odor perception as risk reducer, 559
  protected classes of, 1023t
Employers
  hazard communication responsibilities of, 489–491. *See also* Hazard communication
  right to discharge, 1022–1023, 1023t
Encephalitis, 301t
Energy demands, musculoskeletal injury and, 938
Engineering controls. *See also* Control practices
  OSHA definition of, 540–541
Enteritis, campylobacter, 299t, 304t
Environmental (community) noise assessment, 862–869
Environmental factors, in skin diseases, 706–707
Environmental impairment, 1169–1170
Environmental legislation
  Clean Water Act (P.L. 92-500), 1113–1114
  Clear Air Act (P.L. 101–549), 1113
  Comprehensive Environmental Response, Conservation, and Liability Act (CERCLA, Superfund) (P.L. 96-510), 482, 1115–1116, 1154
  Resource Conservation and Recovery Act (P.L. 94–560), 546, 1115
  Safe Drinking Water Act (P.L. 93-523), 546, 1114
  Toxic Substances Control Act (P.L. 94-469), 1113
Environmental management, 6
Environmental Risk Categories (ERCs), 550, 551t
Environmental testing, 347–348
Environmental tobacco smoke, 783
Environmental transmission, of infectious diseases, 294t–295t
Environmental variability, 243–246
EPA (Environmental Protection Agency). *See also* Regulatory issues
  community right to know regulations, 458
  Design for the Environment Program, 548
  medical waste tracking by, 335–336
  minimization versus elimination policy, 546–549
  Noise Reduction Rating (NRR), 887–894, 889t, 890t, 891t, 893t
  pesticide exposure studies, 753–754
  premanufacturing notification to, 542
Epidemic defined, 343
Epidemiological modeling, of infectious disease transmission, 330–332
Epidermis, 700–703
Epizootic defined, 343
Epoxy paints, 82
Equipment design, for hazard elimination, 440

Equipment safety, 431
Equivalent effective temperature corrected for radiation (ETCR), 801, 821
ERCs (Environmental Risk Categories), 550, 551t
Ergonomics, 5. *See also* Ergonomics job analysis
  defined, 932–934, 933t, 1077
  published documents on, 43t–47t
Ergonomics job analysis, 931–985
  examples of forms, 957–960, 958t
  of manual handling tasks, 950–954
    anthropometric design criteria, 953
    physiological data, 953
    psychophysical data, 953–954
    strength data, 952
  methods for, 946–950
    findings evaluation/normative values, 947–950
    findings summary, 947
    job documentation, 946–947
    work factor identification and evaluation, 947
  OSHA forms, 961–971
    computer model for lifting exposures, 968–970
    hand tool checklist, 967–968
    manual handling checklist, 962–963
    task/work methods checklist, 963–964
    video display unit and keyboard issues checklist, 964–966
    worksite assessment checklist, 961–962
    workstation layout checklist, 966–967
  proactive versus reactive approaches to, 938–939
  solutions, 954–956
    analysis for, 954–955
    engineering controls, 955
    medical management program, 955
    training programs, 956
  steps in, 939–946
    equipment, 940–943
    personnel selection, 940
    videotaping, 943–946
  table of ergonomic studies, 971t–978t
  weight limit calculation, 960
  work factors to consider, 935–938
    cold temperatures, 936–937
    contact stresses, 935
    forces required for task, 935
    glove fit, 937
    muscle-contraction frequency and duration, 936
    obstructions, 937
    physical energy demands/fatigue, 938
    postures assumes, 935–936
    prolonged high visual demands, 938
    standing surfaces, 937–938
    vibration, 936
Erysipeloid diseases, 304t
Erythema defined, 736
Erythema infectiosum, 363t
Escape-only respirators, 652–653, 665
ET (effective temperature) index, 800, 801
Ethical issues
  in biological monitoring, 391
  in reporting of results, 425
2-Ethoxyacetic acid, 393t
2-Ethoxyethanol (EGEE), 393t
2-Ethoxyethyl acetate (EGEEA), 393t
Ethyl benzene, 393t, 418t
Ethylenemine, 388t
Ethylene oxide, 388t
Evacuation drills, 456–457
Evacuation wardens, 455
Evaluations
  health, 239–427. *See also* Health evaluations
  safety, 103–238. *see also* Safety evaluations *and individual topics*
    accident investigation, 103–126
    computer simulations, 179–238
    risk assessment, 127–178
  of safety and health program, 1081–1105. *See also* Program evaluation
Evaporation, 793–794
Exchange (doubling) rate of sound, 923
Exclusions, insurance, 1175–1176
Exfoliative dermatitis, 736
Exhaustion, heat, 798–799
Exhaust systems. *See* Airborne contaminants; Ventilation
Experience modification, 1077
Experience Modification Factor, 1042
Explosions
  building design and, 474–475
  fired heater, 469–470
Exposure, to infectious diseases. *See* Infectious disease(s)
Exposure assessment. *See* Health evaluations; Safety evaluations
Exposure estimate calculation, 249
Exposure periods, 247–249
Exposures, liability. *See under* Liability insurance
Extinguishers, portable fire, 444
Eye and face protection, 358t
  ANSI standards, 625–626
  contact lenses, 626–628
  laser eyewear, 628–629
  selection guidelines, 626t–628t
  sideshields, 625
  in welding, 97–98
  when required, 625
Eye disorders, UV light associated, 1001

Eye irritation, from volatile organic compounds (VOCs), 751

f (frequency), 827
Facepiece respirators, 642–645
loose-fitting, 645
tight-fitting, 642–645
Face protection, 358t, 625–629, 626t–628t. *See also* Eye and face protection; Protective equipment
Facial flush, 725
Facial hair, respirator use and, 663–664
Facilities
designing safety into, 474–475
electrical classification of, 474
fire safety aspects of site selection, 465–469
fire safety issues in, 468–469
layout for contaminant control, 572–578
isolation of workers, 577
material transport, 572–573
process location, 573–574
segregation of operations, 576–577
service and maintenance, 575–576
workstations, 574–575
process design, 542
Facultative pathogens, 322
Failure mode and effect analysis (FMEA), 160–166, 432–433
Fainting (syncope), 798
Fall protection, 635–636
False-positive tests, 340–341
Fatalities. *See also* Accidents; Mortality
reporting of, 106
Fault tree analysis (FTA), 166–177, 432–433
Federal Insecticide, Fungicide, and Rodenticide Act (P.L. 92-516), 1114–1115
Federal legislation, 1112–1117. *See also* OSHA
Clean Air Act (P.L. 101-549), 1113
Clean Water Act (P.L. 92-500), 1113–1114
Comprehensive Environmental Response, Compensation, and Liability Act (CERCLA, Superfund) (P.L. 96-510), 482, 1115–1116, 1154–1155
Consumer Product Safety Act (P.L. 92-573), 1116–1117
Federal Insecticide, Fungicide, and Rodenticide Act (P.L. 92-516), 1114–1115
Hazardous Materials Transportation Act (P.L. 93-633), 1116
OSHA Act, 1117–1146
preemptive effect of, 1150
Resource Conservation and Recovery Act (P.L. 94-580), 1115
Safe Drinking Water Act (P.L. 93-523), 1114
Toxic Substance Control Act (P.L. 94-469), 1113
Federal legislative process, 1110t, 1110–1112, 1112t
FEV_1 (forced expiratory volume), 423
Fibers, skin injury and, 716
Fibrous glass filters, 266t
FIDs (flame ionization detectors), 263t
Fields, electromagnetic. *See* Radiation: nonionizing
"Fifth" disease (parvovirus B19), 315t, 363t
Filters
noise, 857–858
types and common applications, 266t
Filter sampling, 265–268
of aerosols, 265–268, 268t
Fire alarms, 448
Fired heaters, 469–470
Fire hazards
in degreasing, 64
in metal cutting, 75
Fireproofing, 449–451
Fire protection, 440–457
emergency action plan, 454–456
evacuation drills, 456–457
fire detection, 448–449
fire alarms, 448
heat detectors, 448
smoke detectors, 448
firefighting systems, 440–448
fire trucks, 448–449
fixed water sprays, 442–444
foam, 447
halon and carbon extinguishing systems, 447–448
hose, 445–446
live hose reels, 444
mains and hydrants, 441–442
monitors, 442
portable fire extinguishers, 444
pumps, 441
water supply, 440–441
fireproofing, 449–451
life safety principles, 457
Fit factor, 665
Fitness (physical), heat tolerance and, 796
Flame ionization detectors (FIDs), 263t
Flavobacterium, 748, 762
Flexibility, of protective clothing, 615
Fluids, cutting, 73, 74t
Fluorides, 393t, 418t
Fluxes
in gas welding, 96
for soldering, 85, 86t, 87–89
Fluxing, 87–89, 88t
FMEA (failure mode and effect analysis), 160–166, 432–433

Foam, firefighting, 447
Folliculitis, acne, and chloracne, 720t, 720–721, 721t
Food processing, infectious diseases related to, 316–317
Foot protection, 609
  ANSI standards and selection, 632–634
  foot hazard posting, 634
Footwear
  conductive, 633
  electrical hazard, 634
  static-dissipative, 634
Forced expiratory volume (FEV_1), 423
Forced vital capacity (FVC), 423
Force requirements, 935. *See also* Ergonomic job analysis
Formaldehyde, 388t
Fraud, in workers' compensation claims, 1059–1060
Free field, 923–924
Freon solvents, 61. *See also* Degreasing; Solvents
Frequency
  electromagnetic, 990, 991t
  infrasonic, 924
  of sampling, 249–251
  of sound, 829, 924
Frequency analyzers, 857–858
  narrow-band, 858
  octave band, 857–858
Frequency rates, 1035, 1077
Frostbite, 716
FTA (fault tree analysis), 166–177, 432–433
Fumes. *See also* Air contaminants
  metal, 91
    from welding, 93, 94, 96
Functional job requirements, 1078
Fungal infections
  general characteristics and transmission mode, 299t, 309t–310t
  modes of transmission, 291t, 295t
  skin, 717–718
Fungi
  defined, 323
  indoor airborne, 746–748
Fungicides. *See* Pesticides
Furfural, 393t, 418t
FVC (forced vital capacity), 423

Gade v. National Wastes Management Association, 1150
Galvanizing, 96–97
Games and gaming, in computer simulation, 223–224
Gamma ray sources, in industrial testing, 77–78
Gas and vapor removing respirators, 648–649
Gas cabinets, 580t
Gas carburizing, 71
Gas chromatography, filters for, 277–278
Gases
  evaluation of
    by direct reading instruments, 262, 263t–264
    sample collection, 257–262
      by absorbent tubes, 257–258
      by evacuated containers and bags, 260–261, 261t
      by impingers and liquid traps, 259
      by length of stain tubes, 259–260
      by passive badges, 258–259
  from welding, 91
Gas jungles, 580t
Gas metal arc welding, 93–95
Gas nitriding, 71
Gastrointestinal absorption, 382
Gas tungsten arc welding, 92–93
Gas welding, *95,* 95–98
Gauss (unit of radiation measure), 991–992
Geiger Mueller (GM) detector, 1009
Geller, Scott, 6
Gender, heat tolerance and, 795
General duty clause, defined, 1142
General liability insurance, 1177–1180
Genetic predisposition, to skin diseases, 705–708
Genus and species, 324–325
German measles. *See* Rubella
Giardiasis, 299t
Glands
  apocrine, 736
  eccrine, 736
Globalization, 6
Glossaries. *See* Terminology
Glove boxes, 580t
Gloves. *See also* Hand protection
  as skin disease preventive, 733
Glues, as allergens, 712t
Goal setting, for health and safety instruction, 520–530
Goggles. *See* Eye and face protection; Eye protection; Personal protective equipment
Grab (short-term) samples, 256, 260–261
Granulomas, 723–724, 736
Grasping/contact RF currents, 996–997
Grinding, polishing, and buffing, health hazards of, 67–70, 69t
Guaranteed cost workers' compensation, 1044

Hair, facial, respirator use and, 663–664
Hall effect, 992
Halon and carbon fire extinguishing systems, 447–448
Hammurabi Code, 2–3
Hand cleansers, 733–734
Hand protection, 358t. *See also* Personal protective equipment

disadvantages of, 631
latex allergy and, 632
leak checking of, 632
selection criteria for, 629–631
sizing of, 632
Hantavirus, protection against, 359
Hantavirus pulmonary syndrome, 308t
Hardware considerations, in computer selection, 695, 696t
Hats. *See* Head protection
Hazard(s)
defined, 1077
as defined in insurance policies, 1177–1180
health, 53–101. *See also* Health hazards *and specific materials and processes*
identification of, 436
prioritizing of, 438
of protective equipment, 605
published documents on, 36t–39t
recognized, 1144
Hazard and operability (HAZOP) analysis, 152–160
Hazard assessment, 9
components of, 19
Program Evaluation Profile (PEP), 1098t–1099t
as related to protective equipment, 602
accident incident investigation, 602–603
job safety analysis, 603
process hazard review, 603
targeted PPE hazard assessment, 603–604
for respirator programs, 655
Hazard communication, 479–515, 523
foot hazard posting, 634
implementation tools, 503–507
hazardous materials information systems, 500–505
for material labeling, 506–507
for training, 505–506
legislative background of, 480–481
noise hazard, 875
OSHA regulations related to, 513–515
access to employee exposure and medical records (20 CFR 1910,20), 513
air contaminants (29 CFR 1916,1000), 514
hazardous waste operations and emergency response (1910,120), 514–515
occupational exposure to hazardous chemicals in laboratories (29 CFR 1910.1450), 514
process safety management of highly hazardous chemicals (1910,119), 514
program management, 492–499, 492–503
flow of communications, 498–499
policies and procedures, 493
program responsibilities, 494–498
quality control and program improvement, 500
recordkeeping, 500–503
program requirements, 481–492
program unit, 507–513
performance testing, 506–513
program appraisal, 507–508
record review, 506
requirements of
chemical program elements, 483–488
hazard determination, 483–484
labels, 485
materials safety data sheets (MSDS), 484–485
definitions used in, 482–483
employer responsibilities, 489–491
employee information and training, 491
labels, 490–491
materials safety data sheets (MSDS), 490
written hazard communication program, 489–490
trade secrets and, 491–492
Hazard determination, 483–484
Hazard elimination, 440
Hazard impact assessment, 433–434. *See also* Control practices; Risk assessment
Hazardous agents, transfer regulations, 336
Hazardous and Solid Waste Amendments, 546–549
Hazardous Materials Transportation Act (P.L. 93-633), 1116
Hazard reporting, Program Evaluation Profile (PEP), 1100t
Head protection
accessories, 623
ANSI standards, 623
bump caps, 623–624
hard hats, 622–623
when required, 623
Health and safety instruction, 517–537
Health care
infectious diseases related to, 312–314
regulations affecting, 332
Health evaluations, 239–427
biological monitoring in. *See also* Biological monitoring
biological exposure indices, 391–400, 392t–395t, 396t–397t, 398t–399t
case studies
arsenic toxicity, 404–405
cholinesterase-inhibiting (organophosphate) pesticide toxicity, 405–406
tetrachloroethylene toxicity, 408–410
toluene toxicity, 406–408

Health evaluations (*cont'd*)
clinical application of biomarkers, 386–391
definition and background, 374–375
practical considerations in, 377–379
scope and purpose of, 375–377
toxicologic and pharmacokinetic considerations in, 379–386
variables affecting, 401t–403t
biomonitoring in, 373–413
chemical exposure, 239–285. *See also* Chemical exposure
air contaminants
aerosol sampling, 264–271
gas and vapor sampling, 257–264
general considerations, 271–277
sampling strategy, 240–257
surface sampling, 277–278
criteria and forms for, 280–285
infectious agents, 287–371. *See also* Biological monitoring; Infectious disease(s); Inoculation
exposure evaluation, 332–348
overview, 289, 312–321
prevention, 348–364
principles of infectious disease, 321–332
Health hazards, 53–101. *See also specific materials and processes*
of abrasive blasting, 53–56
of brazing, 88–89
of degreasing, 58–64
cold, 59
vapor-phase, 59–64, 62t, 567
of electroplating, 64–67, 68t
air contamination, 65–66
technique-related, 64–65
of grinding, polishing, and buffing, 67–70, 69t
of heat treating, 70–72
annealing, 71–72
surface hardening, 71
of metal cleaning, 56–58, 551–559
alkaline, 57–58
immersion, 57–58
salt baths, 58
of metal machining, 72–76
conventional, 72–75, 74t
electrical discharge, 75–76
electrochemical, 75
of metal thermal spraying, 79–80
nondestructive testing, 76–79
by industrial radiography, 76–78
by liquid penetrant, 78–79
by magnetic particle inspection, 78
of painting, 80–85
of soldering, 85–88, 86t, 88t
of welding
arc, 89–95
gas, 95–98
Hearing, measurement of
by audiometers, 846–847, 848t
audiometric database analysis, 849–850
hearing threshold measurements, 848–849
recordkeeping in, 849
test rooms for, 847–848
Hearing conservation programs, 874–878
Hearing loss. *See also* Noise
causes of, 826–827
conductive and sensorineural, 842–843
noise as cause of, 843–844
nonoccupational, 844–845
occupational noise-induced, 844
Hearing protectors, 878–898, 889t, 890t, 891t, 892t, 893t, 897t. *See also under* Personal protective equipment (PPE)
Hearings
legislative, 1111
workers' compensation, 1060–1061
Hearing tests, baseline, 876
Hearing threshold level associated with age and noise (HTLAN), 845
Hearing threshold level associated with age (HTLA), 845
Heat
indoor, 761–762
skin injury and, 716
Heat balance, 793–794
Heat balance equation, 794
Heat cramps, 798
Heat detectors, 448
Heaters, fired, 469–470
Heat exchange, 793–794
Heat exchangers, 472–473
Heat exhaustion, 798–799. *See also* Heat stress
Heating, ventilation, and air conditioning systems (HVACs). *See* HVACs; Indoor air quality (IAQ)
Heat injury, 603
Heat-reflective shields, 814
Heat stress, 791–823
concepts underlying, 793–794
control measures for, 813–816
disorders related to, 797–799
factors affecting heat tolerance, 794–797
indexes of, 800
management of employee heat exposure, 816–820
physiology of, 792–793
significance of, 792
thermal measurements and instrumentation, 800–813
Heat stress index (HSI), 805–808, 806t
Heat stroke, 799
Heat syncope, 798
Heat tolerance, 794–797, 797t
acclimatization and, 794–795
age and, 795

gender and, 795
obesity and, 795–796
physical fitness and, 796
water and electrolyte balance and, 796–797
wellness programs and, 796
Heat transfer, of protective clothing, 613–615
Heat treating
annealing, 71–72
health hazards of, 70–72
annealing, 71–72
surface hardening, 71
quenching, 72
for surface hardening, 71
carburizing, 71
cyaniding, 71
gas nitriding, 71
Heavy metals. *See also* Specific metals
Heinrich, H. W., 4, 9
Helminth infections, 299t, 308t
Helminths, defined, 324
Hemosiderosis, pulmonary, 746t
HEPA filters, 647–648, 665. *See also* Respirators
HEPA systems, 581t
Hepatitis, 297t
childcare-related risks of, 315t
employment issues in, 338–339
immunization for, 353t
incidence of, 345t
Herpesvirus, 297t, 301t, 718
Herpetic whitlow, 297t
Hertz (Hz), 827–828, 924
n-Hexane, 393t, 418t
High performance liquid chromatography (HPLC), filters for, 277–278
Histoplasmosis, 310t, 359
History
medical, 422
occupational, 422
HIV/AIDS
employment issues in, 338–339
false positives in, 341
general characteristics and transmission of, 296t
incidence of, 345t
infectious disease risk and, 361–362
risk to health care workers and, 313
Hives (urticaria), 724–725
HLA proteins, 736
Hoods. *See* Respirators; Ventilation
Hookworm, 308t
Hose, fire, 445–446
Hose reels, for firefighting, 444
HSI (heat stress index), 800
HTLA (hearing threshold level associated with age), 845
HTLAN (hearing threshold level associated with age and noise), 845
Human contact, infectious diseases transmitted by, 291t, 296t–307t
Human Factors and Ergonomics Society, 2
Human immunodeficiency virus (HIV). *See* HIV/AIDS
Human papillomavirus (HPV), 298t
Humans, strengths as compared with machines, 933t
Humidification and dehumidification, 766–767
Humidifier fever, 746t, 762
Humidity
control of, 816
gender differences in tolerance of, 795
HVAC systems, 764–770. *See also* Indoor air quality (IAQ)
evaluation checklist for, 776–777
Hydatid disease (echinococcus), 302t
Hydrants, fire, 441–442
Hydrochlorofluorocarbon solvents (HCFCs), 553t
Hydrofluorocarbon solvents (HFCs), 553t
Hygiene products, as allergens, 712t
Hygrometers, 811
Hyperpigmentation, 721–722, 722t, 736
Hypersensitivity pneumonitis, 746t, 762
Hyperthermia. *See* Heat stress
Hypopigmentation, 736
Hz (hertz), 827–828, 924

IAQ (indoor air quality), 743–789. *See also* Indoor air quality (IAQ)
IARC (International Agency for Research on Cancer), 484
Ice packet vests, 819
Ichthyosis, 706, 736
ICNIRP, 998
magnetic field exposure standards, 992
IEEE/ANSI maximum permissible radiation exposures, 998t
Illness, occupational, defined, 1078
IME (independent medical exam), 1077
Immunity, types of, 352
Immunization
prior to travel, 319–320
safety issues in, 352
Immunodeficiency
infection risk and, 361–362
secondary (acquired), 362
Immunoglobulins (Ig), 330. *See also* Antibodies
Impact, hazard, 433–434. *See also* Control practices; Risk assessment
Impactors, 268
Impact protection, of hard hats, 622–623
Impetigo, 300t, 315t
Impingers
for aerosols, 269
for gases, 259

Implementation, Program Evaluation Profile (PEP), 1097t
Impregnated paper tape sensors, 263t
Impurities, toxic, 561–562
Incidence rates, 938–939, 1034–1035
Incident investigation resources, 109t. *See also* Accident investigation
Incubation period, 330
Incurred losses, 1045
Indemnification claims, 1154
Indemnity (wage loss compensation), 1032. *See also* Workers' compensation
Independent medical exam (IME), 1077
Index case, 343
Indoor air quality (IAQ), 743–789
  building-related illness, 745–746, 762–764
    economic cost of, 763–764
    humidifier fever, 762
    hypersensitivity pneumonitis, 746t, 762
    legionellosis, 762–763
  combustion products, 754–756
  control methods, 778–785
    combustion products, 783
    communications, 784–785
    environmental tobacco smoke, 783
    microbials, 779–781
    particulates, 783–784
    pesticides, 783
    thermal environmental conditions, 784
    volatile organic compounds (VOCs), 781–783
  environmental tobacco smoke, 756–757
  evaluation and checklists, 770–778
    qualitative evaluation, 770–777, 772t
    quantitative evaluation, 777–778
  historical background, 744
  microbials, 746t, 746–749
  particles, 757–758
  pesticides, 753–754
  problem building studies, 758t, 759–764
    British epidemiologic, 759–760
    Danish epidemiologic, 760
    NIOSH, 759
    of perceived air quality (odors), 760–761
    of thermal environment, 761–762
  representative kinds and sources, 745t
  sick building syndrome, 744–745
  ventilation systems (HVACs) and standards, 764–770
    ASHRAE 62-1989 and revision, 768–770
    CO_2 concentration, 767
    humidification and dehumidification, 766–767
    system description, 764–766, 765t
  volatile organic compounds (VOCs), 749–753, 750t, 753t
Indoor air quality procedure (ASHRAE), 768
Industrial ecology, 548
Industrial hygiene, published documents on, 39t–43t
Industrial radiography, 76
Industrial Revolution, 3, 699
Infection(s)
  airborne microbial, 746–749
  asymptomatic, 329
  defined, 328
  of ear, 841
  process of, 328–330
  skin, 717–718
  subclinical, 329
Infection precautions (universal precautions), 350–352, 351t–352t
Infectious agents
  basic principles of, microorganisms and sources, 322–324. *See also* Microorganisms *and specific agents*
  opportunistic, 322
  parasitic, 321–322
  terroristic use of, 336
  transmission of, 326–328, 327t
Infectious diseases. *See also* Infectious agents *and specific diseases and agents*
  bacterial, 291t, 293t
  basic principles of, 321–332
    epidemiological modeling, 330–332
    nomenclature, 324–326
      diseases, 325–326
      microorganisms, 324–325
    process of infection, 328–330
    transmission, 326–328, 327t
  basic transmission information
    environmental contact or direct agent handling, 308t–310t
    human or direct agent handling, 296t–307t
  communicable, 321
  conclusions about, 364
  exposure evaluation, 332–348. *See also* Biological monitoring
    infectious disease surveillance, 342–346
      investigation and reporting, 343
      notifiable diseases and case definitions, 343–346, 344t, 345t
    inspections and environmental testing, 346–348
    laboratory diagnosis, 339–342
    medical and biological, 336–339
      chronically infected workers, 338–339
      periodic examinations, 337
      postexposure examinations, 337–338, 338t
      preplacement examinations, 336–337
      return-to-work examinations, 338
    regulations and guidelines, 332–336
      CDC, WHO, and professional associations, 333

general federal, state, and local, 332–333
hazardous agent transfer, 336
medical waste handling, disinfectants, and medical devices, 335–336
OSHA, 333–335, 334t
fungal, 291t, 295t
occupational aspects of
animal or arthropod contact, 292t–293t
environmental contact, 294t–295t
human contact, 291t
overview of, 289, 312–321
agriculture, food processing, and animal-related, 316–317
childcare-related, 314–316, 315t
health care-related, 312–314
laboratory-related, 317–318
occupational travel-related, 318–320
personal service work-related, 320
unlisted occupation-related, 320–321
wastewater and sewage treatment-related, 320
parasitic, 321–322. *See also and specific organisms;* Helminth infections; Protozoal infections
potential occupational, 311t–312t
prevention of, 348–363
decontaminants used in, 355t
education, training, and administrative controls in, 357
in health care settings, 350–352, 351t–352t
by immunizations, 352, 353t–354t
in laboratories, 350
by personal protective equipment, 358t, 358–359. *See also* Personal protective equipment
by postexposure prophylaxis, 359–361
risk factors and, 361–363
immunodeficiency, 361–362
pregnancy, 362, 363t
strategies for (levels of), 349t
by substitution and engineering controls, 354–357
containment, 356
decontamination, 354–356
ventilation, 356–457
protozoal, 291t, 292t
viral, 291t, 294t
Infectious (effective) dose, 329
Infectivity, defined, 328
Infertility, heat exposure and, 795
Influenza, 298t, 315t. *See also* Respiratory infections
immunization for, 353t
Information sources, 33–52
electronic data sources, 35, 48–49
on hazards, sources, and properties, 34–35
published documents, 36t–49t
industrial hygiene, 39t–43t
medicine, 47t–49t
safety and ergonomics, 43t–47t
workplace hazards and toxicology, 36t–39t
Information systems, for hazard communication, 503–505
Infrared radiation, 1000–1001
Infrared sensors, 263t
Infrasonic frequency, 924
Initial pleadings, 1156
Injuries. *See also* Workers' Compensation
as compared with accidents, 16–18
compensable, 1077
defined, 1078
minor to serious ratio, 106
reporting of, 106
Inner ear, 841–842
Insecticides. *See* Pesticides
Inspection. *See also* Inspection management
for infectious disease, 346–347
OSHA, 1125–1133. *See also* Inspection management; OSHA
of personal protective equipment, 608
Program Evaluation Profile (PEP), 1099t
of respirators, 660–661, 670–672
Inspection management, 1125–1129. *See also* OSHA
closing conference, 1128
employee complaints, 1126
employer preparedness, 1128-1129
initiation of inspection, 1126
opening conference, 1127
postinspection issues, 1129–1133
employee rights and courses of action, 1132–1133
employer options, 1130–1131
formal contest, 1131–1132
informal conference, 1131
petition for modification of abatement (PMA), 1132
posting requirements, 1130
types of violations, 1129–1130
records review, 1127
types of inspections, 1125–1126
walkaround, 1128
Instrumentation, for chemical sampling
aerosols, 269–271, 270t
calibration and maintenance of, 274–275
certification of, 275
radio frequency effects in, 275
Insulation, 813–814
Insulation value, of clothing, 817–818, 818t, 819t
Insurance
liability, 1163–1194. *See also* Liability insurance
workers' compensation, 1029–1080. *See also* Workers' compensation

Insurers, cooperation with, 1061–1062
Integrated samples, 256–267
Integument defined, 736
International Agency for Research on Cancer (IARC), 484
International Council on Non-Ionizing Radiation Protection (ICNIRP), 992
Internet, 692–695
Interval evaluations, 425–426
Intoxicated employees, 1026
Intradermal skin tests, 342
Ionizing radiation, 1004–1015. *See also under* Radiation
  skin damage related to, 717
IRCs (internet relay charts), 694
Irritant chemical, defined, 484
ISO 9000, 6
ISO 14000, 6, 551–552
ISO 14001, AIHA guidance document to, 11–12
ISO (Insurance Services Office) Standard insurance policy, 1172–1173
ISO (International Standards Organization)
  noise exposure criteria, 872–874
  regulatory guidelines, 1140–1141
  vibration standards, 948
Isolation
  as skin disease preventive, 732
  of workers, 577

Job analysis
  defined, 1078
  ergonomic, 931–985. *See also* Ergonomic job analysis
Job safety analysis, 137–146, 603
  example of completed, 146
  step 1: job or task selection, 143
  step 2: sequential task steps, 143–144
  step 3: hazard identification, 144t, 144–145
  step 4: safety control development, 145
Judge, administrative law, 1142
Judicial review, defined, 1143
Justification formula, in cost-benefit analysis, 1069–1071, 1070t

Keratinocytes, 701, 736. *See also* Skin; Skin diseases
Keratitis
  bacterial, 299t
  viral, 297t
Ketones, as skin irritant, 710t
Koch's postulate, 322

Labeling
  employer responsibilities for, 490–491
  of hazardous chemicals, 486
  of hazardous materials
    ANSI standard, 507
    manufacturer's labels, 506
    in workplace, 506
Laboratories
  biological monitoring provisions for workers in, 389t
  hazardous communications laws and, 483
  infectious diseases related to, 317–318
  OSHA hazard communication regulations, 514
Laboratory testing
  for chemical evaluations, 273–274
  for infectious diseases, 339–342. *See also specific techniques*
  protocols for, 422–425
Labor-Management Relations (Taft-Hartley) Act, 1109
Labor unions, return to work and, 1054–1055
Langerhans cells, 736
Languages, for computer simulation, 219–223
Laser burns, 717
Laser eyewear, 628–629
Lasers, as radiation source, 1002t, 1002–1003
Latex allergy, 632, 712t, 724–725
Law
  common, 1150–1151
  evolution of, 1151
  statutory, 1145, 1150
Layoffs, downsizing, and re-engineering, 1026–1027
LD50 (lethal dose 50), 329, 380, 484
LD100 (lethal dose 100), 380, 484
Lead
  biological exposure index (BEI), 393t, 418t
  in drinking water, 546
Lead poisoning, 377, 381, 389t
Learning objectives, 532–533
Leclanche dry cell batteries, material substitution in, 545
Legal issues. *See also individual subtopics*
  employee relations, 1017–1028
  kinds of laws, 1149–1151
  legal case procedure, 1156–1157
    discovery, 1156–1157
    initial pleadings, 1156
    trial and decision, 1157
  legal claims, 1149–1161
  liability insurance, 1163–1194
  OSHA, 1107–1146. *See also* OSHA; Regulatory issues
  overview of legal claims, 1151–1155. *See also* Legal claims
    contract claims, 1153–1154
    statutory law-based, 1154–1155
    tort claims, 1151–1153
  program evaluation, 1081–1105
  recordkeeping and, 1157–1159
  requiring legal advice, 1159–1161
  workers' compensation, 1029–1080
Legal responsibilities, of management, 21–22

Legal terminology, 1142–1146
Legionella pneumophila, 341, 354, 748, 781
Legionellosis, 310t, 745, 746t, 762–763
  incidence, 345t
  protection against, 359
Length of stain tubes, 259–260
LEPs (local emphasis programs), 1124
Leptospira sp., 340
Leptospirosis, 304t, 345t
Leukemia, magnetic fields and risk of, 994
Level, of sound, 924
Level of risk acceptability, 131
Levels of protection, 615, 616–618t
Liability
  limits of, 1174–1175
  strict products, 1153
Liability insurance, 1163–1194
  coverages, 1172–1187
    general requirements, 1172–1177
      claims procedures, 1176–1177
      definitions, 1176
      exclusions, 1175–1176
      insuring agreements, 1173–1174
      limits and deductibles, 1174–1175
    uniqueness of each, 1177–1187
      business auto, 1184–1187
      general liability, 1177–1180
      workers' compensation, 1180–1184. *See also* Workers' compensation
  exposures, 1164–1172
    customers, 1166–1168
      completed operations, 1167–1168
      on-site visitor (or contractor) accidents, 1168
      product liability, 1166–1167
    employees, 1164–1166
      bodily injury by accident, 1165
      occupational disease, 1165
      off-site accidents, 1165–1166
      vehicular accidents, 1165
    third parties, 1168-1172
      class action suits, 1171–1172
      environmental impairment, 1169–1170
      third party involvement, 1170–1171
  mechanics of, 1187–1193
    broker/agent, 1188-1190
    insurance companies, 1190–1191
    underwriters, 1191–1193
Licensing, for radiation source use, 1004–1005, 1007
Lichenification, 719
Life safety principles, 457
Lifting, 950–954, 951t, 962–963. *See also* Ergonomic job analysis
Light-duty provisions, 1053–1054
Light photoxicity, 713–714
Limits of liability, 1174–1175
D-Limonene, 61
Liquid penetrant, 78–79
Liquid samples, 348
Literature search, by computer, 690
Litigant profile, workers' compensation, 1033–1034
LOAEL (lowest observable adverse effect level), 380
Local emphasis programs (LEPs), 1124
Loss, defined, 1078
Loss control, 1078
Loss control manager, 435, 455
Loss control programs, 434–440. *See also* Fire protection
  control measures, 439–440
    hazard elimination, 440
    human elements, 439
  evaluation, 438–439
  identification of hazards, 436–438
  organization, 434–436
  policy statement, 434
Losses
  incurred, 1045
  paid, 1045
Lost time. *See also* Workers' compensation
  defined, 1078
Low-fuel-pressure shutdown valves, 470
Lump sum settlements, 1061
Lyme disease, 305t, 345t, 718
Lymphocyte transformation test (LTT), 390
Lynch Ryan Shift, 1039–1040

Machines, basic strengths of, 933t
Machining, metal, 72–76. *See also* Metal machining
Mad-cow disease (bovine spongiform encephalopathy), 323
Magnetic fields, 990–992. *See also* Radiation: nonionizing
  exposure standards for, 995t
  leukemia risk and, 994
Magnetic particle inspection, 78–79
Magnetic resonance imaging (MRI), 991. *See also* Radiation: nonionizing
Magnets, 992. *See also* Radiation: nonionizing
Maintenance
  Program Evaluation Profile (PEP), 1102t
  of ventilation systems, 586–587
Maintenance manuals, computer evaluation of, 230–231
Malaria, 303t, 345t
Management
  commitment of, 18-19
  employee relations responsibilities, 1017–1018. *See also* Employee relations
  general principles and background of, 1–12
    history, 2–6
    organizational issues, 7–9

Management (*cont'd*)
resources, 2
system elements, 9–11
leadership Program Evaluation Profile (PEP), 1095t–1098t
program oversight responsibilities of, 1047–1049
role in computer simulation, 182
roles and responsibilities of, 13–32
elements of safety and health program, 18–20
financial impact of accidents, 14–18
hidden costs and, 23–25
legal, 21–22
motivational aspects of, 20–21
personnel/staffing aspects of, 25–26
root causes of accidents and, 22–23
Management systems, in accident prevention, 121–124
Manual handling task assessment, 950–954, 951t, 963–964. *See also* Ergonomic job analysis
Massachusetts Toxic Use Reduction Act of 1989, 549–550
Material Safety Data Sheets (MSDSs), 483, 485–486
employer responsibilities, 490
management and tracking of, 503–504
storage and documentation of, 500
trade secrets and, 492
Materials substitution, 546
Material transport, 572–573
Maximum permissible exposure, for radiation, 1009, 1009t
Maximum use limit, 665
McNamara-O'Hara Public Service Act, 1109
Measles, 298t
immunization for, 353t
in pregnancy, 363t
Measurement. *See* Evaluations; Sampling
Mechanical causes of skin disorders, 716
Mechanical equipment hazards checklist, 138
Medical control, of skin disorders, 735
Medical devices, infection regulations for, 335–336
Medical emergencies, heat stroke as, 799
Medical evaluation, of infectious disease exposure, 336–339, 338t. *See also under* Infectious disease(s)
Medical examinations, in respirator programs, 674–676
Medical help, obtaining, 109–110
Medical history, protocols for, 422
Medical management programs, in ergonomics, 955
Medical provider relationships, 1051–1052
Medical records, access to, 513
Medical screening provision primary health effects, 388t–389t
Medical surveillance, 415–428. *See* Health also evaluations
assessment of
effectiveness, 426
need for, 416–421
worksite, 417–419
decision making in, 419–420
defined, 387
of heat stress control, 816–817
implementation of, 421–426
medical evaluation, 421–426
medical protocol development, 421
notification, 421
personnel and facility selection for, 420
Program Evaluation Profile (PEP), 1102t–1103t
Medical waste handling, regulations for, 335–336
Medical Waste Tracking Act, 335–336
Medications
heat tolerance and, 797, 797t
interfering with biological monitoring, 383t–384t
photoxicity and, 714
prescription in workplace, 1025
use and abuse of, 1024–1026
Medicine, published documents on, 47t–48t
MEDLINE searches, 690
Melanin, 701
Melanocytes, 701–703. *See also* Pigmentation abnormalities; Skin; Skin diseases
Melanoma, 722–723
Menninger Foundation disability study, 1055–1056
Mercury, 377
in batteries, 545
biological exposure index (BEI), 393t
environmental, 546
Metabolic aspects, of skin diseases, 704
Metabolic (body) heat, 793. *See also* Heat stress
Metabolism
aerobic versus anaerobic, 795
of toxicants, 382–385
medicines affecting, 383t–384t
Metal cleaning
acid, 56–57
alkaline, 57–58
immersion, 57–58
salt baths, 58
solvent substitution programs in, 551–559, 553t
Metallized reflective fabric, 819
Metal machining, health hazards of, 72–76
conventional, 72–75, 74t
electrical discharge, 75–76
electrochemical, 75

Metals
 as allergens, 712t
 thermal spraying of, 79–80
Metal salts, primary skin irritant, 709t
Methanol, 393t, 418t
Methemoglobin inducers, 393t, 418t
+2-Methoxyethanol (EGME), 395t
N-Methylacetamide, 393t
Methyl chloroform (1,1,1-trichloroethane), 61, 62t, 95, 393t, 418t
Methyl chloromethyl ether, 389t
Methylene chloride, 62t. *See also* Degreasing; Solvents
Methylethyl ketone (MEK), 393t, 418t
Methylisobutyl ketone (MIBK), 394t
Microbials. *See also* Infectious disease(s)
 evaluation checklist for contamination, 771–772, 772t
 indoor airborne, 746t, 746–749, 779–781
Microcomputers. *See* Computers
Microorganisms. *See also and specific microorganisms;* Infectious agents; Infectious disease(s)
 nomenclature of, 324–325
 physical characteristics of, 322t
 sources of, 324
Microphones, for noise exposure measurement, 861–862
Microscopic examination, 339–340
Microthermal effect, 1000
Microwave band radiation, 999–1000. *See also* Radiation: nonionizing
Microwave ovens, 996. *See also* Radio frequency radiation exposure
Microwaves, skin damage related to, 717
Middle ear, 840–841
Miliaria (prickly heat), 716, 798
Misrepresentation, 1152–1153
Modeling. *See* Computer simulation
Molds, indoor airborne, 746–748
Molecular test modalities, 341–342
Monte Carlo sampling, 193
 simulation algorithms for, 216–219
 from standard distributions, 202–214
Morbidity and Mortality Weekly Report (CDC), 343
Mortality, from accidents, 104–105, 105t
MRP-III, magnetic field exposure standards, 995t
MSDs (musculoskeletal disorders). *See* Ergonomic job analysis
Multiemployee worksite defined, 1143
Mumps, immunization for, 353t
Muscle contraction, 936. *See also* Ergonomic job analysis
Musculoskeletal disorders. *See* Ergonomic job analysis
M values (energy expenditure), 810t
MVC (maximum voluntary contraction), 954
Mycobacterium marinum, 305t
Mycobacterium tuberculosis, 300t, 313

Nail discoloration and dystrophy, 725
Napththylamine, 389t
National Ambient Air Quality Standards, 755–756, 756t
National Council on Compensation Insurance (NCCI), 1037, 1042–1043. *See also* Workers' compensation
National Fire Protection Association (NFPA). *See* NFPA
National Institute for Occupational Safety and Health (NIOSH). *See* NIOSH
National Library of Medicine MEDLINE database, 690
Nationally Recognized Testing Laboratories, 275
National Safety Council (NSC). *See* NSC
National Toxicology Program (NTP), 484, 485–486
Natural wet-bulb temperature (NWB), 800, 821
NCCI (National Council on Compensation Insurance), 1037, 1042–1043. *See also* Workers' compensation
Needs analysis, for health and safety instruction, 518–520
Negative pressure respirator, 665
Negligence, 1151–1152
Neoplasms. *See also* Cancer; Carcinogenicity
 skin, 722–723, 723t
NFPA (National Fire Protection Association), 2, 275, 457, 474
NHZ (nominal hazard zone), 1003
Nickel plating, 66. *See also* Electroplating
Nicotine, 757
NIOSH (National Institute for Occupational Safety and Health), 4. *See also* OSHA
 control technology reports, 591–599
 ergonomics program guidelines, 933–934
 grinding ventilation study of, 70
 lifting equation, 951–952
 noise exposure regulations/criteria, 870
 noise reduction rating method, 887–894, 889t, 890t, 891t, 893t
 problem building studies, 759
 respirator standards, 641, 649, 652–653, 656
 sampling protocols, 256–267
 skin disease statistics of, 699–700
 vibration standards, 948
 videotaping study, 563–564
 Work Practices Guide, 948
NIPTS (noise-induced permanent threshold shift in hearing), 845
Nitrobenzene, 394t, 418t
4-Nitrobiphenyl, 389t

Nitrogen dioxide, in gas welding, 96
N-Nitrosodimethylamine, 389t
NMR (nuclear magnetic resonance, MRI) imaging, 991. *See also* Radiation: nonionizing
No-fault system, workers' compensation as, 1030–1031
Noise, 603, 825–930
  ambient, 922
  community, 918–922, 919t, 922t
  control procedures, 898–918, 913t, 916t, 918t. *See also* Noise control
  defined, 924
  ear and hearing and, 838–861. *See also* Ear(s); Hearing loss; Hearing
  hearing conservation programs, 874–878
  hearing protectors, 878–898, 889t, 890t, 891t, 892t, 893t, 897t. *See also under* Personal protective equipment (PPE)
  measurement of, 851–869. *See also* Noise control; Noise measurement
  nonauditory physiologic effects of, 845–846
  occupational exposure criteria
    ANSI, ACGIH, and ISO, 872–874
    OSHA, 869–872, 871t
    steady-state versus impulsive noise, 874
  physical principles of, 827–838. *See also* Sound
    amplitude, 829–838
    wavelength, 827–828
  primary health effects, 389t
  random, 924
  terminology of, 922–925
  white, 925
Noise absorption coefficients, 914, 915t
Noise barriers, 914–918
Noise control, 898–922
  community noise, 918–922, 919t, 922t
  source control, 899–910
    generated-noise reduction, 900–910
      fans, 902
      fluid turbulence, 902
      general principles, 900–902
      by silencers, 902, 907–908
      vibration isolation, 908, 910
      wind, 902
    structure-borne (in rooms), 910–918
    types of sources, 899–900
  source-path-receiver paradigm for, 898–899
Noise-induced permanent threshold shift (NIPTS) in hearing, 845
Noise measurement
  of exposure
    environmental, 862–869
    occupational, 859–862
  by frequency analyzers, 857–858
  instrument calibration for, 859
  by level recording (data logging), 859
  sound exposures and long-term average sound levels, 855–857, 856t
  by sound level meter, 850–855, 853t
  by tape recording, 858
Noise reduction coefficient (NRC), 924
Noise Reduction Rating (NRR), 887–894, 889t, 890t, 891t, 893t
No-light-duty provisions, 1053–1054
Nominal hazard zone (NHZ), 1003
Nondestructive testing
  health hazards of, 76–79
  by industrial radiography, 76–78
  by liquid penetrant, 78–79
  by magnetic particle inspection, 78
Nonionizing radiation, 988-1004. *See also under* Radiation
Notice of Proposed Rulemaking (NPRM), 1122
Notifiable diseases, 343–346, 344t, 345t
NPA respiratory guidelines, 641
NPRM (Notice of Proposed Rulemaking), 1122
NRC (noise reduction coefficient), 924
NRR (Noise Reduction Rating), 887–894, 889t, 890t, 891t, 893t
NSC (National Safety Council), 2
  incidence and frequency rates, 1035
  pesticide exposure recommendations, 754
  risk assessment guidelines, 132–133
NTP (National Toxicology Program), 484

Obesity, heat tolerance and, 795–796
Obstructions, musculoskeletal injury and, 937
Occupational dermatoses, 697–741. *See also* Skin diseases
Occupational health and safety. *See also* OSHA; OSHA Act
  management principles and background, 1–12
Occupational Health and Safety Act. *See* OSHA Act
Occupational Health and Safety Administration. *See* OSHA
Occupational history, protocols for, 422
Occupational illness defined, 1078
Occupational noise exposure. *See* Noise
Occupational physician, in biological monitoring, 387–388
Occupational Safety and Health Review Commission, 1143
Occupational travel, infectious diseases related to, 318–320
OD (optical density), 1003
Oils, for metal cutting, 74t
Operations, completed, 1167–1168
Operator safety, 431
Opportunistic pathogens, 322
Optical density (OD), 1003
Orf virus disease, 301t

Organizational culture, 2, 7–8
Organizational issues, in management, 7–9
Organophosphates
  biological exposure index (BEI), 394t, 418t
  biological monitoring case study, 405–406
OSHA Act, 4, 1117–1146
  employee rights and responsibilities under, 1018
  General Duty Clause of, 333
OSHA (Occupational Safety and Health), hazard communication regulations, occupational exposure to hazardous chemicals in laboratories (29 CFR 1910.1450), 514
OSHA (Occupational Safety and Health Administration), 1117–1146. *See also* Regulatory issues
  aerosol inhalation standards, 264–265
  audiometer/audiometry standards, 846–848
  bloodborne pathogens standard, 110, 313, 326, 334t
  compliance strategies, 1133
  construction industry training requirements, 527–529
  contact lens policy, 627
  defined, 1143–1144
  ergonomics job analysis forms, 961–971
    computer model for lifting exposures, 968–970
    hand tool checklist, 967–968
    manual handling checklist, 962–963
    task/work methods checklist, 963–964
    video display unit and keyboard issues checklist, 964–966
    worksite assessment checklist, 961–962
    workstation layout checklist, 966–967
  general industry training requirements, 522–526
  hazard communication regulations, 513–515
    access to employee exposure and medical records (20 CFR 1910,20), 513
    air contaminants (29 CFR 1916,1000), 514
    hazardous waste operations and emergency response (1910,120), 514–515
    process safety management of highly hazardous chemicals (1910,119), 514
  hazardous materials compliance review, 508–513
  hazardous materials record review, 508
  history of, 4–5
  infectious disease regulations, 333–335, 334t
  inspection management, 1125–1129
    closing conference, 1128
    employee complaints, 1126
    employer preparedness, 1128-1129
    initiation of inspection, 1126
    opening conference, 1127
    postinspection issues, 1129–1133
      employee rights and courses of action, 1132–1133
      employer options, 1130–1131
      formal contest, 1131–1132
      informal conference, 1131
      petition for modification of abatement (PMA), 1132
      posting requirements, 1130
      types of violations, 1129–1130
    records review, 1127
    types of inspections, 1125–1126
    walkaround, 1128
  inspection targeting, 1123–1124, 1124t, 1125t
  Medical Record Access Standard, 425
  medical screening provision primary health effects, 388t–389t
  noise exposure regulations/criteria, 869–872
  personal protective equipment (PPE) regulations, 608–610, 611t
  personal protective equipment standards, 608–610
  Process Safety Management Standard, 5
  Program Evaluation Profile (PEP)
    accident investigation, 1100t
    contractor safety, 1098t
    data analysis, 1101t
    emergency preparedness, 1103t–1104t
    employee participation, 1096t
    hazard control, 1101t–1102t
    hazard reporting, 1100t
    implementation, 1097t
    inspection, 1099t
    maintenance, 1102t
    management leadership, 1095t–1098t
    medical program, 1102t–1103t
    safety and health training, 1104t–1105t
    survey and hazard analysis, 1098t–1099t
  recordkeeping guidelines, 1048
  respirator regulations, 641
  respirator standards, 646, 649, 652
    for training, 660
  review commission, 1133–1139
    case review, 1135–1139
    defenses before, 1135
    preemption, 1134–1135
  Safety and Health Program Guidelines, 18–20
  Safety and Health Program Standard, 9–10
  standards development, 1117–1119, 1118t
  standards input strategies, 1120–1123
    Advance Notice of Proposed Rulemaking (ANPR), 1119, 1121–1122
    legal challenge, 1123
    national consensus standards, 1120

OSHA (it cont'd)
Notice of Proposed Rulemaking (NPRM), 1122
prepublication drafts, 1121
stakeholder and town meetings, 1121
trade and professional associations, 1120–1121
standards issues as of 1996, 417t
summary of citations and fines, 31–32
Voluntary Protection Program, 6, 10, 1139–1140
website of, 693
OSHA recordkeeping, 1144
Otitis (ear infection), 315t, 841
Outsourcing, 6
Outstanding reserves, 1045
Ovens
microwave, 996
RF induction, 996
Oxygen consumption, energy demand and, 938
Oxygen deficiency, respirators and, 664

Paid losses, 1045
Painting, health hazards of, 80–85
Paints
composition of, 81–82
defined, 81
epoxy, 82
water-borne, 82
Paint spray booth design, 83–84
Paper filters, 266t
PAPRs (powered-air purifying respirators), 647. *See also* Respirators
Papule defined, 736
Parasitic infections. *See also and specific organisms;* Helminth infections; Protozoal infections
skin, 718
Parathion, 394t, 418t. *See also* Organophosphates
Particulate filter respirators, combination cartridges and canisters for, 649
Particulates
indoor air-borne, 757–758
TLVs for, 264–265
Parvovirus B19 ("Fifth disease"), 315t, 363t
Pascals (Pa), 830
Passive immunity, 352
Patch testing, 728–730, 729t
PCBs, skin disease related to, 719–720
PCR (polymerase chain reaction), 341–342
PCs. *See* Computers
Pediculosis (lice), 315t
Pellister, solid-state, 263
Penetrant, liquid, 78–79
Penetration, of protective clothing, 613
Penetration injury, 603, 634
Pentachlorophenol (PCP), 394t, 418t
PEP (postexposure prophylaxis), 359–361. *See also* Prevention: of infectious diseases
PEP (Program evaluation profile). *See* Program Evaluation Profile (PEP)
Perchloroethylene, 62t, 395t. *See also* Degreasing; Solvents
Percutaneous absorption, 703–704. *See also* Skin diseases
Performance-based standards, 4–5
Performance objectives, in injury response, 1052–1053
Performance-type noise codes, 920
Per incident liability, 1175
Permanent disability, 1078
Permeability, of gloves, 631
Permeation, of protective clothing, 612–613
Permissible exposure limit (PEL), 485
Per policy aggregate liability, 1175
Per site aggregate liability, 1174–1175
Personal hygiene, skin diseases and, 707
Personal protective equipment (PPE), 601–638. *See also* Protective equipment
belts
back, 636
seat, 635
chemical protective clothing, 612–622
decontamination of, 615, 622
ensembles, 615, 616t–618t
selection considerations, 612
chemical resistance rating, 612–615
degradation, 613
permeation, 612–613
durability, 615
duration of use, 615
ease of decontamination, 615
flexibility, 615
heat transfer, 613–615
vision, 615
electrical protection, 636
eye and face protection, 624–629
ANSI standards, 625–626
contact lenses, 626–628
laser eyewear, 628–629
selection guidelines, 626t–628t
sideshields, 625
when required, 625
fall protection, 635–636
foot protection, 632–635
ANSI standards and selection, 632–634
foot hazard posting, 634
general considerations, 602–612
employer versus employee payment, 610–612
hazard assessment, 602
accident incident investigation, 602–603
job safety analysis, 603

process hazard review, 603
targeted PPE hazard assessment, 603–604
hazards of PPE itself, 605
inspection, maintenance, and storage, 608
limitations of PPE, 605
program goals, 602
selection and acquisition, 606–607
standards and regulations, 608–610, 611t
user training, 607–608
hand protection, 629–632
disadvantages, 631
glove fit and musculoskeletal injury, 937
latex allergy and, 632
leak checking of, 632
selection criteria, 629–631
sizing, 632
head protection, 622–624
accessories, 623
ANSI standards, 623
bump caps, 623–624
hard hats, 622–623
when required, 623
hearing protectors, 878–898
communication with, 894–898
performance limitations of, 878–879
selection and ratings of, 887–894, 889t, 890t, 891t, 893t
types of, 879–887
concha-seated, 882
insert, 879–882, 885–886
moderate or flat attenuation, 885
muff-type, 882–885.886
levels of protection, 615, 616–618t
limitations with lasers, 1002–1003
for radiation source handling, 1008
as skin disease preventive, 732–734
for thermal protection, 819
Personal sampling, 245
Personal service work, infectious diseases related to, 320
Personnel, 25–26
for ergonomic job analysis, 940
training of, 517–537. *See also* Training
Pertussis (whooping cough), 300t, 353t
Pesticides
cholinesterase-inhibiting (organophosphates)
biological exposure index (BEI), 405–406
biological monitoring case study, 405–406
Federal Insecticide, Fungicide, and Rodenticide Act (P.L. 92–516), 1114–1115
indoor, 753–754, 783
Peterson, Dan, 8, 9
Peterson(,) Dan, 5
Peters, Tom, 8
Petition for Modification of Abatement (PMA), 1132
Petroleum, as skin irritant, 710t
P4SR (predicted 4-hour sweat rate), 800, 801–803
Pharmacodynamic medications, 384t
Pharyngitis, 315t
Phenol, 394t, 418t
Photoionization detectors (PIDs), 263t
Photoresist products, 555, 559
Phototesting, for allergy, 714
Physical causes of skin diseases, 716–718
biological (infection), 717–718
Physical examination, protocols for, 422
Physical hazards
in abrasive blasting, 55–56
in heat treating, 72
Physician(s)
occupational, in biological monitoring, 387–388
provider relationships and workers' compensation, 1051–1052
selection for medical surveillance, 419–420
Pickling, of metals, 56–57
PIDs (photoionization detectors), 263t
Pigmentary abnormalities, 721–722, 722t
Piping, safety design of, 473
Pitch (of sound), 924
Plague *(Yersinia pestis),* 305t, 345t, 353t
Plan, Do, Check, Act (Deming), 9, 10, 104
Plant accident reports. *See* Accident reports
Plants (industrial). *See* Facilities
Plants (vegetation)
infectious diseases transmitted by, 294t–295t, 308t–310t
photosensitizing, 714, 715t
skin irritant, 711–713, 712t
Plasma metal spraying procedures, 79–80
PMA (Petition for Modification of Abatement), 1132
PMN (premanufacturing notification), 585–586
Pneumonia, 315t. *See also* Legionellosis; Respiratory infections
Pneumonitis, hypersensitivity, 746t, 762
Poison ivy, 711
Policy statement, 26
Poliomyelitis, immunization for, 353t
Polishing, 68–70, 69t
Pollution, prevention versus control of, 546–548
Pollution Prevention Act of 1990, 546
Polycarbonate filters, 266t
Polychlorinated biphenyls, skin disease related to, 719–720
Polymerase chain reaction (PCR), 341–342
Polymers, toxic impurities in, 561–562
Polyvinyl chloride filters, 266t
P-100 (HEPA) filters, 647–648

Portal of entry, 326
Positive-pressure respirator, 665
Postexposure examination, for infectious disease, 337–338, 338t
Postexposure prophylaxis, 359–361
Postinjury response systems, 1045–1062. *See also under* Workers' compensation
Powder coating, 555, 556t–557t
Power Engineering Co. v. Envirochem Service, 1153–1154
PPE (personal protective equipment). *See* Protective equipment
PPN (preferred provider network), 1078
Practical points table, for biomonitoring, 396t–397t
Predicted 4-hour sweat rate (P4SR), 800, 801–803
Predictive value, 339
  of tests, 341
Preferred provider network (PPN), 1078
Pregnancy
  hyperthermia and birth defects, 795
  infection risk and, 362–364, 363t
  radiation exposure in, 1009t
Preplacement examination, for infectious disease, 336–337
Presbycusis, 826. *See also* Hearing loss
Prescription medications, 1026
Presentation
  in health and safety instruction, 534–536
  materials for, 535
  styles of, 534–535
Pressure-demand respirator, 665
Pressure relieving systems, 471
Prevention
  of accidents, 119–121
  of infectious disease, 348–363, 349t, 351t–352t
    education, training, and administrative controls, 357
    immunizations, 352, 353t
    personal protective equipment, 358t, 358–359
    postexposure prophylaxis, 359–361
    risk factors and, 361–363
      immunodeficiency, 361–362
      pregnancy, 362–363
    substitution and engineering controls, 353–357
  of skin disorders, 731
Prickly heat (miliaria), 716, 798
Primary containment, 356
Primary health effects, of medical screening toxicants, 388t–389t
Primary prevention, defined, 349t
Prions, 323
Privileged and confidential documents, 1157–1159
Probability calculations, 1066–1068, 1067t
Problem building studies
  of perceived air quality (odors), 760–761
  of thermal environment, 761–762
Process equipment, designing safety into, 469–474
Process hazard review, 542–543, 603
Process Safety Management Standard, 5
Productivity, respirator use and, 658
Product liability, 1166–1167
Professional associations, OSHA input from, 1120–1121
Program evaluation, 1081–1105
  importance of, 1087–1088
  management awareness of, 1086–1087
  management motivation for, 1081–1083
  management roles and responsibilities in, 1085–1086
  methods for, 1083–1084
  OSHA Program Evaluation Profile (PEP), 1089–1105
    accident investigation, 1100t
    contractor safety, 1098t
    data analysis, 1101t
    emergency preparedness, 1103t–1104t
    employee participation, 1096t
    example of evaluation, 1092–1093
    final report compilation, 1093–1094
    hazard control, 1101t–1102t
    hazard reporting, 1100t
    implementation, 1097t
    inspection, 1099t
    maintenance, 1102t
    management involvement in, 1094
    management leadership, 1095t–1098t
    medical program, 1102t–1103t
    PEP evaluation format, 1092
    pre-evaluation preparation/scoring guidelines, 1089–1091
    recordkeeping and documentation, 1091–1092
    safety and health training, 1104t–1105t
    survey and hazard analysis, 1098t–1099t
    value of, 1094–1095
  overview of, 1084–1085
Program Evaluation Profile (PEP). *See under* Program evaluation
Program protection factor, 665
Prophylaxis. *See* Prevention
Propiolactone, 389t
Protected classes of employees, 1023t
Protection factors, for respirators, 657t
Protective equipment
  in electroplating, 67
  personal (PPE), 601–638. *See also* Personal Protective Euipment (PPE) *and specific types*
  respiratory, 639–676. *See also* Respirators
    problems related to, 663–664
    programs, 654–662, 667–671

recordkeeping, 662–663
regulations and guidance documents, 641
sample program, 667–671
terminology, 664–666
types, 641–652
for welding, 97–98
Protocol development, for medical surveillance, 421–425
Protozoa defined, 323–324
Protozoal infections
general chacteristics, 308t
general characteristics and transmission, 299t, 302t–303t
Pseudomonas, 748
Psittacosis, 306t, 345t
Psychometric wet-bulb temperature, 800
Psychophysics, 953–954. *See also* Ergonomic job analysis
Psygrometers, 811
Public relations, in emergency planning, 460
Pulmonary function tests, 423
Pulmonary hemosiderosis, 746t
Pumps
centrifugal, 565–567
fire, 441
safety design of, 472
Pyoderma, 300t

Q fever, 306t, 324, 718
Qualitative evaluation, indoor air quality (IAQ), 770–777, 772t
Qualitative fit test, 665
Quality control, in hazard communication, 500
Quantitative evaluation, indoor air quality (IAQ), 777–778
Quantitative fit test (QNFT), 656, 665, 669
Quenching, 72
Quick returners, 1056
Quick Selection Guide to Protective Clothing, 630

Rabies, 302t, 353t
Radiant heat measurement, 813
Radiation. *See also specific sources*
in arc welding, 92
in gas welding, 96
ionizing, 1004–1015
administration procedures for, 1006–1007
ALARA (as low as reasonably advisable) concept of, 1010
biological effect and risks of, 1010
detection and measurement of, 1008-1009
emergency procedures, 1012–1014
engineering and environmental controls, 1011
exempt source material, 1005
laboratory surveillance and management audits, 1012
licensing for, 1004–1005
machine sources of, 1005
maximum permissible exposure limits, 1009, 1009t
radiation and contamination surveys, 1010–1011
radiation safety liaison, 1006
Radiation Safety Officer, 1005–1006
safe handling and dose reduction techniques, 1008
skin damage related to, 717
transportation of sources, 1012
waste disposal and source deposition, 1012
worker exposure monitoring, 1010
worker training, 1007–1008
nonionizing, 988-1004
extremely low frequency and very low frequency, 993–995
infrared, visible light, and ultraviolet, 999–1001, 1000t
lasers, 1002–1003
physics of, 988–991, 990t, 991t
radio frequency, 995–1000, 999t
static fields, 991–993
optical (light), 603, 627
thermal, 793
from welding generally, 94–95
Radiation dermatitis, 717
Radiation Safety Officer (RSO), 1005–1006
Radiation shielding, 814
Radio frequency effects, on chemical test instruments, 275
Radio frequency radiation exposure, 995–1000, 998t, 999t
Radiography, industrial, 76
Radioisotopes, 77–78
Radiologic evaluation, chest films, 423
Ramazzini, Beardini, 3
Random noise, 924
Random number generation, 193–194
Rashes. *See* Skin disorders
Raynaud's disease, 736
Recognized hazards, 1144
Recordkeeping, 425
for audiometry, 849
of chemical evaluations, 276–277
defined, 1144
in hazard communication, 500–503
for hearing conservation programs, 875–876
as management tool, 1048
OSHA guidelines for accident, 1048
Record retention, 1159
Records, access to medical, 513

Recycling. *See also* Environmental regulations; Toxic use reduction programs
  of automobile parts, 546
  Resource Conservation and Recovery Act (P.L. 94–560), 1115
  state programs in, 549–550
Re-engineering, 1026–1027
Regulations, infectious disease-related, 332–336. *See also under* Infectious disease(s)
Regulatory issues, 1107–1146
  federal legislation, 1112–1117. *See also* OSHA
    Clean Air Act (P.L. 101–549), 1113
    Clean Water Act (P.L. 92–500), 1113–1114
    Comprehensive Environmental Response, Compensation, and Liability Act (CERCLA)(P.L. 96-510), 482, 1115–1116, 1154–1155
    Consumer Product Safety Act (P.L. 92-573), 1116–1117
    Federal Insecticide, Fungicide, and Rodenticide Act (P.L. 92-516), 1114–1115
    Hazardous Materials Transportation Act (P.L. 93-633), 1116
    OSHA Act, 1117–1146
    Resource Conservation and Recovery Act (P.L. 94-580), 1115
    Safe Drinking Water Act (P.L. 93-523), 1114
    Toxic Substance Control Act (P.L. 94-469), 1113
  federal legislative process, 1110t, 1110–1112, 1112t
  ISO guidelines, 1140–1141
  legal terminology, 1142–1146
  OSHA, 1117–1146. *See also* OSHA
    compliance strategies, 1133
    inspection management, 1125–1129
      closing conference, 1128
      employee complaints, 1126
      employer preparedness, 1128-1129
      initiation of inspection, 1126
      opening conference, 1127
      postinspection issues, 1129–1133
        employee rights and courses of action, 1132–1133
        employer options, 1130–1131
        formal contest, 1131–1132
        informal conference, 1131
        Petition for modification of abatement (PMA), 1132
        posting requirements, 1130
        types of violations, 1129–1130
      records review, 1127
      types of inspections, 1125–1126
      walkaround, 1128
    inspection targeting, 1123–1124, 1124t, 1125t
    review commission, 1133–1139
      case review, 1135–1139
      defenses before, 1135
      preemption, 1134–1135
    standards development, 1117–1119, 1118t
    standards input strategies, 1120–1123
      Advance Notice of Proposed Rulemaking (ANPR), 1119, 1121–1122
      legal challenge, 1123
      national consensus standards, 1120
      Notice of Proposed Rulemaking (NPRM), 1122
      prepublication drafts, 1121
      stakeholder and town meetings, 1121
      trade and professional associations, 1120–1121
    voluntary protection programs, 1139–1140
  safety regulation history, 1109–1110
Reinsurance, 1191
Remote damper controls, 470
Replacement programs, for airborne toxicants, 545–559, 548t, 551t, 553t, 554t, 556t–558t
Replications, computation of, 201–202
Reportable diseases, 343–346, 344t, 345t
Reporting, 424–425
  of accidents, 105–107
Reserves, outstanding, 1045
Reservoir defined, 324
Resignation, voluntary, 1023–1024
Resistance to infection, 329
Resonance, 924
Resource assessment, in emergency planning, 459
Resource Conservation and Recovery Act (P.L. 94-580), 546, 1115
Resources. *See also* Information sources
  CD-ROMs, 695
  computers in literature search, 690
Respirator programs, 654–662. *See also* Airborne-contaminants; Respirators
  administration, 654
  equipment selection and cost, 655–659, 657t
  hazard assessment, 655
  maintenance, 660–662
    cleaning and disinfection, 661
    inspection, 660–661
    storage, 662
  recordkeeping, 662–663
  sample, 667–676
  training, 659–660
  written standard operating program (SOP), 654–655

Respirators. *See also* Respirator programs
airline (breathing air systems), 650–651
air-purifying, 645–652
particulate filter, 647–648
regulations and standards, 645–647, 647t
biological monitoring provisions for, 389t
comfort level of, 658–659
cost of, 657–658
effectiveness of, 655–657, 657t
escape-only, 652–653
facepieces, 642–645
loose-fitting, 645
tight-fitting, 642–645
history of, 640–641
maintenance of, 660–662
particulate filter
combination cartridges and canisters for, 649
gas and vapor removing, 648–649
protective, 358–359
regulations regarding, 641
self-contained breathing apparatus (SCBAs), 651–652, 653t
special problems of, 663–664
in cold environments, 664
facial hair, 663–664
oxygen deficiency, 664
vision, 663
voice communications, 664
terminology of, 664–666
Respiratory absorption, 382
Respiratory infections. *See also specific diseases*
childcare-related risks of, 314, 315t
general characteristics and transmission, 296t
hantavirus pulmonary syndrome, 308t
legionellosis, 310t, 345t, 359, 745, 746t, 762–763
Respiratory protection, 358t. *See also* Protective equipment
Responsibility, in workers' injuries, 1046–1047
Results, review of, 424
Retrospective rating workers' compensation, 1044
Return to work, 1053–1057
ADA (Americans with Disabilities Act) and, 1072–1074, 1074t
Return-to-work examinations, for infectious disease, 338
Reverberation, 924
Review commission, 1133–1139. *See also under* OSHA
Review of results, 424
RF induction ovens, 996
RF (radiofrequency) exposure, 995–1000, 998t, 999t
Rickettsial infections
Lyme disease, 305t, 345t, 718
Rocky Mountain spotted fever, 306t, 345t
Right of entry, 1144
Rights, walkaround, 1145
Right to discharge, 1022–1023, 1023t
Right-to-know, worker, 479–515, 1144. *See also* Hazard communication
Ringworm *(Tinea capitis),* 299t
Risk
definition of, 432
quantification of. *See* Risk assessment
Risk assessment, 127–178. *See also individual topics*
benefits of, 128-129
comparative techniques for, 134t–135t
in emergency planning, 459
failure mode and effect analysis (FMEA), 160–166, 432–433
fault tree analysis (FTA), 166–177, 432–433
hazard and operatbility (HAZOP) analysis, 152–160
job safety analysis, 137–146
by modeling (FMEA and FTA), 160–166, 432–433
overview of, 129–133
by past experience, 432
safety checklists in, 133–137
what-if analysis, 147–152
Risk assessment report, 1170
Risk factors, for infectious disease
immunodeficiency, 361–362
pregnancy, 362–364, 363t
Risk management. *See* Control practices; Fire protection; Loss prevention
Risk matrix, 130–131
Risk score formula, 1066–1068, 1067t
Risk score method, 1066
Robotics, 571–572
Rocky Mountain spotted fever, 306t, 345t, 718
Roosevelt, Theodore, 3–4
Root causes of accidents, *14,* 22–23
Root mean square (RMS) evaluation, 924, 990–991
Routes of entry, 380
RSO (Radiation Safety Officer), 1005–1006
Rubber manufacturing
dust control in, 560–561, 561t
solvent use in, 562
Rubber products, as allergens, 712t
Rubella (German measles), 298t, 353t, 363t, 398t
Rubeola (measles), 298t, 353t, 363t, 398t

Safe Drinking Water Act (P.L. 93-523), 546, 1114

Safety. *See also* Control practices; Fire protection; Loss prevention; Safety and Health programs; Safety evaluations
  basic process of, 431
  as defined for workers' compensation, 1078
  equipment, 431
  operator, 431
  published documents on, 43t–47t
Safety and health management. *See* Management
Safety and health policy statement, 26
Safety and health programs
  achievement of effective, 20–21
  evaluation of, 25, 1081–1105. *See also* Program evaluation
  OSHA guidelines for, 18–20. *See also* OSHA; Regulatory issues
  roadblocks to effectiveness of, 27–28
Safety and Health Program Standard, 9–10
Safety and health resources, 2
Safety and health training, 20
Safety audits, 431
Safety checklists, 133–141
Safety evaluations, 102–237
  accident investigation, 103–126
    benefits of, 104–105
    management systems in, 121–124
    need for, 107–108
    overview of, 109–113
    participants in, 108-109
    reporting requirements, 105–107
    step 1: fact determination, 113–115
    step 2: immediate causes (acts and conditions), 115–117
    step 3: underlying factors, 117–119
    step 4: organizational prevention, 119–121
    summary of, 124–125
  computer simulation in, 179–237. *See also* Computer simulation
    applications of, 224–235
    fundamentals of, 193–224
    principles of, 180–181
    as problem-solving methodology, 181–185
    simulation modeling and analysis, 185–193
  risk assessment, 127–178. *See also* Risk assessment *and individual topics*
    benefits of, 128-129
    failure mode and effect analysis (FMEA), 160–166, 432–433
    fault tree analysis (FTA), 166–177, 432–433
    hazard and operatbility (HAZOP) analysis, 152–160
    job safety analysis, 137–152
    overview of, 129–133
    safety checklists in, 133–137
Safety Fundamentals and Comprehensive Practice Examination, 7
Safety standards, 4
Safety statistics, 677–684
Safety valves, 440, 470–471
Salmonellosis, 307t
Salt balance, 796–797
Salt baths, for metal cleaning, 58
Sample(s). *See also* Sampling
  chain of custody for, 273–274
  grab (short-term), 256, 260–261
  integrated, 256–267
  liquid and bulk, 348
  shipping of, 275–276
  surface, 348
Sample values, 196–198
Sampling. *See also* Biological monitoring; Sample(s)
  air
    for chemicals, 562–564, 563t
    for microbials, 748–749
  of chemical contaminants, 257–262
    aerosols, 264–271
      by filter sampling, 265–268, 266t
      by impactors, 268
      by impingers, 269
    gases
      by absorbent tubes, 257–258
      by direct reading instruments, 262, 263t–264
      by evacuated containers and bags, 260–261, 261t
      by impingers and liquid traps, 259
      by length of stain tubes, 259–260
      by passive badges, 258–259
  for chemical contamination, 240–257
    duration of, 247–249
    environmental variability and, 243–246
    frequency of, 249–251
    general principles, 240–242
    location for, 246–247
    method selection, 251, 252t–255t
    purpose of, 243
  grab (short-term), 256
  for infectious disease assessment, 348–349
  integrated, 256–257
  personal, 245
  for volatile organic compounds (VOCs), 751–752
Sampling period, 247–248
Sanitization, 356
SARA (Superfund Amendments and Reauthorization Act), 481, 505, 1116. *See also* CERCLA; Hazard communication
SBS (sick building syndrome), 744–745. *See also* Indoor air quality (IAQ)

Scabies, 315t
SCBAs (self-contained breathing apparatus), 651–652, 653t, 666
Scrotum, percutaneous absorption in, 704
Seat belts, 635
Seated body dimensions, 949t
Secondary containment, 356
Secondary prevention defined, 349t
Security issues, in emergency planning, 459–460
Segregation of operations, 576–577
Self-contained breathing apparatus (SCBAs), 651–652, 653t, 666
Self-evaluative privilege, 1158
Self-insurance, workers' compensation, 1044
Semiconductor industry, ventilation systems used in, 580t–581t
Semiconductors
  employee reaction to substitutions, 555
  as sensors, 263
Semiconductor Safety Association, 579
Sensitivity, of tests, 339, 340–341
Sensitizer chemical defined, 484
Sensors, direct-reading, 262, 283t
Seroconversion, 340
Service and maintenance layout, 575–576
Services installation, 439
Settlement (legal) defined, 1144
Severity rate defined, 939
Sewage treatment, infectious diseases related to, 320
Shielded metal arc welding, 90–92
Shields, heat-reflective, 814
Shigellosis (bacterial dysentery), 300t
Shipping, of chemical samples, 275–276
Shoes. *See* Foot protection
Short-term (grab) samples, 256, 260–261
SIC code, 939
Sick building syndrome, 744–745. *See also* Indoor air quality (IAQ)
Sideshields, 625
Significant threshold shifts (STSs), 849–850, 876–877, 924, 925
Silver filters, 266t
Skin. *See also* Skin diseases
  anatomy of, 700–703
Skin appendages, 736
Skin disorders, 697–741. *See also specific diseases*
  anatomic aspects of, 700–703
  causal factors in, 704–718
  childcare-related infections, 315t
  diagnosis of, 726–730
    approach to, 726–727
    diagnostic tests (patch testing), 728–730
    patient history in, 726–727
    by sites affected, 727–728
    visual inspection in, 727
  direct causes of, 707–718
    allergy (hypersensitization), 708–714
      photoallergy, 714, 715t
      phototesting for, 714
      phototoxicity, 713–714
      to plants and woods, 711–713, 712t
    chemical burns, 708, 710t
    chemical irritants, 707–708, 708t, 709t–710t
    mechanical, 716
    physical, 716–718
      biological (infection), 717–718
      electricity, 717
      heat or cold, 716–717
      ionizing radiation, 717
      microwaves, 717
  history of occupational, 697–699
  incidence of, 699–700
  indirect causes of, 705–707
    age, 706
    environmental and climatic factors, 706–707
    genetic predisposition, 705–708
    personal hygiene, 707
  metabolic aspects of, 704
  percutaneous absorption and, 703–704
  physical findings in occupational, 718–726
    acroosteolysis, 725–726
    contact urticaria (hives), 724–725
    eczematous contact dermatitis, 718–719
      acute, 718–719, 719t
      chronic, 719, 720t
    facial flush, 725
    folliculitis, acne, and chloracne, 720t, 720–721, 721t
    granulomas, 723–724
    nail discoloration and dystrophy, 725
    neoplasms, 722–723, 723t
    pigmentary abnormalities, 721–722, 722t
    sweat-induced reactions, 721
    ulcerations, 723
  prevention of, 731–735
    by control measures, 734–735
    by direct measures, 731
    by indirect measures, 732–734
  signs of systemic intoxication in, 726, 726t
  streptococcal, 300t
  terminology of, 735–737
  treatment of, 730–731
  UV light associated, 1001
Skin tests, intradermal, 342
Skin type, 705
Slurry methods, of dust control, 560–561
Smear sampling. *See* Surface sampling
Smoke
  combustion products in indoor air, 754–755, 754–756
  tobacco, 756–757, 783
Smoke detectors, 448

Snuffing by steam, 470
Sodium hydroxide, in metal cleaning, 58
Soil, infectious diseases transmitted by, 294t–295t, 308t–310t
Soldering and brazing, health hazards of, 85–89, 86t, 88t
Solders, 85–86
  materials substitution in, 546
Solid-state pellister, 263
Solvents. *See also* Metal cleaning; Volatile organic compounds (VOCs)
  characteristics of generic categories, 553
  chlorinated hydrocarbon, decomposition of, 98
  in paints, 81–82
  primary skin irritant, 710t
  in rubber manufacturing, 562
  substitution programs, 551–559, 553t
  toxic impurities in, 562
  for vapor-phase degreasing, 61, 62t, 567
Sound. *See also* Hearing; Noise
  physical principles of, 827–838
Sound-absorbing materials, 910–918
Sound intensity/sound intensity level, 830–831, 925
Sound level meters, 851–855
  exponential time-averaging, 854–855, 856t
  frequency-weighting networks, 852–853, 853t
  microphones and directional characteristics, 851–852
Sound power/sound power level, 831–832, 925
Sound pressure/sound pressure level, 830, 925
  effective, 923
  peak level of, 924
  root mean square, 924, 990–991
Source defined, 324
Species and genus, 324–325
Specificity, of tests, 340–341
Sporotrichosis, 310t
Sports medicine, 1078
Spreadsheet software, 215–216, 688–689
Sprinkler systems, for firefighting, 442–444
Stachybotrys chartarum, 747
Staff, training of, 517–537. *See also* Training
Staffing, 25–26. *See also* Personnel
  for medical surveillance, 419–420
Stakeholder meetings, 1121
Standard 29 CFR, 333, 334t–335t, 335. *See also* Bloodborne pathogens standard; OSHA
Standard operating program (SOP), for respirators, 654–655
Standing body dimensions, 948t
Standing surfaces, 937–938
State laws and workers' compensation, 1031–1032
State Plan, defined, 1144–1145
Static-dissipative footwear, 634
Static fields, 990–992. *See also* Radiation: nonionizing
Statistical analysis, computers in, 677–684
Statistical methods
  for computer simulations, 193–224. *See also* Computer simulation: fundamentals of
  environmental variability and, 243–246
Statutes of repose, 1145
Statutory law, 1145, 1150
  legal claims based on, 1154–1155
Stay (of legal action), 1145
Steam, snuffing by, 470
Sterilization, 356
Storm water systems, 473–474
Streptococcal skin diseases, 300t
Stress
  contact, 935
  employee relations and, 1027–1028
Stressors, diseases of adaption and, 846
Strict products liability, 1153
Stroke, heat, 799
STSs (significant threshold shifts), 849–850, 876–877, 924, 925
Styrene, 394t, 418t
Subclinical infection, 329
Substitution
  for airborne toxicants, 545–559, 548t, 551t, 553t, 554t, 556t–558t
  as skin disease preventive, 732
Sulfur dioxide, 754–756
Superfund Amendments and Reauthorization Act (SARA), 481, 505, 1116. *See also* CERCLA; Hazard communication
Superfund (Comprehensive Environmental Response, Compensation, and Liability Act, CERCLA) (P.L. 96-510), 482, 1115–1116, 1154–1155
Supplied-air hoods, 645–647, 650–652. *See also* Respirators
Surfaces, standing, 937–938
Surface samples, 348
Surface sampling, chemical, 277–278
Surveillance
  defined, 416
  infectious disease, 342–346, 344t, 345t. *See also* Infectious disease(s); Notifiable diseases
  medical, 387, 415–428. *See also* Medical surveillance
Susceptibility, biological markers of, 387
Sweat, characteristics of, 706
Sweat-induced skin reactions, 721
Sweden, videotaping of workstations in, 564
Swipe sampling. *See* Surface sampling

Symptoms survey, 941–942
Syncope, heat, 798

Taft-Hartley (Labor-Management Relations) Act, 1109
Tape recording, in noise measurement, 858
Targeted industries (OSHA), 1124t, 1125t
Target organ effects, 484
Task analysis, 530–531
Task assessment, manual handling, 950–954, 951t. *See also* Ergonomic job analysis
TCA (1,1-trichloroethane), 554.554t
Technological advances, in sampling equipment, 240
Teflon filters, 266t
Telephones, cellular, 996, 998
Temperature
  dry-bulb air, 800
  equivalent effective corrected for radiation (ETCR), 801, 821
  measurement instruments for, 811–813
    amenometers (air velocity), 812–813
    black-globe thermometers, 813
    hygrometers (humidity), 811
    psygrometers (humidity), 811
    thermometers, 811
  respirators and, 664
  skin sensitivity and, 706–707
  wet-bulb, 800, 821
  wet-bulb globe (WBGT) index, 808–809
  wet-globe, 810–811, 822
Termination, 1021–1023
  employer's right to discharge, 1022–1023, 1023t
  guidelines for, 1021–1022
  informing co-workers, 1022
  informing employee, 1022
Terminology
  of noise and sound, 922–925
  of regulatory law, 1142–1146
  of respirator use, 664–666
  of thermal protection, 821–822
  of workers' compensation, 1044–1045, 1077–1078
Terrorism, 336, 459–460
Tertiary prevention defined, 349t
Tesla (unit of radiation measure), 992
Testing
  environmental, 347–348
  of new facilities, 582–586
Tetanus, 307t, 310t, 345t
  immunization for, 353t, 354t
Tetrachloroethylene, biological monitoring case study, 408–410
Textiles, skin injury and, 716
Thermal characteristics
  ergonomics and, 936–937
  of protective clothing, 615
Thermal injury, 603. *See also* Burns; Thermal measurement
  heat stress, 791–823. *See also* Heat stress
    concepts underlying, 793–794
    control measures for, 813–816
    disorders related to, 797–799
    factors affecting heat tolerance, 794–797
    management of employee heat exposure, 816–820
    physiology of, 792–793
    significance of, 792
    thermal measurements and instrumentation, 800–813
  by light radiation, 1000
  to skin, 716–717
Thermal insulation, of clothing, 817–818, 819t
Thermal relaxation time, 717
Thermal spraying, of metals, 79–80
Thermometers, 811, 813
Thermoregulation, by skin, 703
Third-party liability situations, 1168-1169
Threshold level, of sound, 925
Threshold limit values (TLVs), 375, 485. *See also* Biological monitoring
  for aerosols, 264–265
  for dermal exposure, 278
  for heat exposure, 810t
Tickborne infections, 328, 718
Tick-borne infections
  Lyme disease, 305t, 345t, 718
  Rocky Mountain spotted fever, 306t, 345t, 718
Tinea, 299t
Tire manufacture, 571
Tire manufacturing, 560–562, 561t
Titer, antibody, 340
TLVs (threshold limit values), 375, 485. *See also* Biological monitoring
  for aerosols, 264–265
  for dermal exposure, 278
  for heat exposure, 810t
Tobacco smoke, 756–757
  environmental, 783
Toluene
  biological exposure index (BEI), 394t, 419t
  biological monitoring case study, 406–408
Topography, fire safety aspects of, 469
Tort claims, 1151–1153
Total quality management (TQM), accident investigation and, 104
Townmeetings, 1121
Toxic chemical defined, 484
Toxicity
  biological monitoring case studies, 404–411
    arsenic, 404–405
    cholinesterase-inhibiting pesticides (organophosphates), 405–406

Toxicity (*cont'd*)
tetrachloroethylene, 408–410
toluene, 406–408
determinants of, 379–386, 380t
systemic intoxication in skin diseases, 726, 726t
Toxic materials, airborne. *See* Airborne contaminants
Toxicology
defined, 379
published documents on, 36t–39t
Toxic Substance Control Act (P.L. 94-469), 1113
Toxic use reduction programs, 546–552
industry, 549–552
national (U.S.), 546–548
state, 548–549
Toxoplasmosis, 303t
in pregnancy, 363t
Trade associations, OSHA input from, 1120–1121
Trade secrets, 491–492, 500, 1158-1159
Training, 7
in emergency planning, 459
employer hazard communication, 491
in ergonomics, 956
health and safety instruction
evaluation of instruction, 536–537
feedback, 537
goal setting, 520–530
learning audience, 534
learning objectives, 532–533
needs analysis, 518–520
presentation, 534–536
task analysis, 530–531
for heat stress control, 816
in infection prophylaxis, 357
in personal protective equipment use, 607–608
Program Evaluation Profile (PEP), 1104t–1105t
of radiation workers, 1007–1008
record keeping for, 500–502
for respirator use, 659–660, 668
safety and fire prevention, 475–476
safety and health, 20
Training defined, 517
Transient anemic crisis, 363t
Transitional duty, 1078. *See also* Return to work
Transmission
bloodborne, 326
contact, 326
defined, 326
direct-contact, 326
of infectious agents, 326–328, 327t
modes of, 291t–294t
types of, 327–328
Transmission loss, of sound, 925
Transportation, of radiation sources, 1012
Transportation noise, 921
Trauma, cumulative, 5
Travel, infectious diseases related to, 318–320
1,1,1-Trichloroethane (methyl chloroform), 61, 62t, 95, 393t, 418t
1,1-Trichloroethane (TCA), 554.554t
Trichloroethylene, 62t, 395t, 419t. *See also* Degreasing; Solvents
Trichlorotrifluroethane, 62t
Troubleshooting model, 231–235
Trucks, fire, 448–449
Tuberculosis, 300t, 313, 345t, 359
immunization for, 354t
Tuberculosos, 329–330
Tularemia, 307t, 345t
Turpentine, as skin irritant, 710t
TVOC (total volatile organic concentration), 749–750
Two-step testing, 342
Tympanic membrane (eardrum), 840

Ulcerations, skin, 723. *See also* Skin disorders
Ultrasonic defined, 925
Ultrasonic degreasers, 60–61
Ultrasound, in industrial inspection, 79
Ultraviolet analyzers, 263t
Ultraviolet germicidal irradiation (UVGI), 357
Ultraviolet (UV) radiation, 1000–1001
Umbelliferae (phototoxic plants), 714, 715t
Underwriters, insurance, 1191–1193
Uniform resource locators (URLs), 692
Unions, return to work and, 1054–1055
United Kingdom, materials banned in, 546
Unit statistical report, 1078
Universal precautions, 350–352, 351t–352t
Unsafe working conditions defined, 1145
Urine
as index of exposure, 377
medicines affecting, 384t
URLs (uniform resource locators), 692
Urticaria (hives), 724–725
U.S. Centers for Disease Control and Prevention. *See* CDC
U.S. Department of Agriculture (USDA), 336
U.S. Department of Transportation (DOT), 1116
U.S. Environmental Protection Agency. *See* EPA
U.S. Food and Drug Administration. *See* FDA
U.S. Nuclear Regulatory Commission, 77
USDA, hazardous agent regulation by, 336
UV photoxicity, 713–714
UV (ultraviolet) radiation, 1000–1001

Vaccines, safety of, 352
Values, organizational, 8–9

Valves
  battery limit, 473
  depressuring, 470–471
  emergency shut-down, 473
  gas control, 580t
  safety and shutoff, 440
  safety shutdown, 470–471
Vanadium pentoxide, 395t
Vapor(s). *See also* Gases
  defined, 666
  evaluation of chemical, 257–264
Vapor-phase degreasing, 59–64, 567
Vapor-removing respirators, 648–649
Variability, environmental, 243–246
Variance, of regulation, 1145
Varicella (chickenpox), 296t, 315t, 354t, 363t
VDTs (video display terminals), radiation exposure and, 994–995
Vectorborne transmission, 328, 718
Vehicular accidents, 1165
Vehicular insurance, 1184–1187. *See also* Liability insurance
Velocity, of sound, 925
Ventilation. *See also* Airborne contaminants
  acceptance and startup testing of, 582–586
  for contaminant control, 577t, 577–582, 578t, 579t
  dilution, 357
  for heat control, 815–816
  as skin disease preventive, 732
  types used in semiconductor industry, 580t–581t
  for vapor-phase degreasers, 62, 567
Ventilation Manual (ACGIH), 57
Ventilation rate procedure (ASHRAE), 768
Ventilation systems (HVACs) and standards, 764–770
  humidification and dehumidification, 766–767
  system description, 764–766, 765t
Verification, of instruments, 274–275
Vesicle defined, 737
Vibration, 936, 948
Video display terminals (VDTs), radiation exposure and, 994–995
Videotaping
  for ergonomic analysis, 942–948
  in process control, 563–564
Vigilance demands, 938
Vinyl chloride, primary health effects, 389t
Viral infections. *See also specific diseases*
  arthropod-borne, 301t
  childcare-related risks of, 315t
  general characteristics and transmission mode, 296t–298t, 301t–302t, 308t
  immunodeficiency and risk of, 362
  modes of transmission, 291t, 294t
Virulence defined, 328
Viruses defined, 323
Visible light radiation, 1000–1001
Vision
  protective clothing and, 615
  respirator interference with, 663
Vital few analysis, 1037t, 1037–1038. *See also* Workers' compensation
VLF (very low frequency) radiation, 990, 991t, 993–995
Volatile organic compounds (VOCs), 550, 749–753, 750t, 753t. *See also* Air contaminants
  control of, 781–783
  evaluation checklist for, 772–776
Voluntary protection programs, 6, 10, 1139–1140
Voluntary resignation, 1023–1024

Waiting period, 1078
Walkaround inspection, 1128
Walkaround rights, 1145
Wallemia sebi, 747
Warning signs, of employee problems, 1019
Warrant defined, 1146
Warranty, breach of, 1154
Warts, 298t
Waste disposal, of radiation sources, 1012
Waste handling, OSHA regulations for hazardous, 514–515
Waste Reduction Innovation Technology Evaluation (WRITE) Program, 546–547, 547t, 571t
Wastewater treatment, infectious diseases related to, 320
Water. *See also* Environmental transmission
  Clean Water Act (P.L. 92-500), 1113–1114
  drinking, 546
  for firefighting, 440–441
  infectious diseases transmitted by, 294t–295t, 308t–310t
  Safe Drinking Water Act (P.L. 93-523), 1114
  as source of microbials, 748, 766–767
  storm water systems, 473–474
Water and electrolyte balance, 796–797, 798–799
Water-borne paints, 82
Water-cooled clothing, 819
Water draws, in furnaces, 471–472
Waterless cleaners, 733–734
Water sprays, for firefighting, 442–444
Wavelength
  electromagnetic, 988–989
  of sound, 925
WCCC (workers' compensation classification code), 1037, 1041–1043
Weekly Claims Control, 1038
Weight limit calculation, 960
Weil disease (leptospirosis), 304t

Welding, 89–98
arc
gas metal, 93–95
gas tungsten, 92–93
shielded metal, 90–92
ear damage in, 839
gas, 95–98
health hazards of
arc, 89–95
gas, 95–98
protective lenses for, 627, 628t
Wellness programs, 796
Wet-bulb temperature
modified, 803
natural (NWB), 800, 821
psychometric, 800, 821
Wet-globe temperature, 810–811, 822
Wet sinks, 580t
What-if analysis, 147–152, 689
workers' compensation issues, 1075–1076
White noise, 925
WHO, infectious disease regulations of, 333
Whooping cough (pertussis), 353t
Whopping cough (pertussis), 300t
Wood and wood products, as allergens, 712t, 712–713
Work, return to, 1053–1057
Worker right-to-know, 479–515. *See also* Hazard communication
Workers. *See also* Employees; Personnel; Staff
in hazard assessment, 604
Workers' compensation, 4, 1029–1080
Americans with Disabilities Act (ADA) and, 1072–1074
cost-benefit decision making, 1062–1072
cost-benefit analysis, 1063–1064
graphical calculation, 1071
justification formula, 1069–1071
risk-score method, 1065–1068, 1066t
employer situational evaluation, 1034–1041
costs per employee/percent of payroll, 1037
data determination, 1034
incidence rates, 1034–1035, 1036t
injury cause-cost analysis (vital few analysis), 1037
Lynch Ryan shift, 1039–1040
vital few by exposure hours, 1037–1038
weekly claims control, 1038
insurance aspects of, 1180–1184
legal aspects of, 1155
long-term commitment in, 1079
overview of, 1030–1034
current status of, 1031
definition, 1031
functional problems of, 1030–1031
general benefits in, 1031–1033
international, 1031
litigant profile, 1033–1034
as "no-fault" system, 1030
state law differences, 1030–1031
postinury injury response strategy, 1045–1062
accountability for worksite management, 1049–1051
claims management communications, 1057–1059
claims provider involvement, 1061–1062
focus on current claims, 1061
fraud, 1060
hearing and settlement, 1060–1061
management controls of safety programming, 1047–1049
medical provider relationships, 1051–1052, 1052t
performance objectives, 1052–1053
responsibility, 1046–1047
return-to-work opportunities/incentives, 1053–1057
program provisions, 1041–1045
insurance types, 1044
premium calculation, 1041–1044, 1042t
terminology of, 1044–1045
terminology, 1044–1045, 1077–1078
what-if examples, 1075–1076
Workers' compensation classification code (WCCC), 1037, 1041–1043
Work factors, 947. *See also* Ergonomic job analysis
Working position, anthropomentric data, 950t
Work-rest regimens, 820
Worksite, multiemployee, defined, 1143
Worksite assessment, 417–419
Worksite hazards analysis. *see* Hazards analysis
Workstation layout, 574–575
Work task modification, 570–572
World Health Organization (WHO). *See* WHO
World Wide Web, 692–695
Wright v. McDonald's Corporation, 1152
WRITE Program, 546–547, 547t, 571t

XAD sorbent, 259
Xenobiotics. *See also* Biological monitoring; Toxicity
defined, 385
excretion of, 385–386
Xerosis, 706, 737
X-ray machines, 1005. *See also* Radiation: ionizing
X-rays. *See also* Radiologic evaluation
sources in industrial testing, 76–77
Xylene, 382, 395t, 419t

YAG lasers. *See* Lasers
Yersinia pestis, 305t

Z87 designation, 626
Zinc, in galvanizing, 96–97
Zoning, for noise control, 920